[redacted]

[redacted]

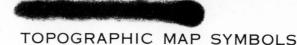

TOPOGRAPHIC MAP SYMBOLS

VARIATIONS WILL BE FOUND ON OLDER MAPS

Hard surface, heavy duty road, four or more lanes

Hard surface, heavy duty road, two or three lanes

Hard surface, medium duty road, four or more lanes

Hard surface, medium duty road, two or three lanes

Improved light duty road

Unimproved dirt road and trail

Dual highway, dividing strip 25 feet or less

Dual highway, dividing strip exceeding 25 feet

Road under construction

Railroad, single track and mult ple track

Railroads in juxtaposition

Narrow gage, single track and multiple track

Railroad in street and carline

Bridge road and railroad

Drawbridge, road and railroad

Footbridge

Tunnel, road and railroad

Overpass and underpass

Important small masonry or earth dam

Dam with lock

Dam with road

Canal with lock

Buildings (dwelling, place of employment, etc.)

School, church, and cemetery — Cem

Buildings (barn, warehouse, etc.)

Power transmission line

Telephone line, pipeline, etc. (labeled as to type)

Wells other than water (labeled as to type) — Oil, Gas

Tanks; oil, water, etc. (labeled as to type) — Water

Located or landmark object; windmill

Open pit, mine, or quarry; prospect

Shaft and tunnel entrance

Horizontal and vertical control station:

Tablet, spirit level elevation — BM△ 5653

Other recoverable mark, spirit level elevation — △ 5455

Horizontal control station: tablet, vertical angle elevation — VABM△ 9519

Any recoverable mark, vertical angle or checked elevation — △*3775*

Vertical control station: tablet, spirit level elevation — BM× 957

Other recoverable mark, spirit level elevation — ×954

Checked spot elevation — ×*4675*

Unchecked spot elevation and water elevation — ×*5657* *870*

Boundary, national

State

County, parish, municipio

Civil township, precinct, town, barrio

Incorporated city, village, town, hamlet

Reservation, national or state

Small park, cemetery, airport, etc.

Land grant

Township or range line, United States land survey

Township or range line, approximate location

Section line, United States land survey

Section line, approximate location

Township line, not United States land survey

Section line, not United States land survey

Section corner, found and indicated

Boundary monument: land grant and other

United States mineral or location monument

Index contour

Intermediate contour

Supplementary contour

Depression contours

Fill

Cut

Levee

Levee with road

Mine dump

Wash

Tailings

Tailings pond

Strip mine

Distorted surface

Sand area

Gravel beach

Perennial streams

Intermittent streams

Elevated aqueduct

Aqueduct tunnel

Water well and spring

Disappearing stream

Small rapids

Small falls

Large rapids

Large falls

Intermittent lake

Dry lake

Foreshore flat

Rock or coral reef

Sounding, depth curve — *10*

Piling or dolphin

Exposed wreck

Sunken wreck

Rock, bare or awash; dangerous to navigation

Marsh (swamp)

Submerged marsh

Wooded marsh

Mangrove

Woods or brushwood

Orchard

Vineyard

Scrub

Inundation area

Urban area

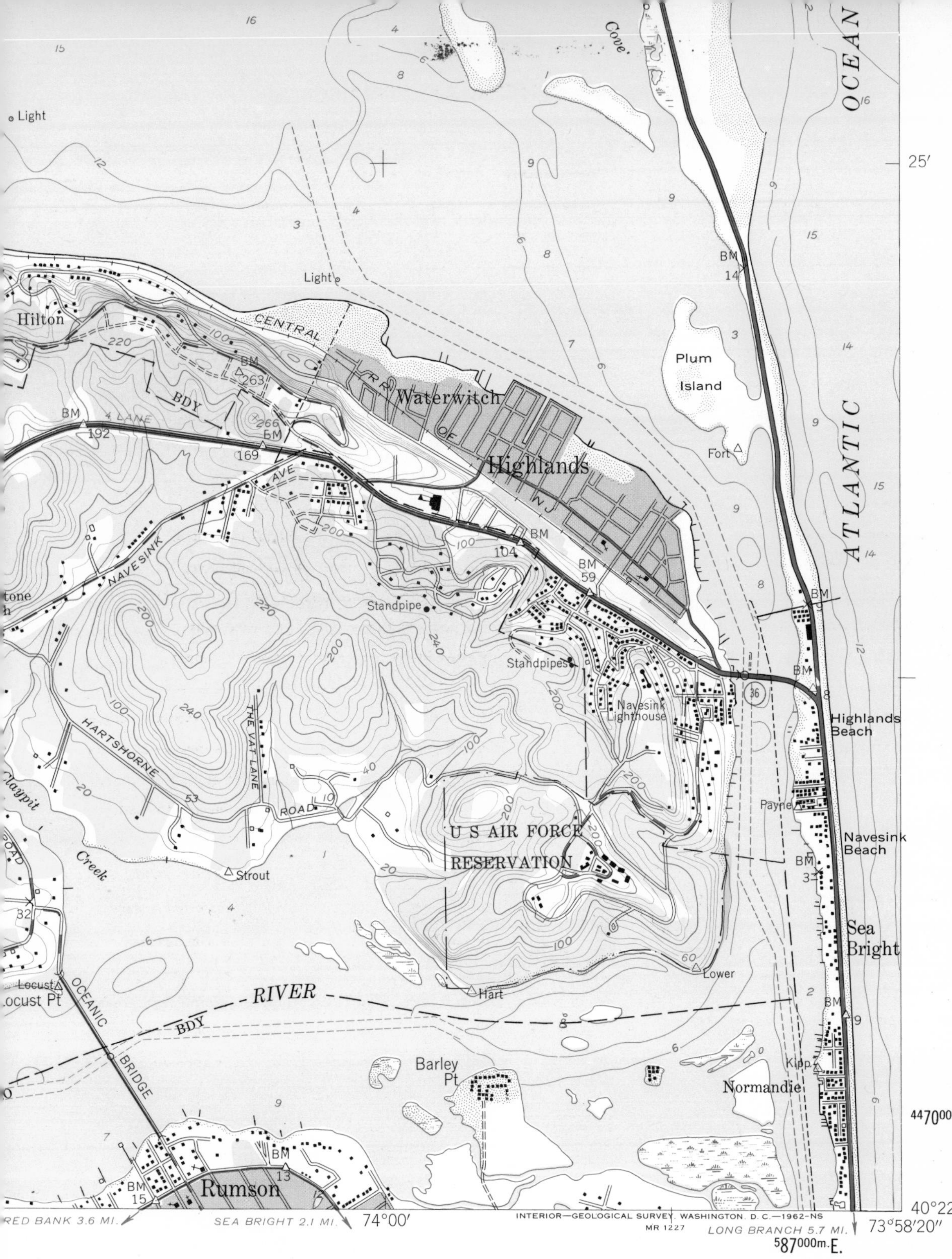

SCALE 1:24000

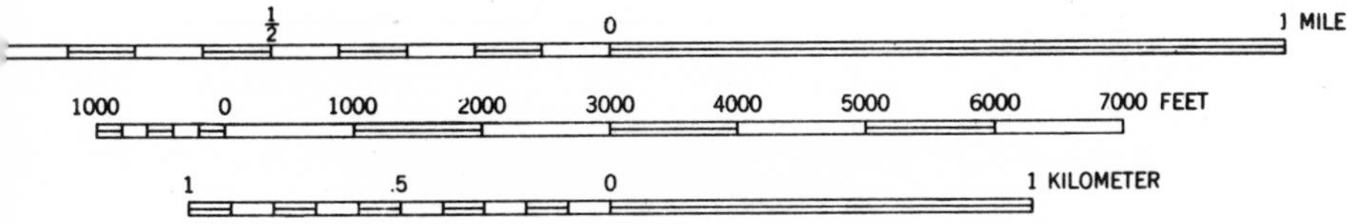

QUADRANGLE LOCATION

CONTOUR INTERVAL 20 FEET
DATUM IS MEAN SEA LEVEL
DEPTH CURVES AND SOUNDINGS IN FEET—DATUM IS MEAN LOW WATER
SHORELINE SHOWN REPRESENTS THE APPROXIMATE LINE OF MEAN HIGH WATER
THE MEAN RANGE OF TIDE IS 3.8 FEET

SANDY HOOK, N. J.—N. Y.
N4022.5—W7358.33/7.5x9.17

1954

PHYSICAL GEOGRAPHY

COVER PHOTOGRAPH

Satellite false-color imagery of Antarctica, showing the extent of sea ice at the end of January, which is midsummer of the Southern Hemisphere. The continental outline is superimposed in black. Data was gathered by Scanning Microwave Radiometer aboard Nimbus-5, a weather satellite. The microwaves, with a wavelength of 1.55 cm, pass readily through the cloud cover. The highest region of the Antarctic Ice Sheet appears as dark blue to the right of the south pole. Black spots are gaps in the data. (Courtesy of the National Aeronautical and Space Administration.)

PHYSICAL GEOGRAPHY
FOURTH EDITION

ARTHUR N. STRAHLER

John Wiley and Sons, Inc., New York London Sydney Toronto

This book was set in Bodoni by York Graphic Services, Inc. It was printed and bound by Vail-Ballou Press, Inc. The cover designer was Eileen Thaxton. The drawings were designed and executed by John Balbalis with the assistance of the Wiley Illustration Department. Regina R. Malone supervised production.

Library of Congress Cataloging in Publication Data

Strahler, Arthur Newell, 1918–
Physical geography.

Bibliography: p.
1. Physical geography—Text-books—1945–
I. Title.
GB56.S75 1974 910′.02 74-9994
ISBN 0-471-83160-3

Printed in the United States of America

10 9 8 7 6 5 4 3 2 1

Preface

THE fourth edition of *Physical Geography* maintains the traditional structure and content of the full introductory college course in physical geography, but at the same time reflects current trends and changing priorities in professional geography. One trend is toward emphasis upon the global balances of radiation, heat, and water. Flow and storage systems of energy and matter in atmosphere and hydrosphere are emerging as dominant concepts in the understanding of man's physical environment. Another trend is toward emphasis upon man's impact on the environment, including pollution and degradation of natural physical systems. Because much of this environmental impact comes through withdrawal, processing, and disposal of water and minerals, including mineral fuels, it is becoming increasingly important to develop the resource aspect of physical geography.

The requirements for a purposeful physical geography have been met by taking advantage of the revised content of the third edition of *Introduction to Physical Geography*, which includes much new text material and many new illustrations. In partial compensation, there have been some deletions and abbreviations of less important topics. Use of a more compact layout has resulted in a volume offering substantially more information in fewer pages, an important consideration in times of rising costs and material shortages.

The principal changes are as follows: Chapter 7 has been expanded to include an overview of the oceans along with the atmosphere. Chapters 8 through 11, covering atmospheric processes, have been extensively revised to give better coverage of the radiation and heat balances, as well as to include material on air pollution and inadvertent climate modification. A major change has been to follow the meteorological chapters with two chapters dealing with the hydrologic cycle and the water balance. The first of these, Chapter 12, is devoted to the global water balance and the soil-moisture balance; the latter topic has been completely rewritten and given new figures to illustrate water-balance regimes. There follows a chapter on ground water and surface water, tracing the flow paths of surplus water to completion of the hydrologic cycle. Water pollution is introduced here. Geologic aspects of runoff and ground water are treated in Chapters 25, 26, and 28, as topics in geomorphology.

Four chapters (Chapters 14 through 17) form a climate group. The introductory chapter on climate classification has been greatly changed by discussions of the role of climate in environmental processes and the importance of radiation and soil-moisture balances as viable bases of climate classification. However, the Köppen system is retained and the three chapters on world climates are virtually unchanged from the previous edition. The subject of climatic regimes and their expression through thermohyet diagrams has been deleted entirely. Chapters 18 through 21, dealing with soils and natural vegetation, remain intact with no changes, except that the *7th Approximation* soils classification system, now known as the *Soil Taxonomy Used by the U.S. National Cooperative Soil Survey*, has been introduced at the end of Chapter 19. This new section describes the orders and suborders of the new soil taxonomy and shows their global distribution on a new map. Thus it is now feasible to present major features of the new soil classification system as an alternative to the 1938 system.

I am particularly indebted to Dr. Roy W. Simonson and Dr. Guy D. Smith, both leaders in development of the new taxonomy, for critical review of the new text material.

Turning to Part IV, a major improvement in the chapter on earth materials (now Chapter 22) has been the rewriting of most of the text on minerals and rocks and the introduction of new material on ore minerals and ore deposits. Orientation has thus been shifted to the resource aspect of the lithosphere. The chapter on the earth's crust and its relief forms (now Chapter 23) has been brought up to date by new text and figures on plate tectonics and continental drift.

The final ten chapters form a geomorphology group, little altered in content from the previous edition, except that new sections have been introduced on environmental impact. Topics include the hydrologic effects of urbanization, man-induced valley aggradation and mass wasting, and man-induced coastal changes. Chapter arrangement has been altered to place the three chapters on structurally-controlled landforms immediately following the chapters on fluvial processes. Thus the final three chapters now deal with the work of glaciers, waves and currents, and wind. Another important change is that the Murphy system of world landform classification has been placed in an appendix, replaced by a new description of the geomorphic provinces of North America. Together with new maps and a large table, this material is placed at the end of Chapter 30, applying concepts of fluvial denudation and structural landform classes covered in Chapters 24 through 30.

An innovation in keeping with modern trends in geography is the review of remote sensing techniques in geographical research, presented as Appendix V. I am grateful to Dr. John E. Estes, University of California at Santa Barbara, for reviewing the text of this appendix and for selecting and describing materials for the color-plate insertion illustrating remote sensing imagery.

Exercises have been retained as in the previous edition, but with minor changes appropriate to the revised text content. The bibliography has been extensively updated, with many new headings for such topics as air pollution, urban climate change, mineral resources, water resources, and plate tectonics.

I am deeply indebted to a group of geography instructors who responded to detailed questionnaires covering both major and minor proposals for revision. The guidance furnished by their collective response has been particularly valuable because it reflects the contemporary experience in the college classroom through a period of changing priorities and practices in geographic education. Particularly gratifying was the almost complete endorsement of all of the author's proposals to introduce new material on environmental impact.

Over a period of nearly a quarter century, many helpful suggestions have been sent in to me by colleagues and students who have used the foregoing editions of my physical geography textbooks. It is a source of regret that I cannot acknowledge individually the aid of these many contributors. Their total effort has played a major role in strengthening this newest work and making it more useful in the teaching of physical geography.

Arthur N. Strahler

Arthur N. Strahler

Arthur N. Strahler (b. 1918) received his B.A. degree in 1938 from the College of Wooster, Ohio, and his Ph.D. degree in geology from Columbia University in 1944. He is a fellow of the Geological Society of America and the Association of American Geographers. He was appointed to the Columbia University faculty in 1941, serving as Professor of Geomorphology from 1958 to 1967 and as Chairman of the Department of Geology from 1959 to 1962. His published research has dealt largely with quantitative and regional geomorphology. At present he is associated with the University of California at Santa Barbara in a research capacity. He is the author of several widely used college textbooks of physical geography, earth science, and environmental science.

Contents

PART IV LANDFORMS OF THE EARTH'S CRUST

APPENDIX

PHYSICAL GEOGRAPHY

Introduction

PHYSICAL geography brings together and unifies several branches of natural science for the purpose of understanding the relationship of man to his physical environment. Man interacts with his environment in a shallow layer—the *life layer*—lying at the contact between the atmosphere and the land surface and between the atmosphere and the ocean surface. These contact zones, which we may call *interfaces*, experience intense activity in the form of exchanges of energy and matter. Man and all other life forms participate in the processes of the life layer. Organisms receive from their physical environment energy and matter necessary for the life processes. Energy and matter are stored in organic tissues, later to be released to the physical environment. Physical geography is thus a study of the workings of an environment that not only nourishes and stimulates life processes, but also places constraints and limitations upon those processes.

The physical geographer is particularly interested in the place-to-place variations in the environment of the life layer. Global patterns of environmental variables are studied with great interest by the physical geographer. He attempts to classify these variables with respect to their distribution. A scientist having related objectives is the ecologist. He studies groups of organisms interacting with their environments; these constitute *ecosystems.* The ecologist deals particularly with the biological processes within ecosystems, but he is also concerned with the global distribution of ecosystems as influenced by variations in the physical environment. To a large extent, physical geography forms the basis for a study of the ecology of man, or *human ecology*.

Interaction implies an exchange or reciprocity between two parts of a system. In this case we find that not only does man respond to the forces of the environment, but man in turn acts upon and modifies environmental processes and forms. In our modern industrial age particularly, man creates many forms of environmental degradation and pollution. With enormous quantities of energy at his disposal, man is capable of severely damaging or destroying natural features of the environment and their ecosystems. Part of the purpose of physical geography is to recognize man's impact upon the environment, and particularly to show how susceptibility of the land to that impact varies from place to place over the globe. Such knowledge can be put to use intelligently for land use planning and resource management.

Broadly speaking, the environment of the life layer includes four realms of matter. One realm is the *atmosphere*, or gaseous envelope surrounding the planet. A second is the *hydrosphere*, encompassing the global reserve of water in gaseous, liquid, and solid states. Third is the *lithosphere*, consisting of mineral matter in the solid state. Fourth is the realm of organic matter itself, the *biosphere.* For an understanding of the atmosphere, we draw upon the sciences of *meteorology* and *climatology*. The sciences of *hydrology* and *oceanography* provide an understanding of the hydrosphere. *Geology* is the source science for an understanding of processes and forms of the lithosphere. Within these large categories of natural science are disciplines that specifically deal with the interactions at the interfaces of the life layer. An example is *soil science*, or *pedology*, which must deal simultaneously with atmosphere, hydrosphere, and lithosphere. Another is *geomorphology*, the study of landforms that comprise the physical platform of the life layer.

Before embarking upon an investigation of the four global realms of matter a number of important global topics must be treated. First and most fundamental are the form of the earth—a concern of the science of *geodesy*—and the relationship between earth and sun—a part of *astronomy*. Much of astronomy is beyond the concern of the geographer, for only two bodies, the sun and the moon, appreciably affect life on earth. Because all energy for sustaining life, all motive power for streams, winds, and ocean currents, comes by radiant emanation from the sun, and because the intensity of this energy changes through daily and annual cycles, an understanding of the motions of the earth in its orbit about the sun is a prime essential. The moon, as the body that controls ocean tides, enters into physical geography in only a minor way.

Because data of the earth sciences are often best represented by maps, and perhaps many are impossible to describe without maps, the science of maps, *cartography*, is an essential ingredient of physical geography. True, cartography is really a science of technique rather than a basic earth science, but it deserves a place early in the list of topics so that it may provide a means for representing the information to follow.

The study of plant forms and their distribution, which we may call *plant geography*, must be included in physical geography because plants are the primary producers of organic matter. In short, man depends upon plants for his food. We are therefore interested in the interaction of plants with their physical environment. Plants constitute important physical features of the earth's land surfaces, and, moreover, the plant cover modifies physical processes that shape the lands.

The professional physical geographer will usually be a specialist in only one of the several fields involved, such as climatology, geomorphology, or soil science. Besides carrying out original research in his chosen specialty, to which he may be making important scientific contributions, the physical geographer attempts to keep informed on important developments as they occur in the other fields of specialization. He is thus able to assemble and integrate pertinent fragments of knowledge into a unified picture of the natural environment of man at any place on the globe at any season of year.

Today increasing emphasis is being placed upon a field of study called *environmental science*, which we recognize as interdisciplinary. Environmental science aims particularly at analysis of environmental problems caused by man. Physical geography has always been at the heart of environmental science, for physical geographers have always concerned themselves with the interaction between man and the environment.

Physical geography is closely involved with the subject of the earth's natural resources. When we study the pathways and storages of fresh water on the lands we are also studying a vital resource. Man and other land animals, as well as all land plants, require fresh water for survival. Minerals furnish the materials for most manufactured products and structures of our industrial society. Mineral fuels—coal, oil, natural gas, and uranium—furnish almost all of the energy for industrial and urban needs. Consequently, as we investigate the processes that form rocks and shape the earth's crust and land surface, we are also investigating the origin of mineral resources. As man withdraws, utilizes, and disposes of water and mineral resources, he makes severe impacts upon the natural environment. We shall be interested in assessing the extent of these impacts, the better to understand what must be done to minimize environmental degradation.

I

THE EARTH AS A GLOBE

CHAPTER 1

Form of the Earth; the Geographic Grid

THE spherical form of the earth is one of the facts of our physical environment, which children learn at an early age, but probably few people give much thought to some of the simple proofs of the earth's sphericity. For example, the evidence that people have repeatedly sailed or flown completely around the globe is tacitly assumed to prove the sphericity of the earth, whereas it means only that the earth is a solid body. Circumnavigation could also be performed on a cubical or cylindrical earth.

Even without optical instruments we can get some inkling that the earth's surface curves downward away from us through the observation that sunlight illuminates the tops of high clouds after sunset and before sunrise.

One proof of the earth's sphericity may be had from observations at sea. As a ship recedes farther and farther into the distance, it appears to sink slowly beneath the water level (Figure 1.1). Seen through binoculars or a telescope the sea surface will appear to rise until the decks are awash, then gradually to submerge the funnel and the masts, leaving finally only smoke visible above the horizon. The explanation obviously lies in the fact that the sea surface curves downward away from us. To prove that this curvature is spherical would require numerous observations in which measurements were made of the amount of apparent sinking of a vessel per unit of distance in many different directions away from the observing point.

A second proof is found in the observation that in all lunar eclipses, at which time the earth's shadow falls on the moon, the edge of the shadow appears as an arc of a circle. It can be shown by geometrical proof that a sphere is the only body that will always cast a circular shadow upon another sphere. Because at the time of these eclipses the earth is rarely turned in just the same position, we may conclude that, no matter what earth profile is cast on the moon, the circular shadows are all alike and the earth must be spherical.

Photographs taken from rockets and earth satellites (such as the *Tiros* weather satellites) at extremely high altitudes show the horizon as a curved line (Figure 1.2). Because the curvature appears to be the same in many widely separated parts of the earth, a series of such photographs provides us with a third proof of the earth's spherical form.

A fourth proof may be had from observation of the position of Polaris, the north star. To the observer at the equator Polaris is on the horizon, but as he travels toward the north pole the star seems to be located higher and higher in the sky, until, at the north pole, it is directly overhead in the sky. It would be found that Polaris rises 1° higher in the sky for every 69 mi (111 km) of northward travel by the observer. A similar condition would hold for travel from the equator to the south pole if a star nearly in line with the earth's axis in the southern sky could be observed. Thus we can prove that all north-south lines drawn from pole to pole (e.g., meridians) are arcs of circles, and that the earth is spherical.

A fifth proof of the earth's sphericity comes from the observation that an object near sea level will weigh very nearly the same amount on a spring type of scales at any place on the globe. Knowing that the weight depends upon the pull of gravity, we conclude that the object weighs the same everywhere because all points on the earth's sur-

Figure 1.1 Seen through a telescope, a distant ship seems to be partly submerged.

Figure 1.2 Curvature of the earth's horizon shows clearly on this photograph of the southwestern United States and northern Mexico, taken from a Navy Viking-12 rocket at an altitude of 143 mi (230 km). On the left are Lower California and the Gulf of California. To the right the view extends as far as the Los Angeles area. (Official U.S. Navy Photograph.)

face are equidistant from its center, hence, the earth is a sphere.

Now, a pendulum clock will also serve to measure the force of gravity. If the pendulum is kept of exactly constant length the clock will keep constant time so long as it is acted upon by a constant gravity. Thus, if a pendulum clock is found to keep good time at all points at sea level over the earth, the spherical form is proved. Extremely precise calculations based upon this principle, however, show slight variations in gravity, which, as will be explained later, led to the discovery that the earth's true form is not a perfect sphere.

A sixth proof is found in surveying operations with precise telescopic instruments. Suppose that an engineer should drive two posts into the ground, one mile apart, to such a depth that when he sights from the top of the first to the top of the second, his line of sight will be perfectly horizontal according to the sensitive level bubble on the telescope (Figure 1.3). Now suppose that he drives a third post in line with the first two, but a mile beyond the second, and that he adjusts the height of the third post so that a telescopic line of sight back to the second post is a perfect horizontal, as shown by his level bubble. Were he not aware of the earth's curvature, the engineer might be surprised to find that if he now sights with his telescope from the top of the first to the top of the third post, the top of the second projects above this new line of sight. This is explained by the fact that a telescopic line of sight does not follow the curve of the earth, but is a tangent straight line which extends out into space. Surveyors, therefore, have to make corrections for the earth's curvature, and because this correction is approximately constant for all places on the earth, we may conclude that the earth is spherical.

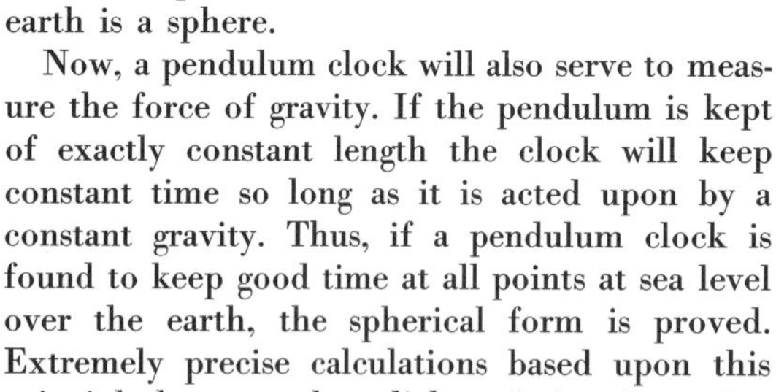

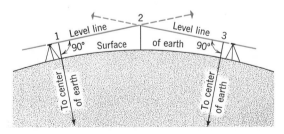

Figure 1.3 Because of the earth's surface curvature, the sight lines of surveying telescopes do not maintain a constant elevation.

As a seventh and last proof, it may be noted that modern navigation methods are based on the assumption that the earth is a sphere. When we consider that, for more than a century, positions of vessels have been correctly determined innumerable times by these methods, it becomes obvious that the correctness of the assumption has been established many times over.

Eratosthenes' measurement of the earth

Although the ancient Greeks, among them Pythagoras (540 B.C.) and associates of Aristotle (384–322 B.C.), believed the earth to be spherical and had speculated upon its circumference, it remained until 200 B.C. for Eratosthenes, librarian at Alexandria, to perform a direct measurement based upon a sound principle of astronomy. He observed that at Syene, Egypt, located on the upper Nile River close to the Tropic of Cancer, $23\frac{1}{2}°$ N, the sun's noon rays on the date of the summer solstice (June 21) shone directly upon the floor of a deep, vertical well. In other words, the sun was then in the zenith point in the sky and its rays were perpendicular to the earth's surface at that latitude (Figure 1.4). At Alexandria, however, on this same date, the rays of the noon sun made an angle of one-fiftieth of a circumference, or 7° 12′, with respect to the vertical.

As we can see from the relations between parallel sun's rays and radial lines from the earth's center, the arc of the earth's surface lying between Alexandria and Syene is also equal to 7° 12′ or one-fiftieth of the earth's circumference. Therefore, it is necessary only to determine the ground distance along the north-south line between the two places, multiply this by 50, and the circumference is known.

Eratosthenes took as the distance between Alexandria and Syene 5000 stadia, but this may have been only a rough estimate. This gave 250,000 stadia for the earth's circumference. Using a stadium equivalent of 185 m, the circumference comes out to 46,250 km or about 26,660 statute miles, which is of the same general order of size as the true value of about 25,000 mi (40,000 km).

From Eratosthenes' classic experiment, it is an easy step to design an astronomical method of measuring the earth's figure. We need only to select a north-south line, whose length can be measured directly on level ground by surveying means. The line should be at least 69 mi (111 km) long to give an arc of about one degree. At the ends of the line the altitude of any selected star can be measured at its highest point above the horizon or with respect to the vertical, using a level bubble or a plumb bob as a means of establishing a true horizontal or vertical reference. The difference in angular position of the star will be the arc of the earth's circumference lying between the ends of the measured line. This very procedure is believed to have been followed by Arabs of the ninth century. Their measurements were probably much more accurate than those of Eratosthenes, but because the units of measure are not known in modern equivalents, their work cannot be checked.

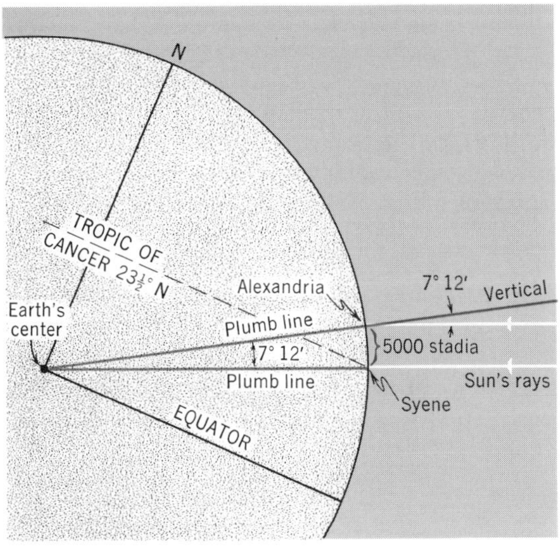

Figure 1.4 Eratosthenes' method of measuring the earth's circumference.

Earth's surface curvature and visibility

The amount of curvature of the earth's surface may be stated in terms of the actual distance between a curved line lying on the earth's surface (as on the calm ocean) and a tangent straight line originating at the same point. This distance we shall call the *divergence* (Figure 1.5). Because the air decreases in density upwards, a light ray will not be a straight line but will bend earthward. The effect of this phenomenon, which is known as *refraction*, is to decrease the divergence by about one-seventh of what it would be if the earth had no atmosphere. A simple rule for finding the divergence in feet between the surface line and the light-ray line is to take three-fifths of the square of the number of miles between the two desired places (points A and B in Figure 1.5). The formula may be written as follows for English units of measure:

$$h = \frac{3}{5}K^2$$

$$\text{(Exact value: } h = 0.574K^2\text{)}$$

where h = number of feet between surface and light-ray lines, and K = distance in miles. For ex-

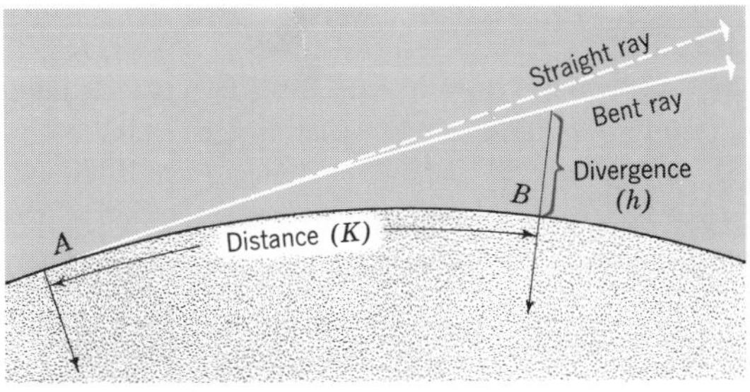

Figure 1.5 Even though a light ray is bent slightly earthward, its divergence from the earth's surface increases with distance.

ample, if the distance is 10 mi, the two lines have diverged to the extent of being approximately 60 ft apart.

If we already know the divergence in feet of the surface line from the light-ray line, we may find the distance in miles separating the two points by the formula

$$K = 1\tfrac{1}{3}\sqrt{h}$$
$$K = 1.317\sqrt{h}$$

That is to say, take the square root of the number of feet of divergence and multiply by $1\tfrac{1}{3}$. If the divergence were 81 ft, for example, the distance in miles would be $9 \times 1\tfrac{1}{3}$, or 12 mi.

In metric units, where h is divergence in meters, K is distance in kilometers, use the following formulas:

$$K = 3.80\sqrt{h}$$
$$h = 0.069K^2$$

In Table 1.1 figures are given for a variety of examples.

TABLE 1.1

ENGLISH UNITS			
Divergence (Feet)	Distance (Miles)	Distance (Miles)	Divergence (Feet)
1	1.32	1	0.6
2	1.86	2	2.3
5	2.94	5	14.4
10	4.16	10	57.4
20	5.89	20	230
50	9.31	50	1440
100	13.2	100	5740
200	18.6	200	23,000
500	29.4	500	144,000
1000	41.6		
5000	93.1		

METRIC UNITS			
Divergence (Meters)	Distance (Kilometers)	Distance (Kilometers)	Divergence (Meters)
1	3.80	1	0.07
2	5.37	2	0.28
5	8.50	5	1.73
10	12.0	10	6.92
20	17.0	20	27.7
50	26.9	50	173
100	38.0	100	692
200	53.7	200	2,770
500	85.0	500	17,500
1000	120		
5000	269		

The amount of curvature becomes of great interest and practical importance in problems of visibility over the open ocean. The expanse of ocean visible from a single point increases greatly with rising elevation above the water surface. Table 1.1 shows that, from a point 5 ft above the surface, as from a small boat, the radius of vision is about 3 mi, whereas from a point 100 ft above sea level, as from the mast of a ship, it increases to 13 mi.

Where both points lie at different elevations above sea level, as for example, a lighthouse 100 ft above sea level and the bridge of a ship 65 ft above sea level (Figure 1.6), the distance of visibility of the light is the sum of the two distances obtained by solving the curvature problem in two parts. For the observer on the ship's bridge, the radius of vision of the horizon is about $10\tfrac{1}{2}$ mi. The light rays from the lighthouse are tangent to the sea surface at a distance of 13 mi. The total is therefore about $23\tfrac{1}{2}$ mi. We have not taken into account other factors that might modify our calculations, such as waves, which would add to the level of the horizon and thus tend to reduce visibility.

The problem of range of visibility has now been extended to indefinitely great heights above the earth's surface by means of orbiting satellites and other space vehicles. Disregarding the effect of the earth's atmosphere in bending and scattering light rays, the problem becomes that of a cone applied to a sphere (Figure 1.7). The observer is considered to be located at the apex of the cone. The observer's horizon (limit of earth visibility) is a circle coinciding with the line of tangency of cone and sphere. The radius of this horizon circle of visibility can be calculated when the equivalent arc in degrees is determined by trigonometry.[1] The number of degrees of arc is multiplied by 69 mi (111 km) to obtain length of radius. Table 1.2 gives distance values for various heights, assuming an earth radius of 4000 mi and 6370 km.

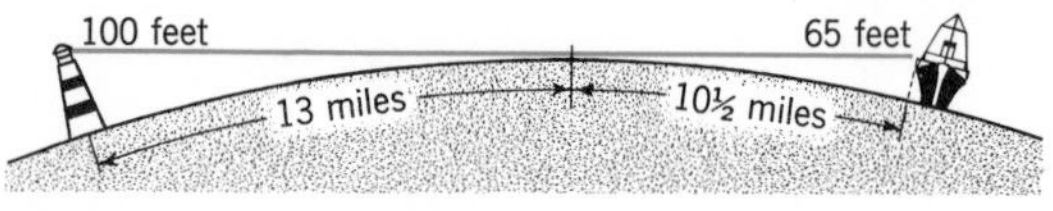

Figure 1.6 Curvature of the earth's surface limits the range of visibility from high points.

The earth as an oblate ellipsoid

In 1671 a French astronomer, Jean Richer, was sent by Louis XIV to the island of Cayenne, French Guiana, to make certain astronomical observations. His clock had been so adjusted that its

[1]The formula is

$$\cos \alpha = \frac{R}{R + h}$$

where α is the angular distance in degrees, R is the earth's radius, and h is height above earth's surface in same units of measure as for the radius.

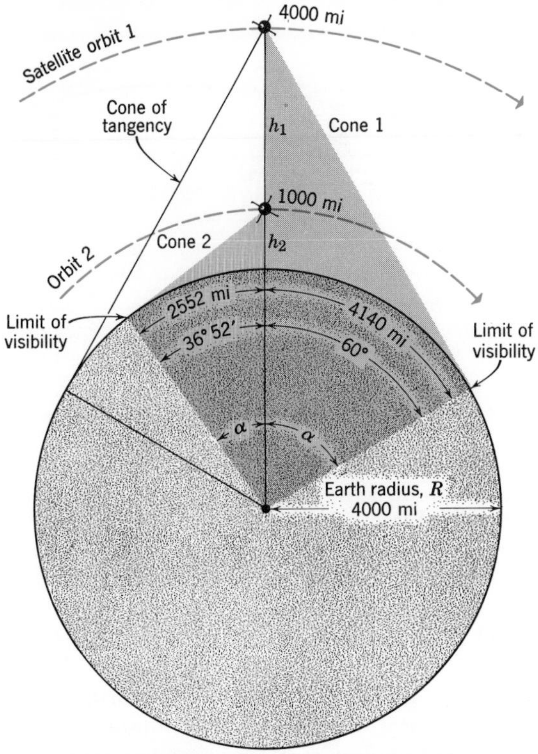

Figure 1.7 Extent of visibility from an orbiting satellite increases with height above earth as indicated by a succession of tangent cones.

pendulum, slightly over 39 in (99.4 cm) long, beat the exact seconds in Paris. Upon arriving on Cayenne, which is near the equator, Richer found the clock to be losing about two and one-half minutes per day. This he correctly attributed to a somewhat lesser force of gravity near the equator, and it was soon realized that this phenomenon could be accounted for only by supposing that the equatorial portions of the earth's surface lie farther from the earth's center than do more northerly places. Refined measurements of a similar type have since revealed that the true form of the earth is like that of a spherical globe, compressed along the polar axis and bulging slightly around the equator (Figure 1.8). This form is known as an *oblate ellipsoid* or *ellipsoid of revolution.* A cross section through the poles gives an ellipse rather than a circle. The equator remains a circle and is the largest possible circumference on the ellipsoid. The earth's oblateness is attributed to the centrifugal force of the earth's rotation, which deforms the somewhat plastic earth into a form in equilibrium with respect to the forces of gravity and rotation.

Confirmation of the earth's oblateness was obtained in the eighteenth century by the work of two scientific expeditions sent out under the auspices of the Royal Academy of Sciences of Paris. One party went to Lapland where, in the years 1736–1737 an arc of 57′ was measured. Finding this arc to be longer than a known equivalent arc at Paris, France, they demonstrated a flattening of the earth toward the poles. Meanwhile, the second party, which had set out for Peru in 1735, actually began measurements near the equator at Quito, Ecuador, completing the measurement of an arc of more than 3° in 1743. The length of a degree of arc there proved to be less than an equivalent arc in France and still less than in Lapland, providing conclusive evidence of the earth's resemblance to an oblate ellipsoid.

TABLE 1.2

ENGLISH UNITS	
Height Above Earth's Surface (Miles)	Radius of Circle of Visibility (Miles)
100	874
200	1225
400	1700
1000	2520
2000	3320
4000	4140

METRIC UNITS	
Height Above Earth's Surface (Kilometers)	Radius of Circle of Visibility (Kilometers)
100	1120
200	1570
400	2200
1000	3350
2000	4500
4000	5780

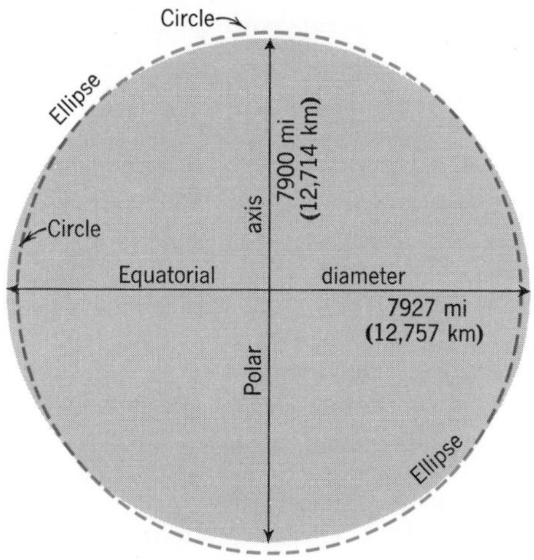

Figure 1.8 The earth is slightly elliptical in a cross section passing through its polar axis. Dimensions are those of the International Ellipsoid of Reference (Hayford, 1909).

Rounded off to the nearest whole mile, the dimension given in Figure 1.8 for the earth's equatorial diameter is 7927 mi (12,757 km), whereas the length of polar axis is 7900 mi (12,714 km), a difference of about 27 mi (43 km). The *oblateness* of the earth ellipsoid, or *flattening of the poles,* is the ratio of this difference to the equatorial diameter, or roughly $^{27}/_{7927}$, which reduces to a fraction only slightly larger than $^{1}/_{300}$. Further details concerning the earth ellipsoid and various calculations of its precise dimensions are given later in this chapter.

Using the above figures, the earth's equatorial circumference is about 24,900 mi (40,075 km). For rough calculations, the figure of 25,000 mi (40,000 km) may be used. The science of *geodesy* (from the Greek words meaning "to divide the earth"), which takes for its goal the determination of the form and dimensions of the earth, developed from the need to ascertain precisely the nature of the oblate ellipsoid that the earth resembles. The scientist who practices geodesy, the *geodesist,* uses extremely precise surveying methods along with delicately refined determinations of the force of gravity to achieve this goal.

The earth as a geoid

Although the oblate ellipsoid is a much better description of the form of the earth than is the sphere, there is need for still further refinement. The earth's figure, which geodesists are trying to measure and describe, is not the configuration of the ground surface, for this rises and falls in a highly irregular way over the sea floor and continents. The surface whose form is sought is the sea-level surface of the oceans extended in an imaginary way under the lands to form a continuous figure known as the *geoid.* If we could crisscross the continents with canals or tunnels at sea level, permitting the ocean water to seek its level in the heart of the continent, the geoid could be established.

Because of the presence of a large rock mass above sea level in a continent, the force of gravity at sea level is somewhat diminished there. Consequently the sea-level surface, or geoid surface, lies somewhat higher under the continents than the ellipsoid, which is used as a surface of reference (Figure 1.9). Under the deep ocean basins, where a large mass of rock is replaced by less dense water, the force of gravity at sea level is greater, depressing the surface of the geoid beneath the surface of the reference ellipsoid. Thus the geoid can be thought of as an undulating surface of irregular form (Figure 1.10). It can be described in terms of its position above or below the imaginary ellipsoid surface, but is too complex a surface to be described by a simple formula.

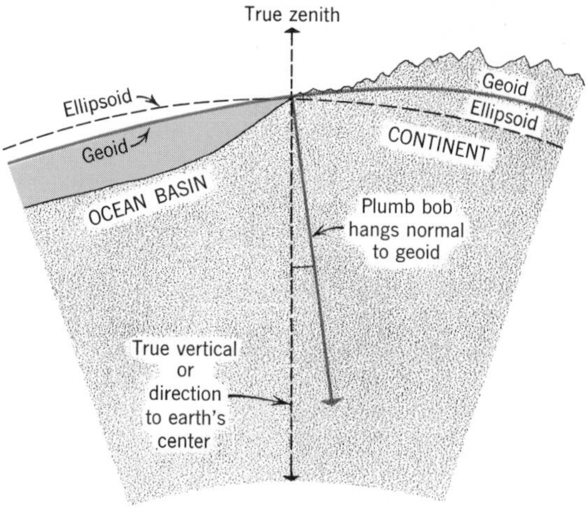

Figure 1.9 Relative positions of geoid and ellipsoid are reversed from ocean basin to continent.

Distances of separation between geoid and ellipsoid typically reach amounts of 65 to 100 ft (20 to 30 m) over the continents. Notice that these values are extremely small compared with the difference of 27 mi (43 km) between earth's polar and equatorial diameters in the ellipsoidal form.

Much of modern research in geodesy is devoted to determining the surface of the geoid. This is important because the direction of downward pull of gravity depends on the form of the geoid surface (Figure 1.9). Because surveying and astronomical observations are based on use of the plumb line or level bubble to give true vertical and horizontal reference directions, the accuracy of observations hinges on knowledge of the geoid.

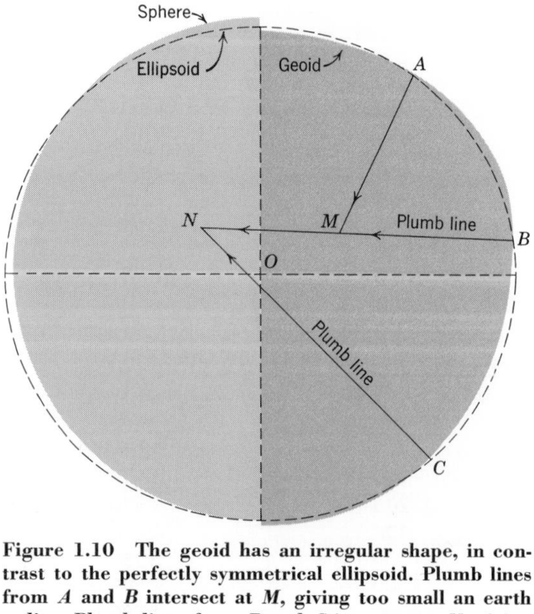

Figure 1.10 The geoid has an irregular shape, in contrast to the perfectly symmetrical ellipsoid. Plumb lines from *A* and *B* intersect at *M*, giving too small an earth radius. Plumb lines from *B* and *C* intersect at *N*, giving too great an earth radius. (After W. A. Heiskanen.)

Great and small circles

If a perfect sphere is divided exactly in half by a plane passed through the center, the intersection of the plane with the sphere is the largest circle that can be drawn on the sphere and is known as a *great circle* (Figure 1.11). Circles produced by planes passing through a sphere anywhere except through the center are smaller than great circles and are designated *small circles.*

It will be of great value to the student of physical geography to be thoroughly familiar with the properties of great circles, because they frequently enter into such global subjects as meridians, navigation, illumination of the globe, and map projections. The following properties may be listed.

1. A great circle results when a plane passes through the center of a sphere, regardless of the attitude of the plane.
2. A great circle is the largest possible circle that can be drawn on the surface of a sphere.
3. An infinite number of great circles can be drawn on a sphere.
4. One and only one great circle can be found that will pass through two given points on the surface of the sphere (unless the two points are at the extremities of the same diameter, in which case an infinite number of great circles can be drawn through them). This follows the geometrical law that three points determine a plane, the third point in this instance being the center of the sphere.
5. An arc of a great circle is the shortest distance, following the surface, between any two points on a sphere.
6. Intersecting great circles bisect each other.

In view of the earlier discussion of the earth's form, in which it was stated that the earth is not a perfect sphere, but an oblate ellipsoid, the student may wonder whether the properties of great circles can be properly applied to the earth. For all ordinary purposes, including the use of great circles, the earth may be treated as a sphere without fear of appreciable error. Throughout most of the subject matter that follows in later chapters, the spherical form will be assumed. An exception is in regard to the exact values for degrees of latitude, a subject requiring use of the ellipsoidal form.

One use of great circles that may be elaborated on here is in navigation. Wherever ships must travel over vast expanses of open ocean between distant ports, or planes must make long flights, it is desirable in the interests of saving fuel and time to follow the great-circle arc between the two points, provided, of course, that there are no obstacles or other deterring factors preventing the use of the great-circle path. Navigators use special types of maps which have the property of always showing great-circle arcs as straight lines. These are known as *great-circle sailing charts* and are discussed more fully under the subject of map projections. To plot the shortest course between any two points, it is necessary only to draw a straight line between the two points on the chart.

Great-circle courses may easily be found by the student using only a small globe and a piece of thin string or a rubber band (Figure 1.12). The string can be held in such a way that it is stretched tightly against the surface of the globe between the two thumbnails, each of which is on one of

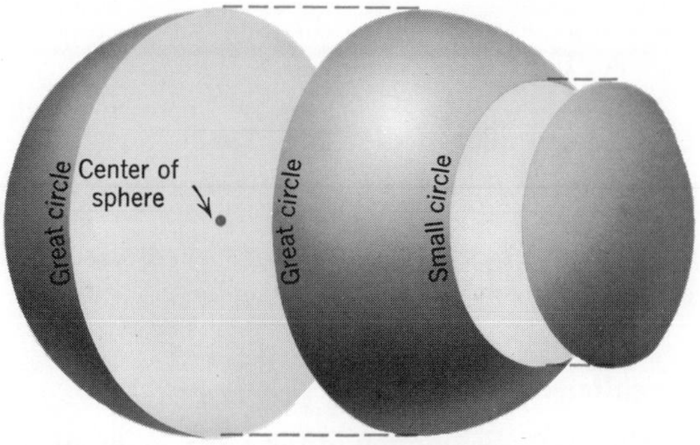

Figure 1.11 A great circle is made by a plane cutting a sphere into two equal halves; a small circle, by a plane cutting the sphere into unequal parts.

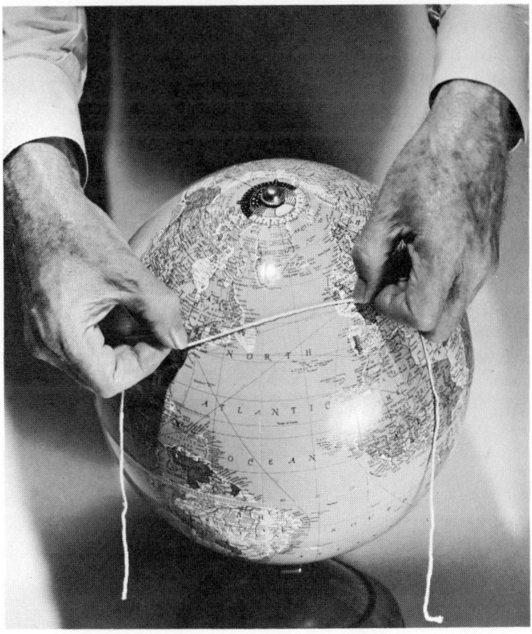

Figure 1.12 A great-circle course may be found by stretching a string between two points on a globe. (Charles Phelps Cushing.)

the points between which the great-circle course is desired. If a rubber band is used, a complete great circle can be shown; it is of special value for points on opposite sides of the globe. Some globes show great-circle routes between distant ports on the Pacific, Atlantic, or Indian oceans. These may readily be checked by the stretched piece of string.

Meridians and parallels

The rotational motion of the earth, spinning on its axis, provides two natural points—the poles—upon which to base the *geographic grid,* a network of intersecting lines inscribed upon the globe for purposes of fixing the location of surface features. It consists of a set of north-south lines connecting the poles—the *meridians*—and a set of east-west lines running parallel with the equator—the *parallels* (Figure 1.13).

All meridians are halves of great circles, whose ends coincide with the earth's north and south poles. Although it is true that opposite meridians taken together comprise a complete great circle, it is well to remember that a single meridian is only half of a great circle and contains 180° of arc. Additional characteristics of meridians are:

1. All meridians run in a true north-south direction.
2. Meridians are spaced farthest apart at the equator and converge to common points at the poles.
3. An infinite number of meridians may be drawn on a globe. Thus a meridian exists for any point selected on the globe. For representation on maps and globes, however, meridians are selected at suitable equal distances apart.

Parallels are entire small circles, produced by passing planes through the earth parallel to the plane of the equator. They possess the following characteristics:

1. Parallels are always parallel to one another. Although they are circular lines, they always remain equal distances apart.
2. Parallels are always true east-west lines.
3. Parallels intersect meridians at right angles. This holds true for any place on the globe, except the two poles, despite the fact that the parallels are strongly curved near the poles.
4. All parallels except the equator are small circles; the equator is a complete great circle.
5. An infinite number of parallels may be drawn on the globe. Therefore, every point on the globe, except the north or south pole, lies on a parallel.

Longitude

The location of points on the earth's surface follows a system in which lengths of arc are measured along meridians and parallels (Figure 1.14). Taking the equator as the starting line, arcs are measured north or south to the desired points. Taking a selected meridian, or *prime meridian,* as a reference line, arcs are measured eastward or westward to the desired points.

The *longitude* of a place may be defined as the arc, measured in degrees, of a parallel between the place and the prime meridian (Figure 1.14). The prime meridian is almost universally accepted as that which passes through the Royal Observatory at Greenwich, near London, England, and is often referred to as the *meridian of Greenwich.* This meridian has the value 0° longitude. The longitude of any given point on the globe is measured eastward or westward from this meridian, whichever is the shorter arc. Longitude may thus range from 0° to 180°, either east or west. It is commonly written in the following form: *long.*

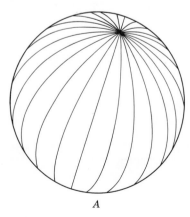

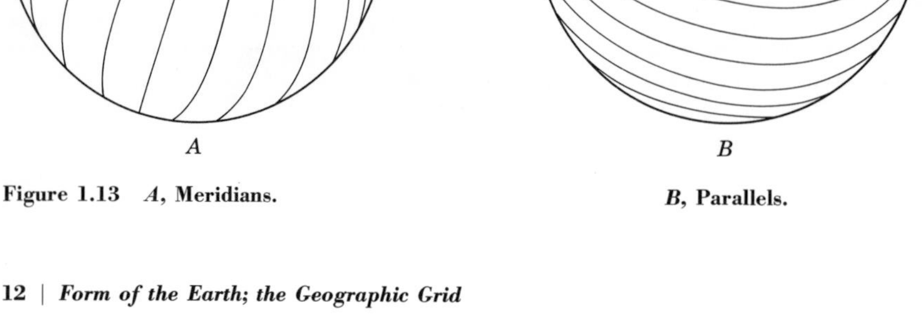

A *B*

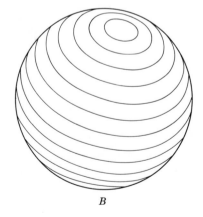

Figure 1.13 ***A*, Meridians.** ***B*, Parallels.**

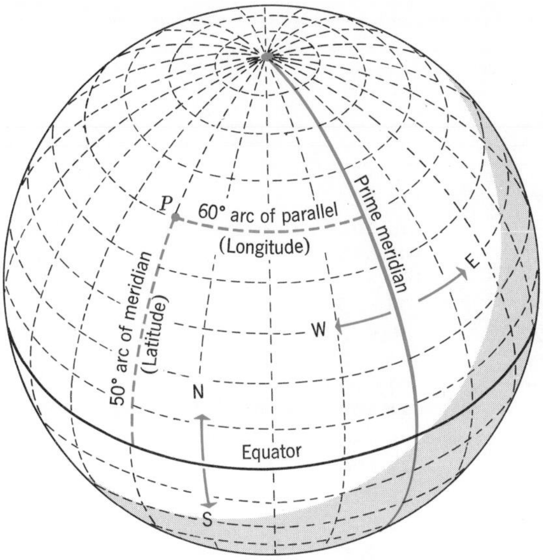

Figure 1.14 The point *P* has a latitude of 50° N, a longitude of 60° W.

77° 03′ 41″ W, which may be read "longitude 77 degrees, 3 minutes, 41 seconds west of Greenwich."

If only the longitude of a point is stated, we cannot tell its precise location because the same arc of measure applies to an entire meridian. For this reason, a meridian might be defined as a line representing all points having the same longitude. This definition explains why the expression "a meridian of longitude" is often used. Confusion may arise in the mind of the student because of the statement that longitude is measured along a parallel of latitude, but this may be clarified by the realization that, in order to measure the arc between a point and the prime meridian, it is necessary to follow eastward or westward along one of the parallels (Figure 1.14).

The actual length, in miles or kilometers, of a degree of longitude will depend upon where it is measured. At the equator this distance may be computed by dividing the earth's circumference by 360°:

$$\frac{24{,}900 \text{ mi}}{360^\circ} = 69 \text{ statute miles (approx.)}$$

$$\frac{40{,}075 \text{ km}}{360^\circ} = 111 \text{ km (approx.)}$$

The student should memorize the value of 1° of longitude at the equator as 69 mi (111 km) because many computations of map distance and scale can be made by converting degrees of longitude into distance. Other figures that hold at the equator are

$$\begin{aligned} 1' \text{ of longitude} &= 1.15 \text{ statute miles (1.85 km)} \\ &= 1 \text{ nautical mile (approx.)} \\ 1'' \text{ of longitude} &= 0.019 \text{ statute mile (0.03 km)} \\ &\quad \text{or about 100 ft (30 m).} \end{aligned}$$

Because of the rapid convergence of the meridians northward or southward, care should be taken not to employ these equivalents inadvertently except close to the equator. It is a further useful item of knowledge that the length of 1° of longitude is reduced to about one-half as much at the 60th parallels, or about $34\frac{1}{2}$ mi ($55\frac{1}{2}$ km).

Latitude

The *latitude* of a place may be defined as the arc, measured in degrees, of a meridian between that place and the equator (Figure 1.14). Latitude may thus range from 0° at the equator to 90° north or south at the poles. The latitude of a place, written as *lat. 34° 10′ 31″ N*, may be read "latitude 34 degrees, 10 minutes, 31 seconds north." When both the latitude and longitude of a place are given, it is accurately and precisely located with respect to the geographic grid.

For almost all practical purposes, we consider the earth to be a sphere, and therefore the parallels of latitude are taken to be exactly equidistantly spaced if they are drawn on a globe for unit amounts of arc, as, for example, every 10°. The length of a degree of latitude is almost the same as the length of 1° of longitude at the equator, slightly over 69 mi (111 km), and so that figure may be used for ordinary purposes.

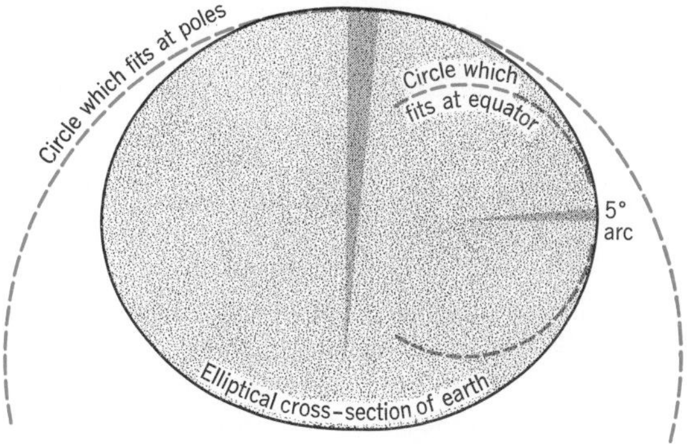

Figure 1.15 Because the earth is an ellipse in cross section, the length of a degree of latitude is very slightly greater at the poles than at the equator.

TABLE 1.3*

Latitude (Degrees)	Length of 1° of Latitude: Statute Miles	Length of 1° of Latitude: Kilometers	Length of 1° of Longitude: Statute Miles	Length of 1° of Longitude: Kilometers
0	68.704	110.569	69.172	111.322
5	68.710	110.578	68.911	110.902
10	68.725	110.603	68.129	109.643
15	68.751	110.644	66.830	107.553
20	68.786	110.701	65.026	104.650
25	68.829	110.770	62.729	100.953
30	68.879	110.850	59.956	96.490
35	68.935	110.941	56.725	91.290
40	68.993	111.034	53.063	85.397
45	69.054	111.132	48.995	78.850
50	69.115	111.230	44.552	71.700
55	69.175	111.327	39.766	63.997
60	69.230	111.415	34.674	55.803
65	69.281	111.497	29.315	47.178
70	69.324	111.567	23.729	38.188
75	69.360	111.625	17.960	28.904
80	69.386	111.666	12.051	19.394
85	69.402	111.692	6.049	9.735
90	69.407	111.700	0.000	0.000

*Based on Clarke ellipsoid of 1866, from *U.S. Geological Survey Bulletin* 650, "Geographic Tables and Formulas," by S. S. Gannett, 1916, pp. 36–37.

To be very precise, and take into account the oblateness of the earth, it must be recognized that a degree of latitude changes slightly in length from equator to poles. Using figures of the Clarke ellipsoid of 1866, the length of 1° of latitude at the equator is 68.704 statute miles (110.569 km); at the poles it is 69.407 mi (111.700 km), or 0.7 mi (1.1 km) longer. One degree at the poles is 1 percent longer than at the equator. The difference is by no means trivial and must be taken into account in construction of large-scale maps.

The explanation of this variation in length of a degree of latitude may be had from a diagram showing how degrees of latitude are determined (Figure 1.15). Because of the earth's oblateness, the surface curvature is less strong near the poles than at the equator. That is to say, a smaller circle can be fitted to the curvature near the equator than at the poles, as shown in Figure 1.15. A single degree on the largest circle has a greater length of arc than a degree on the smallest circle. Hence the length of a single degree of latitude will be greatest near the poles and least near the equator. In order to obtain the correct values for specific latitudes, it is necessary to consult prepared tables. Table 1.3 gives the lengths of single degrees of both latitude and longitude for various latitudes.

Statute mile and nautical mile

Both marine and air navigation use the *nautical mile* as the unit of length or distance. Meteorology (weather science) of the upper atmosphere has also adopted as the unit of wind speed the mariner's *knot*, which is a velocity of one nautical mile per hour. It is therefore worthwhile for the geographer to understand the nautical mile.

On July 1, 1954, the U.S. Department of Defense adopted the *international nautical mile*, defined as exactly equivalent to 1852 international meters, or 6076.103333 ft (the digit 3 is repeated indefinitely). Therefore, dividing this value in feet by 5280, the number of feet per *statute mile*, we arrive at the equivalent: 1 international nautical mile = 1.150777 statute miles. For ordinary calculations, then, the value of 1.15 statute miles (1.85 km) per nautical mile is quite satisfactory.

At what place on the earth does the international nautical mile equal the length of one minute of arc of the earth ellipsoid? This can be computed by first multiplying 1.150777 by 60 to give 69.04663 statute miles per degree of arc. Next, consulting Table 1.3, this figure is seen to be very close to the length of one degree of latitude at 45°, which is given as 69.054 mi according to the Clarke ellipsoid of reference. Furthermore, if all

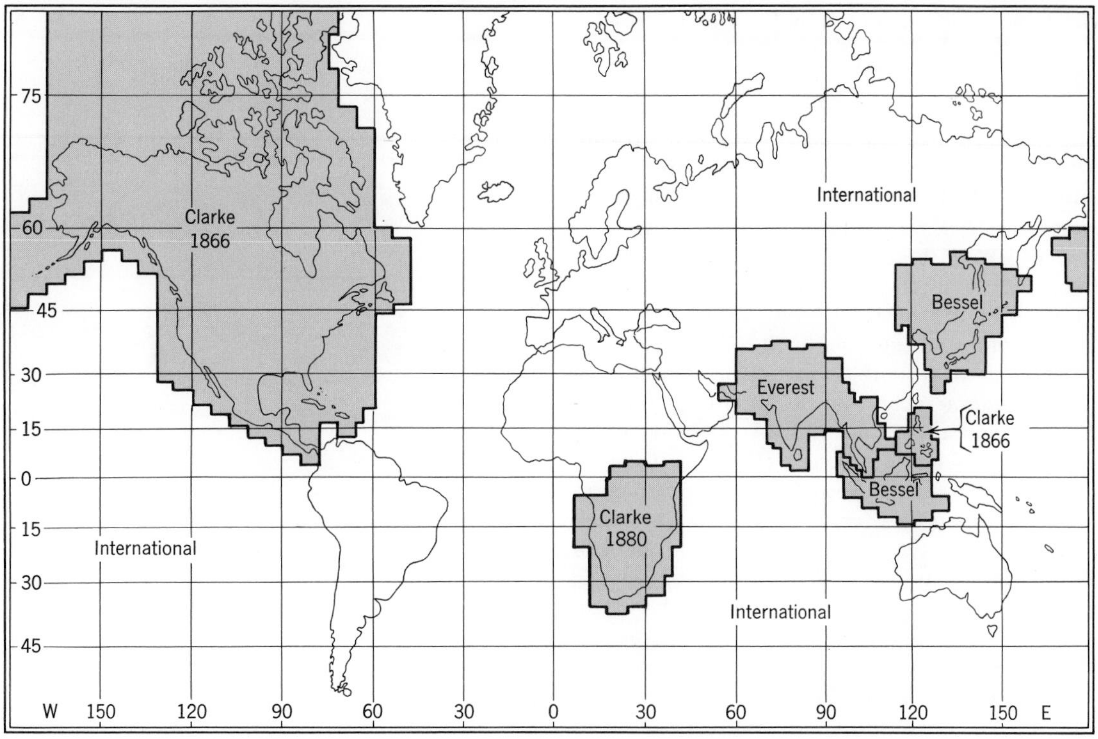

Figure 1.16 World regions are assigned to each of five ellipsoids of reference. (After Department of the Army, TM 5-241.)

of the values of the second column of Table 1.3 are added and the average value computed, it will be found to be 69.055. This leads to the conclusion that the international nautical mile very closely approximates the average length of one minute of latitude, or that it is the $\frac{1}{5400}$ part of the length of a meridian between equator and pole.

Earth ellipsoids

Preparation of maps of the earth's surface requires precise plotting of a network of meridians and parallels that form the framework upon which terrain details are inscribed. Exact lengths of degrees of latitude and longitude can be stated only after the dimensions of the earth ellipsoid are agreed upon. Unfortunately, a single set of dimensions has not been used over the entire earth. Instead, five sets of ellipsoid dimensions have been widely used: (1) the *International ellipsoid*, whose values were computed by J. F. Hayford of the U.S. Coast and Geodetic Survey in 1909 and adopted by the International Geodetic and Geophysical Union in 1924; (2) the *Clarke ellipsoid of 1866* computed by A. R. Clarke, the head of the English Ordnance Survey; (3) the *Clarke ellipsoid of 1880*, a recomputation by General Clark; (4) the

TABLE 1.4*

Ellipsoid	Semimajor Axis a	Semiminor Axis b	Flattening f	Approximate Fraction
Astrogeodetic (Fischer 1960)	6,378,160	6,356,778	0.003,352	1/298
International (Hayford 1909)	6,378,388	6,356,912	0.003,367	1/297
Clarke 1866	6,378,206	6,356,584	0.003,390	1/295
Clarke 1880	6,378,301	6,356,584	0.003,408	1/293
Bessel 1841	6,377,397	6,356,079	0.003,343	1/299
Everest 1830	6,377,276	6,356,075	0.003,324	1/301

*Data from Departments of the Army and the Air Force, TM 5-241, TO 16-1-233.

Bessel ellipsoid computed in 1841 by a Prussian astronomer of that name; and (5) the *Everest ellipsoid* of 1830.

For purposes of a unified system of international military mapping, the world is divided up into areas, each assigned to one of the above five ellipsoids (Figure 1.16). Thus, military maps of North America will be based on the Clarke ellipsoid of 1866; those of Europe on the International ellipsoid; those of central Africa on the Clarke ellipsoid of 1880; those of India on the Everest ellipsoid, and so forth. The reason for assignment of regions to particular ellipsoids is that precise surveying and mapping went forward for many decades under individual governments according to selected reference ellipsoids. To make use of existing map information, it is practical to accept the ellipsoids for those regions for which mapping has been completed and to establish boundaries (heavy lines in Figure 1.16) for the extension of those areas so as to cover the earth.

In order that the student may compare the five ellipsoids, dimensions are given in Table 1.4. The unit of length used in this table is the *international meter*, equal to 1.093611 American yards. The *semimajor axis* of the ellipsoid, designated by the letter a, is the radius of the equator circle (Figure 1.17). The *semiminor axis*, designated by the letter b, is exactly one-half the length of the polar axis.

The oblateness, or flattening of the poles, designated by the letter f, is defined as $f = (a - b)/a$.

In this table is included the *Astrogeodetic ellipsoid* of 1960, whose dimensions were computed from earth-satellite data and all other available geodetic data.

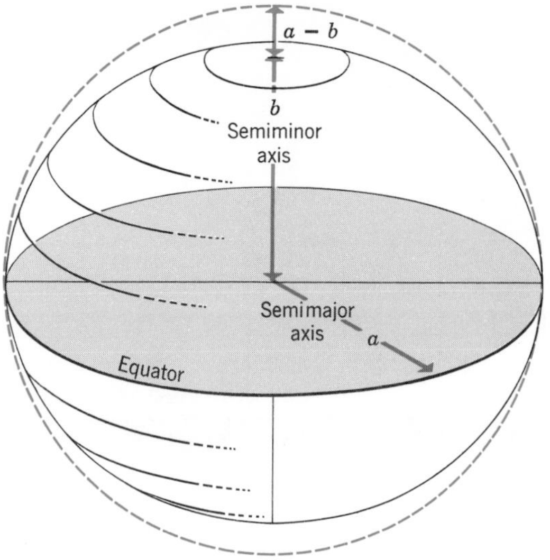

Figure 1.17 Figures of Table 1.4 give dimensions of semimajor and semiminor axes of the ellipsoid as defined here. (After Department of the Army, TM 5-241.)

The figures of Table 1.4 may seem unnecessarily elaborate to the average person, who can make no possible use of such trivial differences in the dimensions of the earth ellipsoid. Nevertheless, these data show something of the degree of precision practiced in geodesy and needed in many scientific applications. The six ellipsoids differ in semimajor axis by as much as 1100 m, which is almost two-thirds of a mile; the semiminor axes by about 850 m, or half a mile.

Review Questions

1. Describe at least six proofs of the earths' approximate sphericity.

2. Explain how Eratosthenes estimated the earth's circumference.

3. How may the amount of the earth's surface curvature be expressed? What effect has the decreasing density of the earth's atmosphere on a horizontal line of sight?

4. Explain how a knowledge of earth's surface curvature can be applied to problems of visibility.

5. How did Richer discover that the earth is an oblate ellipsoid, instead of a true sphere? How and when was this inference confirmed by surveying methods?

6. What geometrical form has a cross section of the earth cutting through the poles? What is the oblateness of the earth? What fraction approximately expresses the oblateness?

7. What is the geoid? How does the surface of the geoid depart from the ellipsoid? Why?

8. What is a great circle? How is it formed? How many great circles can be drawn upon the surface of a sphere? What is a small circle?

9. List six properties of great circles. Of what practical importance are great circles?

10. What is a meridian? How are meridians formed on a globe? List the characteristics of meridians.

11. What is a parallel? How are parallels formed? List the characteristics of parallels.

12. Define and explain longitude. How is longitude written? Give an example. What is a prime meridian? Where is the Greenwich meridian? How long is a degree of longitude, in miles, at the equator? at 60° latitude? at the poles?

13. Define and explain latitude. How is latitude written? Give an example. How long is a degree of latitude, in miles? Does it vary from equator to poles? How much? Why?

14. What is a nautical mile? a knot? At what latitude is the international nautical mile most nearly equal to the length of one minute of arc of the ellipsoid?

15. List five earth ellipsoids used in international military mapping. Why is one ellipsoid not used for the entire earth?

Exercises

1. What is the most distant point on the sea surface visible from an eye point (*a*) 49 ft above sea level, (*b*) 121 ft above sea level, and (*c*) 4900 ft above sea level?

2. A shore battery is firing shells into a floating target 25 ft high and 20 mi distant. Is any part of the target visible to an observer at the battery who is 50 ft above sea level? 400 ft above sea level?

3. At a distance of 12 mi the tips of a ship's funnels are just on the horizon as seen through a submarine periscope at sea level. How high above sea level do the ship's funnels rise?

4. (*a*) Survivors on a life raft are 18 mi from a lighthouse. Taking the light to be 100 ft above sea level, and the maximum eye level of the survivors to be 6 ft above the sea surface, can they see the rays from the lighthouse? (*b*) Calculate the distance at which light would just be visible on the horizon.

5. (*a*) Determine exactly the oblateness of the earth, using the following formula:

$$\text{Oblateness} = \frac{a - b}{a}$$

where a = equatorial diameter, 7926.68 mi; b = polar diameter, 7899.98 mi. (*b*) On a perfectly scaled globe, 10 in. in equatorial diameter, how much shorter would the polar diameter be than the equatorial diameter?

6. Using a small globe and a piece of thin string or a rubber band, make great circle courses between (*a*) Seattle and Tokyo; (*b*) New York and Liverpool; (*c*) New York and Bombay; (*d*) Colombo, Ceylon, and Buenos Aires, Argentina; and (*e*) Miami, Florida, and Capetown, South Africa. For each route, list the principal cities or geographical features lying on or very near the route.

7. Using a small globe, give as closely as possible the latitude and longitude of the following cities: New York, Capetown, Shanghai, Honolulu, London, and Rio de Janeiro.

8. What error has been made in each of the following notations? (*a*) Lat. 5° 08′ 31″ S, long. 191° 33′ 04″ W. (*b*) Lat. 89° 71′ 23″ N, long. 88° 21′ 56″ E. (*c*) Lat. 21° 43′ 59″ E, long. 177° 03′ 00″ E. (*d*) Lat. 94° 21′ 10″ N, long. 103° 42′ 51″ W. (*e*) Lat. 48° 57′ 45″ N, long. 2° 00′ 31″ N.

9. Show, by a geometric construction, that a degree of longitude at the 60th parallel is one-half as long as a degree of longitude at the equator. Label your diagrams, and attach a full explanation.

10. From how many different starting points on the globe would it be possible to travel 100 mi north, then 100 mi east (or west), then 100 mi south and be exactly at the same starting point? (The southern-hemisphere case is simple, but can you solve this problem for the northern hemisphere?)

CHAPTER 2

Map Projections

A map projection is an orderly system of parallels and meridians used as a basis for drawing a map on a flat surface. The fundamental problem is to transfer the geographic grid from its actual spherical form to a flat surface in such a way as to present the earth's surface or some part of it in the most advantageous way possible for the purposes desired.

One way to avoid the map-projection problem is to use only a globe. Unfortunately a globe has shortcomings. First, we can see only one side of a globe at a time. Second, a globe is on too small a scale for many purposes. On globes ranging from a few inches to two or three feet in diameter, only the barest essentials of geography can be shown. The few large globes in existence, those several feet in diameter, may show considerable detail, but they serve also to accentuate a third shortcoming of globes—their lack of portability. Flat maps printed on paper can be folded compactly so that many may be carried in a small pocket, whereas even the smallest globe is a cumbersome and delicate object. Ease of reproduction greatly favors maps over globes. Making a quality globe requires not only that a map be printed but also that the map be trimmed and carefully pasted onto a spherical shell.

The problem of map projection must therefore be squarely faced in an endeavor to learn what types of networks of parallels and meridians are best suited to illustration of various portions of the earth's surface. It is well to point out, however, that no map projection will ever substitute fully for a globe to show general world relations, and use of a globe is to be recommended in conjunction with flat maps.

Developable geometric surfaces

Certain geometric surfaces are said to be *developable* because by cutting along certain lines they can be made to unroll or unfold to make a flat sheet. Two such forms are the *cone* and the *cylinder* (Figure 2.1). Were the earth conical or cylindrical, the map-projection problem would be solved once and for all by using the developed surface. No distortion of surface shapes or areas would occur, although it is true that the surface would be cut apart along certain lines. The earth belongs to a group of geometric forms said to be *undevelopable,* because, no matter how they are cut, they cannot be unrolled or unfolded to lie flat. It is possible to draw a true straight line in one or more directions on the surface of a developable solid, but nowhere can this be done on an undevelopable form such as a spherical surface. In order to make the parts of a spherical surface lie perfectly flat, it must be stretched—more in some places than in others. Thus, it is impossible to make a perfect map projection.

When a map is made of a very small part of the earth's surface—for example, an area four miles across—the map-projection problem can be ignored. If the meridians and parallels are drawn as straight lines, intersecting at right angles and correctly spaced apart, the actual error present is probably so small as to fall within the width of the lines drawn and is not worth correcting. As the area included on the map is increased, however, the problem gains in importance. When an attempt is made to show the whole globe, very serious trouble develops. Only by some compromise can the distortion be reduced to a reasonable degree over important parts of the earth's surface. We should remember that the human eye cannot see the entire surface of a globe at once, and the marginal section within view is greatly foreshortened. Thus, map projections actually improve our ability to perceive the earth's surface.

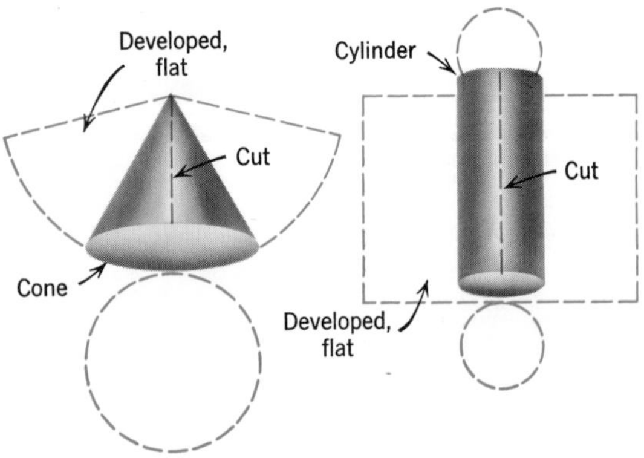

Figure 2.1 The cone and cylinder are developable geometric forms.

Although, for purposes of simplicity, this chapter treats projections of a spherical globe upon a flat map, it is well to point out that in precise plotting of a map projection the earth ellipsoid is the geometrical form that is actually used. But because the earth's oblateness is slight, the earth can be assumed to be a true sphere for an elementary and descriptive study of map projections.

Map scale

All globes and maps depict the earth's features in much smaller size than the true features that they attempt to represent. Globes are intended, in principle, to be perfect models of the earth itself, differing from the earth only in size, but not in shape. The *scale* of a globe is therefore the ratio between the size of the globe and the size of the earth, where size is expressed by some measure of length or distance (but not area or volume). Take, for example, a globe 10 inches in diameter representing the earth, whose diameter is about 8000 miles. The scale of the globe is therefore the ratio between 10 inches and 8000 miles. Dividing both figures by ten, this reduces to a scale stated as *one inch represents 800 miles,* a relationship that holds true for distances between any two points on the globe. (Stated in metric units, the scale of this globe would be the ratio of 25 cm to 12,900 km, or *1 centimeter represents 516 kilometers.*)

Scale is more usefully stated as a simple fraction, termed the *fractional scale,* or *representative fraction* (*R.F.*), which can be obtained by reducing both map and globe distances to the same unit of measure, thus:

$$\frac{1 \text{ in. on globe}}{800 \text{ mi on earth}} = \frac{1 \text{ in.}}{800 \times 63{,}360 \text{ in. (per mile)}}$$

$$= \frac{1 \text{ in.}}{50{,}688{,}000 \text{ in.}} = \frac{1}{50{,}688{,}000}$$

This fraction may be written as 1:50,688,000 for convenience in printing. The advantage of the representative fraction is that it is entirely free of any specified units of measure, such as the foot, mile, meter, or kilometer. Persons of any nationality understand the fraction, regardless of the language or units of measure used in their nation, provided only that they use arabic numerals.

A globe is a *true-scale model* of the earth, in that the representative fraction applies to any distances on the globe, regardless of the latitude or longitude, and regardless of the compass direction of the line whose distance is being considered. This is to say that the scale remains constant over the entire globe. Map projections, however, cannot have the uniform-scale property of a globe, no matter how cleverly devised. In flattening the curved surface of the sphere to conform to a flat plane, all map projections stretch the earth's surface in a nonuniform manner, so that the fractional scale changes from place to place. Thus, we cannot say about a map of the world that "the scale of this map is 1:50,000,000," because the statement is false for any form of projection.

It is quite possible, however, to have the fractional scale of a flat map remain true, or constant, in certain specified directions. For example, one type of projection preserves constant scale along all parallels, but not along the meridians. This condition is illustrated in Figure 2.2*B*, which is a part of the network of the polyconic projection shown complete in Figure 2.17. Another type of projection keeps scale constant along all meridians, but not along parallels. This is shown in Figure 2.2*C*, which is part of the polar position of azimuthal equidistant projection shown in Figure 2.12. Still other projections have changing scale along both meridians and parallels, as illustrated in Figure 2.2*D*, which is part of the gnomonic projection shown in Figure 2.11.

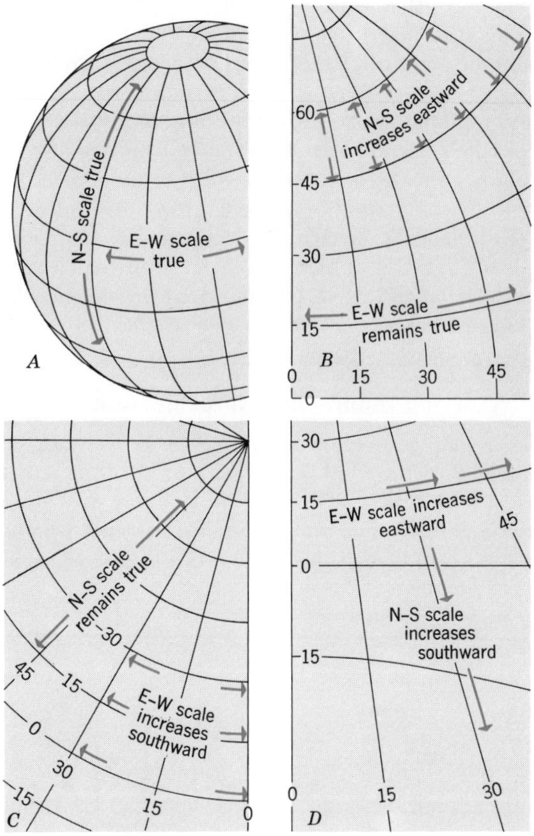

Figure 2.2 Although scale is true in all directions on a globe, *A*, scale changes must occur on all map projections. *B*, Scale is true along all parallels but not along all meridians. *C*, Scale is true along all meridians but not along all parallels. *D*, Scale changes along both parallels and meridians.

Preserving areas on map projections

Because a globe is a true-scale model of the earth, given areas of the earth's surface are shown to correct relative scale everywhere over its surface. The scale of distance is constant in all compass directions. If we should take a small wire ring, say one inch in diameter, and place it anywhere on the surface of the ten-inch globe, the area enclosed will represent an equal amount of area of the earth's surface. But a similar procedure would not enclose constant areas on all parts of most map projections, only on those having the special property of being *equal-area* projections.

At this point a good question arises. If, as stated above, no projection preserves a true, or constant, scale of distances in all directions over the projection, how can circles of equal diameter placed on the map enclose equal amounts of earth area? The answer is suggested in Figure 2.3. The square, one mile on a side, encloses one square mile between two meridians and two parallels. The square can be deformed into rectangles of different shapes, but if the dimensions are changed in an inverse manner, each will still enclose one square mile. The scale has been changed in one direction to compensate for change in another in just the right way to preserve equal areas of map between corresponding parts of intersecting meridians and parallels. Hence, any small square or circle moved about over the map surface will enclose a piece of the map representing a constant quantity of area of the earth's surface. Projections shown in Figures 2.22, 2.23, 2.25, and 2.26 have the equal-area property, but it is also obvious that these networks have had distortions of shape, particularly near the outer edges of the map.

Preserving shapes on map projections

A map projection is said to be *conformal*, or *orthomorphic*, when any small piece of the earth's surface has the same shape on the map as it does on a globe. Thus, the appearance of small islands or countries is faithfully preserved by a conformal map. One characteristic of a conformal projection is that parallels and meridians cross each other at right angles everywhere on the map, just as they do on the globe. However, not all projections whose parallels and meridians cross at right angles are conformal.

Projections shown in Figures 2.10, 2.16, 2.18, and 2.20 are true conformal networks. Examine the intersections of parallels and meridians on each. They will be found to have right-angle intersections in all cases. However, the same statement can be made about the projections shown in Figures 2.14 and 2.15, neither of which is truly conformal.

Another way of saying that parallels and meridians intersect at right angles is that *shearing* of areas does not occur. Figure 2.4 illustrates the meaning of shearing. For projections consisting of straight parallels and meridians, shearing gives parallelograms formed of acute and obtuse angles. For projections with curved meridians and parallels, straight lines are drawn tangent to the curves at the point of intersection. If these tangent lines cross at right angles, the projection is not sheared; but if the tangents form obtuse and acute angles, shearing is present. Conformal maps are not sheared, but not all maps without shearing are conformal. A conformal map cannot have equal-area properties besides, so that some areas are greatly enlarged at the expense of others. Generally speaking, areas near the margin of a conformal map have a much larger scale than central ones.

In all conformal projections the principle of construction prevents coverage of the entire globe within a single orientation of the projection. Instead, as totality of global coverage is approached, dimensions of the map sheet needed to show totality approach infinity.

Whether a conformal or equal-area projection is to be selected depends on what is to be shown.

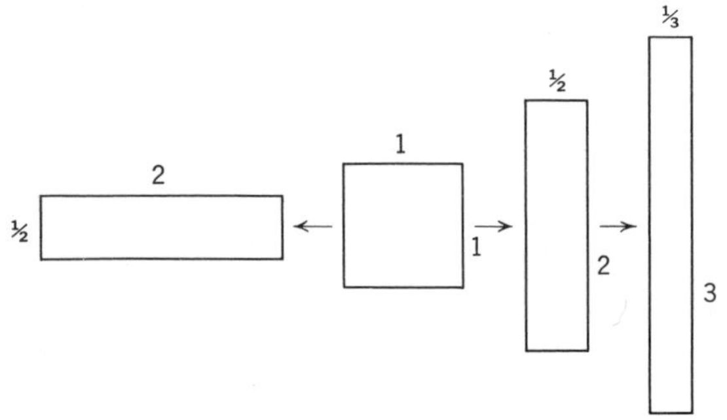

Figure 2.3 Areas can be preserved even though scales and shapes change radically.

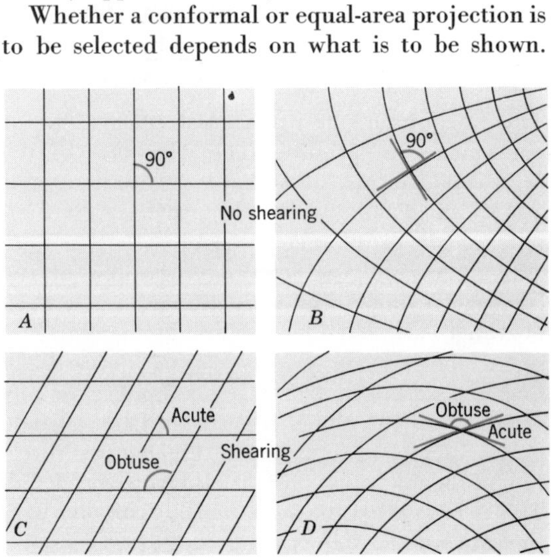

Figure 2.4 Shearing of areas is a defect of many map projections.

Where the areal distribution of something, such as grain crops, or forest-covered lands, is to be shown, an equal-area projection is needed. For most general purposes, a conformal type is preferable because physical features most nearly resemble their true shapes on the globe. Many map projections are neither perfectly conformal nor equal area, but represent a compromise between the two. This compromise may be desired either to achieve a map of more all-around usefulness or because the projection has some other very special property that makes its use essential for certain purposes.

Classification of map projections

Map projections may be classified according to the following groups: (1) *zenithal* (*azimuthal*), (2) *conic*, (3) *cylindric*, and (4) *individual*, or unique types.

The *zenithal*, or *azimuthal*, group of projections includes all types that are centered about a point and have a radial, or wheel-like, symmetry. Some zenithal projections can actually be demonstrated in the laboratory by the following method (see Figure 2.5). A wire replica of the earth, in which the wires represent parallels and meridians, is used. A tiny light source, such as a flashlight bulb or an arc light, is placed at the center of the wire globe (or at any one of several prescribed positions). In a darkened room, the shadow of the wire globe is cast upon a screen or upon the wall or ceiling. This shadow is a true geometric projection. All projections made with this apparatus are of the zenithal type, which are characterized by the following properties (Figure 2.6).

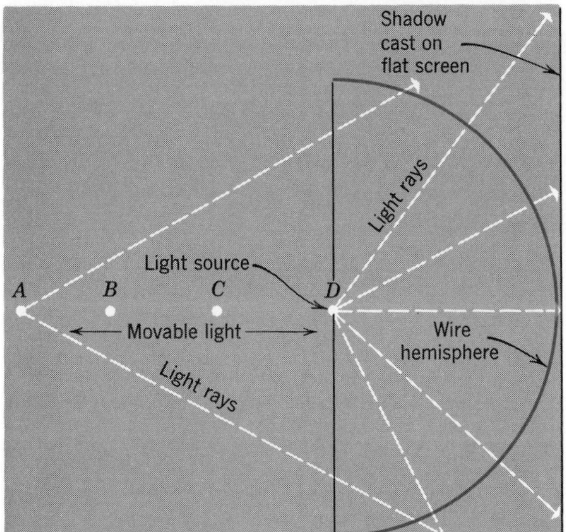

Figure 2.5 Certain zenithal projections may be made by using a light, a wire hemisphere, and a screen.

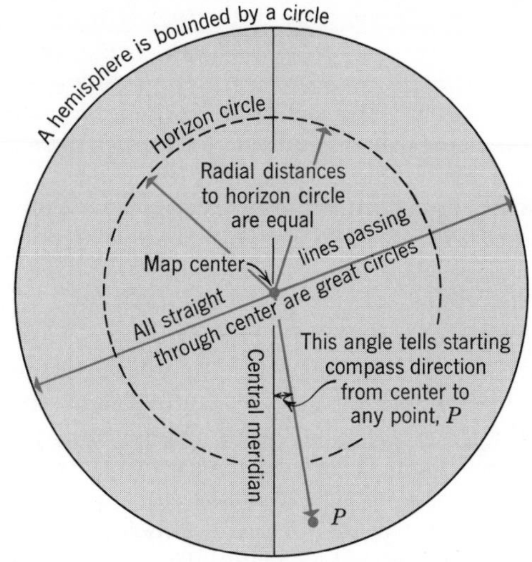

Figure 2.6 Zenithal projections have special properties that hold true regardless of where the map is centered on the globe.

1. A line drawn from center point of the map to any other point gives the true compass direction taken by a great circle as it leaves the center point, headed for the outer point. This direction, or "azimuth," may be measured with respect to the central meridian of the projection. Continual readjustment of course with respect to geographic north will be needed along the route, unless it coincides with a meridian or the equator.

2. When a complete globe or hemisphere is shown, the map is circular in outline. Inasmuch as any map can be trimmed down to have a circular outline, this feature is not a reliable criterion of the azimuthal class.

3. The map possesses a center point around which all its properties are grouped. All changes of scale and distortion of shapes occur uniformly (concentrically) outward from this center.

4. All points equidistant from the center lie on a circle, known as the *horizon circle*. When the entire globe is shown by a zenithal map, the circular edge of the map represents the opposite, or *antipodal*, point on the globe. When a hemisphere is shown, the outer edge of the map represents a great circle, everywhere equidistant from the point on which the projection is centered.

5. All great circles that pass through the center point of the projection appear as straight lines on the map. Likewise, all straight lines drawn through the center point of the map are true great circles.

Zenithal projections appear in three positions, or orientations: (1) *polar*, (2) *equatorial*, and (3) *oblique* or *tilted*, illustrated in Figure 2.9 to Figure 2.13. In the polar position, the center of the projection coincides with the north or south pole; in the equatorial position, the center is somewhere on the equator; in the oblique position, the center is at any desired point intermediate between the equator and poles. Although the equatorial and oblique types may not seem to be radially symmetrical, they nevertheless possess, just as truly as the polar type, the five characteristics described above.

The *conic* group of projections is based on the principle of transferring the geographic grid from a globe to a cone, then developing the cone to a flat map. This principle, too, can be demonstrated in the laboratory with the wire globe and a point source of light (Figure 2.7). Instead of a vertical flat screen, however, a translucent cone of stiff paper is seated on the wire globe, much as a lampshade is seated on a lamp. The shadow of the wires cast upon the conical shade gives a conic projection. If this shadow were traced in pencil or ink and the cone unrolled, a true conic projection would result. Simple conic projections possess the following features (Figures 2.14 and 2.15). All meridians are straight lines, converging to a common point at the north (or south) pole. All parallels are arcs of concentric circles, whose common center lies at the north (or south) pole. A complete conic projection is a sector of a circle, never a complete circle. A conic projection cannot show the whole globe and usually shows little more than the northern (or southern) hemisphere.

Cylindric projections are based on the principle of transferring the geographic grid first onto a cylinder wrapped about the earth, then unrolling the cylinder to make a flat map (Figure 2.8). Simple cylindric projections are easy to draw because they consist of intersecting horizontal and vertical lines (see Figure 2.18). The completed map is rectangular in outline, and the whole circumference of the globe can be shown. When the cylinder is tangent to the equator, meridians are equally spaced vertical lines. Parallels are spaced in various ways, according to the particular projection desired.

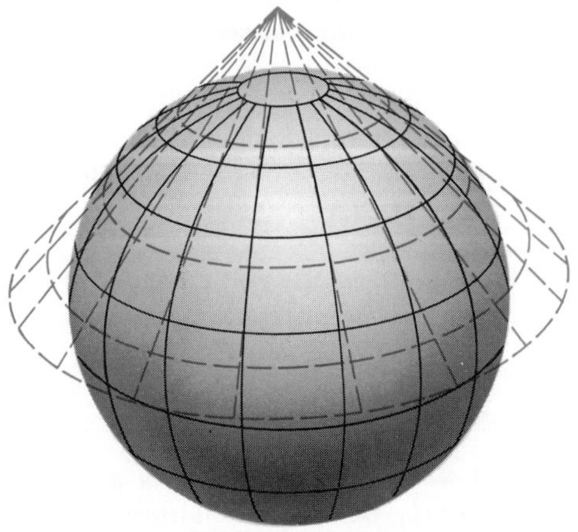

Figure 2.7 Conic projections use the principle of a cone resting on a sphere.

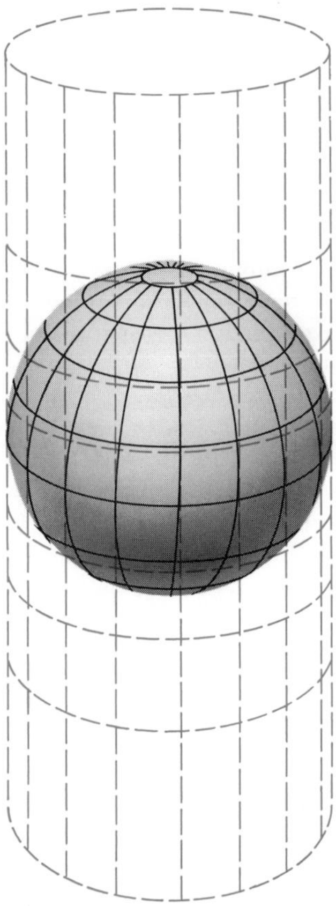

Figure 2.8 Cylindric projections use the principle of a cylinder wrapped around the globe.

Many other kinds of map projections exist, each based upon some unique principle.

Most of the particular types of projections selected for illustration here are important or useful ones, but not all the important map projections in general use are included in the list. Those selected illustrate the principles and classes already explained.

Zenithal (azimuthal) class

1. *Orthographic projection.* The *orthographic* projection employs a principle of construction illustrated in Figure 2.9. Parallel rays, or lines, are used to project the geographic grid of one hemisphere on a tangent plane. It can also be imagined as resulting when the shadow of a hemispherical wire globe is cast on a screen by light rays coming from a very distant source, such as the sun. In the polar projection, the parallels of latitude crowd close together near the outer margin. This form serves to distinguish it from other

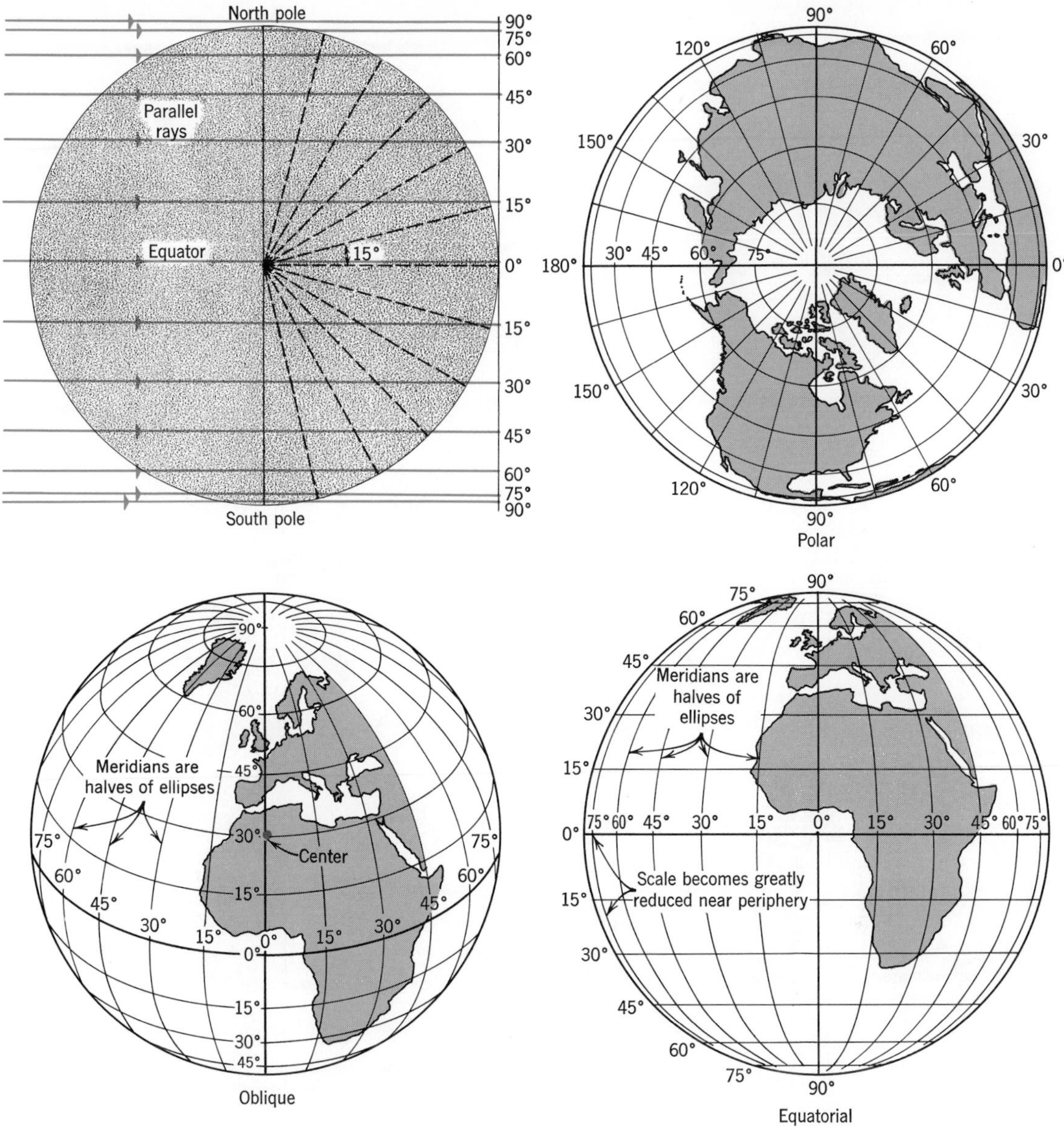

Figure 2.9 The orthographic projection gives the effect of viewing a globe from different vantage points.

polar zenithal projections. In the equatorial projection, the meridians are parts of true ellipses and show close crowding near the outer margin, whereas the parallels are straight, horizontal lines, spaced more closely near the poles. No other zenithal projection in the equatorial position has straight, horizontal parallels. In the oblique position, the closer crowding of meridians and parallels near the outer margin is also noticable.

The largest possible portion of a globe that can be shown on the orthographic projection is one hemisphere. The projection is neither equal-area nor conformal. The scale of miles is much larger near the center than near the outer edges. Use of this projection is quite limited. It gives a visual effect of a globe in three dimensions and is very similar to a photograph taken of a globe. For this reason, it often appears to illustrate articles or books on global political or military strategic problems. It gives a true picture of relations between countries or continents that are located near the central point of the projection. A little shading added to the map accentuates the perspective effect.

2. *Stereographic projection.* In the *stereographic* projection the point from which construction lines, or rays, emanate is located on the globe at a point diametrically opposite to the point

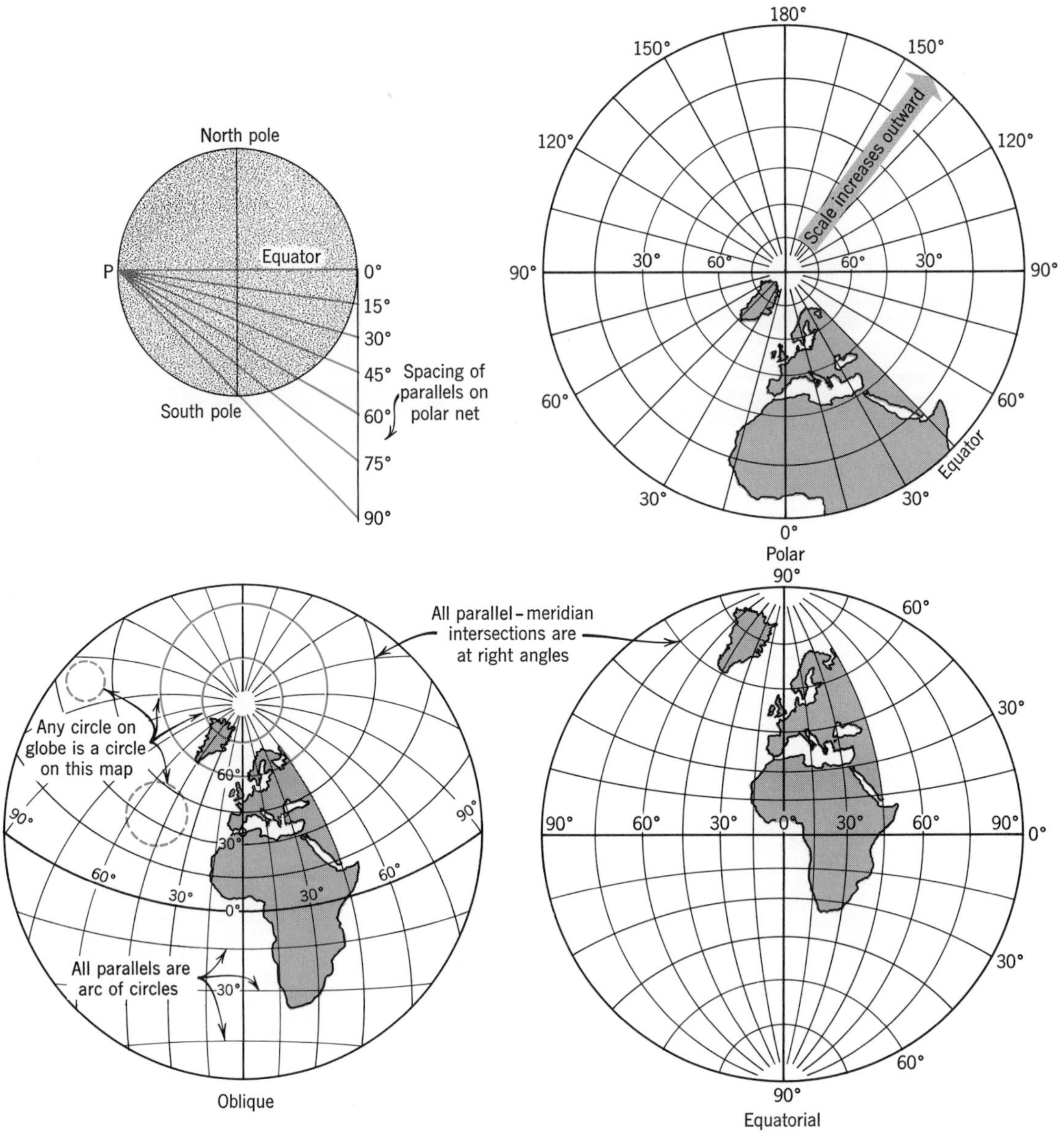

Figure 2.10 **The stereographic projection is the only conformal zenithal projection.**

where the tangent plane touches the globe. Whereas the orthographic projection gives a map of exactly the same diameter as the original globe used, the stereographic projection gives a much larger map than the original globe. Furthermore, the stereographic net can show much more than one hemisphere, although it cannot show the whole globe. This is evident from inspection of the construction diagram shown in Figure 2.10. The principal distinguishing characteristic of this projection is evident in all three of its positions; parallels and meridians show close spacing near the map center and increasingly wider spacing toward the outer margins. On any stereographic projection, the parallels and meridians are either straight lines or arcs of circles. No other kinds of curved lines occur. The reason for this is that the stereographic projection is truly conformal. All lines that are circles on the globe are shown as circles on the map. The scale, however, grows greatly from the map center toward the periphery.

With the enormous growth in importance of polar regions in the age of long-range missile and aircraft operation, the polar sterographic projection has assumed great imporatnce. It forms the base on which the Universal Polar Stereographic Military Grid System is constructed for latitudes between 80° and the poles (Chapter 3).

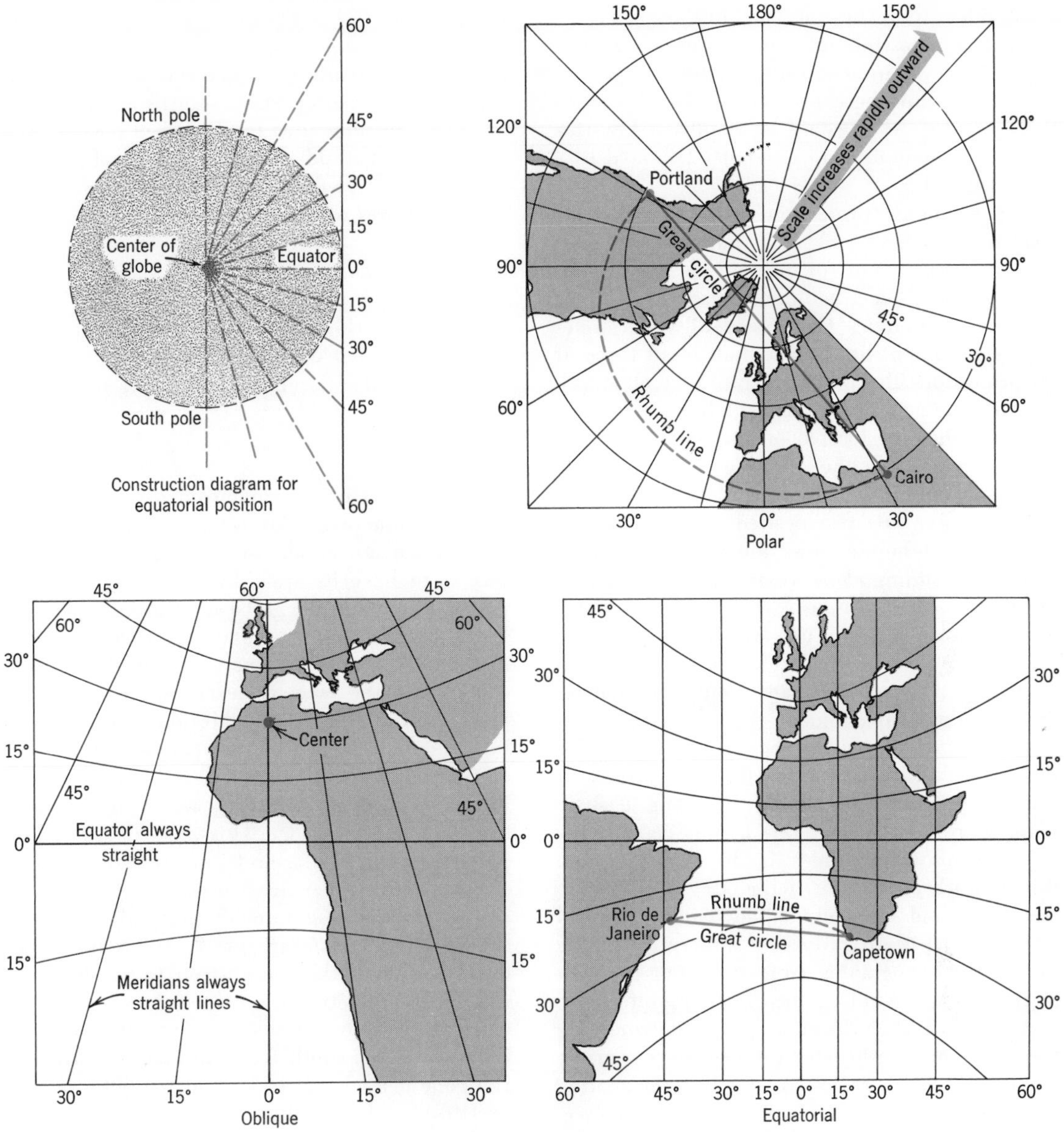

Figure 2.11 The gnomonic projection shows all great circles as straight lines, regardless of where they lie.

World Aeronautical Charts issued by the U.S. Coast and Geodetic Survey on scale of 1 : 1,000,000 are based on a polar stereographic projection for latitudes 80° to 90°. The U.S. National Weather Service daily weather map is printed on a polar stereographic projection.

The stereographic projection also has great importance in certain branches of science. Because of its perfect conformal properties, it is used in the study of mineral crystals, where the crystal faces can be plotted on a blank stereographic net. In solving certain problems of descriptive geometry, in which the intersections of various planes and lines must be determined, this net is valuable.

3. ***Gnomonic projection (great-circle sailing chart).*** The *gnomonic* projection is made by drawing rays from a point at the center of the globe, as illustrated in Figure 2.11. The resulting network is easily distinguished because the spacing of meridians and parallels increases enormously outward from the map center and results in great distortion of shapes of land areas in the outer part of the map. Although the stereographic net also increases in scale outward from the center, the increase is much less and the true shapes of small land areas are preserved. The gnomonic projection comes out with a vastly greater size than the original construction globe. For geometrical reasons evident from the construction lines in Figure 2.11, it is impossible to show a complete hemisphere. It is even impractical to include the greater part of a hemisphere because of the enormous sheet of paper that would be required. For this reason, a gnomonic map is usually trimmed to a rectangular shape.

The gnomonic projection, with its grotesque distortions of both scale and shapes, would find little use were it not for one unique and important property. On a gnomonic map, all straight lines are great circles. Conversely, all great circles appear as straight lines. Note that on all three of the projections illustrated in Figure 2.11, all meridians and the equator are straight lines, regardless of where they are located on the map. For navigational purposes, the plotting of great-circle courses is accomplished by merely connecting with a straight line any two desired points. For this reason, the gnomonic projection goes by the name *great-circle sailing chart* when adapted to navigational uses. Illustrations of great-circle routes plotted on a gnomonic net are shown in Figure 2.11 on the polar and equatorial maps. (Refer to Figure 2.18 to see these same routes plotted on the Mercator projection.)

4. *Azimuthal equidistant projection.* The *azimuthal equidistant* projection (Figure 2.12) cannot be produced optically by using a single point source for lines or rays as was possible for the previous three kinds. As the name implies, this network is made by deliberately spacing the meridians and parallels equidistantly outward from the map center. Moreover, there is nothing to stop the cartographer from extending the map to include the whole globe. The opposite pole, or *antipode*, is then shown as a circle surrounding the map. Thus constructed the map scale remains constant along all radial straight lines emanating from the map center. This gives the map a specialized use for air navigation. When centered on a particular city or airport, great-circle routes can easily be laid off and measured by drawing a line from the central point to any desired point on the map and finding the distance on a graphic scale having equally spaced units. Moreover, the correct compass direction for starting the flight can be measured from the map as the angle between the plotted route and the central meridian of the map.

The azimuthal equidistant net is often used for small-scale hemispherical maps. The polar position is easy to construct, requiring only a compass, protractor, scale, and straightedge, and makes a pleasing map to show grouping of the world's principal land areas about the Arctic Ocean.

5. *Azimuthal equal-area projection.* This projection (Figure 2.13) was designed by J. H. Lambert in 1772 and is therefore often called the *Lambert azimuthal equal-area projection.* It is constructed according to a formula that gives it true equal-area properties, something that none of the projections thus far described possess. The geometrical method of finding spacing of parallels for the polar position is shown in Figure 2.13. Chords of arcs of a semicircle divided into equal parts are used as radii for the projection. Spacing of meridians and parallels becomes slightly closer toward the periphery of the map, but not nearly so close as in the orthographic projection.

The azimuthal equal-area projection is widely used for small-scale maps of general geographical nature.

Conic class

6. *Perspective conic projection.* The perspective conic projection is based on the principle that a cone can be placed over a globe in such a way as to have its apex directly over the north pole and to touch the globe along a single parallel. If the parallels and meridians are then projected upon the cone by drawing rays from the globe's center, and the cone is developed to a flat surface, a conic map results. Meridians are straight lines radiating from the pole; parallels are arcs of concentric circles, centered on the pole. The projection is easy to construct with compass, protractor, and straightedge. This map will always be a portion, or sector, of a circle, and never a complete circle as with the zenithal projections.

The parallel that touches the globe is called the *standard parallel* (Figure 2.14). On this parallel, the scale is the same as on the globe from which the projection was made, whereas the scale is larger everywhere else on the map and increases both north and south from the standard parallel. If a cone is used that touches at the 30th parallel of latitude, the resulting map is exactly half a circle. But, because any other parallel except the equator or pole could be chosen and a cone found to fit, the resulting map might be more or less than a half circle. A perspective conic projection with one standard parallel makes a fairly good global map, without shearing of areas, but can effectively show little more than the northern hemisphere. However, modified conic projections are greatly improved, so that actually the perspective type is rarely used.

One effective modification is made by passing a cone through two parallels of latitude (Figure 2.15). The cone thus cutting the surface of the globe is said to take the *secant* position or, simply, is a *secant cone.* The resulting map now has two standard parallels along which the scale is the same. Scale increase north and south of the standard parallels is considerably reduced. Between the

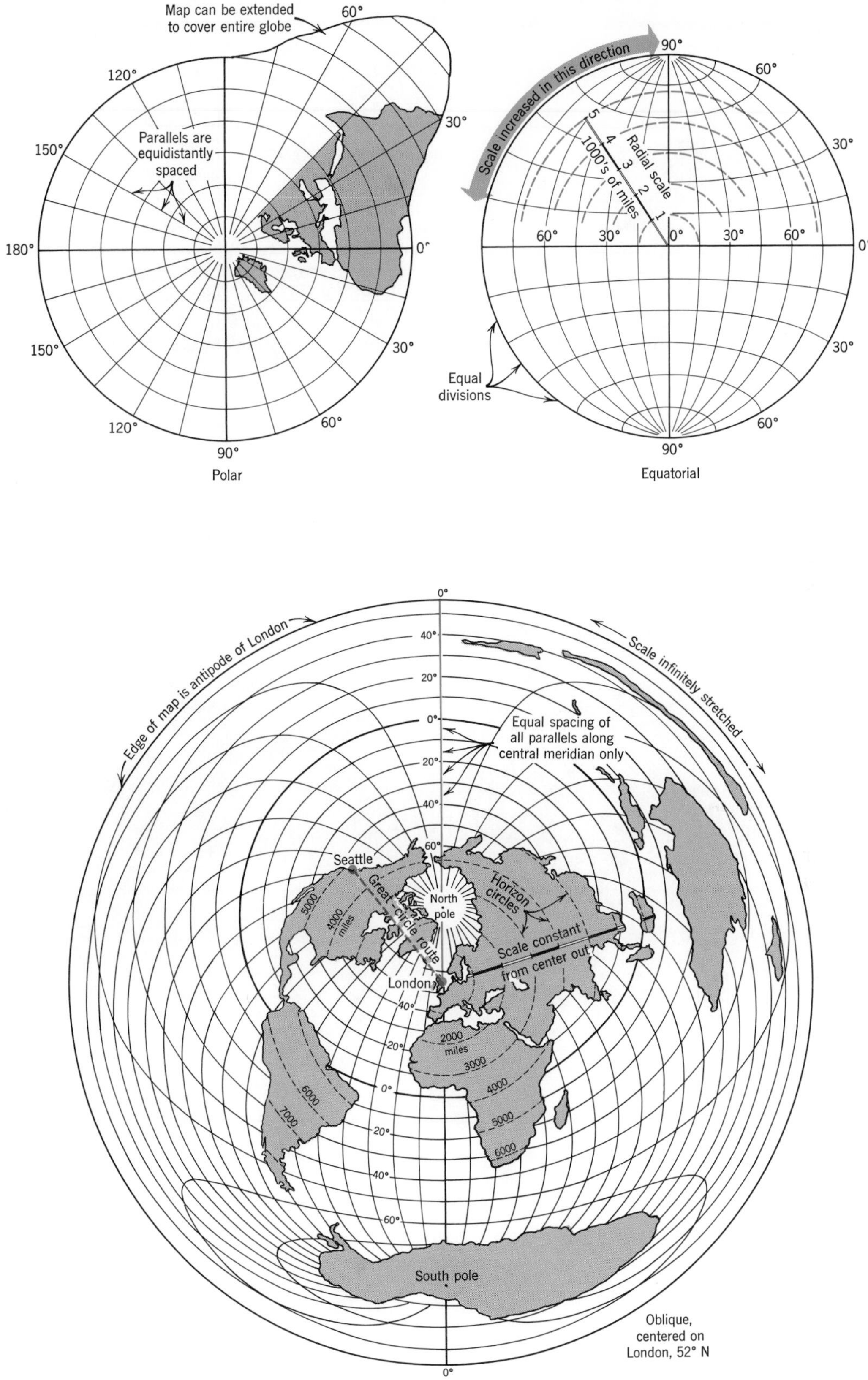

Figure 2.12 The azimuthal equidistant projection is useful in measuring distances from the center to other points.

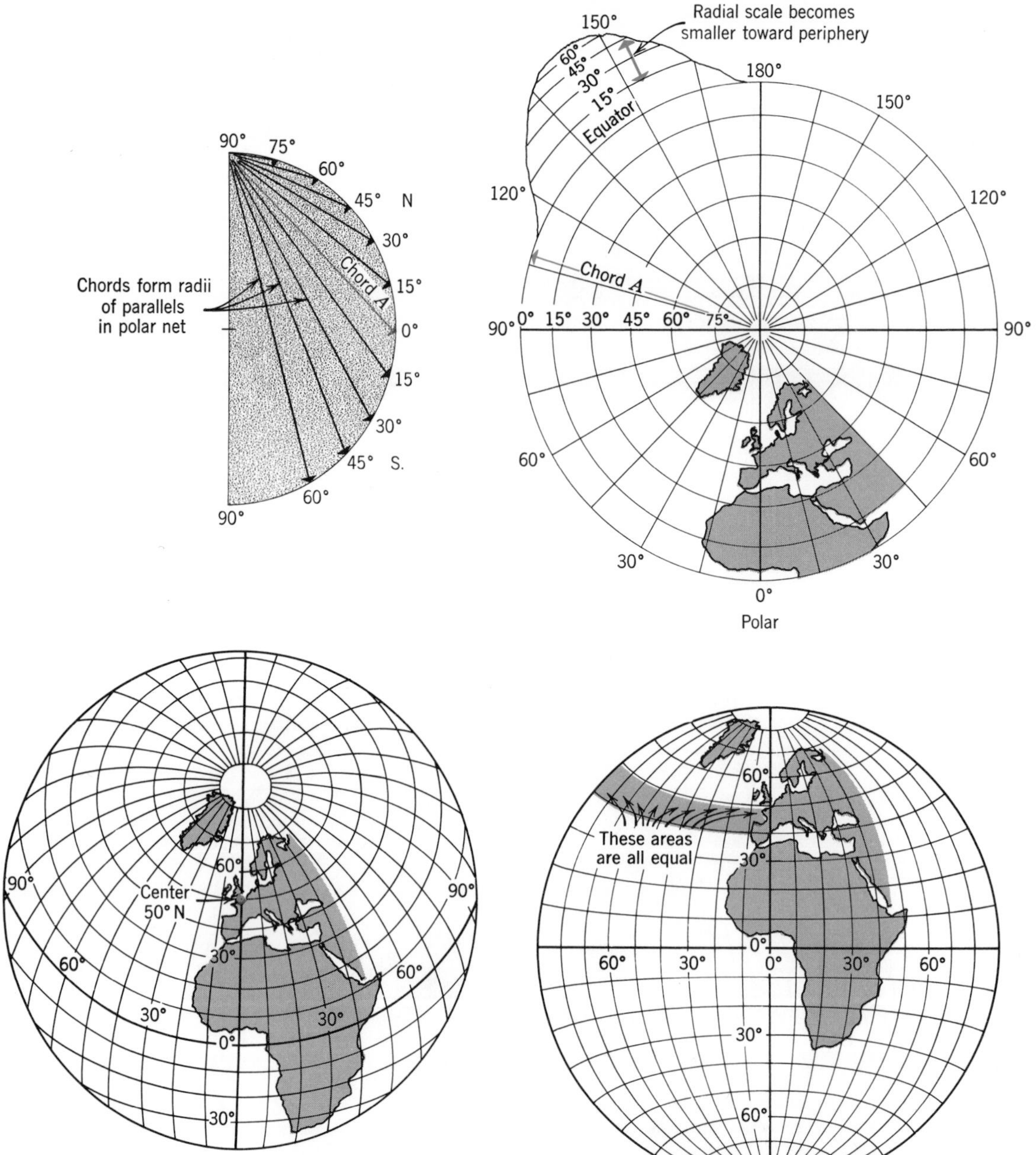

Figure 2.13 The azimuthal equal-area projection gives an excellent base for northern hemisphere maps.

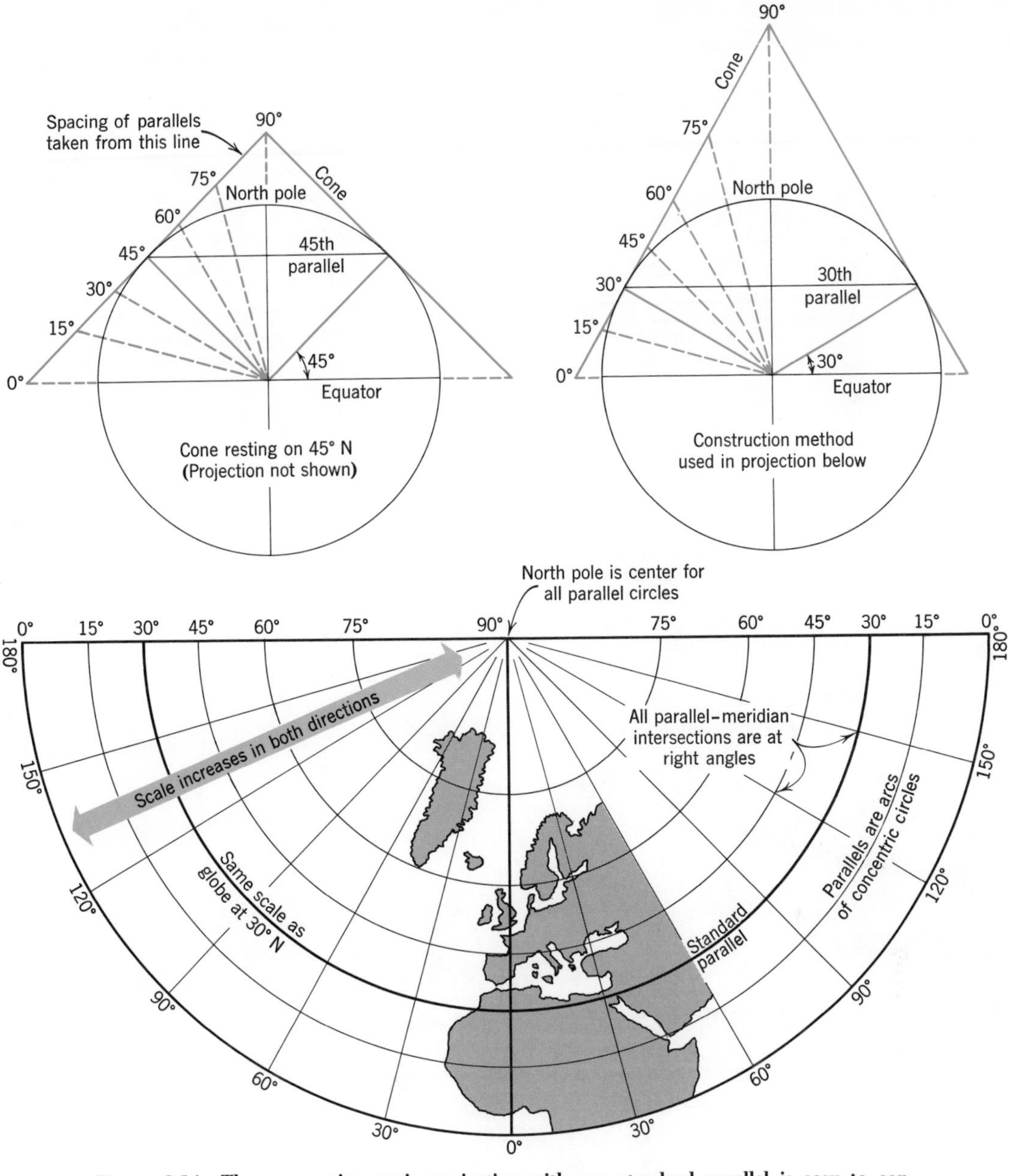

Figure 2.14 The perspective conic projection with one standard parallel is easy to construct.

standard parallels, the scale is less than on these parallels. By selecting the two parallels so as best to fit a particular continent or country in middle latitudes, a highly useful map with small errors of scale and shape results.

7. *Lambert conformal conic projection.* The perspective conic projection with two standard parallels has been further improved by adjusting the spacing of all other parallels in such a way that the map has true conformal properties. This is the *Lambert conformal conic* projection (Figure 2.16). Data for constructing the projection can be found in prepared tables. Because the meridians are straight lines converging to a common point and the parallels are arcs of concentric circles, not only is the projection easy to construct, but also, if it is used as a base of a series of large-scale maps, the individual map sheets will fit perfectly with their neighbors.

The Lambert conformal projection is a highly important type and is in widespread use. When the 33rd and 45th parallels are taken as standard parallels (Figure 2.15), the maximum scale error is about 0.5 percent for nine-tenths of the United States, and a straight line drawn on the map so closely approximates a great circle that the gno-

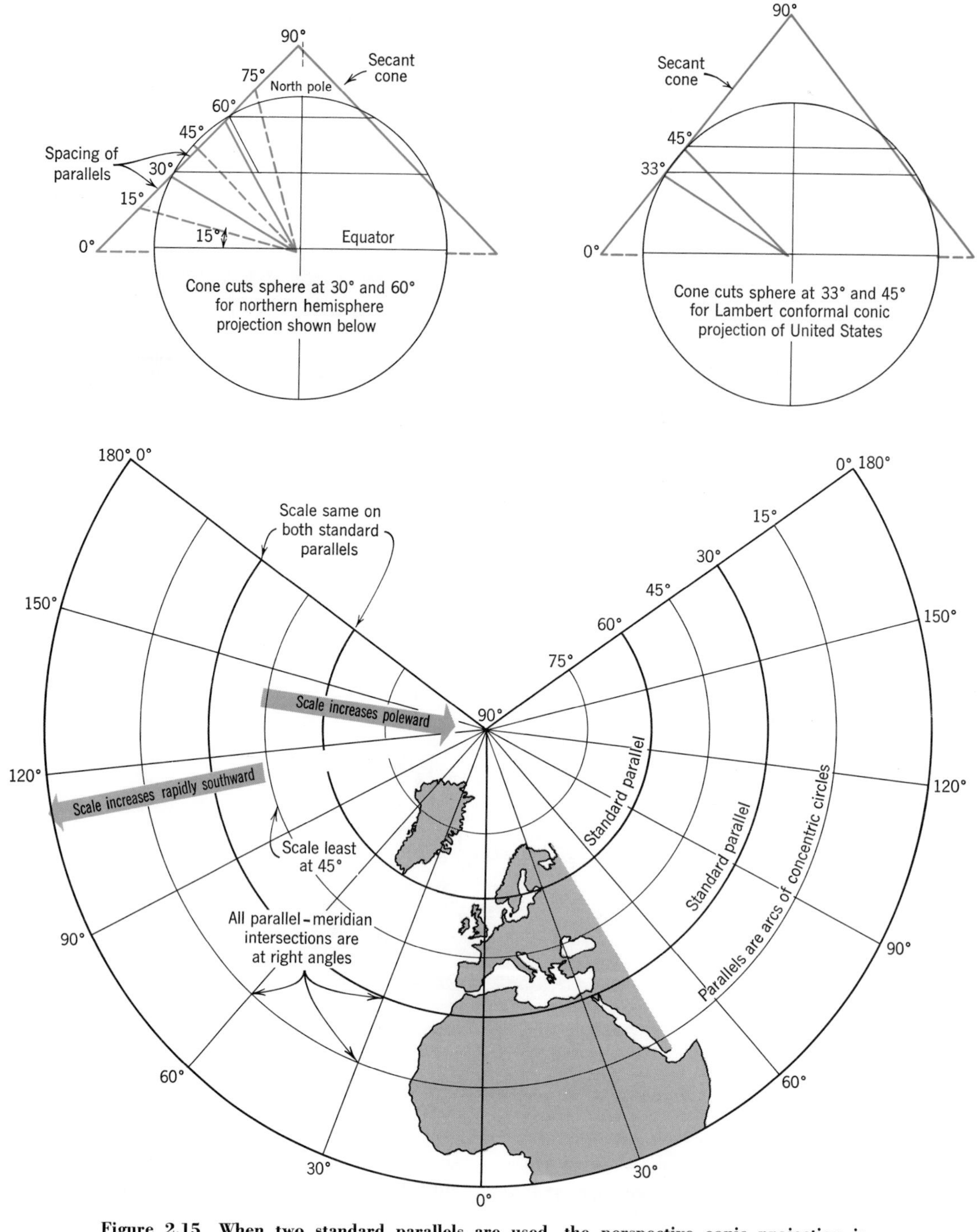

Figure 2.15 When two standard parallels are used, the perspective conic projection is improved in scale qualities.

monic chart is not needed for air navigation. The U.S. Coast and Geodetic Survey sectional aeronautical charts of the United States on a scale of 1:500,000 illustrate this application (Figure 2.16).

Another important use of the Lambert conformal conic projection is for the world aeronautical charts, 1:1,000,000 also issued by the U.S. Coast and Geodetic Survey. From the equator to 80° latitude, 20 different cones, each with its own two standard parallels, are used in belts 4° in width.

8. *Polyconic projection.* If two standard parallels are better than one, why not use three or four, or more, standard parallels? Of course, the same cone cannot pass through more than two parallels, but it is feasible to have several cones,

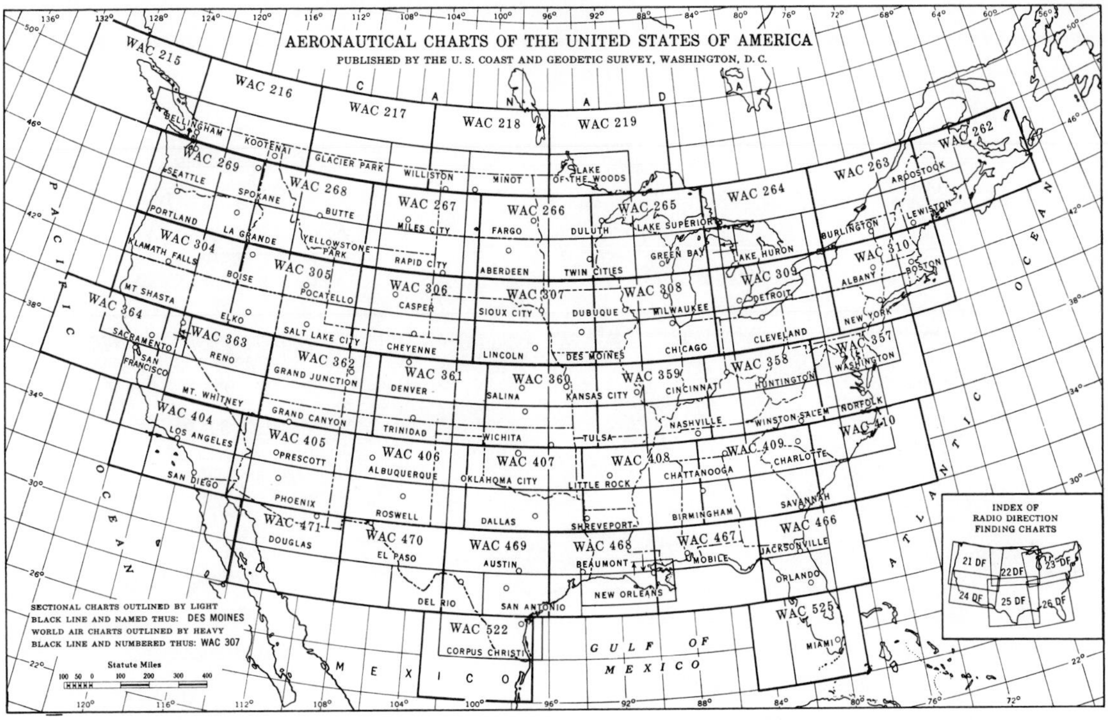

Figure 2.16 A Lambert conformal conic projection of the United States provides an excellent base for a series of large-scale maps. (From U.S. Coast and Geodetic Survey Chart 3060b.)

using only those parts that are near their respective standard parallels, thus obtaining a *polyconic* (from the Greek, meaning "many cones") projection. The flattened cone segments shown in Figure 2.17, left, are separated along their edges, and so it is necessary to stretch the map strips to achieve a continuous map. Furthermore, instead of imagining a specific number of standard parallels, an infinite number are imagined, so that true scale exists along the line of any parallel on the map. Only along the central meridian, which is a vertical straight line, is the scale the same as along all parallels. Every other meridian is a curved line along which the scales become greater toward the outer margins. Notice also that the equator is a straight line at right angles to the central meridian, and that all other parallels are arcs of circles (but not of concentric circles).

The polyconic projection is neither equal-area conformal but, near the middle of the net, distortions of scale and shape are very small. Within 560 mi (900 km) of either side of the central meridian of the map, the scale error does not exceed 1 percent. Tables are available giving data for construction of the projection.

Many sets of large-scale maps published by various agencies have been based on the polyconic projection or a slight modification of it. Ferdinand Hassler, first director of the U.S. Coast and Geodetic Survey, devised the polyconic projection in 1820, and it was formerly used by that organization for many of its maps. The U.S. Geological Survey has used the polyconic net as a base for its topographic maps and various other maps of the United States. The International Map of the World, on a scale of 1:1,000,000, uses a modified polyconic grid. One disadvantage of the polyconic projection is that meridians are curved. When individual sheets of a map series are prepared, they are usually centered on a straight central meridian passing through the map center. Hence the meridians bounding the map on both the left- and right-hand margins curve in toward the top. When adjoining map sheets are trimmed to these meridians, the sheets do not fit together perfectly.

Cylindrical projections

Direct geometrical projection of the geographic grid upon a cylindrical surface results in no useful map projection. While the principle of development of a tangent cylinder remains valid in concept, all valuable forms of cylindrical projections are actually constructed from mathematical formulas, each uniquely designed to impart a useful property to the projection.

9. *Mercator projection.* Perhaps the best known of all map projections is the *Mercator* net, devised by Gerardus Mercator in 1569 and used by him for a world map (Figure 2.18). It is based upon a mathematical formula. The

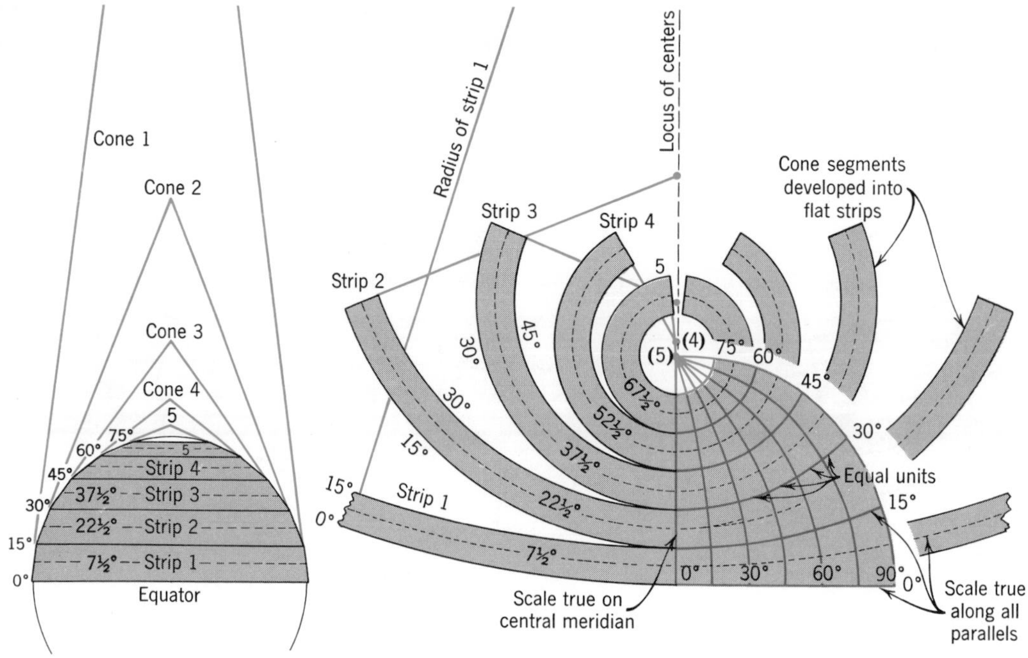

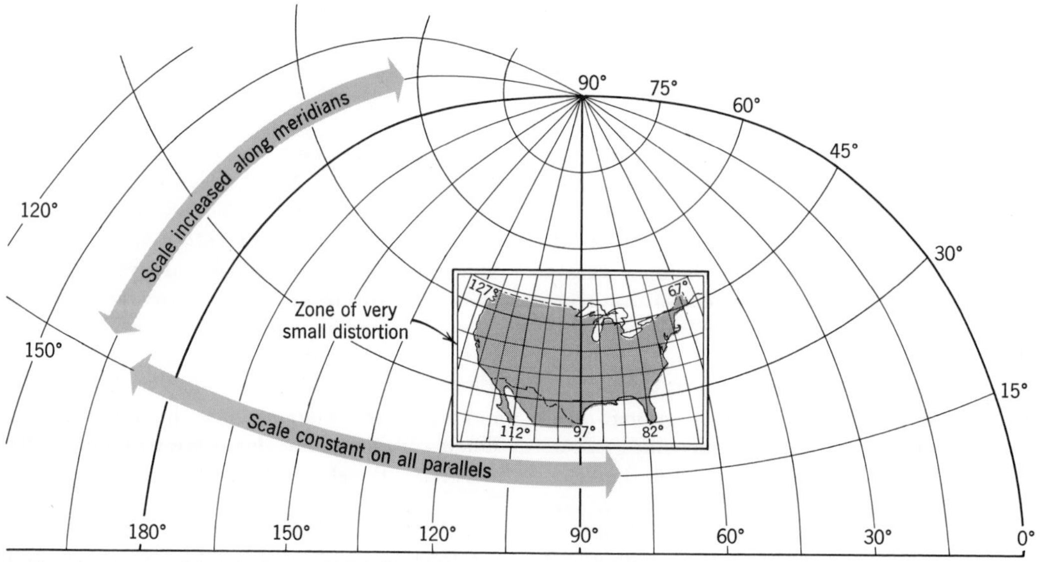

Figure 2.17 The polyconic projection is used for small areas in middle latitudes.

principle, however, can be explained without mathematical expression, as follows. On any cylindrical projection in which the meridians are straight vertical lines, equidistantly spaced, the meridians have had to be spread apart (see right side of Figure 2.18). Only along the equator are they the same distance apart as on a globe of the same equatorial scale. In order to maintain them as parallel lines, the normally converging meridians have had to be spread apart in a greater and greater ratio as the poles are approached. At 60° N and S lat., the meridians are spread apart twice as far as originally, because at that place a degree of longitude is only half what it is at the equator. At the poles the spreading is infinitely greater, because the poles themselves are infinitely tiny points. Now, to maintain the map as a truly conformal map, we must space the paral-

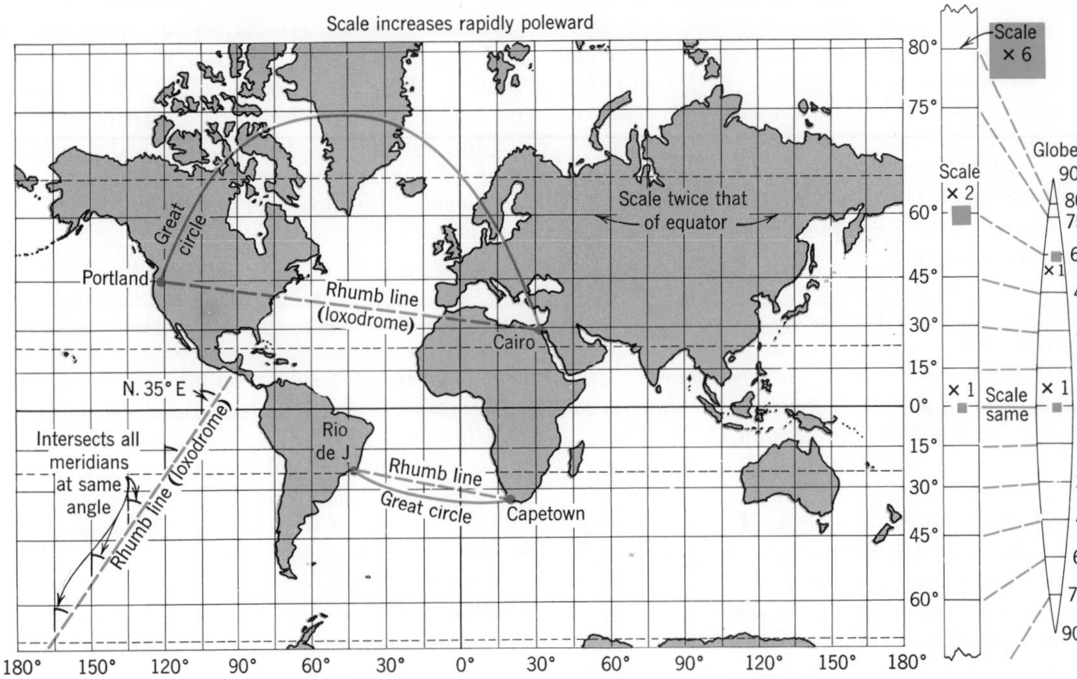

Figure 2.18 The equatorial Mercator projection shows all lines of constant compass direction as straight lines.

lels increasingly far apart toward the poles, using the same ratio of increase that resulted when the meridians were spread to make vertical lines.[1] For example, near the 60th parallel north, parallels must be spread twice as far apart as on the globe because, as explained above, the meridians here are also spread twice as far apart. At 80° latitude the scale is enlarged almost six times. Near the poles, the spacing of parallels increases enormously and rapidly approaches infinity. Because an enormous sheet of paper would be needed to show extreme polar regions, the Mercator map is usually cut off about 80° or 85° N and S lat. The poles can never be shown.

The Mercator chart is a true conformal projection. Any small island or country is shown in its true shape. The scale of the map, however, becomes enormously greater toward the poles. The classical Mercator projection described above is in the equatorial position. That is, the cylinder is tangent to, or touches, the earth's equator, as illustrated in Figure 2.8. The earth's polar axis here coincides with the cylinder axis. Other forms of Mercator projection can be obtained by turning the globe within the cylinder so that the great circle of tangency is a pair of meridians, or any other great circle.

[1]The map scale increases poleward as the secant of the latitude. The secant of 60° is 2; of 70° is 2.9; of 80° is 5.8; of 85° is 11.5; and of 89° is 57.3. The scale is stretched by these factors at the stated latitudes.

The really important, unique feature of an equatorial Mercator projection is that a straight line drawn anywhere on the map, in any direction desired, is a line of constant compass bearing. Such a line is known to navigators as a *rhumb line*, or *loxodrome* (Figure 2.18). If this line is followed, the ship's (or plane's) compass will show that the course is always at a constant angle with respect to geographic north.[2] Once the proper compass bearing is determined, the ship is kept on the same bearing throughout the voyage, if the rhumb line is to be followed. The equatorial Mercator is the only one of all known projections on which all rhumb lines are true straight lines, and vice versa. A protractor can be used with reference to any meridian on the map, and the compass bearing of any straight line can be measured off directly.

The relation of great-circle routes to rhumb lines is shown by two examples on Figures 2.11 and 2.18. Notice that, on the gnomonic map, great circles are straight and rhumb lines curved, whereas on the Mercator chart rhumb lines are straight and great circles curved. Along the equator and all meridians (but only on these lines), rhumb lines and great circles are identical and are straight lines on both charts. In navigation, it is desirable to follow a great-circle course, which is the shortest distance, yet at the same time this is difficult to do because compass bearing constantly changes

[2]Correction would be necessary for changing magnetic declination, explained in Chapter 3.

along the great-circle track. In practice, the course is drawn on a great-circle chart, then transferred as a series of straight segments, or legs, to the Mercator chart. The navigator can now measure the compass bearing of the first leg by means of a protractor. After the ship has traveled the distrance of the first leg, the ship is turned to follow the bearing of the second leg, and so on. Although this is not truly a great-circle route, it fits so closely that the extra distance is negligible.

Aside from its indispensable property for navigational and scientific uses, the equatorial Mercator projection has little to recommend it for unrestricted global use. Except for equatorial regions, for which it provides an excellent grid, distortions of scale are very serious. Because of infinite stretching toward the poles, this map fails completely to show how the land areas of North America, Asia, and Europe are grouped around the polar sea. In the mind of an inexperienced user, it may enhance a false sense of isolation between inhabitants of these lands. In this respect, zenithal or conic projections are preferable for study of continental regions in middle and high latitudes.

On the other hand, certain forms of geographical information are best shown on the Mercator projection. Because of its accurate depiction of the compass directions of lines, the Mercator net is preferred for maps of direction of flow of ocean currents and winds, direction of pointing of the compass needle, or lines of equal value of air pressure and air temperature. Examples of such uses of the Mercator projection will be seen in later chapters.

10. *Transverse Mercator projection.* The principle of a cylinder tangent to the globe along a chosen pair of opposite meridians yields the transverse Mercator projection (Figure 2.19). Figure 2.20 shows a transverse Mercator projection tangent to the globe on the Greenwich (0° longitude) and 180th meridians. This particular form of transverse Mercator net is also known as the *Gauss conformal* projection. The projection extends infinitely far to left and right. Points in line

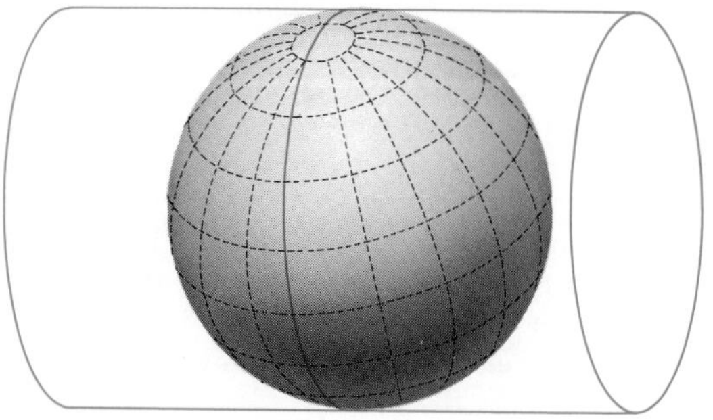

Figure 2.19 The transverse Mercator projection uses the principle of a cylinder tangent on a pair of meridians.

with the cylinder axis, at lat. 0°, long. 90° E and lat. 0°, long. 90° W, can never be shown. Scale of the map is constant only on the central meridian. If the user selects a narrow belt extending but a few degrees of arc to the east and west of the central meridian, scale increase is very small throughout and the advantages of a true conformal projection are enjoyed.

A slight modification has brought the transverse Mercator projection to its peak of usefulness. Instead of having the enclosing cylinder tangent along a meridian, the cylinder cuts through the surface of the globe, intersecting it on two smaller circles, taking what is described as the *secant position* (Figure 2.21). Now the map scale is constant along two straight, parallel lines on the map. These lines are equidistant from the central meridian.

For purposes of the Universal Transverse Mercator military grid system, described in Chapter 3, the two lines of equal scale are separated on the map by a distance of 360,000 m, or 360 km (223.6 mi). Scale changes are extremely small within a 6° belt of longitude. The transverse Mercator projection is thus an excellent base for large-scale topographic maps, for which purpose it has been adopted since World War II by the U.S. Army Map Service. To cover the entire world, the central meridian is moved by 6° intervals, so that 60 central meridians are required in all. Regions lying poleward of 80° latitude, however, are covered by polar stereographic projections.

Other types

11. *Mollweide homolographic projection.* One projection rather widely used to show the entire globe is the *homolographic* projection, invented by Karl B. Mollweide in 1805 (Figure 2.22). "Homolographic" is a word often used to mean "equal area," a property that this projection possesses. One hemisphere is outlined by a circle; the other hemisphere is divided into two parts and added with an elliptical outline to either side of the circle. All other meridians, except the straight central meridian, are halves of ellipses. The equator is twice as long as the central meridian, which is also true on a globe. Parallels are straight, horizontal lines, becoming more closely spaced toward the poles. The spacing of parallels, so adjusted as to give the map equal-area properties, is obtained by an involved method. Tables are available for construction.

The Mollweide projection has distinct advantages as well as disadvantages. Its equal-area property makes it valuable for showing the global areal distribution of geographical or political entities. Severe distortion in the polar regions, however, has hindered its wider use. It can, of course, be centered on any desired meridian so as to re-

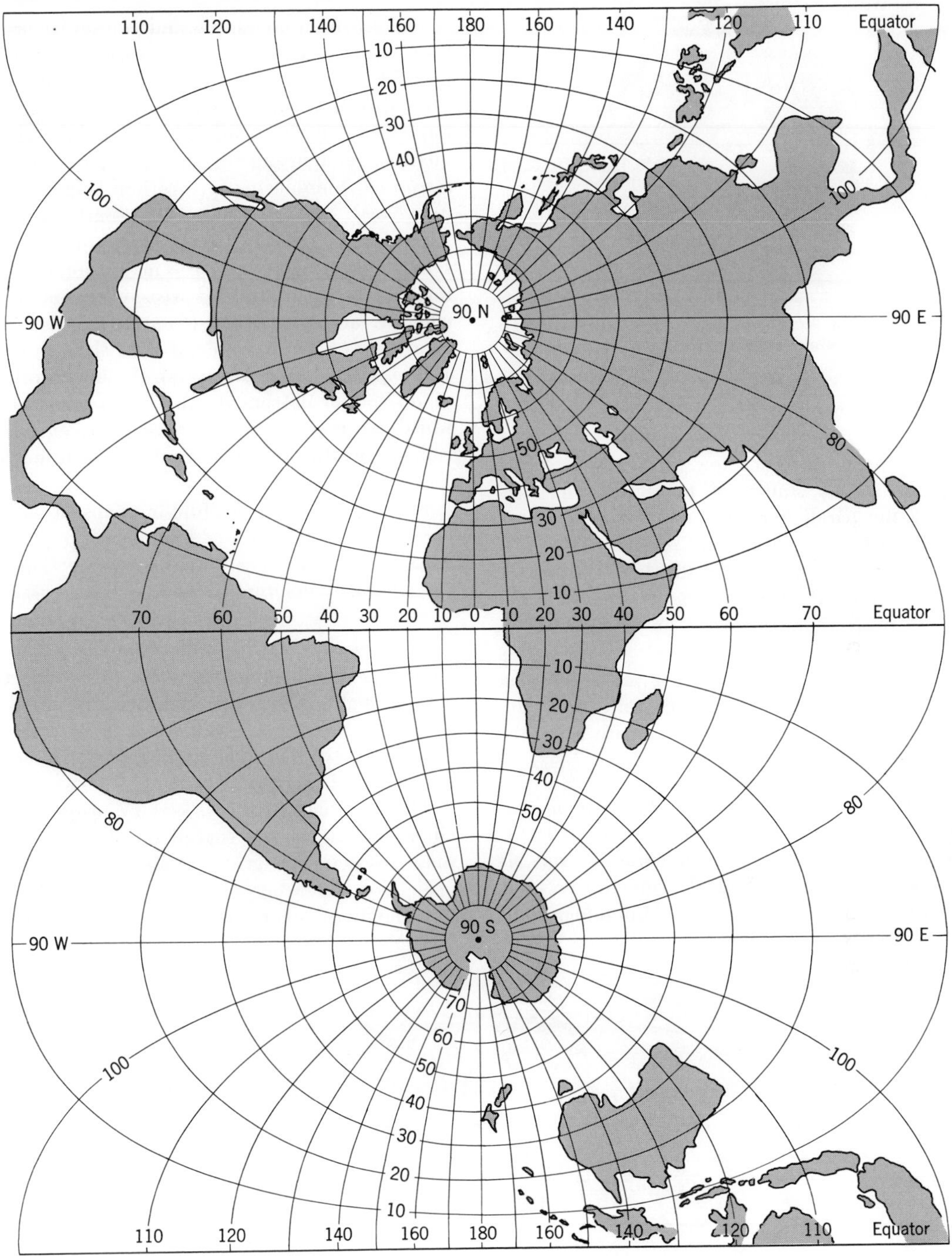

Figure 2.20 A world transverse Mercator projection.

duce distortion for a particular area. It makes a good base for maps of Africa and South America, either of which can be included in the central area of relatively little distortion. Interrupted and tilted forms have proved valuable as world maps.

12. ***Sinusoidal projection.*** In some ways the *sinusoidal* projection (sometimes called the Sanson-Flamsteed projection) is similar to the Mollweide homolographic projection. It is an equal-area projection with straight central meridian and horizontal straight parallels (Figure 2.23). The difference lies in the type of curve used in meridians. Whereas the homolographic net uses ellipses, the sinusoidal net uses families of *sine curves.* Figure 2.24 shows three wave crests of a sine curve superimposed. If we turn Figure 2.24

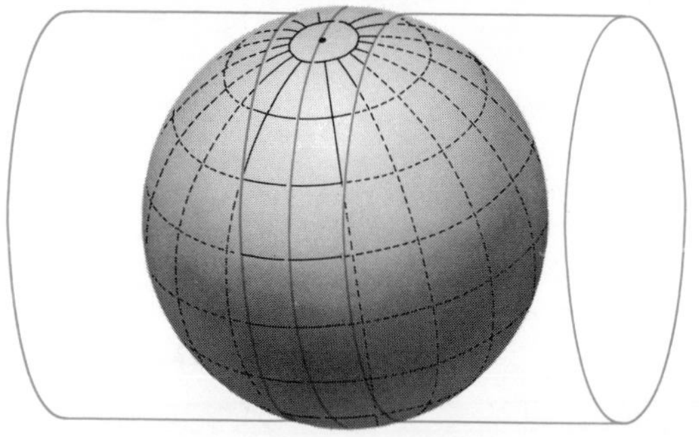

Figure 2.21 The secant position of the transverse Mercator projection minimizes scale changes in a narrow zone.

sidewise and compare it with Figure 2.23, it is evident that a similar set of curves has been used to draw the projection. Now, it is a remarkable property of the sinusoidal projection that, by spacing the parallels equidistantly from equator to poles, a true equal-area net results.

The same advantages and disadvantages apply to the sinusoidal projection as to the homolographic projection. Distortion in polar areas is not quite so great in the sinusoidal net but is nevertheless offensive. If Africa or South America is placed in the center of the projection, the continent is extremely well shown, with little distortion of scale or shape.

13. *Homolosine projection.* The *homolosine* projection, invented by Dr. Paul Goode in 1923, is a combination of the homolographic and sinusoidal types. The sinusoidal projection is used between 40° N and S lat., and the homolographic projection is used for the poleward parts.

In both homolographic and sinusoidal projections, the shearing of polar areas is especially marked to the extreme right and left of the map. This distortion can be reduced by centering each important land area on its own straight central meridian and fitting together the parts. The *interrupted* homolosine projection (Figure 2.25) has North America, Eurasia, South America, Africa, and Australia, each based on the best-suited meridian. Because the map cannot thus fit together between the land areas, except along the equator, large gaps occur. If our interest is in land areas only (as, for example, if we wish to show the areas under wheat cultivation), the interruption is of no special detriment. To show oceans of the world, we may center the oceans on central meridians, making the interruptions occur on land areas.

Plates 1 through 5 (fold-in maps at end of book), use the interrupted homolosine projection. As a further modification, these projections are condensed by deleting parts of ocean areas.

14. *Eckert IV projection.* Popular in Europe among geographers is an equal-area projection with straight, horizontal parallels, invented by Professor Max Eckert (Figure 2.26). Of the six projections that he devised, this one is given the designation *Eckert IV*.

The meridians of the Eckert IV projection are ellipses, equidistantly spaced on all parallels. In these respects, this grid resembles the Mollweide projection, but whereas in the Mollweide projection the meridians converge to common points at the poles, the Eckert IV projection uses for each

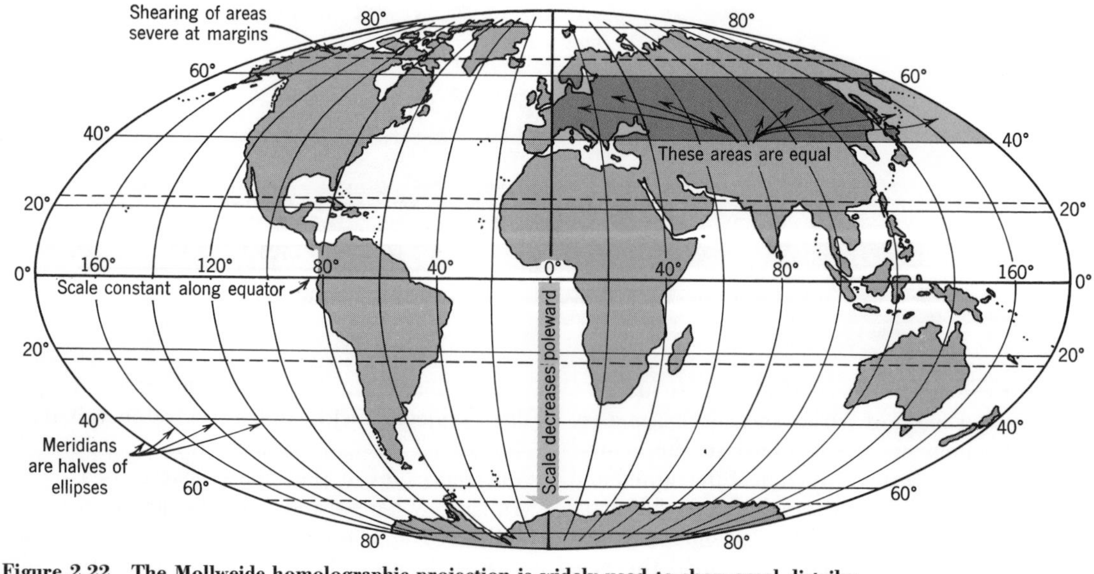

Figure 2.22 The Mollweide homolographic projection is widely used to show areal distributions over the globe.

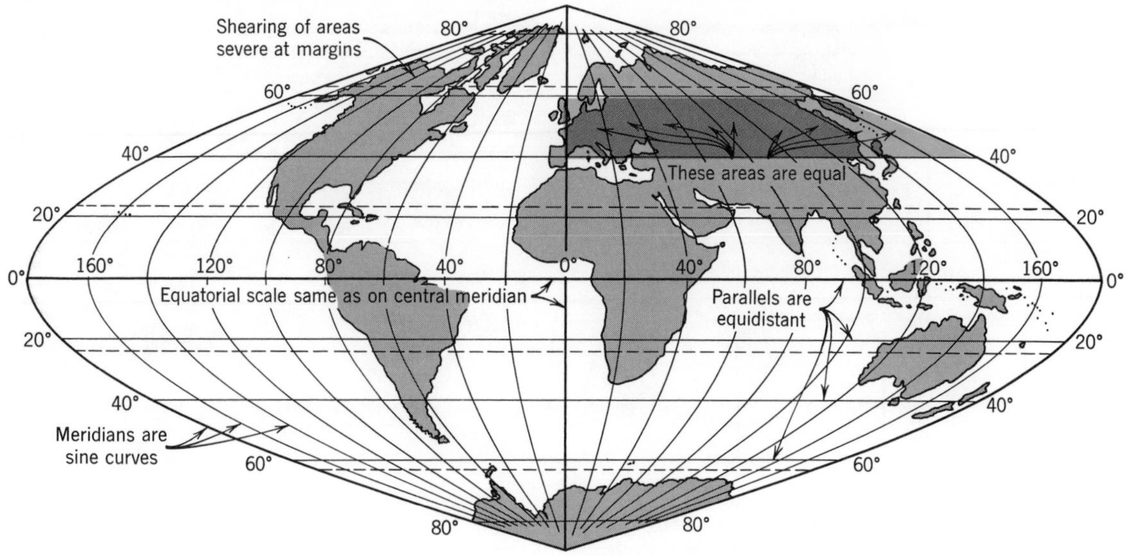

Figure 2.23 The sinusoidal projection is an excellent equal-area projection for lower latitudes.

pole a horizontal line half the length of the equator. In so doing, serious crowding and distortion in high latitudes is relieved. Nevertheless, there is considerable shearing of areas in high latitudes.

To achieve an equal-area map, the spacing of parallels in the Eckert IV projection is adjusted by formula, the spacing being widest close to the equator and diminishing to about one-fourth that amount close to the poles. The Eckert IV projection can readily be interrupted in a manner similar to that seen in the interrupted homolosine projection. Both projections are excellent for world maps showing areal distributions.

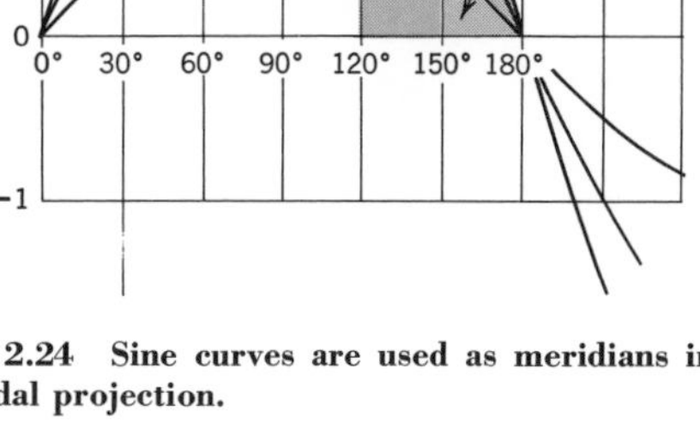

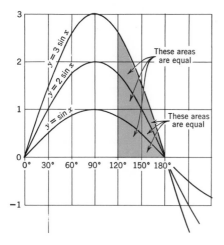

Figure 2.24 Sine curves are used as meridians in the sinusoidal projection.

Summary

Fourteen projections have been explained and illustrated. This should serve to give the geography student a good start into map-projection principles. But, because more than 200 projections have been invented, unfamiliar types will frequently be encountered. The majority of them are closely enough related to one or more of the types discussed in this chapter so that the principle is either obvious from inspection of the map or can be understood from a brief note of explanation.

The following admonitions may serve to point out some of the principles of selecting a projection.

1. Use a good globe whenever possible if points or areas are to be related in distance and direction. There is no substitute for a true-scale model of the earth's surface features.

2. Where a map is required, select the projection best suited to the need. For example, use an equal-area map to show the areal distribution of things. Use conic or azimuthal maps for high latitudes.

3. Before you begin to draw conclusions and analyze geographical factors from a map, be sure that you know the qualities of the projection, whether equal-area, conformal, or neither. If in doubt, compare two or more projections of the same areas and see what possible erroneous concepts each might give. The more of the earth's surface a map shows, the greater should your caution become.

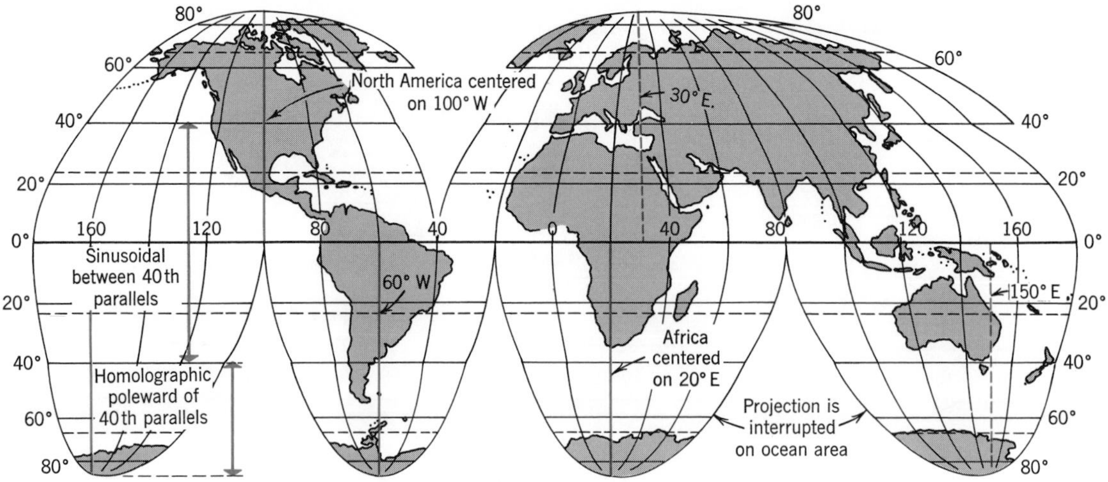

Figure 2.25 Goode's interrupted homolosine projection combines the homolographic and sinusoidal projections. (Based on Goode Base Map. Copyright by the University of Chicago. Used by permission of the University of Chicago Press.)

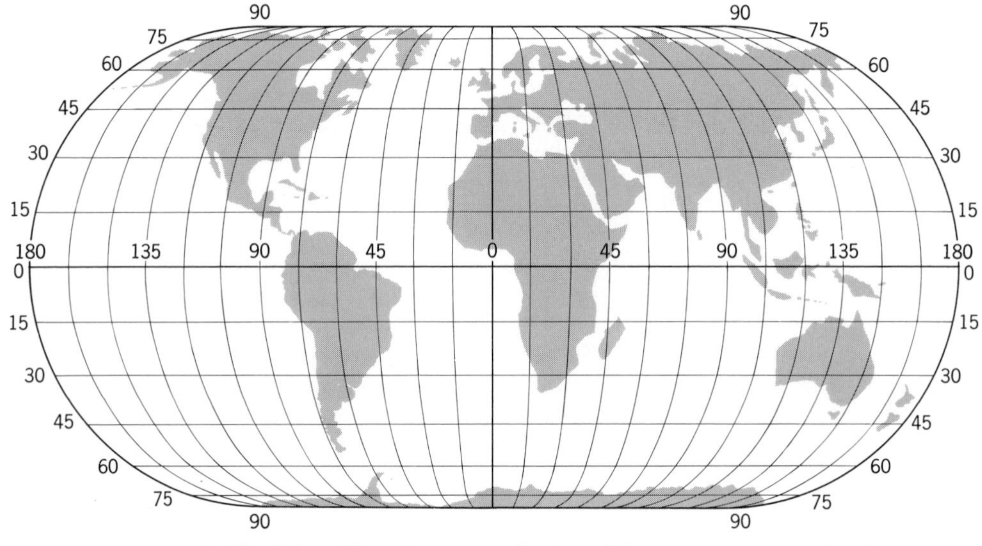

Figure 2.26 The Eckert IV projection uses horizontal lines to represent each pole.

Review Questions

1. What is a map projection? What is the basic problem of map projection?

2. For what purposes is a globe inferior to a flat map? For what purposes is a globe preferable to a flat map?

3. What is a developable solid? an undevelopable solid? Give examples, and explain what this has to do with map projections.

4. Explain the concept of map scale. What is the representative fraction? Which varieties of map projections preserve constant scale in all places and directions over the map?

5. What is an equal-area map projection? Name three equal-area types.

6. What is a conformal map projection? Name three conformal types.

7. Are projections that are neither equal-area nor conformal of any value? Illustrate with specific examples.

8. Into what large groups may map projections be classified?

9. List the properties common to all zenithal (azimuthal) map projections.

10. List the properties common to conic projections. In what way does the polyconic projection differ from other conic types?

11. What advantages do the cylindrical projections have over other groups?

12. For each of the zenithal (azimuthal) projections that you have studied, answer the following questions. (*a*) What principle of construction is used? (*b*) What conditions of scale pertain along a radial line extending outward from the center of the map? (*c*) In the equatorial position, what kind of lines are the parallels? the meridians? (Are they curved or straight; if curved, what kind of curves?) (*d*) Can the projection show an entire hemisphere? an entire globe? (*e*) Is the projection conformal, equal-area, or neither? (*f*) What special value has the projection? For what uses can it be recommended?

13. What advantage is gained by having two standard parallels instead of one on a perspective conic projection? How do scale conditions compare?

14. Explain the basic principle of the Lambert conformal conic projection. What advantages has this projection? What use is made of it? Would a world map be satisfactory on this projection?

15. Explain the principle of the polyconic projection. Along what lines is the scale constant? For what purposes is this a good projection? Name some of the map agencies that have used it. Which is better for sets of large-scale maps, the Lambert conformal conic projection or the polyconic projection? Explain.

16. Explain the principle of the equatorial Mercator projection. Is this an equal-area or conformal projection? What shape has a complete Mercator map showing the entire globe? How does the map scale at 60° latitude compare with the equatorial scale?

17. What is a rhumb line? What is a loxodrome? What projection shows all great circles as straight lines? What use is made of rhumb lines and great circles in navigation? Under what circumstances are these two types of lines identical on a Mercator projection?

18. Describe the transverse Mercator projection. What advantages has this projection? In what way does use of the secant position improve the scale qualities? How is this projection used as the base for the Universal Transverse Mercator Grid?

19. Compare the homolographic, sinusoidal, and Eckert IV projections as to construction principles and properties. Are these projections equal-area, conformal, or neither?

20. What advantages has Goode's interrupted homolosine projection over the homolographic and sinusoidal projections?

Exercises

1. For each of the following statements select from the fourteen projections discussed in this chapter all those to which the statement applies. (List polar, equatorial, and oblique positions of azimuthal projections as separate types.) (*a*) All are equal-area projections. (*b*) All are conformal projections. (*c*) All are neither equal-area nor conformal. (*d*) All the parallels are curved lines. (*e*) All the parallels are straight lines. (*f*) All the meridians, except the central ones, are curved lines. (*g*) All the meridians are straight lines. (*h*) All the curved parallels and meridians are true arcs of circles. (*i*) All the parallels are arcs of circles, but not necessarily concentric. (*j*) All the parallels and meridians intersect at true right angles. (*k*) Neither pole can be shown. (*l*) Only one pole can be shown at a time. (*m*) Map outline is circular when exactly one hemisphere is shown. (*n*) The projection can be constructed by drawing lines or rays from a single point, through a globe, onto a flat surface or a developable surface.

2. Make a table on which are noted the physical properties of the fourteen projections that you have studied. Down the left-hand side of a large sheet of paper list the fourteen projections. Make vertical columns across the paper, using each one to note a certain property of the projection. A suggested list of column headings is as follows: Class to which projection belongs. Name of projection (if azimuthal type, subdivide into polar, equatorial, oblique). Shape of map when showing complete globe. Shape of map when showing one hemisphere. Form of meridians (curved or straight; kind of curves). Form of parallels (curved or straight; kind of curves). Equal-area, conformal, or neither. Special properties (such as "great circles are straight lines"). Best uses (such as "large-scale maps," "small-scale maps," "navigation," and "areal distribution"). Perhaps other columns could be added. A chart of this type is excellent for review purposes.

3. Using only a pencil, compass, ruler, and protractor, construct some of the simpler projections that you have studied. Several easy ones are suggested below. (*a*) Polar orthographic projection. (*b*) Stereographic projection (polar position). (*c*) Azimuthal equidistant projection (polar position). (*d*) Azimuthal equal-area projection (polar position). (*e*) Perspective conic projection, one standard parallel, 45° N.

CHAPTER **3**

Location and Direction on the Globe

THIS chapter deals with the problem of determining locations of points and directions of lines on the earth's surface, in order that they may be correctly shown on maps. Equally important to geographers is a reverse problem, that of reading from a map the location, size, or orientation of features shown on maps and of stating this information in terms of some established system. The geographer relies heavily upon maps. Not only do maps provide him with information that he uses in his studies, but much of the new information that he gathers or synthesizes must be shown on maps. The making of maps, a specialized field combining a mathematical science with a graphic art, is known simply as *cartography*. Although the actual publication of finished maps is best left to the professional cartographer, there are many things that a geographer needs to know about maps in order to gain the maximum amount of information from them and to select base maps best suited to showing the information that he wants to display.

The system of parallels and meridians described in Chapter 1 provides a network of lines to which specific points on the earth's surface are tied. It is the task of the geodesist, geologist, and surveyor to measure the position, size, and shape of the earth's natural and man-made features. The task of the cartographer is to compile this information and to draw it as precisely and effectively as possible for reproduction on printed sheets.

Maps cover a wide range both in the amount of earth's surface shown by a sheet of a given size and in the type of material represented. On a small sheet of paper may be printed a map of the entire globe or a detailed city plan showing but a few city blocks. With respect to the type of information shown we can classify maps in two large groups, *planimetric* and *topographic*. Planimetric maps show the exact location of surface features projected on a single plane, as if no differences of vertical elevation existed. A planimetric map may show such features as shorelines, rivers, lakes, boundary lines, roads, and the precise position of cities and mountain peaks, but it gives no information as to relief forms and slope of the land surface or how high above sea level or surrounding objects the various features lie. On the other hand, topographic maps, treated in Appendix I, are intended to depict the relief features of the land surface and to indicate the degree of slope of the ground.

Relative and absolute locations

Suppose that an explorer has reached a little known land where he finds three prominent mountain peaks, A, B, and C, not previously mapped or described. With suitable surveying equipment he finds that Peak A lies five miles north of Peak B; that Peak C lies seven miles due east of Peak B. Taking a blank sheet of paper and letting one inch on the paper represent one mile of horizontal ground distance, he can draw a map to show the correct relative locations of the three peaks with respect to one another. A triangle connecting peaks A, B, and C is true in shape and is true in orientation with respect to north. The *relative location* of the points is thus determined with respect to one another, but not with respect to any fixed, world-wide system of reference lines, such as the parallels and meridians with a prime meridian passing through Greenwich, England.

If the explorer is equipped with suitable knowledge and instruments, he can determine the approximate location of one of the peaks in terms of latitude and longitude by astronomical methods. This establishes the *absolute location* of the peak, in turn, permitting the absolute location of the remaining points. Absolute location describes the position of a point in such terms that its unique position with respect to all other points on the globe is made clear. In general, three grid systems (sets of interesecting lines) described in this chapter provide absolute location of points: the geographic grid, military grid, and U.S. Land Office grid. Triangulation surveying by itself, and designation of position by compass bearing and distance provide relative locations.

Triangulation and base lines

Planimetric maps show the exact location of features on a map, as if all were reduced to a sea-level surface. Before any major mapping operation is begun, it is essential to locate relative to one another with extreme accuracy several key points within the area. Because all additional points are located with reference to the first set, the highest order of accuracy is required for the original key points. Secondary sets of points are located with somewhat less accurate methods, which take less time and money. From these, still larger numbers of points are determined in more rapid fashion.

Surveying of points begins with the careful measurement of a long, straight *base line*, which may be several miles long. The base line usually follows a highway, railroad, or beach so as to encounter the fewest obstacles. Between points marked by permanent monuments set in the ground, the base line is measured by means of a steel tape. A special nickel-steel alloy, known as *invar*, is used for the tape, which must change length very little with temperature changes. The tape must be stretched with exactly the right force, measured by a spring balance, so that error due to its elasticity is the minimum (Figure 3.1). With extreme care and an elaborate outfit, it is possible to measure the length of a base line so that the actual error will not be more than one part in 300,000. Thus, a line five miles long is measured to within one inch of its true length. Actual errors in first-order surveying are commonly very much smaller, generally about one part in two million.

Electronic distance-measuring instruments have come into widespread use within the past two decades. Based on transmission of either light waves or high-frequency radio waves, these instruments are capable of base-line measurement with accuracy exceeding one part in nine million.

Once the base line has been measured, triangulation is begun. *Triangulation* is the measurement of large triangles on the ground, using the geometric principle that if the length of one side of a triangle (the base line) is known, and the two angles made by the base and adjacent sides are also known, the remaining sides and angle can be calculated (Figure 3.2). Thus, if a surveyor sets up a telescopic instrument, known as a *theodolite*, upon one end of the base line (point A), he can sight both to a flag on the other end of the base line and to a second flag on a third, unknown point (point C) and measure the horizontal angle between the two lines of sight. If he then takes his theodolite to the other end of the base line (point B) and sights back to points A and C, measuring the angle between them, he has the necessary information to determine the lengths of the other two sides of the triangle and thus to locate point C exactly with respect to the base line. The great advantage of triangulation over actual ground measurement is the saving in time and labor, especially where the terrain is rugged. (The Grand Canyon was mapped in this way.) Once the first triangle ABC is measured, either of the other two sides AC or CB can be used as base lines from which to construct more triangles. As more triangles are added, the known points are extended across country as a *system of quadrilaterals* (Figure 3.2). These, in turn, join one another to form an intersecting network (Figure 3.3).

Figure 3.1 Measuring a base line along a railroad track. The closest man applies tension while the nearer of the two kneeling men matches the zero mark of the tape to the mark on the rail. (Official photograph U.S. Coast and Geodetic Survey.)

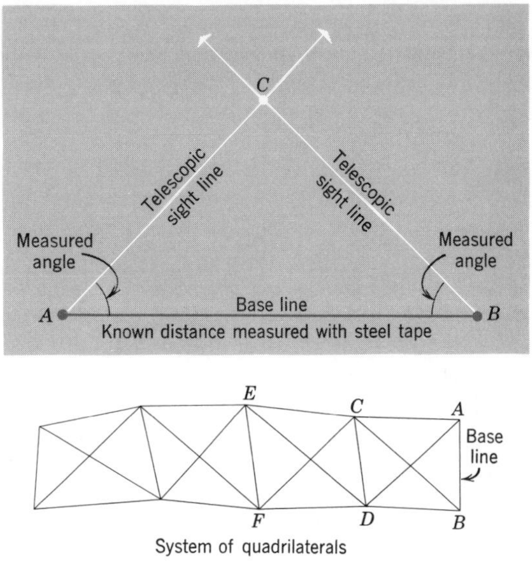

Figure 3.2 The principle of triangulation.

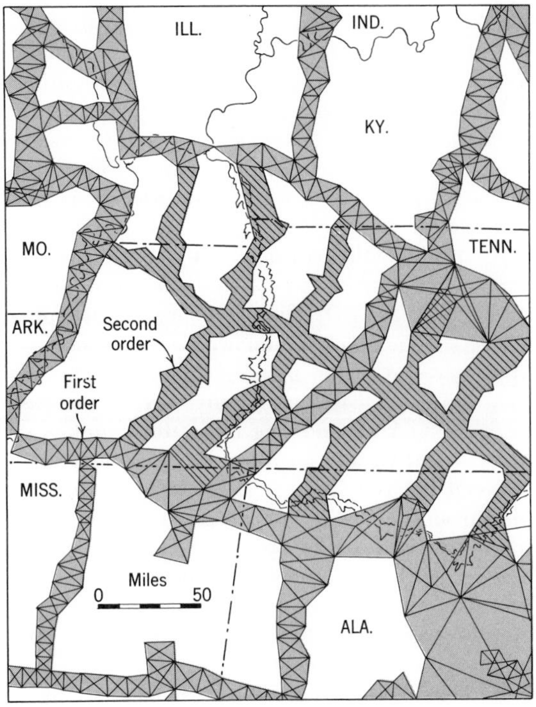

Figure 3.3 A network of triangulation quadrilaterals in the Tennessee Valley region. (After TVA Tech. Report 23, 1951, U.S. Govt. Printing Office.)

Triangulation and measurement of base lines is described as of *first order, second order, third order*, and *fourth order*, depending upon the degree of accuracy attained. First-order and second-order triangulation in the United States is carried out by the U.S. Coast and Geodetic Survey, which achieves a degree of precision almost incredible to the ordinary person. For example, in first-order work, the sum of the three angles of a triangle must approach the theoretical value of 180° to within 1 second of arc. (There are 3600 seconds to 1 degree.) Work is done at night, using a small electric light whose pinpoint image can be picked up by telescope many miles away. During the daytime the image of a distant point shifts about incessantly in the telescope field because the light rays are bent irregularly as unequal heating of the air sets up eddies in the lower atmosphere. Special towers are set up for operation of the theodolite. An inner tower supports the instrument while an independent outer tower supports the observer (Figure 3.4).

Fourth-order triangulation is used by mapping parties to establish a large number of points within the area, and neither requires nor uses the elaborate precision instruments and techniques of the higher orders.

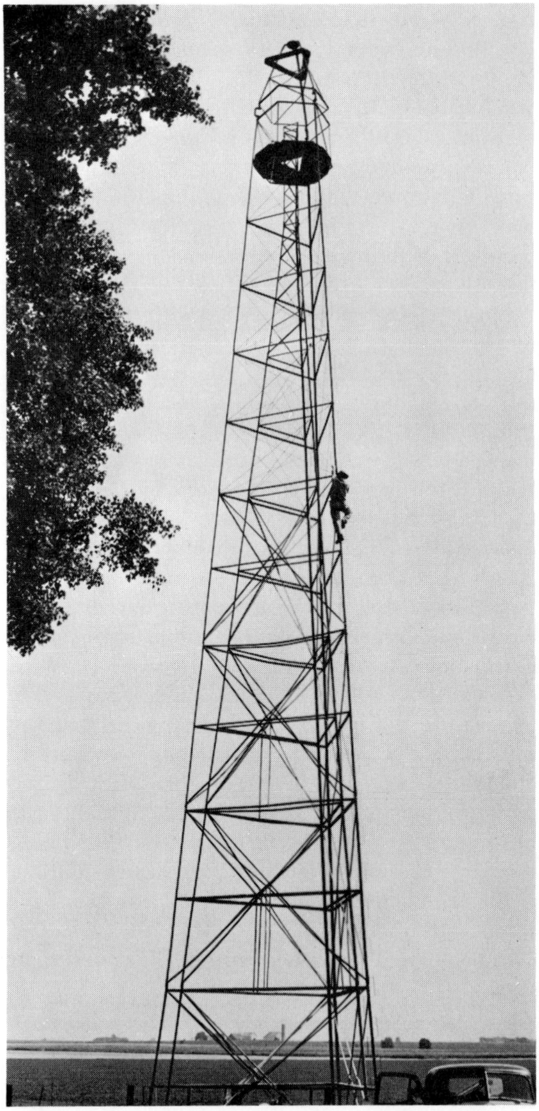

Figure 3.4 A 90-ft steel triangulation tower. (Official photograph, U.S. Coast and Geodetic Survey.)

Although triangulation is fundamentally a system of determining relative location of a given point with respect to other points, a triangulation system can be made one of absolute location by precise determination of the latitude and longitude of one point on the base line, then computing by trigonometry the geographic coordinates of all other points in the network.

In general, the plan of the U.S. Coast and Geodetic Survey was to span the United States with a number of first-order triangulation networks, each being a system of quadrilaterals begun at a coastal point and extended into the heart of the country. Many such traverses, following generally along parallels and meridians, now provide the United States with a control network (Figure 3.5). Triangulation points on this net serve as the reference points for all mapping and surveying opera-

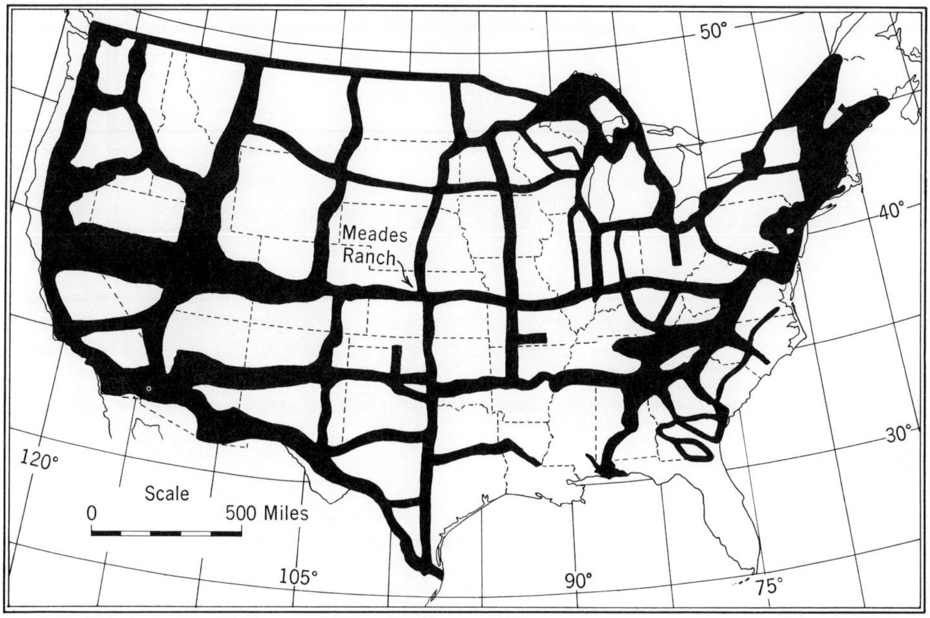

Figure 3.5 The first-order triangulation network of the United States as it was in 1929, shortly after the establishment of the North American Datum of 1927. (After C. V. Hodgson, in *Physics of the Earth*, Vol. II, "Figure of the Earth," National Research Council, 1931.)

tions carried out locally. Information on the precise latitude and longitude of these points is available.

Unfortunately, the most precise determinations of latitude and longitude at various coastal points do not result in exact agreement as to the position of a triangulation station where two or more traverses have met at a common point in the heart of the country. Therefore, after east-west and north-south triangulation surveys were joined in the United States, a single reference station was selected to become the absolute standard of geographical position. This is a station known as "Meades Ranch," located in central Kansas, approximately at the geographical center of the United States. Using the Clarke ellipsoid of 1866 as the reference ellipsoid, latitude and longitude were computed for Meades Ranch station. All other positions of the triangulation networks were adjusted accordingly. This set of absolute locations of control points is known as the *North American Datum of 1927* and all local mapping is tied to it. Both Canada and Mexico decided to adopt the datum based on Meades Ranch. All of North America thus has its triangulation based on the 1927 datum, and is the only continent having all of its triangulation carried out with reference to a common ellipsoid and a single starting point on that ellipsoid.

Vertical control by leveling

Triangulation establishes location of points projected upon the imaginary ellipsoid surface, but does not measure the precise vertical distance between a given point and a reference level. Such information, known as *vertical control,* is supplied by *precise leveling.* The instrument used is known simply as a *level,* or *spirit level.* It is a telescope mounted horizontally with a level bubble astride the telescope tube. By means of thumb screws, the telescope can be adjusted until brought to an exactly level sight line. Sighting horizontally to a point of known elevation, the height of instrument is read on a scaled rod held erect on that point (Figure 3.6). The telescope is then turned to another point, where the scaled rod is again read. Elevation of the second point is determined by adding and subtracting the height readings.

As with triangulation, leveling is described as of first order, second order, third order, and lower order. That of the highest precision, first-order leveling, is carried out by the U.S. Coast and Geodetic Survey in such a way as to form a fundamental level net of which several thousand miles have been completed, largely following railroads and highways. A level line starts at the coast, where the reference level, or datum, is mean sea level. Although easily determined, mean sea level varies from place to place along the coast, so that two precision level lines meeting in a common point may not agree as to elevation. All first-order level nets were adjusted in 1929 to agree with respect to the *Sea-level Datum of 1929.*

Level stations are marked by permanent *benchmarks* for use in local mapping and surveying (see Appendix I).

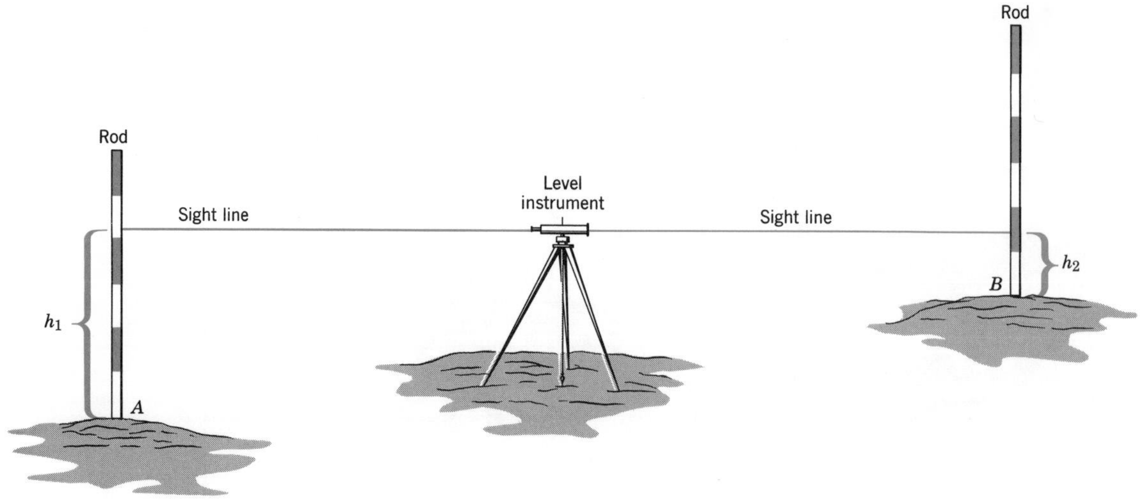

Figure 3.6 Height h_1 first read on the scaled rod at position A is added to the known elevation of point A to determine instrument elevation. Height h_2 is then read and subtracted from instrument elevation to give elevation of point B.

Triangulation by orbiting satellite

Orbiting satellites are providing an entirely new technique for triangulation on a global scale. From two widely separated ground stations laser beams are simultaneously bounced off a single orbiting satellite, enabling the distances and angles from ground stations to the satellite to be calculated with extreme accuracy. Knowing the lengths of these two legs of a triangle, the third leg, which is the ground distance, can be calculated.

Global triangulation by satellite began in the 1960s under the U.S. National Geodetic Satellite Program. Initially, fixed ground cameras were used to photograph the reflected sunlight from a satellite. For the first time triangles were established to link the continents and to establish accurately the locations of isolated islands of the vast ocean areas. Figure 3.7 shows the world satellite triangulation network together with a smaller scale of satellite triangulation for the North American continent.

One of the interesting aspects of laser beam triangulation is that the order of accuracy may become sufficiently high to measure the extremely slow rates of continental separation associated with movements of lithospheric plates (Chapter 23). On the basis of geologic evidence, continents are calculated to be drifting apart at separation speeds on the order of 1 to 2 in. (2 to 5 cm) per year. With sufficient accuracy of satellite triangulation the accumulated separation of several years' duration will become evident through significant changes in length of triangulation base lines.

Direction on the earth's surface

Triangulation, astronomical observation, and leveling establish the relative and absolute positions of points on the ellipsoid of reference as well as the elevations of those points with respect to the same ellipsoid. A third essential form of information for recording geographical information and depicting it on maps is the direction followed by a line from one point to another on the ellipsoid. Direction is essential to the description of many phenomena of physical geography, for example, the trends of mountain ranges and coast lines or the directions of flow of rivers, ocean currents, and winds.

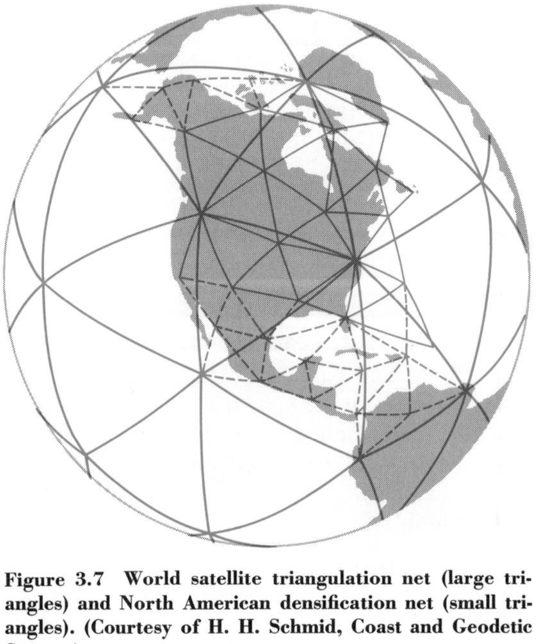

Figure 3.7 World satellite triangulation net (large triangles) and North American densification net (small triangles). (Courtesy of H. H. Schmid, Coast and Geodetic Survey.)

The simplest and most obvious statement of direction is had by measuring the angle between a given line and a meridian of longitude. The geographic meridians define *geographic north.* Because a meridian can be drawn through any point on the globe, geographic direction can be stated for a line originating at any point on the globe.

Geographic direction is the angle between a meridian and the small segment of a great circle that crosses the meridian. For small distances—a few miles—both the meridian and the great-circle segment can be considered as two straight lines on a plane surface. Actually, both lines are curved and follow the surface of the ellipsoid of reference. (Strictly speaking, therefore, neither the meridian nor the great circle are true circles, but are slightly elliptical in shape to conform to the oblate ellipsoid.)

To state the direction of motion of winds or ocean currents, or the trend of a given line at a given point offers no problems because we are measuring the angle between the meridian and a very short (infinitely short) segment of a great circle. However, a serious difficulty arises if we wish to state the direction of a great circle between two distant points, for example between Portland, Oregon, and Cairo, U.A.R. (Refer to Figure 2.11 for a great circle arc between these two cities.) It is impossible to give a single-valued direction of this great circle line with respect to geographic north, because the angle of intersection between the line and the series of meridians that it crosses is everywhere different. In other words, the direction of a great circle constantly changes. (Only on a meridian or on the equator is the great circle a line of constant direction.)

One solution to stating the direction that Cairo occupies with respect to Portland is to establish the rhumb line (loxodrome), which has a constant angle with respect to all meridians that it crosses. (Refer to Figure 2.18 for this rhumb line as it appears on the Mercator projection.) Use of the rhumb line as a basis of measurement of direction is not generally practical, as the rhumb line is not the most direct line between two points on the earth's surface.

Our conclusion concerning the measurement of direction on the earth's surface is that the statement of direction must be limited in practice to very short distances and is most meaningful as indicating the direction of motion of an object or the trend of a line at a particular point.

The earth's magnetic field

The magnetic compass has been the indispensible tool of the geographer following its general adoption as a navigational instrument in the fourteenth century. Since then, the magnetic compass has served in geographical exploration and mapping of the entire globe. Even today, this instrument is essential in civil surveying, large-scale mapping, and in guiding countless travelers on foot and in small boats and aircraft. Some knowledge of the earth's magnetic field and the use of the magnetic compass is therefore desirable in the training of every geographer.

The earth can be thought of as a simple bar magnet, the axis of which approximately coincides with the earth's geographic axis (Figure 3.8). Magnetism is generated within the earth's metallic core, a central spherical body about half of the earth's diameter. The earth's magnetic axis is inclined several degrees with respect to the geographic axis. Hence the north and south *magnetic poles* do not coincide with the geographic north and south poles, nor does the *magnetic equator* coincide with the earth's geographic equator.

Lines of force of the earth's magnetic field, shown in Figure 3.8, pass outward through the earth's surface and into surrounding space. A magnetic compass needle, which is nothing more than a delicately balanced bar magnet, orients itself in a position of rest parallel with the lines of force. Although the force lines are directed downward into the earth, the needle is balanced by placing a counterweight at one end, so as to remain horizontal. Thus the north-seeking end of the compass needle points in the general direction of the north magnetic pole. Because of many irregularities in the magnetic field, and hence in the configuration of the lines of force, the magnetic compass is subject to many local irregularities that can only be corrected by use of a detailed map specially prepared for this purpose.

The major cause of uncertainty in using the magnetic compass as a direction indicator lies in the difference in position between the geographic north pole and the magnetic north pole. The magnetic north pole is located about lat. 70° N, long. 100° W, in the vicinity of the Boothia Peninsula and Prince of Wales Island of the Northwest Territories of Canada (see Figure 3.10). The south magnetic pole is located in Antarctica at about lat. 68° S, long. 143° E. The location of these poles is constantly shifting, a further cause of uncertainty in use of the compass.

Figure 3.9 shows that a great circle through *magnetic north,* the direction taken by the north-seeking pole of compass needle, forms an angle with respect to the meridian, which indicates *true north,* or *geographic north.* At point *A*, the magnetic compass points at an angle west of the meridian; here the *declination of the compass* is described as *west declination.* At *B* the declination is east. At *C*, however, the magnetic pole lies on the meridian of the point *C*; hence there is no

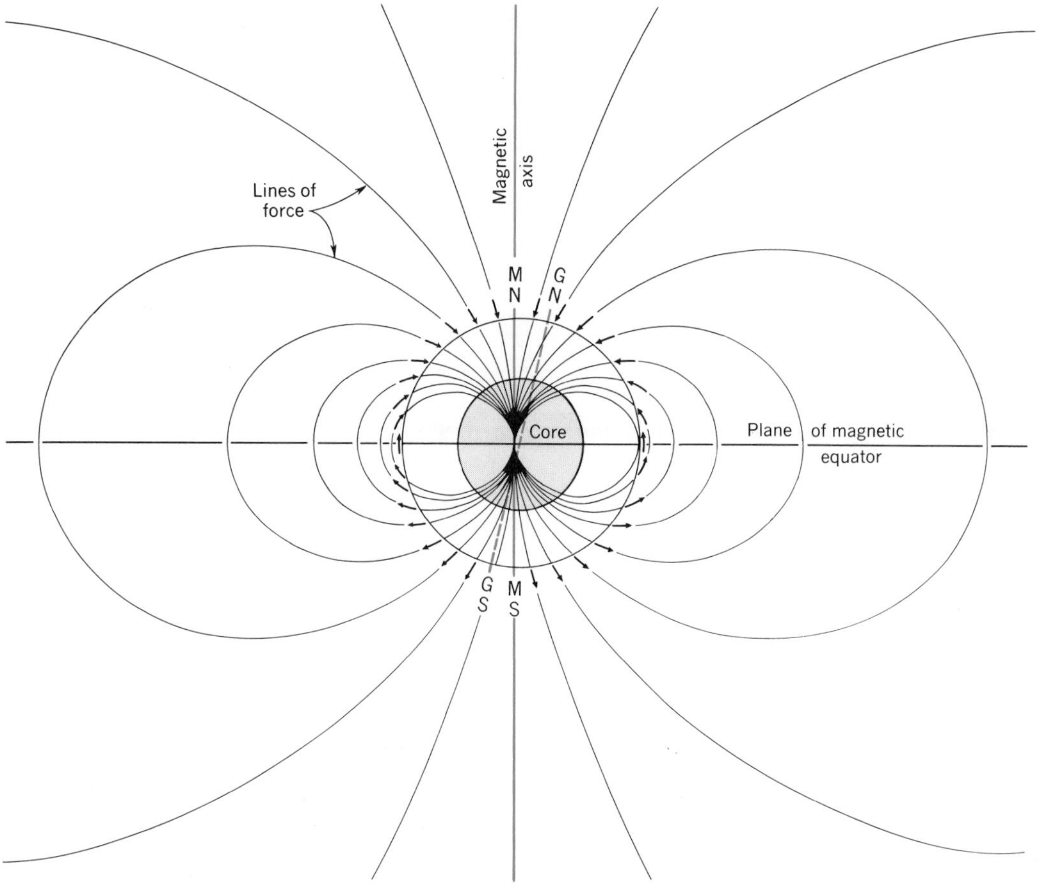

Figure 3.8 Lines of force of the earth's magnetic field are shown diagrammatically in a cross section drawn through the magnetic and geographic poles. The small arrows show the inclination of lines of force at surface points over the globe. (From A. N. Strahler, 1963, *The Earth Sciences*, Harper and Row, Inc., N.Y.)

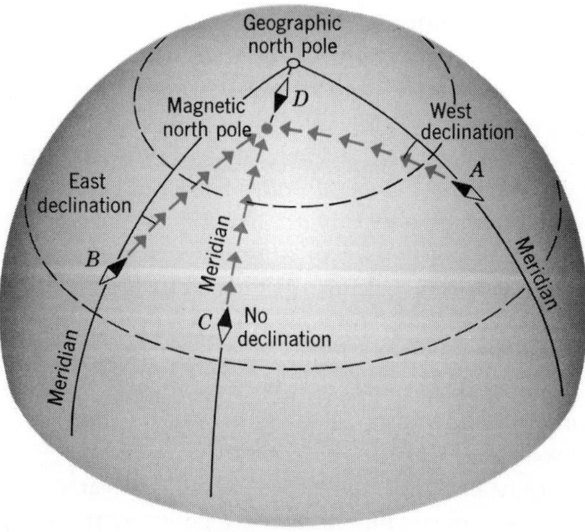

Figure 3.9 Whether declination is east or west depends on the observer's global position with respect to the magnetic and geographic north poles. (From A. N. Strahler, 1963, *The Earth Sciences*, Harper and Row, Inc., N.Y.)

declination and the magnetic compass indicates true north. At position *D*, midway between the two poles, the compass points due south.

Compass declination must be taken into account by adding or subtracting the appropriate number of degrees of declination from the indicated magnetic direction. Information is obtained from an *isogonic map* on which lines of equal declination (*isogonic lines*) are drawn (Figure 3.10). Notice that, on nautical charts, declination is termed *variation of the compass.*

In the region surrounding the north magnetic pole declination ranges from 0° to 180°. Close to the magnetic poles the horizontal intensity of magnetic force is very weak, rendering the ordinary navigation compass unusable. Instead, a magnetized needle balanced upon a horizontal axis points directly downward toward the earth's center (see Figure 3.8).

A detailed isogonic map of part of North America (Figure 3.11) can be read to adjust one's

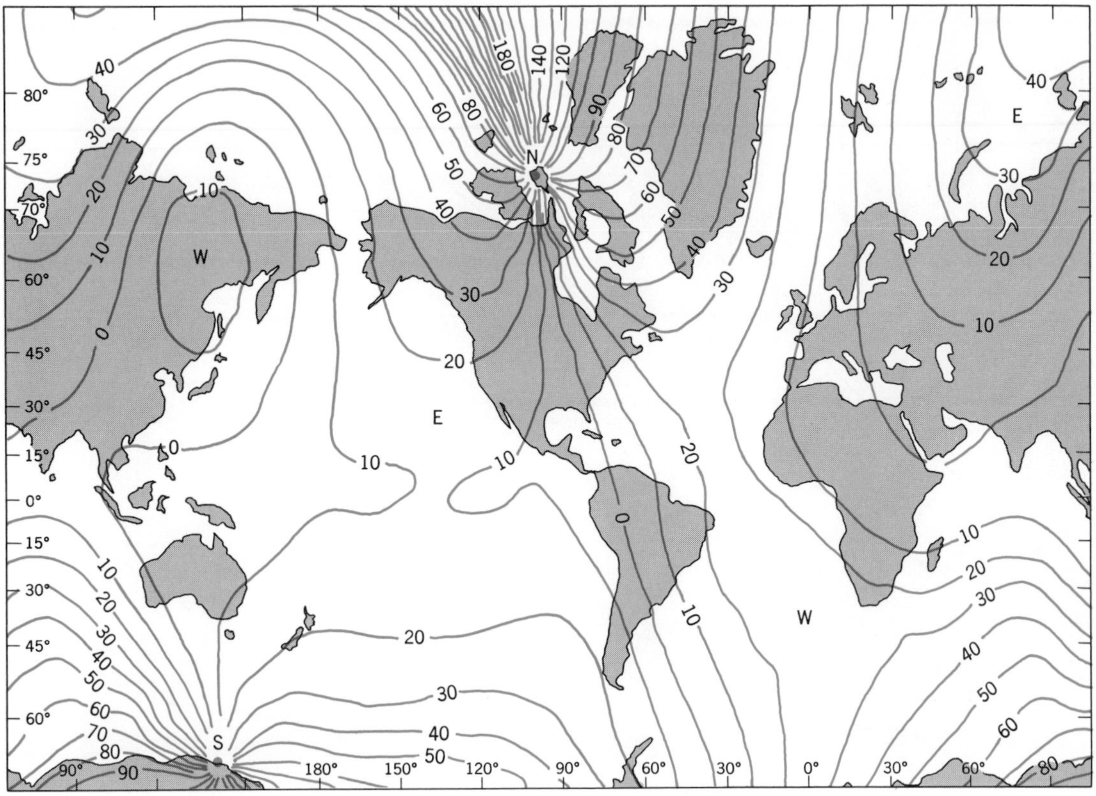

Figure 3.10 On this world isogonic map, declination is given in intervals of ten degrees. (Based on data of U.S. Navy Oceanographic Office. From A. N. Strahler, 1963, *The Earth Sciences*, Harper and Row, Inc., N.Y.)

magnetic compass for use in the field and with maps. The line of zero declination, known as the *agonic line,* passes through Lake Superior and southward along the east coast of Florida.

Because the earth's magnetic field is slowly but constantly shifting, a slight correction must be made in applying declination. For example, in the area of New York City, declination is increasing in value by about 3 minutes of arc annually. Most good large-scale topographic maps will have marginal information giving both the declination in the year in which the map was published and the amount and direction of annual change. The reader can then multiply the annual correction by the number of elapsed years to bring the declination to its correct current value.

Bearings and azimuths

A system of stating direction must be established for a given purpose. For example, in using maps, it is frequently necessary to state the direction followed by a road or stream, or to describe the direction that can be taken to locate a particular object with respect to some known reference point. In air and marine navigation, the direction from one point to another must be stated. All systems measure the angle that the given line makes with a north-south line. The unit of angular measurement most common in map work is the *degree,* 360 of which comprise a complete circle, but other systems of angular measurement, such as the *mil* (of which there are 6400 in a complete circle), are sometimes preferable for special applications.

Two systems of stating directions with respect to north can be used (Figure 3.12). (*a*) *Compass quadrant bearings* are angles measured eastward or westward of either north or south, whichever happens to be the closer. Examples are shown in Figure 3.12*A*. The direction from a given point to some object on the map is thus written as "N 49° E" or "S 70° W." All bearings range between 0° and 90°. Compass bearings may be magnetic bearings, related to magnetic north, or true bearings, related to geographic north. Unless specifically stated otherwise, a bearing should be assumed to be a true bearing. A disadvantage of compass quadrant bearings is that the same number of degrees can be repeated for four different bearings, once for each of the four combinations of north or south, east or west. This may cause confusion or mistakes.

(*b*) *Azimuths* are used by military services and

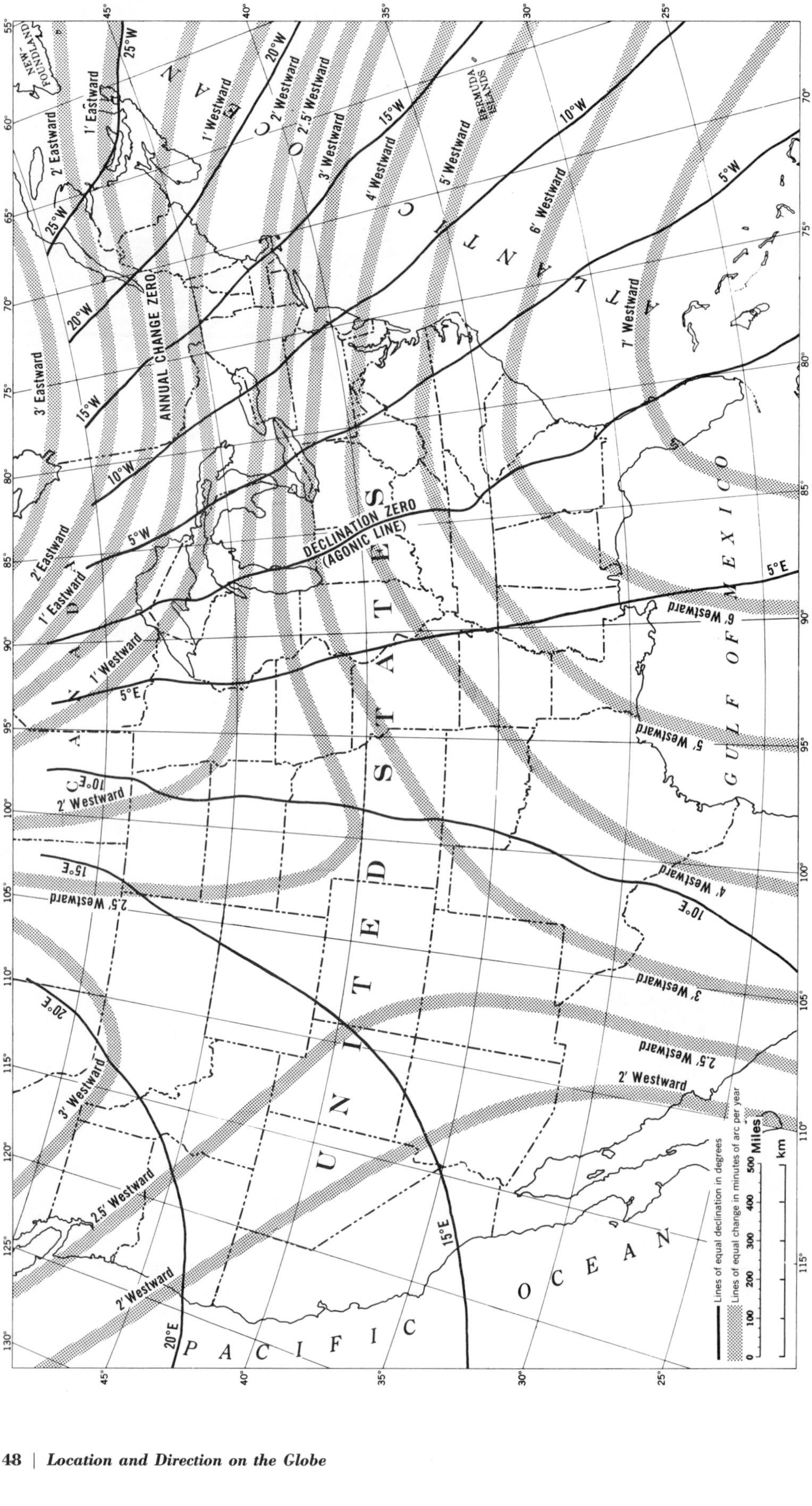

Figure 3.11 Isogonic map of the United States for 1965. (After U.S. Coast and Geodetic Survey.)

in air and marine navigation generally. As shown in Figure 3.12*B*, all azimuths are read in a clockwise direction from north and therefore range between 0° and 360°. There is no repetition of numbers and no use of the words "north," "south," "east," or "west," as in the quadrant bearing system. Azimuths are usually measured from either magnetic north or true north, referred to as *magnetic azimuth* and *true azimuth* respectively. On air navigation charts (Exercise 2), magnetic azimuth (called "bearing")[1] is used. For each radio transmitter, magnetic north at the station serves as the reference line.

With a knowledge of map scales and azimuths (or bearings), we can describe the position of any desired object on a map with reference to a known point. Using the graphic scale, the distance from the known point to the object is measured; then the azimuth or bearing of a line between the two places is read from a protractor laid directly upon the map. For example, a certain farmhouse might be found to lie 512 m from a highway intersection along a true azimuth of 224°. Although this system accurately locates one point with respect to a second point, the location of the second point must still be described in some other manner. We therefore turn next to a study of coordinate systems for the absolute location of points on the earth's surface.

[1]The expression *full-circle bearing* may be used to distinguish this use of "bearing" from the compass-quadrant bearing system.

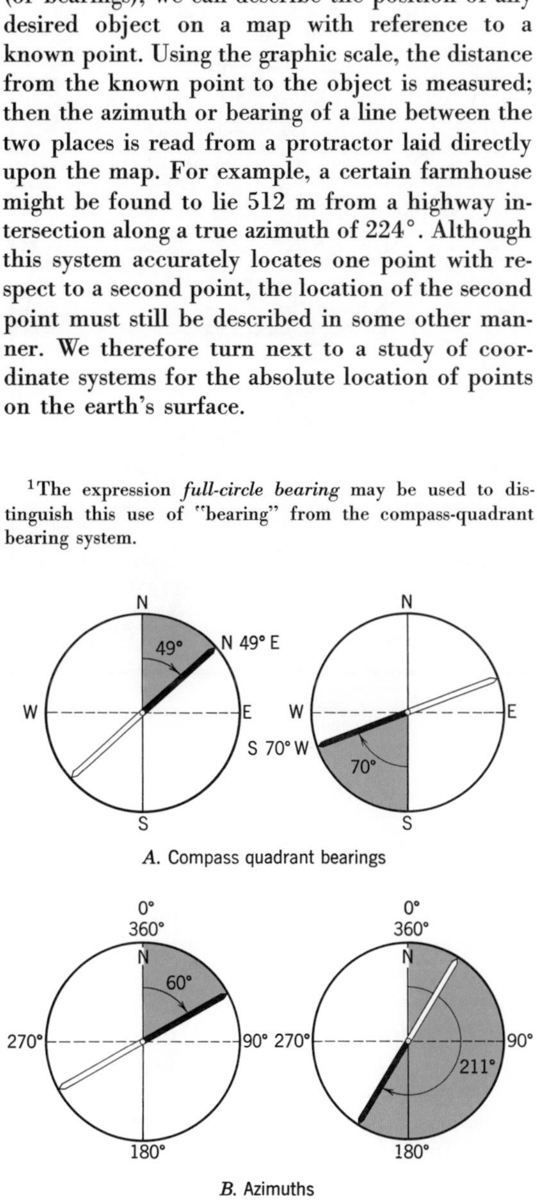

Figure 3.12 Directions are expressed by means of bearings or azimuths.

Coordinate systems used on maps

Any system whereby points on the earth's surface are located with reference to a previously determined set of intersecting lines can be called a *coordinate system*. The student is already familiar with the system of parallels and meridians used throughout the world, and the designation of any point on the earth's surface in terms of latitude and longitude (Chapter 1). This grid constitutes the commonest of the coordinate systems in general use. Two other systems will be included in this discussion: the *military grid* and the *township grid of U.S. Land Survey*.

The geographic grid

A geographer often speaks of the latitude and longitude of a place as *geographic coordinates*, and of the network of parallels and meridians on a globe or map as the *geographic grid* (Chapter 1). Most published sets of large-scale maps use the geographic grid to determine the position and size of individual map sheets in a series. A single map sheet, or *quadrangle*, is bounded on the right- and left-hand margins by meridians, and on the top and bottom by parallels, which are a specified number of minutes or degrees apart. Thus, individual sheets may be fitted together to form unified groups (Figure 3.13). Parallels and meridians are usually labeled at the corners of the map. Often additional meridians and parallels are printed on the map, subdividing the area into smaller rectangles. Where the actual lines are not printed, their positions are sometimes shown by short lines, or *ticks*, at the edges of the map; their intersections are shown by small crosses on the map interior.

Further information on map quadrangles and their relations to scales and areas will be found in Appendix I.

One caution must be taken in the use of foreign maps. Although the system of latitude and longitude based on the prime meridian of Greenwich, England, is widely used, some European countries refer to a meridian of their own choosing, such as the meridian passing through the astronomical observatory in the capital city. For some older European series, such as the German topographic maps on a scale of 1:100,000, a special prime meridian, the *meridian of Ferro*, is used. This meridian passes through Ferro, westernmost island of the Canaries, and has a longitude 17° 14′ W of Greenwich, almost exactly 20° west of the Paris meridian on which French topographic maps are based. It is wise to check the system of a foreign map before determining the location of places in terms of geographic coordinates.

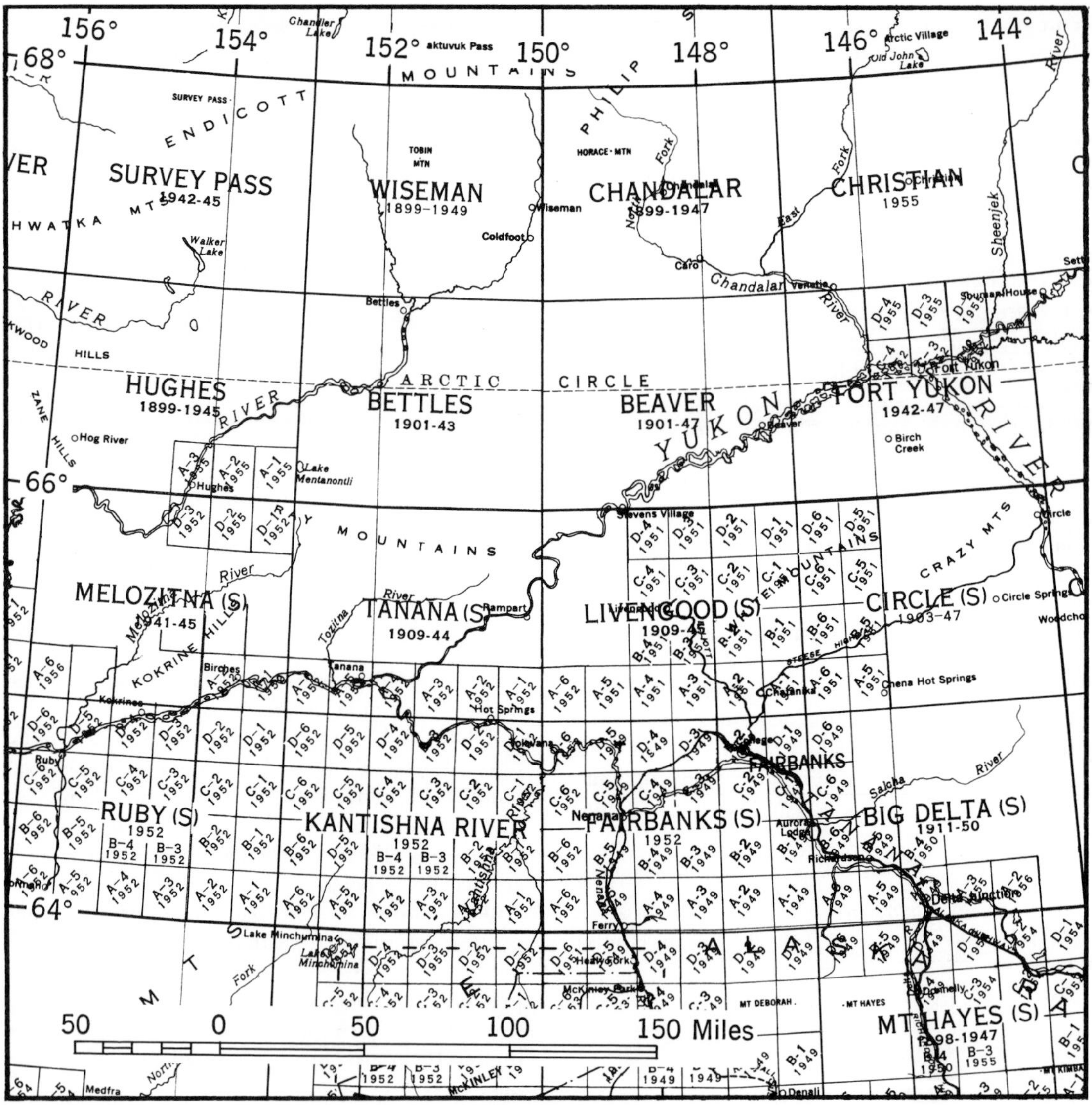

Figure 3.13 This part of the index sheet of Alaska shows the arrangement of quadrangles for scales 1:63,360 and 1:250,000. (U.S. Geological Survey.)

Spherical and plane coordinates

The geographic coordinates may be considered as *spherical coordinates*, because they designate location of points on a spherical (or ellipsoidal) surface. The meridians and parallels are neither straight, equidistantly spaced lines over the globe, nor can they form such a net on any of the useful map projections. Therefore, an entirely different system, that of *plane coordinates*, must be invented to provide a system of straight lines intersecting at perfect right angles on the flat (plane) map sheet, assuming a particular map projection to be used. A grid thus formed consists of true squares on the map and is superimposed upon the geographic grid.

Plane coordinate systems have been set up by individual states of the United States, some of which use the transverse Mercator projection; others use the Lambert conformal conic projection. The grid is scaled in thousands of feet. Many of the large-scale topographic maps of the U.S. Geological Survey indicate the position of 10,000-foot grid lines by ticks on the map margin. Important as these state grids are for surveying and mapping, they will be little used by the geographer. Instead, a single world-wide *military grid system* of plane coordinates will be encountered over a wide range of map scales and geographical locations.

Military grid coordinates

The military grid uses the *meter* as the basic unit of length, although earlier military systems used the yard. The grid is essentially a network of squares, each square 1000 m wide. A portion of a 1000-meter grid is shown in Figure 3.14. Numbers on vertical grid lines increase eastward, toward the right; those on horizontal grid lines increase northward, or upward. Only two digits are printed on most grid lines, those that denote thousands and tens of thousands. In the case of the 1000-meter grid, three zeros have been dropped as well as those digits that denote hundreds of thousands and millions. Full numbers are printed once near the lower left-hand corner of the map. Grid line spacing of 1000 m is used on large-scale maps, 1:100,000 or smaller, whereas a spacing of 10,000 m is used in the grid printed on maps of scale smaller than 1:100,000.

In stating grid coordinates, the number of meters east (right) is given first; then the number of meters north (up). This gives a simple rule: "Read right up." In giving the grid coordinates of a point on the map, the first step is to determine the 1000-meter grid square in which the point lies. A grid square is designated by the grid coordinates of its lower left-hand corner. Thus grid square *A* in Figure 3.14 is designated by the intersection of grid lines *87* east and *80* north. These numbers are written together as *8780*. This is a shorthand notation for coordinates 687,000 m east, 3,800,000 m north.

For a particular point within a grid square, the coordinates may be read to the tenth part of a grid square, which is 100 m. Point *B* of Figure 3.14 lies about four-tenths of the distance from 84 to 85, so that one coordinate is *844* east. It lies about five-tenths of the distance up from 76 to 77, so that the second coordinate is 765. These are written together as a six-digit number: 844765. Should we need to locate the point to the nearest 10 m, still another digit is read by more precise measurements. Thus Point *C* is found to have coordinates *8715* east and *7783* north; written as a single number: 87157783. In all three examples, above, an even number of digits forms the combined number. Therefore, given a coordinate designation, the number is broken in half. The first half is taken as the *easting*; the second half as the *northing*.

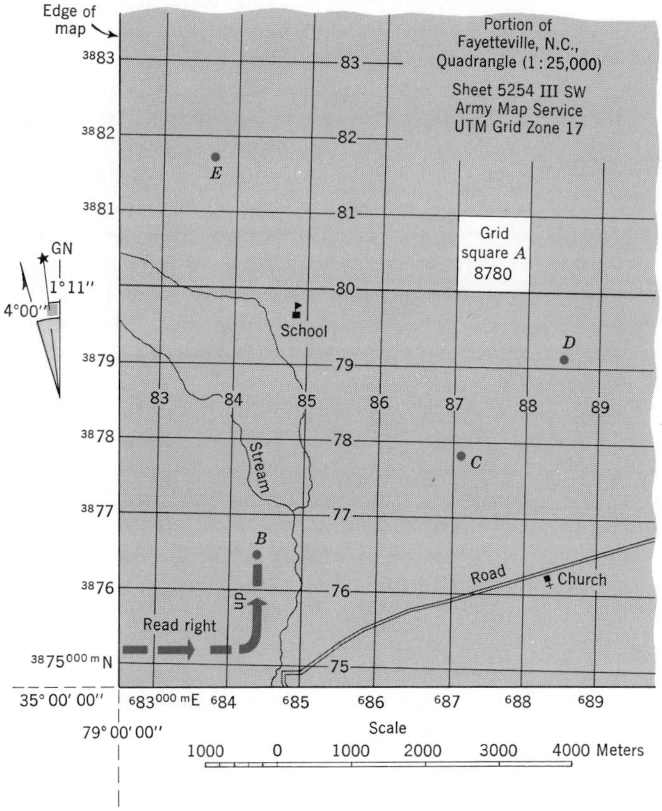

Figure 3.14 Military grid coordinates on the 1000-meter grid.

Grid zones

Grid coordinates of any small area belong to a particular *grid zone.* An international system provides military grid zones for the entire globe. Between lat. 80° S and 80° N the *Universal Transverse Mercator Grid System* is used; poleward of 80°, the *Universal Polar Sterographic Grid System* is used. These systems are named for the map projections on which they are based. (Both projections are explained in Chapter 2.)

The Universal Transverse Mercator grid system, hereafter referred to as the *UTM* grid, consists of 60 grid zones, each 6° of longitude in width (Figure 3.15). An additional one-half degree on each side provides for overlap into the adjacent zone. The *origin* of the grid zone lies at the intersection of the central meridian, which is a straight north-south line, and the equator, which is a straight east-west line. In order to have all eastings increase toward the right across the entire zone, the central meridian is given the arbitrary value of *500,000 meters east.* The equator is given the value of *0 meters north* as the reference line for northings increasing upward to the 80th parallel north. For the southern hemisphere, the equator is given the arbitrary northing of *10 million meters north,* so that northings begin with their lowest value at 80° S lat. and increase northward to attain that figure at the equator.

The relationship of grid lines to parallels and meridians in various parts of a grid zone is shown in Figure 3.15. Near the equator, both sets of lines are approximately parallel to one another. Toward higher latitudes, the two sets of lines diverge increasingly, because the meridians converge whereas the grid lines remain equidistant. The grid zone itself narrows greatly because it is bounded by meridians. This leads to a definition of a second kind of declination, *grid declination,* which is the angle between *grid north* (the direction taken by the vertical grid lines) and true (geographic) north. Grid declination can be read directly with a protractor placed on the angle formed between a grid line and the meridian marking the edge of the map. The angle between grid north and magnetic north is termed the *grid magnetic (GM) angle.* Large-scale military maps, such as those produced by the U.S. Army Map Service, carry in the lower margin a diagram telling the relations among the three norths (Figure 3.16). Three kinds of azimuth are also possible on a military map; *grid azimuth, true azimuth,* and *magnetic azimuth* (Figure 3.17).

The Universal Polar Stereographic Grid system, hereafter designated the *UPS* grid, is superimposed upon a polar stereographic projection within the circle formed by the 80th parallel (Figure 3.18). An additional half degree of latitude is provided for overlap with the UTM grid. The vertical grid lines are parallel with meridians of 0° and 180° longitude; the horizontal grid lines are parallel with meridians of 90° E and 90° W long. Although the origin of the grid is the pole, this point is given an arbitrary easting and northing of 2 million meters. Grid declination therefore ranges from 0° to 180°, depending upon location. The North Zone is shown in Figure 3.18. The South Zone, applied to the Antarctic region, is essentially the same except for a reversal in position of the 0° and 180° meridians which form the central meridian of the projection.

Further details of UTM and UPS grid systems are given in Appendix I.

The United States Land Survey

Students of American geography and history may frequently run across maps showing the division of lands in the central and western United States. The survey lines have exerted a powerful control on the size, shape, and distribution of farmsteads, homes, townships, and counties, and on the location of roads (Figure 3.19).

In 1785, Congress authorized a survey of the territory lying north and west of the Ohio River. To avoid the irregular and unsystematic type of land subdivision that had grown up in seaboard states during colonial times, Congress specified that the new lands should be divided into six-mile squares, now called *congressional townships,* and that the grid of townships should be based upon a carefully surveyed east-west base line, designated the "geographer's line." Meridians and parallels laid off at six-mile intervals from the base line were to form the boundaries of the townships. This general plan, believed to have been proposed by Thomas Jefferson, was subsequently carried out to cover the balance of the central and western states.

The survey began in 1786 under the direction of Thomas Hutchins, who had been appointed *Geographer of the United States.* With thirteen assistants, Hutchins commenced work at the southwest corner of the state of Pennsylvania. From this point he laid off a line due north to the north bank of the Ohio River. From there he ran his "geographer's line" due west for a distance of 42 mi (68 km), when the work was suspended because of danger of attack by hostile Indians. The line was later carried entirely across what is now the state of Ohio.

The *principal meridians* and *base lines,* from which rows of townships were laid off, are shown in Figure 3.20. Principal meridians run north or south, or both, from selected points whose latitude

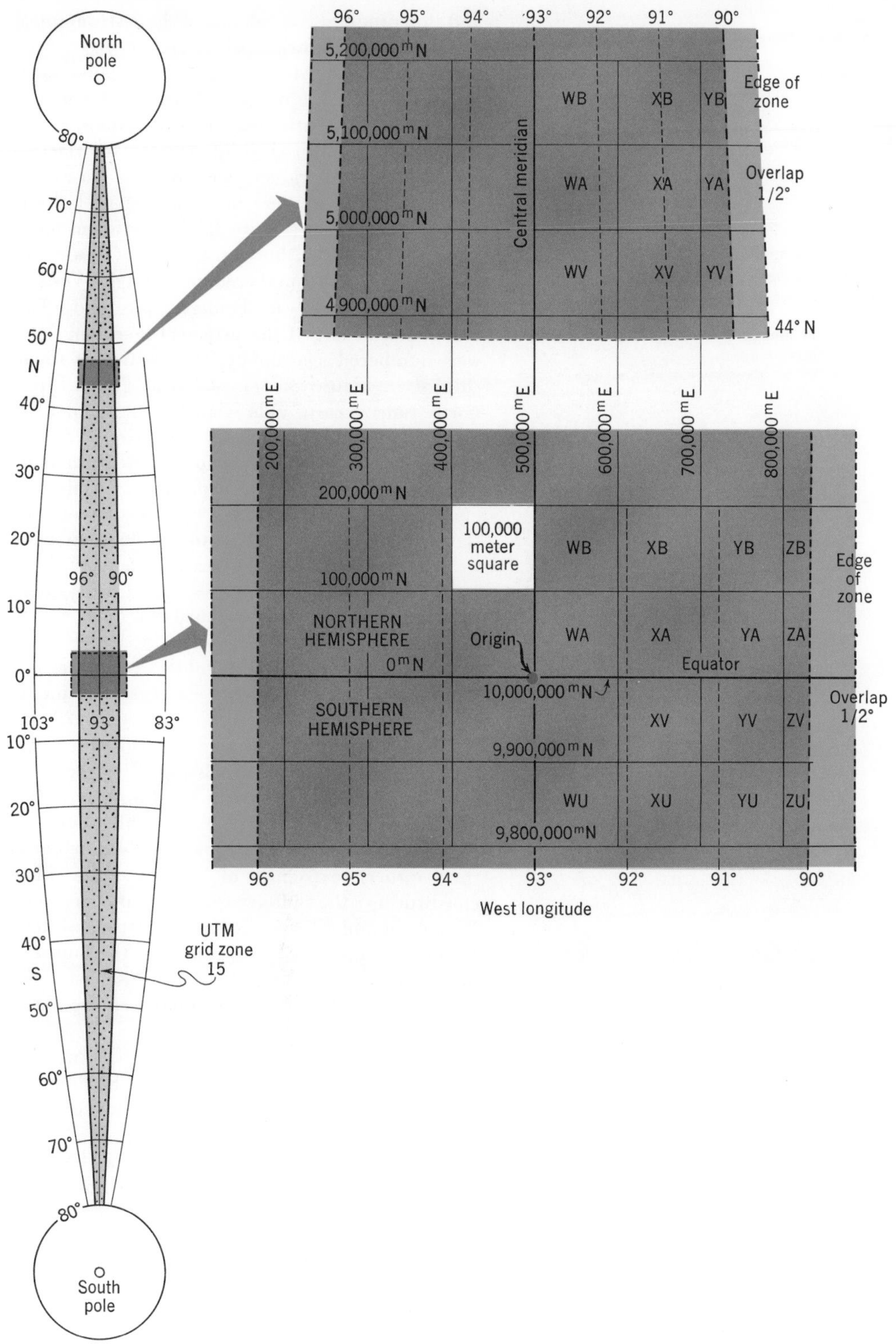

Figure 3.15 The UTM grid zone near the equator and at 45° N. (Data from U.S. Departments of the Army and Air Force, TM, 5-241, 1951.)

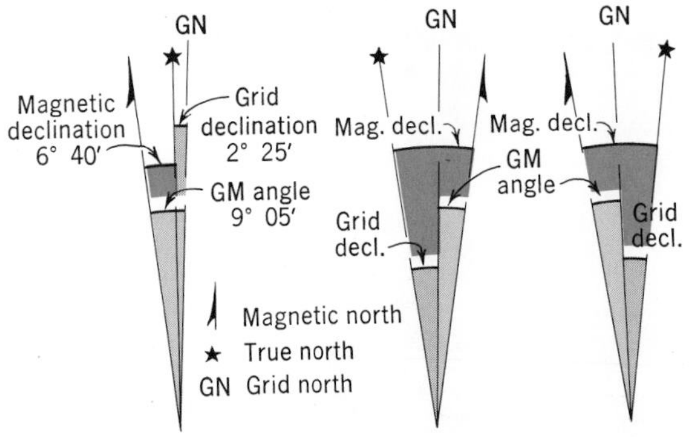

Figure 3.16 A special marginal symbol shows the relations among three kinds of north.

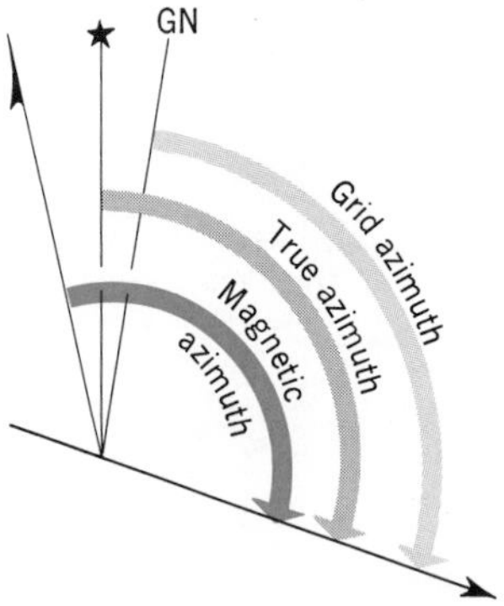

Figure 3.17 Azimuth may be measured with respect to three norths.

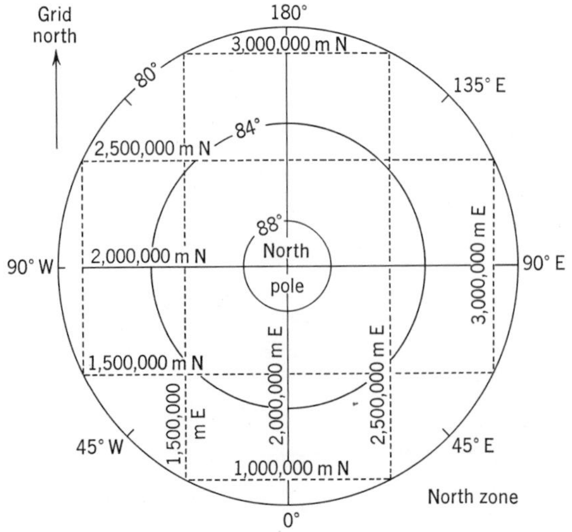

Figure 3.18 The Universal Polar Stereographic Grid, North Zone. (After U.S. Department of the Army and Air Force, TM 5-241, 1951.)

and longitude were calculated by astronomical means. Some 32 principal meridians have been surveyed. Westward from the Ohio-Pennsylvania boundary these are numbered from 1 through 6, after which they are designated by names.

Through the initial point selected for starting the principal meridian, an east-west base line was run, corresponding to a parallel of latitude through that point (Figure 3.21). North and south from the base line, horizontal tiers of townships were laid off and numbered accordingly. Vertical rows of townships, called *ranges*, were laid off to the right and left of the principal meridians, and were numbered accordingly. The position of Township *A* in Figure 3.21 is stated as "township 2 north, range 3 east," abbreviated "T. 2 N., R. 3 E." Township *B* is designated "T. 3 S., R. 6 W." The area governed by one principal meridian and its base line is restricted to a particular section of country, usually about as large as one or two states. Where two systems of townships meet, they do not correspond, because each system was built up independently of the others. Of all the systems, that based on the 5th principal meridian covers the largest area. Its most northerly tier of townships is number 163; the most westerly range, number 104. In all, it has an east-west spread covering 726 mi, a north-south extent of 1122 mi.

Notice that the 4th principal meridian is interrupted by a bend of the Mississippi River between Iowa and Illinois, as well as by the western end of Lake Superior (Figure 3.20). This system is unusual in having two base lines, one situated near the southern extremity of the zone, the other constituting the interstate boundary between Wisconsin and Illinois. Townships within western Illinois are numbered with respect to the southern of the two base lines; those of Wisconsin and northeastern Minnesota are numbered from the northern base line.

A number of small systems, each with its own base line and principal meridian, will be found to lie as enclaves within larger systems. Most of these are for the survey of Indian reservations. Examples shown in Figure 3.20 are the Uinta, Ute, and Navajo meridians with their respective base lines. The Cimarron meridian and base line, covering what is now the panhandle of Oklahoma, was originally applied to a single tract of land, the Cimarron Ranch.

Because the range lines, on eastern and western boundaries of townships, are meridians converging slightly as they are extended northward, the width of townships is progressively diminished in a northward direction. In order to avoid a considerable reduction in township widths in the more northerly tiers, new base-lines, known as *standard parallels*, are surveyed for every four tiers of townships.

Figure 3.19 Iowa farm lands laid out on the Land Office grid (Aero Service, Litton Industries).

They are designated 1st, 2nd, 3rd Standard Parallel N, etc. (Figure 3.21). The ranges will be found to offset at the standard parallels, and in consequence, roads that follow range lines make an offset or jog when crossing standard parallels (Figure 3.22). When the Dakota Territory was divided to enter the Union as two states, the boundary was placed on the 7th standard parallel from the base

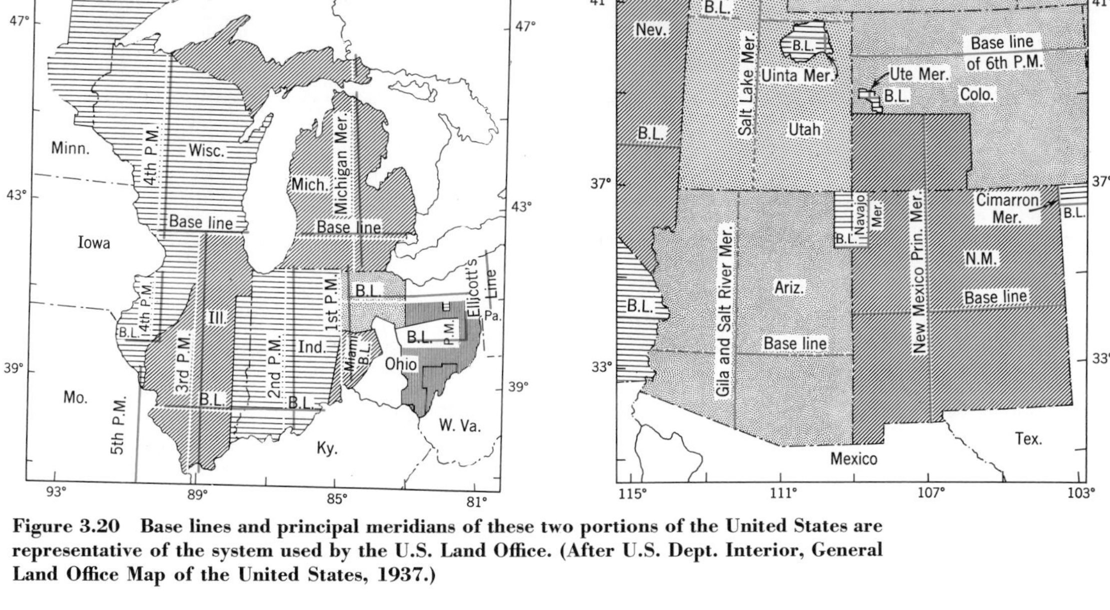

Figure 3.20 Base lines and principal meridians of these two portions of the United States are representative of the system used by the U.S. Land Office. (After U.S. Dept. Interior, General Land Office Map of the United States, 1937.)

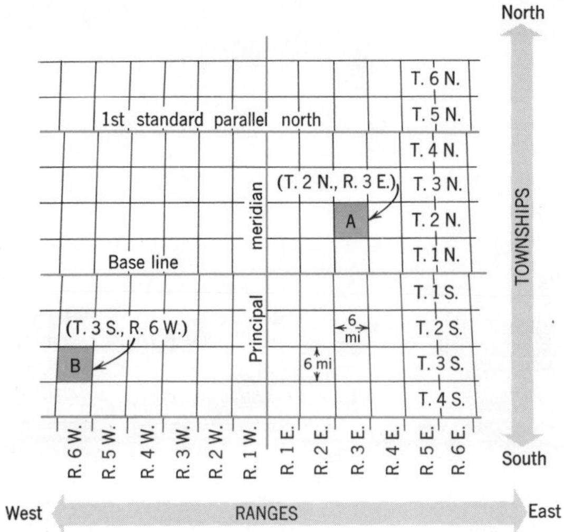

Figure 3.21 Designation of townships and ranges.

line of the 5th principal meridian in preference to the 46th parallel of latitude which lies four miles to the north of it, because the former line was already in use as a boundary between farms, sections, townships, and counties.

Subdivisions of the township are square-mile *sections*, of which there are 36 to the township. These are numbered in the manner illustrated in Figure 3.23. Each section may be subdivided into halves, quarters, and half quarters, or even smaller units. These divisions, together with the number of acres contained in each, are illustrated in Figure 3.24. Abbreviations are read as follows: "N $\frac{1}{2}$ SW $\frac{1}{4}$" as "north half of the southwest quarter"; "SW $\frac{1}{4}$ SW $\frac{1}{4}$ SW $\frac{1}{4}$" as "southwest quarter of the southwest quarter of the southwest quarter." Abbreviation of the location of an area might read as follows: NE $\frac{1}{4}$ of SE $\frac{1}{4}$ of Sec. 24, T. 28 N., R. 6 W., 5 PM meaning "the northeast quarter of the southeast quarter of section No. 24, township 28 north, range 6 west of the 5th principal meridian."

R. 6 W.

6	5	4	3	2	1
7	8	9	10	11	12
18	17	16	15	14	13
19	20	21	22	23	24
30	29	28	27	26	25
31	32	33	34	35	36

T. 2 S.

Figure 3.23 A township is divided into 36 sections, each one a square mile.

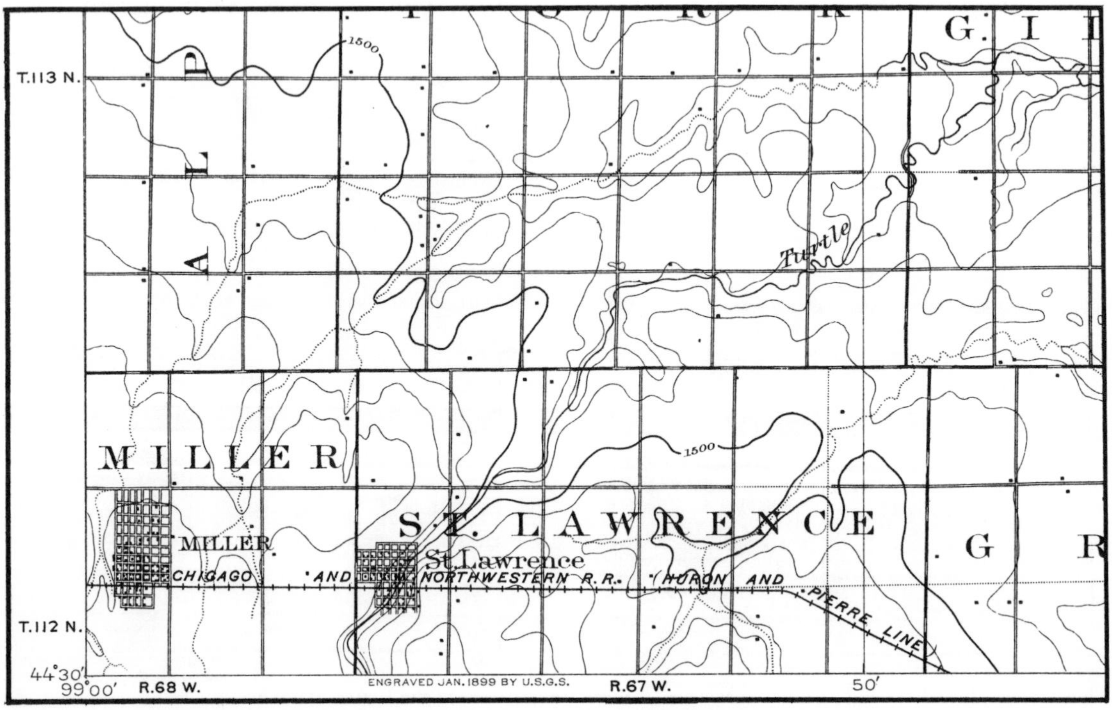

Figure 3.22 This portion of the Redfield, South Dakota, Quadrangle shows the offsetting of ranges along a standard parallel. (U.S. Geological Survey topographic map.)

The unit of length used in the U.S. Land Office survey is the surveyor's *chain*, consisting of 100 *links*. The chain is equal to 66 ft (20.1 m) and there are 80 chains to the statute mile. A single land section, or one square mile, measuring 80 chains on a side, contains 6400 square chains, or 640 acres. The value of 80 is subdivisible by halving into 40, 20, and 10 chains. Thus the smallest land unit shown in Figure 3.24 is a plot of 10 acres, measuring 10 chains on a side.

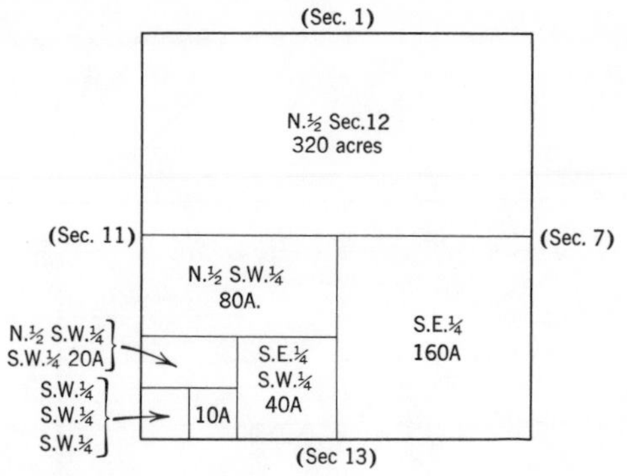

Figure 3.24 **A section may be subdivided into many units. (After Willis E. Johnson.)**

Review Questions

1. What is the scope of cartography? of geodesy? What role does each play in location of earth features on the globe and on maps?

2. Distinguish between relative location and absolute location of points on the earth's surface.

3. What is the principle of triangulation? How and why is a base line measured? What orders of accuracy are defined in triangulation?

4. What is the North American Datum of 1927? Why is it needed? Who uses it?

5. How is precise leveling carried out? What datum is used? How are leveling control points designated and preserved?

6. What problem arises in stating the direction of a line drawn between two distant points on the globe?

7. Describe the earth's magnetic field. Define the magnetic poles and magnetic equator.

8. Explain magnetic declination of the compass. How is declination taken into account when determining geographic north by means of a compass?

9. Explain the compass quadrant bearing system of designating direction. How does it differ from azimuth, or full-circle bearing?

10. Explain the geographic coordinates and show how the geographic grid is used in limiting individual quadrangles of map system. What is the meridian of Ferro?

11. What advantages have plane coordinates over spherical coordinates in map grids? Describe the plane coordinates used by the states.

12. How can the position of a point on a map be designated by military grid coordinates? What abbreviations are used? What units of length are used? What are northing and easting?

13. Describe in detail a grid zone in the UTM grid system. How is it related to the map projection on which it is based?

14. Define grid declination, grid north, and the grid magnetic angle. List three kinds of azimuth and define each.

15. Describe in detail the UPS grid system and its relation to the projection on which it is based.

16. Explain the U.S. Land Office system of subdivision of western lands. Include discussion of principal meridian, base line, township, range, standard parallel, and section. How are townships designated? How are sections subdivided?

Exercises

1. See Figure 3.11. Estimate the magnetic declination at the following places. (*a*) Miami, Florida. (*b*) Boston, Massachusetts. (*c*) San Francisco, California. (*d*) New Orleans, Louisiana. (*e*) Chicago, Illinois. (*f*) Vancouver, British Columbia. (*g*) Halifax, Nova Scotia.

2. See Figure 3.14. (*a*) Give the grid designation of the square in which the letter *D* is printed. (*b*) Give the grid coordinates to the nearest 100 m of the point *E*. (*c*) Using a protractor, measure the grid declination. (*d*) Measure the grid azimuth of a line from Point *C* to the center of grid square *A*. (*e*) Measure the grid azimuth from Point *B* to a point (unmarked) located at 833794.

3. See Figure 3.22. (*a*) On a sheet of thin tracing paper laid over this map, number all sections with their correct numbers according to the system shown in Figure 3.23. (*b*) State accurately and fully the location of the letter "T" of the word "Turtle" on this map, using the Land Office system illustrated in Figures 3.21, 3.23, and 3.24.

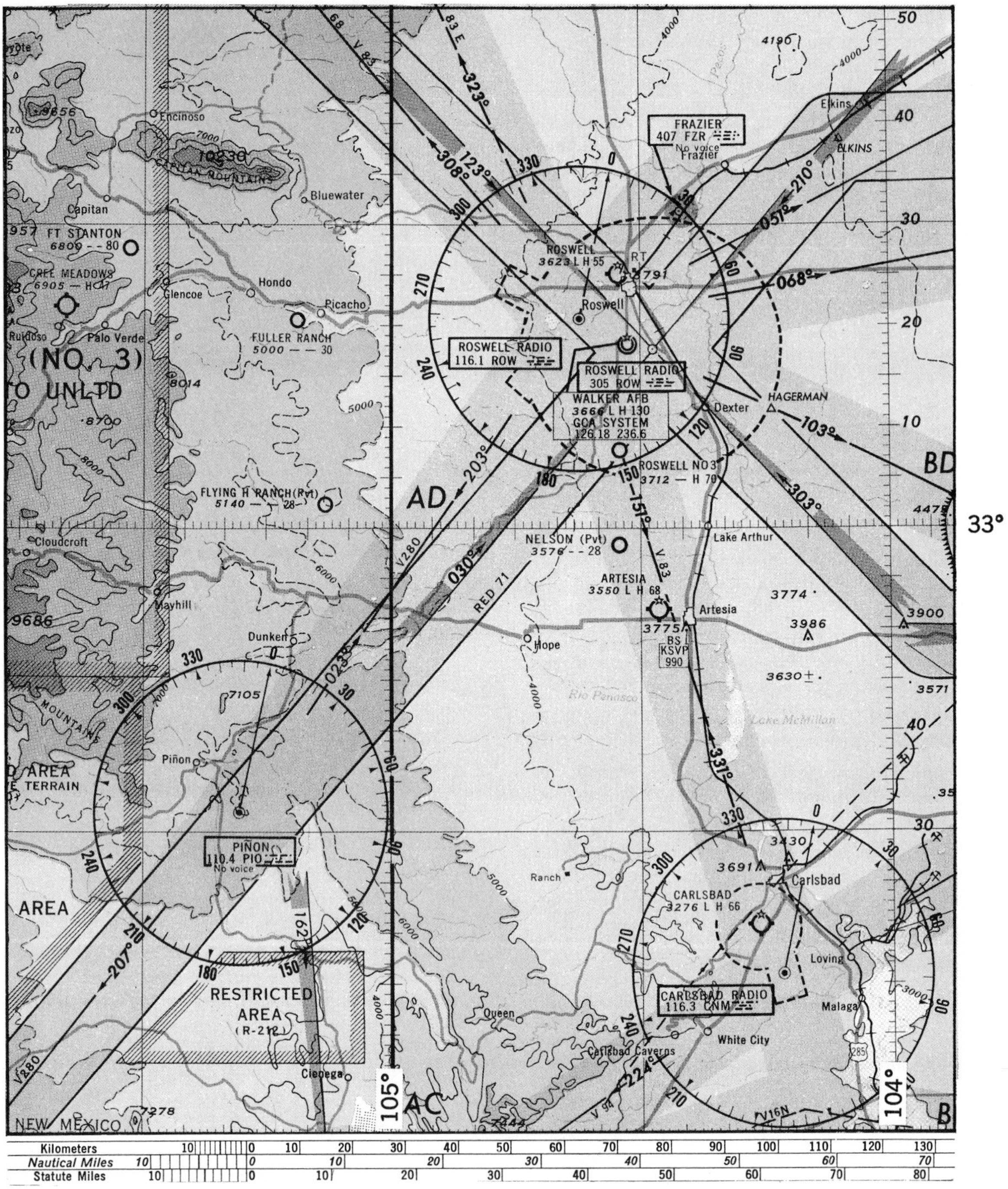

4. Refer to the accompanying air-navigation exercise map. This is a reduced portion of Sheet 406 of the World Aeronautical Chart issued on the scale of 1:1,000,000. (*a*) Give the geographic coordinates to the nearest one minute for Artesia (town), Roswell Radio, and Carlsbad Airport. (*b*) Using a protractor, estimate the magnetic declination in this area. (*c*) What is the magnetic bearing of a line from Carlsbad Airport to Pinon Radio? (*d*) What is the magnetic bearing of a line from Fuller Ranch Airport to Roswell Radio?

CHAPTER 4

Illumination of the Globe

OF the utmost importance to man is the relation between the earth and the sun's rays. The angles at which rays strike the earth at different latitudes and at different times of day and year determine the apparent path of the sun in the sky, the lengths of day and night, and the occurrence of seasons. Rhythmic variations in the rates of receipt of solar energy by different parts of the earth at different times act as fundamental controls of atmospheric temperatures, which in turn have a major effect on air pressure, winds, precipitation, storms, and oceanic circulation—all of which taken together make up the earth's varied climates. That is why one must thoroughly master earth-sun relationships before going ahead to the subjects of weather and climate. Because the earth is turning on its axis at the same time that it is moving in a path about the sun, and because the earth's axis is tilted with respect to the plane of its orbit, these relationships are often difficult to understand. We must learn to think in terms of three dimensions, to imagine ourselves viewing the earth from various vantage points in space; then imagine the same situations as they would appear to an observer standing at various points on the earth.

Rotation of the earth

The spinning of the earth on its polar axis is termed *rotation.* In this study of earth-sun relationships we use the period of rotation, the *mean solar day*, consisting of 24 mean solar hours. This day is the average time required for the earth to make one complete turn in respect to the sun.

Direction of earth rotation can be determined by applying one of the following rules. (*a*) If we imagine ourselves to be looking down upon the north pole of the earth, the direction of turning is counterclockwise. (*b*) If we place a finger upon a point on a globe near the equator and push the finger eastward, it will cause the globe to rotate in the correct direction (Figure 4.1). This demonstrates a common expression, "eastward rotation of the earth." (*c*) Direction of earth rotation is opposite that of the apparent motion of the sun, moon, and stars. Because these bodies appear to travel westward across the sky the earth must be turning in an eastward direction.

The velocity of rotation, defined as rate of travel of a point on the earth's surface in a circular path due to rotation alone, may easily be computed by dividing the length of parallel at the latitude of the point in question by 24, the approximate period of rotation. Thus, at the equator, where the circumference is about 25,000 mi (40,000 km), the velocity of an object on the surface is about 1050 mi (1700 km) per hour. At the 60th parallel the velocity is half this amount, or about 525 mi (850 km) per hour. At the poles it is, of course, zero. We are unaware of this motion because the rotation is at an almost perfectly constant rate.

Two important physical phenomena result from the decrease in rotational velocity with increase in latitude. First, there is a *centrifugal force* generated by the earth's turning which gives surface objects a faint tendency to fly off into space. Because the force of gravity is 289 times greater than this centrifugal force at the equator, objects cannot leave the surface, but the practical effect is to reduce the weight of objects slightly. Near the equator, where the centrifugal force is strongest, this effect is most marked. For example, an object that would weigh 289 pounds at the equator if the earth were not turning actually weighs 1 pound less.

Another effect of the decreasing rotational velocity with increasing latitude is to cause objects in motion to be deflected slightly to the right or left of their paths. This effect will be more fully treated under the subject of the earth's wind systems.

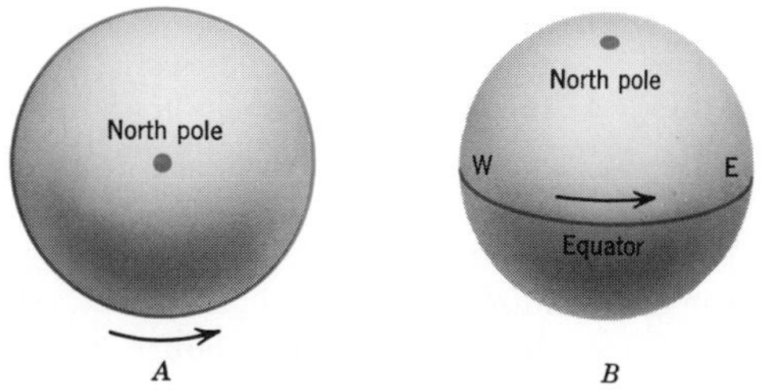

Figure 4.1 The direction of rotation of the earth can be thought of as (*A*) counterclockwise at the north pole, or (*B*) eastward at the equator.

How constant is the rate of the earth's rotation? Is the earth "slowing down"? Astronomers have calculated that the length of day should be increasing by about 0.0016 sec per century because of the braking action of the tides. This small quantity can be disregarded in the study of physical geography. We may confidently assume that the earth's turning with respect to the stars provides an almost perfect timepiece. Other extremely minute variations, both irregular and seasonal, can also be detected in the earth's period of rotation, but these, too, can be disregarded.

Proof of the earth's rotation; the Foucault experiment

The several kinds of proofs offered for the earth's rotation are beyond the needs of the student of physical geography, but one proof in particular, the *Foucault experiment*, is so outstanding as to warrant study. It has been described as follows.[1]

> In 1851, the French physicist, M. Leon Foucault, suspended from the dome of the Pantheon, in Paris, a heavy iron ball by wire 200 feet long. A pin was fastened to the lowest side of the ball so that when swinging it traced a slight mark in a layer of sand placed beneath it. Carefully the long pendulum was set swinging. It was found that the path gradually moved around toward the right. (See Figure 4.2.) Now either the pendulum changed its plane or the building was gradually turned around. By experimenting with a ball suspended from a ruler one can readily see that turning the ruler will not change the plane of the swinging pendulum. If the pendulum swings back and forth in a north and south direction, the ruler can be entirely turned around without changing the direction of the pendulum's swing. If at the north pole a pendulum was set swinging toward a fixed star, say Arcturus, it would continue swinging toward the same star and the earth would thus be seen to turn around in a day. The earth would not seem to turn but the pendulum would seem to deviate toward the right, or clockwise.
>
> At first thought it might seem as though the floor would turn completely around under the pendulum in a day, regardless of the latitude. It will be readily seen, however, that it is only at the pole that the earth would make one complete rotation under the pendulum in one day, or show a deviation of 15° in an hour. At the equator the pendulum will show no deviation, and at intermediate latitudes the rate of deviation varies.

Table 4.1 gives the hourly change in direction of the pendulum's swing and the total time required for the direction to change through 360°.

[1] W. E. Johnson, *Mathematical Geography*, American Book Co., New York, 1907, pp. 54–57.

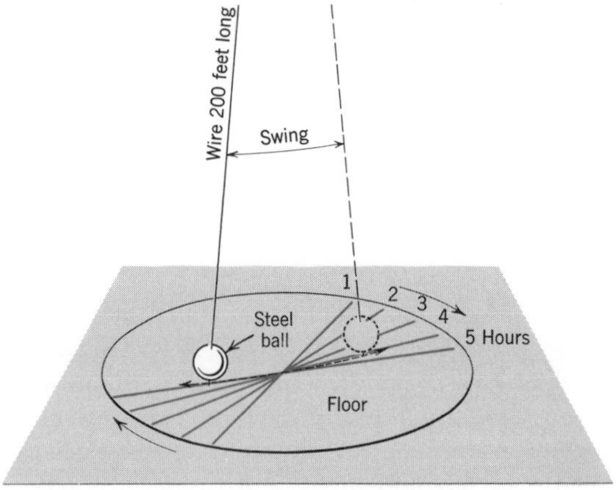

Figure 4.2 The Foucault pendulum experiment proves earth rotation.

For the student familiar with elementary trigonometry, it may be explained that the amount of turning of the pendulum's direction per hour varies according to the *sine of the latitude* and may be obtained by the formula

$$d = 15 \sin L$$

where d = number of degrees of turning per hour and L = latitude.

Another proof of the earth's rotation lies in the oblate ellipsoidal form of the earth. In order to explain the bulging at the equator and shortening of the polar axis, centrifugal force caused by rotation on an axis is required.

TABLE 4.1

Latitude	Hourly change in Pendulum Direction (Degrees)	Total time for 360° change in Direction (Hours)
0°	None	None
5	1.31	275
10	2.60	138
15	3.88	93
20	5.13	70
25	6.34	57
30	7.50	48
35	8.60	42
40	9.64	37
45	10.61	34
50	11.49	31
55	12.29	29
60	12.99	28
65	13.59	26.5
70	14.10	25.6
75	14.49	24.9
80	14.77	24.5
85	14.94	24.1
90	15.00	24.0

Revolution of the earth

The motion of the earth in its orbit around the sun is termed *revolution*. Care should be taken to use the terms rotation and revolution correctly and not to use them interchangeably. The period of revolution, or year, is the time required for the earth to complete one circuit around the sun. This may be measured in different ways. For example, the time required for the earth to return to a given point in its orbit with reference to the fixed stars is called the *sidereal year*. The period of time from one vernal equinox to the next is the *tropical year*, which has a length of $365^d\ 5^h\ 48^m\ 45.68^s$, or approximately $365\frac{1}{4}$ days. Every four years the extra one-fourth day difference between the tropical year and the calendar year of 365 days totals nearly one whole day. By inserting a 29th day in February every leap year, we are able to correct the calendar with respect to the tropical year. Further minor corrections are necessary to perfect this system, but these are beyond the scope of present discussion.

In its orbit the earth moves in such a direction that if we imagine ourselves in space, looking down upon the earth and sun so as to see the north pole of the earth, the earth is traveling counterclockwise around the sun (Figure 4.3). It is further worth noting that nearly all planets and their satellites in our solar system have the same direction of rotation and revolution, suggesting that their motions were imparted to them at a time when they and the sun were originally formed.

Earth's orbit

The earth's orbit is an ellipse, rather than a circle, although the *ellipticity*, or degree of flattening of the ellipse, is very slight. The sun occupies one *focus* of the ellipse. Ellipses of various shapes are easily constructed using only a drawing board, two pins or thumb tacks, a piece of thread or thin string, and a pencil, as illustrated in Figure 4.4. The thread is made into a loop which passes around the two pins and serves to guide the pencil point. By this device we maintain the sum of the distances from pencil point to each pin always the same. Each pin is located at one *focus* of the ellipse. The two *foci* lie on a line which is the maximum diameter of the ellipse and is called the *major axis*. The shortest diameter, drawn at right angles to the major axis, is known as the *minor axis*. The size of the ellipse may be controlled by the length of the loop of thread, whereas the ellipticity may be controlled by changing the distance between the two foci.

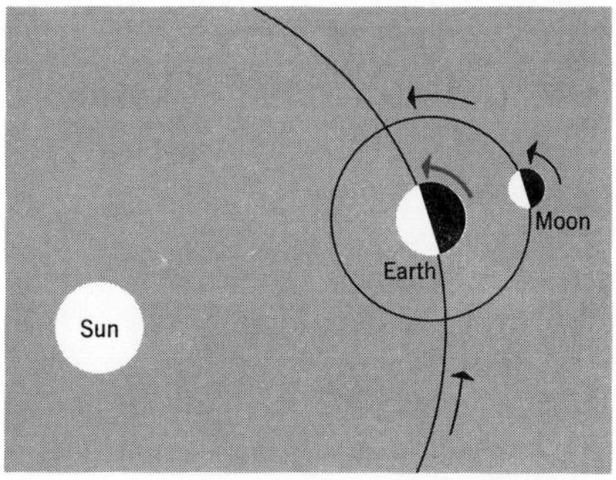

Figure 4.3 Viewed as if from a point over the earth's north pole, the earth both rotates and revolves in a counterclockwise direction.

Perihelion and aphelion

The mean distance between earth and sun is about 93 million mi (150 million km), but because of the ellipticity of the orbit the distance may be 1.5 million mi (2.4 million km) greater or less than this figure (Figure 4.5). The distance is least, or about 91.5 million mi (147 million km), on about January 3, at which time the earth is said to be in *perihelion* (from the Greek *peri*, around or near; and *helios*, the sun). On about July 4, the earth is at its farthest point from the sun, or in *aphelion* (from the Greek *ap*, away from; *helios*, sun), at a distance of 94.5 million mi (152 million km).

These differences in distance do cause some difference in the amount of solar energy received by the earth, but they are not the cause of summer and winter seasons. This is obvious because perihelion, when the earth should receive most heat, falls at the coldest time of year in the northern hemisphere. Moreover, opposite seasons are present simultaneously in the northern and southern hemispheres, proving that another cause exists. It is probable, however, that if all other conditions were considered to be equal,

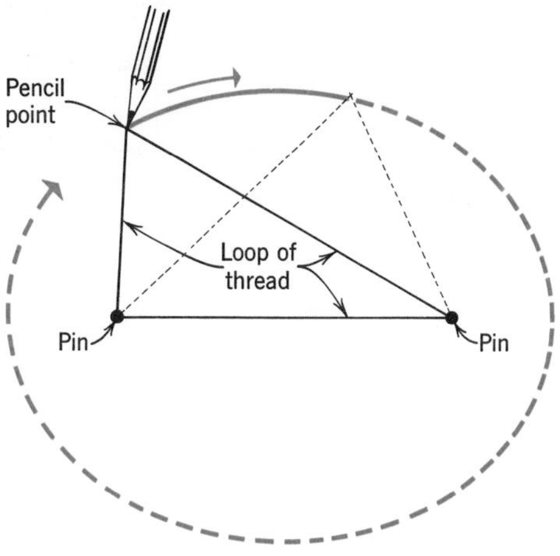

Figure 4.4 An ellipse can be easily constructed.

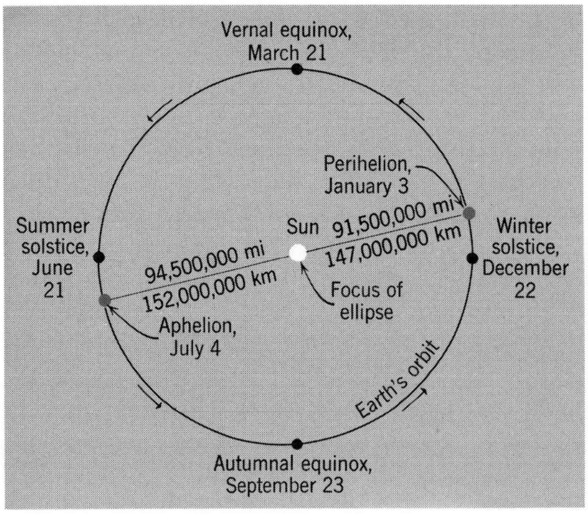

Figure 4.5 The earth's orbit is an ellipse in which the sun is located at one focus.

summers and winters would be slightly intensified in the southern hemisphere and slightly moderated in the northern hemisphere as a result of the relation of the dates of perihelion and aphelion to the summer and winter seasons.

The mean velocity of the earth in its orbit is about 66,600 mi (107,000 km) per hour but varies according to the part of the orbit occupied. The velocity is greatest at perihelion, least at aphelion. The cause and importance of this variation are discussed in Chapter 5.

Inclination of the earth's axis

Most globes that are made to rotate on the polar axis are fixed in a tilted position. So accustomed are we to seeing the earth represented this way that a globe with its axis vertical seems unnatural. For tilted globes the plane in which the earth's orbit and the sun lie, or the *plane of the ecliptic,* is imagined to be horizontal and to pass through the center of the globe (Figure 4.6). The trace of the plane of the ecliptic upon the globe is a great circle. Most globes have an *ecliptic circle* drawn on them. It will be seen to cut across the equator at opposite points on the globe (from the rule that intersecting great circles bisect each other) and to run as far north as $23\frac{1}{2}°$ N lat., and as far south as $23\frac{1}{2}°$ S lat. By rotating the globe slowly, a position will be found in which the ecliptic circle lies in a horizontal plane parallel with the table top. In this position the globe may be used to illustrate the fact that the plane of the equator is inclined $23\frac{1}{2}°$ with the plane of the ecliptic. More exactly, this angle is 23°27′, but the difference between this value and $23\frac{1}{2}°$ can be disregarded. The earth's axis makes an angle of $66\frac{1}{2}°$ with the plane of the ecliptic, and is tilted $23\frac{1}{2}°$ from a line perpendicular to that plane. No other single fact connected with earth-sun relationships is so important as the inclination of the earth's axis.

The earth's axis, although always making the angle $66\frac{1}{2}°$ with the plane of the ecliptic, maintains a fixed orientation with respect to the stars.[2] The earth's axis continues to point to the same spot in the heavens as it makes its yearly circuit

[2]This statement does not take into account the precessional motion of the axis.

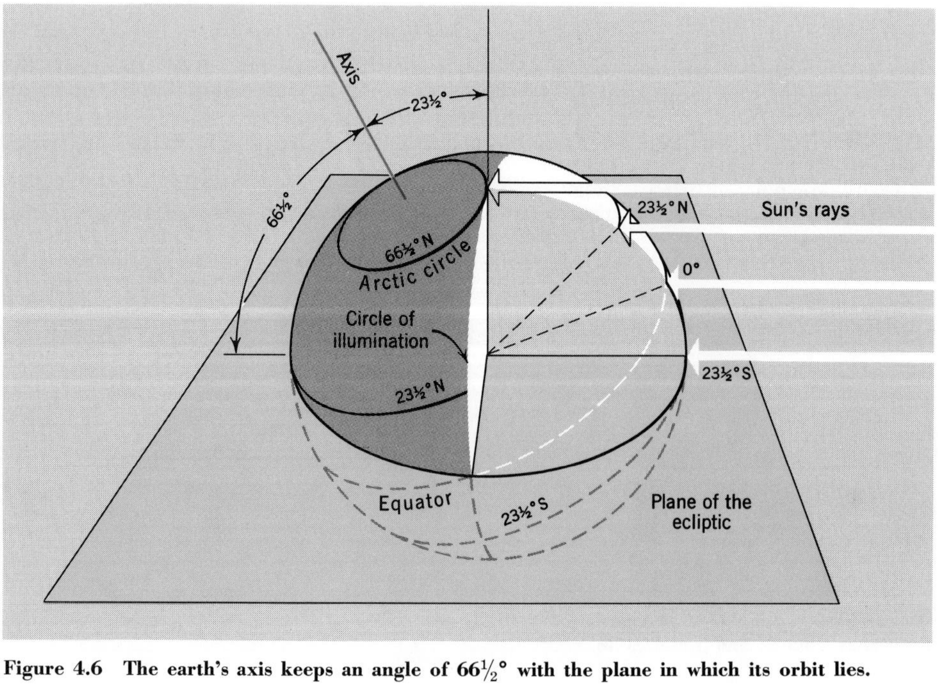

Figure 4.6 The earth's axis keeps an angle of $66\frac{1}{2}°$ with the plane in which its orbit lies.

around the sun. To help in visualizing this movement, hold a globe so as always to keep the axis tilted at $66\frac{1}{2}^\circ$ with horizontal. Move the globe in a small horizontal circle, representing the orbit, at the same time keeping the axis pointed at the same point on the ceiling.

As a direct consequence of the facts (1) that the earth's axis keeps a fixed angle with the plane of the ecliptic, and (2) that the axis always points to the same place among the stars, it will be seen that at one point in its orbit the earth's axis leans toward the sun, that at an opposite point in the orbit the axis leans away from the sun, and at the two intermediate points the axis leans neither toward, nor away from, the sun (Figure 4.7). Here again, a globe or, better still, four globes, may be used to aid in visualizing the facts. The four critical positions will be treated in detail.

Solstice and equinox

On June 21 or 22 the earth is so located in its orbit that the north polar end of its axis leans at the maximum angle $23\frac{1}{2}^\circ$ toward the sun. The northern hemisphere is tipped toward the sun, and the southern hemisphere is tipped away from the sun. This condition is named the *summer solstice.* Six months later, on December 21 or 22, the earth is in an equivalent position on the opposite point in its orbit. At this time, known as the *winter solstice,* the axis again is at a maximum inclination with respect to a line drawn to the sun, but now it is the southern hemisphere that is tipped toward the sun.

Midway between the dates of the solstices occur the *equinoxes,* at which time the earth's axis makes a 90° angle with a line drawn to the sun, and neither the north nor south pole has any inclination toward the sun. The *vernal equinox* occurs on March 20 or 21; the *autumnal equinox* occurs on September 22 or 23.[3] Conditions are identical on the two equinoxes as far as earth-sun relationships are concerned, whereas on the two solstices the conditions of one are the exact reverse of the other. For this reason, it is necessary to consider each solstice separately, whereas the equinoxes can be treated together.

Winter solstice

Conditions at the winter solstice, December 21 or 22, are best studied with the aid of diagrams. Figure 4.8 is a cross-sectional representation to show the angles at which the sun's rays strike the earth. Keep in mind that "winter" applies only to the northern hemisphere and that the southern hemisphere is experiencing its summer season.

The great circle that marks the boundary between sunlit and shadowed halves of the earth

[3]The exact time of solstices and equinoxes ranges into two calendar days because of the one-fourth day difference in the tropical year and the calendar year, which builds up to one whole day every fourth year and is corrected by adding one day in leap year.

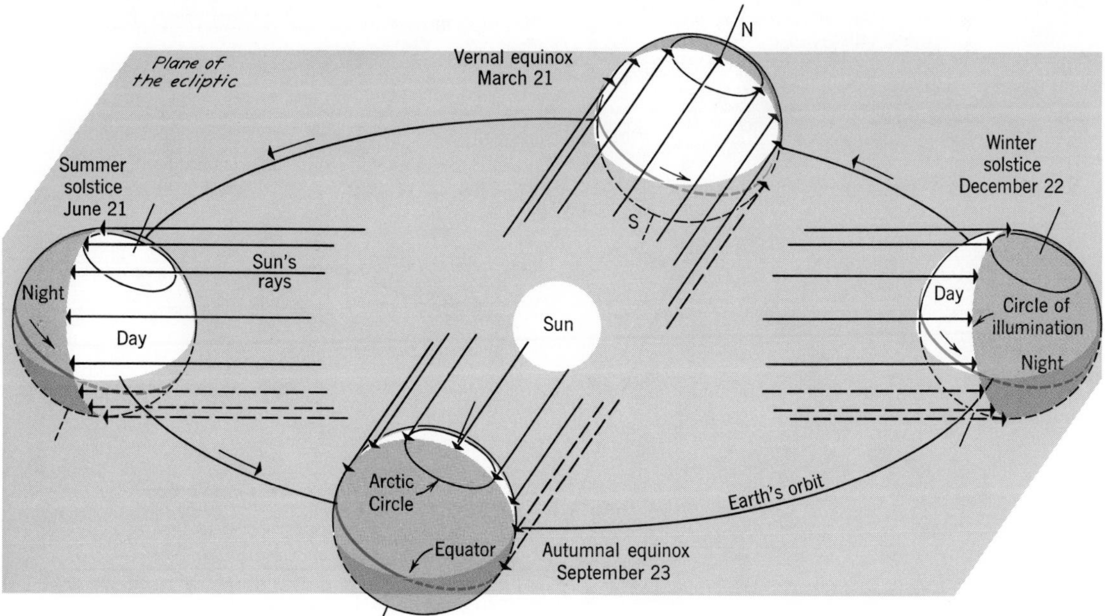

Figure 4.7 The seasons result because the tilted earth's axis keeps a constant orientation in space as the earth revolves about the sun.

is called the *circle of illumination.* At the winter solstice, it divides all parallels of latitude that it crosses (except the equator) into unequal parts. The circle of illumination is tangent to the *Arctic Circle* ($66\frac{1}{2}°$ N lat.) and the *Antarctic Circle* ($66\frac{1}{2}°$ S lat.). This occurrence explains why the two parallels are given special designation on the globe. The circle of illumination bisects the equator in accordance with the law that any two intersecting great circles bisect each other.

Because of the position of the great circle at the winter solstice, day and night are unequal in length over most of the globe. This inequality may be estimated from Figure 4.8 by noting what proportions of a given parallel lie on either side of the circle of illumination. The following facts are evident.

(*a*) Night is longer than day in the northern hemisphere.

(*b*) Day is longer than night in the southern hemisphere.

(*c*) The inequality between day and night increases from the equator poleward.

(*d*) At corresponding latitudes north and south of the equator the relative lengths of day and night are in exact opposite relation.

(*e*) Between the Arctic Circle, $66\frac{1}{2}°$ N, and the north pole, night lasts the entire 24 hours.[4] This is evident from Figure 4.8 because the entire polar area north of the Arctic Circle lies on the shaded side of the circle of illumination and hence will not come into the sun's rays even when the earth turns through 360°.

(*f*) Between the Antarctic Circle, $66\frac{1}{2}°$ S, and the south pole day lasts the entire 24 hours, because the earth's turning fails to bring any part of this area into the zone of darkness.

Altitude of noon sun at winter solstice

At any given instant, all places on the earth where the sun is at its highest point in the sky lie on a single meridian. Thus noon occurs simultaneously at all points having the same longitude. For this reason, noon is often termed the *meridian passage* of the sun. The vertical angle of the sun above the horizon at noon is designated the *altitude.* It may be determined from Figure 4.8 by measuring the angle between a ray from the sun and a line tangent to the globe at a selected latitude. Although the earth's surface is curved, the apparent world in which

[4]This statement does not take into account *twilight,* which provides considerable light near the Arctic Circle.

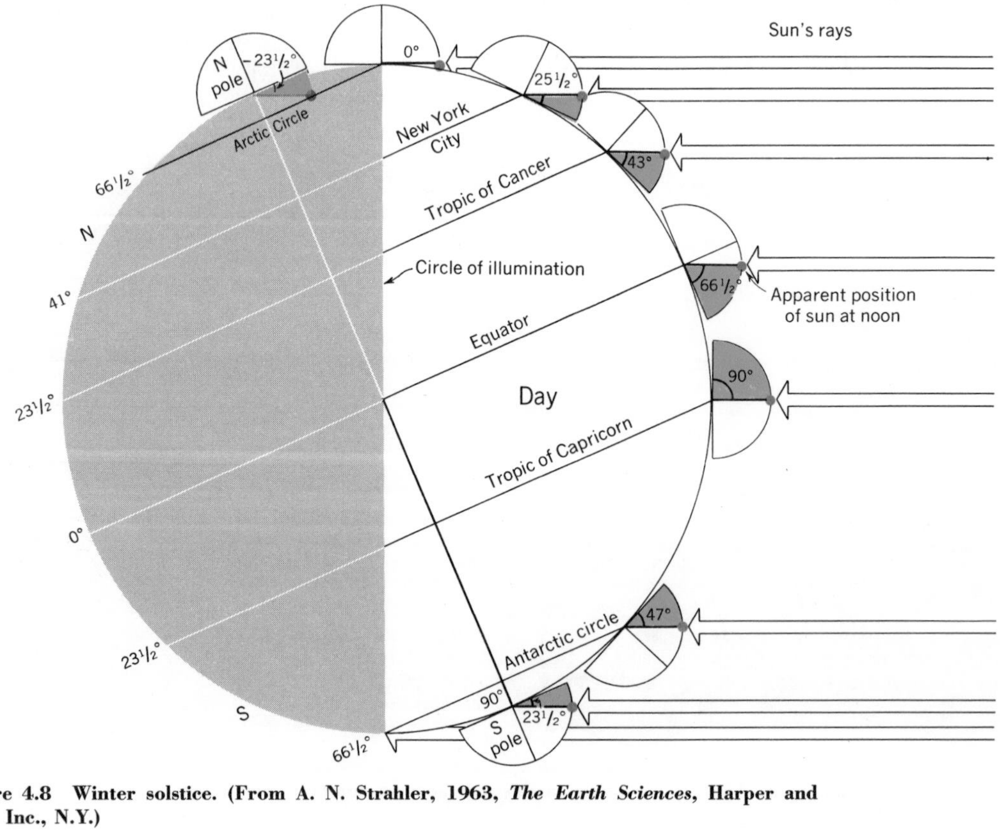

Figure 4.8 Winter solstice. (From A. N. Strahler, 1963, *The Earth Sciences,* Harper and Row, Inc., N.Y.)

we, as tiny individuals, live is a flat world. Within the limits of vision the horizon appears to make a circle on a flat plane. This explains why a straight tangent line may be used for measuring the angle on the diagram. The significant latitudes and altitudes are as follows.

(*a*) At lat. $23\frac{1}{2}°$ S the sun's rays at noon strike the earth at an angle of 90° above the horizon. Thus the sun is exactly in the center of the sky, or *zenith.* The parallel $23\frac{1}{2}°$ S has therefore been designated the *Tropic of Capricorn.*[5] It is the most southerly parallel that the sun's rays can strike vertically. The latitude at which the sun's rays strike with an angle of 90° is the same as a value termed the *sun's declination.* At the winter solstice, therefore, the sun's declination is $23\frac{1}{2}°$ S.

(*b*) At the equator the sun's noon altitude is $66\frac{1}{2}°$ above the southern horizon. Notice that this altitude is equal to 90° minus $23\frac{1}{2}°$.

(*c*) At the Arctic Circle, $66\frac{1}{2}°$ N, the sun at noon is exactly on the horizon.

(*d*) At the Antarctic Circle, $66\frac{1}{2}°$ S, the sun at noon has an altitude of 47° above the northern horizon.

(*e*) At the south pole, the noon sun has an altitude of $23\frac{1}{2}°$ above the horizon.

Careful analysis of the various altitude figures given above will show that there is a systematic relation between latitude and the sun's noon altitude. The general rule, which applies at any selected time of year for any selected latitude, may be stated as follows. *The sun's noon altitude at a place is equal to 90° minus the arc of meridian between the place and the parallel where the sun's rays strike vertically.*

In using this rule, great care must be taken to determine correctly the actual number of degrees of arc separating the desired place from the parallel where the sun's rays strike vertically. This method should be used to determine the sun's noon altitude at the Tropic of Capricorn, equator, Arctic Circle, Antarctic Circle, and south pole, checking the answers against the values cited above for those latitudes. A small globe will prove useful in solving problems of the sun's noon altitude. First, locate that parallel of latitude representing the sun's declination, then measure the arc north or south to the place.

Path of sun in sky at winter solstice

The path of the sun in the sky at the winter solstice is illustrated for various latitudes in Figure 4.9. The horizon is drawn as a circle lying in a horizontal plane, and the sky is visualized as a hemispherical celestial dome. The sun daily completes an entire circle inscribed on the celestial sphere. On certain occasions, as on the equinoxes, these circular paths are great circles on the celestial sphere; otherwise they are small circles.

One general statement about the path of the sun in the sky holds true for any latitude at any time of year: *The plane of the sun's path at all times makes an angle with the plane of the horizon equal to 90° minus the latitude.* Checking this against the diagrams in Figure 4.9, it is seen that at the poles the sun's path lies in a plane parallel with the horizon plane, because the latitude of the poles, 90° N and S, subtracted from 90°, equals 0°. At the equator, the sun's path always lies in a plane perpendicular to the horizon plane, because the latitude, 0°, subtracted from 90°, equals 90°. Do not, however, confuse this angle with the sun's noon altitude, which is measured from the center point of the horizon circle. The two angles are identical only at the equinoxes.

Direction of sunrise and sunset at solstice

The compass direction of sunrise and sunset points on the horizon varies greatly with latitude, as shown in Figure 4.9. At the Antarctic Circle on the December solstice, sunrise and sunset occur at the same instant—midnight—at a point due south on the horizon. At all places between the Antarctic Circle and the Arctic Circle, the sun rises at some point between south and east, and sets at a point between south and west.

Solstice conditions at the poles

At the poles, the path of the sun in the sky is the most extraordinary of all places on the earth (Figure 4.9). Here the sun does not rise and sink in a slanting path with respect to the horizon, as at other latitudes. Instead it follows a horizontal circle, remaining parallel with the horizon throughout the day. In actuality this path is spiral, but so low a spiral that it cannot be detected by ordinary observation. At the December solstice the sun at the north pole remains $23\frac{1}{2}°$ below the horizon throughout the day, while at the south pole it is constantly $23\frac{1}{2}°$ above the horizon. At the south pole we would have no natural way of determining when noon would occur because the sun's altitude remains constant. Moreover, all meridians converge to a point at the pole so that we could not refer the time to any local meridian.

[5]So named because, in ancient times, the sun had a position among the stars of the constellation of Capricorn at the time of the winter solstice.

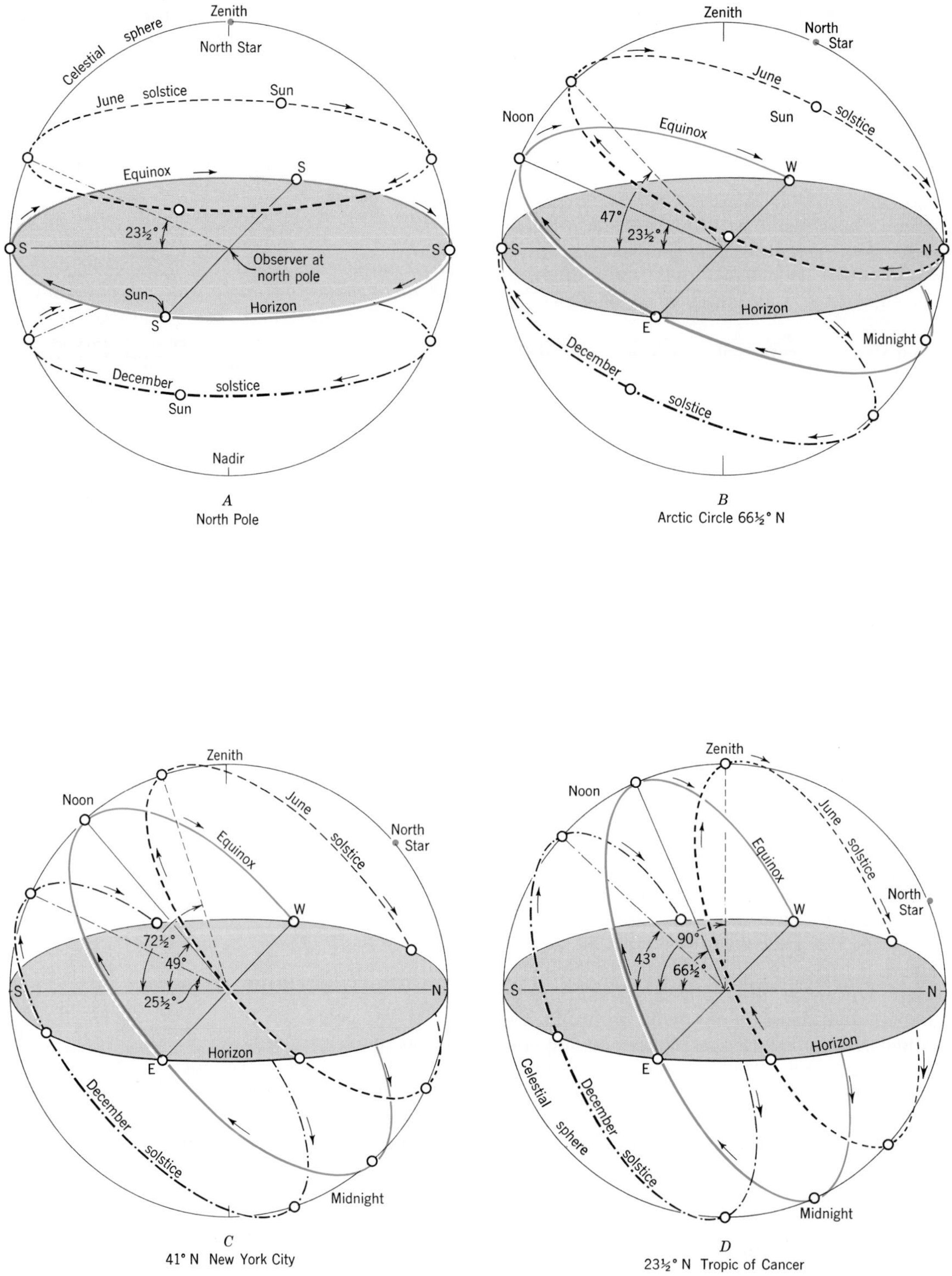

Figure 4.9 To the earth-bound observer the earth's surface is a flat, horizontal disc. The sun, moon, and stars seem to travel on the inner surface of a hemispherical dome above him. The path of the sun in the sky at various latitudes is shown here for the equinox and both solstices.

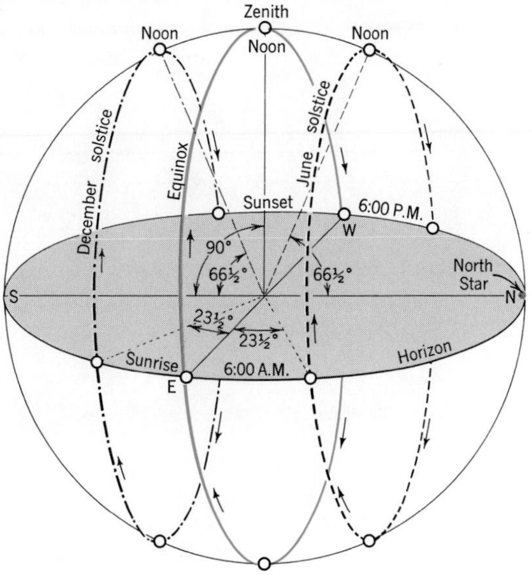

E
Equator
0° Lat.

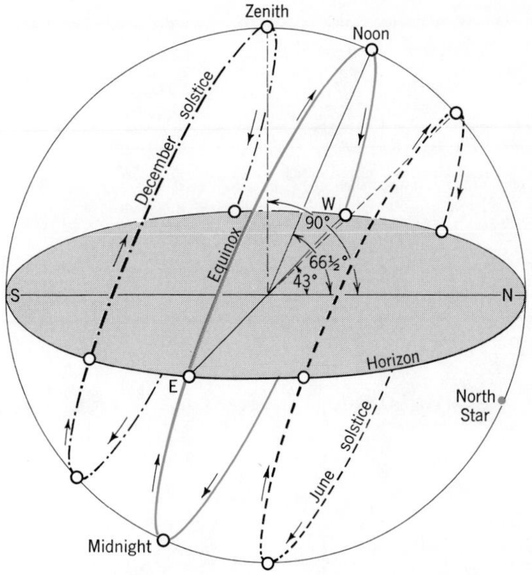

F
23½° S Tropic of Capricorn

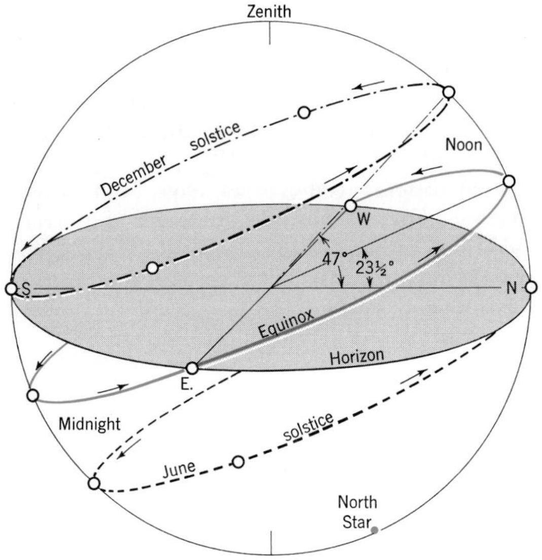

G
66½° S. Antarctic Circle

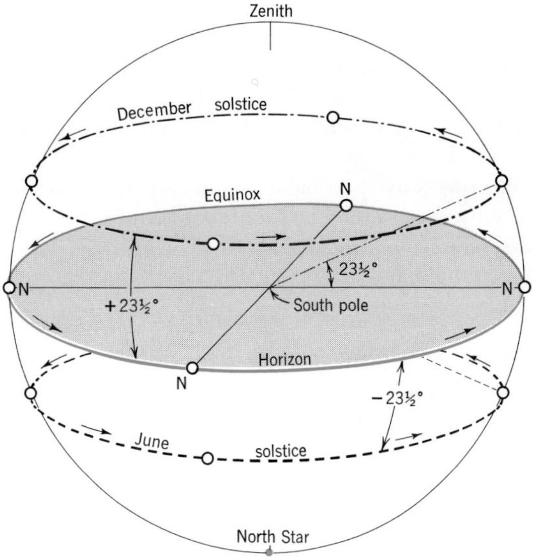

H
South pole
90° S

Figure 4.10 This photograph of the sun was taken at midnight at Hammerfest, Norway, 70°40′ N. lat., during the summer solstice period. The sun has reached its lowest point in the sky. (Photograph by A. Kalland, Hammerfest.)

Another curious fact is that the shadow of an object would always point due north no matter what the hour of the day, because all points of the horizon are north from the south pole. A simple sundial could be made by a perpendicular rod in the center of a horizontal disc whose circumference is divided into 24 equal parts. Selecting any particular point on the circumference as midnight, the hours could be read directly by noting the shadow of the rod on a target held above the edge of the disc.

Summer solstice

In almost every way, the conditions at summer solstice, June 21 or 22, are the exact reverse of winter solstice conditions. At this time, the north polar end of the earth's axis is inclined directly toward the sun and the northern hemisphere enjoys the same conditions of increased sunshine that the southern hemisphere had during the winter solstice (Figure 4.10). Instead of a special diagram to show the relation of sun's rays to the earth, it is suggested that Figure 4.8 be turned upside down, exchanging north for south, Arctic for Antarctic, and Tropic of Cancer for Tropic of Capricorn. The various statements made in the previous pages concerning circle of illumination, length of day and night, and the sun's noon altitude at the winter solstice may reread, with suitable changes to fit the reversed conditions of the summer solstice.

The path of the sun in the sky on June 21 is shown in Figure 4.9. Notice that the sun's noon altitude at each latitude differs from that on December 22 by 47° ($23\frac{1}{2}$° plus $23\frac{1}{2}$°), and that conditions at the poles have been exactly reversed. For all latitudes between the Arctic and Antarctic Circles the sun rises on the northeastern horizon, instead of on the southeastern horizon, and sets on the northwestern horizon instead of on the southwestern horizon.

The equinoxes

On March 20 or 21 and September 22 or 23, vernal and autumnal equinoxes, respectively, the relation of earth to sun's rays is identical and the two dates may be treated jointly. Figure 4.7 shows the general situation. Although the earth's axis, as always, is inclined $66\frac{1}{2}$° to the plane of the ecliptic, the inclination is so oriented that it is neither toward nor away from the sun. The sun's rays make an angle of 90° with the earth's axis. Further details of the equinoctial conditions are shown in Figure 4.11.

The circle of illumination at the equinoxes passes through the poles and hence coincides with the meridians as the earth turns.

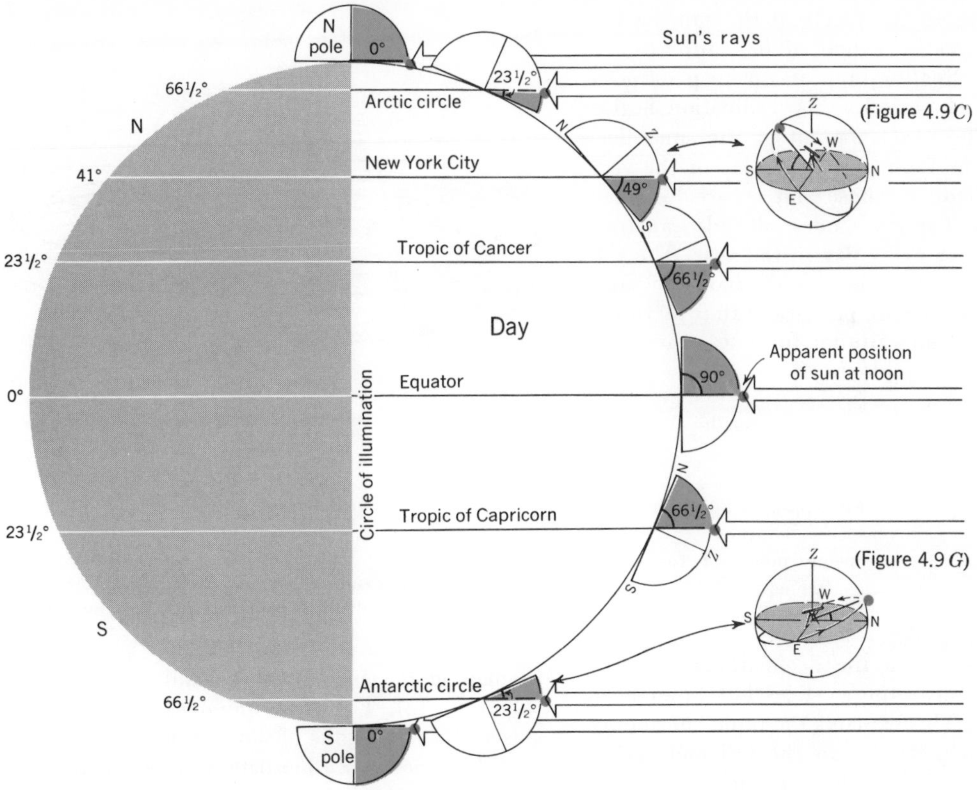

Figure 4.11 Equinox. (From A. N. Strahler, 1963, *The Earth Sciences*, Harper and Row, Inc., N.Y.)

As evident from Figure 4.11, the parallels are divided into equal halves by the circle of illumination. Hence day and night are of exactly equal length, twelve hours each, at all latitudes.[6] Conditions are the same for both northern and southern hemispheres. Sunrise occurs at 6:00 A.M. (local apparent solar time) and sunset at 6:00 P.M. at all places on the globe, except at the poles, where special conditions prevail.

The sun's noon altitude is found by direct measurement of the angle between the sun's parallel rays and tangent lines drawn at selected points (Figure 4.11). A few such measurements should reveal that the altitude is always the *colatitude*, or 90° minus the latitude. Thus, on the equinoxes, but not for any other time of year, the sun's noon altitude may be computed by a single simple subtraction, if only the latitude is given. Although the altitude is the same for similar latitudes both north and south of the equator, it should be remembered that the angle is measured from the southern horizon in the northern hemisphere and from the northern horizon in the southern hemisphere.

[6]This explains the word *equinox*, from the Latin *aequus*, equal; and *nox*, night. We are not taking into account twilight, which extends the period of daylight before sunrise and after sunset.

The path of the sun in the sky at the equinoxes is illustrated in Figure 4.9. In each case, the path is midway between the paths at the solstices. The sun rises at a point due east on the horizon and sets at a point due west on the horizon at all latitudes except at the two poles. At the poles, the sun remains on the horizon all day long, traveling one complete circuit of the horizon in 24 hours. Notice, however, that the direction of apparent movement of the sun is opposite for the two poles.

At the equator the sun has an altitude of 90° at noon; moreover, its path is in a plane perpendicular to the horizon plane. For this reason, the sun changes altitude 15° per hour throughout the day at the equator on the equinoxes. Furthermore, at the equator on these dates, the shadow of any vertical rod will point due west from 6:00 A.M. to noon, will disappear precisely at noon, and will point due east from noon until 6:00 P.M.

Intermediate dates

What may be said of the earth-sun relationships on the days between the equinoxes and solstices? The sun's declination changes continuously from solstice to solstice. The word "solstice," in fact, is derived from the Latin *sol*, sun, and *stare*, to stand, referring to the fact that the sun, having reached the greatest extent of its southward or northward declination, appears to keep its position for a brief period as its declination begins to reverse. The exact amount of the sun's declination may be found for every day of the year in *The Air Almanac,* whose contents are described in Chapter 5. The sun's noon altitude can thus be computed for every day in the year. The following facts regarding the rate of change of sun's declination may be of practical value in rough computation of sun's altitude for dates other than the solstices and equinoxes.

Amount of Change in Declination per Month	Month
$11\frac{3}{4}°$	First months before and after equinox
$8\frac{1}{2}°$	Second months before and after equinox
$3\frac{1}{4}°$	Months adjacent to solstices

It is apparent that the declinational changes are very slow near the solstices but rapid near the equinoxes. This explains the rapid shortening and lengthening of days in the fall and spring months and the apparent persistence of the sun in the very high or very low path in the sky in June-July and December-January periods.

The path of the sun in the sky throughout the year takes intermediate positions from those drawn on Figure 4.9. If all the daily paths from one winter solstice to the next summer solstice were carefully drawn on such a figure, they would be seen to form a spiral of very low pitch. Perhaps this concept will clarify the manner in which the sun behaves at the poles (Figure 4.12). Ascending day after day in this low-pitched spiral from its lowest position below the horizon at the winter solstice, the sun requires three months to attain the level of the horizon at the vernal equinox. After this, it continues to spiral upward, reaching its highest point about June 21, then starts to spiral back down, again reaching the horizon around September 23 and continuing to sink back to its winter-solstice position. This is why the sun rises and sets only once each year at the north pole. Sunrise occurs on March 21, sunset on September 23, separated by six months of sunshine and six months of twilight and total darkness. At the south pole similar, but reversed, conditions prevail.

Finding lengths of day and night and times of sunrise and sunset for all latitudes

To determine approximately the length of day and night and times of sunrise and sunset for various latitudes and at various times of a year,

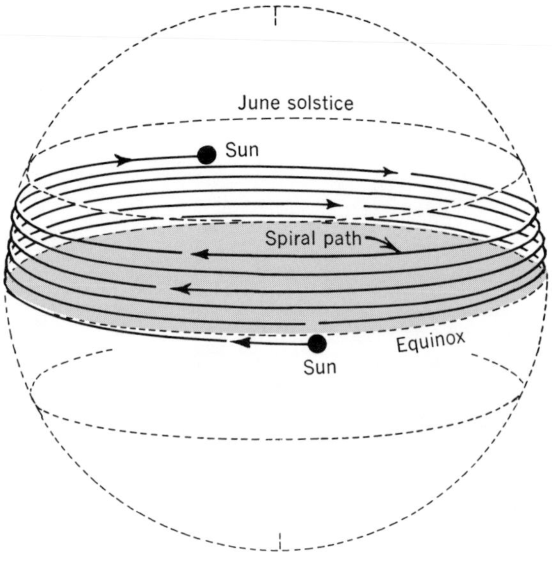

Figure 4.12 The sun's apparent path in the sky throughout the year at the north pole is a spiral of low pitch.

a small globe and a rubber band can be used (Figure 4.13). For the equinoxes, no problem exists because lengths of day and night are equal at all latitudes. For other dates proceed as follows.

1. *Winter solstice.* Place the rubber band so as to represent the circle of illumination (great circle) crossing the equator at the 90° meridians west and east, and just tangent to the Arctic and Antarctic Circles where they cross the Greenwich and 180th meridians. Now select a particular latitude, for example, 40° N. Taking the Greenwich meridian as noon, count 15° of longitude to

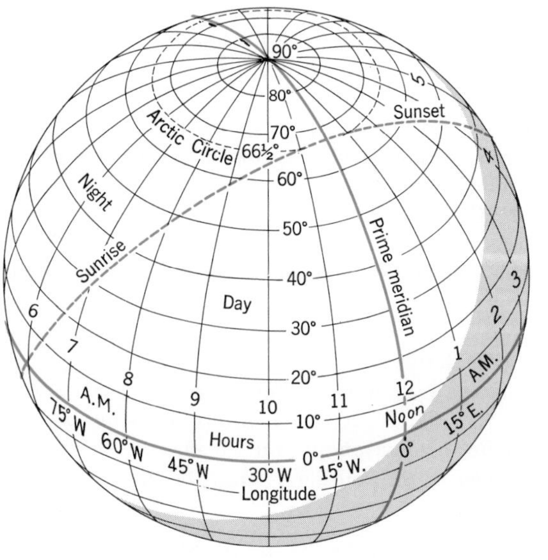

Figure 4.13 A small globe can be used as a graph for determining the lengths of day and night at any latitude.

the hour, westward along the 40th parallel, until the rubber band is reached. This number of hours subtracted from 12:00 noon gives the approximate time of sunrise (local apparent solar time), and added to 12:00 noon gives the time of sunset. The lengths of day and night can be obtained by simple calculation. The shorter arc in this case gives the length of day; the longer arc, by way of the Pacific, gives length of night.

2. *Summer solstice.* Turn the globe around and use the Pacific side, leaving the rubber band in the same position as before.

3. *Other times of year.* Determine the sun's declination from the analemma, Figure 5.7. Then adjust the rubber band so that it is tangent to the parallels having the latitude 90° minus the declination. Be sure that the rubber band maintains a great circle and cuts the equator at the 90th meridians west and east, as before. Proceed with the determinations as described for the solstices.

The method explained here will, if carefully carried out, permit times of sunrise and sunset to be read correctly within one-fourth hour or less for the low and middle latitudes. It is not so satisfactory for high latitudes because of the oblique crossing of the parallels by the circle of illumination. Notice that the globe gives local apparent solar time, which may disagree considerably with standard time systems (Chapter 5). Time of sunrise and sunset can also be read directly from a prepared diagram (Figure 4.14).

Twilight

In order to simplify the treatment of lengths of day and night it was assumed that total darkness comes on instantaneously at setting of the sun and that sunrise is marked by immediate change from total darkness to sunlight. As everyone knows, this is not so. *Twilight,* a diffuse illumination that follows sunset and precedes sunrise, provides important additions to the period of useful daylight, especially in the higher latitudes. Because morning and evening twilight periods are of equal duration and of similar origin, no distinction will be made between them.

Twilight is attributed to scattering action of air molecules and to the presence of minute particles of dust and moisture disseminated throughout the earth's atmosphere. They reflect the sun's rays back to the earth's surface long after the sun has disappeared below the horizon of observers on the ground (Figure 4.15). After sunset, this diffuse light steadily dies out, being finally distinguished only as a faint glow in the western sky.

The duration of twilight depends on the thickness of the earth's atmosphere and the rate at

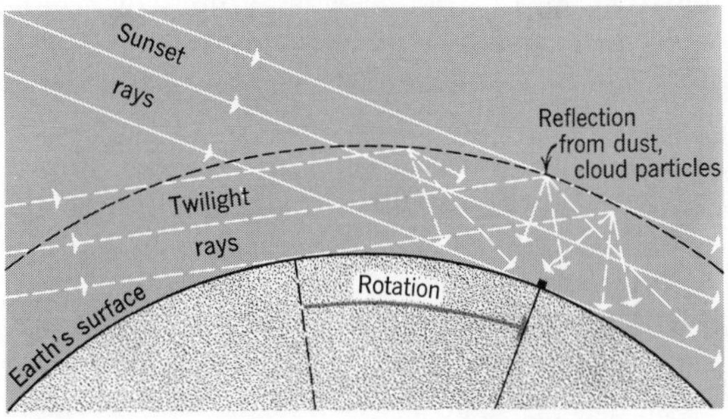

Figure 4.15 Twilight is a diffuse reflection from atmospheric molecules, dust, and moisture particles.

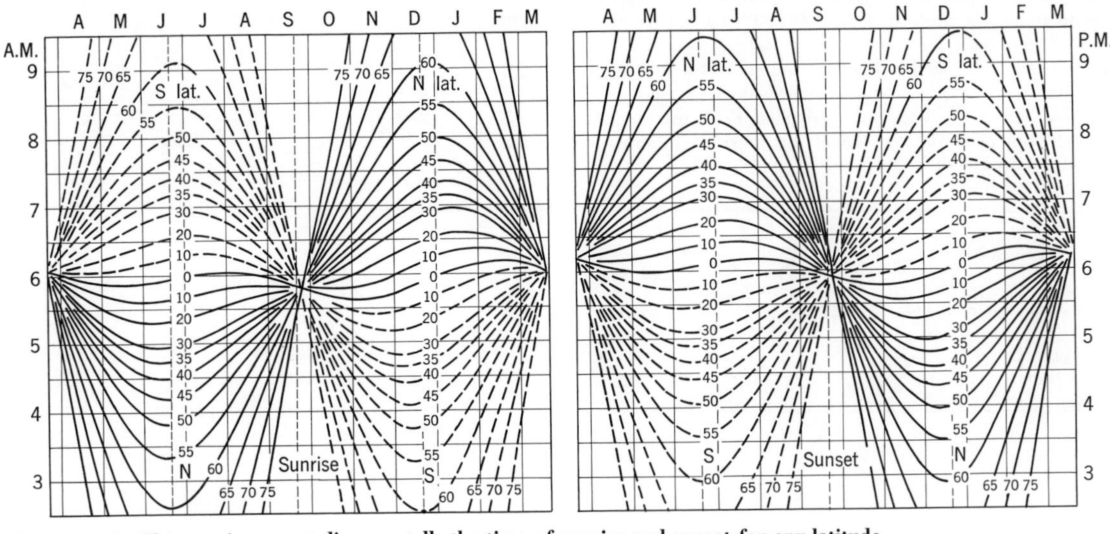

Figure 4.14 This sunrise-sunset diagram tells the time of sunrise and sunset for any latitude on any date of the year. (After U.S. Navy Oceanographic Office No. 5175.)

which the sun sinks below the horizon. Assuming that the atmosphere is of uniform thickness and has a uniform horizontal distribution of dust, the length of twilight will depend on the path of the sun in the sky which, in turn, depends on latitude. This is illustrated in Figure 4.16. Where the plane of the sun's path is vertical, as at the equator, the sun sinks below the horizon most rapidly, that is, at the rate of 15° per hour. At higher latitudes, the slanting path of the sun causes the sun to descend more slowly below the horizon. At 60° N, for example, the sun's path lies in a plane that makes an angle of 30° with the horizon plane (90° − 60° = 30°), and on the equinoxes twilight lasts approximately twice as long as at the equator.[7]

Three kinds of twilight are recognized. *Astronomical twilight* is the period during which any detectable glow exists in the sky, and it is considered to last while the sun is between the horizon and a point 18° below the horizon. *Nautical twilight* endures while the sun is between the horizon and a point 12° below the horizon. At 12° the general outlines of ground objects are visible, although the horizon is probably indistinct and all of the stars used for navigation can be seen. *Civil twilight* is the period during which normal outdoor activities can be carried on without the aid of lights; it is the period during which the sun is between the horizon and a point 6° below the horizon. At 40° latitude, near the time of the equinoxes, civil twilight lasts about 30 minutes, which is the period commonly designated in legal statutes. Duration of civil twilight is given to the nearest one minute for each day of the year and for a wide range of latitudes in the *Air Almanac*, which also carries tables enabling nautical twilight to be calculated.

Figure 4.17 is a twilight diagram prepared by the Oceanographic Office of the U.S. Department of the Navy. On it may be found the length of astronomical twilight for latitude 0° to 70° at any time of year.

Many persons subscribe to the popular idea that, near the equator, twilight is lacking or is extremely short. This supposition has no basis in fact. Astronomical twilight at the equator, it is true, is shorter than anywhere else on the globe, but in order to reach a point 18° below the horizon the sun, descending at a rate of 15° per hour, requires 1 hour and 12 minutes. Civil twilight, based on the sun's being 6° below the horizon, would last one-third as long, or 24 minutes. Carefully conducted observations by scientists have confirmed these facts.

At high latitudes, owing to the low slant of the path of the sun as it goes below the horizon, twilight is greatly lengthened. In northerly regions the twilight period, relative to length of night,

[7]The sun's path at 60° N may be considered to approximate the hypotenuse of a 30° 60° 90° plane triangle, of which the short leg represents the actual distance below the horizon. In such a triangle, the short leg is half the length of the hypotenuse. At the equator, the sun would follow the short leg of the triangle, thus reaching the same number of degrees below the horizon in half the time required at the 60th parallel.

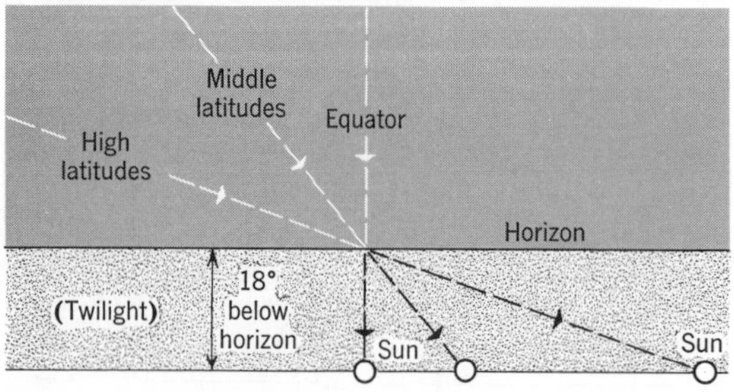

Figure 4.16 Duration of twilight depends on the slope of the sun's path below the horizon.

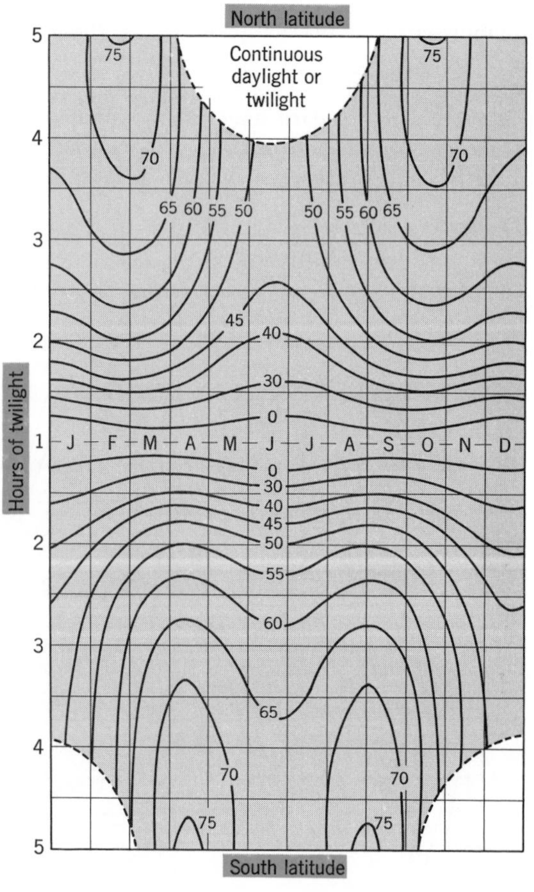

Figure 4.17 Twilight diagram. (After U.S. Navy Oceanographic Office No. 5175.)

is enormously lengthened during June and July; in southerly regions, during December and January. Above a certain critical latitude, which varies from about 48° to 90° throughout the year, the combined lengths of twilight before sunrise and after sunset equal or exceed the total time the sun is below the horizon and twilight lasts all night. Another way of saying this is that the sun's path never reaches a point more than 18° below the horizon. Various latitudes should be examined on the twilight diagram (Figure 4.17) in order to note the wide range in duration throughout the year.

Duration of twilight at the poles is easy to compute. Because the sun's path is a very low-pitched horizontal spiral, twilight persists throughout the whole period that the sun's declination is changing from 0° to 18°. This means that, at the north pole, astronomical twilight lasts from September 23 to November 14, and from January 29 to March 21, each of which is a period of seven weeks.

Review Questions

1. Why are earth-sun relationships important in the study of geography?

2. What is meant by rotation of the earth? What is the period of rotation with reference to the sun? Is this longer or shorter than the period of rotation with reference to the stars?

3. What is the direction of rotation of the earth? Does the earth's rotation have any observable effects on objects at or near the earth's surface aside from the astronomical effects? Is the earth slowing in its rate of rotation?

4. Describe and explain the Foucault pendulum experiment. What does it prove? Could you determine your latitude by means of a Foucault pendulum? How?

5. What is meant by revolution of the earth? In what two ways can the length and starting point of a year be reckoned? Which one is used in our calendar system? How long is this type of year? How must it be periodically corrected to fit our calendar?

6. What is the direction of the earth's revolution? How does this compare with directions of rotation and revolution of the moon and the other members of the solar system?

7. What form has the earth's orbit? What is perihelion? What is aphelion? On what dates do they occur? What distances separate earth and sun at perihelion and at aphelion? What effect does this have on the seasons?

8. Describe the way in which the earth's axis is tilted with respect to the plane of the earth's orbit. What is the angle of tilt? Does this angle change throughout the year?

9. Describe each of the two solstice and two equinox positions of the earth with respect to the sun, giving the dates of each in correct sequence.

10. What is the circle of illumination? Where is the circle of illumination located on the date of equinox? on December solstice? on June solstice?

11. Describe the conditions of global illumination, length of day and night, and path of sun in the sky at June solstice. Give details for the equator, Tropics of Cancer and Capricorn, Arctic and Antarctic Circles, and poles.

12. What rule or formula can be used to determine the sun's noon altitude at any selected date of the year for any desired latitude?

13. Describe the conditions of global illumination, length of day and night, and path of sun in the sky at equinox. Give details for the equator, Tropics of Cancer and Capricorn, Arctic and Antarctic Circles, and poles. How is the sun's noon altitude related to latitude on the date of equinox?

14. Describe the path of the sun in the sky throughout the year at the north pole. How do conditions differ at the south pole?

15. Describe the path of the sun in the sky at the Arctic Circle at equinox and at each solstice.

16. How rapidly does the sun's declination change between one solstice and the next equinox? Give approximate figures.

17. Describe in a general way how the compass direction of sunrise and sunset varies with season of year and with latitude.

18. How can the times of sunrise and sunset be approximated for various latitudes and dates, using only a globe and a rubber band?

19. What is twilight? What causes twilight? What factors determine the duration of twilight? What is the difference between civil and astronomical twilight? How long does astronomical twilight last at the poles? at the equator?

Exercises

1. Practice drawing ellipses of various sizes and degrees of ellipticity, using a drawing board, pins or thumbtacks, a piece of thread or thin string, and a pencil. (Directions are contained in the text, p. 68 and Figure 4.4.) On one of these ellipses draw and label the following: focus, major axis, minor axis, and radius vector. Does the sum of the radius vectors always remain the same for all points on the ellipse? How do you know this?

2. Construct an ellipse to represent the earth's orbit around the sun. Label the following: earth, sun; perihelion and aphelion with their dates and distances. With arrows, show the directions of the earth's rotation and revolution. Find points on the orbit that correspond to the equinoxes and solstices, and label them appropriately.

3. Arrange four small globes in a circle on the laboratory table in such a way that they represent correctly the equinoxes and solstices. Using a piece of chalk, draw on the table top the earth's orbit, and write the correct name and date of the equinox or solstice beside each globe. Indicate by an arrow the direction of earth's revolution. Take care that the axes of all four globes are parallel.

4. To what parallel (or pole) does each of the following statements apply? **(*a*)** Day and night are always of equal length (three answers). **(*b*)** The sun's noon altitude is 90° on June 21. **(*c*)** The sun's noon altitude is 50° on September 23 (two answers). **(*d*)** The sun at midnight on December 22 is exactly on the horizon at a point due south. **(*e*)** The path of the sun in the sky at all times of year lies in a plane perpendicular to the horizon plane. **(*f*)** The sun at noon on December 22 is exactly on the horizon at a point due south. **(*g*)** The path of the sun in the sky at all times of the year lies in a plane inclined 45° with the plane of the horizon. **(*h*)** On March 21 the sun's noon altitude is 66½° (two answers). **(*i*)** Throughout the day, the shadow of a vertical rod sweeps clockwise around the rod at the rate of 15° per hour and remains the same length. **(*j*)** The shadow of a straight east-west wall with a straight horizontal top remains the same width from midmorning to midafternoon and is as wide as the wall is high. **(*k*)** The shadow of a vertical rod points due west during the morning, then due east during the afternoon. **(*l*)** The circle of illumination bisects the parallel on August 9.

5. Given the following information, find the latitude of the place.

Noon Sun Altitude	Latitude at Which Sun's Rays Strike Earth Perpendicularly at Noon	Date	Latitude of Place
(a) 11°	Not given	Sept. 23	______
(b) 44½°	12° N	Not given	______
(c) 90°	Not given	March 21	______
(d) 90°	Not given	June 21	______
(e) 0°	20° S	Not given	______
(f) 8½°	Not given	Dec. 22	______
(g) 81°	9° S	Not given	______

6. **(*a*)** Imagine the earth's axis to be inclined 45° to the plane of the ecliptic. Make a diagram similar to Figures 4.9*A* and 4.9*E* to show the path of the sun in the sky at the equinoxes and solstices. **(*b*)** Imagine the earth's axis to be inclined 0° to the plane of the ecliptic. Make a diagram similar to Figures 4.9*A* and 4.9*E* to show the path of sun in the sky at equinoxes and solstices.

7. Using a small globe and a rubber band according to directions given on p. 77, determine the time of sunrise and sunset and the length of the day for every 10° parallel of latitude from 0° to 60° N. Do this first for the winter solstice, then for the summer solstice. Record your results in a table as follows:

Latitude	Sunrise	Sunset	Length of Day
0°	6:00 A.M.,	6:00 P.M.	12 hours
10°	etc.		
20°			

CHAPTER 5

Time

The need for understanding global time relationships and the various kinds of time in use scarcely needs emphasis in this modern day of instantaneous communication and high-speed travel. Before the coming of the telegraph, problems of time differences were of little or no concern to people who lived most of their lives in one community. Even the traveler was caused only the inconvenience of resetting his watch to the time used by local communities. The amount of time consumed in getting from one place to another was so much more than the difference in watch time between the two places that the difference was of little practical consequence.

When it became possible to transmit messages instantaneously by telegraph, differences in local time resulting from differences in longitudinal position were immediately apparent. With the development of rapid means of travel it became important to correct schedules for the gain or loss of time incurred by passage across the meridians. East-to-west flight in the middle latitudes can now approximate the speed necessary to keep pace with the sun. For example, a plane leaving New York at 12:00 noon, Eastern Standard time, can by traveling about 800 mph (1300 km per hr) arrive in San Francisco at 12:00 noon, Pacific Standard time.

Longitude and time

To avoid confusion and make the study of time relations as simple as possible it is necessary to think of the earth as standing still and of the sun as completing one circuit about the earth every 24 hours. This is a perfectly permissible concept inasmuch as earth-sun relationships are purely relative. Imagine that a meridian is free to sweep westward around the globe and that it constantly maintains such a speed as to be located always where the sun's rays strike the earth's surface at the highest possible angle. This line we shall call the *noon meridian* (Figure 5.1). Directly opposite this meridian, on the other side of the globe, is the *midnight meridian.* It, too, sweeps westward over the globe and remains constantly 180° of longitude apart from the noon meridian. Whereas the noon meridian separates the forenoon and afternoon of the same calendar day, the midnight meridian is the dividing line between one calendar day and the next.

Because the noon meridian sweeps over 360° of longitude every 24 hours, it must cover 15° of longitude every hour, or 1° of longitude every 4 minutes. We therefore find it convenient to state that one hour of time is the equivalent of 15° of longitude. This equality forms the basis for all calculations concerning time belts of the globe. For example, if the noon meridian reaches one place on the globe 4 hours after it leaves another place, the two places are separated by 60° of longitude.

Enlarging this concept of time meridians still further, let it be imagined that, in addition to the noon and midnight meridians, there are 22 hour circles, each one a half great circle, 15° of longitude apart from its neighbor. The hour circles are equidistantly spaced between the noon and midnight meridians (Figure 5.1). Each hour circle will then represent a given hour of the

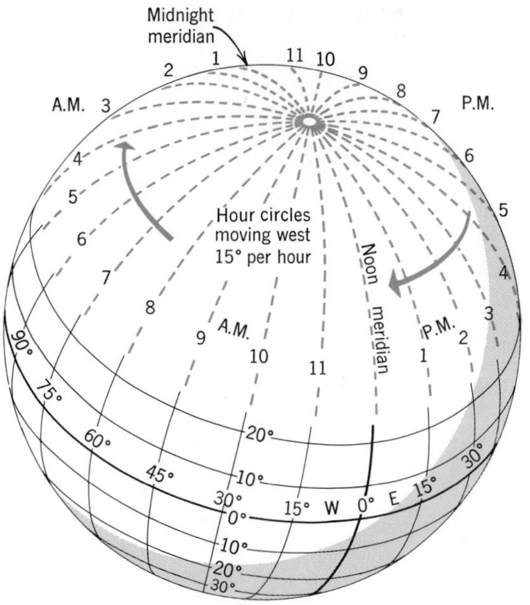

Figure 5.1 The hours may be thought of as meridians spaced 15° apart and moving westward around the globe.

day and can be labeled with a specific hour number which it keeps permanently. Together with the noon and midnight meridians, the hour circles can be imagined to form a birdcagelike net enclosing the globe and attached only at the north and south poles. As a further convenience in analyzing global time relations, it is most helpful if the globe used has meridians drawn for every 15°. If so, wherever the noon meridian of the time net coincides with the Greenwich meridian, all other hour circles coinside with true meridians on the globe.

A working model of global time relations is illustrated in Figure 5.2. If a small globe is available, make a cardboard girdle to fit about the equator and mark upon the cardboard the positions of the time meridians. If no globe is available, make two cardboard disks of different radius and attach them at their centers in such a way that one disk can be turned while the other remains still. On the inner disk, draw radii to represent 15° meridians of a globe seen from a point above the north pole. On the outer disk draw similar radii, but label them in hours to represent the time net. As a further refinement, the inner disk may consist of a hemispherical world map.

People sometimes become confused when trying to decide whether the watch time of places to the east (or west) of them is ahead or behind their own watch time. This is especially evident where the problem is to calculate when a radio or television program will be on the air in different parts of the country, or to decide whether to set one's watch ahead or behind one hour when traveling from one standard time zone to another.

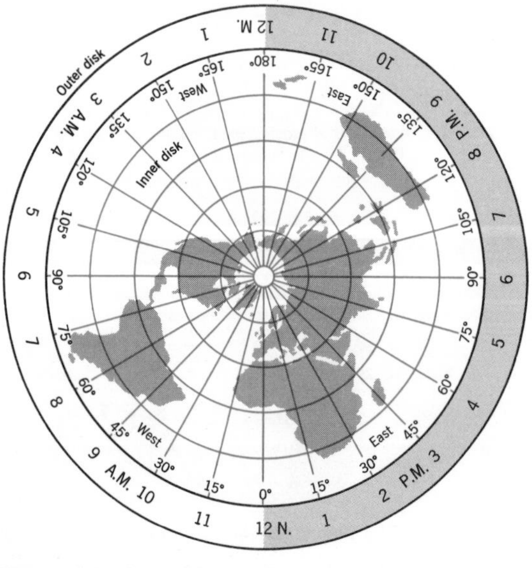

Figure 5.2 A working model helps to clarify global time relationships.

To avoid confusion, let the time meridians be visualized as moving westward around the globe. Then consider, for example, that you are in New York City and the time there is 12:00 noon. The noon meridian which is at New York left Greenwich, England, five hours earlier. Therefore, in England five hours must have elapsed since noon and it must be 5:00 P.M. in that country. From this we get the general rule: *places that lie east of you have a later hour.* Again, consider that the noon meridian is at New York City. It will take that meridian about three hours to travel westward to reach San Francisco. Hence it must be 9:00 A.M. in that city. This gives a counterpart of the rule stated above, namely: *places that lie west of you have an earlier hour.* Both rules are subject to qualifications where the International Date Line lies between the places.

Local time

One means of establishing a time system for a small community is to take the meridian of longitude that passes through some central point in the town or city, as for example the courthouse or a cathedral. All clocks of the community are set to read 12:00 noon when the sun is directly over that meridian; that is to say, when a shadow cast by a vertical rod points due north. As will be explained fully later, the sun is an erratic timekeeper, and it would be necessary to take the mean value of the sun's noon position, but this need not concern us now. The time system thus derived is called *local time* and may be defined as mean solar time based on the local meridian. All places located on the same meridian, regardless of how far apart they may be, have the same local time, whereas all places located on different meridians have unlike local times, differing by four minutes for every degree of longitude.

Standard time

The undesirability of using local time systems in each community of a highly populated country in our modern day is obvious. American railroads, about 1870, introduced a standardized system covering considerable belts of territory, but this system was developed by the railroad companies for their own convenience. Consequently, if several railroads met or passed in a single town, the inhabitants might have had to contend with several different kinds of railroad time in addition to their own local time. It is said that, before 1883, as many as five different time systems were used in a single town and that the railroads of the United States altogether followed 53 different systems of time.

The obvious solution to such problems is *standard time,* based on a *standard meridian,* whose local time is arbitrarily given to wide strips of country on both sides. Thus, all clocks within the belt are set to a single time. It is evident that, if standard meridians are 15° apart, adjacent zones will have standard times differing by exactly one hour. Furthermore, if these meridians represent longitudes that are multiples of 15 (for example, 60°, 75°, 90°, or 105°), each successive standard time zone will differ from the standard time of Greenwich, England, by whole hour units. This is the system of standard time zones employed over most of the globe.

Standard time in the United States

The present system of standard time in the United States was placed in operation on November 18, 1883, but it was not until March 19, 1918, that Congress passed legislation directing the Interstate Commerce Commission to determine time-zone boundaries. The standard meridians and boundaries are show on Figure 5.3. The several United States time zones and their meridians are as follows:

Eastern standard time	75th meridian
Central standard time	90th meridian
Mountain standard time	105th meridian
Pacific standard time	120th meridian
Alaska standard time	150th meridian
Hawaii standard time	150th meridian

If it had been carried out precisely, the system would have resulted in belts extending exactly $7\frac{1}{2}°$ east and west of each standard meridian, but a glance at the map shows that great liberties have been taken in locating the boundaries. Wherever the time-zone boundary could conveniently be located along some already existing and widely recognized line, this was done. Natural physiographic boundaries have been used. For example, the Eastern time-Central time boundary line follows Lake Michigan down its center, and the Mountain time-Pacific time boundary follows a ridge-crest line also used by the Idaho-Montana state boundary. Most frequently, the time-zone boundary follows state and county boundaries. For example, before 1941, the Eastern time-Central time boundary passed north to south through Georgia, but in that year the ICC held hearings in Georgia and officially moved the time-zone boundary westward to the Alabama-Georgia state line, thus bringing all of Georgia into the Eastern standard zone.

The time belts are by no means equally distributed on both sides of the standard meridians, as a glance at the map will show. An extreme example is found in western Texas, where the Central time-Mountain time boundary follows the western boundary of Texas, even crossing west of the standard meridian (105° W) of the Mountain time zone. Although such deviations may seem

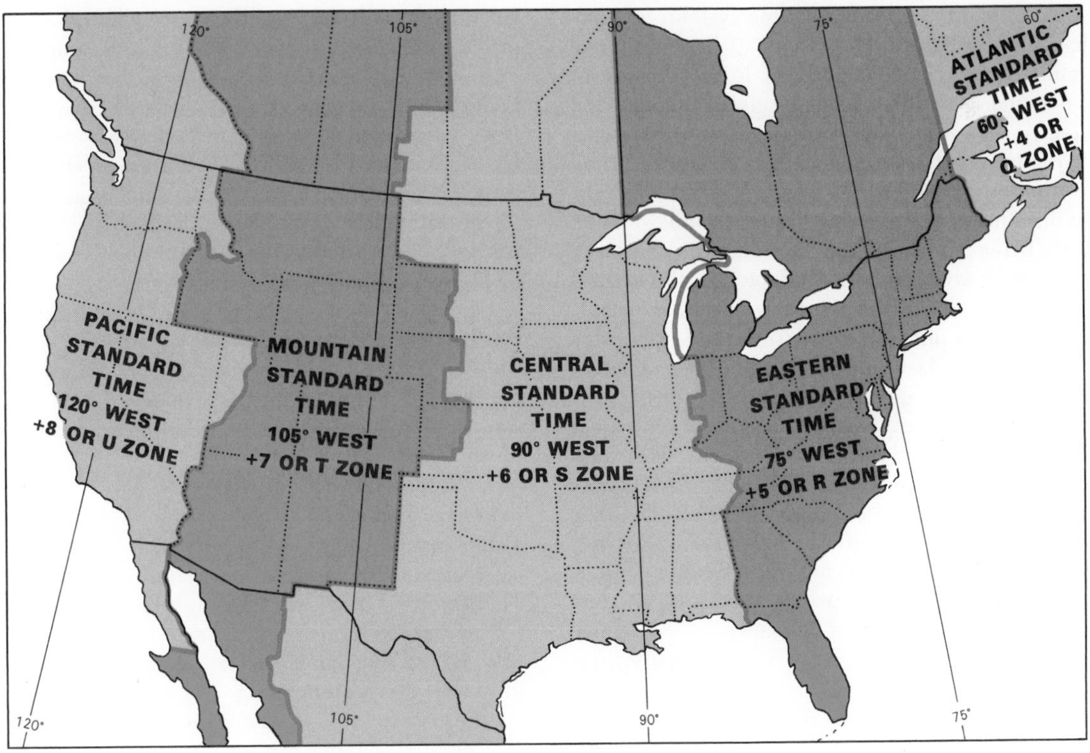

Figure 5.3 Time-zone map of the United States and southern Canada. (Based on data by Interstate Commerce Commission and Department of Transportation.)

odd, they cause little difficulty when the boundaries once are well established. The advantage gained by extreme deviations of the boundaries is in permitting entire states to operate under one kind of time. In some places, the determining factor was a railroad division-point junction or terminal. Where the line was drawn through such a point, as for example at Ogden or Salt Lake City, divisions east and west of the city each ran under a single time.

Daylight saving time and War time

Because many human activities, especially in cities and manufacturing areas, start well after sunrise but continue long after sunset, it would seem desirable to set forward the hours of daylight in order to utilize them to best advantage. A considerable saving in electric power would be made if the early morning daylight period, wasted while people are still in bed and offices and factories are closed, were transferred to the early evening when the large majority of persons are awake and busy. The adjusted time system is known as *daylight saving time* and is obtained by setting ahead all timepieces by one hour. Thus, when the sun is over the standard meridian (i.e., noon by the mean sun), all clocks in that time zone read 1:00 P.M. Sunrise and sunset at the equinoxes or equator, instead of occurring at 6:00 A.M. and 6:00 P.M., would occur at 7:00 A.M. and 7:00 P.M., respectively.

In terms of the standard time system, we may describe daylight saving time as based on the standard meridian lying 15° of longitude east of the standard meridian normally giving the standard time to that zone. For example, Eastern daylight saving time is the same as Atlantic standard time of the meridian 60° W long.; Central daylight saving time is the same as Eastern standard time, both being based on the meridian 75° W long.

Daylight saving time was adopted in the United States during the First World War and, by act of Congress, was put into effect from the last Sunday in April to the last Sunday in September of 1918. After that war, it was used locally throughout the United States where authorized by local legislation. During the Second World War, daylight saving time was used nationally throughout the entire period of February 1942, to October 1945, and was known as War time. England, during the same war period, employed a double daylight saving time in which clocks were running two hours ahead of Greenwich Civil time. This was desirable because of the unusually long summer days that England enjoys as a result of her relatively far northerly latitude. Many European countries normally use daylight saving time, which goes under the name *Summer time*, during a part of the year. Nations whose time is advanced by an hour throughout the entire year are Great Britain, Ireland, Spain, France, the Netherlands, Belgium, Portugal, and the USSR.

In April 1966, The United States Congress passed the *Uniform Time Act*, which requires that daylight saving time be applied uniformly throughout each state, unless the legislature of that state has voted to remain instead on standard time. In the latter case, standard time must be applied uniformly throughout the entire state. Daylight time goes into effect at 2:00 A.M. of the last Sunday of April and continues until 2:00 A.M. of the last Sunday of October. In 1974 year-around daylight saving time was instituted throughout the United States as a means of conserving energy during an acute shortage of petroleum imports.

World time zones

In 1884 an international congress was held in Washington to consider the subject of world standard time. As a result, standard times of countries throughout the world are based on standard meridians, which are multiples of the unit 15° and thus differ from one another by whole hourly amounts. In all global time calculations, the prime meridian of Greenwich, England, is taken as the reference meridian. All time zones of the globe are described in terms of the number of hours' difference between the standard meridian of that zone and the Greenwich meridian. In order to distinguish whether the time zones lie east or west of the Greenwich meridian, the time is designated *fast* for all places east of Greenwich (east longitude), and *slow* for all places west of Greenwich (west longitude). U.S. Eastern standard time, for example, is said to be "5 hours slow."

An alternative system of designating world time zones uses letters of the alphabet, as shown in Figures 5.3 and 5.4.

Figure 5.4 is a world map on which the 24 principal standard time zones of the world are shown. Within each time zone is inscribed the number of hours' difference between the zone time and Greenwich time. Careful examination of this map brings out a number of interesting facts. A few countries or islands lie about midway between 15° meridians. Under such circumstances, a standard meridian is chosen which lies halfway between the two and is thus a multiple of $7\frac{1}{2}$°. The standard time of the country is therefore fast or slow by some multiple of a half hour. Iran ($3\frac{1}{2}$ hours fast) and Surinam ($3\frac{1}{2}$ hours slow) illustrate this point. India, for a large country, is unusual in having the time $5\frac{1}{2}$ hours fast. The country having the greatest east-west extent is the Soviet Union, with eleven standard time

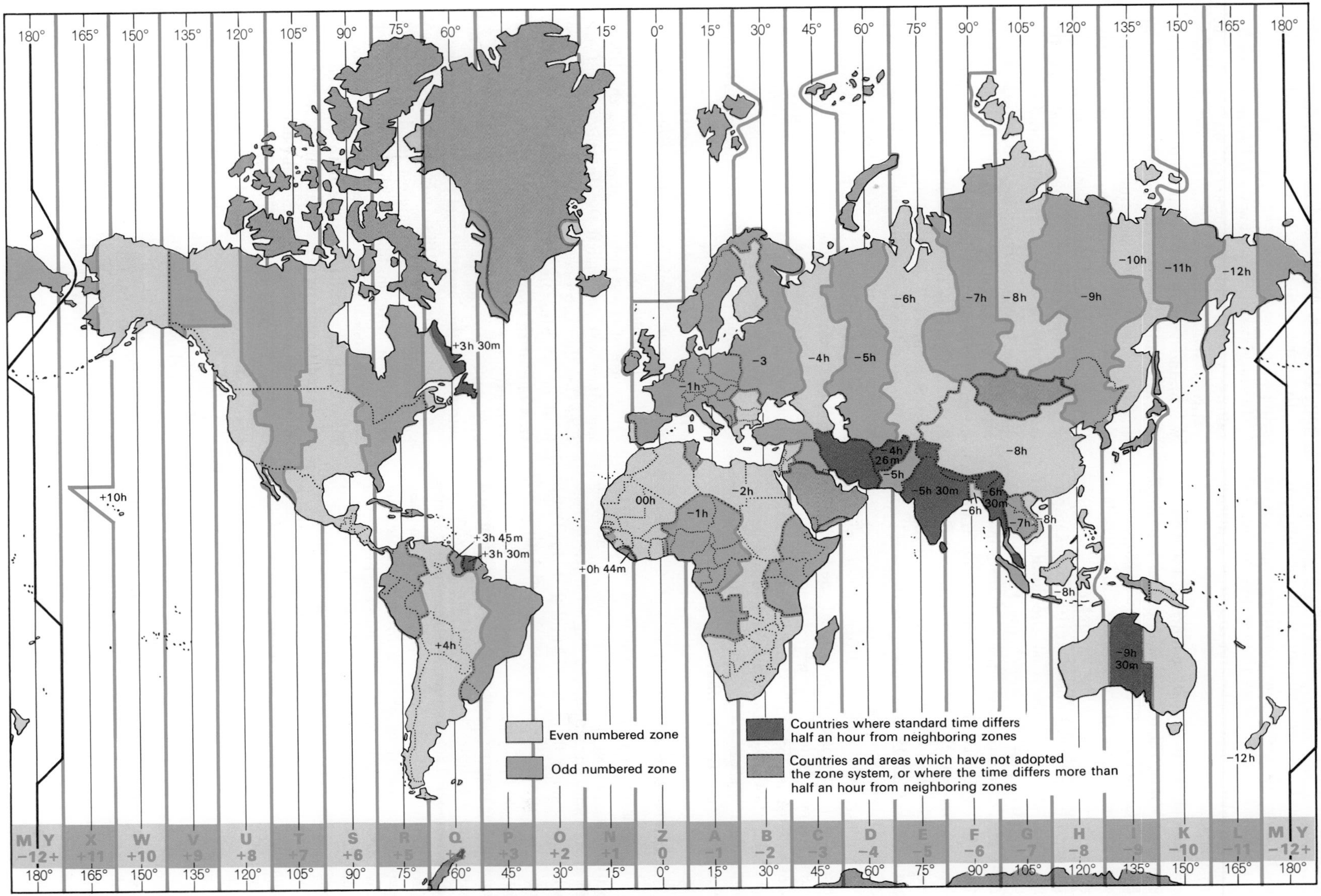

Figure 5.4 Time-zone map of the world. (After U.S. Navy Oceanographic Office, No. 5192.)

zones, but these are all advanced by one hour with respect to the standard time meridians in each zone, to give a perpetual daylight saving time. Canada occupies six time zones. Notice that Newfoundland and Labrador use the time $3\frac{1}{2}$ hours slow.

Although virtually all countries today have standard times related by whole hours or half hours to Greenwich Civil time, it was not so long ago that many countries determined their standard times by meridians of the capital cities or those passing through local observatories. For example, in 1905, all of France used the local time of the meridian passing through the Paris Observatory. This gave a time which was $0^h\ 9^m\ 20.9^s$ fast. India in 1905 used the local time of the Madras Observatory, which is $5^h\ 20^m\ 59.1^s$ fast of Greenwich Civil time. Ireland in 1905 was using the time $0^h\ 25^m\ 21.1^s$ slow, which is the local time of the Dublin meridian.

By 1967, only a few small nations or colonies persisted in the use of odd local meridians for their standard time. These were Liberia ($0^h\ 44\frac{1}{2}^m$ slow), Guyana ($3^h\ 45^m$ slow), Chatham Island ($12^h\ 45^m$ fast), Saudi Arabia, and Mongolia. An up-to-date list of world standard times will be found in *The Air Almanac*.

The International Date Line

If we were to take a world map or globe on which 15° meridians are drawn and count them in an eastward direction, starting with the Greenwich meridian as 0, we would find that the 180th meridian is number 12, and that the time of this meridian is, therefore, 12 hours fast. Counting in a similar manner westward from the Greenwich meridian, we find that the 180th meridian is again number 12, but that the time is 12 hours slow. Both results are, of course, correct; and the explanation becomes obvious when we note that the difference in time between 12 hours fast and 12 hours slow is 24 hours, or a full day. At the precise instant when the noon meridian coincides with the Greenwich meridian, the 180th meridian coincides with the midnight hour meridian. At this instant, and only at this instant, the same calendar day exists on both sides of the meridian. At all other times the calendar day on the west (Asiatic) side of the 180th meridian is one day ahead of that on the east (American) side. For example, if it is Monday on the Asiatic side of the 180th meridian, it is Sunday on the American side. Confusion can be avoided and the correct dates consistently obtained if it is remembered that by counting hours from the Greenwich meridian eastward around the globe by way of Asia, the 180th meridian has the time 12 hours fast, hence that the Asiatic side of the meridian must be a day ahead of the other side.

In the days of very slow trans-Pacific travel by sailing vessel and low-powered steamship, an entire calendar day was merely omitted on a westbound voyage and an entire day repeated on an eastbound voyage. The change was made at any convenient place and time in midocean and usually planned so as to have neither two Sundays in one week nor a week without a Sunday.[1]

It was due to their failure to advance the calendar by a whole day that the crew of Magellan's only surviving ship, reaching Seville after circumnavigating the globe in a westward direction, found that in Spain it was September 8, 1522, whereas by their own reckoning it was only September 7 of that year.

When traveling across the Pacific in today's high-speed jet aircraft, the 24-hour correction is made at the time the 180th meridian is crossed. Suppose that the plane is traveling eastward toward North America and crosses the meridian at 4:00 P.M., (standard time 12 hours fast) on a Tuesday. At the instant of crossing, the time becomes 4:00 P.M. Monday. When traveling westward to the Orient the time moves ahead by a full day. For example, if the meridian is crossed at 9:30 A.M. Wednesday, the time becomes 9:30 A.M. Thursday.

Because of these peculiar properties, the 180th meridian was designated the *International Date Line* by the International Meridian Conference held in Washington, D.C., in 1884 (Figure 5.5). It is one of the fortuitous occurrences of modern civilization that, after the Greenwich meridian had come into widespread use in English-speaking countries as the international basis for the reckoning of longitude, the 180th meridian should have been found to fall in an almost ideal location—squarely in the middle of the world's largest expanse of ocean. Nevertheless, the International Date Line has had to deviate both eastward and westward to permit certain land areas and groups of islands to have the same calendar day (Figure 5.5). By an eastward bulge passing through Bering Strait, the easternmost part of Siberia is included in the Asiatic side, and a westward deflection of the line allows the Aleutian Islands to be included with the Alaskan peninsula. A few degrees south of the equator the date line is shifted eastward $7\frac{1}{2}°$ and thus avoids cutting through the Ellice, Wallis, Fiji, and Tonga island groups, which have the same day as New Zealand.

[1]Willis E. Johnson, 1907, *Mathematical Geography*, American Book Co., New York. See pp. 96–103.

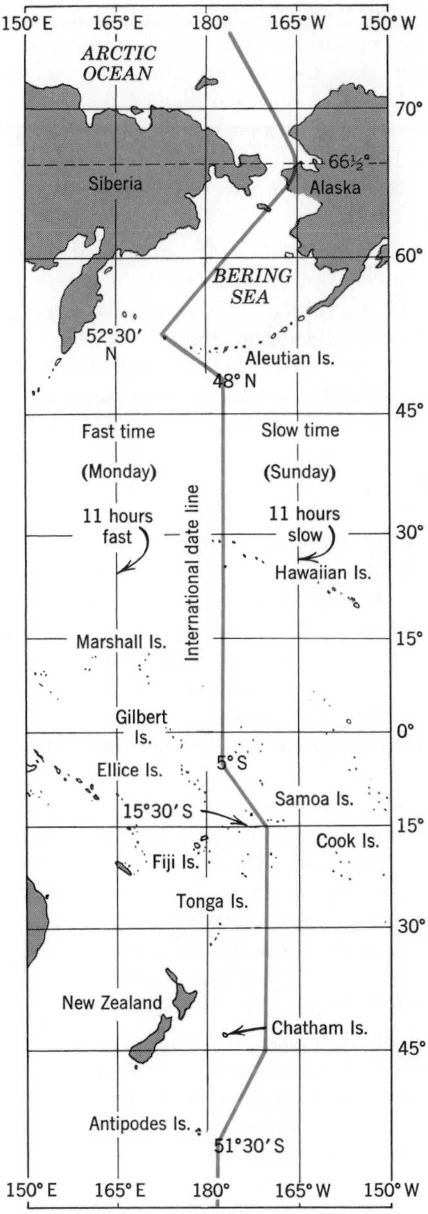

Figure 5.5 The International Date Line.

Duration of days on the globe

One of the most curious aspects of global time is the manner in which calendar days appear and disappear on the globe. If a day lasts 24 hours for a specific place on the earth, and the same series of hours reaches places lying farther west at a later time, it follows that a calendar day exists more than 24 hours for the earth as a whole. To develop this idea fully, use the following visual aids.

Holding a small globe in your hands so that the Pacific Ocean is before you, imagine that the International Date Line is a narrow slit extending from pole to pole on the 180th meridian. (For the time being, let it be assumed that the line has no deviations from the meridian.) Imagine further that calendar days—Monday, Tuesday, etc.—issue from the long slit and spread westward over the globe like a thin film. The leading edge of this film extends from pole to pole and corresponds to the midnight time meridian. The film, which is now issuing from the slit, can be designated Monday, and all global areas that it covers have the calendar day Monday. Traveling at the rate of 15° of longitude per hour, the midnight meridian, or leading edge of the film, will require twelve hours to arrive at the Greenwich meridian. At this precise moment, it is midnight in England and the calendar day Monday covers the eastern hemisphere. The western hemisphere still has the calendar day Sunday, which may be pictured as retreating ahead of Monday. Twelve hours later the imaginary film which we are calling Monday has spread over the entire globe and the edge of the film has reached the slit at the 180th meridian. At this precise instant, Monday envelops the entire earth and is the only calendar day present anywhere.

Because no calendar day can ever cross the date line, we shall have to picture Monday as disappearing into the slit whence it had begun to issue 24 hours earlier. As it does so, the next calendar day, Tuesday, is beginning to issue from the slit and to spread across the Pacific Ocean toward Asia, just as Monday had done earlier. Now we are prepared to answer the intriguing question: What is the total number of hours that the calendar day Monday, whose progress we are observing, will exist on the globe? The answer is that Monday will exist 48 hours. It has required 24 hours for the film representing Monday to spread around and completely cover the earth, and it requires an additional 24 hours for it to disappear back into the slit. Thus Monday is present on the earth's surface for a continuous period of 48 hours, although for any specified place on the globe it can last only 24 hours.

Solar and sidereal time

Unfortunately, the sun is a bad timekeeper, running sometimes slow, sometimes fast, with a total range of more than a half hour from one extreme to the other. The stars, on the other hand, provide a perfect timepiece, but they do not operate according to the conventional system of hours and days that our clocks and calendars follow. Our time system averages out the sun's errors and so is basically controlled by the sun, but the stars are used to check the accuracy of the corrected sun time. The main object of a study of sun time, or *solar time*, is to learn how and why it differs from star time, or *sidereal time*, and what causes the sun to run fast or slow at various times of year.

The interval of time required for 360° of rotation of the earth, causing a given star to return to exactly the same position in the sky, is 23^{h} 56^{m} 4.09^{s} of mean solar time and is known as the *sidereal day*. The interval of time required for successive passages of the sun over a given meridian (i.e., from noon to noon), if averaged throughout the year, is exactly 24 hours and is known as the *mean solar day*. The reason why the value is exactly 24 hours, with no odd number of minutes or seconds more or less, is that our 24-hour-day system was chosen to divide the mean solar day into equal parts.

The solar day is about four minutes longer than the sidereal day. An explanation may be found by study of Figure 5.6, in which the earth's size is enormously exaggerated. Suppose that, when the earth is at *A*, the sun and a particular star are both exactly over the same meridian at the same instant. One day later, after the earth has turned through 360°, the star is again over the meridian, but, because the earth has moved about 1° along in its orbit, a slight amount of additional turning will be required to bring the sun over the meridian again. About 1° more than the full 360° is necessary, and this requires an additional four minutes.

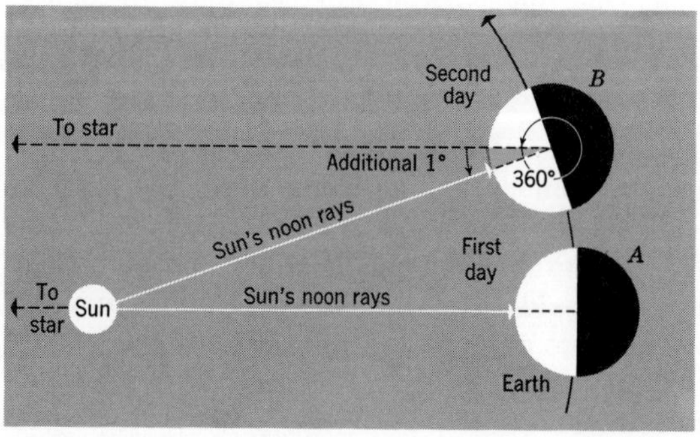

Figure 5.6 The earth must turn slightly more than 360° each day to bring the sun to noon position.

Apparent solar time and mean solar time

In order to determine when solar noon occurs, we might set up a vertical straight rod, from the base of which is drawn a straight line pointing to true north. When the rod's shadow coincides exactly with the north line it is solar noon, and the sun is directly above the meridian passing through the rod.[2] If the time of solar noon by reference to an accurate clock were recorded day after day throughout the year, it would be found that at certain times of the year noon occurs a few minutes early, at other times a few minutes late, and that on only four days in the entire year is the sun over the meridian exactly on time.

Apparent solar time is the system of days and hours which goes strictly by the sun itself and which is therefore continually changing in value from day to day. *Mean solar time* is the system of days and hours mathematically computed in order to give the average value to every hour and day. The imaginary sun that would run on mean solar time is termed by astronomers the *mean sun*. All accurate clocks and watches in general use run on mean solar time. The difference in value between apparent and mean solar time is known as the *equation of time*. The sun is said to be *fast* when it arrives over the meridian before 12:00 noon by mean solar time, and the equation of time is said to be *positive*. When the sun is *slow*, or arrives late over the meridian, the equation of time has a *negative* value. From September through December the sun is fast; from January through March it is slow. During these two periods, the equation of time reaches a value of plus 16 and minus 14 minutes, respectively. During May the sun is again fast; during July and August it is again slow, but in these periods the equation of time does not exceed plus 4 and minus $6\frac{1}{2}$ minutes, respectively.

The analemma

Values of the equation of time for any day in the year can be estimated from a graph known as the *analemma* (Figure 5.7). Two things are shown on the analemma: (1) the equation of time, and (2) the declination of the sun. Values of the equation of time are plotted to the left or right of the vertical center line, depending on whether the sun is fast or slow. Values for the sun's declination are plotted above or below a horizontal center line, and they range from $23\frac{1}{2}$° south to $23\frac{1}{2}$° north. Therefore, for every calendar day, there is a point on the analemma which simultaneously registers the equation of time and the sun's declination. When all these points have been plotted and connected by a curving line, the curious figure-eight graph results. A crude analemma is often found on globes, printed astride the equator in the Pacific Ocean.

[2] In the southern hemisphere, the shadow would point due south. The method described here would be difficult to use in low latitudes where the sun's altitude is great and the shadow short.

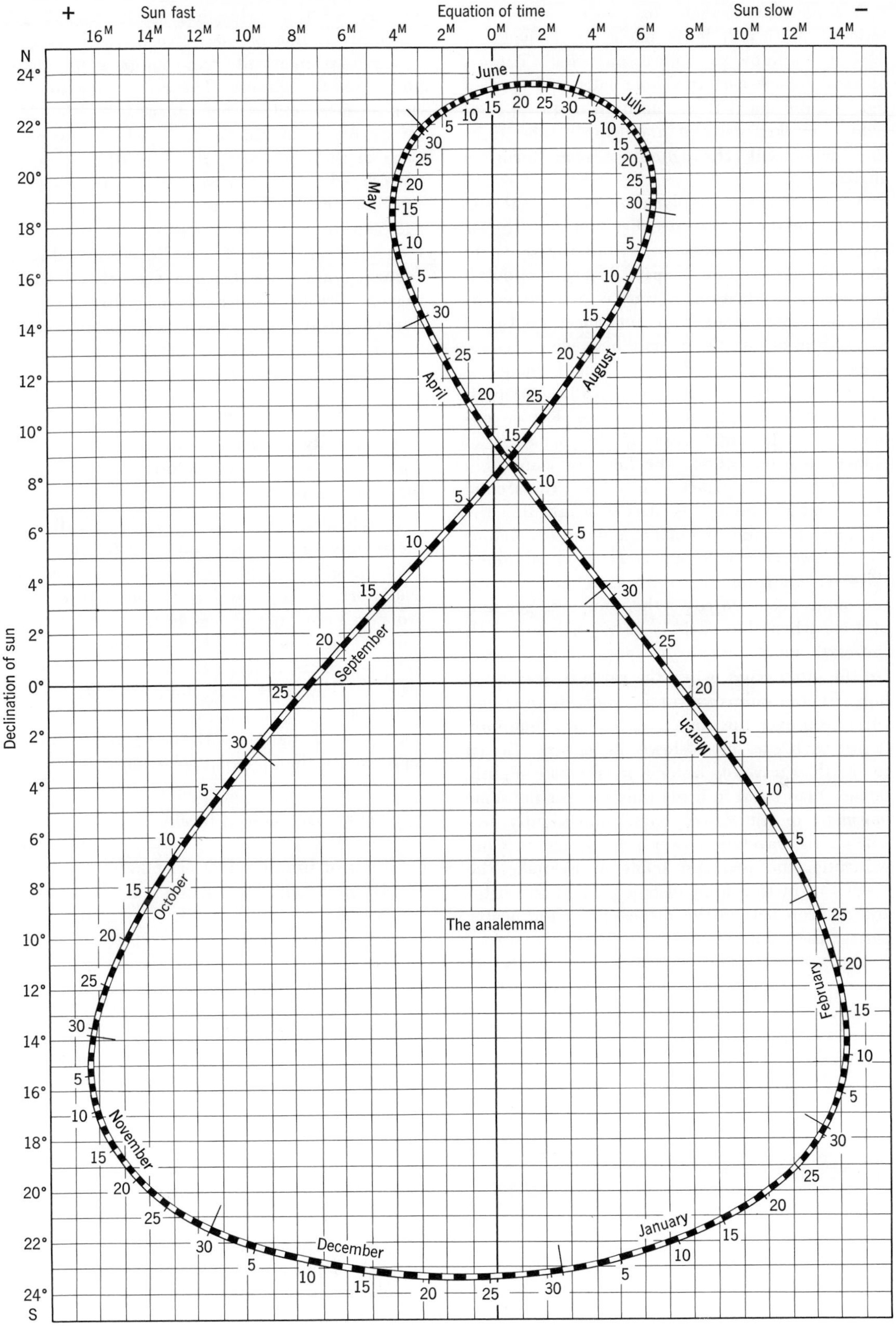

Figure 5.7 The analemma is a graph which gives both the declination of the sun and the equation of time for every day of the year.

A crude natural analemma can be obtained as follows. A tiny hole is made in a shade or darkened window pane on a south wall. The ray of sunlight entering this hole falls as a spot of light on the floor. If the position of the spot of light is marked daily at noon, mean solar time, the marks will, throughout a year, form an analemma.

A comprehensive explanation of why the sun runs fast or slow in the peculiar, but nevertheless systematic, way it does is beyond the scope of this discussion, but a partial explanation is feasible. The equation of time is determined by the combination of two influences, both of which tend to vary the interval between successive meridian passages of the sun. One influence is the varying speed of the earth in different parts of its orbit. The great astronomer Kepler discovered fundamental laws of the behavior of planets in their orbits. The first law states that the orbit of each planet is an ellipse; the second states that a plant moves at such a rate that the straight line connecting planet and sun (the radius vector) sweeps over equal areas in equal times. It is evident from Figure 5.8 that, in order for the radius vector to cover the same area per unit of time when the earth is near perihelion (*A*), the earth must increase its speed of revolution. On the other hand, when the earth is near aphelion (*B*), the radius vector is relatively long and will sweep over the same area per unit time only if the earth goes more slowly in its orbit. When the earth is traveling faster it must rotate slightly farther than usual to bring the sun over the same meridian on successive days, and slightly less when it is traveling slower near aphelion. Consequently, the real sun tends to overtake the mean sun in the more distant part of the orbit and to drop behind the mean sun in the nearer part of the orbit.

The second influence helping to determine the equation of time is less easily understood. Because of the inclination of the earth's axis, the unit of time of successive meridian passages changes systematically from equinox to solstice, and back to equinox. The reason for this will not be clear unless the subject is studied from an astronomical approach, in which the yearly path of the sun among the stars (the ecliptic) is considered in relation to meridian circles on an imaginary celestial sphere. There is a tendency for the sun to run fast twice a year (from May through July and from November through January), and slow twice a year (from August through October and from January through April). Superimposing this tendency upon that caused by changing velocity of revolution gives the peculiar effects shown on the analemma.

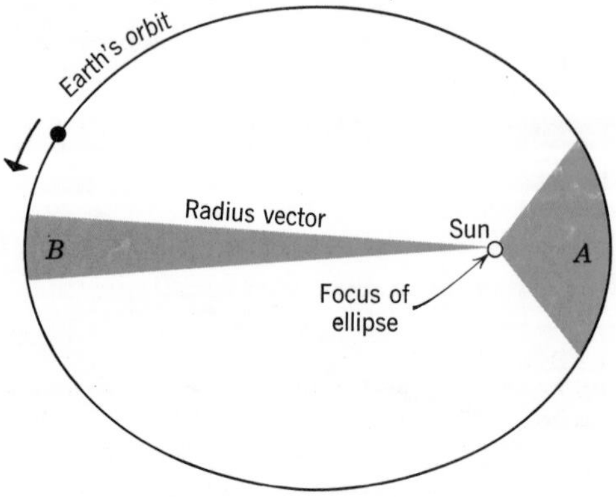

Figure 5.8 The radius vector must sweep over equal areas in equal times.

Use of the analemma

A carefully constructed analemma is a handy tool for the approximate solution of problems of the type suggested below.

1. When will the sun be over the meridian and the shadow of a vertical rod point to true north? This question would need to be answered if a true north-south line was to be drawn, or in order to know at what moment to read the altitude of the noon sun. A systematic solution of the problem is outlined below. Suppose that the place is New York City, long. 74° W, and the date February 25. To avoid confusion, always start your calculations with the figure 12:00 *noon*, representing the apparent solar noon.

Apparent solar noon of local meridian	12:00 noon
Equation of time for February 25 (If *slow*, add. If *fast*, subtract.)	0:13 slow
Mean solar time of local meridian	12:13 P.M.
Correction for difference between local meridian and standard time zone meridian (75° W long.) at rate of 1° = 4 (If standard meridian lies to west, subtract the correction; if to east, add.)	0:04
	12:09 P.M.

Thus, a watch, set to Eastern standard time, will read 12:09 P.M. when the sun arrives over the local meridian on February 25.

2. What will be the altitude of the noon sun for a given place on a given date? Suppose that the place is Capetown, South Africa, lat. 34° S, and that the date is December 10. The following method may be used to solve problems of this type.

Declination of sun on December 10 (read from analemma)	23° S
Latitude of Capetown	34° S
Number of degrees difference between these two parallels	11°
Difference between 11° and 90° (*Answer*)	79°

The sun's noon altitude at Capetown on December 10 is 79° above the northern horizon.

Exact times of sunrise and sunset

Thus far, explanation of lengths of day and night and of times of sunrise and sunset have been oversimplified by assuming that the earth has no atmosphere (hence no refraction of light rays) and that the sun is a tiny pinpoint source of light. Were this so, the length of day on the equinox date would be very nearly 12 hours, with sunrise at 6:00 A.M., local time, and sunset at 6:00 P.M., local time.

Anyone who consults an almanac or newspaper will find that on the equinox date the length of day is about 12 hours and 10 minutes for places at about 40° latitude, such as New York, Chicago, and San Francisco. After correcting for both longitude and equation of time, as explained above, the time of sunrise may still prove to be, say, 5:56 A.M.; the time of sunset 6:05. Why is the length of day some 8 to 10 minutes longer than 12 hours at this latitude? Two factors contribute.

First, because of the earth's atmosphere, light rays are bent so that a sight line is slightly curved, with the convexity upward, as explained in Chapter 1. A horizontal light ray thus bends down over the curve of the earth and our visual horizon is actually lowered by a small amount, approximately 36 minutes of arc (Figure 5.9). For this reason, the sun is actually in sight for a longer time than if the earth had no atmosphere and the day is lengthened accordingly.

Second, the sun is a disk of light, whose average width is equivalent to about 32 minutes of arc as seen from the earth. Sunrise is defined as the instant of appearance of the upper rim (called *upper limb*) of the sun's disk above the horizon, and sunset is defined as the instant of total disappearance of the upper limb below the horizon.

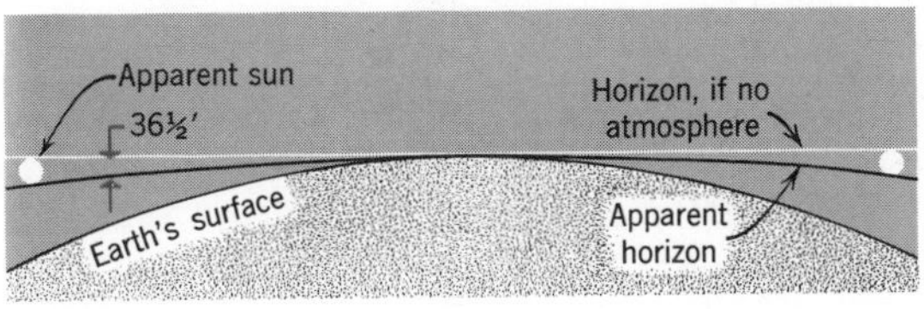

Figure 5.9 Atmospheric refraction lowers the apparent horizon.

At both sunrise and sunset, then, time is added to the day—enough for the sun to rise and set through half of its diameter, or 16′ in each case. If we add the 16′ to 36′, the total is 52′. On a slanting path at 40° latitude, this is equivalent to about $4\frac{1}{2}$ minutes of time (Figure 5.10). Doubling this to include both sunrise and sunset, the total is about 9 minutes, which is quite close to the value given in the almanac or newspaper as the excess over 12 hours. At higher latitudes the excess will be greater, because of the lower angle of the sun's slanting path as it passes below the horizon. For this reason, on the day of equinox the length of day at 72° N lat. is about $12^h\ 21^m$, which is roughly 14 minutes longer than at the equator, where the length is about $12^h\ 07^m$.

The Air Almanac

One authoritative source of astronomical information required for navigation is *The Air Almanac*, produced jointly by Her Majesty's Nautical Almanac Office, Royal Greenwich Observatory, England, and the Nautical Almanac Office of the U.S. Naval Observatory, Washington, D.C. It is printed separately by the two countries, however, and can be obtained in the U.S. through the Superintendent of Documents. The *Air Almanac* is issued in three numbers yearly, covering the periods January–April, May–August, and September–December. Two other authoritative sources are the *Nautical Almanac* and the *American Ephemeris and Nautical Almanac*, also published jointly with Great Britain.

In the *Air Almanac* will be found the sun's declination, as well as time of sunrise and sunset, and duration of civil twilight for a wide range of latitudes. World standard time is given for most countries.

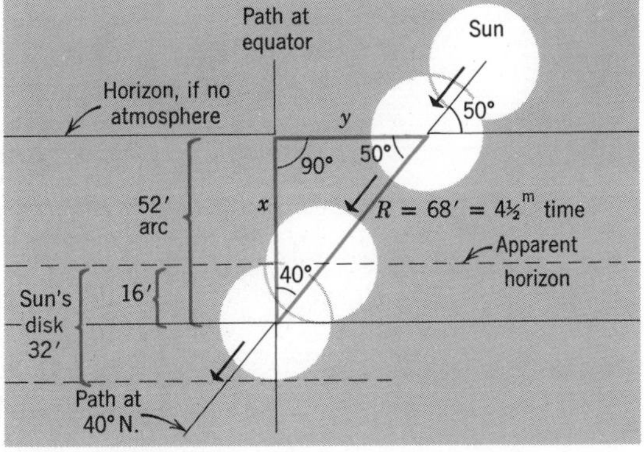

Figure 5.10 The slanting path of the setting sun in middle latitudes lengthens the time needed for it to sink out of sight below the horizon, as compared with a perpendicular path at the equator.

Review Questions

1. Explain how longitude is related to time. How many degrees of longitude are the equivalent of one hour of time?

2. How many hour meridians are there on the globe? In which direction do they travel?

3. Do places located east of you have a time that is earlier or later than your time?

4. What is local time? On what is it based? Can places that differ in longitude have the same local time? Can places that differ in latitude have the same local time?

5. What is standard time? Explain the system of standard time used in the United States. Which meridians are used?

6. How are boundaries between standard time zones determined? Give examples of various kinds of boundaries used in the United States.

7. What is daylight saving time? Why is it used? When is it used?

8. Explain the system of world time zones now in general use. On what prime meridian is it based? What is fast time? slow time? What advantages are there in conforming to the world time zone system?

9. What is the International Date Line? Describe its location and form. Explain the difference in calendar days on the two sides of this line. Which side has the earlier day? Is it possible for the same day to exist simultaneously on both sides of the line?

10. How long does a given calendar day exist on the globe? Describe the life history of one calendar day as it envelops the globe and then disappears.

11. What is the difference between solar time and sidereal time? Which is the more constant time? Which is the basis of our hour system?

12. Explain why the sidereal day is shorter than the mean solar day.

13. How is mean solar time different from apparent solar time? Which time is followed by our clocks? What is the mean sun?

14. What is the equation of time? How is it designated? What range of values does it have? Where can we find out the value of the equation of time?

15. Explain the analemma. What information does it give? How is it constructed? Why does the graph have such a peculiar lopsided figure-eight form? What types of problems can be solved with the aid of an analemma? Give examples.

16. Why, at the equinox date, is the sun actually seen above the horizon for more than 12 hours? Does this excess time over 12 hours change with latitude?

17. What almanacs provide authoritative data for use in navigation? By what agencies are they published? What information of interest in the study of time and global illumination is contained in the *Air Almanac?*

Exercises

1. Using a small globe having 15° meridians or, if no globe is available, the time zone map of the world (Figure 5.4), determine the standard time in use by each of the following places. (State the time in hours and minutes fast or slow.) ***(a)*** Philippines. ***(b)*** Iceland. ***(c)*** India. ***(d)*** Hawaii. ***(e)*** Spain. ***(f)*** Guam. ***(g)*** Ethiopia. ***(h)*** New Zealand.

2. Give the exact longitude of the meridian used for the standard times of the following places. ***(a)*** Rocky Mountain region (7 hours slow). ***(b)*** Maldive Republic (5 hours 5 minutes fast). ***(c)*** Liberia (44 minutes slow). ***(d)*** New York City (5 hours slow). ***(e)*** Cook Islands (10 hours 30 minutes slow).

3. An airlines traveler, delayed at a foreign airfield, notes that the sun is setting just at the moment when his watch, set to Greenwich Civil time when he left London, reads 1:32. The date is March 21. ***(a)*** Assuming the sun to be on time, what is his longitude? (Two answers.) ***(b)*** On March 21, the sun is eight minutes slow. Correct your answers to ***(a)*** so as to take this fact into account. (Disregard the effects of atmospheric refraction and sun's semidiameter.)

4. Construct an analemma by plotting the data of Table 5.1 on cross-section paper. Use an 8½ by 11 inch sheet of paper ruled five squares to the inch. Allow one square for each degree of declination and one square for each minute of time fast or slow. Lay out the page similar to Figure 5.7, and label the analemma fully. After the points have been plotted, each with the date labeled beside it, connect the points with a smooth curve. Because these figures are rounded off to the nearest one-half, they will lie slightly to one side or the other of a smooth curve.

5. Using the analemma, Figure 5.7, as a source of data, solve the following problems. ***(a)*** At what time, according to a clock set for Central Standard time, will the noon sun be over the meridian at Amarillo, Texas (102° W long.), on October 5? ***(b)*** What will be the altitude of the noon sun at New York City (41° N lat.) on February 25? ***(c)*** At what time, according to a clock set for Indian Standard time (5½ hours fast), will the sun rise at Bombay (73° E long.) on September 23? (On the equinox the sun, if on time, would rise at 6:00 A.M. local time.)

TABLE 5.1

Date		Equation of Time	Declination	Date		Equation of Time	Declination
Jan.	1	−3	23° S	July	10	−5	22½
	10	−7	22		20	−6½	21
	20	−11	20		30	−6½	18½
	30	−13½	17½	Aug.	10	−5½	16
Feb.	10	−14	15		20	−4	12½
	20	−14	11		30	−1	9
March	1	−13	8	Sept.	10	+2½	5
	10	−10½	4½		20	+6	1½
	20	−8	½		30	+9½	2½ S
	30	−5	3½ N	Oct.	10	+12½	6½
April	10	−1½	7½		20	+15	10
	20	+1	11		30	+16	13½
	30	+3	14½	Nov.	10	+16	17
May	10	+4	17		20	+14½	19½
	20	+4	20		30	+11½	21½
	30	+3	22	Dec.	10	+7½	23
June	10	+1	23		20	+3	23½
	20	−1	23½				
	30	−3½	23				

CHAPTER 6

Moon and Tides

AN understanding of ocean tides and tidal currents is important to the geographer who is concerned with coastal geography, shoreline landforms, ocean commerce, harbor systems, reclamation of marsh lands, debarkation of armed forces, and many other topics relating to human activity at or near shorelines. Although the tide means little or nothing to inhabitants of midcontinental regions, its influence is continuous and vital to coastal inhabitants.

To understand ocean tides with their seemingly complex variations from time to time and place to place, a knowledge of the moon and its motions is essential. Both the sun and moon exert tide-producing forces upon the earth, but it is the moon, by reason of its closeness, that controls the timing of the tidal rise and fall of ocean level. We turn first, therefore, to a study of the moon's motions, orbit, and phases.

The moon's orbit

The moon, a satellite of the earth, is about 2160 mi (3480 km) in diameter and has a mass of about $\frac{1}{81}$ that of the earth. The moon revolves in an elliptical orbit in which the mean distance between earth and moon is about 240,000 mi (385,000 km). The direction of revolution is similar to the earth's direction of revolution about the sun. If we imagine ourselves to be looking down upon the solar system in such a way that the earth's north pole is below us, the moon's motion is counterclockwise (Figure 6.1). It is also worth noting that the moon rotates upon an axis more or less parallel to the earth's axis, and that both bodies rotate in the same direction, counterclockwise, if we look down upon the north polar end of the earth's axis. This high degree of uniformity in direction of revolution and rotation is found throughout the solar system and strongly suggests that the planets and their satellites originated as condensations in a slowly rotating, flattened cloud of gas and dust. Similarity in process of their formation from a single solar nebula would be expected to produce the uniformity of motion that we observe today.

The moon's orbit is an ellipse, considerably more flattened than the ellipse of the earth's orbit, with the earth located at one focus (Figure 6.2). When at its nearest point to the earth, the moon is said to be in *perigee;* when farthest, in *apogee.* Distances from the earth's center to moon's center are about 221,500 mi (356,000 km) in perigee and about 253,000 mi (407,000 km) in apogee. In accordance with Kepler's law of areas, explained previously in connection with our study of the equation of time, the moon's speed of revolution is somewhat faster near perigee and slower near apogee.

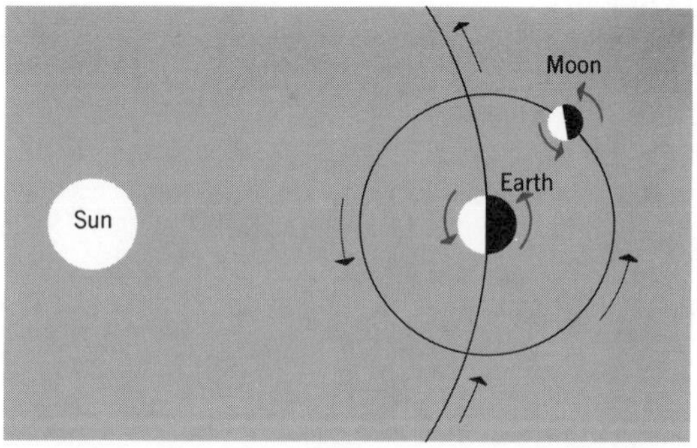

Figure 6.1 Moon and earth revolve and rotate in the same direction.

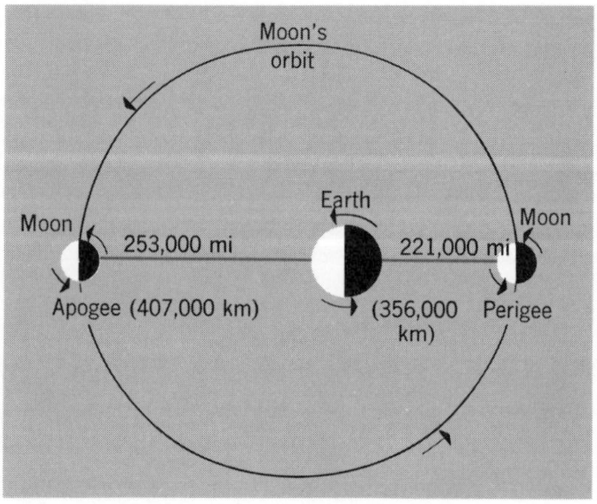

Figure 6.2 The moon's orbit is an ellipse. Distances shown are from center of earth to center of moon.

Period of moon's revolution

If we observe the moon's position with relation to a star very near it in the sky, then observe it again exactly 24 hours later, the moon will be found to be about 13° eastward of the same star. Falling back eastward at this rate, 13.2° per 24 hours, it takes about 27½ days for the moon to be relocated in exactly the same meridional position with respect to the stars. This period, which is 27.32166 days or $27^d\ 7^h\ 43^m\ 11\frac{1}{2}^s$, is called the *sidereal month.* It is the time required by the moon to complete one revolution about the earth.

With reference to the sun, however, the moon's period of revolution is somewhat longer, being about 29½ days. The explanation of this fact is much like that which accounts for the difference between solar and sidereal time. Because the earth is moving in its orbit about the sun, the sun's position is steadily changing with reference to the stars. In order for the moon to complete one whole revolution with respect to the sun, the moon must travel an additional small angular distance beyond 360°. The extra time increases the moon's average time of orbit to 29.53 days, which has been called the *synodic month.* Whereas the sidereal month is always exactly of the same duration, the synodic month may be several hours more or less than 29.53 days, which is only the mean figure. The total possible variation in length of the synodic month is about 13 hours.

It is the synodic month that is of special importance to the physical geographer because the appearance of the moon in the sky and the periods of rise and fall of tides are regulated according to this interval of time.

Inclination of the moon's orbit

The plane containing the moon's orbit is inclined at an angle of 5°09′ to the plane of the ecliptic (Figure 6.3). Thus, during a single revolution, the moon will lie in the plane of the ecliptic only at two places, known as *nodes.* For most purposes, however, the moon may be thought of as moving almost in the plane of the ecliptic, and hence following a path in the sky very similar to that taken by the sun.

Declination of the moon

Just as the sun's declination ranges over a total of 47° from summer solstice to winter solstice, the moon's declination experiences a similar range, but with the possibility of an additional 5°09′ both north and south, or a total possible range of 57°12′. This maximum declination occurs only once every 18½ years.

The moon's entire cycle of declination from maximum south to maximum north, and return, is experienced in 27.2 days, a period known as the *tropical month.* This is much the same as saying that in a month the moon passes through two "equinoxes" and two "solstices" of its own as compared with a similar set of declination changes accomplished by the sun in a whole year. Should we observe the moon's path in the sky on successive nights throughout a month, we would notice that the moon's path is quite low in the sky during one part of the month but becomes fairly high in the sky approximately two weeks later.

Conjunction, opposition, quadrature, and syzygy

When both sun and moon are on the same side of the earth, so that all three bodies lie approximately on a straight line, the moon is said to be in *conjunction* with the sun (Figure 6.4). At this time, the possibility exists for an eclipse of the sun, or *solar eclipse,* but this is a rare occurrence because the moon is so small and the plane of its orbit is tilted about 5° with respect to the plane

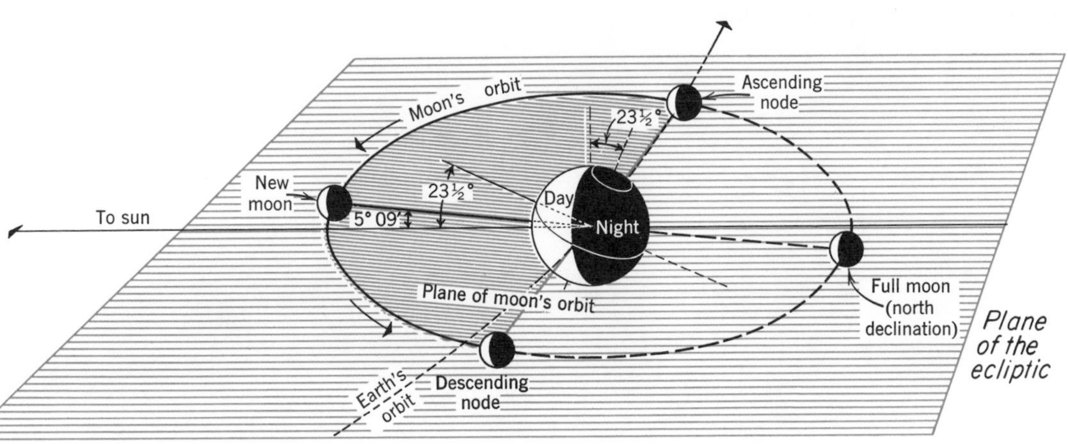

Figure 6.3 The moon's orbit is shown here as it might look at the time of the winter solstice.

of the ecliptic. When the moon and sun are on the opposite sides of the earth they are said to be in *opposition*. Again all three bodies are approximately in a straight line. The possibility now exists for an eclipse of the moon, or *lunar eclipse*, in which the earth's shadow falls on the moon, partly or completely covering it for a short period. The chances of our seeing a lunar eclipse from a given place on earth are much better than for seeing a solar eclipse.

The word *syzygy* combines the meanings of conjunction and opposition. Thus, when we are told that the moon is in syzygy, we know that all three bodies are approximately in a straight line, but we do not know whether sun and moon are on the same side or opposite sides of the earth.

The word *quadrature* means that sun and moon are so situated that rays drawn from each to the earth make an angle of about 90° (Figure 6.4). The moon is thus in quadrature twice every synodic month.

Phases of the moon

Illumination of the moon and earth and the progressive changes in appearance, or *phases*, of the moon throughout the synodic month, are illustrated in Figure 6.5. At the outset, it is important to make clear that one-half of the moon's surface is always illuminated by the sun's rays, just as one-half of the earth's surface is always illuminated. To the earth-bound observer, however, the amount of the illuminated half of the moon that can be seen changes throughout the month and ranges from none visible to the entire illuminated half visible.

The synodic month begins with the phase of *new moon*, when sun and moon are in conjunction (see Figure 6.5). Because the illuminated half of the moon faces entirely away from the earth, the moon would appear entirely dark to the observer on the earth, except for a faint glow of light reflected to it by the earth. There is another reason, however, why we cannot see the moon at this time. As is evident from the diagram, both the sun and moon are approximately in the same position in the sky so that the sun's blinding rays effectively conceal the moon. In this phase the moon and sun both rise at about the same time and move together across the sky. This statement is, of course, generalized, because the moon travels more slowly across the sky and is falling behind at the rate of about 12° every 24 hours.

About $3\frac{3}{4}$ days after conjunction, the moon has traveled one-eighth of the distance around its orbit (Figure 6.5). It is now visible in the sky as a thin crescent whose points are directed away from the sun. This is called the *crescent new moon*. During the preceding $3\frac{3}{4}$ days, the moon has dropped behind the sun in the sky about 45°. Hence the crescent new moon rises in the eastern horizon when the sun has already reached a point in the sky about midway between the horizon and its noon position. The crescent moon follows the same general path as the sun, but is still shining low in the western sky long after the sun has set.

After about $7\frac{1}{2}$ days have elapsed in the synodic month, quadrature is reached (Figure 6.5). The moon is in the phase of *first quarter*, in which it appears as a half circle of light. Roughly speaking, the moon in this phase rises about the time the sun is in its noon position, and reaches its highest point in the sky when the sun is setting. We are here assuming that the sun is rising and setting about 6:00 A.M. and 6:00 P.M., as it would do near the time of equinoxes or near the equator.

By the time the moon has traveled three-eighths of its orbit, and is about $11\frac{1}{4}$ days old, we see it in the sky as about three-quarters illuminated. This is described as a *gibbous moon*.

When the moon is $14\frac{3}{4}$ days old in the synodic month, it is in opposition to the sun and is in

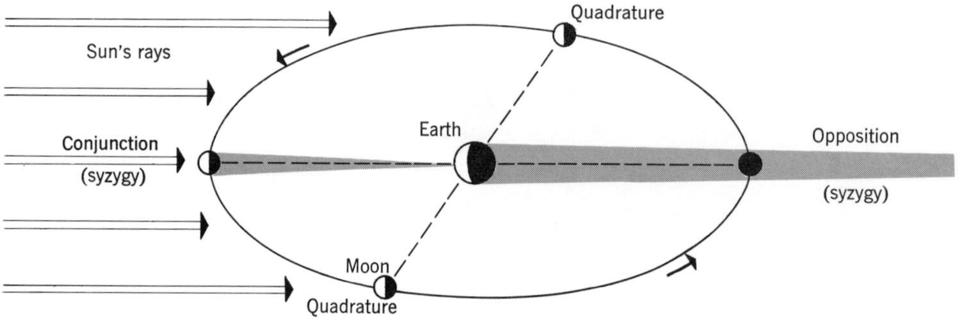

Figure 6.4 These relationships among sun, moon, and earth influence the height of tides.

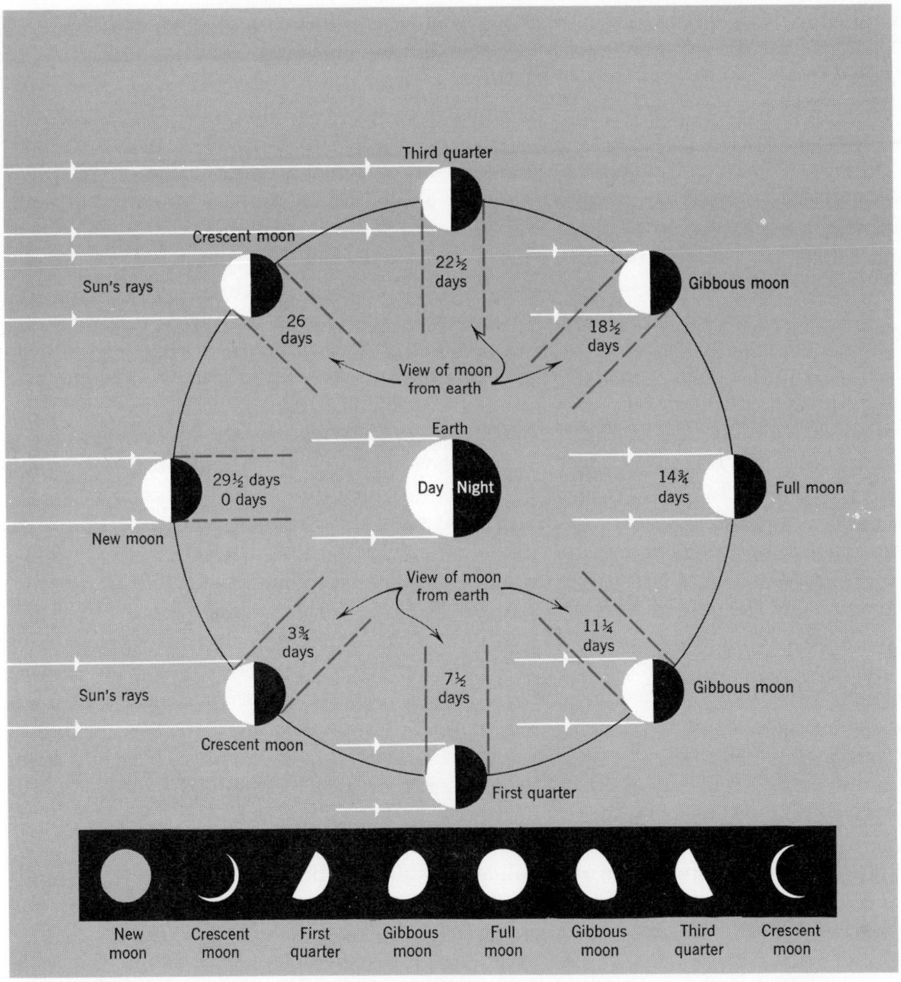

Figure 6.5 Phases of the moon. Diagrams below show outline of moon as seen in the southern half of the sky.

the phase of *full moon* with the entire illuminated half visible from the earth. Because moon and sun are on opposite sides of the earth, the full moon will be highest in the sky at about midnight. If day and night are about equal in length, the moon will rise when the sun is setting and will set when the sun rises.

Now refer back to Figure 6.3, which shows conditions near the winter solstice in late December. When the moon is full, its north declination is at the maximum and its rays strike the earth surface vertically at about the Tropic of Cancer. The moon's path in the sky is then relatively high for observers in the northern hemisphere, which explains the popular observation that the winter full moon "rides high." At the time of summer solstice, however, the path of full moon is low in the sky and the brilliance of moonlight small as compared with the winter full moon.

The remaining phases of the moon are similar to those already described, except that they occur in the reverse order. One important difference is that the moon appears as if it were the mirror image of its corresponding phases of the first half of the synodic month. For example, the horns of the old crescent moon, although pointing away from the sun, will be directed the opposite way in the sky from those of the new crescent moon.

By the time the phase of old crescent moon is reached, 26 days have elapsed in the synodic month, and the moon will have lagged so far behind the sun in the sky that it seems, instead, to be traveling about 45° ahead of the sun. By the 29th day, the moon has fallen back to a place almost coincident with the sun and the synodic month draws to a close.

Rotation of the moon

Should we photograph the moon from earth at many different times and carefully compare the photographs, it would be found that 41 percent of the moon's surface is never seen and that a map of the moon, compiled from the photographs, could show only 59 percent of the moon's total surface. It is therefore evident that the moon at all times keeps the same side toward the earth. This means that the moon rotates on its axis exactly once in each sidereal month of 27.32166 days. It is believed that tidal friction has been responsible for slowing the moon's rotation to the point where it no longer turns with respect to the earth.

Gravitation and tides

Although it was known from the first century A.D. that the *tide*, or periodic rise and fall of ocean level, is controlled in some manner by the sun and moon, it was not until Sir Isaac Newton published the law of gravitation in 1686 that the true explanation became known.

Because the tides depend upon gravitation, which is the mutual attraction between any two masses, it is desirable to restate the law of gravitation: Two bodies attract each other with a force proportional to the product of their masses and inversely proportional to the square of the distance between them.

According to the first part of this law, if one body is twice as massive as another, the more massive one will exert twice as strong an attractive force as the smaller. According to the second part of the law, if the distance between two masses is doubled, the gravitational force is reduced to one-fourth of what it was.

Lunar tides

Now consider Figure 6.6, in which the earth is represented as having a uniform depth of ocean water covering it. That part of the globe at *T* is most strongly attracted by the moon because it is closest. Near *C*, the earth's center, the moon's gravitational attraction is less than at *T*, and at *A* it is least of all. Because the gravitational attraction diminishes from *T* to *A* there is a tendency for the earth to be pulled apart. We might imagine that the ocean water at *T* tries to pull away from the main mass of the earth, centered at *C*, while the main body of the earth tends to pull away from the ocean water at *A*. The effect is much like the stretching of a chain of skaters when they crack the whip. The spherical earth tends to be stretched along the line to the moon and to be deformed into a prolate ellipsoid. So far as the subject of ocean tides is concerned, the solid earth can be considered to remain unaffected by the stretching force, or tide-producing force, although slight responses are detectable in the form of *earth tides*. The oceans, however, are composed of fluids that respond readily to small forces and water will move toward centers at *T* and *A*.

It is beyond the scope of this text to explain in detail the tidal forces. It will suffice here to state that the tide-producing force resulting from decrease in gravitational attraction from points *T* to *A* (Figure 6.6) can be resolved into a *tractive force* (drag force) acting parallel with the earth's surface. Distribution of this tractive force is shown by arrows in Figure 6.7. The tractive force is zero along the great circle passing through the points N and S, but increases to a maximum on small circles lying 45° and 135° of arc from point *T*. From this maximum, the tractive force again diminishes to zero at the points *T* and *A*.

The earth is thus marked off into two hemispheres of tidal influence. The ocean water tends to flow toward the centers *T* and *A*, where the water level will rise, but to flow away from the great circle of zero tractive force, where the water level will sink.

Period of lunar tides

Because the earth rotates eastward upon its axis, the two tidal centers move westward with respect to the surface of the earth. At any particular point on the globe near the equator, the passing of either of these centers causes a rise of water level to a maximum called *high*

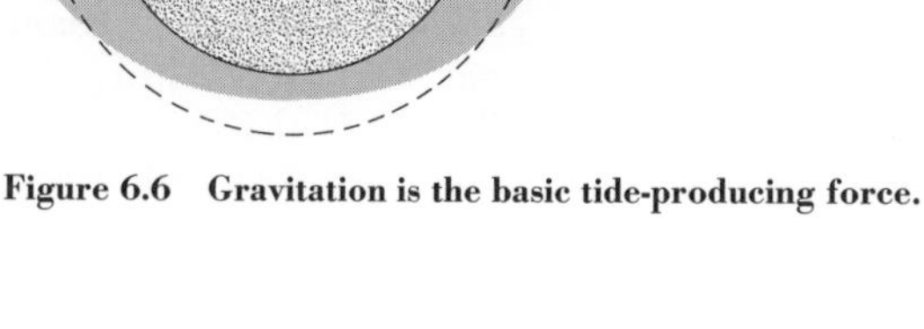

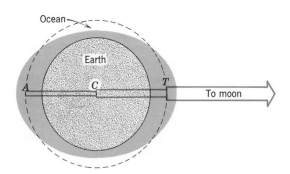

Figure 6.6 Gravitation is the basic tide-producing force.

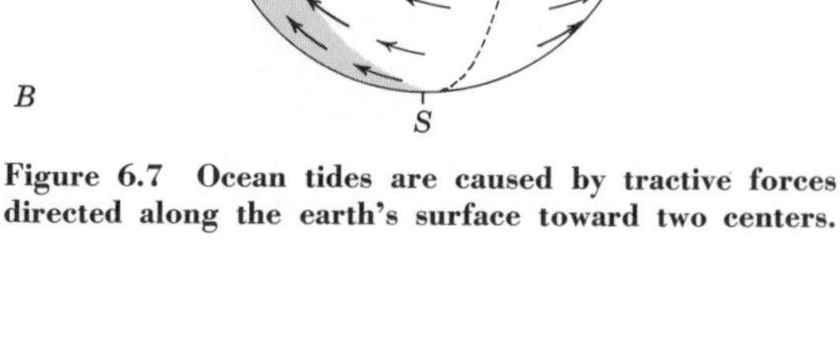

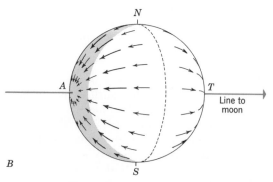

Figure 6.7 Ocean tides are caused by tractive forces directed along the earth's surface toward two centers.

water, whereas halfway between, the ocean level falls to a minimum called *low water*. Between these extremes is mean tide level, which is the average of high and low waters taken over a long period.

Because it takes 24 hours and 50 minutes for the earth to turn once with reference to the moon, two high waters and two low waters will occur during that period. Successive high waters therefore occur about $12\frac{1}{2}$ hours apart, and the interval between high water and the next low water is about $6\frac{1}{4}$ hours. Because our 24-hour calendar day is determined by the mean sun, whereas the tide is governed by the moon, the high or low waters will be found to occur about 50 minutes later on each successive day.

Should we actually compare the time of high water with the time of the moon's meridian passage for a given coastal point, it would be found that the high water may occur several hours after the moon's meridian passage. This lag of the tide behind the moon is known as the *lunitidal interval* or *establishment of the port*. It varies considerably according to the location of a coast and varies at different times of year for the same port. At Fort Hamilton, in New York Harbor, for example, the lunitidal interval is about $7\frac{3}{4}$ hours.

Typical semidaily tide curve

Should we make half-hourly observations of the position of water level against a measuring stick, or *tide staff*, attached to a pier or sea wall, we could plot the changes in water level and thus draw a graph of the tide. Figure 6.8 is such a graph constructed for Boston Harbor during a 24-hour period. In agreement with what has been stated thus far, it can be seen that $12\frac{1}{2}$ hours elapsed between successive high waters and between successive low waters. The time between successive high and low waters was about $6\frac{1}{4}$ hours. This interval, if averaged out over a long period of observations, would be 6 hours and 12 minutes.

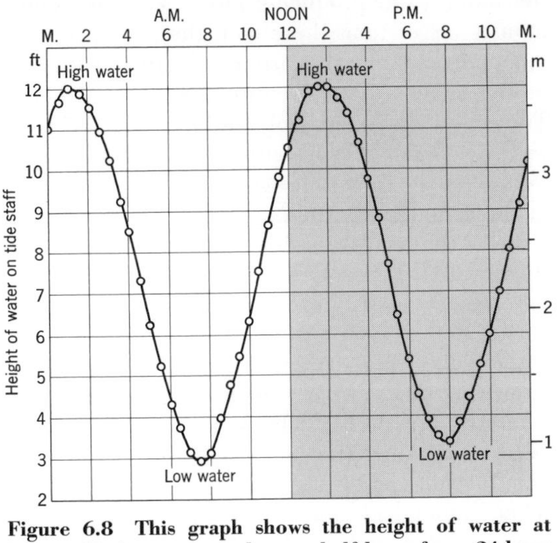

Figure 6.8 **This graph shows the height of water at Boston Harbor measured every half hour for a 24-hour period. (After H. A. Marmer.)**

The *range of tide* on the graph is about 9 ft (2.7m). Although the high waters reached the same mark, the low waters differed by 0.5 ft (0.15 m). Observations taken over a long period would show that the range in Boston Harbor averages about 10 ft (3 m), but may be as great as 14 ft (4.3 m) and may differ greatly from one day to the next.

It is worth noting further that the half-hourly changes of water level are not by any means uniform. Midway between high and low waters the level rises or drops about 2 ft (0.6 m) per hour, whereas close to the high- or low-water points, the change amounts to only 4 to 8 in. (10 to 20 cm) per hour. To the mathematician, a curve of the type shown here is known as a *sine curve*. To an observer at the seashore, the characteristics of this curve are apparent from the fact that high water, once attained, seems to persist for a long time, then is followed by a fairly rapid drop of sea level to low water which, again, seems to persist for a long time.

Diurnal inequality of the tide

It has already been explained that the moon experiences a declination north and south of the equator, approximately equal in amount to the sun's yearly declination but accomplished during a tropical month of 27.2 days. When the moon's declination is farthest north, the tidal center which lies at the point where the moon's rays strike the earth vertically sweeps westward around the earth about on the Tropic of Cancer ($23\frac{1}{2}$ N lat.), whereas the opposite tidal center sweeps around following the Tropic of Capricorn ($23\frac{1}{2}$ S lat.) (Figure 6.9). The importance of this fact is that, for specific places lying north or south of the equator, successive high or low waters are of

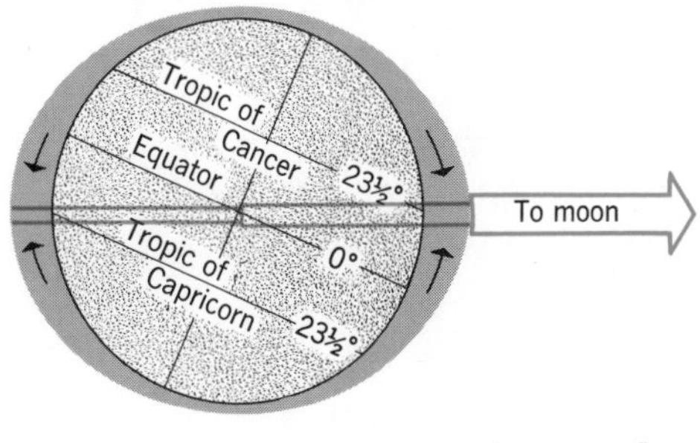

Figure 6.9 **The moon's changing declination influences the tide.**

unequal size, but alternate ones are equal. This phenomenon is known as the *diurnal inequality* of the tide. It is most marked twice a month when the moon's declination reaches a maximum, at which time the tides are designated *tropic tides,* but disappears at the two times during the month when moon rays fall vertically on the equator, when the tides are termed *equatorial tides.* Figure 6.10 shows two tide curves for Portland, Maine. In *A* is shown the equatorial form in which high and low waters repeat the same levels. In *B* is shown the tropic form with about two feet difference in successive high waters, and the same in successive low waters.

Daily, semidaily, and mixed tides

From the foregoing discussion, it can be seen that where diurnal inequality of tides exists, the tidal curve results from the combining of two important component forces, or *constituents:* (1) a *semidaily* constituent, and (2) a *daily* constituent. The semidaily constituent results from the presence of the two tidal centers and, in the pure state, gives a tide curve in which successive high and low waters repeat previous levels. This is illustrated by Figure 6.11*A*. The daily constituent is a result of the moon's declination and gives a tide curve having one high and one low water each lunar day. This is illustrated in Figure 6.11*D*. Most tide curves are combinations of the two constituents and are known as *mixed types.* Where the combination is such that the semidaily constituent is dominant, the curve will show two high waters and two low waters, but either the high waters or low waters will have strong diurnal inequality. One mixed type is represented in Figure 6.11*B*.

When combined in the ratio such that the daily constituent is twice that of the semidaily constituent, illustrated in Figure 6.11*C*, a curious curve results. There is one high and one low water, separated by a *stand* of sea level for several hours. The period of stillstand is known as the *vanishing tide.*

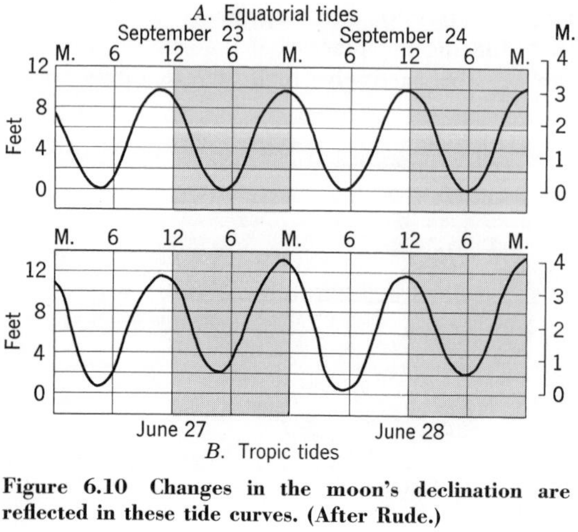

Figure 6.10 Changes in the moon's declination are reflected in these tide curves. (After Rude.)

Figure 6.11*D* is a daily type of tide curve at Manila, Philippine Islands, in which the semidaily constituent is so small as to cause only a slight irregularity in curve form.

Tide curves of the United States coast

Certain general statements may be made about tide curves for the three coasts of the United States. Along the Atlantic coast the typical curve is of the semidaily type, with little diurnal inequality, as illustrated by the curve for Portland, Maine (Figure 6.11*A*). Tide curves of the North Pacific coast are characteristically of the mixed type, exhibiting strong diurnal inequality as shown by a Seattle, Washington, tide curve (Figure 6.11*B*). On the Gulf of Mexico the daily constituent is very strong, so that tide curves alternate between an equatorial form (*A*), which has two high and two low waters (Figure 6.12), and a tropic form (*C*), having only one high and one low water daily. Of the three curves shown for Galveston, Texas, the upper one (*A*) occurs about 25 percent of the time, the middle curve (*B*) about 50 percent, and the lowest one (*C*) about 25 percent, all being directly dependent upon the moon's declination.

Neap and spring tides

Thus far, the tide-producing force of the sun has not been taken into account, although this is an important force, operating in the same way as the moon's tide-producing force. Although enormously larger than the moon, the sun is so very much farther from the earth that its tide-producing power is only five-elevenths that of the moon. The moon always controls the time at which low and high waters occur, whereas the sun's effect is to modify the tidal range greatly at different times in the synodic month.

From the relative position of the sun and moon (Figure 6.4), it becomes evident that, in syzygies, the tide-producing forces of the sun and moon are exerted in such a way as to complement each other. This produces tides of unusually great range, known as *spring tides,* which occur about twice a month (every $14\frac{3}{4}$ days), at new moon and full moon, when the moon and sun are in conjunction and opposition, respectively (Figure 6.13). When moon and sun are in quadrature, in the phases of first and third quarters, the sun's tide-producing force tends to balance out that of the

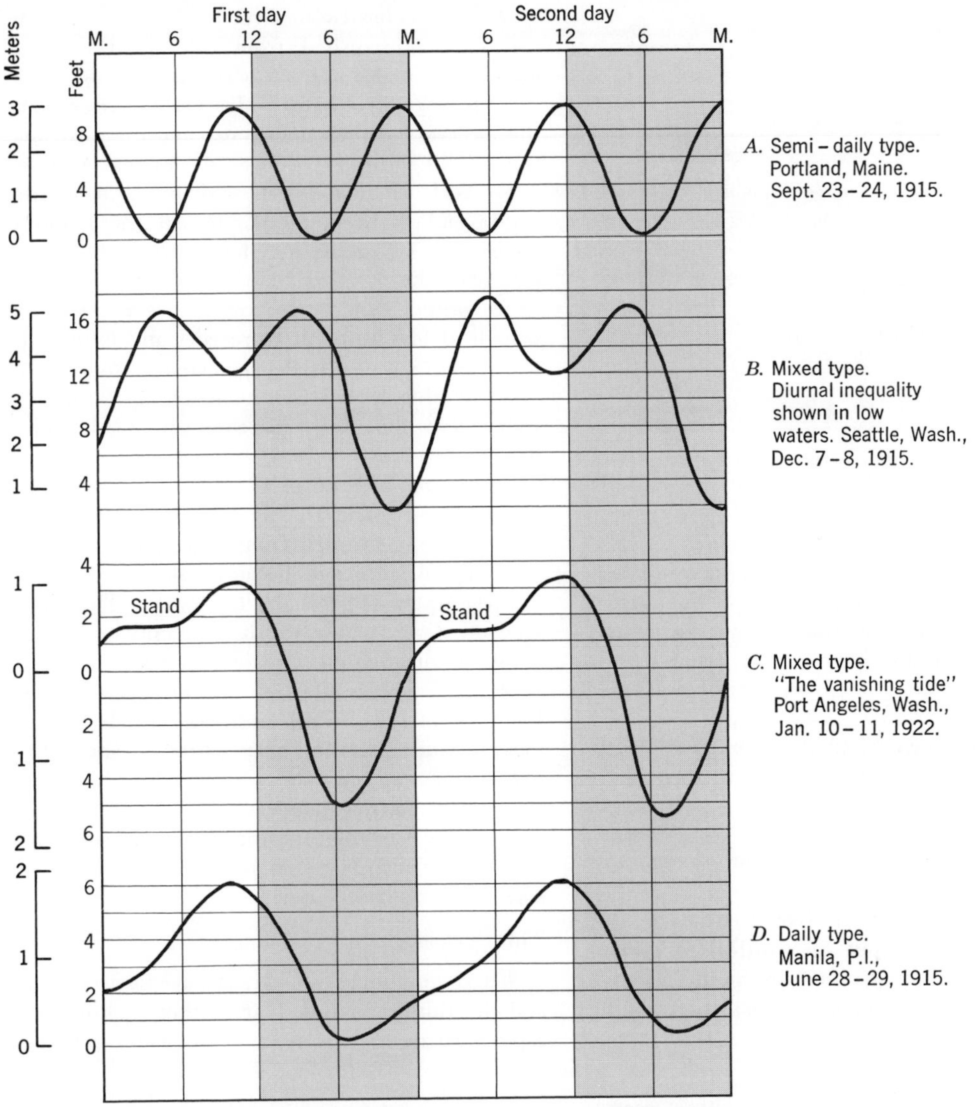

Figure 6.11 Tide curves range from simple semi-daily types, through mixed types, to daily types. (After Rude and Marmer.)

moon, causing tides of unusually small range, known as *neap tides.* Spring tides are about 20 percent greater than the average tide; neap tides are about 20 percent less.

Perigean and apogean tides

Still another important variation occurs in the range of tides. When the moon is at perigee in its orbit, nearest the earth, its tide-producing power is markedly greater than average and results in *perigean tides* which are 15 to 20 percent greater than average. The time interval from perigee to perigee is 27.5 days. When the moon is at apogee, farthest from the earth, tides are about 20 percent less than average and are known as *apogean tides.*

On occasions, when spring tides coincide with perigean tides, the tidal range is, of course, abnormally great, but when neap tides and apogean tides occur together the range is abnormally small.

River tides

Many of the world's great rivers experience tides in their lower parts and are known as *tidal rivers.* This condition has resulted where the coastal area has recently subsided, or the ocean level risen, causing the lower part of the river to be drowned. In a strict sense, such water bodies are not rivers, but are arms of the sea or *estuaries.*

As the tide rises to high water at the seaward mouth of tidal rivers, a wave is generated which

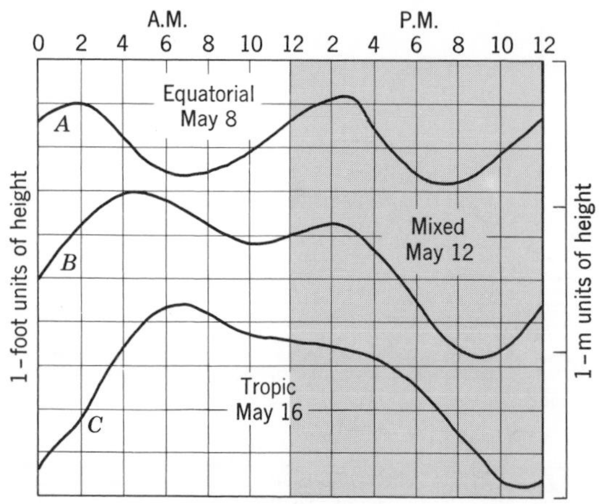

Figure 6.12 At Galveston, Texas, the tide curve varies considerably throughout the lunar month. (After H. A. Marmer.)

runs toward the inner end. Rate of travel of this tidal wave depends upon depth of the water, being faster for deeper water. It is expressed mathematically by the formula

$$v = 3.36\sqrt{d}$$

where v = speed of wave in knots, and d = depth of water in feet. In metric units the formula is

$$v = 3.13\sqrt{d}$$

where v = wave speed in meters per second, and d = depth in meters.

For example, in a tidal river 50 ft deep, the wave would travel about 25 nautical miles per hour. The time of high water for points located 25 nautical miles (29 statute miles) apart would be about one hour different.

One characteristic of river tides, whereby they may be distinguished from tides of the open ocean, is that the interval between one low water and the next high water is distinctly shorter than the interval between high water and the next low water. This is illustrated by a tide curve at Albany, New York (Figure 6.14), which lies near the upper end of the tidal portion of the Hudson River. The inequality may be explained by application of the formula stated above. The crest of an incoming tidal wave travels faster than the trough of low water that precedes and follows it, because the water is deeper. Hence, there is a tendency for high water to catch up with low water, and the effect increases the farther the distance upriver.

In general, the range of river tides decreases toward the head of the tidal river as a result of loss of energy through friction with the channel bottom and sides and because the seaward flow of river water opposes the tidal wave. Thus, the Hudson River, whose width and depth are fairly constant throughout its tidal portion, has an average tidal range of 4.4 ft (1.3 m) at its mouth, but 131 mi (211 km) upstream, at Troy, the head of tide water, this range is reduced to 3.0 ft (0.9 m). Exceptions occur where the tidal river or estuary narrows appreciably landward. The tidal range may then increase inland because the energy of the tidal wave is concentrated into a smaller amount of water.

Tidal bores

Where outgoing river currents are fairly strong, and the tidal river or estuary rather shallow, the rapidly rising high water may advance upstream

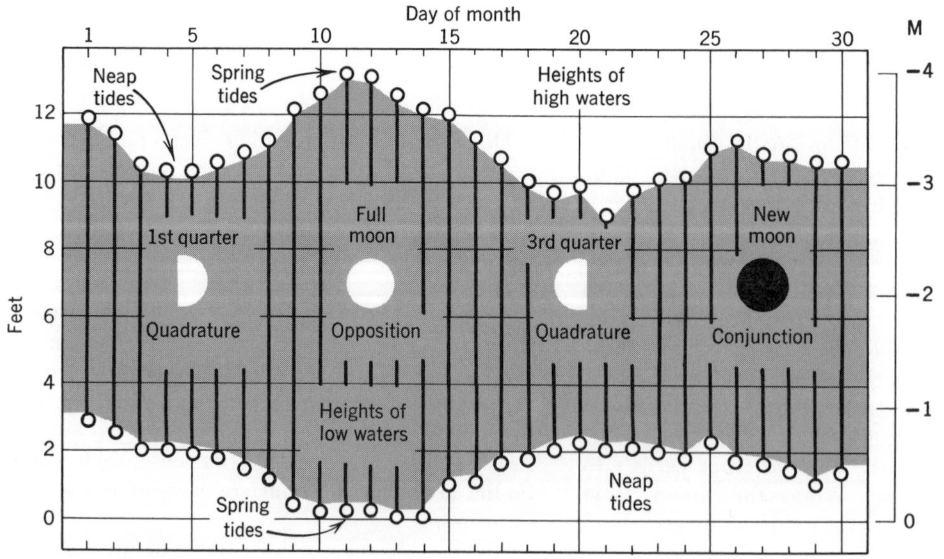

Figure 6.13 Neap and spring tides. (After H. A. Marmer.)

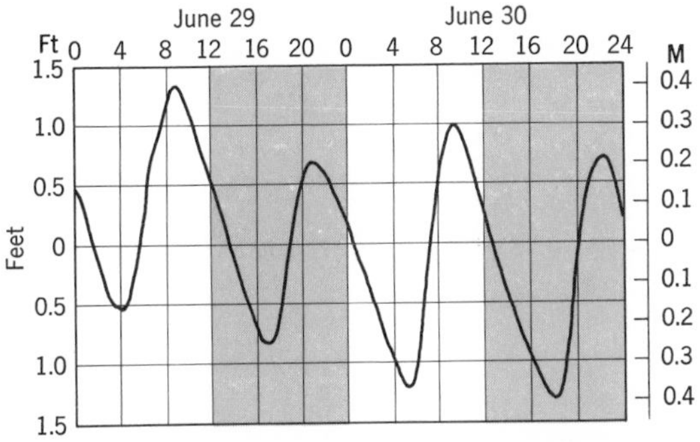

Figure 6.14 A river tide is well shown by this tide curve from the Hudson estuary at Albany, New York, more than 100 mi (160 km) inland. (After H. A. Marmer.)

as a nearly vertical wall several feet high, known as a *tidal bore* (Figure 6.15). Bores are characteristic of certain river mouths, such as those of the Amazon, Colorado, Yangtze, Tsientang, Hooghly, Severn, Elbe, and Weser rivers. Where bores are unusually well developed, as in the Tsientang River mouth at Hangchow, China, the moving wall of water may be 10 to 15 ft (3 to 5 m) high and is reported to be a terrifying and destructive phenomenon.

Theories of tidal behavior in the open oceans

The reader may have gained the impression from the foregoing discussion of tide-producing forces that there are two regions of high water always present on the globe and that these may be likened to two great water waves, each traveling westward around the earth once every 24 hours and 50 minutes. These *tidal waves* would be of such great breadth and small height as to be imperceptible to observers on shipboard in mid-ocean. Because the globe consists of ocean basins separated by vast continents, this simple concept cannot apply, but it gave rise long ago to the *progressive-wave theory* of tides. In the southern hemisphere, between latitudes 40° and 65° S, a broad expanse of almost unbroken ocean encircles the globe. Just as the prevailing westerly winds in this region have free rein to blow over the sea in great gales, so the tide-producing forces were assumed to have freedom to produce two tidal waves sweeping westward around and around the earth. It was further supposed that the progressive waves generated other tidal waves which swept northward up the Atlantic and Pacific oceans. The speed of these secondary waves would be determined by depth of water, rather than by the lunar period, and would reach points progressively farther north along the coasts at later times. Although the progressive-wave theory thus explained certain characteristics of the tide, it proved to be inadequate to explain many tidal peculiarities brought to light by newer and more abundant data. The progressive-wave theory enjoyed widespread popularity because of its simplicity, but it should be regarded as both outmoded and incorrect.

The *oscillation theory*, now followed in explaining tidal behavior in open ocean basins, is based on the principle that a body of water can be set in rhythmic motion, or oscillation, by the tide-producing forces, but that it will experience periodic rises and falls of level in a manner determined by its size and shape. This is roughly illustrated by the back-and-forth movement of water in a shallow tray caused by a slight uptilting of one end of the tray. Where an ocean body is of such a size and shape that its natural period of oscillation is approximately the same as that of the tide-producing forces, it will respond readily; otherwise, it will not have marked tides. The oscillation theory, refined by taking into account the deflective force of the earth's rotation upon flowing water, explains many otherwise anomalous tidal features. For example, the occurrence of a daily tidal curve, with only one high and one low water per day, may result because that particular portion of the ocean has a natural period of oscillation which responds to the daily constituent in the tide-producing force but not to the semidaily constituent. Modern tidal theory is still actively developing and is a highly specialized and mathematical subject of scientific research.

Figure 6.15 A small tidal bore coming up the estuary of the Colorado River. (Godfrey Sykes: Colorado Delta, American Geographical Society.)

Tidal currents

Our discussion so far has treated only the rising and falling of water level. A related subject is the production of *tidal currents,* or streamlike movements of water in and out of bays and tidal rivers, resulting from the tidal changes in ocean level.

The relationships between tidal-current speed and the tide curve are shown in Figure 6.16. When the tide begins to fall, an *ebb current* sets in, reaching maximum speed about at *midtide.* Flow ceases about the time when the tide is at its lowest point, a condition known as *slack water.* As the tide begins to rise, a landward current, the *flood current,* begins to flow and gains in strength to attain a maximum speed at about midtide. Note that the ebb current is stronger than the flood current, a condition explained by the fact that the rivers contribute a considerable discharge of fresh water from the land which must escape to the sea. This stream discharge augments the ebb current but opposes the flood current.

Thus, in the lower Hudson River the ebb velocity at one point may be 2.4 knots (1.2 m/sec); the flood velocity, 1.6 knots (0.6 m/sec), or about two-thirds as great. Such currents are important in harbor navigation. Considerable skill is required in maneuvering large vessels into their berths and guiding ferryboats into their slips where tidal currents are running past the ends of the piers.

Unusually strong tidal currents result where bays connect with the open ocean by narrow inlets. Because water level in the bay cannot rise sufficiently rapidly to maintain the same level as the rising ocean, a marked difference in the two water levels may develop. A strong *hydraulic* current then pours through the narrow inlet. When, on the other hand, the level of the open ocean falls to low water, the surface of the bay will be higher and a strong hydraulic current will flow seaward through the inlet. Tidal currents of this type may develop velocities of 5 to 10 knots (2.5 to 5.0 m/sec), or even as much as 12 knots (6 m/sec) if range of tide is great and configuration of the bay especially favorable. Currents of such magnitude may interfere with navigation to the extent that the passage of vessels in and out of harbors may have to wait upon the occurrence of favorable conditions.

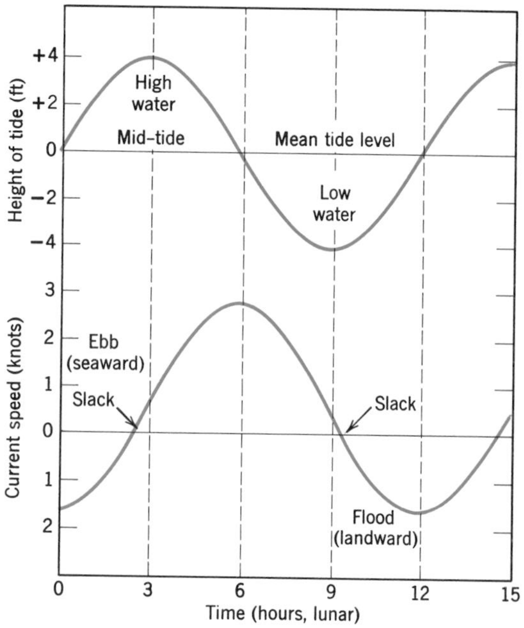

Figure 6.16 Relationship between tidal-current speed and the tide curve.

Review Questions

1. How does the moon compare with the earth in diameter and mass? Approximately how far is the moon from the earth? In what direction does the moon revolve about the earth?

2. What shape has the moon's orbit about the earth? What are perigee and apogee? Compare these conditions with aphelion and perihelion. What can be said about the moon's speed of revolution at perigee and apogee?

3. How does the moon's position in the sky change from one day to the next, if it is observed at the same hour each day?

4. What is the sidereal month? How is it determined? What is the synodic month? How is it determined? Which of these two periods forms the basis for the phases of the moon and the tide periods?

5. How does the plane of the moon's orbit about the earth lie in relation to the plane of the earth's orbit about the sun? What are the nodes? How does inclination of the plane of the moon's orbit influence the likelihood of the occurrence of solar eclipses?

6. Describe the moon's declination. How great a range of declination is possible? How often does the moon pass through "solstice" and "equinox" positions (using an analogy with the sun's declination)?

7. In winter, is full moon associated with high or low path of the moon in the sky? How does this compare with the path of the moon during summer? Explain your answer.

8. Define conjuction, opposition, quadrature, and syzygy. Explain what conditions are required for a solar eclipse; for a lunar eclipse. Which is more commonly seen?

9. Describe the phases of the moon, starting with new moon and continuing through the synodic month. Tell when each phase occurs, and explain the appearance (shape) of the moon. How would you be able to distinguish a new crescent moon from an old crescent moon?

10. How much of the moon's surface is visible from the earth, irrespective of length of time it is observed? Explain your answer.

11. When was the cause of tides first fully understood? State the law of gravitation.

12. Explain how the attraction between moon and earth tends to create a heaping up of ocean waters at two opposite centers. Why is the water not drawn only toward the side nearest the moon?

13. What is meant by high water, low water, and mean sea level? What is the natural period of tides? How does this tide period fit in with our solar time system?

14. What is the lunitidal interval? Is it constant at all times and at all places?

15. Describe the semidaily tide curve in its simplest form. How might this curve be measured? What is meant by range of tide? What kind of mathematical curve is the simple semidaily tide curve? Does the hourly amount of rise or fall of water level remain constant?

16. What is diurnal inequality of the tide? Why does it exist? What is the difference between tropic and equatorial tides?

17. Explain the concept of tide constituents. What kind of tide curve results from the semidaily constituent? from the daily constituent? from various mixtures of both constituents? What is a vanishing tide?

18. Describe the typical tide curves of the Atlantic, Gulf, and Pacific coasts of the United States.

19. What are neap and spring tides? Describe the effect of quadrature and syzygy on tide ranges.

20. What are perigean and apogean tides? When do they occur? What changes in tide range can be expected at apogee and perigee?

21. Explain the movement of tidal waves in tidal rivers. What determines the velocity of wave travel? How does the curve of a river tide differ from the normal tide curve on an open coastline? What is a tidal bore?

22. Explain how the progressive-wave theory of tides differs from the oscillation theory. What features of tides are explained by the oscillation theory that are not explained in the progressive-wave theory?

23. How are tidal currents produced in bays? What is meant by flood and ebb? What is slack water? How can hydraulic tidal currents be produced in bays?

Exercises

1. The figures below give the height of water at San Francisco, California, every hour for one day. Plot these on a graph similar to Figure 6.8, and draw a smooth tide curve through the points. (Data from H. A. Marmer.)

Hour	Height (Feet)	Hour	Height (Feet)
12 midnight	0.0	1 P.M.	−0.3
1 A.M.	−1.5	2 P.M.	−1.5
2 A.M.	−2.3	3 P.M.	−2.4
3 A.M.	−2.6	4 P.M.	−2.5
4 A.M.	−2.3	5 P.M.	−1.9
5 A.M.	−1.5	6 P.M.	−1.0
6 A.M.	−0.3	7 P.M.	0.1
7 A.M.	0.8	8 P.M.	1.2
8 A.M.	2.0	9 P.M.	2.0
9 A.M.	2.6	10 P.M.	2.3
10 A.M.	2.8	11 P.M.	2.2
11 A.M.	2.3	12 midnight	1.2
12 noon	1.2		

2. (*a*) What type of tide curve is illustrated by the graph that you have just drawn for Exercise 1? (*b*) Give as closely as you can the heights of the two high waters and the two low waters. (*c*) What were the ranges between successive high and low waters? (*d*) Were both low waters at the same mark? (*e*) Were both high waters at the same mark? If not, explain. (*f*) Approximately what was the moon's declination on this date? (*g*) How much time elapsed between the first low water and the next high water? (*h*) How much time elapsed between successive low waters? (*i*) between successive high waters?

3. The figures below give the hourly heights of tide at San Francisco for a 24-hour period. Make a graph to show this tide curve, following the same procedure in Exercise 1. (Data from H. A. Marmer.)

Hour	Height (Feet)	Hour	Height (Feet)
12 midnight	0.0	1 P.M.	−2.6
1 A.M.	−0.5	2 P.M.	−3.3
2 A.M.	−0.5	3 P.M.	−3.3
3 A.M.	−0.3	4 P.M.	−2.9
4 A.M.	0.5	5 P.M.	−1.8
5 A.M.	1.4	6 P.M.	−0.9
6 A.M.	2.2	7 P.M.	0.1
7 A.M.	2.7	8 P.M.	1.1
8 A.M.	2.8	9 P.M.	1.6
9 A.M.	2.3	10 P.M.	1.8
10 A.M.	1.3	11 P.M.	1.5
11 A.M.	−0.2	12 midnight	0.9
12 noon	−1.5		

4. (*a*) What type of tide curve is represented by the graph drawn for Exercise 3? (*b*) List the heights of low and high waters as they occurred. (*c*) Explain why the first high water is higher than the second; the second low water much lower than the first. (*d*) What is the maximum range of tide between successive high and low waters on this graph? (*e*) Determine the intervals of time between successive high and low waters. Compare these with the data of Exercise 1.

5. Using the same method as for Exercises 1 and 3, plot the tide curve for St. Michael, Alaska, for which the data are given below. (Data from H. A. Marmer.) Notice that heights are given for every second hour and that a two-day period is covered.

First Day		Second Day	
Hour	Height (Feet)	Hour	Height (Feet)
0	1.0	0	1.1
2	2.0	2	2.3
4	2.3	4	3.0
6	1.7	6	2.9
8	0.4	8	1.9
10	−0.5	10	0.4
12 noon	−1.2	12 noon	−0.5
14	−1.6	14	−1.3
16	−1.8	16	−1.8
18	−1.8	18	−2.2
20	−1.6	20	−2.2
22	−0.5	22	−1.8
24	1.1	24	−0.5

6. (*a*) What type of tide curve is represented by the graph drawn for Exercise 5? (*b*) What interval of time elapsed between successive high waters? (*c*) between successive low waters? (*d*) How does this compare with tide periods shown in the graphs of Exercises 1 and 3? (*e*) Explain the differences in the three tide curves in terms of daily and semidaily constituents.

II

ATMOSPHERE AND HYDROSPHERE

CHAPTER 7

The Earth's Atmosphere and Oceans

MAN lives at the bottom of an ocean of air; he is an air-breather dependent upon favorable conditions of pressure, temperature, and chemical composition of the atmosphere that surrounds him. He also lives on the solid outer surface of the earth, upon which he is dependent for food, clothing, shelter, and means of movement from place to place. But the air and the land are not two entirely separate realms; they constitute an interface across which there is a continual flux of matter and energy. Man's surface environment is a shallow but highly complex zone in which atmospheric conditions exert control upon the land surface, while at the same time the surface of the land exerts an influence upon the properties of the immediately adjacent atmosphere.

Essentially the same statements apply with respect to the surface of the oceans and the atmospheric layer above it. Man utilizes the surface of the sea as a source of food and a means of transportation. There is a continual flux of energy and matter between the sea surface and the lower layer of the atmosphere. Here, again, we find an interface of vital concern to Man. The sea influences the atmosphere above it while the atmosphere influences the sea beneath it.

Our object in this book is to examine the atmosphere and oceans with particular reference to air-land and air-sea interfaces which are so vital to Man. To the geographer, the distributions of physical properties of the ocean and atmosphere are matters of special interest, concerned as he is with spatial relationships on a global scale. The physical geographer seeks to describe and explain the manner in which the environmental elements of weather and climate change with latitude and season, and with geographical position in relation to oceans and continents. The geographer seeks out the broad patterns of similar regions and attempts to define their boundaries and organize them into systems of classification.

States of matter

Study of the atmosphere, ocean, and land requires continual application of the principles relating to three basic states of matter: *gaseous state, liquid state*, and *solid state*. A *gas* is a substance that expands easily to fill any small empty container, is readily compressible, and usually much less dense than liquids and solids of the same chemical composition. While the atmosphere is largely in the gaseous state, it also contains varying amounts of substances in the liquid and solid states.

A *liquid* is a substance that flows freely in response to unequal stresses, but characteristically maintains a free upper surface. Liquids are compressed only slightly with strong stresses. Liquids have densities closely comparable with solids of the same composition.

Although the world ocean is largely composed of water in the liquid state, it also contains substances in the gaseous and solid states. Both gases and liquids belong to the class of *fluids*. Layers of fluids tend to assume positions of equilibrium at rest in which a less dense fluid overlies a more dense fluid.

Solids are substances that resist changes of shape and volume and are typically capable of withstanding large unequal stresses without yielding. When yielding occurs, it is usually by sudden breakage. Although the earth's crust is largely in the solid state, it also contains substances in both gaseous and liquid states.

A further observation concerning states of matter is that a *change of state* is possible and occurs frequently in the world of nature. Most important and widespread is the change of state of water from water vapor (a gas) to liquid water and *vice versa*, and from liquid water to ice (solid state) and *vice versa* (Chapter 10). These changes of state require either an input of heat energy or the disposal of heat energy, depending upon the direction of the change.

The broad generalizations and principles that have been set down here concerning states of matter are further refined, explained, qualified, and applied in the ensuing chapters of this book.

Composition of the atmosphere

The earth's atmosphere consists of a mixture of various gases surrounding the earth to a height of many miles. Held to the earth by gravitational attraction, this envelope of air is densest at sea

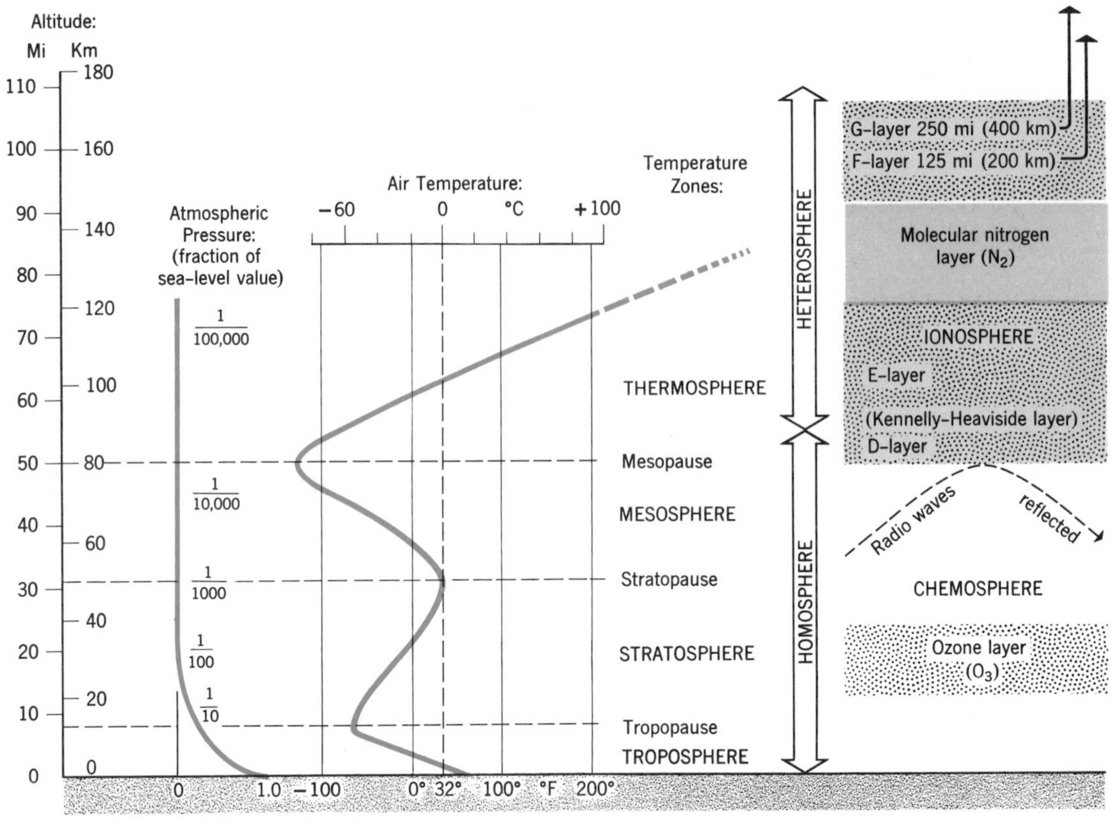

Figure 7.1 Structure of the atmosphere. (After A. N. Strahler, *The Earth Sciences*, Harper and Row, New York.)

level and thins rapidly upward. Although almost all of the atmosphere (97 percent) lies within 18 mi (29 km) of the earth's surface, the upper limit of the atmosphere can be drawn approximately at a height of 6000 mi (10,000 km), a distance approaching the diameter of the earth itself. The science of *meteorology* deals with the physics of this atmosphere.

From the earth's surface upward to an altitude of about 50 mi (80 km) the chemical composition of the atmosphere is highly uniform throughout in terms of the proportions of its component gases. The name *homosphere* has been applied to this lower, uniform layer, in contrast to the overlying *heterosphere*, which is nonuniform in an arrangement of spherical shells.

Pure, dry air of the homosphere consists largely of *nitrogen* (78.084 percent by volume) and *oxygen* (20.946 percent). Nitrogen does not easily enter into chemical union with other substances, and can be thought of as primarily a neutral filler substance. In contrast, oxygen is highly active chemically and combines readily with other elements in the process of *oxidation*. Combustion of fuels represents a rapid form of oxidation, whereas certain forms of rock decay (weathering) represent very slow forms of oxidation.

The remaining 0.970 percent of the air is mostly *argon* (0.934 percent). *Carbon dioxide*, although constituting only about 0.033 percent, is a gas of great importance in atmospheric processes because of its ability to absorb heat and thus to allow the lower atmosphere to be warmed by heat radiation coming from the sun and from the earth's surface.

Green plants, in the process of *photosynthesis*, utilize carbon dioxide from the atmosphere, converting it with water into solid carbohydrate. A pronounced rise in the carbon dioxide content of the atmosphere has been noted since 1900 and is a possible result of Man's combustion of vast quantities of wood, coal, petroleum, and natural gas. In this change, we may find an example of Man's impact upon his environment, for an increase in carbon dioxide, as well as in atmospheric dusts, can result in appreciable increases in average atmospheric temperatures.

The remaining gases of the homosphere are *neon, helium, krypton, xenon, hydrogen, methane*, and *nitrous oxide*. They are present in extremely minute percentages.

Subdivisions of the homosphere

All of the component gases of the homosphere are perfectly diffused among one another, so as to give the pure, dry air a definite set of physical properties, just as if it were a single gas.

However, the homosphere can be subdivided into layers according to temperatures and zones of temperature change. Figure 7.1 shows how temperature is related to altitude. Starting at the earth's surface, air temperature falls steadily with increasing altitude at the fairly uniform average rate of $3\frac{1}{2}$ F° per 1000 ft (6.4 C° per km). This rate of temperature drop is known as the *environmental temperature lapse rate.* Departures from this rate will be observed, depending upon geographical location and season of year. The layer in which the environmental lapse rate is approximately uniform is known as the *troposphere.* Figure 7.2 shows details of a typical atmospheric temperature sounding in middle latitudes.

The uniform lapse rate gives way rather abruptly at a height of 8 to 9 mi (12.5 to 15 km) to a layer, known as the *stratosphere,* in which temperature at first holds essentially constant with increasing height (Figure 7.2). The level at which the troposphere gives way to the stratosphere is termed the *tropopause.* Figure 7.3 shows that the altitude of the tropopause is least at the poles, 5 to 6 mi (8 to 10 km), whereas at the equator, the tropopause is encountered at 10 mi (17 km). If the troposphere is thought of as a complete surface in three dimensions, it resembles an oblate ellipsoid with a polar flattening and an equatorial bulge.

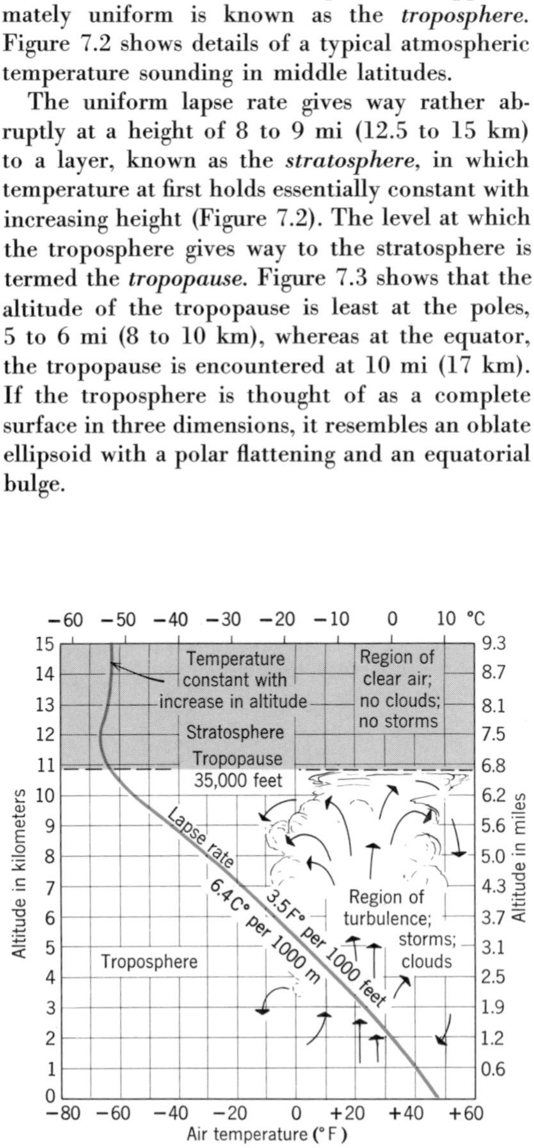

Figure 7.2 A typical environmental lapse rate curve.

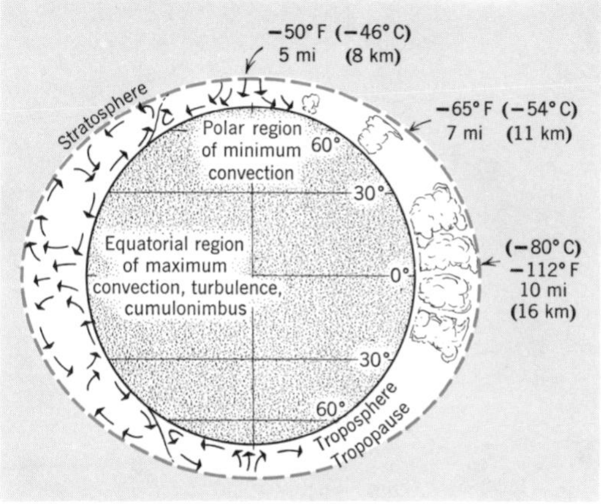

Figure 7.3 Schematic cross section of the troposphere. Figures give height and temperature of the tropopause.

Seasonal changes in the elevation of the tropopause are marked in middle and high latitudes. For example, at 45° latitude the average altitude in January is 8 mi (12.5 km), but rises to 9 mi (15 km) in July. Temperatures at the tropopause are markedly lower at the equator than at the poles, as shown in Figure 7.3. At first glance, this relationship may seem strange, accustomed as we are to considering the equatorial region to be hot and the poles cold. However, with a constant temperature lapse rate assumed, the higher the tropopause, the colder will be the air.

Upward through the stratosphere there sets in a slow rise in temperature until a value of about 32°F (0°C) is reached at about 30 mi (50 km). Here, at the *stratopause,* a reversal to falling temperature sets in. Temperature decreases through the overlying *mesosphere,* a layer extending upward to about 50 mi (80 km), where a low point of −120°F (−83°C) is reached. This level of temperature minimum and reversal is termed the *mesopause.* With further increasing altitude, a steep climb in temperature is observed within the *thermosphere.*

The troposphere and man

It is the lowermost atmospheric layer, the troposphere, that is of most direct importance to man in his environment at the bottom of the atmosphere. Almost all phenomena of weather and climate that physically affect man take place within the troposphere.

In addition to pure dry air, the troposphere contains *water vapor,* a colorless, odorless gaseous form of water which mixes perfectly with the other gases of the air. The degree to which water vapor

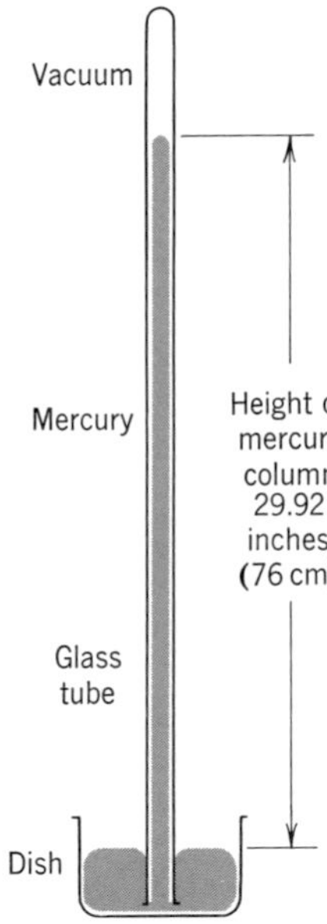

Figure 7.4 Principle of the mercurial barometer.

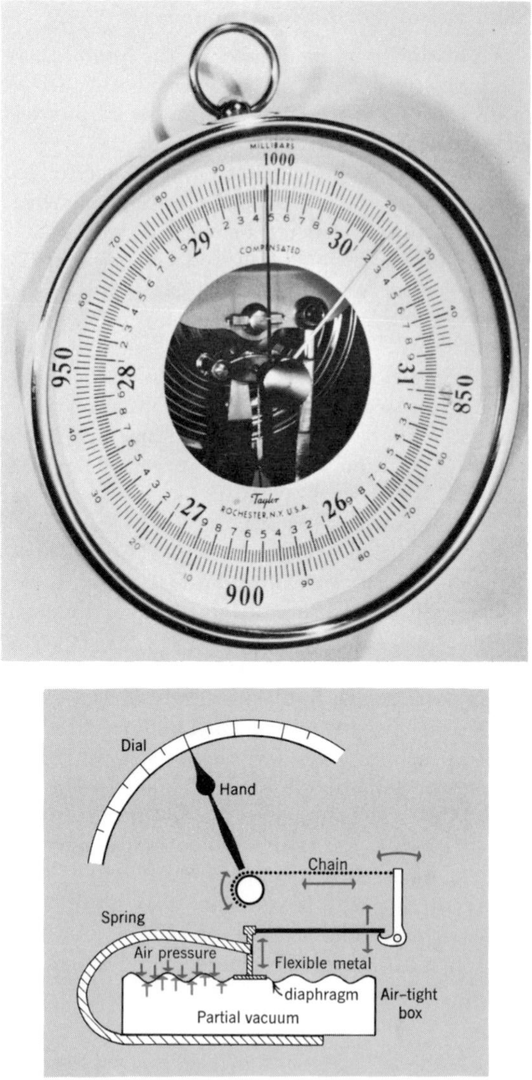

Figure 7.5 Aneroid barometer (above) and schematic diagram of workings (below). (Photograph by Taylor Instrument Co., Rochester, New York, Courtesy of Science Associates, Inc. Diagram after Wold, *College Physics*.)

is present is designated as the *humidity* and is of primary importance in weather phenomena. Water vapor can condense into clouds and fog. If condensation is excessive, rain, snow, hail, or sleet, collectively termed *precipitation*, may result. Where water vapor is present only in small proportions, extreme dryness of air typical of the hot deserts results. There is, in addition, a most important function performed by water vapor. Like carbon dioxide, it is capable of absorbing heat, which penetrates the atmosphere in the form of radiant energy from the sun and earth. Water vapor gives to the troposphere the qualities of an insulating blanket, which prevents the rapid escape of heat from the earth's surface.

The troposphere contains myriads of tiny dust particles, so small and light that the slightest movements of the air keep them aloft. They have been swept into the air from dry desert plains, lake beds and beaches, or explosive volcanoes. Strong winds blowing over the ocean lift droplets of spray into the air. These may dry out, leaving as residues extremely minute crystals of salt which are carried high into the air. Forest and brush fires are yet another important source of atmospheric dust particles. Countless meteors, vaporizing from the heat of friction as they enter the upper layers of air, have contributed dust particles.

Dust in the troposphere contributes to the occurrence of twilight and the red colors of sunrise and sunset, but the most important function of dust particles is not observable and is rarely appreciated. Certain types of dust particles serve as *nuclei*, or centers, around which water vapor condenses to produce cloud particles.

The stratosphere and higher layers are almost free of water vapor and dust. Clouds are rare and storms are absent in the stratosphere, although winds of high speed are observed.

Atmospheric pressure

Although we are not constantly aware of it, air is a tangible, material substance. At sea level, the atmosphere exerts a pressure of about 15 lb per square inch (about 1 kg per square centimeter) on every solid or liquid surface exposed to it. Because this pressure is exactly counterbalanced

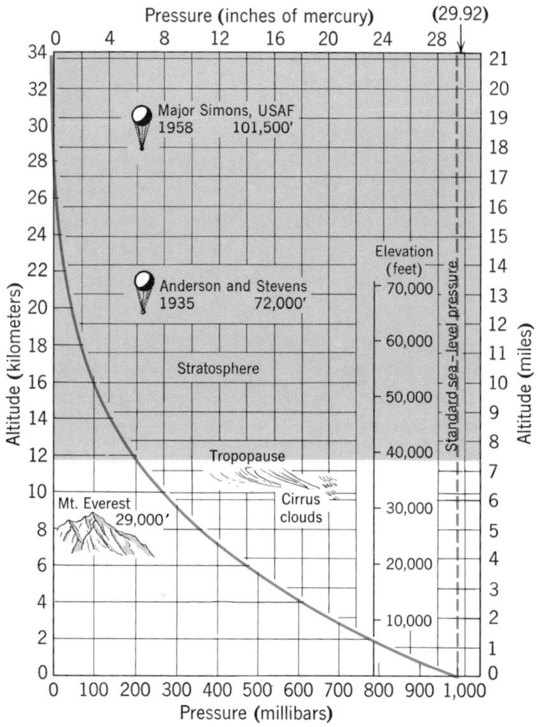

Figure 7.6 Decrease of air pressure with altitude. (Data from Humphreys, *Physics of the Air*.)

by the pressure of air within liquids, hollow objects, or porous substances, its ever-present weight creates no special concern. The pressure on one square inch of surface can be thought of as the actual weight of a column of air one inch in cross section extending upward to the outer limits of the atmosphere. Air is readily compressible. That which lies lowest is most greatly compressed and is, therefore, densest. In an upward direction, both density and pressure of the air fall off rapidly.

The meteorologist uses another method of stating the pressure of the atmosphere, based on a classic experiment of physics first performed by Torricelli in the year 1643. A glass tube about 3 ft (1 m) long, sealed at one end, is completely filled with mercury. The open end is temporarily held closed. Then the tube is inverted and the end is immersed into a dish of mercury. When the opening is uncovered, the mercury in the tube falls a few inches, but then remains fixed at a level about 30 in (76 cm) above the surface of the mercury in the dish (Figure 7.4). Atmospheric pressure now balances the weight of the mercury column. Should the air pressure increase or decrease, the mercury level will rise or fall correspondingly. Here, then, is an instrument for measuring air pressure and its variations.

Any instrument that measures atmospheric pressure is a *barometer*. The type devised by Toricelli is known as the *mercurial barometer*. With various refinements over the original simple device it has become the standard instrument.

Pressure is read in inches or centimeters of mercury, the true measure of the height of the mercury column. Standard sea-level pressure is 29.92 in. on this scale. In metric units this is 76 cm (760 mm).

Another unit has been introduced by meteorologists. This is the *millibar* (mb). One inch of mercury is equivalent to about 33.9 mb. Standard sea-level pressure is 1013.2 mb, and each $\frac{1}{10}$ inch of mercury is equal to about 3 mb (0.1 in. = 3.39 mb). In this book, both systems of stating air pressure will be used.

Another type of barometer is the *aneroid barometer* (Figure 7.5). It consists of a hollow metal chamber partly emptied of air and sealed. The walls of the chamber are flexible, so that the chamber expands and contracts as the outside air pressure varies. These movements operate a hand which is read against a calibrated circular dial. The aneroid is compact and easily carried in a plane or on the person.

Vertical distribution of pressure

Figure 7.6 shows how pressure falls with increasing altitude. For every 900 ft (275 m) of rise in elevation, the mercury column falls $\frac{1}{30}$ of its height. As the graph shows (by a steepening of the curve), the rate of drop of the mercury becomes less and less with increasing altitude until, beyond a height of 30 mi (50 km), decrease is extremely slight.

The effects of decreased air pressure upon human physiology, upon the boiling point of water, and upon the gain and loss of atmospheric heat are treated in Chapter 17.

Phenomena of the outer atmosphere

Man continues to extend his activities into more distant layers of the atmosphere. Of particular interest in environmental science are advances in use of communications satellites and remote sensing instruments which scan the earth's surface. Weather satellites continually orbit the earth, providing photographs that materially assist in weather forecasting and in the early detection of tropical storms. Certain physical phenomena of the outer atmospheric regions are thus of importance in the broad framework of physical geography.

Of particular interest in the development of radio communication on a global scale is a layer known as the *ionosphere*, located in the altitude range of 50 to 250 mi (80 to 400 km).

As shown in Figure 7.1, the ionosphere is essentially identical in position with the lower thermosphere. The ionosphere consists of a number of layers in which the process of *ionization* takes place. Here highly energetic gamma rays and X-rays from the solar radiation spectrum are absorbed by molecules and atoms of nitrogen and oxygen. In the absorption process, each molecule or atom gives up an electron, becoming a positively charged *ion.* The electrons thus released form an electric current that flows freely on a global scale within the ionosphere. Of particular interest in the geography of radio communication is the ability of the layers of ions to reflect radio waves and thus to turn them back toward the earth. Most of the important reflection of long-wave radio waves takes place in the lower part of the ionosphere, which bears the name of *Kennelly-Heaviside layer.* Without such reflection, long-distance radio communication would not be possible. Because the process of ionization requires direct solar radiation, the ionospheric layers, of which there are five, are developed on the sunlight side of the earth (Figure 7.7). On the dark side, under nighttime conditions, the layers tend to weaken and disappear.

Yet another phenomenon, one of vital concern to Man and all other life forms on earth, is the presence of an *ozone layer* largely occurring in the region from 12 to 21 mi (20 to 35 km) elevation, but also extending upward to an elevation of 30 to 35 mi (50 to 55 km) (see Figure 7.1). The ozone layer thus extends from the upper stratosphere into the mesosphere. The ozone layer is a region of concentration of the form of oxygen molecule known as *ozone,* (O_3), in which three oxygen atoms are combined instead of the usual two atoms (O_2). Ozone is produced by the action of ultraviolet rays upon ordinary oxygen atoms. The ozone layer thus serves as a shield, protecting the troposphere and earth's surface from most of the ultraviolet radiation found in the sun's radiation spectrum. If these ultraviolet rays were to reach the earth's surface in full intensity, all exposed bacteria would be destroyed and animal tissues severely burned. Thus the presence of the ozone layer is an essential element in man's environment. It is also interesting to note that the high temperatures of the mesosphere are produced by the absorption of the ultraviolet rays in the upper part of the ozone layer.

The magnetosphere

In Chapter 3 the earth's magnetic field was described as resembling that of a bar magnet oriented on a magnetic polar axis inclined a few degrees from the earth's geographic axis. The lines of force shown in Figure 3.8 also extend out into space, comprising the earth's *external magnetic field.* If we assume, for purposes of comparison,

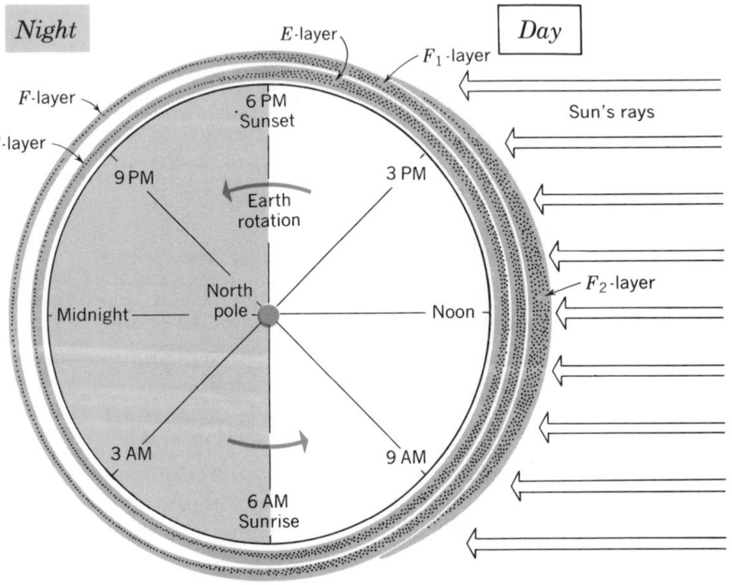

Figure 7.7 This diagram of the ionospheric layers is a cross section through the earth's equator. The observer looks down upon the cross section from a point over the north pole. (Based on a figure by B. F. Howell, Jr., 1959, *Introduction to Geophysics*, McGraw-Hill, N.Y. From A. N. Strahler, *The Earth Sciences*, Harper and Row, N.Y.)

that the earth's atmosphere extends outward to a distance equal to twice its own radius, or 8000 mi (13,000 km) it becomes evident that the magnetic field extends far beyond the outermost limits of the atmosphere. The effective limit of the external magnetic field lies perhaps 40,000 to 80,000 mi (64,000 to 130,000 km) from the earth. All of the region within this limit is described as the *magnetosphere;* its outer boundary, the *magnetopause.*

The simplest geometrical model for the shape of the magnetosphere would be a doughnut-shaped ring surrounding the earth. The plane of the ring would lie in the plane of the magnetic equator, while the earth would occupy the opening in the center of the doughnut. Actually, this ideal shape does not exist because of the action of the *solar wind,* a more or less continual flow of electrons and protons emitted by the sun. Pressure of the solar wind acts to press the magnetopause close to the earth on the side nearest the sun (Figure 7.8). Here the distance to the magnetopause is on the order of 10 earthradii (about 40,000 mi, or 64,000 km). Lines of force in this region are crowded together and the magnetic field is intensified. On the opposite side of the earth, in a line pointing away from the sun, the magnetopause is drawn far out from the earth and the force lines are greatly attenuated. The extent of this "tail" is not known, but the entire shape of the magnetosphere has been described as resembling a comet. Length of the magnetic tail has been estimated to be at least 4 million miles (6,400,000 km) and is possibly vastly longer.

Radiation belts

In 1958, satellites *Explorer I* and *III,* carrying Geiger counters, sent to earth information concerning the existence of a region of intense radioactivity within the magnetosphere. It was soon discovered that two ring-shaped belts of radiation existed, one lying within the other (Figure 7.9). These rings were named the *Van Allen radiation belts,* after the physicist who first described them. An inner belt was found to lie about 2300 mi (2600 km) from the earth's surface; an outer and much more intense belt at about 8000 to 12,000 mi (13,000 to 19,000 km) distance.

The Van Allen radiation belts represent concentrations of charged particles—protons and electrons—trapped within lines of force of the earth's external magnetic field. These highly energetic particles are derived from the sun and are trapped upon entering the magnetopause. Intensity of trapped radiation fluctuates over a wide range. Solar flares from the sun's surface, occurring at irregular intervals, send bursts of ion clouds toward the earth. At such times, the intensity of trapped particle radiation is greatly increased. One manifestation of such events is the *aurora,* which is most intense over arctic and antarctic latitudes. On

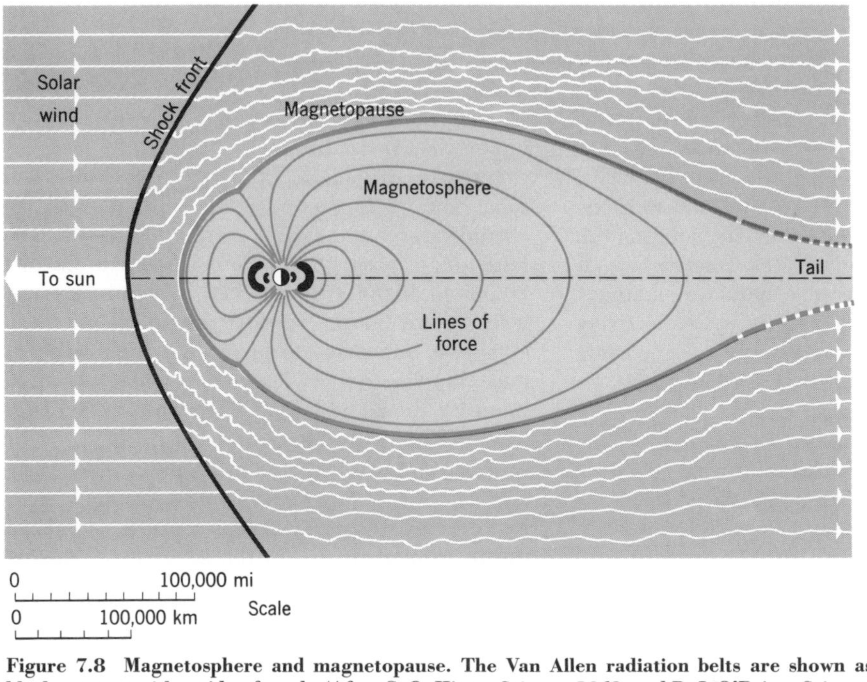

Figure 7.8 Magnetosphere and magnetopause. The Van Allen radiation belts are shown as black areas on either side of earth. (After C. O. Hines, *Science*, 1963, and B. J. O'Brien, *Science*, 1965.)

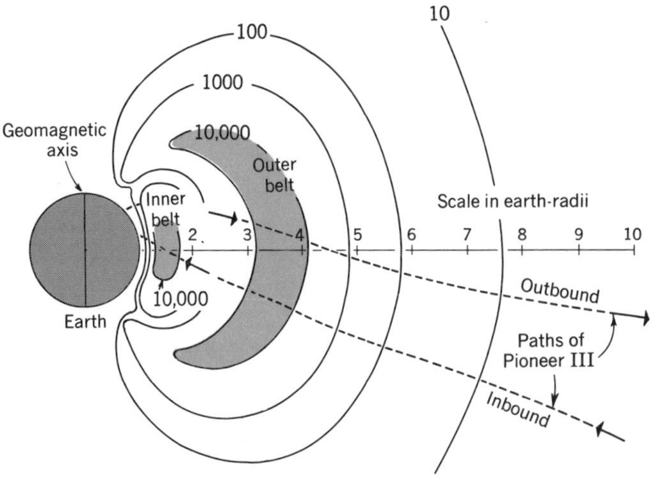

Figure 7.9 Cross section of the Van Allen radiation belts as they were first interpreted from data sent back to earth from satellites *Explorer IV* and *Pioneer III*. Contour lines of equal radiation intensity are scaled in relative values, each quantity being ten times greater than the next lower value. (Based on data of J. A. Van Allen, 1959, Jour. Geophysical Research, Vol. 64. From A. N. Strahler, 1963, *The Earth Sciences*, Harper and Row, N.Y.)

earth, severe disturbances to the magnetic field, known as *magnetic storms*, accompany the arrival of ion clouds from solar flares and seriously disrupt radio communication.

Cosmic particles and ionizing radiation

Quite different from charged particles of the solar wind are *cosmic particles* (often called *cosmic rays*). These are elementary particles traveling through space at speeds approaching that of light; they carry tremendous energy. Cosmic particles are protons—portions of the nucleus of the atom. Most are nuclei of hydrogen atoms, others are of helium, and a very small number are of other heavier atomic nuclei. Their penetrating power is enormous and they can reach the earth's surface.

Cosmic particles arrive from all points in interstellar space. A single cosmic particle, entering the earth's atmosphere, impacts the nucleus of an atom of gas, giving rise to a complex chain of disintegrations and energy dissipations, collectively called a *cosmic shower*. Products of a cosmic shower are neutrons, protons, and gamma rays. As we have already mentioned, gamma rays are also found in the sun's radiation spectrum.

The term *ionizing radiation* is used to cover radiation capable of tearing off electrons from atoms that intercept that radiation. Gamma rays, together with X-rays, are a source of ionizing radiation. We have seen that this is the same process acting to produce ions in the ionosphere. However, since cosmic particles penetrate the atmosphere with much greater energy than X-rays, the effects of cosmic radiation are important at the earth's surface and make up a substantial part of the steady, or *background* ionizing radiation to which life forms are exposed. Terrestrial sources of ionizing radiation include natural radioactivity of elements in crustal rocks and radioactivity from products of nuclear test explosions. Finally, there are important sources from medical X-rays and television tubes.

The biological effect of ionizing radiation is to produce changes in genetic materials within the cells of organisms. Since man is also a producer of ionizing radiation at intensity levels potentially much higher than natural levels, the total production of ionizing radiation is a subject of great importance in environmental science.

Intensity of ionizing radiation from cosmic and solar sources increases greatly with increase in altitude above the earth's surface, because of the lessening of absorption by atmospheric atoms. At 20,000 to 30,000 ft (6 to 9 km) radiation intensity level is some 30 times greater than at ground level. It is also important to know that high latitudes have much stronger levels of cosmic radiation intensity than low latitudes, roughly by a factor of five. This latitudinal effect is explained by the trapping action of the earth's magnetic field: particles pass much more readily through the steeply inclined force lines near either pole than through the horizontal lines near the equator.

From the standpoint of man's environment, the biological effect of cosmic radiation becomes important when humans fly at high altitudes in jet aircraft. The problem becomes particularly acute for passengers in supersonic transport (SST) aircraft, for these newer aircraft encounter radiation at an intensity level more than half again greater than for conventional subsonic jet aircraft. Great-circle routes between Northern Hemisphere cities pass largely over polar regions, where radiation levels are highest. Although for passengers flying only a few trips per year in SST aircraft, the annual dosage of ionizing radiation would not reach dangerous levels, the dosage for crew members flying some 480 hours per year at such altitudes would approach the maximum permissible dosage established by the International Commission on Radiological Protection.

Environmental importance of the magnetosphere to man and all other life forms at the earth's surface lies in the shielding action of the lines of force of that magnetic field. Should the earth's magnetic field weaken or disappear, as it may have done for short periods in the geologic past, the intensity of ionizing radiation would be increased, with possible important effects upon the genetic materials of organisms. It has been speculated that periods of rapid evolutionary change, including extinction of species, have been brought about by weakening of the magnetic field at times when the polarity of the entire field has been reversed.

Thus we see that while physical geography is aimed at understanding man's physical environment close to the earth's surface, it has been necessary to extend our inquiry far into outer space in search of forces that determine that quality of that surface environment.

Man and the oceans

The importance of the oceans to man is felt in a wide range of dimensions and scales. One environmental role played by the oceans is climatic. The huge water mass of the oceans retains a large quantity of heat, which is gained or lost very slowly. As we shall see, the oceans effectively moderate the seasonal extremes of over much of the earth's surface. The oceans supply water vapor to the atmosphere, so that the rain that falls on the lands and is the source of man's vital fresh water supplies originates from the ocean surface by a process of distillation of salt water.

The oceans sustain a vast and complex assemblage of marine life forms, both plant and animal. This organic production provides man with a small share of his food. Throughout history the oceans have served as trackless surfaces of transport of people and the commodities that sustain their civilizations. Winds, waves, currents, sea ice, and fog are environmental factors—sometimes favorable and sometimes hazardous—which the oceans impose upon men and their ships at sea.

The zone of contact between oceans and lands is a unique environment for man. In Chapter 32 we investigate the processes by which ocean waves and currents shape coastal land features. Man uses and modifies the coastal zone in various ways, ranging from ports to recreational facilities.

The world ocean

Using the term *world ocean* to refer to the combined ocean bodies and seas of the globe, let us consider some statistics that emphasize the enormous extent and bulk of this great salt water layer. The world ocean covers about 71 percent of the globe (Figure 7.10); its average depth is about 12,500 ft (3800 m), when shallow seas are included with the deep main ocean basins. For major portions of the Atlantic, Pacific, and Indian Oceans the average depth is about 13,000 ft (4000 m). The total volume of the world ocean is about 317 million cu mi (1.4 billion cu km), which comprises just over 97 percent of the world's free water. Of the small remaining volume, about 2 percent is locked up in the ice sheets of Antarctica and Greenland, and about 1 percent is represented by fresh water of the lands. These figures show that the *hydrosphere*, which is a generalized word for the total free water of the earth (whether as gas, liquid, or solid), is largely represented by the world ocean. To place the masses of the atmosphere and oceans in their proper planetary perspective, compare the following figures (the unit of mass used here is 10^{21} kg):

Entire earth	6000
World ocean	1.4
Atmosphere	0.005

What are the basic differences in properties and behavior between the world ocean and the atmosphere? How do atmosphere and ocean interact in the region of their interface? The answers to these questions are vital to understanding environmental processes, because marine life depends upon the exchanges of matter and energy across the atmosphere-ocean interface. It is also significant that the earliest life forms originated and developed in the

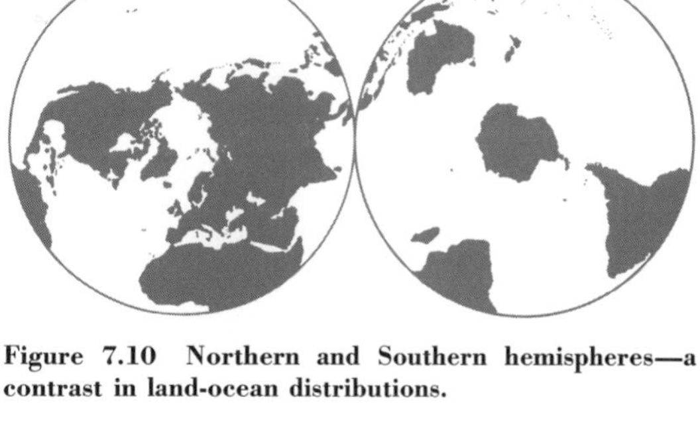

Figure 7.10 Northern and Southern hemispheres—a contrast in land-ocean distributions.

shallow layer of water immediately beneath the interface.

The atmosphere, being composed of a gas that is easily compressed, has no distinct upper boundary; it becomes progressively denser toward its base under the load of the overlying gas. The oceans, composed of liquid water that shows very little change of density under large compressional forces, has a sharply defined upper surface in contact with the densest layer of the overlying atmosphere. Whereas the most active region of the atmosphere is the lowermost layer—the troposphere—the most active region of the ocean is its uppermost layer. At great ocean depths, water moves extremely slowly and maintains a uniformly low temperature. One reason for intense physical and biological activity in the uppermost ocean layer is that the input of energy and matter from the overlying atmosphere drives water motions in the form of waves and currents. The atmosphere is also the source layer of heat and of condensed fresh water entering the ocean. But the ocean surface also returns heat and water (in vapor form) to the lower atmosphere, a phenomenon of primary importance in driving atmospheric motions. Interaction between atmosphere and ocean surfaces is a topic we shall need to explore further in later chapters.

Compartmentation of the oceans by intervening continental masses inhibits the free global interchange of ocean waters, whereas the atmosphere is free to move globally. Another difference in the two bodies is that the atmosphere has little ability to resist stresses and therefore moves easily and rapidly, changing its velocity very quickly from place to place. In contrast, the ocean water can move only sluggishly and is very slow to respond to the changes in force applied by winds.

Composition of sea water

Sea water is a solution of salts—a brine—whose ingredients have maintained approximately fixed proportions over a considerable span of geologic time. The principal constituents of sea water are listed in Table 7.1.

Besides their importance in the chemical environment of marine life, these salts constitute a vast reservoir of mineral matter from which certain constituents may be extracted by man for his use. One way to describe the composition of sea water is to state the principal ingredients that would be required to make an artificial brine approximately like sea water. These are listed in Table 7.1. Of the various elements combined in these salts, chlorine alone makes up 55 percent by weight of all the dissolved matter, and sodium 31 percent. Important, but less abundant than elements of the five salts in Table 7.1, are bromine, carbon, strontium, boron, silicon, and fluorine. At least some trace of half of the known elements can be found in sea water. Sea water also holds in solution small amounts of all of the gases of the atmosphere, principally nitrogen, oxygen, argon, carbon dioxide, and hydrogen.

TABLE 7.1 COMPOSITION OF SEA WATER

Name of Salt	Chemical Formula	Grams of Salt per 1000 g of Water
Sodium chloride	$NaCl$	23
Magnesium chloride	$MgCl_2$	5
Sodium sulfate	Na_2SO_4	4
Calcium chloride	$CaCl_2$	1
Potassium chloride	KCl	0.7
With other minor ingredients, to total		34.5

Layered structure of the oceans

As with the atmosphere, the ocean has a layered structure, the layers being recognized in terms of temperature or chemical composition. In the troposphere, air temperatures are generally highest at ground level and diminish upward. In the oceans, temperatures are generally highest at the sea surface and decline with depth. This trend is to be expected, since the source of heat is from the sun's rays and from heat supplied by the overlying atmosphere.

With respect to temperature, the ocean presents a three-layered structure in cross section, as shown in the left-hand diagram of Figure 3.11. At low latitudes throughout the year and in middle latitudes in the summer there develops a warm surface layer. Here, wave action mixes heated surface

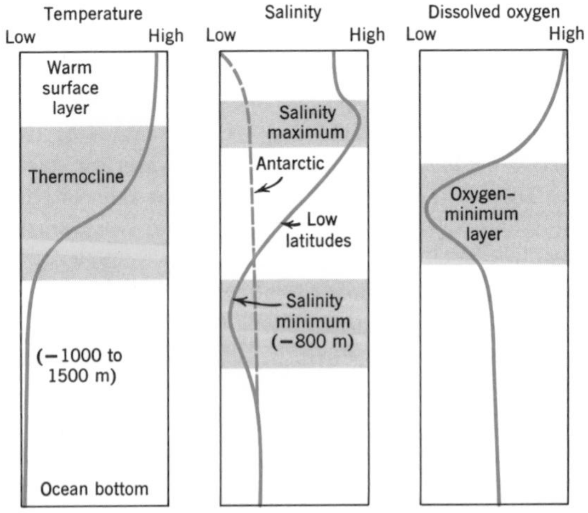

Figure 7.11 Schematic diagram of changes with depth of temperature, salinity, and oxygen content. Depths are not to scale. (After W. E. Yasso, 1965, *Oceanography*, Holt, Rinehart and Winston, New York, Figure 2.4.)

water with the water below it to give a warm layer that may be as thick as 1600 ft (500 m), with a temperature of 70° to 80°F (20° to 25° C) in oceans of the equatorial belt. Below the warm layer temperatures drop rapidly, constituting a second layer known as the *thermocline.* Below the thermocline is a third layer of very cold water extending to the deep ocean floor. Temperatures near the base of the deep layer are in the range of 32° to 40°F (0° to 5°C). In arctic and antarctic regions, the three-layer system is replaced by a single layer of cold water, as shown in the north-south profile of Figure 7.12. Temperature is a prime environmental factor controlling the abundance and variety of marine life, the bulk of which thrives in the shallow upper layer.

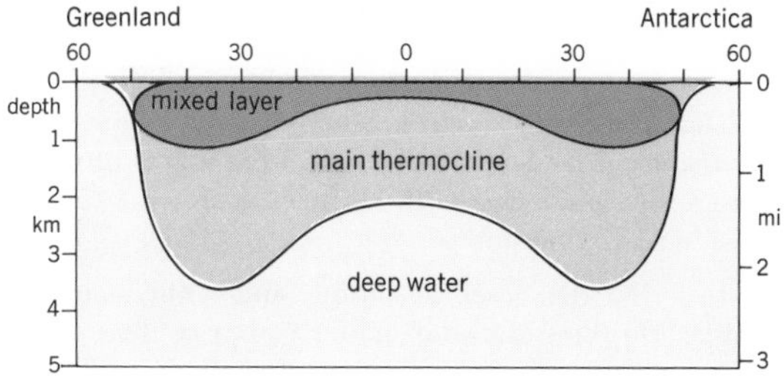

Figure 7.12 Schematic diagram of three-layered structure of the oceans. (After J. Williams, *Oceanography.* Copyright © 1962, Little, Brown, Boston, p. 94, Figure 7.4.)

The content of free oxygen (O_2) dissolved in sea water shows an oxygen-rich surface layer, accounted for by the availability of atmospheric oxygen and the activity of oxygen-releasing plant life in the sea. As shown in the right-hand diagram of Figure 7.11, oxygen content falls rapidly with depth; over large ocean areas there is a distinct minimum zone of low oxygen content. Here, oxygen has been consumed by biological activity. In deep water, the oxygen content holds to a uniform and moderate value down to the ocean floor.

Retrospect and prospect

The broad overview of atmosphere and oceans presented in this chapter has revealed the major elements of physical structure and chemical composition of those two great fluid layers. Most of the information has been about static conditions that one would encounter when probing upward into the atmosphere and downward into the oceans.

In the next four chapters, we turn to the great systems of flow of matter and energy that continually involve the atmosphere and oceans, making them dynamic rather than static bodies. First, we shall trace the course of radiant energy from the sun as it passes through the atmosphere, reaches the earth, and is returned to outer space. The solar radiation system establishes the thermal environment of the biosphere, supplying it with the energy for biological processes. There follow accounts of the vast systems of transport of the atmosphere and oceans, whereby energy and matter are redistributed over the globe to provide more moderate and more favorable life conditions than would otherwise exist on our planet. Accompanying these circulation systems are intense disturbances of the air and sea—storms that constitute environmental stresses and hazards, often of phenomenal proportions.

Looking over the other planets of the solar system, the uniqueness of Earth as a life environment is most striking. Only our planet has both a great world ocean and a comparatively dense oxygen-rich atmosphere combined with favorable temperature range. Mars, our nearest planet, has practically no free water in any form and only a very rarified atmosphere with little oxygen. Venus, matching us closely in size, has a much denser atmosphere than Earth. But while there is some free oxygen in the atmosphere of Venus, water in any form is apparently almost totally lacking, and surface temperatures are very much higher than on Earth. Little Mercury, with no atmosphere or water, roasts in the sun's rays. The great outer planets—Jupiter, Saturn, Uranus, and Neptune—probably have huge quantities of water, but it is frozen solid. Their atmospheres, composed largely of ammonia and methane, would be lethal to life such as ours even if the surface temperatures were not impossibly cold. The lunar environment has no free water or atmosphere to offer. So there is really no other place for man to live but on planet Earth.

REVIEW QUESTIONS

1. Describe man's environment on earth in terms of the realms of land, sea, and air. In what way is physical geography concerned with sciences that treat the atmosphere and oceans?

2. Describe the gaseous, liquid, and solid states of matter. Do these states of matter comprise three complete separate and distinct realms of planet Earth? How are energy changes related to changes of state?

3. Compose definitions of the terms *atmosphere* and *meteorology*. What is the basis for subdividing the atmosphere into the homosphere and the heterosphere?

4. List the nonvarying component gases of the homosphere in order of decreasing proportion by volume, giving approximate percentages for each. Evaluate each gas in terms of its importance to plant and animal life on earth.

5. Describe the subdivisions of the homosphere. What conditions of temperature prevail in each zone? At what altitude is the tropopause encountered over the equator? over the poles? over middle latitudes? What changes in altitude of the tropopause take place throughout the year?

6. Why is the troposphere of special interest to geographers? What are the constituents of the troposphere, in addition to gases of the pure, dry air? What importance has each constituent?

7. What is the cause of atmospheric pressure? How is pressure measured? Describe Torricelli's experiment and explain the principle involved. Describe a mercurial barometer.

8. What value is given to standard sea-level pressure in inches and centimeters of mercury? What is the equivalent value in millibars?

9. Explain the principle of the aneroid barometer. What advantages has this type over the mercurial barometer?

10. At what rate does air pressure diminish with increasing altitude? Is the rate constant? Does the pressure-altitude curve reveal the position of the tropopause? Explain how air density is related to air pressure.

11. Explain the formation of the ionosphere. At what level does the ionosphere occur? What relation exists between intensity of development of ionospheric layers and the exposure of the atmosphere to sunlight? How does the ionosphere promote long-distance radio communication? What is the Kennelly-Heaviside layer?

12. What is the ozone layer? At what altitude is it found? How is ozone produced? How does the ozone layer influence the environment of plant and animal life at the earth's surface?

13. Describe the magnetosphere. What is the magnetopause? How is the shape of the magnetosphere influenced by the solar wind?

14. What are the Van Allen radiation belts? How were they first discovered? What is the source of radioactivity in these belts?

15. Describe the world ocean in terms of surface extent, average depth, and volume. Compare the mass of the world ocean with that of earth and atmosphere.

16. Compare the atmosphere and oceans with respect to their density distribution, temperature structure, and freedom of motion.

17. What are the principal chemical components of sea water? Of what value are sea salts to man?

18. Describe the layered structure of the oceans with respect to temperature and oxygen content. How do these layers relate to distribution of organic life and to organic processes?

19. Compare the earth's planetary environment with environments of the other planets and the moon.

Exercises

1. At Chicago, Illinois, on a day in mid-April, the surface air temperature measures 50°F. A vertical sounding of the atmosphere by balloon reveals a nearly constant temperature lapse rate of 3.6 F° per 1000 ft. At the tropopause a temperature of −58°F is reported. **(*a*)** Calculate the altitude of the tropopause (assume the elevation of Chicago to be sea level). **(*b*)** Convert your answer to kilometers. **(*c*)** Prepare a graph similar to Figure 7.2, showing this lapse rate curve. Label the position of the tropopause. (Data from S. Petterssen.)

2. On the same day that the observations of Exercise 1 are being made, a similar atmospheric sounding is being carried out at Nauru Island, located in mid-Pacific, close to the equator. Here the sea-level temperature is 30°C and the lapse rate is constant at 6.6 C° per km. The tropopause is encountered at 16.8 km. **(*a*)** Calculate the air temperature at the tropopause. **(*b*)** Convert your answer to degrees Fahrenheit. **(*c*)** Plot the data on the same graph as that used in Exercise 1. Attach altitude and temperature scales in both English and metric units. (Data from S. Petterssen.)

3. **(*a*)** Plot a pressure-altitude curve using the information that air pressure decreases by $\frac{1}{30}$ of its value for every 1000 ft of ascent. Use 30.00 in. of mercury as starting sea-level pressure. Thus, at an altitude of 1000 ft, the pressure will be reduced to 29.00 in. At 2000 ft. the pressure will be reduced by $\frac{1}{30}$ of 29.00 or 0.97 in., and will thus measure 28.03 in. Calculate pressures to at least 50,000 ft altitude. Plot the values as points and connect with a smooth curve. **(*b*)** Compare your results with the curve in Figure 7.6. A curve of this type is known as a *negative exponential curve.* Similar curves are observed in the radioactive decay rate of uranium 238, in the rate of radiative cooling of a hot object, and in many other important phenomena of science.

4. Using the graph in Figure 7.6 as a direct source of information, answer the following. **(*a*)** What is the barometric pressure in inches of mercury on Mt. Everest? **(*b*)** What is the pressure in millibars at 72,000 ft, the elevation reached by Anderson and Stevens in 1935? at 100,000 ft, reached by man in a balloon in 1958? **(*c*)** In the region between three and five miles altitude, what is the rate of change of pressure in millibars per kilometer? **(*d*)** Through how many millibars of pressure does the barometer drop from 30,000 to 60,000 ft?

CHAPTER **8**

The Global Radiation and Heat Balances

MAN'S physical environment lies at an interface experiencing intense fluxes of energy and matter from the atmosphere to land and water surfaces, and from land and water surfaces to the atmosphere. These exchanges of energy and matter are not only physical, but organic as well. Plants and animals gain energy and matter from the atmosphere. This they place in storage. Organisms also release stored energy and matter to the atmosphere. The full spectrum of man's environment must therefore include the science of ecology and so goes far beyond the bounds of physical geography. But we must make a beginning, and it lies with a study of the physical and largely inorganic processes and forms of the environment. Understanding of natural processes of the earth's surface is a primary concern of physical geography, but it is not enough. Our major goal is to analyze the global distribution patterns of the processes and structures of man's environment.

The atmosphere and man

We will begin with a study of atmospheric processes, which fall under the general heading of *meteorology*—the physics of the lower atmosphere. We will accompany this analysis of atmospheric processes with a study of the characteristic place-to-place variations in the intensity of those atmospheric processes, which falls under the general heading of the science of *climatology*. For example, we will first examine the process of gain and loss of heat energy by the air layer near the ground, knowing that the basic principles apply at any place on the globe. This is the meteorological aspect of the subject. We then examine the global pattern of the thermal (heat) environment as it is expressed by characteristic values of air temperatures. This is the climatological aspect of the subject. Later, we shall be able to recognize global climatic types and regions in terms of combinations of the basic physical environmental ingredients of life processes. These are heat energy, as expressed in air temperature, and availability of water to plants, as expressed by precipitation and evaporation.

The atmosphere is involved in yet another facet of the environment—that of physical stress and hazard to man and other life forms. Atmospheric phenomena such as high winds (in hurricanes, squalls, and tornadoes), fog, ice (as snow and hail), and torrential rains (causing floods) influence man over a wide scale of intensities, ranging from minor inconvenience to wholesale losses of life and property. We shall therefore be interested in analyzing weather disturbances of various kinds. In so doing we are better able to evaluate global environment in terms of hazard to man, his crops, his cities, and his arteries of transport.

Finally, physical geography must take into account the ways in which man alters his environment. Through enormous expenditures of energy, possible through combustion of fuels accumulated by organic processes through geologic time, man loads the atmosphere with waste heat and many forms of pollutants. Already, significant changes in the climate of cities have been documented, while more subtle man-made changes of climate on a global scale are suspected. So we find that man and environment relationships are reciprocal. The nature of interactions between man and environment must be a central theme of the study of physical geography.

Concept of the radiation and heat balances

All life processes as well as practically all exchanges of matter and energy at the interface between the earth's atmosphere and the surfaces of the oceans and lands are supported with radiant energy supplied by the sun. The planetary circulation systems of atmosphere and oceans are driven by solar energy. Exchanges of water vapor and liquid water from place to place over the globe depend upon this single energy source. It is true that some heat flows upward through the lithosphere to the earth's surface from internal radioactive and volcanic sources, but the amount is trivial in comparison with the energy that the earth intercepts from the sun's rays.

The flow of energy from sun to earth and then out into space is a complex system, since it involves not only electromagnetic radiation, but also energy storage and transport as heat in the gaseous, liquid, and solid matter of the atmosphere, hy-

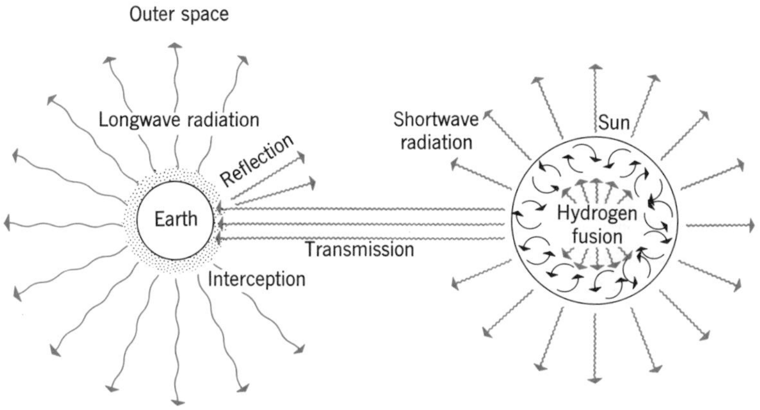

Figure 8.1 Schematic diagram of the sun-earth-space radiation system. (From A. N. Strahler, "The Life Layer," *Jour. Geogr.*, Vol. 69, p. 72 © 1970 by *The Journal of Geography*; reproduced by permission)

drosphere, and lithosphere. However, we can simplify the study of this total system by first examining each of its parts. We will start with the radiation process itself and develop the concept of a *radiation balance*, which is perhaps the most important control system of the earth's surface environment.

Organisms respond directly to the heating and cooling of the air, water, or soil that surrounds them. These temperature changes result from the gain or loss of energy by the absorption or emission of radiant energy. When a substance absorbs radiant energy, the surface temperature of that substance is raised. This process represents a transformation of radiant energy into the energy of *sensible heat*, a physical property measured by the thermometer. Sensible heat, in the case of a gas, represents the kinetic energy of motion of the gas molecules, which are in constant high-speed flight and endlessly colliding with one another. A rise in temperature of the gas represents an increase in the average velocity of the molecules and in the frequency of their collisions.

Many of the biochemical processes taking place within organisms as well as many common inorganic chemical reactions are intensified by an increase in temperature of the solutions in which these reactions are occurring. Severe cold, which is simply the lack of kinetic energy within matter, may greatly reduce, or even completely stop biochemical and inorganic reactions. This is why the vital environmental ingredient of heat—heat in the air, water, and soil—needs to be thoroughly understood.

We are all familiar with the cyclic nature of temperature changes. There is a daily rhythm of rise and fall of temperature as well as a seasonal rhythm. There are also systematic average changes in air temperatures from equatorial to polar latitudes as well as from oceanic to continental surfaces. Correspondingly, the lower atmosphere and the surfaces of the lands and oceans must be receiving and giving up heat energy in daily and seasonal cycles. There must also be great differences in the quantities of heat received and given up in low latitudes as compared to high latitudes.

Despite the existence of thermal cycles and latitudinal contrasts in temperature, human history as well as the geologic record indicate an overall uniformity of the global thermal environment through time. It is apparent that the earth as a planet maintains within fairly narrow limits a certain average *planetary temperature* which has depended on maintenance of approximately the same distance from the sun and approximately the same planetary surface properties. If this were not the case, the gradual drift toward either increasing heat or increasing cold would ultimately render the earth's surface too hot or too cold to support life.

Solar energy is intercepted by our spherical planet and the level of heat energy tends to be raised. At the same time, our planet radiates heat into outer space, a process that tends to diminish the level of heat energy. Incoming and outgoing radiation processes are simultaneously in action (Figure 8.1). In one place and time more heat is being gained than lost; in another place and time more heat is being lost than gained.

There exists in combination with the global radiation balance a global *heat balance*, and the two together constitute the earth's total *energy balance*. Equatorial regions receive through solar radiation much more heat than is lost directly to space, whereas polar regions lose by radiation into space much more heat than is received. So there must be included in the energy system mechanisms of heat transfer adequate to export heat from a region of excess and to carry that heat into a region of deficiency. On our planet, motions of the atmosphere and oceans act as heat-transfer mechanisms. A study of the earth's heat balance will not be complete until the global patterns of air and water circulation are described and explained in Chapter 9.

Storage of heat energy in a latent form is an important part of the earth's heat balance. It was noted in Chapter 7 that changes of state between gaseous, liquid, and solid states are accompanied by the taking up of heat energy or release of heat energy. Water in its three states—as water vapor in the atmosphere and as liquid and solid water in the oceans and over the land surfaces—absorbs and liberates heat as it changes from one state to another. Consequently, a study of the earth's heat balance will not be complete until the processes of change of state of water in the atmosphere and hydrosphere are examined in Chapter 10.

Upon reflection, it becomes apparent that the movement of water through atmosphere, oceans, and upon the lands comprises a system of equal importance to the flow of heat and that the activities of these two systems are closely intermeshed. The concept of a *water budget* with a *water balance* can be developed and takes its place beside the heat budget (Chapters 10 and 12). The heat budget can be thought of as dealing with energy and the water budget with matter. Together, these two great systems of energy and matter form a single, grand planetary system and permit us to relate and explain many of the environmental phenomena of our earth within a single unified framework.

Man's environment as an open system

We can view the flow of energy from sun to earth and back to outer space as constituting a great natural open system. In the broadest sense, an *open system* is any organized configuration of matter open to the inflow or outflow of energy or matter, or both. The life layer, man's environment on earth, can be regarded as an open system with arbitrary boundaries—one in the atmosphere above, and another lying at the base of a shallow layer of soil or water. It is through the upper boundary that the open system of the life layer has its input and output of energy.

Within the life layer energy moves in many complex pathways, involving many subsystems and subcycles. These movements of energy require the presence of matter, both to transport energy from place to place and to store it temporarily. Air and water, as highly mobile fluids, are the principal transporters of energy. Solid earth materials, such as soil, rock, and organic compounds, along with air and water, provide for storage of energy within the open system. Placing energy in storage requires that it be transformed. Thus incoming solar energy in the radiant form goes into storage as sensible heat. In this form the energy may be transported by moving air or water, or it may be further transformed into latent energy.

An essential principle in the operation of open systems is that they tend to attain a dynamic equilibrium, or *steady state,* in which rate of input of energy and matter equals rate of output, while the quantity of stored energy and matter within the system remains constant. In the case of the total global system of atmosphere, hydrosphere, and lithosphere, only energy enters and leaves the system. Matter remains within the system boundary where it is forced to run repeatedly through flow cycles. Matter is thus *recycled* within the system. We can conclude that, with respect to matter only, the global system is closed, rather than open. Energy, on the other hand, is cycled within the system and also enters and leaves the system.

When we refer to a steady state of an open system, such as the life layer, we intend to describe a long-term average state. Because of strong daily and annual rhythms of energy input, the entire system rhythmically changes its rates of energy transformation and its reservoir of stored energy. Most of this rhythmic change, following the astronomical controls, is removed by taking yearly averages of rates of flow of energy and matter. From year to year the total quantity of energy in storage should remain about constant. The global steady state is also subject to readjustments following longer cycles of change having periods of decades or centuries.

The biochemical energy cycle

To the inorganic or physical energy cycle, there must be added an organic phase, the *biochemical energy cycle,* in which a part of the incoming solar energy is used and given up by plants and animals. Briefly, this cycle consists of the absorption of solar energy by plants in the manufacture of carbohydrate compounds, which provide the food for animals. Eventually, after much recycling, most of these compounds are oxidized in a process known as *respiration,* and the energy is returned to the atmosphere. However, some fraction of the carbohydrate compounds may be stored in soil layers, for example, as peat in bogs. In the geologic past, the total accumulation of hydrocarbon compounds in rock strata as coal and petroleum has been enormous, and represents a great storage bank of solar energy. While the biochemical energy cycle involves only a tiny fraction (about one-tenth of a percent) of the solar energy received by the earth, the process is of the first order of importance in the biosphere, since it represents the energy cycle of the whole organic world.

Solar radiation

A systematic approach to the earth's heat budget begins with an examination of the input, or source, of energy from solar radiation. This radiation is traced as it penetrates the earth's atmosphere and is absorbed or transformed. We then turn to the mechanism of output of heat energy by the earth as a secondary radiator.

Our sun, a star of about medium mass and temperature, as compared with the overall range of stars, has a surface temperature of about 11,000°F (6000°C). The highly heated, incandescent gas that comprises the sun's surface emits a form of energy known as *electromagnetic radiation.* This form of energy transfer can be thought of as a collection, or *spectrum,* of waves of a wide range of lengths traveling at the uniform velocity of 186,000 mi (300,000 km) per second. The energy travels in straight lines radially outward from the sun, and requires about $9\frac{1}{3}$ minutes to travel the 93 million miles (150 million kilometers) from sun to earth. Although the solar radiation travels through space without energy loss, the intensity of radiation within a beam of given cross section (such as one square inch) decreases inversely as the square of the distance from the sun. The earth thus intercepts only about one two-billionth of the sun's total energy output.

The solar radiation spectrum consists of (*a*) X-rays, gamma rays, and ultraviolet rays, carrying about 9 percent of the total energy, (*b*) visible light rays, 41 percent, and (*c*) invisible infrared and heat rays, 50 percent. Table 8.1 gives the wavelengths in microns, of the various parts of the spectrum. A micron is equivalent to one ten-thousandth of a centimeter. Hereafter, the term *shortwave* radiation is applied to the visible and ultraviolet portion of the spectrum (wavelengths less than 0.7 microns) as distinct from the *infrared* portion (wavelengths longer than 0.7 microns). We see that the total energy of the radiation spectrum is about equally divided between shortwave and infrared portions.

TABLE 8.1

	Wavelength	Total Energy (%)
(Shortest)		
X-rays and gamma rays	1/2000 to 1/100 micron	9
Ultraviolet rays	0.2 to 0.4 micron	
Visible light rays	0.4 to 0.7 micron	41
(Longest)		
Infrared rays	0.7 to 3000 microns	50

The source of solar energy is in the sun's interior where, under enormous confining pressure and high temperature, hydrogen is converted to helium. In this fusion process, a vast quantity of heat is generated and finds its way by convection and conduction to the sun's surface. Because the rate of production of energy is constant, the output of solar radiation varies only slightly. Hence, at the average distance from the sun, the solar energy which is received upon a unit area of surface held at right angles to the sun's rays is also nearly constant. Known as the *solar constant,* this radiation rate has a value of about 2 gram calories per square centimeter per minute. It is assumed, of course, that the radiation is measured beyond the limits of the earth's atmosphere so that none has been lost. One gram calorie per square centimeter constitutes a unit measure of heat energy termed the *langley.* Therefore, we can say that the solar constant is equal to 2 langleys per minute. In English heat units, the solar constant is equivalent to 430 Btu per square foot per hour. Orbiting space satellites equipped with suitable instruments for measuring electromagnetic radiation intensity have provided precise data on the solar constant. Minor fluctuations in radiation intensity are observed, and these are attributed to variations in output of ultraviolet rays accompanying disturbances of the sun's surface.

Insolation over the globe

Because the earth is a sphere (disregarding oblateness), only one point on earth—that upon which the sun's noon rays are perpendicular—presents a surface at right angles to the sun's rays. In all directions away from this point, the earth's curved surface becomes turned at a decreasing angle with respect to the rays until the circle of illumination is reached. Along that great circle the rays are parallel with the surface. (See Chapter 4 for a discussion of the relation between the earth and the sun's rays throughout the year.)

Let us now assume that the earth is a perfectly uniform sphere with no atmosphere. Only at the subsolar point will solar energy be intercepted at the maximum rate of 2 langleys per minute. (Hereafter the term *insolation* will be used to mean the interception of solar energy by an exposed surface.) At any particular place on the earth, the quantity of insolation received in one day will then depend upon two factors: (1) the angle at which the sun's rays strike the earth, and (2) the length of time of exposure to the rays. These factors are varied by latitude and by the seasonal changes in the path of the sun in the sky.

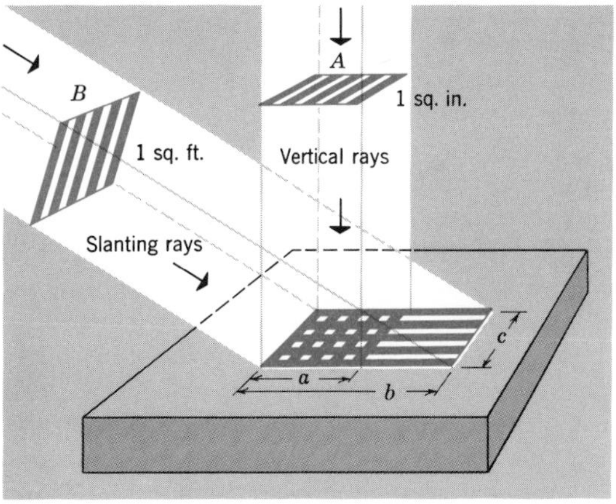

Figure 8.2 The angle of the sun's rays determines the intensity of insolation upon the ground.

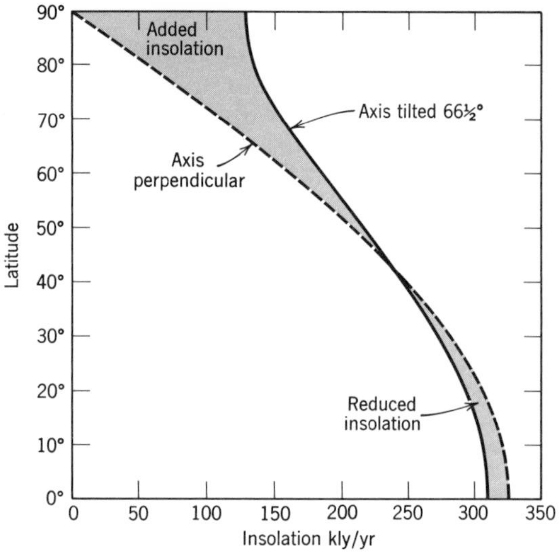

Figure 8.3 Meridional profile of annual total insolation (solid line).

Figure 8.2 shows that intensity of insolation is greatest where the sun's rays strike vertically, as they do at noon at the latitude equal to the sun's declination ranging between the tropics of Cancer and Capricorn. With diminishing angle, the same amount of solar energy spreads over a greater area of ground surface. Hence, on the average, the polar regions receive the least heat per unit area. This fact helps to explain the general distribution of average air temperatures over the globe, from a maximum at low latitudes to a minimum near either pole.

We have seen in Chapter 4 that, because of the inclination of the earth's axis, the angle of the sun's noon rays shifts through a total range 47° from one solstice to the next. This cycle does not make the yearly total of insolation for the entire globe different from an ideal situation in which the earth's axis would not be inclined, but it does cause a great difference in both latitudinal and seasonal distribution of the insolation.

Consider first that, if the earth's axis were perpendicular to the orbital plane, the poles would not receive any insolation, regardless of time of year, whereas the equator would receive an unvarying maximum. In other words, equinox conditions (shown in Figures 4.9 and 4.11) would apply every day of the year. The earth's inclination, by exposing the poles alternately to the sun, redistributes the yearly total insolation toward higher latitudes, but deducts somewhat from the equatorial zone. Figure 8.3 shows the total annual insolation from equator to poles in thousands of langleys per year (solid line), as well as the insolation that would result if the earth's axis had no tilt (dashed line). Notice how much insolation the polar regions actually receive—over 40 percent of the equatorial value.

Second, the earth's inclination produces seasonal differences in insolation at any given latitude, and these differences increase toward the poles, where the ultimate in opposites (six months of day; six of night) is reached. Along with the variation in angle of the sun's rays operates another factor, the duration of daylight. At the season when the sun's path is highest in the sky, the length of time it is above the horizon is correspondingly greater. The two factors thus work hand in hand to intensify the contrast between amounts of insolation at opposite solstices.

A three-dimensional diagram (Figure 8.4) shows how insolation varies with latitude and with season of year. Figure 8.5 shows graphs of insolation at various selected latitudes from equator to north pole. These diagrams show insolation at the outer limits of the atmosphere and thus would apply at the ground surface only for an earth imagined to have no atmosphere to absorb or reflect radiation. Notice that the equator receives two maximum periods (corresponding with the equinoxes, when the sun is overhead at the equator) and two minimum periods (corresponding to the solstices, when the sun's declination is farthest north and south of the equator). At the Arctic Circle, $66\frac{1}{2}°$ N, insolation is reduced to nothing on the day of the winter solstice, and with increasing latitude poleward this period of no insolation becomes longer. All latitudes between the Tropics of Cancer and Capricorn have two maxima and two minima, but one maximum becomes dominant as the tropic is approached. From $23\frac{1}{2}°$ to $66\frac{1}{2}°$ there is a single continuous insolation cycle with maximum at one solstice, minimum at the other.

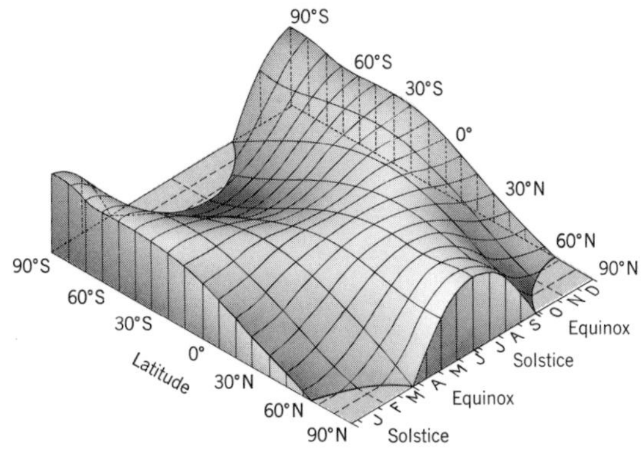

Figure 8.4 Insolation at various latitudes and seasons. (After W. M. Davis.)

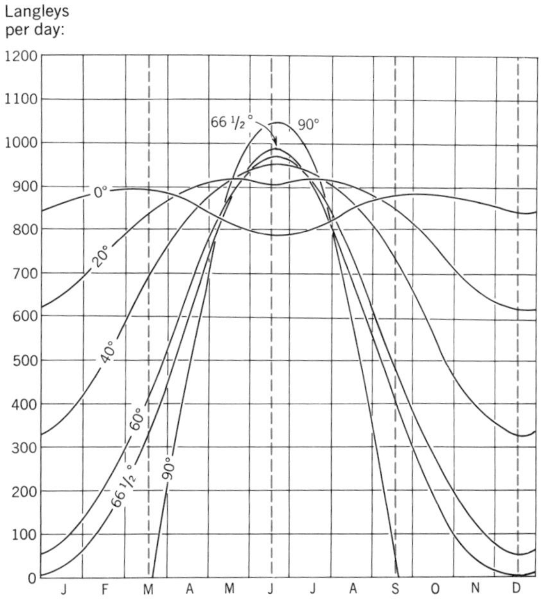

Figure 8.5 Insolation at various latitudes in the northern hemisphere. (From A. N. Strahler, 1971 *The Earth Sciences*, 2nd ed., Harper and Row, New York.)

World latitude zones

The sun's path in the sky, which determines the flow of solar energy reaching the earth's surface—and hence governs the thermal environment of man—provides a basis for dividing the globe into latitude zones (Figure 8.6). It is not intended that the specified zone limits be taken as absolute and binding, but rather that the system be considered as a convenient terminology.

The *equatorial zone* lies astride the equator and extends to 10° latitude north and south. Within this zone, the sun throughout the year provides intense insolation, while day and night are of roughly equal duration. Astride the Tropics of Cancer and Capricorn are the *north tropical zone* and *south tropical zone* respectively, spanning the latitude belts 10° to 25° north and south. In this zone, the sun takes a path close to the zenith at one solstice and is appreciably lower at the opposite solstice. Thus a marked seasonal cycle exists, but is combined with a potentially large total annual insolation. You are, of course, aware that literary usage and some geographical works differ from what is described here, for the word *tropics* has been widely used to denote the entire belt of 47 degrees of latitude between the tropics of Cancer and Capricorn.

Immediately poleward of the tropical zones are transitional regions which have become widely accepted among geographers as the *subtropical zones*. For convenience, these zones are here assigned the latitude belts 25° to 35° north and south, but it is understood that the adjective "subtropical" as applied to environmental regions may extend a few degrees farther poleward or equatorward of these parallels.

The *middle-latitude zones*, lying between 35° and 55° north and south latitude represent regions in which the sun's path shifts through a relatively large range of noon altitudes, so that seasonal contrasts in incoming solar energy are strong. Strong seasonal differences in lengths of day and night exist as compared with the tropical zones.

Bordering the middle-latitude zones on the poleward side are the *subarctic zones*, 55° to 60° north and south latitudes, transitional between middle-latitude and arctic zones.

Astride the Arctic and Antarctic circles, $66\frac{1}{2}$° north and south latitudes, lie the *arctic zones*, which may be further differentiated, if desired, into an *arctic zone* and an *antarctic zone*. The latitudinal extent of the arctic zones is here specified as 60° to 75° north and south, but these limits should not be imposed severely. The arctic zones have an extremely large yearly variation in lengths of day and night, yielding enormous contrasts in incoming solar energy from solstice to solstice. Notice that classical usage, as found in standard dictionaries, considers the "arctic" or "arctic region" as the entire area from arctic circle to pole.

The *polar zones*, north and south, are circular areas between 75° latitude and the poles. Here the polar regime of six months day and six months night yields the ultimate in seasonal range of incoming solar radiation.

Insolation losses in the atmosphere

As the sun's radiation penetrates the earth's atmosphere, a series of selective depletions and diversions of energy take place. At an altitude of 95 mi (150 km), the radiation spectrum possesses almost 100 percent of its original energy, but in penetration to an altitude of 55 mi (88 km) ab-

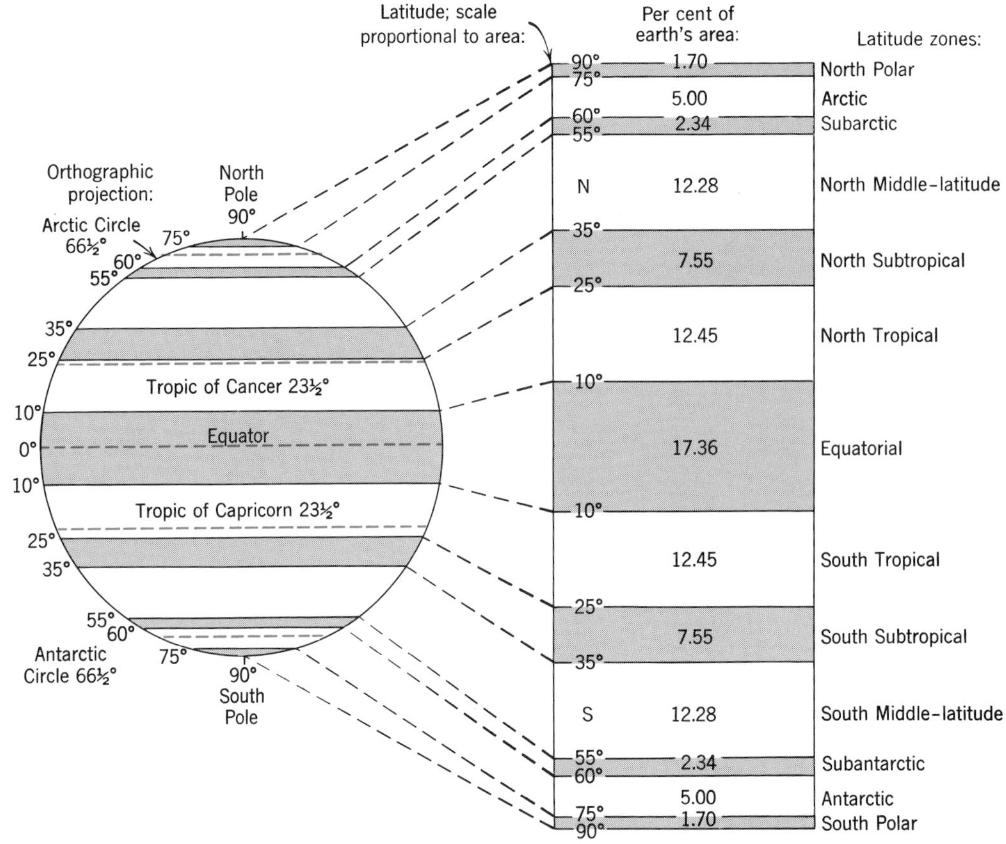

Figure 8.6 A system of environmental latitude zones.

sorption of X-rays is almost complete and some of the ultraviolet radiation has been absorbed as well. As noted in Chapter 7, the ionosphere is developed in this region of the atmosphere by the effect of the highly energetic X-rays, gamma rays, and ultraviolet rays upon molecules and atoms of nitrogen and oxygen.

As solar radiation penetrates into deeper and denser atmospheric layers, gas molecules cause the visible light rays to be turned aside in all possible directions, a process known as *Rayleigh scattering.* Where dust particles are encountered in the troposphere, further scattering occurs. The total process may be described as *diffuse reflection.* That the clear sky is blue in color is explained by Rayleigh scattering of the shorter visible wavelengths. These predominantly blue light waves reach our eyes indirectly from all parts of the sky. The red wavelengths and infrared rays are less subject to scatter and largely continue in a straight-line path toward earth. The setting sun appears red because a part of the red rays escape deflection from the direct line of sight.

As a result of all forms of shortwave scattering, some solar energy is returned to space and forever lost, while at the same time some scattered shortwave energy also is directed earthward. The latter is referred to as *diffuse sky radiation,* or *down scatter.* An additional but minor cause of energy loss is that which occurs in the ozone layer (see Chapter 7), as oxygen molecules are broken into atoms and reformed into ozone molecules.

Another form of energy loss takes place as the sun's rays penetrate the atmosphere. Both carbon dioxide and water vapor are capable of directly absorbing infrared radiation. Absorption results in a rise of sensible temperature of the air. Thus some direct heating of the lower atmosphere takes place during incoming solar radiation. Although carbon dioxide is a constant quantity in the air (0.033 percent by volume) the water vapor content varies greatly from place to place, being as low as 0.02 percent under desert conditions to as high as 1.8 percent in humid equatorial regions. Absorption correspondingly varies from one global environment to another.

All forms of direct absorption listed above, namely X-ray, gamma ray, and ultraviolet absorption in the ionosphere and ozone layer, combined with direct longwave absorption by carbon dioxide, water vapor, and other gas molecules and dust particles, is estimated to average as little as 10 percent for conditions of clear, dry air, to as high as 30 percent when a cloud cover exists.

Figure 8.7 shows in a highly diagrammatic way the range of values of the various forms of reflection and absorption that may occur. When skies are clear, reflection and absorption combined may total about 20 percent, leaving as much as 80 percent to reach the ground.

Yet another form of energy loss must be brought into the picture. The upper surfaces of clouds are extremely good reflectors of shortwave radiation. Air travelers are well aware of how painfully brilliant the sunlit upper surface of a cloud deck can be when seen from above. Cloud reflection can account for a direct turning back into space of from 30 to 60 percent of total incoming radiation (Figure 8.7). Thus we see that, under conditions of a heavy cloud layer, the combined reflection and absorption from clouds alone can account for a loss of from 35 to 80 percent of the incoming radiation and allow from 45 to 0 percent to reach the ground. A world average value for reflection from clouds to space is about 21 percent of the total insolation (Table 8.2).

The surfaces of the land and ocean reflect some shortwave radiation directly back into the atmosphere. This small quantity, about 6% as a world average, may be combined with cloud reflection in evaluating total reflective losses.

Table 8.2 lists the percentages given so far for the energy losses in insolation by reflection and absorption. Altogether the losses to space by reflection total 32 percent of the total insolation. Figure 8.8 shows the same data as in the upper part of Table 8.2, but in graphic form.

The percentage of radiant energy reflected back by a surface is termed the *albedo*. This is an important property of the earth's surface because it determines the relative rate of heating of the surface when exposed to insolation. Albedo of a water surface is very low (2 percent) for nearly vertical rays, but high for low-angle rays. It is also extremely high for snow or ice (45 to 85 percent). For fields, forests, and bare ground the albedos are of intermediate value, ranging from as low as 3 percent to as high as 25 percent. For the earth as a planet, average albedo is on the order of 32 percent (Table 8.2), and is determined largely by the extent of cloud cover.

The radiation balance

Any substance that possesses heat sends out energy of electromagnetic radiation from its surface. The amount of radiant energy thus sent out

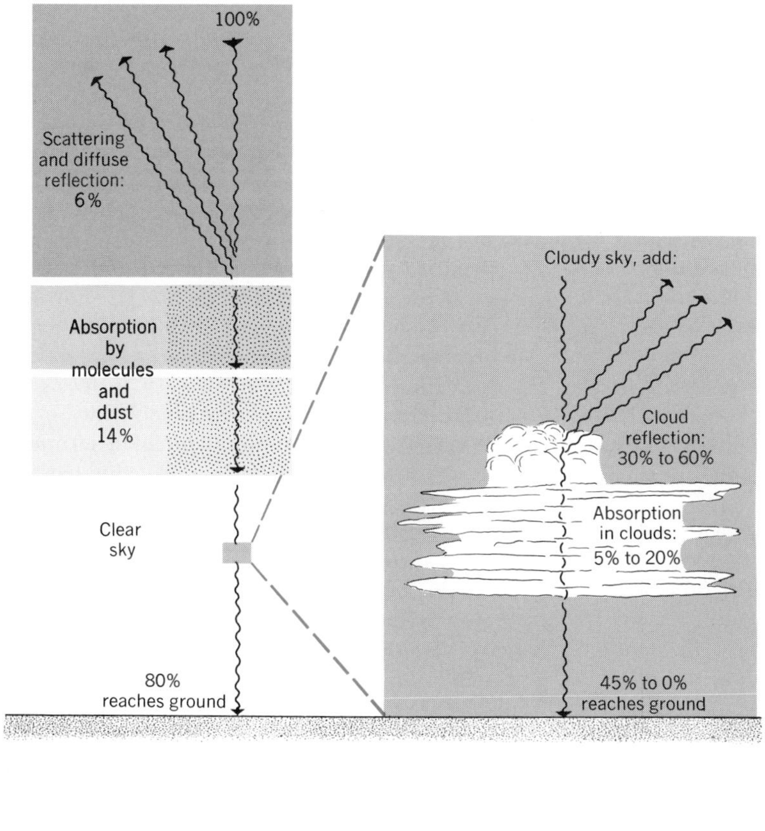

Figure 8.7 Losses of incoming solar energy by scattering, reflection and absorption. (From A. N. Strahler, 1971, *The Earth Sciences*, 2nd ed., Harper and Row, New York.)

TABLE 8.2 THE GLOBAL RADIATION BUDGET

Incoming Solar Radiation (shortwave and infrared)		Percentage
Total at top of atmosphere		100
Diffuse reflection to space (Rayleigh scatter, dust)	5	
Reflection from clouds to space	21	
Direct reflection from earth's surface	6	
Total reflection loss to space by earth-atmosphere system (Earth's albedo)		32
Absorbed by molecules, dust, water vapor, CO_2, clouds	18	
Absorbed by earth's surface	50	
Total absorbed by earth-atmosphere system		68
Sum of absorption and reflection		100
Outgoing Radiation (longwave)		
Infrared radiation from earth's surface	98	
Lost to space	8	
Absorbed by atmosphere	90	
Infrared radiation emitted by atmosphere	137	
Lost to space	60	
Absorbed by earth's surface as counter radiation	77	
Effective (net) outgoing radiation from earth's surface		21
Effective (net) outgoing radiation from atmosphere		47
Effective (net) outgoing radiation from earth-atmosphere system		68

Source: Data from W. D. Sellers (1965), *Physical Climatology*, Chicago and London, Univ. of Chicago Press, Tables 6 and 9. Note: 100% represents a value of 263 kilolangleys per year.

is directly proportional to the fourth power of the absolute temperature of the substance. Also, the lower the temperature of the radiating material, the longer are the wavelengths of the rays emitted.

The ground or ocean surface, possessing heat derived originally from absorption of the sun's rays, continually radiates this energy back into the atmosphere, a process known as *ground radiation* or *terrestrial radiation*. This infrared radiation consists of wavelengths longer than 3 or 4 microns and is referred to here as *longwave radiation*. The atmosphere also radiates energy both toward the earth and outward into space where it is lost. Note that longwave radiation is quite different from reflection, in which the rays are turned back directly without being absorbed. Longwave radiation from both ground and atmosphere continues during the night, when no solar radiation is being received.

Energy radiated from the ground is easily absorbed by the atmosphere because it consists largely of very long wavelengths (4 to 30 microns), in contrast to the visible light rays (0.4 to 0.7 microns) and shorter infrared rays (0.7 to 3.0 microns) which make up almost all of the entering solar radiation. Absorption of longwave radiation by water vapor and carbon dioxide takes place largely in wavelengths from 4 to 8 microns and 12 to 20 microns. However, radiation in the range of wavelengths between 8 and 12 microns passes freely through the earth's atmosphere and into outer space.

Of the longwave energy radiated from the ground, a portion is radiated back to the earth's surface, a process called *counterradiation*. Thus the atmosphere receives heat by an indirect process in which the radiant energy in shortwave form is permitted to pass through, but that in longwave form is delayed in making its escape. For this reason, the lower atmosphere with its water vapor and carbon dioxide acts as a blanket which returns heat to the earth and helps to keep surface temperatures from dropping excessively during the night or in winter at middle and high latitudes. Somewhat the same principle is employed in greenhouses and in homes using the solar-heating method (Figure 8.9). Here the glass permits entry of shortwave energy. Accumulated heat cannot escape by mixing with cooler air outside. The expression *greenhouse effect* has been used by meteorologists to describe the atmospheric heating principle.

The lower half of Table 8.2 shows the components of outgoing or longwave radiation for the

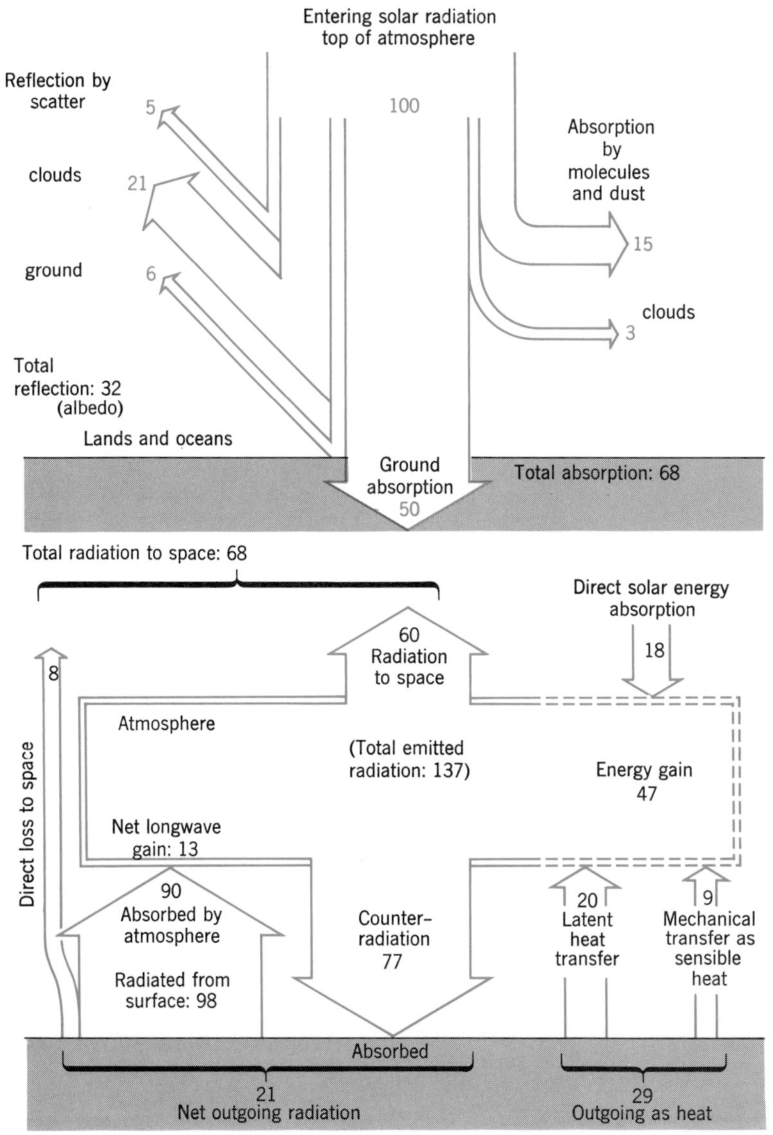

Figure 8.8 Schematic diagram of the global radiation balance. Figures correspond with those in Table 8.2.

planet as a whole, using the same percentage units as for the incoming radiation. Figure 8.8 shows the same data graphically. The total longwave radiation leaving the earth's land and ocean surface is equivalent in amount to 98 percentage units. Of this, 8 units are lost to space, while 90 units are absorbed by the atmosphere. In turn, the atmosphere emits longwave radiation. The total of this radiation is equivalent in amount to 137 percentage units of the insolation at the top of the atmosphere. This figure may at first seem absurd, but the next two lines show that this radiation is divided into two parts, one of which goes out into space (60 units) and the other of which is absorbed by the earth's surface as counterradiation (77 units). The last three lines of the table show the net figures as follows: the earth's surface has a net outgoing longwave radiation of 21 units, while the atmosphere has a net outgoing longwave radiation of 47 units. Combining these two figures gives 68 units for the net outgoing radiation from the entire earth-atmosphere system, which equals the total energy absorbed by that same system (see upper total).

Further study of Table 8.2 will reveal what seems to be a discrepancy. In the upper part of the table, the absorption of energy by the earth's land and water surface is given as 50 percent whereas in the lower part of the table the outgoing radiation from the same surfaces is given as 21

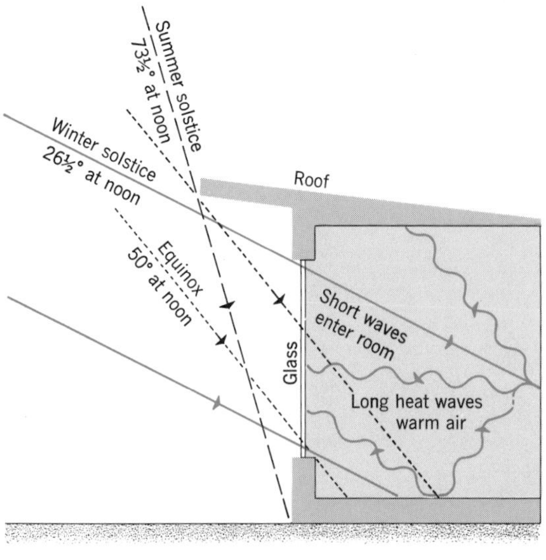

Figure 8.9 Principle of solar heating of a house.

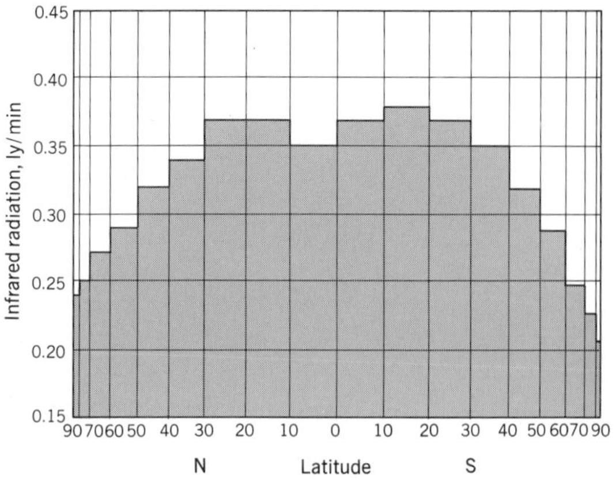

Figure 8.11 Meridional profile of mean longwave radiation from the earth, as determined from satellite data. Data from T. H. Vonder Haar and V. E. Suomi, 1969, *Science*, Vol. 163, p. 667. (From A. N. Strahler, 1971, *The Earth Sciences*, 2nd ed., Harper and Row, New York.)

percentage units, the discrepancy being 29 units. Actually, two other mechanisms transfer the missing energy back into the atmosphere. One is as *latent heat,* through the evaporation of surface water. A second is the mechanical transfer of heat from surface to air. The heat is first conducted as sensible heat from water or soil into the overlying air layer, then carried upward in turbulent eddies. (The reverse flow of heat can also occur in this way.) Evidently about 60 percent of the incoming energy is returned to the atmosphere by the combined processes of evaporation and sensible heat transfer. Of this 60 percent, about $\frac{2}{3}$ returns as latent heat, and about $\frac{1}{3}$ as sensible heat.

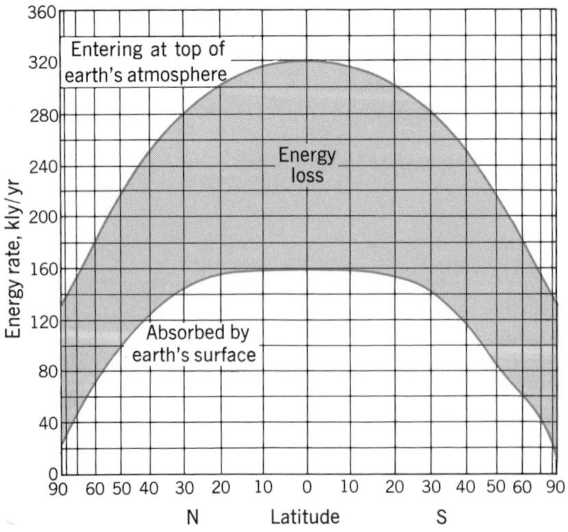

Figure 8.10 Meridional profiles of mean annual values of entering solar radiation at top of atmosphere and radiation absorbed by earth's surface. Area between the two curves represents energy loss in atmosphere. Data from W. D. Sellers, 1956, *Physical Climatology* (From A. N. Strahler, 1971, *The Earth Sciences*, 2nd ed., Harper and Row, New York.)

Latitude and the radiation balance

Earlier in this chapter, in relating latitude to insolation, we showed that the tilt of the earth's axis causes a poleward redistribution of insolation, as compared with conditions that would apply if the axis were perpendicular to the orbital plane (Figure 8.3). Let us now look deeper into the wide range in rates of incoming and outgoing energy in terms of a profile spanning the entire latitude range of 90° N to 90° S. In this analysis, yearly averages are used, so that the effect of seasons is not seen. Figure 8.10 is a graph of solar radiation (insolation) plotted against latitude. Units are kilolangleys (thousands of langleys) per year. The upper curve shows insolation entering the top of the atmosphere; it is essentially the same information as that shown in Figure 8.3. The lower curve shows insolation absorbed at the earth's surface. The intervening area between the two curves represents the loss of energy as insolation penetrates the atmosphere, combined with shortwave reflection from the surface. The horizontal scale is so adjusted that it is proportional to the area of earth's surface between 10-degree parallels of latitude. (It is the same type of latitude scale used in the right-hand part of Figure 8.6.) Notice that near the equator, about 50 percent of the entering insolation reaches the ground, whereas near the poles less than 20 percent is absorbed by the

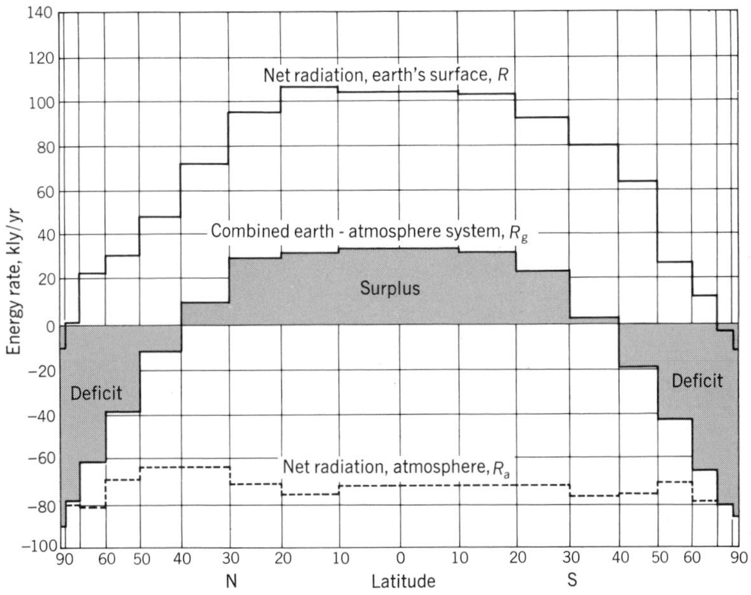

Figure 8.12 Meridional profiles of mean net radiation at earth's surface, from atmosphere, and from combined earth-atmosphere system. Data from W. D. Sellers, 1965, *Physical Climatology*. (From A. N. Strahler, 1971, *The Earth Sciences*, 2nd ed., Harper and Row, New York.)

surface. This difference is understandable in view of the low angle of attack of the sun's rays at high latitudes, causing those rays to pass through a much greater thickness of atmosphere than at the pole and thus to experience proportionately greater losses by reflection and absorption. Also, the surface albedo is much greater in high latitudes, so that a much greater proportion of shortwave energy is reflected from snow-covered surfaces of arctic and polar regions than in low latitudes.

Figure 8.11 shows the average longwave radiation emitted by the earth as a planet. Because the data are recorded from an orbiting satellite, they represent the total radiation of the earth-atmosphere system. Obviously, the radiation is greatest from low latitudes, where the input of radiation is greatest and surface temperatures are highest. But notice that there are two maxima in the graph, one centered over each of the subtropical belts. There is a small but pronounced dip close to the equator. Here we see the effect of two subtropical desert belts that girdle the globe, each representing a zone of reduced cloud cover and high ground surface temperatures. Over the equatorial belt, cloud cover is denser on the average and cuts down on longwave radiation.

To make an assessment of the earth's radiation balance in terms of latitude, all forms of incoming radiation are combined and balanced against all forms of outgoing radiation. This difference is known as the *net all-wave radiation*. As applied to the earth's surface, *incoming* radiation includes both direct and indirect solar radiation as well as downward longwave counterradiation. *Outgoing*

TABLE 8.3* ANNUAL MERIDIONAL HEAT TRANSPORT

Latitude (°N)	Heat Transport kcal/yr × 10^{19}
90	0.00
80	0.35
70	1.25
60	2.40
50	3.40
40	3.91
30	3.56
20	2.54
10	1.21
0	−0.26

*Data from W. D. Sellers, 1965, *Physical Climatology*, University of Chicago Press, Table 12.

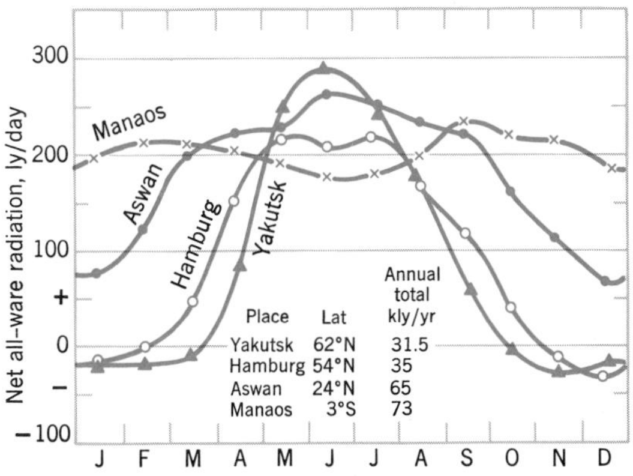

Figure 8.13 Net all-wave radiation throughout the year at four representative stations. (Data by courtesy of David H. Miller.)

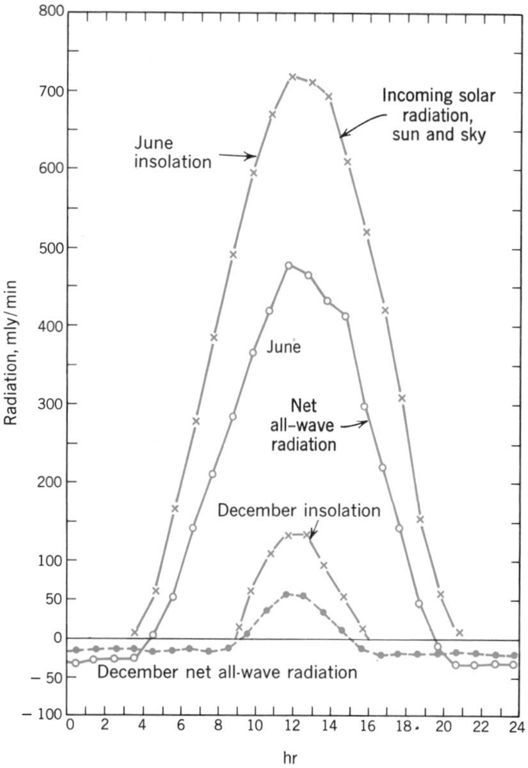

Figure 8.14 Mean daily cycles of June and December incoming and outgoing radiation at Hamburg, Germany. (Data by courtesy of Ernst Frankenberger.)

radiation combines both longwave radiation and reflected shortwave radiation. The net all-wave radiation is therefore the energy flux, whether downward (positive) or upward (negative).

Figure 8.12 shows net all-wave radiation for the

Figure 8.15 Pyranometer for measurement of solar and sky radiation. (Photograph by courtesy of WeatherMeasure Corporation, Sacramento, Calif.)

earth's surface (upper graph), the atmosphere (lower graph), and for the combined earth-atmosphere system (middle graph). Notice that the net radiation is positive for nearly all of the earth's surface, whereas it has a negative value for all of the atmosphere. But when these two graphs are combined, there results a large region of surplus radiation from about 40° N to 30° S, and two high-latitude regions of deficit. On the diagram, the areas labeled "deficit" are together equal to the area labeled "surplus," as the radiation balance requires.

It is obvious from Figure 8.12 that the earth's energy balance can be maintained only if heat is transported from the low-latitude belt of surplus to the two high-latitude regions of deficit. This poleward movement of heat is described as *meridional,* for example, moving north (or south) along the meridians of longitude. We should expect the rate of meridional heat transport to be greatest in middle latitudes, and this fact is shown by the figures in Table 8.3. The unit used is the kilocalorie (1000 calories) multiplied by 10 raised to the nineteenth power (10^{19}).

The meridional flow of heat is carried out by circulation of the atmosphere and oceans. In the atmosphere, heat is transported both as sensible heat and as latent heat. The latter subject is explained in Chapter 10.

Annual and daily cycles of radiation

The effects of changing angles of attack of the sun's rays upon the parallels of latitude from season to season are shown in a series of curves in Figure 8.13. The quantity scaled on the vertical axis is net all-wave radiation in units of langleys per day. Compare these curves with the insolation curves shown in Figure 8.5. There is a general resemblance in curve forms when similar latitudes are compared. Manaos, Brazil, located near the equator, has a large radiation surplus in every month, but with two minor maxima and two minor minima. Aswan, U.A.R. has a strong annual cycle but a large surplus of radiation in all months. For middle latitudes the situation is different. Both Hamburg, Germany, and Yakutsk, U.S.S.R., have winter periods of radiation deficit, that of Yakutsk lasting through six months. But see how the summer peak at Yakutsk exceeds that of even the equatorial station! Later in this chapter, the way in which air temperatures near the ground surface respond to the annual radiation cycle is brought into the picture to complete the annual cycle of the energy balance.

The daily cycle of incoming shortwave radiation, as measured close to the ground surface, is shown in Figure 8.14 for Hamburg, Germany. Data are for June and December, the months of maximum and minimum values. Insolation begins close to sunrise and ceases close to sunset. Not only is insolation vastly greater in intensity in June, but it lasts for nearly three times as long a period of daylight. Near the equator, a similar insolation curve would differ very little from month to month and would always span approximately a 12-hour period.

Incoming shortwave energy from both the direct solar beam and the indirect sky radiation are simultaneously measured by an instrument known as the *pyranometer* (Figure 8.15). A sensing cell enclosed in a glass bulb receives radiation from the entire hemisphere of the sky. This is the standard instrument for measuring solar radiation at observing stations.

Figure 8.14 also shows the daily curves of net all-wave radiation for June and December at Hamburg. During the hours that the sun is in the sky, the curve rises and falls symetrically, peaking at noon. However, during the night hours, a negative value sets in and is held almost constantly while the sun is below the horizon.

Daily and seasonal cycles of radiation and heat at the earth's surface are a vital environmental factor of the biosphere and deserve careful attention. The responses of plants and animals to these rhythms of energy change explain in large part the worldwide ranges in life regions, from the rich life zones of the low latitudes to the sparse life zones of arctic and polar regions.

As we shall find in Chapter 12, the water needs of plants are very large in low-latitude zones of large radiation surpluses. In an equatorial-zone locality, such as Manaos (Figure 8.13) a very large quantity of rain falls throughout the year to furnish the needed water. In contrast, where rainfall is scanty, as at Aswan, desert conditions prevail.

Measurement of air temperature

The remainder of this chapter deals with the surface thermal environment, specifically, with sensible heat expressed in air temperatures. At every recording weather station, the temperature of the air is read at regular intervals from thermometers mounted inside a boxlike shelter built several feet off the ground (Figure 8.16). The instruments are protected from direct sunlight, but air is allowed to circulate freely through the shelter. Standard equipment consists of a pair of *maximum-minimum* thermometers, one of which shows the maximum temperature, the other the minimum, that have occurred in the period since last reset. In addition, an automatic recording thermometer, called a *thermograph*, may be used to draw a continuous temperature record on a piece of graph paper (Figure 8.17).

Figure 8.16 A standard thermometer shelter. Rain gauge at left. (National Weather Service.)

Throughout this book, the *Fahrenheit* temperature scale is used—the common everyday scale among the general public in the United States. Freezing temperature is 32°; the boiling point is 212° on this scale (Figure 8.18). One advantage of this scale is that the general reader can relate Fahrenheit temperatures to degrees of bodily comfort or discomfort. The *Centigrade*, or *Celsius*, temperature scale, in which 0° is freezing point and 100° the boiling point, is now used in all countries of the world except the United States; it is the scale used in physics, chemistry, and meteorology. Fahrenheit and Centigrade (Celsius) measurements are distinguished by the letter "F" or "C" following the figure.

Land and water differences

Before we examine the annual and daily rhythms of air temperature change, we need to introduce the fact that land and water surfaces have quite different properties in absorption and radiation of heat. The general law may be stated as follows. Land surfaces are rapidly and intensely heated under the sun's rays, whereas water surfaces are only slowly and moderately heated. On the other hand, land surfaces cool off more rapidly and reach much lower temperatures than water surfaces when solar radiation is cut off. Temperature contrasts are therefore great over land areas, but only moderate over water areas. It is further true that the larger the mass of land, the greater are seasonal temperature contrasts. Because the

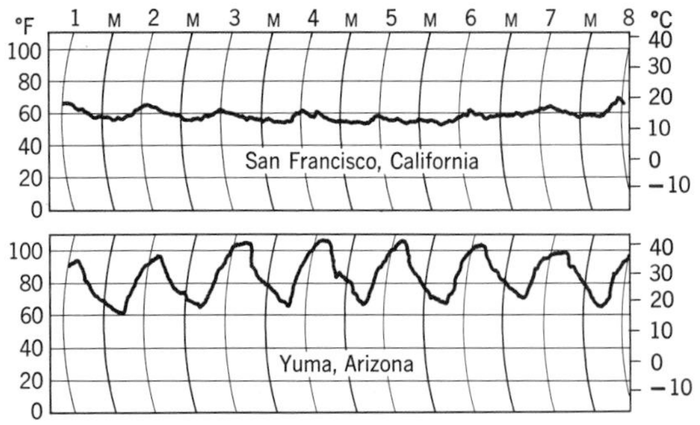

Figure 8.17 Thermograph trace sheets show temperatures for one week. (After Kincer, U.S. Dept. of Agriculture.)

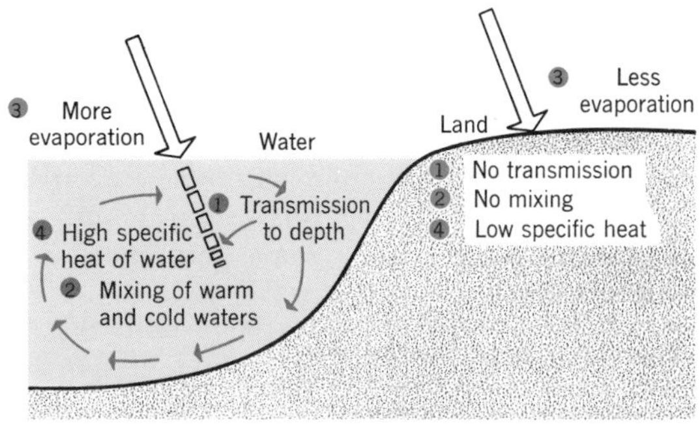

Figure 8.19 Contrasts in heating of land and water.

heating of ground and water surfaces controls heating of the atmosphere above, the same observations apply to air temperature as to surface temperature.

An explanation of the law of land and water contrasts may be found in the application of certain simple principles of physics (Figure 8.19). Water is transparent and permits heat rays to penetrate many feet, thus distributing the heat through a thickness of several feet of water. The ground surface, being opaque, absorbs heat only at the surface, which thus attains a higher temperature than the water surface. Ocean waters are mixed by vertical rising and sinking motions in the surface layer, allowing heat to be distributed and stored through a great mass of water, but no such movement can occur in the ground. Water surfaces permit continual evaporation, which is a cooling process and serves to alleviate the surface heat. Ground surfaces, which are commonly moist and covered by vegetation, also permit cooling by evaporation, but to a lesser degree than ocean surfaces. As a further cause of contrast, water must absorb almost five times as much heat energy in order to rise in temperature the same amount as dry soil or rock. If heat is being applied equally to both substances, the ground will attain a high temperature long before the water will; *specific heat* of the water is said to be great, that of rock or dry soil to be small.

Annual cycle of air temperature

In order to build statistical information about temperatures for longer periods of time than a single day, a unit known as the *mean daily temperature* is used. The National Weather Service follows a very simple method of obtaining the mean daily temperature, using readings made once a day from the maximum-minimum thermometers. The maximum and minimum temperatures for one day are added together and divided by 2. If the mean daily temperatures are collected for many years and averaged for each calendar day or month, then plotted on a graph, a smooth curve of annual temperature is obtained. Figure 8.20 shows such curves for two places at about 50° latitude. Winnipeg, Manitoba, has a midcontinent location; the Scilly Islands, England lie to the lee of the North Atlantic Ocean.

Although insolation reaches a maximum at summer solstice, the hottest part of the year for inland regions is about a month later, since heat energy continues to flow into the ground well into August. Air temperature maximum, closely corre-

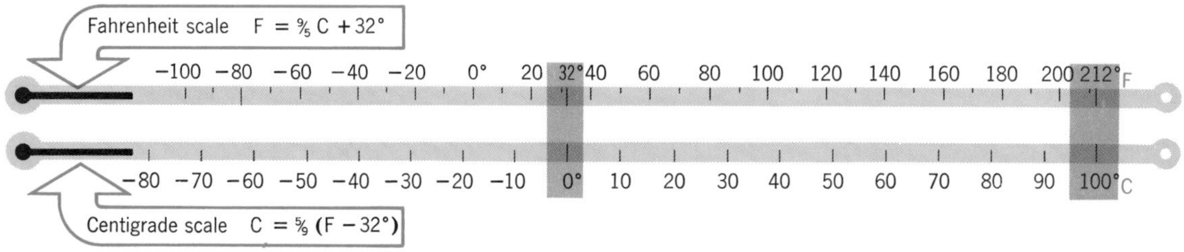

Figure 8.18 Fahrenheit and Centigrade (Celsius) temperature scales.

sponding with maximum ground output of long-wave radiation, is correspondingly delayed. (Bear in mind that this cycle applies to middle and high latitudes, but not to the region between the tropics of Cancer and Capricorn.) Similarly, the coldest time of year for large land areas is January, about a month after winter solstice, because the ground continues to lose heat even after insolation begins to rise.

Over the oceans there are two differences. (1) Maximum and minimum temperatures are reached about a month later than on land—in August and February, respectively—because water bodies heat or cool much more slowly than land areas. (2) The yearly range is less than over land, following the law of temperature differences between land and water surfaces. Coastal regions are usually influenced by the oceans to the extent that maximum and minimum temperatures occur later than in the interior. This principle shows nicely for monthly temperatures of the Scilly Islands, off the Cornwall coast of southeast England. February is slightly colder than January (Figure 8.20).

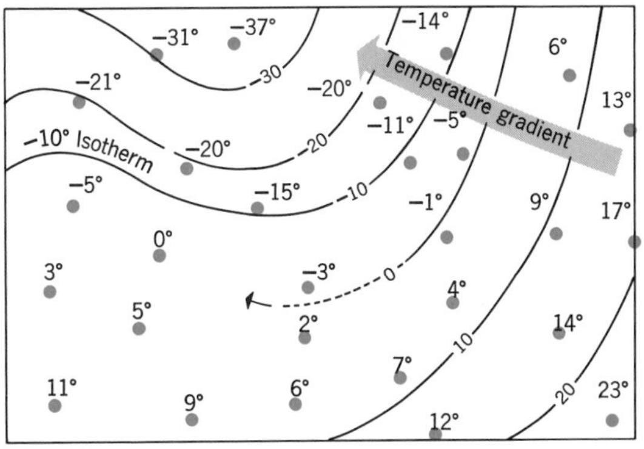

Figure 8.21 Construction of isotherms.

Air temperature maps

The distribution of air temperatures over large areas can best be shown by a map composed of *isotherms*. Similar to topographic contours, whose meaning and construction are discussed at length in Appendix I, isotherms are drawn to connect all points having the same temperature. Figure 8.21 shows a weather map upon which the observed air temperatures have been recorded in the correct places. These may represent single readings taken at the same time everywhere, or they may represent the averages of many years of records for a particular day or month of a year, depending upon the purposes of the map. Usually, isotherms representing 5° or 10° differences are chosen, but they can be drawn for any selected temperatures. The isotherms pass through the observing stations only when the station readings coincide with the value selected for an isotherm. Otherwise it is necessary to draw the isotherms by estimating their proper position between stations. The value of isothermal maps is that they make clearly visible the important characteristics of the prevailing temperatures. Centers of high or low temperature are clearly outlined. Zones of gradation are readily seen. From a mere mass of figures on a map, these features are not easily grasped.

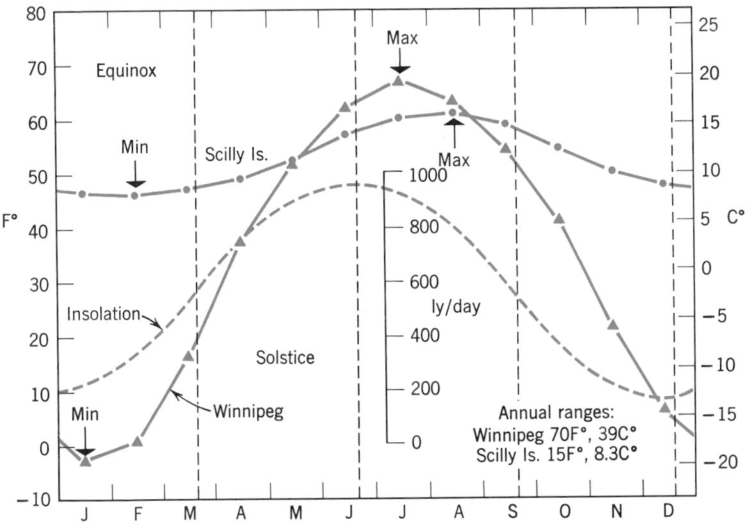

Figure 8.20 Annual cycles of monthly mean air temperature for two stations at 50° N lat.: Winnipeg, Manitoba, Canada, and Scilly Islands, England. (Data of Meteorological Office, Air Ministry, Great Britain.)

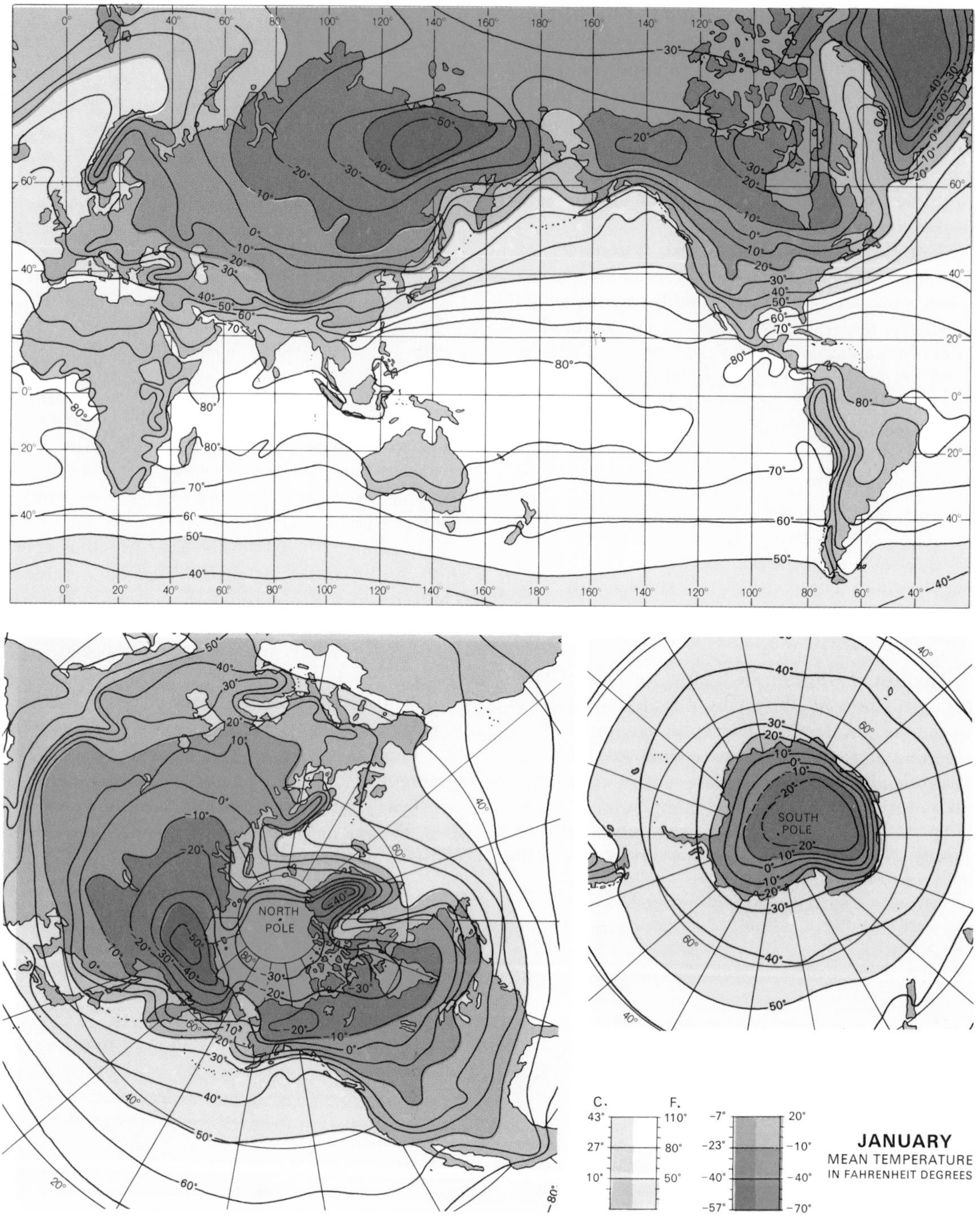

World temperature distribution

Isothermal maps of the world for January and July are shown in Figure 8.22. Annual temperature range is shown in Figure 8.23. Isotherms have a general east-west trend around the earth because of the general decrease of insolation from equator to polar regions. The east-west trend and parallelism of isotherms are best developed in the southern hemisphere, south of the 25th parallel, where land areas are small. In the northern hemisphere, isotherms show wide northward and southward deflections where they pass from a land area to an ocean area, particularly in January, when land and ocean surface temperatures are brought most strongly into contrast.

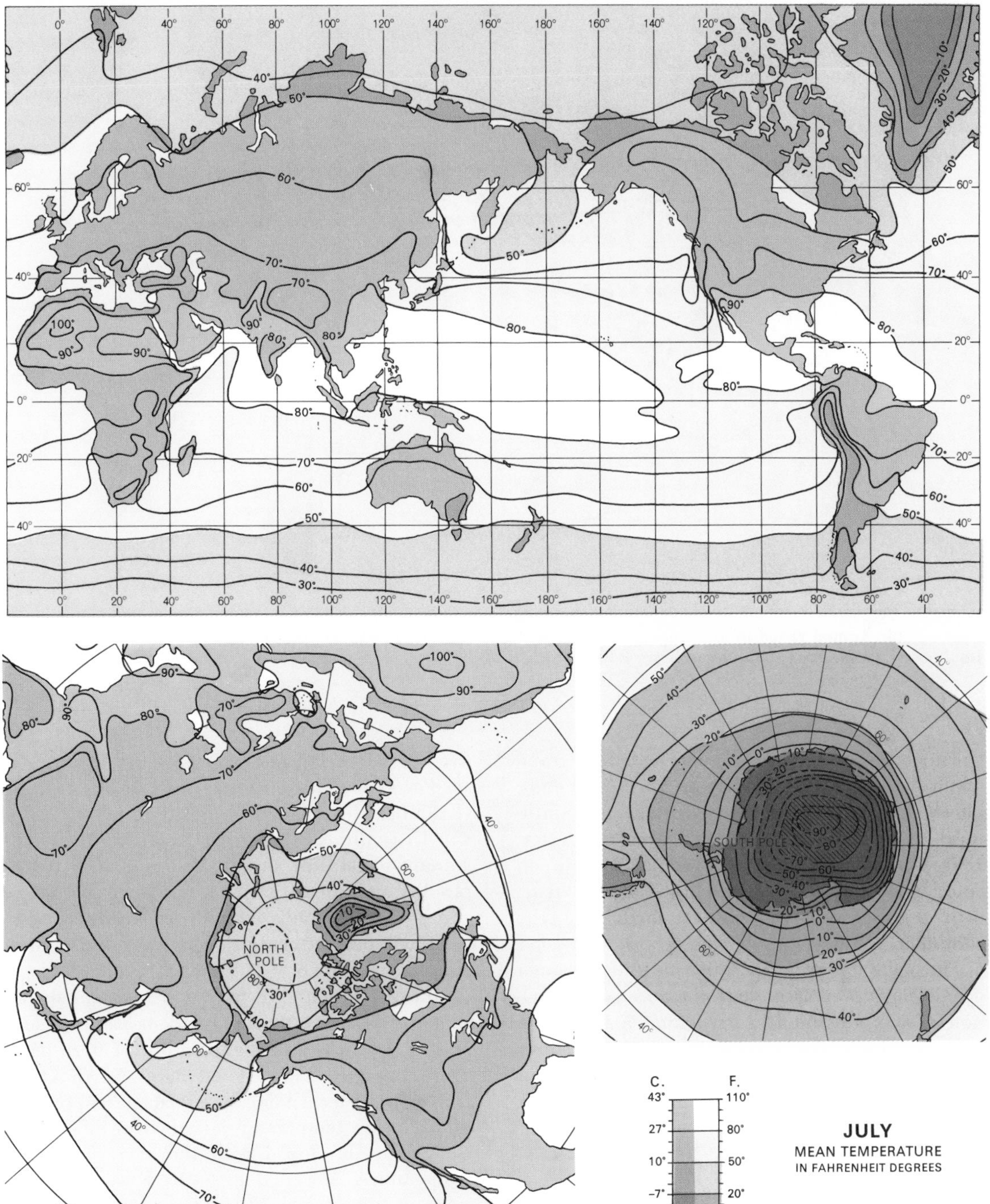

Figure 8.22 ***(above and left)*** **January and July air temperatures in degrees Fahrenheit. (Compiled by John E. Oliver from data by World Climatology Branch, Meteorological Office, *Tables of Temperature*, 1958, Her Majesty's Stationery Office, London; U.S. Navy, 1955, *Marine Climatic Atlas*, Washington, D.C.; and P. C. Dalrymple, 1966, American Geophysical Union. Cartography by John Tremblay.)**

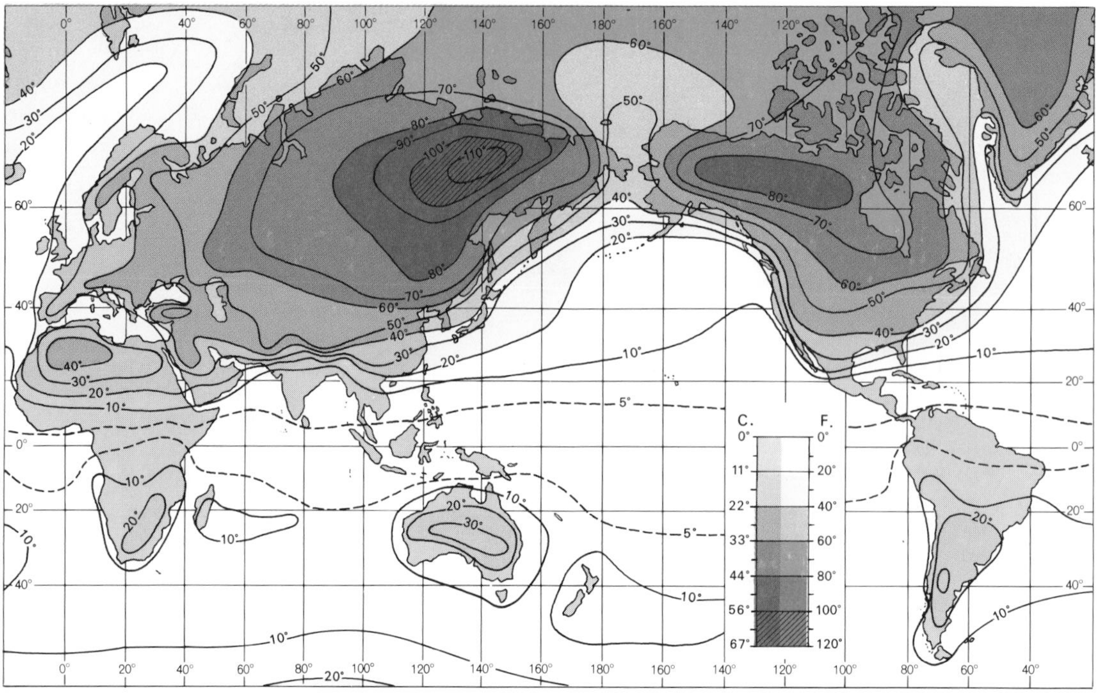

Figure 8.23 Annual range of air temperatures. Data show differences between January and July means. (Same data sources as Figure 8.22.)

The land-water effect is represented diagrammatically in Figure 8.24, for the northern hemisphere. The January isotherm is deflected southward over the land, northward over the water. Temperatures along a single parallel are low on land but high on water. In July the reverse is true, with the isotherm pushed far north over the continent.

Throughout the year, isotherms shift through several degrees of latitude, following the declination of the sun but lagging behind a month or so in time. Over large water areas, such as the south Pacific, the annual shift amounts to only about five degrees of latitude, whereas over landmasses, such as Africa, this shift is as much as 20° latitude. (Examine the change in position of the 70°F (21°C) isotherms over Africa.) This difference in amount of latitude shift is also explained by the rapidity and intensity with which lands are heated and cooled as compared with ocean areas.

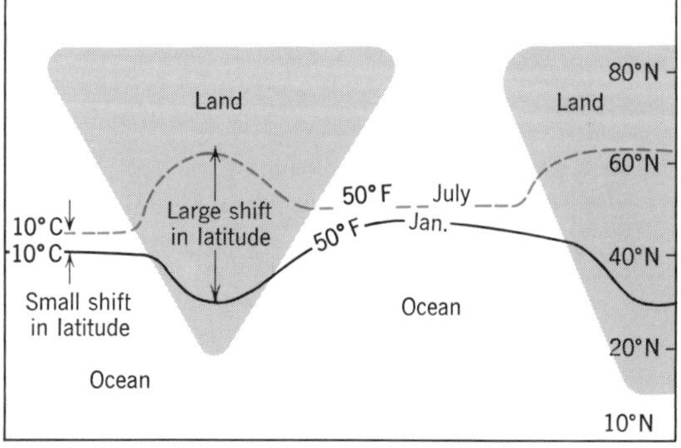

Figure 8.24 Seasonal shift of an isotherm.

Certain definite centers of high and low temperature occur and are shown by isotherms which are completely closed to form oval or irregular-shaped enclosures. Notice that all of them are over landmasses. In July, high-temperature centers occur over the southwestern United States, North Africa, and southwestern Asia. In January, a continental center of low temperature occurs over Siberia and is strongly developed with the average January temperatures lower than −50°F (−46°C). A corresponding region of low temperature, marked off by the closed isotherm of −30°F (−34°C), occurs in northernmost North America. It is not so well developed as that of Asia because of the presence of considerable areas of Arctic Ocean among the islands of the northern fringe of the landmass and the smaller size of the North American landmass.

Permanent centers of low temperature exist over Greenland and Antarctica, the two regions of massive icecaps. Temperatures over Greenland do not, however, reach the extreme low of northern

Siberia in January, although the annual average temperature of the icecap is much lower.

A comparison of winter temperatures at the two poles is instructive, since one lies in a region of deep ocean and the other in the heart of a continent at high elevation. Over the ice-covered sea surface, north-polar January mean temperature is probably about −30°F (−35°C). By contrast, the July average at the south pole is about −75°F (−60°C), because of intense radiation of heat from the elevated plateau.

The annual range of monthly mean temperatures at any desired location may be roughly computed from the January and July maps but is more conveniently analyzed on a world map of annual temperature ranges (Figure 8.23). The lines on this map are drawn through points of equal range and may be called *corange lines.* In northern Siberia, the range is about 110 F° (61 C°), greatest of any place on earth. Next are north-central Canada, just west of Hudson Bay, and Greenland with ranges of 70 F° (40 C°). The Arctic Ocean and surrounding continental fringes have a comparable range. Then follow Africa, South America, and Australia, with maximum ranges of about 30 F° (17 C°). An equatorial belt, about 35° of latitude in width over the oceans and about 10° wide across Africa and South America, has an annual range of 5 F° (2.8 C°) or less.

Annual temperature range has environmental significance, for it is an indicator of thermal stress (or the lack of it) placed upon plants as well as upon man and the other animals. Thermal stress is small in the equatorial zone but severe in the subarctic zone. Climates having a small annual temperature range are described as *equable.*

Daily cycle of air temperature

When hourly air temperature readings are taken throughout a 24-hour period and plotted on a graph, the curve for a clear day typically shows one low point near sunrise and one high point in midafternoon with a fairly smooth curve throughout. This rhythmic rise and fall of air temperature is termed the *daily*, or *diurnal, air temperature cycle.*

Figure 8.25 relates the typical diurnal air temperature curve (bottom) with cycles of both incoming and outgoing heat energy. The diagrams are schematic and apply to a typical middle-latitude location (40° to 45° latitude). We assume equinox conditions (March 21 or September 23), with the time of sunrise and sunset as 6:00 A.M., and 6:00 P.M., respectively. The uppermost graph shows incoming shortwave radiation throughout a clear day. Disregarding small amounts of indirect shortwave energy before sunrise and after sunset, the curve begins at 6:00 A.M., reaches a maximum at noon, and ends at 6:00 P.M.

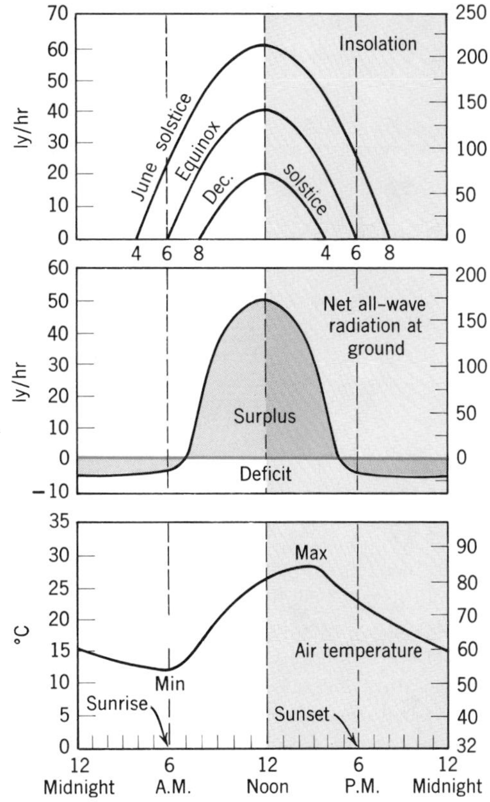

Figure 8.25 Relation of diurnal air temperature curve to insolation and net ground radiation, middle latitudes at equinox.

The middle graph shows the net all-wave radiation at the ground surface. This curve represents the same quantity shown in Figure 8.14 and has been explained earlier in this chapter. Where the curve is above the zero line, excess energy is passing upward from ground to atmosphere; where it is below the zero line, excess energy is passing from atmosphere to ground. These positive and negative values are labeled *surplus* and *deficit,* respectively. When a surplus is present, the temperature of the air layer above the ground tends to rise; when a deficit exists, the air tends to be cooled. The net all-wave radiation curve tends to be symmetrical with respect to noon (the maximum point) and to be nearly flat during hours of darkness. As the graph shows, a surplus normally sets in about one hour after sunrise and thereafter increases rapidly. A deficit sets in about an hour or so before sunset.

The typical diurnal air temperature curve, shown in the lowermost graph of Figure 8.25, is not symmetrical. The minimum point occurs at about sunrise. Temperature rises steeply as the radiation surplus increases rapidly. Air tempera-

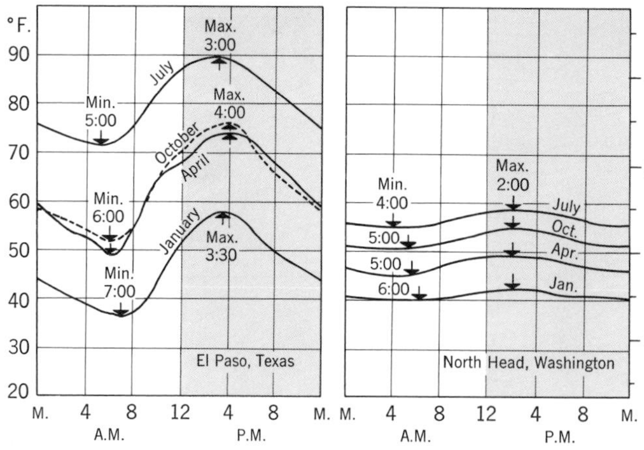

Figure 8.26 The average daily march of air temperatures for two stations. (After J. B. Kincer, U.S. Dept of Agriculture.)

ture continues to rise past noon, for the radiation surplus, although beginning to diminish, is still large. The air temperature maximum typically occurs between 2:00 and 4:00 P.M. Thereafter, the air temperature begins to fall, even though a radiation surplus exists, as the middle curve shows. If air temperature depended only upon ground radiation, the maximum temperature would occur later in the day, perhaps at about 5:00 P.M. under the equinox conditions illustrated. However, another factor has entered the picture. During the early afternoon, mixing of the lower air in turbulent motions increases in intensity, carrying the heated air upward and replacing it with cooler air. The effect of this mixing is to cause the air temperature curve to begin to drop long before the radiation surplus has ended.

Under conditions of June solstice, incoming solar radiation is greatly increased (upper graph in Figure 8.25); the insolation commences much earlier and ends much later. The surplus part of the net ground radiation curve (not shown) is similarly broadened and raised in height. The hour of minimum temperature is set back correspondingly to perhaps 4:00 A.M. However, the hour of maximum temperature remains essentially the same. At December solstice, corresponding reductions of incoming energy and a narrowing and lowering of the part of the surplus radiation curve occur. The time of minimum temperature is advanced correspondingly.

Figure 8.26 shows the average diurnal cycle of air temperature at two places, one of interior continental location in a dry climate, the other very close to the ocean water on a windward coast. Notice that the hour of minimum temperature changes with solstice and equinox, but that the hour of maximum temperature remains fairly constant. These two graphs show the contrasting effects of water and land in controlling the daily and annual temperature ranges.

Temperature inversion and frost

Although air temperature typically falls with increasing elevation, weather conditions at night in the lower air on land are typically such that, instead of falling, the temperature first rises with increasing height above ground before beginning to drop off into the normal lapse rate (Figure 8.27). This condition is termed *temperature inversion* and signifies that warmer air overlies colder air.

Low-level temperature inversion, or *ground inversion*, commonly results at night from rapid heat loss by radiation from the ground surface and basal air layer upward into space. Rapid radiation loss is favored by the presence of clear skies. Over snow-covered surfaces on clear winter nights, inversion is particularly marked. If the heat loss is great during a night in spring or early fall, the

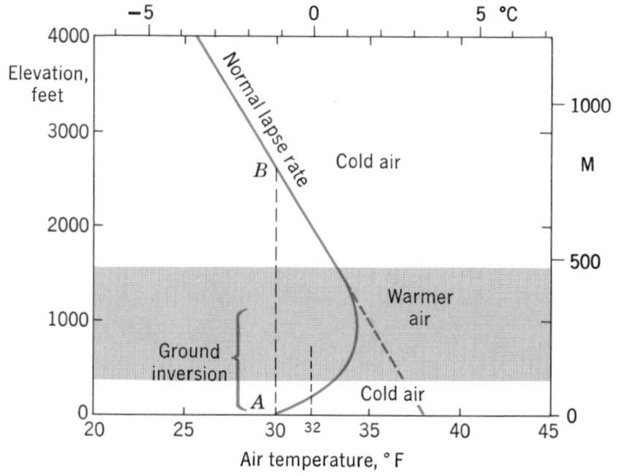

Figure 8.27 A low-level temperature inversion.

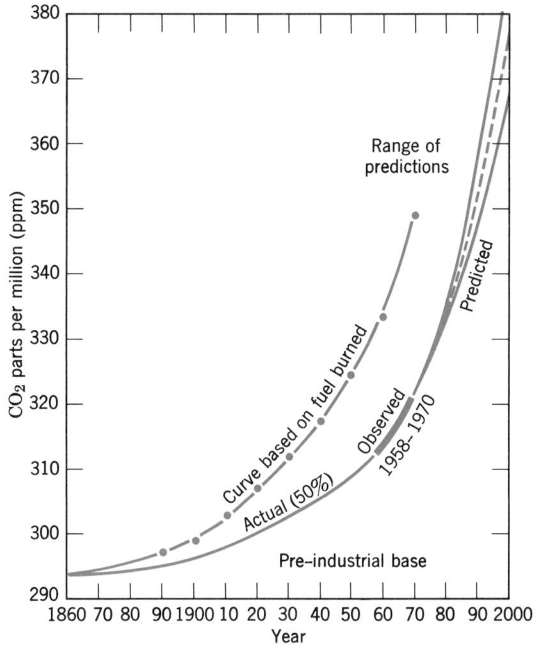

Figure 8.28 Increase in atmospheric carbon dioxide since 1860 with projections to the year 2000. (Data of L. Machta, from *Inadvertent Climate Modification*, SMIC, The MIT Press, Cambridge, Mass., p. 237, Figure 8.13.)

air temperature close to the ground may drop below freezing, resulting in a killing frost which damages sensitive crops. There are other causes and varieties of both temperature inversions and killing frosts, but these involve movements of masses of air and are not phenomena of heat radiation.

Killing frost may be prevented in orchards and citrus groves by causing a circulation of air that mixes the warmer air above with the cold-air layer near the ground. Where the cold-air layer is thin this effect may be accomplished by the use of powerful motor-driven propellors that circulate air much as does a fan in a room.

Carbon dioxide, dust, and global climate change

Atmospheric changes induced by man fall into four categories with respect to basic causes: (1) Changes in concentrations of the natural component gases of the lower atmosphere; (2) Changes in the water vapor content of the troposphere and stratosphere; (3) Alteration of surface characteristics of the lands and oceans in such a way as to change the interaction between the atmosphere and those surfaces; (4) Introduction of finely divided solid substances into the lower atmosphere, along with gases not normally found in substantial amounts in the unpolluted atmosphere. In this and following chapters we will give brief summaries of various facets of the total impact of man upon the atmosphere.

Atmospheric carbon, as carbon dioxide (CO_2), and molecular oxygen (O_2) are both involved in continuous cycles of interchange between atmosphere and oceans, soil and rock layers, and the biosphere. Carbon is linked with oxygen in CO_2 as a free gas in the atmosphere and in solution in ocean water and fresh water of lakes and streams, and in soil water. Both land and marine plants withdraw and use CO_2 to create carbohydrate compounds of plant tissues. Animals consume plants, releasing CO_2 back to the atmosphere in the process of biological oxidation (respiration). Decomposition of plant matter also releases CO_2 to the atmosphere.

From the long-range point of view, both carbon and oxygen (as CO_2 and H_2O) have been added throughout the earth's long history to the atmosphere by outgassing from the earth's interior. If it had not been for the unceasing work of plants in removing CO_2 and storing it as hydrocarbon compounds in the earth (coal, oil, natural gas), and of marine organisms in removing CO_2 and storing it in sediments as carbonate compounds, the earth would have an atmosphere rich in CO_2, perhaps like that of the planet Venus. Under preindustrial conditions of recent centuries, CO_2 added to the atmosphere by outgassing has been removed and stored in mineral form at a comparable rate, keeping the atmospheric content fixed at a level of roughly 0.03 percent by volume, or 300 parts per million (300 ppm). The problem is, of course, that man has recently begun to extract and burn hydrocarbon fuels at a rapid rate, releasing into the atmosphere the combustion products (principally H_2O and CO_2) and a great deal of heat, as well. How does this activity affect our environment?

During the past 110 years, there has been an increase in atmospheric CO_2 from about 295 parts per million (ppm) to a 1970 value of about 320 ppm, an increase of about 10 percent. Moreover, the rate of increase in this period, while slow at first, has become much more rapid toward the end of the period. This increase is shown by the lower curve of Figure 8.28. Now, we also have a fairly good evaluation of the quantity of hydrocarbon fuel burned during the same period, and from this we can calculate the increase in atmospheric CO_2 that would have resulted in the same period, if all of the additional CO_2 had remained in the atmosphere. It is estimated that 40 to 50 percent of the CO_2 produced has remained in the atmosphere. The excess CO_2 has been removed from the atmosphere, most of it going into solution in ocean waters, but some of it perhaps going into increased rates of plant growth. The upper curve

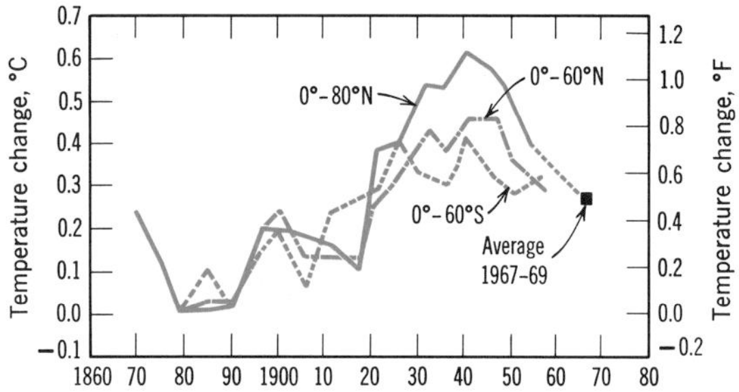

Figure 8.29 Observed changes in mean annual air temperatures between the equator and 80° N. lat. (Generalized from data of J. M. Mitchell, Jr., 1970, National Oceanographic and Atmospheric Administration.)

of Figure 8.28 shows estimated concentration of CO_2 if none had been absorbed by the oceans. It is based on an assumed absorption value of 50 percent of the industrial production.

The lower curve in Figure 8.28 shows one of several recent calculations of the rising atmospheric CO_2 content from 1860 to 1970. It is based in part on observations of actual CO_2 values recorded at Mauna Loa, Hawaii, in the period 1958 to 1970. Projection of this curve into the future shows a CO_2 value of 375 ppm in the year 2000. Other estimates place the 2000 value somewhat higher or lower, as shown by the solid lines in the upper right-hand part of Figure 8.28.

We turn next to consider the environmental effects to be anticipated from an increase of atmospheric CO_2. Because CO_2 is an absorber and emitter of longwave radiation, its presence in larger proportions will raise the level of absorption of both incoming and outgoing radiation, changing the energy balance so as to raise the average level of sensible heat in the atmospheric column. Thus, a general air temperature rise is the anticipated result. So, we shall want to look next at available temperature data to see if such an increase has occurred.

Figure 8.29 shows the observed change in mean hemispherical air temperature for approximately the past century. From 1920 to 1940, when fuel consumption was rising rapidly, the average temperature increased by about 0.6 F° (0.4 C°). This relationship follows the predicted pattern. However, since 1940, the graph shows a drop in temperature, despite the rising rate of fuel combustion. Assuming that the atmospheric warming because of increased CO_2 is a valid effect in principle, some other factor, working in the opposite direction, has entered the picture and its cooling effect has outweighed that of warming by CO_2.

When we calculate in terms of complete combustion of all of the estimated world supplies of hydrocarbon fuels, and make the assumption that 50 percent of the CO_2 thus produced is added to the atmospheric quantity, an average air temperature increase of between 2 F° and 4 F° (1.1 C° and 2.2 C°) is indicated, assuming this cause alone is acting. Recent calculations show that as CO_2 increases, its temperature-raising influence falls off, so that no really serious runaway heating effect need be feared.

Two feedback mechanisms may come into play as temperature rises. First, as seawater temperature increases, more CO_2 is released into the atmosphere. This mechanism would tend to accelerate the air temperature rise. Second, higher temperatures would increase evaporation and raise water vapor content of the atmosphere, in turn increasing cloudiness and the earth's albedo. With more solar radiation turned back into space, atmospheric temperatures would be lowered. These two mechanisms tend to counteract each other's effects in a self-regulatory, thermostatlike mechanism, and it is very difficult to predict the outcome.

Downturn in average air temperatures since about 1940 may be the result of greatly increased input of dust into the upper atmosphere by a number of major volcanic eruptions, the first of which was the 1947 eruption of the Icelandic volcano, Hekla. It is also suspected by some observers that man-made dusts from industrial sources have increased sufficiently to cause, or at least contribute to the temperature downturn. Increased fallout of dust from the lower troposphere is well documented and is related to industrial activities. An increase in volcanic dust at high elevations acts to increase the losses of insolation to space through scattering and diffuse reflection, with the result that less energy arrives at lower atmosphere levels. There follows a reduction in level of sensible heat of the lower atmosphere. At this point in time, it is not possible to decide the extent of man's activity in increasing atmospheric dust content at high levels, but the climatic impact may prove eventually to be important.

Fuel combustion and climate change

Molecular oxygen, O_2, moves in the reverse direction from CO_2 during exchanges between atmosphere and biosphere. Oxygen placed in long-term storage as hydrocarbon compounds in the earth (peat, coal, petroleum) and as carbonate sediments act so as to balance out the rate at which oxygen (as CO_2) is added to the atmosphere by outgassing. However, man's large-scale combustion of hydrocarbon fuels requires that a large quantity of O_2 be withdrawn from the atmosphere, to be converted with hydrocarbon into CO_2 and H_2O. Combustion releases heat as well. Now the question of environmental significance is, "What will be the impact of large-scale withdrawal of free oxygen and addition of heat?"

The possibility of a lowering of O_2 content of the atmosphere to levels deleterious or dangerous to animal life has been the subject of concern to a number of environmentalists in recent years. However, recent analysis of the oxygen balance and storage capacity have satisfactorily shown that there can be no oxygen depletion by these causes of sufficient magnitude to constitute a serious threat to life.

On the other hand, the increased production of heat by fuel combustion is an environmental impact to be reckoned with. At present, the added input of heat from power generation (fuel combustion and nuclear power) is an extremely tiny fraction of the longwave energy radiated from the earth, so that its effect on the planetary heat balance is negligible. When the existing rates of increase in power generation are projected into the future, an increase of about 2 F° (1 C°) in the earth's mean air temperature from the added heat of combustion might be expected in about 90 years. While the hydrocarbon fuels supply lasts, the warming effect of increased CO_2 can be expected to supplement the direct effect of power-generation heat.

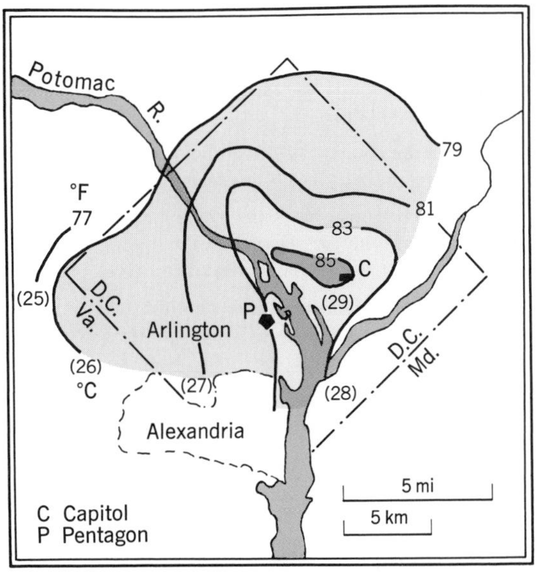

Figure 8.30 Heat island over Washington, D.C. Isotherms of air temperature at 10 P.M. local time on a day in early August. (Data of H. E. Landsberg, 1950, *Weatherwise*, Vol. 3, No. 1.)

Urbanization and the heat island

Applying the principles of the radiation and heat balances at the interface of the atmosphere and the solid ground surface, we can anticipate the impact of man as his cities spread, replacing a richly vegetated countryside with blacktop and concrete. Not only do the thermal properties of the surface change, but also hydrologic factors of evaporation and transpiration.

In the urban environment, the absorption of solar radiation causes higher ground temperatures for two reasons. First, foliage of plants is absent, so that the full quantity of solar energy falls upon the bare ground. Absence of foliage also means absence of transpiration (see Chapter 12), which elsewhere produces a cooling of the lower air layer. A second factor is that roofs and pavements of concrete and asphalt hold no moisture, so that evaporative cooling cannot occur as it would from a moist soil. The thermal effect is that of converting the city into a hot desert. The summer temperature cycle close to the pavement of a city may be almost as extreme as that of the desert floor. This surface heat is conducted into the ground and stored there. The thermal effects within a city are actually more intense than on a sandy desert floor, because the capacity of solid concrete, stone, or asphalt to conduct and hold heat is greater than that of loose, sandy soil. Because of more rapid conductivity, the solid materials absorb heat to a greater depth than loose, dry soil in a given period of heating. An additional thermal factor is that vertical masonry surfaces absorb insolation or reflect it to the ground and to other vertical surfaces. The absorbed heat is then radiated back into the air between buildings.

As a result of these changes in the radiation and heat balances, the central region of a city typically shows summer air temperatures several degrees higher than for the surrounding suburbs and countryside. Figure 8.30 is a map of the Washington, D.C. area, showing isotherms for a typical August afternoon. The isotherms delineate a *heat island.* You might suppose that at night the city air temperatures would fall below those of the surrounding countryside, since longwave radiation from bare, dry surfaces would cause rapid heat loss, as it does in the desert. Instead, we find that the heat island persists through the night because of the availability of a large quantity of heat stored in the ground during the daytime hours. In winter, additional heat is radiated by walls and roofs of buildings, which conduct heat from the inside. Even in summer, man adds to the city heat output through use of air conditioners, which expend enormous amounts of energy at a time when the outside air is at its warmest.

In succeeding chapters, we will make further observations of the impact of urbanization upon climate of the cities. Other weather phenomena, such as wind speeds, fog, clouds, and precipitation show measurable changes as a result of heat output and the injection of pollutants into the air.

REVIEW QUESTIONS

1. Describe the various facets of man's interaction with the atmosphere.

2. Explain the concept of the global radiation balance. How is it related to the global heat balance?

3. From what source does the earth's atmosphere receive its heat energy? Describe the solar radiation spectrum. What is the solar constant?

4. What factors determine the amount of insolation received each day at a given place on the earth? On the average, throughout the year, how is insolation related to latitude?

5. How does the occurrence of the seasons (changing declination of the sun) affect insolation in equatorial latitudes? in middle latitudes (40° to 50°)? in near-polar regions?

6. Name the world latitude zones in order from equator to poles. Give the approximate latitude span and insolation conditions for each zone.

7. Describe the losses in solar radiation in the troposphere. Why should insolation received at the ground surface vary from day to day?

8. How is the earth's atmosphere heated? What has ground radiation to do with atmospheric heating?

9. What is meant by net all-wave radiation? How does the average annual net radiation vary from equator to poles?

10. Describe the annual cycle of net all-wave radiation for locations in equatorial, tropical, middle-latitude, and subarctic zones.

11. Describe the typical daily cycle of net all-wave radiation. How does it vary with the seasons in middle latitudes?

12. Under what conditions is air temperature measured at standard stations? How do Fahrenheit and Centigrade temperature scales compare? Explain how temperature can be converted from one scale to the other.

13. Explain the basic differences of land and water surfaces as regards their properties for absorbing and transmitting insolation. How may these differences be expected to influence air temperatures over continental areas as contrasted with ocean areas?

14. What is the mean daily temperature? How is it computed?

15. How does the annual temperature curve for a middle-latitude place with an inland location differ from that of a place located on a seacoast? Is there a difference in the times at which maximum and minimum temperatures normally occur in these two locations? Why?

16. What are isotherms? How are they drawn? In what way are isothermal maps useful?

17. What is the general trend of isotherms on the globe? What influence has the changing sun's declination upon the isotherms of monthly mean temperatures? Is the latitudinal shift of isotherms greater over land or water areas? Explain your answer.

18. What effect have the land masses of North America and Asia upon the isotherms for January and July? Where is the earth's greatest annual temperature range experienced? Where would you expect to find a minimum annual range of temperature?

19. Describe the normal daily temperature curve. When are maximum and minimum air temperatures normally reached? How is the air temperature cycle related to the cycle of insolation and to net all-wave ground radiation? Describe the effect of changing seasons in middle latitudes upon the time of minimum daily temperature.

20. What is temperature inversion? Explain how a ground inversion occurs. How are killing frosts produced by radiation of heat?

21. What has caused a rise in level of atmospheric carbon dioxide? What climatic effect may be expected to result?

22. How have average global atmospheric temperatures changed in the past century? What cause or causes may have been responsible for these changes?

23. What impact does the combustion of fossil fuels have upon the atmosphere?

24. Explain how urbanization alters the radiation and heat balances of a city. What is the *heat island?*

Exercises

1. **(*a*)** Taking into account only latitude, and considering insolation on a unit area of horizontal ground surface at the equator to represent 100 percent, what percentage of this insolation will be received by a similar unit area of ground surface at 30° N lat. **(*b*)** at 45° S? **(*c*)** at 60° S? **(*d*)** at 90° N? These answers will hold true only for equinox conditions.

2. **(*a*)** Using Angot's insolation in the table below, prepare graphs to show the changes in insolation throughout the year at the equator, 20° N, 40° N, 60° N, and the north pole. Compare with Figure 8.5. **(*b*)** Why does the south pole receive more insolation in January than the north pole receives in July? **(*c*)** Why is the yearly total for 90° S less than for 90° N?

ANGOT'S INSOLATION TABLE

(To accompany Exercise 2)

Insolation received during each month of the year at various latitudes, assuming a completely transparent atmosphere. The value of one unit used in this table is 889 gram calories per square centimeter, which is the quantity of radiation received at the equator in one day at equinox. (Data from Brunt, *Physical and Dynamical Meteorology*, Cambridge University Press, 1939.)

	Latitude	Jan.	Feb.	Mar.	Apr.	May	Jun.	Jul.	Aug.	Sep.	Oct.	Nov.	Dec.	Total for year
	90	...	...	1.9	17.5	31.5	36.4	32.9	21.1	4.6	...	...	...	145.9
	80	...	0.1	5.0	17.5	30.5	35.8	32.4	20.9	7.4	0.6	...	...	150.2
°N	60	3.0	7.4	14.8	23.2	30.2	33.2	31.1	24.9	16.7	9.0	3.8	1.9	199.2
	40	12.5	17.0	23.1	28.6	32.4	33.8	32.8	29.4	24.3	18.4	13.4	11.1	276.8
	20	22.0	25.1	28.6	30.9	31.8	32.0	31.8	30.9	28.9	25.8	22.5	20.9	331.2
Equator		29.4	30.4	30.6	29.6	28.0	27.1	27.6	28.6	30.1	30.2	29.5	28.9	350.0
	20	33.8	32.2	29.0	24.9	21.2	19.6	20.5	23.7	27.7	31.1	33.3	34.1	331.1
°S	40	34.8	30.4	23.9	17.4	12.5	10.4	11.6	15.8	21.9	28.5	33.6	36.0	276.8
	60	33.0	25.3	16.0	8.1	3.3	1.7	2.7	6.5	13.6	22.6	31.1	35.3	199.2
	80	34.2	20.5	6.3	0.3	...	...	...	...	3.8	16.0	31.0	38.1	150.2
	90	34.7	20.7	3.2	...	...	...	...	...	1.0	15.6	31.5	38.7	145.4

3. Convert the following Fahrenheit temperatures to Centigrade: 32°, 0°, 212°, 71°, −22°, 44°.

4. Convert the following Centigrade temperatures to Fahrenheit: 0°, 100°, 32°, 11°, 37°, −40°.

5. The figures given below are mean monthly temperatures (°F) for Buenos Aires, Argentina, located at 35° S lat. (Data from Trewartha.) **(*a*)** Prepare a graph similar to Figure 8.20 to illustrate the annual temperature march at this place. Scale the graph both in Fahrenheit and Centigrade scales. **(*b*)** What is the highest monthly average? **(*c*)** In what month does it fall? **(*d*)** What is the lowest monthly average? **(*e*)** In what month does it fall? **(*f*)** What is the annual range?

January 74°F	May 55°F	September 55°F
February 73°F	June 50°F	October 60°F
March 69°F	July 49°F	November 66°F
April 61°F	August 51°F	December 71°F

6. On a sheet of tracing paper complete the drawing of isotherms on Figure 8.21. Then draw in red pencil 5° isotherms on the Centigrade temperature scale. This may be done by first converting the desired Centigrade temperature (i.e., 5°C) to Fahrenheit and interpolating this isotherm between the lines already drawn.

7. Prepare a temperature graph similar to Figure 8.26 showing temperatures during the 24-hour period for which hourly data are listed in the table. Scale the graph in Fahrenheit units on the left, and in Centigrade units on the right. After plotting the points, draw a smooth temperature curve. Label maximum and minimum points.

BISMARCK, North Dakota, July Hourly Averages

(Data from U.S. Dept. of Agriculture)

12 midnight	66.5°F	1 P.M.	76.5°F
1 A.M.	64.5°F	2 P.M.	77.5°F
2 A.M.	63.0°F	3 P.M.	78.5°F
3 A.M.	61.5°F	4 P.M.	80.0°F
4 A.M.	60.0°F	5 P.M.	80.0°F
5 A.M.	59.0°F	6 P.M.	79.0°F
6 A.M.	58.0°F	7 P.M.	78.0°F
7 A.M.	58.5°F	8 P.M.	77.0°F
8 A.M.	61.0°F	9 P.M.	74.0°F
9 A.M.	64.0°F	10 P.M.	70.5°F
10 A.M.	67.0°F	11 P.M.	68.0°F
11 A.M.	72.0°F	12 midnight	66.5°F
12 noon	74.0°F		

CHAPTER 9

Winds and the Global Circulation

MAN'S environment at the earth's surface depends for its quality as much upon atmospheric motions as it does upon the flow of heat by radiation and conduction. Air in motion represents a form of kinetic energy. In the form of very strong winds—as in hurricanes and tornadoes—air in motion is a severe environmental hazard. Winds also transfer energy to the surface of the sea, as wind-driven waves. Wave energy in turn travels to the shores of continents, where it is transformed into surf and coastal currents capable of reshaping the coastline.

But air in motion has another, more basic role to play in the planetary environment. Large-scale air circulation transports heat, both as sensible heat and as latent heat present in water vapor. Because of the global energy imbalance—a surplus in low latitudes and a deficit in high latitudes—atmospheric circulation must transport heat across the parallels of latitude from the region of deficit to the regions of surplus. Figure 9.1 shows this transport in schematic form. Notice that circulation of ocean waters is also involved in transport of sensible heat. But this oceanic circulation is a secondary mechanism, driven largely by surface winds.

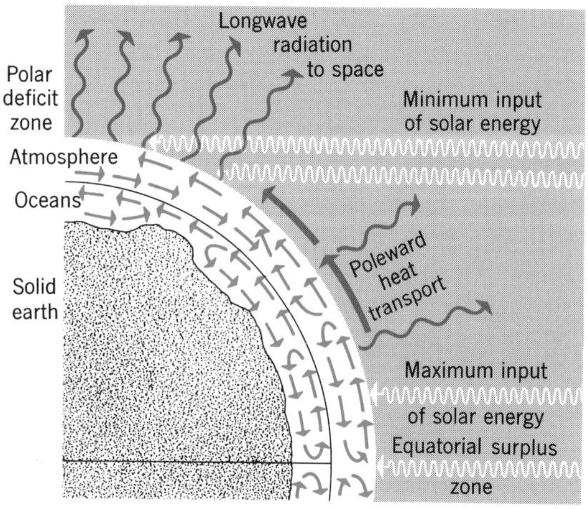

Figure 9.1 The global heat balance.

Equally important to man's environment are the rising and sinking motions of air. We shall find in Chapter 10 that precipitation, the source of all fresh water on the lands, requires massive lifting of large bodies of air heavily charged with water vapor. Conversely, large-scale sinking motions in the atmosphere lead to aridity and the occurrence of deserts. In this way the earth's surface becomes differentiated into regions of ample fresh water and regions of water scarcity.

With these broad generalizations in mind, we turn to examine the forces that set the atmosphere in motion and drive the ceaseless global circuits of air circulation.

Winds and the pressure gradient force

Winds are dominantly horizontal air motions with respect to the earth's surface. (Dominantly vertical air motions are referred to by other terms, such as *updrafts* or *downdrafts*.) To explain winds, we must first consider barometric pressure and its variations from place to place.

In Chapter 7, we found that barometric pressure falls with increasing altitude above the earth's surface, the rate of pressure decrease becoming less as we ascend. For an atmosphere at rest, the barometric pressure will be the same within a given horizontal surface at any chosen altitude above sea level. Thus, the surfaces of equal barometric pressure, or *isobaric surfaces*, are horizontal; in a cross section of a portion of the atmosphere at rest, these surfaces appear as horizontal lines, as shown in Figure 9.2*A*. (For convenience, we have shown surface pressure as 1000 mb.)

Suppose, now, that the rate of upward pressure decrease is more rapid in one place than another, as shown in Figure 9.2*B*. As we proceed from left to right across the diagram, the upward rate of pressure decrease is more rapid. The isobaric surfaces now slope down toward the right. At a selected altitude, say 1000 m (color line), barometric pressure declines from left to right. As shown in Diagram *C*, which is a type of map, the 1000-m horizontal surface cuts across successive isobaric surfaces. The trace of each isobaric surface is a

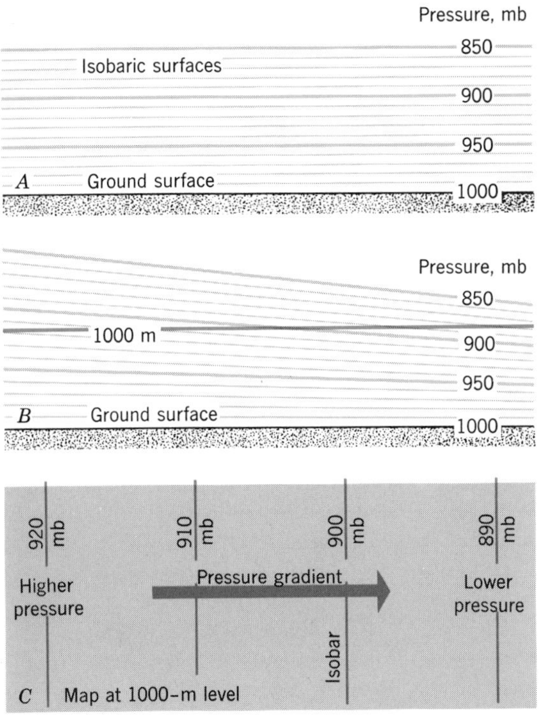

Figure 9.2 Isobaric surfaces and the pressure gradient.

line, known as an *isobar.* The isobar is thus a line showing the location on a map of all points having the same barometric pressure.

The change in barometric pressure across a horizontal surface constitutes a *pressure gradient,* its direction indicated by the broad arrow in Diagram *C*. The gradient is in the direction from higher pressure (at the left) to lower pressure (at the right). One can think of the sloping barometric surface as a sloping hillside; the downward slope of the ground surface is analogous to the pressure gradient.

Where a pressure gradient exists, air molecules tend to drift in the same direction as that gradient. This tendency for mass movement of the air is referred to as the *pressure gradient force.* The magnitude of the force is directly proportional to the steepness of the gradient. Wind is thus the horizontal motion of air in response to the pressure gradient force.

Sea and land breezes

Perhaps the simplest example of the relationship of wind to the pressure gradient force is a common phenomenon of coasts—the sea breeze and land breeze, illustrated in Figure 9.3. Diagram *A* shows the initial situation in which no pressure gradient exists. During the daytime, more rapid heating of the lower air layer over the land than over the ocean causes a pressure gradient from sea to land (Diagram *B*). Air moving landward in response to this gradient from higher to lower pressure constitutes the *sea breeze.* At higher levels a reverse flow sets in. Together with weak rising and sinking air motions, a complete flow circuit is formed. During the night, when radiational cooling of the land is rapid, the lower air becomes colder over land than over the water. Higher pressure now develops over land and the barometric gradient is reversed. Air now moves from land to sea as a *land breeze* (Diagram *C*).

This illustration shows that a pressure gradient can be developed through unequal heating or cooling of a layer of the atmosphere. Air that is warmed also expands and becomes less dense. Air that is cooled contracts and becomes denser. The upward change in barometric pressure is then more rapid within the cooler air layer than within the warmer layer. Heat energy drives the circulation system by changing air densities and setting up barometric pressure gradients. The entire mechanism is often described as a *heat engine,* since kinetic energy of air motion is derived from the input of heat. This type of air circulation is also important on a large scale in the atmosphere.

The Coriolis force and its effect on winds

If the earth did not rotate upon its axis, winds would follow the direction of pressure gradient. Instead, earth rotation produces another force, the *Coriolis force,* which tends to turn the flow of air. The direction of action of the Coriolis force is stated in Ferrel's law: Any object or fluid moving horizontally in the northern hemisphere tends to be deflected to the right of its path of motion, regardless of the compass direction of the path.

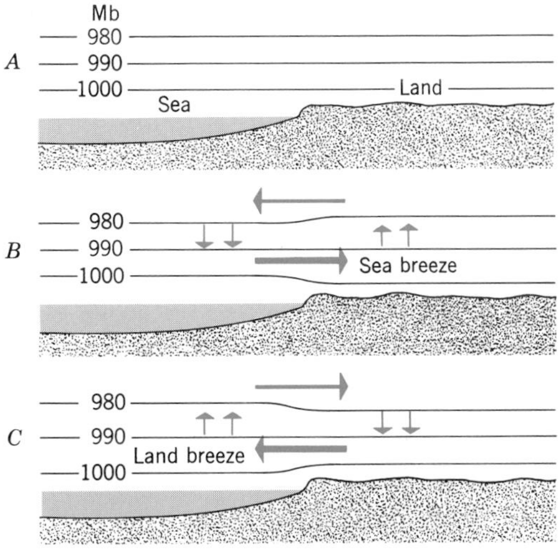

Figure 9.3 Sea breeze and land breeze.

In the southern hemisphere a similar deflection is toward the left of the path of motion. The Coriolis force is absent at the equator but increases progressively poleward.

On Figure 9.4 the various arrows show how an initial straight line of motion is modified by the deflective force. Note especially that the compass direction is not of any consequence. If we face down the direction of motion, turning will always be toward the right hand in the northern hemisphere. Because the deflective force is very weak, it is normally apparent only in freely moving fluids such as air or water. Ocean current patterns are greatly affected by it, and streams occasionally will show a tendency to undercut their right-hand banks in the northern hemisphere. Driftwood floating in rivers at high northerly latitudes concentrates along the right-hand edge of the stream. Rifle bullets are slightly deflected over long ranges.

Applying these principles to the relation of winds to pressure (Figure 9.5), the gradient force (acting in the direction of the pressure gradient) and the Coriolis force (acting to the right of the path of flow) quickly reach a balance, or equilibrium, when the wind has been turned to the point that it flows in a direction at right angles to the pressure gradient, that is, parallel with the isobars. The ideal wind in this state of balance with respect to the two forces is termed the *geostrophic wind* for cases in which the isobars are straight. Where isobars are curved, centrifugal force must also be taken into account, but, in general, air flow at high altitudes parallels the isobars (Figure 9.5). The rule for the relation of wind to pressure in the northern hemisphere is known as *Ballot's law;* it states: Stand with your back to the wind and the low pressure will be on your left, the high on your right.

Near the earth's surface, at levels from the ground upward to about 2000 or 3000 ft (600 or 900 m), yet another force modifies the wind direction. This is the force of friction of the air with the ground. It acts in such a way as to counteract in part the Coriolis force and to prevent the wind from being deflected until parallel with the isobars. Instead, the wind blows obliquely across the isobars, the angle being from 20° to 45°. Figure 9.6 illustrates surface winds and is typical of conditions found on the surface weather map. The angle is large for rugged terrain; small for smooth surfaces, such as water or a flat plain. Wind speed is reduced in proportion to ground friction.

Figure 9.6 also illustrates the principle that wind speeds are, in general, proportional to steepness of the pressure gradient. Where the isobars are closely spaced, winds are strong; where isobars are far apart, winds are weak. (This principle can be verified by examining the daily surface weather

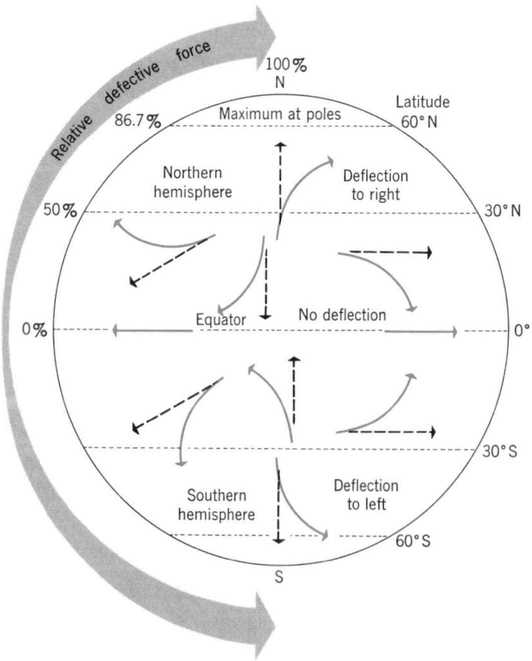

Figure 9.4 Deflective force of the earth's rotation.

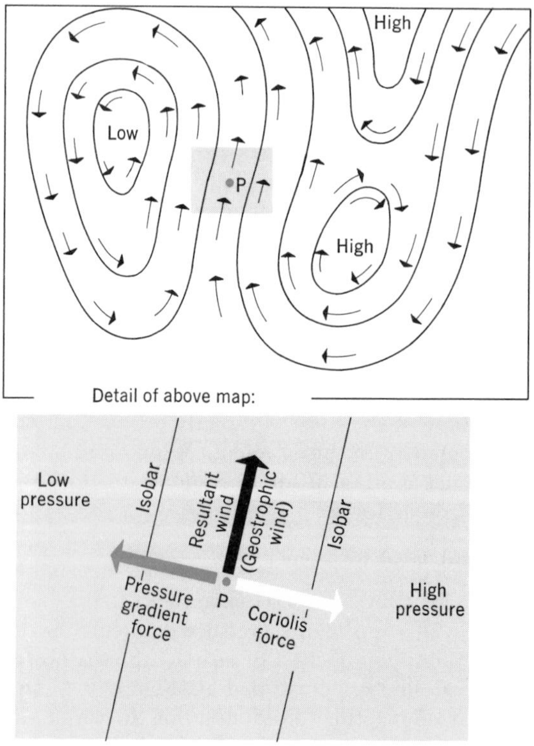

Figure 9.5 Wind follows isobars at high levels.

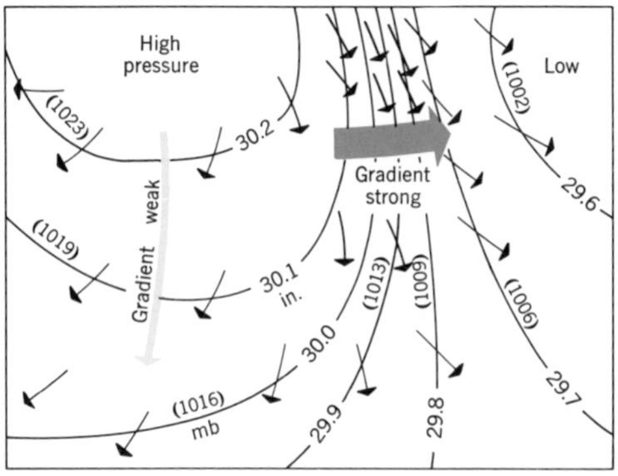

Figure 9.6 Relation of winds to isobars.

map, Figure 11.8*B*. An explanation of the wind symbols is given in Figure 11.9.) Notice that for curved isobars, pressure gradient follows a curved line as well, because direction of the pressure gradient is always at right angles to the isobars.

Cyclones and anticyclones

In the language of meteorology and climatology, a center of low pressure is designated a *cyclone;* a center of high pressure is an *anticyclone.* Cyclones and anticyclones may be of the stationary, or semipermanent type, or they may be rapidly moving pressure centers such as create the weather disturbances described in Chapter 11.

As shown in Figure 9.5 air motion at high altitudes, where surface friction does not act, parallels the isobars. We can thus give the rule that in the northern hemisphere winds move anticlockwise (counterclockwise) about a cyclone; clockwise about an anticyclone. In the southern hemisphere these patterns are exactly reverse: clockwise about a cyclone; anticlockwise about an anticyclone. The simple upper-air map in Figure 9.7 shows how these rules apply.

For surface winds, which move obliquely across the isobars, the systems for cyclones and anticyclones in both hemispheres are shown in Figure 9.8. Winds in a cyclone in the northern hemisphere show an anticlockwise inspiral; in an anticyclone, a clockwise outspiral. Note the reversal between the labels "anticlockwise" and "clockwise" in the southern hemisphere. In both hemispheres the surface winds spiral inward upon the center of the cyclone, hence the air is *converging* upon the center and must also rise to be disposed of. For the anticyclone, by contrast, surface winds spiral out from the center, which represents a *diverging* of air flow and must be accompanied by a sinking (subsidence) of air in the center of the anticyclone to replace the outmoving air.

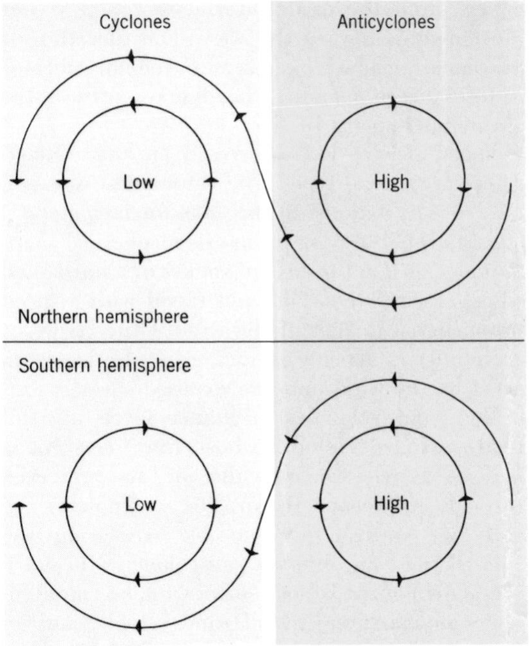

Figure 9.7 Winds at high levels around cyclones and anticyclones.

Measurement of winds

A description of winds requires measurement of two quantities: direction and speed. Direction is easily determined by a *wind vane,* one of the commonest of the amateur weather instruments. Wind direction is stated in terms of the direction from which the wind is coming. Thus an east wind

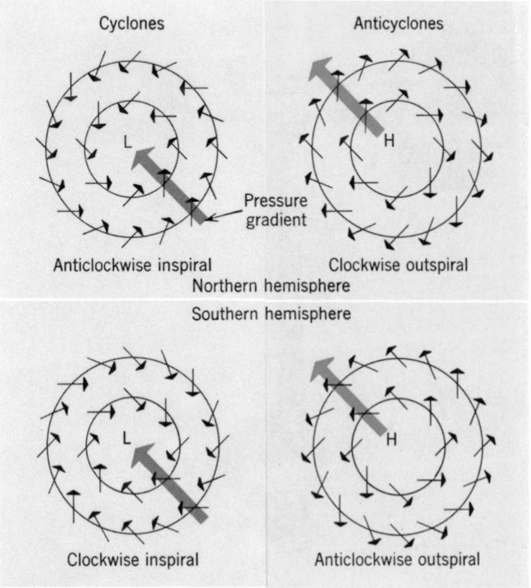

Figure 9.8 Surface winds within cyclones and anticyclones.

comes from the east, but the direction of air movement is toward the west. The direction of movement of low clouds is an excellent indicator of wind direction and can be observed without the aid of instruments.

Speed of wind is measured by an *anemometer.* There are several types. The commonest one seen at weather stations is the *cup anemometer.* It consists of three or four hemispherical cups mounted as if at the ends of spokes of a horizontal wheel (Figure 9.9). The cups travel with a speed proportional to that of the wind. Other types of anemometers depend on measuring the force exerted by the wind upon an exposed surface.

For wind velocities at higher levels a small hydrogen-filled balloon, whose rate of ascent is known, is released into the air and observed through a telescope (Figure 9.10). Knowing the balloon's vertical position by measuring the elapsed time, an observer can calculate the horizontal drift of the balloon downwind. For modern, upper-air measurements of wind velocity and direction the balloon carries a target that reflects radar waves and can thus be followed when the sky is overcast.

For wind observations collected over a long period of time, a device known as a *wind rose* is constructed. One is illustrated in Figure 9.11. Wind directions are reduced to eight compass sectors shown by lines radiating from the central point. The percentage of total length of time during which wind blows from these sectors of the compass is indicated by length of the lines. On charts showing wind roses, there is printed

Figure 9.9 A cup-type anemometer. (National Weather Service.)

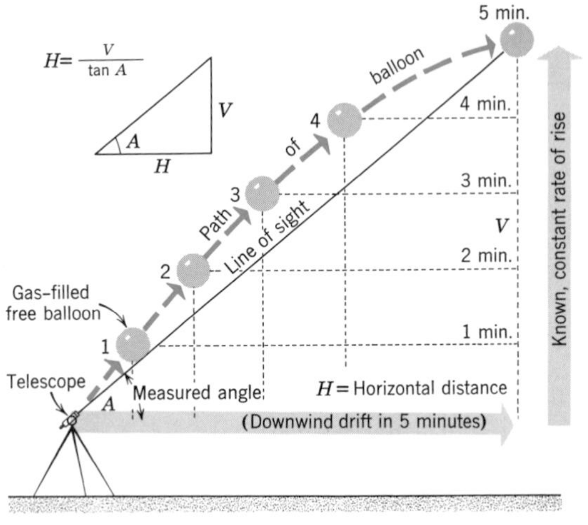

Figure 9.10 The velocity of winds aloft can be calculated by following the path of a free gas-filled balloon with a telescopic instrument.

on the margin a scale from which the percentages may be measured. Small feather lines attached to the ends of the radial lines are used to show the average wind speed during the period for which the wind rose is constructed.

A scale of numbers was long used in English-speaking countries to state the strength of wind. This is the *Beaufort scale,* devised by Admiral Sir Francis Beaufort of the British navy in 1806. It consists of numbers ranging from 0 to 12. Table 9.1 gives a description of this scale, together with common evidences of the various wind strengths and the velocities in miles per hour.

The Beaufort force scale of numbers has now been largely replaced by a direct statement of wind velocity in knots.

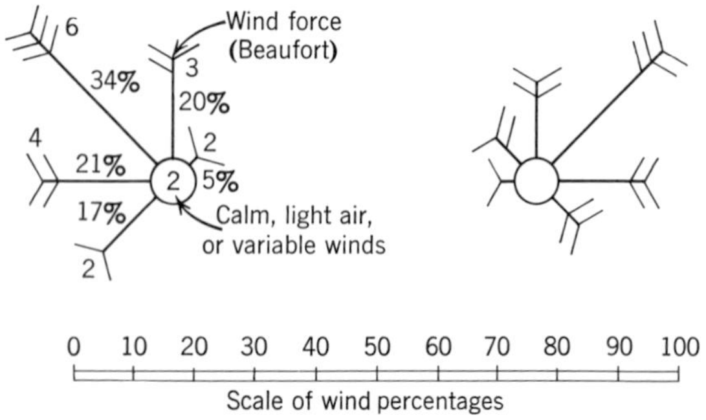

Figure 9.11 On U.S. Navy Oceanographic Office pilot charts of the oceans, a wind rose is used to show the average of wind durations and strengths for each 5° square of latitude and longitude.

TABLE 9.1 THE BEAUFORT SCALE OF WINDS*

Beaufort Number	Name of Wind	Observable Features	Velocity in Miles per Hour, 20 ft Above Ground	Equivalents in Kilometers per Hour
0	Calm	Smoke rises vertically.	Less than 1	Less than 1.6
1	Light air	Smoke drifts downwind. Wind does not move wind vane.	1 to 3	1.6 to 4.8
2	Light breeze	Wind felt on face; leaves rustle. Vane moved by wind.	4 to 7	6.4 to 11.3
3	Gentle breeze	Leaves and twigs in constant motion; wind extends light flag.	8 to 12	12.9 to 19.3
4	Moderate breeze	Raises dust and loose paper; small branches are moved.	13 to 18	20.9 to 29.0
5	Fresh breeze	Small trees in leaf begin to sway; crested wavelets form on inland waters.	19 to 24	30.6 to 38.6
6	Strong breeze	Large branches in motion; whistling heard in telegraph wires; umbrellas used with difficulty.	25 to 31	40.2 to 49.9
7	Moderate gale	Whole trees in motion; inconvenience felt in walking against wind.	32 to 38	51.5 to 61.1
8	Fresh gale	Twigs break off trees; progress generally impeded.	39 to 46	62.8 to 74.0
9	Strong gale	Slight structural damage occurs (chimney pots and slate removed).	47 to 54	75.6 to 86.9
10	Whole gale	Seldom experienced inland; trees uprooted; considerable structural damage occurs.	55 to 63	88.5 to 101.4
11	Storm	Very rarely experienced; accompanied by widespread destruction.	64 to 75	103.0 to 120.7
12	Hurricane		Above 75	Above 120.7

*After U.S. National Weather Service.

Global distribution of surface pressure systems

To understand the earth's surface wind systems, we must study the global system of barometric pressure distribution. Once we grasp the patterns of isobars and pressure gradients, the prevailing or average winds can be predicted.

World isobaric maps are constructed to show average pressures and winds for the two months of seasonal temperature extremes over large landmasses—January and July (Figures 9.12 and 9.13). Because observing stations lie at various altitudes above sea level, their barometric readings must be reduced to sea level equivalents, using the standard rate of pressure change with altitude, explained in Chapter 7. When this has been done, and the daily readings are averaged over long periods of time, small but distinct pressure differences remain from place to place.

If 1013 mb (29.92 in., 76 cm) is taken as standard sea-level pressure, readings higher than this will frequently be observed in middle latitudes, occasionally up to 1040 mb (30.7 in., 78 cm) or higher. These pressures are designated as *high*. Pressures ranging down to 982 mb (29 in., 74 cm) or below are *low*.

Figure 9.12 (*following pages*) Average January and July barometric pressures (millibars, reduced to sea level) and surface winds. Wind arrows in polar regions largely inferred from isobars. (Compiled by John E. Oliver from data by Y. Mintz, G. Dean, R. Geiger, and J. Blüthagen. Cartography by John Tremblay.)

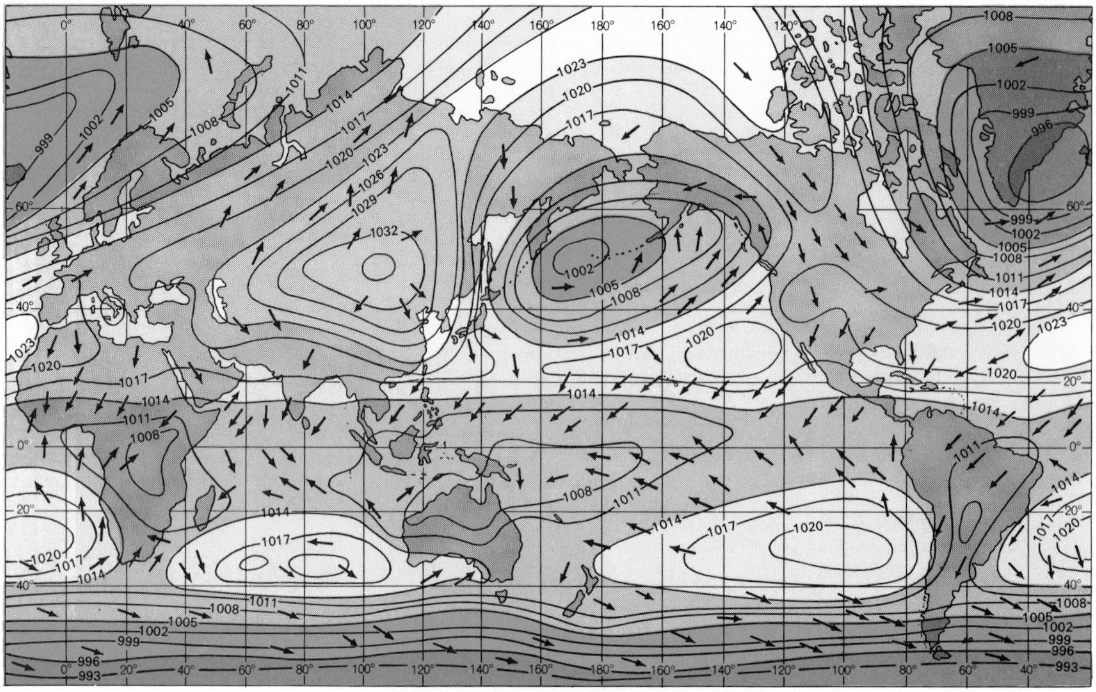

Figure 9.12, ***left***

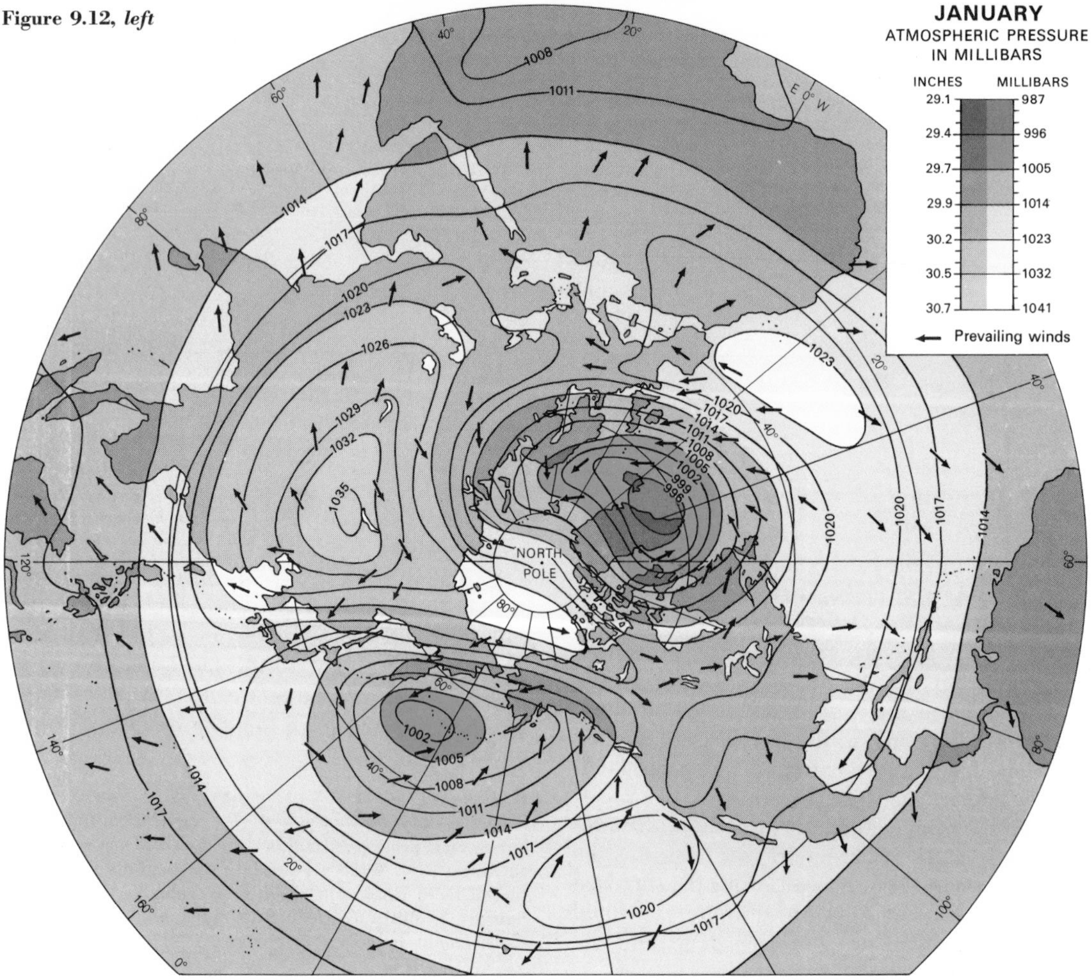

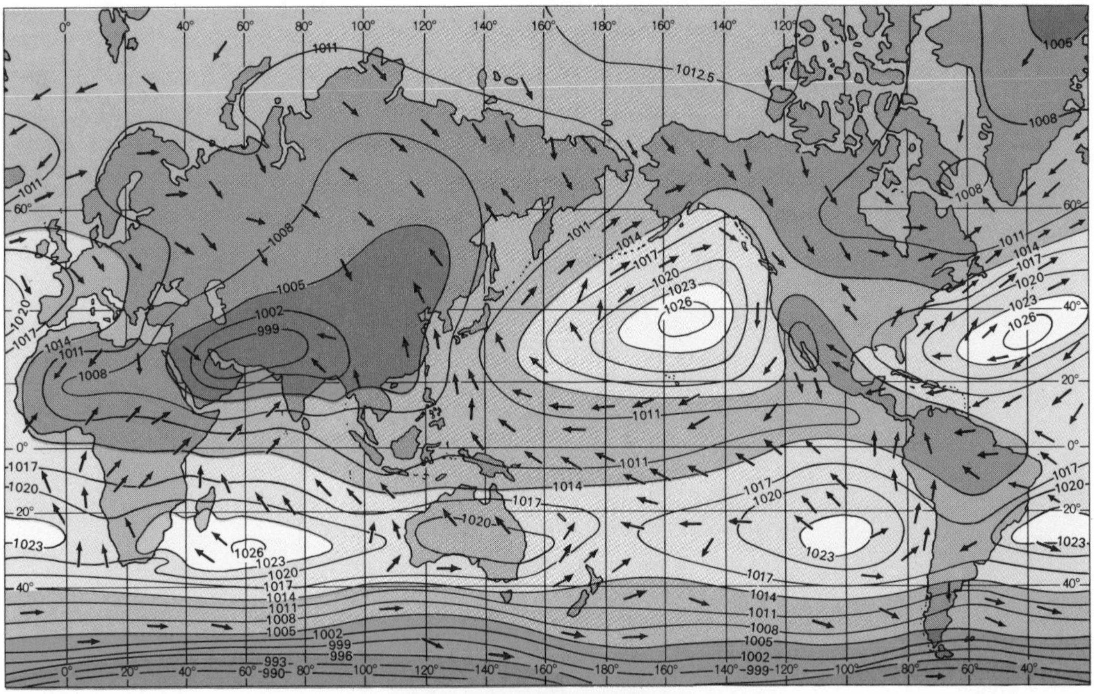

Figure 9.12, *right*

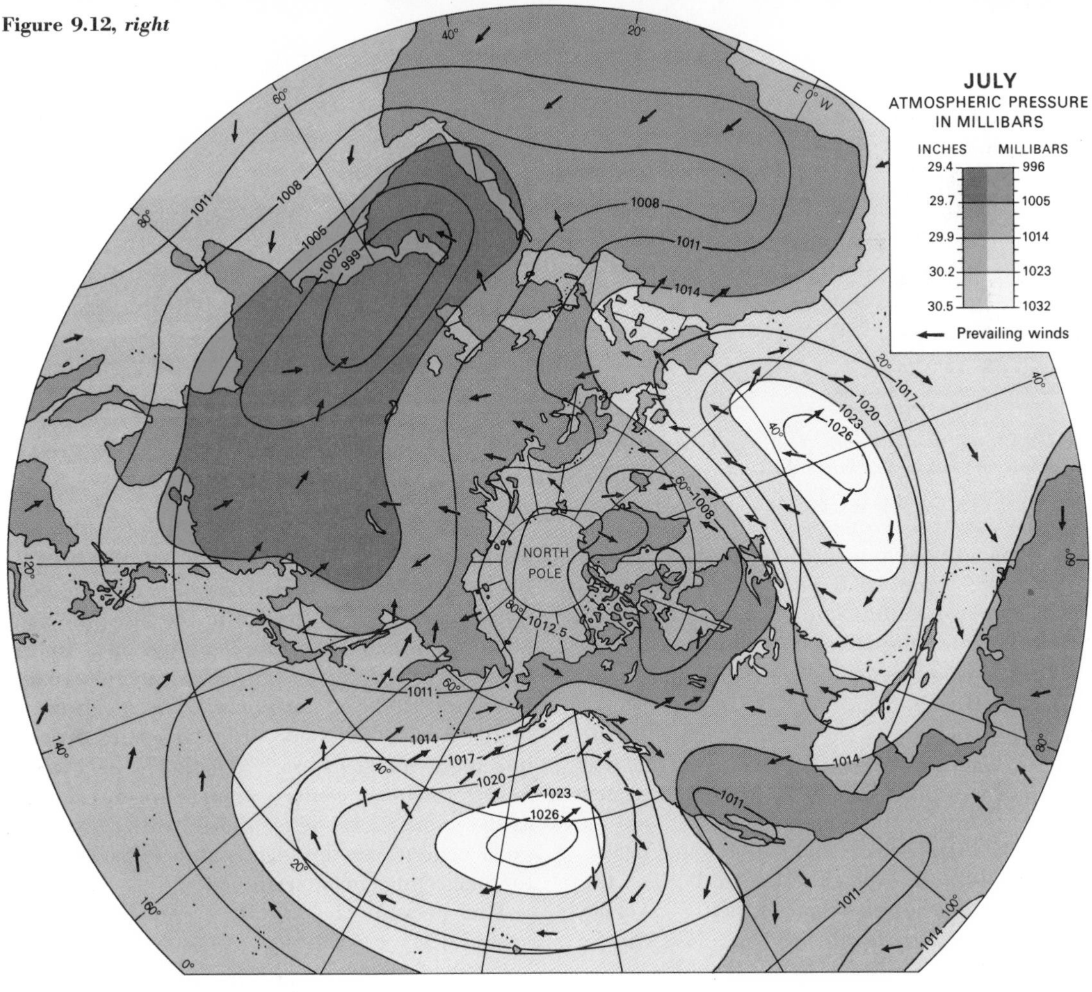

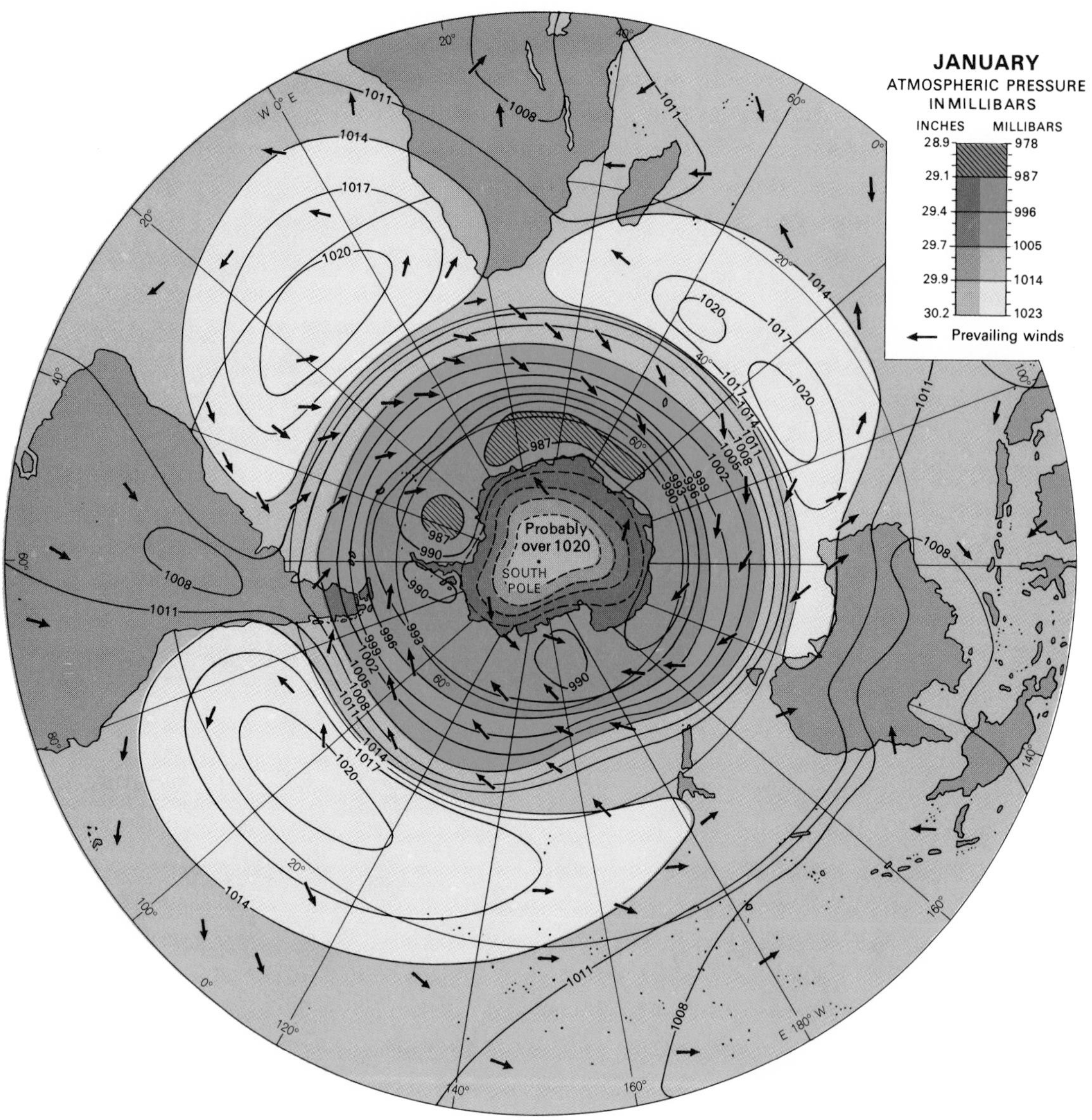

Figure 9.13 (*above and right*) Southern hemisphere pressures and winds, January and July. (Same data sources as in Figure 9.12)

In the equatorial zone is a belt of somewhat lower than normal pressure, between 1011 and 1008 mb (29.9 and 29.8 in., 76 and 75.7 cm), which is known as the *equatorial trough.* Lower pressure is made conspicuous by contrast with belts of higher pressure lying to the north and south and centered on about latitudes 30° N and S. These are the *subtropical belts of high pressure.* In the southern hemisphere this belt is clearly defined but contains centers of high pressure, termed *pressure cells.* In the northern hemisphere in summer the high-pressure belt is dominated by two oceanic cells, one over the eastern Pacific, the other over the eastern North Atlantic. Average pressures exceed 1026 mb (30.3 in., 77.0 cm) in the centers of the cells.

Poleward of the subtropical high-pressure belts are broad belts of low pressure, extending roughly from the middle-latitude zone to the arctic zone but centered and intensified in the subarctic zone at about the 60th parallels of latitude. In the southern hemisphere, over the continuous expanse of southern ocean the *subantarctic low-pressure belt* is especially defined with average pressures as low as 984 mb (29.1 in., 73.9 cm). The polar zones have permanent centers of high pressure known as the *polar highs,* better illustrated by the south polar zone where the high contrasts strongly with the encircling subantarctic low.

The pressure belts shift seasonally through several degrees of latitude, just as do the isotherm belts that accompany them.

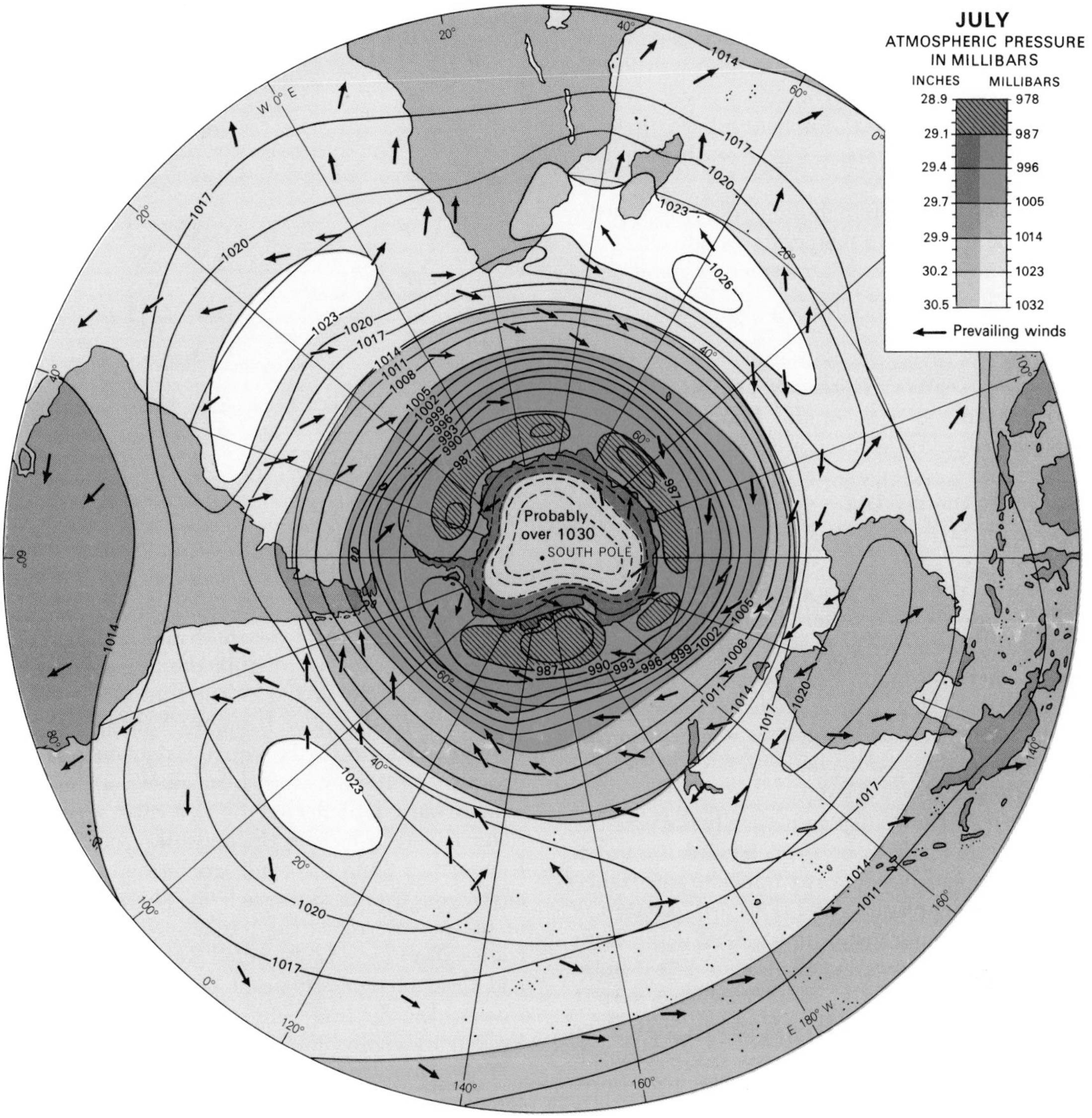

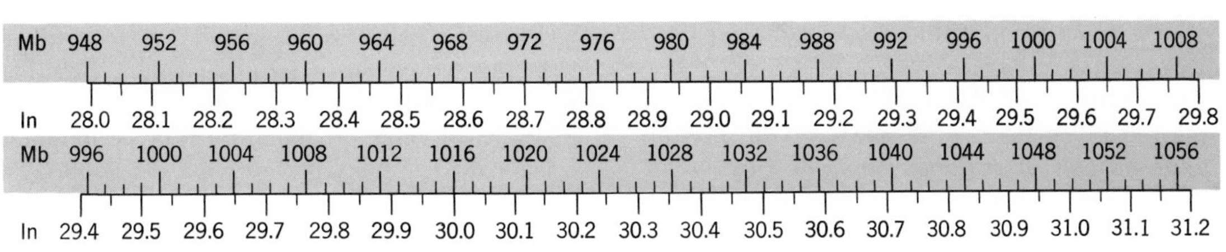

Figure 9.14 Scale for conversion of millibars to inches.

Northern hemisphere pressure centers

The vast landmasses of North America and Asia, separated by the North Atlantic and North Pacific oceans, exert such a powerful control over pressure conditions in the northern hemisphere that the belted arrangement typical of the southern hemisphere is absent.

Land areas develop high-pressure centers at the same time that winter temperatures fall far below those of adjacent oceans. In summer, land areas develop low-pressure centers, at which season land-surface temperatures rise sharply above temperatures over the adjoining oceans. Ocean areas show centers of pressure opposite to those on the lands, as seen in the January and July isobaric maps. In winter, pressure contrasts are greater, just as temperature contrasts are greater. Over north central Asia is developed the *Siberian high,* with pressure average exceeding 1035 mb (30.6 in., 77.7 cm). Over central North America is a clearly defined, but much less intense, ridge of high pressure, called the *Canadian high.* Over the oceans are the *Aleutian low* and the *Icelandic low,* named after the localities over which they are centered. These two low-pressure areas have much cloudy, stormy weather in winter, whereas the continental highs characteristically have a large proportion of clear, dry days.

Figure 9.15 shows diagrammatically the pressure centers as they appear grouped around the north pole. Highs and lows occupy opposite quadrants.

In summer, pressure conditions are exactly the opposite of winter conditions. Asia and North America develop lows, but the low in Asia is more intense. It is centered in southern Asia where it is fused with the equatorial low-pressure belt. Over the Atlantic and Pacific oceans are two well-developed cells of the subtropical belt of high pressure, shifted northward of their winter position and considerably expanded. These are termed the *Azores* (or *Bermuda*) *high* and the *Hawaiian high* respectively.

Earth's surface wind systems

Prevailing surface winds during the months of January and July are suggested by arrows on the pressure maps, Figures 9.12 and 9.13. A highly diagrammatic representation of the wind systems in Figure 9.16 shows the earth as if no land areas existed to modify the belted arrangement of pressure zones.

Over the equatorial trough of low pressure, lying roughly between 5° S and 5° N lat., is the *equatorial belt of variable winds and calms,* or the *doldrums.* There are no prevailing surface winds here, but a fair distribution of directions around the compass. Calms prevail as much as a third of the time. Centrally located on a belt of low pressure, this zone has no strong pressure gradients to induce a persistent flow of wind.

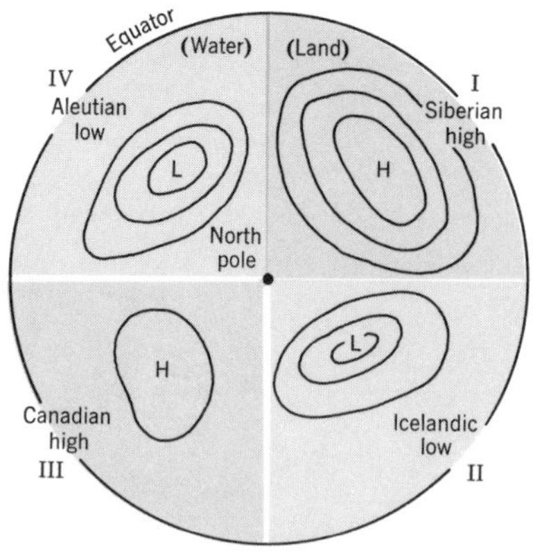

Figure 9.15 **Northern hemisphere pressure centers in January.**

North and south of the doldrums are the *trade wind* belts, covering roughly the zones lying between 5° and 30° N and S. The trades are a result of a pressure gradient from the subtropical belt of high pressure to the equatorial trough of low pressure. In the northern hemisphere, air moving equatorward is deflected by the earth's rotation to turn westward. Thus, the prevailing wind is from the northeast and the winds are termed the *northeast trades.* In the southern hemisphere, deflection of the moving air to the left causes the *southeast trades.* Trade winds are noted for their steadiness and directional persistence. Figure 9.17 shows wind roses within the trade-wind belts. Notice that most winds come from one quarter of the compass.

The system of doldrums and trades shifts seasonally north and south, through several degrees of latitude, as do the pressure belts that cause them. Because of the large land areas of the northern hemisphere, there is a tendency for these belts to be shifted farther north in the northern hemisphere in summer (July) than they are shifted south in winter (January). The trades are best developed over the Pacific and Atlantic oceans, but are upset in the Indian Ocean region by the proximity of the great Asiatic landmass.

The trade winds provided a splendid avenue for westward travel in the days of sailing vessels. Steadiness of wind and generally clear weather made this a favorite zone of mariners. Crossing of the doldrums was hazardous because of the possibility of being becalmed for long periods and because of the uncertainty of wind direction. The trade wind belts are not altogether favorable for navigation and flying, however, because over certain oceanic portions, at certain seasons of the year,

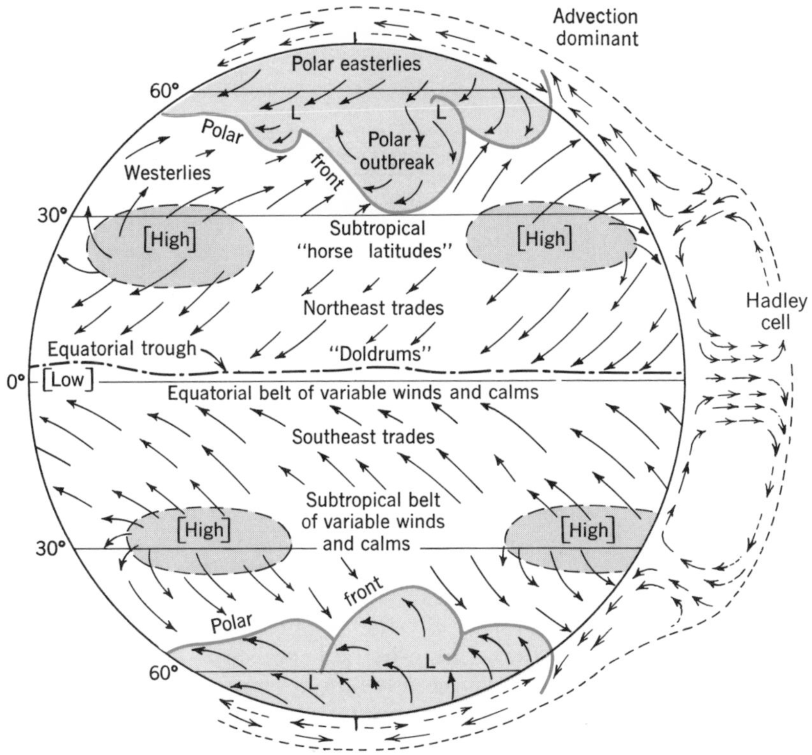

Figure 9.16 The general scheme of atmospheric circulation.

terrible tropical storms known as hurricanes or typhoons occur (Chapter 11).

Between latitudes 25° and 35° are what have long been called the *subtropical belts of variable winds and calms,* or *horse latitudes,* coinciding with the subtropical high-pressure belt. Instead of being continuous even belts, however, the high-pressure areas are concentrated into distinct anticyclones or cells, located over the oceans. Figure 9.18 shows anticyclones in the northern and southern hemisphere and the resultant surface winds. The apparent outward spiraling movement of air is directed equatorward into the easterly trade wind system; poleward into the westerly wind system. The cells of high pressure are most strongly developed in the summer (January in the southern hemisphere, July in the northern). There is also a latitudinal shifting following the sun's declination. This amounts to less than 5° in the southern hemisphere, but it is about 8° for the strong Hawaiian high located in the northeastern Pacific.

Wind roses for the horse latitudes are shown in Figure 9.19. Winds in the high pressure cells are distributed around a considerable range of compass directions. Calms prevail as much as a quarter of the time. The cells have generally fair, clear weather, with a strong tendency to dryness. Most of the world's great deserts lie in this zone and in the adjacent trade wind belt. An explanation of the dry, clear weather lies in the fact that the anticyclonic cells are centers of descending air, settling from higher levels of the atmosphere and spreading out near the earth's surface. Descending air, as explained more fully in the next chapter, becomes increasingly dry.

Between latitudes 35° and 60°, both N and S, is the belt of the *westerlies,* or *prevailing westerly winds.* Moving from the subtropical anticyclones toward the subarctic lows, these surface winds are shown on Figure 9.16 to blow from a southwesterly quarter in the northern hemisphere, from a northwesterly quarter in the southern hemisphere. This generalization is somewhat misleading, however, because winds from polar directions are frequent and strong. It is more accurate to say that within the westerly wind belt, winds blow from any direction of the compass, but that the westerly components are definitely predominant. Storm winds are common in this belt, as are frequent cloudy days with continued precipitation. Weather is highly changeable.

In the northern hemisphere, landmasses cause considerable disruption of the westerly wind belt, but in the southern hemisphere, between the latitudes 40° and 60° S, there is an almost unbroken belt of ocean. Here the westerlies gain great

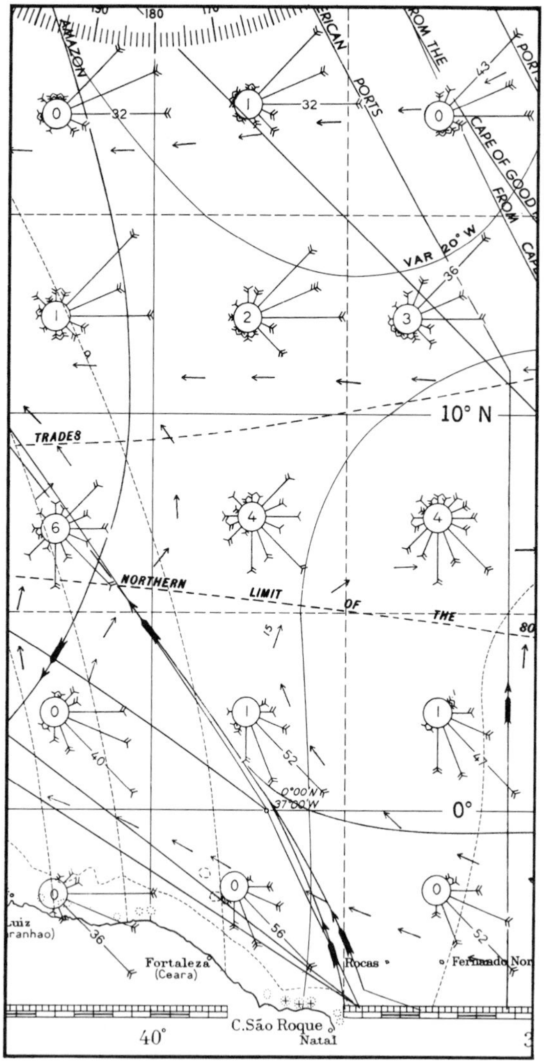

Figure 9.17 This portion of an Oceanographic Office pilot chart of the North Atlantic for July shows parts of both the northeast and southeast trade wind belts, with the narrow doldrum belt lying between. The area shown here extends from 5° S lat. to 20° N lat. Wind roses occupy 5° squares. (U.S. Navy Oceanographic Office, Chart 1400.)

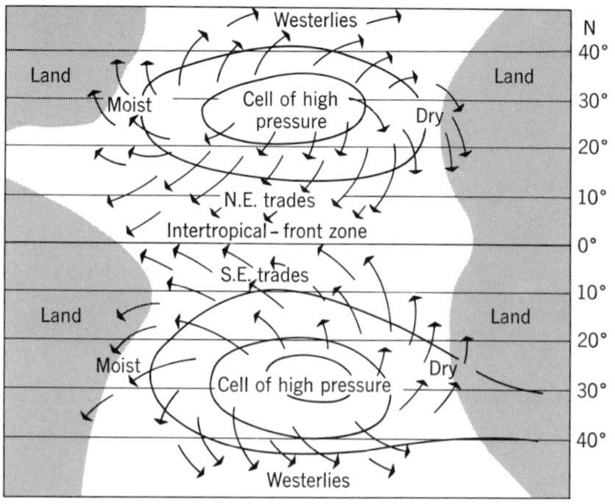

Figure 9.18 Semi-permanent centers of high pressure and surface winds.

strength and persistence, giving rise to the mariner's expressions, "the roaring forties," "the furious fifties," and "the screaming sixties." This belt was extensively used for sailing vessels traveling eastward from the South Atlantic Ocean to Australia, Tasmania, New Zealand, and the southern Pacific islands. From these places it was then easier to continue eastward around the world to return to European ports. Rounding Cape Horn was relatively easy on an eastward voyage, but in the opposite direction, in the face of prevailing stormy westerly winds, was fraught with great danger.

Although the westerly wind belts no longer exert a strong influence over the routes of modern ocean vessels, they are important in long-distance flying. Transoceanic and transcontinental flights in the easterly direction require less fuel and a shorter time. On westward flights, strong head winds may eat dangerously into the fuel supply on the plane and in any event necessitate reduced pay loads.

A wind system often termed the *polar easterlies* has been described as characteristic of the arctic and polar zones (Figure 9.16). The concept is greatly oversimplified, if not actually erroneous, for winds in these regions take a variety of directions, as dictated by local weather disturbances. Perhaps in Antarctica, where an icecapped landmass rests squarely upon the pole and is surrounded by a vast oceanic expanse, the outward spiraling flow of polar easterlies is a valid concept. Deflected to the left in the southern hemisphere, the radial winds would spiral counterclockwise, producing a system of southeasterly winds.

Monsoon winds of Asia and North America

Frequent reference has been made to the powerful control which Asia and North America exert upon conditions of temperature and pressure in the northern hemisphere. Because pressure conditions control winds, it is obvious that these areas must develop wind systems relatively independent of the belted system of earth winds so well illustrated in the southern hemisphere.

In summer, southern Asia develops a cyclone into which there is a considerable flow of air. This may be a *heat low*, or *thermal low*, limited to the lower levels of the atmosphere (Figure 9.20). From the Indian Ocean and the southwestern Pacific warm, humid air moves northward and northwestward into Asia, passing over India, Indochina, and China. This air flow constitutes the *summer monsoon*, which is accompanied by heavy rainfall in southeastern Asia.

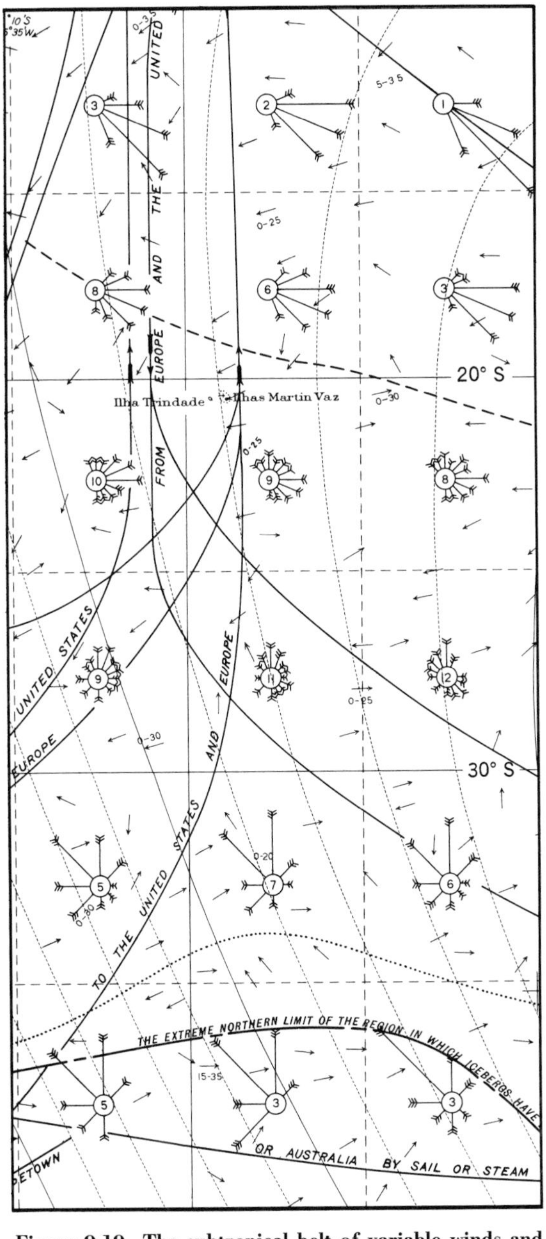

Figure 9.19 The subtropical belt of variable winds and calms (horse latitudes) lies in the center of this portion of an Oceanographic Office pilot chart of the South Atlantic for June–July–August. To the north are the steady southeast trades; to the south, the variable northwesterlies. The area shown here extends from 10° to 40° S lat. Wind roses occupy 5° squares. (U.S. Navy Oceanographic Office, Chart 2600.)

In winter, Asia is dominated by a strong center of high pressure, from which there is an outward flow of air reversing that of the summer monsoon. Blowing southward and southeastward toward the equatorial oceans, this *winter monsoon* brings dry, clear weather for a period of several months.

North America does not have the remarkable extremes of monsoon winds experienced by southeastern Asia, but there is nevertheless an alternation of temperature and pressure conditions between winter and summer. Wind records show that in summer there is a prevailing tendency for air originating in the Gulf of Mexico to move northward across the central and eastern part of the United States, whereas in winter there is a prevailing tendency for air to move southward from sources in Canada. Australia, too, shows a monsoon effect, but being south of the equator it reverses the conditions of Asia.

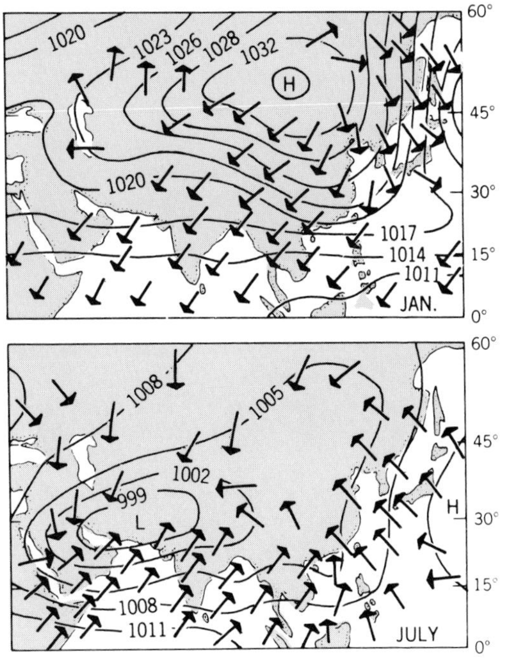

Figure 9.20 Surface maps of pressure and winds over southeastern Asia in January and July. Pressure in millibars. (Same data source as Figure 9.12.)

Global circulation systems

The surface wind systems thus far described represent only a shallow basal air layer a few thousands of feet thick, whereas the troposphere is from 5 to 12 mi (8 to 20 km) thick. What is the nature of air flow at these higher levels? Slowly moving high- and low-pressure systems are found aloft, but that these are generally simple in pattern with smoothly curved isobars. Winds, which may be extremely strong and follow the isobars closely, move counterclockwise around the lows (northern hemisphere), but clockwise around the highs, as shown in Figure 9.7.

The general or average pattern of upper air flow is sketched in Figure 9.21. Two systems dominate. One is the system of *westerlies* blowing in a complete circuit about the earth from about latitude 25° almost to the poles. At high latitudes these

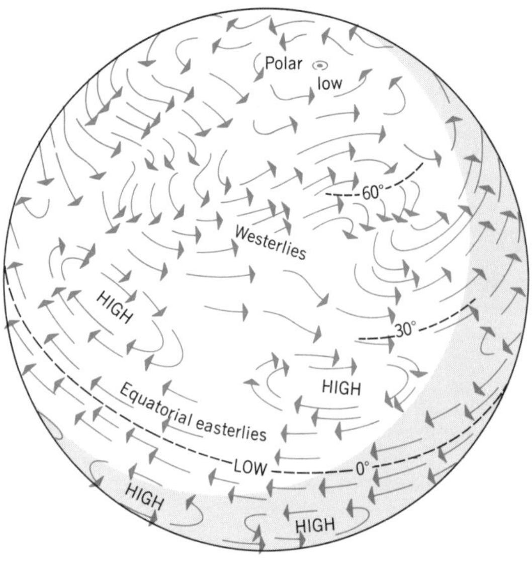

Figure 9.21 Schematic representation of circulation in the upper part of the troposphere, 20,000 to 40,000 ft (6 to 12 km).

westerlies constitute a circumpolar whirl, coinciding with a great polar low-pressure center. Toward low latitudes the pressure rises steadily at a given altitude, to form two high-pressure ridges at latitudes 15° to 20° N and S. These are the high-altitude parts of the subtropical highs, but are shifted somewhat equatorward. In the high-pressure zones, wind velocities are low, just as in the horse latitudes at sea level. Between the high-pressure ridges is a trough of weak low pressure, in which the winds are easterly, comprising the second major circulation system of the globe, termed the *equatorial easterlies.* At lower elevation their influence spreads into somewhat higher latitudes as the trade winds.

Seen in meridional cross section (Figure 9.16) circulation in equatorial and tropical latitude zones resolves itself into two circuits, one in each hemisphere. Air rises over the equatorial zone but subsides in the subtropical cells, forming the *Hadley cell.* Some of the subsiding air escapes poleward, into the westerlies. Thus the circulation of low latitudes is a heat engine of the type described earlier in the explanation of sea and land breezes.

Upper-air waves and the jet stream

The uniform flow of the upper-air westerlies is frequently disturbed by the formation of large undulations, called *upper-air waves* (or *Rossby waves*). As shown in Figure 9.22, these waves grow in amplitude and ultimately become cut off. The waves develop in a zone of contact between cold, polar air and warm, tropical air.

It is by means of the upper air waves that warm air of low latitudes is carried far north at the same time that cold air of polar regions is brought equatorward. In this way horizontal mixing, or *advection,* develops on a gigantic scale and serves to provide heat exchange between the equatorial region of energy surplus and the polar regions of energy deficit.

Associated with the development of such upper air waves at altitudes of 30,000 to 40,000 ft (10,000 to 12,000 m) are narrow zones in which wind streams attain velocities up to 200 to 250 knots (350–450 km per hr). This phenomenon, named the *jet stream,* consists of pulselike movements of air following a broadly curving track (Figure 9.23). In cross section the jet may be likened to a stream of water moving through a hose, the center line of highest velocity being surrounded by concentric zones of less rapidly moving fluid, as pictured in Figure 9.24.

The jet stream is an important factor in the operation of jet aircraft in the range of their normal cruising altitudes. In addition to strongly increasing or decreasing the ground speed of the aircraft, there is a form of air turbulence that at times reaches hazardous levels. This is *clear air turbulence* (CAT); it is avoided when known to be severe in intensity.

Local winds

In certain favorable localities, *local winds* are generated by immediate influences of the surrounding terrain, rather than by the large-scale pressure systems that produce global winds and large traveling storms. Local winds are of environmental importance in various ways. They may exert a powerful stress on animals and plants, when the winds are dry and extremely hot or cold. They are also important in affecting the movement of atmospheric pollutants and in some localities carry pollution plumes far downwind from the sources.

One class of local winds—sea breezes and land breezes—has been explained earlier in this chapter. The cooling sea breeze (or lake breeze) of summer is an important environmental resource of coastal communities, for it adds to the attraction of the shore zone as a recreation facility.

Mountain and *valley winds* are local winds following a daily alternation of direction in a manner similar to the land and sea breezes. The air moves from valleys, upward over rising mountain slopes, toward the summits during the day, when slopes are intensely heated by the sun. The air then moves valleyward, down the ground slopes, when the same slopes have been cooled at night by radiation of heat from ground to air. These winds are therefore responding to local pressure

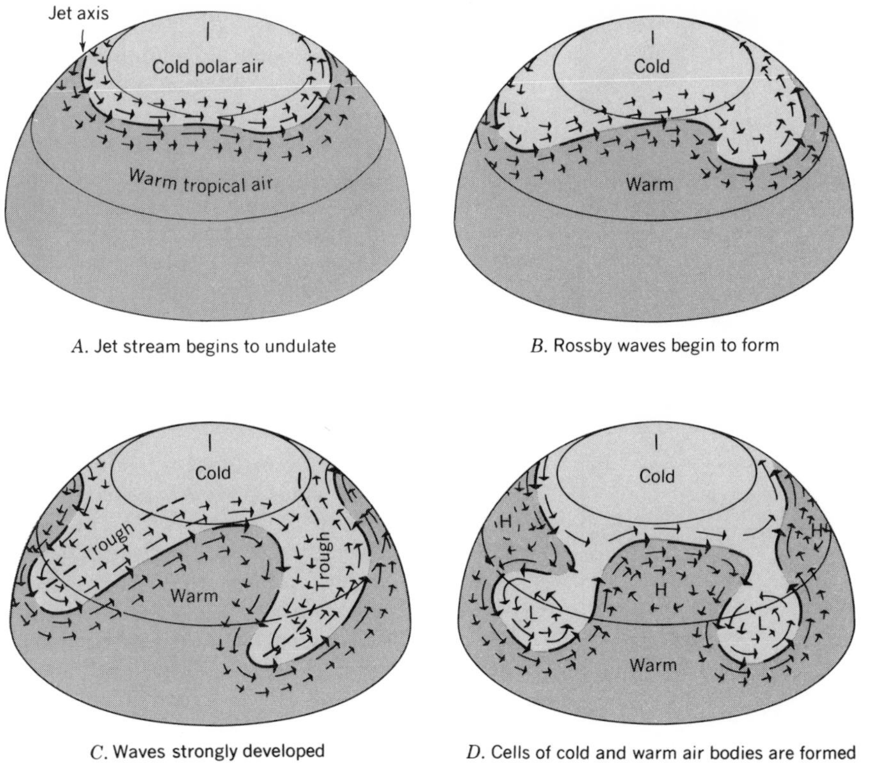

Figure 9.22 Development of upper-air waves in the westerlies. Modified from diagrams by J. Namias, NOAA National Weather Service. (From A. N. Strahler, 1971, *The Earth Sciences*, 2nd ed., Harper and Row, New York.)

gradients set up by heating or cooling of the lower air.

Still another group of local winds are known as *drainage winds*, or *katabatic winds*, in which cold air flows under the influence of gravity from higher to lower regions. Such cold, dense air may accumulate in winter over a high plateau or high interior valley. When general weather conditions are favorable some of this cold air spills over low divides or through passes to flow out upon adjacent lowlands as a strong, cold wind. Drainage winds occur in many mountainous regions of the world and go by various local names. The *bora* of the northern Adriatic coast and the *mistral* of southern France are well-known examples. In southern California there issues on occasion from the Santa Ana Valley a strong, dry east wind, the *Santa Ana*, which blows across the coastal lowland. This air is of desert origin and may carry much dust and silt in suspension. On the icecaps of Greenland and Antarctica, powerful katabatic winds move down the gradient of the ice surface and are funneled through coastal valleys to produce powerful blizzards lasting for days at a time.

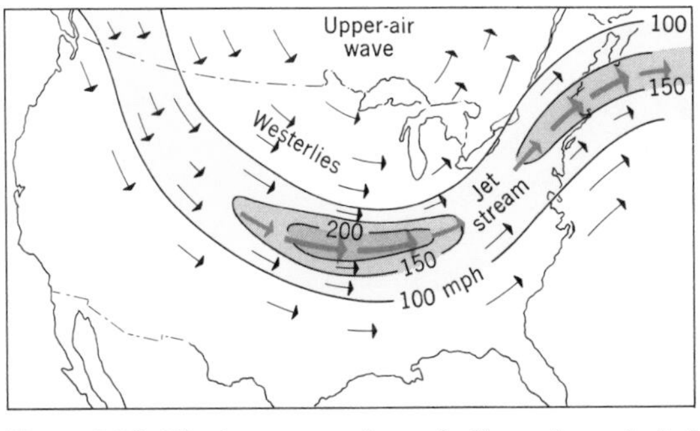

Figure 9.23 The jet stream, shown by lines of equal wind speed. (After National Weather Service.)

Still other types of local winds, bearing such names as *foehn* and *chinook*, result when strong regional winds passing over a mountain range are forced to descend on the lee side with the result that the air is heated and dried. These winds are explained in Chapter 10.

Ocean currents

Ocean currents provide a high degree of regulation to thermal environments of the earth's surface. On a global scale, the vast current systems aid in exchange of heat between low and high latitudes and are thus essential in sustaining the heat balance. On a local scale, warm water currents bring a moderating influence to coasts in arctic

Figure 9.24 The jet stream. (From National Weather Service.)

Generalized scheme of ocean currents

To illustrate surface water circulation an idealized ocean extending across the equator to latitudes of 60° or 70° on either side may be taken for illustration (Figure 9.25). Perhaps the most outstanding features are the circular movements, or *gyres,* around the subtropical highs, centered about 25° to 30° N and S. An *equatorial current* marks the belt of the trades. Whereas the trades blow to the southwest and northwest, obliquely across the parallels of latitude, the water movement follows the parallels. Thus, ocean currents trend at an angle of about 45° with the prevailing surface winds, because of the deflective force of the earth's rotation.

A slow eastward movement of water over the zone of the westerly winds is named the *west-wind drift.* It covers a broad belt between 35° and 45° in the northern hemisphere, and between 30° or 35° and 70° in the southern where open ocean exists in the higher latitudes.

The equatorial currents are separated by an *equatorial countercurrent.* This is well developed in the Pacific, Atlantic, and Indian oceans (Figure 9.26).

Along the west sides of the oceans in low latitudes the equatorial current turns poleward, forming a warm current paralleling the coast. Examples

latitudes; cool currents greatly alleviate the heat of tropical deserts along narrow coastal belts.

Practically all of the important surface currents of the oceans are set in motion by prevailing surface winds. Energy is transferred from wind to water by the frictional drag of the air blowing over the water surface. Because of the Coriolis force, the water drift is impelled toward the right of its path of motion (northern hemisphere), and therefore the current at the water surface is in a direction about 45° to the right of the wind direction. Under the influence of winds, currents may tend to bank up the water close to the coast of a continent, in which case the force of gravity, tending to equalize the water level, will cause other currents to be set up.

Density differences may also cause flow of ocean water. Such differences arise from greater heating by insolation, or greater cooling by radiation, in one place than another. Thus, surface water chilled in the arctic and polar seas will sink to the ocean floor, spreading equatorward and displacing upward the less dense, warmer water.

Still another controlling influence upon water movements is the configuration of the ocean basins and coasts. Currents initially caused by winds impinge upon a coast and are locally deflected.

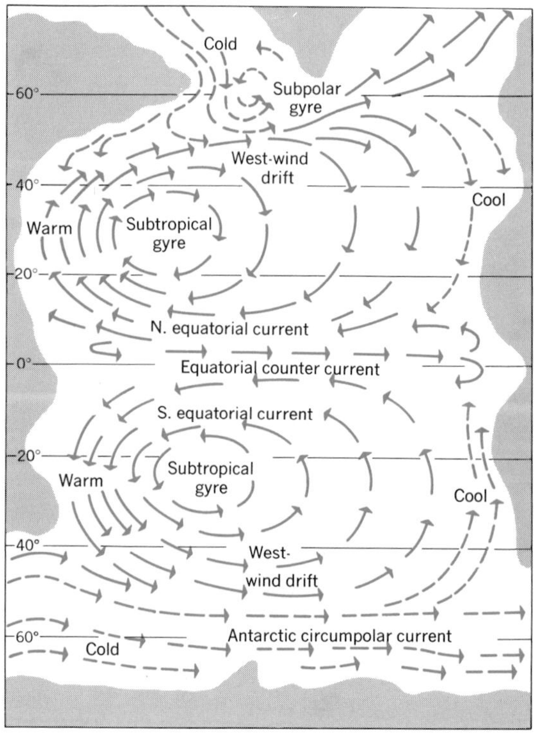

Figure 9.25 Schematic map of system of ocean currents.

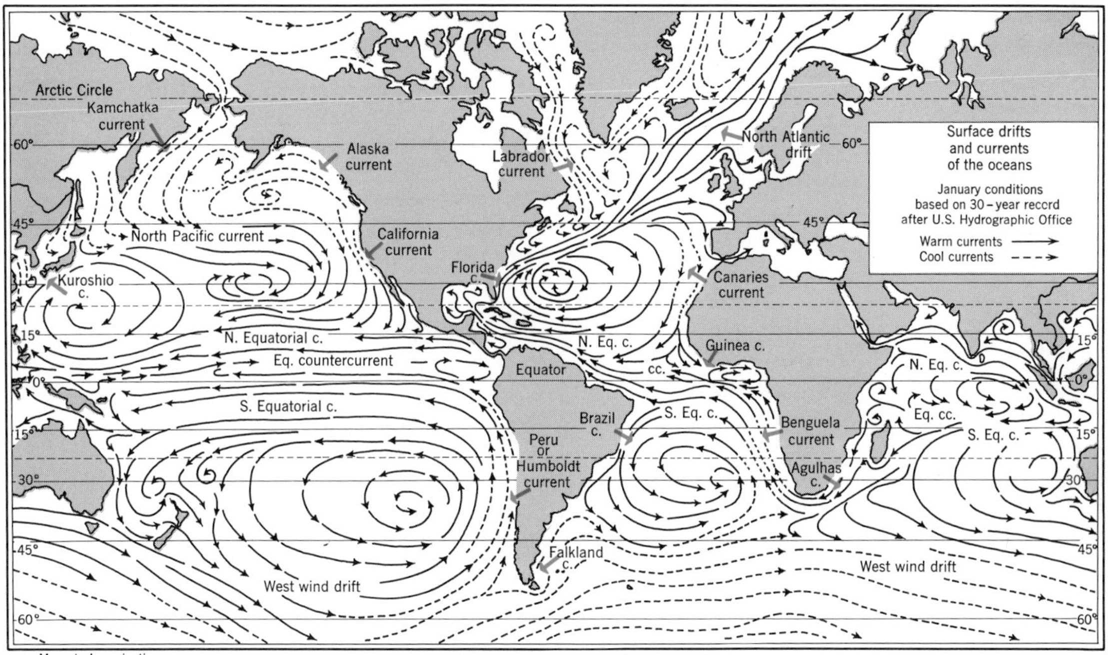

Figure 9.26 Surface drifts and currents of the oceans (January). (After U.S. Navy Oceanographic Office.)

are the *Gulf Stream* (*Florida* or *Caribbean* stream), the *Japan current* (*Kuroshio*) and the *Brazil current*, which bring higher than average temperatures along these coasts. This is illustrated in Figure 8.22 by the southward bulge of the 60°F (16°C) July isotherm along the west side of the South Atlantic.

The west-wind drift, upon approaching the east side of the ocean, is deflected both south and north along the coast. The equatorward flow is a cool current, produced by upwelling of colder water from greater depths. It is well illustrated by the *Humboldt current* (Peru current) off the coast of Chile and Peru; by the *Benguela current* off the southwest African coast; by the *California current* off the west coast of the United States; and by the *Canaries current*, off the Spanish and North African coast. Note that this cold upwelling causes a marked equatorward deflection of the isotherms, illustrated in Figure 8.22 by the northward bend in the 70°F (21°C) January isotherm along the east side of the South Pacific and South Atlantic oceans.

In the northern eastern Atlantic Ocean, the westwind drift is deflected poleward as a relatively warm current. This is the *North Atlantic current*, which spreads around the British Isles, into the North Sea, and along the Norwegian coast. The port of Murmansk, on the Arctic Circle, has year-round navigability by way of this coast. Note that in Figure 8.22 isotherms are deflected northward where they cross this current. In winter this effect is much more pronounced than in summer.

In the northern hemisphere, where the polar sea is largely landlocked, cold water flows equatorward along the west side of the large straits connecting the Arctic Ocean with the Atlantic basin. Three principal cold currents are the *Kamchatka current*, flowing southward along the Kamchatka Peninsula and Kurile Islands; the *Greenland current*, flowing south along the east Greenland coast through the Denmark Strait; and the *Labrador current*, moving south from the Baffin Bay area through Davis Strait to reach the coasts of Newfoundland, Nova Scotia, and New England.

In both the north Atlantic and Pacific oceans the Icelandic and Aleutian lows in a very rough way coincide with two centers of counterclockwise circulation involving the cold arctic currents and the west-wind drifts.

The antarctic region has a relatively simple current scheme consisting of a single *antarctic circumpolar current* moving clockwise around the antarctic continent in latitudes 50° to 65° S, where a continuous expanse of open ocean occurs.

Oceanographers today recognize that oceanic circulation involves the complex motions of water masses of different temperature and salinity characteristics. Sinking and upwelling are both important motions in certain areas of the oceans.

REVIEW QUESTIONS

1. What environmental role is played by atmospheric motions?

2. Explain the pressure gradient force. In what direction does it act?

3. Explain the workings of the sea breeze and the land breeze. In what respect are these circulation systems heat engines?

4. What is the Coriolis force? What causes this force? In what direction does it act? How does it vary with latitude and hemisphere?

5. What is the relationship between isobars and wind direction at high altitudes? What is Ballot's law? What is the geostrophic wind?

6. Explain how friction with the earth's surface modifies wind direction and strength.

7. Describe the flow of winds around cyclones and anticyclones at high altitudes and near the earth's surface. How are convergence and divergence related to vertical air motions?

8. With what instruments are winds measured? How is the direction of a wind designated? How can wind direction and speed in the upper air levels be determined?

9. Describe the principal pressure belts of the globe, giving the latitude and approximate pressures for each.

10. Why do the pressure belts shift in latitude throughout the year? Is this shift over as great a latitude as the sun's declination?

11. Why do the landmasses of North America and Asia disrupt the belted pressure pattern of the globe? What pressures occur on these land areas in the winter solstice season? In the summer solstice season?

12. What pressure centers dominate the North Atlantic and North Pacific oceans during the summer months? During the winter months?

13. What conditions of winds and calms prevail in the equatorial trough? Why are steady, prevailing winds absent here?

14. Describe the trade winds. Which way do they blow in the northern hemisphere? In the southern hemisphere? Explain the direction of the trade winds. How were the trade winds used by mariners?

15. What are the subtropical belts of variable winds and calms? Describe the cells of high pressure of which this belt is composed. How do they shift in latitude seasonally?

16. Describe the westerly wind belts. How do the westerlies compare with the trades for constancy of direction and strength? How does the westerly wind belt influence transoceanic sailing and flying?

17. Describe the monsoon wind systems of southeastern Asia. What general type of weather condition is associated with the summer monsoon? With the winter monsoon?

18. Describe the general global circulation in the middle and upper troposphere. What are the westerlies? The equatorial easterlies? What is the Hadley cell? How does it operate?

19. Describe the jet stream and its relation to upper air waves of the westerlies.

20. What are local winds? Explain the mountain and valley breeze. Where are katabatic (drainage) winds found? Give examples.

21. Discuss the causes of surface ocean currents. Show how the Coriolis force modifies the direction of flow.

22. Sketch a hypothetical ocean modeled after the Pacific or Atlantic and indicate the general scheme of surface currents. Label fully.

23. List three warm currents and three cool currents, giving the location and direction of flow of each. What effect have these currents upon isotherms of air temperatures?

24. What current systems conduct very cold water into the northern Atlantic Ocean? What current prevails in the southern ocean, in the region surrounding the continent of Antarctica?

Exercises

1. The following air pressures are given in one of three units of measurement: inches, millimeters, or millibars. Convert the given figure into the two other units.

Inches	Millimeters	Millibars
30.12	______	______
______	710	______
______	______	1006
29.60	______	______
______	______	984.5

2. Draw wind arrows of the type used on National Weather Service daily maps (Figure 11.9) for each of the following conditions:

	Direction	Velocity Knots
(a)	SSW	15
(b)	N	35
(c)	ESE	10
(d)	290°	75
(e)	260°	110

3. Using the scale of wind percentages given, estimate the percentages of time during which the wind blows from each of the sectors on the wind rose on the right-hand side of Figure 9.11. Total the percentages and subtract from 100 percent in order to determine the figure that should go in the circle representing light air and calm. (Measure the lengths of shafts from the center of the circle.)

4. Prepare a wind rose for the following data. Use the same method as in Exercise 3. (Data from Hydrographic Office, U.S. Navy.) From what belt of winds was this wind rose taken? Compare with Figures 9.17 and 9.19, and find this wind rose.

Sector	Percent of Time	Average Force (Beaufort)
N	21	5
NE	12	5
E	6	5
SE	6	5
S	8	5
SW	10	5
W	12	5
NW	20	5

5. The map below shows barometric pressure conditions at high levels by means of contours drawn upon the 500-millibar pressure surface. Contours are labeled in thousands of feet. The map shows conditions at 7:00 P.M., E.S.T. on a day in early April. (Data of National Weather Service.)

Treating the contours as if they were isobars, draw numerous arrow points on the contours to show the direction of air motion. Draw broad, sweeping arrows to show the probable position of the jet stream in two places on the map. Label a *cyclone*; label an *anticyclone*.

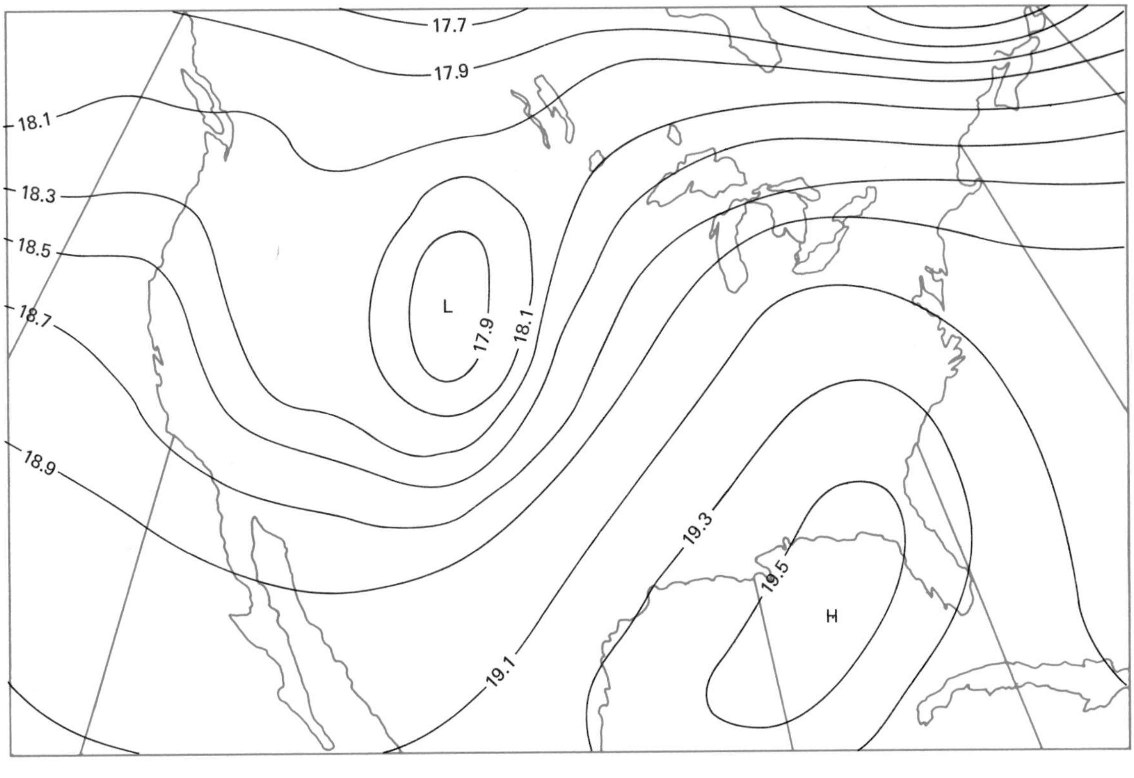

6. The weather map on this page shows barometric pressures observed simultaneously at many United States and Canadian observing stations. Pressures are in millibars. Only the last two digits are given. Hence, *16* designates *1016*; *96* designates *996*. Station location is shown by a dot beside the numeral.

Draw isobars for the entire map, using an interval of 4 mb. Isobars should run thus: 992, 996, 1000, 1004, 1008, 1012, 1016, 1020, 1024, 1028. Label isobars. Label lows and highs. Use pencil lightly at first; then draw final lines as smooth flowing curves, except where abrupt pressure changes (fronts) occur. Finally, draw many short, straight arrows across the isobars to show the directions of surface winds.

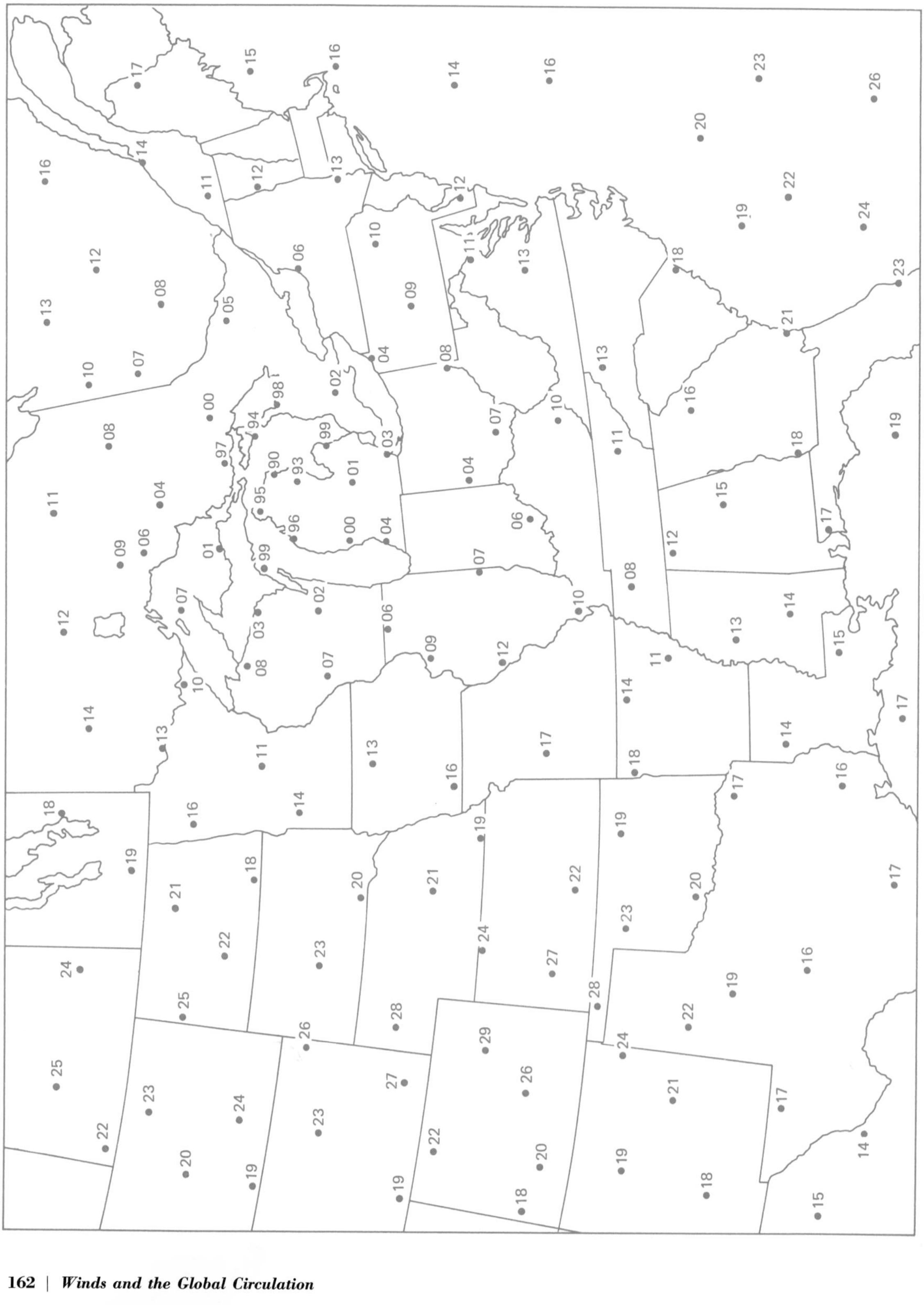

CHAPTER **10**

Atmospheric Moisture and Precipitation

WE have pointed out that heat and water are vital ingredients of the environment of the biosphere, or life layer. Plant and animal life of the lands, upon which man depends for much of his food, require fresh water. Man himself uses fresh water in many ways. The only basic source of fresh water is from the atmosphere through condensation of water vapor. In this chapter, we are concerned mostly with water in the vapor state in the atmosphere and the processes by which it passes into the liquid (or solid) state and ultimatelv arrives at the surface of the ocean and the lands through the process of precipitation.

Water also leaves the land and ocean surfaces by evaporation and so returns to the atmosphere. Evidently, the global pathways of movement of water form a complex network. We have already mentioned that there is a global water balance, just as there is an energy balance, and that the water balance deals with flow of matter and so compliments the energy balance.

Let us first review basic terms and processes involved in an understanding of the water balance.

Water states and heat

Water occurs in three states, (1) frozen as ice, a crystalline *solid*, (2) *liquid* as water, and (3) *gaseous* as water vapor (Figure 10.1). From the gaseous vapor state, molecules may pass into the liquid state by *condensation*, or, if temperatures are below the freezing point, they can pass by *sublimation* directly into the solid state to form ice crystals. By *evaporation*, molecules can leave a water surface to become gas molecules in water vapor. The analogous change from ice directly into water vapor is also designated sublimation. Then, of course, water may pass from liquid to solid state by *freezing*, and from solid state to liquid state by *melting*. All of this can be represented by a triangle in which the three states of water form the corners. Arrows show the six possible changes of state.

Of great importance in weather science are the exchanges of heat energy accompanying changes of state. For example, when water evaporates, sensible heat, which we can feel and measure by thermometer, passes into a hidden form held by the water vapor and known as the *latent heat of vaporization*. This results in a drop in temperature of the remaining liquid. The cooling effect produced by evaporation of perspiration from the skin is perhaps the most obvious example. For every gram of water that is evaporated, about 600 calories change into the latent form. In the reverse process of condensation, an equal amount of energy is released to become sensible heat and the temperature rises correspondingly. Similarly, the freezing process releases heat energy in the amount of about 80 calories per gram of water, whereas melting absorbs an equal quantity of heat. This is referred to as the *latent heat of fusion*. When sublimation occurs, the heat absorbed by vaporization, or released by crystallization, is still greater for each gram of water, for the latent heats of vaporization and fusion are added together.

Humidity

The amount of water vapor that may be present in the air at a given time varies widely from place to place. It ranges from almost nothing in the cold, dry air of arctic regions in winter to as much as 4 or 5 percent of a given volume of the atmosphere in the humid equatorial zone.

The term *humidity* simply refers to the degree to which water vapor is present in the air. For any specified temperature there is a definite limit to

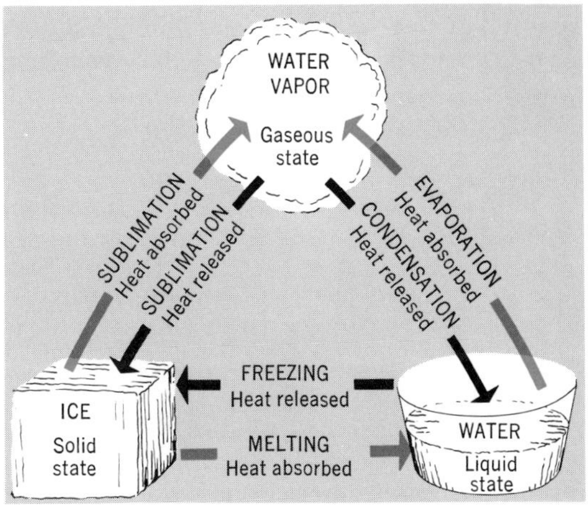

Figure 10.1 **Three states of water. (After C.A.A., U.S. Dept. Commerce.)**

the quantity of moisture that can be held by the air. This limit is known as the *saturation point.* The proportion of water vapor present relative to the maximum quantity is the *relative humidity,* expressed as a percentage. At the saturation point, relative humidity is 100 percent; when half of the total possible quantity of vapor is present, relative humidity is 50 percent, and so on.

A change in relative humidity of the atmosphere can be caused in one of two ways. If an exposed water surface is present, the humidity can be increased by evaporation. This is a slow process, requiring that the water vapor diffuse upward through the air. The other way is through a change of temperature. Even though no water vapor is added, a lowering of temperature results in a rise of relative humidity. This is automatic and is a logical consequence of the fact that the capacity of the air to hold water vapor has been lowered by cooling; thus the existing amount of vapor represents a higher percentage of the total capacity of the air. Similarly, a rise of air temperature results in decreased relative humidity, even though no water vapor has been taken away. The principle of relative humidity change caused by temperature change is illustrated by a graph of these two properties throughout the day (Figure 10.2). As air temperature rises, relative humidity falls, and vice versa.

A simple example may be given to illustrate these principles. At a certain place the temperature of the air is 60°F (16°C), the relative humidity 50 percent. Should the air become warmed by the radiant energy from the sun and ground surface to 90°F (32°C), the relative humidity automatically drops to 20 percent, which is very dry air. Should the air become chilled during the night and its temperature fall to 40°F (5°C), the relative humidity will automatically rise to 100 percent, the saturation point. Any further cooling will cause condensation of the excess vapor into liquid form. As the air temperature continues to fall, the humidity remains at 100 percent, but condensation continues. This may take the form of minute droplets of dew or fog. If the temperature falls below freezing, condensation occurs as frost upon exposed surfaces.

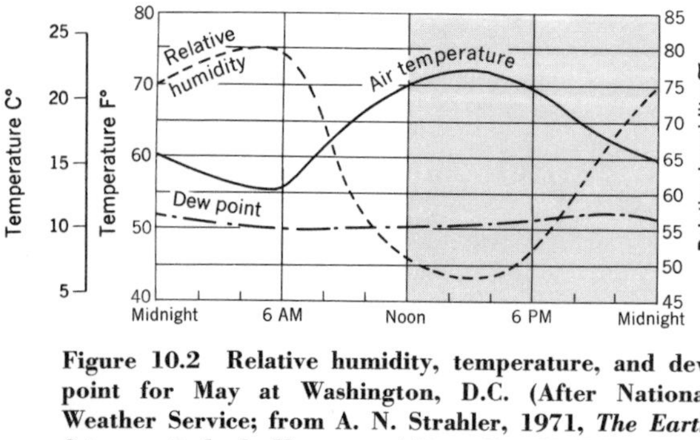

Figure 10.2 Relative humidity, temperature, and dew point for May at Washington, D.C. (After National Weather Service; from A. N. Strahler, 1971, *The Earth Sciences*, 2nd ed., Harper and Row, New York.)

The term *dew point* is applied to the critical temperature at which the air is fully saturated, and below which condensation normally occurs. An excellent illustration of condensation due to cooling is seen in summertime when beads of moisture form on the outside surface of a pitcher or glass filled with ice water. Air immediately adjacent to the cold glass or metal surface is sufficiently chilled to fall below the dew point temperature, causing moisture to condense on the surface of the glass.

In understanding the relationships between atmospheric temperature and relative humidity, a very homely analogy may prove helpful. An ordinary sponge, if left to soak up water, will take up moisture to its full capacity. This is analogous to the manner in which air will gradually increase its humidity to the saturation point, if allowed to stand over a water surface and to maintain a constant temperature. If the sponge is now lifted out of the water and held carefully in the hand it will continue to hold the absorbed water. Suppose now the sponge is slowly squeezed. Water is expelled. In a like manner, the lowering of temperature below dew point of the saturated air expels moisture by condensation of the excess water vapor. After most of the water has been squeezed from the sponge it may be released, but not permitted to touch the water. This is analogous to a rise of air temperature back to the previous starting point, but without being permitted to take up water. This condition would prevail over an interior desert region. The air, like the sponge, is now holding only a small fraction of its total possible moisture content. If it is again cooled no condensation will result from the air until a temperature below the previous minimum is reached, just as no water will be released from the sponge until it is squeezed even harder than previously.

Absolute humidity

Although relative humidity is an important indicator of the state of water vapor in the air, it is a statement only of the relative quantity with respect to a saturation quantity. The actual quantity of moisture present is denoted by *absolute humidity,* defined as the weight of water vapor contained in a given volume of air. Weight is stated in grams, volume in cubic meters. For any specified air temperature, there is a maximum weight of water vapor that a cubic meter of air can hold (the saturation quantity). Figure 10.3 is a graph showing this maximum moisture content of air for a wide range of temperatures.

In a sense, the absolute humidity is a geographer's yardstick of a basic natural resource—water—to be applied from equatorial to polar regions. It is a measure of the quantity of water that can be extracted from the atmosphere as precipitation. Cold air can supply only a small quantity of rain or snow; warm air is capable of supplying huge quantities.

One disadvantage of using absolute humidity in the study of atmospheric moisture is that when air rises or sinks in elevation, it undergoes corresponding volume changes of expansion or compression. Thus the absolute humidity cannot remain a constant figure for the same body of air. Modern meteorology therefore makes use of another measure of moisture content, *specific humidity*, which is the ratio of weight of water vapor to weight of moist air (including the water vapor). This ratio is stated in units of grams of water vapor per kilogram of moist air. When a given parcel of air is lifted to higher elevations without gain or loss of moisture the specific humidity remains constant, despite volume increase.

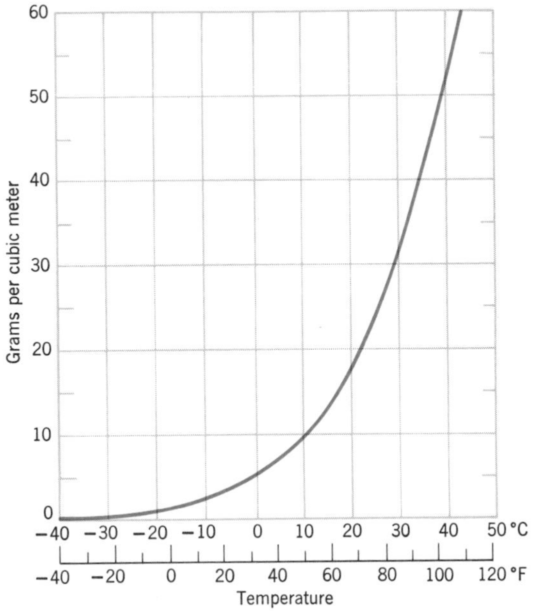

Figure 10.3 Maximum absolute humidity for a wide range of temperatures.

Specific humidity is often used to describe the moisture characteristics of a large mass of air. For example, extremely cold, dry air over arctic regions in winter may have a specific humidity of as low as 0.2 grams per kilogram, whereas extremely warm moist air of tropical regions may hold as

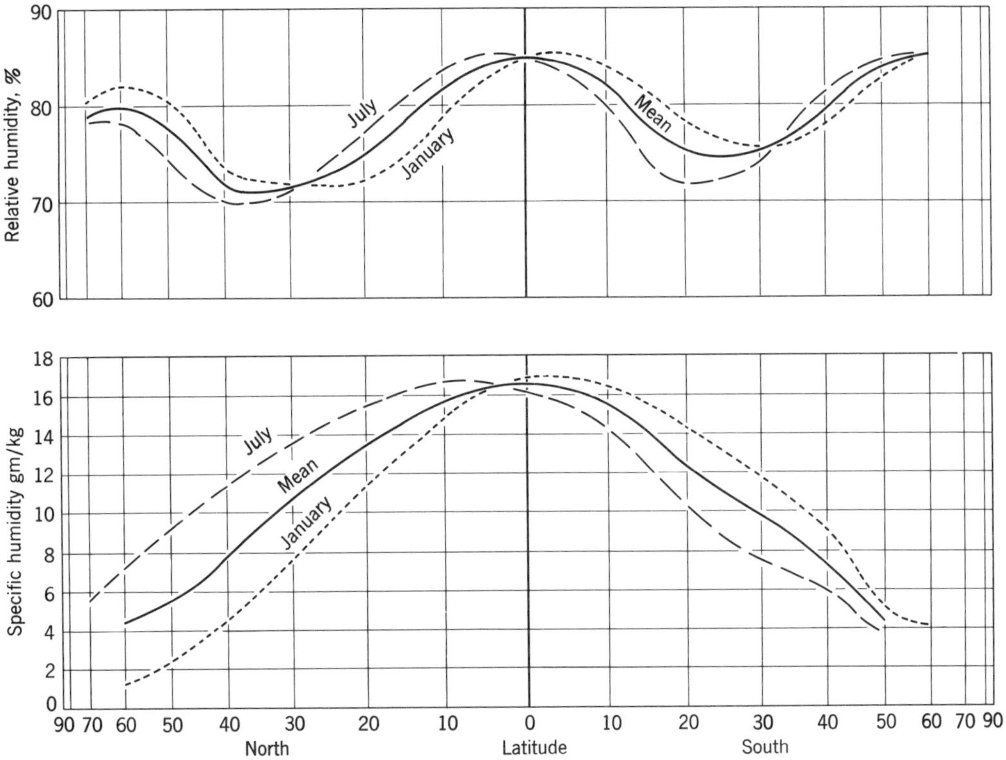

Figure 10.4 Relation of relative humidity to latitude (above) and of specific humidity to latitude (below). (After Haurwitz and Austin, 1944, *Climatology*.)

much as 18 grams per kilogram. The total natural range on a world-wide basis is such that the largest values of specific humidity are from 100 to 200 times as great as the least.

Figure 10.4 is a graph of average specific humidity and latitude, showing that the warm air of equatorial regions normally holds vastly more water vapor than cold air of arctic and polar regions.

How humidity is measured

Humidity of the air can be measured in two ways. An instrument known as a *hygrometer* indicates relative humidity on a calibrated dial. One simple type uses a strand of human hair which lengthens and shortens according to the relative humidity and thereby activates the dial (Figure 10.5). A continuous record of humidity can be obtained by means of a *hygrograph*. Using the same basic mechanism as the hygrometer, a continuous, automatic record is drawn by a pen on a sheet of paper attached to a rotating drum.

A different principle is applied in the *sling psychrometer*. This instrument is simply a pair of thermometers mounted side by side (Figure 10.6). One is of the ordinary type; the other has a piece of wet cloth around the bulb. If the air is fully saturated (relative humidity 100 percent), there will be no evaporation from the wet cloth and both thermometers will read the same. If, however, the air is not fully saturated, evaporation will occur, cooling the cloth-covered thermometer below the temperature shown on the ordinary thermometer.

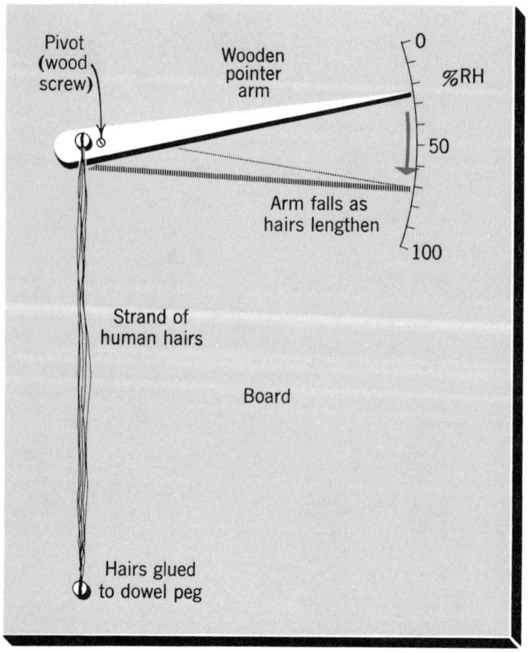

Figure 10.5 A simple hygrometer.

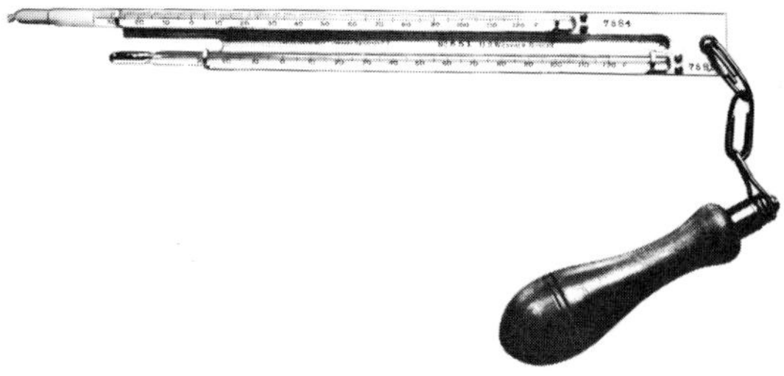

Figure 10.6 Sling psychrometer. (National Weather Service.)

Because the rate of evaporation depends on dryness of the air, the difference in temperature shown by the two thermometers will increase as relative humidity decreases.

Standard tables are available to show the relative humidity for a given combination of wet- and dry-bulb temperatures. In order to be sure that maximum possible evaporation is taking place, the two thermometers are attached by a swivel joint to a handle by which the thermometers can be swung around in a circle by hand. Other types have a fan to blow air past the wet thermometer bulb.

How condensation occurs

Falling rain, snow, sleet, or hail, referred to collectively as *precipitation*, can result only where large masses of air are experiencing a steady drop in temperature below the dew point. This condition cannot be brought about by the simple process of chilling of the air through loss of heat by long-wave radiation during the night. Instead, it is necessary that the large mass of air be rising to higher elevations. This statement requires that a new principle of weather science be explained.

One of the most important laws of meteorology is that rising air experiences a drop in temperature, even though no heat energy is lost to the outside (Figure 10.7). The drop of temperature is a result of the decrease in air pressure at higher elevations, permitting the rising air to expand. Because individual molecules of the gas are more widely diffused and do not strike one another so frequently, the sensible temperature of the gas is lowered. When no condensation is occurring, the rate of drop of temperature, termed the *dry adiabatic rate*, is about $5\frac{1}{2}$ F° per 1000 feet of vertical rise of air. In metric units the rate is 1 C° per 100 meters. The dew point also declines with rise of air; the rate is 1 F° per 1000 ft (0.2 C° per 100 m).

If water vapor in the air is condensing, the adiabatic rate is less, about 3.2 F° per 1000 ft (0.6 C° per 100 m), owing to the partial counteraction of temperature loss through the liberation of latent heat during the condensation process. This modified rate is referred to as the *wet adiabatic*, or *saturation adiabatic* rate (Figure 10.7). Adiabatic cooling rate should not be confused with the environmental lapse rate, explained in Chapter 7. The environmental lapse rate applies only to still air whose temperature is measured at successively higher levels.

Where condensation is occurring directly in the form of snow (ice crystals), the adiabatic rate is intermediate in value between dry and saturated rates.

The various ways in which large masses of air are induced to rise to higher elevations are treated more fully in a later paragraph. Because the actual fall of rain, snow, or other forms of precipitation is preceded by the formation of clouds, we must first consider the various types of clouds and their significance.

Clouds

Clouds consist of extremely tiny droplets of water, 0.0008 to 0.0024 in (0.02 to 0.06 mm) in diameter, or minute crystals of ice. These are sustained by the slightest upward movements of air. In order for cloud droplets to form, it is necessary that microscopic dust particles serve as centers, or *nuclei*, of condensation. Dusts with a high affinity for water are abundant throughout the atmosphere.

Where the air temperature is well below freezing, clouds may form of tiny ice crystals. However, water in such minute quantities can remain liquid far below normal freezing temperatures; the liquid is said to be *supercooled*. Thus, water droplets exist at temperatures down to 10°F (−12°C); a mixture of water droplets and ice crystals from 10° to −20°F (−12 to −30°C) or even lower; and predominantly ice crystals below −20°F (−30°C). Below −40°F (−40°C) all of the cloud is ice. Clouds appear white when thin or when the sun is shining upon the outer surface. When dense and thick, clouds appear gray or black underneath simply because this is the shaded side.

Cloud types may be classified on the basis of two characteristics: general form and altitude. On the basis of form there are two major groups: *stratiform* or layered types, and *cumuliform* or massive, globular types (Figure 10.8).

The stratiform clouds are blanketlike, often covering vast areas, but are fairly thin in comparison to horizontal dimensions. Stratiform clouds are subdivided according to the level of elevation at which they lie. The highest type is the *cirrus* cloud and its related forms, *cirrostratus* and *cirrocumulus* (Figures 10.8 and 10.9). These are roughly within the altitude range of 20,000 to 40,000 ft (6000 to 12,000 m) and are composed of ice crystals. Cirrus is a delicate, wispy cloud, often forming streaks or stringers across the sky. It does not interfere with the passage of sunlight or moonlight and appears to the ground observer to be moving very slowly, if at all. Cirrus bands often indicate the presence of the jet stream aloft. It is possible for the observer on the ground to estimate direction of upper air flow by means of fibrous cirrus formations. *Cirrostratus* is a more complete layer of cloud, producing a *halo* about the sun or moon. Where the layer consists of closely packed globular pieces of cloud, arranged in groups or lines, the name cirrocumulus is given. This is the *mackerel sky* of popular description.

At intermediate height range, from 6500 to 20,000 ft (2000 to 6000 m), are the *altostratus* and *altocumulus* clouds. Altostratus is a blanket layer, often smoothly distributed over the entire sky. It is grayish in appearance, usually has a smooth underside, and will often show the sun as a bright spot in the cloud. Altocumulus is a layer of individual cloud masses, fitted closely together in geometric pattern. The masses appear white, or somewhat gray on the shaded sides, and blue sky is seen between individual patches or rows. Altostratus is commonly associated with the development of bad weather, whereas altocumulus is usually characteristic of generally fair conditions.

In the low cloud group, from ground level to 6500 ft (2000 m), are *stratus, nimbostratus*, and *stratocumulus* clouds. Stratus is a dense, low-lying dark-gray layer (Figure 10.9*C*). If rain or snow is falling from this cloud, it is termed nimbostratus, the prefix *nimbo* merely meaning that precipitation is coming from the cloud. Stratocumulus is a

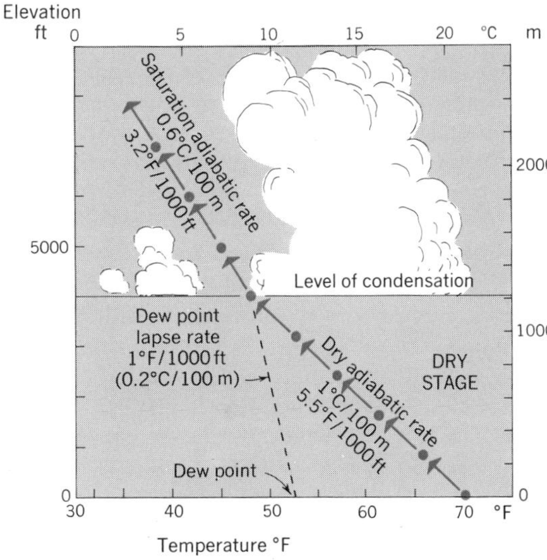

Figure 10.7 Adiabatic changes of temperature in a rising air mass. (From A. N. Strahler, 1971, *The Earth Sciences*, 2nd ed., Harper and Row, New York.)

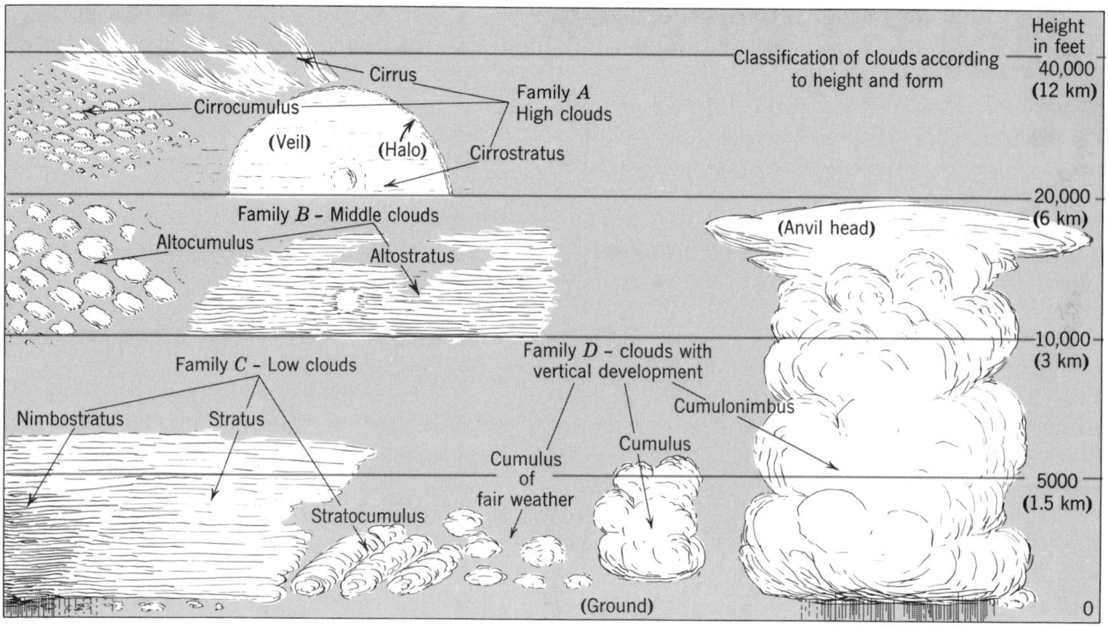

Figure 10.8 Cloud types are grouped into *families* according to height range and form.

low-lying cloud layer consisting of distinct grayish masses of cloud between which is open sky. The individual masses often take on the form of long rolls of cloud, oriented at right angles to the direction of wind and cloud motion. Stratocumulus is generally associated with fair or clearing weather, but sometimes rain or snow flurries issue from individual cloud masses.

Fog is simply a form of stratus cloud lying very close to the ground. One type, known as a *radiation fog*, is formed at night when temperature of the basal air falls below the dew point. Another type, *advection fog*, results from the movement of warm, moist air over a cold or snow-covered ground surface. Losing heat to the ground, the air layer undergoes a drop of temperature below the dew point, and condensation sets in. A similar type of advection fog is formed over oceans where air from over a warm current blows across the cold surface of an adjacent cold current. Fogs of the Grand Banks off Newfoundland are largely of this origin because here the cold Labrador current comes in contact with warm waters of Gulf Stream origin.

The cumuliform clouds tend to display a height as great as, or greater than, their horizontal dimensions. *Cumulus* is a white, woolpack cloud mass, often showing a flat base and a bumpy upper surface somewhat resembling a head of cauliflower (Figure 10.9*E*). These clouds look pure white on the side illuminated by the sun, but may be gray or black on the shaded or underneath side. Small cumulus clouds are associated with fair weather. Under different conditions, discussed below, individual masses grow into *cumulonimbus*, the thunderstorm cloud mass of enormous size which brings heavy rainfall, thunder and lightning, and gusty winds (Figure 10.9*F*). A large cumulonimbus cloud may extend from a height of 1000 to 2000 ft (300 to 600 m) at the base up to 30,000 or 40,000 ft (9000 to 12,000 m). When seen from a great distance, the top of the cumulonimbus cloud is pure white, but to observers beneath, the sky may be darkened to almost nighttime blackness. More will be said of this cloud in connection with thunderstorms.

Forms of precipitation

Precipitation results when condensation is occurring rapidly within a cloud. *Rain* is formed when cloud droplets in large numbers are caused to coalesce into drops too large to remain suspended in the air. The drops may then grow by colliding with other drops and joining with them to become as large as 0.25 in (7 mm) in diameter; but above this size they are unstable and break into smaller drops. Falling droplets less than 0.02 in (0.5 mm) in diameter make a *drizzle*.

Sleet, as the term is used in the United States, consists of pellets of ice produced from freezing of rain. The raindrops form in an upper, warmer layer, but fall into an underlying cold air layer. (Elsewhere in the English-speaking world sleet means a mixture of rain and snow.)

Snow consists of masses of crystals of ice, grown directly from the water vapor of the air, where

A. Cirrus in parallel trails and small patches.

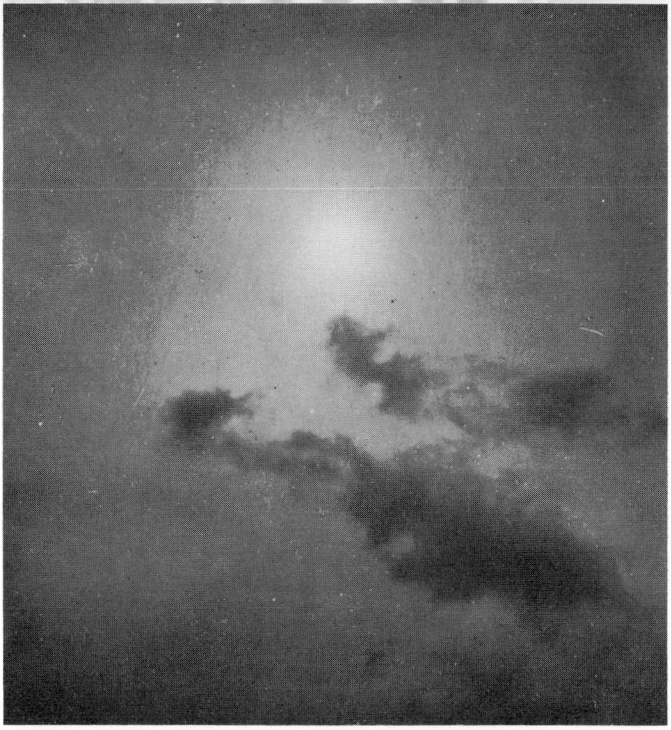

B. Thin altostratus with fractostratus patches below.

C. Stratus, a uniform layer extending below the level of the hilltop, with shreds of fractostratus.

D. Stratocumulus in irregular horizontal rolls.

E. Cumulus of fair weather.

F. Cumulonimbus, an isolated thunderstorm showing rain falling from base.

Figure 10.9 Cloud types. (Photographs courtesy of National Weather Service and U.S. Navy.)

Figure 10.10 Snowflakes. (Ewing Galloway.)

air temperature is below freezing. Individual snow crystals, which can be carefully caught upon a black surface and examined with a strong magnifying glass, develop in six-sided, flat crystals, or as prisms. They display infinite variations in their beautiful symmetrical patterns (Figure 10.10).

Hail consists of rounded lumps of ice, having an internal structure of concentric layers, much like an onion. Ordinarily the ice is not clear but has a frosted appearance. Hailstones range from 0.2 to 2 in (0.5 to 5 cm) in diameter and may be extremely destructive to crops and light buildings (Figure 10.11). Hail occurs only from the cumulonimbus cloud type, inside of which are extremely strong updrafts of air. Raindrops are carried up to high altitudes, are frozen into ice pellets, then fall again through the cloud. Suspended in powerful updrafts, the hailstone grows by the attachment and freezing of droplets, much as ice accumulates on the leading edge of an airplane wing. Eventually the hailstone escapes from the updrafts and falls to earth.

When rain falls upon a ground surface that is covered by an air layer of below-freezing temperature, the water freezes into clear ice after striking the ground or other surfaces such as trees, houses, or wires (Figure 10.12). The coating of ice that results is called a *glaze*, and an *icing storm* is said to have occurred. Actually no ice falls, so that ice glaze is not a form of precipitation. Icing storms cause great damage, especially to telephone and power wires and to tree limbs. Roads and sidewalks are made extremely hazardous.

How precipitation is measured

Precipitation is generally stated in units of inches or centimeters that fall per unit of time. One inch of rainfall, for example, is a quantity sufficient to cover the ground to a depth of one inch, provided that none is lost by runoff, evaporation, or sinking into the ground. A simple form

Figure 10.11 These hailstones, larger than hens' eggs (arrow), fell at Girard, Illinois, on August 13, 1929. (National Weather Service.)

Figure 10.12 Heavily coated wires and branches caused heavy damage in eastern New York State in January 1943, as a result of this icing storm. (Courtesy of National Weather Service, and New York Power and Light Co., Albany, N.Y.)

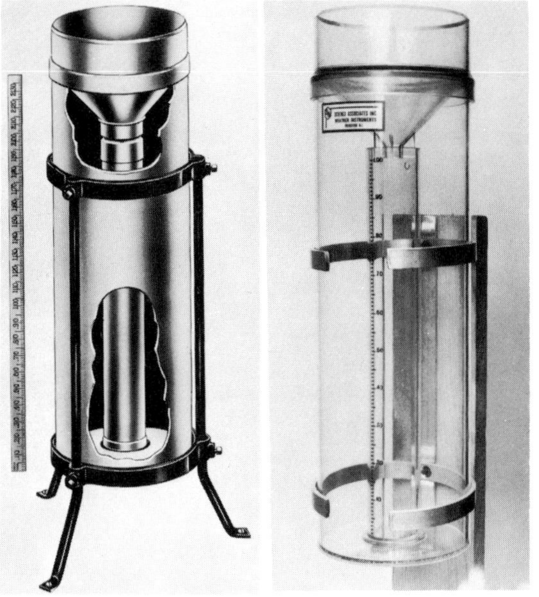

Figure 10.13 Standard 8-in. rain gauge used by the National Weather Service (left). A plastic 4-in. gauge showing funnel and graduated tube inside (right). (Photographs courtesy of Science Associates, Inc., Princeton, N.J.)

of *rain gauge* can be operated merely by setting out a straight-sided, flat-bottomed pan and measuring the depth to which water accumulates during a particular period. Unless this period is short, however, evaporation seriously upsets the results. Furthermore, very small amounts of rainfall, such as 0.1 in. (0.25 cm), would make too thin a layer to be accurately measured. To avoid this difficulty, as well as to reduce evaporation loss, good rain gauges are made in the form of a cylinder whose base is a funnel leading into a narrow tube (Figure 10.13). A small amount of rainfall will fill the narrow pipe to a considerable height, thus making it easy to read accurately, once a simple scale has been provided for the pipe. This gauge requires frequent emptying unless it is equipped with automatic devices for this purpose.

Snowfall is measured by melting a sample column of snow and reducing it to an equivalent in water. Thus, rainfall and snowfall records may be combined for purposes of comparison. Ordinarily, a ten-inch layer of snow is assumed to be equivalent to one inch of rainfall, but this ratio may range from 30 to 1 in very loose snow to 2 to 1 in old, partly melted snow.

Precipitation-producing conditions

Thus far we have observed that precipitation results when air rises and is adiabatically cooled below the dew point so rapidly that not only do clouds form, but rain, snow, or hail is produced as well. Consider, then, how large masses of air are actually induced to rise to higher elevations. The three possible ways are (1) *convectional,* (2) *orographic,* and (3) *cyclonic* or *frontal.*

Convectional precipitation results from a *convection cell,* which is simply an updraft of warmer air, seeking higher altitude because it is lighter than surrounding air (Figure 10.14). Suppose that on a clear, warm summer morning the sun is shining upon a landscape consisting of patches of open fields and woodlands. Certain of these types of surfaces, such as the bare ground, heat more rapidly and transmit radiant heat to the overlying air. Air over a warmer patch is thus warmed more than adjacent air and begins to rise as a bubble, much as a hot-air balloon after being released. Vertical movements of this type are often called *thermals* by sailplane pilots who use them to obtain lift.

As the air rises, it is cooled adiabatically so that eventually it will reach the same temperature as the surrounding air and come to rest. Before this happens, however, it may be cooled below the dew point. At once condensation begins, and the rising air column appears as a cumulus cloud whose flat base shows the critical level above which condensation is occurring (Figure 10.14). The bulging "cauliflower" top of the cloud represents the top of the rising warm air bubble, pushing into higher levels of the atmosphere. Usually the small cumulus cloud dissolves after drifting some distance downwind. However, should this convection continue to develop, the cloud may grow to a cumulonimbus mass, or thunderstorm, from which heavy rain will fall. Why, you may ask, does such spontaneous cloud growth take place and continue beyond the initial cumulus stage, long after the original input of heat energy is gone?

Actually, the unequal heating of the ground served only as a trigger effect to release a spontaneous updraft, fed by latent heat energy liberated

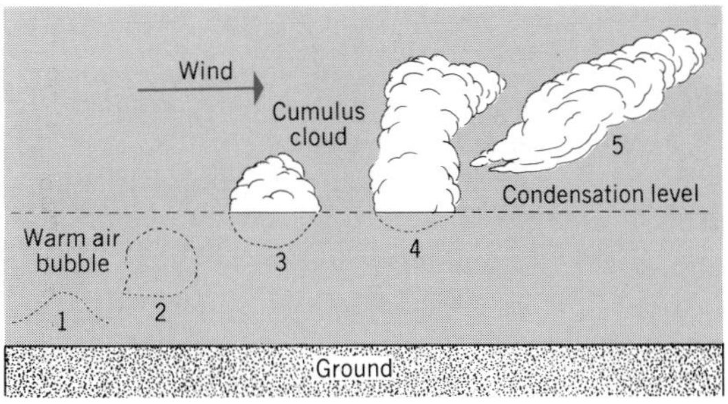

Figure 10.14 Rise of a bubble of heated air to form a cumulus cloud. (From A. N. Strahler, 1971, *The Earth Sciences*, 2nd ed., Harper and Row, New York.)

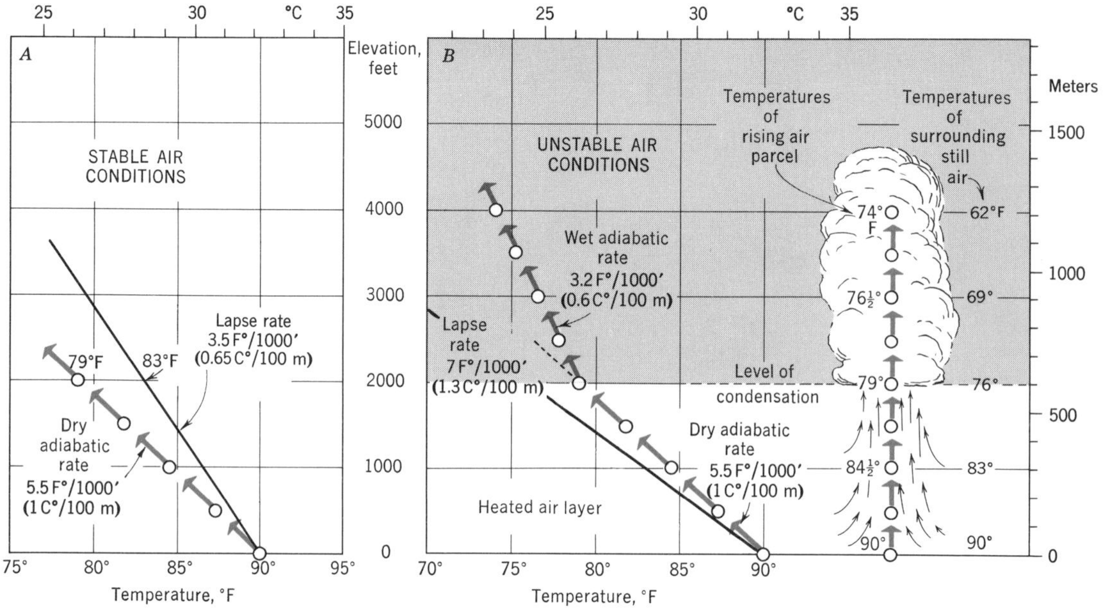

Figure 10.15 *A*, **Forced ascent of stable air.** *B*, **Spontaneous rise of unstable air.**

from the condensing water vapor. Recall that for every gram of water formed by condensation 600 calories of heat are released.

The graph of Figure 10.15*A* is a plot of altitude against air temperature. The small circles represent a small parcel of air being forced to rise steadily higher, following the same dry adiabatic rate of cooling shown in Figure 10.7. To the right of this line is a solid line showing the temperature of the undisturbed surrounding air; it is the environmental lapse rate, shown in Figure 7.2. Suppose that the air parcel is lifted from a point near the ground, where its temperature is 90°F (32°C). After the air parcel has been carried up 2000 ft (600 m), its temperature has fallen about 11 F° (6 C°) and is now 79°F (26°C); whereas the surrounding air is cooler by only about 7 F° (4 C°), and has a temperature of 83°F (28°C). The air parcel would thus be cooler than the surrounding air at 2000 ft (600 m), and if no longer forcibly carried upward, would tend to sink back to the ground. These conditions represent *stable* air, not likely to produce convectional rise, because the air would resist lifting.

When the air layer near the ground is excessively heated by the sun, the environmental lapse rate is increased (solid line has lower slope, Figure 10.15*B*). The air parcel near the ground begins to rise spontaneously because it is lighter than air over adjacent, less intensely heated ground areas. Although cooled adiabatically while rising, the air parcel at 1000 ft (300 m) has a temperature of 85°F (29°C), but this is well above the temperature of the surrounding still air. The air parcel, therefore, is lighter than the surrounding air and continues its rise. At 2000 ft (600 m), the dew point is reached and condensation sets in. Now the rising air parcel is cooled at the reduced wet adiabatic rate of 3.2 F° per 1000 feet (0.6 C° per 100 m), because the latent heat liberated in condensation offsets the rate of drop due to expansion. At 3000 ft (900 m), the rising air parcel is still several degrees warmer than the surrounding air, and therefore continues its spontaneous rise.

The air described here as spontaneously rising during condensation is *unstable* in properties. In such air the updraft tends to increase in intensity as time goes on, much as a bonfire blazes with increasing ferocity as the updraft draws in greater supplies of oxygen. Of course, at very high altitudes, the bulk of the water vapor having condensed and fallen as precipitation, the energy source is gone; the convection column then weakens and air rise finally ceases.

Unstable air, given to spontaneous convection in the form of heavy showers and thunderstorms, is most likely to be found in warm, humid areas such as the equatorial and tropical oceans and their bordering lands throughout the year, and the middle-latitude regions during the summer season.

The second precipitation-producing mechanism is described as *orographic*, which means "related to mountains." Prevailing winds or other moving masses of air may be forced to flow over mountain ranges (Figure 10.16). As the air rises on the windward side of the range, it is cooled at the adiabatic

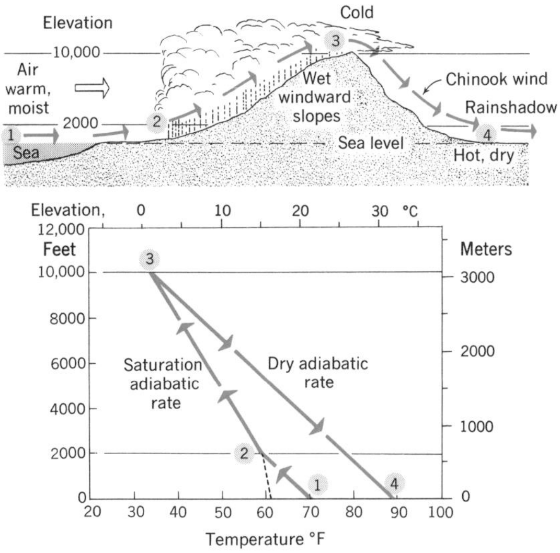

Figure 10.16 Forced ascent of oceanic air masses, producing precipitation and a rainshadow desert. (From A. N. Strahler, 1971, *The Earth Sciences*, 2nd ed., Harper and Row, New York.)

rate. If cooling is sufficient, precipitation will result. After passing over the mountain summit, the air will begin to descend the lee side of the range. Now it will undergo a warming through the same adiabatic process and, having no source from which to draw up moisture, will become very dry. A belt of dry climate, often called a *rainshadow*, may exist on the lee side of the range. Several of the important dry deserts of the earth are of this type.

Dry, warm *foehn winds* (Europe) and *chinook winds* (northwestern North America), which occur on the lee side of a mountain range, may cause extremely rapid evaporation of snow or soil moisture. These winds result from turbulent mixing of lower and upper air in the lee of the range. The upper air, which has little moisture to begin with, is greatly dried and heated when swept down to low levels.

An excellent illustration of orographic precipitation and rainshadow occurs in the far west of the United States. Prevailing westerly winds bring moist air from the Pacific Ocean over the coast ranges of central and northern California and the great Sierra Nevada Range, whose summits rise to 14,000 ft (4000 m) above sea level. (See Figure 16.15.) Heavy rainfall is experienced on the windward slopes of these ranges, nourishing great forests. Passing down the steep eastern face of the Sierras, air must descend nearly to sea level, even below sea level in Death Valley. The adiabatic heating thus caused, and a consequent drop in humidity, produces part of America's great desert zone, covering a strip of eastern California and all of Nevada.

Much orographic rainfall is actually of the convectional type, in that it takes the form of heavy convectional showers and thunderstorms. The storms are induced, however, by the forced ascent of unstable air as it passes over the mountain barrier.

A third type of precipitation is *cyclonic*. This topic cannot be fully understood until the entire subject of cyclonic storms and fronts has been developed. It will suffice here to note that in middle and high latitudes much of the precipitation occurs in cyclonic storms or eastward-moving centers of low pressure into which air is converging and being forced to rise.

Thunderstorms

A *thunderstorm* is an intense local storm associated with a large, dense cumulonimbus cloud in which there are very strong updrafts of air (Figures 10.8 and 10.9*F*). Thunder and lightning normally accompany the storm, and rainfall is heavy, often of cloudburst intensity, for a short period. Violent surface winds may occur at the onset of the storm.

Recent studies of cumulonimbus clouds show that a single thunderstorm consists of individual parts, called *cells*. Air rises within each cell as a succession of bubble-like air bodies, rather than as a continuous updraft from bottom to top (Figure 10.17). As each bubble rises, air in its wake is brought in from the surrounding region, a process called *entrainment*. Rising air in the thunderstorm cell can reach vertical speeds up to 3000 ft (900 m) per minute. Rapid condensation will

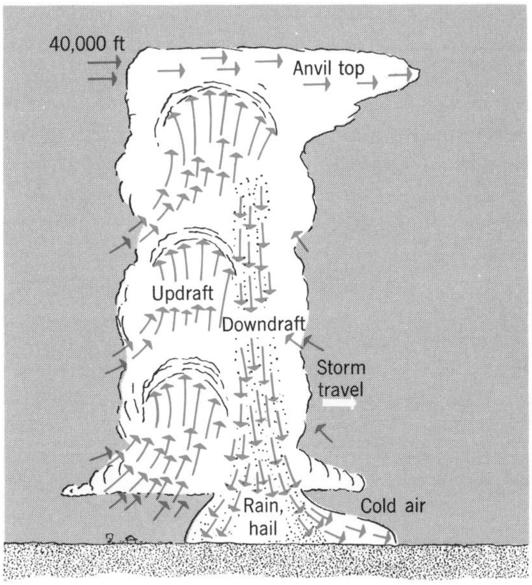

Figure 10.17 Schematic diagram of interior of a thunderstorm cell.

be in the form of rain in the lower levels, mixed water and snow at intermediate levels, and snow at high levels.

Upon reaching high levels, which may be on the order of 20,000 to 40,000 ft (6 to 12 km) or even higher, the rising rate diminishes and the cloud top is dragged downwind to form an *anvil top* (Figure 10.17). Ice particles falling from the cloud top act as nuclei for condensation at lower levels, a process called *seeding*. The rapid fall of condensation adjacent to the rising air bubbles exerts a frictional drag upon the air and sets in motion a downdraft. Striking the ground where precipitation is heaviest, this downdraft forms a local squall wind, which is sometimes strong enough to fell trees and do severe structural damage to buildings. We have already noted that hail is produced in the updrafts of a thunderstorm and that it can fall with damaging intensity in the thunderstorm downdraft.

Thunderstorms may be classified into several types, based on the mechanism or cause of initial lift of the air column that sets off the spontaneous growth of the storm. One common type is the *thermal*, or *air mass thunderstorm*, which is set off by thermal convection caused by solar heating of the ground and lower air layer. Figure 10.14 illustrates the process. Storms of this type are often widely scattered over a large region covered by warm, moist air. The time of occurrence is typically in the late afternoon when air temperatures near the ground reach their highest levels.

In the *orographic thunderstorm*, air is forced to rise over a mountain barrier, as shown in Figure 10.16. If the air is warm and moist, with unstable properties, it readily breaks into heavy showers and thunderstorms. The torrential monsoon rains of the Asiatic and East Indian mountain ranges are largely of this type. For example, Cherrapunji, a hill station facing the summer monsoon air drift in northeast India, averages 426 in (1082 cm) of rainfall annually. In the arid southwestern United States, isolated mountain ranges and high plateaus receive abundant summer rain in the form of thunderstorms orographically induced. Here rich forests grow while the surrounding lower areas are barren or sparsely vegetated deserts.

Another type of thunderstorm is set off by the forced rise of a layer of warm air over a layer of cold air. Such *frontal thunderstorms*, are explained in the following chapter.

Latent heat and the global balances of energy and water

An understanding of the processes of condensation and precipitation allows us now to take another look at the global energy balance, and to include the mechanism of latent heat transport.

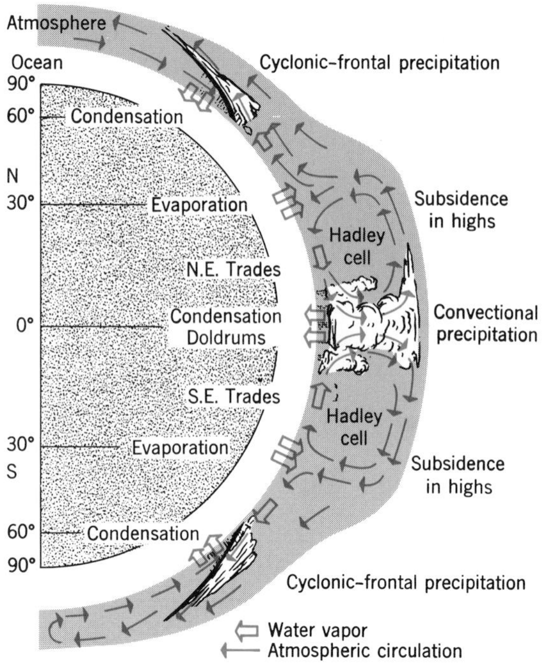

Figure 10.18 Schematic diagram of exchanges of water vapor along a meridional profile. (After A. N. Strahler, 1970, *Jour. Geography*, Vol. 69, p. 74, Figure 3.)

Figure 10.18 is a schematic diagram to show how evaporation and condensation are involved in energy exchange in each of the major latitude zones. We can now also summarize the pattern of transport of water vapor across the parallels of latitude; this is a transport of matter and is part of the global water balance. Figure 10.19 shows the average annual rate of vapor transport, in units of 10^{15} kilograms of water per year. Let us review the energy exchanges and water transports for each latitude zone.

The equatorial zone is characterized by a rise of moist, warm air in innumerable convectional cells reaching to the upper limits of the troposphere. As condensation and precipitation occur here, enormous amounts of latent heat energy are liberated. This zone has been called the "fire box" of the globe, in recognition of this intense production of sensible heat by condensation. Moving equatorward to replace the rising air are the tropical easterlies, or trades, within the Hadley cell circulation (Figure 10.18). Water evaporates from the ocean surface over the subtropical highs and is carried equatorward in vapor form. This transport shows in Figure 10.19 by peaks in the transport curve at about 10° N and 10° S lat. (Where the line of the graph lies to right of the center line, the movement is northward; where left, movement is southward.)

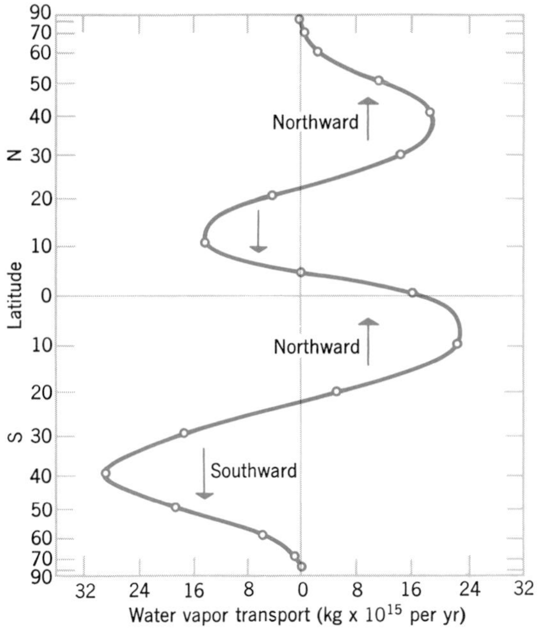

Figure 10.19 Graph of mean annual meridional transport of water vapor. (Simplified from data of W. D. Sellers, 1965, *Physical Climatology*, Univ. of Chicago Press, p. 94, Figure 29.)

Next, we move into middle latitudes, where upper-air waves constantly form and dissolve in the westerlies. Here, cyclones and anticyclones exchange cold, polar air for warm, tropical air across the parallels of latitude. Water vapor is also transported poleward and reaches a peak rate of flow at about 40° N and 40° S lat. as the graph in Figure 10.19 shows. Condensation in cyclonic storms removes water vapor from the troposphere in increasing quantities into high latitudes, so that the movement of water vapor declines and reaches zero at the poles.

Air pollution

We make good use of an understanding of the processes of condensation and precipitation when probing into man-made changes in the atmospheric environment. These are inadvertent changes, largely the result of urbanization and the growth of industrial processes, which have enormously increased the rate of combustion of hydrocarbon (fossil) fuels in the past half century or so. We have already considered the thermal effects of man's use of energy. Now we turn to consider the kinds of foreign matter, or *pollutants,* injected by man into the lower atmosphere, and the effects of pollution upon air quality and urban climate.

We recognize two classes of atmospheric pollutants. First, there are solid and liquid particles, which are designated collectively as *particulate matter.* Dusts found in smoke of combustion, as well as droplets naturally occurring as cloud and fog, fall into the category of particles. Second, there are compounds in the gaseous state, included under the general term *chemical pollutants* in the sense that they are not normally present in measurable quantities in clean air remote from densely populated, industrialized regions. (Excess CO_2 produced by combustion is not usually classed as a pollutant.)

One group of chemical pollutants of industrial and urban areas is of primary origin, that is, produced directly from a source on the ground. Gases included in this group are *carbon monoxide* (CO), *sulfur dioxide* (SO_2), *oxides of nitrogen* (NO, NO_2, NO_3), and hydrocarbon compounds. However, these chemical pollutants cannot be treated separately from particulate matter, since they are often combined within a single suspended particle. Certain dusts are said to be *hygroscopic* because they have an affinity for water and easily take on a covering water film (ordinary table salt is an example). The water film in turn absorbs the chemical pollutants. Hygroscopic particles of salt, derived from the sea surface, are normally present in great numbers in the atmosphere.

When particles and chemical pollutants are present in considerable density over an urban area, the resultant mixture is known as *smog.* Almost everyone living in middle latitudes is familiar with smog through its irritating effects upon the eyes and respiratory system and its ability to obscure distant objects. When concentrations of suspended matter are less dense, obscuring visibility of very distant objects, but not otherwise objectionable, the atmospheric condition is referred to as *haze.* Atmospheric haze builds up quite naturally in stagnant air masses as a result of the infusion of various surface materials. Haze is normally present whenever air reaches high relative humidity because water films grow on hygroscopic nuclei. Nuclei of natural atmospheric haze particles consist of mineral dusts from the soil, crystals of salt blown from the sea surface, hydrocarbon compounds (pollens and terpenes) exuded by plant foliage, and smoke from forest and grass fires. Dusts from volcanoes may, on occasion, add to atmospheric haze.

It is evident at this point that what we are calling atmospheric *pollutants* are of both natural and man-made origin, and that man's activities can supplement the quantities of natural pollutants present. Table 10.1 illustrates this complexity by listing the primary pollutants according to sources.

Not all man-made pollution comes from the cities. Isolated industrial activities can produce

TABLE 10.1 SOURCES OF PRIMARY ATMOSPHERIC POLLUTANTS[a]

Pollutants from Natural Sources	Sources of Man-made Pollutants
Volcanic dusts	Fuel combustion (CO_2, SO_2, lead)
Sea salts from breaking waves	Chemical processes
Pollens and terpenes from plants	Nuclear fusion and fission
(*Aggravated by man's activities:*)	Smelting and refining of ores
Smoke of forest and grass fires	Mining, quarrying
Blowing dust	Farming
Bacteria, viruses	

[a] After Assn. of Amer. Geographers, 1968, *Air Pollution*, Commission on College Geography, Resource Paper No. 2, Figure 3, p. 9.

pollutants far from urban areas. Particularly important are smelters and manufacturing plants in small towns and rural areas. Sulfide ores (metals in combination with sulfur compounds) are processed by heating in smelters close to the mine. Here, sulfur compounds are sent into the air in enormous concentrations from smokestacks. Fallout over the surrounding area is destructive to vegetation. An example is the Ducktown, Tennessee smelter, located in a richly forested area of the southern Appalachians. An area of several square miles surrounding the smelter has been denuded of vegetation and severely eroded; it resembles a patch of badlands in the arid lands of the West. Mining and quarrying operations send mineral dusts into the air. For example, asbestos mines (together with asbestos processing and manufacturing plants) send into the air countless threadlike mineral particles, some of which are so small that they can be seen only with the electron microscope. These particles travel widely and are inhaled by humans, lodging permanently in the lung tissue. Nuclear test explosions inject into the atmosphere a wide range of particles, including many radioactive substances capable of traveling thousands of miles in the atmospheric circulation. Certain of these pollutants belong to a very special and dangerous class, for they are sources of ionizing radiation, discussed in Chapter 7.

Man-induced forest and grass fires add greatly to smoke palls in certain seasons of the year. Plowing, grazing, and vehicular traffic raise large amounts of mineral dusts from surfaces of deserts and steppes. Bacteria and viruses, which we have not as yet mentioned, are borne aloft in air turbulence when winds blow over contaminated surfaces, such as farmlands, grazing lands, city streets, and waste disposal sites.

Figure 10.20 shows major atmospheric pollutants in percentage form as to both component matter and source, for 1968. The proportions will, of course, change somewhat from one year to the next. Much of the carbon monoxide, half of the hydrocarbons, and about a third of the oxides of nitrogen comes from exhausts of gasoline and diesel engines in vehicular traffic. Generation of electricity and various industrial processes contribute most of the sulfur oxides, because the coal and

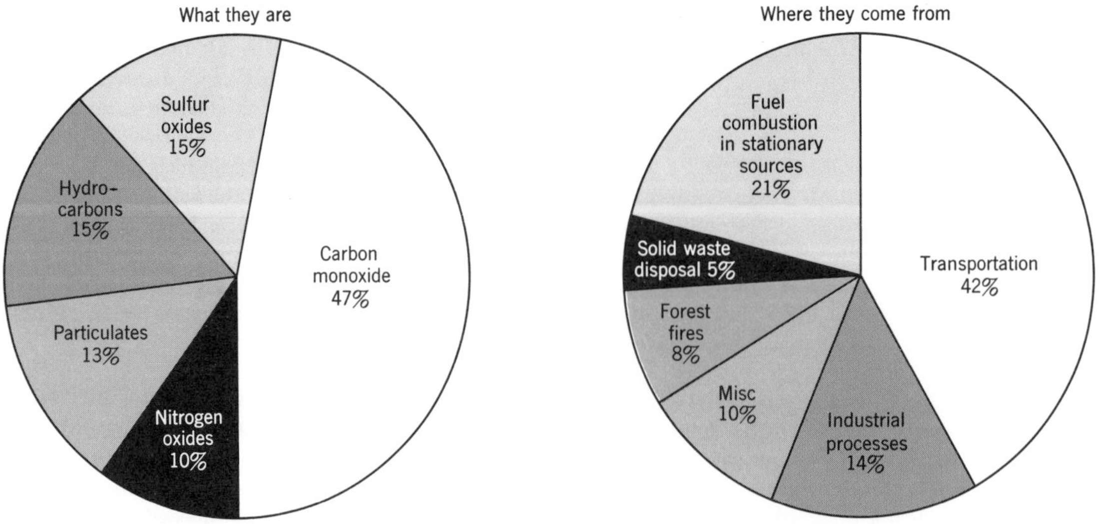

Figure 10.20 Air pollution emissions in the United States, 1968. Percentage by weight. (Data from National Air Pollution Control Administration, HEW.)

lower-grade fuel oil used for these purposes are comparatively rich in sulfur. These same sources also supply most of the particulate matter. *Fly ash* is the term applied to the coarser grades of soot particles emitted from smokestacks of generating plants. These particles settle out quite quickly within close range of the source. Combustion used for heating of buildings is a comparatively minor contributor to pollution, because the higher grades of fuel oil are low in sulfur and are usually efficiently burned. Very finely divided carbon comprises much of the smoke of combustion and is capable of remaining in suspension almost indefinitely because of its colloidal size. Forest fires comprise a secondary contributor of particles. Burning of refuse is a minor contributor in all categories of pollutants.

In the smog of cities there are, in addition to the ingredients mentioned above, certain chemical elements contained in particles contributed by exhausts of automobiles and trucks. Included are particles that contain lead, chlorine, and bromine. About half of the particles in automobile exhaust contain lead.

Primary pollutants are conducted upward from the emission sources by rising air currents that are a part of the normal convective process. The larger particles settle under gravity and return to the surface as *fallout.* Particles too small to settle out are later swept down to earth by precipitation, a process called *washout.* By a combination of fallout and washout the atmosphere tends to be cleansed of pollutants. In the long run a balance is achieved between input and output of pollutants, but there are large fluctuations in the quantities stored in the air at a given time. Pollutants are also eliminated from the air over their source areas by winds that disperse the particles into large volumes of cleaner air in the downwind direction. Strong winds can quickly sweep away most pollutants from an urban area, but during periods when a stagnant anticyclone is present, the concentrations rise to high values.

In polluted air, certain chemical reactions take place among the components injected into the atmosphere, generating a secondary group of pollutants. For example, sulfur dioxide (SO_2) may combine with oxygen to produce *sulfur trioxide* (SO_3), which in turn reacts with water of suspended droplets to yield *sulfuric acid* (H_2SO_4). This acid is both irritating to organic tissues and corrosive to many inorganic materials. In another typical reaction, the action of sunlight upon nitrogen oxides and organic compounds produces *ozone* (O_3), a toxic and destructive gas. Reactions brought about by the presence of sunlight are described as *photochemical.* One toxic product of photochemical action is *ethylene,* produced from hydrocarbon compounds.

Inversion and smog

Concentration of pollutants over a source area rises to its highest levels when the vertical mixing (convection) of the air is inhibited by a stable configuration of the vertical temperature profile of the air mass. The principles of stable versus unstable air conditions were covered earlier in this chapter and we shall now apply those principles to the problem of air pollution over cities.

When the normal environmental lapse rate of 3.5°F per 1000 ft (0.6°C per 100 m) is present, there is resistance to mixing by vertical movements, as we have already shown (Figure 10.21). Consider next that the environmental lapse rate is steepened by heating of an air layer near the ground, because of excess heat radiated and conducted from hot pavements and roof tops (Figure 10.21, right). When the temperature gradient (lapse rate) of the heated air becomes greater than the dry adiabatic rate of 5.5 F° per 1000 ft (1.0 C° per 100 m), a condition of instability exists and a bubble of warm air can begin to rise, like a

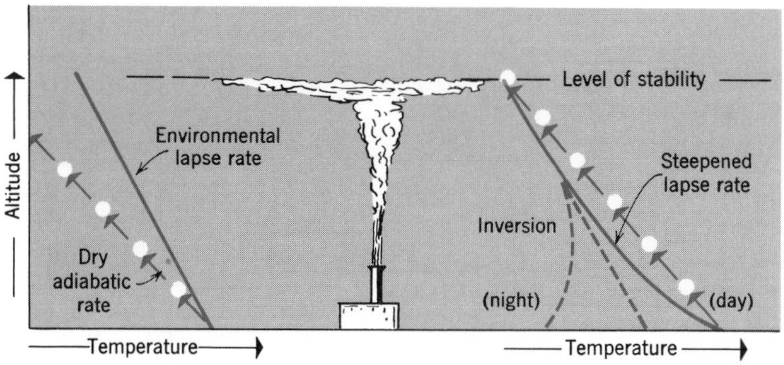

Figure 10.21 Relation of dry adiabatic lapse rate (circles) to various environmental lapse rates.

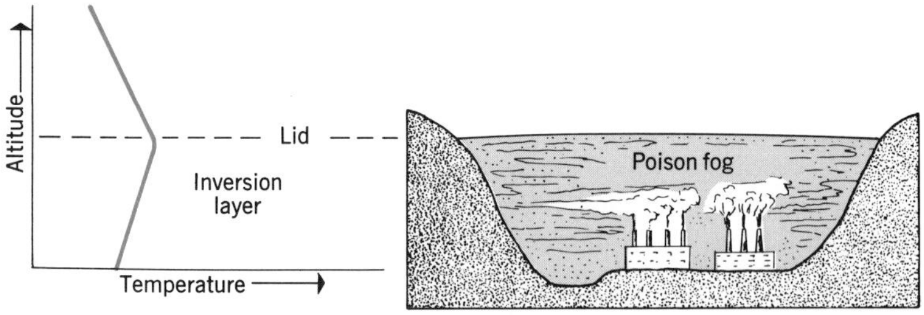

Figure 10.22 A low-level inversion (*left*) was the predisposing condition for poison fog accumulation at Donora, Pennsylvania, in October 1948. (From A. N. Strahler, 1972, *Planet Earth; Its Physical Systems Through Geologic Time*, Harper and Row, New York.)

helium-filled balloon. Assume that the lapse rate lessens with increased altitude, as shown by the curved line in Figure 10.21. Cooled at the dry adiabatic rate, the temperature of the rising bubble then falls faster than does the temperature of the surrounding air. When the bubble has reached an altitude at which its temperature (and therefore also its density) matches that of the surrounding air, it can rise no further and convection ceases.

Now suppose that instead of the bubble of warm air we substitute the hot air from a smokestack (Figure 10.21). The resultant rise follows essentially the same pattern, although initially faster and in the form of a vertical jet. Carrying up with it the pollutants of combustion, the rising hot air gradually cools and reaches a level of stability, where it spreads laterally. Cooling by longwave radiation and mixing with the surrounding air will reinforce the adiabatic cooling, since a truly adiabatic system would not be realistic in nature.

Recall that at night when the air is calm and the sky clear, rapid cooling of the ground surface typically produces a *low-level temperature inversion*, illustrated in Figure 8.27. In cold air, the reversal of the temperature gradient may extend hundreds of feet into the air. A low-level temperature inversion represents an unusually stable air structure. When this type of inversion develops over an urban area, conditions are particularly favorable for entrapment of pollutants to the degree that heavy smog or highly toxic fog can develop, as shown in Figure 10.22. The upper limit of the inversion layer coincides with the *cap*, or *lid*, below which pollutants are held. The lid may be situated at a height of perhaps 500 to 1000 ft (150 to 300 m) above the ground.

Although situations dangerous to health during prolonged low-level inversion have occurred a number of times over European cities since the Industrial Revolution began, the first major tragedy of this kind in the United States occurred at Donora, Pennsylvania, in late October 1948. The city occupies a valley floor hemmed in by steeply rising valley walls which prevent the free mixing of the lower air layer with that of the surrounding region (Figure 10.22). Industrial smoke and gases from factories poured into the inversion layer for five days, increasing the pollution level. Since humidity was high, a poisonous fog formed and began to take its toll. In all, 20 persons died and several thousand persons were made ill before a change in weather pattern dispersed the smog layer.

Related to the low-level inversion, but caused in a somewhat different manner, is the *upper-level inversion*, illustrated in Figure 10.23. Recall that anticyclones are cells of subsiding air that diverges at low levels. Within the center of the cell, winds

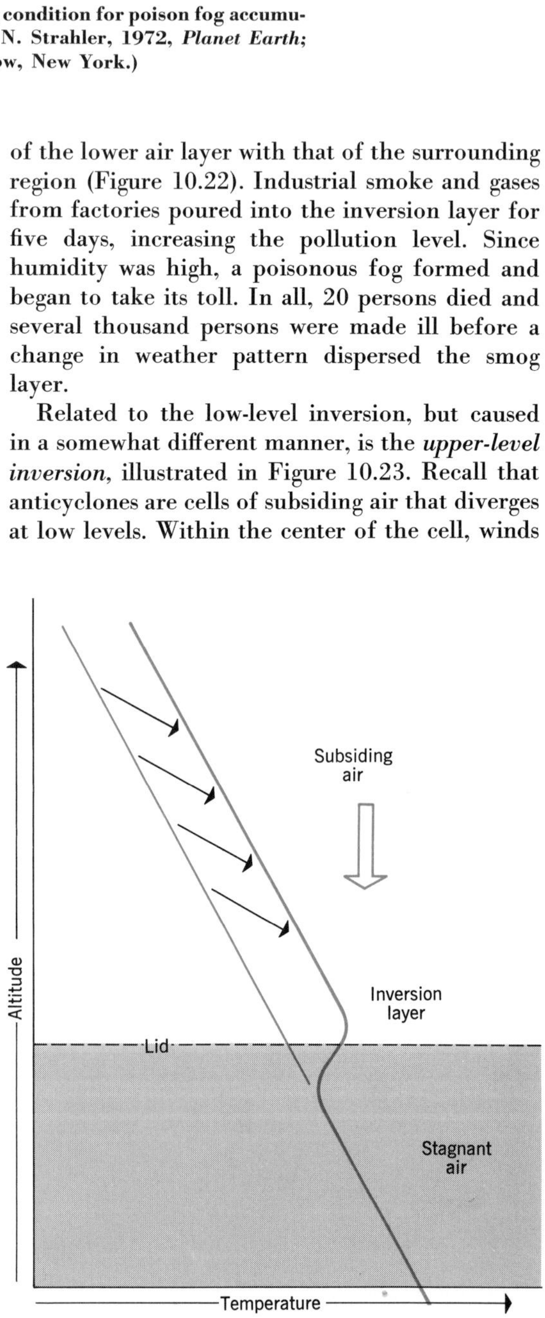

Figure 10.23 An upper-air inversion caused by subsidence. (From A. N. Strahler, 1972, *Planet Earth; Its Physical Systems Through Geologic Time*, Harper and Row, New York.)

are calm or very gentle. As the air subsides, it is adiabatically warmed, so that the normal temperature lapse rate is displaced to the right in the temperature-altitude graph, as shown in Figure 10.23 by the diagonal arrows. Below the level at which subsidence is occurring, the air layer remains stagnant. The temperature curve consequently develops a kink in which a part of the curve shows an inversion. For reasons already explained, the layer of inverted temperature structure strongly resists mixing and acts as a lid to prevent the continued upward movement and dispersal of pollutants. Upper-level inversions develop occasionally over various parts of the United States when an anticyclone stagnates for several days at a stretch.

For the Los Angeles Basin of southern California, and to a lesser degree the San Francisco Bay area and over other west coasts in these latitudes generally, special climatic conditions produce prolonged upper-air inversions favorable to persistent smog accumulation. The Los Angeles Basin is a low, sloping plain, lying between the Pacific Ocean and a massive mountain barrier on the north and east sides. Cool air is carried inland over the basin on weak winds from the south and southwest, but cannot move farther inland because of the mountain barrier. It is a characteristic of these latitudes that the air on the eastern sides of the subtropical high pressure cells is continually subsiding, creating a more-or-less permanent upper-air inversion that dominates the western coasts of the continents and extends far out to sea (see Figure 15.15). The effect is particularly marked in the summer season, when the Azores and Bermuda highs are at their largest and strongest. The subsiding air over the Los Angeles Basin is warmed adiabatically as well as heated by direct absorption of solar radiation during the day, so that it is markedly warmer than the cool stagnant air below the inversion lid, which lies at an elevation of about 2000 ft (600 m). Pollutants accumulate in the cool air layer, producing the characteristic smog that first became noticeable for its irritating qualities in the early 1940's. Because water vapor content is generally low, Los Angeles smog is better described as a dense haze than as a fog, for it does not shut out the sun or cause a reduction in visibility to a degree that interferes with vehicular operation or the landing and takeoff of aircraft (Figure 10.24). Close to the coast, true fogs are frequent, but this condition normally is prevalent on dry west coasts adjacent to cold currents. The upper limit of the smog stands out sharply in contrast to the clear air above it, filling the basin like a lake and extending into valleys in the bordering mountains.

One important physical effect of urban air pollution is that it reduces visibility and illumination. A smog layer can cut illumination by 10 percent in summer and 20 percent in winter. Ultraviolet radiation is absorbed by smog, which at times completely prevents these wavelengths from reaching the ground. Reduced ultraviolet radiation may prove to be of importance in permitting increased bacterial activity at ground level. City

Figure 10.24 Smog layer in downtown Los Angeles. Base of the temperature inversion is located immediately above the top of the smog layer, at a height of about 300 ft (100 m) above the ground. (Photograph by Los Angeles County Air Pollution Control District.)

TABLE 10.2 CLIMATE OF A CITY AS COMPARED WITH THAT OF THE SURROUNDING COUNTRYSIDE[a]

Radiation	
Total insolation	15 to 20 percent less
Ultraviolet (winter)	30 percent less
Ultraviolet (summer)	5 percent less
Sunshine duration	5 to 15 percent less
Temperature	
Annual mean	0.9 to 1.8 F° (0.5 to 1.0 C°) higher
Winter minimum	1.8 to 3.6 F° (1.0 to 2.0 C°) higher
Relative humidity	2 to 3 percent less
Cloudiness	
Cloud cover	5 to 10 percent more
Fog in winter	100 percent more
Fog in summer	30 percent more
Precipitation	
Total quantity	5 to 10 percent more
Snowfall	5 percent less
Particulate matter	10 times more
Gaseous pollutants	5 to 25 times more
Wind speed	
Annual mean	20 to 30 percent lower
Extreme gusts	10 to 20 percent lower
Calms	5 to 20 percent more frequent

[a] Data of H. E. Landsberg, 1970, *Meteorological Monographs,* Vol. 11, p. 91, Table 1.

smog cuts horizontal visibility to some one-fifth to one-tenth of the distance normal for clean air. Where atmospheric moisture is sufficient, the hygroscopic particles acquire water films and can cause formation of true fog with near-zero visibility. Over cities, winter fogs are much more frequent than over the surrounding countryside. Coastal airports, such as La Guardia (New York), Newark (New Jersey), and Boston suffer severely from a high incidence of fogs augmented by urban air pollution.

A related effect of the urban heat island (Chapter 8) is the general increase in cloudiness and precipitation over a city, as compared with the surrounding countryside. This increase results from intensified convection generated by heating of the lower air. For example, it has been found that thunderstorms over the city of London produce 30 percent more rainfall over the city than over the surrounding country. Increased precipitation over an urban area is estimated to average from 5 to 10 percent over the normal for the region in which it lies.

Table 10.2 summarizes the main climatic differences between a city and the surrounding countryside. Keep in mind that this summary is a generalization applied to highly industrialized nations in middle latitudes, and there are differences among cities with respect to magnitude of a given effect.

As we found in Chapter 8, a large city produces a heat island. Within this warm air layer, pollutants are trapped beneath an inversion lid. The layer of polluted air takes the form of a broad *pollution dome* centered over the city when winds are very light or near calm (Figure 10.25). However, when there is general air movement in response to the pressure gradient field, the pollutants are carried far downwind to form a *pollution plume.* Figure 10.26 has two maps showing plumes from the major cities of the Atlantic seaboard. The *V*-lines show zones of fallout beneath the plumes, while the small dot at the open end of the *V* shows the distance traveled by the air in one day from the source. The left-hand map shows the effects of weak southerly winds on a day in June. In this situation pollution from one city affects another and generally the pollutants remain over the land, contaminating suburban and rural areas over a wide zone. The right-hand map, for a day in February, shows the effect of strong westerly winds, causing the pollution to be carried directly to sea.

Harmful effects of atmospheric pollution

A list of the harmful effects of atmospheric pollutants upon a plant and animal life and upon inorganic substances would be a very long one if fully developed. We can only suggest some of these effects. For humans in cities, both sulfur dioxide and hydrocarbon compounds, altered by photochemical reaction to produce sulfuric acid and ethylene, respectively, are irritants to the eyes and to the respiratory system. Nitrogen dioxide is also an eye and lung irritant when present in sufficient quantities. Photochemical alteration of nitrogen oxides leads to production of ozone, which acts as an irritant in smog and would be lethal if it occurred in large concentrations.

For persons suffering from respiratory ailments, such as bronchitis and emphysema, the breathing of heavily polluted air can bring on disability and even death, as statistics clearly show. During the London fog of December 1952, the death rate

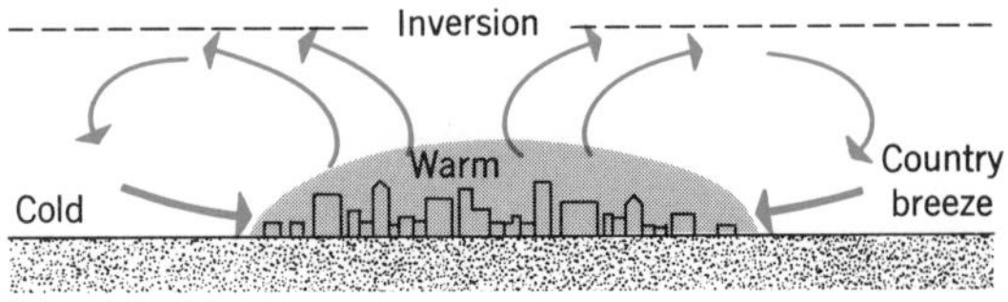

Figure 10.25 Air flow at night within an urban heat island. (After H. E. Landsberg, 1970, *Science*, Vol. 170, p. 1271, Figure 5.)

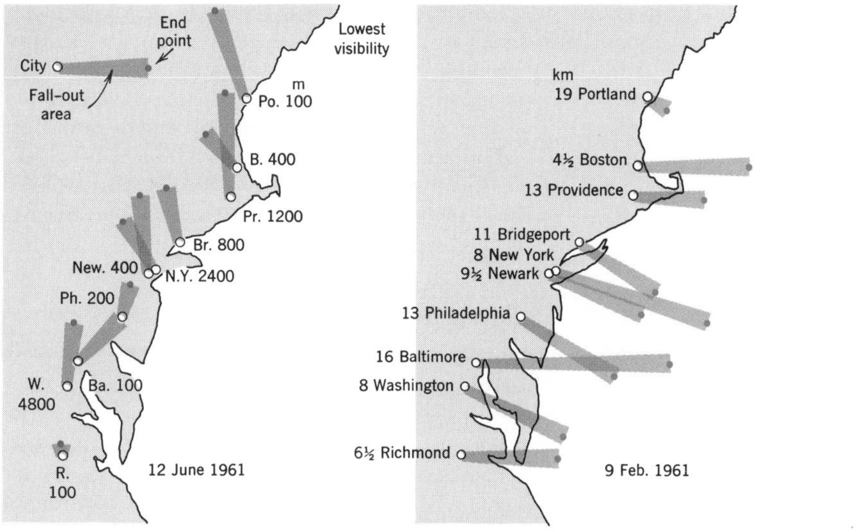

Figure 10.26 Pollution plumes from cities on the eastern seaboard under conditions of weak, southerly winds (*left*) and strong westerly winds (*right*). The dot at end of each plume represents approximate distance traveled by pollutants at the end of a 24-hour period. (After H. E. Landsberg, 1962, in *Symposium—Air Over Cities*, Sanitary Eng. Center Tech. Rept. A62-5, Cincinnati, Ohio.)

approximately doubled, while an increased death rate over the average rate persisted for weeks after the event. In a more recent London fog, that of December 1962, more than 300 deaths were attributed to the breathing of polluted air. Particularly hard hit were the very young and the old. There is also a suspected linkage between breathing of atmospheric pollutants and lung cancer, since the incidence of that disease is higher in cities than in other areas. It is thought that the accumulation of atmospheric hydrocarbon compounds on lung tissues may predispose to the onset of lung cancer.

Carbon monoxide is a cause of death when inhaled in sufficient quantities. Everyone knows that the carbon monoxide from automobile exhaust will kill in a short time when breathed in a closed garage. Carbon monoxide levels are a general indicator of the degree of air pollution from vehicular exhausts, but concentrations rarely reach sufficient levels in the open air to be a threat to life. Nevertheless, the long-continued inhalation of small amounts of carbon monoxide is suspected of harmful effects, not as yet evaluated.

Ozone in urban smog has a most deleterious effect upon plant tissues and in some cases has caused the death or severe damage of ornamental trees and shrubs. Sulfur dioxide is injurious to certain plants, and is a cause of loss of productivity in truck gardens and orchards in polluted air. Atmospheric sulfuric acid in cities has in some places largely wiped out lichen growth.

Although secondary in the sense that the loss is in dollars, rather than in lives and health of animals and plants, the deterioration of various materials subjected to polluted air forms an important category of harmful effects. Building stones and masonry are susceptible to the corrosive action of sulfuric acid derived from the atmosphere. Metals, fabrics, leather, rubber, and paint deteriorate and discolor under the impact of exposure to urban air pollutants. Ozone, in particular, is deleterious to natural rubber, causing it to harden and crack. The sulfuric acid produced from sulfur dioxide corrodes exposed metals, particularly steel and copper. Not the least of the economic losses from pollution are simply from soilage of clothing, automobiles, furniture, and interior floors, walls, and ceilings. The cleaning bill totaled for a large city is truly staggering, when calculated to include the labor and cleaning agents expended by householders.

Lead and other toxic metals in the polluted atmosphere are a particular source of concern for human health in the future. The lead-bearing particles from auto exhausts tend to concentrate in the grass, leaves, and soil near major highways. There is good reason to suppose that humans ingest lead particles directly from the air and that these may prove to be a health hazard. Although lead poisoning from atmospheric sources has not yet been documented in humans, there is now evidence that it has caused the deaths of animals in city zoos. Tests have shown high levels of lead in the tissues of the dead animals and no source other than atmospheric particles has been found for the ingested lead. Animals in outdoor cages showed higher lead levels than animals kept indoors. These findings are ominous in tone and tend to reinforce the conclusion of the Air Pollution

Control Office of the Environmental Protection Agency that atmospheric lead pollution is a possible health hazard.

Radioactive substances in the atmosphere are a special form of environmental hazard because of the genetic damage that is done to plant and animal tissues exposed to ionizing radiation. This radiation is also important in causing mutations.

Late in 1970, the federal Clean Air Act was signed into law, giving the Environmental Protection Agency authority to set national standards for tolerable limits of pollutants in the air. Included in the act is provision for an emergency alert system that will signal the need to reduce fuel combustion and curtail the use of automobiles when the danger point is reached in a given city.

REVIEW QUESTIONS

1. In what ways is the presence of water vapor in the air important in man's environment? What are the sources of atmospheric moisture?

2. Describe the three states in which water may exist. How, and in what amounts, is heat liberated or absorbed as water passes from one state to another?

3. Define relative humidity. What effect has a change of air temperature upon the relative humidity of the air? What is the dew point temperature? Explain the phenomenon of condensation of water vapor.

4. What is absolute humidity? Specific humidity? What use have these measures of water vapor? What ranges of specific humidity may be expected from equatorial to polar regions?

5. How is humidity of the air measured? Explain the principle of the hygrometer and psychrometer.

6. Explain the process whereby condensation and precipitation occur as a result of the lifting of a mass of air. What is the adiabatic rate? How does this differ from the environmental lapse rate?

7. Of what are clouds composed? How are clouds classified? Name the common cloud types, give their height ranges, and describe their forms. What are stratiform and cumuliform clouds?

8. Name and describe the various forms of precipitation. Is an icing storm a form of precipitation? How does hail form?

9. In what units is rainfall measured? Describe a rain gauge. How is snowfall measured? How does 1 foot of snow compare with 1 foot of rainfall for relative amounts of water?

10. What are the three ways in which precipitation can occur on a large scale? Describe how a convection cell operates to produce rainfall. What keeps the updraft in operation during a convectional storm? What is an unstable air mass?

11. Explain how orographic precipitation occurs. Why is there often a dry area, or rainshadow, on the lee side of a mountain chain? Is the air which has reached a rainshadow area normally warmer or cooler than it was at the same elevation over the coastal zone from which it came? Explain.

12. Describe the internal meteorological conditions of a thunderstorm. What size and height do these storms commonly attain? How is hail formed in a thunderstorm? What keeps a thunderstorm active? What is the source of energy?

13. Name three common types of thunderstorms. What can be said about time of day and geographical regions of occurrence of thunderstorms?

14. Describe the transport of water vapor across the parallels of latitude from one zone to another. What changes of state occur in the Hadley cell circulation? In middle and high latitudes?

15. What classes of atmospheric pollutants are recognized? Name the component substances and particles in each class. Distinguish between primary and secondary air pollutants.

16. To what extent are air pollutants manmade and to what extent natural? Name the natural sources.

17. Describe the fallout, washout, and dispersal of air pollutants. What effect do winds have on urban air pollution?

18. Explain how inversions are conducive to smog formation and persistence. What two basic types of inversions are important in incurring air pollution? What is a pollution dome? A pollution plume?

19. Describe the harmful effects of air pollution.

Exercises

1. A mass of rising air has a temperature of 65°F at 1000 ft elevation. If the air temperature is falling at the dry adiabatic rate, what will the temperature be at 4000 ft? Assuming that this air initially has a dew-point temperature of 36°F, at what level will condensation set in? Refer to Figure 10.7 for guidance.

2. The graph in Figure 10.16 shows the changes in air temperature accompanying the passage of a mass of air over a mountain range and down the other side. Construct a similar graph of this type to show what happens during the following sequence of changing conditions.

Air at 84°F at sea level rises up mountain slopes;

reaches dew-point temperature of 42°F; condensation sets in. Air continues to rise (at wet-adiabatic rate) until, at the summit, temperature has dropped to 35°. Air descends lee slope of range to floor of an interior basin, 500 ft elevation. (Follows dry adiabatic rate down.)

(*a*) At what elevation was dew-point temperature reached? (*b*) How high was the summit of the range? (*c*) What was the air temperature on arrival at the floor of the interior basin? (*d*) How does the humidity of the air compare at the start and at the finish of this series of changes? (*e*) Suppose the air were next forced to pass over another range of the same elevation farther inland; would the same series of changes happen all over again? Explain.

3. During a torrential summer thunderstorm, an automatic recording rain gauge showed that the rainfall in each five-minute interval was as given below. The first period began at 3:40 P.M.

(*a*) Show this information by a simple bar graph. Let each bar, from left to right, represent one five-minute period. Let the height of each bar be proportional to the amount of rain, using a scale of one inch on the graph equals 0.10 inch of rainfall. (*b*) How much rain, measured in inches, fell during the entire storm? (*c*) Was the heaviest rain near the beginning or end of the rain period? (*d*) How long did the rainfall period last? This exercise illustrates a typical burst of rainfall from a moving thunderstorm.

4. The accompanying map of the lower Ohio Valley region shows the total rainfall during a four-day rain storm in October. Lines have been drawn through stations that have received the same amount of rainfall. Lines of this type, termed *isohyetal lines*, are drawn like isobars, isotherms, or topographic contours. Complete the isohyetal lines on this map. Draw these on a sheet of thin tracing paper laid over the map. Label the five- and ten-inch lines. (Based on data of D. W. Mead.)

Five-minute Period	1	2	3	4	5	6	7	8	9
Rainfall, inches	0.22	0.30	0.42	0.09	0.04	0.02	0.01	0.01	0.00

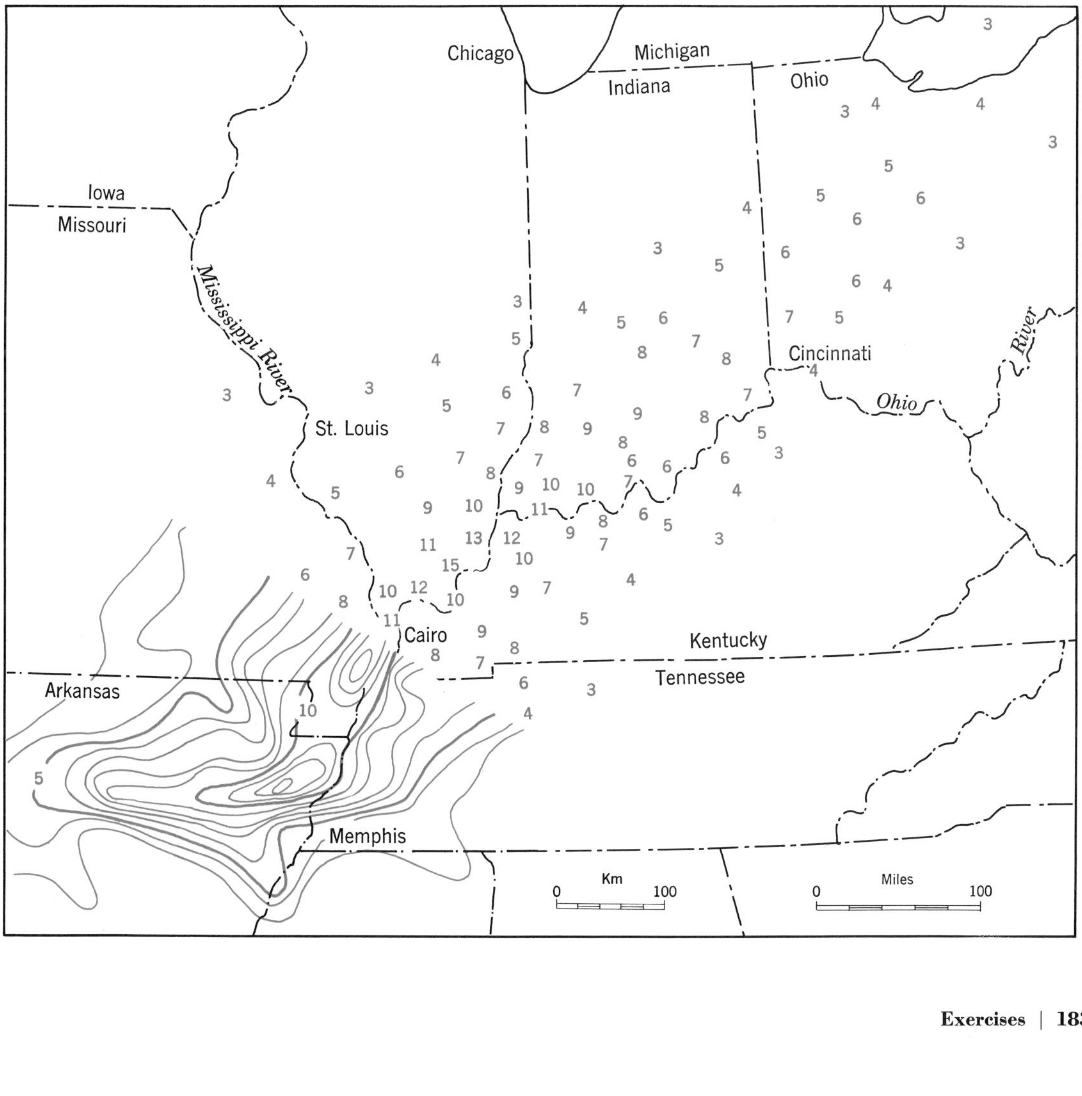

CHAPTER 11

Air Masses, Weather Fronts, Cyclonic Storms

THE atmosphere exerts stress—severe at times—upon man and other life forms through weather disturbances involving extremes of wind speeds, cold, and precipitation. Under the general category of storms, these phenomena constitute environmental hazards, not only directly, but indirectly through their attendant phenomena—storm waves and storm surges of the seas and river floods, mudflows, and landslides of the lands. Weather disturbances of lesser magnitude are among the beneficial environmental phenomena, for they bring precipitation to the land surfaces, and thus recharge the vital supplies of fresh water upon which man and all other terrestrial life forms depend.

An understanding of weather disturbances of all intensities enables man to predict their times and places of occurrence, and thus to give warnings and allow protective measures to be taken. This function of the atmosphere scientist can be placed under the heading of *environmental protection.* It is also possible to a limited degree for man to modify atmospheric processes in a deliberate way, in order to ameliorate the intensity of wind stresses and precipitation intensities, in the case of storms, and to increase precipitation over drought areas. These activities come under the heading of *planned weather modification.* Here, again, we find that the relationship of man with the atmosphere is one of interaction.

Traveling cyclones

Much unsettled, cloudy weather experienced in middle and high latitudes is associated with traveling cyclones. The convergence of masses of air toward these centers is accompanied by lift of air and adiabatic cooling, which, in turn, produces cloudiness and precipitation. By contrast, much fair, sunny weather is associated with traveling anticyclones in which the air tends to subside and spread outward, causing adiabatic warming, a process that is unfavorable to the development of clouds and precipitation.

Cyclones may be very mild in intensity, passing with little more than a period of cloud cover and light rain or snow. On the other hand, if the pressure gradient is strong, winds ranging in strength from moderate to gale force may accompany the cyclone. In such a case, the disturbance may be called a *cyclonic storm.*

Moving cyclones fall into three general classes. (1) The *wave cyclone* of middle and high latitudes (also called *extratropical cyclone*). It ranges in severity from a weak disturbance to a powerful storm. (2) The *tropical cyclone* of low latitudes over ocean areas. It ranges from a mild disturbance to the terribly destructive *hurricane,* or *typhoon.* (3) The *tornado.* Although a very small storm, it is an intense cyclonic vortex of enormously powerful winds. It is on a very much smaller scale of magnitude than other types of cyclones and must be treated separately.

Air masses

Cyclones of middle and high latitudes depend for their development and structure upon the coming together of large bodies of air of contrasting physical properties. A body of air in which the upward gradients of temperature and moisture are fairly uniform over a large area is known as an *air mass.* In areal extent, a single air mass may be of subcontinental proportions; in vertical dimension it may extend through the troposphere. A given air mass is characterized by a distinctive combination of temperature, environmental lapse rate, and specific humidity. Thus we find air masses differing widely in temperature—from very warm to very cold—and in moisture content—from very dry to very moist.

A given air mass may have a rather sharply defined boundary between itself and a neighboring air mass. This discontinuity is termed a *front.* We found an example of a front in the contact between polar and tropical air masses below the axis of the jet in upper air waves, as shown in Figure 9.22. This feature is called the *polar front*; it represents the highest degree of global generalization. Fronts may be nearly vertical, as in the case of air masses having little motion relative to one another, or they may be inclined at an angle not far from the horizontal, in cases where one air mass is sliding over another. A front may be al-

TABLE 11.1 PROPERTIES OF TYPICAL AIR MASSES

Air Mass	Symbol	Properties	Temperature °F	(°C)	Specific Humidity g/kg
Continental-arctic (and continental antarctic)	cA (cAA)	Very cold, very dry (winter)	−50°	(−46°)	0.1
Continental polar	cP	Cold, dry (winter)	12°	(−11°)	1.4
Maritime polar	mP	Cool, moist (winter)	39°	(4°)	4.4
Continental tropical	cT	Warm, dry	75°	(24°)	11
Maritime tropical	mT	Warm, moist	75°	(24°)	17
Maritime equatorial	mE	Warm, very moist	80°	(27°)	19

most stationary with respect to the earth's surface, but nevertheless the adjacent air masses may be in relatively rapid motion with respect to each other along the front.

The properties of an air mass are derived in part from the regions over which it passes. Because the entire troposphere is in more or less continuous motion, the particular air-mass properties at a given place reflect the composite influence of trajectories covering thousands of miles and passing alternately over oceans and continents. This complexity of influences is particularly important in middle and high latitudes in the northern hemisphere, within the flow of the global westerlies. However, there are vast tropical and equatorial areas over which air masses reflect quite simply the properties of the ocean and land surfaces over which they move slowly or tend to stagnate. Thus, over a warm equatorial ocean surface the lower levels of the overlying air mass may develop a high water vapor content combined with a steep environmental temperature lapse rate. Over a large tropical desert, slowly subsiding air forms an air mass of high temperatures and low relative humidities. Over cold, snow-covered land surfaces in arctic latitudes in winter, the lower layer of the air mass remains very cold with a very low water vapor content. Meteorologists have designated as *source regions* those land or ocean surfaces that strongly impress their temperature and moisture characteristics upon overlying air masses moving over them.

Air masses move from one region to another following the patterns of barometric pressure. During such migration, lower levels of the air mass undergo gradual modification, taking up or losing heat to the surface beneath, and perhaps taking up or losing water vapor as well.

Air masses are classified according to two categories of generalized source regions: (1) latitudinal position on the globe, which primarily determines thermal properties, and (2) underlying surface—continents or oceans—determining the moisture content. With respect to latitudinal position, five types of air masses are as follows:

Air Mass	Symbol	Source Region
Arctic	A	Arctic ocean and fringing lands
Antarctic	AA	Antarctica
Polar	P	Continents and oceans, 50°–60° N. and S.
Tropical	T	Continents and oceans, 20°–35° N. and S.
Equatorial	E	Oceans close to equator

With respect to type of underlying surface, two further subdivisions are imposed on the preceding types as follows:

Air Mass	Symbol	Source Region
Maritime	*m*	Oceans
Continental	c	Continents

By combining types based on latitudinal position with those based on underlying surface a list of six important air masses results; these are listed in Table 11.1. (Figure 14.4 shows global distribution of source regions of these air masses.) Table 11.1 gives typical values of temperature and specific humidity at the surface, although a wide range in these properties may be expected, depending on season. The equatorial air mass, mE, holds about 200 times as much water vapor as the extremely cold arctic and antarctic air masses, cA and cAA. The maritime tropical air mass mT and maritime equatorial mE air mass are quite similar in temperature and water vapor content. Both are capable of very heavy yields of precipitation. The continental tropical air mass cT has its source

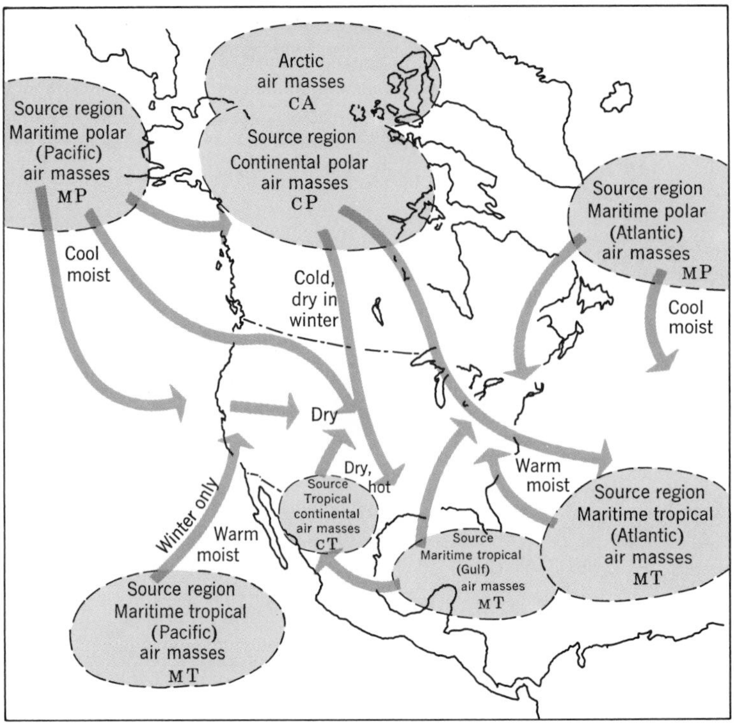

Figure 11.1 North American air mass source regions and trajectories. (After Haynes, U.S. Dept. Commerce.)

region over subtropical deserts of the continents. Although typically it has a substantial water vapor content, it has low relative humidity when highly heated during the daytime. The polar maritime air mass mP, with specific humidity of 4 to 5 g/kg in winter, originates over middle-latitude oceans. It has a high relative humidity and can yield substantial precipitation.

North American air masses

The continental polar air masses of North America (symbol, cP) originate over north-central Canada and are characterized by low temperature and low moisture content (Figure 11.1). These air masses form tongues of cold air, which periodically extend south and east from the source region to produce anticyclones accompanied, in winter, by low temperatures and clear skies. Over the Arctic Ocean and its bordering lands of the arctic zone there develops the arctic air mass (symbol, A) which is extremely cold and stable. When this air mass invades the United States, it produces a severe cold wave.

Over the North Pacific and Bering Straits originate the maritime polar air masses (symbol, mP). With ample opportunity to absorb moisture both over the source region and throughout their travel southeastward to the west coast of North America, these air masses are characteristically cool and moist, with a tendency in winter to become unstable, giving heavy precipitation over coastal ranges.

Another maritime polar air mass of the North American region originates over the North Atlantic Ocean. It, too, is cool and moist.

Note that the polar air masses (mP, cP) originate in the subarctic latitude zone (Figure 8.6), not in the polar latitude zone. The meteorological definition of the word "polar" for air masses has long been in use and has international acceptance, hence cannot be changed to conform with the latitude zones defined in Chapter 8.

Of the tropical air masses the commonest visitor to the central and eastern states is the maritime tropical air mass (symbol, mT) from the Gulf of Mexico. It moves northward, bringing warm, moist, unstable air over the eastern part of the country. In the summer, particularly, this air mass brings hot, sultry weather to the central and eastern United States. This air mass gives frequent thunderstorms. Closely related is a maritime tropical air mass from the Atlantic Ocean east of Florida, over the Bahamas.

During the summer there originates over northern Mexico, western Texas, New Mexico, and Arizona a tropical continental air mass (symbol, cT) which is hot and dry. This air mass does not travel widely, but governs weather conditions over the source region.

Over the Pacific Ocean in the cell of high pressure located to the southwest of Lower California is a source region of another maritime tropical air mass. It visits the United States only in winter and affects only the California coast.

Cold and warm fronts

The structure of a frontal contact zone in which cold air is invading the warm-air zone is shown in Figure 11.2. A front of this type is termed a *cold front.* The colder air mass, being heavier, remains in contact with the ground and forces the warmer air mass to rise over it. The slope of the cold front surface is greatly exaggerated in the figure, being actually of the order of slope of 1 in 40 to 1 in 80 (meaning that the slope rises 1 foot vertically for every 80 feet of horizontal distance). Cold fronts are associated with strong atmospheric disturbance; the warm air thus lifted often breaks out in violent thunderstorms. These may also occur along a line well in advance of the cold front, a *squall line.* Thunderstorms can be seen on the radar screen (Figure 11.3).

Figure 11.4 illustrates a *warm front* in which warm air is moving into a region of colder air. Here, again, the cold air mass remains in contact with the ground, and the warm air mass is forced to rise as if ascending a long ramp. Warm fronts have lower slopes than cold fronts, being of the order of 1 in 80 to as low as 1 in 200. Moreover, warm fronts are commonly attended by stable atmospheric conditions and lack the turbulent air motions of the cold front. Of course, if the warm air is unstable, it will develop convection cells and there will be heavy showers and thunderstorms.

Cold fronts normally move along the ground at a faster rate than warm fronts. Hence, when both types are in the same neighborhood, as they are in the cyclonic storm, the cold front may overtake the warm front. An *occluded front* then results (Figure 11.5). The colder air of the fast-moving cold front remains next to the ground, forcing both the warm air and the less cold air to rise over it. The warm air mass is lifted completely free of the ground.

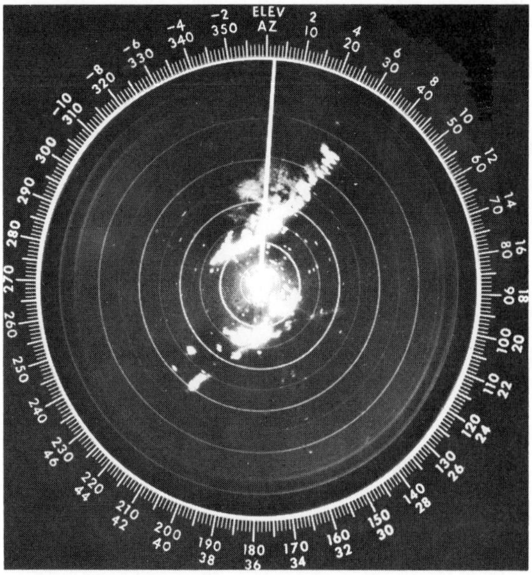

Figure 11.3 Photograph of a radar screen on which lines of thunderstorms show as bright patches. The heavy circles are spaced 50 nautical miles (80 km) apart. (National Weather Service.)

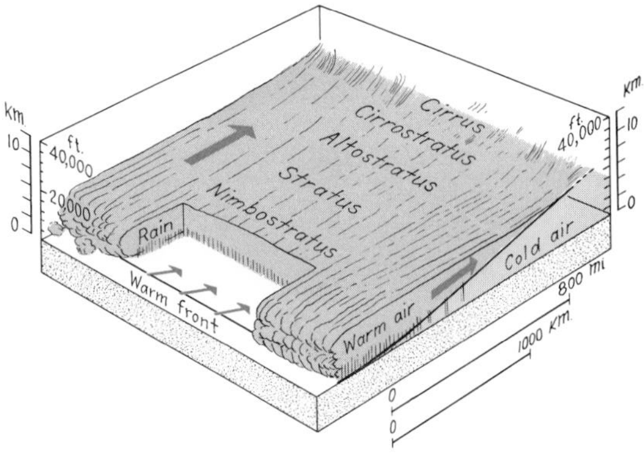

Figure 11.4 A warm front.

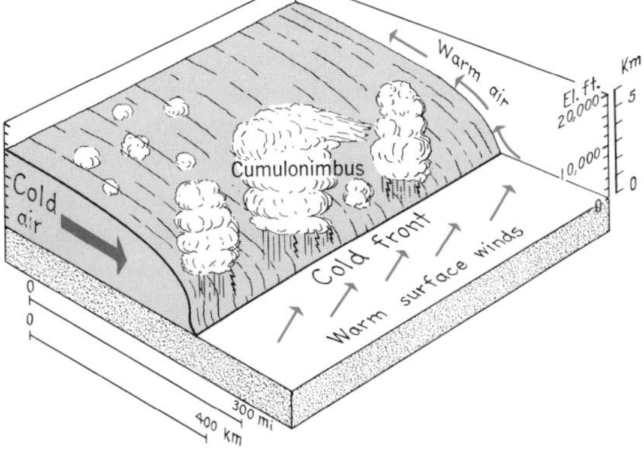

Figure 11.2 A cold front.

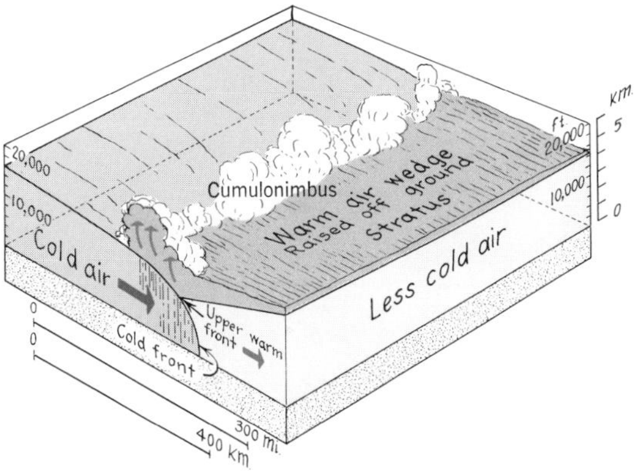

Figure 11.5 An occluded warm front.

Wave cyclones

The dominant type of weather disturbance of middle and high latitudes is the *wave cyclone*, a vortex that repeatedly forms, intensifies, and dissolves along the frontal zone between cold and warm air masses. The Norwegian meteorologist, J. Bjerknes, at the time of World War I, recognized the existence of atmospheric fronts and developed his *wave theory* of cyclones.

The term *front*, used by Bjerknes, was particularly apt because of the resemblance of this feature to the fighting fronts in western Europe, then active. Just as vast armies met along a sharply defined front which moved back and forth, so masses of cold polar air meet in conflict with warm, moist air from the subtropical regions. Instead of mixing freely, these unlike air masses remain clearly defined, but interact along the polar front in great whorls whose structure is not unlike the form of an ocean wave seen in cross section.

A meteorological situation favorable to the formation of a wave cyclone is shown by a surface weather map, Figure 11.6. A trough of low pressure lies between two large cyclones (highs); one is made up of a cold, dry, polar air mass, the other of a warm, moist, maritime air mass. Air flow is in opposite directions on the two sides of the front, setting up a *shear zone*, and producing an unstable situation.

A series of individual blocks, Figure 11.7 shows the various stages in the life history of a wave cyclone. At the start of the cycle the polar front is a smooth boundary along which air is moving in opposite directions. In Block *A* of Figure 11.7, the polar front shows a bulge, or *wave*, beginning to form. Cold air is turned in a southerly direction, warm air in a northerly direction, as if each would invade the domain of the other.

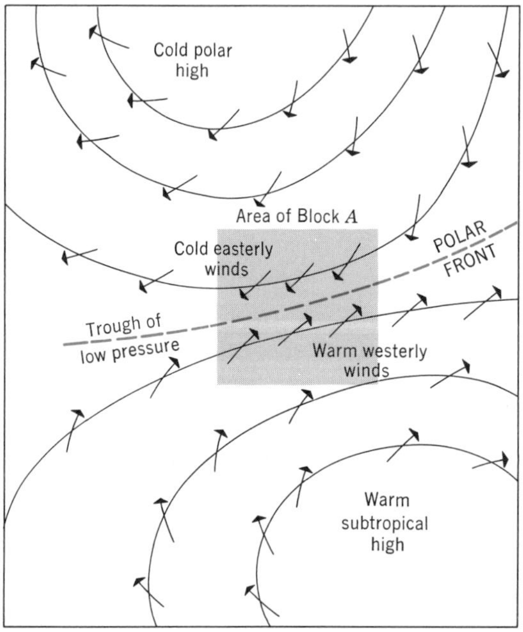

Figure 11.6 The trough between two high-pressure regions is a likely zone for development of a wave cyclone.

In Block *B* the wavelike disturbance along the polar front has deepened and intensified. Cold air is now actively pushing southward along a cold front; warm air is actively moving northeastward along a warm front. Each front is convex in the direction of motion. The zone of precipitation is now considerable, but wider along the warm front than along the cold front. In a still later stage the more rapidly moving cold front has reduced the zone of warm air to a narrow sector. In Block *C*, the cold front has overtaken the warm front, producing an occluded front and forcing the warm air mass off the ground, isolating it from the parent region of warm air to the south. The source of moisture and energy thus cut off, the cyclonic storm gradually dies out and the polar front is reestablished as originally (Block *D*).

Cyclones on the daily weather map

Further details of a wave cyclone are illustrated by surface weather maps for two successive days (Figure 11.8). Standard map symbols for wind directions and speeds are shown in Figure 11.9. Isobars are in millibars; the interval is 4 mb.

Map *A* shows a cyclone in a stage approximately equivalent to Block *B* of Figure 11.7. The storm is centered over western Minnesota and is moving northeastward. Note the following points: (*a*) Isobars of the low are closed to form an oval-shaped pattern. (*b*) Isobars make a sharp V where crossing cold and warm fronts. (*c*) Wind directions, indicated by arrows, are at an angle to the trend of the isobars and form a pattern of counterclockwise inspiraling. (*d*) In the warm-air sector there is northward flow of warm, moist tropical air toward the direction of the warm front. (*e*) There is a sudden shift of wind direction accompanying the passage of the cold front, as indicated by the widely different wind directions at stations close to the cold front, but on opposite sides. (*f*) There is a severe drop in temperature accompanying the passage of the cold front, as shown by differences in temperature readings at stations on either side of the cold front. (*g*) Precipitation, shown by shading, is occurring over a broad zone near the warm front and in the central area of the cyclone, but extends as a thin band down the length of the cold front. (*h*) Cloudiness, shown by degree of blackness of station circles, is great over the entire cyclone. (*i*) The low is followed on the west by a high (anticyclone) in which low temperatures and clear skies prevail. (*j*) The 32°F (0°C) isotherm crosses the cyclone diagonally from northeast to southwest, showing that the southeastern part is warmer than the northwestern part.

A cross section through Map *A* along the line AA′ shows how the fronts and clouds are related. Along the warm front is a broad area of stratiform clouds. These take the form of a wedge with a thin leading edge of cirrus. Westward this thickens to altostratus, then to stratus, and finally to nimbostratus with steady rain. Within the warm air mass sector the sky may partially clear with scattered cumulus. Along the cold front are violent thunderstorms with heavy rains, but this is along a narrow belt that passes quickly.

The second weather map, Map *B*, shows conditions 24 hours later. The cyclone has moved rapidly northeastward into Canada, its path shown by the line labeled *storm track*. The center has moved about 800 mi (1300 km) in 24 hours, a speed of just over 40 mi (65 km) per hour. The cyclone has occluded. An occluded front replaces the separate warm and cold fronts in the central part of the disturbance. The high-pressure area, or tongue of cold polar air, has moved in to the west and south of the cyclone, and the cold front is passing over the eastern and Gulf Coast states. Within closed isobars around the anticyclone the skies are clear and winds are weak. In another day the entire storm will have passed out to sea, leaving the eastern United States in the grip of cold but clear weather. A cross section below the map shows conditions along the line BB′, cutting through the occluded part of the storm. Note that the warm

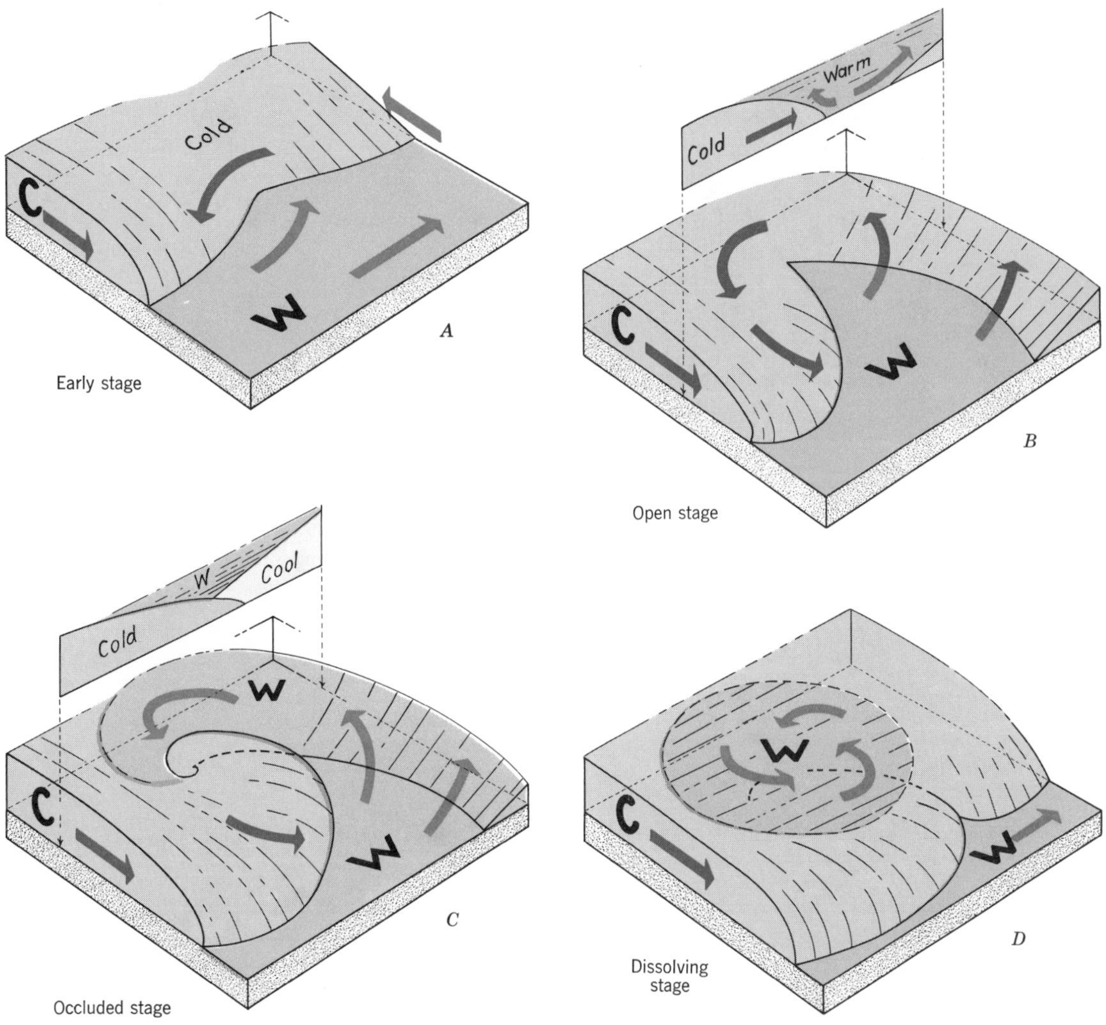

Figure 11.7 The development of a wave cyclone.

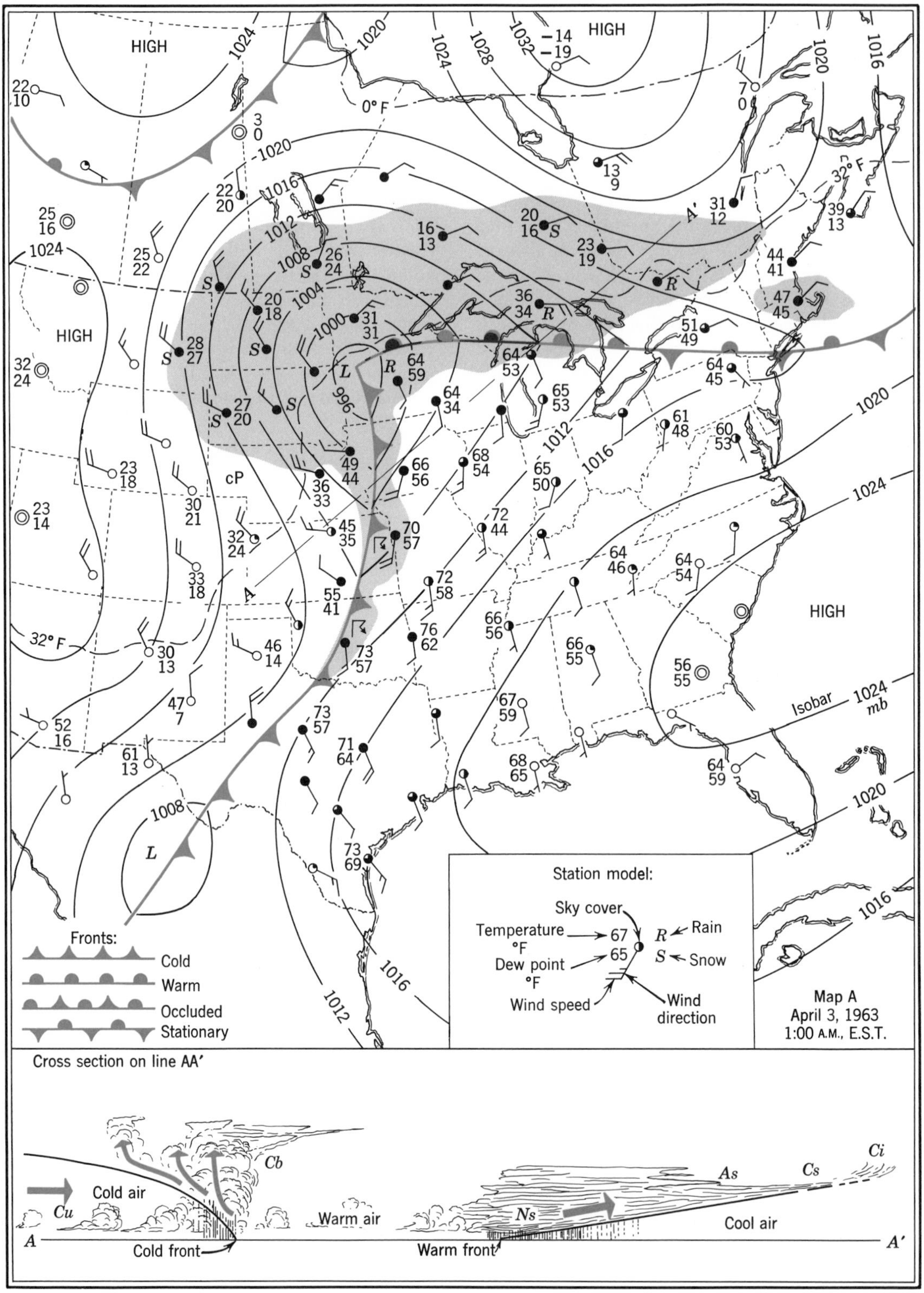

Figure 11.8 A wave cyclone shown on surface weather maps for successive days. Pressures are given in millibars; temperatures in degrees Fahrenheit. Areas experiencing precipitation

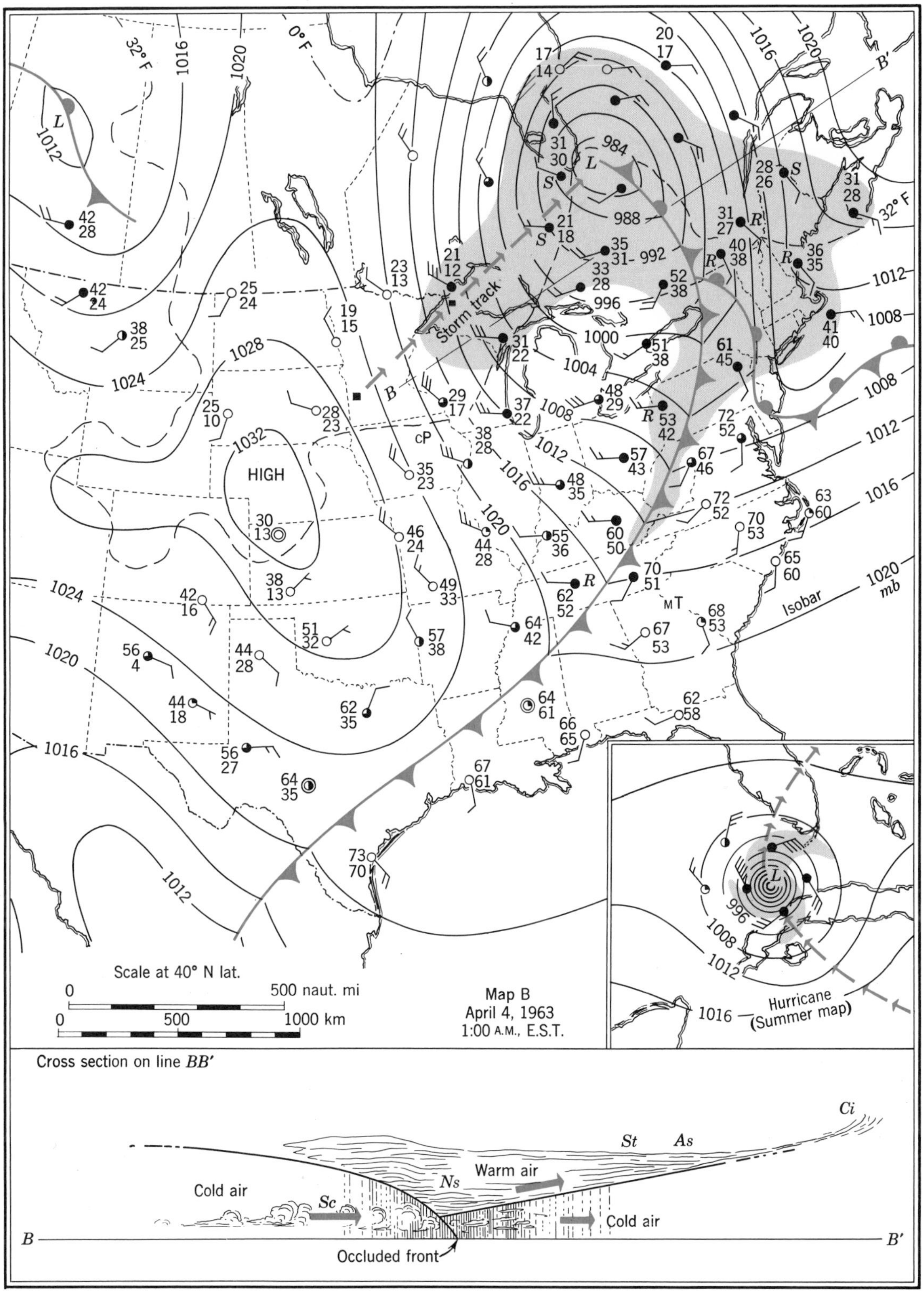

are shaded. See Figure 11.9 for explanation of wind symbols. (Modified and simplified from Daily Weather Map of the National Weather Service.)

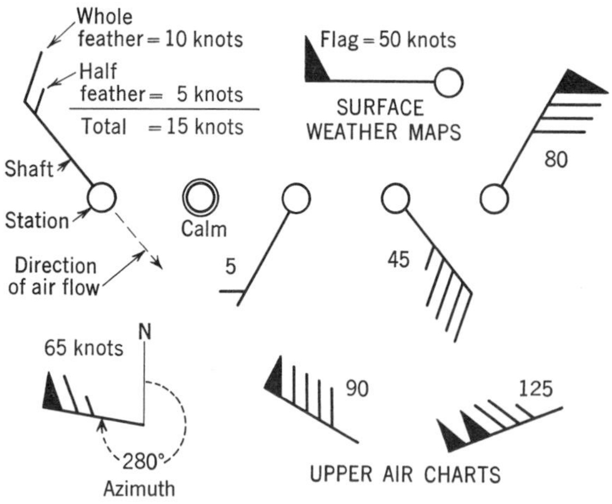

Figure 11.9 Standard weather map symbols for winds.

air mass is being lifted higher off the ground and is giving heavy precipitation.

Long observation on the movements of cyclones and anticyclones has revealed that certain tracks are most commonly followed. Figure 11.10 is a map of the United States showing these common paths. Notice that whereas some cyclonic storms travel across the entire United States from places of origin in the North Pacific, such as the Aleutian low, others originate in the Rocky Mountain region, the central states, or the Gulf Coast. Most tracks converge toward the northeastern United States and pass out into the North Atlantic, where they tend to concentrate in the region of the Icelandic low. General world distribution of paths of cyclonic storms is shown in Figure 11.11. Notice the heavy concentration in the neighborhood of the Aleutian and Icelandic lows. Wave cyclones commonly form in succession to travel in a chain across the north Atlantic and north Pacific oceans. Figure 11.12, a world weather map, shows several such *cyclone families.* As each cyclone moves northeastward it deepens and occludes, becoming an intense upper-air vortex. For this reason, cyclones arriving at the western coasts of North America and Europe are typically occluded.

In the southern hemisphere, storm tracks are more nearly along a single lane, following the parallels of latitude. This appears to be the result of uniform ocean surface throughout the middle latitudes, only the southern tip of South America breaking the monotonous oceanic expanse. Furthermore, the polar-centered icecap of Antarctica provides a centralized source of polar air.

Tropical and equatorial weather disturbances

Weather systems of tropical and equatorial zones show some basic differences from those of middle and high latitudes. Weakness of the Coriolis force close to the equator and lack of contrast between air masses conspire to prevent development of strong, clearly defined fronts and intense wave cyclones of the types that characterize higher latitudes. Nevertheless, there is no lack of intense atmospheric activity in the form of convective cells, for the high specific humidity of maritime air masses in low latitudes guarantees an enormous reservoir of energy in the latent form. This same energy source powers the most formidable of all storms—the tropical cyclone.

Figure 11.12 shows a typical set of weather conditions in the subtropical and equatorial zones, in addition to showing the general scheme of traveling wave cyclones in middle and high latitudes.

Fundamentally the arrangement consists of two rows of high-pressure cells, one or two cells to each land or ocean body. The northern row lies approximately along the Tropic of Cancer; the southern row, along the Tropic of Capricorn. Between the subtropical highs lies the equatorial trough of low pressure. Toward this trough converge the northeast and southeast trade winds. For this reason the trough is called the *Intertropical convergence zone.* At higher levels in the troposphere, the air flow

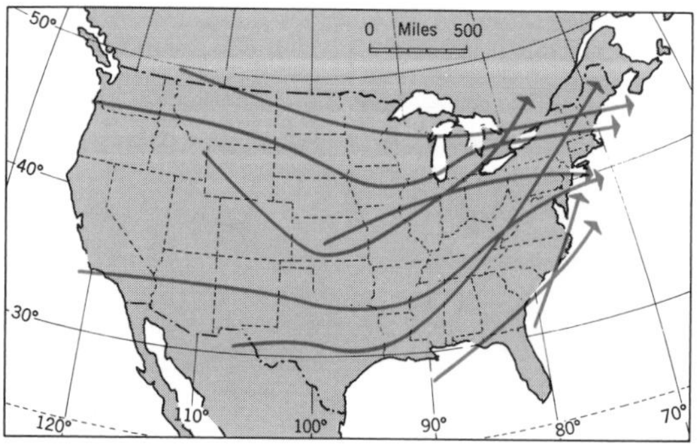

Figure 11.10 Common tracks taken by wave cyclones passing across the United States. (After Bowie and Weightman.)

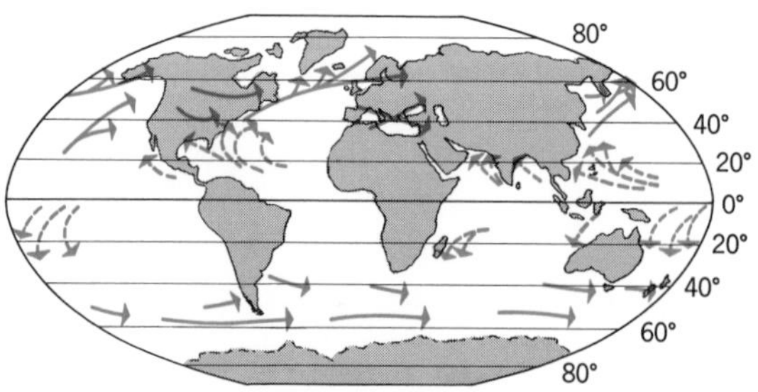

Figure 11.11 Principal tracks of wave cyclones are shown by solid lines; those of tropical cyclones by dashed lines. (After Petterssen, *Introduction to Meteorology.*)

Figure 11.12 A daily weather map of the world for a given day during July or August might look like this map, which is a composite of typical weather conditions. (After M. A. Garbell.)

is almost directly from east to west in the form of persistent *tropical easterlies*, described in Chapter 8.

One of the simplest forms of weather disturbance is an *easterly wave*, a slowly moving trough of low pressure within the belt of tropical easterlies. These waves occur in a zone 5° to 30° latitude over oceans both north and south of the equator, but not over the equator itself. Figure 11.13 is a simplified weather map of an easterly wave, showing isobars, winds, and the zone of rain. The wave is simply a series of indentations in the isobars to form a shallow pressure trough. Note that it travels westward, perhaps 200 to 300 mi (325 to 500 km) per day. Air flow tends to converge on the eastern, or rear, side of the wave axis. This causes the moist air to be lifted and to break out into scattered showers and thunderstorms. The rainy period may last for a day or two.

Another related disturbance is an *equatorial wave*, or *weak equatorial low*, which forms in the center of the equatorial trough. Although the normal air flow is from east to west, there develops an eddy in the tropical easterlies. Here air flow is locally reversed and tends to run opposite to the main stream. The result is a weak low-pressure center into which moist equatorial air masses converge, causing rainfall from many individual convectional storms within the low. Several such weak lows are shown on the world weather map (Figure 11.12), lying along the equatorial trough. Because the map is for a day in July or August, the trough is shifted well north of the equator.

Another distinctive feature of tropical weather is the occasional penetration of powerful tongues of cold polar air from the middle latitudes into very low latitudes. These are known as *polar outbreaks* and bring unusually cool, clear weather with strong, steady winds moving behind a cold front with squalls. The polar outbreak is best developed in the Americas. Outbreaks which move southward from the United States over the Carib-

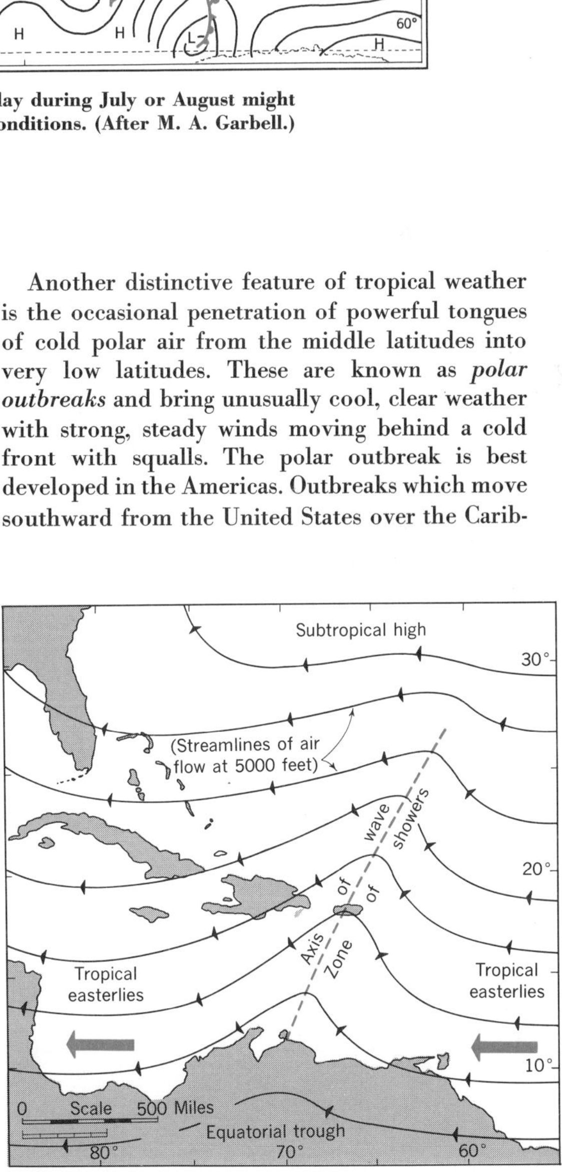

Figure 11.13 An easterly wave passing over the West Indies. (After Riehl.)

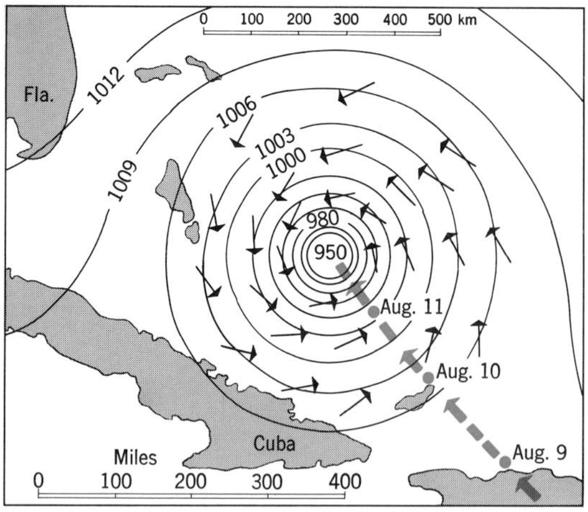

Figure 11.14 A hurricane (tropical cyclone) of the West Indies.

bean Sea and Central America are called *northers* or *nortes;* those which move north from Patagonia into tropical South America are called *pamperos* or *friagems.* One such outbreak is shown over South America on the world weather map (Figure 11.12).

Tropical cyclones

One of the most powerful and destructive types of cyclonic storms is the *tropical cyclone,* otherwise known as the *hurricane* or *typhoon.* The storm develops over oceans in latitudes 8° to 15° N and S, but not close to the equator, where the Coriolis force is extremely weak. In many cases an easterly wave simply deepens and intensifies, growing into a deep, circular low. High sea-surface temperatures which are over 80°F (27°C) in these latitudes, are of basic importance in the environment of storm origin. Warming of air at low level creates instability and predisposes toward storm formation. Once formed, the storm moves westward through the trade wind belt. It may then curve northwest and north, finally penetrating well into the belt of westerly winds. The tropical cyclone is an almost circular storm center of extremely low pressure into which winds are spiraling with great speed, accompanied by very heavy rainfall (Figures 11.14 and 11.15). Storm diameter may be 100 to 300 mi (150 to 500 km); the wind velocities range from 75 to 125 mi (120 to 200 km) per hour, sometimes much more; and the barometric pressure in the center commonly falls to 965 mb (28.5 in, 72.4 cm) or lower (Figure 11.16).

A brief description of the passage of a tropical cyclone at sea might be as follows. During the day preceding the storm the air is generally calm, the pressure somewhat above normal, and the sky shows cirrus clouds in long streamers, seeming to originate from a distant point on the horizon. The cirrus may be veil-like, giving a halo to the sun or moon and producing a red sunset. A long swell is felt on the sea, this being the train of dying storm waves that have outrun the slowly moving storm center. As the storm approaches, the barometer begins to fall. Wind springs up. A great dark wall of cloud approaches. When this envelopes the ship, torrential rainfall begins. The wind rises quickly to terrifying intensity. Great waves break over the vessel; the spray blows in continuous sheets which reduce visibility almost to zero.

This terrible storm continues for several hours and is abruptly followed by total calm and clearing skies, and sometimes by a sharp rise in temperature. The barometer has now reached its lowest point and the vessel is in the calm *central eye* of the storm. This is merely a hollow vortex produced by the rapid spiraling of air in the storm, comparable to the funnel-like air hole in the center of a vortex of water passing down a drain. Although the air is clear and calm, seas are mountainous and rise in great peaklike masses which are of gravest peril to the vessel. The period of calm may last a half hour. Then a great dark wall of cloud strikes the vessel and winds of high velocity again set in, but this time, in reverse direction to those of the first half of the storm. For several hours more the full fury of the storm

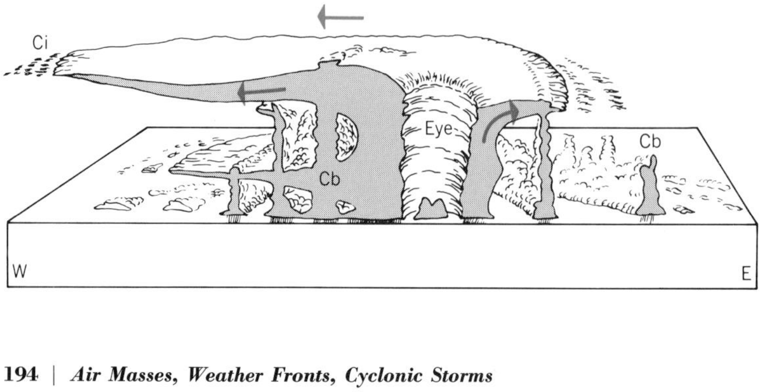

Figure 11.15 Schematic diagram of a hurricane. Cumulonimbus clouds in concentric rings rise through dense stratiform clouds in which the air spirals upward. Width of the diagram represents about 600 mi (1000 km). Highest clouds are at elevations often over 30,000 ft (9 km). (Redrawn from NOAA, National Weather Service, R. C. Gentry, 1964, *Weatherwise,* Vol. 17, p. 182.)

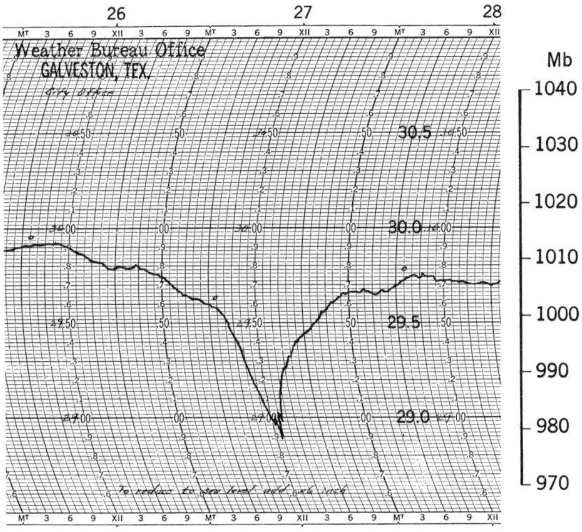

Figure 11.16 Trace sheet from a barograph at Galveston, Texas, during the hurricane of July 27, 1943. (U.S. Weather Bureau.)

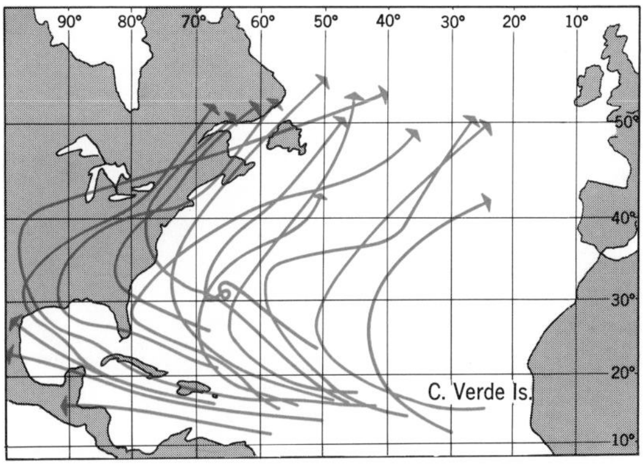

Figure 11.18 Tracks of some typical hurricanes occurring during August. (After U.S. Navy Oceanographic Office.)

rages, then gradually the winds abate, the clouds break, and fair weather returns.

World distribution of tropical cyclones is limited to six regions, all of them over tropical and subtropical oceans (Figure 11.17): (1) West Indies, Gulf of Mexico, and Caribbean Sea; (2) western North Pacific, including the Philippine Islands, China Sea, and Japanese Islands; (3) Arabian Sea and Bay of Bengal; (4) eastern Pacific coastal region off Mexico and Central America; (5) south Indian Ocean, off Madagascar; and (6) western South Pacific, in the region of Samoa and Fiji Islands and the east coast of Australia, Curiously enough, these storms are unknown in the South Atlantic. Tropical cyclones never originate over land, although they often penetrate far into the margins of continents.

Paths, or tracks, of tropical cyclones of the North Atlantic (Figure 11.18) show that most of the storms originate at 10° to 20° latitude, travel westward and northwestward through the trades, then turn northeast at about 30° to 35° latitude into the zone of the westerlies. Here the intensity lessens and the storms change into typical middle-latitude cyclones. In the trade wind belt the cyclones travel some 6 to 12 mi (10 to 20 km) per hour, in the westerlies, from 20 to 40 mi (30 to 60 km) per hour.

The occurrence of tropical cyclones is restricted to certain seasons of year, depending on the global location of the storm region. Those of the West Indies, and off the western coast of Mexico, occur largely from May through November, with maximum frequency in late summer or early autumn. Those of the western North Pacific, Bay of Bengal, and Arabian Sea are spread widely through the year but are dominant from May through November. Those of the South Pacific and south Indian oceans occur from October through April. Thus, they are restricted to the warm season in each hemisphere.

The environmental importance of tropical cyclones lies in their tremendously destructive effect upon island and coastal habitations (Figure 11.19).

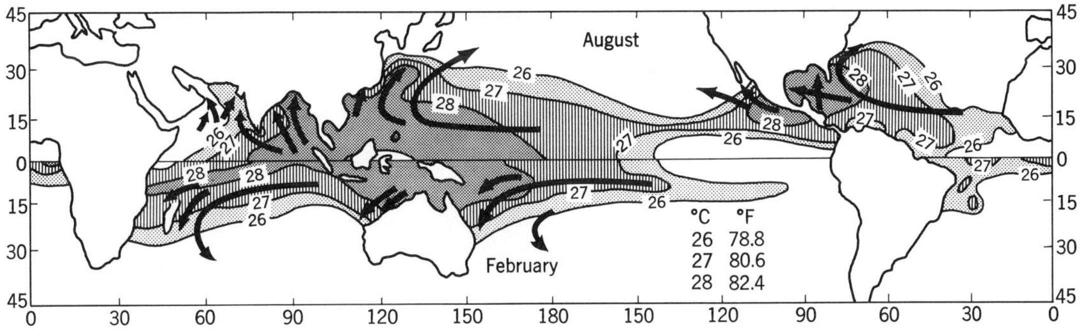

Figure 11.17 Typical paths of tropical cyclones in relation to sea surface temperatures (°C) in summer of the respective hemisphere. (After Palmen, 1948.)

Figure 11.19 This devastation along the south coast of Haiti was caused by Hurricane Flora on October 3, 1963. (Miami News photo.)

Wholesale destruction of cities and their inhabitants has been reported on several occasions. A terrible hurricane which struck Barbados in the West Indies in 1780 is reported to have torn stone buildings from their foundations, destroyed forts, and carried cannon more than a hundred feet from their locations. Trees were torn up and stripped of their bark. More than 6000 persons perished there.

Coastal destruction by storm waves and greatly raised sea level is perhaps the most serious effect of tropical cyclones. Where water level is raised by strong wind pressure, great breaking storm waves attack ground ordinarily far inland of the limits of wave action. A sudden rise of water level, known as a *storm surge,* may take place as the hurricane moves over a coastline. Ships are lifted bodily and carried inland to become stranded. If high tide accompanies the storm the limits reached by inundation are even higher. The terrible hurricane disaster at Galveston, Texas, in 1900 was wrought largely by a sudden storm surge inundating the low coastal city and drowning about 6000 persons. At the mouth of the Hooghly River on the Bay of Bengal, 300,000 persons died as a result of inundation by a 40-foot (12-meter) storm surge which accompanied a severe tropical cyclone in 1737. Low-lying coral atolls of the western Pacific may be entirely swept over by wind-driven sea water, washing away palm trees and houses and drowning the inhabitants.

Of importance, too, is the large quantity of rainfall produced by tropical cyclones. A considerable part of the summer rainfall of certain coastal regions can often be traced to a few such storms.

Tornadoes

The smallest but most violent of all known storms is the *tornado.* It seems to be a typically American storm, being most frequent and violent in the United States, although occurring in Australia in substantial numbers and reported occasionally in other places in middle latitudes. Tornadoes are also known throughout tropical and subtropical regions of the globe.

The tornado is a small, intense cyclone in which the air is spiraling at tremendous velocity. It appears as a dark *funnel cloud* (Figure 11.20), hanging from a large cumulonimbus cloud. At its lower end the funnel may be from 300 to 1500 ft (90 to 460 m) in diameter. The funnel appears dark because of the density of condensing moisture, dust, and debris swept up by the wind.

Wind speeds in a tornado exceed anything known in other storms. Estimates of wind speed run to as high as 500 mi (800 km) per hour. There is, in addition, a violent updraft in the funnel. As the tornado moves across the country the funnel writhes and twists. The end of the funnel cloud may alternately sweep the ground, causing complete destruction of anything in its path, and rise in the air to leave the ground below unharmed. Tornado destruction occurs both from the great

Figure 11.20 This funnel cloud was photographed by William L. Males at Cheyenne, Oklahoma, on May 4, 1961. The tornado was less than one mile from the observer when the picture was taken.

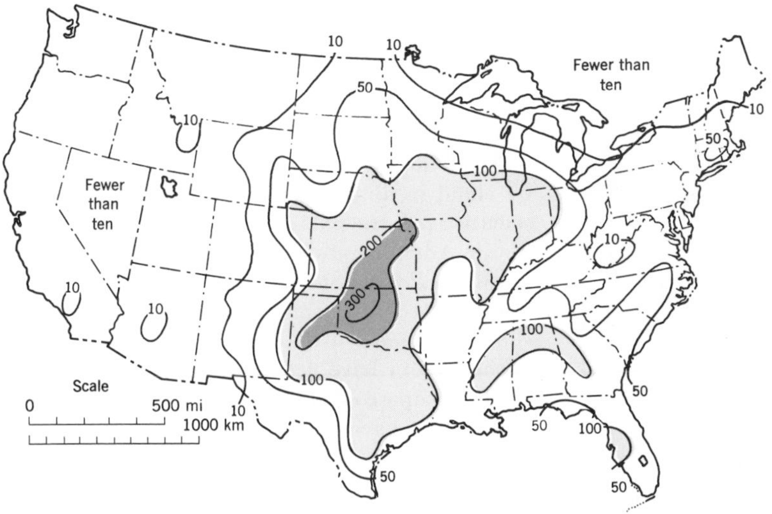

Figure 11.21 Distribution of tornadoes in the United States, 1955 to 1967. Figures tell number of tornadoes observed within 2-degree squares of latitude and longitude. (Data of M. E. Pautz, 1969, ESSA Tech. Memo WBTM FCST 12, Office of Meteorological Operations, Silver Spring, Md.)

wind stress and from the sudden reduction of air pressure in the vortex of the cyclonic spiral. Closed houses literally explode. It is even reported that the corks will pop out of empty bottles, so great is the difference in air pressure.

Tornadoes occur as parts of powerful cumulonimbus clouds in the squall line that travels in advance of a cold front. They seem to originate where turbulence is greatest. They are commonest in the spring and summer but occur in any month. Where maritime polar (mP) air lifts warm, moist tropical (mT) air on a cold front, conditions may become favorable for tornadoes. They occur in greatest numbers in the central and southeastern states and are rare over mountainous and forested regions. They are almost unknown from the Rocky Mountains westward and are relatively fewer on the eastern seaboard (Figures 11.21 and 11.22).

Devastation from a tornado is complete within the narrow limits of its path (Figure 11.23), but fortunately the storms are uncommon and the total danger small. Even in states having the most tornadoes, deaths from automobile and other accidents are a very much more serious danger. Storm cellars built completely below ground provide satisfactory protection if they can be reached in time. Although a tornado can often be seen or heard approaching, a cold front passing during the hours of darkness, as it often does, may present no warning. The National Weather Service maintains a tornado forecasting and warning system. Whenever weather conditions conspire to favor tornado development, the danger area is alerted and systems for observing and reporting a tornado are set in readiness. Communities in the paths of tornadoes may thus be warned in time for inhabitants to take shelter.

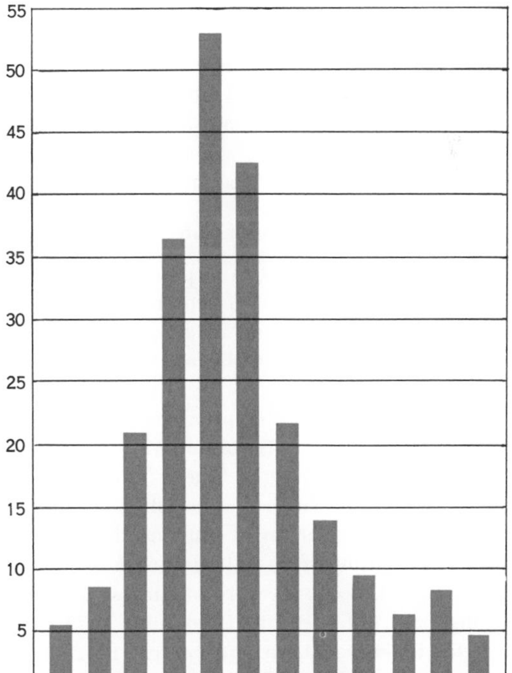

Figure 11.22 Average number of tornadoes reported in each month in the United States for a period 1916–1960. (From National Weather Service.)

Figure 11.23 Tornado devastation at Ionia, Iowa, on April 23, 1948, included this store in which two persons were killed. (National Weather Service and *Des Moines Register*.)

Waterspouts are similar in structure to tornadoes but form at sea under cumulonimbus clouds. They are smaller and less powerful than tornadoes. Sea water may be lifted 10 ft (3 m) above the sea surface, and the spray is carried higher. Waterspouts are commonly found in subtropical waters of the Gulf of Mexico and off the southeastern coast of the United States and seem to result from air turbulence occurring when continental air masses spread out over these oceans.

Planned weather modification

Recent years have seen a sharp increase in man's efforts to modify weather phenomena. One goal is to increase precipitation in areas experiencing drought; another is to ameliorate the severity of storms.

One method of inducing precipitation is known as *cloud seeding*, the introduction of minute particles into dense cumulus clouds. The particles, which may be of silver iodide smoke, serve as nuclei for added intensity of condensation and can lead to formation of cumulonimbus clouds with heavy rainfall. Some measure of success by cloud seeding was obtained in 1970 and 1971 over drought-ridden southern and central Florida. In another region, *Project Skywater*, begun in 1969, is aimed at increasing winter snowfall over mountains of southwestern Colorado. Using silver iodide smoke, a snowfall increase of 16 percent is anticipated, and will augment runoff of streams in the area.

An attempt to reduce the severity of a hurricane by cloud seeding methods was made in 1969 by scientists of the National Oceanic and Atmospheric Administration (NOAA). The experiment, known as *Project Stormfury*, apparently produced significant results. A 30 percent reduction in wind speed is claimed to have been achieved. Seeding is thought to have acted to cause rapid condensation of supercooled liquid particles in the storm, and thus to have quickly drained off latent heat.

Yet another aspect of planned weather modification is that of reducing the severity of hailstorms. Annual losses from crop destruction by hailstorms are estimated to approach $300 million. Damage to wheat crops is particularly severe in a north-south belt of the High Plains, running through Nebraska, Kansas, and Oklahoma. A much larger region of somewhat less hailstorm frequency extends eastward generally from the Rockies to the Ohio Valley. Corn is a major crop over much of this area. Research begun in 1971 by the National Center for Atmospheric research (NCAR) is studying the structure and dynamics of the isolated cumulonimbus clouds that produce hail. Research will lead to development of cloud-seeding techniques by means of which the severity of hail storms can be reduced. The Soviet Union, which has agricultural regions experiencing heavy crop losses through hail, claims to have been successful in developing the seeding technique to reduce hail damage.

Fog dispersal is another form of weather modification that has invited research and experimentation because of its great potential use at airports. Seeding experiments have shown that fog consisting of supercooled droplets can be cleared by seeding, using liquid propane or dry ice. Seeding causes rapid transformation of water droplets into ice particles. The very cold fogs to which this method applies are only a small percentage of all fogs that occur in middle and high latitudes. Warm fogs require other methods for dispersal, and these have met with some success.

It is well to keep in mind that the forms of planned weather modification described here apply to very small areas for short periods of time. No means yet exists to change precipitation appreciably over large areas on a sustained basis.

REVIEW QUESTIONS

1. What are the varieties of cyclonic storms? Is the thunderstorm a cyclonic storm?

2. What is an air mass? What meteorological variables determine the characteristics of an air mass?

3. How are air masses classified and designated? Discuss the air masses of North America, giving their source regions, paths of movement, and generally associated weather characteristics. What air masses are found over polar regions? over the equatorial belt?

4. Describe the structure of a cold front, a warm front, and an occluded front. What types of clouds and precipitation are characteristic of each of these fronts?

5. Explain the wave theory of middle-latitude cyclones as originated by J. Bjerknes. What is the polar front?

6. How does a cyclone develop and change along the polar front? How are cold and warm fronts involved in development of the wave cyclone?

7. Describe the changes in wind direction and strength, cloudiness and precipitation, and temperature that an observer experiences when a middle-latitude cyclone passes with its center north of the observer. Do the same for a cyclone in which the center passes south of the observer.

8. Describe the weather conditions associated with an anticyclone, or high, such as might normally follow the cyclone mentioned in the previous question.

9. What is the prevailing direction of movement of cyclonic storms in the United States? Do storm tracks show any tendency to concentrate at particular places in the northern hemisphere? If so, where?

10. What is an easterly wave? What weather does it bring? What is an equatorial wave? What weather does it bring?

11. Describe polar outbreaks of tropical regions. What names do these outbreaks have locally in the Central American area? in South America?

12. Describe the weather conditions in a tropical cyclone, or hurricane. Where do these storms originate? What paths do they normally follow? Explain the calm central eye of a tropical cyclone.

13. Name the regions of the world where tropical cyclones occur. Tell the seasons of occurrence of storms in each locality. How is season of storms related to position of the equatorial trough?

14. Describe some of the destructive effects of tropical cyclones along low-lying coasts. How is ocean level affected by these storms? What is a storm surge?

15. What is a tornado? Where and how does it occur? Why are tornadoes in the United States commonest in the spring? What is a waterspout?

16. What are the basic purposes of planned weather modification? In what ways has cloud seeding been used to modify weather? How can hailstorms be reduced in severity? How can fog be dispersed?

Exercises

1. **(*a*)** On both weather maps shown in Figure 11.9 draw isotherms for every 5°F (5°, 10°, 15°, 20°, etc.). Draw the isotherms on a sheet of thin tracing paper placed over the page. Include a tracing of shorelines and national boudaries. Superimpose the two tracing sheets and locate the areas of greatest temperature rise or drop. **(*b*)** Describe the pattern of isobars and winds within the deep low on map *A*. **(*c*)** Compare temperatures in the high pressure centers located over Labrador, the Rocky Mountains, and the southeastern United States. Why do such large differences in temperature exist within areas having similar pressure conditions? **(*d*)** Draw a pressure profile across map *A* from Idaho to Georgia. The method of drawing this profile is exactly the same as that illustrated in Figure I 20 of Appendix I, except that barometric pressure is substituted for altitude. **(*e*)** Study the distribution of dew-point temperatures over map *A*. Identify the air masses present on the map and label them on the same tracing sheet used in **(*a*)**.

2. Using the information in Figure 11.14, imagine that this hurricane turns slightly westward so as to head toward Miami, Florida. Describe conditions of winds, pressures, clouds, rainfall, surf, and tides at Miami for each successive day, beginning with August 12. Assume that the center of the storm passes over the city and that the rate of progress of the storm center remains the same as it was between August 10 and 11. (*Hint.* Trace the isobars and wind arrows on transparent paper; then move the tracing to successive positions along the storm track.)

3. Refer to Figure 11.12 and answer the following questions. **(*a*)** How many low-pressure centers are shown on this map? **(*b*)** How many high-pressure centers are shown? **(*c*)** How many lows fall into each of the following classes? (1) Extratropical (middle-latitude) cyclones; (2) tropical cyclones, and (3) weak lows of the equatorial trough. **(*d*)** At what approximate latitude or latitude belt does each of the following lie? (1) Equatorial trough; (2) subtropical cells of high pressure, northern hemisphere; (3) subtropical cells of high pressure, southern hemisphere; (4) middle-latitude cyclone centers, northern hemisphere; and (5) middle-latitude cyclone centers, southern hemisphere. **(*e*)** Of the middle-latitude cyclones shown on this map, how many are occluded; how many are not?

CHAPTER 12

The Soil-Moisture Balance

A FUNDAMENTAL concept of physical geography is that the availability of water to plants and animals is a more important factor in the environment than precipitation itself, which is only the amount of water received from the sky. Much of the water received as precipitation is lost in a variety of ways and is not usable for plants and animals. Just as in a fiscal budget, when the monthly or yearly loss of moisture exceeds the precipitation a budgetary deficit results; when precipitation exceeds the losses, a budgetary surplus results. An analysis of the water balance, or water budget, of a given area of the earth's surface is actually arrived at in much the same way as the understanding of a fiscal budget and requires only additions and subtractions of amounts within fixed periods of time, such as the calendar month or year.

On the other hand, to understand why such gains and losses occur involves study of the physical processes affecting water in its vapor, liquid, and solid states not only in the atmosphere but also in the soil and rock and in exposed water of streams, lakes, and glaciers. The science of *hydrology* treats such relationships of water as a complex but unified system on the earth. In this chapter, our primary concern is with the hydrology of the soil zone.

Surface and subsurface water

We may classify water according to whether it is *surface water*, flowing exposed or ponded upon the land, or *subsurface water*, occupying openings in the soil, overburden, or bedrock.* That which is held in the soil within a few feet of the surface is termed *soil moisture*, and is the particular concern of the botanist, soil scientist, and agricultural engineer. That which is held in the openings of the bedrock is usually referred to as *ground water* and is studied by the geologist. Surface water and ground water are the subjects of Chapter 13. The geologic aspects of ground water are considered in later chapters.

*See Chapter 18 and Figure 18.1 for explanations of these terms.

The hydrologic cycle and the global water balance

Water of oceans, atmosphere, and lands moves in a great series of continuous interchanges of both geographic position and physical state, known as the *hydrologic cycle* (Figure 12.1). A particular molecule of water might, if we could trace it continuously, travel through any one of a number of possible circuits involving alternately the water vapor state and the liquid or solid state.

The pictorial diagram of the hydrologic cycle given in Figure 12.1 can be quantified for the earth as a whole. Figure 12.2 is a mass-flow diagram relating the principal pathways of the water circuit. We can start with the oceans, which are the basic reservoir of free water. Evaporation from the ocean surfaces totals about 109,000 cu mi per year. (Metric equivalents shown in table in Figure 12.2.) At the same time, evaporation from soil, plants, and water surfaces of the continents totals about 15,000 cu mi. Thus the total evaporation term is 124,000 cu mi; it represents the quantity of water that must be returned annually to the liquid or

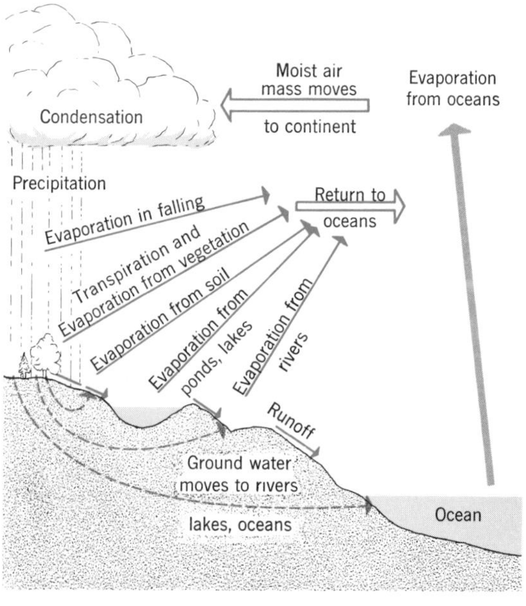

Figure 12.1 The hydrologic cycle. (After Holtzman.)

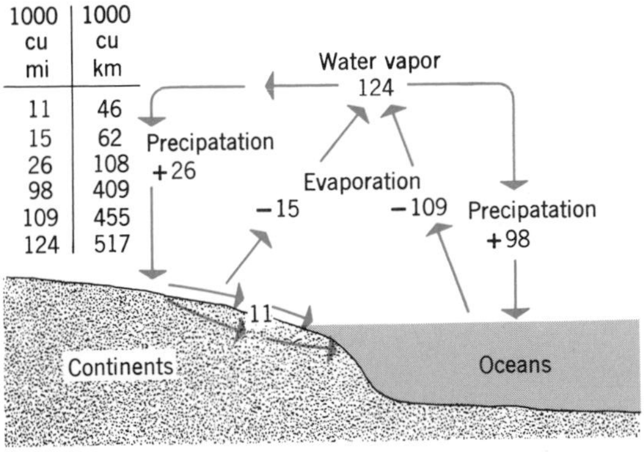

Figure 12.2 The global water balance. Figures give average annual water flux in and out of world land areas and world oceans. (Data of M. I. Budyko, 1971.)

solid state. Precipitation is unevenly divided between continents and oceans; 26,000 cu mi is received by the land surfaces, and 98,000 cu mi by the ocean surfaces. Notice that the continents receive about 11,000 cu mi more water as precipitation than they lose by evaporation. This excess quantity flows over or under the ground surface to reach the sea; it is collectively termed *runoff*.

We can state the global water balance, as follows.

$$P = E + G + R$$

where P = precipitation
E = evaporation
G = net gain or loss of water in the system, a storage term
R = runoff (positive when out of the continents, negative when into the oceans)

All terms have the dimensions of mass per unit time (for example, metric tons per day), or volume per unit time (for example, cubic miles per year). When applied over the span of a year, and averaged over many years, the storage term G can be neglected, since the system is essentially closed so far as matter is concerned. The quantities of water in storage in the atmosphere, on the lands, and in the oceans will remain about constant from year to year.

The equation then simplifies to

$$P = E + R$$

Using the preceding figures given for the continents

$$26{,}000 = 15{,}000 + 11{,}000$$

and for the oceans

$$98{,}000 = 109{,}000 - 11{,}000$$

For the globe as a whole, combining continents and ocean basins, the runoff terms cancel out:

$$26{,}000 + 98{,}000 = 15{,}000 + 109{,}000$$

$$124{,}000 = 124{,}000$$

From a global water balance in terms of water volume per unit of time (cu mi/yr; cu km/yr), we now turn to water balance data in terms of water depth per unit of time (in/yr; cm/yr). Whereas the first set of data tell us the total quantity of water transported into or out of a specified area in one year, the second set will tell us the intensity of water flow, independent of total surface area and total water quantity.

The water balance can be estimated for latitude belts of 10-degree width to reveal the response to the latitudinal changes in radiation and heat balances from equatorial zone to polar zones. Figure 12.3 shows the average annual values of precipitation, evaporation, and runoff. The runoff term R in this case includes import or export of water in or out of the belt by ocean currents as well as by stream flow. Notice the equatorial zone of water surplus with positive values of R, in contrast to the subtropical belts, with an excess of evaporation. The runoff term R, which is here negative in sign, represents the importation of water by ocean currents to furnish the quantity needed for evaporation. Water surpluses occur poleward of the 40th parallel but the values of all three terms decline rapidly at high latitudes, going nearly to zero at the poles. This graph should be compared with Figure 10.19, showing water vapor transport across parallels of latitude. Notice that the precip-

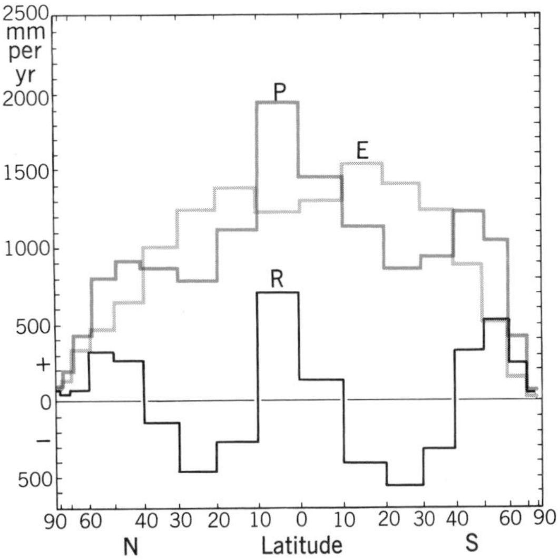

Figure 12.3 The water balance for 10-degree latitude zones. Data from several sources compiled by W. D. Sellers, 1965. (From A. N. Strahler, 1971, *The Earth Sciences*, 2nd ed., Harper and Row, New York.)

itation surpluses are sustained by importation of water vapor, and evaporation surpluses by export of water vapor.

Great inequalities exist in global amounts of water stored in the gaseous, liquid, and solid states. Table 12.1 gives a breakdown of storage quantities. Water of the oceans constitutes over 97 percent of the total, as we would expect. Next comes water in storage in glaciers, a little over 2 percent. The remaining quantity, about one-third of 1 percent, is mostly held as subsurface water, so that surface water in lakes and streams is a very small quantity, indeed. Yet it is this surface water, together with the very small quantity of soil water, that sustains all life of the lands. Indeed, some environmentalists consider that fresh surface water will prove to be the limiting factor in the capacity of our planet to support the rapidly expanding human population. The quantity of water vapor held in the atmosphere is also very small—only about 10 times greater than that held in streams.

Long-term changes in water balance quantities are associated with atmospheric environmental changes. For example, atmospheric cooling on a global scale would bring a reduction of water vapor storage and would reduce precipitation and runoff generally. But at the same time, a greater proportion of that precipitation would be in the form of snow, so that the storage in ice accumulations would rise and the storage in ocean waters would fall. These changes describe the changing water balance associated with onset of an *ice age*, or *glaciation*, and constitute a major environmental change already experienced by our planet at least four times in the past two million years.

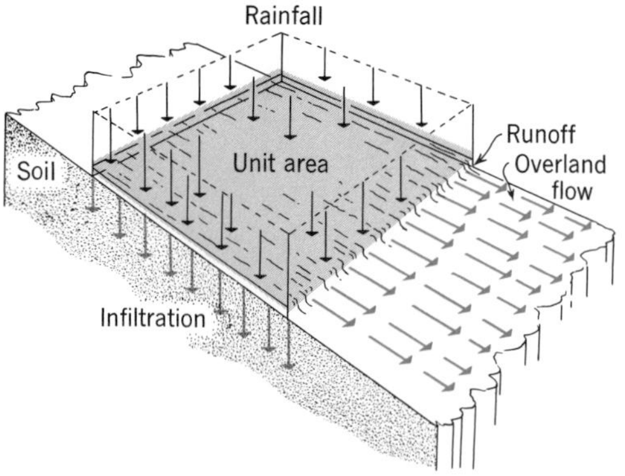

Figure 12.4 Rainfall, infiltration, and overland flow. (From A. N. Strahler, 1971, *The Earth Sciences*, 2nd ed., Harper and Row, New York.)

Infiltration and runoff

Most soil surfaces in their undisturbed, natural states are capable of absorbing the water from light or moderate rains, a process known as *infiltration.* Such soils have natural passageways between poorly fitting soil particles, as well as larger openings, such as earth cracks resulting from soil drying, borings of worms and animals, cavities left from decay of plant roots, or openings made by heaving and collapse of soil as frost crystals alternately grow and melt. A mat of decaying leaves and stems breaks the force of falling drops and helps to keep these openings clear. If rain falls too rapidly to be passed downward through these soil openings the excess amount flows as a surface water film or sheet down the direction of ground slope, a runoff process termed *overland flow* (Figure 12.4).

TABLE 12.1 DISTRIBUTION OF THE WORLD'S ESTIMATED WATER[a]

Location	Surface Area		Water Volume		Percent of Total
	Sq Mi	Sq Km	Cu Mi	Cu Km	
SURFACE WATER					
Fresh-water lakes	330,000	860,000	30,000	125,000	0.009
Saline lakes and inland seas	270,000	700,000	25,000	104,000	0.008
Average in stream channels	—	—	300	1,250	0.0001
SUBSURFACE WATER	50,000,000	130,000,000			
Soil moisture and intermediate-zone (vadose) water			16,000	67,000	0.005
Ground water within 0.5 mi (0.8 km) depth			1,000,000	4,170,000	0.31
Ground water, deep-lying			1,000,000	4,170,000	0.31
Total liquid water in land areas			2,070,000	8,637,000	0.635
ICECAPS AND GLACIERS	6,900,000	18,000,000	7,000,000	29,200,000	2.15
ATMOSPHERE	197,000,000	510,000,000	3,100	13,000	0.001
WORLD OCEAN	139,500,000	360,000,000	317,000,000	1,322,000,000	97.2
Totals (rounded)			326,000,000	1,360,000,000	100

[a] Data from Dr. Raymond L. Nace, U.S. Geological Survey, 1964.

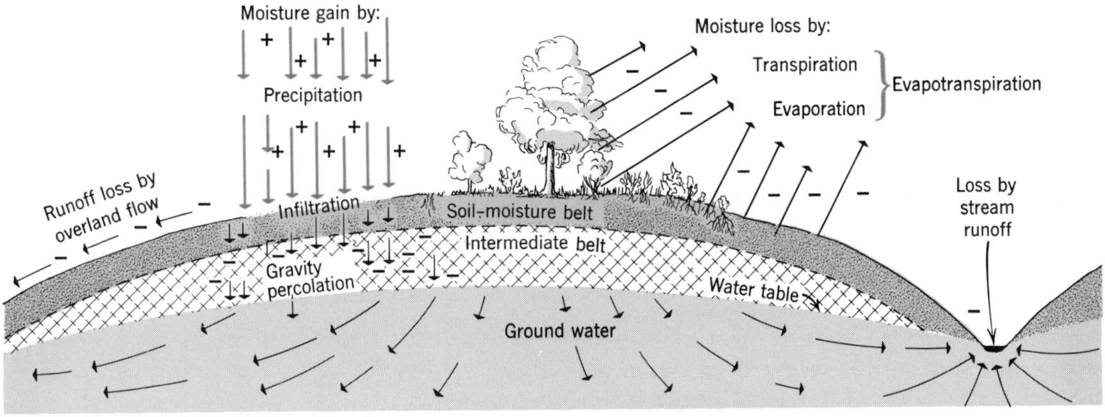

Figure 12.5 The soil-moisture belt occupies an important position in the hydrologic cycle.

As stated in Chapter 10, rainfall is measured in units of inches or centimeters per hour. This is the depth to which water will accumulate in each hour if rain is caught in a flat-bottomed, straight-sided container, assuming none to be lost by evaporation or splashing out. Similarly, infiltration is stated in inches or centimeters per hour and might be thought of as the rate at which the water level in the same container might drop if the water were leaking through a porous base. Runoff, also stated in inches (cm) per hour, may be thought of as the amount of overflow of the container per hour when rain falls too fast to be disposed of by leaking through the base.

Evaporation and transpiration

Between periods of rain, water held in the soil is gradually given up by a twofold drying process. First, direct evaporation into the open air occurs at the soil surface and progresses downward. Air also enters the soil freely and may actually be forced alternately in and out of the soil by atmospheric pressure changes. Even if the soil did not "breathe" in this way, there would be a slow diffusion of water vapor surfaceward through the open soil pores. Ordinarily only the first foot (30 cm) of soil is dried by evaporation in a single dry season, but in the prolonged drought of deserts, drying will extend to depths of many feet. Second, plants draw the soil water into their systems through vast networks of tiny rootlets. This water, after being carried upward through the trunk and branches into the leaves, is discharged in the form of water vapor, through leaf pores into the atmosphere, a process termed *transpiration.* Further details of plant transpiration are given in Chapter 20.

In studies of climatology and hydrology it is convenient to use the term *evapotranspiration* to cover the combined moisture loss from direct evaporation and the transpiration of plants. The rate of evapotranspiration slows down as soil moisture supply becomes depleted during a dry summer period because plants employ various devices to reduce transpiration. In general, the less moisture remaining, the slower is the loss through evapotranspiration.

Figure 12.5 shows diagrammatically the various terms explained up to this point and serves to give a more detailed picture of that part of the hydrologic cycle involving the soil. As the plus signs show, the *soil-moisture belt* gains water through precipitation and infiltration. As the minus signs show, the soil loses water through transpiration, evaporation, and overland flow, and by *gravity percolation* downward through the soil to the ground water zone below. Between the soil-moisture belt and the ground water zone is an *intermediate belt,* holding moisture but at a depth too great to be returned to the surface by evapotranspiration.

Moisture in the soil

When infiltration occurs during heavy and prolonged rains (or when a snow cover is melting) the water is drawn downward by gravity through the soil pores, wetting successively lower layers. Soon the soil openings are filled with water moving downward, except for some air entrapped in the form of bubbles. Then the percolation continues downward into the bedrock. Suppose now that the rain stops and a period of several days of dry weather follows. The excess soil water continues to drain downward, but some water clings to the soil particles and completely resists the pull of gravity through the force of *capillary tension.* We are all familiar with the way in which a water droplet seems to be enclosed in a "skin" of surface molecules, drawing the droplet together into a spherical shape, so that it clings to the side of a glass indefinitely without flowing down. Similarly, tiny films of water adhere to the soil grains, particularly at the points of grain contacts, and will stay until disposed of by evaporation or by absorption into plant rootlets.

When a soil has first been saturated by water, then allowed to drain under gravity until no more water moves downward, the soil is said to be holding its *field capacity* of water. This takes no more than two or three days for most soils. Most is drained out within one day. Field capacity is measured in units of depth, usually inches or centimeters, just as with precipitation. This means that for a given cube of soil, say 12 in on a side (1 cu ft), if we were to extract all of the field moisture, it might form a layer of water 3-in deep in a pan 1-ft square. This would be equivalent to complete absorption of a 3-in rainfall by a completely dry 12-in layer of soil. (Metric units of volume and depth may be substituted.)

Field capacity of a given soil depends largely on its texture. Sandy soil has a very low field capacity; clay soil has high field capacity. This effect is shown in Figure 12.6, a graph in which field capacity is plotted against soil texture, from coarse to fine. It should also be noted that sandy soils reach their field capacity very quickly, both because of the ease with which the water penetrates and the low quantity required. Clay soils take long rain periods to reach field capacity because the infiltration is slow and the total quantity required to be absorbed is great.

Agricultural scientists also use a measure of soil moisture termed the *wilting point*. This is the quantity of soil moisture below which plants are unable to extract further moisture from the soil and the foliage will wilt. As Figure 12.6 shows, the wilting point also depends upon particle size.

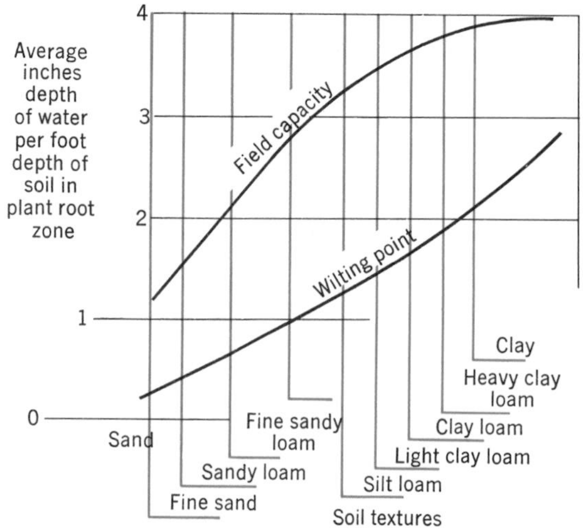

Figure 12.6 Field capacity and wilting point. See Figure 18.2 for definitions of terms. (After Smith and Ruhe, *Yearbook of Agriculture*, 1955.)

The soil moisture cycle

Equipped with the foregoing explanations of processes and terms relating to water gains and losses in the soil, we can turn next to consider the annual water budget of the soil, involving principles of great concern not only in plant geography and agriculture, but in the further study of ground water, surface runoff, stream flow, and therefore of the sculpturing of land slopes.

Figure 12.7 shows the annual cycle of soil moisture for the year 1944 at an agricultural experiment station in Coshocton, Ohio. If we follow the

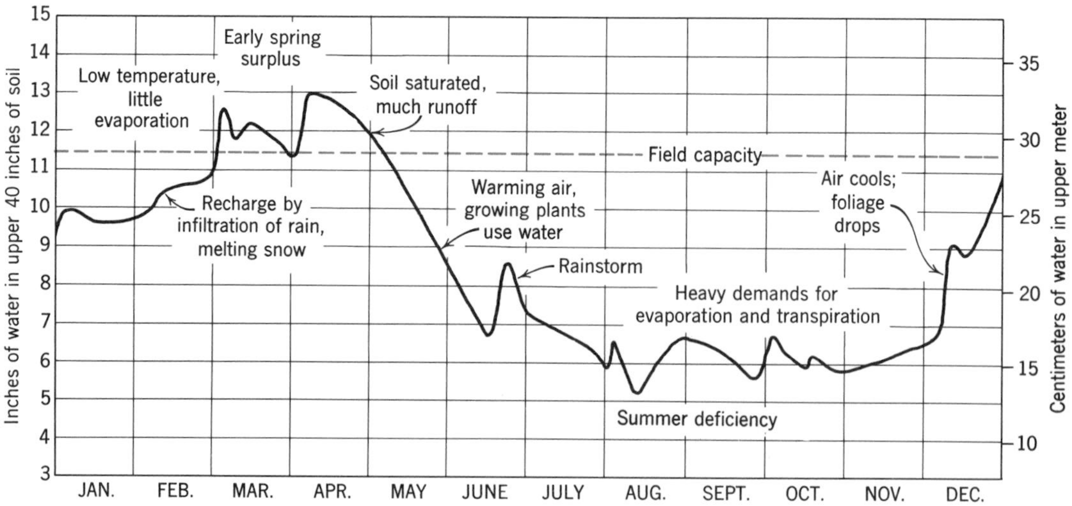

Figure 12.7 Annual cycle of soil moisture. (After Thornthwaite and Mather, *The Water Balance*, 1955.)

changes in this example, the cycle it shows can be considered generally representative of conditions in humid, middle-latitude climates where there is a strong temperature contrast between winter and summer. Let us start with the early spring (March). At this time the evaporation rate is low, because of low temperatures. The abundance of melting snows and rains has restored the soil moisture to a surplus quantity. For two months the quantity of water percolating through the soil and entering the ground water keeps the soil pores nearly filled with water. This is the time of year when one encounters soft, muddy ground conditions, whether driving on dirt roads or walking across country. This, too, is the season when runoff is heavy and major floods may be expected on larger streams and rivers. In terms of the soil water budget, a *moisture surplus* exists.

By May, the rising air temperatures, increasing evaporation, and the full growth of plant foliage, bringing on heavy transpiration, have reduced the soil moisture to a quantity below field capacity, although it may be restored temporarily by unusually heavy rains in some years. By midsummer, a state of heavy *moisture deficiency* exists in the water budget. Even the occasional heavy thunderstorm rains of summer cannot restore the water lost by steady and heavy evapotranspiration. Small springs and streams dry up, the soil becomes firm and dry. By November (and sometimes in September), however, the soil moisture again begins to increase. This is because the plants go into a dormant state, sharply reducing transpiration losses, while, at the same time, falling air temperatures reduce evaporation. By late winter, usually in February at this location, the field capacity of the soil is again restored.

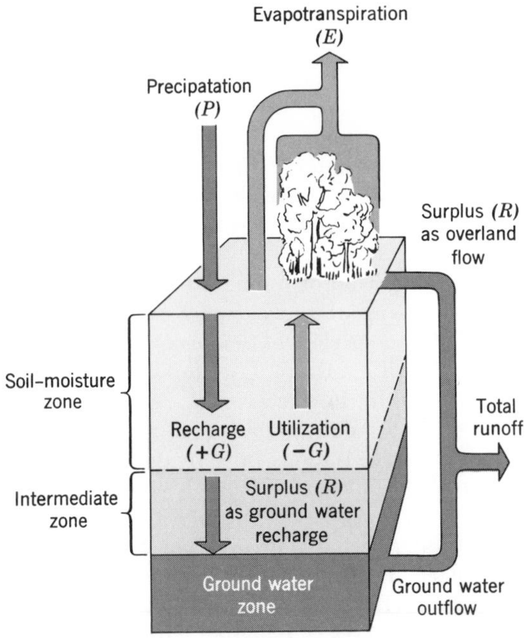

Figure 12.8 Schematic diagram of the soil-moisture balance in a soil column of unit cross section.

The soil-moisture budget[1]

From the foregoing example of soil moisture changes throughout a single year's time, we go on to a more generalized concept of the soil-moisture budget. Let us first return to the basic water-balance equation, modifying it to apply specifically to the soil-moisture zone. This zone has a depth equal to the lower limit to which plants send their roots; it is the storage zone for moisture available to plants. Figure 12.8 is a schematic diagram of the terms in the moisture-balance equation. The soil column is assumed to have a unit cross-sectional area, and we can therefore use flow units of depth-per-unit-time (for example, mm/day or mm/month), as in previous statements of the general water-balance equation. The equation is as follows.

$$P = E + G + R$$

where P = precipitation
E = evapotranspiration
G = change in soil-moisture storage
R = runoff (by overland flow, or by infiltration to the ground water zone beneath)

To proceed further, we must recognize two ways to define evapotranspiration. First is *actual evapotranspiration*, which is the true or real rate of water vapor return to the atmosphere from the ground and its plant cover. Second is *potential evapotranspiration*, representing the water vapor flux under an ideal set of conditions. One condition is that there be present a complete (or closed) cover of homogeneous vegetation consisting of fresh green leaves, and no bare ground exposed through that cover. The leaf cover is assumed to have a uniform height above ground—whether the plants be trees, shrubs, or grasses. A second condition is that there be an adequate water supply, such that the field capacity of the soil is maintained at all times. This condition can be fulfilled naturally by abundant and frequent precipitation, or artificially by irrigation.

Several rigorous methods have been devised for estimating the monthly potential evapotranspiration at any given location on the globe. The

[1] In this chapter we refer to water held in the soil as *moisture*, because that is the term most commonly used by scientists who study this subject from the agronomist's point of view. However, *soil water* is a fully acceptable alternative term for *soil moisture*.

method used in this chapter is based on air temperature, as well as latitude and date. The latter variables determine intensity and duration of solar radiation received at the ground. Put another way, potential evapotranspiration is a measure of the maximum capability of a land surface to return energy from the surface to the atmosphere by the mechanism of latent heat flux under the defined conditions.

Beyond this point, we shall use the symbol E_p to designate potential evapotranspiration and the symbol E_a to designate actual evapotranspiration. Note that when P exceeds E_p, or when the soil is at field capacity, E_p and E_a will have the same numerical value, but at all other times E_a will be a smaller value than E_p.

Suppose now that mean daily values of precipitation P and both forms of evapotranspiration, E_p and E_a, are plotted on a graph to show a full year's cycle for a place in the northern hemisphere (Figure 12.9). For the sake of simplicity we have used smooth curves, rather than a sequence of 365 points, to depict the changes in the variable quantities. We shall assume that precipitation is uniformly distributed throughout the entire year, and shows as a horizontal line on the graph. On the other hand, potential evapotranspiration, E_p, is assumed to show a strong annual cycle with a low value in winter and a high value in summer, as is typical in middle latitudes.

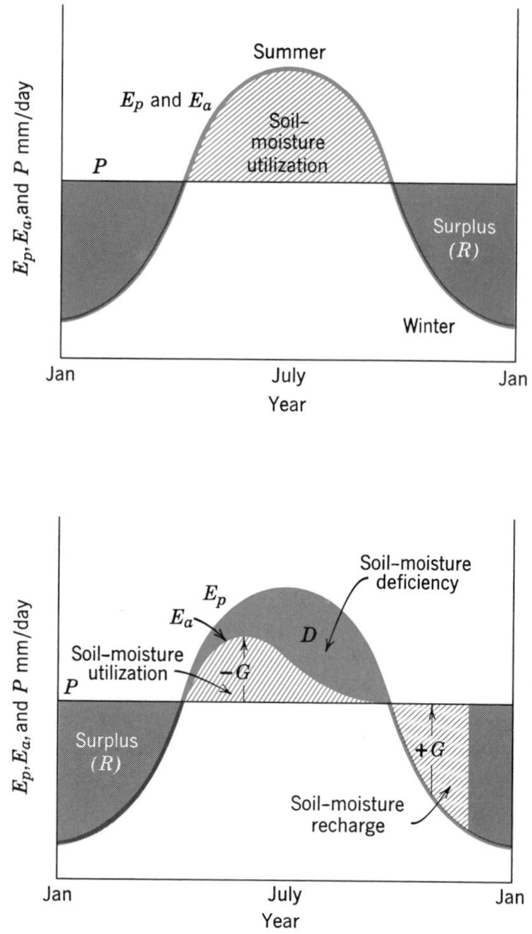

Figure 12.9 Schematic graphs of the soil-moisture budget.

Consider two idealized cases. First, as shown in Diagram *A*, full irrigation is supplied during the summer, so that E_a and E_p remain identical in value. During the period when P is greater than E_p, there is a *water surplus* and runoff (R) is generated. In the second case, shown in Diagram *B*, no irrigation water is available in the summer. When E_p exceeds P, the plants must draw upon moisture stored in the soil in an attempt to sustain as rapid growth as possible at all times. There is, of course, a continuing supply of precipitation constituting an input of moisture into the soil zone, but the plants are capable of using more than this amount. Under these circumstances, the curve of actual evapotranspiration, E_a, drops off to a lower value than the curve of E_p. As we have previously stated, it is assumed that the rate at which soil moisture is depleted falls off at a rate proportional to the quantity of soil moisture remaining. The curve of E_a falls off rapidly at first, then begins to level off. The store of soil moisture is never fully depleted, but may approach a negligible quantity after a long season of depletion. Daily change in soil-moisture storage, G, is represented by the vertical distance between the curves of P and E_a (G has a negative sign).

The difference between the daily values of E_p and E_a (or $E_p - E_a$), when accumulated from the time of onset of soil moisture depletion until the depletion period is ended, is referred to as the *soil-moisture deficiency* (D); it is stated in units of water depth (mm or in). (In other words, D is a quantity of water, rather than a flow rate of water.) The soil-moisture deficiency is represented graphically in Diagram B as the area between the curves of E_p and E_a.

Once the period of soil-moisture deficiency has ended (crossing of the curves E_p and P), there must follow a period of *soil moisture recharge*, in which soil moisture is restored to the field capacity. All P in excess of E_p now goes into recharge. The daily rate of increase in soil-moisture storage is represented by the term G with a positive sign. When recharge is complete, a period of *water surplus* sets in and runoff (including ground water recharge or overland flow, or both) takes place. The period of soil-moisture recharge is shown in Diagram B by a distinctive pattern between the curves of P and E_a. This area is equal to the area on the same graph representing soil-moisture utilization.

TABLE 12.2 SIMPLIFIED EXAMPLE OF A SOIL-MOISTURE BUDGET

Equation:	P	=	E_a	$+G$		$+R$	E_p	$(E_p - E_a) = D$
January	110	=	10			+100	10	0
February	90	=	20			+ 70	20	0
March	60	=	35			+ 25	35	0
April	30	=	60	−30			60	0
May	25	=	70	−45			85	15
June	20	=	60	−40			95	35
July	25	=	50	−25			90	40
August	40	=	45	− 5			70	25
September	70	=	45		+25		45	0
October	90	=	30		+60		30	0
November	105	=	15		+60	+ 30	15	0
December	120	=	15			+105	15	0
Totals	785	=	455	−145	+145	+330	570	115
	785	=	785					

We are now ready to turn to the practical application of the soil-moisture balance equation to the simplified data of an observing station. Monthly, rather than daily means of P are used, and these represent long-term averages. Figure 12.10 is an idealized yearly moisture budget. Each month is represented by two vertical bars, whose heights are proportional to flow rate in units of millimeters per month. Each bar is divided up into segments showing the several terms of the water balance equation. The left-hand bar shows P, part of which matches an equal quantity of moisture lost by evapotranspiration, and part of which may represent recharge ($+G$) or surplus (R). The right-hand bar shows E_p, part of which may represent moisture deficiency (D) or soil-moisture utilization ($-G$). Numerical values can now be assigned to each term of the water balance equation for each month, as in Table 12.2 (all values in this example are multiples of 5 mm).

By means of simple arithmetic, we have calculated the water balance for each month singly and for the year as a whole. Tallied separately at the right are the monthly differences between E_p and E_a, giving a total soil-moisture deficiency, D, of 115 mm for the year. This is the quantity of water

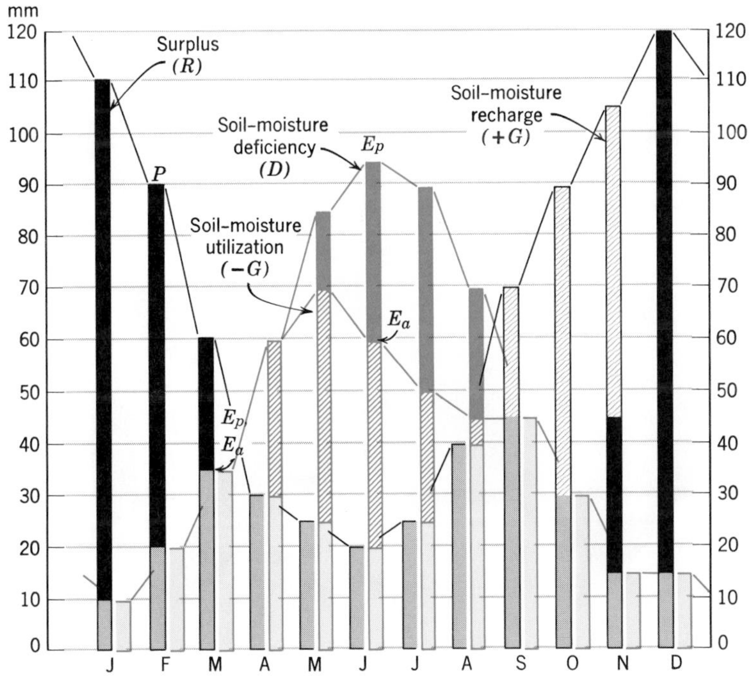

Figure 12.10 A model soil-moisture budget.

that would have to be supplied by irrigation to sustain the full value of E_p. The importance of the water balance calculation in estimating the need for irrigation of crops should be obvious. We have also been able to calculate the annual runoff, 330 mm, and can use this information to estimate the recharge of ground water and the runoff into streams. In this way an assessment of the water resource potential of a region can be made and is a vital consideration in planning regional economic development and resource management.

Soil-moisture regimes

C. Warren Thornthwaite, a distinguished climatologist who was concerned with practical problems of crop irrigation, developed the foregoing principles of the soil-water balance and proposed a system of classification of world climates based on those principles. Associates and students of Thornthwaite extended his work and collected hydrologic data for a vast network of observing stations on all land areas of the globe. Moisture deficiencies were computed on the basis of a field capacity of 300 mm, considered as the best single representative value for general use.

Figure 12.11 is a map of the United States showing isopleths of equal total annual potential evapotranspiration. Notice that the highest values occur over the desert basins of the southwest. Here, insolation is intense and air temperatures are generally high. High values also prevail over a belt bordering the Gulf Coast from Texas to Florida. As you would expect, values fall steadily as one progresses northward because of lowered total annual insolation and colder mean annual air temperatures. The world range in values of total annual E_p is much larger than this map indicates. Over large areas of the southern Sahara Desert, for example, the values are over 1500 mm/yr, while values of 1400 to 1750 mm/yr are prevalent over equatorial Africa. In arctic latitudes, annual E_p is reduced in some localities to as little as 200 to 250 mm/yr. Not only does the annual total E_p grade from high values in low latitudes to low values in high latitudes, but the pattern of seasonal distribution changes radically. As with monthly values of net radiation (Figure 12.13), monthly E_p values in the equatorial zone show nearly uniform high values throughout the year. With increasing latitude, an annual cycle becomes stronger, and, where ground is frozen and plants are dormant in a winter season, there are several consecutive months of virtually zero E_p. These global trends will show conspicuously in the examples to follow (see Figure 12.12).

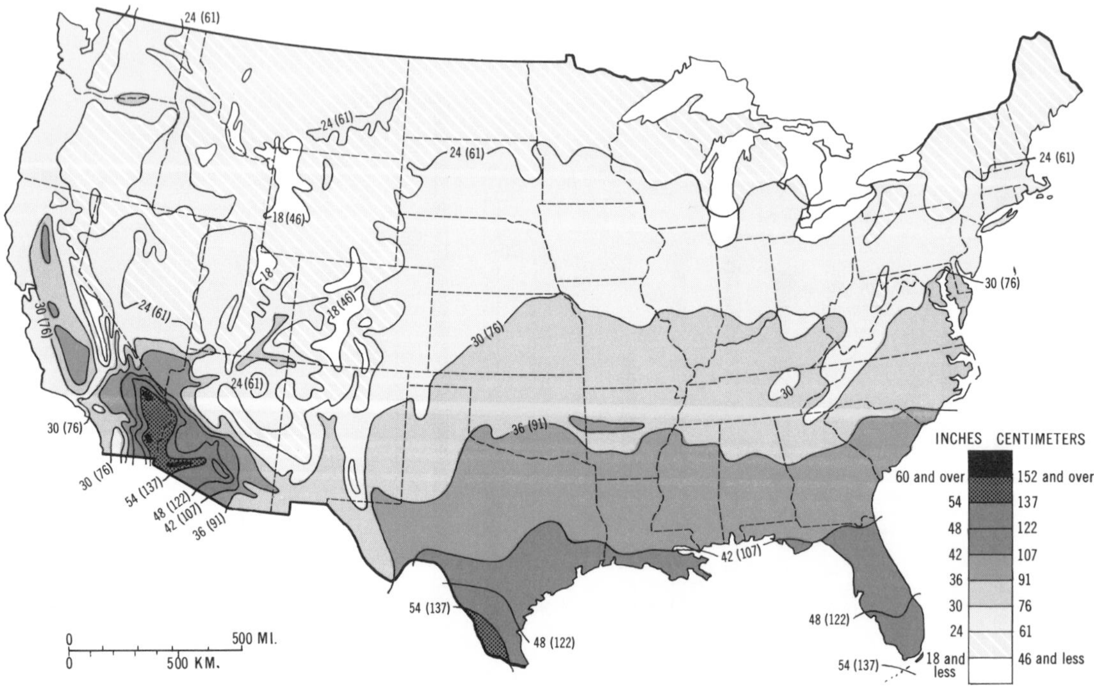

Figure 12.11 Average annual potential evapotranspiration in the United States. (Courtesy of *The Geographical Review*, vol. 38, 1948. Copyrighted by the American Geographical Society of New York.)

We offer here some examples of the wide global range in *soil-moisture regimes.* By "regime" we mean "prevailing tendency," or "dominant pattern" with respect to the annual rhythms of change in soil-moisture deficiencies and water surpluses. Figure 12.12 is a collection of moisture balance graphs for stations showing a wide range of regimes; they are arranged in order of increasing latitude. Let us start with the *Mediterranean regime,* illustrated by the soil-moisture budget for Los Angeles, California (Graph *D*). The regime name derives from its prevalence in lands bordering the Mediterranean Sea, but it is found generally on western continental coasts in the latitude range 30° to 50°. Notice that the graph for Los Angeles is very similar to the idealized example we used to show calculation of the terms of the water-balance equation (Figure 12.10). This regime is one of a large soil-moisture deficiency. Because P declines to low values at the very time when E_p rises to its maximum, the large summer deficiency is accentuated. Here is a region of dry climate in which crop irrigation will be extremely valuable; in fact, summer crop cultivation will not be possible here without irrigation. The native vegetation must be adapted to a long, hot summer with little or no rainfall and a long period of severely depleted soil moisture. The winter excess of P over E_p, on the other hand, gives a period of soil moisture recharge, but it is insufficient to generate a surplus.

An example of a regime with an enormous water surplus is the *equatorial regime,* illustrated in Graph *A*, showing data of Bougainville, a Pacific island of the Solomon group, lying close to the equator. Both P and E_p are remarkably uniform throughout the year, but P is much the greater in every month. As a result, there is a tremendously large annual total water surplus. Soil moisture storage remains at capacity—300 mm—in all months. Plant growth can proceed at the maximum rate in all months; the equatorial rainforest is the typical natural vegetation. Stream flow in this regime is copious throughout the year and leads to the possibility of development of hydroelectric power where mountainous relief is present and a demand for power exists.

The *tropical wet-dry regime* is illustrated by Graph *B*, showing data for Calcutta, India, lying close to the Tropic of Cancer. Here is a regime of great moisture contrasts. Precipitation is slight in the low-sun months (November to February), but rises to huge values in the rainy season (June to September). The annual curve of E_p also shows a strong seasonality, being moderately high in coincidence with the rainy season. As a result, there is a large water surplus (June to September), but a severe soil-moisture deficiency preceding the rains (March to May). Agriculture depends on these monsoon rains; should they not materialize, the moisture deficiency can extend through the year, bringing crop failure and famine.

The *tropical desert regime* is illustrated by data for Alice Springs, which lies in the heart of the Great Australian Desert (Graph *C*). While P is measurable in small amounts in all months, the values of E_p are always the larger. Consequently a soil-moisture deficiency prevails throughout all months of the year and racks up a very large total. Soil moisture storage is close to zero at all times. It is easy to understand why vegetation is sparse in a desert such as this. Notice that because this station is in the southern hemisphere, the annual cycle seems inverted in phase. Actually, the cycles of P and E_p, which are closely in phase, have their peak when the sun is highest in the sky.

Passing up Los Angeles, Graph *D*, which we have already analyzed, we arrive in middle latitudes. Graph *E* shows a station in the heart of the United States and illustrates the *continental humid regime.* The annual cycles of both P and E_p are closely in phase at Manhattan, Kansas; both show a strong summer maximum. However, values of E_p are the larger from June through October, so that a small soil-moisture deficiency develops in summer. But, because the deficiency does not make serious inroads into the soil moisture storage (300 mm), crops thrive without irrigation and yields are high in most years. In winter, E_p falls to zero in three months (December to February), and recharge sets in. By spring, water surplus is produced, but it is not a large amount.

We now travel north, still in the heart of the North American continent, to reach Medicine Hat, Alberta, a station in the Great Plains region of Canada (Graph *F*). The regime is much like that of Manhattan, Kansas, but P is much less than E_p in summer months, generating a large total soil-moisture deficiency. We can call this pattern the *continental arid regime.* Recharge in winter is insufficient to restore the field capacity of the soil, so that no surplus occurs.

At about the same latitude as Medicine Hat, but located on the Pacific Coast, is a station showing the *wet west coastal regime* (Graph *G*). Although the regime at Prince Rupert, British Columbia, may seem totally unlike that for Los Angeles (Graph *D*), the annual cycles of P and E_p show the same out-of-phase relationship. Notice the reduced values of P in the summer. However, since P is always larger than E_p, a surplus occurs in every month and yields a total annual surplus —even greater than for Bougainville (Graph *A*). This huge water surplus might be put to man's use both for hydroelectric power and to be transported to regions of water deficiency.

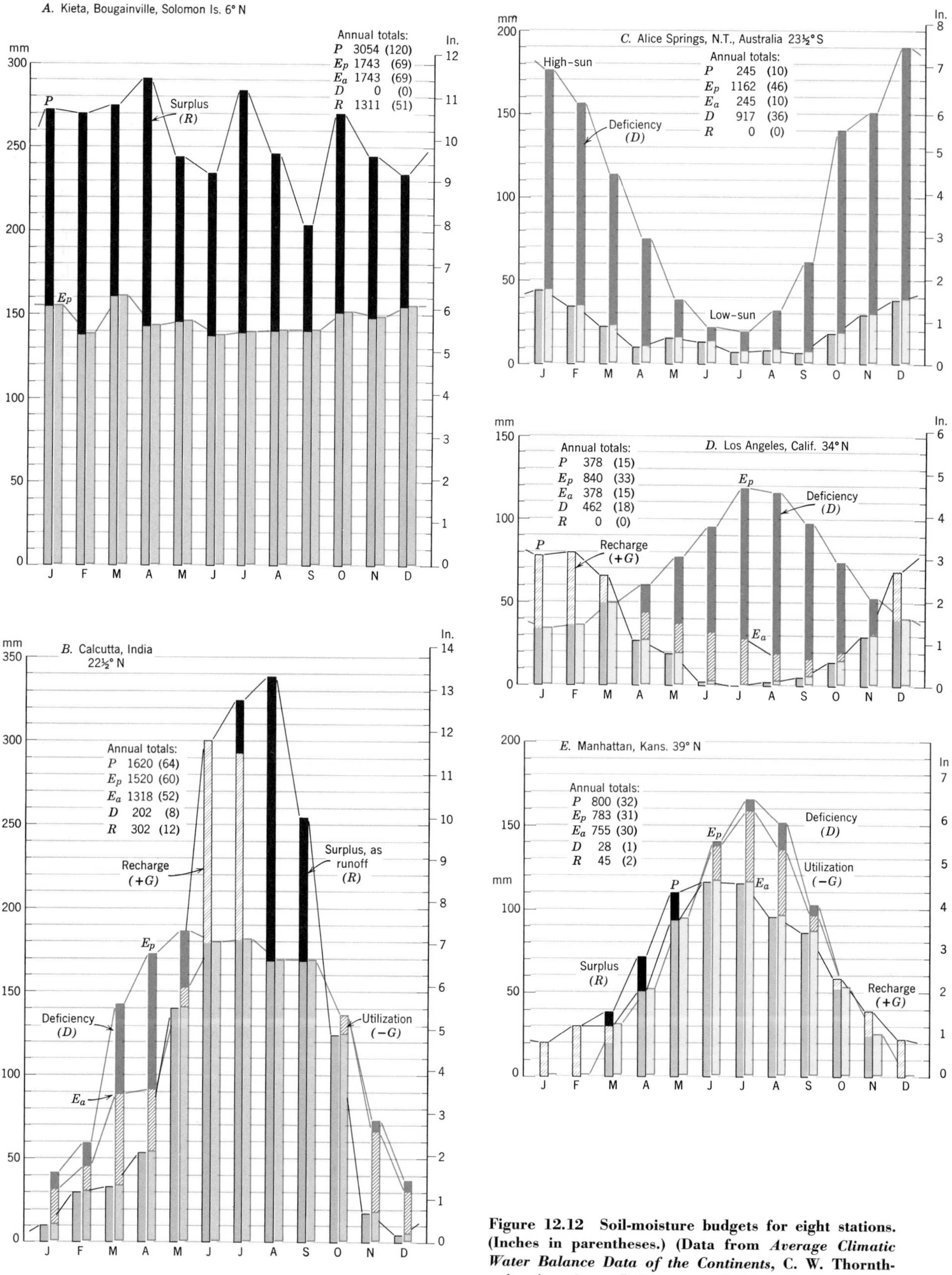

Figure 12.12 Soil-moisture budgets for eight stations. (Inches in parentheses.) (Data from *Average Climatic Water Balance Data of the Continents*, C. W. Thornthwaite Associates, Laboratory of Climatology, *Publ. in Climatology*, 1962–1965, Centerton, N.J.)

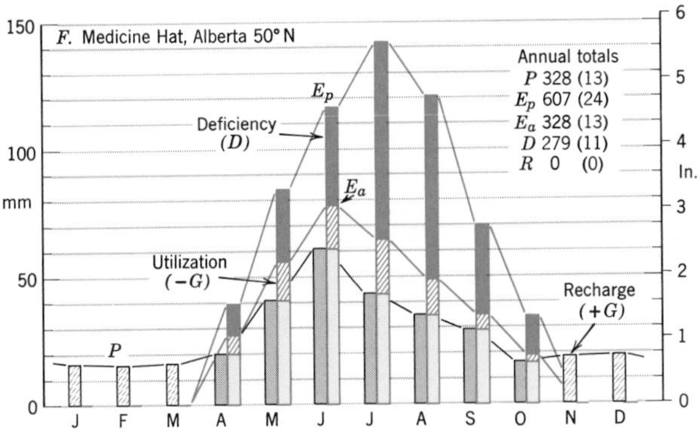

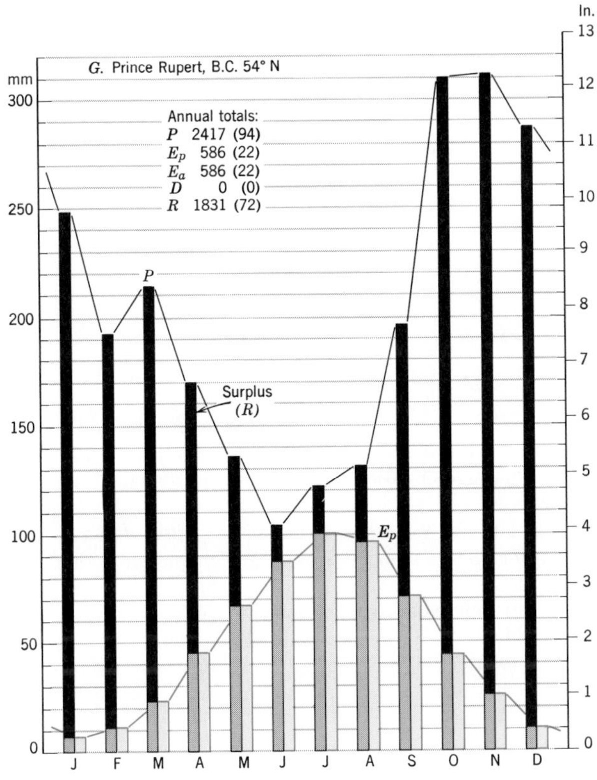

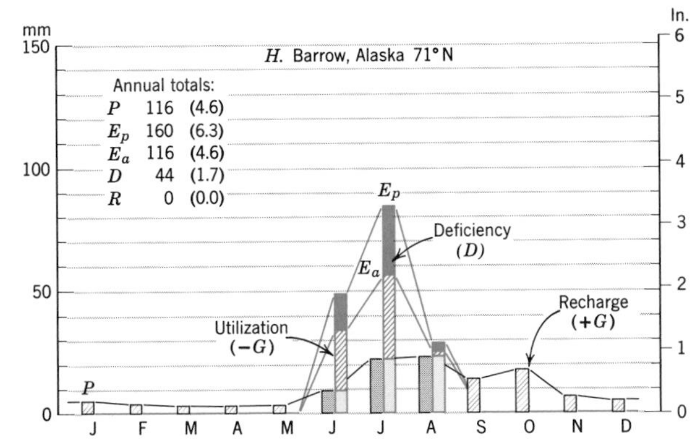

Finally, we travel to Barrow, Alaska, a station on the shores of the Arctic Ocean, latitude 71° N, lying poleward of the Arctic Circle (Graph *H*). Because the ground is frozen from September through May, no E_p occurs for nine consecutive months. Under an *arctic regime*, E_p rises to a sharp peak in summer, and even exceeds P to the extent that a soil-moisture deficiency develops in three months.

This very brief investigation of only a few examples of soil-moisture budgets gives an appreciation of the great global range in qualities of the environments of the life layer. Considerations of soil-moisture deficiency and water surplus are vital, not only in understanding the adaptation of plants and animals to their environments, but also in understanding the opportunities and limits of man's use of plant resources. Intelligent planning for environmental management depends heavily upon an understanding of all phases of the water balance.

REVIEW QUESTIONS

1. Distinguish between surface water and subsurface water; between soil water and ground water.

2. Give a complete general account of the hydrologic cycle. What quantities of water are evaporated annually from the oceans? From the lands? How is this balanced by precipitation and runoff?

3. Where is most of the world's water held in storage? What environmental effects would result from changes in quantity of water held in storage as glacial ice?

4. What is infiltration? What is overland flow? In what units is each stated?

5. How is soil water held in place? What is the source of soil moisture? What is field capacity? How does it vary with soil texture? What is the wilting point?

6. To what depth does soil moisture evaporate? How do atmospheric pressure changes assist in this process?

7. Explain the process of transpiration by plants. What quantities of soil water are thus lost? To what depth does this loss extend? Define evapotranspiration.

8. Describe the annual cycle of soil water in humid, middle-latitude climates. When and why do water surpluses and deficits occur?

9. State the soil-moisture balance equation and define the terms used. Distinguish between actual and potential evapotranspiration.

10. Under what circumstances does a soil-moisture deficiency occur? When does soil moisture recharge take place?

11. Explain how the soil-moisture balance is calculated, using monthly values.

12. How is annual potential evapotranspiration related to latitude? To air temperature? Illustrate with examples from the United States.

13. What is meant by a soil-moisture regime? Name six important regimes and describe the distinctive annual and seasonal features of each.

14. Of what importance is the soil-moisture balance in agriculture? In water resource development?

Exercises

Tables on the opposite page give soil-moisture balance data for eight stations, representing each of the eight soil-moisture regimes described in Chapter 12. For each station prepare a graph similar to graphs in Figure 12.12. However, instead of drawing bars, plot each number as a point on the graph. Use the following symbols: P, solid dot; E_p, open dot (circle); E_a, triangle. Connect the points with straight line segments. Use a vertical line to mark the change from soil-moisture recharge ($+G$) to surplus (R). Color the areas between curves, using the following code: D, red; R, blue; $-G$, yellow; $+G$, diagonal blue bars.

Use graph paper ruled 10 mm to the centimeter. Use a vertical scale such that 5 mm represents 10 mm of P, E_p, and E_a. On the horizontal scale allow 10 mm for each calendar month. Plot the monthly mean values along the midline of the month.

As each graph is completed, identify the soil-moisture regime it illustrates and label the graph accordingly. (Stations are listed in the table in alphabetical order.)

Calculate the values of D and S for those stations having a moisture deficiency or a water surplus or both. Enter these values along with total annual P, E_p, and E_a in any convenient space on the graph.

(Data from C. W. Thornthwaite Associates, *Average Climatic Water Balance Data of the Continents*, Laboratory of Climatology, Publications in Climatology, Vols. 15–18, 1962–1965.)

	J	F	M	A	M	J	J	A	S	O	N	D	Year
Hyderabad, Pakistan ($25\frac{1}{2}°$ N, $68\frac{1}{2}°$ E)													
P	5	8	5	2	5	10	79	53	15	0	2	2	186
E_p	25	33	118	175	204	205	203	184	165	153	72	26	1563
E_a	5	8	5	2	5	10	79	53	15	0	2	2	186
Irkutsk, U.S.S.R. ($52\frac{1}{2}°$ N, $104\frac{1}{2}°$ E)													
P	10	9	7	16	31	58	77	68	44	17	16	16	369
E_p	0	0	0	5	65	109	129	104	52	0	0	0	464
E_a	0	0	0	5	51	85	99	82	47	0	0	0	369
Januarete, Brazil ($\frac{1}{2}°$ N, 69° W)													
P	244	217	259	362	358	351	299	272	264	232	213	204	3275
E_p	112	110	119	115	112	103	103	106	109	122	118	119	1348
E_a	112	110	119	115	112	103	103	106	109	122	118	119	1348
Kaduna, Nigeria ($10\frac{1}{2}°$ N, $7\frac{1}{2}°$ E)													
P	0	2	13	69	147	180	218	312	279	76	5	0	1301
E_p	84	100	143	157	156	119	100	93	100	114	94	82	1342
E_a	36	34	42	82	148	119	100	93	100	112	73	47	986
Leipzig, East Germany ($51\frac{1}{2}°$ N, $12\frac{1}{2}°$ E)													
P	40	33	41	47	60	67	85	67	49	50	40	42	621
E_p	0	2	18	45	89	114	126	105	72	39	14	4	628
E_a	0	2	18	45	88	107	115	91	62	39	14	4	585
New Plymouth, New Zealand (39° S, 174° E)													
P	112	104	91	119	152	155	160	142	127	140	122	112	1536
E_p	93	81	76	59	42	30	28	33	42	59	70	87	700
E_a	93	81	76	59	42	30	28	33	42	59	70	87	700
Sevilla, Spain ($37\frac{1}{2}°$ N, 6° W)													
P	44	65	70	50	36	27	2	2	27	70	99	67	559
E_p	17	22	39	57	98	140	180	176	121	71	33	19	973
E_a	17	22	39	55	81	89	61	35	38	70	33	19	559
Sian, China (34° N, 109° E)													
P	4	8	17	40	52	56	91	106	105	59	15	5	558
E_p	0	3	24	57	112	156	176	151	89	50	14	1	833
E_a	0	3	19	43	62	69	99	109	89	50	14	1	558

CHAPTER 13

Ground Water and Surface Water

WE have now followed the hydrologic cycle through the atmospheric phase and the soil-moisture phase. When a soil-moisture deficiency exists, water returns directly to the atmospheric phase by evapotranspiration, and thus a subcircuit is completed. On the other hand, when a water surplus exists in the soil moisture budget, the hydrologic cycle is continued by the runoff phase, and this may take one of two paths.

First, surplus water may percolate through the soil, traveling downward under the force of gravity to become a part of the underlying ground water body. After following subterranean flow paths, this water emerges to become surface water, or it may emerge directly in the shore zone of the ocean (Figure 12.1).

Second, surplus water may flow over the ground surface as runoff to progressively lower levels. In so doing, the dispersed flow becomes collected into streams, which eventually conduct the runoff to the ocean. In this chapter, we trace both the subsurface and surface pathways of flow of surplus water, and thus complete the hydrologic cycle by means of which the total global water balance is maintained.

Surplus water, as runoff, is a vital part of the environment of terrestrial life forms and of man in particular. Surface water in the form of streams, rivers, ponds, and lakes constitutes one of the distinctive environments of plants and animals. Like the oceans, fresh waters of the lands support primary producers of carbohydrate compounds. Consumers in elaborate food chains depend upon this primary product. Man, like other air-breathing land animals, finds a substantial food resource in the lesser members of this food chain.

Our heavily industrialized society requires enormous supplies of fresh water for its sustained operation. Urban dwellers consume water in their homes at rates on the order of 50 to 100 gallons per person per day. Huge quantities of water are used for cooling purposes in air-conditioning units and power plants.

In view of projections based upon existing rates of increase in water demands, we shall be hard put in the future to develop the needed supplies of pure fresh water. Water pollution also tends to increase as populations grow and urbanization advances over broader areas. Thus the available resource of pure fresh water is shrinking while demands are rising.

An understanding of the processes of runoff is basic to the study of man's impact upon natural hydrologic systems. Knowledge of hydrologic processes also enables us to evaluate the total water resource and to plan for its management. To this end, we need to learn not only the nature and pathways of the flow of water, but also how flow quantities are measured.

Ground water

Ground water is that part of subsurface water which fully saturates the pore spaces of the rock or its overburden and which behaves in response to gravitational force. The ground water occupies the *zone of saturation* (Figure 13.1). Above it is the *zone of aeration*, in which water does not fully saturate the pores. We have seen that the soil-

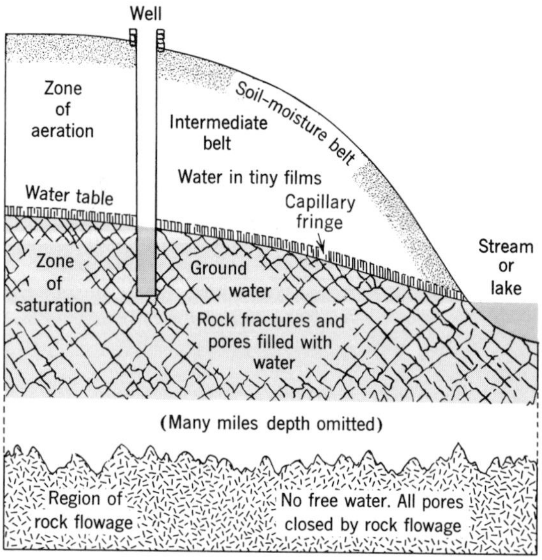

Figure 13.1 Subsurface water zones. (After Ackerman, Colman and Ogrosky.)

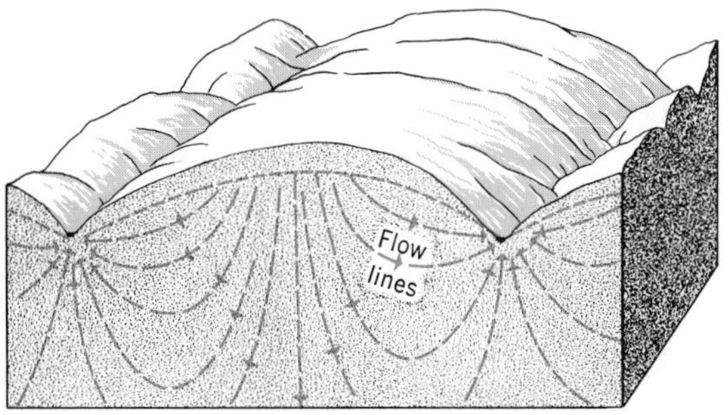

Figure 13.2 Theoretical paths of ground-water movement under divides and valleys. (After M. K. Hubbert.)

moisture belt is the uppermost layer of the zone of aeration; moisture is held in this belt by capillary force in tiny films adhering to the soil particles. A similar condition prevails through the underlying intermediate belt. The sole basis for distinguishing these two belts is that the soil belt represents a shallow zone of moisture usable by plants, whereas the intermediate belt is too deep for capillary water to be returned to the atmosphere by either direct evaporation or transpiration. The depth of the zone of aeration may be very shallow or missing (when the ground water is close to the surface in low, flat regions), or up to several hundred feet thick in hilly or mountainous regions with low ground water level.

At the base of the zone of aeration is the *capillary fringe*, a thin layer in which the water has been drawn upward from the ground water body through capillary force. The action is much like the rise of kerosene in a lamp wick, or of water in a blotter whose edge is immersed. Water in the capillary fringe largely fills the soil pores, hence, is continuous with the ground water body. Thickness of the capillary fringe depends on the soil texture, because capillary rise is higher when the openings are smaller. Thus, in a silty material the capillary fringe may be 2 ft (0.6 m) thick, but only a fraction of an inch (1 cm) thick in coarse sand or fine gravel with large pore spaces.

Natural discharge of ground water

The subsurface phase of the hydrologic cycle is completed when the ground water emerges along lines or zones where the water table intersects the ground surface. Such places are the channels of streams and the floors of marshes and lakes. By slow seepage and spring flow the water must emerge in sufficient quantity to balance that which enters the ground water table by percolation through the zone of aeration.

Figure 13.1 shows the theoretical paths of flow of ground water as calculated by use of basic principles of the physics of fluids. Water follows paths curved concavely upward. Water entering the slope midway between divide and stream flows rather directly. Close to the divide point on the water table, however, the flow lines go almost straight down to great depths in the earth from which they recurve upward to points under the streams. Progress along these deep paths would be incredibly slow; that near the surface would be faster. The most rapid flow is encountered close to the line of discharge in the stream, where the arrows are shown to converge.

The water table

Ground water is extracted for man's use from wells dug or drilled to reach the ground water zone. In the ordinary shallow well, water rises to the same height as the *water table*, or upper boundary of the zone of saturation.

If wells are numerous in an area, the position of the water table can be mapped in detail by plotting the water heights and noting the trends in elevation from one well to the other. When this is done it is usually seen that the water table is highest under the highest areas of surface, namely, hilltops and divides, but descends toward the valleys where it appears at the surface close to streams, lakes, or marshes (Figure 13.3). The reason for such a configuration of the water table is that water percolating down through the zone of aeration tends to raise the water table, whereas seepage into streams, swamps, and lakes tends to draw off ground water and to lower its level.

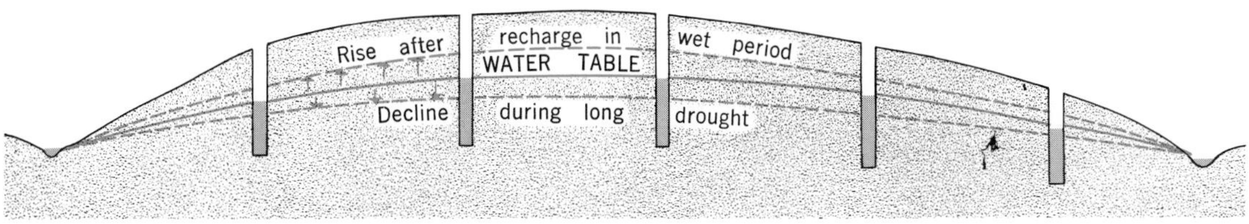

Figure 13.3 The water table conforms to surface topography.

Because ground water moves extremely slowly, a difference in water table level, or *head*, is built up and maintained between areas of high elevation and those of low elevation. In periods of water surplus accompanying abnormally high precipitation, this head is increased by a rise in the water table elevation under divide areas; in periods of water deficit, occasioned by drought, the water table falls (Figure 13.3).

In humid climates having a strong seasonal cycle of precipitation, and in climates in which soil moisture is frozen for several months of the year, a seasonal cycle of rise and fall of the water table is produced. This cycle is illustrated by a graph of water table fluctuations in an observation well on Cape Cod, Massachusetts (Figure 13.4). Percolating water reaches the water table in abundant quantities from late winter to late spring, a phenomenon known as *ground water recharge.* This recharge period, which reflects the period of water surplus in the soil-moisture budget, causes a seasonal rise in the water table. Recharge declines to very small amounts from midsummer to early winter, and during this time, the water table steadily declines as water moves to lower levels under the force of gravity. Effects of a major drought—1964 to 1966—show in the graph by a downward trend of the average elevation.

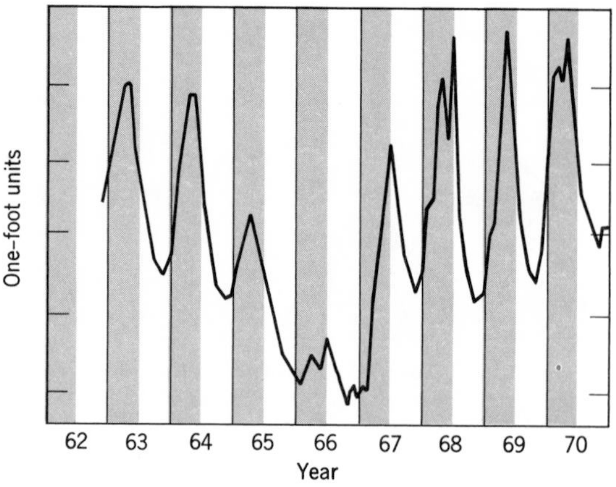

Figure 13.4 Hydrograph of an observation well on Cape Cod, Massachusetts, showing the characteristic annual cycle of rise and fall of the water table. (Data of U.S. Geological Survey.)

Water-table lakes, marshes, and bogs

In humid climates with a large annual surplus in the water budget and with generally high water tables, the various events of geologic history, such as erosion and deposition by wind and by ice sheets of the most recent glacial stage, have created natural depressions with floors at or below the water table elevation. Seepage of ground water, as well as direct runoff of precipitation, maintains these free water surfaces permanently throughout the year. Examples of fresh water-table ponds are found widely distributed in North America and Europe where plains of glacial sand and gravel contain natural pits and hollows left by the melting of stagnant ice masses. (See Chapter 31.) Figure 13.5 is a block diagram showing small fresh-water ponds on Cape Cod. The surface elevation of these ponds coincides closely with the level of the surrounding water table. Pond levels rise and fall seasonally in a rhythm similar to that of the water table (Figure 13.4).

Many former fresh-water water-table ponds have, since they were formed by glacial ice, become partially or entirely filled by the organic matter from growth and decay of water-loving plants. The result is a *bog* whose surface lies close to water table (Chapter 20).

Marshes and swamps, where water stands at or close to the ground surface over a broad area, represent the appearance of the water table at the surface. Such areas of poor surface drainage have a variety of origins. For example, the broad, shallow fresh-water swamps of the Atlantic and Gulf coastal plain represent regions only recently emerged from the sea. Others are created by the shifting of river channels on floodplains.

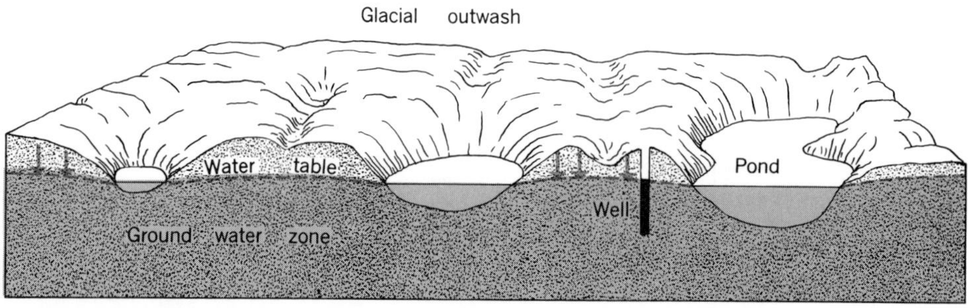

Figure 13.5 Water table ponds in glacial deposits, largely sand, on Cape Cod, Massachusetts. (From A. N. Strahler, *A Geologist's View of Cape Cod*, Copyright © 1966 by Arthur N. Strahler. Reproduced by permission of Doubleday & Company.)

Pumping, drawdown, and recharge of wells

Of increasing importance in environmental science is the effect on the water table of man's withdrawals of ground water. The drilling of vast numbers of wells, from which water is forced out in great volumes by powerful pumps, has profoundly altered nature's balance of ground water recharge and discharge. Increased urban populations and industrial developments require larger water supplies, needs that cannot always be met from construction of new surface water reservoirs.

In agricultural lands of the semi-arid and desert climates, heavy dependence is placed upon irrigation water from pumped wells, especially since many of the major river systems have already been fully developed for irrigation from surface supplies. Wells can be drilled within the limits of a given agricultural or industrial property, and hence can provide immediate supplies of water without need to construct expensive canals or aqueducts. A few of the physical principles of water wells are treated here to aid in understanding the basis of complex economic and legal problems arising from ground water development.

Formerly the small well needed to supply domestic and livestock needs of a home or farmstead was actually dug by hand as a large cylindrical hole, lined with masonry where required. By contrast, the modern well put down to supply irrigation and industrial water is drilled by powerful machinery which may bore a hole 12 to 16 in (30 to 40 cm) or more in diameter to depths of 1000 ft (300 m) or more, although much smaller-scaled wells and well-boring machines suffice for domestic purposes. Drilled wells are sealed off by metal casings which exclude impure near-surface water and prevent clogging of the tube by caving of the walls. Near the lower end of the hole, where it enters the aquifer, the casing is perforated so as to admit the water through a considerable surface area. Rate of flow of a well or spring is stated in units of gallons or liters per minute or per day. The yields of single wells range from as low as a few gallons per day in a domestic well to many millions of gallons per day for large, deep industrial or irrigation wells.

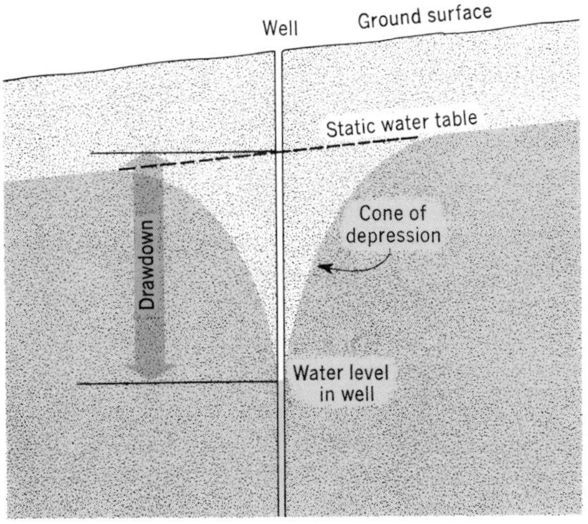

Figure 13.6 Drawdown and cone of depression in a pumped well.

As water is pumped from a well, the level of water in the well drops and the surrounding water table is lowered in the shape of a conical surface, termed the *cone of depression*, the height of which is termed the *drawdown* (Figure 13.6). By producing a steeper gradient of the water table, the flow of ground water toward the well is also increased, so that the well will yield more water. This holds only for a limited amount of drawdown, beyond which the yield fails to increase. The cone of depression may extend as far out as 8 to 10 mi (13 to 16 km) or more from a well where very heavy pumping is continued. Where many wells are in operation, their intersecting cones produce a general lowering of the water table.

Depletion often greatly exceeds the rate at which the ground water of the area is recharged by percolation from rain or from the beds of streams. In an arid region, much of the ground water for irrigation is from wells driven into thick sands and gravels which are lowland deposits of transported overburden of a type termed *alluvium*. (These features are described in Chapter 25.) Recharge of such deposits depends on the seasonal flows of water from streams heading high in adjacent mountain ranges. Where such highly permeable materials exist, the extraction of ground water by pumping can greatly exceed the recharge by stream flow. Cones of depression deepen and widen; deeper wells and more powerful pumps are then required. Overdrafts of water accumulate and the result is exhaustion of a natural resource not renewable except by long lapses of time. In humid areas where a large annual water surplus exists natural recharge is by general percolation over the ground area surrounding the well. Here the prospects of achieving a balance of recharge and withdrawal are highly favorable through the control of pumping and the return of waste waters or stream waters to the ground water table by means of *recharge wells* in which water flows down, rather than up. Use of waste water and stream water for ground-water recharge is not always a wholly satisfactory procedure, for the possibilities of polluting the ground water are always present. It is to this subject that we turn our attention next.

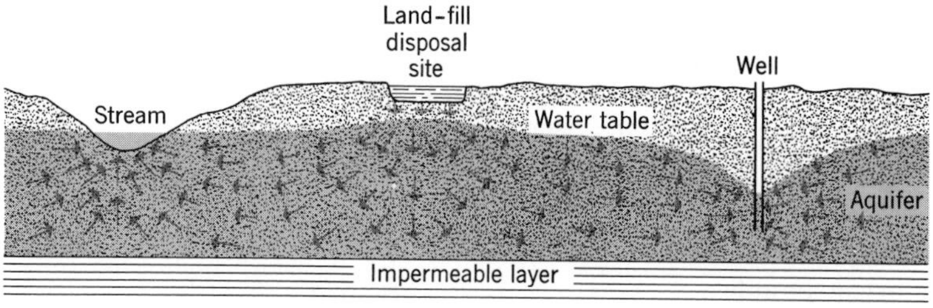

Figure 13.7 Leachate from a waste disposal site moves toward a supply well (right) and a stream (left). (From A. N. Strahler, 1972, *Planet Earth; Its Physical Systems Through Geologic Time*, Harper and Row, New York.)

Pollution of ground water

Disposal of solid wastes poses a major environmental problem in the United States because our advanced industrial economy produces an endless source of garbage and trash. Traditionally, these waste products have been trucked to the town or city dump, and there burned in continually smoldering fires that emit foul smoke and gases. The partially consumed residual waste is then buried under earth. In recent years, a major effort has been made to improve solid-waste disposal methods. One method is high-temperature incineration. Another is the *sanitary land-fill* method in which waste is not allowed to burn, but is continually covered by protective overburden, usually sand or clay available on the land-fill site. The waste is thus buried in the zone of aeration and is subject to reaction with percolating rainwater infiltrating the ground surface. This water picks up a wide variety of ions from the waste body and carries these down as *leachate* to the water table. Once in the water table, the leachate follows the flow paths of the ground water.

As shown in Figure 13.7, there normally develops a *mound* in the water table beneath the disposal site. Loose soil of the disposal area facilitates infiltration of precipitation, while lack of vegetation reduces the evapotranspiration. Consequently, the recharge here is greater than elsewhere and the mound is maintained. After leachate has moved vertically down by gravity percolation to the water table mound, it moves radially outward from the mound to surrounding lower points on the water table. As shown in Figure 13.7, a supply well with its cone of depression draws ground water from the surrounding area. Linkage between outward flow from the waste disposal site and inward flow to the well can bring leachate into the well, polluting the ground water supply. (The term *aquifer* in Figure 13.7 designates a rock layer through which ground water moves freely.)

An important step in guarding against this form of pollution is to place a monitor well (or several monitor wells) on a line between the disposal site and the well. Chemical tests for presence of leachate are made regularly, while the slope of the water table can also be determined. Movement of leachate toward the supply well may be blocked by placement of a recharge well (or wells), building a fresh-water accumulation (actually an inverted cone), which will oppose the movement of the leachate.

Pollution of supply wells by partially treated effluent infiltrating the ground at sewage disposal plants can occur in a basically similar manner.

It is not necessary for a ground-water mound to be present for pollutants to travel to distant points. Where the water table has a pronounced slope, as it generally does everywhere except near the summit of a broad ground-water divide, leachate or any pollutant introduced at a given point migrates as a *pollution plume* along the flow paths of the ground water.

Yet another potential source of pollution of ground water supplies is from highways and streets, through spillage of chemicals and from deicing salt applied during the winter months. Spillage of large volumes of liquids from tank trucks and tank cars as a result of highway and railroad accidents poses a serious threat because a large slug of pollutant can be injected into the ground-water recharge system. The commonest pollutants to be feared are automobile fuel and heating oil, but many toxic industrial chemicals are transported in tank trucks and cars. Leakage of fuels from underground storage tanks, used in all gasoline stations, is a related source of possible pollution.

Salt water intrusion

A serious consequence of sustained, heavy ground water withdrawal in coastal zones is that wells near the shore eventually draw salt ground water and must be abandoned. To understand how this happens, we must examine the relationship between salt and fresh ground water.

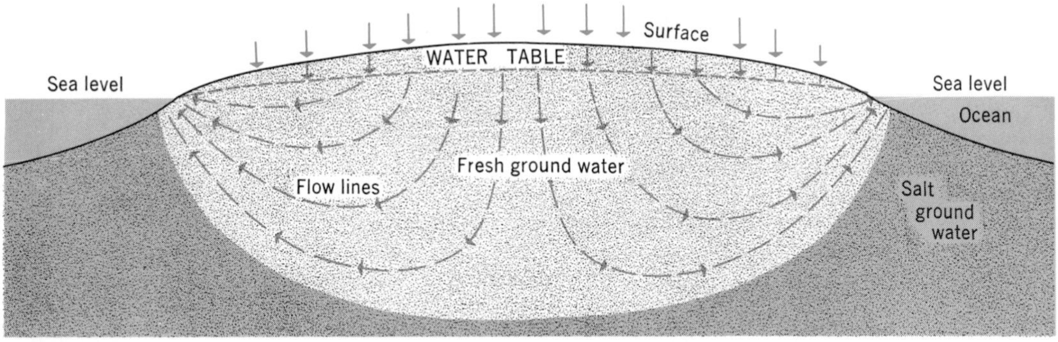

Figure 13.8 Fresh-water and salt-water relations in an island or peninsula. (After G. Parker.)

Figure 13.8 shows an idealized diagram through an island or a narrow peninsula. The body of fresh ground water takes the shape of gigantic lens with convex faces, except that the upper surface has only a broad curvature whereas the lower surface, in contact with the salt ground water, bulges deeply downward. Because fresh water is less dense than salt, we can think of this fresh water lens as floating upon the salt water, pushing it down much as the hull of an ocean liner pushes aside the surrounding water. The ratio of densities of fresh water to salt water is as 40 to 41. Hence, if the water table is, say, 10 feet above sea level, the bottom of the fresh water lens will be located 400 feet below sea level, or forty times as deep as the water table is high with respect to sea level.

The fresh ground water extends seaward some distance beyond the shoreline. Although the salt ground water is stagnant, the fresh water travels in the curved paths shown by arrows in Figure 13.8. If water is pumped excessively from wells close to the coast, the contact of salt with fresh ground water shifts landward, where it may intersect the wells and contaminate the fresh water.

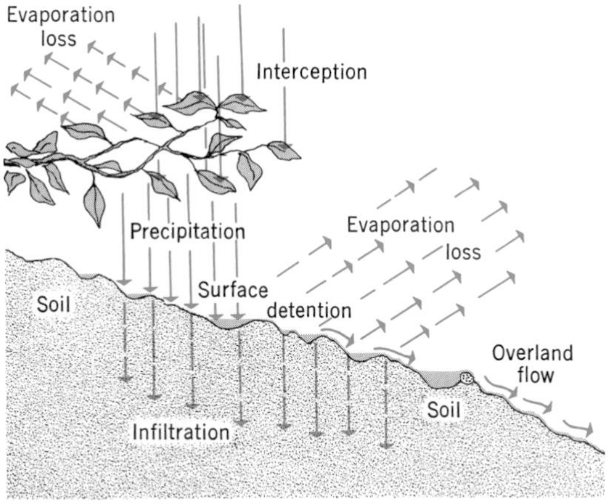

Figure 13.9 Precipitation and overland flow.

When such a condition exists, the only cure is to cease pumping and allow the fresh water slowly to push the contact back toward its original seaward position, or to create a fresh water recharge barrier between the wells and the coastline. Resumed pumping must then be regulated to a lower rate.

We have now traced the subsurface movements of surplus water beneath the lands. We turn next to trace the surface flow paths of surplus water.

Forms of overland flow

Runoff that flows down the slopes of the land in more or less broadly distributed films, sheets, or rills is referred to as *overland flow* in distinction with *channel flow*, or *stream flow*, in which the water occupies a narrow trough confined by lateral banks. Within this broad definition, overland flow can take many forms. It may be a continuous thin film, called *sheet flow*, where the soil or rock surface is extremely smooth, or a series of tiny rivulets connecting one water-filled hollow with another, where the ground is rough or pitted. On a grass-covered slope, overland flow is subdivided into countless tiny threads of water, passing around the stems. Even in a heavy and prolonged rain, overland flow in full progress on a sloping lawn may not be visible to the casual observer. On heavily forested slopes bearing a thick mat of decaying leaves and many fallen branches and tree trunks, overland flow may pass almost entirely concealed beneath this cover.

Progress of overland flow

Imagine that a hillside slope which has been thoroughly drained of moisture in a period of drought is subjected to a period of rain. What are the successive stages in the production of overland flow? If heavy vegetation such as forest is present, much of the rain at the beginning is held in droplets on the leaves and plant stems, a process termed *interception*. This water may be returned directly to the atmosphere by evaporation, so that if the rainfall is brief, little water reaches the ground.

The ground surface is capable of absorbing by infiltration even a heavy rain in the early stages of a rain period. Therefore, unless the rain continues for an hour or so, no overland flow may be expected. Soon, however, the soil passages are sealed or obstructed, dropping the infiltration rate to a low, but constant value. Any excessive precipitation now remains on the ground surface, first accumulating in small puddles or pools which occupy natural hollows in the rough ground surface or are held behind tiny check dams formed by fallen leaves and twigs (Figure 13.9). *Surface detention* is the term applied to the holding of water on a slope by such small natural containers. Assuming that the rain continues to fall with sufficient intensity, water then overflows from hollow to hollow, becoming true overland flow.

Because any given square unit of ground on a hill side must receive the overland flow from the entire strip of ground of that width lying upslope of it, we may expect the rate of discharge (volume of water passing across a given line in a given unit of time) to increase in direct proportion to the length of the total path of flow. Depth of the flowing layer might therefore be expected to increase the farther downslope it progresses, but this increase may be small because the flow velocity will also be increasing down the slope.

At the base of a hill slope, overland flow is disposed of by passing into a stream channel or lake, or by sinking into the ground, should a highly permeable layer of sand, gravel, or blocky slide rock be encountered.

Overland flow is measured in inches or centimeters of water per hour, just as for precipitation and infiltration. Therefore, a simple formula expresses the rate at which overland flow will be produced by a given unit of ground surface as follows:

Rate of production of overland flow = rate of precipitation − rate of infiltration

For example, if the rate of infiltration became constant at a value of 0.4 inches per hour, and the rate of rainfall was a steady 0.6 inches per hour (a heavy rain), the runoff would be produced at a rate of 0.2 inches per hour, assuming none to be returned to the atmosphere by evaporation.

Drainage systems

Seeking to escape to progressively lower levels and eventually to the sea, runoff becomes organized into *drainage systems,* which we may describe as more or less pear-shaped areas bounded by divides, within which ground slopes and branching stream networks are adjusted to dispose as efficiently as possible of the runoff and its contained load of mineral particles.

A typical stream network contributing to a single outlet is shown in Figure 13.10. Note that each fingertip tributary receives runoff from a small area of land surface surrounding the channel. This area may be regarded as the *unit cell* of the drainage system. The entire surface within the outer divide of the drainage basin constitutes the *watershed* for overland flow. Thus, a drainage system is a converging mechanism funneling and integrating the weaker and more diffuse forms of runoff into progressively deeper and more intense paths of activity.

The study of drainage systems brings us in contact with two fields of science, hydrology and geology. Much of the study of the water itself, particularly as to the quantities of water involved

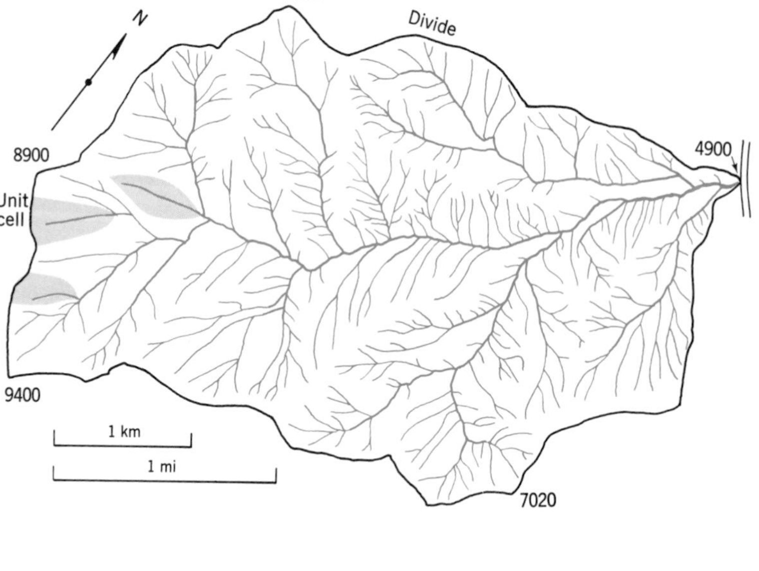

Figure 13.10 Channel network of the basin of Pole Canyon, Spanish Fork Peak Quadrangle, Utah. Data of U.S. Geological Survey and Mark A. Melton. (From A. N. Strahler, 1972, *Planet Earth; Its Physical Systems Through Geologic Time,* Harper and Row, New York.)

in runoff and their variations in response to precipitation, is done by hydrologists, who are affiliated with the profession of civil engineering. The study of streams in eroding and transporting rock materials, so as to shape the landforms of drainage systems, is done by geologists. In the United States, both groups pool their efforts in the analysis and solution of problems of both runoff and ground water under the Water Resources Division of the U.S. Geological Survey, to which is given the responsibility for assessing the surface and ground water resources of the nation.

The Forest Service studies problems of runoff in the nation's forests; the Soil Conservation Service is concerned with the effects of runoff in causing land erosion and related agricultural problems. Runoff problems concerned with the improvement of irrigation works and navigable waterways are dealt with by engineers and scientists of the Bureau of Reclamation and the U.S. Army Corps of Engineers.

Stream channels

The channel of a stream may be thought of as a long, narrow trough, shaped by the forces of flowing water to be most effective in moving the quantities of water and sediment supplied from the drainage basin or watershed. Channels may be so narrow that a person can jump across them, or as wide as one mile (1.5 km) for great rivers such as the Mississippi. Taking the entire range of natural channel widths as between one foot and one mile, a 5000-fold difference in size can exist.

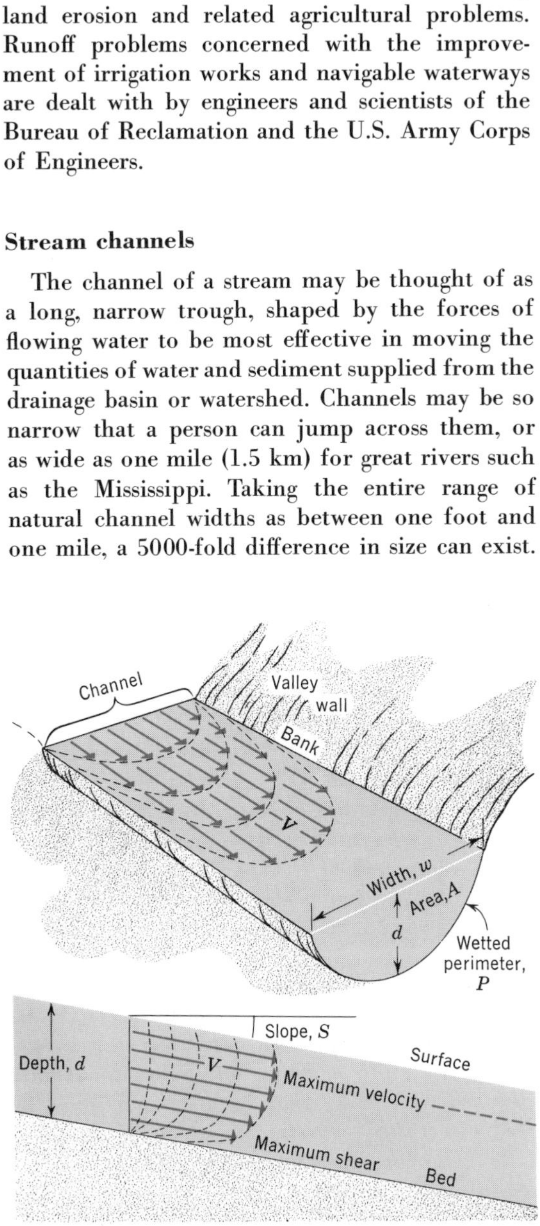

Figure 13.11 Geometry of a stream channel.

Hydraulic engineers who must measure stream dimensions and flow rates have adopted a set of terms to describe channel geometry (Figure 13.11). *Depth*, in feet or meters, is measured at any specified point in the stream as the vertical distance from surface to bed. *Width* is the distance across the stream from one water's edge to the other. *Cross-sectional area, A*, is the area in square feet or square meters of a vertical slice across the stream at any specified place. *Wetted perimeter, P*, is the length of the line of contact between the water and the channel, as measured from the cross section.

Channel *slope*, *S* (or *gradient*), is the angle between the water surface and the horizontal plane. Slope can be stated in feet per mile or meters per kilometer. Thus a slope of five feet per mile means that the stream surface undergoes a vertical drop of five feet for each mile of horizontal distance downstream. Slope can also be given in terms of *per cent grade*, a common practice in engineering. A grade of 3 percent, or 0.03, means that the stream drops 3 feet for every 100 feet of horizontal distance.

Stream flow

Gravity acts upon the water of a stream to exert pressure against the confining walls. A small part of the gravitational force is aimed downstream parallel with the surface and bed, causing flow. Resisting the force of downstream flow is the force of resistance, or friction, between the water and the floor and sides of the channel. As a result, water close to the bed and banks moves slowly; that in the deepest and most centrally located zone flows fastest. Figure 13.11 indicates by dotted lines the manner in which flow takes place, or the *velocity distribution*. We can imagine that each dot is a given drop of water and that we observe its subsequent positions at equal time intervals. The single line of highest velocity is located in midstream, if the channel is straight and symmetrical, but about one-third of the distance down from surface to bed.

The above statements about velocity need to be qualified. Actually, in all but the most sluggish streams, the water is affected by *turbulence*, a system of innumerable eddies that are continually forming and dissolving. Therefore, a particular molecule of water, if we could keep track of it, would actually describe a highly irregular, corkscrew path as it is swept downstream. Motions would include upward, downward, and sideward directions. Turbulence in streams is extremely important because of the upward elements of flow that lift and support fine particles of sediment. The murky, turbid appearance of streams in flood is ample evidence of turbulence, without which sediment would remain near the bed. Only if we measure the water velocity at a certain fixed point for a long period of time, say several minutes, will the average motion at that point be downstream and in a line parallel with the surface and bed. It is such average values that are shown by the arrows in Figure 13.10.

Because the average velocity at a given point in a stream differs greatly according to whether it is being measured close to the banks and bed, or out in the middle line, a single figure, the *mean velocity*, is computed for the entire cross section to express the activity of the stream as a whole. Mean velocity in streams is commonly equal to about six-tenths of the maximum velocity, but depends on the relative depth of the stream.

The last and most important measure of stream flow is *discharge*, Q, defined as the volume of water passing through a given cross section of the stream in a given unit of time. Commonly, discharge is stated in cubic feet per second, abbreviated to *cfs*. Sometimes the hydraulic engineer simply states this quantity as *second feet*. In metric units discharge is stated in cubic meters per second (cms). Discharge may be obtained by taking the mean velocity V, and multiplying it by cross-sectional area, A. This relationship is stated by the important equation, $Q = AV$, sometimes referred to as "the equation of continuity of flow."

As a stream flows downgrade, potential energy of elevation is transformed into kinetic energy of motion. If there were no resistance to flow, the water body of the stream would accelerate as if it were an object falling in a vacuum. Instead, as flow velocity increases, resistance builds up, withdrawing more energy as friction. The result is that a *terminal velocity* is quickly reached and the mean velocity becomes constant. Sensible heat generated by the resistance is conducted through the channel walls and the upper stream surface and is lost at a rate equal to its production rate.

Stream channels differ in the amount of resistance offered to flow. Resistance is large in a broad but shallow channel, and is much less in a deep, narrow channel. The optimum channel would be semicircular in cross section. However, streams are often required to have broad, shallow channels because of the load of mineral matter that must be carried, or because the banks are weak and will not hold a steep attitude.

Velocity, depth, and gradient

It is intuitively obvious that water will flow faster in a channel of steep gradient than one of low gradient, since the component of gravity acting parallel with the bed is stronger for the steeper gradient. As shown in Figure 13.11, velocity increases quickly where a stream passes from a pool of low gradient to a steep stretch of rapids. As velocity V increases, cross-sectional area A must also decrease, for otherwise their product, AV, would not be held constant. In the pool, where velocity is low, cross-sectional area is correspondingly increased.

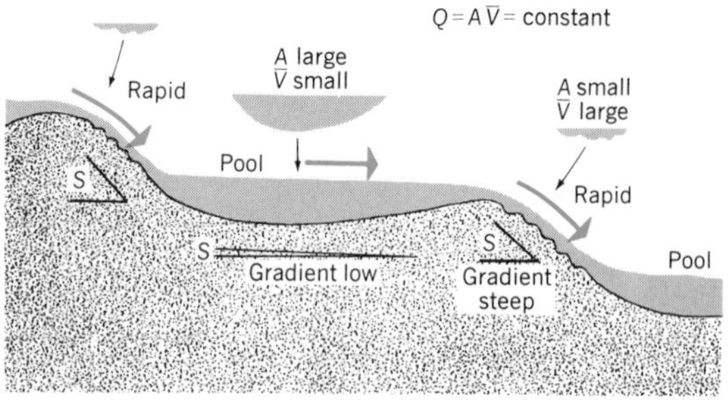

Figure 13.12 Schematic diagram of relationships among cross-sectional area, mean velocity, and gradient. (From A. N. Strahler, 1971, *The Earth Sciences*, 2nd ed., Harper and Row, New York.)

Streams of water and of ice (glaciers) represent gravity-flow systems. The potential energy contained in the stream at its headwaters is derived indirectly from the solar radiation that initially evaporates the ocean water and sets in motion the atmospheric circulation. Although condensation of water vapor transforms most of the latent heat back into sensible heat, a small fraction remains with the precipitation as potential energy.

Stream gauging

An important activity of the U.S. Geological Survey is the measurement, or gauging, of stream flow in the United States. In cooperation with states and municipalities this organization maintains over 6000 river-measurement stations on principal streams and their tributaries. The figures on discharge thus obtained are published by the Geological Survey in a series of *Water-Supply Papers*. Information on daily discharge and flood discharges is essential for planning the distribution and development of surface waters as well as for design of flood-protection structures and for the prediction of floods as they progress down a river system.

A stream gauging station requires a device for measuring the height of the water surface, or *stage* of the stream. Simplest to install is a *staff gauge*, which is simply a graduated stick permanently attached to a post or bridge pier. This must be read directly by an observer whenever the stage is to be recorded. More useful is an automatic-recording gauge, which is mounted in a *stilling tower* built beside the river bank (Figure 13.13). The tower is simply a hollow masonry shaft filled

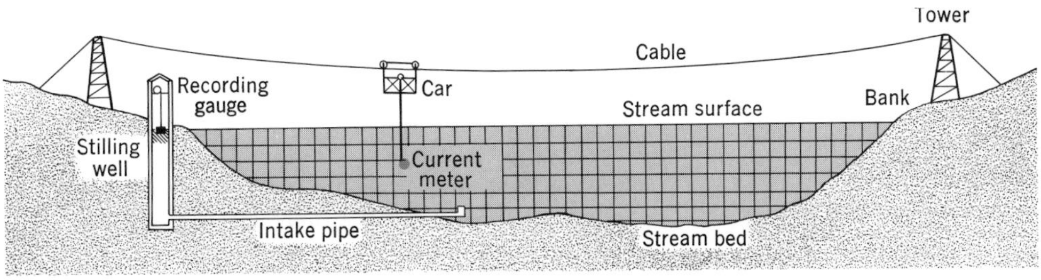

Figure 13.13 Idealized diagram of stream gauging installation.

by water which enters through a pipe at the base. By means of a float connected by cable to a recording mechanism above, a continuous ink-line record of the stream stage is made on a graph paper attached to a slowly rotating drum.

To measure stream discharge, it is necessary to determine both the area of cross section of the stream and the mean velocity. This requires that a *current meter* (Figure 13.14) be lowered into the stream at closely spaced intervals so that the velocity can be read at a large number of points evenly distributed in a grid pattern through the stream's cross section (Figure 13.13). A bridge often serves as a convenient means of crossing over the stream; otherwise a cable car or small boat is used. The current meter has a set of revolving cups whose rate of turning is proportional to current velocity. The Price current meter, pictured in Figure 13.14, is in general use by the Geological Survey and will measure velocities from 0.2 to 20 ft (0.06 to 6 m) per second. As the velocities are being measured from point to point, a profile of the river bed is also made by sounding the depth. Thus, a profile is drawn and the cross-sectional area is measured from the profile. Mean velocity is computed by summing all individual velocity readings and dividing by the number of readings. Discharge can then be computed using the formula, $Q = AV$.

Figure 13.14 In gauging large rivers, the current meter is lowered on a cable by a power winch. Earphones, connected by wires to the meter, receive a series of clicks whose frequency indicates water velocity. (U.S. Geological Survey photograph.)

In practice, the number of velocity readings at each point of sounding is reduced to two: one at 0.2 the depth; one at 0.8 the depth. The average of these two readings gives a close approximation to the true mean velocity. For shallow streams, a single velocity reading at 0.6 the depth suffices at each point of sounding.

Because of the time and labor required to measure discharge repeatedly by current readings, it is practical to take instead only a limited set of such measurements over a wide range of discharges. From these measured discharges is constructed a *rating curve,* or *stage-discharge curve,* permitting discharge to be estimated directly from gauge height. For the sample curve in Figure 13.15, eight points of discharge were actually measured by the current-meter method. These were plotted on the graph against gauge height in feet. A smooth curve was then drawn through the points. Thus, if the gauge height is known to be 20 ft, we can estimate that the discharge is occurring at a rate of about 16,500 cfs. Rating curves enable estimates of total discharge to be computed from stage records alone, despite wide fluctuations in stage.

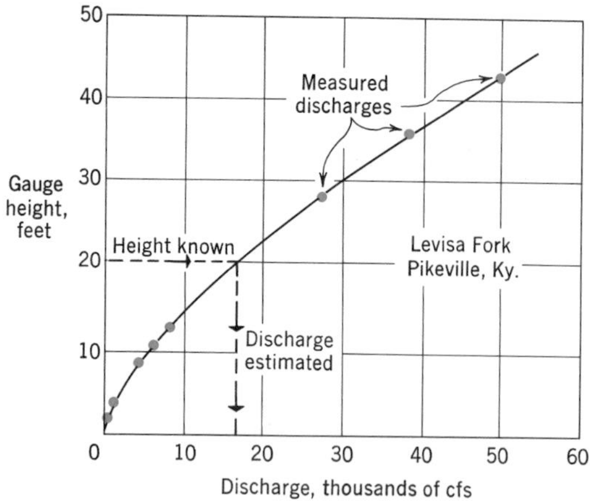

Figure 13.15 This rating curve applied to Levisa Fork, Kentucky, during the period October 1945 to January 1946. (After Hoyt and Langbein, *Floods*.)

A single rating curve may be useful only for limited periods of time because of changes in form of the river channel. Such changes may take place by channel erosion in floods. The rating curve is therefore recomputed and corrected as necessity demands.

Stream flow and precipitation

By studying the records of stream discharge in relation to precipitation on a given watershed, the hydrologist has developed a set of basic principles applying to the variations in stream discharge with different lengths and intensities of storms and with different sizes of watersheds.

Consider first a very small watershed, just over one acre (0.4 hectare) in area, upon which a heavy rain fell in a total period of about an hour. Figure 13.16 is a graph showing what happened to the water from beginning to end of the storm. Rainfall was measured with the rain gauge and is shown in terms of the intensity of rainfall, or quantity falling in each 5- or 10-minute period. Rain began at 4:21 P.M. and was extremely heavy for nearly 40 minutes, after which it let up rapidly and ceased entirely by 5:40. Discharge, measured at the outlet point of the small watershed, is shown in Figure 13.16 by a smooth curve scaled in cubic feet per second. Because of the high initial infiltration capacity of the soil, all of the rain was at first absorbed by the soil or was detained in surface irregularities. About 6 minutes after the rain began, discharge set in and rose rapidly for a half hour, reached the peak just after 4:50, then declined again and became zero by 5:50 P.M.

The lower part of Figure 13.16 is another form of representation in which the quantities of rainfall and runoff are accumulated from beginning to end. Here both discharge and rainfall are scaled in terms of inches of water depth, so that the values can be directly subtracted. At the end of the storm, about 5:40 P.M., a total of 1.2 in (3 cm) of rain had fallen, but only about 1 in (2.5 cm) of water had been disposed of by runoff. This leaves 0.2 in (0.5 cm) of loss through combined evaporation and infiltration. An important principle of this water graph, or *hydrograph*, as it is generally called, is that for a small watershed, the response of runoff is rapid, with little time lag between. Let us now look at the hydrographs of larger areas over longer periods of time to see the effect of watershed size and storm duration on stream flow.

Figure 13.17 shows the hydrograph of Sugar Creek, Ohio, with a watershed area of 310 sq mi (805 sq km). Sugar Creek basin, a part of the much larger Muskingum River watershed, is outlined in Figure 13.18, a map showing by isohyetal lines the rainfall during the 12 hr storm of August 6 and 7, 1935, for which the hydrograph was constructed. Over the area of Sugar Creek, the average total rainfall was 6.3 in (16 cm) for the entire storm, but the total quantity discharged by Sugar Creek

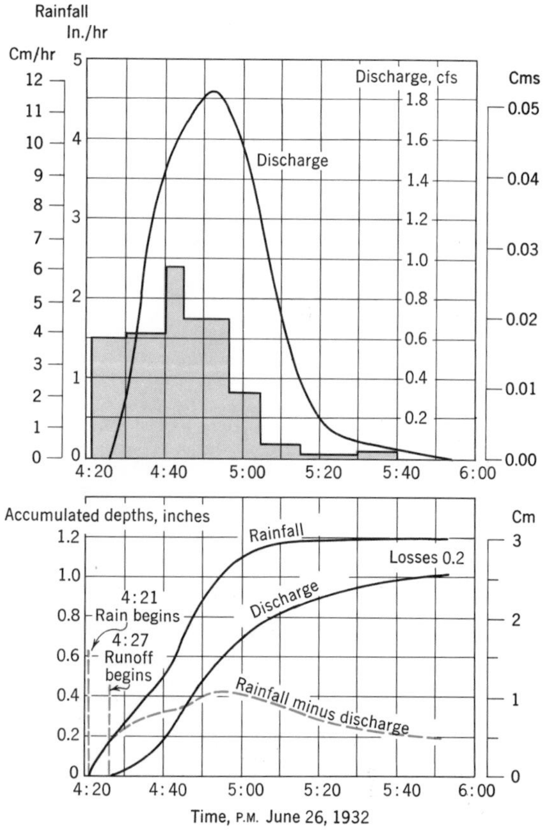

Figure 13.16 This hydrograph accounts for the receipt and outflow of water for a very small drainage area of about one acre (0.4 hectare) near Hays, Kansas, during a rainstorm on June 26, 1932. (After E. E. Foster, *Rainfall and Runoff*.)

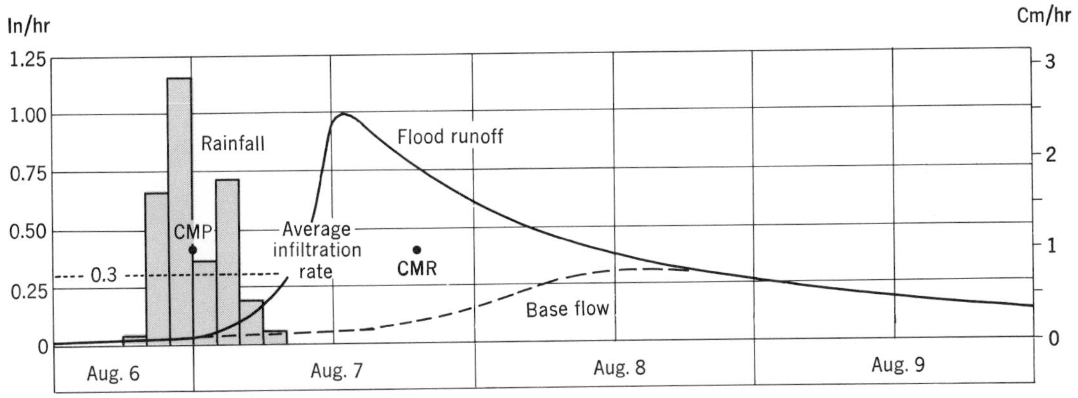

Figure 13.17 Four days of flow of Sugar Creek, Ohio. (After Hoyt and Langbein, *Floods*.)

was only 3 in (7.5 cm). This means that 3.3 in (8.5 cm), or more than half of the rainfall, was retained on the watershed, having infiltrated to become part of the soil and ground water, or had evaporated.

Studying the rainfall and runoff graphs in Figure 13.17, we see that prior to the onset of the storm, Sugar Creek was carrying a small discharge. This was being supplied by the seepage of ground water into the channel and is termed *base flow*. After the heavy rainfall began several hours elapsed before the stream gauge at the basin mouth began to show a rise in discharge. This interval is known as the *lag time* and indicated that the branching system of channels was acting as a temporary reservoir, receiving inflow more rapidly than it could be passed down the channel system to the stream gauge. The term *channel storage* is applied to runoff delayed in this manner during the early period of a storm.

Lag time is measured as the difference between center of mass of precipitation (CMP) and center of mass of runoff (CMR), as labeled in Figure 13.17. The peak flow of Sugar Creek was reached almost 24 hours after the rain began; the lag time was about 18 hours. Note also that the rate of decline in discharge was much slower than the rate of rise. In general, the larger a watershed, the longer is the lag time between peak rainfall and peak discharge; the more gradual is the rate of decline of discharge after the peak has passed. Because much rainfall had entered the ground and had reached the water table, a slow but distinct rise is seen in the amount of discharge contributed by base flow.

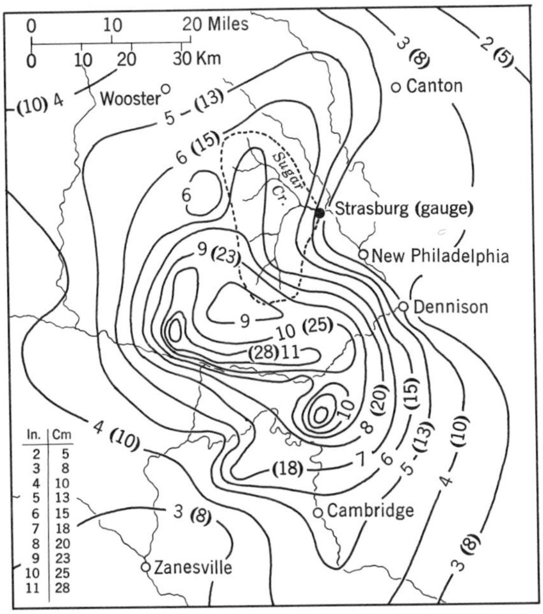

Figure 13.18 Isohyets of Sugar Creek Watershed, Ohio. Centimeter equivalents in parentheses. (After Hoyt and Langbein, *Floods*.)

Base flow and surface water flow

In regions of humid climates, where the water table is high and normally intersects the important stream channels, the hydrographs of larger streams will show clearly the effects of two sources of water: (*a*) *base flow* and (*b*) *surface water flow*. Figure 13.19 is a hydrograph of the Chattahoochee River, Georgia, a large river draining a watershed of some 3350 sq mi (8700 sq km), much of it in the humid southern Appalachian Mountains. The sharp, abrupt fluctuations in discharge are produced by overland flow following rain periods of one to three days duration. These are each similar to the hydrograph of Figure 13.17, except that they are here shown much compressed by the time scale.

After each rain period the discharge falls off rapidly, but if another storm occurs within a few days, the discharge rises to another peak. The enlarged inset graph, showing details of the month of January, reveals how this effect occurs. Where a long period intervenes between storms, the dis-

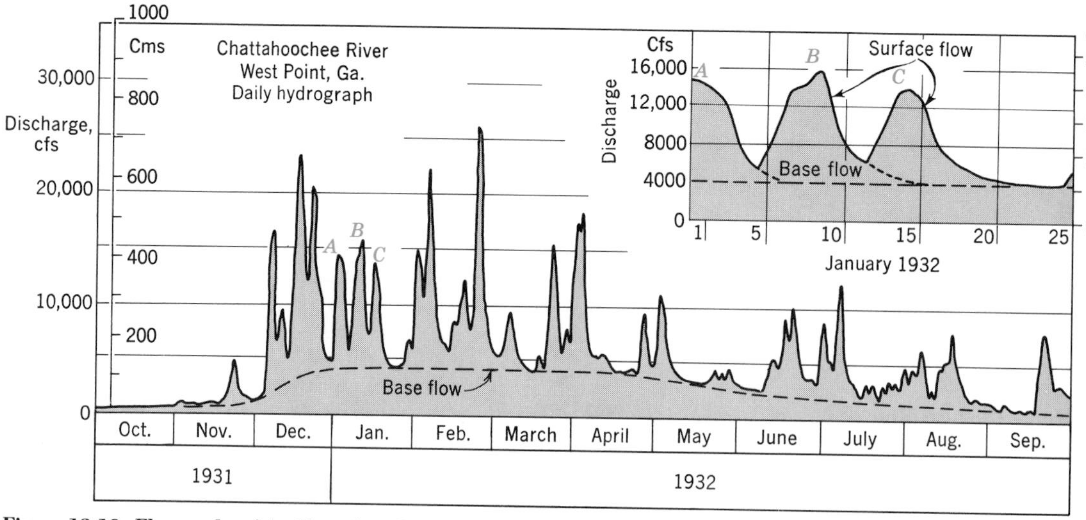

Figure 13.19 Flow peaks of the Chattahoochee River. (After E. E. Foster, *Rainfall and Runoff*.)

charge falls to a low value, the base flow, where it levels off. Throughout the year the base flow, which represents ground water inflow into the stream, undergoes a marked annual cycle. During the period of recharge (winter and early spring), water table levels are raised and the rate of inflow into streams is increased. For the Chattahoochee River, the rate of base flow during January, February, March, and April holds uniform at about 4000 cfs (110 cms). As the heavy evapotranspiration losses of spring reduce soil water, and therefore cut off the recharge of ground water by downward percolation (see Chapter 12), the base flow falls steadily. The decline continues through the summer, reaching by the end of October a low of about 1000 cfs (30 cms) supplied entirely from base flow.

With a knowledge of soil-moisture regimes to guide our reasoning, we can anticipate that base flow will be important in regions having a substantial water surplus, but will be unimportant or absent in regions having a substantial soil-moisture deficiency and no surplus.

Figure 13.20 shows comparative hydrographs for one-year periods for three watersheds of approximately the same areas. That of Ecofina Creek, Florida, is unusual in showing a large proportion of base flow and lack of strong discharge peaks. The explanation may lie partly in the low relief and gentle slopes of the watershed, but more particularly in the presence of cavernous limestone beneath the surface. Most of the excess rainfall enters the ground-water system and is discharged copiously through solution passages. Potato Creek, Georgia, in a region of steep slopes, has occasional extreme peak flows that occur in winter and early spring when soil moisture is excessive and the proportion of surface runoff is high. Antelope Creek, gauged at Red Bluff, California, in the northern part of the state, likewise shows the extreme discharges typical of mountain watersheds, but the complete absence of peaks from June through October reflects the long summer drought of the Mediterranean soil-moisture regime, during which only base flow can be maintained.

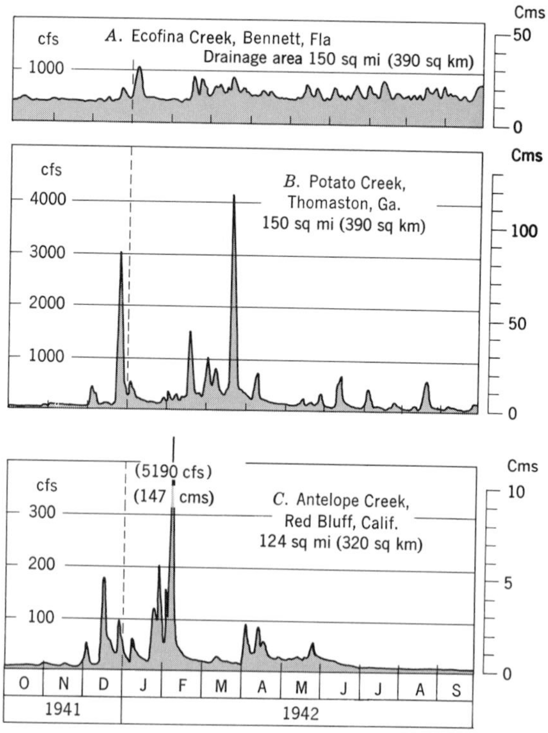

Figure 13.20 Hydrographs of three streams differ because of differences in climate, relief, and rock type. (After E. E. Foster, *Rainfall and Runoff*.)

Finally, examine the hydrograph of the Missouri River at Omaha, Nebraska, from October 1940, to September 1942 (Figure 13.21). This great river, draining 322,800 sq mi (840,000 sq km) of watershed, is a major tributary of the Mississippi River. Note that the discharge, ranging from 10,000 to 100,000 cfs (280 to 2800 cms), is many times greater than the discharges of smaller streams considered thus far. High rates of flow are chiefly from snowmelt, which occurs on the High Plains in spring and in the Rocky Mountain headwater areas in early summer. This event explains the sudden high discharges from April through June. During midwinter, when soil moisture is frozen and total precipitation small over the watershed as a whole, the discharge rises little above the base flow. Ground water recharge occurring in the spring raises summer levels of base flow to about 20,000 cfs (570 cms), or two to three times the winter base flow.

Floods

Everyone has seen enough news photographs of river floods to have a good idea of the appearance of flood waters and the havoc wrought by their erosive power and by the silt and clay that they leave behind. Nevertheless, even the hydraulic engineer may not be fully satisfied that he can exactly define the term *flood.* Perhaps it is enough to say that a condition of flood exists when the discharge of a river cannot be accommodated within the margins of its normal channel, so that the water spreads over adjoining ground upon which crops or forests are able to flourish.

Most larger streams of humid climates have a *floodplain,* a belt of low flat ground bordering the channel on one or both sides inundated by stream waters about once a year, at the season when abundant supplies of surface water combine with effects of a high water table and ample soil moisture to supply more runoff than can stay within the heavily scoured troughlike channel (Chapter 25). Such annual inundation is considered a flood, even though its occurrence is expected and does not prevent the cultivation of crops after the flood has subsided, or does it interfere with the growth of dense forests which are widely distributed over low, marshy floodplains in all humid regions of the world. Still higher discharges of water, the rare and disastrous floods which may occur as infrequently as three to five decades, inundate ground lying above the floodplain, principally affecting broad steplike expanses of ground known as *terraces* (Figure 13.22).

For practical purposes, the National Weather Service, which provides a flood-warning service, designates a particular stage of gauge height at a given place as the *flood stage,* implying that the critical level has been reached above which overbank flooding may be expected to set in. Immediately at or below flood stage the river may be described as being in the *bank-full stage,* the flow being entirely within the limits of the heavily scoured channel.

Downstream progress of a flood wave

The rise of a river stage to its maximum height, or *crest,* followed by a gradual lowering of stage, is termed the *flood wave.* The flood wave is simply a large-sized rise and fall of river discharge of the type already analyzed in earlier paragraphs, and follows the same principles. Figure 13.23*A* shows the downstream progress of a flood on the Chattooga-Savannah river system. In the Chattooga River near Clayton, Georgia, the flood peak or crest was quickly reached—one day after the storm—and quickly subsided. On the Savannah River, 65 mi (105 km) downstream at Calhoun Falls, South Carolina, the peak flow occurred a day later, but the discharge was very much larger because of the larger area of watershed involved. Downstream another 95 mi (153 km), near Clyo, Georgia, the Savannah River crested five days after the initial storm with a discharge of over 60,000 cfs (1700 cms). This set of three hydrographs shows that (*a*) the lag time in occurrence of the crest increases downstream, (*b*) the entire period of rise and fall of flood wave becomes longer downstream, and (*c*) the discharge increases greatly downstream as watershed area increases.

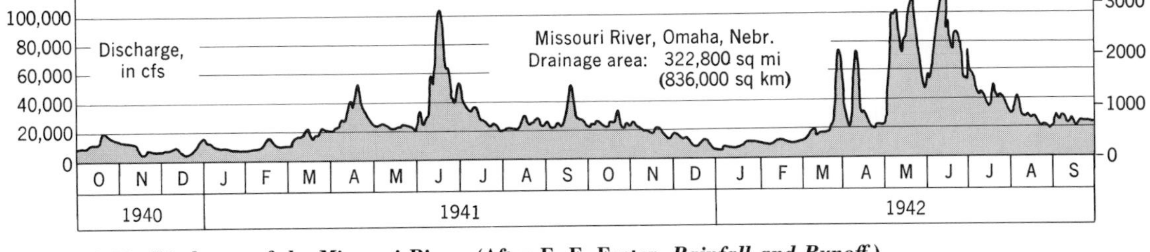

Figure 13.21 Discharge of the Missouri River. (After E. E. Foster, *Rainfall and Runoff.*)

Figure 13.22 The city of Hartford was partly inundated by the Connecticut River flood of March 1936. The river channel is to the left, its banks marked by a line of trees. (Official Photograph, 8th Photo Section, A. C., U.S. Army.)

Figure 13.23*B* is a somewhat different presentation of the same flood data, in that the discharge is given in terms of a common unit of area, the square mile, thus eliminating the effect of increase in discharge downstream and showing us only the shape or form of the flood crest.

Flood prediction

The National Weather Service operates a River and Flood Forecasting Service through 85 offices located at strategic points along major river systems of the United States. Each office issues river and flood forecasts to the communities within the associated district, which is laid out to cover one or more large watersheds. Flood warnings are publicized by every possible means. Close cooperation is maintained with such agencies as the American Red Cross, the U.S. Army Corps of Engineers, and the U.S. Coast Guard, in order to plan evacuation of threatened areas, and the removal or protection of vulnerable property.

Stream flow data are the basis of graphs of flood stages telling the likelihood of occurrence of given stages of high water for each month of the year. Figure 13.24 shows expectancy graphs for four selected stations. The meaning of the strange looking bar symbols is explained in the key. The Mississippi River at Vicksburg illustrates a great river corresponding largely to spring floods so as to yield a simple annual cycle. The Colorado River at Austin, Texas, is chosen to illustrate a river draining largely semi-arid plains. Summer floods

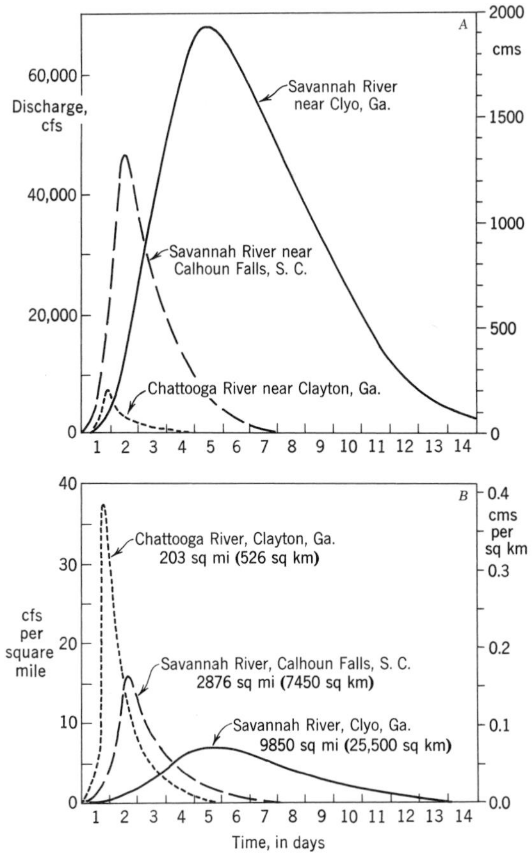

Figure 13.23 The downstream progress of a flood wave. (After Hoyt and Langbein, *Floods*.)

are produced directly by torrential rains from invading moist tropical air masses. Floods of the late summer and fall are often attributable to tropical storms (hurricanes) moving inland from the Gulf of Mexico. The Sacramento River at Red Bluff, California, has a winter flood season when rains are heavy, but a sharp dip to low stages in late summer, which is the very dry period for the California coastal belt. The flood expectancy graph for the Connecticut River at Hartford shows two seasons of floods. The more reliable of the two is the early spring, when snowmelt is rapid over the mountainous New England terrain; the second, in the fall, when rare but heavy rainstorms, some of hurricane origin, bring exceptional high stages.

Regulation of floods of large rivers is taken up in Chapter 25, in connection with landforms of floodplains bordering alluvial rivers.

Hydrologic effects of urbanization

Hydrologic characteristics of a watershed are altered in two ways by urbanization. First, an increasing percentage of the surface is rendered impervious to infiltration by construction of roofs, driveways, walks, pavements, and parking lots. It has been estimated that in residential areas, for a lot size of 15,000 sq ft (0.34 acre; 1400 sq m) the impervious area amounts to about 25 percent, whereas for a lot size of 6000 sq ft it rises to about 80 percent. An increase in proportion of impervious surface reduces infiltration and increases overland flow generally from the urbanized area. In addition to increasing flood peaks during heavy storms, there is a reduction of recharge to the ground body beneath, and this reduction in turn decreases the base flow contribution to channels in the area. Thus the full range of stream discharges, from low stages in dry periods to flood stages, is made greater by urbanization.

A second change caused by urbanization is the introduction of storm sewers that allow storm runoff from paved areas to be taken directly to stream channels for discharge. Runoff travel time to channels is thus being shortened at the same time that the proportion of runoff is being increased by expansion in impervious surfaces. The two changes together conspire to reduce the lag time, as shown by the schematic hydrographs in Figure 13.24. Figure 13.25 shows how combined effect of increase in sewered area and impervious area is to increase the frequency of occurrence of overbank floods as compared with frequency that existed prior to urbanization. The ratio given on the vertical axis is the increase in yearly number of overbank flows based on a drainage area of one square mile. Thus, for 50 percent impervious area and 50 percent sewered area, the number of overbank flows yearly has increased by a factor of about $3\frac{1}{2}$.

Increase in flood peaks results in inundation of areas previously above flood limits except during rare flood events, with attendant water damage to properties adjacent to the channel. If the stream should be free to adjust its dimensions by deepening and widening its channel to accommodate higher flood peaks, there would be a serious problem arising from the resulting load of sediment, a factor that is discussed in Chapter 25.

One partial solution to the problem created by storm sewering is to return storm runoff to the ground water body by means of infiltrating basins. This program has been adopted on Long Island, where infiltration rates are high in sandy glacial materials. Another method of disposal of storm runoff is by recharge wells. At Orlando, Florida, storm runoff enters wells that penetrate cavernous limestone. The capacity of the system to absorb runoff without clogging appears to be adequate. In Fresno, California, a large number of gravel-packed wells of 30-in (76-cm) diameter receive runoff from streets. The system has proved successful in disposing of storm drainage.

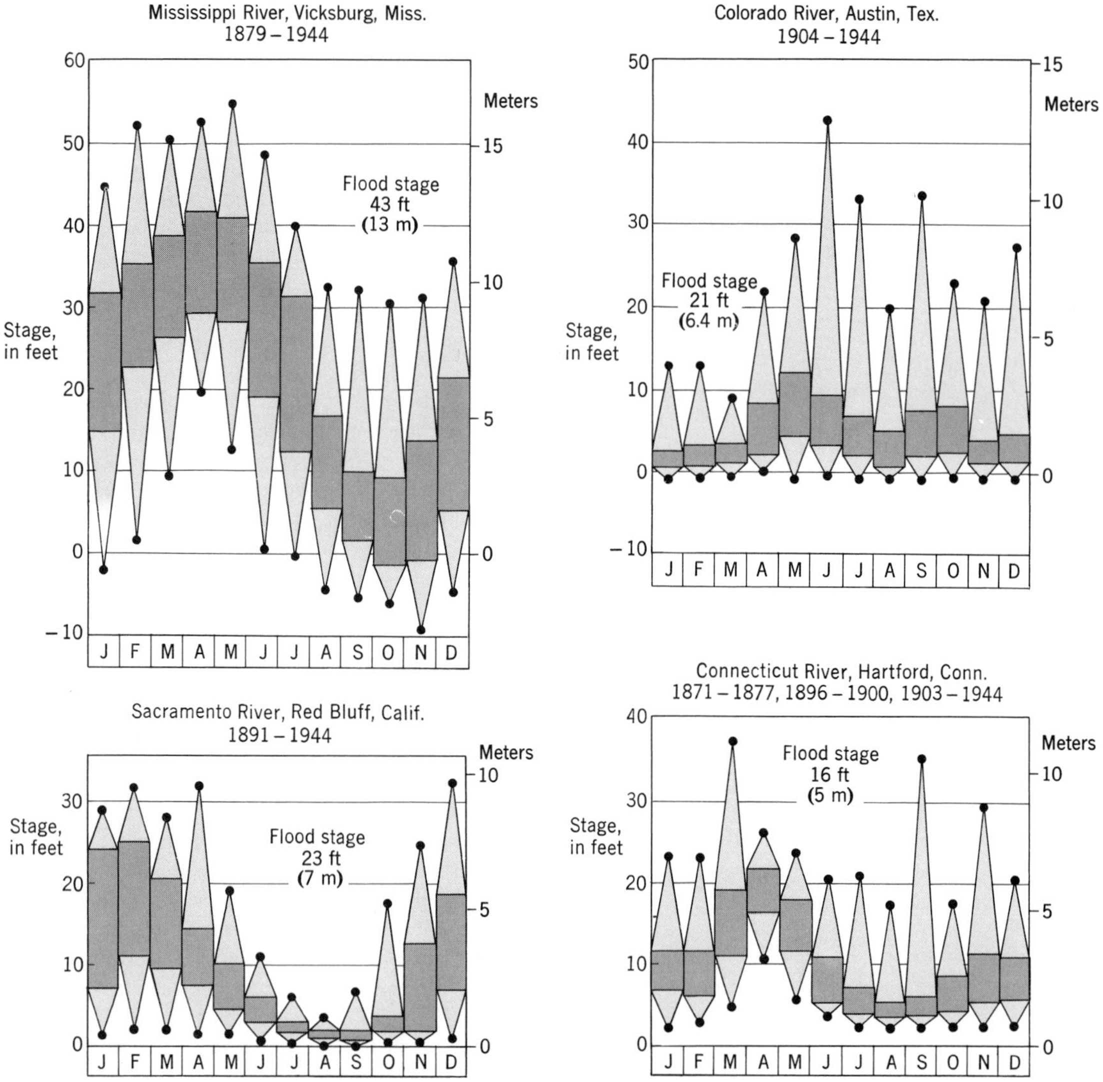

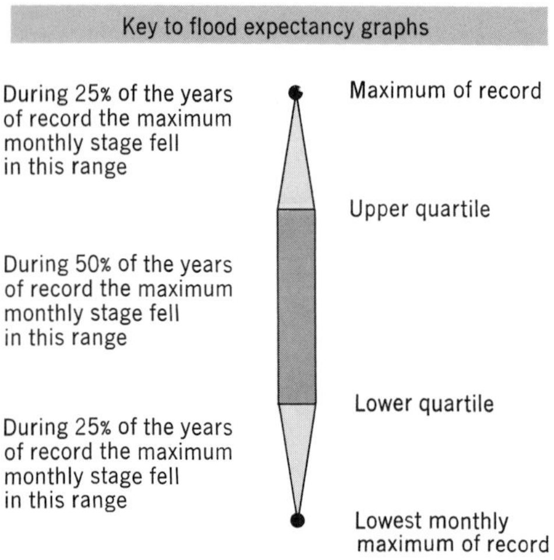

Figure 13.24 The highest water stage that occurred in each month is given in terms of percentages on these graphs of four rivers. (After National Weather Service.)

Water pollution

In this chapter we have examined the flow of water itself—largely from the hydrologist's point of view—leaving geological considerations for later chapters. Streams, lakes, bogs, and marshes are specialized habitats of plants and animals; their ecosystems are particularly sensitive to changes induced by man in the water balance and in water chemistry. Not only does our industrial society make radical physical changes in water flow by construction of engineering works (dams, irrigation systems, canals, dredged channels), but we also pollute and contaminate our surface waters with a large variety of wastes. Some of these wastes are in the form of ions in solution. The subject of water quality is appropriate to discuss in this chapter, and we shall want to examine the sources of pollutants introduced into the natural flow systems by man. Introduction of mineral sediment

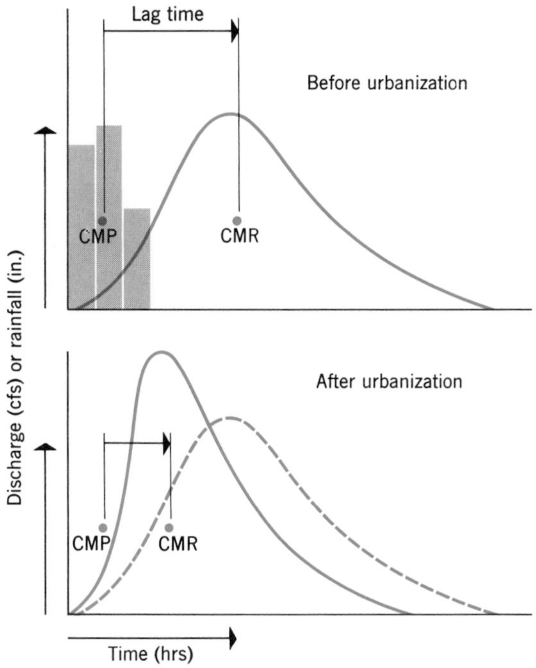

Figure 13.25 Schematic hydrographs showing the effect of urbanization upon lag time and peak discharge. Points *CMP* and *CMR* are centers of mass of rainfall and runoff, respectively, as in Figure 13.17. (After L. B. Leopold, 1968, U.S. Geological Survey, Circular 554.)

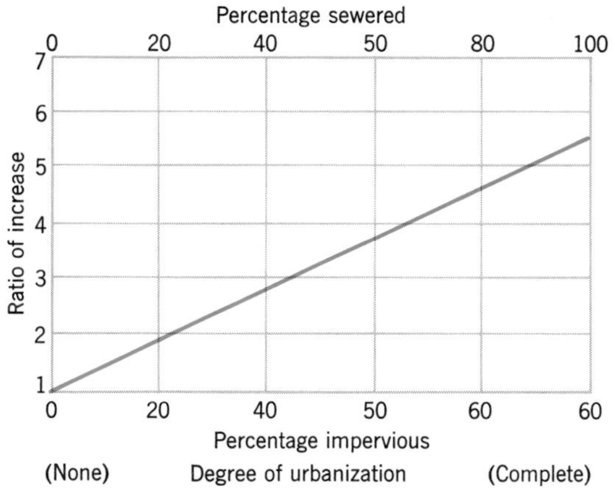

Figure 13.26 Increase in frequency of overbank flows with increase in degree of urbanization. Data are adjusted to a drainage area of one square mile. (After L. B. Leopold, 1968, U.S. Geological Survey, Circular 554.)

into surface waters as a result of land disturbances is a suitable topic for Chapter 25.

Chemical pollution by direct disposal into streams and lakes of wastes generated in industrial plants is a phenomenon well known to the general public and can be seen first-hand in almost any industrial community in the United States. Direct outfall of sewage, whether raw (untreated) or partially treated, is another form of direct pollution of streams and lakes that does not need to be elaborated upon, for it, too, is a commonplace phenomenon for all to see (and smell).

In urban and suburban areas pollutant matter entering streams and lakes includes deicing salt, lawn conditioners (lime and fertilizers), and sewage effluent. In agricultural regions, important sources of pollutants are fertilizers and the body wastes of livestock. Major sources of water pollution are associated with mining and processing of mineral deposits. In addition to chemical pollution, there is thermal pollution from discharge of coolant waters from nuclear generating plants. There exists also the possibility of contamination from radioactive substances from nuclear processing plants.

Among the common chemical pollutants of both surface water and ground water are sulfate, nitrate, phosphate, chloride, sodium, and calcium ions. Sulfate ions enter runoff both by fallout from polluted urban air and as sewage effluent. Important sources of nitrate ions are fertilizers and sewage effluent. Excessive concentrations of nitrate in water supplies are highly toxic, and at the same time their removal is difficult and expensive. Phosphate ions are contributed in part by fertilizers and by detergents in sewage effluent. Phosphate and nitrate are plant nutrients and can lead to excessive growth of algae in streams and lakes—a process included in the term *eutrophication.* Chloride and sodium ions are contributed both by fallout from polluted air and by deicing salts used on highways. Instances are recorded in which water supply wells close to highways have become polluted from deicing salts.

A particular form of chemical pollution of surface water goes under the name of *acid mine drainage* and is an important form of environmental degradation in parts of Appalachia where abandoned coal mines and strip mine workings are concentrated (Chapter 28). Ground water emerging from abandoned mines, as well as soil water percolating through strip mine spoil banks, is charged with sulfuric acid and various salts of metals, particularly of iron. The sulfuric acid is formed by reaction of water with iron sulfides, particularly the mineral *pyrite* (Fe_2S), which is a common constituent of coal seams. Acid of this origin in stream waters can have adverse effects upon animal life. In sufficient concentrations it is lethal to certain species of fish and has at times caused massive fish kills. The acid waters also cause corrosion to boats and piers. Government sources have estimated that nearly 6000 mi (10,000 km) of streams in this country, together with almost 40 sq mi (100 sq km) of reservoirs and lakes are seriously affected by acid mine drainage. One particularly undesirable by-product of acid mine drainage is precipitation of iron to form slimy red and yellow deposits in stream channels.

Sulfuric acid may also be produced from drainage of mines from which sulfide ores are being extracted, and from tailings produced in plants where the ores are processed. Such chemical pollution also includes salts of various toxic metals, among them zinc, lead, arsenic, copper, and aluminum. Yet another source of chemical pollution of streams is by radium derived from the tailings of uranium ore processing plants.

Toxic metals, among them mercury, along with pesticides and a host of other industrial chemicals, are introduced into streams and lakes in quantities that are locally damaging or lethal to plant and animal communities. In addition, sewage introduces live bacteria and viruses that are classed as biological pollutants; these pose a threat to health of man and animals.

Thermal pollution of surface waters

Thermal pollution is a term applied generally to the discharge of heat into the environment from combustion of fuels and from nuclear energy conversion into electric power. We have described thermal pollution of the atmosphere and its effects in Chapter 8. Thermal pollution of water is different in its environmental effects because it takes the form of heavy discharges of heated water locally into streams, estuaries, and lakes. The thermal environmental impact may thus be quite drastic in a small area.

Nuclear power plants require the flow of large quantities of water for cooling. Other industrial processes also rely upon water as a coolant. Large air-conditioning plants also use water to cool the condensing coils. A cooling system may be designed to recycle the coolant, causing the heat to be dissipated into the atmosphere, the principle being the same as the water-cooled automobile engine and most small air conditioners. In this case the heat load is placed on the atmosphere by radiation and conduction. Evaporation of water may be used as a means of cooling in recycling systems, but this requires an input of water to replace that lost by evaporation. A third method is to withdraw cold water from a lake or stream or from the ground water body, pass it through the cooling system, and discharge the heated water into the same source body. For example, a large air-conditioning plant may use cold ground water and dispose of the heated water into recharge wells. Commonly, such disposal by well recharge is required by law to conserve ground water.

Nuclear power plants have large cooling requirements, with the result that they may be sited near a large body of cold water. For example, a proposed nuclear generating plant on Lake Cayuga, a deep finger lake in New York State, will require the pumping of 750 million gallons (4 million liters) per day from the cold lower depths of the lake, where water temperature is on the order of 43°F (6°C). This water will be discharged into the surface layer of the lake at a temperature of about 70°F (21°C). Environmentalists fear that this infusion of warm water may delay the annual fall cooling of the surface layer, which in turn will delay the annual overturn of lake water and lengthen the period of time that the lower water remains stagnant. It is also feared that the bringing to the surface of nutrients present in the deep water will increase plant growth in the surface layer and lead to an oxygen deficiency.

It is appropriate here to point out that the forms of pollution we have listed here often do severe damage to the esthetic qualities of landscapes of streams and lakes.

Fresh water as a natural resource

Fresh water is a basic natural resource essential to man in his varied and intense agricultural and industrial activities. Runoff held in reservoirs behind dams provides water supplies for great urban centers, such as New York City and Los Angeles; diverted from large rivers, it provides irrigation water for highly productive lowlands in arid lands, such as the Imperial Valley of California and the Nile Valley of Egypt. To these uses of runoff are added hydroelectric power, where the grade of a river is steep, or routes of inland navigation, where the grade is gentle.

Unlike ground water, which represents a large water storage body that can be withdrawn much as a mineral is mined for exploitation, fresh surface water in the liquid state is stored only in small quantities. (An exception is the Great Lakes system.) Referring back to Table 12.1, note that the quantity of available ground water is about 30 times as large as that stored in fresh-water lakes, while the water held in streams is only about 1/100 that in lakes. Both surface water and ground water resources are renewable, but because of small natural storage capacities, surface water can be drawn only at a rate comparable with its annual renewal through precipitation. Dams are built to develop useful storage capacity for surplus runoff that would otherwise escape to the sea, but once the reservoir has been filled, water use must be scaled to match the natural supply rate averaged over the year. Development of surface water supplies brings on many environmental changes, both physical and biological, and these must be taken into account in planning for future water developments.

Studies of the total United States water resource have led to the conclusion that for the nation as a whole there is an adequate fresh water supply for at least several decades into the future. For individual regions of the country, however, we can anticipate severe shortages that can be remedied only by transfer of water from regions of surplus. The enormous concentration of persons and industry into a small area, as for example Los Angeles County and the New York City region, require development of surface water facilities in distant uplands, but it is only by continual and costly expansion of these facilities that shortages can be avoided. In other words, the real problem of supply is not in the total quantity of fresh water available but in its distribution. Compounding distribution problems are those of water pollution. If the quality of our fresh water supplies continues to be degraded it will be necessary to resort increasingly to expensive water treatment procedures necessary to keep the water usable.

In recent years, much emphasis has been placed upon alternatives to massive water transfers to meet rising demands. The basic alternative is a reduction in water use, particularly reductions in wasteful use in urban systems and in the huge demands for irrigation in arid lands. It is obvious that problems of water resource management involve a broad spectrum of economic, political, and cultural factors.

REVIEW QUESTIONS

1. What is the environmental importance of runoff? How can a knowledge of hydrology be useful in water resource management?

2. Define ground water. How does one distinguish between zones of saturation and aeration? Between soil water and intermediate water belts? What is the capillary fringe? How does its thickness vary with soil texture? What is the water table?

3. Where is ground water discharged from the ground to become runoff? Describe the paths of ground water flow from divide to stream.

4. Describe the form of the water table under a hilly topography. What fluctuations may be expected in the water table? What is ground water recharge and how does it take place?

5. What are water-table lakes? Do they show great seasonal changes of water level? What is a bog? A marsh?

6. How does pumping cause drawdown of water table? What is the cone of depression? How far may it extend from a well? How rapidly is ground water recharged from alluvial deposits of arid regions?

7. What are the major sources of ground water pollution? Describe the movement of pollutants in the ground water zone.

8. Describe and explain the relation of a fresh ground water body to the salt ground water beneath an island. What physical principle governs the relation of water table elevation to depth of fresh water? How is salt water contamination of wells caused?

9. Distinguish between overland flow and channel flow. To what extent is overland flow observable?

10. Describe the development of overland flow, beginning with the onset of rainfall. What is interception? Surface detention? What equation relates overland flow to precipitation and infiltration?

11. What are the general characteristics of a natural drainage system? What is the shape of the drainage basin? Define a watershed.

12. What branches of earth science deal with runoff and drainage systems? What governmental agencies take responsibility for various phases of this study?

13. List the various geometrical elements of a stream channel, state how each is defined and measured. Explain the meaning of "percent of grade."

14. Describe the manner in which velocity varies throughout a stream, both across the channel and in vertical section. Why is the velocity not the same in all parts of the stream? What is stream turbulence?

15. What is the mean velocity of a stream? State and explain in words the equation relating discharge to area and mean velocity.

16. What physical phenomenon limits the flow velocity of a stream? How is stream velocity related to depth? To gradient?

17. Describe the methods and equipment used in stream gauging. How is mean velocity actually measured and computed?

18. What is a hydrograph? Describe the manner in which runoff occurs following a rainstorm on a watershed. Why is there a time lag between rainfall and runoff? For larger drainage basins, how do these relationships differ?

19. What is base flow? What supplies base flow? What is channel storage and how does it affect the hydrograph of a large watershed?

20. Describe the annual changes in base flow and surface water contributions in a watershed typical of the humid southeastern United States.

21. How is a flood defined? What is the bank-full stage of a stream flow?

22. Describe the progress of a flood crest down a large river. How does discharge change with increasing distance downstream? How does the form of the flood wave change?

23. What government agency supplies river and flood forecasts? How are its activities organized and what is their function?

24. How do flood expectancies differ for the Mississippi River at Vicksburg, Mississippi, and the Colorado River at Austin, Texas? Explain.

25. What are the principal pollutants of surface and ground water? From what sources do they come? What is acid mine drainage? What is thermal pollution?

26. Comment on the adequacy of surface water supplies for meeting the nation's future needs. What alternatives exist to construction of massive water transfer systems? Should further development of irrigation systems in arid parts of the nation be encouraged or discouraged? Give your reasons.

Exercises

1. The accompanying table gives U.S. Geological Survey data for the measurement of discharge of the Wabash River at Montezuma, Indiana, on March 21, 1960. Figures in the first column give horizontal distance from a control point on the left bank and represent the points at which the current meter was lowered. The second column gives the width of section to which the sounding applies. Notice that these widths vary because measurements are made from a bridge whose several piers prohibit uniformly spaced soundings. The fourth column gives mean velocity of each section, obtained by averaging current meter velocity readings at 0.2 and 0.8 of depth. The fifth and sixth columns give area of the section and its discharge.

(*a*) Using a sheet of graph paper, plot the cross section of the river. Let 1 in. on the horizontal scale equal 100 ft; let 1 in. on the vertical scale equal 20 ft to give a vertical exaggeration of 5 times. Draw vertical lines to show locations of soundings. Above this cross section, plot a similar cross section to natural scale (no vertical exaggeration).

(*b*) Complete the computations of mean velocity, area, and discharge for all blank spaces in the table. Sum the area and discharge columns to obtain total cross-section area and total discharge.

(*c*) Using the totals obtained in (*b*), compute the mean velocity of the entire stream.

(*d*) Prepare a graph on which mean velocity is scaled on the vertical axis; depth on the horizontal axis. Plot a point on the graph for each of the 30 pairs of measurements of mean velocity and depth. Does there seem to be a significant relationship between mean velocity and depth? Describe this relationship.

1. Horizontal Distance (Feet)	2. Width of Section (Feet)	3. Depth (Feet)	4. Mean Velocity (Feet per Second)	5. Area of Section (Square Feet)	6. Discharge (cfs)
10	Left edge of water				
30	25	6.3	____	158	171
60	30	20.8	2.50	____	1560
90	30	27.2	3.25	816	____
120	30	33.5	____	1000	4080
150	30	37.8	4.16	____	4700
180	23	34.8	3.52	800	____
200	17	37.9	____	644	2540
220	24	43.8	4.02	____	4220
250	30	42.1	5.59	1260	____
280	30	37.5	____	1120	6900
310	30	38.4	5.72	____	6580
340	30	35.5	6.34	1060	____
370	29	30.3	____	879	4220
400	27	28.9	5.11	____	3980
430	28	23.7	3.36	664	____
460	35	14.6	____	511	552
500	45	11.8	1.24	____	658
550	50	11.0	0.94	550	____
600	52.5	11.8	____	618	531
660	47.5	11.2	1.09	____	580
700	45	10.0	1.34	450	____
750	51	11.5	____	586	1490
805	52.5	13.0	2.36	____	1610
860	46.5	16.2	1.43	753	____
900	45	16.7	____	752	541
950	52	15.8	1.46	____	1200
1005	42.5	21.8	1.98	926	____
1040	35.5	15.8	____	561	1160
1080	45	12.1	2.12	____	1150
1130	55	6.2	1.22	341	____
1190	Right edge of water				

2. Given below are observed values of gauge height (stage) and corresponding measured discharges for a permanent gauging station on Little Ossipee River in Maine. (Data from U.S. Geological Survey.)

Gauge Height (Feet)	Discharge, cfs
0.38	11
0.80	55
1.5	220
2.2	650
3.0	1400
3.8	2100
4.4	2800

(*a*) Plot these data on a graph similar to that in Figure 13.15. Fit a smooth curve to the points.

(*b*) Using the rating curve that you have drawn, estimate discharge for the following gauge heights: 0.5, 1.0, 2.6, 4.1.

(*c*) Plot the same stage-discharge data on 3-cycle logarithmic paper (logarithmic vertical and horizontal scales). Note that a straight line fits the points very well, indicating that discharge increases as some fixed power (exponent) of the gauge height.

3. Given below are precipitation and runoff data for an early autumn rainstorm on the Delaware River Watershed. The stream gauge is at Port Jervis, N.Y. (Data from U.S. Geological Survey.)

Rainfall	Inches
6 P.M., Sept. 28 to 6 A.M., Sept. 29:	0.1
6 A.M., Sept. 29 to 6 P.M., Sept. 29:	0.9
6 P.M., Sept. 29 to 6 P.M., Sept. 30:	3.7
6 P.M., to midnight, Sept. 30:	0.1
Total	*4.8*

Runoff	cfs
Sept. 28	1,000 (base flow)
Sept. 29	1,000 (base flow)
Sept. 30	12,000
Oct. 1	74,000
Oct. 2	47,000
Oct. 3	21,000
Oct. 4	13,000
Oct. 5	9,000
Oct. 6	8,000
Oct. 7	7,000
Oct. 8	6,000
Oct. 9	5,000 (base flow)

(*a*) Prepare a graph, similar to Figure 13.17, to show precipitation and runoff for each day from September 28 through October 9. Suggested vertical scale: let one-half inch equal 0.5 in. precipitation and 10,000 cfs runoff. Let one-half inch on the horizontal scale equal one day. Label the graph fully. Sketch the position of the rising base flow curve throughout the entire period.

(*b*) Estimate the positions of the centers of mass of precipitation and of runoff; label as in Figure 13.17. Estimate the lag time.

(*c*) For each day, subtract the base flow from the total daily discharge. Sum these quantities to determine the total surface runoff from the storm.

III

CLIMATE, SOILS, AND VEGETATION

CHAPTER 14

Classification of Global Climates

CLIMATE, a basic environmental factor, is perhaps the keystone of physical geography. In the broadest sense, *climate* is the characteristic condition of the atmosphere near the earth's surface at a given place or over a given region. Components that enter into the description and classification of climate are mostly the same as weather components used to describe the state of the atmosphere at a given instant. If weather information deals with the specific event, then climate represents a generalization of weather. A statement of the climate of a given observing station, or of a designated region, is described through the medium of weather observations accumulated over many years time. Not only are mean, or average, values taken into account, but also the departures from those means and the probabilities that such departures will occur.

In foregoing chapters we have presented much information that falls within the definition of climate. For example, world maps of average January and July barometric pressures, winds, and air temperatures are expressions of climate. The physical components of climate are many; they include such measurable quantities as net radiation, sensible heat, barometric pressure, winds, relative and specific humidity, dew point, cloud cover and type, fog, precipitation type and intensity, evaporation and transpiration, incidence of cyclones and anticyclones, and frequency of frontal passages. Shall we try to include all of these components in our analysis of climate? The need to be selective in approaching *climatology*, the science of climate, requires us to examine our purposes.

Climate and life

A goal of physical geography is to recognize and define environmental regions of significance to man and to all life forms generally. Which are the vital components of climate with respect to man in particular, and to the biosphere in general? Keep in mind that the basic food supply of animals is organic matter synthesized by plants. Plants are the *primary producers* of organic matter, which sustains life. Carbohydrate compounds produced by plants are consumed by animals. Animals are *consumers;* they derive food either from plants or from other animals or from both. The consumers are organized into *food chains* by means of which stored energy and matter in the form of organic molecules are passed from one individual to another. Ultimately, of course, plants and animals die and their tissues revert by oxidation into the primary components of life, largely carbon dioxide and water. Organisms that play this role are classed as *decomposers.*

In view of man's dependence upon the primary producers, it seems logical that we should select those components of climate that are vital to plant growth. Plants occupy both marine and terrestrial environments. Because plant life of the lands is most directly exposed to the atmosphere for exchanges of energy and matter, we will want to concentrate on climate of the land-atmosphere interface, letting the climate of the ocean-atmosphere interface take a secondary place.

Climate classification

Knowing what our broad goals are to be in climatology is an important first step. Our concern is with those atmospheric components essential to life on the lands. Basically there are two vital components: energy and water, both in available forms. Plants require exposure to solar radiation in order to carry on *photosynthesis* (the process by which carbohydrate molecules are synthesized from CO_2 and H_2O); plants require sensible heat as measured by air and soil temperatures within specified limits; plants require water in the capillary water form in the root zone of the soil. Although land animals are consumers, they too require fresh water and a tolerable range of surrounding (ambient) temperature.

Our problem in climatology is therefore to select categories of available information that correlate closely with the needs of life on the lands. We can use the same basic information that enters into the global balances of radiation, heat, and water. The data must then be viewed in terms of global distribution patterns. Regional classes must be defined and delineated. Two or more categories of information may then be combined and refined into a finished climate classification system.

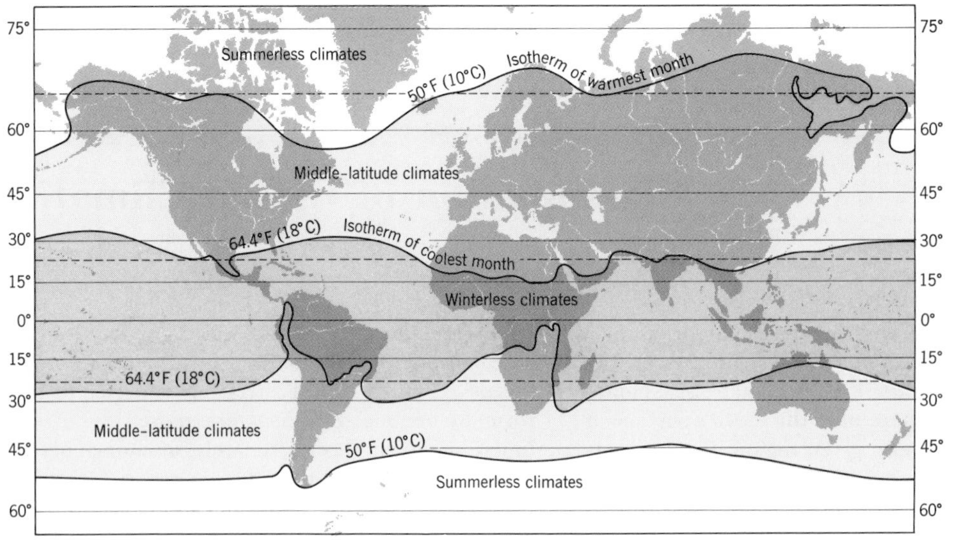

Figure 14.1 On the basis of temperature alone, the earth's surface can be divided into three major climatic groups.

If we have done our work well, the regional units within the climate system will strongly reflect the control of the atmosphere upon terrestrial life and will give some indication of the opportunities and constraints that the atmospheric environment imposes on man as he seeks to increase his food supplies and water resources at the same time that he extends his areas of urban and industrial land use. The climate classification we seek must have utility in guiding land-use planning and population growth, as well as describing the environment.

Let us now review some of the categories of information that may prove useful in producing a meaningful system of climate classification.

Net radiation as a basis of climate classification

Among geographers the proposal to use insolation or net radiation as the basis of a global classification of climates is only a very recent event. A preliminary attempt to set up a classification system using insolation was made by Werner H. Terjung in 1970; it was followed in the same year by his preliminary system of climate classification using net radiation.

Recall from Chapter 8 that net radiation is an expression of the difference between incoming and outgoing energy in both shortwave and longwave forms. Referring back to Figure 8.13, we see that the annual cycle of net radiation shows striking differences from equatorial to subarctic zones. Terjung analyzes the annual curves of net radiation with respect to the maximum values, the annual range of values, and the form of the curve. Combining these parameters he derives a system of climate types and a world map showing their distribution.

Net radiation is regarded as the best indicator of energy available for plant growth. In combination with a measure of available water in the form of soil moisture, net radiation may succeed as the basis of classifying world climates. At present, however, more conventional systems using other climatic variables are dominant in physical geography; it is to these we turn next.

Temperature as a basis of climate classification

Temperature of the lower air layer, as measured in the standard thermometer shelter (Chapter 8) has long provided the first variable quantity in leading systems of climate classification. Monthly data based on daily readings of the maximum-minimum thermometer have been accumulated for many decades at thousands of observing stations the world over. Thus, availability has been an important factor in favoring the use of air temperature in climate classification.

Using monthly mean air temperature, three major climate groups can be defined: (1) winterless climates of low latitudes, (2) climates of middle latitudes with both a summer and a winter season, and (3) summerless climates of high latitudes. A *winterless* climate is commonly defined as one in which no month of the year has a monthly mean temperature lower than 64.4°F (18°C). The approximate position of the appropriate isotherm is shown on a world map (Figure 10.1). Note that the isotherm shows a considerable latitude range, bending equatorward over the cool west-coast ocean currents and over the land in north Africa and Australia.

A *summerless* climate is commonly defined as one in which no month has a monthly mean temperature higher than 50°F (10°C). The appropriate isotherm, shown in Figure 14.1, has a wide latitude range in the northern hemisphere, being farthest poleward over the landmasses of North America and Eurasia, but dipping to middle latitudes over the intervening oceans. The 50°F (10°C) isotherm of the warmest month closely coincides with the northernmost limit of tree growth, hence it separates the regions of boreal forests from the treeless arctic tundra. Here then, is an example of a thermal climatic boundary selected to coincide with a natural vegetation boundary.

Climates having both a summer and a winter season lie in the regions between the two boundary isotherms described above and constitute a *middle-latitude* group.

Air temperature is an important environmental factor in plant physiology and reproduction; it enters into many activities of animal life as well (hibernation, migration). For man, air temperature is an important physiological factor and relates directly to the quantity of energy expended in space heating of buildings. Nevertheless, air temperature alone does not define meaningful climate classes, since the ingredient of water availability is missing. The winterless climates, for example, include extremes of dry and moist environments.

Precipitation as a basis for climate classification

Precipitation data, obtained by the simple rain gauge (Chapter 10) are abundantly available for long periods of record for thousands of observing stations widely distributed over the globe. Small wonder, then that monthly and annual precipitation data form the cornerstone of most of the widely used climate classifications.

At this point it is important to obtain an overview of the global patterns of precipitation and their relationship to air mass source regions and prevailing movements of air masses.

Average annual precipitation is shown on a world map by means of *isohyets*, lines drawn through all points having the same annual quantity. A detailed world precipitation map will be found in Plate 1 (insert color map at end of book). For regions where all or most of the precipitation is rain, we use the word "rainfall," for regions where snow is an important part of the annual total, we use the word "precipitation."

Seven *precipitation regions* can be recognized in terms of annual total in combination with location (Table 14.1). As you read the following descriptions, locate the regions on Plate 1.

TABLE 14.1 WORLD PRECIPITATION REGIONS

Name	Latitude Range	Continental Location	Prevailing Air Masses	Annual Precipitation: Inches	Annual Precipitation: Centimeters
1. Wet equatorial belt	10° N to 10° S	Interiors, coasts	mE	Over 80	Over 200
2. Trade-wind coasts (windward tropical coasts)	5–30° N and S	Narrow coastal zones	mT	Over 60	Over 150
3. Tropical deserts	10–35° N and S	Interiors, west coasts	cT	Under 10	Under 25
4. Middle-latitude deserts and steppes	30–50° N and S	Interiors	cT,cP	4–20	10–50
5. Humid subtropical regions	25–45° N and S	Interiors, coasts	mT (summer)	40–60	100–150
6. Middle-latitude west coasts	35–65° N and S	West coasts	mP	Over 40	Over 100
7. Arctic and polar deserts	60–90° N and S	Interiors, coasts	cP,cA	Under 12	Under 30

1. The *wet equatorial belt* of heavy rainfall, over 80 in (200 cm) annually, straddles the equator and includes the Amazon River basin in South America, the Congo River basin of equatorial Africa, much of the African coast from Nigeria west to Guinea, and the East Indies. Here the prevailingly warm temperatures and high moisture content of the mE air masses favor abundant convective rainfall. Thunderstorms are frequent year around.

2. Narrow coastal belts of high rainfall, 60–80+ in (150–200 cm+) per year, extend from near the equator to latitudes of about 25° to 30° N. and S. on the eastern sides of every continent or large island. For examples, see the eastern coasts of Brazil, Central America, Madagascar, and northeastern Australia. These are the *trade-wind coasts*, or *windward tropical coasts*, where moist mT air masses from warm oceans are brought over the land by the trades. Encountering coastal hills, escarpments, or mountains, these air masses produce heavy orographic rainfall.

3. In striking contrast to the wet equatorial belt astride the equator are the two zones of huge *tropical deserts* lying approximately upon the tropics of Cancer and Capricorn. These hot, barren deserts, with less than 10 in (25 cm) of rainfall annually and in many places with less than 2 in (5 cm) are located under and caused by the subtropical cells of high pressure where the subsiding cT air mass is adiabatically warmed and dried. Note that these deserts extend off the west coasts of the lands and out over the oceans. Such rain as these areas experience is largely convective and extremely unreliable.

4. Farther northward, in the interiors of Asia and North America between lat. 30° and lat. 50°, are great continental *middle-latitude deserts* and expanses of semiarid grasslands known as *steppes*. Annual precipitation ranges from less than 4 in (10 cm) in the driest areas to 20 in (50 cm) in the moister steppes. Dryness here results from remoteness from ocean sources of moisture. Located in a region of prevailing westerly winds, these arid lands occupy the position of rain shadows in the lee of coastal mountains and highlands. Thus the Cordilleran Ranges of Oregon, Washington, British Columbia, and Alaska shield the interior of North America from moist mP air masses originating in the Pacific. Upon descending into the intermontane basins and interior plains, the mP air masses are warmed and dried.

Similarly, mountains of Europe and the Scandinavian peninsula serve to obstruct the flow of moist mP air masses from the North Atlantic into western Asia. The great southern Asiatic ranges likewise prevent the entry of moist mT and mE air masses from the Indian Ocean.

The southern hemisphere has too little land in the middle latitudes to produce a true continental desert, but the dry steppes of Patagonia lying on the lee side of the Andean chain are roughly the counterpart of the North American deserts and steppes of Oregon and northern Nevada, which lie in the rain shadow of the Sierra Nevada and the Cascade Range.

5. On the southeastern sides of the continents of North America and Asia, in lat. 25° to 45°, and to a less marked degree in these same latitudes in the southern hemisphere in Uruguay, Argentina, and southeastern Australia, are the *humid subtropical regions*, with 40–60 in (100–150 cm) of rainfall annually. These regions lie on the moist western sides of the subtropical high-pressure centers in such a position that humid mT air masses from the tropical ocean are carried poleward over the adjoining land. Commonly too, these areas receive heavy rains from tropical cyclones.

6. Still another distinctive wet location is on *middle-latitude west coasts* of all continents and large islands lying between about 35° and 65° in the region of prevailing westerly winds. These zones have been mentioned in Chapter 10 as good examples of coasts on which abundant orographic precipitation falls as a result of forced ascent of mP air masses. Where the coasts are mountainous, as in Alaska and British Columbia, Patagonia, Scotland, Norway, and South Island of New Zealand, the annual precipitation is over 80 in (200 cm). Small wonder that these coasts formerly supported great valley glaciers that carved the deep bays (fiords,) so typically a part of their scenery.

7. The seventh precipitation region is formed by the *arctic* and *polar deserts*. Northward of the 60th parallel, annual precipitation is largely under 12 in (30 cm), except for the west-coast belts. Cold cP and cA air masses cannot hold much moisture, consequently they do not yield large amounts of precipitation. At the same time, however, the relative humidity is high and evaporation rates are low. Consequently these arctic and polar regions have abundant moisture in the air and soil and are not to be considered as dry in the same sense as the tropical deserts.

As would be expected, zones of gradation lie between the precipitation regions listed above. Moreover, this list of regions does not recognize the fact that seasonal patterns of precipitation differ from region to region.

The soil-moisture balance as a basis for climate classification

In recent years, geographers have focused attention upon the water balance as a basis for a climate classification. The soil-moisture balance, particularly, has been favored as the foundation of a climate system because it represents availability of water for plants, as well as an assessment of the availability of surplus water to supply stream flow and ground water.

The soil-moisture balance has been developed fully in Chapter 12. The important concept involved is that precipitation by itself does not indicate the amount of water actually available to plants. Evapotranspiration must be subtracted from precipitation to reveal the net quantity of water as a surplus or a deficit.

Climate classification based upon the soil-moisture balance uses two variables: (1) precipitation, and (2) evapotranspiration. Temperature is by no means disregarded in this classification, for temperature enters into the calculation of evapotranspiration. Insolation is also used in calculating evapotranspiration.

It is interesting to note that both precipitation and evapotranspiration are measured in the same quantities, namely, water depth or its equivalent per month or per year. Thus it is possible to place the classification system on a strictly quantitative basis with minimal difficulty. A system using temperature and precipitation must operate with two unlike quantities: degrees F or C, and inches (or cm) per month or year.

C. Warren Thornthwaite, the climatologist who developed the concepts of the soil-moisture balance (Chapter 12), also devised a climate classification system using precipitation and evapotranspiration. The Thornthwaite system uses an elaborate code to designate climate types. Although important in advanced work in climatology, the Thornthwaite system is rarely used in an introductory course.

Vegetation and soils as bases of climate classifications

Botanists and geographers have long been aware that plants are highly responsive to differences in climate. With each plant species is associated a particular combination of climatic factors most favorable to its growth, as well as certain extremes of heat, cold, or drought beyond which it cannot survive. Plants tend to adapt their physical forms to meet the stresses of climate, hence we find that there is a wide range of forms taken by assemblages of the dominant plant species, and that the overall form patterns or habits closely reflect climate. These subjects are treated in Chapters 20 and 21.

There is much to recommend a climate classification based on the different forms taken by groups of plants. On the other hand, such plant forms represent responses to climate, rather than causes of global climate variation. A fundamental principle of scientific classification is that the setting up of classes of things is better done according to the causes of the class differences than according to the effects that differences produce.

Since the late nineteenth century, soil scientists have recognized that the several fundamental classes of mature soils, seen on a world-wide scope, are more strongly controlled by climatic elements than by any other single factor. (See Chapters 18 and 19.) But plants also contribute to determining the properties of the very soils upon which they depend. Thus, as in the case of vegetation, soils reflect differences in climate but do not cause those differences. A climate classification based upon soils would be highly meaningful, but not scientifically as desirable as a climate classification based upon causes.

The Köppen climate classification system

The foregoing concepts of devising climate classes that combine temperature and precipitation characteristics, but of setting limits and boundaries fitted into known vegetation and soil distributions, were actually carried out in 1918 by Dr. Wladimir Köppen of the University of Graz in Austria. The classification was subsequently revised and extended by Köppen and his students to become the most widely used of climatic classifications for geographical purposes.

The Köppen system is strictly empirical. This is to say that each climate is defined according to fixed values of temperature and precipitation, computed according to the averages of the year or of individual months. In such a classification, no concern whatever need be given to the causes of the climate in terms of pressure and wind belts, air masses, fronts, or storms. It is possible to assign a given place to a particular climate subgroup solely on the basis of the records of temperature and precipitation of that place, provided, of course, that the period of record is long enough to yield meaningful averages. Air temperature and precipitation are the most easily obtainable surface weather data, requiring only simple equipment and a very elementary observer education. A climate system based on these data has a great advantage, in that the areas covered by each subtype of climate can be delineated for large regions of the world.

The Köppen system features a shorthand code of letters designating major climate groups, subgroups within the major groups, and further subdivisions to distinguish particular seasonal characteristics of temperature and precipitation.

Five major climate groups are designated by capital letters as follows (Figure 14.2). Groups *A*, *C*, and *D* have sufficient heat and precipitation for growth of high-trunk trees (e.g., forest and woodland vegetation).

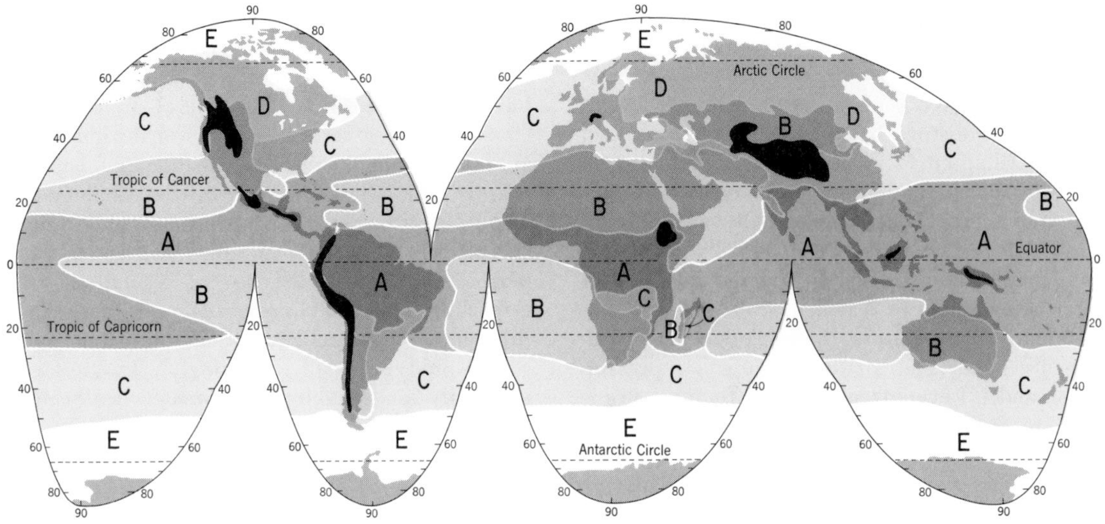

Figure 14.2 Highly generalized world map of major climatic regions according to the Köppen classification. Highland areas in black. (Based on Goode Base Map. Copyright by the University of Chicago. Used by permission of the Geography Department.)

A *Tropical climates.* Average temperature of every month is above 64.4°F (18°C). These climates have no winter season. Annual rainfall is large and exceeds annual evaporation.

B *Dry climates.* Potential evaporation exceeds precipitation on the average throughout the year. No water surplus; hence no permanent streams originate in *B* climate zones.

C *Warm temperate (mesothermal)*[1] *climates.* Coldest month has an average temperature under 64.4°F (18°C), but above 26.6°F (−3°C); at least one month has an average temperature above 50°F (10°C). The *C* climates thus have both a summer and a winter season.

D *Snow (microthermal)*[2] *climates.* Coldest month average temperature under 26.6°F (−3°C). Average temperature of warmest month above 50°F (10°C), that isotherm coinciding approximately with poleward limit of forest growth.

E *Ice climates.* Average temperature of warmest month below 50°F (10°C). These climates have no true summer.

Note that four of these five groups (*A, C, D,* and *E*) are defined by temperature averages, whereas one (*B*) is defined by precipitation-to-evaporation ratios. This procedure may seem to be a fundamental inconsistency.

[1]The middle latitudes are anything but "temperate" in seasonal and day-to-day weather. *Mesothermal* is widely substituted for *temperate* to imply intermediate temperatures as compared with the extreme heat of dry deserts and the extreme cold of polar and arctic climates.

[2]The term *microthermal,* meaning "small heat," is widely substituted in the United States, along with *mesothermal,* in the titles of Köppen groups *D* and *C,* respectively.

Subgroups within the five major groups are designated by a second letter according to the following code:

S *Steppe climate:* A semi-arid climate with about 15 to 30 in (38 to 76 cm) of rainfall annually at low latitudes. Exact rainfall boundary determined by formula taking temperature into account.

W *Desert climate:* Arid climate. Most regions included have less than 10 in (25 cm) of rainfall annually. Exact boundary with steppe climate determined by formula.

(The letters *S* and *W* are applied only to the dry *B* climates, yielding two combinations, *BS* and *BW*.)

f Moist. Adequate precipitation in all months. No dry season. This modifier is applied to *A, C,* and *D* groups.

w Dry season in winter of the respective hemisphere (low-sun season).

s Dry season in summer of the respective hemisphere (high-sun season).

m Rainforest climate despite short, dry season in monsoon type of precipitation cycle. Applies only to *A* climates.

From combinations of the two letter groups, eleven distinct climates emerge as follows:

Af Tropical rainforest (Also *Am,* a variant of *Af*).
Aw Tropical savanna.
BS Steppe climate.
BW Desert climate.
Cw Temperate rainy (humid mesothermal) climate with dry winter.

Cf Temperate rainy (humid mesothermal) climate moist all seasons.

Cs Temperate rainy (humid mesothermal) climate with dry summer.

Df Cold snowy forest (humid microthermal) climate moist in all seasons.

Dw Cold snowy forest (humid microthermal) climate with dry winter.

ET Tundra climate.

EF Climates of perpetual frost (icecaps).

To differentiate still more variations in temperature or other weather elements. Köppen added a third letter to the code group. Meanings are as follows:

a With hot summer; warmest month over 71.6°F (22°C) (*C* and *D* climates).

b With warm summer; warmest month below 71.6°F (22°C) (*C* and *D* climates).

c With cool, short summer; less than four months over 50°F (10°C) (*C* and *D* climates).

d With very cold winter; coldest month below −36.4°F (−38°C) (*D* climates only).

h Dry-hot; mean annual temperature over 64.4°F (18°C) (*B* climates only).

k Dry-cold; mean annual temperature under 64.4°F (18°C) (*B* climates only).

As an example of a complete Köppen climate code, *BWk* would refer to *cool desert climate*, *Dfc* would refer to *cold, snowy forest climate with cool, short summer*. The complete list of climate code designations is included in the summary table of climates to follow. A world map of climates following a later revision of the Köppen system (the Köppen-Geiger system) appears as Plate 2.

An explanatory-descriptive climate system

Earlier chapters on weather processes, global circulation, air mass properties and source regions, and cyclonic storms have armed the reader with many principles concerning the causes of weather patterns and their seasonal variations. It would be a pity if this knowledge were left unused with no application in the systematic study of world climates, in favor of a purely empirical system based on temperature and precipitation averages expressed by code letters.

The most satisfying systems of natural science classification are those described as *genetic*, meaning that the genesis, or origin, of the phenomena is put first in designing the classification units. Because a genetic system gives an explanation of

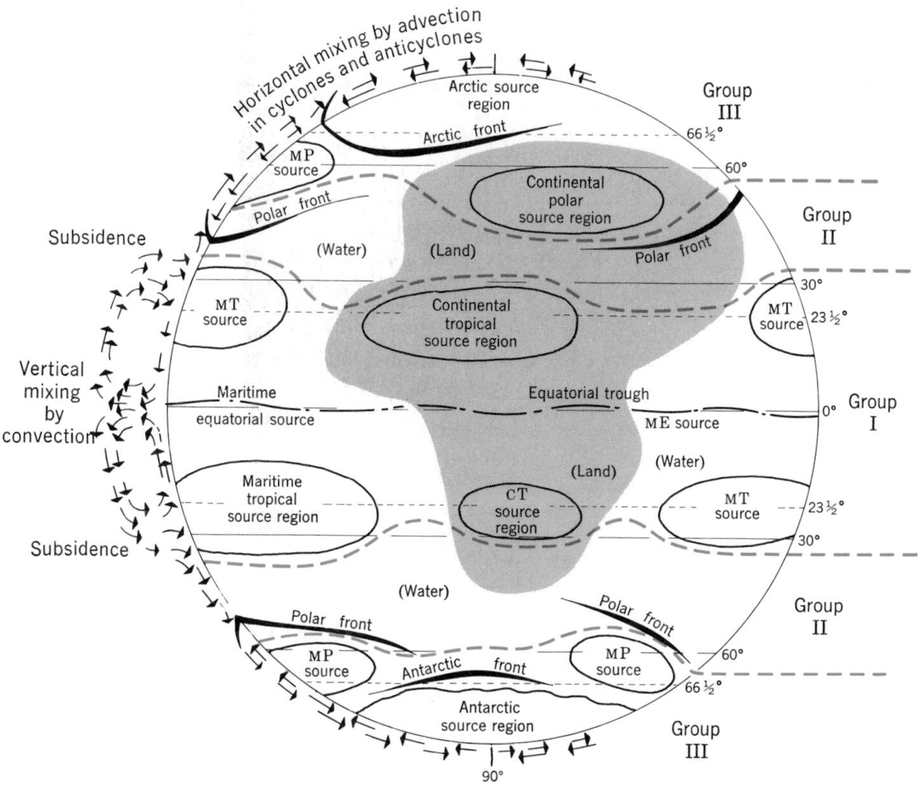

Figure 14.3 Global diagram of major climate groups. (After Petterssen and others.)

the things classified, it may be considered as *explanatory*. If the explanation is carried out largely in verbal statements (as distinct from numerical or mathematical statements) it may be labeled as *descriptive*. This general approach to classification can therefore be designated as *explanatory-descriptive* in contrast to the *empirical-quantitative* approach followed in the Köppen system.

In the chapters to follow, world climates are treated by an explanatory-descriptive system based on causes and effects. It will not be difficult to introduce the Köppen code symbols into such a system by way of interrelating the two. In fact, it will soon become apparent that the explanatory-descriptive system is simply providing a reasonable scientific explanation for the existence of Köppen's climate groups and subgroups.

Air mass source regions and frontal zones as a basis of classification

The system of climate analysis used in the succeeding chapters is based on the location of air mass source regions and the nature and movement of air masses, fronts, and cyclonic storms. A schematic global diagram, Figure 14.3, shows the principles of the classification. A world map, Figure 14.4, shows the general locations of each of fourteen fundamental types.

Three major climate groups are shown in Figure 14.3. Group I includes the tropical air mass source regions and the equatorial trough or convergence zone between. *Equatorial* is here defined as pertaining to the latitude zone within a few degrees north and south of the equator; *tropical* is defined as referring to the two latitude zones included within a few degrees north and south of the tropics of Cancer and Capricorn (Figure 8.6).

Climates of Group I are controlled by the dynamic subtropical high-pressure cells, or anticyclones, which are regions of air subsidence and are basically dry, and by the great equatorial trough of convergence that lies between them. Though it is true that air of polar origin occasionally invades the tropical and equatorial latitudes, it may be said of the climates of Group I that they are almost wholly dominated by tropical and equatorial air masses.

Climates of Group II are in a zone of intense interaction between unlike air masses: the *polar front zone*. Tropical air masses moving poleward and polar air masses moving equatorward are in conflict in this zone, which contains a procession of eastmoving cyclonic storms. Locally and seasonally either tropical or polar air masses may dominate in these regions, but neither has exclusive control.

Climates of Group III are dominated by polar and arctic (including antarctic) air masses. The two polar continental air mass source regions of northern Canada and Siberia fall into this group, but there is no southern hemisphere counterpart to these continental centers. In the belt of the 60th to 70th parallels, air masses of arctic origin meet polar continental air masses along an *arctic front zone*, creating a series of east-moving cyclones.

Fourteen major climate types are listed below. Where a compound climate type is formed by the alternation of two simple types, it follows them on the list. Simple names are applied to these climate types following common usage among geographers. Numbers correspond with those in Figure 14.4. Köppen symbols are added to the table to show the close correspondence of these fourteen types with his eleven climates and their subtypes.

In Chapters 15–17 we examine in detail the characteristics and distributions of the fourteen climate types tabulated here. An important synthesis will be to identify the soil moisture regime associated with each climate type.

GROUP I: LOW-LATITUDE CLIMATES (CONTROLLED BY EQUATORIAL AND TROPICAL AIR MASSES)

Climate Name		Köppen Symbol	Air Mass Source Regions and Frontal Zones, General Climate Characteristics
1. Wet Equatorial Climate 10° N–10° S lat. (Asia 10°–20° N)	Af Am	Tropical rainforest climate, and Tropical rainforest climate, monsoon type	Equatorial trough (convergence zone) climates are dominated by warm, moist tropical maritime (mT) and equatorial (mE) air masses yielding heavy rainfall through convectional storms. Remarkably uniform temperatures prevail throughout the year.
2. Trade Wind Littoral Climate 10°–25° N and S lat.	Af-Am	Included in climates	Tropical easterlies (trades) bring maritime tropical (mT) air masses from moist western sides of oceanic subtropical high-pressure cells to give narrow east-coast zones of heavy rainfall and uniformly high temperatures. Rainfall shows strong seasonal variation.

3. Tropical Desert and Steppe Climates 15° – 35° N and S lat.	*BWh* *BSh*	Desert climate, hot, and Steppe climate, hot	Source regions of continental-tropical (cT_s) air masses in high-pressure cells at high level over lands astride the Tropics of Cancer and Capricorn give arid to semi-arid climate with very high maximum temperatures and moderate annual range.
4. West Coast Desert Climate 15° – 30° N and S lat.	*BWk* *BWh*	Desert climate, cool, and Desert climate, hot (*BWn* in earlier versions, *n* meaning frequent fog)	On west coasts bordering the oceanic subtropical high-pressure cells, subsiding maritime tropical (mT_s) air masses are stable and dry. Extremely dry, but relatively cool, foggy desert climates prevail in narrow coastal belts. Annual temperature range is small.
5. Tropical Wet-Dry Climate 5°–25° N and S lat.	*Aw* *Cwa*	Tropical rainy climate, savanna; also Temperate rainy (Humid mesothermal) climate, dry winter, hot summer	Seasonal alternation of moist mT or mE air masses with dry cT air masses gives climate with wet season at time of high sun, dry season at time of low sun.

GROUP II: MIDDLE-LATITUDE CLIMATES (CONTROLLED BY BOTH TROPICAL AND POLAR AIR MASSES)

6. Humid Subtropical Climate 20°–35° N and S lat.	*Cfa*	Temperate rainy (Humid mesothermal) climate, hot summers	Subtropical, eastern continental margins dominated by moist maritime (mT) air masses flowing from the western sides of oceanic high-pressure cells. In high-sun season, rainfall is copious and temperatures warm. Winters are cool with frequent continental polar (cP) air mass invasions. Frequent cyclonic storms.
7. Marine West Coast Climate 40°–60° N and S lat.	*Cfb* *Cfc*	Temperate rainy (Humid mesothermal) climate, warm summers, and same but cool, short summers	Windward, middle-latitude west coasts receive frequent cyclonic storms with cool, moist maritime polar (mP) air masses. These bring much cloudiness and well-distributed precipitation, but the winter maximum. Annual temperature range is small for middle latitudes.
8. Mediterranean Climate 30°–45° N and S lat.	*Csa* *Csb*	Temperate rainy (Humid mesothermal) climate, dry, hot summer, and same, but dry, warm summer	This wet-winter, dry-summer climate results from seasonal alternation of conditions causing climates 4 and 7; mP air masses dominate in winter with cyclonic storms and ample rainfall, mT_s air masses dominate in summer with extreme drought. Moderate annual temperature range.
9. Middle-latitude Desert and Steppe Climates 35°–50° N and S lat.	*BWk* *BWk'* *BSk* *BSk'*	Desert climate, cool same, but cold; and Steppe climate, cool same, but cold	Interior, middle-latitude deserts and steppes of regions shut off by mountains from invasions of maritime air masses (mT or mP), but dominated by continental tropical (cT) air masses in summer and continental polar (cP) air masses in winter. Great annual temperature range; hot summers, cold winters.

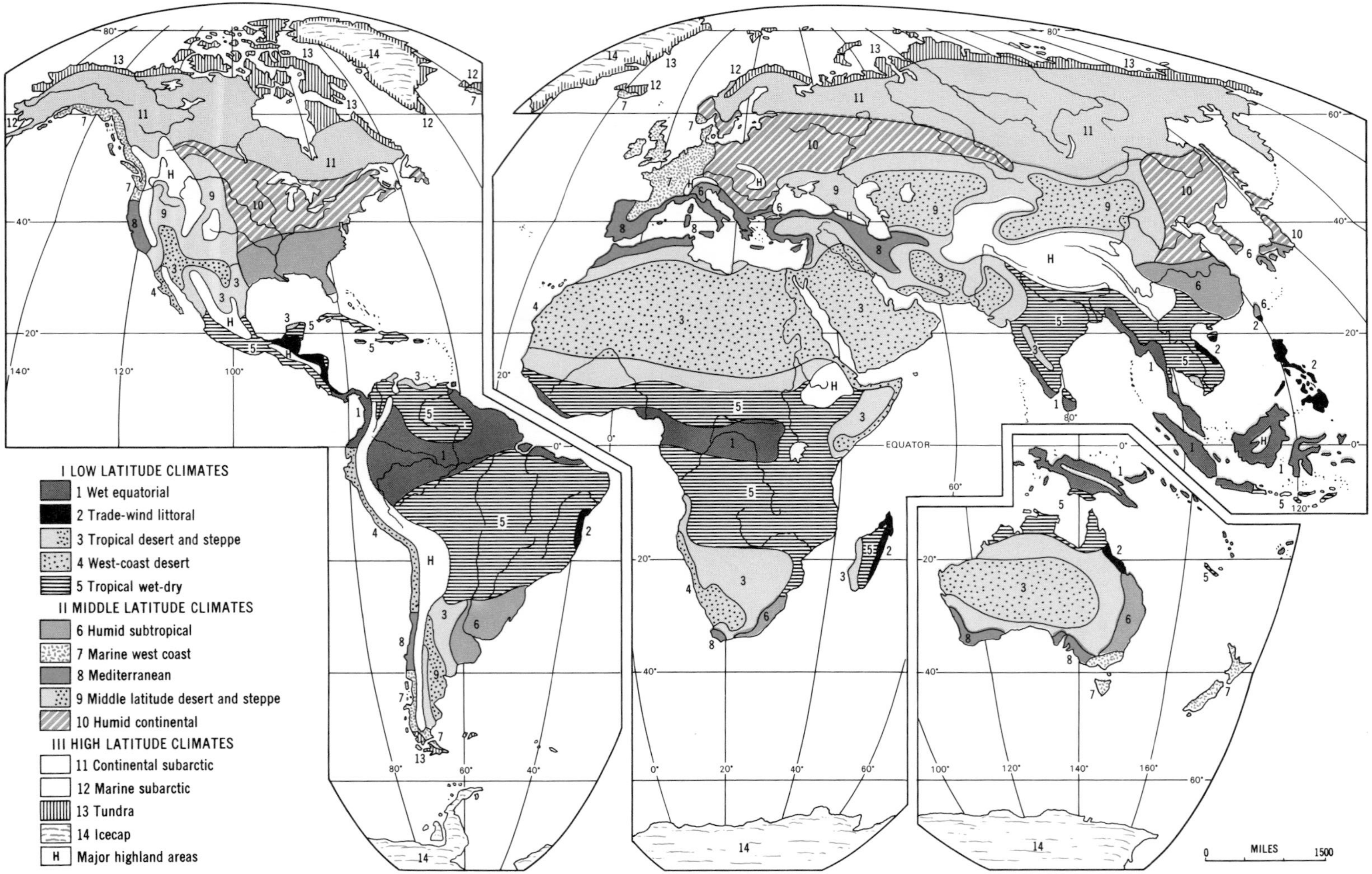

Figure 14.4 Generalized and simplified world map showing distribution of fourteen climates. In many respects, these climate regions correspond with regions as defined by G. T. Trewartha. (Aitoff's equal-area projection adapted by V. C. Finch.)

10. Humid Continental Climate 35°–60° N lat.	*Dfa*	Cold, snowy forest (Humid microthermal) climate, moist all year, hot summers, and	Located in central and eastern parts of continents of middle latitudes, these climates are in the polar front zone, the battle ground of polar and tropical air masses. Seasonal contrasts are strong and weather highly variable. Ample precipitation throughout the year is increased in summer by invading maritime tropical (mT) air masses. Cold winters are dominated by continental polar (cP) air masses invading frequently from northern source regions.
	Dfb	same, but warm summers; also	
	Dwa	Cold, snowy forest (Humid microthermal) climate, dry winters, hot summers, and	
	Dwb	same, but warm summers	

GROUP III: HIGH-LATITUDE CLIMATES (CONTROLLED BY POLAR AND ARCTIC AIR MASSES)

11. Continental Subarctic Climate 50°–70° N lat.	*Dfc*	Cold, snowy forest (Humid microthermal) climate, moist all year, cool summers, and	This climate lies in source region of continental polar (cP) air masses, which in winter are stable and very cold. Summers are short and cool. Annual temperature range is enormous. Cyclonic storms, into which maritime polar (mP) air is drawn, supply light precipitation, but evaporation is small and the climate is therefore effectively moist.
	Dfd	Same, but very cold winters, also	
	Dwc	Cold, snowy forest (Humid microthermal) climate, dry winter, cool summer, and	
	Dwd	same, but very cold winter	
12. Marine Subarctic Climate 50°–60° N and 45°–60° S lat.	*ET*	Polar, tundra climate	Located in the arctic frontal zones of the winter season, these windward coasts and islands of subarctic latitudes are dominated by cool mP air masses. Precipitation is relatively large and annual temperature range small for so high a latitude.
13. Tundra Climate North of 55° N South of 50°S	*ET*	Polar, tundra climate	The arctic coastal fringes lie along a frontal zone, in which polar (mP, cP) air masses interact with arctic (*A*) air masses in cyclonic storms. Climate is humid and severely cold with no warm season or summer. Moderating influence of ocean water prevents extreme winter severity as in Climate 11.
14. Icecap Climate (Greenland, Antarctica)	*EF*	Polar climate, perpetual frost	Source regions of arctic (*A*) and antarctic (*AA*) air masses situated upon the great continental icecaps have climate with annual temperature average far below all other climates and no above-freezing monthly average. High altitudes of ice plateaus intensify air mass cold.
Highland Climates			Cool to cold moist climates, occupying high-altitude zones of the world's mountain ranges, are localized in extent and not included in classification system.

REVIEW QUESTIONS

1. In what ways is climate different from weather? Name the important climatic variables.

2. Explain how life processes of the biosphere are dependent upon the atmospheric environment. What are the vital components of the atmospheric environment with respect to plant life on the lands?

3. How can net radiation be used as the basis of a system of climate classification? What vital component of climate is not included in such a system?

4. Explain how temperature can be used as a basis for a climate classification. Why is this element alone unsatisfactory as a basis for a climate system?

5. How can annual precipitation be used as a basis for a climate classification? Why does the factor of evaporation make this classification of limited value?

6. Name and describe the seven basic precipitation regions. How is each related to prevailing air masses?

7. How can the soil-moisture balance be made the basis of a system of climate classification? What advantage has the use of evapotranspiration over air temperature, in combination with precipitation?

8. Why is it that the distribution of natural vegetation types forms a workable basis for distinguishing climate zones?

9. Explain the basic principles underlying the Köppen climate system. What are the five major climate groups? What code system is used to designate eleven climates? Name these and give symbols for each.

10. What do each of the following letters mean in the Köppen system: *a, b, c, d, h, k?*

11. Explain how a climate classification system can be based on the air mass source regions, patterns of air mass flow, and frontal zones. What advantage has a natural science classification that is based on origin of the things classified as against one that is based on numerical values?

12. How may the climates of the globe be classified into three major groups on the basis of air mass types?

Exercises

Given below are monthly means of temperature and precipitation for ten climate stations. Figures in final column are mean annual temperature and mean total yearly precipitation. Determine the Köppen climate group to which each station belongs and give the first two letters of the Köppen letter code. If possible, determine the third letter as well. State the criteria by which each station is assigned to its group. Refer to Plate 2 for further details of the Köppen system. **(*a*)** Stanley, Falkland Is. **(*b*)** Madang, New Guinea. **(*c*)** Cochin, India. **(*d*)** Port Darwin, N.T., Australia. **(*e*)** Sacramento, California. **(*f*)** Winnipeg, Manitoba, Canada. **(*g*)** Lima, Peru. **(*h*)** Port Nolloth, Union of South Africa. **(*i*)** Kazalinsk, Kazak, U.S.S.R. **(*j*)** Dawson, Yukon Terr., Canada.

		J	F	M	A	M	J	J	A	S	O	N	D	Year
(*a*)	°F	13.3	13.6	27.0	43.9	53.8	62.2	69.1	65.8	56.1	43.9	30.9	20.1	41.5
	In.	0.7	0.6	0.5	1.0	2.0	2.9	1.8	1.2	1.3	0.7	0.6	0.6	13.9
(*b*)	°F	−14.4	7.4	13.3	33.8	46.8	58.5	62.6	58.5	46.0	30.0	6.8	−9.2	27.1
	In.	0.1	0.1	0.1	0.4	0.9	1.7	2.4	3.0	1.3	0.3	0.2	0.2	10.3
(*c*)	°F	71.1	73.6	83.1	91.6	94.5	93.7	89.2	86.5	89.2	88.9	80.8	71.1	84.4
	In.	0	0	0.1	0	0.3	0.9	3.5	2.8	1.1	0.4	0	0	9.0
(*d*)	°F	33.4	35.2	41.9	53.2	64.0	72.1	76.6	75.0	68.0	57.7	46.2	36.9	55.0
	In.	3.4	3.7	3.6	3.4	3.5	3.9	4.6	4.3	3.4	2.8	2.6	3.3	42.5
(*e*)	°F	47.8	49.5	52.5	58.3	66.4	74.1	79.7	79.5	73.2	66.2	57.0	57.8	63.0
	In.	2.2	1.8	1.3	0.9	0.8	0.6	0.3	0.6	0.7	1.4	2.9	2.5	16.0
(*f*)	°F	77.9	78.4	79.3	79.9	80.6	79.9	80.2	79.7	79.5	79.7	79.0	78.3	79.3
	In.	9.7	7.1	7.3	7.8	6.5	7.0	6.7	7.8	6.9	7.9	10.1	10.5	95.1
(*g*)	°F	80.8	80.6	80.8	80.4	78.4	75.0	74.8	78.1	81.3	82.6	81.7	81.0	79.7
	In.	9.6	8.9	8.1	4.1	2.0	0.2	0.1	1.1	2.0	4.4	6.0	7.9	54.5
(*h*)	°F	−14.3	−12.8	−2.0	12.0	28.6	38.7	45.7	44.2	36.3	26.1	12.4	−3.8	17.6
	In.	1.1	1.3	0.8	1.5	1.5	1.1	2.6	2.0	1.2	1.5	2.2	1.5	18.3
(*i*)	°F	63.3	63.0	62.2	58.3	54.1	51.6	50.4	53.4	60.1	63.5	64.6	63.9	59.0
	In.	10.6	8.4	8.3	2.2	0.9	0.6	0.8	0.3	0.3	1.3	3.5	8.5	45.7
(*j*)	°F	22.6	23.9	31.6	42.4	52.5	61.9	67.5	65.7	59.2	49.6	37.9	27.7	45.1
	In.	3.9	4.3	3.8	3.4	3.3	3.3	3.2	3.1	3.1	3.1	3.5	3.9	41.9

CHAPTER 15

Equatorial and Tropical Climates

CLIMATES of group I are those of low latitudes. They are controlled largely by equatorial and tropical air masses.

1. Wet equatorial climate (Af, Am)

Bringing together the information on global pattern of weather elements contained in previous chapters and focusing attention on the equatorial belt of the earth, it will be seen that within 5° of latitude north and south of the equator the following conditions prevail:

(*a*) Temperatures average close to 80°F (27°C) for every month (Figure 8.22), so that the annual range is extremely small.

(*b*) Air pressures average between 1009 and 1012 mb, or slightly less than normal sea-level pressure (Figure 9.12).

(*c*) The general flow of air is from east to west in the tropical easterlies at high altitude, but somewhat more equatorward in the surface trade winds, which originate in the subtropical high-pressure cells. Here, then, is a region of converging warm, unstable air masses meeting along the equatorial trough (Chapter 11).

(*d*) Examination of the world rainfall map (Plate 1) shows this to be a belt of heavy rainfall, exceeding 80 in (200 cm) annually for the most part. This belt has been explained in Chapter 14 as convectional rainfall from great cumulonimbus masses in weak, slowly moving easterly waves and equatorial lows.

From the above information we can compile a description of the wet equatorial climate. Average annual temperatures are close to the 80°F(27°C) mark; seasonal range of temperature is so slight as to be imperceptible because the sun is high throughout the year. Rainfall is heavy during the entire year, but with considerable differences in monthly averages because of the seasonal shifting of the equatorial convergence zone and a consequent variation in air mass characteristics.

Figure 15.1 is a monthly average temperature-rainfall graph for Iquitos, Peru, a typical wet equatorial station located at about 3° S lat. in the broad basin of the Amazon River. Note that the annual range in temperature is only 4 F° (2.2 C°)

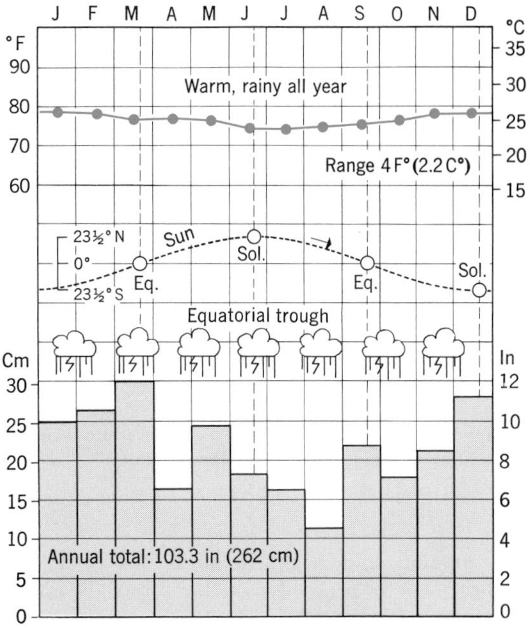

Figure 15.1 Climograph for Iquitos, Peru, $3\frac{1}{2}$° S. lat., upper Amazon River basin. (Data from Blair.)

and that the annual rainfall total is more than 100 in (250 cm). In all but one month the monthly rainfall averages more than 6 in (15 cm). According to Köppen's definition of this *Af* climate, no month averages less than 2.4 in (6 cm) of rainfall.

Some idea of the extreme monotony of the daily temperature cycle in this climate can be had from Figure 15.2, which shows minimum and maximum daily temperatures for two months at Panama, 9° N lat. An especially noteworthy feature is that the daily range is normally from 15 to 20 F° (8 to 11 C°), a vastly greater range than the annual range of monthly mean temperatures. In other words, daily variations far exceed seasonal variations in the wet equatorial climate.

An illustration of month-to-month differences in average monthly rainfall in the wet equatorial climate is given in Figure 15.3. All the stations shown here lie within a few degrees of the equator and are basically similar in climate. The fact that some stations show one maximum and one minimum in the rainfall graph, whereas others have

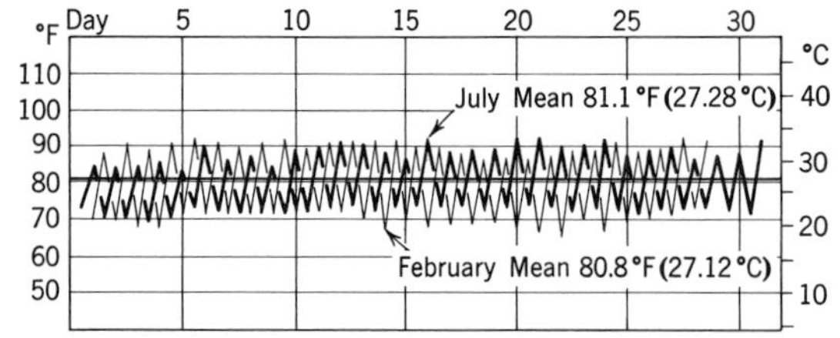

Figure 15.2 July and February temperatures at Panama, 9° N lat. (After Mark Jefferson, *Geographical Review*.)

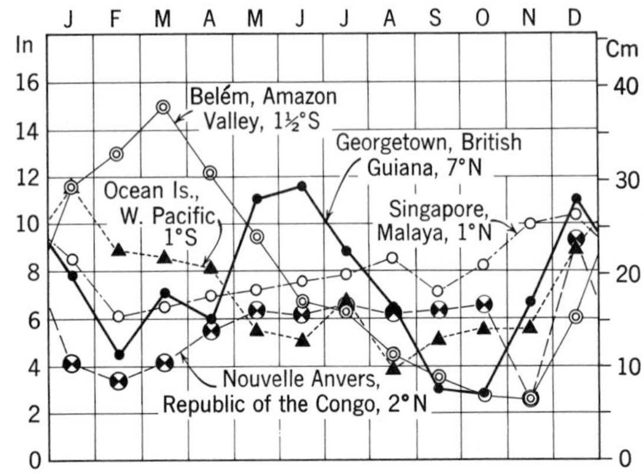

Figure 15.3 Monthly rainfall averages for stations in wet equatorial climates. (Data from Trewartha.)

two maxima and two minima is not easily explained, but is probably attributable to a rather complex series of seasonal changes in air mass movements and positions of the equatorial trough.

The soil-moisture regime is of the equatorial type (Figure 12.12*A*) in which large water surpluses are generated in most months of the year.

Because of the abundant rainfall and prevailingly warm temperatures, the equatorial region is characterized by growth of *rainforest*, or *selva*, described in detail in Chapter 21, a vegetation type unexcelled for luxuriance of tree growth and numbers of species (Figures 21.2, 21.3, 21.4). Broadleaf trees rise to heights of 100 to 150 ft (30–45 m), forming a dense leaf canopy through which little sunlight can reach the ground. Giant *lianas* (woody vines) hang from the trees. The forest is evergreen, although individual species have a rhythm of leaf-shedding.

Rainforest is the home of the small forest animals, of which the monkeys are perhaps the best representatives, taking advantage of the continuous forest canopy for living and traveling. Birds, too, are numerous in species and are spectacularly plumaged.

With a large water surplus and prevailingly high temperatures, chemical processes are active on the rocks and soils in the wet equatorial regions. Leaching out of all soluble constituents of the deeply decayed rock results in a distinctive type of soil, termed a *latosol* (color plate). Reddish or yellowish, and often containing irregular nodules of reddish iron hydroxides, this soil is especially rich in hydroxides of iron, manganese, and aluminum. These have been left behind in the soil after the soluble minerals (including silica) have been carried down through the soil and into the streams and rivers. Large concentrations of the iron, manganese, or aluminum minerals occurring as lenses or layers in the soil are termed *laterite*. These minerals may be extracted as ores of commercial value. *Bauxite* is the foremost ore of aluminum in use today. Manganese ores are likewise of great value, but lateritic iron ore is as yet not so widely exploited because of iron ore abundance in other forms.

Vegetation and climate work hand in hand to make lateritic soils and ore bodies. The soil-forming regime of laterization is discussed in Chapter 18. At the prevailingly warm temperatures, bacteria in the upper soil layer are unusually vigorous and consume virtually all dead vegetation. Thus *humus*, the black, partially decomposed plant matter present in most soils of middle-latitude and arctic climates, is almost entirely absent on well-drained sites. Rotting and decay of bedrock may be very deep in wet equatorial regions of low topographic relief. It is reported that even to depths as great as 300 ft (90 m) the rock has been found to be soft and crumbly from chemical action.

Stream flow tends to be fairly constant and extremely copious because a large water surplus exists throughout much of the year and provides ample runoff (Figure 12.12*A*). River channels are lined along the banks with dense vegetation. Sand bars or sand banks are not so conspicuous as in drier regions. Floodplains have meanders and many swampy sloughs where the river channels have shifted their courses (Figure 15.4). Although water is abundant, river systems such as the Amazon carry relatively little material in chemical solution. This has been explained as resulting from the thorough leaching of soils, which have little soluble mineral matter left to contribute.

Transportation is by rivers whose broad courses can be utilized by dugout canoes or shallow-draft river craft. The use of light aircraft has, however, greatly simplified long-distance transportation. Villages are characteristically built close to the river banks.

Hilly or mountainous belts have very steep slopes, on which flows, slides, and avalanches of soil and rock frequently occur, stripping away all forest and soil down to the bedrock.

Figure 15.4 This air view over the rainforest of the Amazon Basin shows the Rio Negro, a tributary of the Amazon River. The view is southward from a point about 0°23′ S, 64°05′ W. (Photograph by Colonel Richmond, courtesy of the American Geographical Society.)

Of economic importance are several forest products. Rainforest lumber, such as mahogany, ebony, or balsawood, is a valuable tropical product. Quinine, cocaine, and other drugs come from the bark and leaves of tropical plants; cocoa comes from the seed kernel of the cacao plant. Rubber, made from the sap of the rubber tree, is now largely an economic product of Malaya, Sumatra, and Ceylon, although the tree comes from South America, where it was first exploited.

Although most of the wet equatorial climate lies in a belt 10° of latitude on either side of the equator, the Malabar coast of India and the coasts of Burma and Thailand, located between 10° and 25° N lat., have a warm, wet climate with large annual total of rainfall, which also supports rainforest. This Asiatic climate may be considered as a special wet monsoon type. Under the Köppen system it is designated by the symbol *Am*, with the stipulation that the average rainfall of the driest month shall be less than 2.4 in (6 cm). (See Figure 15.20.) Figure 15.5 shows the temperature-

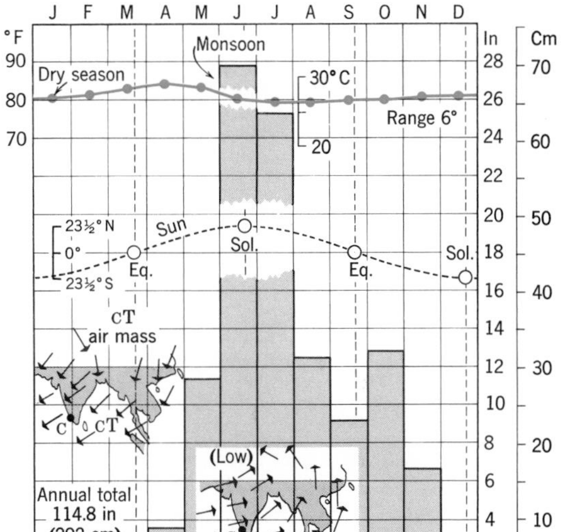

Figure 15.5 Climograph for Cochin, India, 10° N. lat. (Data from Clayton, Smithsonian Institution.)

precipitation graph of a wet monsoon climate. A short dry season occurs in the low-sun season, when the winter monsoon is in effect and continental air masses move southward from interior Asia. This dry period is too brief to deplete soil moisture and ground water reserves. Rainforest therefore thrives. Note that the summer (rainy) monsoon brings enormous quantities of rainfall in June and July. This rain is derived from convectional storms within moist maritime equatorial air masses moving northward from the Indian Ocean.

2. Trade-wind littoral climates (var. of Af and Am)

A study of the world rainfall map, Plate 1, shows that along the east coasts of Central and South America, Madagascar, Indochina, the Philippines, and northeastern Australia, in latitudes 10° to 25°, there are narrow belts of heavy rainfall. These are coasts exposed to moist maritime tropical air masses brought by the tropical easterlies, or trade winds, from the oceanic subtropical highs. These air masses are abundantly supplied with moisture, as is typical of maritime tropical air on the western sides of the high-pressure cells. When these air masses encounter the hill and mountain slopes of the coasts, a heavy orographic precipitation results. Easterly waves cause periods of rainy weather. In addition, the high-sun solstice period sees the onset of tropical cyclones, to which these coasts are vulnerable.

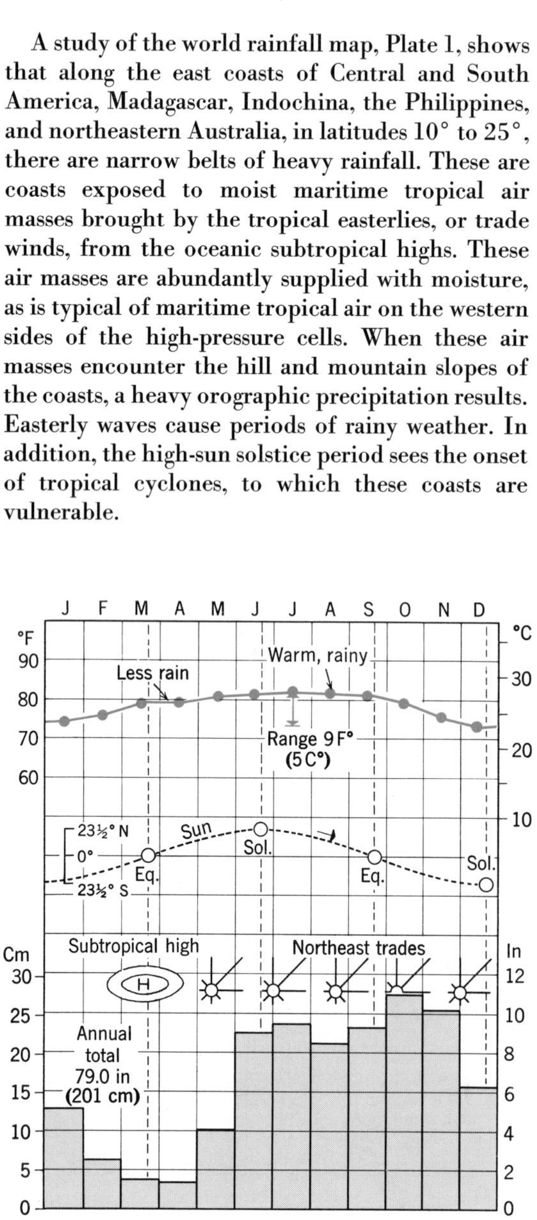

Figure 15.6 Climograph for Belize, British Honduras, 17° N. lat. (Data from Trewartha.)

Because of its coastal position with respect to the easterlies, this climate may be called a *trade-wind littoral climate.* Under the Köppen system it is included with the rainforest climates *Af* and *Am.* It is treated as a separate climate here because the precipitation characteristics differ from the wet equatorial climate, and the temperatures show a somewhat more pronounced annual range.

The temperature and rainfall graph for a representative station of this climate is shown in Figure 15.6. Belize, British Honduras, is located in Central America at 17° N lat. The total rainfall is great, almost 80 in (200 cm) annually, and rain is abundant in most months. A tendency toward dryness, typical of the tropical wet-dry climates (which generally elsewhere lie at this same latitude), is seen in the low rainfall of February, March, and April. The temperature cycle has a range of only 9 F° (5 C°) because of the moderating influence of the nearby ocean but this range is conspicuously greater than that of the wet equatorial climate.

The warm, wet climates of trade-wind coasts support a tropical rainforest vegetation, somewhat similar to the equatorial rainforest.

3. Tropical desert and steppe climates (BWh, BSh)

In strong contrast to the wet equatorial climate are the very dry climates controlled by subsiding, outwardly moving air of the continental high-pressure cells which dominate much of the earth's land areas in the latitude belts 15° to 35°, roughly centered on the Tropics of Cancer and Capricorn. Here lie the source regions of the continental tropical (*cTs*) air masses. The vast deserts of north Africa, Arabia, Iran, and West Pakistan exemplify this climate type, as do also the Sonoran Desert of the southwestern United States and northern Mexico, the interior Kalahari Desert of South Africa, and the Australian Desert.

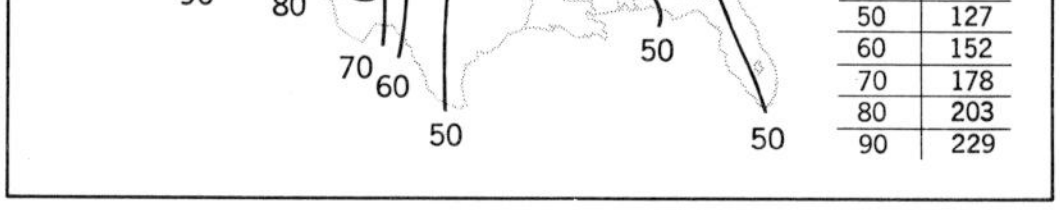

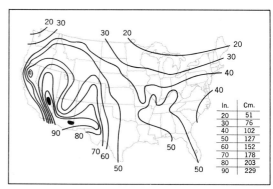

Figure 15.7 Annual evaporation (inches) from a free water surface in the United States. (After Mead.)

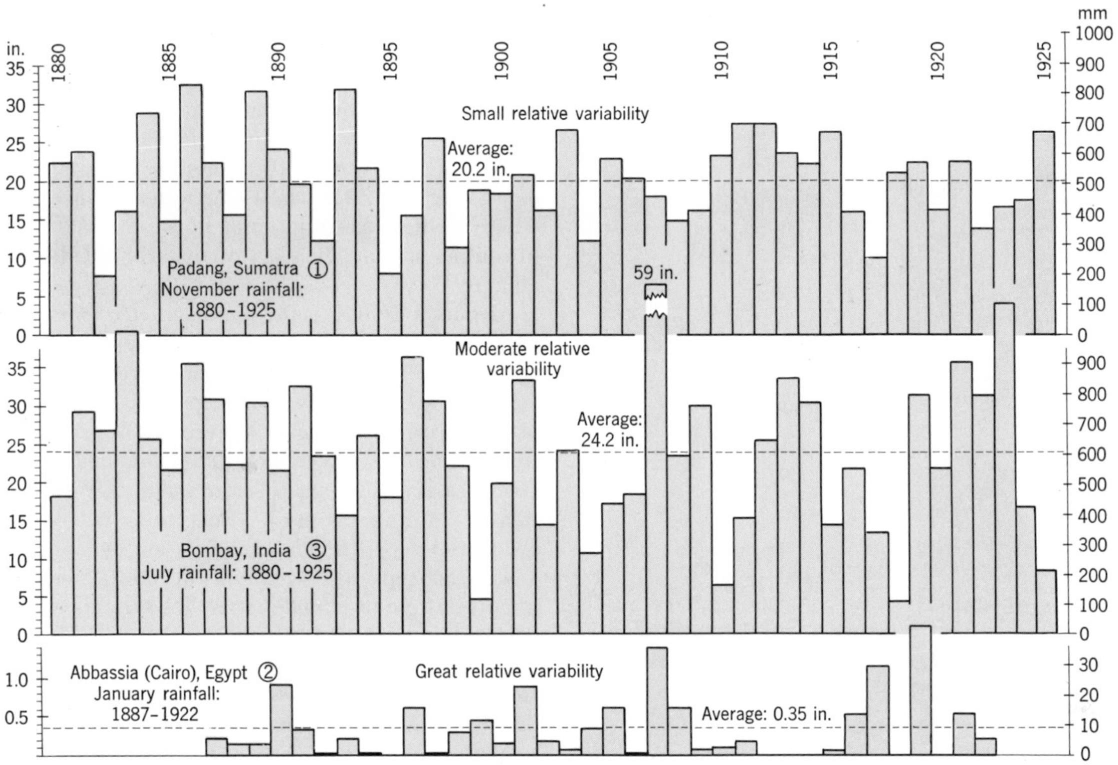

Figure 15.8 These three graphs illustrate the concept of variability of rainfall by showing the actual amount of rain received each year during the month which, on the average, is the rainiest month at that place. (Data from H. H. Clayton, Smithsonian Institution.)

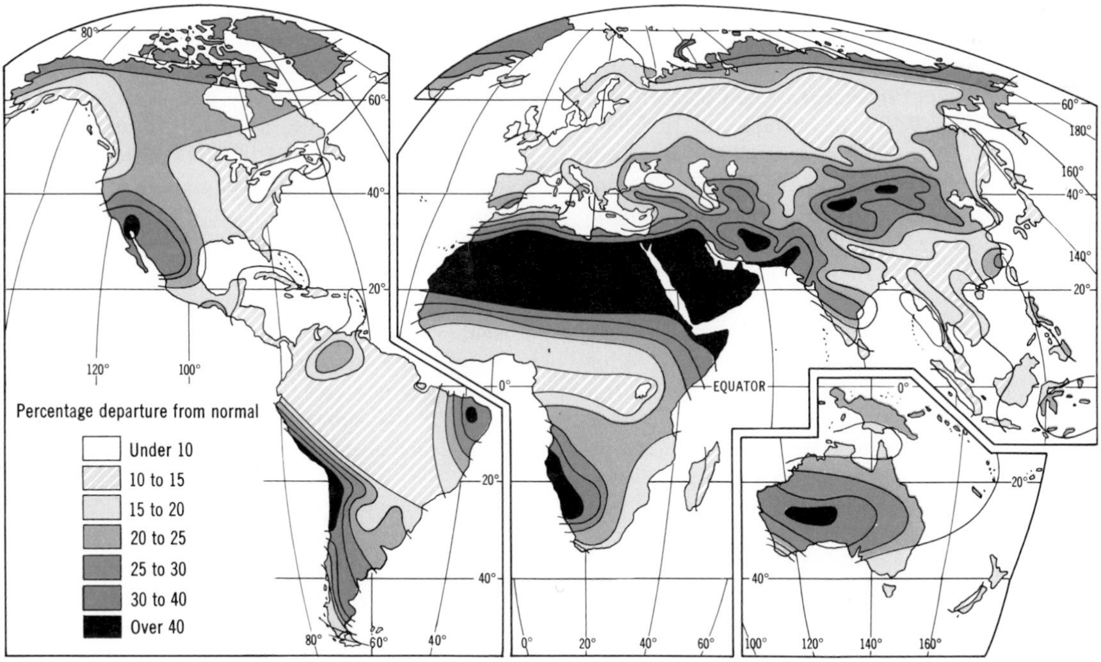

Figure 15.9 Precipitation variability map of the world. (After William Van Royen, 1954, *Atlas of the World's Agricultural Resources*, Prentice-Hall, Englewood Cliffs, N.J. Based on Goode Base Map. Copyright by the University of Chicago. Used by permission of the Geography Department.)

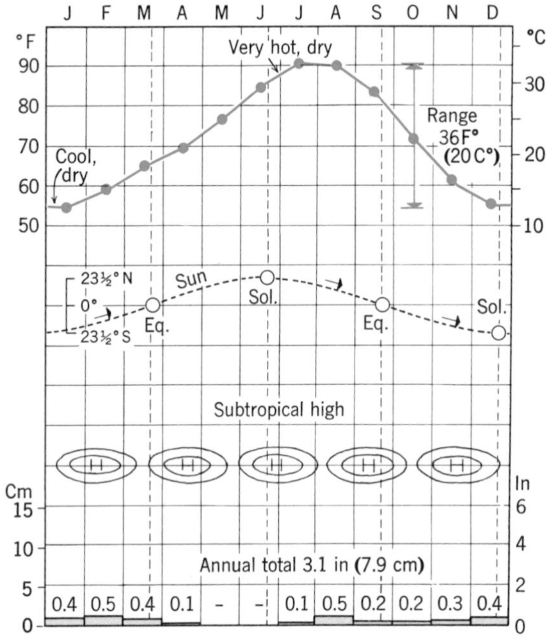

Figure 15.10 Climograph for Yuma, Arizona, 33° N. lat. (Data from Trewartha.)

Within this region of general aridity we can recognize the truly arid zones, or *deserts* (*BWh*), in which annual rainfall is less than 10 in (4 cm), and the semi-arid zones, or *steppes* (*BSh*), in which rainfall is from 10 to 30 in (4 to 12 cm) annually.

In the continental interiors in these tropical regions, far removed from oceanic sources of moisture, extreme aridity prevails. Average annual rainfall is often less than 5 in (2 cm); at some localities a period of several years may pass without measurable rainfall. As an example, at In Salah, Algeria, the average annual rainfall for a 15-year record was 0.6 in (1.5 cm). Other Algerian stations had even less. Because these are regions of subsiding air in the general global scheme of circulation, adiabatic heating reduces the relative humidity of the air to low levels much of the time. At the drier stations in the Sahara Desert, relative humidity at 1:00 P.M. averages 25 to 30 percent for the year, but from 15 to 20 percent in the hottest months. Although in the summer low pressure due to ground-surface heating develops over the tropical landmasses (as seen in Figures 9.12 and 9.13), this is only a low-level condition, with the permanent highs persisting at higher levels of the atmosphere.

The capacity of tropical desert air for evaporation of exposed water surfaces is enormous (Figure 15.7). In the Sonoran Desert, annual evaporation from a free water surface exceeds 90 to 100 in (230 or 250 cm) annually, or about twenty times as much as falls in rain. It is obvious that the full amount of evaporation that is possible does not take place in the tropical deserts. Once the stream channels and soil have become dry after a rain, further evaporation is limited to a small amount of moisture slowly brought to the surface locally by capillary movement from moist soil or rock at depth. Figure 12.12*C* shows the water balance for Alice Springs, Australia, and illustrates well the prevailing water deficiency of the tropical desert climate.

Although dryness is the dominant characteristic of the continental tropical air mass source regions, sporadic heavy rainfall does occur from violent convectional storms. Penetration of maritime tropical or equatorial air may be responsible for such storms. During a single cloudburst confined to an area of a few square miles, the major portion of rainfall of one or more year's total may fall, producing debris floods in the stream channels.

The concept of variability in rainfall is an important one in climate study. By variability we mean the degree to which the rainfall of the individual years differs from the average value computed over a long period of years. Figure 15.8 illustrates this principle. The yearly amounts of rainfall are shown by bars for three stations for a period of many years. At the top the record for Padang, Sumatra, gives data for a wet equatorial climate; the bottom graph gives rainfall for Abbassia, Egypt (near Cairo), a tropical desert station. The middle graph is of a combined climate type (tropical wet-dry climate). We say that the rainfall variability of the wet equatorial climate is small because the individual yearly amounts were, in this case, never more than twice the average or less than a third of the average. On the other hand, Abbassia shows great variability because, although several years had no rainfall at all, five of the years had more than twice the average amount and two of the years had four or more times the average.

Rainfall variability has been estimated for the world in terms of percentage departure from the normal value (Figure 15.9). As would be expected, the tropical desert areas lying near the tropics of Cancer and Capricorn have the highest variability; the equatorial belt of heavy rainfall, the least variability of the low-latitude part of the globe.

Consider next the temperature conditions of the dry continental tropical air mass source regions. Figure 15.10 shows the yearly march of temperatures at Yuma, Arizona, a representative North American station of the tropical desert (*BWh*) climate type. Two points are noteworthy: (*a*) temperatures are very high during the period of high sun; (*b*) annual range is moderately strong. Normally the annual range of temperature (hottest month average minus coolest month average) in these climates is 30 to 40 F° (17 to 22 C°) and is directly related to the height of the sun in the sky.

For those who consider southern Arizona to be a hot desert, the data of Bou-Bernous, Algeria, may be enlightening (Figure 15.11). Here the mean daily temperature averaged over 100°F (38°C) in July—a full 10 F° (5.5 C°) higher than Yuma, Arizona. Notice that the mean value of the highest temperature observed in July is 120°F (48.9°C).

More interesting, perhaps, than annual range is the normal daily range of temperatures in the tropical deserts. Figure 15.12 shows maximum and minimum daily readings for the months of January and July at Phoenix, Arizona. Note that the daily range is often as much as 35 F° (22 C°), the average daily range about 30 F° (17 C°). In no other climate does so great a daily range occur. This is explained by the rapid nightly heat loss from the ground and lower air layers because of the low water-vapor content of the air. On the other hand, insolation during the day is extremely intense and air temperatures soar to great heights. An all-time record range was recorded in Bir Milrha in the Sahara Desert, south of Tripoli, where 31°F and 99°F (−0.6°C and 37.2°C) were recorded on the same day, a range of 68 F° (37.8 C°).[1]

The world's highest temperatures are recorded in certain parts of the continental tropical deserts. A world record maximum of 136.4°F (58°C) was officially observed in the shade under standard shelter at Azizia, Tripoli. Figure 15.13 shows the highest temperatures ever observed in the western United States. The Sonoran Desert region of the lower Colorado River valley clearly surpasses all other areas for high temperatures.

In the extremely dry deserts (*BWh*), virtually the entire land surface appears to be free of vegetation and consists of bare rock, stream gravels and sands, or drifting dune sands. This does not mean that vegetation is wholly absent, rather, that the plants are thinly scattered over the surface and lack foliage to protect or obscure the bare ground. Desert plants are adapted to long drought periods by thick, fleshy leaves and stems, which store water for long periods of time but which do not permit loss of water through the surface. Representative of desert vegetation are the cacti and other shrubs seen in the Sonoran Desert of southwestern United States and northern Mexico (Figure 21.29). Chapter 16 treats the adaptation of plants to a dry (xerophytic) environment; Chapter 21 describes desert vegetation and its distribution.

[1]G. T. Trewartha, *An Introduction to Weather and Climate*, McGraw-Hill Book Co., New York, 1943, p. 367.

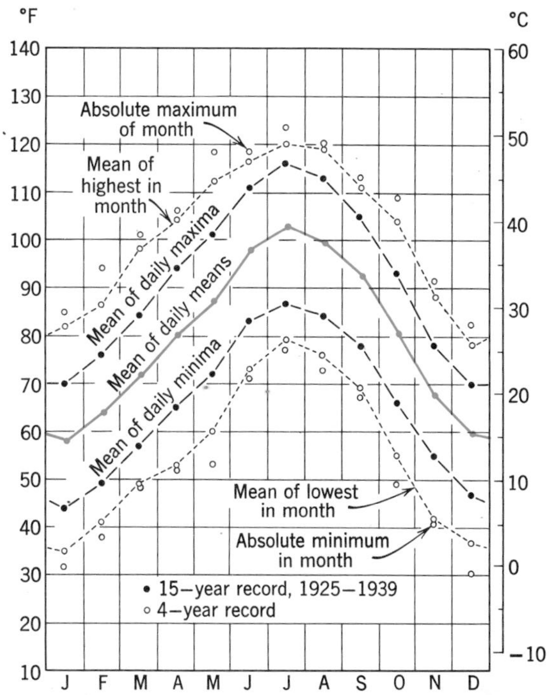

Figure 15.11 Air temperature data for Bou-Bernous, Algeria, 27°17′ N, 02°53′ W, elevation 1509 ft (460m). (Data from Meteorological Office of Great Britain.)

Soils of the deserts are lacking in humus and are of a grayish or reddish color, depending upon the type of iron compound present to produce staining. These soils contain excessive amounts of calcium carbonate and other salts, which are left near the ground surface by evaporating water. In the centers of shallow lakes, the salts concentrate to form white saltflats, entirely sterile and almost perfectly smooth. Soil-forming processes of deserts are discussed in Chapter 18, the great soil groups of deserts in Chapter 19.

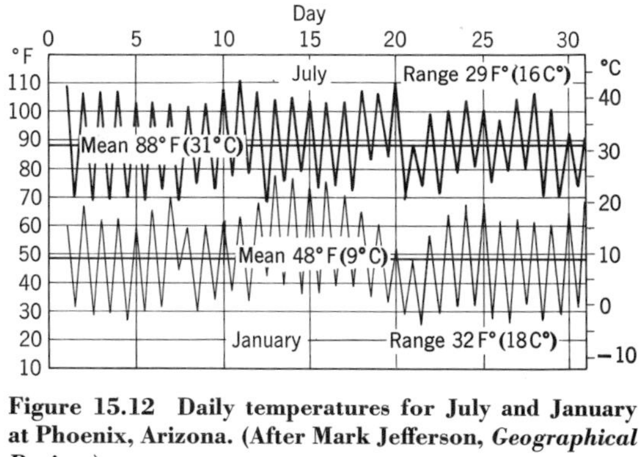

Figure 15.12 Daily temperatures for July and January at Phoenix, Arizona. (After Mark Jefferson, *Geographical Review*.)

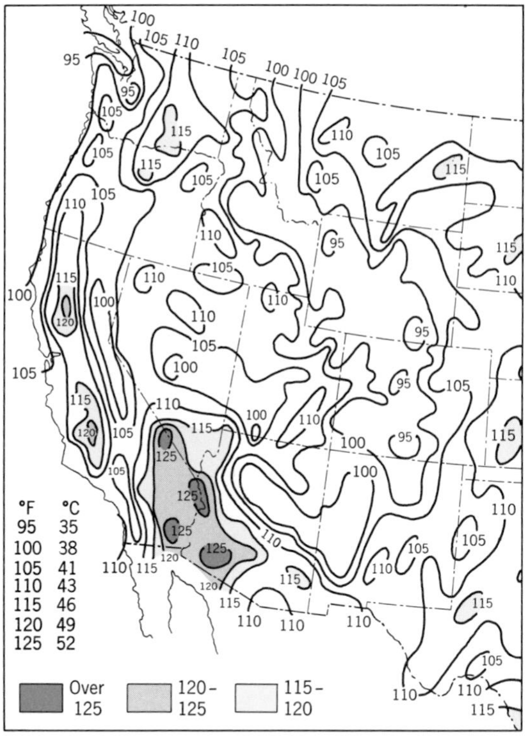

Figure 15.13 The highest temperatures (Fahrenheit) ever observed during the period 1899 to 1938. (After J. B. Kincer, U.S. Dept. of Agriculture, *Yearbook of American Agriculture*, 1941.)

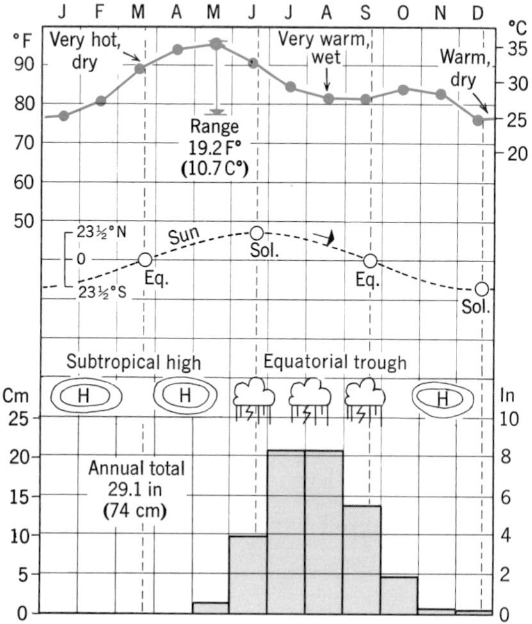

Figure 15.14 Climograph for Kayes, Mali, 14½° N lat. (Data from Trewartha.)

Landforms of mountainous deserts are described in Chapter 26, along with an explanation of stream action and ground water in arid climates.

Tropical steppe climate (*BSh*) borders the tropical deserts on both north and south, and in places on the east as well. Locally because of altitude, plateaus and high plains within what would otherwise be desert have the semi-arid steppe climate. Steppe zones lying equatorward of the deserts are transitional into the tropical wet-dry climate (*Aw*) and resemble it in many ways (Figure 15.14). Steppes on the poleward fringes of the tropical deserts grade into the Mediterranean climate (*Cs*) in many places. Steppes typically are grasslands of short grasses and other herbs, and with locally developed shrub and woodland (see Chapter 21). These areas are able to support limited numbers of grazing animals but are not generally moist enough for crop cultivation without irrigation. Soils are commonly of the *brown soil* and *chestnut soil* groups, containing some humus (Chapter 19).

4. West coast desert climate (BWk, BWh)

Again referring to the world rainfall map, Plate 1, it will be seen that all west coasts in latitudes 15° to 30° are extremely dry, generally with less than 10 in (25 cm) of rainfall annually. The Atacama desert of Chile and the Namib Desert of coastal southwest Africa are perhaps the most celebrated of these deserts, but they exist also in Lower California, the Morrocan coast of Africa, and the west coast of Australia. The arid belt extends continuously eastward to inland continental tropical deserts.

Does it not seem strange that extreme dryness exists immediately along the shores of the oceans, close to possible sources of moist maritime air masses? The interior tropical deserts, already discussed, are logically explained by land-centered high-pressure cells into which moist air masses cannot easily drift, but the dry west coast strips are located between the oceanic and continental high-pressure cells where we might expect to find some development of fronts and convergence of air masses. The key to this problem seems to lie in the fact that the oceanic subtropical high-pressure cells are inherently dry on their east sides. The circulation in these cells is thought to be such that the air on the east sides is subsiding as it moves outward, hence, is adiabatically heated and its humidity reduced. The result is dry, stable air masses that bring an arid zone not only to the coast but extending far seaward as well. Strangely enough, there are dry deserts over the oceans in these tropical latitudes (Figure 15.15).

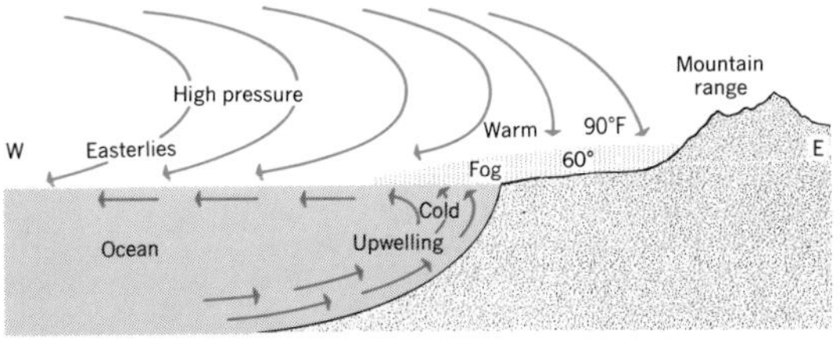

Figure 15.15 Subsiding air above a coastal temperature inversion. (From A. N. Strahler, 1971, *The Earth Sciences*, 2nd ed., Harper and Row, New York.)

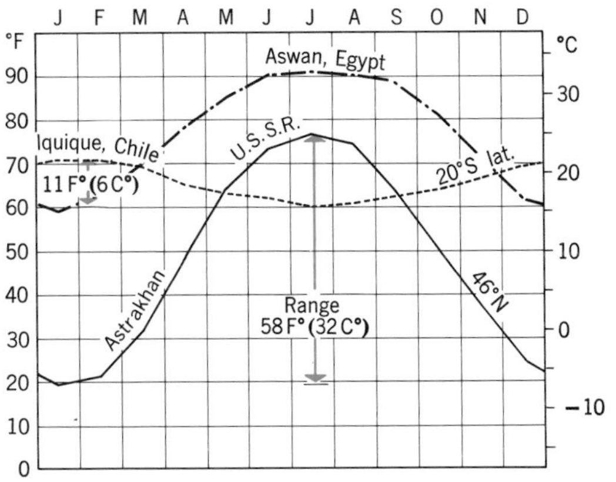

Figure 15.16 Temperature conditions compared for coastal and interior tropical desert locations. (Data from H. H. Clayton, Smithsonian Institution, and Trewartha.)

In what respects do the dry west coasts differ in climate from the interior continental deserts which they adjoin? The principal difference is in temperature. The coastal deserts are relatively cool, with annual average temperatures around 65°F (18°C), whereas the interior continental deserts average some 10 F° (5 C°) higher. The presence of cool upwelling and equatorward-flowing currents, such as the Humboldt and Benguela currents, explains the lowered temperatures. (See Chapter 9.) The annual range of temperature in the coastal deserts is very low, as illustrated in Figure 15.16 by the temperature graph of Iquique, Chile (20° S), on the dry Atacama Desert coast of South America. The annual range is only 11 F° (6 C°). In contrast, the temperature cycle of Aswan, Egypt (24° N), shows both a higher average and a greater range.

Although the cool west coast desert climate is treated here as a separate climate from the tropical desert climate, not all climate classifications make the distinction. In the original Köppen classification, the cooler desert west coasts are designated by the symbol *BWn*, in which *n* means *frequent fog* (from the German, *nebel*, meaning fog). Persistent coastal fog banks form in the cool lower air layer overlying the cool ocean water (Figure 15.15). In the latest versions of the Köppen climate system, the symbols *BWh* and *BWk* are applied to the west coast deserts. The *BWk* climate is limited to the South American and southwest African coasts in latitudes 20° to 32° S, where the cool ocean currents are most influential.

Vegetation and soils of the cool west coast deserts are essentially similar to those of the interior deserts. An unusually high incidence of fog on these coasts leads to growth of some specialized plants which can exist on condensed moisture close to the shore.

5. Tropical wet-dry climate (Aw, also Cwa)

We have thus far considered two extreme climate types. One, a wet climate lying in the equatorial zone, the other a desert climate arranged in two belts along the Tropics of Cancer and Capricorn. What of the intermediate zones where these two climate types come together? Knowing from previous study that the pressure and wind belts of the globe migrate northward at June solstice season and southward at December solstice season, we can infer that the intermediate zones in question will have climates that combine the characteristics of the first two types in a seasonal alternation. This results in the tropical wet-dry (*Aw*) climate, which

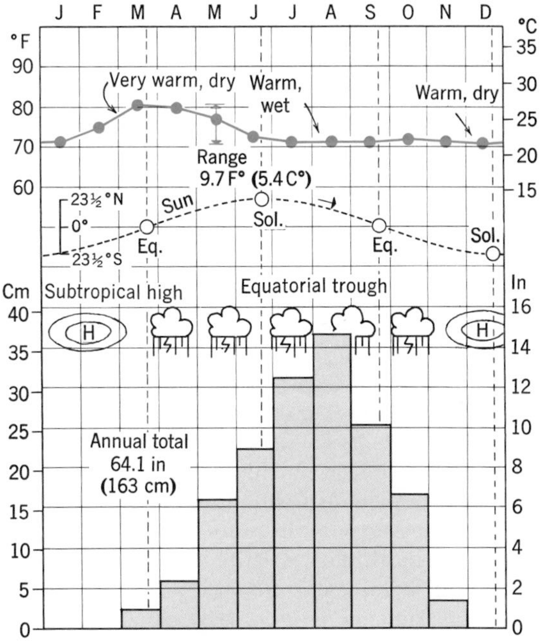

Figure 15.17 Climograph for Timbo, Republic of Guinea, 10°40′ N lat. (Data from Trewartha.)

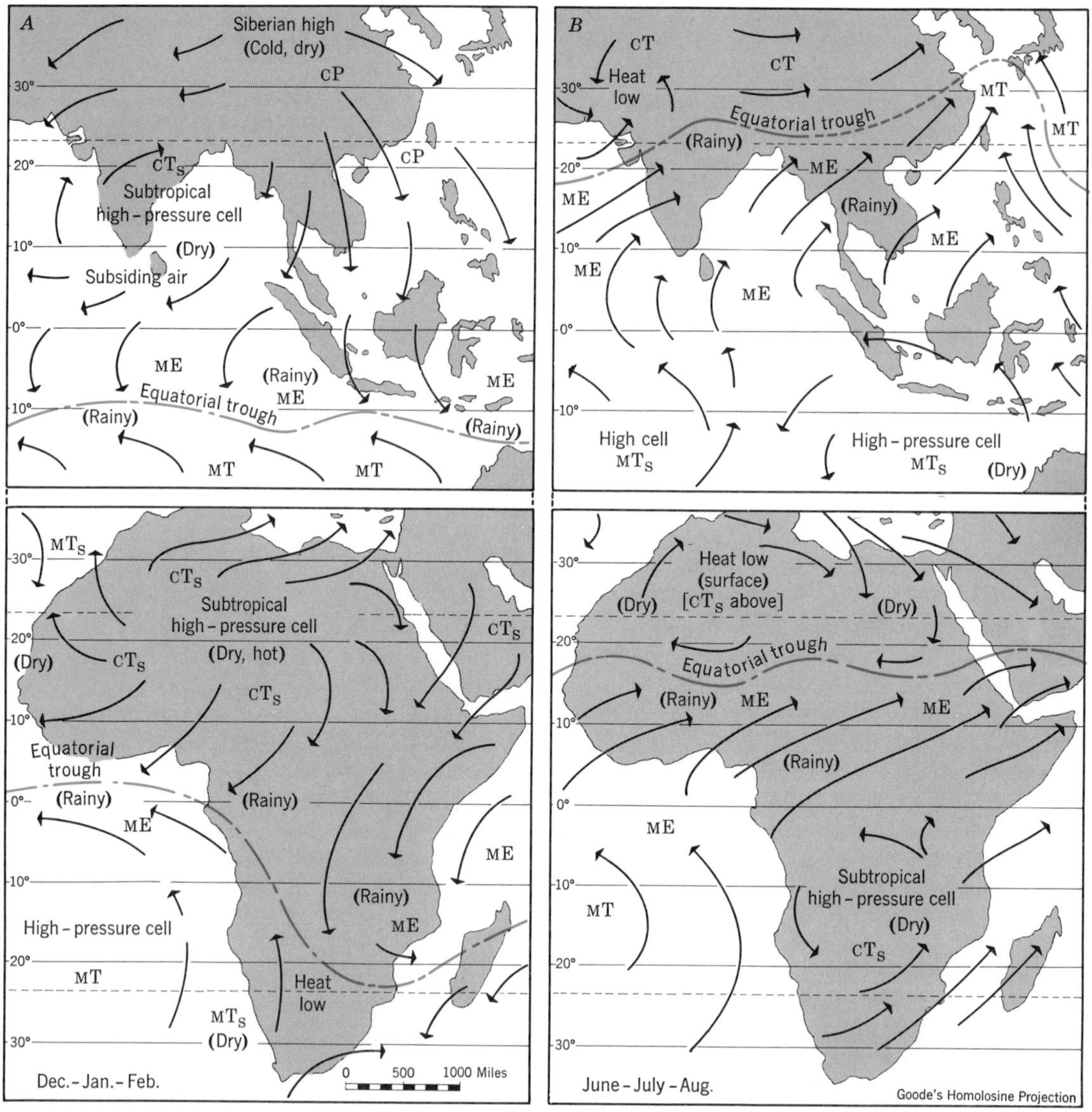

Figure 15.18 These maps of Africa and southern Asia illustrate one interpretation of the air-mass source regions and circulation patterns that govern the equatorial and tropical climates. (Modified after M. A. Garbell, *Tropical and Equatorial Meteorology*, New York, 1947. Based on a Goode Base Map. Copyright by the University of Chicago. Used by permission of the Department of Geography.)

has a wet season controlled by moist, warm equatorial and maritime tropical air masses at time of high sun and a dry season controlled by the continental tropical air masses at time of low sun.

The latitude belts in which tropical wet-dry climate is found lie roughly between 5° and 25° latitude, throughout Central and South America, Africa, and Australia. In southeast Asia this zone is pushed northward to latitudes 10° to 30° because the continental tropical air mass source region in summer is necessarily situated farther north to conform to the Asiatic landmass.

Climate characteristics of the tropical wet-dry climate can be judged from Figure 15.17, which gives temperature and rainfall data for a representative west African station. This can be studied in conjunction with Figure 15.18, which shows seasonal air mass and flow patterns. Timbo lies closer to the equator than Kayes, representing the tropical steppe (*BSh*) climate (Figure 15.14), hence has a longer wet season and a more even temperature cycle. Note especially the fact that maximum temperatures occur in March, April, or May, rather than in July, because the onset of the rains brings a cloud cover and cooler air temperatures. The soil-moisture regime is illustrated by Calcutta, India (Figure 12.12*B*). Seasonal alternation of moisture deficiency and water surplus is striking.

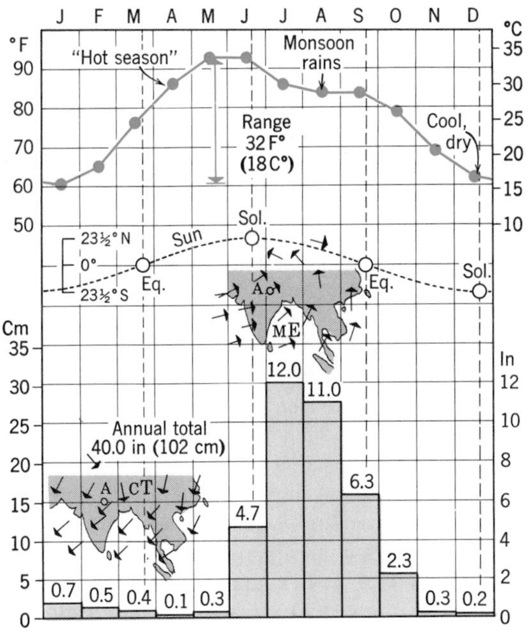

Figure 15.19 Climograph for Allahabad, India, 25° N lat. (Data from H. H. Clayton, Smithsonian Institution.)

The Asiatic tropical wet-dry climate is somewhat different in that the monsoon controls are strong and bring an extreme contrast of wet and dry conditions. If we now look at the record of Allahabad, India, 25° N lat. (Figure 15.19), the basic similarity to Kayes and Timbo in western Africa (Figures 15.14 and 15.17), is very marked. The higher latitude of Allahabad results in much cooler January temperatures, and a greater annual range, but otherwise the temperature curves show the same dip during the rainy season. Köppen places Allahabad and a large belt of southeastern Asia in latitudes 20° to 25° in the *Cwa* climate (temperate, rainy climate, dry winter, hot summer). The *Cw* climate is a poleward extension of the *Aw* climate both here and in Africa and South and Central America. The two are here treated under the tropical wet-dry climate.

The Köppen system provides an exact basis for the distinction between *Am* and *Aw* climates, both of which have a dry season. The *Am-Aw* boundary varies according to both total annual rainfall and driest-month rainfall, as shown by a graph (Figure 15.20). In the annual rainfall range between 40 in (100 cm) and 100 in (250 cm) the *Am* climate can exist with progressively lower values of driest-month rainfall, beginning with 2.4 in (6 cm) and declining to zero. Thus an *Am* station might have one month with no rainfall whatsoever, provided it has an annual total of more than 100 in (250 cm).

The contrast in wet and dry season rainfall in the tropical wet-dry climate is further illustrated in Figure 15.21. Here the rainfall of the one rainest month is compared with the total for the three consecutive driest months for a number of typical stations. Rainfall is not so reliable in the tropical wet-dry climate as in the wet equatorial climate but is more reliable than in the very dry tropical deserts. This can be seen from the record of Bombay, India, in Figure 15.8, and is also apparent from the world variability map, Figure 15.9.

Alternation of wet and dry seasons results in the growth of a distinctive vegetation known generally as the *tropical savanna.* This is characterized by open expanses of tall grasses, interspersed with hardy, drought-resistant shrubs and trees (Figure 21.20). Other areas have savanna woodland, monsoon forest, thornbush, and tropical scrub, all of which are formation classes grouped on Plate 4 under the heading "Raingreen." These formation

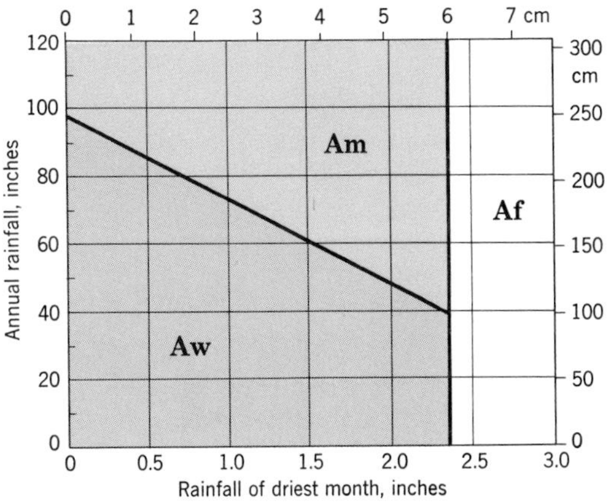

Figure 15.20 This graph shows how *Aw* climates are distinguished from *Am* climates in the Köppen system. (After Haurwitz and Austin.)

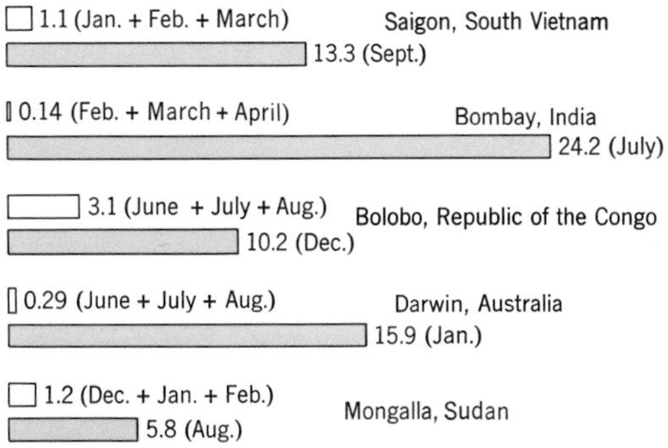

Figure 15.21 Rainfall (inches) of the one wettest month is shown contrasted with the total rainfall of the three consecutive driest months for five wet-dry tropical climate stations. (Data from Clayton and Trewartha.)

classes are described in Chapter 21. Grasses in this climate are dried to straw and many tree species shed their leaves in the dry season. Other trees and shrubs have thorns and small leaves or hard, leathery leaves that resist water loss.

Soils of the wet-dry tropical climates are mostly yellowish or reddish latosols, similar to those described in connection with the wet equatorial climate. (See Chapter 19.) Excessive leaching is again the result of the heavy rainfall and high temperatures. In general, these residual soils of uplands are not fertile and are little cultivated in South America and Africa. Locally, rich floodplain alluvium is highly productive. Stream flow in these regions contrasts greatly with that of the wet equatorial climate, in that the former has a very strong seasonal fluctuation. From flood conditions with extensive low-lying areas under water in the rainy season, the streams pass to a regime of little or no flow in the dry season, when channel bottoms of sand and gravel are exposed and mud flats dry.

Closely related to vegetation and climate is the natural animal life of the savanna grasslands and woodlands. These are the regions of the carnivorous game animals and a vast multitude of grazing animals on which they feed. The grasslands of Africa are the natural home of such herbivores as wildebeest, gazelle, deer, antelope, buffalo, rhinoceros, zebra, giraffe, and elephant. On them feed the lion, leopard, hyena, and jackal. Some of the herbivores depend upon fleetness of foot to escape the predators. Others, such as the rhinoceros, buffalo, and elephant, defend themselves by their size, strength, or armor-thick hide. The giraffe is a peculiar adaptation to savanna woodlands; his long neck permits browsing upon the higher foliage of scattered trees.

The dry season brings a severe struggle for existence to animals of the African savanna. As streams and hollows dry up, the few muddy waterholes must supply all drinking water. Danger of attack by carnivores is greatly increased.

The Indian savanna, woodland, and thornforest have a somewhat similar assemblage. Deer and antelope are especially abundant, with some water buffalo and a few rhinoceroses. The tiger replaces the lion as the principal carnivore. The Indian elephant, however, is largely restricted in natural habitat to the rainforest coastal strips of Burma, the Malabar coast, and Ceylon.

REVIEW QUESTIONS

1. What conditions of temperature, pressure, and rainfall prevail in the equatorial belt? What air masses are present? Under what conditions does heavy precipitation occur? How does the equatorial trough change position throughout the year?

2. Describe the annual temperature curve for a typical station in the wet equatorial (*Af*) climate zone. What range is usually found? How does the daily temperature range compare with the annual range?

3. Approximately how great is the average annual rainfall of the wet equatorial climate? How is rainfall distributed throughout the year? Is variability of rainfall great or small?

4. What is rainforest? Of what types of vegetation is it composed? (See Chapter 21.)

5. What is the lateritic soil (latosol)? What processes are responsible for development of lateritic soils? Of what economic importance are laterites? (See Chapter 19.)

6. Is there much humus in lateritic soils (latosols)? Explain how temperature and bacterial activity affect the accumulation of decomposed vegetation in the hot equatorial regions as contrasted with cold regions.

7. Is the chemical decay of rock relatively small or great in the wet equatorial climate? Explain.

8. What can be said about stream flow in the wet equatorial climate? In what way are rivers of importance to native inhabitants of these regions?

9. In what way does the wet, monsoon-dominated climate of the southern Asiatic coasts (*Am*) differ from that close to the equator (*af*)?

10. Why is there heavy precipitation on east coast belts in latitudes 10° to 25°? What climate occurs here? What type of climate occurs at inland locations in this same latitude belt?

11. What causes the tropical deserts in latitudes 15° to 35°? Explain these deserts in terms of the global pattern of air circulation. What kind of air mass has its source over these areas?

12. Distinguish between desert and steppe climates. What Köppen symbols denote these climates?

13. Discuss the evaporation of moisture in the tropical deserts. How great is evaporation from a free water surface? What influence has this factor upon streams and the ground water table?

14. What type of rainfall occurs in the tropical deserts? Is it widespread or localized? Is rainfall variability in these deserts great or small?

15. What is the nature of the annual temperature cycle in the tropical deserts? Explain these characteristics. What is the nature of the daily cycle of temperature? How does it compare with that of the wet equatorial climate?

16. What kind of vegetation is native to the tropical deserts? To what extent does it afford cover to the ground? What kinds of soils are typical of these deserts?

17. Explain the west coast deserts in latitudes 15° to 30°. How are these related to the subtropical high pressure cells? Why is the air dry in these locations?

18. What is the outstanding feature of the annual temperature cycle of the west coast deserts? Why is the range small? Why are the temperatures abnormally low for this latitude? Why is fog abundant along these coasts?

19. Explain the characteristic seasonal features of the wet-dry tropical climate in terms of air masses and seasonal shifting of the pressure and wind belts of the globe.

20. How do annual precipitation and its distribution throughout the year change if we start with the wet equatorial climate zone and examine the rainfall records for stations progressively farther north until the Sahara Desert is reached?

21. Describe the annual temperature cycle of the tropical wet-dry climate. When is the maximum reached? Why does this not coincide with the time of highest sun?

22. In what way does the Asiatic wet-dry tropical climate differ from that of Africa or South America? Explain. How does Köppen distinguish the climate of north India from the climate of peninsular India?

23. What is the natural vegetation of the wet-dry tropical climate? What are the characteristics of the wild animal life of these regions? What type of soil is generally present? What effect has the seasonal contrast of dry and wet periods upon stream flow?

Exercises

1. Prepare a rainfall-temperature graph (similar to Figure 15.1) for each of the following stations. (Data from Trewartha and Meteorological Office of Great Britain.)

(*a*) *Belém, Brazil,* 1° S.

	J	F	M	A	M	J	J	A	S	O	N	D
Temperature, °F	77.7	77.0	77.5	77.7	78.4	78.3	78.1	78.3	78.6	79.0	79.7	79.0
Precipitation, in.	11.6	12.9	14.9	12.1	9.4	6.7	6.2	4.5	3.5	2.8	2.6	6.0

(*b*) *Cairo, Egypt, U.A.R.,* 30° N.

	J	F	M	A	M	J	J	A	S	O	N	D
Temperature, °F	56	58.5	63.5	70	77	81.5	83	83	79	75.5	68	59
Precipitation, in.	0.2	0.2	0.2	0.1	0.1	$<$0.1	0.0	0.0	$<$0.1	$<$0.1	0.1	0.2

2. For each of the above stations calculate: **(*a*)** Average annual temperature (add monthly mean temperatures and divide by 12); **(*b*)** annual temperature range (difference between means of warmest and coolest months); and **(*c*)** average annual total rainfall (add monthly rainfall figures).

3. (*a*) To what climate does each of the stations in Exercise 1 belong? **(*b*)** Study the data of Belém. If maximum rainfall occurs at times when the sun is highest in the sky, what months should show a maximum rainfall? **(*c*)** Why does this principle not hold true for Belém? **(*d*)** Explain the marked maximum of rainfall in January–April and a minimum period April–November.

4. (*a*) How does it happen that Cairo, a great city of over three million population, lies in an extremely dry, tropical desert? **(*b*)** Name other great cities that lie in true desert climates.

5. What soil-moisture regime is associated with each of the stations in Exercise 1? (See Chapter 12.) Belém has a water surplus of about 50 in (1270 mm); Cairo has a soil-moisture deficiency of about 45 in (1150 mm).

6. Prepare a rainfall-temperature graph similar to Figure 15.1 for each of the following stations. (Data from Trewartha and Meteorological Office of Great Britain.)

(*a*) *Akyab, Burma,* 20° N.

	J	F	M	A	M	J	J	A	S	O	N	D
Temperature, °F	70	72.5	78	82.5	84	81.5	80.5	80.5	81.5	81.5	78	72
Precipitation, in.	0.1	0.2	0.4	2.0	15.4	45.3	55.1	44.6	22.7	11.3	5.1	0.7

(*b*) *Cuyaba, Brazil,* 15½° S.

	J	F	M	A	M	J	J	A	S	O	N	D
Temperature, °F	81	81	81	80	78	75	76	78	82	82	82	81
Precipitation, in.	9.8	8.3	8.3	4.0	2.1	0.3	0.2	1.1	2.0	4.5	5.9	8.1

(*c*) *Benares, India,* 25° N.

	J	F	M	A	M	J	J	A	S	O	N	D
Temperature, °F	60	65	77	87	91	89	84	83	83	78	68	60
Precipitation, in.	0.7	0.6	0.4	0.2	0.6	4.8	12.1	11.6	7.1	2.1	0.2	0.2

7. **(*a*)** Determine average annual temperature, annual temperature range, and annual precipitation for each of the stations in Exercise 6. **(*b*)** To what climate does each station belong?

8. **(*a*)** In what way does the temperature cycle at Akyab, Burma, differ from that of Cuyaba, Brazil? **(*b*)** Explain this difference. **(*c*)** What is the reason that Akyab receives so much more rainfall than Cochin, India (Figure 15.5), although the two are both west-coast Asiatic stations subject to similar monsoon conditions?

9. **(*a*)** Comparing the two tropical wet-dry climates for which data are given, why does the rainfall of Cuyaba, Brazil (Exercise 6), show a minimum in the period June–August, whereas the minimum for Timbo (Figure 15.17), occurs in the period December–March? **(*b*)** Of these two stations, which has the drier dry season? Which has the wetter wet season? **(*c*)** Of these two stations, does the one with greater seasonal rainfall contrast also have the greater annual temperature range? Explain.

10. What soil-moisture regime is associated with each station in Exercise 6? (See Chapter 12.) Akyab has a water surplus of about 150 in (3800 mm) and a soil-moisture deficiency of about 7 in (175 mm). Cuyaba has a water surplus of about 4 in (105 mm) and a soil-moisture deficiency of about 7 in (170 mm). Benares has a soil-moisture deficiency of 17 in (440 mm) and no water surplus.

11. Prepare a rainfall-temperature graph similar to Figure 15.1 for the following station. (Data from Meteorological Office of Great Britain.)

Port Nolloth, S. Africa, 29° S.

	J	F	M	A	M	J	J	A	S	O	N	D
Temperature, °F	60	60.5	60	58	57	55.5	53.5	54	55	56.5	58.5	59.5
Precipitation, in.	0.1	0.1	0.2	0.2	0.3	0.3	0.3	0.3	0.2	0.1	0.1	0.1

12. **(*a*)** Compare the data of the desert stations in Exercises 1 and 11. **(*b*)** What two varieties of tropical desert do they represent? **(*c*)** Why has Port Nolloth such a small annual temperature range? **(*d*)** Is rainfall greater or less during the warmer months than in the cooler months for these two stations? Explain.

13. What soil-moisture regime is associated with the climate of Port Nolloth? This station has a soil-moisture deficiency of about 25 in (640 mm).

CHAPTER **16**

Middle-Latitude Climates

CLIMATES of group II are those of middle latitudes occupying the polar front zone in which both tropical and polar air masses play an important part. The latitude belt in which these climates lie is subject to cyclonic storms; most of the precipitation in these climates occurs along fronts within these cyclones.

6. Humid subtropical climate (Cfa)

The moist nature of the west sides of the oceanic subtropical high-pressure cells has already been discussed in Chapter 14. Maritime tropical (mT) air in this part of the anticyclonic cell is engaged in a slight ascending motion combined with its flow toward higher latitudes. The air mass has a steep lapse rate. Moisture that enters the air mass by evaporation from the warm ocean surface is distributed to great heights. As this moist, unstable maritime tropical air mass moves over the eastern continental coasts in latitudes 25° to 35° and drifts inland, it brings the necessary moisture and latent heat energy for heavy precipitation. Lifting occurs along warm and cold fronts where the tropical air encounters polar air. This general pattern of climate, here called the *humid subtropical climate*, is exemplified by the southern Atlantic and Gulf Coast states of the United States; corresponding regions are found in the Argentina-Uruguay-southern Brazil area of South America, in eastern China and southern Japan, along a small part of the southeastern coast of Africa, and on the east Australian coast.

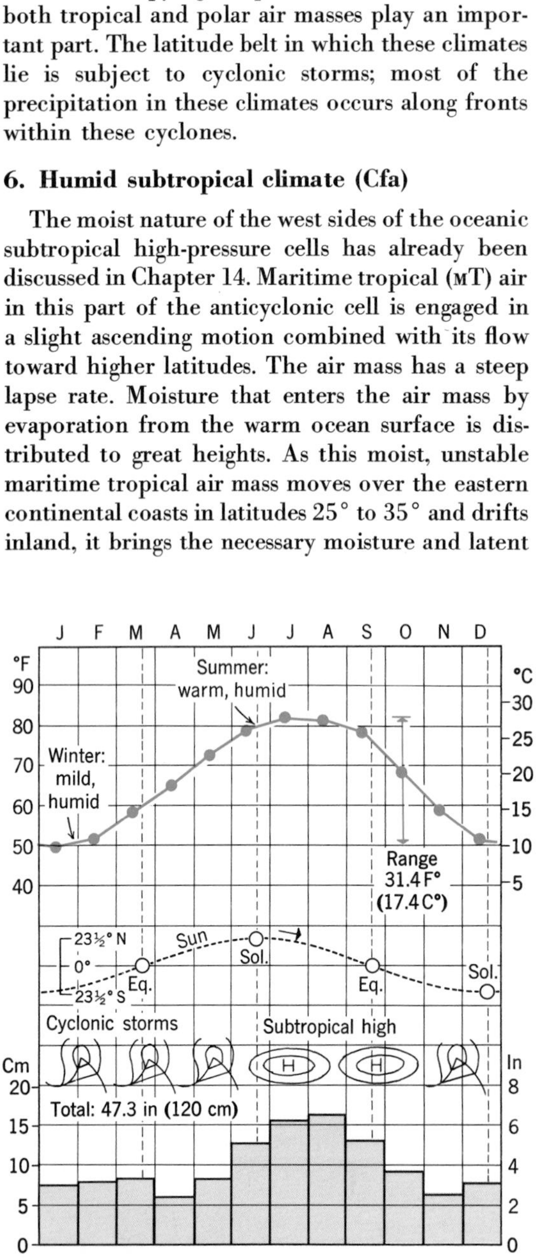

Figure 16.1 Climograph for Charleston, South Carolina, 33° N lat. (Data from Trewartha.)

In the Köppen system, these areas lie in the *Cfa* climate, described as a temperate rainy climate with hot summers. Temperature specifications of the *C* climates have been stated in Chapter 14. The *Cfa* climate has no dry season, and even the driest summer month receives more than 1.2 in (3 cm) of rain. The hot summer specified by the letter *a* is one in which the average temperature of the warmest month is over 71.6°F (22°C).

For an analysis of the temperature and precipitation characteristics of this moist, subtropical climate, we may refer to the record for Charleston, South Carolina (Figure 16.1). Rainfall is ample at all times of the year, but is distinctly greater during the summer when the oceanic high strengthens and the flow of maritime tropical air is increased. Thunderstorms are especially frequent in summer (Figure 16.2). They may be of the thermal (heat) type, or of squall-line or cold front origin. An occasional tropical cyclone may strike the coastal area, bringing very heavy rains. Winter precipitation, some of it in the form of snow, is of frontal type in the frequent middle-latitude cyclones that sweep over these regions. Not all stations in the humid subtropical climate show a summer maximum. For example, Atlanta, Georgia, has a winter maximum, but a secondary peak in midsummer. Typically the annual precipitation totals more than 40 in (100 cm).

Temperatures show a moderately strong range, of very much the same magnitude as in the tropical deserts, but without the extreme heat in summer. Humidity is, of course, very high, in marked contrast to the deserts, and summer climate on the humid east coasts is at times similar to the wet equatorial climate in temperature, rainfall, and

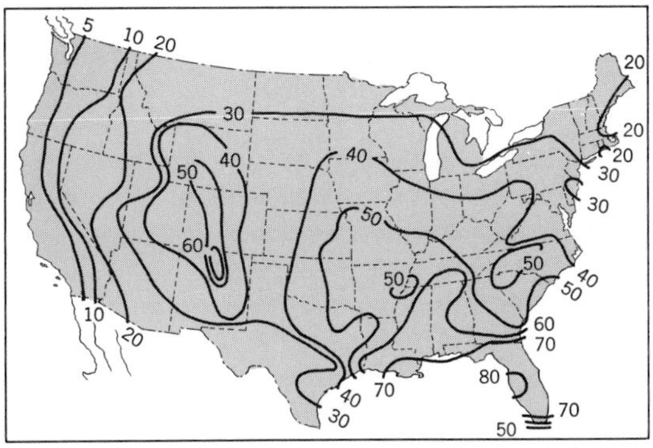

Figure 16.2 Average annual number of days with thunderstorms, based on records of the period 1899 to 1938. (After J. B. Kincer, U.S. Dept. of Agriculture.)

humidity. Winters show the influence of outbursts of polar air masses, which frequently penetrate into subtropical latitudes and bring below-freezing weather with killing frosts. We might say that this climate type is shared to some extent by both tropical and polar air masses, but that the tropical air masses prevail most of the time and dominate in summer.

In southeast Asia the humid subtropical climate is somewhat modified by intensive monsoon development. Winter air masses from interior Asia are very dry, and a winter scarcity of precipitation develops, whereas maritime air masses in summer, together with occasional typhoons, cause a strongly accentuated maximum in the summer. This shows clearly in the precipitation cycle of Shanghai, China (Figure 16.3), which has a winter both drier and cooler than that of Charleston.

A large winter water surplus characterizes the soil-moisture balance of the humid subtropical climate, while by comparison, the summer moisture deficiency is typically very small.

Soils of the moister, warmer parts of the humid subtropical regions are strongly leached red-yellow soils related to the latosols of the humid tropical and equatorial climates. Rich in iron and aluminum oxides, these soils are poor in many of the plant nutrients essential for successful agricultural production.

Forest is the natural vegetation of most of the areas having the humid subtropical climate. Plate 4 shows temperate rainforest to be the dominant type, but with some areas of tropical rainforest occupying the coastal belts in lower latitudes. Much of the sandy coastal region of the southeastern United States today has a second growth forest of longleaf, loblolly, and slash pine, whereas the inland region has summer-green deciduous forest. (These forest types are treated in Chapter 21.) Toward higher latitudes forest gives way to tall-grass prairie such as the *Pampa* of Argentina and Uruguay, and the prairies of Oklahoma and Missouri. Here the soils are of dark prairie and chernozem groups, typical of a dry continental climate. These grasslands occupy regions that are *subhumid*, i.e., best considered transitional to the semi-arid steppes, for they have less than 40 in (100 cm) annual precipitation and a marked dryness of the winter season.

7. Marine west coast climate (Cfb, Cfc)

Shifting attention now toward higher latitudes, and using the information about air masses, fronts, and cyclones presented in Chapter 11, we might expect to find that western coasts in the belt of cyclonic storms would receive ample rainfall from maritime polar air masses and would have rather moderate temperature variations because of the proximity to the oceans from which the air masses tend to drift landward.

These conditions are fulfilled in west coasts of landmasses lying between 40° or 45° and 60° latitude. Situated too far poleward to be dominated by the dry influence of the subtropical oceanic high-pressure cells, these climates lack a dry season. Because continental polar air masses tend to drift eastward, they rarely move westward to visit the west coasts; hence severe dry-cold conditions are uncommon.

Köppen classifies the marine west coast climate as *Cfb*, a temperate rainy climate with warm summers. The average temperature of the warmest month is under 71.6°F (22°C) and at least four

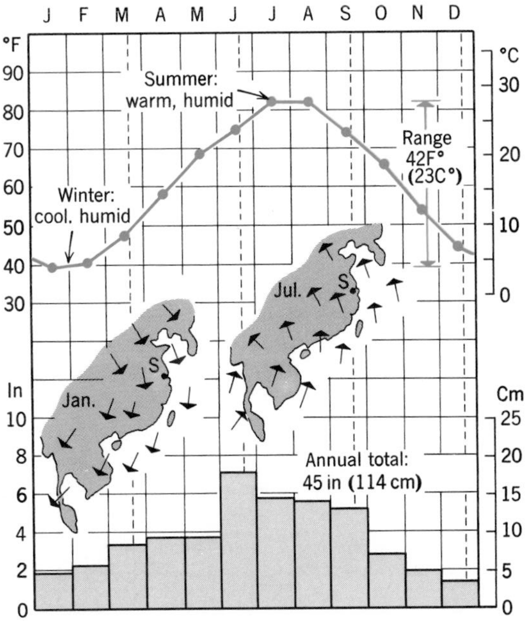

Figure 16.3 Climograph for Shanghai, China, 31° N lat. (Data from Meteorological Office of Great Britain.)

months average 50°F (10°C) or more. The cool summer variety *Cfc* has fewer than four months with averages over 50°F (10°C).

Climate of the middle-latitude west coasts is illustrated by the graphs for Brest, France (Figure 16.4), located at 49° N lat. on the Brittany coast. Precipitation is well distributed throughout the year, but shows a distinct reduction during the summer months. This feature is also found on rainfall graphs for coastal stations on the Pacific northwest coast of the United States. Why should this reduction occur? In summer, the oceanic subtropical high is most strongly developed and moves farthest north, bringing its arid influence to bear just enough to make a distinct decrease in summer rainfall. In other words, this is a manifestation of the same mechanism that causes a west coast desert in low latitudes.

Although the total rainfall for Brest and similar marine west coast stations over much of Europe is not great, judged by tropical or equatorial standards, the cooler air temperatures reduce evaporation and produce a very damp, humid climate with much cloud cover. Nearness to the ocean makes for a small annual temperature range, just as it does in lower latitudes on the west coast deserts. Mild winters and relatively cool summers are characteristic. Winters which are severely cold at the same latitudes in mid-continent and east-continent positions are by contrast surprisingly mild on the west coasts.

The soil-moisture budget of the marine west coast climate shows a strong tendency to follow

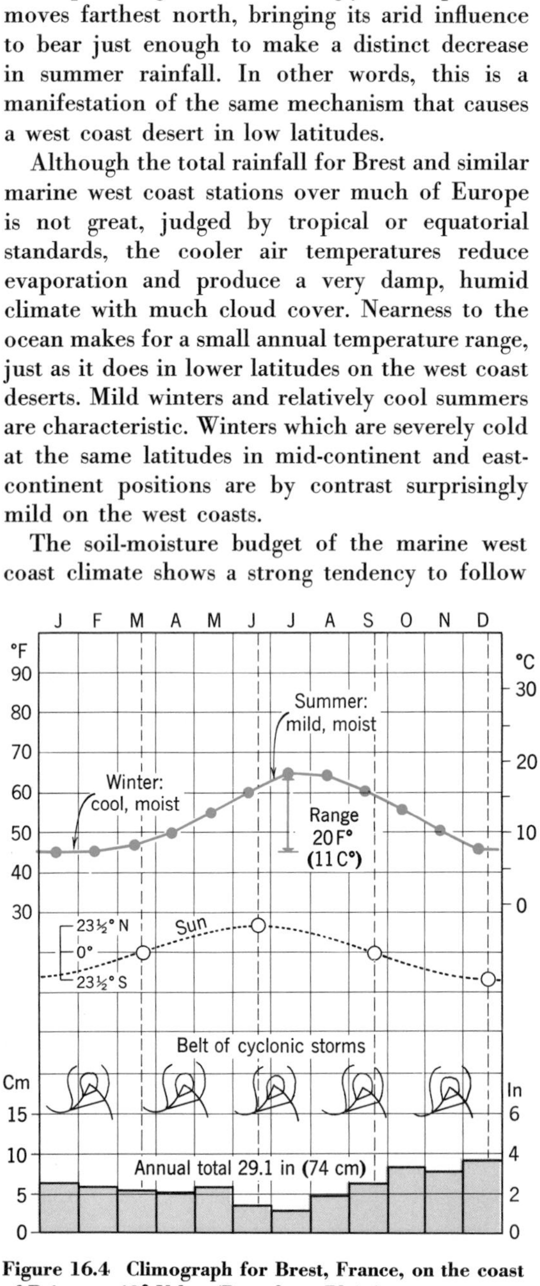

Figure 16.4 Climograph for Brest, France, on the coast of Brittany, 49° N lat. (Data from Blair.)

the Mediterranean regime (Figure 12.12*D*), in that the cycles of precipitation and evapotranspiration are exactly in opposite phase. The diagram for Prince Rupert, B. C. (Figure 12.12*G*) shows this relationship clearly, but because precipitation is ample throughout the summer months, no deficit appears. Instead, there is a very large water surplus.

The influence of coastal ranges upon rainfall is extremely marked in middle latitudes. Whereas low-lying coasts, such as in northern France or southern England, receive only 30 to 40 in (75–100 cm) of precipitation annually, mountainous coasts of British Columbia, Alaska, Norway, and Chile get 60 to 80 in (150–200 cm) and over. This has been a major factor in the development of fiords along the seacoasts. (See Chapter 31.) Heavy snows in the glacial period nourished vigorous valley glaciers which descended to the sea, scouring deep troughs below sea level at their lower ends.

Natural vegetation of the marine west coast climate is forest, but of widely different classes from one world locality to another. Western Europe has a summer-green deciduous forest whereas that of west coast North America is of the needleleaf class. On coast ranges of the Pacific northwest, Douglas fir, red cedar, and spruce grow in magnificent forests (Figure 21.14). Forests of the marine west coast climate belts of Chile and New Zealand are of the temperate rainforest class (Chapter 21).

Soils of the marine west coast climate regions bearing needleleaf forests are of a strongly leached type, the *podzols*, and are acid in nature. Under cool temperatures, bacterial activity is slow, in contrast to the warm tropics, so that vegetative matter is not consumed and forms a heavy surface deposit. Organic acids from the decomposing vegetation react with the soil compounds, resulting in removal of such bases as calcium, sodium, and potassium. Soils of western Europe are of the *gray-brown podzolic* group, typically associated with deciduous forests in middle latitudes. This soil is not so strongly leached and is suited to diversified forms of agriculture.

8. Mediterranean climate (dry summer subtropical climate) (Csa, Csb)

Situated on west coasts between latitudes 30° and 45° is a zone subject to alternate wet and dry seasons because it is located in the transitional zone between the dry west coast tropical desert (on the equatorward side) and the wet west coast climate (on the poleward side). Because these two climate types have been analyzed and described in previous pages, we need only substitute the information to derive a distinctive compound climate type.

Consider the climate of Monterey, California, as an example (Figure 16.5). In summer, when the oceanic subtropical high is most powerfully developed and farthest north, the same desert conditions that prevail permanently farther south take over control of the climate and bring a severe drought. However, the proximity of the ocean with its cool current keeps summer temperatures to a mild 60°F (16°C) average. In winter, the humid regime of middle-latitude cyclones and moist maritime polar (mP) air masses is felt in the ample precipitation.

The dry-summer subtropical climate is particularly extensive in the Mediterranean lands. Hence, the name *Mediterranean climate* is commonly used for this climate. The soil-moisture regime of the Mediterranean climate has been described in Chapter 12, and is well illustrated by the graph for Los Angeles (Figure 12.12*D*).

Köppen classifies the climate of the Mediterranean lands as *Csa*, a temperate rainy climate with dry, hot summers. Those narrow coastal belts directly bordering the cool currents of the open Atlantic, Pacific, and Indian oceans, Köppen designates as *Csb*, distinguished by a markedly cooler summer than the *Csa* climate. Thus Monterey, California, represents a *Csb* climate.

An example of a *Csa* station in the Mediterranean lands is Naples, Italy (Figure 16.6). The annual temperature range here is 28.5 F° (16 C°), which is over twice that at Monterey. The summer is not rainless, but monthly averages are very low in June and July. Annual precipitation is double that of Monterey. In short, the equable influence of the cool west coast desert is replaced by a continental influence.

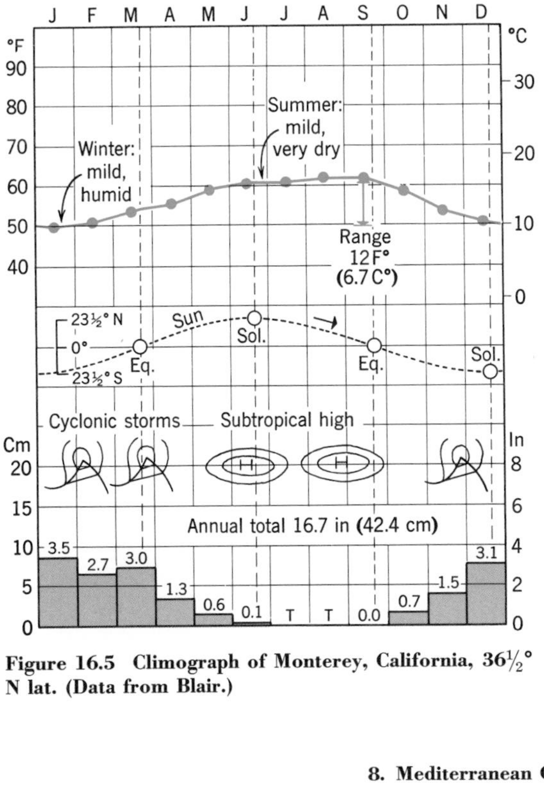

Figure 16.5 Climograph of Monterey, California, 36½° N lat. (Data from Blair.)

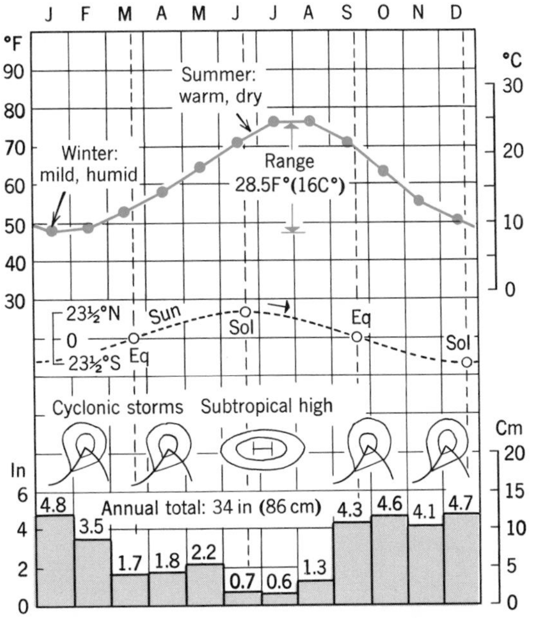

Figure 16.6 Climograph for Naples, Italy, 40½° N lat. (Data from Meteorological Office of Great Britain.)

It is apparent that if one travels from the Mediterranean shore into North Africa, he will pass from the Mediterranean climate into the tropical steppe climate of the Sahara. A transition climate type intermediate between the two is nicely shown by the record for Benghasi, Tripoli, at lat. 32° N (Figure 16.7). Note that the temperature record resembles that of the tropical dry climates, although not so hot, but the rainfall distribution is distinctly of the same type as the Mediterranean rainfall, though not so copious.

The Mediterranean climate, its drought coinciding with the high temperatures of summer, incurs a large water deficiency in middle and late summer (Figure 12.12*D*). Winter rains quickly restore the moisture deficiency and a surplus is developed by early spring.

The occurrence of a wet winter and a dry summer is unique among climate types and results in a distinctive natural vegetation of hardleaved evergreen trees and shrubs, known as *sclerophyll forest* (Chapter 21). Various forms of sclerophyll woodland and scrub are also typical (Figures 21.16, 21.18). Trees and shrubs must withstand the severe summer drought of two to four rainless months and intense evaporation.

Soils of the Mediterranean climate are not readily subject to simple classification. Reddish-chestnut and reddish-brown soils typical of semi-

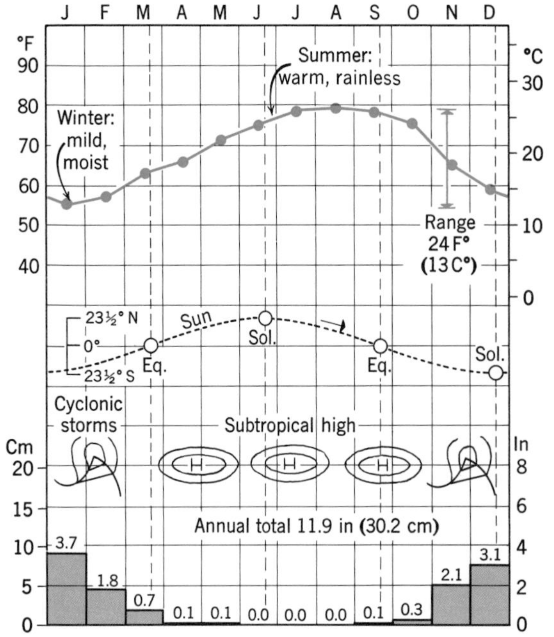

Figure 16.7 Climograph for Benghasi, Libya, 32° N lat., on the North African coast. (Data from Trewartha.)

arid climates are generally present (Plate 3). In the Mediterranean lands *terra rossa,* a red soil formed on limestone, occurs in various locations.

9. Middle-latitude desert and steppe climates (BWk, BSk)

The interiors of North America and Asia, in latitudes 35° to 50°, and, to a limited extent, that of southernmost South America have desert and steppe climates of somewhat complex origin. Three basic air mass controls operate. (1) In summer, when pressure and wind belts are shifted poleward, these regions temporarily become the source regions for continental tropical air masses, developed because of intense heating of the large continental interiors. (2) In winter, the intense development of the Siberian and Canadian highs, which are the source areas for continental polar air masses, causes frequent invasions of relatively dry continental air. (3) Mountain ranges separate these deserts from moist maritime polar and maritime tropical air masses which supply abundant rainfall to the west and southeast coasts. Through forced ascent over these ranges, followed by adiabatic heating upon descending the lee slopes (see Chapter 10), the maritime air masses are deprived of their moisture and raised in temperature as well. Thus the regions in question are poorly situated for obtaining precipitation. Because of the prevailing westerly winds in these latitudes, maritime tropical air masses cannot easily reach these areas from the east. In southern Asia, the northward flow of moist tropical air from the Indian Ocean is blocked by the great Himalayan chain.

Under the Köppen system, two climate varieties are recognized, the semi-arid steppes, *BSk,* and the true desert, *BWk* (Plate 2). The letter *k* signifies a cool climate, with average annual temperature below 64.4°F (18°C). Where the letter *k′* is used, a cold climate is indicated, with the warmest month average below 64.4°F (18°C).

Only a small proportion of the area covered by dry middle-latitude climates is extremely dry: the Turkestan and Gobi deserts of central Asia and parts of the Great Basin in Nevada and Utah (Plate 1). The principal respect in which these deserts differ from the tropical deserts of lower latitudes is that their annual temperature range is much greater and the winter temperatures much lower. Astrakhan, USSR, at 46° N lat., illustrates this feature nicely (Figure 15.16). The enormous annual range of 58 F° (32 C°) is almost double that of Aswan, Egypt; the January mean temperature is a severe 20°F (−7°C), as compared with 60°F (16°C) for Aswan; and the July maximum is only 15 F° (8 C°) less than for Aswan.

Of considerably greater importance geographically than the deserts are the vast semi-arid steppe lands of the middle-latitude dry climates. Partly because of higher elevation, which increases the precipitation, or because of location nearer to maritime air mass invasions, these regions receive from 10 to 20 in (25 to 50 cm) of precipitation annually (Plate 1).

When the semi-arid steppes are followed from subtropical highlands in Mexico to middle lati-

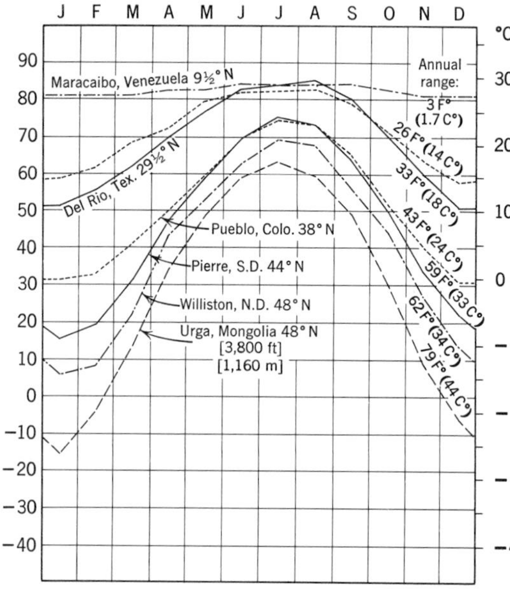

Figure 16.8 Temperature regimes of highland steppe climates. (Data from Trewartha.)

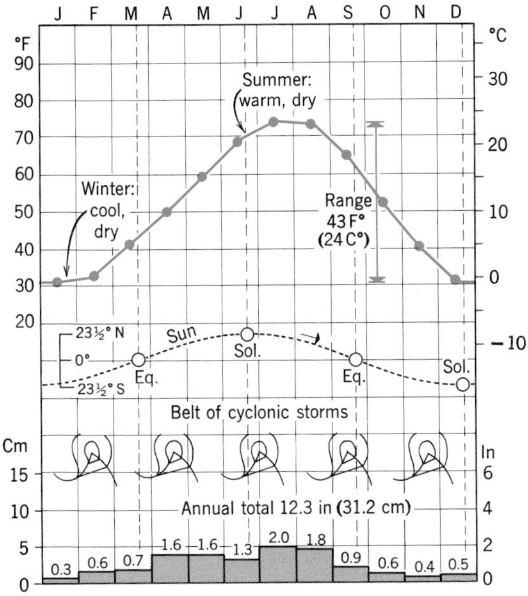

Figure 16.9 Climograph for Pueblo, Colorado, 38° N lat. (Data from Blair.)

tudes, the annual temperature cycles show not only progressively lower temperatures but also greatly increased annual ranges (Figure 16.8). A typical steppe climate of intermediate position is that of Pueblo, Colorado (Figure 16.9). Compared with the Mediterranean climate of Monterey, California (Figure 16.5), which lies at about the same latitude, the annual temperature range of Pueblo is very much greater, the precipitation cycle is just reversed, with the summer maximum clearly marked. In summer large highs drifting over the central United States produce a northerly return flow of air on their western sides, spreading moist tropical air from the Gulf of Mexico far northward and westward into the continental interior. This air is unstable and readily produces thunderstorms when lifted over a mountain range or along a cold front.

The continental arid regime of the soil-moisture budget is illustrated by data for Medicine Hat, Alberta, a station with steppe climate (Figure 12.12*F*). The summer moisture deficiency persists through seven months. Recharge in the winter months is not sufficient to generate a surplus as runoff.

Steppes support a short-grass vegetation cover suited to grazing of cattle and sheep, but inadequate for farming without irrigation or special dry-farming methods. Special attention is given by geographers and climatologists to the boundary between dry and humid climates, particularly because the limits of agriculture without irrigation are drawn by nature as a fluctuating line over a hazardous frontier zone. Here, productivity may in some years be high; in others, drought brings disastrous failures. Figure 16.10 is a map showing precipitation and temperature data for the boundary between dry and humid climates in the Great Plains region of the United States. Let us define a humid climate as one in which the precipitation on the average exceeds evaporation, so as to give permanently flowing streams and a generally moist soil; a dry climate as one in which evaporation on the average exceeds precipitation, so as to give ephemeral streams and a generally dry soil. The humid-arid boundary cannot be permanently located, but its average position runs somewhere along the dashed north-south line in Figure 16.10. Note that in North Dakota, the boundary cuts the 15-in (38-cm) annual isohyet but in southern Texas runs close to the 25-in (64-cm) isohyet. This shift is explained by the temperatures, which are higher

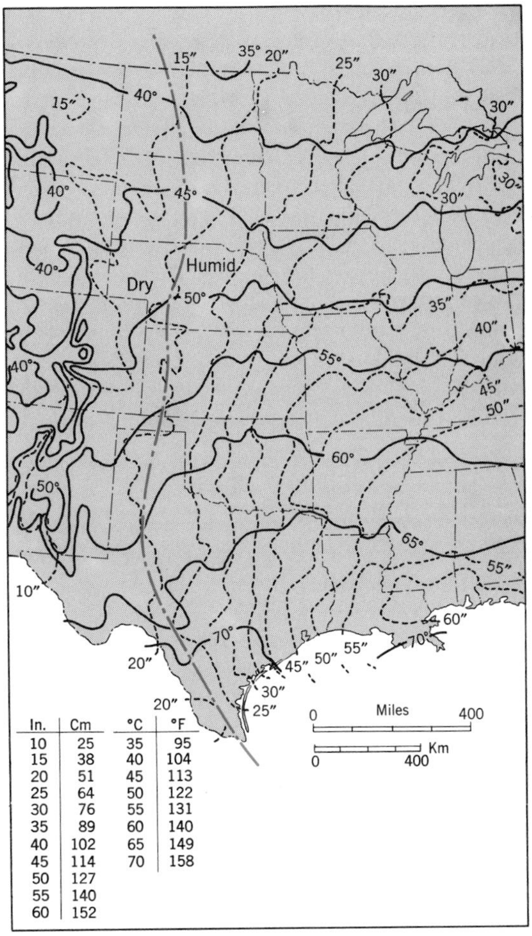

Figure 17.10 On this map, the boundary zone between semiarid and subhumid climates runs north-south. Superimposed are isotherms of average annual Fahrenheit temperature (solid lines) and lines of equal average annual precipitation in inches (broken lines). (After J. B. Kincer, U.S. Dept. of Agriculture.)

in the south and result in greater evaporation losses.

Middle-latitude steppes are characterized by short-grass prairie and by local occurrences of woodland and semi-desert shrubs (Plate 4, Figure 21.27). (This vegetation is described in Chapter 21.) Middle-latitude steppes now constitute the great sheep and cattle ranges of the world. On the vast expanses of the High Plains, the American bison lived in great numbers until almost exterminated by hunters. Likewise, the short-grass veldt of South Africa supported much game at one time. Steppe grasses do not form a complete sod cover, and loose, bare soil is exposed between grass clumps. For this reason, overgrazing or a series of dry years often reduces the hold of grasses enough to permit destructive soil erosion and gullying from heavy local downpours.

Soils of the steppe lands are deficient in humus. They belong to the group of brown soils in the less dry parts; elsewhere they are reddish or gray desert soils. Calcium carbonate is present in excess quantities and may form nodules in the soil. Encrustations of calcium carbonate make a hard, whitish crust termed *caliche* in the southwestern United States. These calcium-rich soils are highly fertile in terms of bases favorable to grain crops, such as wheat, but unless irrigation is employed the advantage is lost. Traced toward humid-climate zones which adjoin the steppes, the soils become darker brown in color, denoting the increasing amounts of humus from heavier growth of grasses. Thus the gray or red desert and steppe soils grade into the rich, dark brown and black chestnut and chernozem soils of the prairies (Plate 3).

10. Humid continental climate (Dfa, Dfb, Dwa, Dwb)

Consider a vast region located in a middle latitude, say between 40° and 55° N, extending from the continental interior to the east coast. North America and Asia are, of course, the areas that we have in mind. The location is intermediate between the source region of polar continental air masses on the north and maritime or continental tropical air masses on the south and southeast. In this polar-front zone, maximum interaction between polar and tropical air masses can be expected along warm and cold fronts associated with east-moving cyclones. In winter, the continental polar air masses dominate and much cold weather prevails; in summer, tropical air masses dominate and high temperatures prevail. Thus, strong seasonal temperature contrasts must be expected in this region. Precipitation is ample throughout the year, because the region lies in a frontal zone. Strong contrasts in air masses result in strong frontal activity and highly changeable weather. This climate may be described as both humid and continental in properties.

The climate picture drawn above applies well to the north-central and northeastern United States and southeastern Canada as well as to northern China, southern Manchuria, Korea, and northern Japan. Very similar climatic conditions also hold for much of central and eastern Europe, the Balkan countries, and Russia, but this third region differs from the first two in that it is influenced by a great source region of continental-tropical air masses lying to the south and southeast, instead of the oceanic source regions of maritime tropical air masses.

Four varieties of climate, according to Köppen, are found in the regions that we are grouping together as having the humid continental climate. *Dfa* and *Dfb* are cold, snowy, forest climates, the first with a hot summer, the second with a warm summer. The precise definitions of *a* and *b* are as previously given. *Dwa* and *Dwb* climates, found in eastern Siberia, Manchuria, and northern Korea, are cold, snowy forest climates with a dry winter. Letters *a* and *b* refer to hot and warm summers, respectively. The winter dryness, typical of all climates of eastern Asia, is explained below.

We are also including here under the humid continental climate the northern part of Köppen's *Cf* climate region in the eastern United States. (Compare Plate 2 and Figure 14.5.) Recall that Köppen used as his *C-D* boundary the isotherm

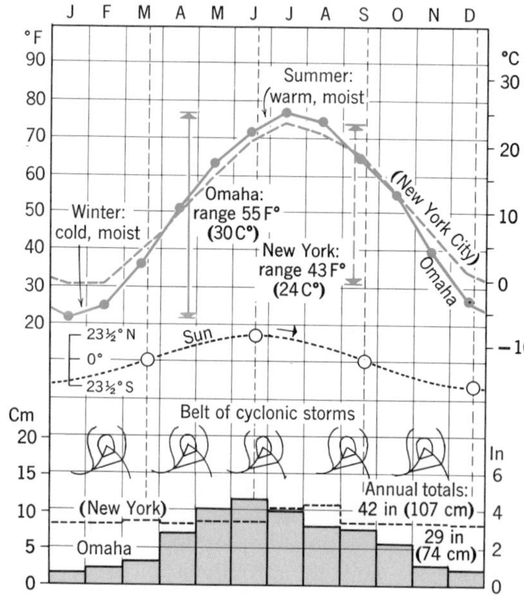

Figure 16.11 A comparison of climates at New York and Omaha, both in the humid continental climate zone. (Data from H. H. Clayton, Smithsonian Institution.)

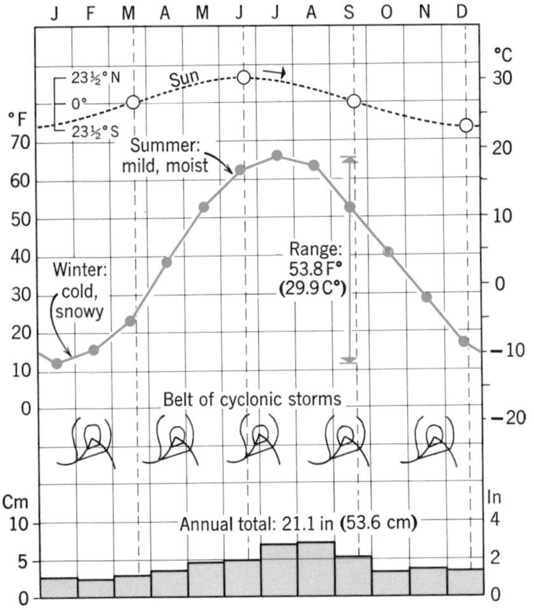

Figure 16.12 Climograph for Moscow, USSR, 56° N lat. (Data from Blair.)

of 26.6°F (−3°C) of the coldest month. Thus, for example, Köppen places New Haven and Cleveland in the same *Cfa* climate as New Orleans and Tampa, despite obvious contrasts in January mean temperatures, soil groups, and natural vegetation between these northern and southern zones.

Some characteristic features of the humid continental climate can be seen from the temperature-precipitation graphs of four stations: Omaha and New York City (Figure 16.11), Moscow, USSR (Figure 16.12), and Peiping, China (Figure 16.13). All four stations have similar temperature curves. Yearly range is near 55 F° (33 C°) for all but New York City, which is moderated by its near-ocean location. Omaha and Peiping, located at the same latitude (40° N), have almost identical temperatures; Moscow, being more than 1000 mi (620 km) farther north 56° N), has a temperature cycle which runs 10 F° (5.5 C°) lower than that of Omaha or Peiping.

Precipitation records of the four stations illustrated show certain marked dissimilarities which require explanation. A summer maximum is apparent in all four but is very weakly defined in New York City because maritime air masses, both polar and tropical, have ready access to the eastern seaboard at all times of the year. Omaha and Moscow have well-defined summer maxima and winter minima, reflecting the predominance of tropical air masses in summer and continental-polar air masses in the winter. Moscow, being farther north, near to the source region of polar continental air masses, has less precipitation than the other stations. Peiping shows a very strong summer maximum and a winter drought. This contrast is characteristic of the east Asiatic middle-latitude stations and reflects the powerful monsoon control, whereby dry continental air masses dominate in winter.

The soil-moisture regime associated with the humid continental climate is well illustrated by data for Manhattan, Kansas (Figure 12.12*E*). The summer moisture deficiency is small, while winter recharge leads only to a small spring surplus.

The general temperature pattern of middle-latitude climates is well illustrated by daily temperature graphs of three stations at about the same latitude across North America (Figure 16.14). Victoria, British Columbia, on the west coast, occupies a windward position with respect to maritime polar air masses from the Pacific. Consequently, the contrasts between July and February (the two extreme months of the year) are small. Moreover, the daily ranges are small, especially in winter. The graph of Winnipeg, Manitoba, shows maximum continental influence with strong annual and daily contrasts. Note especially the wide fluctuations in winter, with outbursts of polar and arctic air bringing very low temperatures. St. Johns, Newfoundland, occupies a leeward position with respect to the continent and is accessible to maritime air masses from the Atlantic. It therefore has only a moderate annual temperature range, but the continental influence is nevertheless clearly marked in the sharply fluctuating daily temperature record.

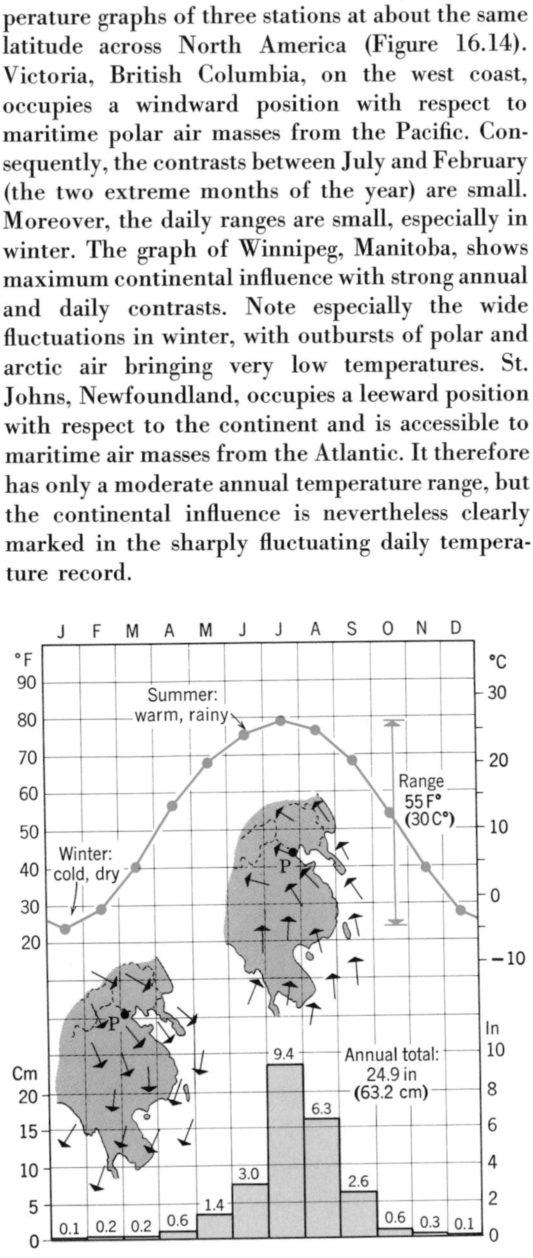

Figure 16.13 Climograph for Peiping, China, 40° N lat. (Data from Trewartha.)

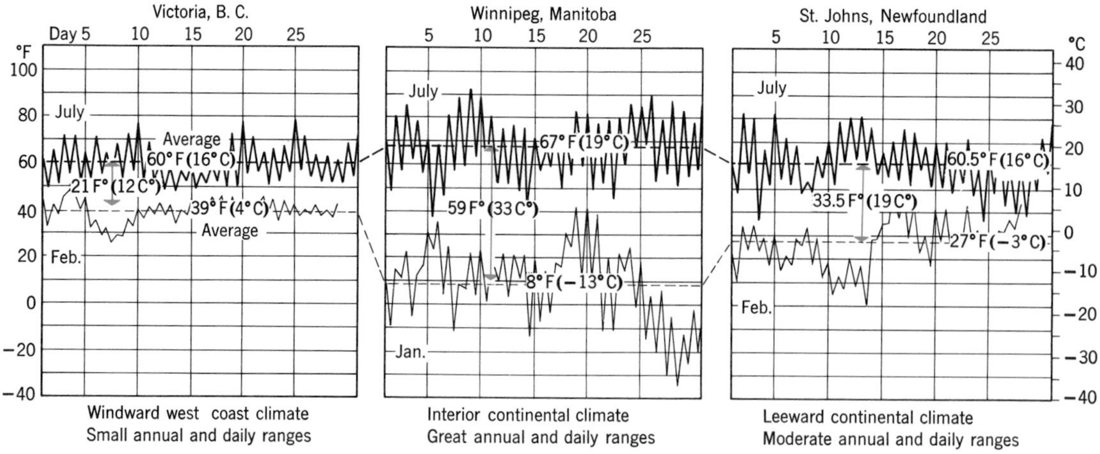

Figure 16.14 Along the 50th parallel in North America lie widely differing climate types. Averages are for months actually shown on graphs. (After Mark Jefferson, *Geographical Review*.)

Three major types of natural vegetation and their associated soils may be recognized in the humid continental climate. The distribution is well illustrated in North America. (See Plates 3 and 4.)

In the more humid eastern sections, including the warmer parts of the humid continental climate zone (*Dfa*), natural vegetation is summergreen deciduous forest (Figure 21.11). Here soils are of the *brown forest* and *gray-brown podzolic* types, rich in humus and moderately leached so as to have a distinct light-colored leached zone under the upper dark layer. In this region, diversified farming and dairying are the most successful uses of the land where topography is favorable.

A northern belt of needleleaf evergreen forest (Figure 21.12) extends along the entire length of the colder northern parts of the humid continental climate zone (*Dfb*). To this may be added the mountain regions of the Adirondacks and northern New England. Here soils are of the *podzol* type, strongly leached, but with an upper layer of humus (Chapter 19). Cool temperatures inhibit bacterial activity which would destroy this organic matter in tropical regions. Podzols are deficient in calcium, potassium, and magnesium, and are, in general, acid in chemical nature. Thus they are not highly productive for crop farming, even though adequate rainfall is generally assured. The podzols are, however, well suited to the growth of conifers.

The drier plains areas of the humid continental climate support a natural tall-grass prairie (Figure 21.26), which grades into the drier steppe regions of short grasses to the west. The prairie soils and chernozem soils, two major soil groups of these grasslands, are typically dark in color and consist of a single, thick upper layer grading into the parent soil material below (Chapter 19). These soils contain abundant calcium, magnesium, and potassium because rainfall is here distinctly less than farther eastward and leaching is less active as a soil-forming process.

REVIEW QUESTIONS

1. Explain the occurrence of a moist climate in subtropical latitudes (25° to 35°) along the eastern continental margins. What air masses prevail over this region? What is the nature of these air masses?

2. Describe the annual cycle of temperature and precipitation at a typical humid subtropical climate (*Cfa*) station on the southeastern Atlantic Coast or eastern Gulf Coast of the United States. When is precipitation at a maximum? Why? Are thunderstorms common or rare?

3. What are the characteristics of climate on the southeastern Asiatic coast in the subtropical latitudes? How does the precipitation cycle differ from that in comparable localities in the southeastern United States? Why is there a difference?

4. What is the natural vegetation of the humid subtropical climate of east coasts in latitudes 25° to 35°? What is a characteristic feature of the soils in this climate zone? To what other groups of soils are they most closely related?

5. What are the important characteristics of the climates of west coasts in middle latitudes (40° to 60°)? What air masses are dominant? From what direction do most air masses and cyclonic storms approach these coasts? How is this reflected in the annual temperature cycle?

6. Why does the marine west coast climate (*Cfb*) show a marked reduction in rainfall during the summer, especially between 40° and 50° latitude?

7. What type of forest vegetation is developed on windward mountain slopes in the marine west coast belts. In what way is orographic rainfall related to the development of these forests and their associated soils?

8. Describe and explain the Mediterranean climate (*Csb*) which prevails on west coasts between 30° and 40° latitude. What is unique about the annual precipitation graph of this climate? How does it compare with graphs of the tropical savanna climate (*Aw*)?

9. What is the natural vegetation of the Mediterranean climate (*Csa, Csb*)?

10. Why are dry climates developed in middle latitudes in interior continental locations? Are they related to the presence of mountain barriers? Give examples. Explain the principle of the rainshadow desert.

11. What distinction is made between extremely arid regions and semi-arid regions or steppes? What is the natural vegetation of steppes? Of the very dry middle-latitude deserts (*BWk*)?

12. How does altitude influence the degree of aridity in the dry middle-latitude regions?

13. What is noteworthy about the annual temperature cycle of the middle-latitude steppes (*BSk*) and deserts (*BWk*)? How does this cycle differ from that of the tropical deserts?

14. How may the boundary between dry and humid climates be determined? Is it a fixed line, or does it fluctuate? Of what economic importance is this boundary?

15. Of what economic value are the middle-latitude steppe lands of the world? What types of soils underlie these regions? Is the soil suitable for grain crops? What is caliche? How does it form?

16. Describe the climate located in middle latitudes in the central and eastern parts of the continents. What air masses are involved in this climate? How does season of year determine which air masses are dominant?

17. Describe the annual temperature cycle of a typical station in the humid continental climate zone (*Dfa, Dfb*). Is the range great or small? Explain. Is the average temperature of the coldest month below freezing?

18. What can be said about the day-to-day temperature fluctuations in the humid continental climate? How do these variations reflect the conflict in air masses?

19. Discuss the natural vegetation and soils of the humid continental climate (*Dfa, Dfb*) in middle latitudes. How do southerly and northerly locations differ in this respect? How do easterly and westerly locations differ?

Exercises

1. Prepare temperature-precipitation graphs similar to Figure 16.1 for each of the following stations. (Data from Trewartha and Meteorological Office of Great Britain.)

(*a*) *Nagasaki, Japan*, 33° N.

	J	F	M	A	M	J	J	A	S	O	N	D
Temperature, °F	42	43	49	58	65	71	79	81	74	65	56	46
Precipitation, in.	2.8	3.3	4.9	7.3	6.7	12.3	10.1	6.9	9.8	4.5	3.7	3.2

(*b*) *Seattle, Washington*, 47½° N.

	J	F	M	A	M	J	J	A	S	O	N	D
Temperature, °F	40	42	45	50	55	60	64	64	59	52	46	42
Precipitation, in.	4.9	3.8	3.1	2.4	1.8	1.3	0.6	0.7	1.7	2.8	4.8	5.5

2. For each of the above stations calculate the mean annual temperature, annual temperature range, and annual rainfall total.

3. (*a*) Compare the record of Charleston, South Carolina (Figure 16.1), with that of Nagasaki, Japan (Exercise 1). Both are located at approximately the same latitude. (*b*) In what climate do they belong? (*c*) Why has Nagasaki a colder winter than Charleston? (*d*) Why has Nagasaki a rainier summer than Charleston?

4. Seattle, Washington, and Brest, France (Figure 16.4), both belong in the humid, marine west-coast climate and are located at about the same latitude. (*a*) Compare the temperature and precipitation figures of the two stations. (*b*) Why does Seattle show greater precipitation in winter but less in summer? (*c*) In what way does Seattle climate resemble the Mediterranean climate of California?

5. The soil-moisture budget of Shanghai, Figure 16.3, is similar to that of Manhattan, Kansas, Figure 12.12*E*, except that at Shanghai the water surplus is much larger (12 in; 300 mm) and the soil-moisture deficiency is close to zero. To what climatic influence do you attribute this difference in the two soil-moisture budgets? Seattle is located between Los Angeles (Figure 12.12*D*) and Prince Rupert (Figure 12.12*G*) on the Pacific coast of North America. The soil-moisture budget for Seattle belongs to the Mediterranean regime. At Seattle the water surplus is about 11 in (280 mm); the soil-moisture deficiency is about 5 in (130 mm). Compare these figures with corresponding values for both Los Angeles and Prince Rupert.

6. Prepare temperature-precipitation graphs similar to Figure 16.1 for each of the following stations. (Data from Trewartha.)

(*a*) *Valparaiso, Chile*, 33° S.

	J	F	M	A	M	J	J	A	S	O	N	D
Temperature, °F	67	66	65	61	59	56	55	56	58	59	62	64
Precipitation, in.	0.0	0.0	0.6	0.2	3.5	5.8	4.8	3.2	0.8	0.4	0.1	0.3

(*b*) *Rome, Italy*, 42° N.

	J	F	M	A	M	J	J	A	S	O	N	D
Temperature, °F	45	47	51	57	64	71	76	76	70	62	53	46
Precipitation, in.	3.2	2.7	2.9	2.6	2.2	1.6	0.7	1.0	2.5	5.0	4.4	3.9

7. (*a*) For each of the stations in Exercise 6 compute mean annual temperature, annual temperature range, and annual precipitation. (*b*) To what type of climate do these stations belong?

8. The soil-moisture budget of Rome shows almost equal values of soil-moisture deficiency and water surplus (7 in; 180 mm). In what respect does the budget of Rome differ from that of Los Angeles (Figure 12.12*D*)? Can the difference be explained in terms of monthly and annual precipitation values?

9. Prepare temperature-precipitation graphs similar to Figure 16.1 for each of the following stations. (Data from Trewartha.)

(*a*) *Lovelock, Nevada,* 40° N.

	J	F	M	A	M	J	J	A	S	O	N	D
Temperature, °F	30	36	43	50	58	66	74	72	62	51	40	31
Precipitation, in.	0.7	0.5	0.4	0.4	0.4	0.3	0.2	0.2	0.3	0.4	0.3	0.4

(*b*) *Williston, North Dakota,* 48° N.

	J	F	M	A	M	J	J	A	S	O	N	D
Temperature, °F	6	8	22	43	53	63	69	67	56	44	27	14
Precipitation, in.	0.5	0.4	0.9	1.1	2.1	3.2	1.7	1.7	1.0	0.7	0.6	0.5

10. (*a*) Compute the mean annual temperature, temperature range, and total annual precipitation for the stations in Exercise 9. (*b*) To what climates do these stations belong? (*c*) How do the climates of Lovelock, Nevada, and Williston, North Dakota, differ as to temperature cycle and precipitation cycle? Explain these differences.

11. Identify the soil-moisture regime of Williston by comparison with the graphs in Figure 12.12. Williston has a soil-moisture deficiency of about 7 in (180 mm) and there is no water surplus. At Williston the potential evapotranspiration is zero from November through March; the total annual evapotranspiration is about 22 in (550 mm).

12. Prepare temperature and precipitation graphs similar to Figure 16.1 for each of the following stations. (Data from Trewartha.)

(*a*) *Bucharest, Rumania,* 43½° N.

	J	F	M	A	M	J	J	A	S	O	N	D
Temperature, °F	26	29	40	52	61	68	73	71	64	54	41	30
Precipitation, in.	1.2	1.1	1.7	2.0	2.5	3.3	2.8	1.9	1.5	1.5	1.9	1.7

(*b*) *Marquette, Michigan,* 46½° N.

	J	F	M	A	M	J	J	A	S	O	N	D
Temperature, °F	16	16	25	38	49	59	65	63	57	46	33	23
Precipitation, in.	2.2	1.8	2.1	2.3	3.1	3.5	3.1	2.8	3.2	3.0	3.0	2.5

13. For each station in Exercise 12 compute mean annual temperature, temperature range, and annual precipitation total.

14. (*a*) Compare the climate of Bucharest, Rumania, with that of New York City (Figure 16.11), both of which belong to the humid continental climate. Note that summer temperatures are similar but that Bucharest has a colder winter. Why is this so? (*b*) Precipitation is very much less in Bucharest than in New York, especially in the winter months. Why? Explain in terms of air masses.

15. (*a*) Compare the climates of Marquette, Michigan, and Moscow, USSR (Figure 16.12). (*b*) Moscow lies almost 10° latitude (700 mi, 1100 km) farther north than Marquette. Does this show up in the temperature figures? Explain. (*c*) Compare the precipitation of these stations. Which has more? Why?

16. The soil-moisture budget of Bucharest shows a deficiency of about 4.5 in (115 mm) and no water surplus. The potential evapotranspiration totals about 28 in (700 mm) for the year. Using these figures, compare the budget of Bucharest with budgets illustrated in Figure 12.12. With which regime is Bucharest most closely identified?

CHAPTER 17

Polar, Arctic, and Highland Climates

CLIMATES of group III, located at high latitudes, are controlled largely by polar and arctic air masses. They have low temperatures, usually low precipitation, and low evapotranspiration.

11. Continental subarctic climate (Dfc, Dfd, Dwc, Dwd)

In the two great landmasses of North America and Eurasia, a vast expanse of interior continental area lies between 50° and 70° N lat. Here are the source regions for the continental polar air masses.

In winter, when excessive heat loss by radiation has resulted in the formation of the prevailing Siberian and Canadian highs, severely cold air temperatures develop over snow-covered surfaces, forming a cold, dense air mass. Typically, a severe temperature inversion prevails in winter, so that the air at, say, 5000 ft (1500 m) may be 10 F° (5 C°) warmer than air at the ground level. Low in moisture content, this air mass is stable and normally clear.

In summer, the source region is shifted farther north; air mass temperatures rise to moderately high levels, but the moisture content, although much greater in summer, is still small by comparison with that of maritime tropical air masses.

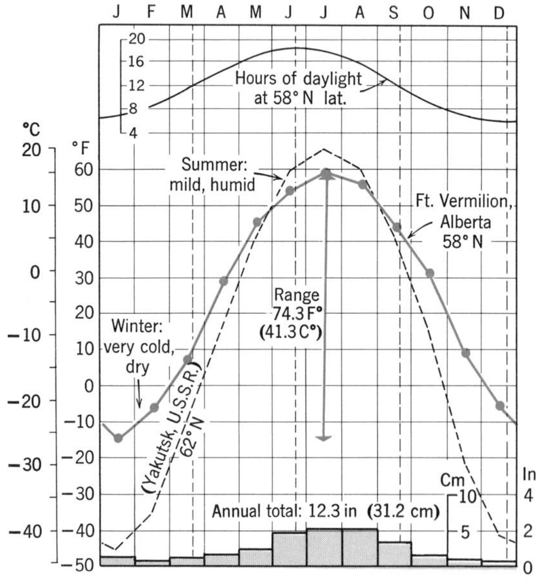

Figure 17.1 Climographs for Ft. Vermilion, Alberta, and Yakutsk, USSR. (Data from Trewartha.)

We might expect, therefore, a climate type showing very great seasonal temperature range, with extremely severe winters and a small annual total precipitation concentrated in the warm months. This climate, called here the *continental subarctic climate,* includes four of Köppen's climate types. The largest area, including belts from Alaska to Labrador and from Scandinavia to Siberia, is classified as *Dfc,* a cold, snowy forest climate, moist all year, with cool, short summers. Less than four months of the year have averages over 50°F (10°C). A still colder climate, *Dfd,* found in northern Siberia only, has very cold winters in which the average of the coldest months is below −36.4°F (−38°C). Climates *Dwc* and *Dwd* are also cold, snowy forest climates, but with a dry winter. They are found only in northeastern Asia. The letters *c* and *d* denote a cool summer and a very cold winter, respectively.

The continental subarctic climate (*Dfc*) is well illustrated by the graph of Ft. Vermilion, Alberta, at 58° N lat. (Figure 17.1). The annual range of 74 F° (41 C°) is remarkable enough, but it is greatly exceeded by that of Yakutsk, USSR, lying in the *Dwd* climate type. The annual temperature range of the continental subarctic climate is the greatest of any on earth, reaching 110 F° (61 C°) in Siberia. Even in central Antarctica, which has the lowest temperatures on earth, the annual range is not over 70 F° (39 C°).

Absolute minimum temperatures of −70° to −80°F (−57° to −62°C) are recorded in northwestern Canada, the lowest being −81.4°F (−63°C) at Snag, Yukon on February 3, 1947. Perhaps the coldest place in the northern hemisphere is Verkhoyansk, USSR, which lies in this climate zone and has a January mean of −59°F (−51°C) and an absolute minimum recorded temperature of −93°F (−69°C).

Summer in the subarctic climate regions is very short. The warmest month average may not greatly exceed 50°F (10°C), and frosts can occur at any time during the summer. Daily maximum temperatures, however, commonly reach 70°F (21°C). At these high latitudes the sun is above the horizon

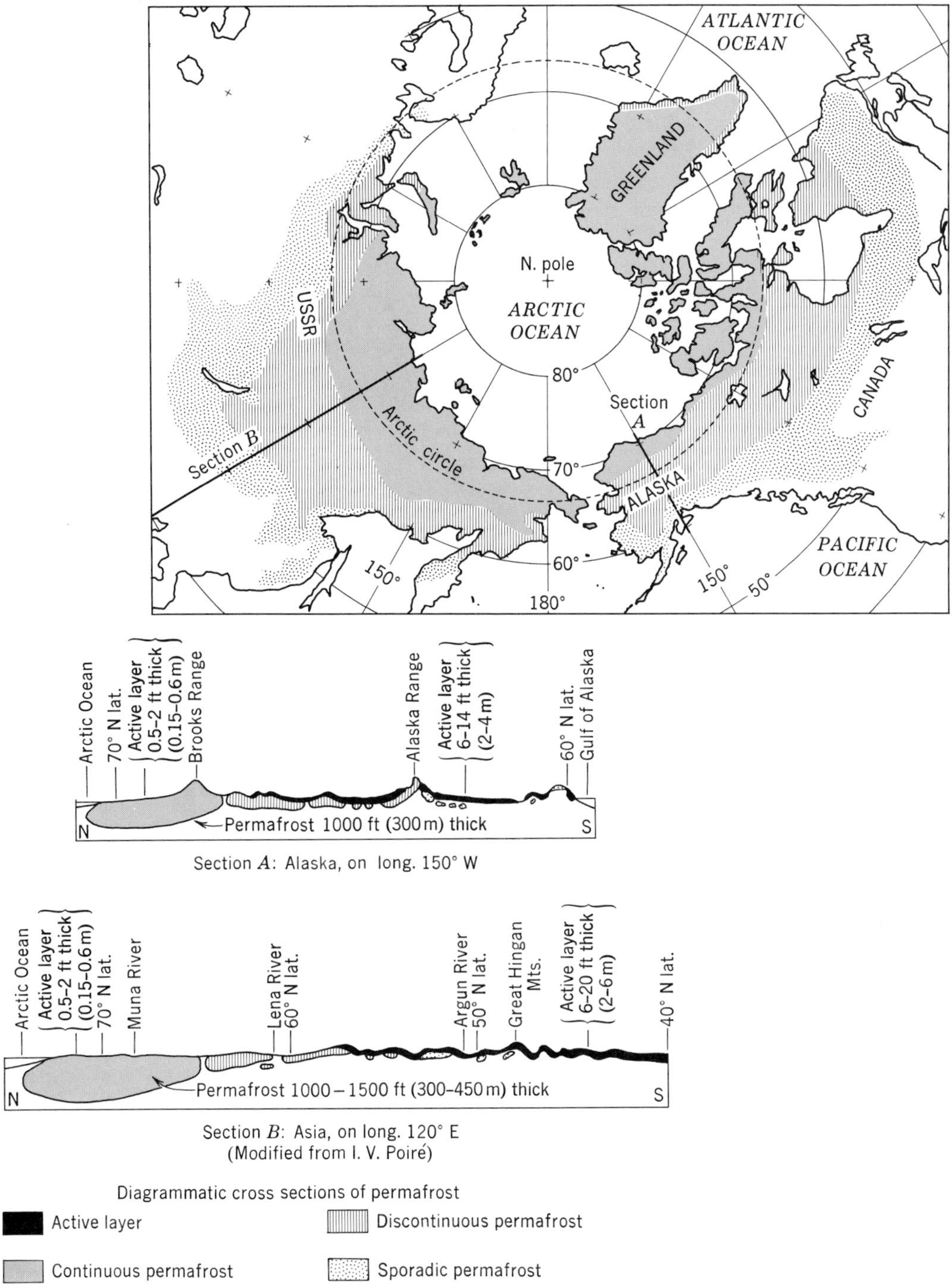

Figure 17.2 Distribution of permafrost in the northern hemisphere, and representative cross sections in Alaska and Asia. (From Robert F. Black, "Permafrost," Chapter 14 of P. D. Trask's *Applied Sedimentation*, John Wiley and Sons, New York, 1950.)

for sixteen to eighteen hours from May through August. (See graph on Figure 17.1.) The long hours of sunshine so accelerate the growth of plants that agriculture may be possible despite a very short growing season.

Winter is the dominant season of the continental subarctic climate. Because subfreezing monthly average temperatures occur for six or seven consecutive months, all moisture in the soil and subsoil is solidly frozen to depths of many feet. (See Figure 17.6.) Summer warmth is insufficient to thaw more than the upper few feet so that a condition of perennially frozen ground, or *permafrost,* prevails over large parts of this and the tundra regions to the north. Seasonal thaw penetrates from 2 to 14 ft (0.6–4m), depending on location and nature of the ground. This shallow zone of alternate freeze and thaw is termed the *active zone.*

The distribution of permafrost in the northern hemisphere is shown in Figure 17.2. Three zones are recognized. Continuous permafrost, which extends without gaps or interruptions under all topographic features, coincides largely with the tundra climate (*ET*), but also includes a large part of the continental subarctic climate (*Dfc, Dfd, Dwd*) in Siberia. Discontinuous permafrost, which occurs in patches separated by frost-free zones under lakes and rivers, occupies much of the continental subarctic climate (*Dfc*) zone of North America and Eurasia. Sporadic occurrence of permafrost in small patches extends into the southern limits of the continental subarctic climate.

Depth of permafrost reaches 1000 to 1500 ft (300 to 450 m) in the continuous zone near latitude 70° (Figure 17.2). Much of this permanent frost is an inheritance from more severe conditions of the last ice age, but some permafrost bodies may be growing under existing climate conditions.

Permafrost presents problems of great concern in engineering and building construction in these cold regions. Buildings must be insulated underneath to protect the ground moisture beneath from melting; otherwise, the building might literally be engulfed in mud. Another serious problem is in the behavior of streams in winter. As the surface of streams or springs freezes over, the water beneath bursts out from place to place, freezing into huge accumulations of ice. Highways are thus made impassable. Scraping of insulating peat, forest litter, and vegetative cover from the frozen ground to make roads and air fields may result in dire consequences. The summer sun thaws the bare ground, which turns into a liquid mud, often growing into sizable lake basins by melting and sapping around the edges of the exposed areas.

Most of the source areas of the continental polar air masses are in regions of less than 20 in (50 cm) of precipitation annually, whereas the northerly portions have less than 10 in (25 cm).

Precipitation is largely cyclonic in type and shows a very definite maximum in the summer months. Snowfall, although conspicuous in winter because it remains upon the ground, accounts for only a fraction of an inch of precipitation per month in the coldest months. Cyclonic storms crossing these areas bring little precipitation at this season. In summer, cyclonic rains are frequent, although thunderstorms are few.

The soil-moisture budget of the continental subarctic climate resembles the arctic regime illustrated by the graph for Barrow, Alaska (Figure 12.12*H*). Evapotranspiration is effectively zero over a seven or eight month period when soil moisture is solidly frozen. However, evapotranspiration peaks sharply in summer and exceeds precipitation by substantial amounts in June, July, and August. Thus a summer moisture deficiency may develop.

The subarctic climate zone coincides with a great belt of needleleaf forest, often referred to as *boreal forest* (Figure 21.12), and open lichen woodland, the *taiga* (Figure 21.25). Trees tend to be small, so that they are economically of less value for lumber than for pulpwood.

Soils of the *podzol* group are associated with the arctic needleleaf forest. As explained before, these soils are strongly leached and of acid type. They are light gray and have a very distinct leached layer beneath the uppermost layer of humus and forest litter. These soils are extremely poor from the standpoint of agriculture. Added to the natural inadequacy of soils of this region is a great prevalence of swamps and lakes left by the departed ice sheets (Figure 31.17). Some rock surfaces were scoured by ice which stripped off the soil entirely. Elsewhere rock basins were formed, or previous stream courses dammed, making countless lakes. Insufficient geologic time has elapsed for good drainage even to begin to be reestablished over large areas.

12. Marine subarctic climate (EM)

In middle latitudes, the west coasts of continents and their offshore islands enjoy a remarkably equable climate, since both temperature and precipitation are highly uniform throughout the year. Followed poleward into latitudes 50° to 70° this same uniformity can be recognized in the *marine subarctic climate,* marked by unusually large total annual precipitation and an unusually small annual temperature range for such high latitudes.

The marine subarctic climate is dominated by maritime polar (mP) air masses throughout the year. The climate distribution actually coincides rather well with the mP air mass source regions in the North Atlantic, North Pacific, and the Southern Ocean. Important in defining the characteristics of this climate is the persistence of cloudy skies and strong winds, and the high percentage of days with precipitation.

The marine subarctic climate is not designated as a separate variety in the Köppen system, which includes it in the tundra climate (*ET*). Recent recognition by geographers of the marine subarctic climate has been coupled with the suggestion that it be entered in the Köppen system with the symbol *EM.*

TABLE 17.1 REPRESENTATIVE MARINE SUBARCTIC STATIONS*

	Mean Annual Temperature		Mean Temperature of Warmest Month		Mean Temperature of Coldest Month		Annual Temperature Range		Total Annual Precipitation	
	°F	°C	°F	°C	°F	°C	F°	C°	Inches	Centimeters
Nanortalik, Greenland, 60°08′N, 45°11′W	34.5	(1.5)	45	(7.2)	26	(−3.3)	19	(10.5)	35	(89)
Vardø, Norway, 70°22′N, 31°06′E	34	(1.1)	48.5	(9.2)	22	(−5.6)	26.5	(14.8)	23.5	(60)
Cumberland Bay, South Georgia, 54°16′S, 36°30′W	35	(1.7)	42	(5.6)	28.5	(−2.0)	13.5	(7.6)	52	(132)
Stanley, Falkland Islands, 51°42′S, 57°51′W	42	(5.6)	49	(9.5)	35.5	(2.0)	13.5	(7.5)	27	(69)

* Data from Air Ministry, Meteorological Office of Great Britain.

The marine subarctic climate is found on limited stretches of windward coasts, on islands, and over wide expanses of ocean in latitudes 50° to 60° in the Bering Sea, and latitudes 55° to 75° in the North Atlantic, touching points on the south Greenland coast, north Iceland, and extreme northern Norway. In the southern hemisphere, the marine subarctic climate occupies almost no land. The few representative stations are found in the southernmost tip of South America, the Falkland Islands, South Georgia Island, and a number of other small islands. Table 17.1 gives temperature and precipitation data for four stations.

Representative stations in the marine subarctic climate qualify as belonging in the *E* climates of Köppen, for the warmest-month means are all below 50°F (10°C). In contrast to the typical tundra climate (*ET*) the annual ranges are all much smaller, the winters much milder, and the annual precipitation totals much larger. These contrasts are brought out strikingly in the comparison of two stations in Figure 17.3. Coldest month temperature mean is not below 20°F (−7°C).

Soils and vegetation of the marine subarctic climate are essentially of the same tundra varieties associated with the tundra climate, treated below. The landscape is treeless and gives the impression of great barrenness, but it is nevertheless surprisingly rich in the variety of plant species that exist in the layer close to the ground.

13. Tundra climate (ET)

The northern continental fringes of North America and Eurasia from the Arctic Circle to about the 75th parallel lie within the outer zone of control of arctic-type air masses, whose source region covers the Arctic Ocean and Greenland. To the south lie the continental polar (cP) and maritime polar (mP) air mass source regions. The land fringes are thus in a frontal zone, which has been known as the *arctic front*, but which may be more simply identified as belonging to a widely shifting polar front, well developed over the northern Pacific and Atlantic oceans. Here many intense east-moving cyclonic storms are developed, and much bad weather may be expected.

These conditions produce a *tundra climate*, described by Köppen with the symbol *ET*. This is a polar climate in which the average temperature of the warmest month is below 50°F (10°C), but above 32°F (0°C).

The tundra climate is illustrated by the temperature-precipitation graph of Upernivik at 73° N lat.

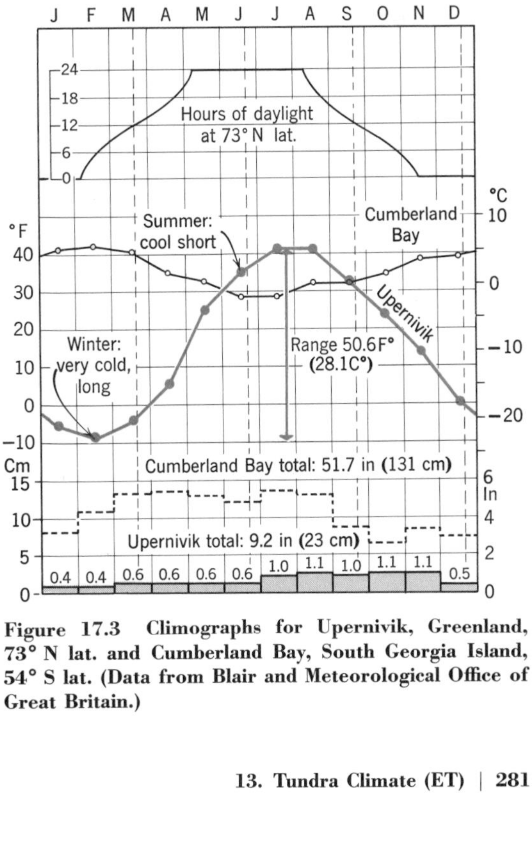

Figure 17.3 Climographs for Upernivik, Greenland, 73° N lat. and Cumberland Bay, South Georgia Island, 54° S lat. (Data from Blair and Meteorological Office of Great Britain.)

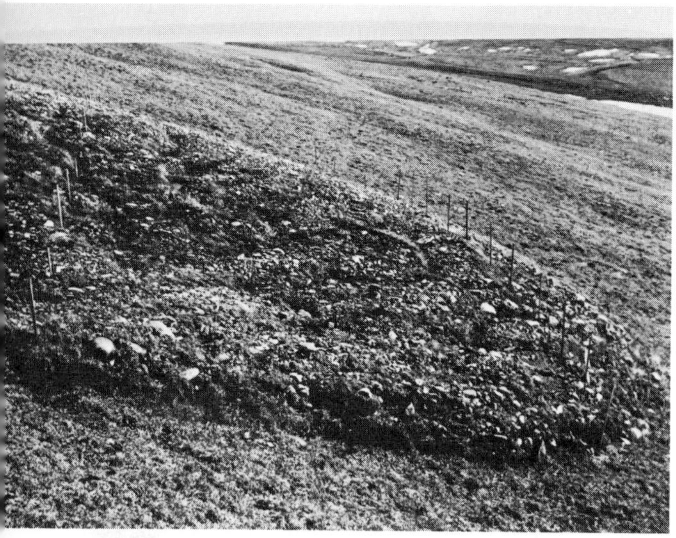

Figure 17.4 Solifluction, a slow soil flowage, has produced this lobate form on a tundra slope on Mt. Pelly, Victoria Island (lat. 69° N, long. 104° W). Surface of lobe stands about 6 ft (1.8 m) above surrounding slope. (Photograph by A. L. Washburn, Arctic Institute of North America.)

on the west Greenland coast (Figure 17.3). Note that (*a*) temperature range is large, but not nearly as great as in the continental subarctic climate; (*b*) the warmest month averages are just over 40°F (4°C) and the coldest month average is well below 0°F (−18°C); and (*c*) precipitation total is less than 10 in (25 cm), with an increased amount falling during and after the warm season. Nearness to the Arctic Ocean explains the somewhat more moderate temperature range and minimum, compared to the continental centers. Coolness of the summer is explained by the nearness to the large ocean body, keeping air temperatures down despite large receipts of solar energy at this latitude near summer solstice. Possibly a more persistent cloud cover is also a factor.

The soil-moisture budget of the tundra climate follows the arctic regime, illustrated by the graph for Barrow, Alaska (Figure 12.12*H*). Evapotranspiration is effectively zero for an eight to nine month period when soil moisture is solidly frozen. However, evapotranspiration rises sharply in summer, exceeding precipitation and giving rise to a moisture deficiency in June, July, and August.

Vegetation of the treeless tundra consists of grasses, sedges, and lichens, along with shrubs of willow (Figure 21.28). Traced southward the vegetation changes into birch-lichen woodland, then into the needleleaf forest. In some places a distinct tree line separates the forest and tundra. Coinciding approximately with the 50° isotherm of the warmest month, this line has been used by Köppen as a boundary between *Df* and *ET* climates.

Tundra soils are noteworthy in that the soil particles are produced almost entirely by mechanical breakup of the parent rock and have suffered little or no chemical alteration. Grayish loam and blue-gray clay layers are present with much peat.

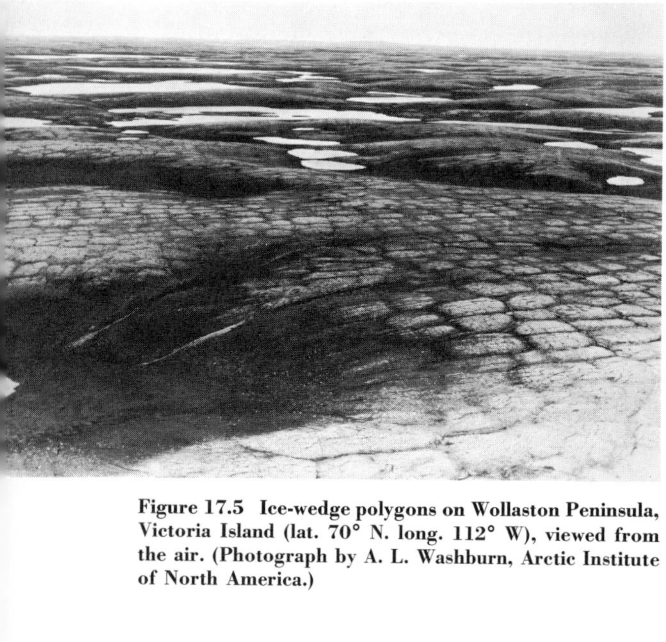

Figure 17.5 Ice-wedge polygons on Wollaston Peninsula, Victoria Island (lat. 70° N. long. 112° W), viewed from the air. (Photograph by A. L. Washburn, Arctic Institute of North America.)

Figure 17.6 This vertical river-bank exposure near Livengood, Alaska, in the subarctic climate region reveals a V-shaped ice wedge surrounded by layered silt of alluvial origin. (Photograph by T. L. Péwé, U.S. Geological Survey.)

Continual freezing and thawing of soil moisture has been responsible for disintegration of the soil particles. Like the soils of the northern continental interiors, soils of the tundra are affected by the permanently frozen, or permafrost, condition (Figure 17.2). The permafrost layer is more than 1000 ft (300 m) thick over most of this region; seasonal thaw reaches only 4 to 24 in (10–60 cm) below the surface.

Geomorphic processes have a somewhat distinctive pattern in the tundra regions, and a variety of curious landforms results. Under a protective sod of small plants, the soil water melts in summer, producing a thick mud which may flow downslope to create bulges and terraces without breaking through the surface. This process is known as *solifluction*, or *sludging*, and forms *solifluction terraces* and *lobes* on slopes (Figures 17.4 and 24.21). In the desert tundra, solifluction occurs without the confining cover and may be described as a layer of thick mud creeping down the slopes. Observers who have studied this flowage find that it is most rapid at midday, the mud flowing at a rate of several feet per hour and carrying along large blocks of rock.

The freeze and thaw of water in the soil gives rise to a curious system of polygonal cracks in flat ground (Figure 17.5). They may result from the shrinkage of the clay as water is withdrawn to form ice crystals in lenses or layers within the soil (Figure 17.6). The resulting pattern is termed *polygonal ground.* On hill and mountain summits the cracks are filled with stones, which seem to be gradually sorted and pushed to the sides of each polygon during alternate freeze and thaw of soil water. These forms are called *stone rings*, or *stone polygons.* On slopes, the stone rings are drawn downslope to produce *stone stripes*, appearing from the air like giant hachures on the ground.

14. Icecap climate (EF)

Three vast regions of ice exist on the earth. These are the Greenland and Antarctic continental icecaps and the large area of floating sea ice in the Arctic Ocean (Figures 31.9, 31.10, and 17.9). The continental icecaps differ in various ways, both physically and climatically, from the polar sea ice and can be separately treated. Glacial and topographic features of Greenland and Antarctica are treated in Chapter 31.

Icecap climate has by far the lowest average annual temperature of all global climates. Designated by Köppen *EF*, or polar climate of perpetual frost, this climate has no monthly average above freezing (32°F, 0°C). Available records indicate mean annual temperatures of −20° to −30°F (−30° to −35°C) for the Greenland icecap, but

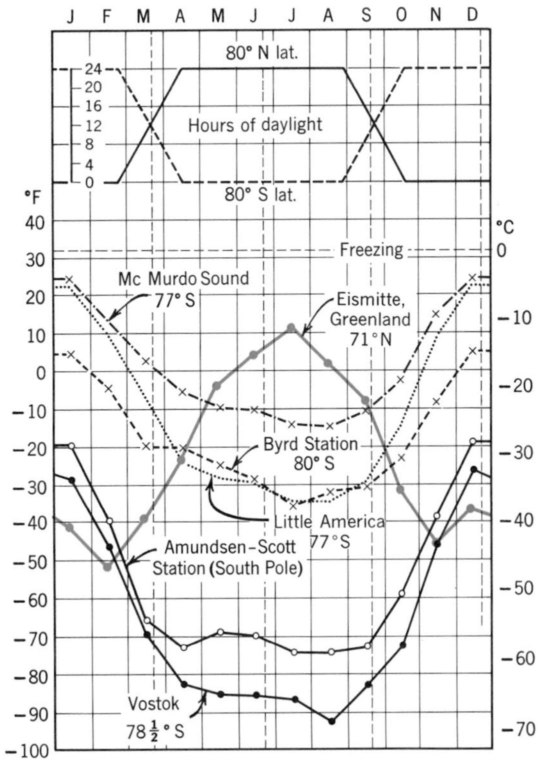

Figure 17.7 Temperature graphs for five icecap stations. (Data from Trewartha and *Encyclopedia of Atmospheric Sciences*, R. W. Fairbridge, ed., Reinhold, New York, 1967, p. 23.)

about −9°F (−23°C) for the Arctic Ocean. Compare this with the 10° to 25°F (−12° to −4°C) mean annual temperature of the tundra climate. The higher Arctic Ocean temperature is due to its sea-level altitude.

The annual temperature graph for a Greenland icecap station, "Eismitte," is shown in Figure 17.7. This information was gathered by a German expedition under the direction of Alfred Wegener in the years 1929 through 1931. Note that only in three months of the year did the mean monthly temperatures rise above 0°F (−18°C), and then only to a July maximum of 12°F (−11°C). The coldest month was February, with −53°F (−47°C), making an annual range of 65F° (36C°).

A shallow layer of air over the ice sheet is chilled severely and at times flows downslope under the influence of gravity toward the margins as a severe blizzard wind. It is reported that so great is the chill of the ice upon the air that tiny ice crystals sometimes form within a few feet of the ground. These may make a "snowstorm," above which the head and shoulders of a man rise clear. Driving blizzard winds pack the sandlike snow into a hard, smooth pavement.

Figure 17.8 The U.S. Coast Guard icebreaker Northwind forces a passage through McClure Strait, Banks Island, Canadian Northwest Territory, in mid-August. To the left is an open lead. Cutting across from left to right is a rugged zone of pressure ridges. (Official U.S. Coast Guard photo.)

Cyclonic storms frequently penetrate Greenland, bringing precipitation to the icecap. The principal nourishment of the ice sheet is from this source.

Climate of the Antarctic icecap was little understood until a number of weather stations were maintained in the International Geophysical Year of 1957–1958. Temperatures in the interior have proved to be far lower than any place on earth. The Russian meteorological station "Vostok," located about 800 mi (1300 km) from the south pole at an elevation of 11,440 ft (3488 m), may be the world's coldest spot (Figure 17.7). Here a record low of −126.9°F (−88.3°C) was observed. Note that this minimum value occurred near the end of the long polar night. At the pole itself (Amundsen-Scott Station), July, August, and September have averages of about −76°F (−60°C) (Figure 17.7). Temperatures run roughly 50 F° (28 C°) higher, month for month, at McMurdo Sound because it is located close to the Ross Sea and is at low elevation.

A remarkable feature of the high, antarctic interior is the intense chilling of air close to the snow surface. A strong temperature inversion develops in winter, so that the air near the surface may be 50 to 60 F° (28 to 33 C°) colder than air a few hundred feet higher. Downslope flow of this heavy, cold air layer causes blizzard winds to develop in favorable valley locations.

Sea ice

Greatly increased utilization and study of high latitudes by both military forces and civilian scientist research groups has brought to attention the phenomenon of floating sea ice. Supply of outposts by ship, maintenance of observing stations on floating ice masses, and submarine operation in the polar sea are activities influenced by sea ice.

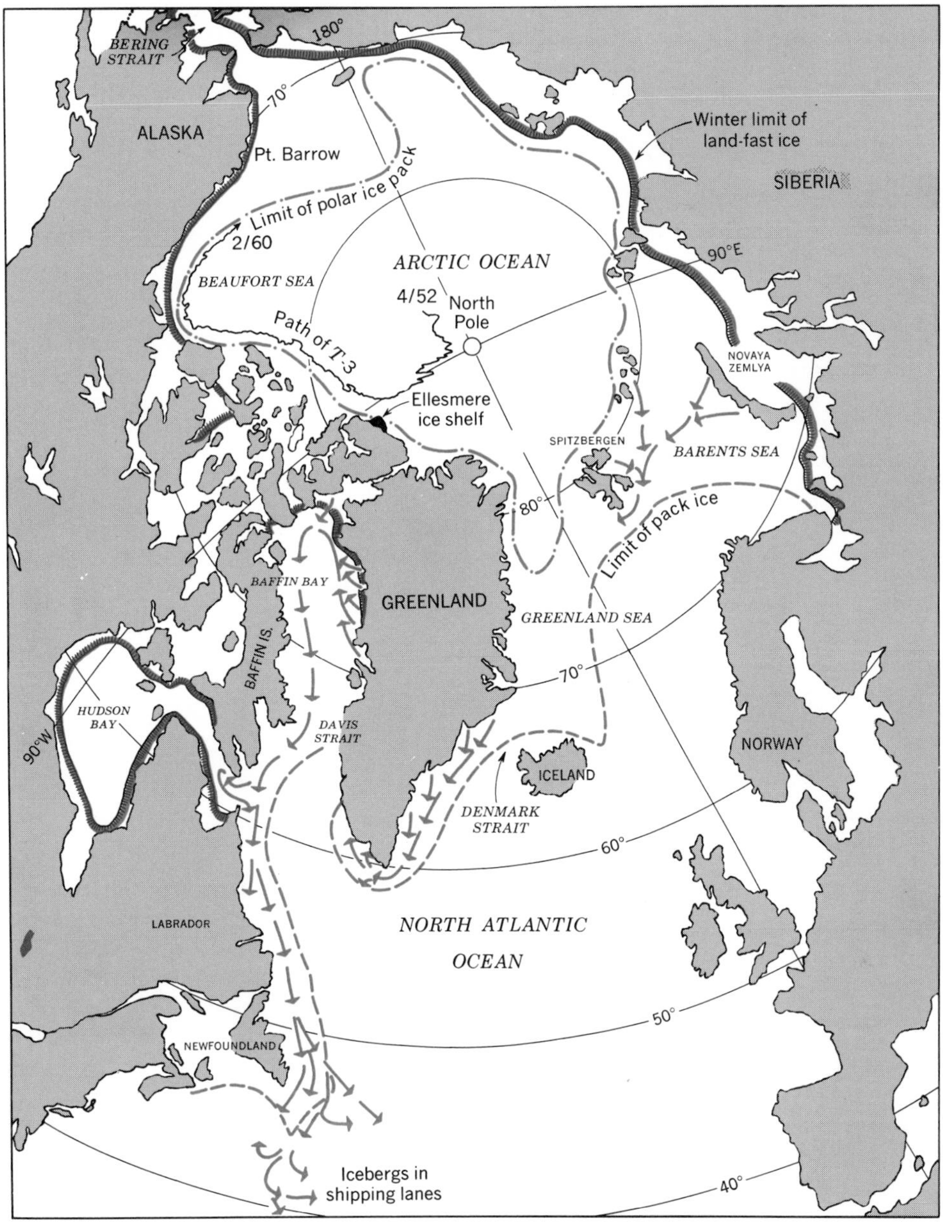

Figure 17.9 Sea ice of the Arctic Ocean. Common tracks of icebergs are shown by arrows. A tilted homolographic projection is used for this map. (Based on data of the National Research Council. From A. N. Strahler, *The Earth Sciences*, Harper and Row, New York.)

The oceanographer distinguishes *sea ice*, formed by direct freezing of ocean water, from *icebergs*, and *ice islands*, which are bodies of land ice broken free from tide-level glaciers and continental ice shelves. Aside from differences in origin, a major difference between sea ice and floating masses of land ice is in thickness. Sea ice, which begins to form when the surface water is cooled to temperatures of about $28\frac{1}{2}$°F (−2°C) is limited in thickness to about 15 ft (5 m).

Pack ice is the name given to ice that completely covers the sea surface (Figure 17.8). Under the forces of wind and currents, pack ice breaks up into individual patches, termed *ice floes*. The narrow strips of open water between such floes are *leads*. Where ice floes are forcibly brought together by winds, the ice margins buckle and turn upward into pressure ridges resembling walls or irregular hummocks (Figure 17.8). The difficulties of travel on foot across the polar sea ice are made extreme by the presence of such obstacles. The surface zone of sea ice is composed of fresh water, the salt

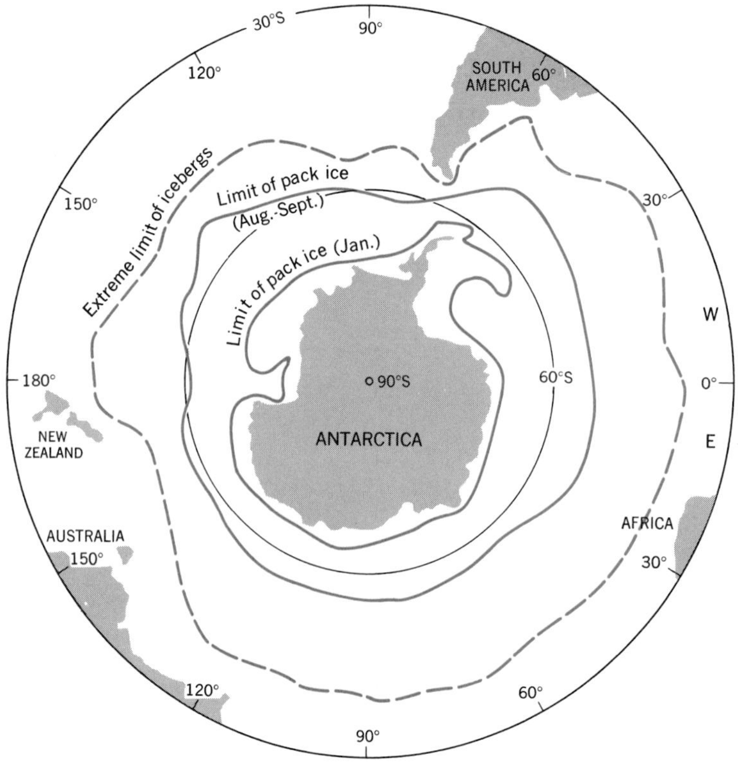

Figure 17.10 Sea ice of the antarctic region. (Data from National Academy of Sciences and American Geographical Society. From A. N. Strahler, *The Earth Sciences*, Harper and Row, New York.)

Figure 17.11 This great iceberg in the east Arctic Ocean dwarfs the U.S. Coast Guard icebreaker Eastwind. (Official U.S. Coast Guard photo.)

having been excluded in the process of freezing.

The Arctic Ocean, which is surrounded by landmasses, is normally covered by pack ice throughout the year, although open leads are numerous in the summer (Figure 17.9). The relatively warm North Atlantic drift maintains an ice-free zone off the northern coast of Norway. The situation is quite different in the antarctic, where a vast open ocean bounds the sea ice zone on the equatorward margin (Figure 17.10). Because the ice floes can drift freely north into warmer waters, the antarctic ice pack does not spread far beyond 60° S lat. in the cold season. In March, close to the end of the warm season, the ice margin shrinks to a narrow zone bordering the Antarctic continent.

Icebergs and ice islands

Icebergs, formed by the breaking off, or *calving*, of blocks from a valley glacier or tongue of an icecap, may be as thick as several hundred feet. Being only slightly less dense than sea water, the iceberg floats very low in the water, about five-sixths of its bulk lying below water level (Figure 17.11). The ice is fresh, of course, since it is formed of compacted and recrystallized snow.

In the northern hemisphere, icebergs are derived largely from glacier tongues of the Greenland

Figure 17.12 This tabular iceberg was observed near the Bay of Whales, Little America, Antarctica, in January 1947. (Official U.S. Coast Guard photo.)

icecap (Figure 17.9). They drift slowly south with the Labrador and Greenland currents and may find their way into the North Atlantic in the vicinity of the Grand Banks of Newfoundland. Icebergs of the antarctic are distinctly different. Whereas those of the North Atlantic are irregular in shape and therefore present rather peaked outlines above water, the antarctic icebergs are commonly *tabular* in form, with flat tops and steep clifflike sides (Figure 17.12). This is because tabular bergs are parts of ice shelves, the great, floating platelike extensions of the continental icecap (Chapter 31). In dimensions, a large tabular berg of the antarctic may be tens of miles broad and over 2000 ft (600 m) thick, with an ice wall rising 200 to 300 ft (60 to 90 m) above sea level.

Somewhat related in origin to the tabular bergs of the antarctic are *ice islands* of the North Polar Sea. These huge plates of floating ice may be 20 mi (32 km) across and have an area of 300 sq mi (800 sq km). The bordering ice cliff, 20 to 30 ft (6 to 10 m) above the surrounding pack ice, indicates an ice thickness of 200 ft (60 m) or more. The few ice islands known are probably derived from a shelf of land-fast glacial ice attached to Ellesmere Island, about 83° N lat. (Figure 17.9). The ice islands move slowly with the water drift of the Arctic Ocean and a charting of their tracks reveals much about circulation in that ocean. (See path of T-3 in Figure 17.9). As permanent and sturdy platforms, ice islands serve as bases of scientific researches from which observations of oceanography, meteorology, and geophysics can be carried out over long periods.

Highland climates

As explained in earlier chapters, increasing elevation brings a great reduction in both pressure and temperature. Thus, climates change greatly within a vertical range of a few thousand feet. In a general way, a rise of altitude is equivalent to an increase in latitude, so that the tundra and icecap climate equivalents can be found among the glaciers of a mountain mass above timber line. In one major respect, however, the analogy is inadequate. Whereas intensity of insolation is progressively less toward the poles, it is increased at higher altitudes. Thus, daily temperature ranges are excessive at high altitudes in middle and low latitudes, but are much less pronounced in equivalent arctic climates.

Pressure and temperature

As stated in Chapter 7, for every 900 ft (275 m) of increased elevation the barometric pressure falls $\frac{1}{30}$ of its value (Figure 7.6).

The physiological effects of a pressure decrease are well known from the experiences of flying and mountain climbing. The principal influence is through an insufficient amount of oxygen to supply the blood through the lungs. At altitudes of 10,000 to 15,000 ft (3000 to 4500 m) mountain sickness (altitude sickness) occurs, characterized by weakness, headache, nosebleed, or nausea. Persons who remain at these altitudes for a day or two normally adjust to the conditions, but physical exertion is always accompanied by shortness of breath.

At reduced pressures the boiling point of water or other liquids is reduced so that cooking time of various foods is greatly lengthened.

The table gives some data on pressure and boiling point relationships. From these figures it is evident that the use of pressure cookers will be of great value about 5000 ft (1500 m) wherever the cooking involves boiling of water.

Elevation		Pressure		Boiling Temperature	
Ft	M	In	Cm	°F	°C
Sea level	0	29.9	76	212	100
1000	300	28.8	73	210	99
3000	900	26.8	68	206	97
5000	1500	24.9	63	203	95
10,000	3000	20.7	53	194	90

From the standpoint of weather and climate, reduced atmospheric pressure is principally effec-

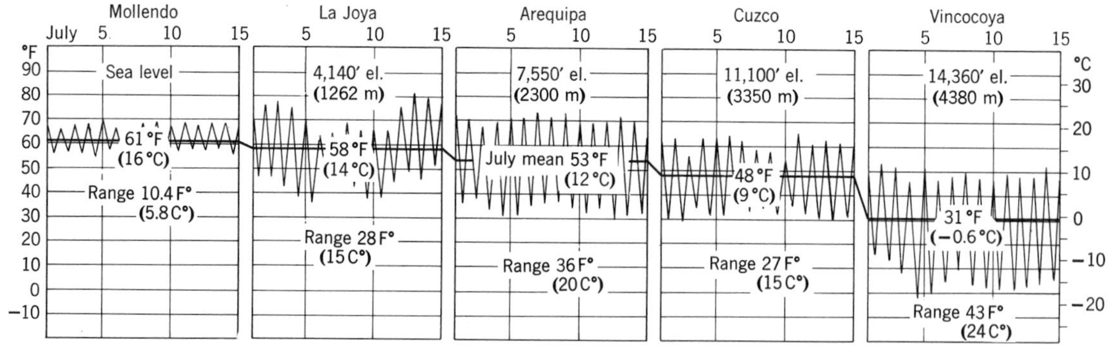

Figure 17.13 Daily maximum and minimum temperatures for mountain stations in Peru, 15° S lat. (After Mark Jefferson, *Geographical Review*.)

tive in that the thinner atmosphere, with relatively less carbon dioxide, water vapor, and dust, absorbs and deflects less solar energy and thus permits a high intensity of insolation at the ground.

Increasing intensity of insolation at higher elevation has a profound influence upon temperature relations. Surfaces exposed to sunlight heat rapidly and intensely, shaded surfaces are quickly and severely cooled. This results in rapid air heating during the day and rapid cooling at night at high-mountain locations (Figure 17.13). Thus at mountain vacation resorts not only is the air pure and the landscape features sharply outlined as if washed clean, but the cool nights and warm days are stimulating physically.

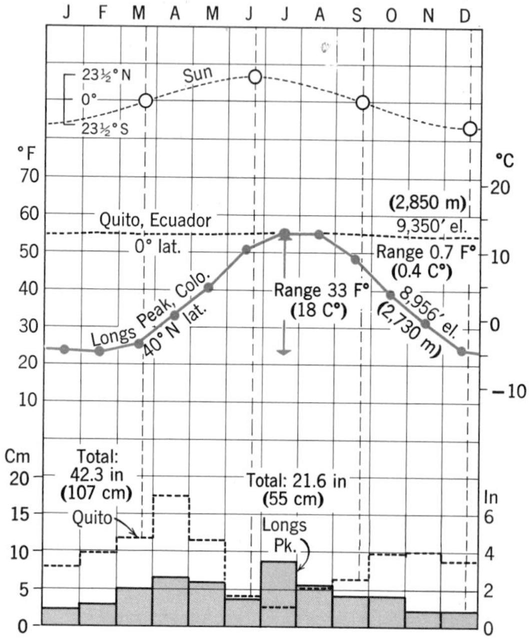

Figure 17.14 A comparison of rainfall and temperature characteristics of two mountain stations. (Data from Trewartha.)

The contrast between exposed and shaded surfaces is particularly noteworthy at high altitudes. It is said that temperatures of objects in the sun and in the shade differ by as much as 40 to 50 F° (22 to 28 C°).

Increased intensity of insolation is accompanied by an increase in intensity of violet and ultraviolet rays. Sunburn is very much more rapid above 5000 ft (1500 m) than at sea level, as many a person has learned by unfortunate experience. The red and infrared rays of the spectrum, on the other hand, are relatively less intensified by increased elevation because they are better able to pass through a dense atmosphere.

The general decrease in air temperature with elevation follows the environmental lapse rate of $3\frac{1}{2}$ F° per 1000 ft (2 C° per 300 m). Thus, we might expect a station 10,000 ft (3000 m) in elevation to have a temperature about 35 F° (20 C°) below that of a nearby sea-level station. Actually the difference is somewhat less than this.

At high altitudes in the equatorial regions, annual range of temperature is very small, much as in the wet equatorial rainforest climates which surround these mountains at low elevations (Figure 17.14). The range at Quito, Ecuador, is thus only 0.7 F° (0.4 C°) for the year. Compare this with Figure 15.1.

In middle and high latitudes, on the other hand, a wide annual temperature variation is to be expected, following the marked variations of insolation from summer to winter, as the Longs Peak graph shows (Figure 17.14).

Precipitation

The general influence of increased elevation is first to bring an increase in precipitation, at least for the first few thousand feet of elevation. This is due to the production of orographic rainfall, generated by the forced ascent of air masses and the resultant cooling of the air. (See Chapter 10.)

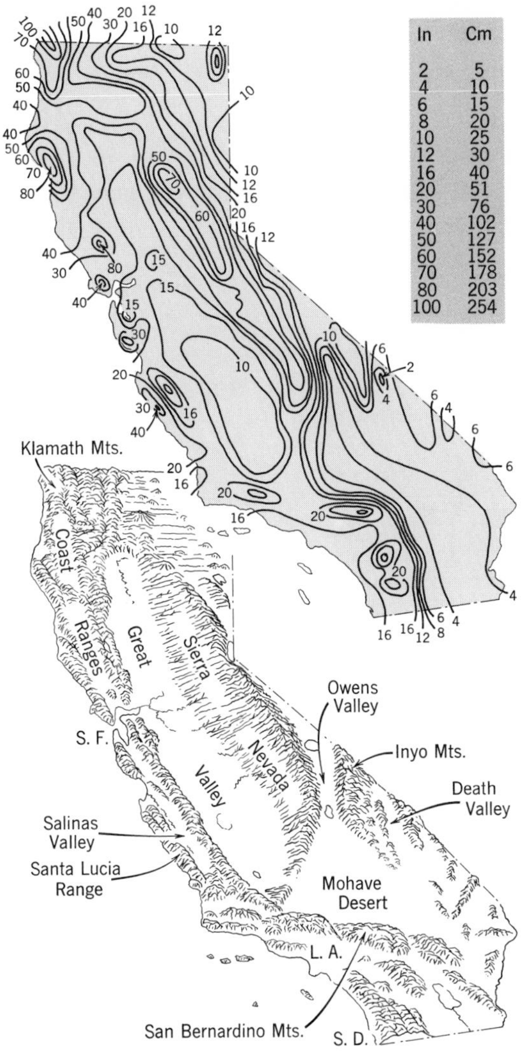

Figure 17.15 The effect of mountain topography on rainfall is well shown by the state of California. Isohyets in inches. (Rainfall map after U.S. Dept. of Agriculture, *Yearbook of American Agriculture*, 1941.)

Above elevations of 6000 to 10,000 ft (1800 to 3000 m), which form the zone of heaviest rainfall in low latitudes, precipitation increase begins to slacken with elevation, owing to the inability of air at the lower temperatures to hold, and therefore to give up, as much moisture.

Generally speaking, mountains and plateaus are humid climate zones. This is particularly striking in an arid or semi-arid region, where the mountains form islands of humid climate surrounded by desert or steppe. Reduction in temperatures results in reduced evaporation, so that a humid condition prevails.

The influence of mountains upon precipitation is strikingly illustrated by the state of California (Figure 17.15), whose topography is greatly diversified. Note that in the Great Valley and Death Valley precipitation is less than 10 in (25 cm), but on the west slopes of the Sierra Nevada and Klamath Mountains it is over 70 in (175 cm).

From the standpoint of river flow and floods, mountain climates are of greatest importance in middle latitudes. The higher ranges serve as snow storage areas, keeping back the precipitation until early or midsummer, releasing it slowly through melting, and thus aiding in the maintenance of continuous river flow. As melting proceeds to successively higher levels, the meltwater is supplied to the drainage basin. Among the snow-fed rivers of the United States are the Columbia, Snake, Missouri, Platte, Arkansas, and Colorado. In the eastern United States, the Appalachians serve in a similar but less pronounced fashion for the Ohio and other rivers. Here, however, the snow is melted in late spring and has no influence in the middle and late summer.

Vegetation and life zones

It has been observed that increased elevation brings climatic conditions approximately equivalent to those encountered by increase in latitude. Consequently there exist distinct zones of natural vegetation which, in a general way, recapitulate the vegetation types of increasingly high latitudes.

Effects of increasing altitude on vegetation in wet equatorial lands are illustrated by an example from southern Peru (Figure 17.16*A*). Two other important regions besides the Andes of South America are the Ruwenzori Range of central Africa and the central mountain range of New Guinea. Where exposed to prevailing winds (trades) or to seasonal winds (wet monsoon) which bring moisture of maritime air masses (mT, mE) precipitation increases greatly with elevation. As a result, rainforest extends high up the mountain slopes. Between 4000 and 6000 ft (1200 and 1800 m) the rainforest gradually changes into *montane forest*, which resembles the temperate rainforest found at low elevations farther poleward (Chapter 21). Montane forest is lower and less dense than equatorial rainforest and becomes still lower with increasing altitude. Tree ferns and bamboos are conspicuous. Epiphytes (air plants) are abundant. Mosses increase with altitude, giving what is termed *mossy forest* in the zone of persistent mists and high relative humidities. Near its upper limit the montane forest trees become dwarfed, constituting *elfin forest*, and are densely festooned with mosses. Above the forest limit, which occurs about at 12,000 ft (3600 m), there sets in a treeless vegetation which may be alpine tundra, scrub, grassland, or heath. (See Chapter 21 for a description of these vegetation forms.) Still higher, the

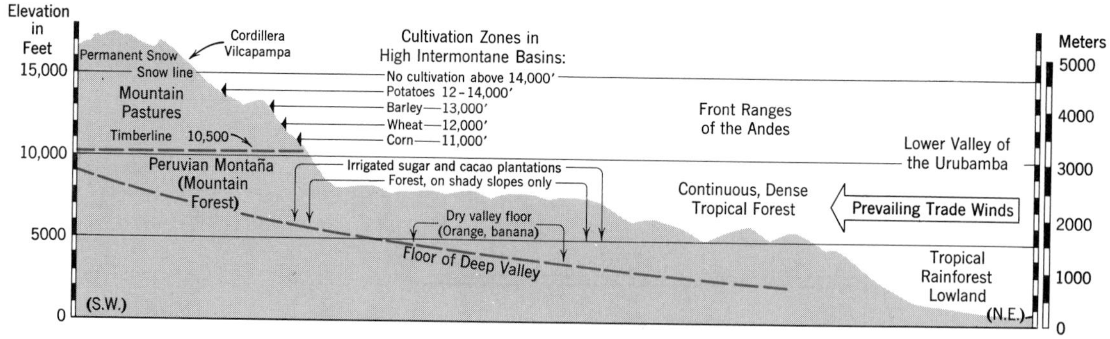

Figure 17.16*A* Altitude zoning of climates in the equatorial Andes of southern Peru, 10°–15° S lat. (After Isaiah Bowman, *The Andes of Southern Peru*, 1916.)

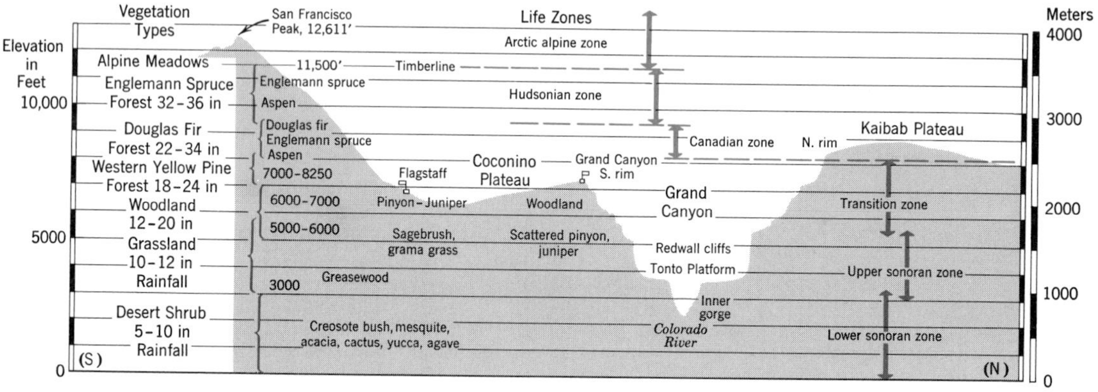

Figure 17.16*B* Altitude zoning of mountain and plateau climates in the arid southwestern United States. Grand Canyon-San Francisco Mountain district of northern Arizona. (After G. A. Pearson, C. H. Merriam, and A. N. Strahler.)

alpine zone gives way to the zone of perpetual snow (Figure 17.16*A*). The snow line lies at about 16,000 ft (5000 m) in central Peru; about at 15,000 ft (4600 m) in the Ruwenzori Range.

In middle latitudes, where steppe or desert exist at low elevations, the zonation is particularly striking. Figure 17.16*B* shows the vegetation zones of the Colorado Plateau region of northern Arizona and adjacent states. Zone names, elevations, dominant forest trees, and annual precipitation data are given in the figure. Ecologists have set up a series of *life zones*, whose names suggest the similarities of these zones with latitude zones encountered in poleward travel on a meridian. The Hudsonian zone, 9500 to 11,500 ft (2900 to 3500 m), bears a needleleaf forest essentially similar to subarctic needleleaf (boreal) forest. Soils are similarly of podzolic type. As the limit of forest, or tree line, is approached the coniferous trees take on a stunted appearance and decrease in height to low shrublike forms.

A vegetation zone of *alpine meadows* (alpine tundra) lies above tree line and resembles in many ways the arctic tundra. The snow line is encountered at about 9000 to 10,000 ft (2750 to 3000 m) in these middle latitudes, which is of course much lower than at the equator. Poleward the snow line decreases in altitude, eventually reaching sea level in the vicinity of the Arctic Circle.

REVIEW QUESTIONS

1. Summarize and explain the characteristics of the continental subarctic climate. How do the Köppen symbols *Dfc, Dfd, Dwc,* and *Dwd* distinguish varieties of this climate? What is particularly noteworthy about the annual temperature cycle? Explain the excessively low winter temperatures. Where is the coldest place in the northern hemisphere?

2. What is perennially frozen ground, or permafrost? What is the origin of permafrost? How great a depth of ground is thawed during the summers in the continental subarctic climate? What are some of the engineering problems associated with permafrost?

3. How does the annual total of precipitation in the northern continental centers compare with that in the tropical deserts? How does evaporation compare in these two regions?

4. Describe the boreal forest and lichen woodland. What is the taiga? What types of soils are found in the continental subarctic climate regions? How has glaciation affected the surfaces of these regions?

5. What are the characteristics of the marine subarctic climate? How is the equability of this climate explained? Compare the marine subarctic climate with the tundra climate.

6. Describe the tundra climate along continental fringes between 65° and 75° N lat. What air masses interact in this zone? In what two regions is cyclonic activity strongly concentrated?

7. Compare the annual temperature cycle of the tundra climate with that of the continental subarctic climate? Compare ranges, warmest month averages, and coldest month averages.

8. What is the tundra? What types of plants are found on the tundra? What kinds of soils are developed here? Describe permafrost conditions in the tundra regions. To what depth is the soil moisture frozen over most of this area? To what depth does the annual summer thaw extend?

9. What is the tree line of the arctic regions? With what monthly isotherm is it approximately identified?

10. What are some of the unusual geomorphic processes active in the tundra regions? Explain the following terms: solifluction, solifluction terraces, polygonal ground, stone rings, stone stripes.

11. What are the essential characteristics of icecap climate? Describe temperature conditions throughout the year. With what air mass source regions is this climate associated?

12. Describe briefly the Greenland and Antarctic icecaps. How is the Greenland icecap nourished, that is, where does the moisture for the snow fields come from?

13. How is sea ice different from ice of icebergs and ice islands? What is the thickness of sea ice? Why is it limited? Describe pack ice, ice floes, leads, and pressure ridges.

14. Explain how the distribution of sea ice differs between the Arctic Ocean and the Antarctic Ocean.

15. How are icebergs formed? What proportion is submerged? Compare the form and source of bergs found in the North Atlantic with those of the antarctic region.

16. What is an ice island? What is its origin? How may it serve the purposes of scientific research?

17. Discuss the influence of increasing altitude upon air pressure. Describe the effects of high altitudes upon physiological processes and upon the boiling point of liquids.

18. What is the effect of increasing altitude upon air temperature? How are maximum and minimum daily temperatures affected? How is the mean annual temperature affected? Is the annual range significantly changed by increased elevation?

19. What influence has increased elevation upon precipitation? Explain this effect. Cite examples of variation of precipitation with elevation.

20. Explain how snow storage in high mountains affects the flow of rivers. Name some snow-fed rivers of the United States.

21. Describe the influence of increased altitude upon natural vegetation and crop cultivation in equatorial regions, using the Andes of Peru as an illustration. What is montane forest? Mossy forest?

22. Describe the influence of increased altitude upon vegetation in the southwestern United States, using the Grand Canyon-San Francisco Mountains region as an illustration. How does the elevation of the alpine meadow zone here compare with that of the upper limit of pastures in the Andes?

23. At what elevation is the snow line encountered near the equator? In middle latitudes? In high latitudes?

Exercises

1. Prepare temperature-precipitation graphs similar to Figure 17.1 for each of the following stations. (Data from Trewartha and Meteorological Office of Great Britain.)

(*a*) *Verkhoyansk, USSR,* 67½° N.

	J	F	M	A	M	J	J	A	S	O	N	D
Temperature, °F	−58.5	−48.5	−26	4.5	32.5	54	56.5	49	35	4.5	−35.5	−54
Precipitation, in.	0.2	0.1	0.1	0.2	0.3	0.9	1.0	1.0	0.5	0.4	0.3	0.1

(*b*) *Point Barrow, Alaska,* 71° N.

	J	F	M	A	M	J	J	A	S	O	N	D
Temperature, °F	−19	−13	−14	−2	21	35	40	39	31	16	0	−15
Precipitation, in.	0.3	0.2	0.2	0.3	0.3	0.3	1.1	0.8	0.5	0.8	0.4	0.4

(*c*) *South Orkneys,* 61° S.

	J	F	M	A	M	J	J	A	S	O	N	D
Temperature, °F	32	33	31	27	19	15	13	15	20	25	28	31
Precipitation, in.	1.5	1.5	1.8	1.7	1.3	1.2	1.2	1.4	1.0	1.0	1.4	0.9

2. (*a*) For each of the above stations, compute the mean annual temperature, temperature range, and annual precipitation total. (*b*) To what climate type does each station belong?

3. Verkhoyansk, Siberia (Exercise 1), represents the region of coldest winter temperature of the northern hemisphere climates. (*a*) Compare the minimum month temperature of Verkhoyansk with Ft. Vermilion, Alberta, and Yakutsk, Siberia (Figure 17.1). (*b*) Study the rate of change of monthly temperature means for successive months in the spring and fall at these stations. What is the greatest change shown for any two successive months?

4. (*a*) Compare the climate of Point Barrow, Alaska (Exercise 1), with that of Upernivik, Greenland (Figure 17.3). (*b*) Which has the colder winter? (*c*) Is January the coldest month for both stations? (*d*) At what time of year does most precipitation occur? (*e*) Compare the precipitation regime of Point Barrow with that of the tropical desert climates in Chapter 15, Exercises 1 and 10. How do the totals compare? (*f*) The South Orkneys (Exercise 1) are a small island group in the South Atlantic. How does the marine location of the Orkneys influence the temperature regime? (*g*) Do the Orkneys receive more or less precipitation than Point Barrow (Exercise 1) and Upernivik (Figure 17.3)?

5. Refer to the soil-moisture budget for Barrow, Alaska, Figure 12.12*H*. Potential evapotranspiration should be zero for those months having an average temperature below the freezing point. Check this statement against the temperature data in Exercise 1.

6. Prepare a temperature-precipitation graph similar to Figure 17.1 for the following station. (Data from Trewartha.)

Kodaikanal, India, 10° N, elevation 7700 ft (2350 m).

	J	F	M	A	M	J	J	A	S	O	N	D
Temperature, °F	55	56	59	61	62	59	58	58	58	57	55	55
Precipitation, in.	2.9	1.4	2.0	4.3	6.0	4.1	5.0	7.0	7.3	9.7	8.2	4.4

7. Kodaikanal, a hill station in southern India popular during the hot season, lies in the Palni Hills, a mountain mass surrounded by lowlands with a tropical wet-dry climate. (*a*) Madras, a nearby station on the east coast of India, has its highest monthly mean temperature (90°F) in May; its lowest (76°F) in January. How do these monthly average temperatures compare with the same months at Kodaikanal? (*b*) Using the lapse rate of 3½°F per 1000 ft, what difference in temperatures might we expect for these two stations? Is this figure roughly comparable to actual temperature differences? (*c*) Does the rainfall regime of Kodaikanal follow the usual monsoon pattern (see Cochin, India, Figure 15.5)?

CHAPTER 18

Soils and Soil-Forming Processes

AN understanding of fundamental principles of soil science, or *pedology*, is indispensable to a geographer. Soils constitute a major environmental factor, influencing by their fertility and special qualities, not only whether a population can be fed, clothed, and housed but also the particular types of food and fiber or lumber products that can be obtained from a region. The systematic study of soils logically follows a study of climates because climate is a primary factor in soil making.

The soil as a dynamic body

Many persons think of the soil as a lifeless, residual layer, which has somehow accumulated over a long period of time and which merely holds a supply of things necessary for plant growth. As soil science has developed, however, it has become known that the soil is a dynamic layer in which many complex chemical, physical, and biological activities are going on constantly. Far from being a static, lifeless zone, it is a changing and developing body. We know now that soils become adjusted to conditions of climate, landform, and vegetation and will change internally when those controlling conditions change.

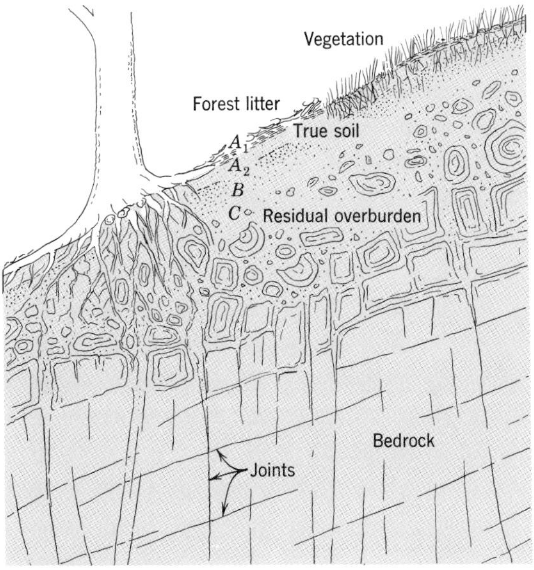

Figure 18.1 Bedrock, residual overburden, and true soil.

The soil scientist restricts the word *soil*, or *solum*, to the surface material, which, over a long period of years, has come to have distinctive layers, or horizons (Figure 18.1). A soil has certain distinctive physical, chemical, and biological qualities, which permit it to support plant growth and which set it off from the infertile substratum, which may consist of *overburden* or solid *bedrock* lying beneath. (See Chapter 22.) The true soil is composed both of mineral and organic particles, whereas the underlying material may be, and usually is, wholly mineral matter.

Soil is made up of substances existing in three states: solid, liquid, and gaseous. For plant growth a proper balance of all three states of matter is necessary.

The solid portion of soil is both inorganic and organic. Weathering of rock produces the inorganic particles that give a soil the main part of its weight and volume (Figure 18.1). These fragments range from gravel and sand down to tiny colloidal particles too small to be seen by an optical microscope. The organic solids consist of both living and decayed plant and animal materials, most being plant roots, fungi, bacteria, worms, insects, and rodents. Colloidal particles of organic matter share with inorganic colloidal particles an important function in soil chemistry.

The liquid portion of soil, the *soil solution*, is a complex chemical solution necessary for many important activities that go on in the soil. Soil without water cannot have these chemical reactions, or can it support life. Gases in the open pore spaces of the soil form the third essential component. They are principally the gases of the atmosphere, together with gases liberated by biological and chemical activity in the soil.

For an understanding of soils, information is needed about (1) the physical-chemical properties and materials of soils, and (2) processes that make and maintain soils.

Physical-chemical make-up of soils

Although a minor factor in itself, *soil color* is perhaps the characteristic that is first noticed about a soil. Color can tell much about how a soil is formed and what it is made of. Soil horizons are

TABLE 18.1 SOIL TEXTURE GRADES

(U.S. Department of Agriculture)

Name of Grade	Diameter, In	Diameter, Mm
Coarse gravel	Above 0.08	Above 2
Fine gravel	0.04–0.08	1–2
Coarse sand	0.02–0.04	0.5–1
Medium sand	0.01–0.02	0.25–0.5
Fine sand	0.004–0.01	0.1–0.25
Very fine sand	0.002–0.004	0.05–0.1
Silt	0.000,08–0.002	0.002–0.05
Clay	Below 0.000,08	Below 0.002

usually distinguishable by color differences. One sequence of colors ranges from white, through brown, to black as a result of an increasing content of *humus* which is finely divided, partially decomposed organic matter. Abundance of humus depends in a general way on luxuriance of vegetation and upon intensity of microbial activity, which in turn depend on climate. Thus we find that in middle latitudes, soils range from black or dark brown in the cool, humid areas to light brown or gray in the semi-arid steppe lands and deserts.

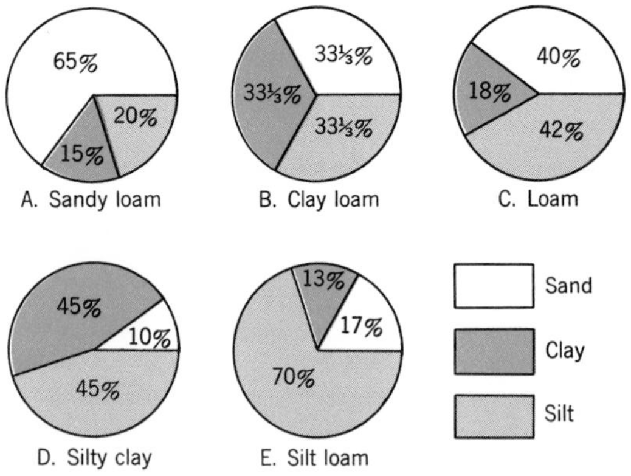

Figure 18.3 Typical compositions of five soil texture classes. These examples are shown as lettered points on Figure 18.2. (After U.S. Dept. of Agriculture, *Yearbook of American Agriculture*, 1938.)

Desert soils have little humus.

Reds and yellows are common colors in soils and are the result of small quantities of iron compounds. The red color is particularly associated with *sesquioxide of iron* (Fe_2O_3), whereas the yellow color may indicate the presence of this same iron compound combined with water (*hydrated iron oxide*). Red color indicates that the soil is well drained, but locally the color may be

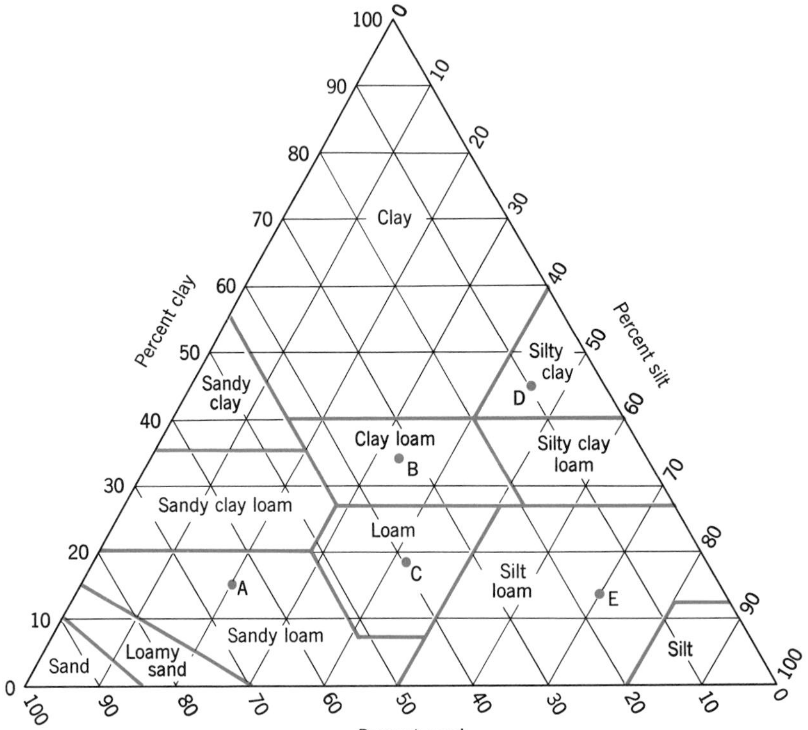

Figure 18.2 Texture classes shown as areas bounded by heavy lines on a triangular graph. (See also Figure 18.3.) (From U.S. Department of Agriculture and Millar, Turk, and Foth, *Fundamentals of Soil Science*, John Wiley and Sons, New York.)

derived from a red source rock such as a red shale or sandstone.

Grayish and bluish colors in soils of humid climates often mean the presence of reduced iron compounds (such as FeO) in the soil and indicate poor drainage or bog conditions. Grayish soils in dry climates mean a meager amount of humus; a white color may be a result of the deposit of salts in the soil. Although some recently formed soils retain the color of the parent overburden or bedrock, the color of fully developed soils is independent of what lies beneath.

Soil texture, a major characteristic of the soil, refers to particle sizes composing the soil. Particles are classified as various grades of gravel, sand, silt, and clay, in decreasing order of size (Table 18.1).

The U.S. Department of Agriculture has set up standard definitions of soil-texture classes in which the proportions of sand, silt and clay are given in percentages. Rather than to attempt to list these classes and give the limiting percentages for each, the information is given in a triangular diagram (Figure 18.2), which enables the percentages of all three components to be shown simultaneously. The corners of the triangle represent 100% of each of the three grades of particles—sand, silt, or clay. The word *loam* refers to a mixture in which no one of the three grades dominates over the other two. Loams therefore appear in the central region of the triangle. A particular soil whose components give it a position at Point *A* in the triangle has 65% sand, 20% silt, and 15% clay; it falls into a texture class known as *sandy loam.* Another soil, whose texture is represented by Point *B* has $33\frac{1}{3}\%$ sand, $33\frac{1}{3}\%$ silt, and $33\frac{1}{3}\%$ clay; it falls into the class of a *clay loam.* Figure 18.3 gives five examples of soil textures; their positions are shown on Figure 18.2.

Texture is important because it largely determines the water retention and transmission properties of the soil as explained later in this chapter. Sand may drain too rapidly; in a clay soil the individual pore spaces are too small for adequate drainage. Where clay and silt proportions are high, root penetration is difficult. Generally speaking, the loam textures are best for plant growth.

Included in the clay grade, smaller than 0.002 mm (0.000,08 in), are mineral soil colloids. Colloid particles are so small that they cannot be seen by optical microscope and will remain suspended indefinitely in water. Individual particles are in the form of thin flakes in the size range 0.001 to 1.0 microns[1] diameter. Finely divided humus provides another class of soil colloids, which may be referred to as *humus colloids*, or *organic colloids.*

Unusual chemical properties of colloids result from their very vast surface area for a given weight. Colloids have the property of being electrically charged and can therefore attract and hold *ions*, the unit chemical particles of dissolved substances. Ions of calcium, magnesium, and potassium are known in soil science as *bases.* These bases may be given up by the colloids to plants, which require them for growth, by a process known as *base*

[1]One micron = 0.001 mm = 0.00004 in.

TABLE 18.2 SOIL ACIDITY AND ALKALINITY[a]

<table>
<tr><td>pH</td><td colspan="10">4.0 4.5 5.0 5.5 6.0 6.5 6.7 7.0 8.0 9.0 10.0 11.0</td></tr>
<tr><td>Acidity</td><td colspan="2">Very strongly acid</td><td>Strongly acid</td><td>Moderately acid</td><td>Slightly acid</td><td>Neutral</td><td>Weakly alkaline</td><td>Alkaline</td><td>Strongly alkaline</td><td>Excessively alkaline</td></tr>
<tr><td>Lime requirements</td><td colspan="2">Lime needed except for crops requiring acid soil</td><td colspan="2">Lime needed for all but acid-tolerant crops</td><td>Lime generally not required</td><td colspan="5">No lime needed</td></tr>
<tr><td>Occurrence</td><td>Rare</td><td>Frequent</td><td colspan="2">Very common in cultivated soils of humid climates</td><td colspan="2"></td><td colspan="2">Common in subhumid and arid climates</td><td colspan="2">Limited areas in deserts</td></tr>
<tr><td>Soil groups</td><td></td><td>Podzols</td><td colspan="2">Gray-brown podzolic soils
Tundra soils</td><td colspan="2">Brown forest soils
Prairie soils
Latosols</td><td colspan="2">Chestnut and brown soils</td><td colspan="2">Black alkali soils</td></tr>
<tr><td></td><td colspan="4"></td><td colspan="4">Tropical black earths</td><td colspan="2"></td></tr>
</table>

[a]Based on data of C. E. Millar, L. M. Turk, and H. D. Foth (1958), *Fundamentals of soil science*, third edition, John Wiley and Sons, New York, 526 pp. See Chart 4.

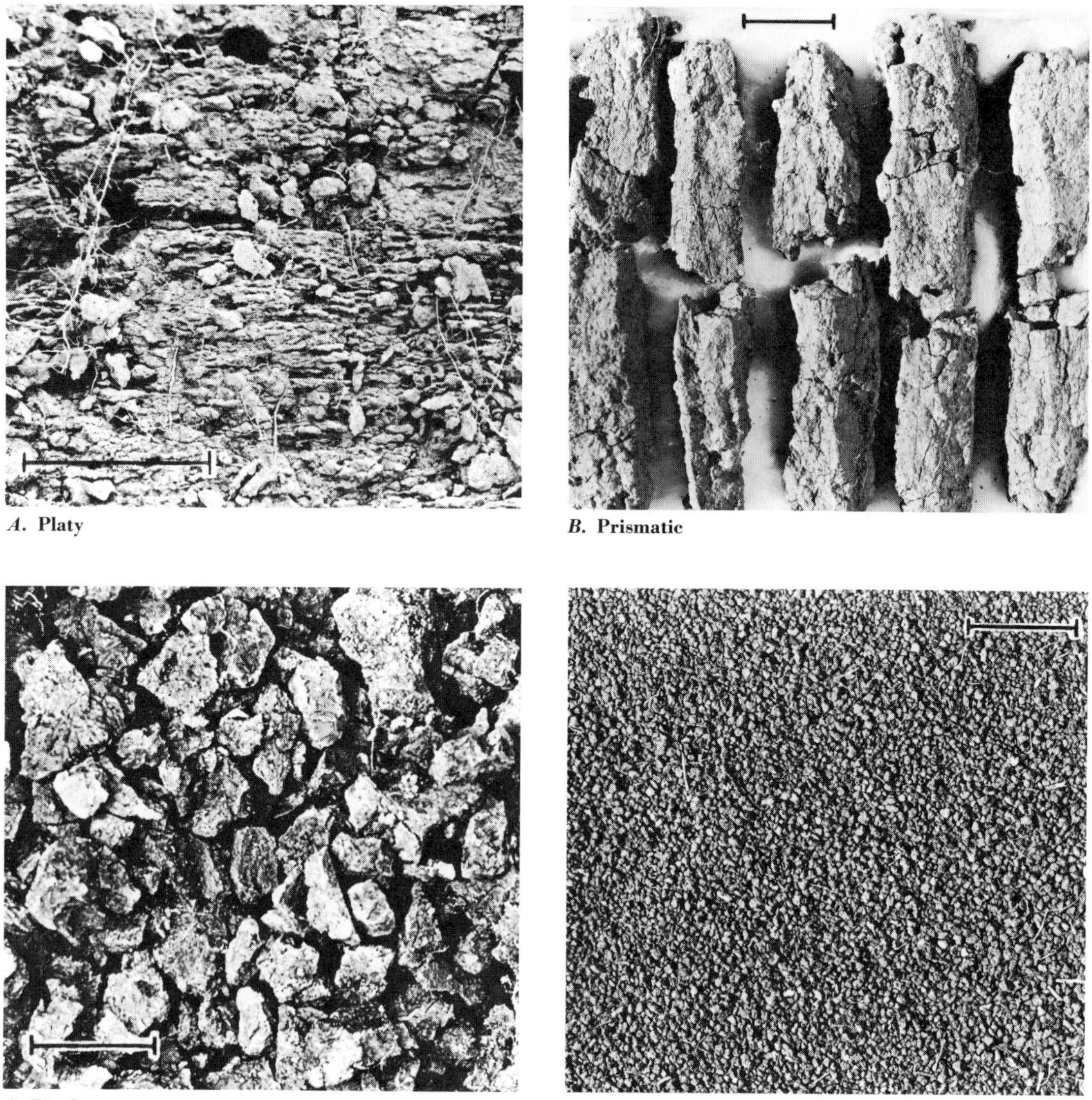

A. **Platy**

B. **Prismatic**

C. **Blocky**

D. **Granular**

Figure 18.4 Four basic soil structures are illustrated here. The black bar on each photograph represents one inch. (Photographs *A* and *C* by Roy W. Simonson; *B* and *D* by C. C. Nikiforoff. Courtesy of Division of Soil Survey, U.S. Dept. of Agriculture.)

exchange. On the other hand, the *hydrogen ion* in the soil solution makes for an *acid* condition. The concentration of hydrogen ions in the soil solution, relative to hydroxyl ions, is known as the pH of the soil, and is a measure of soil acidity or alkalinity. Hydrogen ions are held also by the soil colloids in exchange positions. In soils a value of 7.0 in the pH scale is *neutral,* whereas values are below 7.0 (4.0 to 7.0) in *acid soils* but above 7.0 (7.0 to 10.0) in *alkaline soils.* Table 18.2 will help to explain the scale of acidity and alkalinity as applied by agronomists to soils.

Soil colloids also are useful in holding water in the soil. When present in large quantities colloids may make the soil sticky and tough so that it is difficult to cultivate.

Soil structure refers to the way in which soil grains are grouped together into larger pieces held together by soil colloids (Figure 18.4). Irregular pieces with sharp corners and edges give a *blocky* or nutlike structure. More or less spherical pieces make *granular* and *crumb* structure. Some soils have *columnar* and *prismatic* structure, made up of vertical columns or prisms 0.2 to 4 in (0.5 to 10 cm) across. *Platy* soil structure consists of plates, or flat pieces, in a horizontal position. Soil structure influences the rate at which water is absorbed by the soil, the susceptibility of the soil

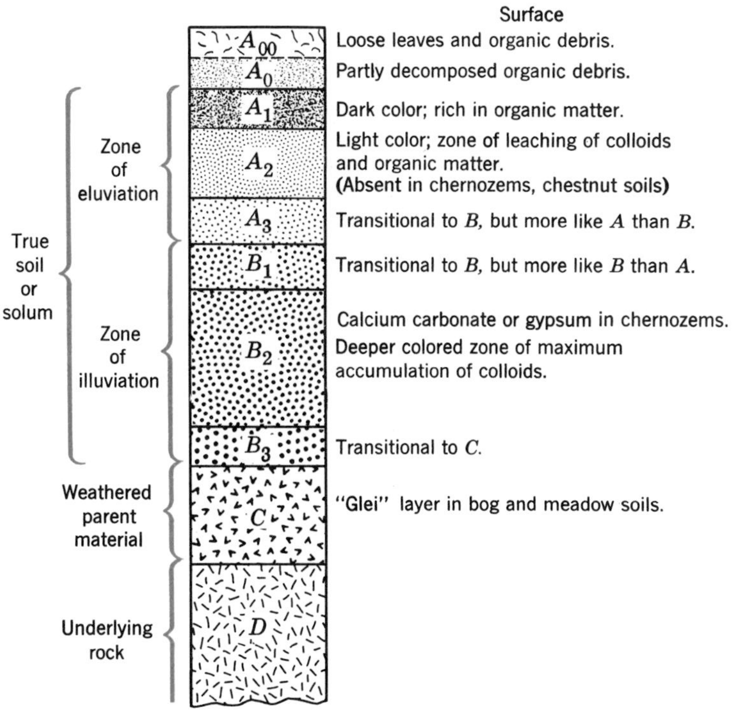

Figure 18.5 Horizons of soil profile. (After U.S. Dept. of Agriculture, 1938.)

to erosion, and the ease of soil cultivation.

Another constituent of the soil is *soil air*, which occupies the pore spaces of the soil when it is not saturated with water. Soil air has been analyzed and found to contain an excess of carbon dioxide, but a deficiency of oxygen and nitrogen.

Soil water, the water temporarily held in the soil, is in reality a complex chemical solution. It is a dilute solution of such substances as bicarbonates, sulfates, chlorides, nitrates, phosphates, and silicates of calcium, magnesium, potassium, sodium, and iron.

The term *soil profile* denotes the arrangement of the soil into layerlike *horizons* of differing texture, color, and consistency (Figure 18.5). Soils are recognized and classified into broad groups on the basis of the parts of the profile that are present. Basically there are three parts to the soil profile. Horizons *A* and *B* represent the true soil, or *solum;* horizon *C* is the subsoil, or weathered parent body. Below this is the parent bedrock or other underlying rock, designated as horizon *D*. The *A* horizon in humid climates is composed of two very different parts. The upper, or A_1, horizon is rich in organic matter and is dark colored. The lower, or A_2, horizon is a zone of leaching. The *B* horizon is usually a zone of accumulation of soil colloids and is dark in contrast to the A_2 horizon above it. These processes and the horizons they produce are discussed more fully in later pages.

Soil-forming processes and factors

Many types of processes and influences, known altogether as *soil formers*, act together to develop a soil. Some of these are *passive* conditions; others are *active* agents. Five principal soil formers are (1) parent material, (2) landform, (3) time, (4) climate, and (5) biological activity.

The first of the passive soil formers is *parent material*, the residual or transported overburden of disintegrated rock making up the bulk of the soil. Certain of the original rock forming minerals have been thoroughly changed chemically into new compounds and reduced to colloidal size. Although many people think that the type of parent material alone determines the kind of soil that is present, this is an inaccurate concept. For example, the same granite forms the bedrock in the Piedmont region in both Maryland and Georgia, but because of climatic differences the soils of these two states are somewhat different. On the other hand, soils of the same major groups may be found to overlie two different types of overburden or bedrock.

An exception to the general rule that soil type is independent of parent material origin is found in young soils that have not had enough time to develop, and in some limestone areas where the influence of the rock is especially strong. Parent material may locally exert a strong control over the soil texture. For example, on the sandy cuestas of the New Jersey coastal plain, the soil consists largely of quartz sand and is very porous. On glacial clays of the Hudson Valley the soil is unusually sticky and dense, deterring rainfall from seeping through.

Another passive soil former is *landform*, or ground-surface configuration (Figure 18.6). Where slope is steep, surface erosion by runoff is more rapid and water penetration is less than on gentle slopes. As a result, the soil will be thinner on steeper slopes. Flat upland areas accumulate a thick soil that has a thick layer of dense clay (clay pan) and is excessively leached. There the products of weathering tend to remain in place. Flat bottom lands likewise have thick soils, but they are poorly drained and dark colored. Here, constant saturation retards decay of vegetation and allows organic matter to accumulate. Gentle slopes where drainage is good but erosion is slow are considered the norm for soil formation. Slow, continuous erosion is a normal geologic soil process whereby the removal balances the formation of new soil from the parent material. Only when this erosion is greatly accelerated does it become harmful to the soil.

Another influence of landform is the slope aspect or direction of exposure of the surface to the slanting rays of the sun. In middle latitudes it is common to find that south-facing slopes, exposed to the warming and drying effects of sunlight, have different conditions of vegetation and soils from north-facing slopes, which retain cold and moisture longer.

A third passive factor in soil formation is *time*. A soil is said to become *mature* when it has been acted upon by all soil-forming processes for a sufficient long time to have developed a profile that changes only imperceptibly with further passage of time. Soils that are evolving from recently deposited river alluvium or glacial till, for example, are considered *young*. In young soils the characteristic horizons are absent or poorly developed. No age in terms of years can be given to all mature soils because the rate at which a mature soil is developed depends on many other factors. Some soils of humid regions in sandy localities may require 100 to 200 years to develop, whereas more commonly, several thousand years may be needed to produce a mature soil. Some soils of tropical and equatorial regions are thought to be as old as one to six million years, or of Pliocene geologic age. As with development of landforms, age of soils is purely relative. Another way of defining a mature soil might be to say that it is in equilibrium with the many processes and forces acting upon it.

Climate and soil

Of the active soil formers, climate is perhaps the most important. Climatic elements involved in soil development are (1) *moisture conditions* affecting the soil (precipitation, evaporation, and humidity), (2) *temperature*, and (3) *wind*.

Precipitation provides the soil water, without which chemical and biological activities are not possible. When soluble chemicals are dissolved in water they ionize, or dissociate into positively and negatively charged particles. Without ionization the many complex chemical interchanges of elements necessary to soil development and plant growth cannot take place. An excess of precipita-

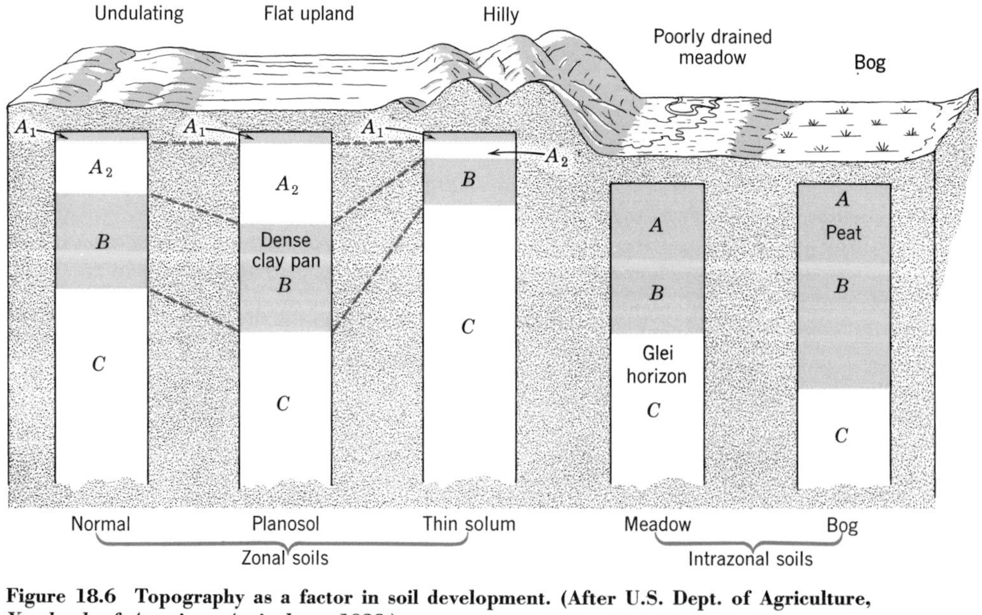

Figure 18.6 Topography as a factor in soil development. (After U.S. Dept. of Agriculture, *Yearbook of American Agriculture*, 1938.)

tion, however, tends to leach away the colloids and ions. This process of downward migration of soil components by waters percolating through the soil is known as *eluviation.* A distinctive leached horizon of the soil, the A_2 horizon, results from this process. (See Figure 18.5). The deposition of colloids and bases in the underlying B horizon is a process known as *illuviation.*

In warm climates where rainfall is extremely heavy, silica (SiO_2) is largely removed from the soil and carried off in streams. This process is termed *desilication.* Thus soils in the wet equatorial rain-forest belts are deficient in silica as well as in such bases as calcium, sodium, magnesium, and potassium, and are generally low in fertility.

In dry climates, evaporation exceeds precipitation and the soil is dry for long periods. Ground water is slowly brought to the surface by capillary attraction and evaporates in the soil, leaving behind the salts that were dissolved in the water. Calcium carbonate, the commonest of these deposits, forms a whitish crust, or *hardpan,* in the soil. In the southwestern United States this material is called *caliche,* and in places it makes the soil as hard and resistant to erosion as if it were a limestone. Gypsum (hydrous calcium sulfate) forms similar encrustations. In intermediate precipitation zones, such as the humid eastern border of the middle-latitude steppes, calcium carbonate appears as small nodules in the soil.

Rainfall and evaporation controls result in the formation of two major groups of soils (Figure 18.7). (1) *Pedalfer soils* show pronounced leaching and occur in the eastern United States where the rainfall is more than 25 in (60 cm) annually. (2) *Pedocal soils* have an excess of calcium carbonate and occur in the western United States where rainfall is less than 25 in (60 cm). The names are coined from the chemical content of the soil. The syllables *al* and *fer* in pedalfer refer to aluminum and iron, respectively, and were chosen because

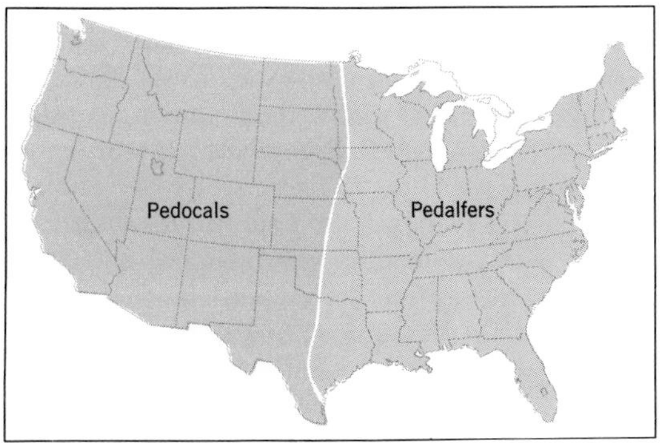

Figure 18.7 Major soil classes of the United States. (After C. F. Marbut.)

excessive leaching leaves behind the aluminum and iron oxides as residual substances which thus become important in quantity. The syllable *cal* in pedocal refers, of course, to calcium, which is present in the carbonate form in all pedocal soils.

Temperature is another important climatic factor in soil formation. It acts in two ways. (1) Chemical activity is generally increased by higher temperatures but reduced by cold,[2] and it ceases when soil water is frozen. Thus, tropical soils have a parent material which is thoroughly altered chemically, whereas soils of the frozen tundra have a parent material which is composed largely of mechanically broken minerals. (2) Bacterial activity is increased by warmer soil temperatures. Where bacteria thrive, as in the humid tropics, they consume all dead plants that lie upon the ground. Thus there is no layer of decomposing vegetation on the ground and little humus within the soils of the humid tropics. In cold continental climates, bacterial action is reduced, and a generous layer of decomposing vegetation (leaf mold) covers the ground under forests. Hence, raw humus is preserved at the soil surface and is important in the upper part of the mature soil profile.

Wind is of minor importance as a climatic factor in soil development. Winds may increase the evaporation from soil surfaces and may remove surface soil in arid regions lacking a plant cover. Windblown dust may accumulate and thereby provide the parent material of a soil.

Biological soil formers

Both plants and animals profoundly influence soil development. The plant kingdom consists of the *macroflora* (trees, shrubs, and herbs) and *microflora* (bacteria and fungi).

Grasses and trees require somewhat different chemical substances for growth. Trees, particularly the conifers, use little calcium and magnesium. Hence they thrive well in the pedalfer soils from which these substances have been leached and which are usually acid. Grasses and small grains (wheat, oats, barley) need abundant calcium and magnesium and do well in the pedocal soils of the semi-arid and marginal lands. For grasses to grow well in acid pedalfers, calcium must be added to the soil in the form of lime or crushed limestone. Plants tend to maintain the fertility of soil by bringing the bases (calcium, magnesium, potassium) from lower layers of the soil into the plant stems and leaves, then releasing them to the soil surfaces as the plant decomposes.

[2]The process of *carbonation,* or reaction of carbonic acid (H_2CO_3) upon minerals, is believed to be quite active at low temperatures because the concentration of carbonic acid is greater in cold water than in warm.

Dead plants provide *humus*, the finely-divided organic matter of the soil. Humus gives a dark brown or black color to the soil, as already noted. Humus particles of colloidal size act in the same way as mineral colloids in holding ions in the soil. The process of humus development, or *humification*, is essentially the slow oxidation, or burning, of the vegetative matter. Acids, known as the *organic acids*, are formed during humification. They aid in decomposing the minerals of the parent soil material. The hydrogen ions of the acid solution tend to replace the ions of potassium, calcium, magnesium, and sodium, which are removed in the leaching process. Soils of the cold humid climates are therefore deficient in the bases and are consequently of low fertility for crop farming. The deficiency of nutrients is remedied by applications of fertilizers rich in nitrogen, phosphorus, and potassium.

Turning now to the microflora, or bacteria and fungi, we find that bacteria consume humus. In cold climates bacterial growth is slow, hence, humus may accumulate on and in the soil. Soils of the subarctic and tundra climates have much undecomposed organic matter, which locally forms layers of peat, but in humid tropical and equatorial climates, bacterial action is intense and most dead vegetation is rapidly oxidized by bacteria. Here humus content of the soil is low. The organic acids formed by humus are therefore also lacking, and certain bases such as aluminum, iron, and manganese accumulate in a large proportion relative to silica. In this way the fundamental differences in soils of cold and warm climates can be traced back to intensity of bacterial activity.

Another function of some bacteria and other soil organisms (algae) is to take gaseous nitrogen from the air and convert it into a chemical form that can be used by plants. This process is known as *nitrogen fixation*. One kind of bacteria (*Rhizobium*) lives in the root nodules of leguminous plants and there fixes nitrogen beneficial to the plant host.

The influence of animals in the soil is largely mechanical, but nevertheless important. Earthworms are a particularly important agent in humid regions. They not only continually rework the soil by burrowing, but also change the texture and chemical composition of the soil as it passes through their digestive systems. Ants and termites bring large quantities of soil from lower horizons to the surface. Such burrowing animals as prairie dogs, gophers, ground squirrels, moles, and field mice disturb and rearrange the soil. Digging of burrows brings soil of lower horizons to the surface; collapse of burrows carries surface soil into lower horizons.

The pedogenic regimes

The foregoing analysis of soil-forming processes has been piecemeal in the sense that each factor and activity has been considered separately and in serial order. If the topics seemed in some cases to be out of order and some repetition occurred, it was because the soil-forming processes act in concert and affect the action of one another. To bring the study of soils into a more unified perspective we may concentrate upon several basic trends in soil development, each leading to formation of a distinctive major soil group under the control of a particular climatic regime. Such basic trends may be referred to as *pedogenic regimes*.

The regime of *podzolization* dominates in climates having sufficient cold to inhibit bacterial action, but sufficient moisture to permit larger green plants (macroflora) to thrive (Figure 18.8). Such conditions exist only in middle and high latitudes, and at high altitudes. The corresponding climatic regime may be equable, provided that it is prevailingly cool (marine west coast climates, poleward of 40° latitude), or it may be a continental regime which has cold winters and adequate precipitation distributed throughout the year (humid continental climate; continental subarctic climate). In its extreme development podzolization is associated with coniferous trees (spruce, fir, hemlock, pine). These plants do not require the bases (calcium, magnesium, and potassium) and hence do not restore them to the soil surface. The result is that humic acids, produced from the abundant leaf mold and humus, leach the upper soil strongly of bases, colloids, and the oxides of iron and aluminum, leaving a characteristic ashgray A_2 soil horizon composed largely of silica (SiO_2) (Figure 18.1). Colloids, humus, and oxides of iron carried out of the A_2 horizon accumulate in the *B* horizon, which may be dark in color, dense in structure, and in some cases hardened to rocklike consistency (*ortstein*).

The pedogenic regime of *laterization* is in some respects a warm-climate relative of podzolization, in that both are associated with climatic regimes of ample precipitation and with forests. Laterization takes place in a warm climate having copious rainfall well distributed throughout the year (equatorial rainforest climate; tropical wet-dry climate with long wet season; humid subtropical climate). A high mean annual temperature and a lack of severe winter season permit sustained bacterial action which destroys dead vegetation as rapidly as it is produced. Consequently little or no humus is found upon or in the soil (Figure 18.8*B*). In the absence of humic acids the sesquioxides of iron (Fe_2O_3) are insoluble and accumulate in the soil

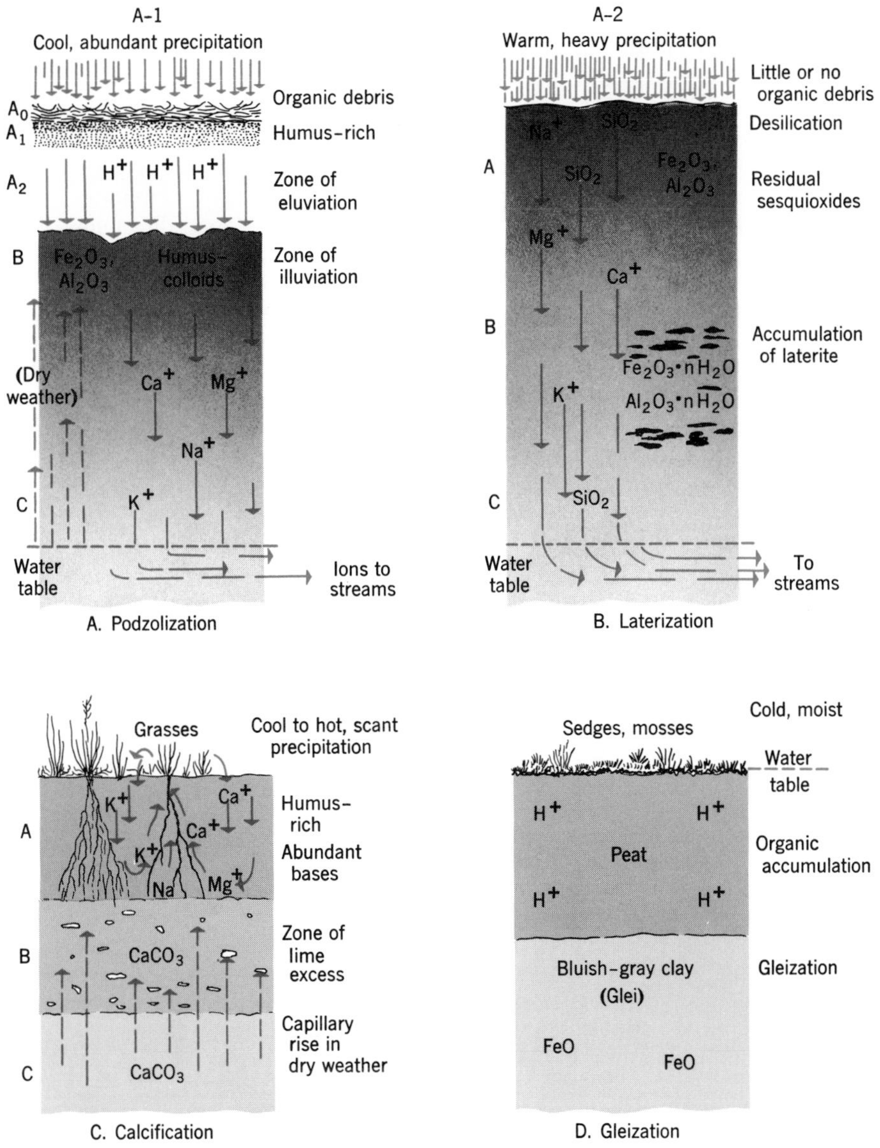

Figure 18.8 Soil development under four pedogenic regimes.

as red clays, nodules, and rocklike layers (*laterite*). Silica, on the other hand, is leached out of the soil and disposed of eventually by stream flow in the process of *desilication*. No distinctive soil horizons are developed. In the absence of silicate colloids the soil tends to be firm and porous rather than sticky and plastic, and will transmit water readily. Laterization results in very low soil fertility because bases are not held in the soil and humus is lacking.

Calcification is a pedogenic regime of climates in which evaporation on the average exceeds precipitation. Calcification is associated with a continental climatic regime with low total annual precipitation (middle-latitude steppe climate) and with a tropical wet-dry climatic regime with a short wet season (tropical steppe climate). Rainfall is not enough to leach out the bases, so that calcium and magnesium ions remain in the soil (Figure 18.8*C*). Grasses, which use these bases, restore them to the soil surface. Colloids remain essentially in place and are not leached out, but are in a dense (flocculated) state and hold the soil into aggregate structures. Calcium carbonate, brought upward by capillary water films and evaporated in dry periods, is precipitated in the *B* horizon of the soil in the form of nodules, slabs, and even dense stony layers (caliche). Microbial activity is restricted and humus may be abundantly distributed throughout the *A* and *B* horizons. Humus occurs in progressively smaller amounts as one traces the soil into climate zones of increasing aridity. Calcification is characteristically associated with grasslands—the steppes and semi-deserts.

The pedogenic regime of *gleization* is characteristic of poorly drained (but not saline) environments under a moist and cool or cold climate. Gleization is thus associated with the climatic regime of polarization (tundra climate) but is also effective in bog environments of continental climates with cold winters. Low temperatures permit heavy accumulations of organic matter to form a surface layer of peaty material (Figure 18.8*D*). Beneath this is the *glei horizon,* a thick layer of compact, sticky, structureless clay of bluish-gray color. The glei horizon lies generally within the zone of ground water saturation; consequently the iron is in a partially reduced condition and imparts the bluish-gray color.

Finally, there is the pedogenic regime of *salinization,* or accumulation of highly soluble salts in the soil. Salinization is associated with the desert climatic regime and takes place in poorly drained locations where surface runoff evaporates. Such locations are typically low-lying valley floors, flats, and basins in the continental interiors; and coastal flats in arid climates. Sulfates and chorides of calcium and sodium are common salts in such soils.

The pedogenic regimes form the basis for classifying the soils of the world into a number of great soil groups, discussed in Chapter 19. The pedogenic regimes also underlie, or are closely interrelated with the geography of plants.

REVIEW QUESTIONS

1. In what way is the soil a dynamic, rather than a lifeless, static body? To what general influences does the soil respond?

2. How is the term *soil* defined by the soil scientist? Of what three states of matter is soil made up? Are all three necessary for plant growth? Describe the types of substances that make up the soil.

3. What is the significance of soil color? What are some of the common soil colors, and what do they mean? In general, what relationship does soil color bear to climate?

4. What is meant by soil texture? What are the various soil textures recognized by the U.S. Department of Agriculture? What influence has soil texture upon agricultural use of the land?

5. What are some of the common soil structures? Describe four structures. How does soil structure influence the rate at which water is absorbed and transmitted by the soil?

6. What are soil colloids? Explain how colloids hold ions. What are the bases commonly present in soils? What is base exchange? What is the relation of hydrogen ion concentration to acidity of the soil? What is meant by pH of the soil?

7. What is soil air? In what way is it different from normal air of the free atmosphere? What are some of the substances commonly found in soil water?

8. Describe an idealized soil profile containing *A*, *B*, and *C* horizons. Which of these horizons constitute the true soil?

9. List the five principal soil formers, including both passive conditions and active agents.

10. From what types of parent material can soils be derived? Does the type of parent material strongly control the characteristics of a fully developed soil profile? Give examples to illustrate. How may young soils strongly reflect the nature of the parent materials? Give examples.

11. How does landform influence development of the soil profile? Explain how various degrees of slope influence thickness of the profile. What conditions of slope are best for the development of soils suited to agriculture?

12. What is a mature soil? Does a mature soil profile continue to evolve through a series of changing forms? How long does it take for a mature soil to develop?

13. What influence has precipitation upon the soil profile? What is eluviation? What is illuviation? What horizons of the soil are affected by these processes? What is meant by desilication? Where is this process most intensively developed?

14. How do soils of dry climates differ fundamentally from soils of humid climates? What two major soil classes have been recognized on this basis?

15. How does temperature affect soil formation? Explain the control exerted by temperature over the type of weathering of the parent material and upon the accumulation of organic matter.

16. How are the needs of different plant groups suited to particular groups of soils? Are plants essential in maintaining the profile of a mature soil?

17. Explain the importance of humus in soil development. What is humification? What acids are formed during humification? What is the action of these acids in the soil? How does bacterial activity determine the amount of humus in the soil? How does this factor determine the fundamental difference between soils of equatorial and arctic regions?

18. What is meant by nitrogen fixation? How is it accomplished?

19. What important influence have earthworms, ants, termites, and burrowing animals upon the soil?

20. Name the basic pedogenic regimes and give the climatic regime in which each develops.

21. Describe and explain the soil profile characteristics associated with each pedogenic regime.

Exercises

1. Refer to Figure 18.2. Estimate the limiting percentages of sand, silt, and clay that define each of the twelve texture classes. Example: *silty clay;* sand 0 to 20 percent, silt 40 to 60 percent, clay 40 to 60 percent.

2. Given below are percentages of sand, silt, and clay in a series of five soil samples. ***(a)*** Refer to Figure 18.2 and locate each sample on the diagram. (Place a sheet of tracing paper over the page; trace the outline of the triangle, and indicate the location of each point by number.) ***(b)*** Name the texture class to which each sample belongs. ***(c)*** Construct a pie-diagram, similar to those in Figure 18.3, for each sample.

	Sand (%)	Silt (%)	Clay (%)
(1)	15	51	34
(2)	72	14	14
(3)	10	85	5
(4)	18	32	50
(5)	47	32	21

CHAPTER 19

The Classification of World Soils

THE soil scientist recognizes that all soils can be subdivided into three orders, known as *zonal, intrazonal,* and *azonal* orders. Zonal soils, formed under conditions of good soil drainage through the prolonged action of climate and vegetation, are by far the most important and widespread of the three orders. Intrazonal soils are simply those formed under conditions of very poor drainage (such as in bogs, flood-plain meadows, or in the playa lake basins of the deserts) or upon limestones whose influence is dominant.

Azonal soils have no well-developed profile characteristics, either because they have had insufficient time to develop or because they are on slopes too steep to allow profile development. Azonal soils include thin, stony mountain soils of the earth's mountain regions (*lithosols*), freshly laid alluvial materials, or dune sands (*regosols*). These usually have poorly developed profiles and cannot be easily classified, whereas the zonal and intrazonal soils have distinctive profile characteristics as the result of long development.

The great soil groups

A widely used classification of soils is illustrated in Table 19.1. The U.S. Department of Agriculture in 1938 recognized about 30 *great soil groups,* but these can be simplified into about 18 principal varieties, including both zonal and intrazonal orders, that have world-wide distributions under similar climatic and geomorphic conditions. Various zonal soil groups have already been mentioned and briefly discussed in earlier chapters describing climates. Combining knowledge of the pedogenic processes and regimes gained in Chapter 18 with an understanding of climatic elements, we can proceed to a systematic study of each of the great soil groups. Some of the soil groups bear Russian names, applied by soil scientists of Russia who pioneered in this field.

The founder of modern theories of soil origin and classification was V. V. Dokuchaiev, a Russian geologist; his studies between 1882 and 1900 led him to the concept that soil is an independent body whose character is determined primarily by climate and vegetation. A Russian follower of Dokuchaiev, K. D. Glinka, expanded the concepts of horizons in the soil profile. For the development of modern soil science in the United States during the 1920's and 1930's, much credit is given to C. F. Marbut, who served for many years as chief of the Soil Survey Division of the United States Department of Agriculture. Marbut translated Glinka's work into English, borrowed and modified the Russian pedologic views and ultimately created a comprehensive system of soil classification for the United States. It is Marbut's system, with further modifications, that is presented here.

Podzol soils

Of the zonal soils of cool humid climates the most widely distributed are the *podzol soils* (or simply, the *podzols*), found closely associated with the subarctic climate, the more northerly parts of the humid continental climate, and the cooler parts of the marine west coast climate. Podzol soils require a cold winter and adequate precipitation distributed throughout the year. (Plate 3 is a world soils map which may be referred to in conjunction with the world maps of climate and vegetation, Plates 2 and 4.) The pedogenic regime is that of podzolization (Chapter 18).

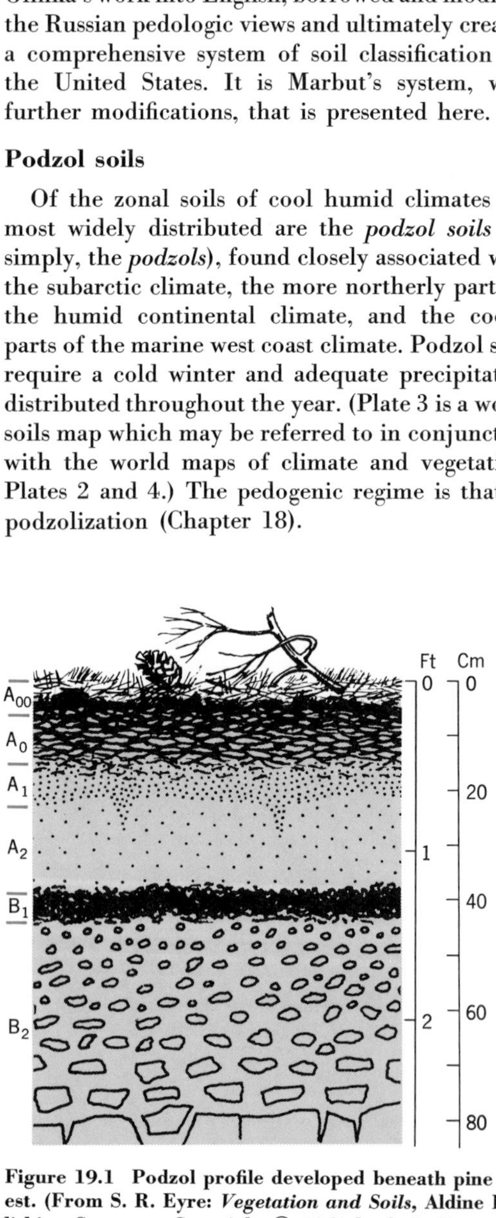

Figure 19.1 Podzol profile developed beneath pine forest. (From S. R. Eyre: *Vegetation and Soils,* Aldine Publishing Company. Copyright © 1968 by S. R. Eyre.)

The podzol profile (Figure 19.1 and color plate, *A*) is the soil profile from which the various soil horizons were originally named. At the very top, lying on the soil surface, is a layer of leaf mold and acid humus termed the A_0 horizon. Below this is the first soil layer, designated as the A_1 horizon. It is a thin, acid layer, rich in humus and varying in color from gray through yellowish brown to a reddish brown. The A_1 horizon is rich in colloids and is a zone of interaction between acids and bases.

Below the A_1 horizon is a distinctive light-colored zone called the A_2 horizon. This is the strongly leached horizon from which colloids and bases have been carried down. It sometimes has a bleached ash-gray or whitish-gray color because coloring agents such as the iron oxides and colloidal humus have been removed. Leaching such as

TABLE 19.1 CLASSIFICATION OF SOILS[a]

Suborders	Great Soil Groups
ZONAL ORDER	
Light-colored podzolized soils of forested regions	PODZOL SOILS Brown podzolic soils GRAY-BROWN PODZOLIC SOILS RED-YELLOW PODZOLIC SOILS (Incl. *terra rossa*)
Lateritic soils of warm, moist subtropical, tropical, and equatorial regions	LATOSOLS Reddish-brown lateritic soils Black and dark-gray tropical soils
Soils of the forest-grassland transition	Degraded chernozem soils
Dark-colored soils of the semi-arid, subhumid, and humid grasslands	PRAIRIE SOILS (BRUNIZEM SOILS) Reddish prairie soils CHERNOZEM SOILS CHESTNUT SOILS REDDISH-CHESTNUT AND REDDISH-BROWN SOILS
Light-colored soils of arid regions	BROWN SOILS GRAY DESERT SOILS (SIEROZEM SOILS) RED DESERT SOILS
Soils of the cold zone	TUNDRA SOILS Arctic brown forest soils
INTRAZONAL ORDER	
Hydromorphic soils of marshes, swamps, bogs, and flat uplands	BOG SOILS MEADOW SOILS (WIESENBÖDEN) Alpine meadow soils PLANOSOLS
Halomorphic soils of poorly-drained arid regions and coastal deposits	SALINE SOILS (SOLONCHAK) ALKALI SOILS (SOLONETZ) Soloth
Calcimorphic soils	RENDZINA SOILS
AZONAL ORDER	
LITHOSOLS	
REGOSOLS	Alluvial soils Sands (dry)

[a]Based on U.S. Department of Agriculture (1938), *Soils and men*; *Yearbook of Agriculture 1938*, U.S. Govt. Printing Office, Washington, D.C. See pp. 993–995.

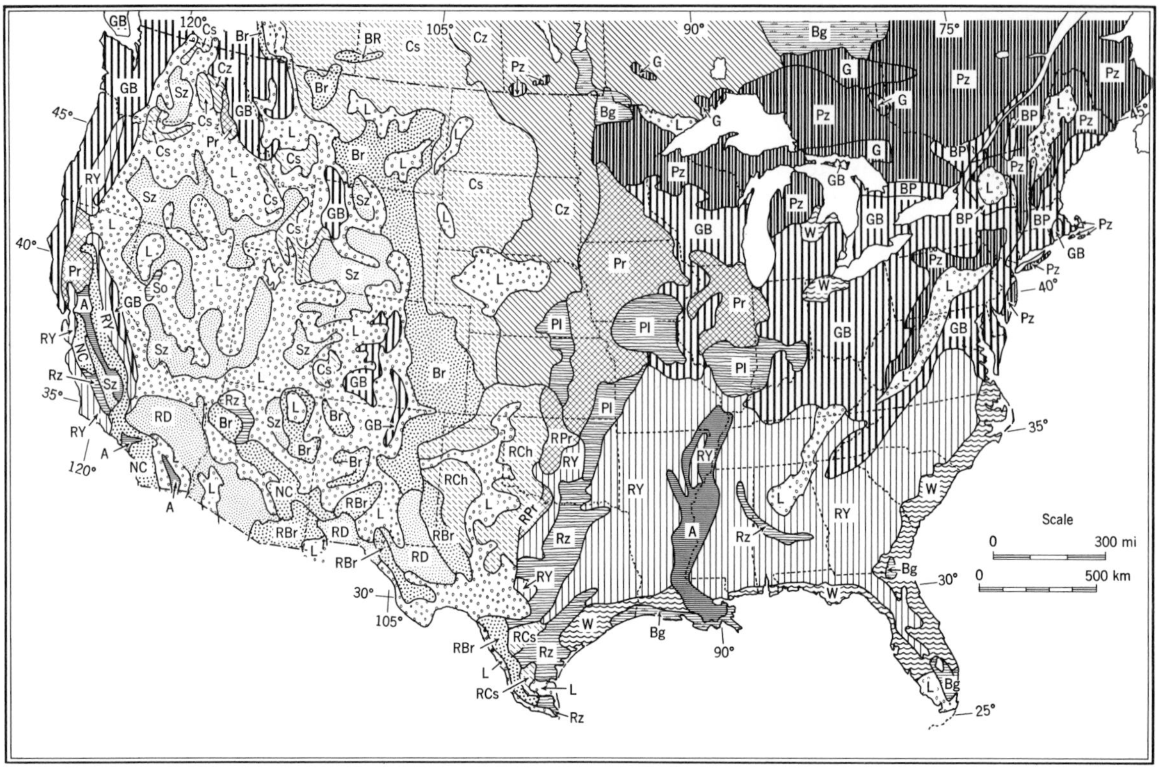

ZONAL SOILS

- Pz Podzol soils
- G Gray wooded soils (Canada)
- BP Brown podzolic soils
- GB Gray-brown podzolic soils
- RY Red and yellow podzolic soils
- Pr Prairie soils
- RPr Reddish prairie soils
- Cz Chernozem soils
- Cs Chestnut soils
- RCh Reddish chestnut soils
- Br Brown soils
- RBr Reddish brown soils
- NC Noncalcic brown (Shantung brown) soils
- Sz Sierozem (Gray desert) soils
- RD Red desert soils

INTRAZONAL SOILS

- Pl Planosols
- Rz Rendzina soils
- So Solonchak and solonetz soils
- W Wiesenböden, ground-water podzol, and half bog soils
- Bg Bog soils

AZONAL SOILS

- L Lithosols and shallow soils, sands, lava beds
- A Alluvial soils

Figure 19.2 Soils of the contiguous 48 United States and southern Canada. (Modified and simplified from map of soil associations of the United States by U.S. Department of Agriculture, Soil Survey Division, in *Yearbook of American Agriculture, 1938,* U.S. Government Printing Office, Washington, D.C., 1938, and *Atlas of Canada,* 1957, Plate No. 35, Canada Dept. of Mines and Technical Surveys.)

occurs in the A_2 horizon constitutes eluviation, explained in Chapter 18.

Below the A_2 horizon of the podzol profile is the *B* horizon, a brownish zone that is enriched by the colloids and bases brought down from the A_2 horizon. The process of accumulation is one of illuviation (Chapter 18). The colloids give a heavy, clayey consistency to the *B* horizon. Excessive deposit of oxides may cause the soil particles in this horizon to become strongly cemented into stony material known as *hardpan,* or, in European soil terminology, *ortstein.* Nodules of soil in the *B* horizon formed by the same process as hardpan are termed *concretions.* They may be composed of clay cemented by *limonite,* a hydrous iron oxide compound. Altogether the *A* and *B* horizons of the podzol profile total less than 3 ft (1 m) in thickness.

In the United States (Figure 19.2), the podzols are found in the northern Great Lakes states, the Adirondacks, and mountainous parts of New England. Here the ample precipitation combined with long, cold winters favors surface accumulation of decomposed vegetation. Organic acids cause a

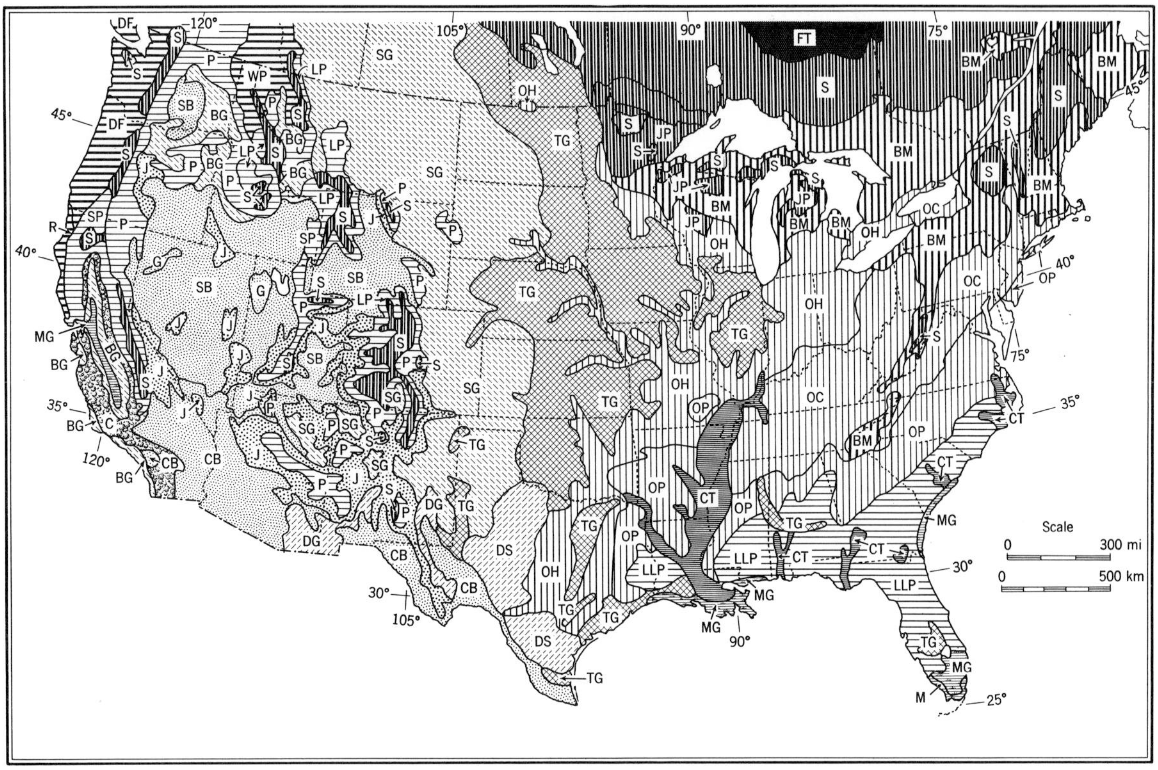

EASTERN FOREST VEGETATION

- FT Subarctic forest-tundra transition (Canada)
- S Spruce-fir (Northern coniferous forest)
- JP Jack, red, and white pines (Northeastern pine forest)
- BM Birch-beech-maple-hemlock (Northeastern hardwoods)
- Oak forest (Southern hardwood forest):
  - OC Chestnut-chestnut oak-yellow poplar
  - OH Oak-hickory
  - OP Oak-pine
- CT Cypress-tupelo-red gum (River bottom forest)
- LLP Longleaf-loblolly-slash pines (Southeastern pine forest)
- M Mangrove (Subtropical forest)

WESTERN FOREST VEGETATION

- S Spruce-fir (Northern coniferous forest)
- Cedar-hemlock (Northwestern coniferous forest):
  - WP Western larch-western white pine
  - DF Pacific Douglas fir
  - R Redwood
- Yellow pine-Douglas fir (Western pine forest)
  - Sp Yellow pine-sugar pine
  - P Yellow pine-Douglas fir
  - LP Lodgepole pine
- J Pinon-Juniper (Southwestern coniferous woodland)
- C Chaparral (Southwestern broad-leaved woodland)

DESERT SHRUB VEGETATION

- SB Sagebrush (Northern desert shrub)
- CB Creosote bush (Southern desert shrub)
- G Greasewood (Salt desert shrub)

GRASS VEGETATION

- TG Tall grass (Prairie grassland)
- SG Short grass (Plains grassland)
- DG Mesquite-grass (Desert grassland)
- DS Mesquite and desert grass savanna (Desert savanna)
- BG Bunch grass (Pacific grassland)
- MG Marsh grass (Marsh grassland)
- Alpine meadow (Not shown)

Figure 19.3 Vegetation of the contiguous 48 United States and southern Canada. (Modified and simplified from maps of H. L. Shantz and Raphael Zon, in *Atlas of American Agriculture*, 1929, U.S. Government Printing Office, Washington, D.C. and Canada Department of Forestry, Bulletin 123, 1963.)

strong leaching of the A_2 horizon. Islands of podzolic soil are found in the Allegheny Mountains, where high altitude gives a cooler climate, and in the sandy New Jersey and Long Island coastal plain belt, where leaching is intense in the porous sand. Podzols of the United States are but the small southern fringe of a great podzol belt corresponding with the needleleaf evergreen forest of the subarctic climate zones in Canada and Eurasia (Figure 19.3 and Plate 3). This forest is dominated by conifers which promote the process of podzolization.

Podzol soils are low in fertility. Leaching of important plant constituents is shown by the association of coniferous forest with this soil. Conifers need little of the calcium, magnesium, potassium, and phosphorus which many other plants require. Consequently the trees do not bring these bases to the ground surface from which they can be restored to the upper soil horizons. Certain species of oaks have a similar lack of use of the bases. The Pine Barrens on podzol soils of the New Jersey coastal plain are examples of adaptation of pine and oak forest to podzol soils.

The podzols cannot produce the crops to feed a large population. Addition of lime and fertilizers to the soil will largely correct soil acidity and replenish the leached bases, but the favorable areas for such treatment are limited by the effects of continental glaciation. Bouldery morainal topography interspersed with swamps and lakes still renders much of the podzol soil area unfit for farming.

Gray-brown podzolic soils

The second major great soil group of humid climates consists of the *gray-brown podzolic soils.* They differ from the podzols in that leaching is less intense and the soil color is brownish. The various soil horizons correspond with those of the podzols (Figures 19.4 and color plate, *B*). The A_1 horizon is a moderately acid humus layer. The A_2 horizon is a grayish-brown leached zone. It is less intensely leached than in the podzols and, consequently, is neither so light colored nor so distinctly limited. The *B* horizon is thick and yellowish brown to light reddish brown in color. Like the podzols, it has concentrated colloids and bases.

The gray-brown podzolic soils contain more of the important bases than the podzols but are nevertheless somewhat acid. Deciduous forests (maple, beech, oak) grow luxuriantly on the forest soils (Figure 19.3). These trees bring the bases up from the *B* horizon, returning them to the surface as dead leaves and branches. Thus the soil is replenished by these bases in a way not found in the podzols.

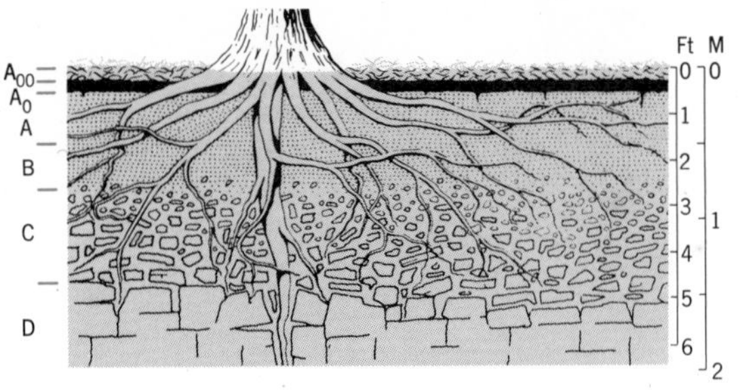

Figure 19.4 Brown forest soil profile formed beneath oak forest. (From S. R. Eyre: *Vegetation and Soils*, Aldine Publishing Company. Copyright © 1968 by S. R. Eyre.)

When treated with lime and fertilizers, the gray-brown podzolic soils make highly productive farms on which diversified crop farming and dairying are well developed. This is seen from the distribution of the gray-brown podzolic soils in the eastern-central United States where rainfall is 35 to 40 in (90 to 100 cm) yearly in the humid continental climate. Note that southern Wisconsin, southern Michigan, Indiana, Ohio, Kentucky, New York, Pennsylvania, and Maryland, and southern New England are largely underlain by these soils (Figure 19.2). These are states noted for the value of their diversified crop production. Gray-brown podzolic soils are also found in the Pacific northwest and on summit areas in the Rocky Mountains.

The gray-brown podzolic soils are also found over much of western Europe in both the marine west coast climate and humid continental climate. Smaller areas occur in the humid continental climate of northern China and northern Japan. Summer-green deciduous forest is associated with these areas of gray-brown podzolic soils.

Of relatively minor extent are the *brown podzolic soils* (Table 19.1), found in southern New England (Figure 19.2). These are transitional between podzols and gray-brown podzolic soils. The brown podzolic soils have only a very thin leached A_2 horizon. Although acid, they can be highly productive when heavily limed and fertilized.

Red-yellow podzolic soils

Farther south than the gray-brown podzolic soils, in a zone of increasingly warmer climate but with equally abundant precipitation, lies a great area of *red-yellow podzolic soils* (Figure 19.2). These soils occupy the southern United States from Texas to the Atlantic, and coincide fairly well with the extent of the humid subtropical climate. A similar geographic relation holds in Japan, while in southern Brazil and southeastern Paraguay there is also a substantial area of these soils at comparable latitudes (Plate 3). Smaller coastal zones of

REPRESENTATIVE EXAMPLES OF PROFILES OF THE GREAT SOIL GROUPS

Terminology and classification are according to the U.S. Department of Agriculture, 1938. Given in parentheses below are approximate equivalents from the U.S. Department of Agriculture, Soil Conservation Service, *Soil Classification, A Comprehensive System, 7th Approximation*, 1960 and 1967. Scale units are in feet. (Photographs were provided by the Soil Conservation Service, U.S. Department of Agriculture, courtesy of Dr. Guy D. Smith, Director, Soil Survey Investigations, Soil Conservation Service.)

ZONAL ORDER

A *Podzol* (*Humod*) Humus podzol, Fontainbleu, France.
B *Gray-Brown Podzolic Soil* (*Udalf*), Minnesota.
C *Yellow Podzolic Soil* (*Udult*), Puerto Rico.
D *Terra Rossa* (*Xeralf*), Portugal.
E *Latosol* (*Orthox*), Puerto Rico.
F *Prairie Soil* or *Brunizem Soil* (*Udoll*), Story County, Iowa.
G *Chernozem Soil* (*Udic Boroll*), Spink County, South Dakota.
H *Chestnut Soil* (*Typic Boroll*), Raymon, Saskatchewan.
I *Reddish Brown Soil* (*Ustalf*), Vernon, Texas.
J *Brown Soil* (*Aridic Boroll*), Scobey, Montana.
K *Sierozem* or *Gray Desert Soil* (*Calciorthid*), Escalante, Utah.
L *Red Desert Soil* (*Argid*), Tucson, Arizona.

INTRAZONAL ORDER

M *Bog Soil* (*Fibrist*), Sphagnum peat, Minnesota.
N *Meadow Soil* or *Humic Glei Soil* (*Aquoll*), Union County, Iowa.
O *Solonchak* or *Saline Soil* (*Salorthid*), Virgin River, Nevada. White horizon is salt.
P *Rendzina* (*Rendoll*), Puerto Rico. Developed on marl.

SOIL PROFILES

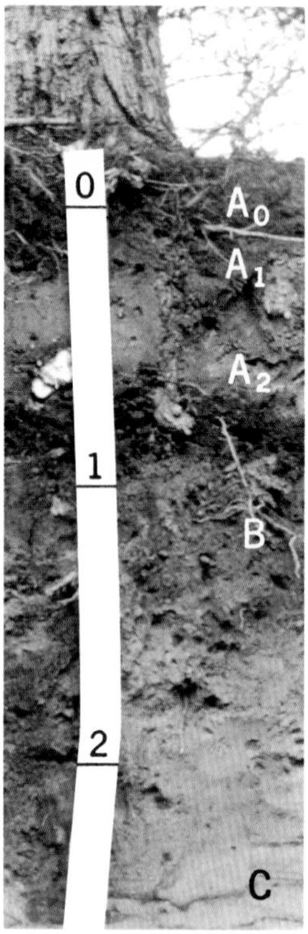

A Podzol

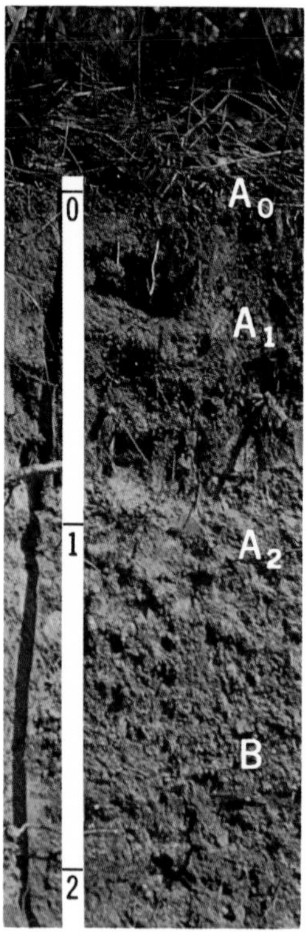

B Gray-brown podzolic soil

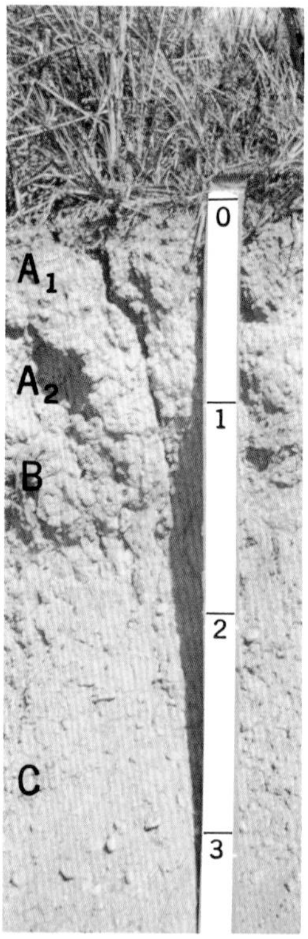

C Yellow podzolic soil

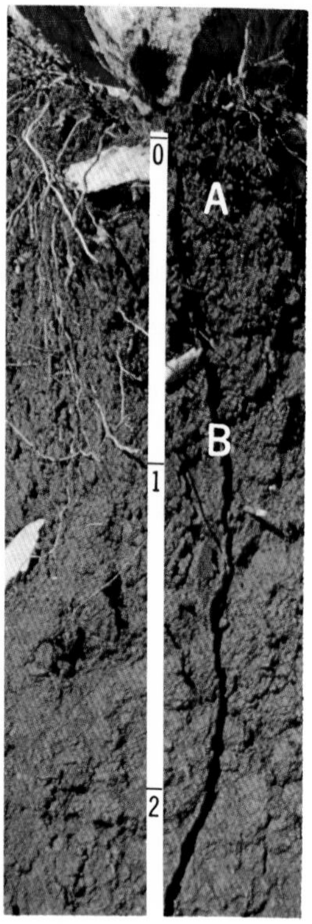

D Terra rossa

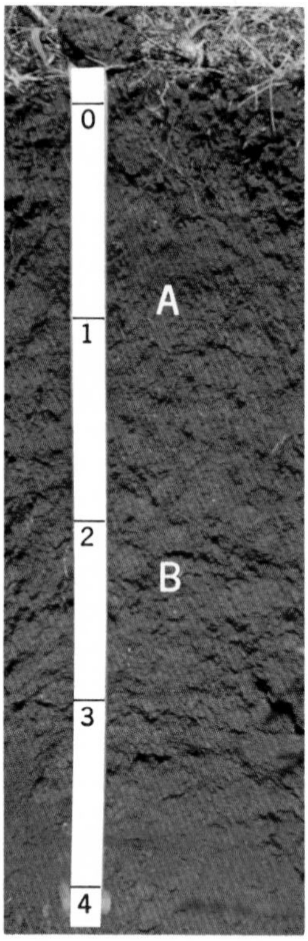

E Latosol

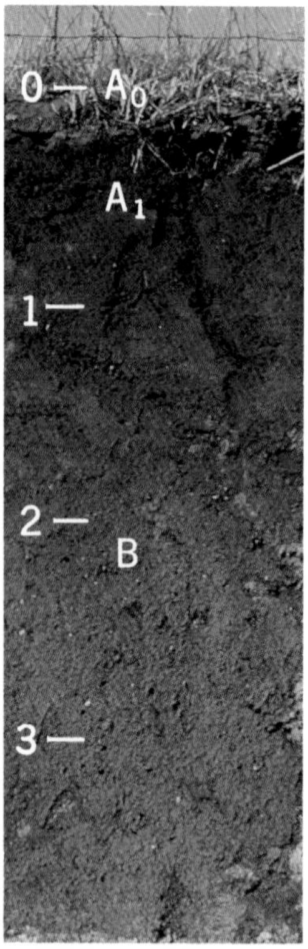

F Prairie soil

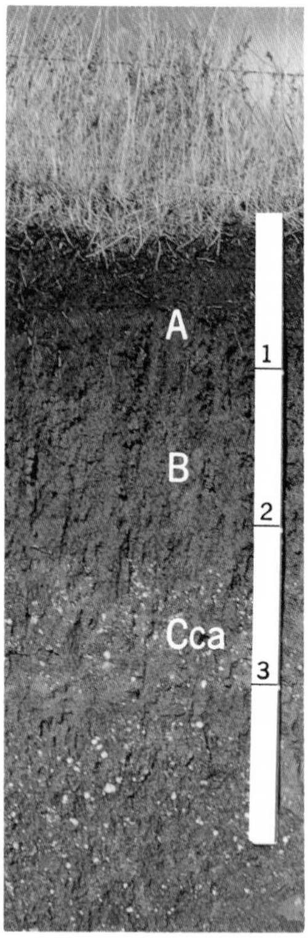

G Chernozem

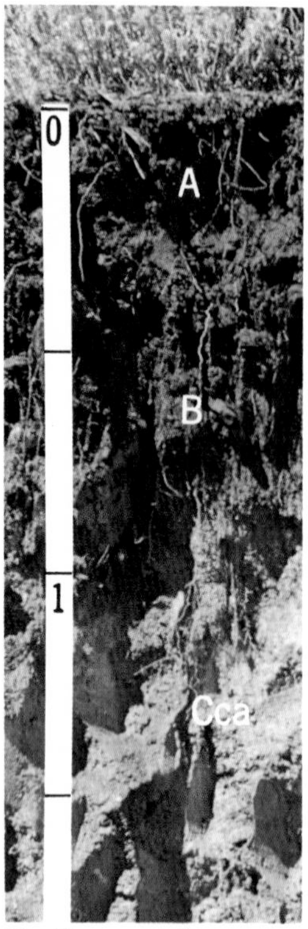

H Chestnut soil

SOIL PROFILES

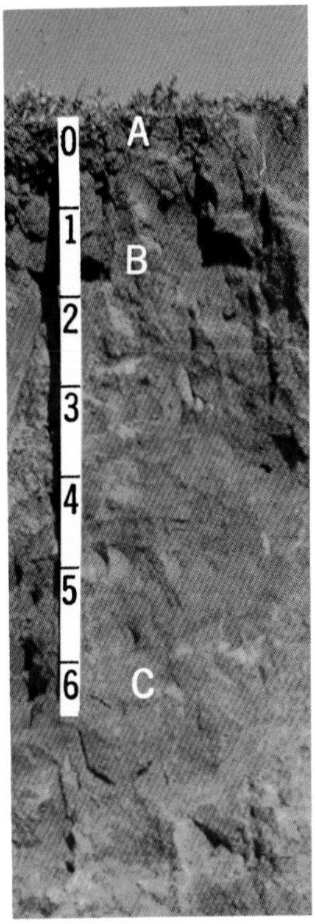

I Reddish brown soil

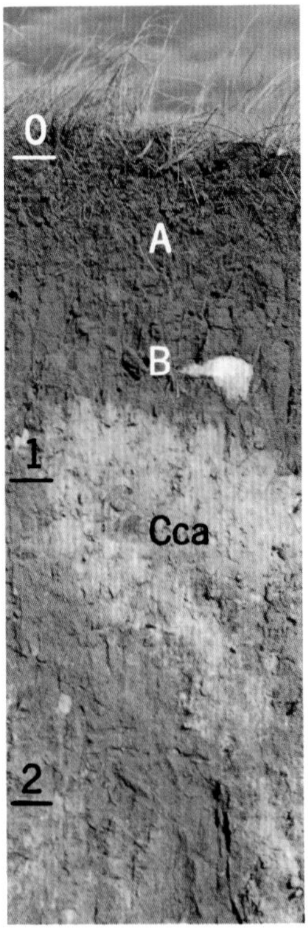

J Brown soil

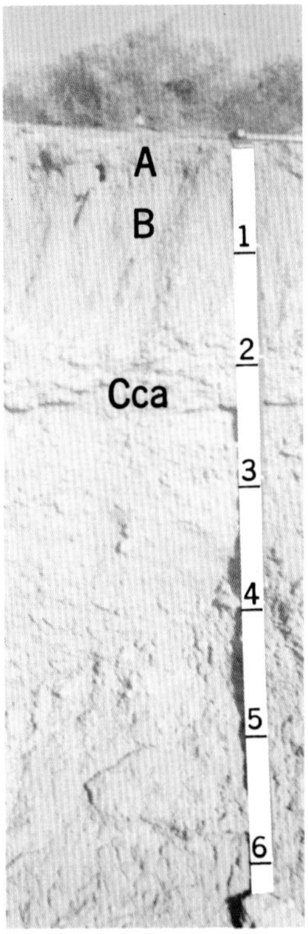

K Sierozem (gray desert soil)

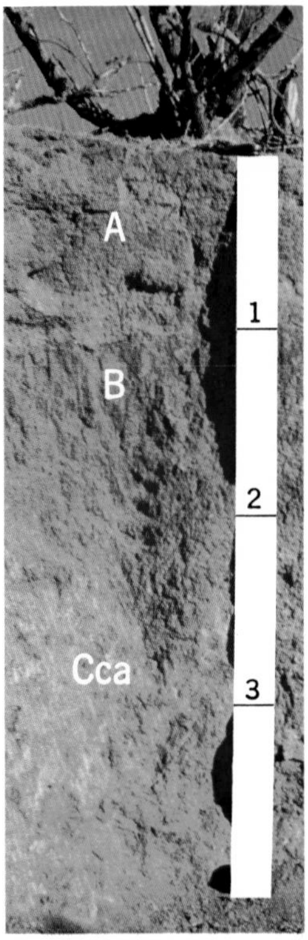

L Red desert soil

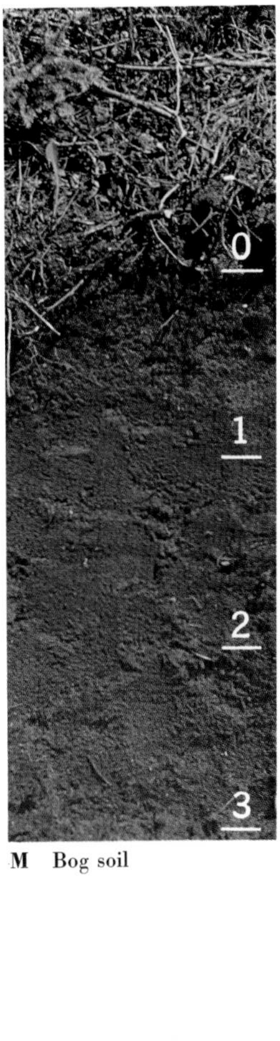

M Bog soil

N Meadow soil (Humic glei soil)

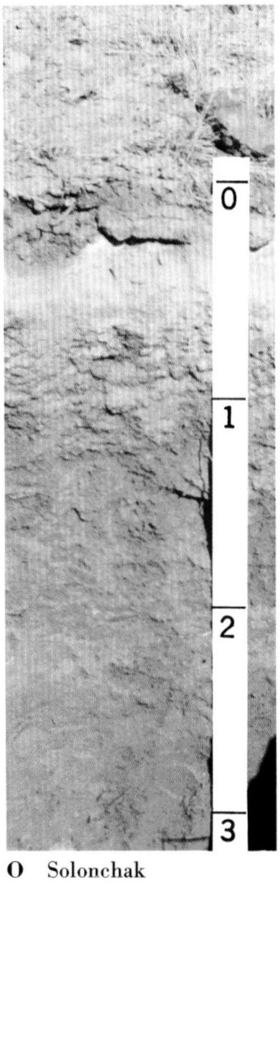

O Solonchak

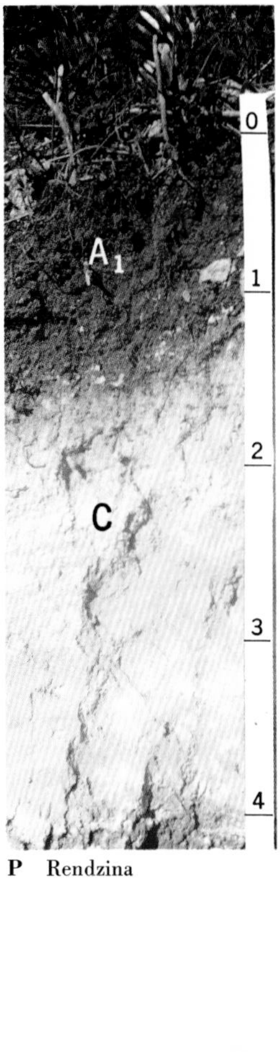

P Rendzina

red-yellow podzolic soils are found in South Africa, Australia, and New Zealand.

The red-yellow soils are of the podzolic type and show the same characteristic leaching of the A_2 horizon (color plate, *C*). Warm summers and mild winters favor bacterial action. Humus content is low. Thus, both podzolization and laterization act in concert as pedogenic processes. The typical red and yellow colors are a staining in the form of hydroxides of iron. The yellow soils are the more strongly leached of the two and are found on sandy coastal plain belts. Aluminum hydroxides are also abundant in these soils, a condition typical of tropical soils in warm, humid regions.

Deciduous forest was the natural vegetation of the northern part of the red soil belt in the United States. Soils of these southern states, though low in plant nutrients, respond well to fertilizers and have been important producers of tobacco, cotton, peanuts, soybeans, corn, sweet potatoes, cowpeas, and many other crops. Yellow soils, owing to their strong leaching, support forests of longleaf, loblolly, and slash pine (Figure 19.3). The pitch pine of the coastal belt is an important source of turpentine and resin; the slash pine is a good pulpwood tree.

In the other world regions of red-yellow podzolic soils noted above, natural vegetation is dominantly rainforest, of both tropical and temperate classes (Plate 4).

A great soil group related to the red podzolic soils is the *terra rossa* of Mediterranean lands (Plate 3). This red soil, poor in humus, is rich in sesquioxide of iron (Fe_2O_3) to which it largely owes its color. The origin of terra rossa is the source of much speculation. Perhaps it represents a soil that was once much richer in humus, but which has lost that humus because of the destruction of forest cover by man and his grazing animals over the course of many centuries. In some areas of its occurrence, terra rossa is associated with limestone as the parent matter, suggesting that its properties are inherited from a calcareous subsoil.

Latosols

Soils of the humid tropical and equatorial zones are called *latosols,* or *lateritic soils* (color plate, *E*). They are characterized as follows: (1) Chemical and mechanical decomposition of the parent rock is complete, owing to the favorable conditions of moisture and heat. (2) Silica has been almost entirely leached from the soil. (3) Sesquioxides of iron and aluminum have accumulated in the soil as abundant and permanent residual materials. (4) Humus is almost or entirely lacking because of the rapidity of bacterial action in the prevailingly warm temperatures. (5) The soil is distinctively reddish because of the presence of sesquioxides of iron. The reasons for the development of these conditions have been explained in Chapter 18 under the pedogenic regime of *laterization.* Loss of silicate clay minerals renders the latosols relatively low in plasticity (stickiness) and remarkably porous. Consequently rainfall sinks readily into these soils.

True latosols are found only in warm, humid regions and, hence, correspond closely with the wet equatorial climate and the tropical wet-dry climate. Though the red-yellow podzolic soils show the effects of laterization, they are not to be classified as true latosols.

Latosols quickly lose their fertility under crop cultivation because excessive leaching has removed the plant nutrients in all but a thin surface layer. However, the soil is favorable for the luxuriant growth of broadleaf evergreen rainforest (Plate 4). Other large areas have the raingreen forest and woodland associated with the tropical wet-dry climatic regime.

An interesting feature of latosols is the local development of accumulations of iron and aluminum sesquioxides into layers that can be cut out as building bricks. The material is termed *laterite.* On exposure to the drying effects of the air, these blocks become very hard. In Indochina, particularly, laterite bricks have been much used as a building material.

Valuable mineral deposits occur as laterites. These are thick layers of such minerals as *bauxite* (hydrous aluminum oxide), *limonite* (hydrous iron oxide), and *manganite* (manganese oxide). They are known as *residual ores* because they are not soluble in soil water and have continued to accumulate as the parent rock has weathered away and the silica and other soluble constituents have been removed. Important bauxite deposits of the Guianas of northern South America and of western India are of this type. Valuable resources of manganese are in laterite deposits.

By no means all soils of the humid tropical and equatorial climates are latosols of the type described above. Variations of the latosols have been described as *yellowish-brown lateritic* and *reddish-brown lateritic* soils. Large upland areas of Africa and India within the tropical wet-dry climate regime have *black* and *dark-gray tropical soils.* The extent of these is given in Plate 3. These dark soils may be related to special conditions of underlying bedrock. For example, the dark soils of peninsular India shown in Plate 3 coincide rather well with the Deccan Plateau, underlain by basalts.

Hydromorphic soils (intrazonal)

Hydromorphic soils are soils associated with marshes, swamps, bogs, or poorly drained flat uplands. All are considered intrazonal soils because of poor drainage.

Bog soils are formed under bog vegetation in regions of cool, moist continental climates. Continental glaciation in North America and Europe left countless basins that have since been largely filled by a succession of water-loving plants (Chapter 20). Here the soil is saturated most of the time and plant decay is greatly retarded. The partly decomposed plants of the bog therefore accumulate into an upper peat layer 3 to 4 ft (1 m) thick (color plate, *M*). Below this is a horizon of sticky, structureless clay, known to soil scientists as the *glei* (also *gley*) horizon. It is of gray-blue color and is largely impervious to water seepage. The process of *gleization* was discussed in Chapter 18.

Meadow soils (*wiesenböden*) are formed on the flood plains of streams where drainage is somewhat better than in the bogs, but is nevertheless poor. These areas are extensively used in the humid middle-latitude climates as pastures because the grass grows rapidly and densely. A thick, humus-rich layer is developed, overlying a sticky gley horizon (color plate, *N*). A similar process of soil development takes place on poorly drained sites at the bases of hill slopes where there is a tendency for water to collect.

The name *humic-glei soils* has been proposed for the meadow soils in combination with half-bog soils.

At high altitudes, where the alpine tundra climate prevails, *alpine meadow soils* are found. Dark in color, these soils support a growth of grasses, sedges, and flowering plants.

Strongly leached soils developed on flat or gently sloping, elevated surfaces are known as *planosols.* The soil horizons are abnormally thick because of slow removal by erosion. Planosols have a thick, dense clay horizon in humid climates. In subhumid climates planosols have a dense, cemented horizon.

Tundra soils

Soils of the arctic tundra are so widespread (Plate 3) that they may be considered a zonal type along with the podzols, gray-brown forest soils, red-yellow soils, and latosols, but because they are poorly drained they are sometimes classified as intrazonal.

Climate conditions favorable for tundra soils have been described under tundra climates of the northern continental fringes, Chapter 17. Intensely cold, long winters cause soil moisture to be frozen during many months of the year. Under these cold conditions chemical alteration of the minerals is slow, and much of the parent material of the soil consists of mechanically broken particles. The slow rate of plant decomposition results in the presence of much raw humus, or peat. The tundra soils do not have simple, distinctive soil profiles, but consist of thin layers of sandy clay and raw humus. The surface may be covered with a sod of lichens, mosses, and herbaceous plants.

In the tundra regions of Siberia and North America, the condition of permanently frozen ground, or *permafrost*, is widespread. (See Chapter 17.) Curious lenses, layers, and vertical wedgelike bodies of ice are present under the soil.

In parts of central Alaska, principally in the valleys of the Yukon and Tanana rivers, are dark-colored soils which have been given great-soil group status as the *arctic brown forest soils* (Plate 3). The profiles of these soils have a thick, dark A_1 horizon rich in organic matter. Downward through lower horizons the soils grade into lighter brown colors and finally to gray in the *C* horizon. These soils may have originated in a surface layer of loess (wind-blown silt).

Chernozem soils

Of the zonal soils in a semi-arid climate among the most distinctive and widely distributed types are the *chernozem soils*, or *black earths* (color plate, *G*). A typical chernozem profile appears to consist essentially of two layers. Immediately beneath a grass sod is a black layer, the *A* horizon, 2 to 3 ft (0.6 to 1 m) thick and rich in humus. In this layer the structure is a crumb or nut. The *A* horizon grades downward into a *B* horizon of brown or yellowish-brown color, then, with a sharp line of demarcation, into a light-colored *C* horizon. As in the podzolic soils, in the *B* horizon colloids and bases accumulate by downward percolation through the *A* horizon, but unlike the podzols, the chernozem soils have no leached A_2 horizon.

Chernozem soils are rich in calcium, which appears in excess as calcium carbonate precipitated in the lower *B* horizon or just beneath the *B* horizon. It has been noted that chernozem soils develop in parent material rich in calcium carbonate. The origin and distribution of chernozem soils have long attracted the soil scientist. Russian investigators in particular have been intensely interested in chernozem soils because they occupy a large area in the Ukraine, surrounding the Black Sea on its west, north, and east sides (Plate 3) and continuing in a great belt eastward along the 55th parallel into the heart of Asia.

Chernozem soils are important in the United States and Canada, where they form a north-south belt starting in Alberta and Saskatchewan and running through the Great Plains of the United States to central Texas. A similar area lies in Argentina. Other areas are mapped in Australia and Manchuria (Plate 3).

Climate has long been thought to be a determining factor in the development of chernozem soils. Comparison of soil and climate maps shows that the middle-latitude chernozems, in the Americas and Europe, lie on the more arid western side of the humid continental climates and with decreasing latitude extend over into the middle-latitude steppe climates. Aridity is therefore a definite contributing cause. The continental location of chernozem areas makes for hot summers and cold winters. Drought periods with strong evaporation dry out the soil, and forests cannot exist. Instead, grasses, which can withstand drought readily and which are tolerant to soils with excesses of mineral salts, flourish on chernozem soils. Steppe grasslands and prairies are the natural vegetation of the middle-latitude chernozem soils.

An important factor in development of the middle-latitude chernozem soils is their occurrence on *loess*, the wind-transported dust so extensively deposited during the glacial period. (See Chapter 33.) Though chernozem soils are not limited to such areas, the texture and lime content of loess, together with the plains topography, have been especially favorable to chernozem development.

Geographically, perhaps the outstanding point of importance regarding the chernozem soils is their productivity for small grain crops—wheat, oats, barley, and rye. Great grain surpluses are exported from chernozem areas in the United States, Canada, and the Ukraine and Argentine, causing them to be described as breadbaskets of the world.

Where forests invade the chernozem soil grasslands, some influence of podzolization is felt and there develops a faint A_2 horizon. Such soils are known as *degraded chernozem soils.* They are transitional to the gray-brown podzolic soils and occupy a geographical position adjacent to them. One particularly large region of degraded chernozem soils lies northwest of the Black Sea and extends from the Danube River of Rumania to the southern Ukraine (Plate 3).

Prairie soils (brunizem soils)

Examination of the soil map (Figure 19.2) will show that between the chernozems and gray-brown podzolic soils in the United States lies a zone of *prairie soils* (or *brunizem soils*). Rainfall is 25 to 40 in (60 to 100 cm) and diminishes greatly across this belt (Figure 16.10). This soil group is similar to the chernozems in general profile and appearance, but differs in that it lacks the excess calcium carbonate of the chernozems (color plate, *F*). The prairie soil is, therefore, a transitional type between the major soil divisions: pedocal and pedalfer.

Of special interest in the United States has been the origin of natural tall-grass prairies of the upper Mississippi Valley and the Great Plains states: Illinois, Iowa, eastern Nebraska, southern Minnesota, northern Missouri, and eastern Kansas (Figure 19.3). Here forests were lacking over vast expanses at the time of the coming of white men. Although many explanations have been offered for the existence of the prairies, including the possibility of deforestation by burning, it seems likely that a major contributing factor has been that the prairie soils become almost completely dry between summer rains down to a depth of a foot or so (0.3 m) as a result of the dryness and heat of summer air masses. Although the prairie grasses can survive these conditions, deciduous forests, which border the prairies on the east, cannot.

Prairie soils are extremely productive, combining the fertility of the chernozems with somewhat moister climate. Perhaps the outstanding crop associated with the prairie-soil belt is corn. The corn belt is practically identical with the prairie-soil belt, although the corn belt actually extends eastward into the gray-brown forest soils of Indiana and Ohio. Corn requires not only high temperatures during the growing season but ample moisture as well. This last requirement is met by periods of summer thunderstorms interspersed with hot, dry periods.

South of the prairie soils in the United States lies a small area of *reddish prairie soils,* located south of the Arkansas River in Kansas, Oklahoma, and Texas (Figure 19.2). The reddish prairie soils are essentially like the prairie soils, but of reddish-brown color. They support a prairie grassland vegetation.

Chestnut soils and brown soils

To the arid side of the chernozem belt lies the belt of *chestnut soils* or *dark brown soils.* They occupy the semi-arid middle-latitude steppe lands, in North America and Asia (Plate 3). The chestnut-soil profile is generally similar to the chernozem, but contains less humus and hence is not so dark in color. The soil structure tends to be prismatic in the *B* horizon.

The chestnut soils are fertile under conditions of adequate rainfall or irrigation, but lie in a hazardous marginal belt in which years of drought and adequate rainfall are alternated. Bordering on the productive chernozem wheat belt of the Great Plains, the chestnut-soil belt of the United States offers a temptation for expanded wheat cultivation. Under special cultivation practices that conserve soil moisture, a period of moist years brings high grain yields to these marginal belts; but a series of drought years can bring repeated crop failures and poverty.

Toward still more arid regions the chestnut soils give way to the *brown soils,* generally similar, but with still less humus and consequently a lighter color (Figure 19.5 and color plate, *J*). In the western United States, the brown soils occupy basins in the central Rockies of Wyoming, the Colorado Piedmont, and parts of the Colorado Plateaus in Colorado, Utah, Arizona, and New Mexico.

The brown soils are typical of the middle-latitude steppes and support a light growth of grasses suitable for livestock grazing. With irrigation they are productive, but farming is not attempted without.

Reddish-chestnut and reddish-brown soils

Widespread in semi-arid and subhumid tropical and subtropical regions of the world are soils of reddish brown surface color which grade downward into a heavier dull-red or reddish-brown subsoil, then into a zone of lime accumulation. These are classified as *reddish-chestnut* and *reddish-brown soils* and grouped with the reddish prairie soils on the world soils map (Plate 3).

Comparison of the distribution of these reddish soils with the distribution of climate and natural vegetation (Plates 2 and 4) suggest that these soils are found in regions of tropical wet-dry climates with a short wet season and with their bordering tropical steppes. They also occupy regions of Asiatic monsoon wet-dry climate and regions of Mediterranean climates in Australia, North Africa, and the Middle East. Vegetation associated with reddish-chestnut and reddish-brown soils is varied, being tall-grass prairie, short-grass steppe, raingreen savanna and scrub, or monsoon forest. All of the areas of reddish-chestnut and reddish-brown soils have in common a marked degree of aridity of climate, whether associated with an annual cycle or drought in the winter (low-sun season) or in the summer (high-sun season). Very high air temperatures and a severe water deficiency occur at one season of the year. Otherwise, in terms of latitude, these reddish soils occupy the same global positions as areas of latosols and lateritic soils, into which the reddish soils grade in Central and South America, Africa, southern Asia, and Australia.

Aridity in the seasonal climatic cycle explains the occurrence of calcium carbonate in lower layers of the reddish-brown and reddish-chestnut soils, for this is a characteristic of the pedocal soils. Red color is evidence of an abundance of the sesquioxides of iron, which accumulate in warm climates where organic acids are not produced in quantity.

Gray desert soils (sierozem) and red desert soils

Soils of the middle-latitude deserts and tropical deserts fall into two great soil groups on the basis of color. (1) *gray desert soils*, or *sierozem*, and (2) *red desert soils*.

Figure 19.5 Profile of a brown soil formed from loess in Colorado. The *B* horizon has good prismatic structure. Depths in feet. (Photograph by C. C. Nikiforoff. Courtesy of Soil Conservation Service, U.S. Dept. of Agriculture.)

The gray desert soils, or sierozem, are well developed in Wyoming and in the deserts of Nevada, western Utah, and southern portions of Oregon and Idaho (Figure 19.2). This latter region is sometimes referred to as the Great Basin because of the prevalence of interior drainage systems ending in evaporating basins. It corresponds approximately with the middle-latitude desert climate and has northward extensions into the middle-latitude steppe climate.

The gray desert soils contain little humus because of the sparse growth of vegetation, such as sagebrush and bunchgrass. Color ranges from light gray to grayish brown (color plate, *K*). Horizons are present, but being only slightly differentiated are not conspicuous. Excessive amounts of calcium carbonate are present at depths of less than one foot (0.3 m) in the form of *lime crust*, or *caliche*, a deposit of calcium carbonate or hydrous calcium sulfate. In places this deposit has the appearance of a hard rock layer and may even resist erosion in such a way as to produce small mesas or platforms. Gravels deposited by streams are often thus cemented into a conglomerate rock. Lime crust forms during prolonged dry periods when ground waters rise slowly surfaceword by capillary attraction and evaporate near the surface, leaving the salts behind in the soil.

In the more arid, hotter tropical deserts are found the *red desert soils* (Plate 3). These range from a pale reddish gray to a pronounced deep red (color plate, *L*). Humus is reduced to the minimum, there being only a scattered growth of desert shrubs. Thus the activity of plants as soil formers reaches its lowest point in the red desert soils, as does also the activity of animals. Color is derived from small amounts of oxides of iron. Horizons are poorly developed; texture is often coarse, with many fragments of parent rock throughout the soil. Deposits of lime carbonate are present, as with the gray desert soils.

The gray and red desert soils are suitable for cultivation only where they are fine textured, as along the floodplain terraces of exotic streams and on the outer slopes of alluvial fans. Irrigation is essential, whether water is diverted from a river or obtained from wells penetrating the ground water reserves in alluvial fans.

Altitude and soils

As explained in Chapter 17, climate zoning in mountains tends to reproduce world climates in a series of vertical zones, the effect of increasing altitude being much the same as that of increasing latitude. Because climate is a major determining factor in soils, we might also expect increasing altitude to result in a series of great soil groups. Such a series is illustrated by soils of the Big Horn Mountains in Wyoming (Figure 19.6). Starting with gray desert soils at the lowest elevation, the series progresses through the zonal soils of dry climates through prairie soils to podzols at the high elevations.

Compare these altitude changes with Figure 19.7, which summarizes the profile changes along an imaginary traverse line from the southwestern desert of the United States, eastward across the high plains region, then north into the Lake Superior region.

Halomorphic soils (intrazonal)

In the steppes and deserts, evaporation on the average greatly exceeds precipitation and there are many topographic depressions that have no outlet to external drainage systems. Here the products of rock weathering are brought by intermittent flowing streams in times of flood and left behind on the basin floor. Along with clay, silt and sand are the dissolved mineral salts that crystallize out as the water evaporates.

Shallow lake basins, or *playas*, of extreme flatness are often covered with only a thin film of water which evaporates rapidly, leaving its salts on the surface. Most persons are familiar with the salt flats of Great Salt Lake in Utah, on which so many automobile speed records have been set. Salts found in the various playa lakes of the southwestern United States are *soda* (Na_2CO_3), *borax* ($Na_2B_4O_7$), *calcium carbonate* ($CaCO_3$), various sulfates (Na_2SO_4, $MgSO_4$, K_2SO_4), chlorides ($NaCl$, $CaCl_2$, $MgCl_2$), and many others. Where salts are thick and pure in the inner parts of the playas there is no soil in the true sense of the word. The term *halomorphic soil* is properly applied to marginal areas where silts and clays make up a large proportion of the body of the soil.

The pedogenic process in the development of halomorphic soils is salinization (Chapter 18). Two major groups of halomorphic soils are recognized: *saline soils* (*solonchak*) and *alkali soils* (*solonetz*). Halomorphic soils are classed as intrazonal because of their poor drainage and limited extent.

Saline soils (solonchak, or *white alkali soils*) contain chlorides, sulfates, carbonates, and bicar-

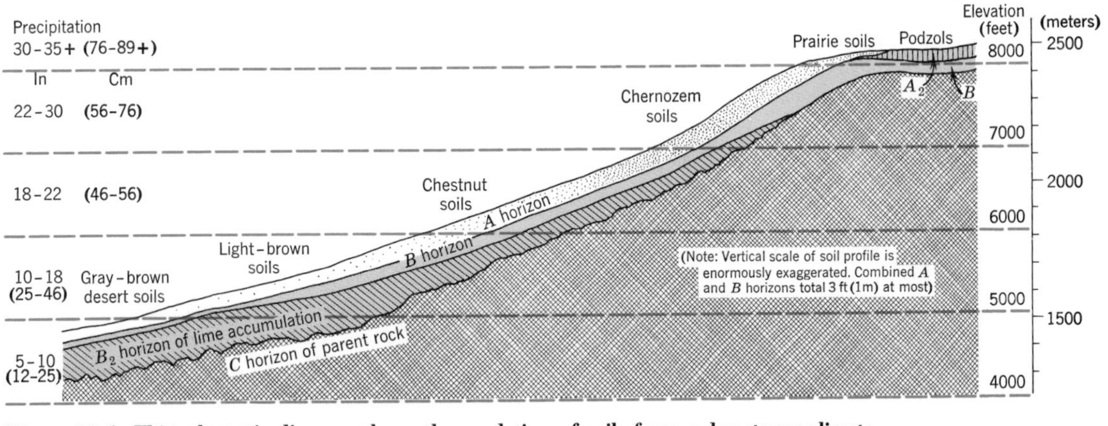

Figure 19.6 This schematic diagram shows the gradation of soils from a dry steppe-climate basin (left) to a cool, humid climate (right) as one ascends the west slopes of the Bighorn Mountains, Wyoming. (After J. Thorp, 1931.)

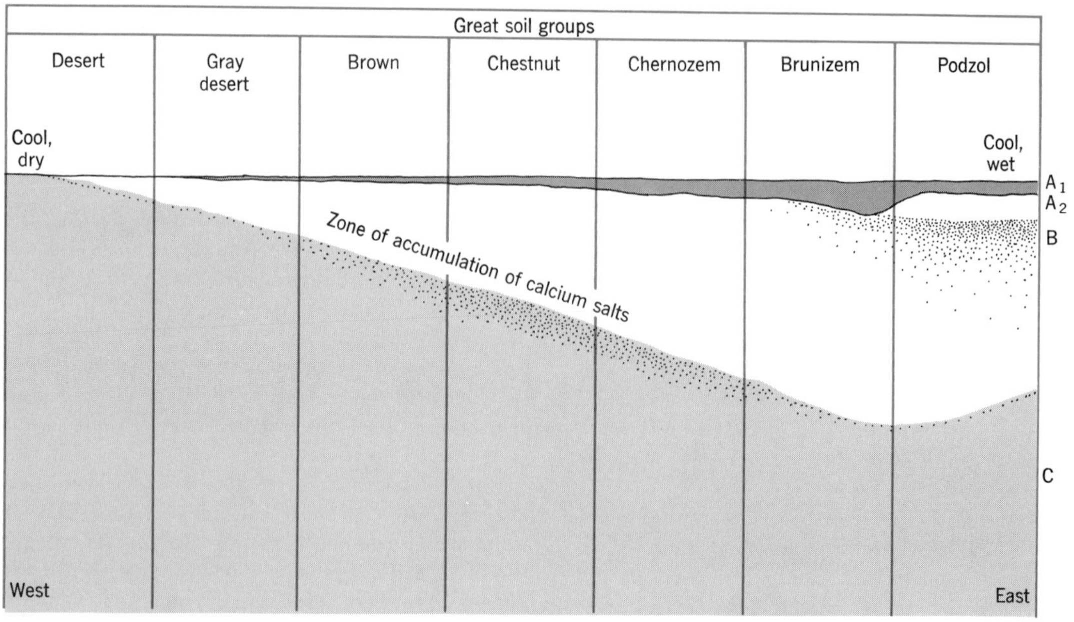

Figure 19.7 Schematic diagram of the relationships among profiles of the great soil groups, traced from dry to humid climates. (After C. E. Millar, L. M. Turk, and H. D. Foth, *Fundamentals of Soil Science*, John Wiley and Sons, New York, 1958.)

bonates of sodium, calcium, magnesium, and potassium. These soils are light colored and show poorly developed horizons (color plate, *O*). Although there are many species of plants adapted to saline soils the plant cover is at best sparse. Plants which are salt-tolerant are known as *halophytic* plants. They include grasses, shrubs, and some trees. Agriculture is not possible on saline soils unless they are flushed out with large quantities of irrigation water to remove the salts. This has been done to advantage to reclaim much good agricultural land in the southwestern United States.

In the second major group of halomorphic soils, the alkali soils (solonetz or *black alkali soils*), sodium salts predominate, especially sodium carbonate ($NaCO_3$). The soil profile is characterized by the presence of a dark, hard, columnar layer, i.e., a layer with prismatic structure (Figure 18.4*B*). Although salts of the alkali soils are of somewhat different chemical properties from salts of the saline soils, both soils occur in the same areas. Alkali soils occupy areas of somewhat better drainage than the saline soils and may be derived from the latter by the removal of more soluble salts (desalinization). Grass and shrub vegetation is found on alkali soils, but the species are particularly alkali-resistant.

With improved drainage—whether by man's intervention or by natural processes—solonetz is transformed by leaching of salts into *soloth*, a halomorphic soil of light-colored, slightly acid surface layer and a heavy-textured, dark brown *B* horizon with columnar structure.

Irrigation and the salinization of soils

Irrigation systems in arid lands divert the discharge of a large river, such as the Nile, Indus, or Colorado, into a distributary system that allows the water to infiltrate the soil of areas under crop cultivation. Ultimately, such irrigation projects suffer from two undesirable side effects—salinization and waterlogging of the soil.

The irrigated area is subject to very heavy water losses through evapotranspiration. Salts contained in the irrigation water remain in the soil and increase in concentration—the process of *salinization.* Ultimately, when salinity of the soil reaches the limit of tolerance of the plants, the land must be abandoned. Prevention or cure of salinization may be possible by flushing down soil salts to lower levels by use of more water. This remedy requires more water use than for crop growth alone. Furthermore, when salts are flushed down to lower levels, they may enter the ground water body and contaminate it with toxic ions, such as nitrates.

Infiltration of large volumes of water causes a rise in the water table, and may ultimately bring the zone of saturation close to the surface—a phenomenon known as *waterlogging.* Crops cannot grow in perpetually saturated soils. Furthermore, when the water table rises to the point that upward movement under capillary action can bring water to the surface, evaporation is increased and salinization is intensified.

One of the largest of the modern irrigation

projects affected adversely by salinization and waterlogging lies within the basins of the lower Indus River in West Pakistan. Here the annual rate of rise of the water table has averaged about one foot (0.3 m), while the annual increase in land area adversely affected is on the order of 50,000 to 100,000 acres (20,000 to 40,000 hectares).

Calcimorphic soils

Another class of intrazonal soils consists of the *calcimorphic soils.* These are soils whose characteristics are strongly related to the presence of lime-rich parent material. The process of *calcification* (introduction of calcium) is dominant in the formation of calcimorphic soils. An important example of the calcimorphic soils are the *rendzina soils.*

Rendzina soils have dark-gray or black surface layers overlying soft, light-gray or white material which is highly calcareous (lime-rich) (color plate, *P*). The parent matter of rendzina soils may be marl (a lime mud), soft limestone, or chalk, all of which are forms of calcium-carbonate. The soil profile is regarded as immature. The natural vegetation is typically grassland, which results in a well-distributed humus in the upper layers and imparts the dark color to the soils, just as in the chernozem.

Distribution of rendzina soils in the United States is fragmented into a number of geologically favorable areas. Substantial areas of these soils are found in undulating prairies of central and northeastern Texas and central-southern Oklahoma (Blackland Prairie and Grand Prairie). Another geographically important area is the Black Belt of Alabama and Mississippi, a lowland underlain by soft limestone. These areas are in the humid subtropical climate. Other rendzina areas occur on a high limestone plateau in northwestern Arizona, and in the Mediterranean climate regime of southern California.

Rendzina soils of the humid subtropical climate are productive agriculturally and yield cotton, corn, and alfalfa. In more arid grassland environments, uses are for grazing and some dry-farming.

The Soil Taxonomy used by the U.S. National-Cooperative Soil Survey

Since its publication in 1938, the soil classification system outlined in Table 19.1 underwent a revision, published in 1949, but the changes were minor. Inadequacy of the older system was then becoming more and more apparent, particularly as new information on soils the world over could not be accommodated within the classes of the older system. Thus a national cooperative effort was launched in the early 1950s to develop a completely new scheme. After it had progressed through a succession of stages over a period of several years, the new scheme was ready for presentation by American pedologists to the Seventh International Congress of Soil Science in 1960. Known at the time as the *Seventh Approximation,* because it was the seventh in the series of revisions, the new system was launched as a tentative model for international study. The description of the newer system was prepared by the Soil Survey Staff of the Soil Conservation Service and published in 1960.[1] Supplements appeared in 1964, 1967, and 1968; use of the system in the United States began in 1965. It is appropriate now to refer to the newer system as the *Soil Taxonomy,* a short title for "A Basic System of Soil Classification for Making and Interpreting Soil Surveys." For those ready to examine the Soil Taxonomy used in the United States and a number of other countries, we offer here a description of the major classes of soils it embodies and the general patterns of their world distribution. Because the terminology is entirely new and the number of new names for classes is very large, the newer system will seem strange to those who have learned and used the older system.

Understanding of the newer system rests upon knowledge of the same basic principles of soil science covered in Chapter 18, and not upon the older system it replaces. However, for a study in depth of the newer system, it would be necessary to master the properties of some 20 diagnostic soil horizons and some 36 diagnostic features, most of which bear new names and new definitions. Instead of introducing these new concepts and greatly lengthening the presentation, we shall refer to horizons in terms already used in Chapter 18.

The newer classification system differs in important ways from the older system. The newer system defines its classes, strictly in terms of the morphology and composition of the soils, that is, in terms of the soil characteristics themselves. Moreover, the definitions are made as nearly quantitative as possible. Every effort has been made to use definitions in terms of features that can be observed or inferred, so that subjective or arbitrary decisions as to classification of a given soil will be avoided.

The newer classification does not recognize a distinction between zonal soils (mostly well-drained, upland soils) and intrazonal soils (mostly poorly drained) in the highest category of classification, which is the soil order. Thus, within a

[1]Soil Survey Staff, Soil Conservation Service, U.S. Department of Agriculture, *Soil Classification; A Comprehensive System (7th Approximation),* U.S. Government Printing Office, Washington, D.C., 265 pp., 1960.

particular soil order we find soils of wet places grouped with soils of well-drained upland surfaces. No attempt is made in the newer classification to assign more importance to one class of soils than another, particularly with respect to the amount of surface area it occupies.

The newer classification recognizes and gives equal importance to classes of soils deriving their characteristics from man's activities, such as long-continued cultivation and applications of lime and fertilizers, or from the accumulation of manmade wastes. Because man's occupance and agricultural exploitation of large expanses of soils have gone on for centuries in various parts of the world, recognition of such man-modified soils is desirable and realistic in a classification system.

A new terminology using coined words gives the system a major advantage, because the syllables can be selected to convey precisely the intended meaning with respect to the properties or genetic factors relating to the soil class. For example, the soil order *Histosol* derives its meaning from the Greek word *histos*, meaning "tissue." These soils are indeed formed of the tissue of plants in various degrees of decomposition. *Aridisols* comprise a soil order owing its properties to aridity of climate; *Oxisols*, to the important proportions of oxides, and so forth.

The soil orders and suborders

The Soil Taxonomy is based on a heirarchy of six categories or levels of classification. From top to bottom they are listed below with approximate numbers of classes recognized in the United States within each category.

Orders	10
Suborders	47
Great groups	180
Subgroups	960
Families	4700
Series	11,000

Numbers given for the lowest three categories refer only to soils of the United States. When soils of all continents are eventually included, the numbers of at least the two lowest categories will be greatly increased. On the other hand, numbers given for the first three categories are expected to remain approximately the same, since they are designed to provide for classification of all known soils of the world.

The following factors serve to distinguish the ten soil orders, each one from the others: (1) gross composition, whether organic or mineral or both, and texture; (2) degree of development of horizons; (3) presence or absence of certain horizons; (4) an index of the degree of weathering of the soil minerals. Generally, the estimation of the first and last factors must be based on comparisons with soil samples that have been studied in a laboratory. Horizons and other gross structures, as well as color, are assessed by field inspection in the walls of test pits. It was the intent of those who prepared the classification that the characteristics selected to distinguish a given order should reflect major differences in the histories of horizon development. In other words, classification of soils according to the results of their genesis remains a governing concept.

Although the soil orders are capable of presentation in a variety of sequences, including alphabetical sequence, we have chosen to arrange the ten orders as follows. We begin with an order that is widespread in the cold tundra regions of the arctic and subarctic zones, then proceed through orders typically widespread in successively lower latitude zones, but largely within moist climates, to the equatorial zone. We then take up two orders that are characteristic of semiarid and arid climates; these are largely situated in tropical, subtropical, and middle-latitude zones. Finally, we recognize an order that occurs throughout all latitude zones and is characterized by the absence of most soil properties (especially the soil horizons) required in the definitions of the other orders.

The ten soil orders are listed below, together with the meanings of the roots used in their names. The suffix *sol* derives from the Latin word *solum*, meaning "soil."

Inceptisols	L.: *inceptum*, "inception" or "beginning."
Histosols	Gr.: *histos*, "tissue."
Spodosols	Gr.: *spodos*, "wood ash."
Alfisols	The syllable *alf* is coined from *al* and *fe*, chemical symbols for "aluminum" and "iron."
Ultisols	L.: *ultimus*, "ultimate."
Oxisols	The syllable *oxi* is short for "oxide."
Vertisols	L.: *verto*, "to turn."
Mollisols	L.: *mollis*, "soft."
Aridisols	L.: *aridus*, "dry."
Entisols	The syllable *ent* has no root meaning, but can be associated with "recent."

Within each order there are suborders ranging in number from as few as two to as many as seven, the total being 47. In pages to follow the suborders are named and defined for each order. Suborders are defined in various ways. For some orders, the list of suborders begins with soils of wet places and moves in sequence to well-drained and increasingly dry environments. In other cases, suborders are defined in terms of humus or mineral content. Meaning of root syllables used in suborder names is given in the legend for Figure 19.9.

Inceptisols[2]

Inceptisols are uniquely defined by combination of the following properties. (1) Soil moisture is available to plants more than half the year or for more than three consecutive months during a warm season. (2) There are present one or more pedogenic horizons (horizons developed by soil-forming processes) formed by alteration or concentration of matter, but without accumulation of translocated materials other than carbonate minerals or amorphous (non-crystalline) silica. (3) Soil textures are finer than loamy sand. (4) The soil contains some weatherable minerals. "Weatherable" means "capable of further chemical alteration" (see Chapter 22 and Table 22.3). (5) The clay fraction of the soil has a moderate to high capacity to retain cations. (*Cations* are principally the positively charged ions of calcium, magnesium, and potassium, referred to in Chapter 18 as *bases*, but also include aluminum, important in strongly acid soils.)

Although the inceptisols are found in a wide range of latitudes having the requisite soil-moisture budgets, it is most meaningful to focus attention upon the widespread occurrence of inceptisols in the regions of tundra climate (Chapter 17). Figure 19.8 is a generalized world map of dominant regions of occurrence of the soil orders. Compare the subarctic and arctic regions of inceptisol occurrence on this map with the regions of tundra climate in Figure 14.5 and Plate 2. The correspondence is generally quite good. Inceptisols are also found in high mountains associated with alpine tundra climate; such locations are found at all latitudes, ranging down to the equatorial zone. Figure 19.9, a map of soil orders and suborders of the United States, shows inceptisols in Alaska and coinciding with the Cordilleran mountain ranges of the northwestern United States.

It is interesting to note also that the inceptisols are mostly found on relatively young geomorphic surfaces, for example, surfaces shaped by glacial ice of the last great ice advance, or Wisconsin glaciation (Chapter 31). Because of their youth, these inceptisols usually lack distinct horizons. Under the cold tundra climate with its permafrost, chemical decomposition of minerals is greatly inhibited.

Inceptisols also occur on recently accumulated alluvial sediments of floodplains; these occurrences are shown on the world map (Figure 19.8) as ranging from equatorial to middle-latitude zones. Examples are the floodplains of the Mississippi, Amazon, Congo, and Ganges-Brahmaputra rivers.

Suborders. Definitions of the five suborders of inceptisols make clear the point that they can occur over the full geographical range from the arctic zone to the equator.

Andepts commonly have a dark surface horizon, are rich in humus and amorphous clay, and have a high cation exchange capacity. Andepts are formed chiefly from recent volcanic deposits of ash and tuff (Chapter 30). An example is the occurrence of andepts on the Island of Hawaii, where volcanic activity is in progress (Figure 19.9).

Aquepts are inceptisols of wet places; they have a gray to black surface horizon, a gray subsoil, and are seasonally saturated with water. The aquepts include some soils formerly called *tundra soils* and some of the *humic gley soils* and *low humic gley soils* in United States usage.

Ochrepts have a brownish horizon of altered materials at or close to the surface and are found in middle-latitude and high-latitude climates. They include some soils formerly called *brown forest soils* in United States usage.

Plaggepts are inceptisols with thick surface horizons of materials added by man through long habitation or long-continued manuring.

Tropepts are inceptisols of low latitudes. They have a brownish or reddish horizon of altered material at or close to the surface. The tropepts include some soils formerly called *latosols* in United States usage.

Umbrepts are inceptisols with a dark acid surface horizon more than 10 in (25 cm) thick. They contain crystalline clay minerals and are brownish in color. They occur in moist middle-latitude and high-latitude climates. The umbrepts include some soils formerly referred to as *brown forest soils*, as *alpine meadow soils*, and as *tundra soils* in United States usage.

Histosols

Histosols are unique in having a very high content of organic matter in the upper 32 in (80 cm). More than half of this thickness has from 20 to 30 percent of organic matter or more, or the organic horizon rests on bedrock or rock rubble. Most histosols are peats or mucks consisting of more or less decomposed plant remains that accumulated in water, but some have formed from forest litter or moss or both in a cool, continuously humid climate, but are freely drained.

[2]Most of the description of soil orders and suborders in the remainder of this chapter is selectively quoted or paraphrased from one of the following two sources: (1) Guy D. Smith, *The Soil Orders of the Classification Used in the U.S.A.*, Romanian Geological Institute, Technological and Economic Bulletin, Series C—Pedology, No. 18, pp. 509–531, 1970. Dr. Guy D. Smith is the former Director, Soil Survey Investigations, Soil Conservation Service, U.S. Department of Agriculture. (2) Roy W. Simonson and Darrell L. Gallup, *Soils of the World*, unpublished report, 1972. Dr. Roy W. Simonson is the former Director of Soil Classification and Correlation, Soil Survey, Soil Conservation Service, U.S. Department of Agriculture. This material has been used with permission of the authors.

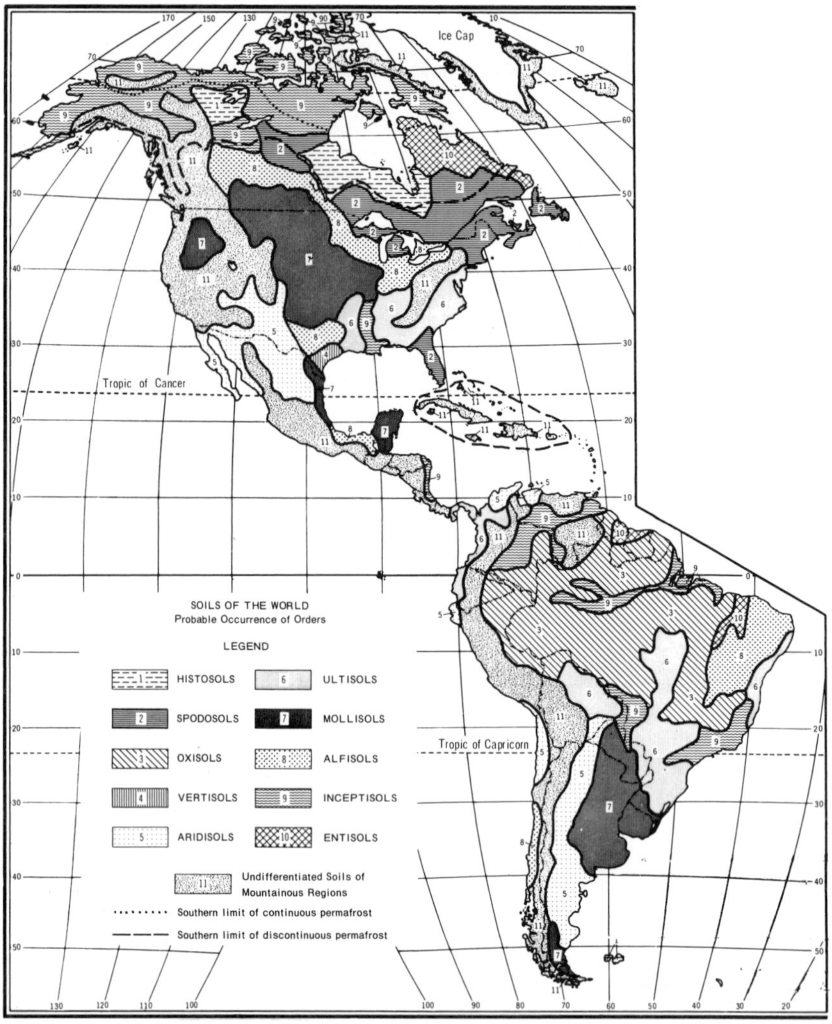

Figure 19.8 World map of soil orders, greatly generalized to show probable regions of dominance. (From Soil Conservation Service, U.S. Department of Agriculture, World Soil Geography Unit, Beltsville, Maryland, 1970.)

Histosols of cool climates are commonly acid in reaction and low in plant nutrients. Important areas of such histosols are shown in Figure 19.8 in the Northwest Territories of Canada and in an area lying to the south of Hudson Bay in Manitoba, Ontario, and Quebec. Histosols are found in low latitudes as well, where conditions of poor drainage have favored thick accumulations of plant matter.

Histosols that are mucks are agriculturally valuable in middle latitudes, where they occur in the beds of countless former lakes in the glaciated zone. After appropriate drainage and application of lime and fertilizers, these mucks are highly productive for truck garden vegetables. Peat bogs are extensively used for cultivation of cranberries (cranberry bogs); *Sphagnum* peat is dried and baled for sale as a mulch for use on suburban lawns and shrubbery beds. For centuries in Europe dried peat from bogs of glacial origin has been used as a low-grade fuel.

Suborders. Histosols are subdivided into four suborders on the basis of soil-moisture regime and degree of decomposition of the plant remains.

Fibrists are histosols composed mostly of *Sphagnum* moss or are histosols that are saturated with water most of the year or artificially drained, and that consist mainly of recognizable plant remains so little decomposed that rubbing does not destroy the bulk of the fibers. Fibrists were formerly classed as *peats* in United States usage.

Folists are histosols formed with free drainage and consisting of forest litter resting on bedrock or rock rubble. (No equivalent in older system.)

Hemists are histosols saturated with water most of the year or artificially drained, and with between 10 and 40 percent of the volume remaining as fibers after

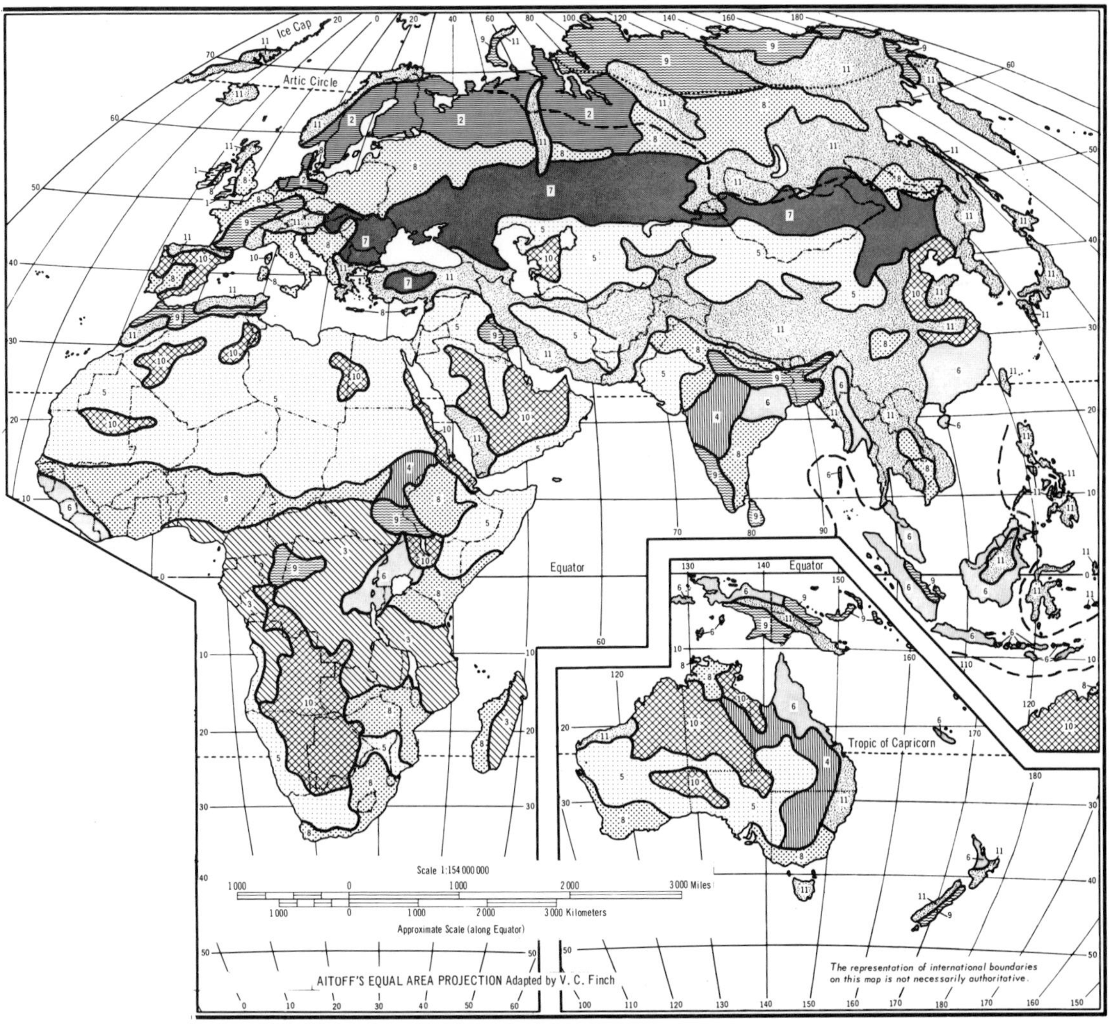

rubbing. Hemists were formerly classed as *peaty mucks* in United States usage.

Saprists are histosols saturated with water most of the year or artificially drained, and so decomposed that after rubbing less than 10 percent of the volume remains as fibers. Saprists were formerly called *mucks* in United States usage.

Spodosols

Spodosols have a unique property: a *B* horizon of accumulation of reddish amorphous materials having a high cation (base) exchange capacity. The amorphous materials contain organic material, compounds of aluminum, and commonly iron, all brought downward from overlying *A* horizons. (Recall that the process of removal of matter from the *A* horizons is *eluviation;* of deposition in the *B* horizon, *illuviation.*) In undisturbed soils (virgin soils) a bleached gray to white horizon overlies the *B* horizon. Spodosols are strongly acid, low in plant nutrients, such as calcium and magnesium, and low in humus. Many spodosols are of sandy texture and have small water-holding capacities.

Spodosols, with some exceptions, were formed under forest cover in a middle-latitude climate that is both cool and moist. In comparing the world distribution of spodosols with that of climates (compare Figure 19.8 with Figure 14.5 and Plate 2), the correspondence with the humid subarctic climate (Dfc) is striking. Spodosols also occupy northerly portions of the humid continental climate (Dfb). Correspondence with needleleaf (boreal) forest is also striking (Plate 4).

As in the case of the inceptisols, regions of spodosols are largely those that experienced the most recent continental glaciation; the soils are therefore very young. Nevertheless, spodosols can be recognized as occurring in latitudes down to the equatorial zone, but in these lower latitudes are limited to sandy parent materials and are of small extent. Soils of parts of Florida are classed as spodosols on the United States map (Figure 19.9).

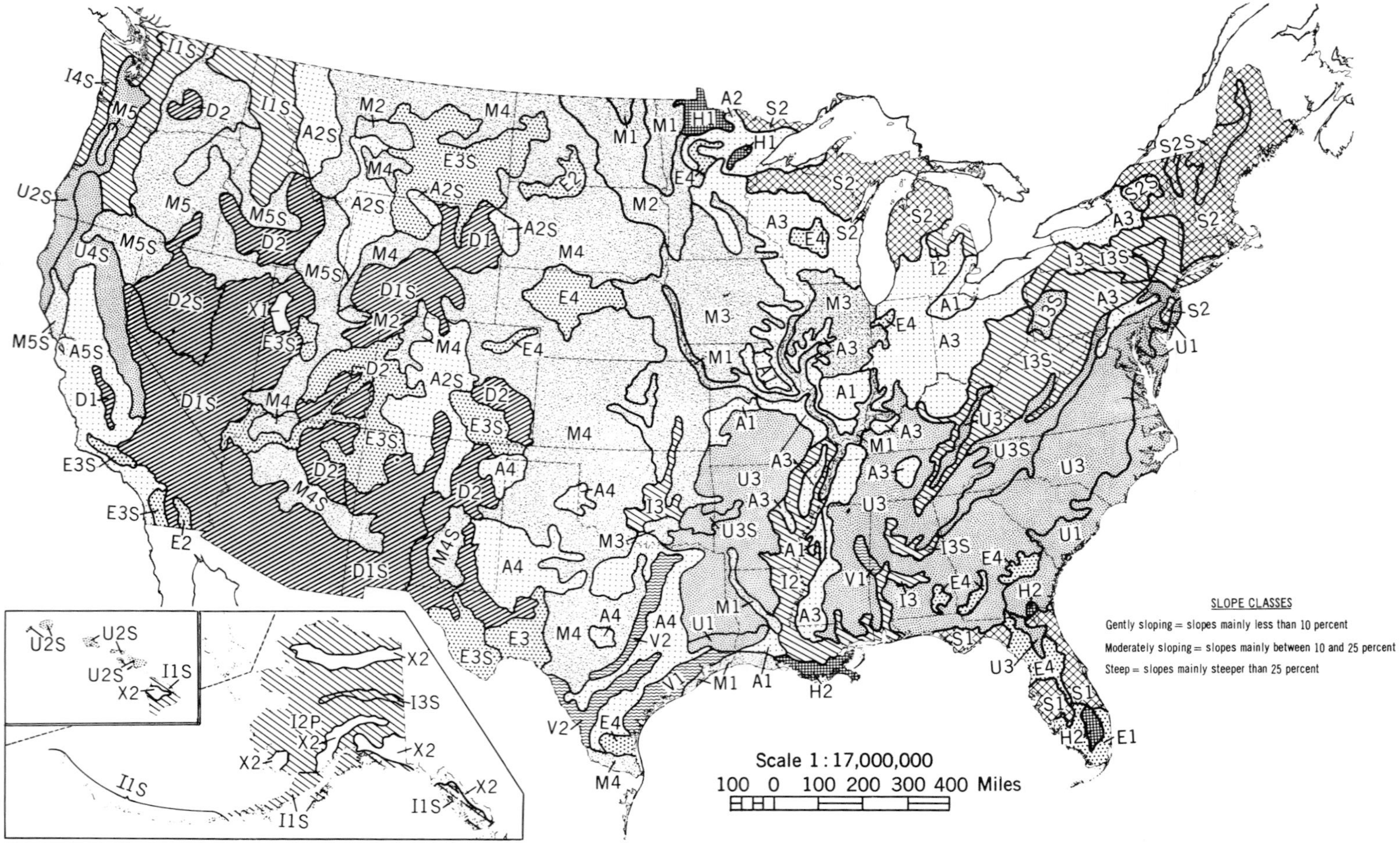

Figure 19.9 Patterns of soil orders and suborders of the United States. (Soil Conservation Service, U.S. Department of Agriculture, Hyattsville, Maryland, 1967.)

LEGEND

Only the dominant orders and suborders are shown. Each delineation has many inclusions of other kinds of soil. General definitions for the orders and suborders follow. For complete definitions see Soil Survey Staff, Soil Classification, A Comprehensive System, 7th Approximation, Soil Conservation Service, U. S. Department of Agriculture, 1960 (for sale by U. S. Government Printing Office) and the March 1967 supplement (available from Soil Conservation Service, U. S. Department of Agriculture). Approximate equivalents in the modified 1938 soil classification system are indicated for each suborder.

ALFISOLS . . . Soils with gray to brown surface horizons, medium to high base supply, and subsurface horizons of clay accumulation; usually moist but may be dry during warm season

A1 AQUALFS (seasonally saturated with water) gently sloping; general crops if drained, pasture and woodland if undrained (Some Low–Humic Gley soils and Planosols)

A2 BORALFS (cool or cold) gently sloping; mostly woodland, pasture, and some small grain (Gray Wooded soils)

A2S BORALFS steep; mostly woodland

A3 UDALFS (temperate or warm, and moist) gently or moderately sloping; mostly farmed, corn, soybeans, small grain, and pasture (Gray–Brown Podzolic soils)

A4 USTALFS (warm and intermittently dry for long periods) gently or moderately sloping; range, small grain, and irrigated crops (Some Reddish Chestnut and Red–Yellow Podzolic soils)

A5S XERALFS (warm and continuously dry in summer for long periods, moist in winter) gently sloping to steep; mostly range, small grain, and irrigated crops (Noncalcic Brown soils)

ARIDISOLS . . . Soils with pedogenic horizons, low in organic matter, and dry more than 6 months of the year in all horizons

D1 ARGIDS (with horizon of clay accumulation) gently or moderately sloping; mostly range, some irrigated crops (Some Desert, Reddish Desert, Reddish- Brown, and Brown soils and associated Solonetz soils)

D1S ARGIDS gently sloping to steep

D2 ORTHIDS (without horizon of clay accumulation) gently or moderately sloping; mostly range and some irrigated crops (Some Desert, Reddish Desert, Sierozem, and Brown soils, and some Calcisols and Solonchak soils)

D2S ORTHIDS gently sloping to steep

ENTISOLS . . . Soils without pedogenic horizons

E1 AQUENTS (seasonally saturated with water) gently sloping; some grazing

E2 ORTHENTS (loamy or clayey textures) deep to hard rock; gently to moderately sloping; range or irrigated farming (Regosols)

E3 ORTHENTS shallow to hard rock; gently to moderately sloping; mostly range (Lithosols)

E3S ORTHENTS shallow to hard rock; steep; mostly range

E4 PSAMMENTS (sand or loamy sand textures) gently to moderately sloping; mostly range in dry climates, woodland or cropland in humid climates (Regosols)

HISTOSOLS . . . Organic soils

H1 FIBRISTS (fibrous or woody peats, largely undecomposed) mostly wooded or idle (Peats)

H2 SAPRISTS (decomposed mucks) truck crops if drained, idle if undrained (Mucks)

INCEPTISOLS . . . Soils that are usually moist, with pedogenic horizons of alteration of parent materials but not of accumulation

I1S ANDEPTS (with amorphous clay or vitric volcanic ash and pumice) gently sloping to steep; mostly woodland; in Hawaii mostly sugar cane, pineapple, and range (Ando soils, some Tundra soils)

I2 AQUEPTS (seasonally saturated with water) gently sloping; if drained, mostly row crops, corn, soybeans, and cotton; if undrained, mostly woodland or pasture (Some Low–Humic Gley soils and Alluvial soils)

I2P AQUEPTS (with continuous or sporadic permafrost) gently sloping to steep; woodland or idle (Tundra soils)

I3 OCHREPTS (with thin or light–colored surface horizons and little organic matter) gently to moderately sloping; mostly pasture, small grain, and hay (Sols Bruns Acides and some Alluvial soils)

I3S OCHREPTS gently sloping to steep; woodland, pasture, small grains

I4S UMBREPTS (with thick dark–colored surface horizons rich in organic matter) moderately sloping to steep; mostly woodland (Some Regosols)

MOLLISOLS . . . Soils with nearly black, organic–rich surface horizons and high base supply

M1 AQUOLLS (seasonally saturated with water) gently sloping; mostly drained and farmed (Humic Gley soils)

M2 BOROLLS (cool or cold) gently or moderately sloping, some steep slopes in Utah; mostly small grain in North Central States, range and woodland in Western States (Some Chernozems)

M3 UDOLLS (temperate or warm, and moist) gently or moderately sloping; mostly corn, soybeans, and small grains (Some Brunizems)

M4 USTOLLS (intermittently dry for long periods during summer) gently to moderately sloping; mostly wheat and range in western part, wheat and corn or sorghum in eastern part, some irrigated crops (Chestnut soils and some Chernozems and Brown soils)

M4S USTOLLS moderately sloping to steep; mostly range or woodland

M5 XEROLLS (continuously dry in summer for long periods, moist in winter) gently to moderately sloping; mostly wheat, range, and irrigated crops (Some Brunizems, Chestnut, and Brown soils)

M5S XEROLLS moderately sloping to steep; mostly range

SPODOSOLS . . . Soils with accumulations of amorphous materials in subsurface horizons

S1 AQUODS (seasonally saturated with water) gently sloping; mostly range or woodland; where drained in Florida, citrus and special crops (Ground–Water Podzols)

S2 ORTHODS (with subsurface accumulations of iron, aluminum, and organic matter) gently to moderately sloping; woodland, pasture, small grains, special crops (Podzols, Brown Podzolic soils)

S2S ORTHODS steep; mostly woodland

ULTISOLS . . . Soils that are usually moist with horizon of clay accumulation and a low base supply

U1 AQUULTS (seasonally saturated with water) gently sloping; woodland and pasture if undrained, feed and truck crops if drained (Some Low–Humic Gley soils)

U2S HUMULTS (with high or very high organic–matter content) moderately sloping to steep; woodland and pasture if steep, sugar cane and pineapple in Hawaii, truck and seed crops in Western States (Some Reddish–Brown Lateritic soils)

U3 UDULTS (with low organic– matter content; temperate or warm, and moist) gently to moderately sloping; woodland, pasture, feed crops, tobacco, and cotton (Red–Yellow Podzolic soils, some Reddish–Brown Lateritic soils)

U3S UDULTS moderately sloping to steep; woodland, pasture

U4S XERULTS (with low to moderate organic–matter content, continuously dry for long periods in summer) range and woodland (Some Reddish–Brown Lateritic soils)

VERTISOLS . . . Soils with high content of swelling clays and wide deep cracks at some season

V1 UDERTS (cracks open for only short periods, less than 3 months in a year) gently sloping; cotton, corn, pasture, and some rice (Some Grumusols)

V2 USTERTS (cracks open and close twice a year and remain open more than 3 months); general crops, range, and some irrigated crops (Some Grumusols)

AREAS with little soil . . .

X1 Salt flats

X2 Rockland, ice fields

NOMENCLATURE

The nomenclature is systematic. Names of soil orders end in *sol* (L. *solum*, soil), e. g., ALFISOL, and contain a formative element used as the final syllable in names of taxa in suborders, great groups, and sub-groups.

Names of suborders consist of two syllables, e. g., AQUALF. Formative elements in the legend for this map and their connotations are as follows:

and	– Modified from Ando soils; soils from vitreous parent materials
aqu	– L. *aqua*, water; soils that are wet for long periods
arg	– Modified from L. *argilla*, clay; soils with a horizon of clay accumulation
bor	– Gr. *boreas*, northern; cool
fibr	– L. *fibra*, fiber; least decomposed
hum	– L. *humus*, earth; presence of organic matter
ochr	– Gr. base of ochros, pale; soils with little organic matter
orth	– Gr. *orthos*, true; the common or typical
psamm	– Gr. *psammos*, sand; sandy soils
sapr	– Gr. *sapros*, rotten; most decomposed
ud	– L. *udus*, humid; of humid climates
umbr	– L. *umbra*, shade; dark colors reflecting much organic matter
ust	– L. *ustus*, burnt; of dry climates with summer rains
xer	– Gr. *xeros*, dry; of dry climates with winter rains

Spodosols of middle and high latitudes are naturally poor soils in terms of agricultural productivity. Because they are acid, the application of lime is a necessity. Because they are poor in the nutrients required by most crops, application of fertilizers is required. With proper management and the input of required industrial products, the spodosols can be highly productive. An example is the high yield of potatoes from spodosols in Maine and New Brunswick. Another unfavorable factor is shortness of the growing season; this factor is limiting to agriculture in the more northerly areas of spodosols.

Suborders. Four suborders of the spodosols are defined in terms of soil moisture regimes and composition.

Aquods are spodosols of wet places, seasonally saturated with water. The surface horizon may be either black or white and lacks free iron oxides. The *B* horizon is brownish or reddish and is often firm or cemented. Aquods were formerly called *ground water podzols* in United States usage.

Ferrods are freely drained spodosols having *B* horizons of accumulation of free iron oxides. This suborder is of rare occurrence.

Humods are freely drained spodosols in which at least the upper part of the *B* horizon has accumulated humus and aluminum, but not iron. Humods were formerly classed with the *podzols* in United States usage.

Orthods are freely drained spodosols with *B* horizons in which iron, aluminum, and humus have accumulated. Orthods were formerly identified with the *podzols* and *brown podzolic soils* in United States usage.

Alfisols

Alfisols are uniquely defined by the following combination of properties. (1) A gray, brownish, or reddish horizon not darkened by humus at or close to the surface. (2) A *B* horizon of clay accumulation. (3) A medium to high supply of bases in the soil. (4) Soil-moisture available to plants more than half of the year or more than three consecutive months during a warm season. The horizon referred to in the second property listed above is a *B* horizon with accumulated silicate clay minerals and is moderately saturated with exchangeable bases such as calcium and magnesium. The A_2 horizon above it is marked by some loss of bases, silicate clay minerals, and sesquioxides.

On the world map (Figure 19.8) dominant areas of alfisols are shown in central North America, western Europe, central Siberia, and north China. These areas are identified with the humid continental climate (Cfa), the marine west-coast climate (Cfb), and the Mediterranean climate (Csb) (see Figure 14.5 and Plate 2). Other important areas of alfisols shown in Figure 19.8 are located in eastern Brazil, central, eastern and southern Africa, western Madagascar, southern Australia, eastern India, and Indochina.

On the whole, alfisols are agriculturally fairly productive under simple management because soil-moisture is usually adequate and bases are not depleted. In western Europe and China alfisols, along with the inceptisols occurring within the same regions, have supported dense populations for centuries.

Suborders. Five suborders of the alfisols are defined in terms of soil-moisture and temperature regimes.

Aqualfs are alfisols of wet places. They are dominantly gray throughout and are seasonally saturated with water. Aqualfs were formerly classed as *low-humic gley soils* and *planosols* in United States usage.

Boralfs are alfisols of boreal forests or high mountains. They have a gray surface horizon, a brownish subsoil, and are associated with a mean annual soil temperature less than 47°F (8°C). The boralfs were formerly called *gray wooded soils* in United States usage.

Udalfs are brownish alfisols of the humid continental climate. They occur in regions having no periods or only short periods (fewer than 90 days in most years) when part or all of the soil is dry. Udalfs were formerly classed as *gray-brown podzolic soils* in United States usage.

Ustalfs are alfisols of warm climates with long periods (more than 90 cumulative days in most years) when the soil is dry. They are brownish to reddish throughout. Formerly the ustalfs were designated as *reddish-brown soils* and *reddish-chestnut soils* in United States usage.

Xeralfs are alfisols of the Mediterranean climates with cool, moist winters and rainless summers. They are brownish to reddish throughout and were formerly designated as *noncalcic brown soils* in United States usage. A typical occurrence is in coastal and inland valleys of central and southern California (Figure 19.9).

Ultisols

Ultisols are unique through the combination of three properties. (1) There is a *B* horizon of clay accumulation. (2) The supply of bases is low, particularly in the lower horizons. (3) The mean annual soil temperature is greater than 47°F (8°C). A point of importance is that the degree of base saturation (degree to which colloids are holding base cations) diminishes rapidly with depth. The highest base saturation is within the upper few inches, reflecting the recycling of bases by plants or additions of fertilizer.

Ultisols characteristically form under forest vegetation in climates with slight to a pronounced seasonal moisture deficit alternating with a surplus. At one season evapotranspiration exceeds precipitation but, at another season, there is a surplus of precipitation so that some water passes through the soil into the substratum. Thus, bases not held in the plant tissue are depleted. The *B* horizon of well-drained ultisols is characteristically red or yellowish brown in color due to concentration of iron oxides.

Looking at the world map (Figure 19.8) we find one large area of dominantly ultisols in the southeastern United States and Caribbean lands, another in southern China. There are also important areas of ultisols in Bolivia, southern Brazil, western and central Africa, India, the East Indies, and northern Australia.

Reference to Figure 14.5 and Plate 2 shows that the climate in areas of ultisols ranges from the humid subtropical climate (Cfa) in the higher latitudes to the tropical wet-dry and trade-wind littoral climates (Cwa, Aw, Am) in the lower latitudes. These climates have large water surpluses in one season, although they cover a large spread in temperature characteristics. The land surfaces in these areas have been subjected to prolonged denudation and weathering, so that crystalline bedrock beneath is often deeply decayed.

Ultisols were largely used under systems of shifting cultivation prior to the development of modern technology, as they still are where soil amendments are not accessible to cultivators. The soils are low enough in plant nutrients so that growth of cultivated plants is poor without general use of lime and fertilizers. However, given essential industrial products and adequate management skills, the ultisols can be highly productive. In fact, many of these soils have large unrealized potentials at the present time.

Suborders. The five suborders of ultisols are defined in terms of their moisture regimes and humus content.

Aquults are ultisols of wet places. They are seasonally saturated with water and are dominantly gray throughout. A few have black surface horizons on gray subsoils. The aquults include some soils formerly called *low humic-gley soils* and *humic-gley soils* in United States usage.

Humults are ultisols with large accumulations of organic matter formed under relatively high, well-distributed rainfall in middle-latitude and tropical zones. Humults include some soils formerly called *reddish brown lateritic soils* and *humic latosols* in United States usage.

Udults are ultisols of humid middle-latitude to tropical zones. They have moderate to small amounts of organic matter, reddish or yellowish *B* horizons, and no periods or only short periods when some part of the soil is dry. Udults include some soils formerly called *red-yellow podzolic* and *reddish-brown lateritic soils* in the United States usage.

Ustults are ultisols of warm climates with long periods when the soil is dry. They are brownish to reddish throughout. Ustults include some soils formerly called *red-yellow podzolic soils* in United States usage.

Xerults are ultisols of Mediterranean climates with cool moist winters and rainless summers. They are brownish to reddish throughout. Xerults include some soils formerly called *noncalcic brown soils* or *red-yellow podzolic soils* in United States usage.

Oxisols

Oxisols are unique through the combination of three properties. (1) Extreme weathering of most minerals other than quartz to kaolin and free oxides. (2) Very low cation exchange capacity of the clay fraction. (3) A loamy or clayey texture (sandy loam or finer). Oxisols characteristically develop in equatorial, tropical, and subtropical zones on land surfaces that have been stable over long periods of time. Most of these surfaces are of early Pleistocene age or much older. During development the climate must be humid. In some areas, however, the oxisols now occupy seasonally arid environments because a climate change has occurred since the time of their formation.

Oxisols of low latitudes lack distinct horizons, except for darkened surface layers, even though they are formed in strongly weathered parent matter. Red, yellow, and yellowish-brown colors are normal to the well-drained oxisols because of the presence of iron oxides. Supplies of plant nutrients are commonly low, and the capacity of the soil to fix phosphorus in unavailable forms is high. The clay fraction consists largely of silicate clay minerals, but with an important proportion of iron and aluminum oxides. However, despite the large clay fraction the soils are friable and are permeable to water and roots.

The world map of soil orders (Figure 19.8) shows oxisols to be dominant in vast areas of equatorial and tropical South America and Africa. These are regions of wet equatorial and tropical wet-dry climates (Af, Am, Aw) shown in Figure 14.5 and Plate 2.

Like the ultisols of warm climates, the oxisols of low latitudes were used under systems of shifting cultivation prior to the advent of modern technology; a large proportion of these soils are still used in that way. The levels of plant nutrients under natural conditions are so low that the yields obtainable under hand tillage are very low, especially after a garden patch has been used for a year or two. Substantial use of lime, fertilizers, and other products of heavy industry is necessary for high yields and sustained production; unrealized agricultural potentials are thought to be high.

Suborders. Five suborders of oxisols are recognized on the basis of their moisture and temperature regime, or other properties very closely associated with climate.

Aquox are oxisols of wet places. They are dominantly gray, or mottled dark red and gray, and are seasonally saturated with water. Aquox include some soils formerly called *ground-water laterite soils* in the United States usage.

Humox are oxisols of relatively cool moist regions, with very large accumulations of organic carbon. Humox occur on high plateaus; they include some soils formerly called *black-red latosols* and some *humic latosols* in United States usage.
Orthox are oxisols of warm, humid regions with short dry seasons or no dry season. These soils are never dry in any horizon for as long as 60 consecutive days. Orthox include soils formerly called *red latosols,* and *red-yellow latosols* in United States usage.
Torrox are oxisols that are dry in all horizons for more than six months of the year, are never continuously moist for as long as three consecutive months, and are not appreciably darkened by humus. (No equivalent in older system.)
Ustox are oxisols that are moist in some horizon more than six months (cumulative) or more than three consecutive months, but are dry in some horizon for more than 60 consecutive days (roughly equivalent to a three-month dry season). Ustox include some soils formerly called *low-humic latosols* in United States usage.

Vertisols

Vertisols are uniquely defined through combination of the following properties. (1) A high content of clay that shrinks and swells with changes in moisture. (2) Deep, wide cracks at some season. (3) Evidence of soil movement in the form of slickensides, microrelief (gilgai), or structural aggregates tilted between horizontal and vertical. (Slickensides are soil surfaces with grooves or striations made by the movement of one mass of soil against another while in a plastic state.) The cation exchange capacity of vertisols is usually high. Horizons are normally weakly expressed.

Vertisols typically form under grass or savanna vegetation in climates with moderate or pronounced seasonal soil-moisture deficiencies. At some season the soils dry out enough that the soil cracks (Figure 19.10). When rains come, some of the surface soil commonly drops into the cracks before they close, so that the soil slowly "swallows itself" (Figure 19.11). Vertisols over the world have had more names than any other kind of soil. Examples are *black cotton soils, tropical black clays,* and *regur.* In the older United States terminology most vertisols were called *rendzinas* for some years and then renamed *grumusols.*

Vertisols are found only in the latitude range of the tropical and subtropical zones. The world map (Figure 19.8) shows a small area of vertisols on the coastal plain of Texas. One large tropical area is in the Deccan region of India, where the vertisols are formed on weathered basaltic rocks. Other areas shown on the map are the Sudan region in east central Africa and a large zone in eastern Australia.

Vertisols are high in exchangeable bases such as calcium and magnesium, most are near neutral (neither alkaline nor acid), and they have intermediate amounts of organic matter. The soils retain large amounts of water because of their fine texture, but much of this moisture is not available to plants. If cultivation is dependent upon human or animal power, yields are low. Problems of using these soils are especially difficult; for example, the soil becomes highly plastic and difficult to till when wet. Many areas of vertisols have thus been left in grass or savanna, which may be grazed. However, with use of modern technology and its energy resources, production of food and fiber from vertisols could be substantial.

Suborders. There are four suborders of vertisols, defined on the basis of soil-moisture regimes. The suborders of vertisols have no equivalents in former United States usage.
Torrerts are vertisols of arid regions, having cracks that remain open throughout the year in most years.
Uderts are vertisols of relatively humid regions, having cracks that remain open for only short periods in most years, less than 90 cumulative days and less than 60 consecutive days.
Usterts are vertisols of tropical regions with distinct wet and dry seasons, or of regions with two dry seasons. The cracks remain open more than 90 cumulative days in most years.
Xererts are vertisols of Mediterranean climates, with cracks that close in winter and open in the summer every year for more than 60 consecutive days.

Mollisols

Mollisols are uniquely defined through a combination of the following properties. (1) A very dark brown to black surface horizon that is more than one-third of the combined thickness of the *A* and *B* horizons or is more than 10 in (25 cm) thick and that has a structure or soft consistence when dry. (2) A dominance of calcium among the extractable cations (bases) in the A_1 and *B* horizons. (3) A dominance of crystalline clay minerals of moderate or high cation exchange capacity. (4) Less than 30 percent of clay in some horizon if the soil has deep wide cracks at some season.

Mollisols characteristically form under grass in climates with a moderate to pronounced seasonal moisture deficiency. A few form in marshes or on marls (compacted muds of calcium-carbonate composition) in humid climates. Comparison of mollisol distribution on the world map (Figure 19.8) with distribution of climates in Figure 14.5 and Plate 2 shows that the mollisols are closely associated with middle-latitude steppes (BSk), with the subhumid parts of the humid continental climate (Dfa, Dfb), and with the less-humid parts of the humid subtropical climate (Cfa). In North America mollisols dominate the Great Basin and Great Plains regions, but there are smaller areas along the Gulf coastal plain of Mexico and the

Figure 19.10 Soil cracks in a Texas vertisol, the Houston black clay, during an extremely dry period. The crop is cotton. (U.S.D.A.—Soil Conservation Service.)

Yucatan Peninsula. In South America there is a large area of mollisols over the pampas of Argentina and Uruguay. In Eurasia a great belt of mollisols stretches from Rumania eastward across the steppes of Russia, Siberia, and Mongolia, into Manchuria.

Mollisols are among the naturally most fertile soils in the world. They now produce the bulk of the grain that moves in commercial trade channels. Most of these soils have not been widely used for crop production except during the last century. Prior to that time they were used mainly for grazing by nomadic herds. The mollisols have favorable characteristics for growing of cereals in large-scale farming and are relatively easy to manage. Production varies considerably from one year to the next because seasonal rainfall is highly variable and soil moisture is the limiting factor in productivity.

Suborders. Seven suborders of mollisols are defined primarily in terms of their moisture and temperature regimes, and in one suborder by the carbonate content.

Albolls are mollisols with a bleached gray to white horizon above a slowly permeable horizon of clay accumulation, saturated with water in at least the surface horizons at some season. Albolls include some soils formerly called *planosols* and *solodized solonetz* in United States usage.

Aquolls are mollisols of wet places. They have a nearly black surface horizon resting on a mottled grayish *B* horizon and are seasonally saturated with water. Aquolls include soils formerly called *humic-gley soils, calcium carbonate solonchak,* and a few *solonetz* in United States usage.

Borolls are mollisols of cold-winter steppes or high mountains. They have a mean annual soil temperature lower than 47°F (8°C). Borolls include some soils formerly called *chernozems, chestnut soils,* and *brown soils* in United States usage.

Rendolls are mollisols formed on highly calcareous parent materials and with more than 40 percent calcium carbonate in some horizon within 20 in (50 mm) of the surface, but without a horizon of accumulation

Figure 19.11 Sidewalk slabs tilted by movements in vertisol (Houston black clay), Travis County, Texas. (U.S.D.A.—Soil Conservation Service.)

of calcium carbonate. Rendolls are mostly formed in humid climates under forest vegetation. They were formerly called *rendzinas* in United States usage.

Udolls are mollisols of humid middle-latitude and subtropical zones. They have brownish hues throughout and no horizon of accumulation of calcium carbonate in soft powdery forms. Udolls include soils formerly called *prairie soils* and *brunizems* in United States usage.

Ustolls are mollisols of middle-latitude and subtropical zones with strongly continental climate characteristics and with long periods (more than 90 cumulative days in most years) in which the soil is dry. Most ustolls have a horizon of accumulation of soft, powdery calcium carbonate starting at a depth of about 20 to 40 in (50 to 100 cm). Ustolls include some soils formerly called *chestnut soils* and *brown soils* in United States usage.

Xerolls are mollisols of Mediterranean climates with cool moist winters and rainless summers. Xerolls include some soils formerly called *brunizems, chestnut soils,* and *brown soils* in United States usage.

Aridisols

Aridisols are uniquely defined through a combination of the following properties. (1) A lack of water available for most plants for very extended periods. (2) One or more pedogenic horizons. (3) A surface horizon or horizons not significantly darkened by humus. (4) Absence of deep wide cracks. The aridisols have practically no available soil moisture most of the time that the soil is warm enough for plant growth, for example, higher than 41°F (5°C). Soil moisture, in most years, is not available for as long as 90 consecutive days when the soil temperature is above 47°F (8°C). Carbonate accumulations are often large at depth in the profile.

Aridisols are primarily soils of arid climates in tropical, subtropical, and middle-latitude zones. They form under conditions that preclude entry of much water into the soil—either extremely scanty rainfall or slight rainfall that for one reason or another does not enter the soil. The vegetation consists of scattered plants, ephemeral grasses and forbs, cacti, and xerophytic shrubs.

Study of the world map of soil orders (Figure 19.8) shows that aridisols occupy regions shown on Figure 14.5 and Plate 2 to have either tropical or middle-latitude desert and steppe climates (BWh, BSh, BWk, BSk). Note that substantial parts of these same arid regions are dominated by entisols, occupying areas of dune sand or rocky desert floor.

Most aridisols are used, as they have been through the ages, for nomadic grazing. This use is dictated by the limited rainfall, which is inadequate for crops without irrigation. Locally, where water supplies from exotic streams or ground water permit, aridisols can be highly productive for a wide variety of crops under irrigation. Great irrigation systems, such as those of the Imperial Valley of the United States, the Nile Valley of Egypt, and the Indus Valley of Pakistan, have rendered aridisols highly productive, but not without attendant problems of salinization and waterlogging (see Chapter 19).

Suborders. Two suborders of aridisols are defined by the presence or absence of a *B* horizon of clay accumulation.

Argids are aridisols with a *B* horizon of clay accumulation. These are largely surfaces of late Wisconsin age or older, and consequently have developed throughout one or more of the Pleistocene pluvial (moist) periods. Argids include some soils formerly called *desert* and *red desert soils* and associated *solonetz* in United States usage.

Orthids are soils without a *B* horizon of clay accumulation but with one or more pedogenic horizons that extend at least 10 in (25 cm) below the surface so that some horizon persists in part if the soil is plowed. Orthids include some soils formerly called *desert* and *red desert soils, calcisols,* and *sierozems* in United States usage.

Entisols

Entisols have in common the combination of a mineral soil with absence of distinct pedogenic horizons that would persist after normal plowing. Entisols are soils in the sense that they support plants, but they may be found in any climate and under any vegetation. Lack of distinct horizons in the entisols may be the result of a parent material, such as quartz sand, in which horizons do not readily form, or of lack of time for horizons to form in recent deposits of volcanic ash or alluvium, on actively eroding slopes, and in soils recently disturbed by plowing to a depth of several feet.

Entisols occur throughout the full global range from equatorial to arctic latitude zones. Areas identified on the world map (Figure 19.8) as dominantly entisols include the heavily glaciated Ungava region of Quebec and Labrador and a number of large dune-sand areas of the North African-Arabian desert. Other desert regions mapped as entisols are the Kalahari Desert of southwestern Africa and parts of the central and northern desert of Australia. A very different type of region having entisols is indicated in central and northern China; this is an expanse of floodplain and deltaic plain of the Yellow and Yangtze rivers.

Entisols of the subarctic zone and tropical deserts are (along with many areas of inceptisols) the poorest of all soils from the standpoint of potential agricultural productivity. In contrast, entisols and inceptisols of floodplains and deltaic plains are among the most highly productive agricultural soils in the world because of their fine texture, ample nutrient content, and large available soil moisture. Dense agricultural populations in central China and in the Ganges-Brahmaputra plain of India and Bangladesh bear witness to this fact.

Suborders. To some extent, the following definitions of the five suborders of entisols allow us to sort out the productive soils from the nonproductive soils.

Aquents are entisols of wet places. They are dominantly gray throughout and stratified or are saturated with water at all times. Aquents include some soils formerly called *low-humic gley soils* in United States usage.

Arents are entisols with recognizable fragments of pedogenic horizons that have been mixed by deep plowing, spading, or other disturbance by man. (No equivalent in former system.)

Fluvents are entisols from recent alluvium, rarely saturated with water, usually stratified, and brownish to reddish throughout at least in the upper 20 in (50 cm), and with loamy or clayey texture in some part of the upper 40 in (1 m). Fluvents include some soils formerly called *alluvial soils* in United States usage.

Orthents are entisols that have a shallow depth to consolidated rock or have unconsolidated materials such as loess, and with skeletal, loamy, or clayey texture in some part of the upper 40 in (1 m). Orthents include some soils formerly called *lithosols* and *regosols* in United States usage.

Psamments are entisols that have sandy texture (sands and loamy sands) throughout the upper 40 in (1 m). They include some soils formerly called *regosols* in United States usage.

REVIEW QUESTIONS

1. What are the three orders into which all soils can be classified? Under what conditions is each of these orders developed? What kinds of soils are included under the azonal order?

2. How did our present system of soil classification develop? What basic concepts underlie the classification? Name persons who made important contributions to this field.

3. Describe the profile of a typical podzol soil. What climatic conditions are required for the development of soils of this group? What is the natural vegetation of this great soil group? How do these soils rate in terms of agricultural productivity? In what ways do they require treatment?

4. How do the gray-brown podzolic soils differ from the podzol soils? How do the climatic factors differ? What geographic positions do the gray-brown podzolic soils have with respect to the podzols? How does natural vegetation of the two soil groups differ? Which soil group is better for agricultural purposes? Why?

5. What are the essential characteristics of the red-yellow soils? With what climate are they associated? Explain how temperature and precipitation determine the features of these soils. What types of forest are found on the red-yellow soils?

6. What is terra rossa? Where is it found? What ideas have been advanced as to the origin of terra rossa?

7. What are latosols (lateritic soils)? What are their essential characteristics? What is meant by laterization? How do latosols rate in terms of agricultural productivity?

8. How can laterite be used as a building material? What are some of the residual ores obtained from laterites?

9. Describe other soils of humid tropical and equatorial climates.

10. What are hydromorphic soils? Why are the bog and meadow soils classed as intrazonal? What is the appearance of the profiles of these soils? What is the glei horizon? How do meadow soils differ from bog soils?

11. What are planosols? To what order do they belong? Where do planosols develop, and what is different about their profiles as compared with associated zonal soils?

12. Describe a tundra soil. In what respect does it differ from the podzol soils? What effect has excessive cold upon the composition of the parent material of the soil? Where are the arctic brown forest soils found?

13. What are the chernozem soils? Where were these soils first studied? Where are they best developed? Describe the chernozem profile.

14. What is the essential difference between the chernozems and the podzolic soils? With what climatic characteristics are the chernozem regions associated?

15. What is the natural vegetation of chernozem soils? How is loess associated with the chernozem?

16. In what way are the prairie soils related to both the chernozems and the gray-brown podzolic soils? What is the nature of the geographic and climatic location of the prairie soils?

17. What are the tall-grass prairies? What theories have been advanced for the absence of forests in these areas? With what crops are the prairie regions now associated in the central United States? Where are reddish prairie soils found?

18. Where are the chestnut soils (dark-brown soils) located in relation to the chernozems? With what climatic conditions are they associated?

19. How do the brown soils differ from the chestnut soils? What is distinctive about the *B* horizon of a brown soil?

20. Describe reddish-chestnut and reddish-brown soils. Where are they found, and with what climates and natural vegetation types are they associated?

21. Describe the gray-desert soils and red-desert soils. How do these two groups differ in geographic location and in climatic controls? Are horizons well developed in these soils? What forms do excessive calcium carbonate and hydrous calcium sulfate take in these soils?

22. How does altitude influence the development of soil types? In an arid region in middle latitudes, what succession of soil groups might be encountered in passing from an intermontane basin to a lofty mountain summit?

23. What are halomorphic soils? Where do saline soils form? What are some of the salts found in playa lakes of the dry deserts? How do the alkali soils differ from the saline soils?

24. Explain the process of salinization of soils in irrigated arid lands. How does waterlogging occur and what are its undesirable effects?

25. What are calcimorphic soils? Describe rendzina soils and give locations of three occurrences. Why are rendzina soils regarded as intrazonal?

26. Describe the basic characteristics of the *U.S. National Soil Taxonomy* system. What advantages has this system over the older system, published in 1938?

27. What is the basis for distinguishing among the ten soil orders of the newer classification system? Which soil factors are recognizable in the field? Which require laboratory analysis?

28. For each of the soil orders, give a brief statement of the natural agricultural productivity of the soil. Which orders require substantial inputs of lime, fertilizers, and other industrial products in order to give satisfactory crop yields?

CHAPTER **20**

Structure and Environment of Vegetation

VEGETATION on the lands of the earth is of prime concern to the geographer. Plants, as stationary objects possessing distinctive physical properties, are an element of the landscape just as fully as landforms, soils, and hydrographic features. That the physical forms of individual plants and of their assemblages vary in some systematic way with latitude, elevation, and continental position excites the interest of the geographer and causes him to inquire more closely into the interrelations of vegetation, soils, landforms, and climate. Plants, as consumable and renewable sources of food, fuel, clothing, shelter and many other categories of life-essentials, form a great natural resource base essential to man. The ways in which man has used these resources to his advantage—or has been hindered by plants in his progress—throughout human history forms a vital part of historical geography; the present and future development and management of plant resources is a part of the fabric of many studies of economic and political geography.

Our concern here is with *natural vegetation*, that is, vegetation which develops without appreciable interference and modification by man. The widespread activities of man in agriculture, herding, forest-cropping, and urbanization guarantee that a considerable proportion of the earth's land surfaces does not bear a natural vegetation. Nevertheless, from the study of islandlike remnants of protected plant life and historical records, the plant geographer can achieve some success in the global mapping of broad units of natural vegetation. (See Chapter 21.) Such maps represent the vegetation that might be expected eventually to return through a series of stages, if man's interference with the natural regimen were to be suspended.

Floristic or structural approach?

Our selection of study topics is extremely limited in terms of the scope of the science of botany and even of plant geography as a specialized field within botany. Emphasis here is not upon a *floristic* approach, that is an approach primarily concerned with the *flora*. Should we describe the flora of a region, we would produce a list of the plant species found there. But such a list would tell us little about the relative abundance of the species; their physical forms; and their characteristic arrangements as landscape elements.

To be sure, where the trees of a forest consist largely of two or three dominant species, it will be desirable to name these species and to designate the forest accordingly. For example, one finds in the plateaus of Arizona and Utah a woodland in which the tree population consists almost exclusively of pinyon pine (*Pinus edulis*) and juniper (*Juniperus utahensis*). Through our knowledge of the characteristic height, branching habit, and foliage of these species we gain a true picture of the physical form of the woodland. Such a terse forest description by floristic means would be impossible in an equatorial forest composed of scores of different species. Instead, the forest must be described according to its *structure*, that is, by the way in which the living parts of the plants are shaped and distributed in space. Generally, in the classification of the world's natural vegetation, the structural approach is most meaningful to follow.

Bioclimatology and ecology

Our study of plant geography will not be concerned with the historical aspects, that is, with the evolution of plants and the spread and decline of floras through geologic time. On the other hand, we will be greatly interested in the *bioclimatological* aspects of plant geography in which the responses of plants to light, heat, and moisture are studied and the knowledge of climate elements and their global distribution is brought into play. If we choose, the bioclimatological approach can be narrowed to the response of a single species to a single meteorological element.

A quite different approach to the question of a plant's relation to its environmental influences is to consider together all the organisms—plant and animal—living in one place at one time in terms of their total environment and of their adjustments to both the environment and to each other. Such a whole dynamic system is referred to as an *ecosystem;* the study of an ecosystem is *plant ecology*. The ecosystem represents the combined use by plants and animals of the resources provided, in one place at one time, by the surroundings. Air, water, nutrients, heat, and light are variously utilized, transformed, stored, and returned to the nonliving state.

The biosphere and its subdivision

To fix the place of natural vegetation units in the scheme of life on earth we begin with the total system, or *biosphere*, which encompasses the entire near-surface zone of the earth favorable to life in one form or another. The biosphere may be subdivided into three environmental divisions, or *biocycles:* (1) *salt water* (oceans), (2) *fresh water* (streams, lakes, ponds), and (3) *land* (soil and the air in contact with it). We are interested primarily in the land biocycle, although the geographer's concern extends into the fresh water and salt water biocycles as well.

The land biocycle is divided into ecological systems of decreasing size and complexity, schematically represented in Figure 20.1*A*. Four great *biochores* comprise the land biocycle: (*a*) *forest,* (*b*) *savanna,* (*c*) *grassland,* and (*d*) *desert.* (See Figure 21.1). These four classes of vegetation are separated on the basis of the structure of the plant assemblages and represent the fundamental response of vegetation to global climate controls, principally available moisture (precipitation and evaporation), but secondarily light, heat, and winds. The four biochores comprise systems too vast in area and having too wide a variation within each biochore to use in the mapping of world vegetation. Therefore, within the biochores are *formation classes*—altogether from 15 to 20 or more in number, depending upon the judgment of the plant geographer whose classification system is used. Thus there are clearly several varieties of forests (e.g., rainforest, deciduous forest, needleleaf forest) each responding to the climate regime to which it is subjected. Definitions of such terms as "forest," "woodland," and "grassland" are given in the discussion of the individual formation classes in Chapter 21. The distinction among the several biochores and their included formation classes is made on the basis of the plant structure, rather than species. Thus classification is structural, rather than floristic.

Plant habitats and communities

As we all know from direct experience, vegetation is strongly influenced by landforms and soil. Vegetation on an upland, that is, on relatively high ground of moderate surface slope and thick soil,

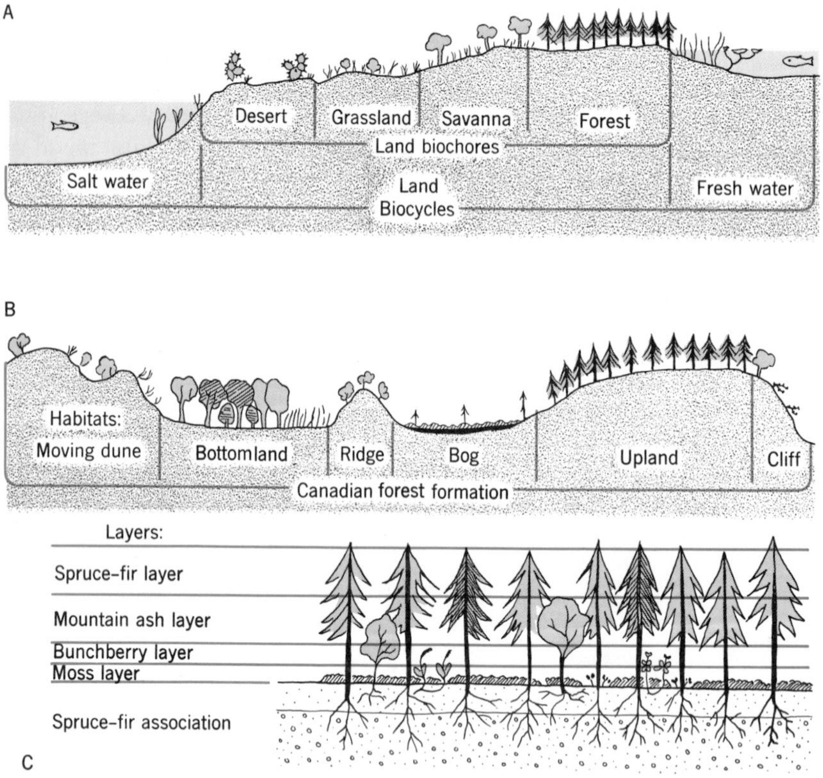

Figure 20.1 Dimensions of the environment. *A*, the biocycles and biochores. *B*, Habitats in the Canadian forest formation. *C*, Detail of spruce-fir upland association. (From Pierre Dansereau, 1951, *Ecology*, Vol. 32.)

is quite different from that on an adjacent valley floor where water lies near the surface much of the time. Vegetation is also strikingly different in form on rocky ridges and on steep cliffs where water drains away rapidly and soil is thin or largely absent.

Thus, each region which we assign to a given formation class is actually a mosaic of smaller units reflecting the inequalities in conditions of slope, elevation, and soil type. Such subdivisions of the plant environment are described as *habitats* (Figure 20.1*B*). In the example shown, the Canadian needleleaf forest (a formation class) actually comprises at least six habitats: upland, bog, bottomland, ridge, cliff, and active dune. Just where each habitat is located and how large an area it occupies depends largely upon the geologic history of the region, e.g., what processes of erosion and deposition have acted in the past to shape the landforms. The patterns of landforms from such processes are in turn influenced by the ancient geologic history which brought about varied combinations of arrangements of weak and resistant rocks. In the final part of this book, the evolution of landforms is explained in some detail.

Closely tied in with the effect of landform upon habitat is the distribution of water in the soil and rock, a topic that was treated in Chapter 12. Finally, each habitat has characteristic soil properties which are determined not only by the conditions of slope and water, but in part by the plants themselves.

Adjacent habitats may possess strikingly different vegetation structures so that the question arises as to which habitat shall be selected as that for which the formation class is named. For this purpose the upland habitat, with its well-drained surface, moderate surface slopes, and well-developed soil, is usually taken as the reference standard. The vegetation structure found in the upland habitat is thus what we have in mind when naming and describing a formation class (Figure 20.1*C*).

Within a given habitat will be found a smaller unit of ecosystem, the *plant community*, composed of rather definite proportions of organisms which are more or less interdependent and which use the resources of their habitat in such a way as to either maintain the habitat or modify it. No particular dimensions and boundary limits can be set for plant communities; these properties will depend on the changing habitat patterns. Within a given plant community can be recognized a structure of vegetation which is distinctive and a certain floristic composition, which may seem more or less fixed, or which may be gradually changing with time.

Structural description of vegetation

If the plant geographer is to describe vegetation according to its structure, or physical properties and outward forms, he must set up a number of categories of information, each designed to contribute an essential element to the description of the vegetation. Six categories are used in a system set up by Dr. Pierre Dansereau.[1] These six categories of information tell us about the growth-form of the plants, their size and stratification, the degree to which they cover the ground, their time functions, and their leaf forms.

1. *Life-form.* Plants can be classified according to their life-forms. The first two forms, *trees* and *shrubs* are erect, woody plants (Figure 20.2). The word "tree" means a perennial plant having a single upright main trunk, often with few branches

[1]Dansereau, Pierre (1957), *Biogeography, an ecological perspective*, The Ronald Press Co., New York. See pp. 147–154.

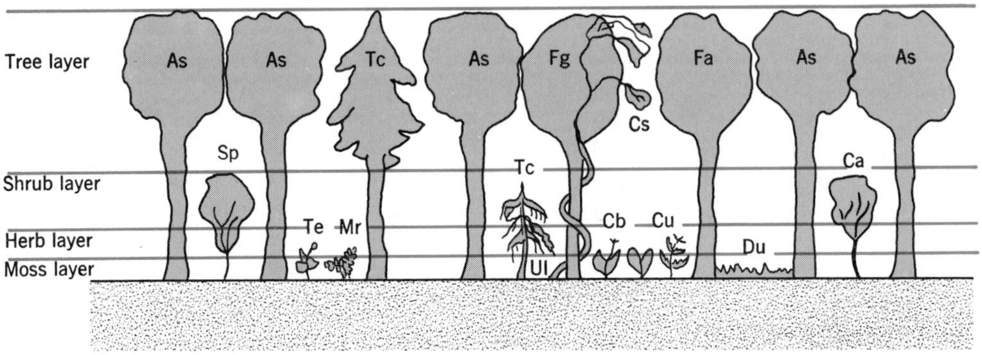

Figure 20.2 Schematic diagram of life forms in a beech-maple-hemlock forest. The tree layer consists of sugar maple (As), ash (Fa), beech (Fg), and hemlock (Tc) and includes a liana (Cs). The shrub layer includes elder (Sp), dogwood (Ca), and a young hemlock (Tc). An epiphyte (Ul) grows upon the hemlock. Plants designated Te, Mr, Cb, and Cu form the herb layer. Moss (Du) forms the lowest layer. (From Pierre Dansereau, 1951, *Ecology*, Vol. 32.)

Figure 20.3 A stand of mature sugar maples, 160 to 200 years old, in the Allegheny National Forest, Pennsylvania, 1939. (U.S. Forest Service photograph.)

in the lower part, but branching in the upper part to a crown which, in the mature individual, contributes to one of the upper layers of vegetation (Figure 20.3). The word "shrub" generally refers to a woody plant having several stems branching near the ground, so as to place the foliage in a mass starting close to ground level. Next are recognized woody vines which climb up on trees: the *lianas* (Figure 20.4). These may climb upon mature plants, or may be lifted during the growth of young trees to which they have become attached. Although the term "liana" is usually associated with woody vines of tropical forests, the meaning may be broadened to include woody vines of middle-latitude forests, such as the poison ivy and Virginia creeper. Thus it is apparent that the life-form classes cut across the taxonomic categories of the plant world. For example, palms and tree ferns look much alike outwardly but are of entirely different plant orders.

Fourth of the life-forms is the plant group known as the *herbs*. These are usually small tender plants, lacking woody stems. They occur in a wide variety of forms and leaf structures; including annuals and perennials, and broad-leaved plants as well as the grasses. Broad-leaved herbs are termed *forbs* in distinction with grasses. The adjective *herbaceous* is applied to this life-form. The herbaceous layer will normally occupy a low position in a layered or stratified plant community. Smaller, and lying in close contact with the ground or attached to tree trunks are the *bryoids*, a life-form named for a plant phylum which includes the mosses and liverworts.

Figure 20.4 Lianas of the equatorial rainforest, near Belém, Brazil. (Photograph by Otto Penner. Courtesy of Instituto Agronómico do Norte.)

Finally, among the life-forms are the *epiphytes*, plants which use other plants as supporting structures and thus live above the ground level, out of contact with soil (Figure 20.5). Familiar to all are the tropical orchids that live high upon tree limbs and are sometimes referred to as "air plants." Ferns also commonly exist as epiphytes (see Figure 21.5).

Not included in our list of life-forms are lower forms of plants, grouped under the name of *thallophytes*. These include bacteria, algae, molds, and fungi—all of which are plants lacking true roots, stems, and leaves. *Lichens*, plant forms in which algae and fungi live together to form a single plant structure, are included within the life-form classification of bryoids. Lichens occur as tough, leathery coatings or crusts and as leaflike masses attached to rocks and tree trunks (Figure 20.6). In arctic and alpine environments lichens may grow in profusion and dominate the vegetation in the near absence of more conspicuous plant forms.

Figure 20.5 An epiphyte, or air plant. This *Bromeliad* is attached to a limb of the host tree by a mass of fibrous roots. The plant leaves form an urn-shaped structure that collects and holds rainfall. About 1/7 natural size. (After W. H. Brown, 1935, *The Plant Kingdom*, courtesy of Blaisdell Publishing Co., a Division of Ginn and Co.)

Figure 20.6 A lichen (*Stereocaulon vulcani*) growing on a rough lava surface, Waiakea Forest Reserve, Maui, Hawaii. (Photograph by Pierre Dansereau.)

2. *Size and stratification.* Each of the life-forms described above may be classified according to size. The words "tall," "medium," and "low" may be given definite limits for each life-form. For example, a tree higher than 25 meters (82 feet) is "tall"; from 10 to 25 meters (33 to 82 feet) is "medium"; 8 to 10 meters (26 to 33 feet), "low." For the smaller life-forms different limits are set (Table 20.1). Such standardization of size enables the plant geographer to make precise descriptions of vegetation. Further standardization is achieved by setting height limits for a series of layers, numbered successively from the ground up (Table 20.1).

3. *Coverage.* The degree to which the foliage of individual plants of a given life-form cover the ground beneath them is designated the *coverage.* We may use four terms descriptive of the coverage: barren or very sparse; discontinuous; in tufts or groups; and continuous. For example, the trees may be of discontinuous coverage whereas the herb layer is continuous, or vice versa.

4. *Function, or periodicity.* Of primary importance in the classification of forms of natural vegetation is the response of the plant foliage to the annual climatic cycle. *Deciduous* plants shed their leaves and become dormant in an unfavorable season, which is either too cold or too dry to permit growth. *Evergreen* plants retain green foliage year-around, although in some cases becoming almost dormant in a cold or dry season. Where the climate is equable (that is, moist and not cold throughout the year) evergreen plants grow continuously. A third class, the *semideciduous* plants, are those which shed their leaves at intervals not in phase with a season. Thus a semideciduous forest would not at any one time have all of its individuals devoid of foliage. As a fourth class we recognize *evergreen-succulent* and *evergreen-leafless* plants, those with very thick fleshy leaves, which retain their foliage year-around, and those with fleshy stems but no functional leaves, such as the cacti.

5. *Leaf shape and size.* Recognition of the leaf form of a plant constitutes an essential part of the structural description (Figure 20.7). One form is the *broadleaf,* familiar to us in such common trees as the maple, beech, and rhododendron. In contrast is the *needleleaf,* also familiar in the pine, spruce, fir, and hemlock. A similar form is the *spine,* which in some plants represents the transformation of the entire leaf. The slender, tapering leaves of grass are referred to as *graminoid* in form. We may also recognize the *small leaf-form,* as in the birch or holly; and the *compound leaf,* as in the hickory and ash.

6. *Leaf texture.* Leaf textures range widely according to the climate and habitat, because of the different degrees to which the water loss from the leaf into the air must be controlled. Leaves of average thickness are described as *membranous;* those which are thin and delicate (as in the maiden-hair fern) are described as *filmy.* Leaves which are hard, thick, and leathery are *sclerophyllous;* a forest dominated by trees and shrubs having such

TABLE 20.1 PLANT SIZE AND STRATIFICATION OF VEGETATION[a]

Plant Size

TALL:	Tree	Over 82 ft (25 m)
	Shrub	6.6 to 26 ft (2 to 8 m)
	Herb	Over 6.6 ft (2 m)
MEDIUM:	Tree	33 to 82 ft (10 to 25 m)
	Shrub	1.6 to 6.6 ft (0.5 to 2 m)
	Herb	1.6 to 6.6 ft (0.5 to 2 m)
	Bryoid	Over 4 in (10 cm)
LOW:	Tree	26 to 33 ft (8 to 10 m)
	Shrub	Under 1.6 ft (0.5 m)
	Herb	Under 1.6 ft (0.5 m)
	Bryoid	Under 4 in (10 cm)

Stratification

LAYER NUMBER	RANGE
7	Over 82 ft (25 m)
6	33 to 82 ft (10 to 25 m)
5	26 to 33 ft (8 to 10 m)
4	6.6 to 26 ft (2 to 8 m)
3	1.6 to 6.6 ft (0.5 to 2 m)
2	4 in to 1.6 ft (10 cm to 0.5 m)
1	0 to 4 in (0 to 10 cm)

[a] After Pierre Dansereau, 1958, Botanical Institute of the University of Montreal, *Contributions* No. 72, pp. 31–32.

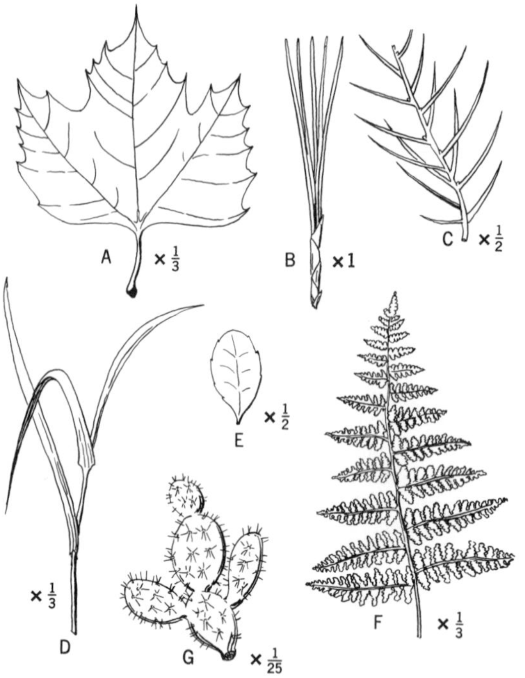

Figure 20.7 Leaf shapes. ***A*****, Large thin leaf (sycamore).** ***B*****, Needle leaf (pine).** ***C*****, Spine.** ***D*****, Graminoid leaf (grass).** ***E*****, Small leaf.** ***F*****, Compound leaf (fern).** ***G*****, Succulent stem (cactus).**

leaves is termed a *sclerophyll forest.* Very greatly thickened leaves, capable of holding much water in their spongy structure, are described as *succulent.*

Environmental factors in plant ecology

With some information now at our disposal concerning the structure of vegetation and its organization into plant communities, associations, formation-classes, and biochores, we can study the environmental factors which require vegetation to assume such varied forms.

The four main classes of environmental factors are (1) *climatic,* (2) *geomorphic* (related to landform), (3) *edaphic* (related to soil), and (4) *biotic* (related to living organisms). Although we treat each of these classes of factors in turn and examine each factor separately, it must be kept in mind that the basic concept of ecology is that many factors act simultaneously and that factors affect one another in a most complex interrelationship. The very plants that are affected by environmental factors may react in such a way as to modify the environment and thus change the factors themselves.

In considering how various factors of the physical environment influence plant structures and distributions, two scale ranges can be treated. One is essentially the global scale and consists of such climatic factors as the seasonal and latitudinal patterns of insolation, light and darkness, temperature, precipitation, and prevailing winds. The other scale of consideration is that of variations of the physical environment found within a relatively small habitat and between adjacent habitats. Thus, although there are vast global climate patterns of deserts and humid regions, we may also find within a region of generally humid climate a few small habitats (such as a dune or cliff) which are extremely dry places for plants to live. We may also find in a large desert some small habitats which are extremely wet most of the time (such as a seep or spring).

Water needs of plants

The need for water is perhaps the dominating consideration in analyzing the physical environment of plants. In conjunction with their growth processes, leaf-bearing plants give off large quantities of water into the atmosphere, a process termed *transpiration,* which is essentially a form of evaporation from water films upon the exposed surfaces of plant cells. A relatively small amount of water is also required by green plants in the process of *photosynthesis* in which the energy of light is used together with carbon dioxide and water to produce carbohydrates. The dominant source of water needed for transpiration and photosynthesis is from the soil, where it is taken up into the plant roots.

The rate of transpiration varies greatly according to the type of plant and the prevailing atmospheric conditions. High temperatures, low humidities, and winds favor high rates of transpiration. The plant structure, particularly of the leaf, largely determines the rate of water loss. Plants with large total foliage surfaces composed of broad, thin leaves have higher rates of loss than plants bearing needle leaves, spines, or small thick leaves (sclerophylls). Under conditions of critical water supply but high rates of evaporation, only those plants can survive that minimize transpiration losses by their special leaf structures and by their small size.

The adaptation of plant structures to water budgets with large water deficiencies is of particular interest to the plant geographer. Transpiration occurs largely from specialized leaf pores, called *stomata*, which are openings in the *epidermis* (outer cell layer) and *cuticle* (an outermost protective layer) through which water vapor and other gases can pass into and out of the leaf (Figure 20.8). Surrounding the openings of the stomata are *guard cells* which can open and close the openings and thus to some extent regulate the flow of water vapor and other gases. Although most of the transpiration occurs through the stomata, some may pass through the cuticle. This latter form of loss is reduced in some plants by thickening of the outer layers of cells or by the deposition of wax or waxlike material on or near the leaf surface. Thus many desert plants have thickened cuticle or wax-coated leaves, stems, or branches.

Another means of reducing transpiration is the development of stomata deeply sunken into the leaf surface that retard outward diffusion of water vapor into dry air, and the restriction in location of stomata to the shaded undersurfaces of leaves. A plant may also adapt to a desert environment by greatly reducing the leaf area, or by bearing no leaves at all. Thus needle-leaves and spines representing leaves greatly reduce loss from transpiration. In cacti the foliage leaf is not present and transpiration is limited to fleshy stems.

In addition to developing leaf structures that reduce water loss by transpiration, plants in a water-scarce environment improve their means of obtaining water and of storing it. Roots become greatly extended to reach soil moisture at increased depth. In cases where the roots reach to the ground water table, a steady supply of water is assured. Plants drawing from such a source are termed *phreatophytes* and may be found along dry channels (draws) and alluvial valley floors in desert regions. Other desert plants produce a widespread but shallow root system enabling them to absorb the maximum quantity of water from sporadic desert downpours which saturate only the uppermost soil layer. Stems of desert plants are commonly greatly thickened by a spongy tissue in which much water can be stored. As already noted, such plants would be described as *succulents*.

A quite different adaptation to extreme aridity is seen in many species of small desert plants that complete a very short cycle of germination, leafing,

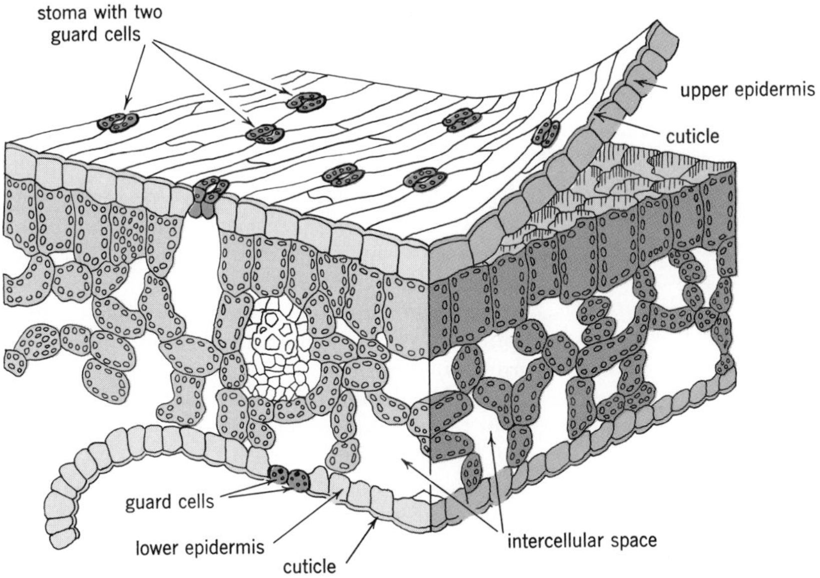

Figure 20.8 Cell structure of a foliage leaf. (After W. W. Robbins and T. E. Weier, 1950, *Botany*, John Wiley and Sons.)

flowering, fruiting, and seed dispersal immediately following a desert downpour.

Classification of plants by water need

Habitats of the land biocycle are differentiated primarily with reference to the degree of saturation of the soil by water. So important is this factor that plants may be classified according to their water requirements. The terminology associated with the water factor is built upon three simple prefixes of Greek roots: *xero-*, dry; *hygro-*(*hydro*), wet; and *meso-*, intermediate or middle. Thus a habitat may be prevailing wet (*hygric*); prevailingly dry (*xeric*), or of intermediate degree of wetness (*mesic*). Those plants which grow in dry habitats are *xerophytes;* those which grow in water or in wet habitats are *hygrophytes* (*hydrophytes*); those of habitats of an intermediate degree of wetness and relatively uniform water availability are *mesophytes.*

The xerophytes are highly tolerant of drought and can survive in habitats which dry quickly following rapid drainage of precipitation, for example, on sand dunes, beaches, and bare rock surfaces. The plants typical of dry climates (deserts) are also xerophytes; cactus is an example. The hygrophytes are tolerant of excessive water and may be found in shallow streams, lakes, marshes, swamps, and bogs; an example is the water lily. The mesophytes are found in upland habitats in regions of ample rainfall. Here the drainage of precipitation is good and moisture penetrates deeply where it can later be used by the plants. On such upland locations the soil may be of intermediate texture—not too coarse or too fine—and is usually thickly and uniformly present.

Certain of the climatic regimes, such as the tropical wet-dry regime (including monsoon climates) and the continental regime where moist but not too cold, have a yearly cycle with one season in which water is unavailable to plants because of lack of precipitation or because the soil water is frozen. This season alternates with one in which there is abundant water. Plants adapted to such regimes are termed *tropophytes,* from the Greek word *trophos,* meaning change, or turn. Tropophytes may meet the impact of the season of unavailable water by dropping their leaves and becoming dormant. When water is again available, they leaf out and grow at a rapid rate. Trees and shrubs which seasonally shed their leaves are said to be *deciduous;* in distinction with *evergreen* trees which retain most of their leaves in a green state throughout the year.

Other climatic factors

The factor of light is of importance in plant ecology. Within the habitat of a given plant association or community the degree of light available will depend in large part upon position of the plant. Tree crowns of the upper layer receive maximum light, but correspondingly reduce the amount available to lower layers. In extreme cases forest trees so effectively cut off light that the forest floor is almost free of shrubs and herbaceous plants. In certain deciduous forests of middle latitudes, the period of early spring, before the trees are in leaf, is one of high light intensity at ground level, permitting certain herbaceous plants to go through a rapid growth cycle. In summer these plants will largely disappear as the leaf canopy is completed. Other herbaceous plants in the same habitat require shade and do not appear until later in the summer.

Treated on a global basis, the factor of light available for plant growth is varied by latitude. Duration of daylight in summer increases rapidly with higher latitude and reaches its maximum poleward of the arctic and antarctic circles, where the sun may be above the horizon for 24 hours (Chapter 4.) Thus, although the growing season for plants is greatly shortened at high latitudes by frost, the rate of plant growth in the short frost-free summer is greatly accelerated by the prolonged daylight. But in still higher, subarctic latitudes plant growth is greatly slowed by the low heat budget, despite perpetual summer daylight.

In middle latitudes, where vegetation is of a deciduous type, the annual rhythm of increasing and decreasing periods of daylight determines the timing of budding, flowering, fruiting, leaf-shedding, and other vegetation activities. As to the importance of light intensity itself, it is generally believed that even on overcast days there is sufficient light to permit plants to carry out photosynthesis at their maximum rates and that direct sunlight constitutes a considerable excess of light.

Temperature, another of the important climatic factors in plant ecology, acts directly upon plants through its influence upon the rates at which the physiological processes take place. In general, we can say that each plant species has an optimum temperature associated with each of its functions, such as photosynthesis, flowering, fruiting, or seed germination, and that there exist some overall optimum yearly temperature conditions for its growth in terms of size and numbers of individuals. There are also limiting lower and upper temperatures for the individual functions of the plant as well as for its total survival. Temperature acts as an indirect factor in many other ways. Higher air temperatures increase the water-vapor capacity of the air and thus induce greater plant transpiration as well as greater rates of evaporative loss of moisture from the soil.

In general, the colder the climate, the fewer

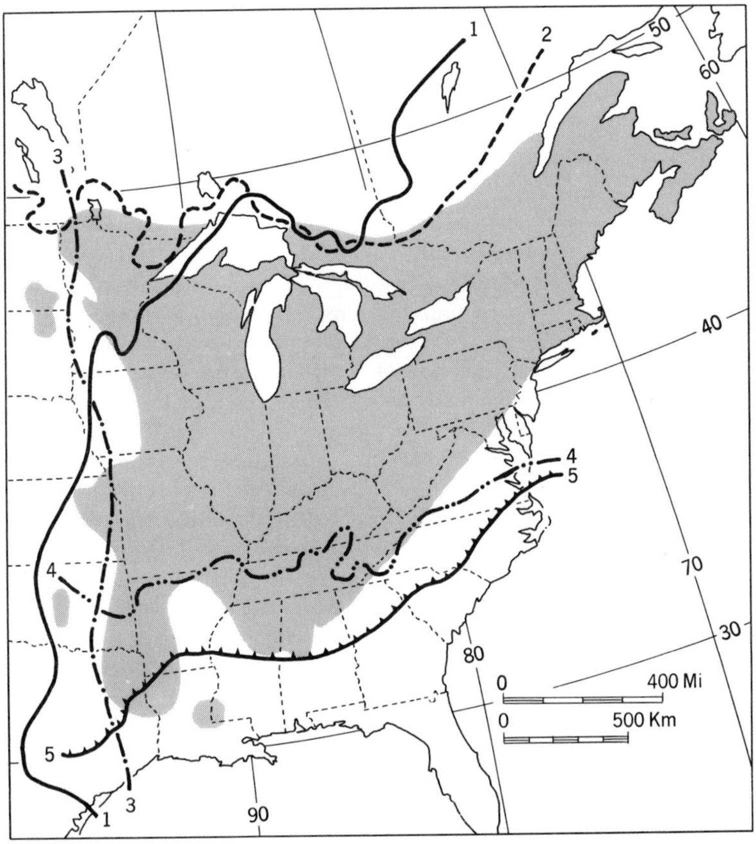

Figure 20.9 Bioclimatic limits of the sugar maple (*Acer saccharophorum*) in eastern North America. The shaded area represents the distribution of the sugar maple. Line 1, 30 in (76 cm) annual precipitation. Line 2, −40°F (−40°C) mean annual minimum temperature. Line 3, eastern limit of yearly boundary between arid and humid climates. Line 4, 10 in (25 cm) mean annual snowfall. Line 5, 16°F (−10°C) mean annual minimum temperature. (Based on *Biogeography—An Ecological Perspective*, by Pierre Dansereau. Copyright © 1957, The Ronald Press Company.)

numbers of species that are capable of surviving. A large number of tropical plant species cannot survive below-freezing temperatures. In the severely cold arctic and alpine environments of high latitudes and high altitudes only a few species can survive. Application of this principle explains why a forest in the equatorial zone has many species of trees, whereas a forest of the subarctic zone may be dominantly of one, two, or three tree species. Tolerance to cold is closely tied up with the ability of the plant to withstand the physical disruption that accompanies the freezing of water. If the plant has no means of disposing of the excess water in its tissues, the freezing of that water will damage the cell tissue.

On the basis of preference for, or tolerance to, temperatures, plant geographers distinguish the following: *megatherms*, plants favoring warm regions; *microtherms*, favoring cold regions; and *mesotherms*, favoring regions of intermediate temperatures.

It is a law of bioclimatology that there is a critical level of climatic stress beyond which a plant species cannot survive; hence that there will exist a geographical boundary that will mark the limits of its distribution. Such a boundary may also be referred to as a *frontier*. Although the frontier is determined by a complex of climatic elements, it is sometimes possible to single out one climatological element that coincides with the plant frontier. An example is seen in the limits of growth of the sugar maple (*Acer saccharophorum*) in North America (Figure 20.9). Here the boundaries on the north, west, and south are found to coincide roughly with selected values of annual precipitation, mean annual minimum temperature, and mean annual snowfall. Another example is seen in the distribution of the yellow pine (*Pinus ponderosa*) of western North America (Figure 20.10). In this mountainous region annual rainfall varies sharply with elevation. The 20 in (50 cm) isohyet of annual total precipitation encloses most of the upland areas having the yellow pine. It is the parallelism of the isohyet with forest boundary that is significant, rather than actual degree of coincidence.

Wind is seen as an important environmental factor in the structure of vegetation in highly exposed positions. Close to timber line in high mountains and along the northern limits of tree growth in the arctic zone trees will be found to be deformed so that the branches project from the lee side of the trunk only (flag shape), or the trunk and branches are bent to near-horizontal attitude, facing away from the prevailing wind direction (Figure 20.11). In such habitats the effect of wind is to cause excessive drying on the exposed side of the plant. The tree limit on mountainsides thus varies in elevation with degree of exposure of the slope to strong prevailing winds and will extend higher on lee slopes and in sheltered pockets.

Geomorphic factors

Geomorphic, or landform, factors influencing plant forms include such elements as *slope steepness* (the angle which the ground surface makes with respect to the horizontal), *slope aspect* (the orientation of a sloping ground surface with respect to geographic north), and *relief* (the difference in elevation of divides and adjacent valley bottoms). In a much broader sense, the geomorphic factor includes the entire sculpturing of the landforms of a region by processes of erosion, transportation and deposition by streams, waves, wind, and ice, and by forces of vulcanism and mountain building. The unique landform assemblage found in any one region is understood in terms of geomorphic processes; these are treated in Chapters 24 through 33. An infinite variety of plant habitats can be ascribed to the geomorphic processes and their individual landforms.

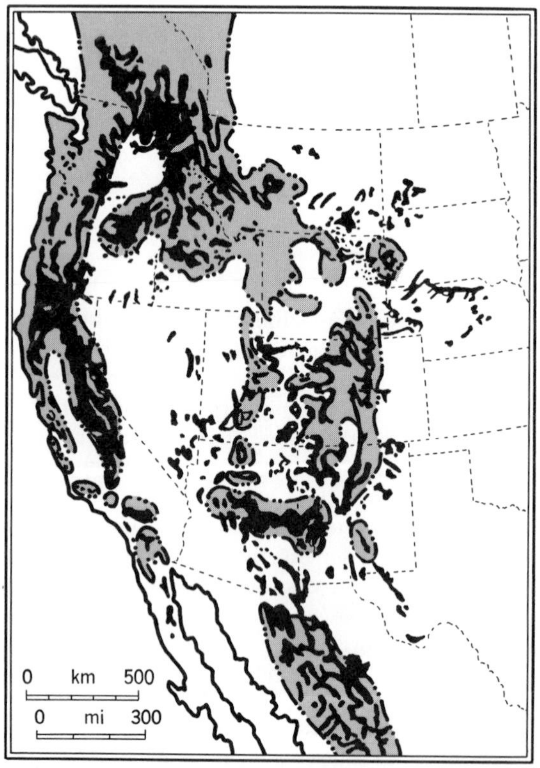

Figure 20.10 Areas of yellow pine (*Pinus ponderosa*) in western North America are shown in solid black. Edge of the shaded area represents the isohyet of 20 in (50 cm) annual precipitation. (Based on *Biogeography—An Ecological Perspective*, by Pierre Dansereau. Copyright © 1957, The Ronald Press Company.)

Slope steepness acts indirectly by its influence upon the rate at which precipitation is drained from the surface. On steep slopes surface runoff is rapid and the water does not long remain available to plants. On gentle slopes much of the precipitation can penetrate the soil and become available for prolonged plant use. More rapid erosion on steep slopes may result in thin soil, whereas that on gentler slopes is thicker. Slope aspect has a direct influence upon plants by increasing or decreasing the exposure to sunlight and to prevailing winds. Slopes facing the sun have a warmer, drier environment than slopes facing away from the sun and therefore lying in shade for much longer periods of the day. In middle latitudes these slope-aspect contrasts may be so strong as to produce quite different plant formations on north-facing and south-facing slopes (Figure 20.12).

Geomorphic factors are in part responsible for the dryness or wetness of the plant habitat within a region having essentially the same overall climate. Each plant community has its own microclimate. Upon divides, peaks, and ridge crests the soil tends to dryness because of rapid drainage of water away from such places and because the surfaces are more exposed to sunlight and to drying winds. By contrast, the valley floors tend to wetness because surface runoff over the ground and into streams causes water to converge there. In humid climates the ground water table may lie close to the ground in the valley floors or may actually coincide with the ground surface to produce marshes and swamps. Watertable ponds and bogs may also occur. Hygrophytic plants thus form distinctive plant communities in the valley floors in humid climates while at the same time mesophytic or xerophytic communities occupy the intervalley surfaces.

Edaphic factors

Edaphic factors are those related to the soil. In Chapter 18 principles of soil development (pedogenesis) have been taken up systematically. In

Figure 20.11 Trees deformed by the effects of cold, dry winds. Timberline scene in Arapaho National Forest, Colorado, 1946. (U.S. Forest Service photograph.)

terms of plant geography we can look at soils in two perspectives. One views broad patterns of the great soil groups, reflecting the pedogenic regimes of podzolization, laterization, calcification, gleization, and salinization. These pedogenic regimes are largely controlled by the climatic regimes and will be found closely correlated with the global patterns of the vegetation formation classes. These broad relationships will be taken up together in Chapter 21. A second view point is in terms of plant habitats—the small-scale mosaic of place-to-place variations of the earth's surface. Here the edaphic factors also act as important controls.

Among the edaphic factors treated in Chapter 18 are: soil texture and structure, humus content; presence or absence of soil horizons; soil alkalinity, acidity, or salinity; and the activity of bacteria and animals in the soil.

Although this book treats the systematic principles of soil science ahead of those of natural vegetation, a good argument might be made for reversing this order of treatment on the grounds that vegetation plays a leading role in the development of soil characteristics. Given a barren habitat, recently formed by some geologic event such as the outpouring of lava or the emergence of a

Figure 20.12 Contrasts in vegetation on opposing valley-side slopes. The heavily wooded slope on the left faces northeast and is in shade during the afternoon. The sparsely wooded slope on the right, facing southwest, receives intense insolation when air temperatures are highest. Long Canyon of the Rio Hondo, Carson National Forest, New Mexico. (U.S. Forest Service photograph.)

coastal zone from beneath the sea, the gradual evolution of a soil profile goes hand in hand with the occupance of the habitat by a succession of plant communities. The plants profoundly alter the soil by such processes as contributing organic matter, or by producing acids which act upon the mineral matter. Animal life, feeding upon the plant life, also makes its contribution to physical and chemical processes of soil evolution. We will give further attention to this topic under the subject of dynamics of vegetation.

Biotic factors

The sustained activity within a particular plant community functioning as a stable ecosystem and the gradual modification of the vegetation at a given place with time throughout a succession of stages require that the plants and animals within that community contribute to its operation through their own physical and chemical processes and through their life cycles of growth and subsequent decay. So vast and complex are these biotic factors that we can do no more here than to suggest by examples what is meant by biotic influence.

Among the biotic factors may be mentioned the activity of bacteria in consuming the dead tissue of larger plants; of earthworms in altering and aerating the soil; of insects and herbivorous animals which attack and consume plants; and of insects and birds which perform such functions as pollination of plants and the dispersal of seeds. A whole realm of biotic influence lies in the diseases of plants, the spread of which may drastically alter the plant associations over wide areas. The geographer is particularly interested in the way in which large herds of grazing and browsing animals, whether wild or domesticated, have contributed to the development of certain formation classes of vegetation. To what extent, say, have herds of American bison been responsible for the maintenance of grasslands against the spread of woodland and forest?

Man is himself perhaps the most potent biotic factor influencing vegetation over the globe today. His role has been dominantly destructive with respect to the plant associations and formation classes that might otherwise be expected to cover the lands in response to the various factors of climate, soil, and geomorphology that we have thus far reviewed.

Dynamics of vegetation

One of the main themes of plant geography is that the vegetation at a particular place evolves with time, usually starting with very simple plant communities, then leading gradually to more complex communities, and ultimately to the establishment of a relatively stable plant community—the *climax*. Starting with a newly formed ground surface, or one that has been denuded of vegetation, the process of *succession* takes place, in which one plant community invades the area and is followed in turn by other plant communities in an orderly sequence, or *sere*, culminating in the vegetation climax.

A new site for the development of vegetation may have one of several origins: a sand dune, a sand beach, the surface of a new lava flow or of a freshly fallen layer of volcanic ash, or the deposits of silt on the inside of a river bend which is gradually shifting. Such a site will not have a true soil with horizons; rather it may be a lithosol—perhaps little more than a deposit of coarse mineral fragments. In other cases, such as that of the floodplain silt deposits, the surface layer may represent redeposited soil endowed with substantial proportions of soil colloids and bases. Ground surfaces from which vegetation has been destroyed by fire will have the soil profile largely intact.

The first stage of a succession is a *pioneer stage*, consisting of a few plant species unusually well adapted to adverse conditions of rapid water drainage and drying of soil, and to excessive exposure to sunlight, wind, and extreme ground and lower air temperatures. As these plants grow, their roots penetrate the soil; their subsequent decay adds humus to the soil. Fallen leaves and stems add an organic layer to the ground surface. Bacteria and animals begin to live in the soil in large numbers. Soon conditions are favorable for other plant species which *invade* the area and displace the pioneers. The new arrivals may be larger plant forms providing more extensive cover of foliage over the ground. In this case the climate near the ground, or *microclimate*, is considerably altered toward one of less extreme air and soil temperatures, higher humidities, and less intense insolation. Now still other species can invade and thrive in the modified environment.

When the succession has finally run its course, there will exist a stable community consisting of certain definite proportions of the various species, each of which contributes to the total structure of the vegetation. This climax stage represents an ideal model for the so-called natural vegetation of a region. Doubt exists as to whether a climax, unchanging with time, can truly be maintained. Some plant geographers consider that the climax must be followed by disturbance leading to a *regression*, or return to one of the earlier stages of the succession, so that a type of self-repeating cycle is the rule. There is also the possibility that climatic change prevents the climax from being maintained in one place.

One type of succession is that in which the

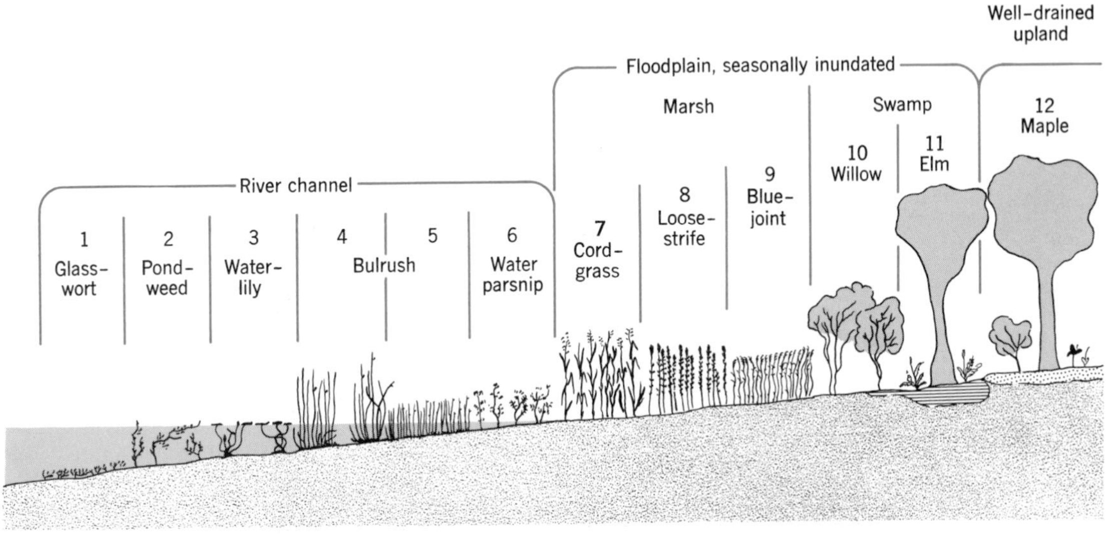

Figure 20.13 Allogenic succession on the bank of the St. Lawrence River. Twelve vegetation belts, each characterized by a plant association, shift gradually from right to left as bank deposition continues. (After Pierre Dansereau, 1956, *Revue Canadienne de Biologie*, Vol. 15.)

normal geomorphic processes build new ground continuously, as on a floodplain, delta, or sandspit. The succession associated with such continuous accretion of ground is described as *allogenic.* An example is shown in Figure 20.13, representing the zonation of vegetation along the St. Lawrence River. The twelve zones shift gradually to the left as silting by the river raises the level of the ground surface and shifts the water line to the left. Here we see the hygrophytes, living largely submerged, giving way to plants which thrive under conditions of intense sunlight on ground that is alternately exposed and inundated. Higher upon the bank are successive forest zones: willow, elm, and finally maple, representing the climax of mesophytic trees.

Figure 20.14 Bog succession in Emmet County, Michigan. A bog mat lies at the edge of the lake, a black spruce forest in background. (Photograph by Pierre Dansereau.)

Another form of succession, described as *autogenic,* results from the alteration of the environment by the plants themselves, and not by outside agencies (such as the silt accumulation from flood stages of a river). The gradual covering of a sand dune by small plants, then by forest, would illustrate autogenic succession. Another illustration is provided by the evolution of a shallow pond or lake of glacial origin in the regions of cold continental climate such as prevail in Canada and northern Europe (Figure 20.14). This *bog succession* is illustrated by stages beginning with the lake as it was left following disappearance of glacial ice, perhaps 10,000 to 15,000 years ago. A remarkable feature of the bog succession is that the organic matter produced by the growth and partial decay of the plants accumulates to such thickness that the open water is actually replaced by an organic mass of a substance known as *peat.* At the water's edge is a zone of sedges, followed by rushes (Figure 20.15). These construct a floating layer that encroaches upon the open water. There follows a zone of sphagnum (peat moss), which eventually completely fills the lake. Now the peat deposit supports hygrophytic trees (largely spruce) which produce a woody peat. This community may in turn be replaced by mesophytic trees, marking the climax stage. In the shallower upland ponds, shown on either side of the profile, the mesophytic growth is achieved much earlier than over the larger water body.

In terms of the vast scope and depth of the

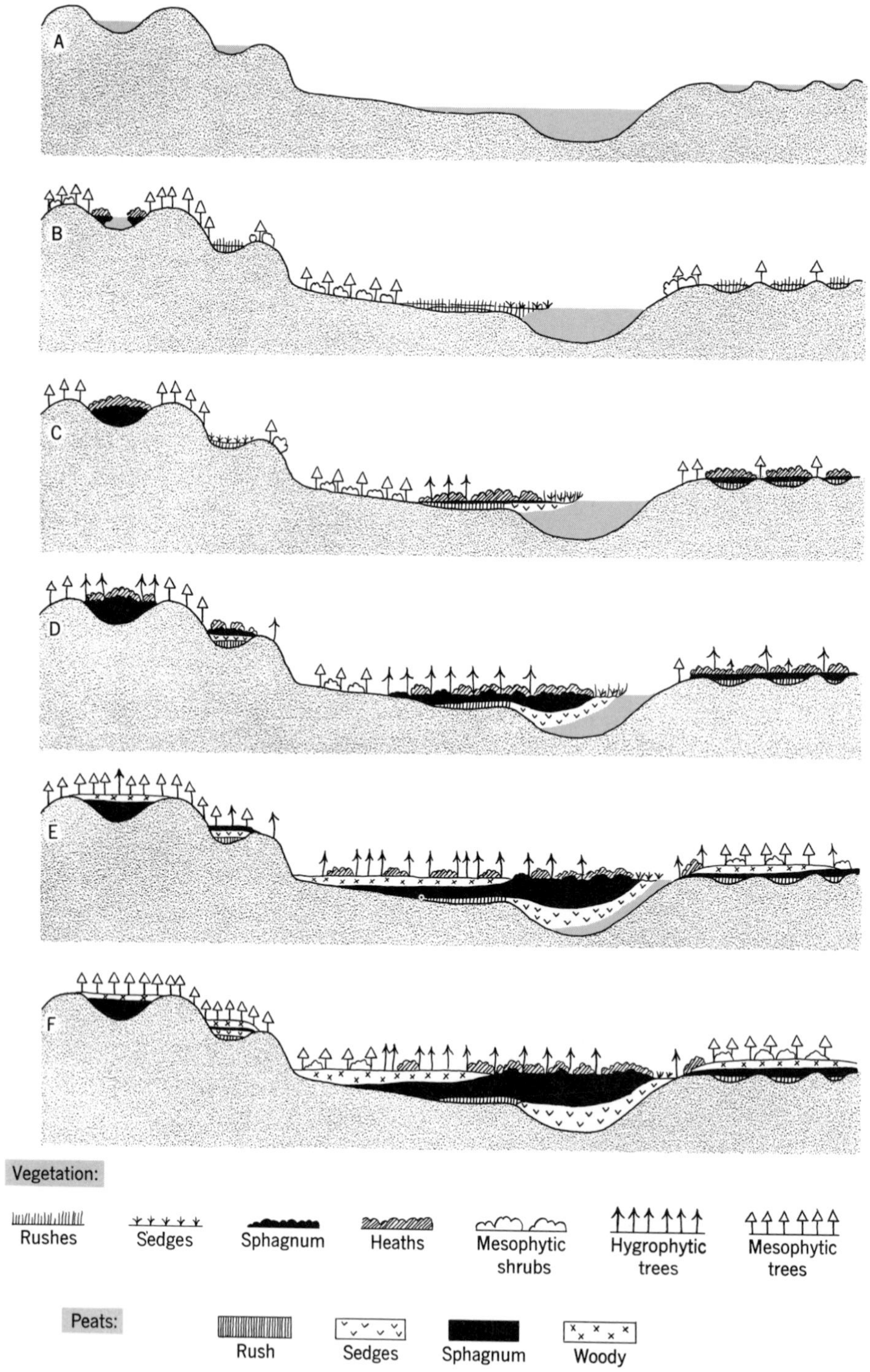

Figure 20.15 Autogenic bog succession typical of the Laurentian shield area of Canada. (After Dansereau and Segadas-Vianna, 1952, *Canadian Journal of Botany*, Vol. 30.)

science of plant geography, this chapter must be taken as a greatly simplified review touching upon only a few essential topics. As with all other subjects which we call collectively "physical geography," plant geography can be seriously approached only through intensive study of principles of botany, including the classification of plants (taxonomy), their evolution and floristic distribution, plant morphology, and the chemistry and physics of plant physiology.

REVIEW QUESTIONS

1. How does vegetation act as a geographic factor? To what other branches of physical geography is plant geography most closely related?

2. What is meant by natural vegetation? What does a map of natural vegetation attempt to portray?

3. Compare floristic with structural approach in the study of natural vegetation. Which approach is most useful in physical geography and why?

4. What is bioclimatology? Define an ecosystem. What is the subject of plant ecology?

5. Name the three environmental divisions of the biosphere. Into what four great biochores is the land biocycle divided? Explain the basis of this subdivision.

6. What are formation-classes? How many formation classes are commonly recognized? On what basis are formation classes set up?

7. How does habitat influence the structure of vegetation? What six habitats can be recognized within the Canadian needleleaf forest? What is the cause of differing habitats? On the basis of what type of habitat are the formation classes defined? What is a plant community?

8. What six categories have been used to form a system of description of the structure of vegetation? Define each category and give the units of subdivision within each.

9. How is a tree distinguished from a shrub? What are lianas? What kind of plants fall within the category of herbs? What are forbs? Bryoids? Epiphytes? Lichens? What kinds of plants are included within the thallophytes?

10. Distinguish deciduous from evergreen plants. What are semideciduous plants? Name the principal leaf forms and give examples from common plants.

11. What are the varieties of leaf textures? What is a sclerophyll forest? What are succulent leaves?

12. List the four main classes of environmental factors in plant ecology. What two scale ranges of phenomena are included in the environment?

13. How do plants use water? What factors affect the rate of transpiration? How are plants adapted to reduce water loss? How do some plants improve their means of obtaining and storing water? What are phreatophytes?

14. Explain how plants are classified according to water need. Name and define the three plant groups so classified. What are tropophytes?

15. Discuss the climatic factors of light, air temperature, and wind as they apply to plant ecology. How do numbers of plant species in a region depend upon temperature conditions? Distinguish megatherms, microtherms, and mesotherms.

16. What bioclimatological law relates a plant species to the level of climatic stress imposed upon it? Give an example of a plant frontier.

17. Discuss the geomorphic factors in plant ecology and show how each factor acts to influence vegetation.

18. What are edaphic factors? What two points of view can be taken in treating the edaphic controls? How do plants themselves modify the soils upon which they grow?

19. Discuss the biotic factors in plant ecology. Give specific examples of the way in which animal life affects vegetation. What is the role of man as a biotic factor in our present day?

20. What is the process of succession in development of vegetation? Define sere and climax. What kinds of new sites may become available for plant successions? What is the pioneer stage? What is meant by a regression following climax?

21. Distinguish allogenic from autogenic succession and give one example of each type.

CHAPTER 21

Distribution of Natural Vegetation

THE principles of description of vegetation in terms of its structure, and the organization of vegetation into plant assemblages of various orders of magnitude (biochore, formation class, association, community), have been treated in Chapter 20. Using these principles in combination with our understanding of the climatic types, pedogenic regimes, and the soil-moisture regimes, we are now prepared to analyze the world-wide distribution of vegetation and to explain its variations with latitude, continental position, and altitude.

The great biochores

All natural vegetation of the lands falls into four major structural subdivisions, the *biochores*, illustrated schematically in Figure 21.1. First is the *forest biochore.* We define a *forest* as a plant formation consisting of trees growing close together and forming a layer of foliage that largely shades the ground. Forests often show stratification, with more than one layer. Shading of the ground gives a distinctly different microclimate than that which would be found over open ground. Forests require a relatively large annual precipitation, but it does not need to be uniformly distributed throughout the year. No single value of precipitation can be stated because the effectiveness of the precipitation depends upon the water loss by evapotranspiration, and this in turn depends upon air temperature and humidity. Consequently, the forest biochore spans a great climatic range, from wet equatorial to cold subarctic.

The *savanna biochore* is a formation consisting of a combination of trees and grassland in various proportions. The appearance of the vegetation can be described as parklike, with trees spaced singly or in small groups and surrounded by, or interspersed with, surfaces covered by grasses, or by some other plant life form, such as shrubs or annuals in a low layer. The savanna biochore indicates a climate of limited total annual precipitation with an uneven distribution throughout the year.

The *grassland biochore* consists of an upland vegetation largely or entirely of herbs, which may include grasses, grasslike plants, and forbs (broadleaf herbs). Degree of coverage may range from continuous to discontinuous and there may be stratification. The grassland biochore may include trees in moister habitats of valley floors and along stream courses where ground water is available. The grassland biochore is typical of a climate which has small total annual precipitation, but otherwise the climate may range from one of extreme heat to one of extreme cold.

The *desert biochore,* associated with climates of extreme aridity, has thinly dispersed plants and hence a high percentage of bare ground exposed to direct insolation and to the forces of wind and water erosion or to freeze-thaw action. Although essentially treeless, the desert biochore may have scattered woody plants. Typically, however, the plants are small, e.g., herbs, bryoids, lichens. Because the desert biochore includes climates ranging from extremely hot tropical desert to extremely cold arctic desert, a great range in plant communities and habitats is spanned by the biochore.

In describing the four great biochores, emphasis has been placed on the vast range of climates spanned by each. Essentially, the biochores are determined by the degree to which moisture is available to plants in a scale ranging from abundant (forest biochore) to almost none (desert biochore). But within each biochore conditions of temperature are vastly different from low to high latitudes and from low to high altitudes. Consequently, there is need to subdivide each biochore into a number of formation classes.

The formation classes

The description of world vegetation in terms of formation classes was first developed fully by Professor A. F. W. Schimper whose monumental work in two volumes entitled *Plant Geography on a Physiological Basis* was first published in German in 1903. This work was subsequently revised by Professor F. C. von Faber and published in 1935. A somewhat similar approach to structural classification of world vegetation was taken by Professor Eduard Rübel who published a major volume in 1930. The classification system described below follows in many respects the recent work of Professor Pierre Dansereau[1] and is based

[1]Dansereau, Pierre (1957), *Biogeography, an ecological perspective,* The Ronald Press Co., New York, 394 pp.

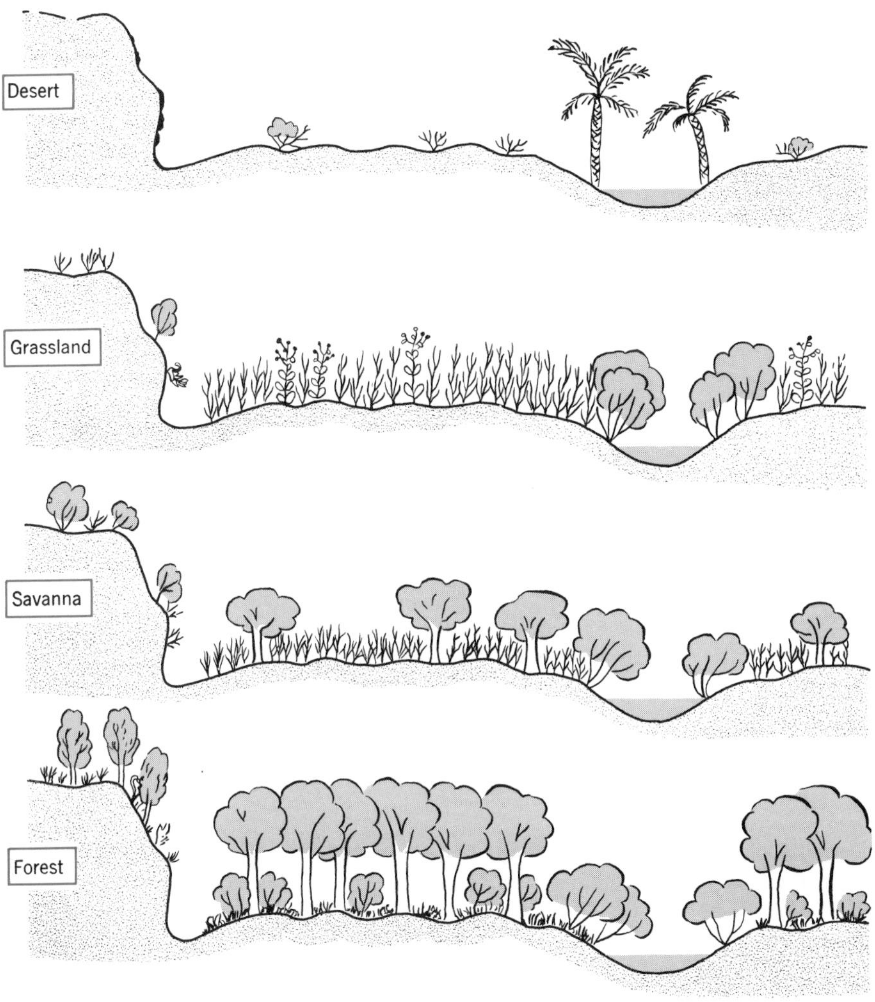

Figure 21.1 Diagrammatic representation of the four great biochores. Well-drained upland is shown in the middle of each profile; steep cliffs on the left; wet or poorly drained ground on the right. Note that the vertical scale of the grassland biochore is exaggerated in comparison with the other three. (From *Biogeography—An Ecological Perspective*, by Pierre Dansereau. Copyright © 1957, The Ronald Press Company.)

essentially on Schimper's and Rübel's principles. The world map of vegetation, Plate 4, follows Rübel's system in showing the distribution of ten map units, designated here by the letters *A* through *J*. The eighteen formation classes listed in this chapter are all included within Rübel's ten map units so that in certain cases two to four formation classes are combined into a single map unit. The map is thus simplified in appearance and brings out broad world patterns that are obviously related to the climate regimes.

Table 21.1 lists eighteen formation classes within the four biochores and gives the corresponding world map units of Plate 4. On the map legend is a similar table in reverse, giving the ten map units and listing the formation classes which are included in each.

Forest biochore

1. *Equatorial rainforest*[2] consists of tall, closely set trees whose crowns form a continuous canopy of foliage and provide dense shade for the ground and lower layers (Figure 21.2). The trees are characteristically smooth barked and unbranched in the lower two-thirds. Trunks commonly are buttressed at the base by radiating, wall-like roots (Figure 21.3). Tree leaves are large and evergreen; from this characteristic the equatorial rainforest is often described as "broadleaf evergreen forest." Crowns of the trees tend to form into two or three layers, or strata, of which the highest layer consists of scattered emergent crowns rising to 130 ft (40 m) and protruding conspicuously above a second

[2] Called *tropical rainforest* by Dansereau and others.

TABLE 21.1 THE FORMATION CLASSES

I Forest Biochore	Equivalent map units (Plate 4)
1. Equatorial rainforest 2. Tropical rainforest	A. Equatorial and tropical rainforest
3. Monsoon forest	C. Raingreen forest, woodland, scrub, and savanna.
4. Temperate rainforest	*B.* Temperate rainforest or Laurel forest
5. Summergreen deciduous forest	*E.* Summergreen deciduous forest
6. Needleleaf forest	*F.* Needleleaf forest
7. Evergreen-hardwood forest (Sclerophyll forest)	*D.* Evergreen-hardwood forest (Sclerophyll forest)
II Savanna Biochore	
8. Savanna woodland 9. Thornbush and tropical scrub 10. Savanna	C. Raingreen forest, woodland, scrub, and savanna
11. Semidesert	*H.* Dry desert and semidesert
12. Heath	(Heath not shown on map)
13. Cold woodland	*I.* Tundra
III Grassland Biochore	
14. Prairie 15. Steppe	G. Steppe and Prairie grasslands
16. Grassy tundra	*I.* Tundra
IV Desert Biochore	
17. Dry desert	*H.* Dry desert and Semidesert
18. Arctic fell field	*I.* Tundra

layer, 50 to 100 ft (15 to 30 m) which is continuous (Figure 21.4). A third, lower layer consists of small, slender trees 15 to 50 ft (5 to 15 m) high with narrow crowns (Figure 21.4).

Typical of the equatorial rainforest are lianas, thick woody vines supported by the trunks and branches of trees. Some are slender, like ropes, others reach thicknesses of 8 in (20 cm). They rise to heights of the upper tree levels where light is available and may have profusely branched crowns. Lianas may depend upon a growing tree to be carried upward, where the liana has no devices with which to climb by itself. Other woody climbers rise by winding about the tree trunk. Epiphytes are numerous in the equatorial rainforest. These plants are attached to the trunk, branches and foliage of trees and lianas, using the "host" solely as a means of physical support. Epiphytes are of many plant classes and include ferns, orchids, mosses, and lichens (Figure 21.5). Some epiphytes are *stranglers.* These woody vines send down their roots to the soil and may eventually surround the tree, perhaps ultimately replacing it. The *strangling fig* (*Ficus*) is an example (Figure 21.6). Other stranglers begin as lianas.

A particularly important botanical characteristic of the equatorial rainforest is the large number of species of trees that coexist. It is said that as many as 3000 species may be found in a square mile. Individuals of a species are thus widely separated. Consequently, if a particular tree species is to be extracted from the forest for commercial uses, considerable labor is involved in seeking out the trees and transporting them from their isolated positions. Representative trees of the rainforest of the Amazon valley are the Brazilnut (*Bertholletia excelsa*) and the silk-cotton tree (species of *Bombax*).

The floor of the equatorial rainforest is usually so densely shaded that plant foliage is sparse close to the ground and gives the forest an open aspect, making it easy to traverse. The ground surface is covered only by a thin litter of leaves. Rapid consumption of dead plant matter by bacterial action results in the absence of humus upon the soil surface and within the soil profile. As explained in Chapter 18, these conditions are typical of the pedogenic process of laterization, with which the rainforest is identified.

Equatorial rainforest is a response to an equable, moist climate which is continuously warm, frost-free, and has abundant precipitation in all months of the year (or, at most, only one or two dry months). A large water surplus characterizes the annual water budget (see Figure 12.12*A*), so that soil moisture is adequate at all times and the export of large amounts of stream runoff allows permanent removal of bases and silica from the soils of the region. In the absence of a cold or dry season, plant growth goes on continuously throughout the year. Individual species have their own seasons of leaf shedding, possibly caused by slight changes in the light period.

Variations in the equatorial rainforest structure are found in specialized habitats and where man has disturbed the vegetation. Where the forest has been cleared by cutting and burning (as for small-plot agriculture or highways), the returning plant growth is low and dense and may be described as *jungle*. Jungle can consist of a tangled growth of lianas, bamboo scrub, thorny palms, and thickly branching shrubs, constituting an impenetrable barrier to travel, in contrast to the open floor of the climax rainforest.

Figure 21.3 Buttress roots at the base of a large tree (*Bombacopsis fendleri*) of the rainforest on Barro Coloradi Island, Canal Zone. (Courtesy of the American Museum of Natural History.)

Figure 21.2 Equatorial rainforest near Belém, Brazil. This forest of tall broad-leaved evergreen trees and numerous lianas is on the relatively high ground (*terra firme*) of the Amazon lowland. (Photograph by Otto Penner, courtesy of Instituto Agronómico do Norte.)

Coastal vegetation in areas of equatorial rainforest is highly specialized. Coasts which receive suspended sediment (mud) from river mouths, and where water depths are shallow, typically have *mangrove swamp forest*, consisting of stilted trees (Figure 21.7). The mangrove prop-roots serve to trap sediment from ebb and flood tidal currents, so that the land is gradually extended seaward. Mangroves commonly consist of several shoreward belts of the red (*Rhizophora*), the black (*Avicennia*), and the white (*Laguncularia*) mangrove. Another common coastal salt-marsh plant of the humid tropics is the screw-pine (*Pandanus*). Typical of recently formed coastal deposits are belts of palms, such as the cocoanut palm (*Cocos nucifera*) (Figure 21.8).

World distribution of equatorial rainforest is shown on the vegetation map, Plate 4, by those areas of rainforest (Class *A*) located close to the equator. The principal world areas are: the Amazon lowland of South America; the Congo lowland of Africa and a coastal zone extending westward from Nigeria to Guinea; and the East Indian region, from Sumatra on the west to the islands of the western Pacific on the east. Poleward borders of these equatorial rainforest regions are, of course, transitional into rainforest of higher latitudes, particularly that found on tropical windward coasts and that of coastal monsoon rainforest belts of south and southeast Asia.

2. *Tropical rainforest* is in many respects structurally similar to equatorial rainforest but has distinct differences imposed upon it by its location, which is on windward coasts, roughly from 10° latitude to the tropics of Cancer and Capricorn

Figure 21.4 Diagram of the structure of tropical rainforest in Trinidad, British West Indies. A representative species of the tallest trees is *Mora exelsa.* (After J. S. Beard, 1946, *The Natural Vegetation of Trinidad,* Clarendon Press, Oxford.)

Figure 21.5 A tree limb covered by epiphytic ferns. Upland rainforest of Kenya, Africa. (Courtesy of the American Museum of Natural History.)

Figure 21.6 This strangler fig has surrounded the trunk of a cabbage palm in southernmost Florida. (Photograph by Bob Haugen, Everglades National Park.)

($23\frac{1}{2}°$). Here we find a distinct annual precipitation cycle, consisting of a long wet season alternating with a season of reduced rainfall, if not actual drought. There is also a marked annual temperature cycle resulting from the variations in height of the sun's path in the sky in tropical latitudes. The cooler temperatures, coinciding approximately with the period of reduced rainfall, impose some stress upon the plants. As a result, there are fewer species and fewer lianas. Epiphytes are, however, abundant because of continued exposure to humid air and cloudiness of the maritime tropical air masses which impinge upon the coastal hill and mountain slopes.

In terms of global distribution, the tropical rainforest is represented on the world map (Plate 4) by those areas of rainforest (Class *A*) which lie between 10° and 25° latitude. The Carribean lands represent one important area of tropical rainforest, although the rainforest is predominantly limited to windward locations. Of interest to American students is the fact that the Everglades of southernmost Florida contains small isolated communities of tropical rainforest, termed *hammocks*, in which one finds the mahogany tree and strangler fig, along with abundant epiphytes. The coast of this same area has extensive mangrove forest and scrub.

In southern and southeastern Asia tropical rainforest is extensive in coastal zones and highlands which have heavy monsoon rainfall and a very short dry season. The Western Ghats of India and the coastal zone of Burma have tropical rainforest supported by orographic rains of the southwest monsoon. In the zone of combined northeast trades and Asiatic summer monsoon are rainforests of the coasts of Vietnam and the Philippine Islands. In the southern hemisphere belts of tropical rainforest extend down the eastern Brazilian coast, the Madagascar coast, and the coast of northeastern Australia.

As is the equatorial rainforest, the tropical rainforest is associated with a pedogenic regime of laterization under an equable climatic regime. It is common practice among plant geographers to combine the equatorial and tropical rainforests into a single formation class, because their points of similarity greatly outweigh their points of difference.

Figure 21.7 Mangrove growing in salt water, Harney's River, Florida. (Courtesy of the American Museum of Natural History.)

Figure 21.8 A grove of cocoanut palms (*Cocos nucifera*) along a sandy beach in the Solomon Islands. Wave action is undermining the trees. (Courtesy of the American Museum of Natural History.)

3. *Monsoon forest* presents a more open tree growth than the equatorial and tropical rainforests. Consequently, there is less competition among trees for light but a greater development of vegetation in the lower layers. Maximum tree heights range from 40 to 100 ft (12 to 35 m), which is less than in the equatorial rainforest. Many tree species are present and may number 30 to 40 species in a small tract. Tree trunks are massive; the bark is often thick and rough. Branching starts at comparatively low level and produces large round crowns. Perhaps the most important feature of the monsoon forest is the deciduousness of most of the tree species present, e.g., the abundance of tropophytes. The shedding of leaves results from the stress of a long dry season which occurs at time of low sun and cooler temperatures. Thus the forest in the dry season has somewhat the dormant winter aspect of deciduous forests of middle latitudes (Figure 21.9). Some writers use the name "tropical deciduous forest" for monsoon forest, emphasizing the deciduousness rather than the climate regime. A representative example of a monsoon forest tree is the teakwood tree. (*Tectona grandis*).

Lianas and epiphytes are locally abundant in monsoon rainforest but are fewer and smaller than in the equatorial rainforest. Undergrowth is often a dense shrub thicket. Where second-growth vegetation has formed, it is typically jungle. Clumps of bamboo are an important part of the vegetation in climax teakwood forest.

As already noted, monsoon forest is a response to a wet-dry tropical climate regime in which a long rainy season with a large total precipitation alternates with a dry, rather cool season. Such conditions are most strongly developed in the Asiatic monsoon climate, but are not limited to that area. Perhaps the type regions of monsoon forest are in Burma (inland from the coastal tropical rainforest belt) and in Thailand and Cambodia. Large areas of deciduous and semideciduous tropical forest occur in west Africa and in central and South America, bordering the equatorial and tropical rainforests into which there is a gradation. Areas of monsoon forest or related types are also described in Indonesia (especially Java and Celebes), in northern Australia, and in western Madagascar. World vegetation maps and standard reference works do not agree closely on the location and extent of the monsoon, or deciduous tropical forest areas. In Plate 4, monsoon forests are included in Class *C*, the raingreen vegetation of the wet-dry tropics, which also represents other formation classes.

The prevailing pedogenic regime of monsoon forest areas is that of laterization. Despite the dry season, a substantial water surplus is developed during the warm rainy season. Humus does not accumulate; leaching of bases and silica is the dominant soil-forming process.

4. *Temperate rainforest*, also referred to as *temperate evergreen forest* and *laurel forest*, differs from the equatorial and tropical rainforests in having relatively few species of trees, and hence large populations of individuals of a species. Trees are not as tall as in the low-latitude rainforests; the leaves tend to be smaller and more leathery; the leaf canopy less dense. Among the trees commonly found in the temperate rainforests of southern Japan and the southeastern United States are evergreen oaks (such as *Quercus virginiana*) and members of the laurel and magnolia families (such as *Magnolia grandiflora*). A quite different temperate rainforest flora is found in New Zealand and consists of large tree ferns, large conifers such as the kauri tree (*Agathis australis*), podocarp trees (*Podocarpus*), and small-leaved southern beeches (*Nothofagus*) (Figure 21.10). Another important type of temperate rainforest, found in the Azores and Canary island groups, is the Canary laurel forest, which formerly covered Europe in the Miocene geological epoch.

Temperate rainforests tend to have a well-developed lower stratum of vegetation that, in different places, may include tree ferns, small palms, bamboos, shrubs, and herbaceous plants. Lianas and epiphytes are abundant. Particularly striking at higher elevations where fog and cloud are persistent is the sheathing of tree trunks and branches by mosses. An example of conspicuous epiphyte accumulation at low elevation is the Spanish "moss" (*Tillandsia usneoides*) which festoons the Evangeline oak, bald cypress, and other trees of the gulf coast of the southeastern United States.

Temperate rainforest represents a response to an equable climatic regime in which the annual range of temperature is small or moderate and rainfall is abundant and well distributed throughout the year. Such conditions are met in three quite different geographical locations: (1) at higher altitudes in the equatorial and tropical zones; (2) along eastern continental margins and on islands in the latitude belt 25° to 35° or 40°; (3) on west coasts from 35° to 55°. In the first location the effect of higher elevation is to reduce temperatures and evaporation and thereby to increase the moisture available to plants.

Figure 21.9 Monsoon forest in Chieng Mai Province, northern Thailand. Scale is indicated by a line of porters crossing the clearing. (Photograph by Robert L. Pendleton, courtesy of the American Geographical Society.)

Figure 21.10 This podocarp forest illustrates a variety of temperate rainforest in a marine west-coast climate, Hari Hari, western coast of South Island of New Zealand. (Photograph by Pierre Dansereau.)

A word about the vegetation of the coastal region of the southeastern United States may save misunderstanding. On forest maps of the United States (Figure 19.3), and on a number of maps of world vegetation, this coastal zone is shown as having needleleaf evergreen or coniferous forest, whereas on Plate 4, it is shown as temperate rainforest. It is true that large areas of sandy upland bear forests of loblolly and slash pine and that bald-cypress is a dominant tree in swamps, but such vegetation represents xerophytic and hydrophytic forms in excessively dry or wet habitats, or the second-growth forest following fire and deforestation. The climax vegetation of mesophytic habitats is nevertheless the evergreen-oak and magnolia forest.

Temperate rainforest spans two pedogenic regimes: laterization and podzolization. In lower latitudes laterization is characteristic, and this grades through regimes in which both laterization and podzolization are effective (red-yellow soils), to the cool higher latitude regions of podzolization.

Figure 21.11 This stand of beech and hemlock in the Allegheny National Forest, Pennsylvania, illustrates mixed summergreen deciduous forest and needleleaf forest in the northeastern United States. (U.S. Forest Service photograph.)

5. *Summergreen deciduous forest*, sometimes called *temperate deciduous forest*, is familiar to inhabitants of eastern North America and western Europe as a native forest type. It is dominated by tall, broadleaf trees which provide a continuous and dense canopy in summer but shed their leaves completely in the winter (Figure 21.11). Lower layers of small trees and shrubs are weakly developed. In spring a luxuriant low layer of herbs quickly develops, but this is greatly reduced after the trees have reached full foliage and shaded the ground.

Summergreen deciduous forest is almost entirely limited to the middle-latitude landmasses of the northern Hemisphere (Plate 4, Class *E*). Common trees of the deciduous forests of eastern North America, southeastern Europe, and eastern Asia (all in the humid-continental climate) are oak (*Quercus*), beech (*Fagus*), birch (*Betula*), hickory (*Carya*), walnut (*Juglans*), maple (*Acer*), basswood (*Tilia*), elm (*Ulmus*), ash (*Fraximus*), tulip tree (*Liriodendron*), sweet chestnut (*Castanea*), and hornbeam (*Carpinus*). In western and central Europe, under a marine west coast climate dominant trees are mostly oak and ash, with beech (*Fagus sylvatica*) in cooler and moister areas.

In poorly drained habitats, the deciduous forest consists of such trees as alder, willow, ash, and elm, and many hygrophytic shrubs. Where the deciduous forests have been cleared in lumbering, pines readily develop as second-growth vegetation.

The summergreen deciduous forest represents a response to a continental climatic regime which at the same time receives adequate precipitation in all months. There is a strong annual temperature cycle with a cold winter season and a warm summer. Precipitation is markedly greater in the summer months (especially so in eastern Asia) and thus increases at the time of year when evapotranspiration is great and the moisture demands are high. Only a small water deficit is incurred in the summer, while a large surplus normally develops in spring. In eastern Asia the winter is exceptionally dry, but this factor is compensated for by cold.

The pedogenic process associated with summergreen deciduous forest is that of podzolization, but moderated by the warm wet summers. As a result, the soils are characteristically of the gray-brown forest group. Toward lower latitudes the tendency to laterization becomes stronger and red-yellow podzolic soils are encountered; toward the continental interiors the tendency to calcification sets in and the deciduous forest extends into regions of darker soils of the grasslands (prairie and chernozem soils). In the summergreen deciduous forests a thick layer of fallen leaves covers the ground and there is abundant humus in the soil.

Because the regions of summergreen deciduous forest have for centuries been the areas of dense populations, few remnants of a primeval forest have survived. Instead, most existing forests are modified by practices of tree farming. Large areas are completely and permanently removed from forest growth by farming, urban development, and roadways.

6. *Needleleaf forest* is composed largely of straight-trunked, conical trees with relatively short branches, and small, narrow, needlelike leaves. These trees are conifers (Figure 21.12). Where evergreen, the needleleaf forest provides continuous and deep shade to the ground so that lower layers of vegetation are sparse or absent, except for a thick carpet of mosses in many places. Species are few and large tracts of forest consist almost entirely of but one or two species.

Reference to the world vegetation map (Plate 4, Class *F*) will show that needleleaf forest is predominantly in two great continental belts, one in North America and one in Eurasia, which span the landmasses from west to east in latitudes 45° to 75°. The needleleaf forest of North America, Europe, and western Siberia is composed of evergreen conifers, such as spruce (*Picea*), fir (*Abies*), and pine (*Pinus*), whereas that of north-central and eastern Siberia is dominantly of larch (*Larix*) which sheds its needles in winter and comprises a deciduous forest (Figure 21.13). Associated with the needleleaf trees is mountain ash. Aspen and balsam poplar (*Populus*), willow, and birch tend to take over rapidly in areas of needleleaf forest

Figure 21.12 Spruce forest, Upper Peribonka River, Quebec, Canada. (Photograph by Pierre Dansereau.)

Figure 21.13 Woodland of larch (*Larix dahurica*), Tompo River region, about latitude 64° N, Yakutsk, U.S.S.R. This scene shows the larch tree near the northern limit of its growth. (Photograph by I. D. Kild'ushevsky, courtesy of Professor B. A. Tikhomirov, Komarov Botanical Institute, Leningrad.)

which have been burned over, or can be found bordering streams and in open places.

Needleleaf evergreen forest extends into lower latitudes wherever mountain ranges and high plateaus exist. Thus in western North America this formation class extends southward into the United States on the Cascade, Sierra Nevada, and Rocky Mountain ranges and over parts of the higher plateaus of the southwestern states. In Europe, needleleaf evergreen forests flourish on all of the higher mountain ranges, and well into Scandinavia.

The needleleaf evergreen forests of British Columbia and California are of particular interest. Here, under a regime of heavy orographic precipitation and prevailing high humidities, are perhaps the densest of all coniferous forests, with the world's largest trees. Forests of coastal redwood (*Sequoia sempervirens*), big tree (*Sequoiadendron giganteum*) and Douglas fir (*Pseudotsuga taxifolia*) are particularly remarkable (Figure 21.14). Individuals of redwood and big tree attain heights of over 325 ft (100 m) and girths of over 65 ft (20 m) (Figure 21.15).

As much of the area of needleleaf evergreen forest in North America and Europe was subjected to glaciation in the Wisconsin stage of the Pleistocene epoch, there abound lakes and poorly drained depressions. These bear the hygrophytic vegetation, described in Chapter 20 as forming a bog succession, and leading to large, thick peat accumulations known in Canada as *muskeg*.

Podzolization, the pedogenic regime of the needleleaf forests, is a direct result of a climatic regime of continentality with a strong polar influence. As explained in Chapter 18, the conifers require little of the bases and these are leached from the soil, which is characteristically acidic.

7. *Evergreen-hardwood forest,* also termed *sclerophyll forest,* consists of low trees with small, hard, leathery leaves. Typically the trees are low-branched and gnarled, with thick bark. The formation class includes much *woodland,* an open forest in which the canopy coverage is only 25 to 60%. Also included are extensive areas of *scrub,* a plant formation type consisting of shrubs having

Figure 21.14 This mixed stand of white pine, Douglas fir, and Englemann spruce is on the western slopes of the Cascade Range in Washington. (U.S. Forest Service photograph.)

Figure 21.15 A grove of Sequoia trees, Sequoia National Park, California. The largest tree measures 51 ft (15m) in circumference. A man can be seen standing at the right of this tree (look for hat). (Courtesy of the American Museum of Natural History.)

a canopy coverage of perhaps 50% (Figure 21.16). The trees and shrubs are evergreen, their thickened leaves being retained despite a severe annual drought. There is little stratification in the sclerophyll forest and scrub, although there may be a spring herb layer.

Evergreen hardwood forest is closely associated with the dry-summer subtropical (Mediterranean) climate and hence quite narrowly limited in geographical extent—primarily to west coasts between 30° and 40° or 45° latitude (Plate 4, Class *D*). In the Mediterranean lands the hardwood forest forms a narrow peripheral coastal belt. Here the woodland consists of such trees as cork oak (*Quercus suber*), live oak (*Quercus ilex*), Aleppo pine (*Pinus halepensis*), stone pine (*Pinus pinea*), and olive (*Olea europaea*). What may have once been luxuriant forests of such trees were greatly disturbed by man over the centuries and reduced to woodland or entirely destroyed. Instead, large areas consist of dense scrub termed *maquis* which includes many species, some of them very spiny. The other northern hemisphere region of evergreen hardwood forest is that of the California coast ranges. Some of this is a woodland composed largely of the live oak (*Quercus agrifolia*) and white oak (*Quercus lobata*). Much of the vegetation is a scrub or "dwarf forest" known as *chaparral*, which varies in composition with elevation and exposure (Figure 21.17). Chaparral may contain wild lilac (*Ceanothus*), manzanita (*Arctostaphylos*), mountain mahogany (*Cercocarpus*), "poison oak" (*Rhus diversiloba*), and live oak. The evergreen hardwood forest is represented in central Chile and in the Cape region of South Africa by a maquislike scrub vegetation which is, however, of quite different flora from those of the northern hemisphere. Important areas of sclerophyll forest, woodland, and scrub are found in southeast, southcentral, and southwest Australia, including several species of eucalypts and acacias (Figure 21.18).

The dry-summer subtropical (Mediterranean) climate of the evergreen hardwood forest is one of great environmental stress because the severe drought season coincides with high air temperatures. Thus a large water deficit occurs in the summer (see Figure 12.12*D*). The wet, mild winter is, by contrast, highly favorable to rapid plant growth. The pedogenic regime is one associated with semi-aridity, namely calcification, and results in soils with some precipitated calcium carbonate in the *B* horizon. Much of the area of the evergreen hardwood forest, woodland, and scrub is classified as having reddish-chestnut, reddish-prairie, and reddish-brown soils, and in the Mediterranean lands, *terra rossa*.

Figure 21.16 Chaparral on steep mountain slopes of the San Dimas Experimental Forest, near Glendora, California, 1953. This vegetation has been protected from burning for many years. (Photograph by A. N. Strahler.)

Savanna biochore

8. *Savanna woodland* consists of trees spaced rather widely apart, permitting development of a dense lower layer, which may be of grasses or shrubs. This formation class is sometimes referred to as "parkland" because of the open, parklike appearance of the vegetation. Savanna woodland is associated with a climate regime in which aridity is sufficiently developed to prevent the tree growth from forming a closed canopy.

Many geographers associate savanna woodland closely with the tropical wet-dry climate. It is this tropical variety of woodland that is implied in the world vegetation map (Plate 4) by the inclusion of savanna woodland with Class *C* vegetation. However, from the standpoint of the vegetation structure itself, the definition of savanna woodland can be broadened to include woodlands of middle latitudes, such as the open stands of yellow pine and of pinyon pine and juniper which, in the western United States, occur in an elevation zone above that of the sagebrush scrub and below that of the needleleaf forests. One might also include in this formation class the Eucalyptus woodlands of southern Australia, which we have elsewhere placed in the formation class of evergreen-hardwood forest.

Figure 21.17 Detail of chaparral vegetation, San Dimas Experimental Forest (see Figure 21.16). A tilted rain gauge, 8 in (20 cm) in diameter, rests on bare ground at the left. (Photograph by A. N. Strahler.)

Figure 21.18 Eucalyptus woodland, Darling Range, Western Australia, about 1930. (Photograph by Agent General for Western Australia, courtesy of the American Geographical Society.)

Referring now to the tropical savanna woodland, the trees are of medium height, the crowns flattened or umbrella-shaped, and the trunks have thick, rough bark (see Figure 21.20). The trees tend toward xerophytic forms with small leaves and thorns, or may be deciduous, shedding their leaves in the dry season. In this respect, tropical savanna woodland is closely akin to monsoon forest, into which it grades. The trees are of species capable of withstanding fires which sweep through the lower layer in dry season; thus many rainforest tree species which might otherwise grow in the wet-dry climate regime are prevented by fires from invading.

Tropical savanna woodland is found widely throughout Africa, South America, southeast Asia, northern Australia, and in Central America and the Carribean Islands (Plate 4, Class *C*). The pedogenic process usually dominant is laterization.

9. *Thornbush and tropical scrub* (which may be treated as two different formation classes) consist of xerophytic trees and shrubs responding to a climate with a very long dry season and only a short, but intense, rainy season. Thornbush, also referred to as "thorn forest," and "thornwoods," consists of tall, closely spaced woody shrubs commonly bearing thorns and largely deciduous (Figure 21.19). Cacti may also be present. The lower layer of herbs may consist of annuals which largely disappear in the dry season, or of grasses. In the drier areas the lower layer may consist only of scattered grass clumps with much bare ground between. One example of the thornbush is the *caatinga* of northeastern Brazil, an open thorn forest of the dry highlands. Another is the *dornveld* of South Africa.

Tropical scrub is a dense growth of low woody shrubs which may occur in patches or clumps separated by barren ground. Scrub may develop in stony, sandy, or gravelly sites in areas of thornbush.

Thornbush and tropical scrub have been included in Class *C* of the world vegetation map (Plate 4) as formation classes of the wet-dry tropical climate regime. Thornbush and tropical scrub are found in many parts of the world where the wet-dry tropical climatic regime is transitional into the tropical desert regime. Soils show the influence of aridity and are subject to calcification, and (in poorly drained sites) of salinization. Reddish-brown and reddish-chestnut soils occur in much of the area of these formation classes.

10. *Savanna* is a vegetation consisting of widely scattered trees rising from a more or less continuous lower layer dominated typically by grasses. Although the term "savanna" most commonly refers to a vegetation in tropical and subtropical latitudes, structurally similar vegetation can be found in the subarctic lands of the northern hemisphere—the *taiga*. (See "cold woodland.") As discussed here, savanna is that of low latitudes, included in Class *C* in the world vegetation map (Plate 4) as a relative of the tropical savanna woodland, monsoon forest, and thornbush and tropical scrub. Like its relatives, the tropical sa-

Figure 21.19 Thornbush in the region of the Tona River, central Africa. (Photograph of the Akeley Expedition, courtesy of the American Museum of Natural History.)

vanna is a response to a wet-dry tropical climate regime in which the severe drought period is one of relatively cooler temperature but which experiences great heat just preceding the onset of the rains.

As with the savanna woodland, the tropical savanna supports trees and shrubs which are xerophytic or deciduous. In some localities the trees are palms, giving a "palm savanna." Grasses in the tropical savanna are characteristically tall, with stiff coarse blades, commonly higher than the height of a man, and even up to 12 ft (4 m) high. In the dry season these grasses form a yellowish straw mat that is highly flammable and subject to periodic burning. It is widely held that, to a greater or lesser degree, periodic burning of the savanna grasses is responsible for the maintenance of the grassland against the invasion of forest. Fire does not kill the underground parts of grass plants, but limits tree growth to a few individuals of fire-resistant species. The browsing of animals, which kills many young trees, is also a factor in maintaining grassland at the expense of forest.

The African savanna is perhaps the most celebrated of the tropical grasslands, spanning the continent from west to east in two great belts, centered about on the 10th parallels of latitude north and south, and connected in east Africa by a broad north-south belt (Figure 21.20). In equatorial latitudes the African savanna replaces the rainforest because of the aridity associated with highlands in Sudan, Kenya, and Tanzania. Characteristic are the flat-topped acacia trees (*Acacia*), and the grotesque baobab (*Adansonia digitata*), which has a large barrel-shaped water-storing trunk (Figure 21.21). Elephant grass (*Pennisetum purpureum*), forming almost inpenetrable thickets, may grow to heights of 16 ft (5 m).

Figure 21.21 Baobab tree (*Adansonia digitata*), near Nairobi, Kenya, Africa. (Photograph of the Akeley Expedition, courtesy of the American Museum of Natural History.)

Savanna in South America is exemplified in the *campo cerrado* of the interior Brazilian Highlands (Figure 21.22). Here the trees are largely evergreen, deep-rooted, and capable of tapping lower levels of soil moisture not available to the grasses during the dry season.

Other important savanna areas occur in northern Australia, India, and southeast Asia. The mesquite savanna of Texas provides an American example. Of interest to American students is the small patch

Figure 21.20 African tall-grass savanna in Kenya. (Photograph by Richard U. Light, courtesy of American Geographical Society.)

Figure 21.22 The *campo cerrado* at Pirassununga, Sao Paulo, Brazil. (Photograph by Pierre Dansereau.)

of rather specialized (edaphic) tropical savanna in southern Florida—the Everglades. Underlain by a limestone formation, the extremely low, flat plain of the Everglades is flooded by a shallow layer of runoff from summer rains and becomes a swamp. In winter the area becomes extremely dry. Coarse saw grass (*Cladium effusum*) covers much of the plain. Scattered trees are represented by palms and, on higher ground, by pines.

The pedogenic process most closely associated with tropical savanna is laterization, promoted by the high temperatures associated with the rainy season. However, laterization gives way to calcification as the savanna is traced toward higher latitudes where thornbush and, ultimately, steppe grasslands are encountered.

11. *Semidesert* (also called *half desert*) is a xerophytic shrub vegetation with a poorly developed herbaceous lower layer. Trees are generally absent. Semidesert shrub vegetation is well developed in subtropical and middle-latitude dry climates having a small annual total rainfall and high summer temperatures. An example is the sagebrush (*Artemisia tridentata*) vegetation of the middle and southern Rocky Mountain region and Colorado Plateau (Figure 21.23). Semidesert shrub vegetation seems recently to have expanded widely into areas of the western United States that were formerly steppe grasslands, as a result of overgrazing and trampling by livestock.

On the world map of vegetation, Plate 4, semidesert is included with dry-desert vegetation. The xerophytic shrub vegetation is to be expected along the less arid margins of the deserts and in favorable highland locations within the desert. Pedogenic processes are calcification, producing brown soils,

Figure 21.23 Sage-brush semidesert, Vermilion Cliffs, near Kanab, Utah, 1906. (Photograph by Douglas Johnson.)

Figure 21.24 Lowland heath, mostly heather (*Calluna vulgaris*), on moraine, North Norfolk. (Photograph by S. R. Eyre.)

and gray desert soils. Salinization is found in poorly drained sites.

12. *Heath* is a low, dense layer of shrubs commonly not more than 10 in (25 cm) in height, which occurs in regions of highly equable but cool climate in middle and high latitudes (Figure 21.24). The shrubs are usually dominated by members of the Heath family (*Ericaceae*); mosses are also important. A common plant of heath vegetation is heather (*Calluna vulgaris*). In very cold climates heath includes small shrubby birches and willows.

Distribution of heath is not shown on the world vegetation map (Plate 4) but is found in small areas in Classes *E* and *I*. A cool, marine west coast climate is favorable to development of heath. Well-distributed precipitation and a temperature regime with small annual range are required. The pedogenic process is podzolization and soils are acidic. Examples of heath vegetation are seen in Ireland and other exposed west coasts of the British Isles (*moors*) and in western and north-central Europe.

13. *Cold woodland*, the last on our list of formation classes in the savanna biochore, is a form of vegetation limited to very cold subarctic and tundra climates. Trees are low in height and well spaced apart; a shrub layer may be well developed. The ground cover of lichens and mosses is distinctive. Cold woodland is essentially equivalent to what is widely referred to as *taiga*,[3] and occurs along the northern fringes of the high-latitude, needleleaf forests. Cold woodland is thus transitional into the treeless tundra and arctic heath.

In North America representative trees of the

[3]The term "taiga" is commonly used to include also needleleaf forest, needleleaf woodland, and needleleaf savanna.

cold woodland are black spruce (*Picea mariana*) and tamarack (*Larix laricina*). In northern Scandinavia a scrubby birch (*Betula odorata*) forms a woodland with a low layer of lichens, such as reindeer "moss" (*Cladonia rangiferina*) (Figure 21.25). In Siberia larch (*Larix*) is the woodland tree. Birch-lichen woodland is also found in the subarctic regions of northwestern Canada and Alaska.

Cold woodland is dominated by the cold continental climate with its severe winters in which all soil moisture is frozen for many months of the year. A shallow depth of thaw accompanies the short "summer," when insolation continues throughout much of the day. The pedogenic process is toward gleization in the poorly drained sites; toward podzolization on upland sites.

Cold woodland is not shown separately on the world vegetation map (Plate 4) but can be found in the tundra areas (Class *I*) along the boundary with needleleaf forest (Class *F*).

Grassland biochore

14. *Prairie* consists of tall grasses, comprising the dominant herbs, and subdominant forbs (broad-leaved herbs) (Figure 21.26). Trees and shrubs are almost totally absent but may occur in the same region as forest or woodland patches in valleys and other topographic depressions. The grasses are deeply rooted and form a continuous and dense sward. The grasses flower in spring and early summer; the forbs, in late summer. In Iowa, a representative region of tall-grass prairie, typical grasses are big bluestem (*Andropogon gerardi*) and little bluestem (*A. scoparius*), a typical forb is black-eyed Susan (*Rudbeckia nitida*).

Figure 21.26 Kalsow Prairie, Iowa, has been set aside as an example of virgin prairie. (Photograph by courtesy of State Conservation Commission of Iowa.)

Figure 21.25 Lichen woodland (open boreal forest) on a sand plain of the Hamilton River delta, Labrador. Black spruce and a few small white birch are widely spaced. The floor is covered by lichen (*Cladonia*). (Photograph by F. Kenneth Hare, courtesy of the American Geographical Society.)

The tall-grass prairies are typically associated with continental, middle-latitude climates described as *subhumid*, that is, in which evapotranspiration and precipitation are almost balanced on an average yearly basis and range between 20 and 40 in (50–100 cm). In summers air and soil temperatures are high, so that on uplands soil moisture is not adequate for tree growth and deeper water sources are beyond reach of tree roots. The North American Prairies are found in a broad belt extending from Illinois, northwestward to southern Alberta and Saskatchewan. On the world map of vegetation (Plate 4) the North American prairie includes the eastern and northern parts of Class *G*. Areas of forest are mixed with areas of prairies in a transitional belt between forest and prairie regions.

Because the tall-grass prairies grade into short-grass prairies and eventually into steppe grasslands in the direction of increasing aridity, it is not practical to try to list all world regions of tall-grass prairies. In Europe, a typical region of tall-grass prairie is the *puszta* of Hungary. The Argentine *pampa* is often cited as a region of prairie vegetation, as are areas in north China.

The pedogenic process associated with prairie vegetation is calcification and results in development of prairie and chernozem soil profiles.

15. *Steppe*, sometimes called *short-grass prairie*, is a formation class consisting of short grasses tending to be bunched and sparsely distributed (Figure 21.27). Scattered shrubs and low trees may be found in the steppe and there exist all grada-

Figure 21.27 Short-grass vegetation of the Great Plains, South Dakota. (Photograph by Douglas Johnson.)

tions into semidesert and woodland formation classes. Ground coverage is small and much bare soil is exposed. Many species of grasses and other herbs occur; a typical grass of the American steppe is buffalo grass (*Buchloe dactyloides*), other typical plants are the sunflower (*Helianthus rigidus*) and loco weed (*Oxytropis lambertii*).

The world distribution of steppe vegetation is extremely wide in terms of latitude, ranging in location from the equator to 55° N and 45° S (Plate 4, Class *G*). Steppes of low latitudes are transitional from vegetation of the wet-dry tropical climates into the dry deserts. Steppes of North Africa are transitional from Mediterranean climate, with its sclerophyll forest, to the African desert. Steppes of middle latitudes are associated with a semi-arid continental climatic regime in which, despite a summer rainfall maximum, evaporation exceeds precipitation on the average. Winters in the middle-latitude steppes are cold and dry; the summers warm to hot. In this climatic regime the dominant pedogenic process is calcification, with salinization in poorly drained sites. Soils contain a large excess of precipitated calcium carbonate and are very rich in bases. Brown soils are typical. Humus content is relatively small because of the sparseness of the vegetation.

16. *Grassy tundra* (including *alpine meadow*) is a grassland biochore formation limited to very cold climates having ample available moisture and often saturated soils. Grassy tundra of the arctic regions flourishes under a regime of long summer days during which time the ground ice melts only in a shallow surface layer. The frozen ground beneath (permafrost) remains impermeable and melt water cannot readily escape. Consequently, in summer a marshy condition prevails for at least a short time over wide areas. Humus accumulates in a well-developed layer.

Plants of the arctic grassy tundra are low and mostly herbaceous, although dwarf willow (*Salix herbacea*) occurs in places. Sedges, grasses, mosses, and lichens dominate the tundra in a low layer (Figure 21.28). Typical species are ridge sedge (*Carex bigelowii*), arctic meadow grass (*Poa arctica*), cottongrasses (*Eriophorum*), and snow lichen (*Cetraria nivalis*). There are also many species of forbs which flower brightly in the summer. Considerable variations in composition of the tundra are seen in the range from wet to well-drained habitats. One form of tundra consists of sturdy hummocks of plants with low, water-covered ground between. In the regions of grassy tundra will also be found areas of arctic scrub vegetation composed of willows and birches.

Under the subarctic climate and a condition of poor soil-water drainage, the pedogenic regime tends toward gleization. Size of plants is in part limited by the mechanical rupture of roots during freeze and thaw of the surface layer of soil, producing shallow-rooted plants. In winter drying winds and mechanical abrasion by wind-driven snow tend to reduce any portions of a plant that project above the snow.

In all latitudes, where altitude is sufficiently high, an alpine tundra is developed above the limit of tree growth and below the vegetation-free zone of barren rock and perpetual snow. Alpine tundra resembles arctic tundra in many physical respects.

Desert biochore

17. *Dry desert* is a formation class of xerophytic plants widely dispersed and providing almost negligible ground cover. In dry periods (which are the rule) the visible vegetation consists of small hard-leaved or spiny shrubs, succulent plants (cacti), or hard grasses. Many species of small annuals may be present, but appear only after a rare but heavy rain has saturated the soil.

Desert floras differ greatly from one part of the

Figure 21.28 Cotton-grass meadows, Arctic coastal plain of Alaska. (Photograph by William R. Farrand.)

Figure 21.29 Shrub vegetation of the Sonoran Desert, southwestern Arizona. Ocotillo plant in the left foreground; saguaro cactus plants in left distance. (Photograph by A. N. Strahler.)

world to another. In the Mohave-Sonoran deserts of the southwestern United States, plants are often large and in places give a near-woodland appearance (Figure 21.29). Well known are the treelike saguaro cactus (*Carnegiea gigantea*), the prickly-pear cactus (*Opuntia imbricata*), the ocotillo (*Fouquiera splendens*), creosote bush (*Larrea tridentata*) and smoke tree (*Dalea spinosa*). Much of the "desert" of the southwestern United States is in fact scrub, thorn scrub, savanna, or steppe grassland (Figure 19.2). In the Sahara Desert (most of it very much drier than in the American desert) a typical plant is a hard grass (*Stipa*); another, found along the dry beds of watercourses, is tamarisk (*Tamarix*). The coastal desert of southwest Africa is known for the strange tumboa plant (*Welwitschia mirabilis*) with strap-shaped leaves radiating from a central tap root which penetrates deep into the ground (Figure 21.30).

Much of the area assigned to dry-desert vegetation has no plants of visible dimensions, being composed of shifting dune sands, or almost sterile salt flats.

Distribution of dry deserts is shown in the world vegetation map, Plate 4, as Class *H*. As a study of climate will show, three variations of the desert trend exist. The true tropical-continental deserts have not only extreme aridity but also extremely high air and soil temperatures; the middle-latitude deserts (30° to 35° lat.) have both aridity and a great annual temperature range including extremes of winter cold; the tropical west coast deserts have remarkable uniformity and relative coolness of temperature along with persistent coastal fog. A dominant pedogenic process of dry deserts is salinization, locally producing areas of salt crust where only salt-loving (halophytic) plants can survive. Calcification is conspicuous on well-drained uplands; encrustations and deposits of calcium carbonate (caliche) are commonplace. Humus is lacking and soils are pale gray or red in color, e.g., sierozems and red-desert soils.

18. *Arctic fell field* is the arctic equivalent of the dry desert, being found in the extremely cold tundra and icecap climates. The arctic fell field consists of rocky ground surfaces, produced by

Figure 21.30 The tumboa plant (*Weltwitschia mirabilis*) on a sandy plain in the Kalahari Desert of southwestern Africa. (Photograph by Robert J. Rodin.)

Figure 21.31 Arctic fell-field vegetation, Etah, Greenland. On the left is *Dryas integrifolia*; on the right, *Cerastium alpinium*. (Photograph of Crockerland Expedition, courtesy of the American Museum of Natural History.)

intense frost shattering and having only small patches of finetextured mineral soil. Vegetation is very sparse and consists mostly of lichens, mosses, and a few small shrubs (Figure 21.31). For example, in northern Baffin Island, typical plants are a small polar willow (*Salix polaris*), purple saxifrage (*Saxifraga oppositifolia*), and woolly moss (*Rhacomitrium lanuginosum*).

Arctic fell field can be found as far north as the extreme limits of land, namely up to 84° N latitude and southward to ice-free parts of Antarctica. Soil-forming processes consist of little more than rock disintegration by frost action, manifested in the formation of stone polygons in which finer particles are sorted from the coarse into patches. On the world vegetation map, Plate 4, arctic fell field is included in Class *I* along with grassy tundra and arctic scrub.

This attempt to classify world vegetation into a number of formation classes and to show their distribution by means of a map can be only partly successful, at best. The infinite variations in structure of vegetation defy a simple categorical system. Moreover, no two authorities in the field of plant geography will set up the same array of classes or produce closely corresponding maps. An appreciation of broad systems of plant formation classes in response to the restraints and freedoms of the spectrum of climatic regimes should be foremost in the mind of the geographer and he must stress the underlying principles and distributional aspects that most systems show in common, rather than be diverted by the seeming contradictions and complexities he frequently encounters.

REVIEW QUESTIONS

1. What are the four great biochores? Describe each briefly and give its general climatic associations.

2. What investigators were responsible for developing the system of vegetation formation classes commonly followed today? Discuss some of the problems that might be connected with preparing a useful world map of natural vegetation.

3. Describe the equatorial rainforest in terms of structure. What is the importance of lianas in this rainforest? Of epiphytes? Comment on the numbers of species of trees found in the equatorial rainforest. Name two representative trees. What climatic and pedogenic regimes are associated with the equatorial rainforest?

4. What coastal vegetation forms may be expected in regions of equatorial rainforest? Describe particularly the mangrove coasts.

5. How does tropical rainforest differ from equatorial rainforest and in what respects is it similar? In what geographic regions is tropical rainforest found? Explain.

6. What are the unique characteristics of the monsoon forest as compared with other low-latitude forests? What accounts for the deciduousness of the trees? Relate the monsoon forest to a particular climatic regime and describe its world distribution pattern.

7. How does temperate rainforest differ in structure and geographic location from equatorial and tropical rainforests? Describe temperate rainforests of America, Asia, and New Zealand. What is particularly noteworthy of the development of epiphytes in the temperate rainforest?

8. What three types of world locations favor temperate rainforest? What climatic and pedogenic regimes are associated with temperate rainforest?

9. Describe the summergreen deciduous forest as to structure and seasonal development. Name common trees of the deciduous forests of eastern North America and Europe. In what ways does the summergreen deciduous forest represent a response to the continental climatic regime? What is the associated pedogenic process?

10. Describe a needleleaf forest in terms of vegetation structure. Define the major world regions of occurrence of needleleaf forest and name common trees of these belts.

11. How does topography influence the extension of needleleaf forests into lower latitudes? Describe the needleleaf forests of the Pacific coastal regions of North America. What is the dominant pedogenic regime in areas of needleleaf forest?

12. Describe the evergreen-hardwood (sclerophyll) forest in terms of vegetation structure. With what climatic regime is this formation class closely associated and how is this association seen in the world geographical distribution of sclerophyll forest, woodland, and scrub? Name some representative sclerophyll forest plants of the Mediterranean region, California, and Australia.

13. What is savanna woodland? With what climatic regime is it most commonly associated? What are the life-form characteristics of trees and shrubs of the tropical savanna woodland?

14. Describe thornbush and tropical scrub as vegetation classes. What is the *caatinga* of Brazil? The *dornveld* of South Africa?

15. How does savanna differ from savanna woodland? Describe the vegetation soils, climate, and animal life associated with tropical savanna, with particular reference to Africa. What is the *campo cerrado* of Brazil?

16. What is semidesert vegetation? Describe semidesert in the southwestern United States.

17. What is heath? What common plants are associated with heath? Describe the conditions of climate and geographic location favorable to occurrence of heath.

18. Describe cold-woodland vegetation. Is *taiga* an equivalent designation for this formation class? What are some representative plants of the cold woodland of North America? Of Scandinavia? Of Siberia?

19. What plants comprise prairie as a vegetation formation class? Name representative examples from the tall-grass prairie of Iowa. With what climatic and pedogenic regimes is prairie vegetation associated? Explain. What is the *puszta* of Hungary? The *pampa* of Argentina?

20. Compare steppe (short-grass prairie) with tall-grass prairie in terms of vegetation structure and climatic regime.

21. Describe the grassy tundra of cold climates, giving names of typical plants. With what conditions of climate, drainage, and pedogenic process is the grassy tundra associated? Compare alpine tundra with grassy tundra of low elevations.

22. Describe dry desert as a plant formation class. Name some typical desert plants of the southwestern United States, the Sahara, and southwest Africa. What is the dominant pedogenic process of the dry desert?

23. What relation does arctic fell field bear to other formation classes of the cold lands in high latitudes? Explain how freeze and thaw of soil water plays a dominant role in the landscape associated with arctic fell field.

IV

LANDFORMS OF THE EARTH'S CRUST

CHAPTER **22**

Earth Materials and Mineral Resources

AT first thought, the dense, hard rock that lies beneath the soil layer might seem to play no active role in man's environment. A highly polished granite slab shows no observable change after decades of exposure to sunlight, rain, frost, and winds. On second thought, we realize that slow changes in exposed rock surfaces are in some instances observable. For example, a century-old tombstone of marble proves to be wasted to the degree that the inscription is no longer legible. Evidently, tightly bonded mineral matter of the marble has yielded to environmental processes of the atmosphere and hydrosphere and has dissolved, releasing its substance as ions in solution. Recall that calcium and magnesium ions are vital ingredients in the soil. Marble is composed in large part of calcium and magnesium; it is thus a source of base ions (Chapter 18).

Release of large numbers of ions from mineral matter of rock must be preceded by fragmentation of the rock into particles ranging in size from sand through silt, to clay composed of particles of colloidal dimension. As we have noted in the study of soil-forming processes, this mass of fragments is the parent mineral matter of the soil. Life on the lands could not have been possible without formation of soil. Thus one major environmental role of geologic processes is to furnish the parent matter of soil and with it essential nutrients of plant growth. Even the floating plants of streams, lakes, and oceans (algae, for example), while not requiring soil for physical support, depend upon ions supplied by the breakdown of rock and transported in solution by streams and currents.

For modern man, the lithosphere takes on yet another vital role, that of supplying him with mineral resources without which industrial development would have been impossible. Mineral resources include hydrocarbon fuels, which are man's principal energy source, and the essential substances of civilization—metals, nonmetallic structural materials, and chemicals for all kinds of industrial processes and synthetic products. To gain a full appreciation of the limitations of these nonrenewable resources, we must know how rocks are formed and how very rare elements dispersed throughout rocks are concentrated into ores.

Composition of the earth's crust

From the standpoint of the environment of man, the really significant zone of the solid earth is the thin outermost layer—the earth's crust. This mineral skin, averaging about 10 mi (17 km) thickness for the globe as a whole, contains the continents and ocean basins and is the source of soil and other sediment vital to life, of salts of the sea, of gases of the atmosphere, and of all free water of the oceans, atmosphere and lands.

Table 22.1 lists in order the eight most abundant elements of the earth's crust, in terms of percentage by weight. Oxygen, the predominant element, accounts for about half the total weight. It occurs in combination with silicon, the second most abundant element.

Aluminum and iron are third and fourth on the list. These metals are of primary importance in man's industrial civilization and it is fortunate, indeed, that they are comparatively abundant elements. There follow four metallic elements which we have already identified as base ions in soils: calcium, sodium, potassium, and magnesium. All are on the same order of abundance, from 2 to 4 percent. Their importance in soil fertility has already been stressed in Chapter 18.

If we were to extend Table 22.1, the ninth-place element would prove to be titanium, which would

TABLE 22.1 ABUNDANT ELEMENTS IN THE EARTH'S CRUST[a]

Element	Symbol	Percent by Weight
Oxygen	O	46.6
Silicon	Si	27.7
Aluminum	Al	8.1
Iron	Fe	5.0
Calcium	Ca	3.6
Sodium	Na	2.8
Potassium	K	2.6
Magnesium	Mg	2.1
	Total	98.5

[a]Data from Brian Mason, 1966, *Principles of Geochemistry*, 3rd. Edition, John Wiley & Sons, New York. Table 3.4, p. 48. Figures have been rounded to nearest one-tenth.

be followed in order by hydrogen, phosphorus, barium, and strontium. Phosphorus is one of the essential elements in plant growth.

A number of metallic elements that we think of as abundant in manufactured products—copper, lead, zinc, nickel, and tin—are present only in extremely low percentages and are very scarce elements, indeed.

The silicate minerals

Rock, which is composed of mineral matter in the solid state, comes in a very wide range of compositions, physical characteristics, and ages. A given rock is usually composed of two or more minerals and usually many minerals are present; however, a few rock varieties consist almost entirely of one mineral. Most rock of the earth's crust is extremely old in terms of human standards, the times of formation ranging back many millions of years. But rock is also being formed at this very hour, as a volcano emits lava that solidifies upon contact with the atmosphere.

A *mineral* is perhaps easier to define; it is a naturally occurring, inorganic substance, usually possessing a definite chemical composition and a characteristic atomic structure. The vast number of known minerals, together with the great number of their combinations into rocks, require that we must generalize and simplify this discussion to a high degree, referring to rocks and minerals in a way meaningful in terms of their environmental properties and value as natural resources.

Most of the rock of the crust is *igneous* in origin, meaning that the crystalline or glassy solid rock has solidified from a high-temperature molten state, or *magma*. Most of the bulk of igneous rock consists of *silicate minerals*, or simply *silicates*, which are compounds containing a combination of *silicon* (Si) and oxygen, usually together with one or more metallic elements. A basic understanding of igneous rocks can be had by considering the various common combinations of only seven silicate minerals or mineral groups; these are shown in Figure 22.1.

Among the most common minerals of all rock types is *quartz;* its composition is *silicon dioxide* (SiO_2). There follow five mineral groups, collectively forming the *aluminosilicates* because all contain aluminum. Two groups of *feldspars* are set apart; the *potash feldspars* contain potassium (K) as the dominant metallic ion, but sodium (Na) is commonly present in various proportions. The *plagioclase feldspars* form a continuous series, beginning with the *sodic*, or sodium-rich varieties, and grading through with increasing proportions of calcium to the *calcic*, or calcium-rich varieties.

Belonging to the *mica group*, which is familar because of its property of splitting into very thin, flexible layers, is *biotite*, a dark-colored mica with a complex chemical formula. Potassium, magnesium, and iron are present in biotite, along with some water. The *amphibole group*, of which the common black mineral *hornblende* is a common representative, is a complex aluminosilicate containing calcium, magnesium, and iron. Similar in outward appearance and having essentially the same component elements is the *pyroxene group*, with the mineral *augite* as a representative. Last on the list is *olivine*, a dense greenish mineral which is a silicate of magnesium and iron, but without aluminum.

Looking down the list of density[1] values of these silicate minerals, you will notice that there is a progressive increase from the least dense, quartz, to the most dense, olivine. This change reflects the decreasing proportion of aluminum and sodium, elements of low atomic weights, and the increasing proportion of calcium and iron, elements of considerably greater atomic weights.

The list as a whole is conveniently divided into two major groups of silicate minerals: a *felsic group*, consisting of quartz and the feldspars, and a *mafic group*, consisting of the silicates rich in magnesium and iron. The coined word "felsic" is easily recognized as a combination of *fel*, for feldspar, and *si* for silica, while the syllable *ma* in "mafic" stands for magnesium, the letter *f* for iron (symbol, *Fe*). The felsic minerals are light in color, as well as of comparatively low density; the mafic minerals are dark in color and of comparatively high density.

Two important mafic minerals that are not silicates occur in many igneous rocks. These are *magnetite*, an iron oxide (Fe_3O_4) and *ilmenite*, an oxide of iron and titanium ($Fe\,TiO_3$). Both of these minerals are black and have high densities —$4\frac{1}{2}$ to $5\frac{1}{2}$ gm/cc.

Keep in mind that the minerals listed are important as sources of ions of primary environmental significance in the life layer. Ions in seawater include silicon, calcium, sodium, potassium, and magnesium. The last four named are major constituents of sea salts (see Table 7.1). Bases important in soils include ions of calcium, sodium, potassium, and magnesium; they are also the principal bases found in stream water derived from areas of igneous rocks. Oxides of aluminum and iron, derived from chemical alteration of the silicate minerals, are important components of soils and provide essential ores of iron and aluminum. From the geologic standpoint, the silicate minerals can be viewed as the basic materials out of which

[1] *Density* of a substance is defined as the *mass* of matter per unit of *volume*. The unit of mass is the gram; the unit of volume is the cubic centimeter. Density of pure water is very close to 1.0 gm/cc.

other rock groups (sedimentary and metamorphic) are created.

Silicate magmas and volatiles

About 99 percent of the bulk of the igneous rocks of the earth's crust consists of the seven silicate minerals or mineral groups listed in Figure 22.1. The remainder consists of minerals of secondary importance in bulk, although their number is very large. Fortunately, the eight silicate minerals or groups combine to form only a dozen or so igneous rock varieties having widespread occurrence. We shall simplify the list to only five representative rock types, with the purpose in mind of illustrating the concepts behind the formation of igneous rocks.

Igneous rocks are derived from *silicate magmas*, formed at considerable depths in the earth under conditions of very high temperatures and pressures. Since these magmas cannot be studied under the conditions of their occurrence at depth, inferences as to their properties depend upon laboratory studies in which minerals are melted and allowed to solidify, and upon direct observations of magmas issuing from active volcanoes. At a depth of several miles silicate magmas probably have temperatures in the range from 900° to 2200°F (500° to 1200°C) and are under pressures 6000 to 12,000 times as great as atmospheric pressure at sea level.

Very important in the behavior of magmas is the presence of substances over and above the elements that comprise the solidified silicate rock. These substances are known as *volatiles*, because they remain in a gaseous or liquid state at much lower temperatures than the silicate compounds. Consequently, the volatiles are separated from the magma as it cools and solidifies into a crystalline state. From analysis of gas samples collected from volcanic emissions, we know that water is the preponderant constituent of the volatile group. Table 22.2 lists the volatiles found in gases emanating from magma of active Hawaiian volcanoes. For comparison, the table lists the proportions of these volatiles in the atmosphere and hydrosphere. Notice that the abundances of the various constituents are of the same general order of magnitude in both lists.

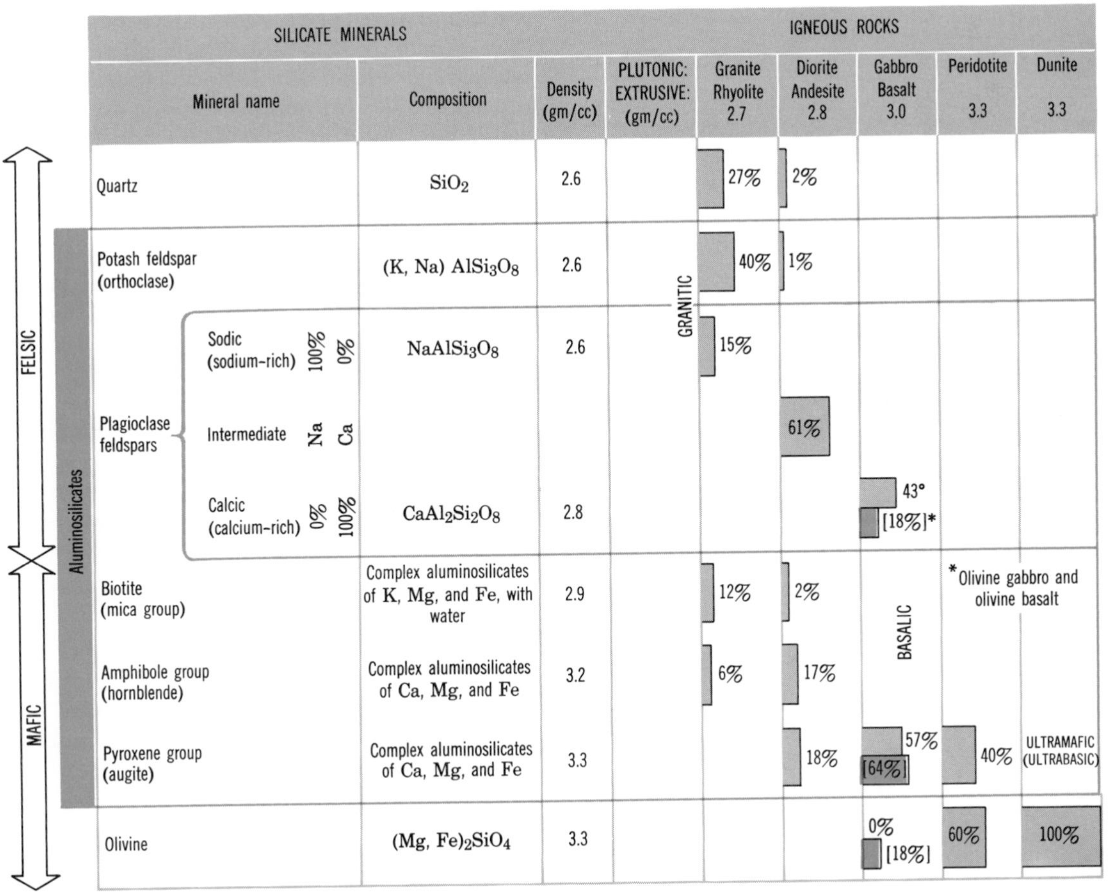

Figure 22.1 Simplified chart of common silicate minerals and abundant igneous rocks. (From A. N. Strahler, 1972, *Planet Earth; Its Physical Systems Through Geologic Time*, Harper and Row, New York.)

TABLE 22.2 VOLATILES IN GASES OF MAGMAS COMPARED WITH VOLATILES IN ATMOSPHERE AND HYDROSPHERE[a]

	Lava Gases from Mauna Loa and Kilauea Volcanoes (Percent by Weight)	Volatiles Free in Earth's Atmosphere and Hydrosphere (Percent by Weight)
Water, H_2O	60	93
Carbon, as CO_2 gas	24	5.1
Sulfur, S_2	13	0.13
Nitrogen, N_2	5.7	0.24
Argon, A	0.3	Trace
Chlorine, Cl_2	0.1	1.7
Hydrogen, H_2	0.04	0.07
Fluorine, F_2	—	Trace

[a]Data from W. W. Rubey, 1952, *Geol. Soc. Amer., Bull.*, vol. 62, p. 1137. Figures have been rounded off.

Figure 22.2 Seen close up, this granite, a coarse-grained intrusive rock, proves to be made of tightly interlocking crystals of a few kinds of minerals.

The emanation of volatiles from the earth's crust is called *outgassing* and is the source of free water of the earth's hydrosphere as well as of the atmospheric gases carbon dioxide, nitrogen, argon, and hydrogen. Chlorine and sulfur compounds of seawater have derived their chlorine and sulfur from outgassing. We can conclude, then, that silicate magmas, with their enclosed volatiles, have through geologic time supplied almost all of the essential components of the atmosphere, hydrosphere, and lithosphere. As life evolved, its processes and forms were adapted to the chemical ingredients supplied by fresh magmas and volatiles, and not from an original lithosphere and atmosphere inherited from the time of accretion of the earth (about 4½ billion years ago). But, as life evolved and spread in the shallow seas and later over the lands, organic processes began to change the composition of atmosphere, the dissolved solids in the oceans, and the mineral deposits on the sea floors. Thus, molecular oxygen (O_2), which is not found in gases of magmas, was released by photosynthesis of plants and eventually built up to a large volume (21 percent) of the atmosphere. The much lower proportion of carbon (as CO_2) in the hydrosphere today than in magmatic gases largely reflects storage of carbon as a mineral compound in solid layers on the bottom of the sea, and to a lesser degree as hydrocarbon compounds in fuels such as coal and petroleum contained in rocks. Sulfur has similarly been removed from the atmosphere and hydrosphere and placed in storage in rocks. The important point is that processes of the biosphere, or life layer, have contributed to an inorganic evolution of the earth's crust.

The presence of water in silicate magmas is most important in the physical behavior of the magma as it works its way from the place of origin to shallower levels in the crust. Even a small amount of water greatly lowers the minimum temperature at which the magma can remain in the liquid state. Consequently, as the magma rises and becomes progressively cooler, the presence of water allows the magma to reach positions much shallower in depth than would otherwise be possible. Extended fluidity of the magma allows it to reach the surface in places, where it can pour out in enormous volumes in fluid layers and tongues as *lava*.

At a certain critical temperature, in combination with a critical pressure, crystallization begins to take place in a slowly cooling magma. Through a complex series of interactions the eight elements listed in Table 22.1 are gathered into compounds as individual crystals of the various silicate minerals. The igneous rocks resulting from crystallization of magmas differ greatly in mineral components and their proportions, either because the original magmas had different compositions, or because mineral groups within a common magma became segregated during crystallization.

The igneous rocks

Before going further with classification of igneous rocks, it will be necessary to recognize that they are classified not only on the basis of mineral composition, but also upon grades of sizes of the component crystals. The term *texture* applies both to the grain size and to the arrangements of crystals of various sizes.

Generally speaking, very gradual cooling of a magma that lies deep within the crust results in formation of large mineral crystals and results in a rock with *coarse-grained* texture. Huge masses

Figure 22.3 A frothy, gaseous lava solidifies into a light, porous scoria (left). Rapidly cooled lava may form a dark volcanic glass (right).

of coarse-grained igneous rock, solidifying at depth in the crust are described by the adjective *plutonic.* Granite is a typical coarse-grained plutonic rock (Figure 22.2). In contrast, rapid cooling of magma as it comes close to the surface or pours out as lava gives very small mineral crystals, or *fine-grained* texture. The lavas, which are classed as *extrusive* in contrast to the *intrusive* rocks solidified beneath the surface, typically are formed of crystals too small to be distinguished with the unaided eye, or even with aid of a magnifying lens. Sudden cooling of magma yields a natural volcanic glass, known as *obsidian* (Figure 22.3 right), while the frothing of a magma as its gases expand near the surface gives a porous rock, known as *scoria* (Figure 22.3 left). Forms taken by various kinds of igneous rock masses are pictured in Figure 30.1.

To simplify rock classification, we shall recognize five rock varieties according to mineral composition, and for each of these classes there will be coarse-grained, plutonic varieties as well as fine-grained or glassy extrusive varieties. As shown in the right half of Figure 22.1, there are five named plutonic varieties and three extrusive varieties (the extrusive varieties of the last two on the list are unimportant). This boils down the list of rocks named to seven.

Bars of varying width, with attached percentage figures, show the typical mineral compositions of these igneous rocks. *Granite* and its extrusive equivalent, *rhyolite,* are rich in quartz and potash feldspar, with lesser amounts of sodic plagioclase, biotite, and amphibole. *Diorite* and its extrusive equivalent, *andesite,* are almost totally lacking in quartz and potash feldspar, but consist dominantly of intermediate plagioclase and lesser amounts of the mafic minerals. Going progressively in the direction of domination by mafic minerals, we come to *gabbro,* and its lava equivalent, *basalt.* Here the plagioclase feldspar is of the calcic type, making up about half the rock, while pyroxene makes up the other half. In a common variety of gabbro and basalt olivine is present at the expense of part of the feldspar. The next rock, *peridotite* is not abundant in the crust, but probably makes up the bulk of the next lower layer, or mantle. It is composed mostly of pyroxene and olivine. Finally, we list *dunite* a rare rock of nearly 100 percent olivine as an example of the extreme mafic end of the mineral series.

Looking back over the five igneous rock classes defined by composition, granite and diorite and related types rich in felsic minerals can be collectively described as *granitic* rocks, while gabbro and basalt and related types can be described as *basaltic* rocks; the extreme mafic types can be described as *ultramafic* rocks.

Densities of the igneous rocks are proportional to the densities of the component minerals. Thus granite has a density of about 2.7; gabbro and basalt, about 3.0; peridotite and dunite, 3.3. Reasoning that geologic processes tend to cause the gross arrangement of igneous rocks to be in layers in order of their densities, granitic rocks will be found in the top layer of the crust, basaltic rocks in the next lower layer, and ultramafic rocks at the bottom. As we shall find in Chapter 23, this proves to be a correct deduction as a first order of generalization.

Mineral alteration

At the interface of the atmosphere and hydrosphere with the lithosphere, oxygen and water are brought into contact with igneous rock. New minerals are formed—compounds that are more stable in the low temperatures and pressures of the surface environment than are the igneous minerals formed at high temperatures and pressures. In this process, known as *mineral alteration,* the solid rock is softened and fragmented, yielding particles of many sizes. When transported in a fluid medium—air, water, or ice—these particles are known collectively as *sediment.* Used in its broadest sense, the term sediment includes both inorganic and organic matter. Dissolved mineral matter in the form of ions in solution may also be included.

Streams carry sediment to lower levels and to locations where accumulation is possible. (Wind and glacial ice also transport sediment, but not necessarily to lower elevations or to places suitable for accumulation.) Usually these sites of accumulation are in shallow seas bordering the continent, but they may also be inland seas and lakes. Thick accumulations of sediment may become deeply buried. Over long spans of time the sediments undergo physical or chemical changes, or both, to become compacted and hardened, producing *sedimentary rock.* The process of compaction and hardening is referred to as *lithification. Diagenesis* is a more general term for complex changes with time, usually involving chemical change and mineral replacement.

Sediments and sedimentary rocks contain essential resources for man's industrial society. First, they provide materials needed to build structures such as highways and buildings. Second, they supply compounds for industrial processes that yield chemicals such as fertilizers and acids. Third, they are the source of hydrocarbon compounds—coal, petroleum, and natural gas—that turn the industrial wheels and provide us with transportation and heat.

The surface environment of the lands is poorly suited to the preservation of a plutonic rock formed under conditions of high pressure and high temperature. Most silicate minerals do not last long, geologically speaking, in the low temperatures and pressures of atmospheric exposure, particularly since free oxygen, carbon dioxide, and water are abundantly available. Rock surfaces are also acted upon by physical forces of disintegration, tending to break up the igneous rock into small fragments and to separate the component minerals, grain from grain. Fragmentation is essential for the chemical reactions of rock alteration, since it results in a great increase in mineral surface area exposed to chemically active solutions. The processes of physical disintegration of rocks are discussed in Chapter 24. Our concern here is with chemical aspects of mineral alteration as a process leading to production of sediment.

Returning now to mineral alteration, the presence of dissolved oxygen in water in contact with mineral surfaces in the soil leads to *oxidation,* which is the combination of oxygen ions with metallic ions, such as calcium, sodium, potassium, magnesium, and iron, abundant in the silicate minerals. At the same time, CO_2 in solution forms a weak acid, *carbonic acid,* capable of reaction with certain minerals. In addition, where decaying vegetation is present, soil water contains complex organic acids, capable of interaction with mineral compounds. Certain common minerals, such as rock salt (NaCl) dissolve directly in water, but simple solution is not particularly effective for the silicate minerals.

Water itself combines with certain mineral compounds in a reaction known as *hydrolysis.* This process is not merely a soaking or wetting of the mineral, but a true chemical change producing a different compound and a different mineral. The reaction is not readily reversible under atmospheric conditions, so that the products of hydrolysis are stable and long-lasting, as are the products of oxidation. In other words, these changes represent a permanent adjustment of mineral matter to a new environment of pressures and temperatures.

TABLE 22.3 COMMON ALTERATION PRODUCTS OF SILICATE MINERALS

Parent Silicate Minerals	Alteration Products	
Potash feldspar and the plagioclase feldspars	Kaolinite and Bauxite	Hydrous aluminosilicates
Muscovite mica and feldspars	Illite	Hydrous aluminosilicate of potassium
Feldspars, mafic minerals, volcanic ash (glass of silicate composition)	Montmorillonite	Complex hydrous aluminosilicate of iron, magnesium, and sodium
Mafic minerals (biotite, amphiboles, pyroxenes, olivine)	Limonite	Hydrous oxide of iron

Table 22.3 lists some important alteration products of the common silicate minerals and mineral groups. Certain of these products are clay minerals. A *clay mineral* is one that has plastic properties when moist, because it consists of minute thin flakes lubricated by layers of water molecules. Potash feldspar undergoes hydrolysis to become *kaolinite,* a white mineral with a greasy feel. Kaolinite becomes plastic when moistened. It is an important ceramic mineral used to make chinaware, porcelain, and tile.

Bauxite is an important alteration product of feldspars, occurring typically in warm climates of tropical and equatorial zones where rainfall is abundant year-round or in a rainy season. Bauxite is actually a mixture of minerals, the dominant constituent being *diaspore,* with the formula $Al_2O_3 \cdot 2\,H_2O$. The presence of *sesquioxide of aluminum,* Al_2O_3, shows that full oxidation has taken place, yielding an unusually stable compound. Unlike kaolinite, which is a true clay with plastic properties, bauxite forms massive rocklike layers below the soil surface. *Laterite* is the name given to this material (see Chapter 18).

Another important clay mineral is *illite,* formed as an alteration product of feldspars and muscovite mica. Illite is a hydrous aluminosilicate of potassium and is an abundant mineral in sedimentary rocks. It occurs as minute thin flakes of colloidal dimensions and is carried long distances in streams (Figure 22.4). *Montmorillonite* is another common clay mineral (more correctly a group of minerals) derived from alteration of feldspar, mafic minerals, or volcanic ash (Figure 22.4).

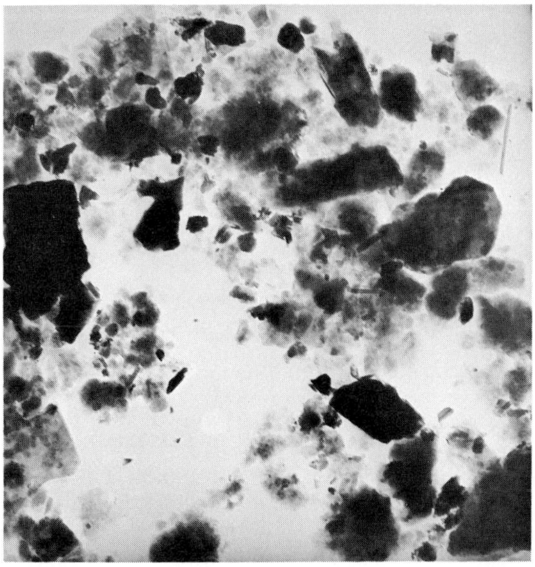

Figure 22.4 **Seen here enlarged about 20,000 times are tiny flakes of the clay minerals *illite* (sharp outlines) and *montmorillonite* (fuzzy outlines). These particles have settled from suspension in San Francisco Bay. (Photograph by Harry Gold. Courtesy of R. B. Krone, San Francisco District Corps of Engineers, U.S. Army.)**

A most important alteration product of the mafic minerals is *limonite*, a hydrous iron compound with the formula $2\,Fe_2O_3 \cdot H_2O$. *Iron sesquioxide* (Fe_2O_3) in limonite is a stable form of iron oxide and is found widely distributed in rocks and soils. It is closely associated with bauxite in laterites of low latitudes. Limonite supplies the typical reddish to chocolate-brown colors of soils and rocks. Some shallow accumulations of limonite were formerly mined as a source of iron.

Silicate minerals differ in their susceptibility to alteration. Olivine is the most susceptible to alteration, followed by the pyroxenes, amphiboles, biotite, and sodic plagioclase feldspar. Potash feldspars are somewhat less susceptible. Muscovite mica is comparatively resistant to alteration, while quartz is immune to hydrolysis and to further oxidation. As a result of the differing susceptibility of silicate minerals to alteration, the igneous rocks also reflect these differences. The mafic rocks are the more easily decomposed, whereas the felsic rocks are the more resistant.

Size grades of sediment particles

Besides mineral composition, the other essential parameter in classifying sediments is particle size, including mixtures of sizes, which relates to the general aspect of rock texture. Table 22.4 is an abbreviated list of size grades, giving the names and dimensional ranges of each grade. The full scale, including subdivisions of each grade into

TABLE 22.4 SIMPLIFIED WENTWORTH SCALE OF GRADE SIZES

	Limits		
Grade Name	Millimeters	Inches	Microns
Boulders			
	256	10	
Cobbles			
	64	2.5	
Pebbles			
	2	0.08	
Sand			
	0.0625		62
Silt			
	0.004		4
Clay (noncolloidal)			
	0.001		1
Clay (colloidal)	down to 0.001 micron		

coarse, medium, and fine categories, is known as the *Wentworth scale*, and is of standard use in geology. Soil scientists use a somewhat different grade scale (see Table 18.1). Clay can be subdivided with respect to its behavior. Particles below about 1 micron (0.001 mm) behave as *colloids*, particles that remain indefinitely in aqueous suspension. Because most natural colloidal clays are thin, platelike bodies or irregularly shaped objects, rather than spherical, grain diameter is not a particularly meaningful term to apply to these very fine clays.

Clastic sediments and sedimentary rocks

The environmental importance of sediments is widespread in the life layer. Sediment (including the soil) provides the physical base, or *substrate*, for practically all forms of life, at least in one phase or another of their life cycles or in food-gathering practices. Water-saturated sands, silts, clays, and muds form the life environments of thousands of species of aquatic organisms under the shallow seas, in tidal estuaries, in lake bottoms and stream beds, and in swamps and marshes. A principle of paramount importance is that organisms modify the sediment in which they live and feed; they also create sediment through life processes, for example, as shells and skeletons. Consequently, a major class of sediments is organically derived, in contrast to the chemically derived products of rock alteration and the physically derived particles of rock disintegration.

Clastic and nonclastic are the two major divisions of sediments. *Clastic sediments* are those derived directly as particles broken from a parent rock source, in contrast to the *nonclastic sediments*, which are of newly created mineral matter precipitated from chemical solutions or from organic activity. The clastic sediments include particles that issue directly from volcanoes in the form of cinder, ash, and volcanic dust. These sediments are described in Chapter 30.

Clastic sediments are derived from any one of the rock groups—igneous, sedimentary, and metamorphic—which gives a very wide range of parent minerals. One sediment source is from the silicate minerals and the alteration products of those minerals. Because the highly susceptible minerals—mostly mafic—are altered prior to transportation, whereas quartz is immune to such alteration, the most important single component of the clastic sediments generally is quartz (Figure 22.5). Second in abundance are fragments of unaltered fine-grained parent rocks. Feldspar and mica are also commonly present. Clay minerals, particularly kaolinite, illite, and montmorillonite, are major constituents of the very fine clastic sediments.

Also important in the clastic sediments are the *heavy minerals.* By "heavy" is meant a mineral of high density. Two good examples of heavy minerals are magnetite, with a density of about 5, and ilmenite; both were mentioned earlier in this chapter. The heavy minerals are derived from any one of a variety of preexisting rocks. They are extremely resistant to abrasion, and, like quartz, travel long distances in streams and along beaches. Concentrations of these minerals can be seen in most beach sands and stream bars. A magnet dragged through the dry sand will accumulate a thick coating of the magnetite grains. While the heavy minerals are not of direct importance to life processes, they are locally significant in large concentrations as metallic mineral sources. For example, titanium is obtained from ilmenite. Another example is the heavy mineral *cassiterite,* an oxide of tin (SnO_2), which is mined from concentrations in gravel, where it is known as *stream-tin.*

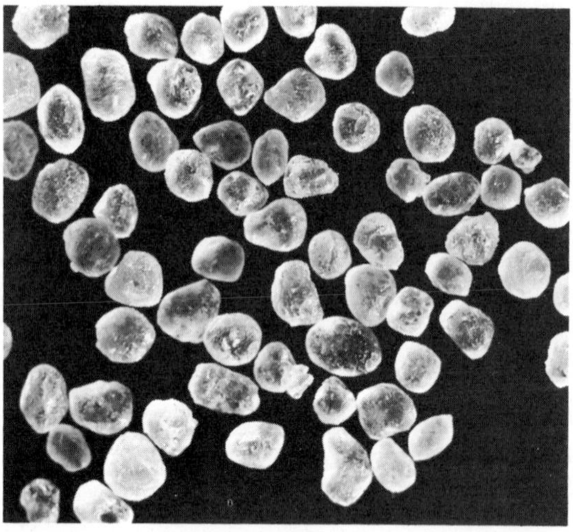

Figure 22.5 Rounded quartz grains from an ancient sandstone. The grains average about 1 mm (0.04 in) in diameter. (Photograph by Andrew McIntyre, Columbia University.)

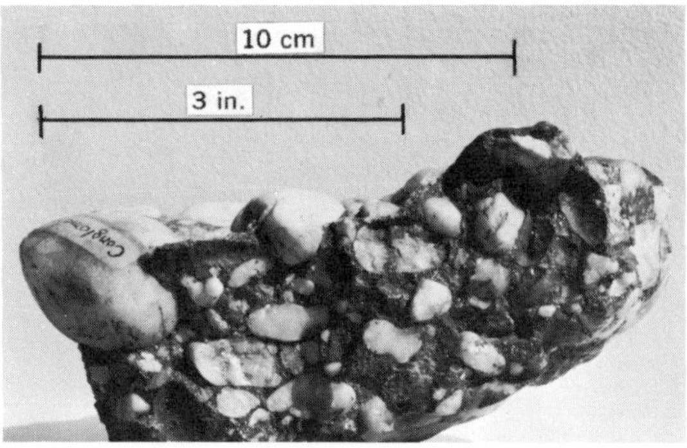

Figure 22.6 Conglomerate is a mixture of pebbles and sand cemented into a hard rock.

The natural range of particle grades in clastic sediment determines the ease and distance of travel of particles in transport by water currents. Obviously, the finer the particles, the more easily they are held in suspension in the fluid; the coarser particles tend to settle to the bottom of the fluid layer. In this way a separation of grades known as *sorting* occurs and determines the texture of the sediment deposit and of the sedimentary rock derived from that sediment. Colloidal clays do not settle out unless they are made to clot together into larger groups, a process of *flocculation.*

Compaction and cementation of layers of sediment leads to formation of sedimentary rocks. The following are important varieties. *Conglomerate* consists of pebbles or cobbles, usually of a very durable rock, set in a matrix of sand (Figure 22.6). Conglomerate layers represent lithified beaches or stream-bed deposits. *Sandstone* is formed of the sand grades of sediment, cemented into a solid rock by silica (SiO_2) or calcium carbonate ($CaCO_3$). The sand grains are commonly of quartz, as shown in Figure 22.5, and sometimes of feldspar, but in other cases are sand-sized fragments of fine-grained rock containing several minerals. *Siltstone* and *shale* are lithified layers of silt and clay particles, respectively, while mixtures of silt and clay, often with some sand, produce *mudstone.*

Shale is the most abundant of the sedimentary rocks. It is formed largely of the clay minerals, kaolinite, illite, and montmorillonite. The compaction of the original clay and mud involves a considerable loss of volume as water is driven out.

A characteristic feature of sedimentary rocks is their layered arrangement, the layers being collectively called *strata.* Typically, layers of different textural composition are alternated or interlay-

Figure 22.7 These limestone strata in Oklahoma were steeply tilted long after being laid down. (Photograph by Lofman. Courtesy of Standard Oil Co., New Jersey.)

ered. The planes of separation between layers are known as *bedding planes* (see Figure 22.7).

Nonclastic sediments and sedimentary rocks

The nonclastic sediments are of particular importance in the environment of the life layer because they represent enormous storages of carbon, obtained from atmospheric carbon dioxide and changed into carbonate and hydrocarbon compounds by both inorganic and organic processes. We recognize two major divisions of the nonclastic sediments. (1) *Chemical precipitates* are compounds precipitated directly from water in which the ions are transported; these sediments are described as *hydrogenic.* (2) *Organically-derived sediments* are created by the life processes of plants and animals; these are described by the adjective *biogenic.*

Chemical precipitates are of major importance in sediments of the sea floor, but a second important environment of deposition is in inland salt lakes of desert regions, where evaporation exceeds precipitation by a wide margin.

The most important hydrogenic and biogenic minerals (hydrocarbon compounds are not included) are listed in Table 22.5, together with their compositions. The first three—*calcite, aragonite,* and *dolomite*—are classed as *carbonates.* Calcite is the dominant carbonate mineral and occurs in many forms. Aragonite is of the same composition as calcite, but of different crystal structure; it is secreted in the shells of certain invertebrate animals. Dolomite contains magnesium as well as calcium. All three carbonate minerals are soft substances, as compared with the silicate minerals. The second group, the *evaporites,* are typically formed where seawater is evaporated in shallow bays and gulfs. *Anhydrite* and *gypsum* are composed of calcium sulfate, the latter in combination with water. Gypsum is an economically important mineral used in the manufacture of structural materials. The third evaporite, *halite,* is commonly known as rock salt, with the composition sodium chloride (NaCl). In refined form it is the table salt we use in cooking and flavoring.

Hematite, a sesquioxide of iron (Fe_2O_3) is a common mineral in sedimentary rocks; it is a major ore of iron. No less important than the carbonates and evaporites is *chalcedony,* the mineral name for silica (SiO_2) in a very fine-grained crystalline form. Chemically, it is not different from the mineral quartz.

The minerals listed in Table 22.5 can be chemically precipitated from seawater, or the water of saline lakes, or can be secreted by organisms to produce the nonclastic sedimentary rocks. Most common of the *carbonate rocks* is *limestone,* of which there are many varieties (Figure 22.7). One source is in reefs built by corals and algae; another form is *chalk,* made up of the skeletons of a marine form of the algae. Other limestones are formed of shell fragments or other broken carbonate matter. Some limestones are densely crystalline. *Dolomite,* a rock of the same name as the mineral that composes it, may have been derived through the alteration of limestone, as magnesium ions of seawater gradually replaced calcium ions.

TABLE 22.5 COMMON HYDROGENIC AND BIOGENIC MINERALS

	Mineral Name	Composition	Density gm/cc
CARBONATES	Calcite	Calcium carbonate $CaCO_3$	2.72
	Aragonite	Calcium carbonate $CaCO_3$	2.9–3
	Dolomite	Calcium-magnesium carbonate $CaMg(CO_3)_2$	2.9
EVAPORITES	Anhydrite	Calcium sulfate $CaSO_4$	2.7–3
	Gypsum	Hydrous calcium sulfate $CaSO_4 \cdot 2\ H_2O$	2
	Halite	Sodium chloride NaCl	2.1–2.3
	Hematite	Sesquioxide of iron (ferric) Fe_2O_3	4.9–5.3
	Chalcedony (chert, flint)	Silica SiO_2	2.6

Chert, composed of chalcedony, is an important siliceous sedimentary rock. It occurs as nodules in limestone, and in some cases as solid rock layers referred to as *bedded cherts*. *Gypsum*, *anhydrite*, and *halite* are layered rocks of their respective mineral compositions and occur in association wyth clastic sedimentary rocks.

Hydrocarbon compounds in sedimentary rocks

Hydrocarbon compounds form a second group of biogenic sediments. These organic substances occur both as solids (peat and coal) and as liquids and gases (petroleum and natural gas) but only coal qualifies physically for designation as a rock.

Peat, a soft, fibrous substance of brown to black color, accumulates in a bog environment where the continual presence of water inhibits decay and oxidation of plant remains. One form of peat is of fresh water origin and represents the filling of shallow lakes (Chapter 20). Thousands of such peat bogs are found in North America and Europe; they occur in depressions remaining after recession of the great ice sheets of the Pleistocene Epoch (Chapter 31). This peat has been used for centuries as a low-grade fuel. Peat of a different sort is formed in the salt-water environment of tidal marshes (Chapter 32).

At various times and places in the geologic past, conditions were favorable for the large-scale accumulation of plant remains, accompanied by subsidence of the area and burial of the compacted organic matter under thick layers of inorganic sediments. Thus *coal seams* interbedded with shale, sandstone, and limestone strata came into existence (Figure 22.8). Groups of strata containing coal seams are referred to as *coal measures;* the individual seams 'range in thickness from a fraction of an inch to as great as 50 ft (12 m) in the exceptional case. Coal consists mostly of compounds of carbon, hydrogen, and oxygen, and usually contains a small proportion of sulfur compounds. Water and hydrocarbon volatiles make up a large part of lower-grade coals, but fixed carbon is the principal constituent of the highest grades of coal.

Figure 22.8 A coal seam, 8 ft (2.4 m) thick exposed in a river bank, Dawson County, Montana. (Photograph by M. R. Campbell, U.S. Geological Survey.)

Petroleum, or *crude oil*, as the liquid form is often called, includes many hydrocarbon compounds. Typically, a sample of crude oil might consist of 82 percent carbon, 15 percent hydrogen, and 3 percent oxygen and nitrogen. *Natural gas*, found in close association with accumulations of petroleum, is a mixture of gases. The principal gas is *methane* (*marsh gas*), CH_4, and there are minor amounts of ethane, propane, and butane, all of which are hydrocarbon compounds. Small amounts of carbon dioxide, nitrogen, and oxygen are also present, and sometimes helium. Petroleum and natural gas are not classed as minerals, but they originated as organic compounds in sediments. The occurrence of accumulations of petroleum and natural gas in sedimentary strata is explained in Chapter 28.

While it is generally agreed that petroleum and natural gas are of organic origin, the nature of the process is hypothetical. A favored explanation for petroleum is that the oil originated within microscopic floating marine plants (phytoplankton) such as the *diatoms*. As each diatom died, it released a minute droplet of oil, which became enclosed in muddy bottom sediment. Eventually, the mud became a shale formation in which the oil was disseminated. Today we have *oil shales* holding petroleum in a dispersed state, and these give support to the organic hypothesis we have stated. However, the occurrence of petroleum in heavy concentrations in porous rock such as sandstone requires that the oil in some manner was forced to migrate from its source region to a reservoir rock. The volatile hydrocarbons and other gases were then segregated to a position above the liquid petroleum.

From the standpoint of man's energy resources, the outstanding concept relating to the hydrocarbon compounds within the earth—*fossil fuels*, they are collectively called—is that they have required hundreds of millions of years to accumulate, whereas they are being consumed at a prodigious rate by our industrial society. These fuels are nonrenewable resources. Once they are gone there will be no more, since the quantity produced in a thousand years by geologic processes is scarcely measurable in comparison to the quantity stored through geologic time.

Figure 22.9 This fragment of schist, 6 in (15 cm) long, has a glistening, undulating surface (above) consisting largely of mica flakes. An edgewise view (below) of the same specimen shows the wavy foliation planes.

Metamorphic rocks

Any of the types of igneous or sedimentary rocks may be altered by the tremendous pressures and high temperatures that accompany mountain-building movements of the earth's crust. The result is a rock so greatly changed in appearance and composition as to be classified as a *metamorphic rock.* Generally speaking, metamorphic rocks are harder and more compact than their original types, except when the latter are igneous rocks. Moreover, the kneading action and baking that metamorphic rocks have undergone produces new structures and even new minerals. Each sedimentary and igneous rock has an equivalent metamorphic rock. The term *metasediment* conveniently covers all metamorphic rocks derived from sedimentary strata.

Shale, on being squeezed and sheared under mountain-making forces, turns into *slate,* a gray or brick-red rock that splits neatly into thin plates so familiar to all as roofing shingles and flagstones of patios and walks. The planes of splitting form a structure called *slaty cleavage.* This is a new structure imposed upon the rock during the process of internal slippage during metamorphism, not merely stratification or bedding. Slate is fine textured and of rather dull surface texture.

Figure 22.10 Gneiss shows a banded surface.

With continued application of pressure and internal shearing, slate changes into *schist,* the most advanced grade of metamorphic rock. Schist has a structure termed *foliation,* consisting of thin but rough and irregularly curved planes of parting in the rock. Schist is set apart from slate and phyllite by the coarse texture of the mineral grains, the abundance of mica, and the presence of scattered large crystals of new minerals such as *garnet* and *staurolite,* which have grown during the process of internal shearing of the rock (Figure 22.9).

The metamorphic equivalent of conglomerate, sandstone, and siltstone is *metaquartzite,* formed by addition of silica (SiO_2) to fill completely the interstices between grains, most of which are normally quartz (also silica). This process is carried out by the slow movement of underground waters carrying the silica into the sandstone, where it is deposited. Pressure and kneading of the rock is not essential in producing a quartzite, but may deform the quartz grains. When a quartzite is fractured, as with a hammer blow, the break will cut across sand grains and pebbles in the rock. In this way, quartzite can be distinguished from a sandstone, which usually breaks around the grains, leaving them mostly intact.

Limestone, upon metamorphism, becomes *marble,* commonly a white rock of sugary texture when freshly broken. During the process of internal shearing, the calcite mineral of the limestone has reformed into larger, more uniform crystals than before. Bedding planes are obscured and masses of mineral impurities are drawn out into swirling streaks and bands.

Finally, the important metamorphic rock, *gneiss,* may be formed either from intrusive igneous rocks, or as a metasediment from strata that have been in close contact with intrusive magmas.

A single description will not fit all gneisses, which vary considerably in appearance, mineral composition, and structure. One common variety is *granite gneiss,* formed directly by flowage of granite in a somewhat plastic state. Granite gneiss resembles granite in its massiveness, general texture, and mineral components, but possesses a streaked appearance, called *lineation,* produced by parallelism of dark minerals which have been drawn out into long, pencil-like shapes in the direction of flow.

Still other gneisses are strongly banded into light and dark layers or lenses (Figure 22.10), which may be contorted into wavy folds. It is possible that in some instances these bands, which have differing mineral compositions, may be the relics of sedimentary strata such as shale and sandstone to which new mineral matter has been added from nearby intrusive rocks.

Metalliferous deposits

Among the nonrenewable mineral resources essential to our modern industrial civilization are *metalliferous deposits*—concentrations of metallic compounds in rocks of the crust. When the degree of concentration of metals is such as to be economically profitable to extract and refine, the deposit is termed an *ore.*[1]

To obtain an appreciation of the scarcity of certain of the essential metals, study the figures given in Table 22.6. The table gives estimated abundance, as percentage by weight, of each metal in average rock of the earth's crust. Notice that whereas aluminum and iron are comparatively abundant elements, many of the industrial metals which we regard as quite commonplace—zinc, copper, and lead, for example—are present in very small percentages in terms of the average rock composition. In the case of mercury, for which there is no adequate substitute at present, the abundance is on the order of about one part in 10 million.

[1]Much of the remainder of this chapter is taken from *Planet Earth; Its Physical Systems Through Geologic Time,* Harper and Row, Publishers, New York. © 1972 by Arthur N. Strahler. Used by permission of the publisher.

TABLE 22.6 SELECTED METALLIC ABUNDANCES IN AVERAGE CRUSTAL ROCK[a]

Symbol	Element Name	Abundance (Percent by Weight)	Annual World Consumption (Tons)[b]
Al	Aluminum	8.1	6,100,000
Fe	Iron	5.0	310,000,000
Mg	Magnesium	2.1	150,000
Ti	Titanium	0.44	10,000
Mn	Manganese	0.10	6,000,000
V	Vanadium	0.014	7,000
Cr	Chromium	0.010	1,400,000
Ni	Nickel	0.0075	400,000
Zn	Zinc	0.0070	3,800,000
Cu	Copper	0.0055	5,400,000
Co	Cobalt	0.0025	13,000
Pb	Lead	0.0013	2,800,000
Sn	Tin	0.00020	190,000
U	Uranium	0.00018	30,000[c]
Mo	Molybdenum	0.00015	45,000
W	Tungsten	0.00015	30,000
Sb	Antimony	0.00002	60,000
Hg	Mercury	0.000008	9,000
Ag	Silver	0.000007	8,000
Pt	Platinum	0.000001	30
Au	Gold	0.0000004	1,600

[a]Data from Brian Mason, 1966, *Principles of Geochemistry,* 3rd. ed., New York, John Wiley & Sons, p. 45–46, Table 3.3 and Appendix III.

[b]New metal only; does not include recycled metal.

[c]As U_3O_8.

Ores represent concentrations of metals in the form of compounds having a high proportion of the metal. Moreover, the compounds must usually form a substantial proportion of the rock itself if it is to qualify as an ore. Take for example, lead, which must comprise approximately 4 percent of the ore for profitable extraction. To have acquired this degree of concentration, lead must be present in approximately 2500 times as great abundance as in the average crustal rock.

Our concern in this chapter is with a brief review of some of the geologic processes by which metals have come to be concentrated into ores. Table 22.7 lists some important ore minerals of industrially important metals. Certain minerals have been described in earlier paragraphs—bauxite, magnetite, hematite, limonite, and ilmenite. Notice that except for the *native metals,* of which *native copper* is an example, all of these ore minerals are either oxides or sulfides. A sulfide mineral consists of one or more metallic elements in combination with sulfur (S). Sulfides are particularly important ore minerals of nickel, zinc, copper, cobalt, lead, mercury, and silver.

The sulfides have considerable environmental importance, quite apart from their value as a resource. They can contribute heavily to air pollution during the smelting process required to drive off the sulfur. Sulfides emitted from the stacks of smelters have caused destruction of vegetation in the area of fallout. Moreover, the percolation of precipitation through abandoned mines and piles of mine wastes (tailings) can pollute streams. (See *acid mine drainage,* Chapter 13.)

Principles of magma crystallization, rock metamorphism, mineral alteration, and deposition of hydrogenic sediments, treated in earlier paragraphs can be put to good use here in a brief sketch of the origin and classification of metalliferous mineral deposits.

One major class of ore deposits is formed within magmas by direct segregation in which mineral grains of greater density sink slowly through the fluid magma while crystallization is still in progress. Masses or layers of a single mineral can accumulate in this way. One example is *chromite,* the principal ore of chromium with a density of about 4.4 (Table 22.7). Bands of chromite ore are sometimes found near the base of the igneous

TABLE 22.7 REPRESENTATIVE METALLIC ORE MINERALS

Metal	Symbol	Mineral Name	Composition
Aluminum	Al	Bauxite (not a single mineral)	Hydrous aluminum oxide
Iron	Fe	Magnetite	$Fe(FeO_2)_2$
		Hematite	Fe_2O_3
		Limonite (not a single mineral)	$Fe_2O_3 \cdot nH_2O$
		Pyrite	FeS_2
Titanium	Ti	Ilmenite	$FeTiO_3$
Manganese	Mn	Pyrolusite	MnO_2
		Manganite	$Mn_2O_3 \cdot H_2O$
Chromium	Cr	Chromite	$FeCr_2O_4$ (with Al, Mg, Fe)
Zinc	Zn	Sphalerite	(SN,Fe)S
Copper	Cu	Native copper	Cu
		Chalcopyrite	$CuFeS_2$
		Chalcocite	Cu_2S
Lead	Pb	Galena	PbS
Tin	Sn	Cassiterite	SnO_2
Molybdenum	Mo	Molybdenite	MoS_2
Tungsten	W	Wolframite	$(Fe,Mn)WO_4$
Uranium	U	Pitchblende (uraninite)	Hydrous uranium oxide
Mercury	Hg	Cinnabar	HgS
Silver	Ag	Native silver	Ag
		Argentite	Ag_2S
Gold	Au	Native gold	Au,Ag'
Platinum	Pt	Platinum	Pt

body. Another example is seen in the nickel ores of Sudbury, Ontario. These sulfides of nickel apparently became segregated from a saucer-shaped magma body and were concentrated in a basal layer (Figure 22.11). *Magnetite* is another ore mineral that has been segregated from a magma to result in an ore body of major importance.

An important process associated with igneous intrusion is *contact metamorphism,* in which the invading magma and the highly active chemical solutions it contains altered the surrounding rock (country rock). In this process, ore minerals were introduced into the country rock in exchange for existing components in the rock. For example, a limestone layer may have been replaced by iron ore consisting of hematite and magnetite (Figure

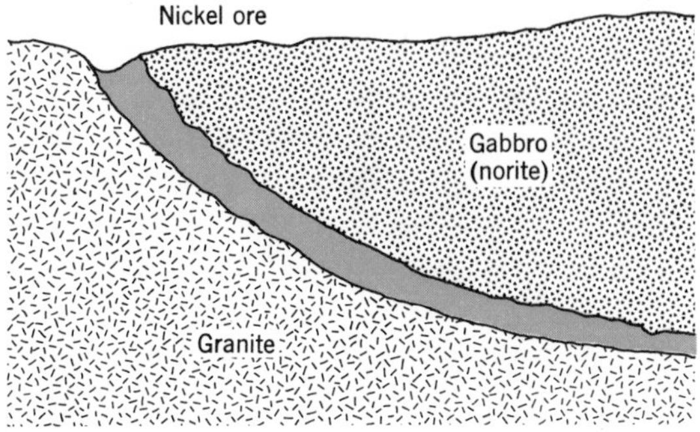

Figure 22.11 Cross section through a nickel ore deposit lying at the base of a body of gabbro and overlying a basement of older granite. (After A. P. Coleman, 1913, Canada Dept. of Mines, *Monograph 170,* p. 34.)

22.12). Ores of copper, zinc, and lead have also been produced in this manner.

Another type of ore deposit is also associated with igneous intrusion. In the final stages of crystallization of a granitic magma there remains a watery magma rich in silica. Under high pressure, this solution leaves the main body of solidifying magma and penetrates the surrounding rock mass in small chambers and narrow passageways to crystallize in the form of *pegmatite* bodies. Pegmatite consists mostly of large mineral crystals. Pegmatite bodies take the form of irregular masses within the parent plutonic rock, and occur as dikes (wall-like bodies), and as veins (Figure 22.13).

While the bulk of all pegmatites consists of quartz, feldspar, micas, and other common minerals of granitic rocks, there occur unusual concentrations of rarer minerals, both metallic and nonmetallic. For example, in certain pegmatites of the Black Hills of South Dakota there occur enormous crystals of the mineral *spodumene,* an aluminosilicate of lithium (Figure 22.14). A single crystal of record size measured 47 ft (14 m) in length and 6 ft (2 m) in diameter; several lesser ones weighed over 30 tons apiece. This and other

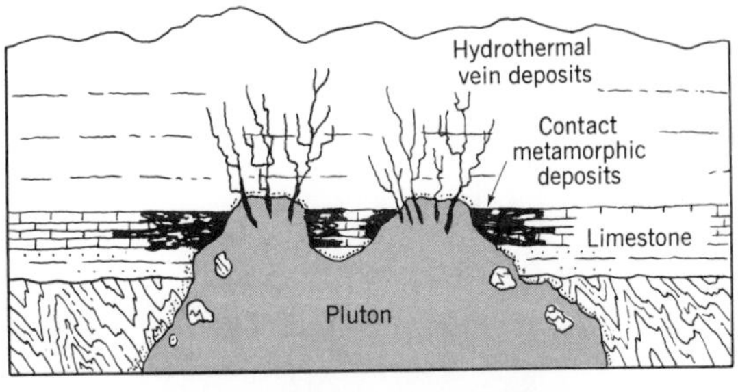

Figure 22.12 Schematic cross section through an intrusive igneous body and the overlying country rock, showing veins and contact metamorphic deposits. (From A. N. Strahler, 1972, *Planet Earth; Its Physical Systems Through Geologic Time,* Harper & Row, New York.)

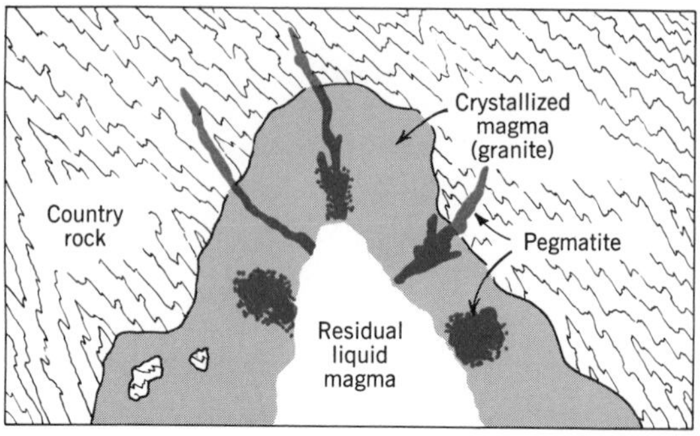

Figure 22.13 Schematic cross section showing the relationship of pegmatite bodies to an intrusive igneous body and its surrounding country rock. (From A. N. Strahler, 1972, *Planet Earth; Its Physical Systems Through Geologic Time*, Harper and Row, New York.)

pegmatite localities are a principal source of the light metal, *lithium*, which has many important uses in industry. Another metal, *beryllium*, is found in pegmatites in the form of the mineral *beryl*, an aluminosilicate of beryllium. Beryllium is an important component in high-strength alloys of copper, cobalt, nickel, and aluminum. Two other metals found in pegmatites, both very rare but essential in industry, are *tantalum* and *columbium*. Because of the large size of pegmatite crystals—from several inches to as much as a few feet in diameter—they are important commerical sources of certain common nonmetallic minerals, principally the feldspars, used in manufacture of pottery, tile, porcelain, and glass. Another example is *muscovite mica*, needed in large sheets and

Figure 22.14 Large spodumene crystals in the Etta pegmatite, Pennington County, South Dakota. The hammer rests upon a single large spodumene crystal. (Photograph by J. J. Norton, U.S. Geological Survey.)

plates for electrical insulation and related uses; it occurs as large sheets only in pegmatites.

Another type of ore deposit is produced by the effects of high-temperature solutions, known as *hydrothermal solutions*, that leave a magma during the final stages of its crystallization and are deposited in fractures to produce mineral veins. Some veins are sharply defined and evidently represent the filling of open cracks with layers of minerals. Other veins seem to be the result of replacement of the country rock by the hydrothermal solution. Where veins occur in exceptional thicknesses, they may constitute a *lode*.

Hydrothermal solutions produce yet another important type of ore accumulation, the *disseminated deposit*, in which the ore is distributed throughout a very large mass of rock. Certain great copper deposits of this type are referred to as *porphyry copper* deposits because the ore has in some cases entered a large body of igneous rocks of a texture class known as a *porphyry*, which had in some manner been shattered into small joint blocks that permitted entry of the solutions. One of the most celebrated of these is at Bingham Canyon, Utah (Figure 22.15).

Hydrothermal solutions rise toward the surface, making vein deposits in a shallow zone and even emerging as hot springs. Many valuable ores of gold and silver are deposits of the shallow type. Particularly interesting is the occurrence of mercury ore in the form of the mineral *cinnabar* (see Table 22.7) as a shallow hydrothermal deposit. Most renowned are the deposits of the Almaden district in Spain, where mercury has been mined for centuries and has provided most of the world's supply of that metal.

A quite different category of ore deposits embodies the effects of downward moving solutions in the zone of aeration and the ground water zone (see Chapter 13). Enrichment of mineral deposits to produce ores in this manner is described as a *secondary* process. Consider first a vein containing primary minerals of magmatic origin (Figure 22.16). These minerals, mostly sulfides of copper, lead, zinc, and silver, along with native gold, are originally disseminated through the vein rock and may not exist in concentrations sufficient to qualify as ores. (See Table 22.7 for sulfide minerals associated with each metal.) Through long continued denudation of the region, the ground surface truncates the vein, which was formerly deeply buried. Assuming a humid climate, there will exist a water table and a ground water zone, above which is the zone of aeration. Water, arriving as rain or snowmelt, moves down through the zone of aeration. The geologist refers to this water as *meteoric*, which is perfectly acceptable from the standpoint of atmospheric science. The meteoric water becomes a weak acid, since it contains dissolved carbon dioxide (carbonic acid) and will also gain sulfuric acid by reactions involving iron sulfide (mineral *pyrite*, Table 22.7).

The result of downward percolation of meteoric water is to cause three forms of enrichment and thus to yield ore bodies. First, in the zone closest to the surface, as soluble waste minerals are removed, there may accumulate certain insoluble minerals, among them gold and compounds of silver or lead, in sufficient concentration to form an ore. This type of ore deposit is known as a *gossan* (Figure 22.16). Iron oxide and quartz will also accumulate in the gossan. In colonial times, iron-rich gossans constituted minable iron ores, but they have been exhausted. Leaching of other minerals carries them down into a *zone of oxidation*. In this second zone there may accumulate a number of oxides of zinc, copper, iron, and lead, along with native silver, copper, and gold. A third zone is that of *sulfide enrichment* within the upper part of the ground water zone, just beneath the water table (Figure 22.16). Sulfides of iron, copper, lead, and zinc may be heavily concentrated in this zone. (Mineral examples are pyrite, chalcopyrite, chalcocite, galena, and sphalerite, described in Table 22.7.) Sulfide enrichment may also effect large primary ore bodies of the disseminated type, such as the porphyry copper of Bingham Canyon, Utah, referred to previously. Here, the enriched layer has already been removed and mining has progressed into low-grade primary ore beneath.

Also in the category of secondary ores is *bauxite* (Table 22.7). We have already explained how this principal ore of aluminum accumulates as a near-surface deposit in warm, wet low-latitude zones where the soil-forming regime of laterization prevails (Chapter 18). Bauxite, a mixture of hydrous oxides of aluminum derived from the alteration of aluminosilicate minerals, is practically insoluble under the prevailing climatic conditions and can accumulate indefinitely as the denudation of the land surface progresses. Produced under similar environmental conditions are residual ores of manganese (mineral: *manganite*), and of iron (mineral: *limonite*) (Table 22.7). The geologist applies the term *laterite* to these residual deposits.

Another category of ore deposit is that in which concentration has occurred through fluid agents of transportation: streams and waves. Certain of the insoluble heavy minerals derived from weathering of rock are swept as small fragments into stream channels and carried downvalley with the sand and gravel as bed materials. Because of their greater density, these heavy minerals become concentrated in layers and lenses of gravel to become *placer deposits*. Native gold is one of the minerals extensively extracted from placer deposits; platinum is another. A third is an oxide of tin, the mineral *cassiterite* (Table 22.7), which forms important placer deposits. Diamonds, too, are concentrated in placer deposits, as are certain other gem stones. Transported by streams to the ocean, gravels bearing the heavy minerals are spread along the coast in beaches, forming a second type of placer deposits, the *marine placers*.

Figure 22.15 Open-pit copper mine at Bingham Canyon, Utah. (Photograph by courtesy of Kennecott Copper Corporation.)

Finally, we can recognize a group of ore deposits in the hydrogenic category of sediments. For the most part, sediment deposition is the principal source of nonmetallic mineral deposits, but some important metalliferous deposits are of this origin. Iron, particularly, occurs as sedimentary ores in enormous quantities. Sedimentary iron ores are oxides of iron—usually *hematite* (Table 22.7). A particularly striking example is iron ore of the Clinton formation of Silurian age, widespread in

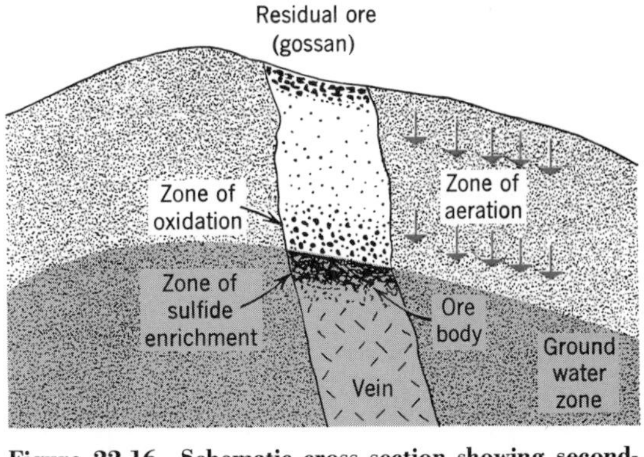

Figure 22.16 Schematic cross section showing secondary ore deposits formed by enrichment of minerals of a vein. (From A. N. Strahler, 1972, *Planet Earth; Its Physical Systems Through Geologic Time*, Harper and Row, New York.)

the Appalachian region. For reasons not well understood, unusually large quantities of iron oxides, derived by weathering of mafic minerals in rocks exposed in bordering lands, were brought to the sea floor and were precipitated as hematite. Another metal, manganese, has been concentrated by depositional processes into important sedimentary ores.

As in the case of the fossil fuels, man is rapidly consuming metallic earth resources that required geologic spans of time to be created and concentrated. So slowly do the geological processes operate that rates of replenishment are infinitesimally small in comparison with the present rates of consumption. Mineral resources are therefore finite, and once we know approximately the world extent of a particular resource, we can predict its expiration according to a given use schedule.

REVIEW QUESTIONS

1. Describe the environmental role played by minerals and rocks of the earth's crust.

2. Name the first eight most abundant elements in the earth's crust. What environmental or industrial importance is attached to each?

3. Name the important rock-forming silicate minerals and give an approximate chemical composition for each. Which are felsic? Which are mafic?

4. What are the physical properties of a magma? What are the volatile constituents of a magma? Why are they important in the environment?

5. What are igneous rocks? How are they classified by texture? Name and describe the important varieties of igneous rocks. Which are felsic? Which are mafic? Which are ultramafic? How do their densities compare?

6. Describe the basic processes of mineral alteration. What is sediment and how is it formed? How are sedimentary rocks formed?

7. What are the clay minerals? Name and describe three clay minerals. What is the environmental and economic importance of the products of mineral alteration?

8. What is the Wentworth scale? Describe colloids produced by mineral alteration.

9. Differentiate between the clastic sediments and the nonclastic sediments. Which minerals are abundant in clastic sediments? What are the heavy minerals? Name two.

10. Name and describe the important clastic sedimentary rocks. What structures do they acquire during deposition?

11. Into what two major groups are the nonclastic sediments divided? Name the important hydrogenic minerals. Under what circumstances are they deposited?

12. What are the common biogenic minerals? How are they formed? What sedimentary rocks are composed of biogenic minerals?

13. What hydrocarbon compounds are found in sedimentary rocks? Distinguish peat from coal. Compare the composition of coal with that of petroleum and natural gas. What is the origin of petroleum?

14. What are the fossil fuels? In what sense are they a nonrenewable resource?

15. How are metamorphic rocks formed? Describe several common kinds of metamorphic rocks.

16. What are metalliferous deposits? What is an ore? Give a general statement about the relative abundances of important metals in the earth's crust. What general classes of chemical compounds make up most of the metallic ores? Give examples.

17. What are pegmatite bodies? How do they form? What important ores occur as pegmatites?

18. How do hydrothermal solutions produce ore deposits? Name an important class of ores occurring as a disseminated deposit.

19. Describe the process of secondary enrichment of mineral deposits. Name several important ores of this origin. What is a placer deposit?

20. What ores are formed as hydrogenic sedimentary deposits?

21. In what sense are metalliferous deposits classed as nonrenewable resources?

CHAPTER 23

The Earth's Crust and Its Relief Forms

A person whose concern lies largely with man's occupation of the earth's surface may justifiably feel little need to study those branches of geology dealing with the earth's interior, the subsurface structures of the continents, the relief features of the ocean floor, and the events of ancient geologic history. It is true that the remoteness of these things—whether in place or in time—relegate them to indirect or remote roles in man's economic, social, and cultural processes, because few immediate or direct environmental effects can be established. Nevertheless, it is worthwhile for the sake of perspective for us to place the surficial features of the lands, atmosphere, and oceans in the broader setting of global geology.

Internal structure of the earth

Man's direct observation of the composition and physical properties of the earth's interior is limited by mining and drilling operations to depths of a few miles at best, so that he must turn to indirect means of obtaining information. The science of *geophysics* is concerned largely with obtaining information about the physical properties of the earth by means of instruments that measure earthquake waves, earth magnetism, and the force of gravity. Interpretation of these data through established laws of physics has yielded surprisingly detailed knowledge of the earth's structure and properties.

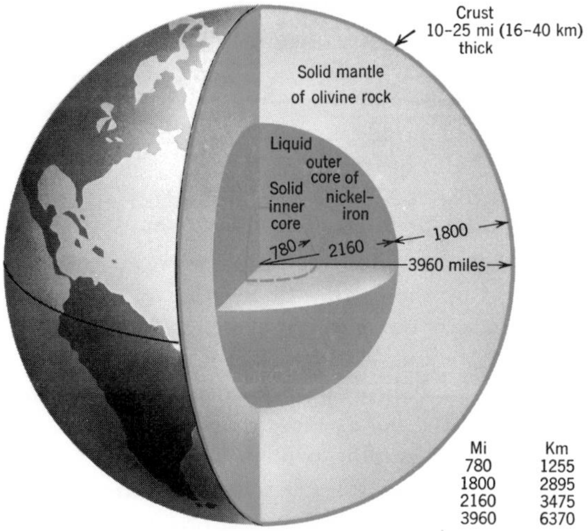

Figure 23.1 Concentric zones make up the earth's interior. The crust is not drawn to true scale.

Figure 23.1 is a cutaway diagram of the earth to show its major parts. The earth is an almost spherical body approximately 3960 mi (6370 km) in equatorial radius. The center is occupied by the *core*, a spherical zone about 2160 mi (3475 km) in radius. Because of the sudden change in behavior of earthquake waves upon reaching this zone, it has been concluded that the outer core has the properties of a liquid, in abrupt contrast to a solid mass which surrounds it. However, the innermost part of the core with a radius of 780 mi (1255 km) may be solid, or crystalline.

Through astronomical calculations it can be shown that the earth has an average density of about $5\frac{1}{2}$, whereas the surface rocks average 3 or less (Chapter 22). This observation must mean that the rock density increases greatly toward the interior, where it may be about from 10 to 15. Iron, with a small proportion of nickel, is considered as the substance probably comprising the liquid core. This conclusion is supported by the fact that many meteorites, representing disrupted fragments of our solar system, are of iron-nickel composition. Temperatures in the earth's core may lie between 4000° and 5000°F (2200° and 2750°C); pressures are as high as three to four million times the pressure of the atmosphere at sea level.

Outside of the core lies the *mantle*, a layer about 1800 mi (2895 km) thick, composed of mineral matter in a solid state. Judging from the behavior of earthquake waves, the mantle is probably composed largely of the mineral olivine (magnesium iron silicate), which comprises an ultramafic rock called *dunite*. This rock, which may be in a glassy state, exhibits qualities of great rigidity and high density in response to sudden stresses of earthquake waves which pass through it. On the other hand, the mantle rock can also adjust by slow flowage to unequal forces which act over great periods of time. In this respect it is somewhat like cold tar, which is hard and shatters easily if struck, but which will slowly flow downhill if left undisturbed for a long time.

The crust

Outermost and thinnest of the earth zones is the *crust,* a layer some 5 to 25 mi (8 to 40 km) thick, formed largely of igneous rocks. The base of the crust, where it contacts the mantle, is sharply defined, a fact known because earthquake waves change velocity abruptly at that level (Figure 23.2*A*). The surface of separation between crust and mantle is called the *Moho,* a simplification of Mohorovičić, the name of the seismologist who discovered it.

From a study of earthquake waves it is concluded that the crust consists of two layers: (1) a lower, continuous layer of basaltic (mafic) rock (Chapter 22); (2) an upper layer of granitic (felsic) rock, which constitutes the bulk of continents. The granitic layer is therefore discontinuous in areal extent, being absent over the ocean basins. Figure 23.2*B* shows schematically a small part of the crust near the margin of a continent. The sedimentary strata of the continents are on the average merely a thin skin, not shown on this diagram, although locally they are several thousand feet thick.

Those parts of the crust forming the continents are much thicker than the crust under the ocean basins, and may be likened to vast icebergs floating in the sea with only a small part visible above the water, but with a great bulk deeply submerged. The glassy rock of the earth's mantle has yielded by slow flowage, much like a very viscous fluid; this has permitted the lighter, rigid, continental plates of the crust to come to rest in the manner of the floating iceberg.

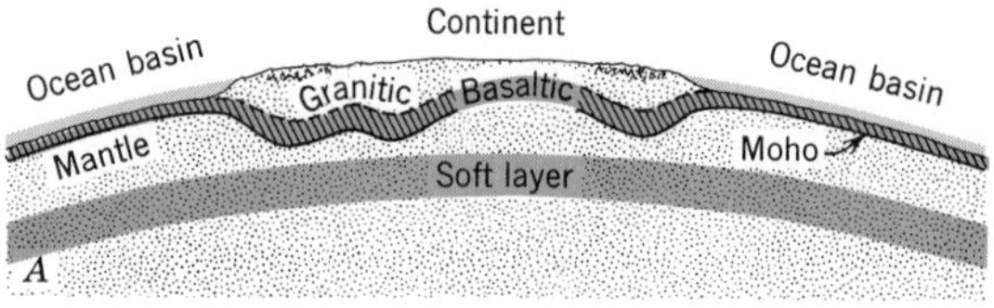

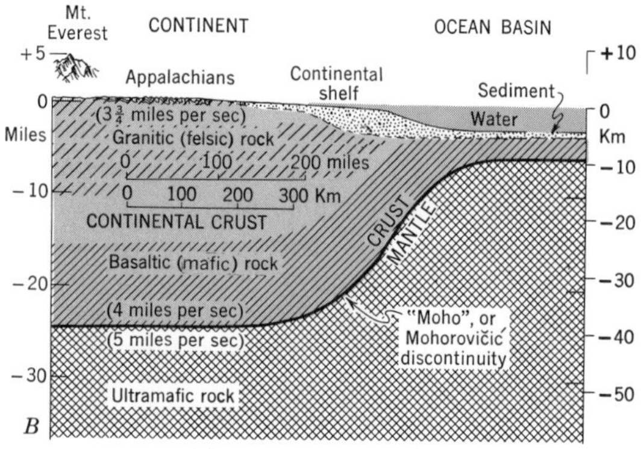

Figure 23.2 The earth's crust is much thicker under continents than beneath the ocean basins.

Distribution of continents and ocean basins

The first-order relief features of the earth are the continents and oceans basins. From a globe or atlas we can compute that about 29% of the globe is land; 71% oceans. If, however, the seas were to drain away, it would become apparent that broad areas lying close to the continental shores are actually covered by shallow water, less than 600 ft (180 m) deep. From these relatively shallow *continental shelves* the ocean floor drops rapidly to depths of thousands of feet. In a sense, then, the ocean basins are brimful of water and have even spread over the margins of ground that would more reasonably be assigned to the continents. If the ocean level were to drop by 600 ft (180 m) the surface area of continents would increase to 35%; the ocean basins decrease to 65%; figures which we may regard as representative of the true relative proportions.

Figure 23.3 shows graphically the percentage distribution of the earth's surface area with respect to elevation both above and below sea level. Note that most of the land surface of the continents is less than 3300 ft (1 km) above sea level. There is a rapid drop off from about −3000 to −10,000 ft (−1 to −3 km) until the ocean floor is reached. A predominant part of the ocean floor lies between 10,000 and 20,000 ft (3 and 6 km) below sea level. Disregarding the earth's curvature, the continents can be visualized as platformlike masses; the oceans as broad, flat-floored basins.

Scale of the earth's relief features

Before turning to a description of the major subdivisions of the continents and ocean basins, it is revealing to consider the true scale of the earth's relief features in comparison with the earth as a sphere. Most relief globes and pictorial relief maps are greatly exaggerated in vertical scale. For a true-scale profile around the earth we might draw a chalk-line circle 21 ft (6.4 m) in diameter, representing the earth's circumference on a scale of 1:2,000,000. A chalk line $^3/_8$ in (0.15 cm) wide would include within its limits not only the highest point on the earth, Mt. Everest, +29,000 feet (+8840 m), but also the deepest known ocean trenches, below −35,000 feet (−10,700 m).

Figure 23.4 shows profiles correctly curved and scaled to fit a globe whose diameter is 21 ft (6.4 m). The topographic profile is drawn to natural scale, without vertical exaggeration. Although the most imposing landscape features of Asia and North America are shown, they seem little more than trivial irregularities on the great global circle.

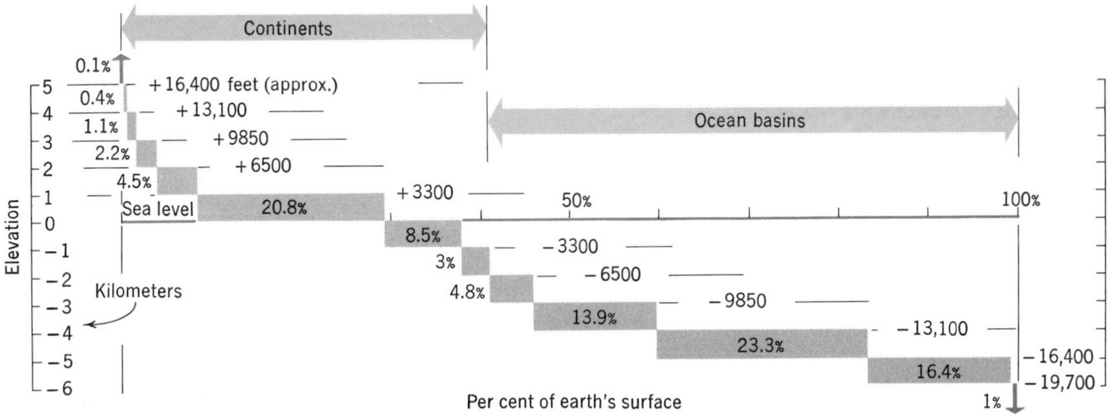

Figure 23.3 **Distribution of the earth's solid surface in sucessively lower altitude zones.**

Second-order relief features of the ocean basins

The North Atlantic Ocean illustrates certain typical features of the ocean basins and the continental margins. Three major units of the North Atlantic are (1) the *continental margin*, (2) the *ocean basin floor*, and (3) the *Mid-Oceanic Ridge* (Figure 23.5). These units may be regarded as second-order relief features of the earth's surface.

Along the eastern margin of North America lies the *continental shelf*, a fairly smooth, sloping plain 75 to 100 mi (120 to 160 km) wide and reaching a depth of 600 ft (180 m) at the outer edge. This shelf, a part of the continental margin, is essentially a zone of deposition of sedimentary rock layers built of material brought from the eastern United States by streams and spread over the sea floor by currents. At its outer edge, the shelf abruptly gives way to a descending *continental slope*, which gives way at the base to a gentler slope, the *continental rise*. The rise merges imperceptibly with the true ocean basin floor at a depth of about 12,000 ft (3700 m).

The slope is scored by many *submarine canyons* (Figure 23.6), whose origin has been strongly debated. The canyons seem to be the work of eroding streams, very likely flows of muddy water, called *turbidity currents*, which are produced when storms or earthquake shocks disturb soft sediment at the canyon heads. These flows travel swiftly down the continental slope because their density is greater than that of the surrounding sea water. Spreading out upon the deep sea floor, turbidity currents come slowly to rest. The sediment is spread in broad layers which have accumulated over millions of years, gradually burying the irregular topographic features of the sea floor and

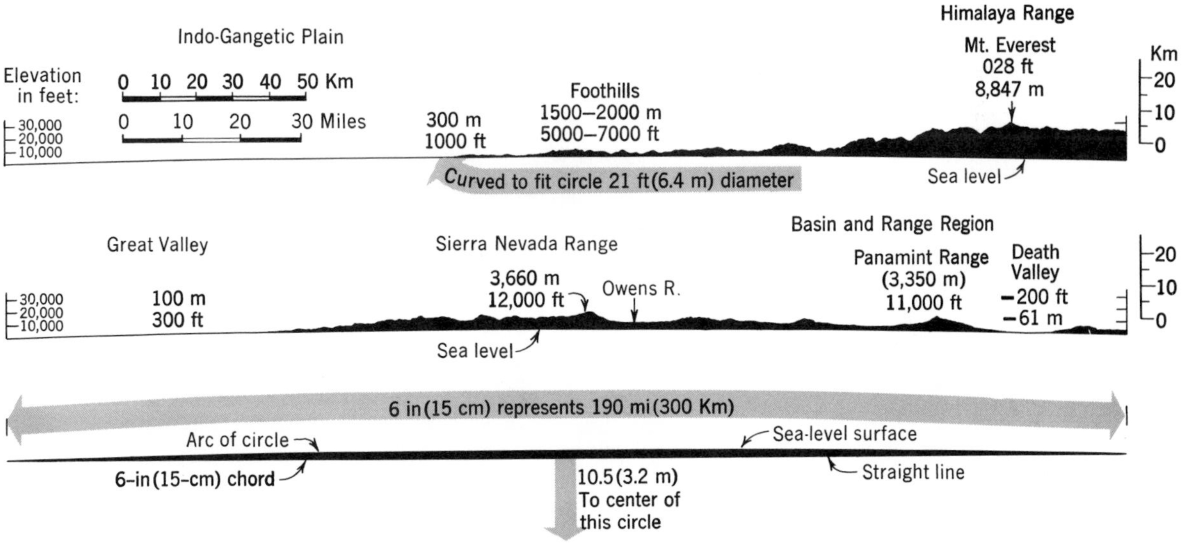

Figure 23.4 **These profiles show the earth's great relief features in true scale with sea-level curvature fitted to a globe 21 ft (6.4 m) in diameter.**

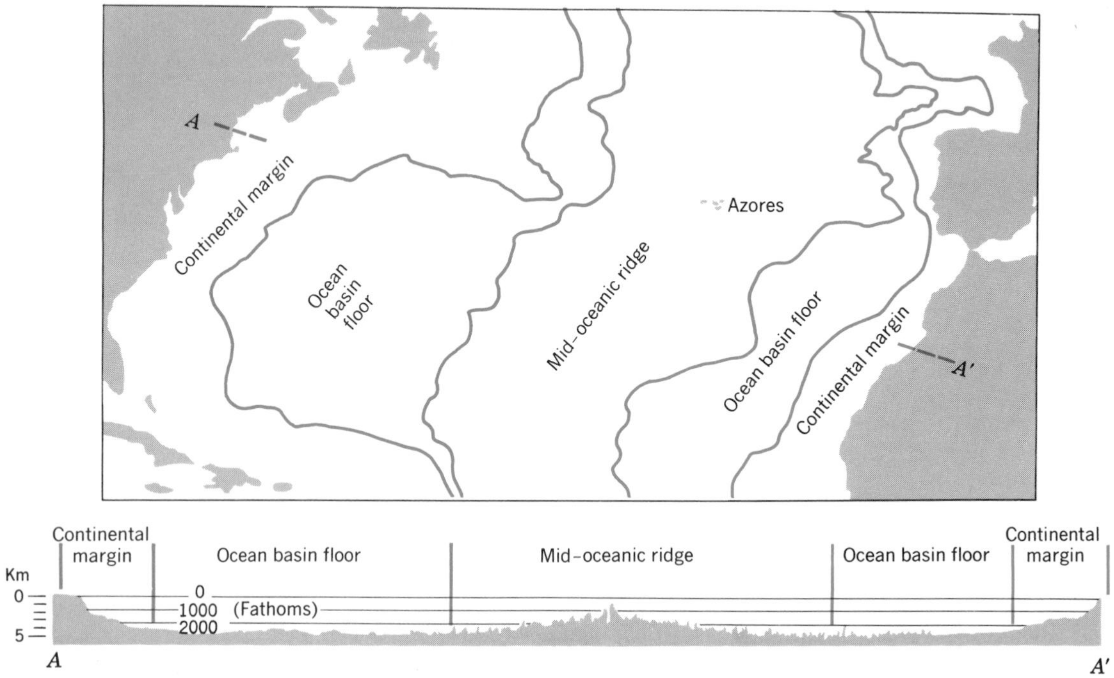

Figure 23.5 Major divisions of the North Atlantic Ocean basin (above), and a representative profile from New England to the coast of Africa (below). Profile exaggeration is about 40 times. (Data of B. C. Heezen, M. Tharp, and M. Ewing, 1959, from A. N. Strahler, 1971, *The Earth Sciences*, 2nd ed., Harper and Row, New York.)

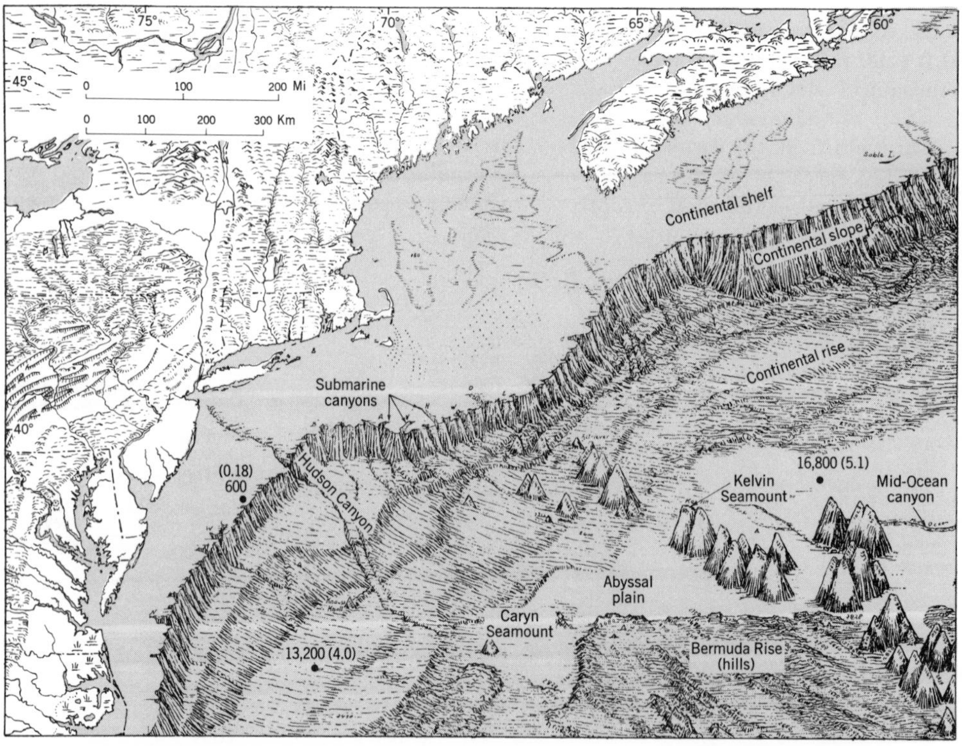

Figure 23.6 Features of the continental margin and ocean-basin floor off the coast of the northeastern United States. Depth in ft; km in parentheses. (Portion of *Physiographic Diagram of the North Atlantic Ocean*, 1968, revised, by B. C. Heezen and M. Tharp, Boulder, Colo., Geol. Soc. of Amer., reproduced by permission.)

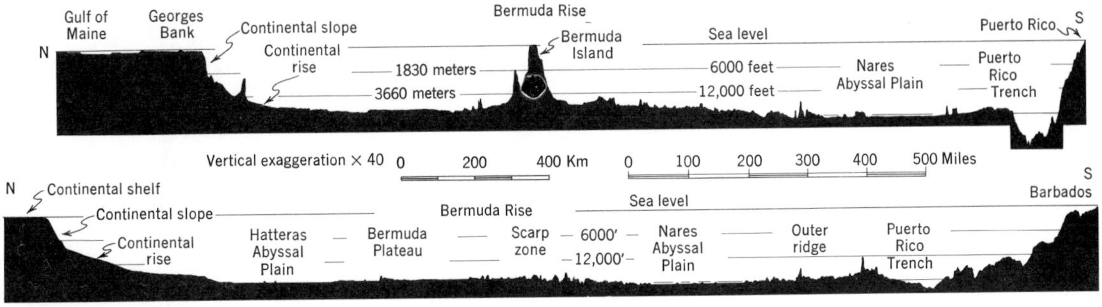

Figure 23.7 Profiles across the Atlantic Ocean basin reveal continental shelf and slope, abyssal plain, trench, and various minor irregularities. (After Bruce C. Heezen, Lamont-Doherty Geological Observatory of Columbia University.)

producing vast, flat *abyssal plains* lying within the ocean basin floor (Figure 23.7). The broad basin of the North Atlantic thus has a remarkably smooth floor over large areas at a depth of about 18,000 ft (5500 m). Rising abruptly from the floor are isolated submarine mountains, named *seamounts,* some of which may be ancient volcanoes. A group of seamounts is shown in the lower right corner of Figure 23.6.

In the center of the Atlantic lies a great submarine mountain range, the *Mid-Atlantic Ridge,* comparable with the Rocky Mountain chain in size and relief but entirely submerged except for the Azores Islands. This ridge is merely a part of a single, continuous Mid-Oceanic Ridge traced through the South Atlantic, Indian, South and East Pacific, and Arctic oceans. It is the expression of a major fracture system of the earth's crust (Figure 23.8). The oceanic crust is in process of being pulled apart along the Mid-Oceanic Ridge, causing a *rift valley* to lie along the center line of the ridge (see profile of Figure 23.5). At the same time mantle rock is being pushed up to elevate the ridge. Evidence in support of crustal spreading in the Mid-Oceanic Ridge zone has come from studies of magnetic properties of the basalt rock in belts adjacent to the ridge. Rate of separation of the crustal masses on either side of the Mid-Oceanic Ridge has been estimated to be about 0.8 in (2 cm) per year in the North Atlantic.

Other features of the ocean floor are *trenches* or *foredeeps*—long, narrow depressions whose

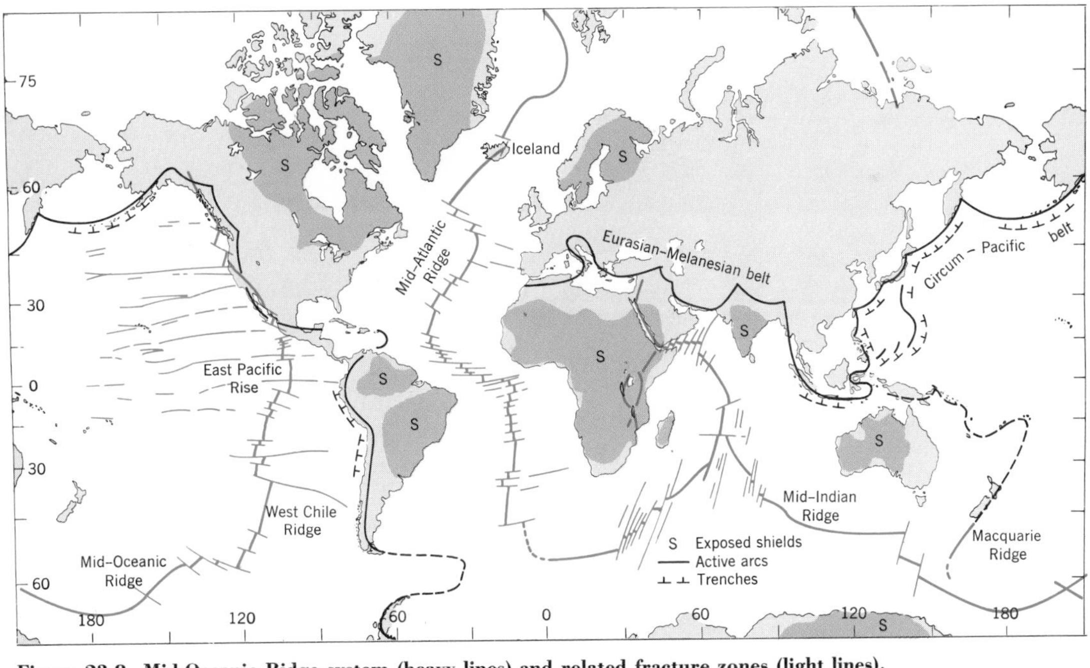

Figure 23.8 Mid-Oceanic Ridge system (heavy lines) and related fracture zones (light lines). (After L. R. Sykes, 1969, in *The Earth's Crust and Upper Mantle,* Geophysical Monograph 13, Washington, D.C., Amer. Geophys. Union, p. 149, Figure 1. Based on data of B. C. Heezen, M. Tharp, H. W. Menard, and other sources.)

bottoms reach depths of 24,000 to 30,000 ft (7500 to 10,000 m) or more (Figure 23.7 and 23.9). The trenches represent downwarped zones associated with recent mountain building movements. Accumulation of sediment on the trench floors has been so slow that these depressions are not filled in, as they would be if above sea level.

Present-day mountain building near the margins of the ocean basins is distributed along long, narrow, curving zones, known as *island arcs* (Figure 23.10). Each arc represents a zone of crustal movement and is associated with chains of active volcanoes and earthquakes. The great ocean deeps generally lie along the outer side of the island arcs.

Second-order relief features of the continents

Just as with the ocean basins, the continents can be subdivided into second-order relief units. Essentially, the continents consist of two fundamental kinds of geologic units: (1) *shields* and (2) *mountain belts,* or *orogenic belts.* The shields are the heartlands of the continental crust and consist of extremely ancient rocks. The age of the shield rocks is *Precambrian,* that is, older than 600 million years. Most of the shield rock is much older than one billion years and some has been dated as $3\frac{1}{2}$ billion years old. Several periods of mountain-making, or *orogeny,* affected the shields throughout Precambrian time and there was much

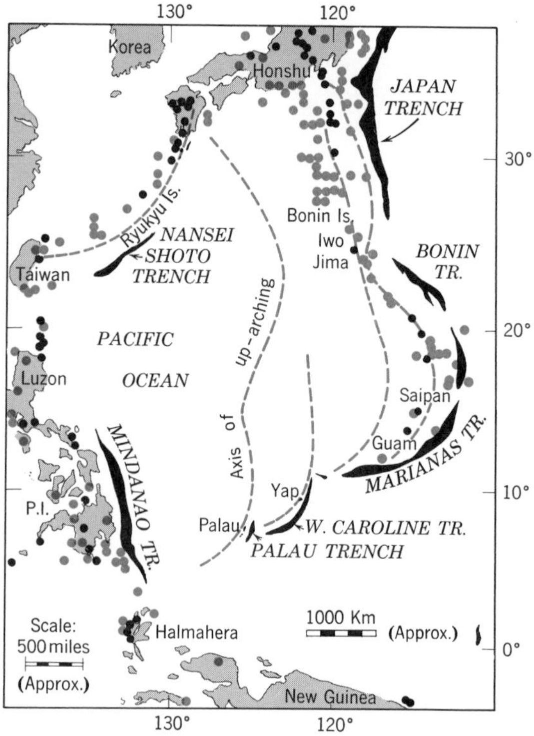

Figure 23.10 This map of the western Pacific shows trenches (solid black), island arcs (dashed lines), active volcanoes (black dots), and deep-focus earthquake centers (circles). (After H. H. Hess.)

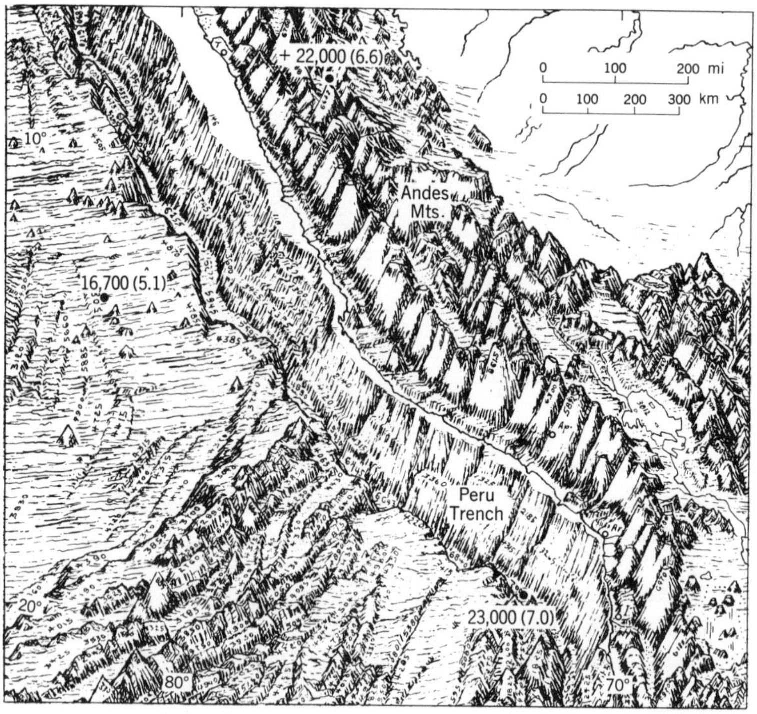

Figure 23.9 The Peru-Chile Trench, off the west coast of South America. (Portion of *Physiographic Diagram of the South Atlantic Ocean,* 1961, by B. C. Heezen and M. Tharp, Boulder, Colo., Geol. Soc. of Amer., reproduced by permission.)

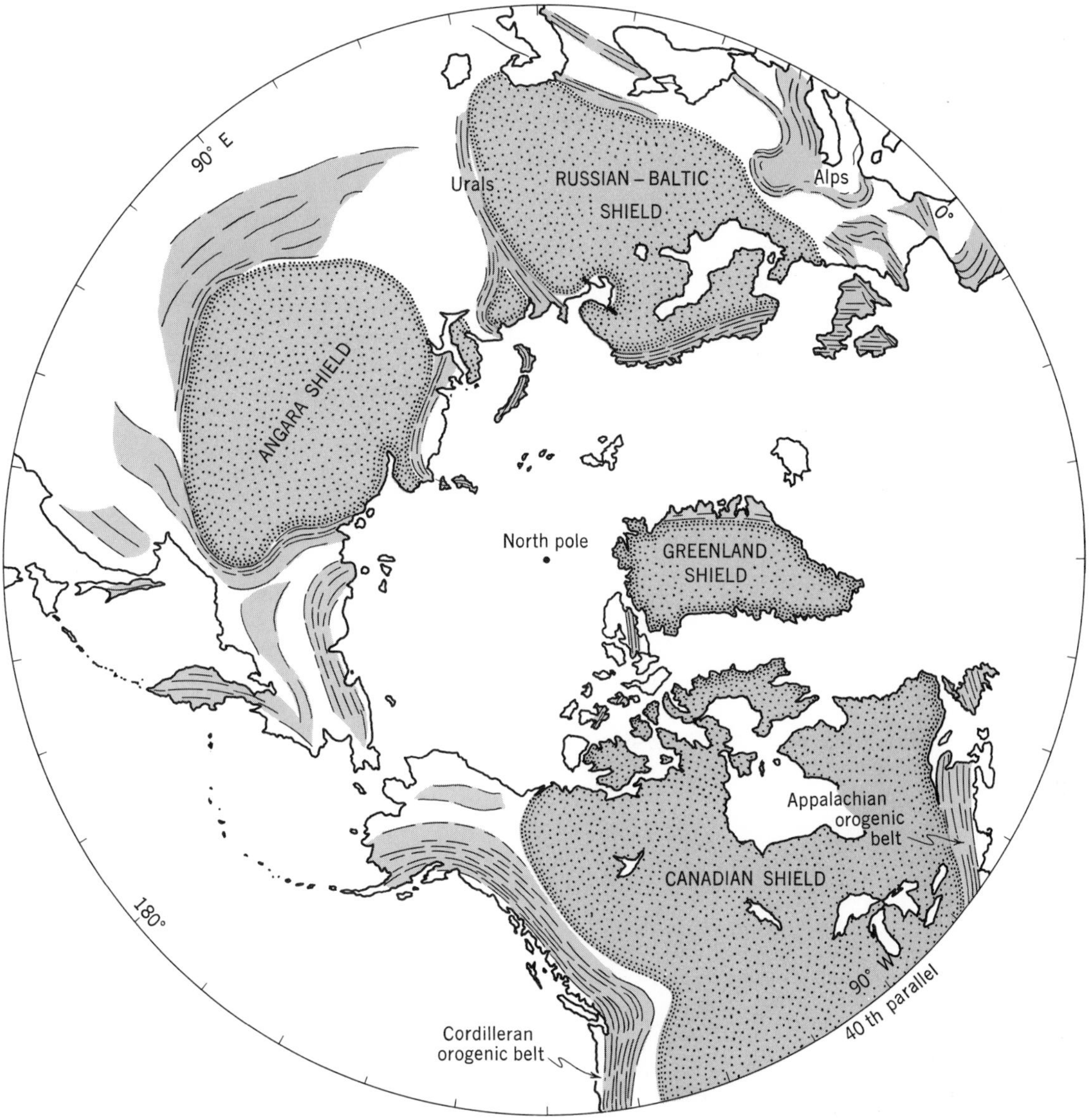

Figure 23.11 Shields are bounded by mountain belts (dashed lines) of Paleozoic and younger age. Shield areas include sedimentary covers. (After A. J. Eardley, *Structural Geology of North America*.)

metamorphism of ancient sedimentary rocks as well as many large intrusions of granitic rock in the form of batholiths (Chapter 21). Whereas great mountain ranges existed at various times during Precambrian time, processes of erosion effectively reduced those mountains to their roots and leveled the ancient continents to nearly featureless plains. The shields today are therefore largely plains and low plateaus and are highly stable parts of the earth's continental crust.

Crustal movements of the shields in later geologic time have been of a type known as *epeirogenic* movements; that is, rising or sinking of the crust over broad areas without appreciably breaking or bending the rocks. Epeirogenic movements reflect crustal stability generally, in contrast to compressional and tensional orogenic movements affecting unstable zones.

When epeirogenic movements were of a negative, or sinking, type, large parts of the shields became submerged as shallow seas and continental shelves. This inundation gave opportunity for sedimentary strata to be deposited on the old metamorphic and igneous rocks. Later, positive, or rising, epeirogenic movements brought the sedimentary cover above sea level where it has since been carved up by streams into hills and plateaus.

Shields of the northern hemisphere are shown in Figure 23.11. North America's geological heart-

TABLE 23.1 THE GEOLOGIC TIME SCALE*

Era	Period		Duration m.y.	Age m.y.	Orogenies
CENOZOIC	Quaternary		2.5	2.5	Cascadian
	Tertiary		62.5	65	
MESOZOIC	Cretaceous		71	136	Laramian
	Jurassic		54	190	Nevadian
	Triassic		35	225	
PALEOZOIC	Permian		55	280	Appalachian (Hercynian)
	Carboniferous	Pennsylvanian	45	325	
		Mississippian	20	345	Acadian
	Devonian		50	395	(Caledonian)
	Silurian		35	430	Taconian
	Ordovician		70	500	
	Cambrian		70	570	
			Duration b.y.	Age b.y.	
PRECAMBRIAN or PROTOZOIC	Upper Precambrian		0.3–0.4	0.9–1.0	Grenville
	Middle Precambrian		0.6–0.8	1.6–1.7	Hudsonian
			0.7–0.9	2.4–2.5	Kenoran
	Lower Precambrian		0.9–1.0		
	Oldest dated rocks			3.4±0.1	
	Earth accretion completed			4.6–4.7	
	Age of universe			17–18	

*Data sources: D. Eicher (1968), *Geologic Time*, Englewood Cliffs, N.J., Prentice-Hall, end paper; M. Kay and E. H. Colbert (1965), *Stratigraphy and Earth History*, New York, Wiley, p. 74.

land is the *Canadian Shield;* that of Europe is the *Russian-Baltic Shield,* also called the *Fenno-Scandian Shield.* Rocks in both of these regions date back to the oldest era of geologic time, from one to three billion years ago. South of the 30th parallel north, and in the Southern Hemisphere, similar shield areas occupy parts of Australia, Africa, South America, India, and Antarctica. These shields, which are also composed of Precambrian rock, are shown in Figure 23.8.

The mountain belts of the continents are long, narrow orogenic zones in which the crust has been compressed and forced to buckle into tight folds and at the same time to be strongly elevated.

The older, inactive, mountain ranges in some places still possess rugged relief and moderately high elevations; elsewhere, they are reduced to hill belts. The youngest ranges are in many places still active zones of crustal deformation and volcanic activity; they rise as spectacular *alpine mountains.*

The geologic time scale

To discuss the development of relief features of the continents requires reference to events of the geologic past. The geographer interested in the occurrence and distribution of ores and mineral fuels will find a knowledge of historical geology extremely helpful in explaining the occurrence of economic mineral deposits in various parts of the world.

Table 23.1 lists the important divisions and subdivisions of geologic time.

Absolute ages given in the table have been verified by means of chemical analyses of radioactive minerals and are generally accepted by geologists, subject to a small percentage of error. In every period of geologic time, there were widespread accumulations of sedimentary strata; in fact, the strata comprise the record itself and constitute sole evidence of the environmental conditions of the time, as well as containing the fossil remains of plant and animal life.

Generally speaking, each major time unit was brought to a close by orogeny, also called *revolution*, disrupting the sequence of sediment deposition. The largest time unit is the *era* of which the last three are *Cenozoic*, *Mesozoic*, and *Paleozoic*, in order of increasing age. All time before this is designated as belonging to the *Precambrian* time. Eras have been recognized within the Precambrian, but the record tends to be fragmentary and confused. The second order of time unit is the *period*, of which the Paleozoic Era has seven; the Mesozoic Era, three; and the Cenozoic Era, two.

An example of crustal evolution—the Appalachians

By reviewing the geologic history of a particular mountain range we can gain insight into a general scheme of geologic events which has been repeated many times in the past. The diagrams of Figure 23.12 represent inferred events in the development of the Hudson Valley region throughout the Paleozoic, Mesozoic, and Cenozoic eras encompassing approximately the last 600 million years.

In block *A* the region is shown as a Paleozoic seaway in which thousands of feet of sedimentary strata had accumulated. An inland seaway deposit of this nature is called a *geosyncline*. The source of sediment was in large part from a chain of volcanic islands lying to the east. The Paleozoic Era was brought to a close by a great orogeny, the *Appalachian Revolution*. Sedimentary strata of the geosyncline were severely crumpled, as well as broken into slices which slid over one another (block *B*). As a result, a great mountain range stood where formerly there had been a seaway. The bending of the strata is generally referred to as *folding*, and the corrugated structures thus produced are simply termed *folds*. The slanting surfaces upon which sliding occurred are termed *overthrust faults*, the process being known as *overthrusting*. The agents of denudation are shown to be dissecting the initial mountain forms even before the folding and thrusting have ceased.

After the Appalachian Revolution, the region shown here remained essentially stable and quiescent for many millions of years. The mountains were reduced by the denudational processes to a land surface of very faint relief termed a *peneplain*. The peneplain is shown in block *C*. Notice that the oldest rock, a gneiss of Precambrian age, is exposed in the core of the mountain belt, whereas the youngest sedimentary layers, of Paleozoic age, remain in the zone of least disturbance, at the left end of the diagram.

In block *D*, the region is shown to have been again subjected to crustal movement, but of a quite different nature from previous movements. This was the *Palisadian disturbance*, which began in the Triassic period. The region has been broken into a series of blocks, each tilted with respect to its neighbor. The fractures are a type of fault, but are different from overthrust faults by having nearly vertical inclination of the breaks and by the absence of any pronounced crustal compression and shortening. The entire breakage scheme may be termed *block faulting*. The peneplain made before faulting forms the smooth, sloping surfaces of the fault blocks.

In block *E*, the region has again been reduced to a peneplain, indicating another prolonged period of relative stability of the earth's crust lasting into the Tertiary period. The more resistant gneiss, granite, conglomerate, and sandstone rise as low hills between broad, flat lowlands underlain by shale.

The region next experienced still another crustal movement, of epeirogenic nature. This was a very simple rising, or upwarping. No faulting or folding occurred. Consequently, the peneplain of block *E* was merely uplifted about 2000 ft (600 m). Streams and other agents of land denudation again set to work and excavated the weaker rocks to form valleys. The resistant sandstones, conglomerates, granites, and gneisses were left as ridges and mountain masses, as shown in block *F*, representing the present. The topography of today is solely the result of different rates of downwasting of the ground surface upon complex rock structures formed through a long series of geologic events.

Chapters 28 to 30 treat the individual peculiarities of landscape features, or landforms, developed on flat-lying strata, folded strata, domed strata, block faults, and the extrusive and intrusive igneous rocks.

First, however, it will be necessary to study the processes of rock weathering and mass wasting of slopes, and the action of streams, waves and currents, glaciers, and wind, because these are the agents that carve landforms from bedrock.

Plate tectonics and continental drift

Obviously, deep-seated forces of enormous magnitude are required to deform crustal rocks on the scale observed in the Appalachian Mountains and other great mountain arcs such as the Alps and Andes. In recent years, geologists and geophysicists

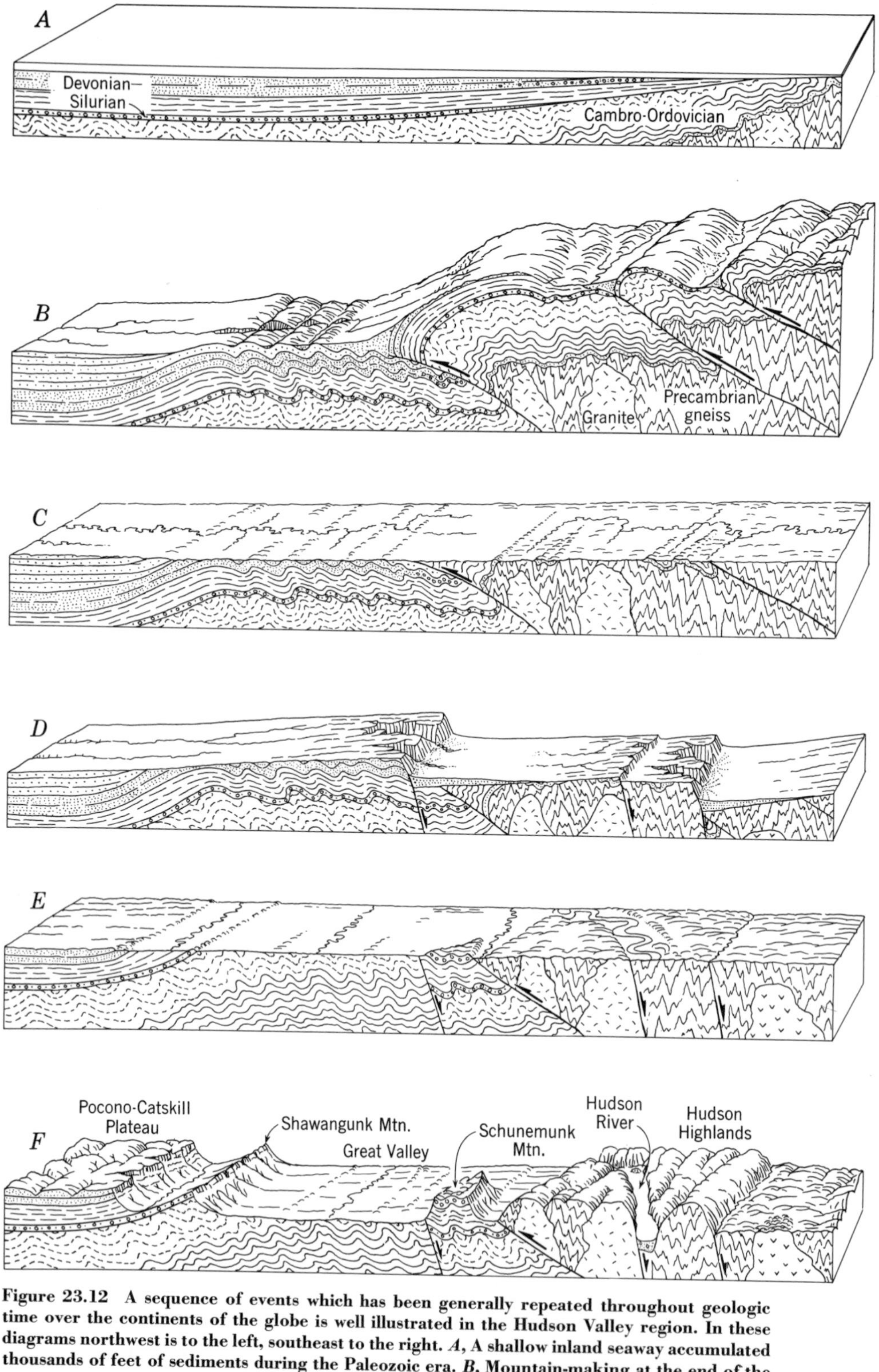

Figure 23.12 A sequence of events which has been generally repeated throughout geologic time over the continents of the globe is well illustrated in the Hudson Valley region. In these diagrams northwest is to the left, southeast to the right. *A,* A shallow inland seaway accumulated thousands of feet of sediments during the Paleozoic era. *B,* Mountain-making at the end of the Paleozoic era produced a series of folds and thrust faults. A general uplifting brought a large mass above sea level. *C,* Following a long period of erosion, a peneplain was produced. *D,* Faulting in the Triassic period produced these gently tilted blocks. *E,* A second long period of erosion resulted in another peneplain. *F,* The region today owes its relief to different rates of removal of the various kinds of rocks (From A. N. Strahler, 1971, *The Earth Sciences,* 2nd ed., Harper and Row, New York.)

have directed their attention toward a zone in the mantle located well below the Moho. Evidence from the study of earthquake waves gives good reason to conclude that there exists in the upper mantle a *soft layer* having little strength (Figure 23.2). This condition of softness, or weakness, sets in at a depth of about 40 mi (60 km). Strength declines with depth to a minimum in a zone near 125 mi (200 km), below which strength again increases. Loss of strength in this soft zone is perhaps caused by accumulated heat of radioactivity, which brings the mantle rock temperature close to its melting point. Like iron heated white hot in a furnace, the rock loses strength and can move slowly by flowage in response to unequally applied stresses.

The crust in combination with the uppermost mantle comprises a strong, rigid earth shell called the *lithosphere.* (This geological usage is somewhat more restrictive than the general usage of the word to designate the entire solid earth realm.) Beneath the lithosphere lies the soft, weak layer, or *asthenosphere.* Recognition of these two layers based upon strength properties is extremely important in understanding crustal separation along the Mid-Oceanic Ridge.

Breaking and bending of crustal rock is included under the general term *tectonic activity*, while the study of these activities and the structures they produce is referred to as *tectonics.* We turn now to an overview of a global system of tectonics in which the strong, brittle lithosphere moves over a soft asthenosphere, impelled by forces that are little understood.

Within the past two decades, exploration of the ocean floors and oceanic crust has brought evidence of lithospheric motions on a colossal scale. The new findings constitute a major revolution in geology, and have led to a unified theory of large-scale geological processes. It all began shortly after World War II with discovery of the Mid-Oceanic Ridge. Later, in the 1960s came evidence that crustal separation is occurring on this line. Further research rapidly led to a general theory of *plate tectonics,* illustrated schematically in Figure 23.13. The entire lithosphere of the earth is considered to be divided up into *lithospheric plates,* of which there are six major plates and many smaller blocks within or between the major plates. The Mid-Oceanic Ridge represents the zone of separation of plates that are drifting apart. A lithospheric plate moves slowly over the soft asthenosphere, much as a slab of butter moves over the surface of a warm skillet. Where two lithospheric plates are moving toward each other they are in collision and one of the two plates is bent down to pass beneath the other, as Figure 23.13 shows. The descent of the plate into the mantle is termed *subduction.* Zones of collision are the orogenic belts already described.

Figure 23.14, a block diagram, shows the relationship between an orogenic belt, such as the Appalachians, and the zone of subduction beneath. As one plate moves down, pressing against the edge of the opposed plate, sediments of the geosyncline are crumpled and metamorphosed. At greater depth the sediment is melted to become magma; this rises to form batholiths in the core of the orogenic belt.

Through geologic time, as lithospheric plates have drifted apart, vast areas of ocean basins with a thin basaltic crust have come into existence. At the same time, the processes of sedimentation, orogeny, and igneous intrusion acting over the subduction zones have gradually created thick continental crust with its felsic (granitic) upper zone. Thus the continents have evolved and increased in size over a 3- to 4-billion year period. We find within the continental shields small core areas which consist of rock from 2.7 to 3.5 billion years (b.y.) in age; these are the *nuclei* of the continents. Nuclei are surrounded by broad zones of somewhat younger rock, 0.8 to 2.7 b.y. in age.

Because the oceanic crust is produced by up-

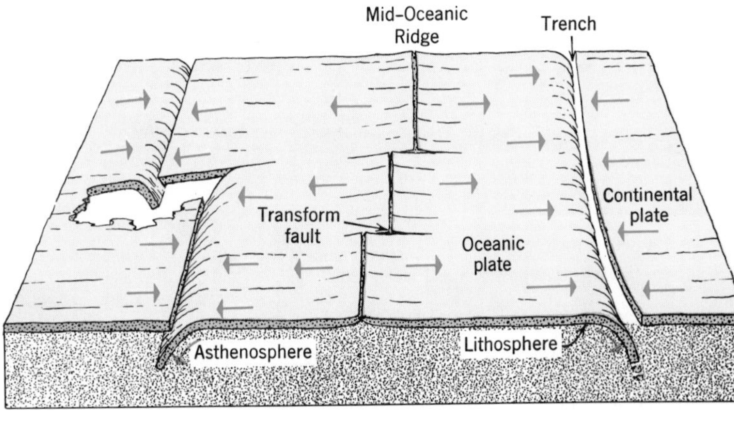

Figure 23.13 Schematic diagram showing major features of plate tectonics. Earth curvature has been removed. (From A. N. Strahler, 1971, *The Earth Sciences*, 2nd ed., Harper and Row, New York.)

welling of basaltic magma along the Mid-Oceanic Ridge, its age is very young as compared with the continental crust. The oldest dated oceanic crust is in the western part of the Pacific Ocean basin; its age is about 75 million years, which is only about one-fortieth the age of rock in the continental nuclei.

As early as the close of the 19th century, the proposal had been made that the continents of North and South America, Europe, Africa, Australia, and Antarctica, and the subcontinent of India (south of the Himalayas) along with Madagascar, had originated as a single supercontinent, named *Pangaea* (Figure 23.15). Pangaea began to break apart about 200 million years ago (Triassic Period). The fragments slowly separated and underwent some horizontal rotation as well. Thus the Atlantic Ocean is envisioned as the area opened up by drifting apart of the Americas from Africa and Europe. This hypothesis of *continental drift* was generally rejected or skeptically regarded by most of the geological fraternity, particularly the Americans, until about 1960, when sound evidence of crustal spreading came to light. Now a majority of geologists have accepted continental drift, much as it was originally outlined, and have fitted it into the global theory of plate tectonics. Figure 23.15 shows postulated stages in the separation of the continents, according to a recent version. (See Table 23.1 for ages of periods.)

Under plate tectonic theory, North and South America are part of the *American Plate*, moving westward with respect to the Mid-Atlantic Ridge. Consequently, the western edge of this plate is being pushed beneath adjoining plates on the west. This zone of subduction is responsible for the great Cordilleran and Andean mountain chains. A huge submarine trench lies off the South American west coast and another off the Aleutian Island chain of Alaska. These trenches are interpreted as expressions of subduction of the plate lying to the west, which is being forced into the asthenosphere as the American Plate overrides it.

Looking at the Pacific Ocean as a whole, we find that it is a single lithospheric plate capped entirely by oceanic crust. Around the margins of this Pacific Plate is a great tectonic ring—the *Circum Pacific Belt* of arcs of mountains or volcanic islands, bordered by deep trenches. It is a belt of crustal unrest, manifested by numerous intense earthquakes as well as volcanic activity, for this is a more-or-less continuous line of subduction of the Pacific Plate.

While the driving force for plate motions is not known, a leading hypothesis attributes the motion to convection currents deep in the mantle (Figure 23.16). In some respects this model convection current system in the mantle resembles the atmospheric circulation at low latitudes within the Hadley cell (Chapter 9). Unequal heating of mantle rock leads to differences in rock density, and this in turn may set in motion very slow rising and sinking motions. It is postulated that horizontal mantle motion between the zone of rising mantle rock (Mid-Oceanic Ridge) and sinking rock (beneath the island and mountain arcs) exerts a dragging force on the lithospheric plate. Many variations of the convection hypothesis have been set forth and various other energy sources have been proposed as well. Thus, while the reality of plate motions seems well established, we are far from understanding the impelling forces.

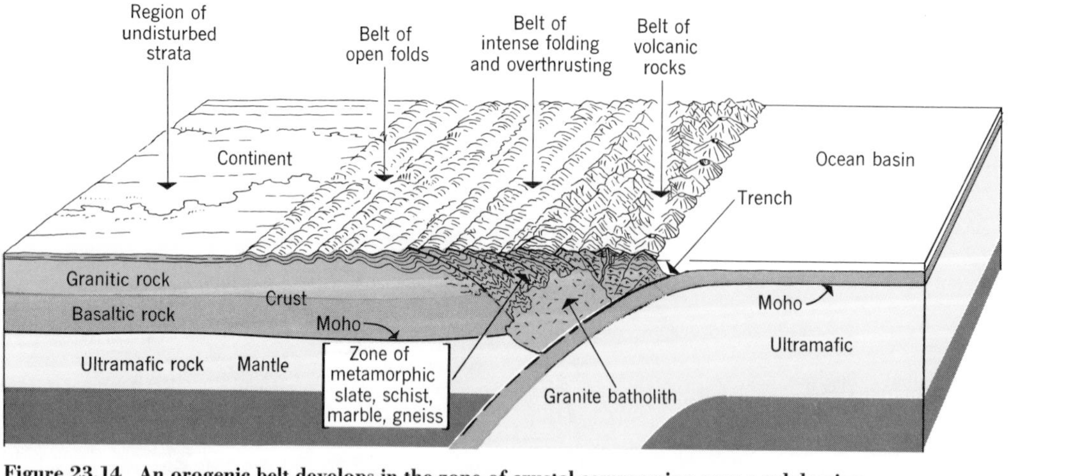

Figure 23.14 An orogenic belt develops in the zone of crustal compression over a subduction zone where lithospheric plates are in collision. (From A. N. Strahler, 1971, *The Earth Sciences*, 2nd ed., Harper and Row, New York.)

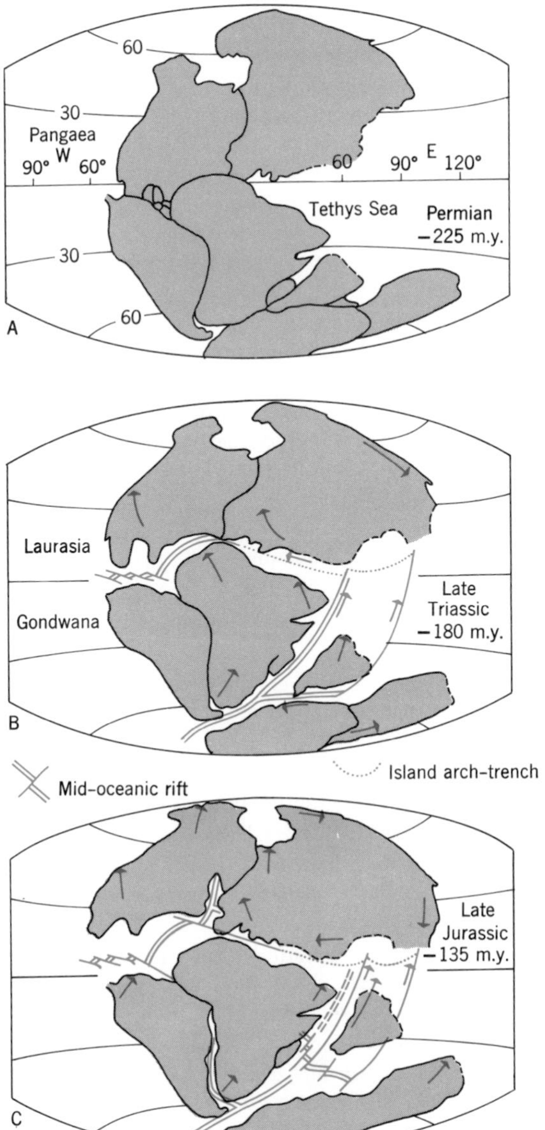

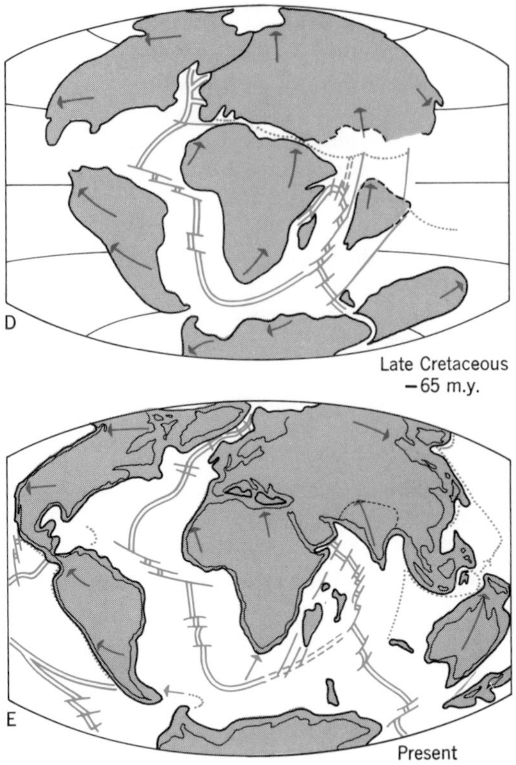

Figure 23.15 The breakup of Pangaea is shown in five stages. Inferred motion of lithospheric plates is indicated by arrows. (Redrawn and simplified from maps by R. S. Dietz and J. C. Holden, 1970, *Jour. Geophys. Research*, Vol. 75, pp. 4943–4951, Figures 2 to 6.)

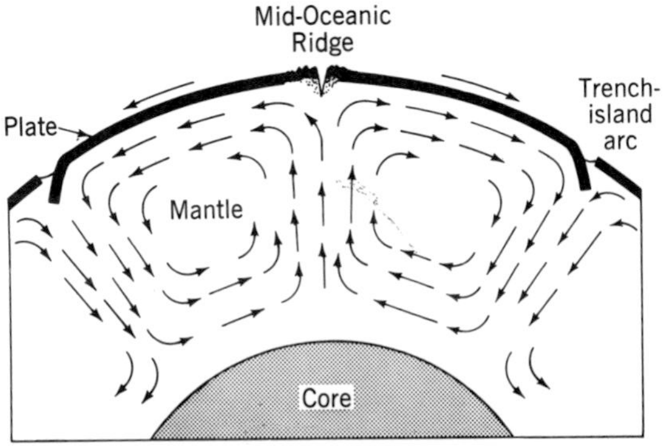

Figure 23.16 Simplified model of a convection system in the mantle in relation to overlying lithospheric plates. (From A. N. Strahler, 1971, *The Earth Sciences*, 2nd ed., Harper and Row, New York.)

Following this brief overview of large-scale forms and processes of the earth's crust, we turn to a study of features of a smaller order of magnitude. These are the landscape elements of the continental surfaces produced by processes acting at the interface of solid earth and atmosphere. In this interface the energy for change is that of solar radiation acting through atmospheric and hydrologic processes.

REVIEW QUESTIONS

1. Describe the earth's core and mantle, giving dimensions, mineral composition, and physical properties. What type of evidence is used to obtain this information? What temperatures and pressures may be expected at the earth's center?

2. What is the earth's crust? How thick is it? How can it be distinguished in properties from the underlying mantle? Of what two rock layers does the crust consist? Compare the continental crust with the oceanic crust as to thickness and composition.

3. Describe the general form of the continents and ocean basins as regards the levels of concentration of surface areas. If sea level were lowered by 600 ft (180 m), what percentage of the earth would be land?

4. On a globe 21 ft (6.4 m) in diameter, how far would the greatest relief features of the earth depart from a perfect circle drawn to represent sea level?

5. What are the principal second-order relief features of the ocean basins?

6. What is a continental shelf? By what type of rock material is it underlain? What is a continental slope? A continental rise? What relation do submarine canyons bear to these features?

7. What are turbidity currents? What work do they perform? What deposits do they build?

8. What kinds of relief features are found on the floors of the ocean basins? Describe seamounts, trenches (foredeeps), and submarine mountain ridges.

9. Describe the Mid-Oceanic Ridge system. What feature marks the central line of the ridge? How do geologists account for the Mid-Oceanic Ridge system?

10. With what sort of crustal deformation are the island arcs and deep trenches of the ocean basins associated?

11. What are the second-order relief features of the continents? What are shields? What are orogenic belts?

12. What are epeirogenic crustal movements? Contrast orogenic movements with epeirogenic movements.

13. Name the eras of geologic time and give the total duration of each in years. For each of the eras, name the periods of geologic time. What great events occurred during the Pleistocene epoch?

14. What type of geologic event has brought to a close each era and many of the periods? What is the known duration of Precambrian time? How much longer is Precambrian time than all of post-Cambrian time?

15. Explain the general scheme of geologic events in which sedimentary rocks are deposited, then deformed, and finally reduced by erosion. What is a geosyncline? What is a revolution?

16. What is meant by folding and overthrusting of strata? Do these deformations require compression or tension of the earth's outer crust?

17. What is a peneplain? At what level do peneplains form? How long a period of time is required to reduce a mountain range to a peneplain?

18. What is faulting? How does block faulting differ from overthrust faulting? Why would you not expect both types to occur simultaneously in the same region?

19. When a peneplain is upwarped, what type of landscape development follows? What happens to the peneplain?

20. What is the soft layer of the mantle? Why does it have low strength? Distinguish between lithosphere and asthenosphere.

21. What is meant by tectonic activity? By tectonics? Give a general account of the theory of plate tectonics. How many major plates are there?

22. What is subduction? How is it related to orogenic belts and their deformation? Where is subduction in progress today?

23. How is plate separation related to forms of the Mid-Oceanic Ridge? What form of igneous activity occurs in the rift zone?

24. How have the continents evolved through geologic time? How is continental crust formed? Compare the age of continental crust with that of oceanic crust.

25. What was Pangaea? Describe the process of continental drift. When did continental separation begin?

26. What mechanism has been postulated as the driving force of lithospheric plate motion? What is the energy source?

CHAPTER 24

Landforms and the Wasting of Slopes

LANDFORMS, the distinctive geometric configurations of the earth's land surface, are of primary concern to the geographer because they exert far-reaching influences on the patterns of human activity. The direct environmental influences are obvious to any thoughtful person. A mountain chain is an effective barrier between groups of people who live in adjacent lowlands. A plain, on the other hand, may be densely populated, rich in agricultural resources, and unified culturally and politically by a network of good roads and railroads that permit people with common interests to intercommunicate freely. One coast line, deeply indented with excellent natural harbors but bordered by a rocky rugged coastal belt, may favor a community of seafaring humans, adapted to fishing, ocean commerce, and shipbuilding. Another coast line, whose simple plan and shallow bottom provide not a single good natural harbor, may be bordered by a low, fertile coastal plain. Here human activity turns naturally to agriculture.

Examples of the direct environmental influences of landforms on man could be cited almost without end, but there are, in addition, indirect environmental influences to be considered. As we have seen in the study of rainfall distribution, a high mountain range profoundly influences climate in adjacent areas. If it shields a nearby lowland from prevailing moisture-bearing winds, a desert results. On mountain slopes, climates become cooler and moister with increased altitude, bringing a changing succession of agricultural and forest conditions, which, in turn, determine the extent of the natural resource. Steepness of land surfaces (slopes) is closely associated with fertility of soils. Hill and mountain slopes have thin, relatively poor soils, readily subject to devastating soil erosion when exposed by axe or plow. Plains may have thick, fertile soils, not easily eroded even under poor farming practices.

Geomorphology

The systematic study of landforms and their origin is known as *geomorphology* (*geo*, earth; *morph*, form; *ology*, science). Landforms must be sorted out into classes or groups of those essentially similar both in outward form and in origin. The geomorphologist is equally interested in forms and in the processes and stages of development of those forms. We will treat landforms in terms of how they came about. Landforms pass through an orderly series of changes, just as do human beings in their life span. Once these life stages are known, any landform can be related to a particular event in a characteristic life cycle.

Genetic landform description

It is possible, of course, to describe landforms by tabulating their size, shape, angles of slope, and orientation without thought to their origin or development. This is an *empirical* approach to natural science. Volumes of figures and other factual data would be required to convey an adequate description of even the simplest assemblages of landforms.

On the other hand, when we examine the development of landforms, we find that the same series of forms is repeated with remarkable similarity over and over again in nature. In order to classify and describe a complex assemblage of forms in terms of orderly sequences of development, we need only a single brief statement giving (*a*) the *structure* of the rock mass beneath, (*b*) the *process* which sculptured the landform and (*c*) the *stage* of its development. Such a description is *genetic* because it emphasizes genesis, or origin. The person who hears or reads such a description, knowing what the ideal forms are like, can relate any particular landform to its proper place in the natural scheme.

The systematic study of landforms according to their origin and stage of development was introduced by Professor William Morris Davis, of Harvard University, about 1890. His influence has been profound and many English-speaking geomorphologists follow the basic plans that he laid down for describing and classifying landforms.

Initial and sequential landforms

In terms of the grand scale of geological processes there are two fundamental classes of landforms. First, there are original crustal features

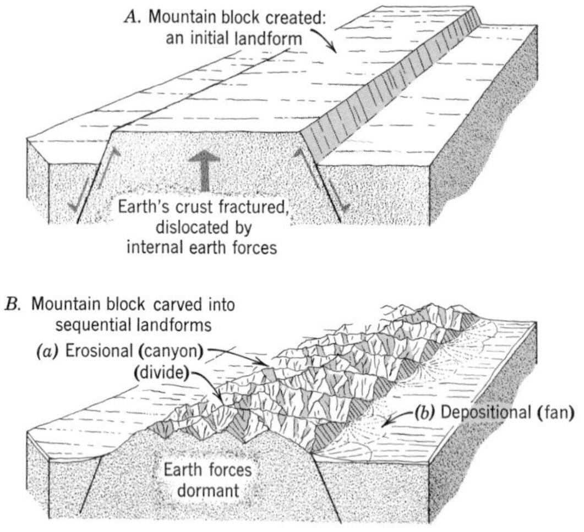

Figure 24.1 Initial and sequential landforms.

produced by tectonic and volcanic activity powered by internal sources of energy. These comprise the *initial landforms* (Figure 24.1*A*). Second, there are landforms made by agents of land surface reduction. Because these follow the initial forms and occur in orderly sequences, they are called collectively the *sequential landforms* (Figure 24.1*B*).

Any landscape is really nothing more than the existing stage in a great struggle or contest. The internal earth forces spasmodically elevate parts of the crust to create initial landforms. The external agents persistently keep wearing these masses down and carving them into vast numbers of smaller sequential landforms.

All stages of this struggle can be seen in various parts of the world. Where high, rugged mountains exist, the internal earth forces have recently dominated. Where certain kinds of low plains exist today, agents of surface reduction have finally triumphed. All intermediate stages can be found. Because the internal earth forces act repeatedly, new landmasses keep coming into existence as old ones are subdued. Judging from conditions in various periods of the geologic past, we are now in a time when continents stand relatively high above sea level. This observation suggests that internal forces were active relatively recently, geologically speaking.

Agents of land sculpture

Sequential landforms are products of one or more of the land-sculpturing agents, running water, waves, ice, and wind. These erosional agents are aided by processes of rock decay and downslope movements of soil and rock under gravitational force. The total process of land surface reduction is termed *denudation*. Denudation affects all continental rock masses that become elevated by orogenic and epeirogenic movements of the earth's solid crust. No part of the land surface is immune from attack. As soon as any rock mass comes to be exposed to the air or to wave attack it is set upon by the denudational agents and processes. They work to one ultimate goal—wearing away the landmass until it becomes a low plain, which is then slowly consumed by waves and perhaps finally covered by ocean waters. The disintegration products are spread over the sea floors surrounding the continents. The processes act with extreme slowness, to be sure, but geologic time is enormously great. Streams and waves seen in action today have had millions of years to do their work. Geomorphologists explain all sequential landforms as results of processes that can be seen acting at the present time.

In the wearing down of landmasses, a great variety of sequential landforms results. Where rock is eroded away, valleys are formed. Between the valleys are ridges, hills, or mountains representing unconsumed parts of landmasses. All such sequential landforms shaped by progressive removal of the bedrock mass are designated *erosional landforms.*

Rock and soil fragments that were removed are deposited elsewhere to make an entirely different set of features, the *depositional landforms.* Figure 24.2 illustrates these two groups of landforms. The ravine, canyon, peak, spur, col and bluffs are erosional landforms; the fan, built of rock fragments below the mouth of the ravine, is a depositional landform. The floodplain, built of material transported by a stream, is also a depositional landform. Once formed, depositional landforms can be carved up to yield a second generation of erosional landforms. The bluffs shown in Figure 24.2 are an example.

Bedrock, soil, and residual overburden

Examination of a freshly cut cliff, such as that in a new highway cut or quarry wall, will reveal several kinds of earth materials, shown in Figure 24.3. Solid, hard rock which is still in place and

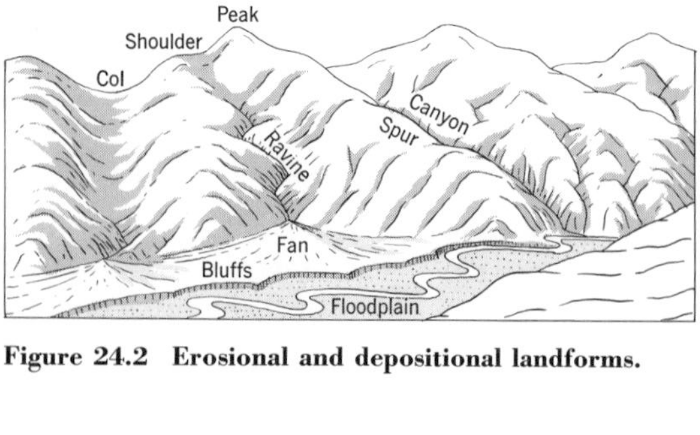

Figure 24.2 Erosional and depositional landforms.

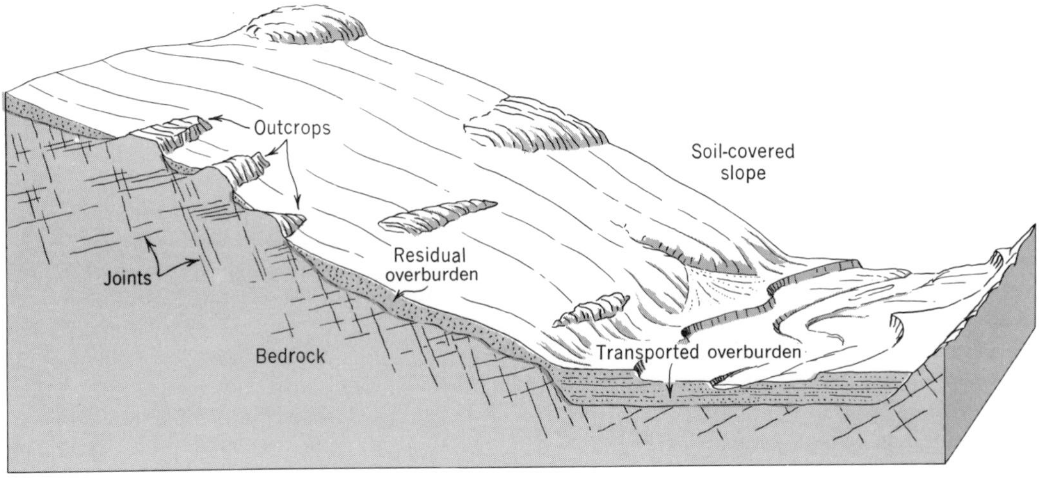

Figure 24.3 Residual and transported overburden.

relatively unchanged is called *bedrock.* It grades upward into a zone where the rock has become decayed and has disintegrated into clay, silt, and sand particles. This may be called the *weathered overburden,* or *residual overburden.* At the top is a layer of true *soil,* often called *topsoil* by farmers and gardeners. Soil properties, soil-forming processes, and soil classification were treated in Chapters 18 and 19. Over the soil may be a protective layer of grass, shrubs, or trees.

One or more of these zones may be missing. Sometimes everything is stripped off down to the bedrock, which then appears at the surface as an *outcrop.* Sometimes following cultivation or forest fires the true soil only is eroded away, leaving exposed the overburden, which is infertile and may become scored by deep gullies. The thickness of soil and overburden is quite variable. Although the true soil is rarely more than a few feet thick, the residual overburden of decayed and fragmented rock may extend down tens or even hundreds of feet. Formation of the overburden is greatly aided by the presence of innumerable bedrock cracks termed *joints* (Figure 24.3), along which water can move easily to promote rock decay.

Transported overburden

Another variety of overburden that may be found covering the bedrock is *transported overburden.* It consists of such materials as stream-laid gravels and sands, floodplain silts, clays of lake bottoms, beach and dune sands, or rubble left by a melting glacier. All types have in common a history of having been transported by streams, ice, waves, or wind.

Whereas residual overburden, formed in place by disintegration of bedrock below it, is of local origin, transported overburden consists of rock and mineral varieties from distant sources and may be quite unlike the underlying minerals and rocks. Figure 24.3 shows stream valley deposits, called *alluvium,* which would be designated as transported overburden in contrast to residual overburden of the adjacent hill slope. Once deposited, transported overburden may remain undisturbed for many thousands of years, in which case a true soil is formed in its uppermost layer.

In a broad sense all of the materials of which the depositional landforms are built up are of transported overburden. In later chapters many kinds of depositional landforms are described and explained in terms of the processes by which the particles are carried.

Influence of bedrock on landforms

Bedrock strongly influences shape, size, and development of the erosional landforms. In some places, rock takes the form of thin layers, lying horizontally, tilted, folded, or broken, as the case may be. Elsewhere it consists of thick, irregular masses extending down to great depths. Some varieties of rock are soft and are readily washed away by streams and waves; others are extremely resistant to all agents of weathering and erosion. To a considerable degree the weakness or resistance of rocks is determined by their origin and age. When varied rocks lie side by side near the surface of the earth's crust the agents of denudation etch them out according to their degree of resistance, the weak rock tending to form valleys or other types of depressions, the resistant ones standing out in bold relief as hills, mountains, or plateaus. Consequently, landforms reflect closely the shape and arrangement of the original rock masses. The study of controls exerted by rock bodies upon landforms will follow logically after a systematic study of the action of the agents of denudation.

The wasting of slopes

The term *slope,* as used in geomorphology, designates some small element or area of the land surface which is inclined from the horizontal. Thus, we speak of "mountain slopes," "hill slopes," or "valley-side slopes" with reference to the inclined ground surfaces extending from divides and summits down to valley bottoms.[1]

Slopes are required for the flow of surface water under the influence of gravity. Therefore, slopes are fitted together to form drainage systems in which surface water flow converges into stream channels, these, in turn, conduct the water and rock waste to the oceans to complete the hydrologic cycle. Nature has so completely provided the earth's land surfaces with slopes that perfectly horizontal or vertical surfaces are extremely rare.

Our concern in this chapter is with the wasting of land slopes under the influence of atmospheric processes in conjunction with gravity. Emphasis is on the slow processes whereby bedrock is transformed into residual overburden. This material in turn moves down the slopes to channels where it can be taken by stream flow to still more distant, lower areas. Slopes are also shaped by other processes—glaciers, winds, and waves—and these will be explained in later chapters.

Weathering and mass wasting

Weathering is the combined action of all processes whereby rock is decomposed and disintegrated because of exposure at or near the earth's surface. Weathering normally changes hard, massive rock into finely fragmented, soft residual overburden, the parent matter of the soil. For this reason, weathering is often described as the preparation of rock materials for transportation by the agents of land erosion—flowing water, glacial ice, waves, and wind. Because gravity exerts its force on all matter, both bedrock and the products of weathering tend to slide, roll, flow, or creep down all slopes in a variety of types of earth and rock movements grouped under the term *mass wasting.*

Weathering processes may be subdivided into two large groups, *physical (mechanical) weathering* and *chemical weathering.* Although these processes are extremely complex and act in combinations that are hard to separate into simple concepts, we shall attempt to identify the most important individual changes and to show what landforms or surface features of the rock and soil are caused by each.

[1] *Slope* is also used to mean inclination from the horizontal, measured as in *dip* (Chapter 28); or it is used as a verb *to slope,* meaning *to incline.*

Geometry of rock breakup

Before examining weathering processes, it is well to introduce four terms applied to the geometrical manner in which bedrock breaks into smaller pieces. In so doing we are not considering the possible forces involved, merely the shapes of the rock fragments as they appear to the eye.

Rocks composed of rather coarse mineral grains (intrusive igneous rocks of granitoid texture and coarse clastic sedimentary rocks) commonly fall apart grain by grain, a form of breakup termed *granular disintegration* (Figure 24.4). The product is a gravel or sand in which each grain consists of a single mineral particle separated from its fellows along the original crystal or grain boundaries. *Exfoliation* is the formation of curved rock shells which separate in succession from the original rock mass, leaving behind successively smaller spheroidal bodies (Figure 24.13). This type of breakup is also called *spalling.*

Where a rock has numerous joints produced previously by mountain-making pressures or by shrinkage during cooling from a magma, the common form of breakup is by *block separation* (Figure 24.4). Obviously, comparatively weak forces can separate such blocks, whereas great forces are required to make fresh fractures through

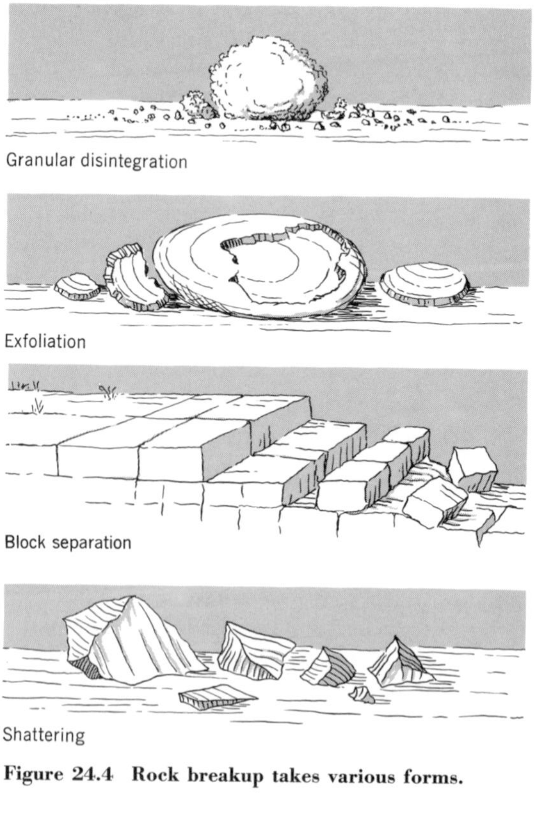

Figure 24.4 Rock breakup takes various forms.

Figure 24.5 A felsenmeer atop Medicine Bow Peak, Snowy Range, Wyoming, at 12,000 ft (3650 m) elevation. The rock is quartzite. (Photograph by A. N. Strahler.)

solid rock. In sedimentary rocks the planes of stratification, or bedding planes, comprise one set of planes of weakness commonly cutting at right angles to the joints. Figures 24.10 and 24.12 show joint blocks being separated by weathering forces. Of course, it is quite possible that a single, solid joint block will later break up either by granular disintegration or by exfoliation.

Shattering is the disintegration of rock along new surfaces of breakage in otherwise massive, strong rock, to produce highly angular pieces with sharp corners and edges (Figure 24.4). The surface of fracture may pass between individual mineral crystals or grains, or may cut through them. Blocks seen in Figure 24.5 are joint blocks, many of which have been shattered into smaller pieces.

Physical weathering processes and forms

The physical, or mechanical, processes of weathering produce fine particles from massive rock by the exertion of stresses sufficient to fracture the rock, but do not change its chemical composition. One of the most important physical weathering processes in cold climates is *frost action*, the repeated growth and melting of ice crystals in the pore spaces or fractures of soil and rock. As water in joints freezes, it forms needlelike ice crystals extending across the openings. As these ice needles grow, they exert tremendous force against the confining walls and can easily pry apart the joint blocks. Even massive rocks can be shattered by the growth of ice crystals created from water that has previously soaked into the rock. Where soil water freezes, it tends to form ice layers parallel with the ground surface, *heaving* the soil upward in an uneven manner.

Freezing water strongly affects soil and rock in all middle- and high-latitude regions having a cold winter season, but its effects are most striking in high mountains, above the timberline. Here the separation and shattering of joint blocks may produce an extensive ground surface littered with angular blocks (Figure 24.5). Such a surface is termed a *felsenmeer* (rock sea), or *boulder field.* Where cliffs of bare rock exist at high altitudes, fragments fall from the cliff face, building up piles of loose blocks into conical forms, termed *talus cones* (Figure 24.15).

Closely related to the growth of ice crystals is the weathering process of rock disintegration by growth of salt crystals. This process operates extensively in dry climates and is responsible for many of the niches, shallow caves, rock arches, and pits in sandstone formations. During long drought periods, ground water is drawn to the surface of the rock by capillary force. As evaporation of the water takes place in the porous outer zone of the sandstone, tiny crystals of salts are left behind. The growth force of these crystals is capable of producing granular disintegration of the sandstone, which crumbles into a sand and is swept away by wind and rain. Especially susceptible are zones of rock lying close to the base of a cliff, for here the ground water tends to seep outward, perhaps prevented from further downward percolation by impervious layers below (Figure 24.6). In the southwestern United States, many of the deep niches thus formed were occupied by Indians, whose cliff dwellings obtained protection from the elements as well as safety from armed attack (Figure 24.7).

An important but little appreciated process of physical weathering is the continual swelling and

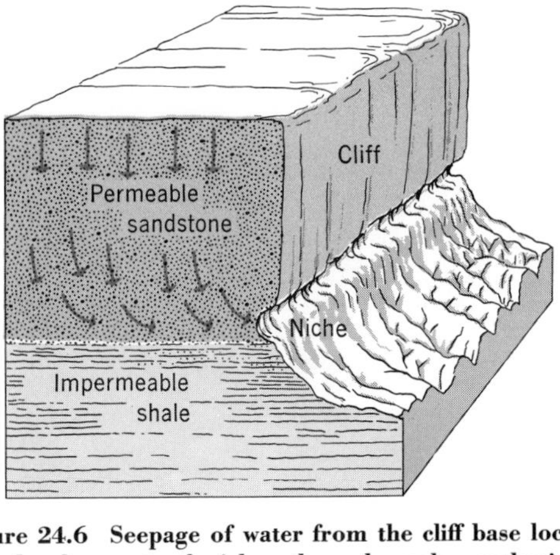

Figure 24.6 Seepage of water from the cliff base localizes development of niches through rock weathering.

Figure 24.7 White House Ruin occupies a deep niche in the sandstone wall of Canyon de Chelly, Arizona. (Photograph by Ray Atkeson.)

shrinking of soils as the particles of fine silt and clay absorb or give up soil water in alternate periods of rain and drought. Shrinkage forms soil cracks in dry periods, making the infiltration of rainfall much more rapid in early stages of an ensuing rain. In clay-rich sedimentary rocks such as shales, the swelling is largely responsible for a spontaneous breakup known as *slaking*, in which

Figure 24.8 Sheeting of granite, a large scale form of exfoliation, facilitates quarrying operations. (Photograph by courtesy of Light Quarry Division of Rock of Ages Corporation, Barre, Vermont.)

the shale crumbles into small chips or pencil-like fragments when exposed to the air.

Most crystalline solids, such as the minerals of rocks, tend to expand when heated and to contract when cooled. Where rock surfaces are exposed daily to the intense heating of the sun alternating with nightly cooling, the resulting expansion and contraction exerts powerful forces upon the rock. Given sufficient time (tens of thousands of such daily alternations), even the strongest rocks may develop fractures. Breakage can take the form of exfoliation or granular disintegration.

A curious but widespread process related to physical weathering results from *unloading*, the relief of confining pressure, as rock is brought nearer to the earth's surface through the erosional removal of overlying rock. Geologists think that rock formed at great depth beneath the earth's surface (particularly igneous and metamorphic rock) is in a slightly contracted state because of the tremendous pressures applied during mountain-making crustal deformations. On being brought to the surface, the rock expands slightly in volume and, in so doing, great shells of rock break free from the parent mass below. The new surfaces of fracture are a form of joint termed *sheeting structure* and show best in massive rocks such as granite and marble, because in a closely jointed rock the expansion would be taken up among the blocks. The rock sheets or shells produced by unloading generally parallel the ground surface and therefore tend to dip valleyward. On granite coasts the shells are found to dip seaward at all points along the shore. Sheeting structure is well seen in quarries, where it greatly facilitates the removal of rock (Figure 24.8).

Where sheeting structure has formed over the top of a single large body of massive rock, an *exfoliation dome* is produced (Figure 24.9). These are among the largest of the landforms due primarily to weathering. In the Yosemite Valley region, California, where domes are spectacularly displayed, the individual rock shells may be as thick as 20 to 50 ft (6 to 15 m).

Other large, smooth-sided rock domes lacking in shells are not true exfoliation domes, but are formed by granular disintegration of a single body of hard, coarse-grained intrusive igneous rock lacking in joints. Examples are the Sugar Loaf of Rio de Janeiro and Stone Mountain, Georgia (Figure 24.4), which rise prominently above surrounding areas of weaker rock.

Finally, in this list of physical weathering processes, the wedging of plant roots deserves consideration as a possible mechanism whereby joint blocks may be separated. We have all seen at one time or another a tree whose lower trunk and roots are firmly wedged between two great joint blocks of

Figure 24.9 North Dome and Basket Dome in Yosemite National Park, California, are exfoliation domes developed from huge masses of solid igneous rock. (Photograph by Douglas Johnson.)

massive rock (Figure 24.10). Whether the tree has actually been able to spread the blocks farther apart is doubtful at best. However, it is certain that the growth of tiny rootlets in joint fractures must be of great importance in loosening countless small rock scales and grains, particularly when a rock has already been softened by decay or fractured by frost action.

Chemical weathering processes and forms

The processes of chemical weathering were explained in Chapter 22, under the essentially synonomous heading of mineral alteration. Recall that oxidation, hydrolysis, and carbonic acid reaction are dominant chemical processes affecting rocks. The stable products of alteration are largely hydrous oxides of aluminum and iron, and various hydrous aluminosilicates. Alteration greatly softens a rock, while the adsorption of water causes volume expansion, and this in turn leads to spontaneous disruption.

The hydrolysis of granite, with accompanying granular disintegration and some exfoliation of thin scales, produces many interesting boulder and pinnacle forms by rounding of angular joint blocks (Figures 22.11 and 22.12). These forms are particularly conspicuous in arid regions because of the absence of any thick cover of soil and vegetation. There is ample moisture in most deserts for hydrolysis to act, given sufficient time. Hydrolysis in fine-grained mafic igneous rocks, such as basalt, commonly gives small-scale exfoliation of a type called *spheroidal weathering* (Figure 24.13).

Figure 24.10 Jointing in sandstone resembles pavement blocks at Artists View, Catskill Mountains, New York. (Photograph by A. N. Strahler.)

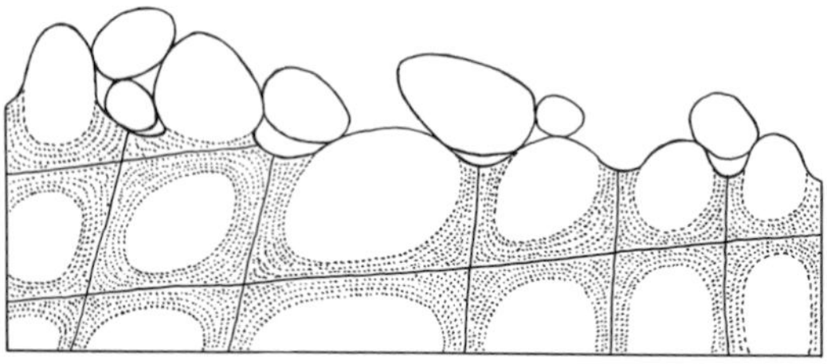

Figure 24.11 Stages in the development of egg-shaped boulders from rectangular joint blocks. (After W. M. Davis.)

In warm, humid climates, hydrolysis of susceptible rocks goes on below the soil and may result in the deep decay or rotting of igneous and metamorphic rocks to depths as much as 100 to 300 ft (30 to 90 m). Geologists who first studied this deep rock decay in the Southern Appalachian region termed the rotted layer *saprolite.* To the engineer, occurrences of deep weathering are of major importance in construction of highways, dams, or other heavy structures. Although saprolite can be excavated by power shovels with little blasting, there is serious danger in the weakness of the material in bearing heavy loads, as well as undesirable plastic properties because of a high content of clay minerals.

Figure 24.12 Egg-shaped granite boulders are produced from joint blocks by granular disintegration in a semiarid climate near Prescott, Arizona. (Photograph by A. N. Strahler.)

Reaction of carbonic acid with carbonate minerals is a major process of chemical weathering in areas underlain by limestone and dolomite. Surfaces of limestones are commonly deeply pitted and grooved (Figure 24.14). More important is the removal of vast amounts of rock in underground locations to produce cavern systems into which surface water disappears (Chapter 28).

It has been estimated that in a humid climate, such as that of the eastern United States, the ground surface of a limestone region may be lowered at the average rate of 1 ft (0.3 m) in 10,000 years through carbonic acid action alone.

In any soil rich in decaying plant matter, a variety of organic acids is formed in the soil solution, and these also react with mineral surfaces to produce chemical weathering. The salts that are products of such reactions are carried down through the soil into the ground water zone, then eventually to streams.

The weathering processes reviewed above, both physical and chemical, work universally but produce few distinctive large landforms or spectular activities that would draw the attention of the average person. Nevertheless, these processes are of enormous importance in slope development in that they prepare the bedrock for soil formation and for erosional removal by the agents of land sculpture. Without the weathering processes, vegetation could not thrive as we know it today, nor could the great continental land masses be easily reduced by the agents of denudation.

Figure 24.13 Spheroidal weathering, shown here, has produced many thin concentric shells in a basaltic igneous rock. Lucchetti Dam, Puerto Rico. (Photograph by C. A. Kaye, U.S. Geological Survey.)

Figure 24.14 Solution rills in limestone, west of Las Vegas, Nevada. Scale is indicated by pocket knife in center. (Photograph by John S. Shelton.)

Figure 24.15 Talus cones at the base of a frost-shattered cirque headwall. Moraine Lake in the Canadian Rockies. (Photograph by Ray Atkeson.)

Mass wasting

Everywhere on the earth's surface, gravity pulls continually downward on all materials. Bedrock is usually so strong and well supported that it remains fixed in place, but should a mountain slope become too steep through removal of rock at the base, bedrock masses break free, falling or sliding to new positions of rest. In cases where huge masses of bedrock are involved, the result may be catastrophic in loss to life and property in towns and villages in the path of the slide. Soil and overburden, being poorly held together, are much more susceptible to gravity movements. There is abundant evidence that on most slopes at least a small amount of downhill movement is going on at all times. Much of this is imperceptible, but sometimes the overburden slides or flows rapidly.

Taken altogether, the various kinds of downslope movements occurring under the pull of gravity, which we have collectively termed mass wasting, constitute an important process in slope wasting and denudation of the lands. A few of the commoner kinds of gravity movements and resulting landforms are described here.

Talus cones

Steep rock walls of gorges and high mountains shed countless rock particles under the attack of physical weathering processes, particularly frost action. These accumulate in a distinctive landform, the *talus cone* (Figure 24.15). A talus slope, or *scree slope*, as it is often called, has a remarkably constant slope angle of about 34° or 35°. So long as the talus slope is freshly formed and contains little very fine material mixed in with the coarse, the angle is constant within one or two degrees of variation, regardless of the rock type or the shape of the blocks.

Most cliffs are notched by narrow ravines which funnel the fragments into individual tracks, so as to produce conelike talus bodies arranged side by side along the cliff. Where a large range of sizes of particles is supplied, the larger pieces, by reason of their greater momentum and ease of rolling, travel to the base of the cone, whereas the tiny grains lodge in the apex. This tends to sort the fragments by size, progressively finer from base to apex (Figure 24.16).

Most fresh talus slopes are unstable, so that the disturbance created by walking across the slope, or dropping of a large rock fragment from the cliff above, will easily set off a sliding of the surface layer of particles. The upper limiting angle to which coarse, hard, well-sorted rock fragments will stand is termed the *angle of repose.* Unstable slopes at the angle of repose are also found on the lee sides of sand dunes.

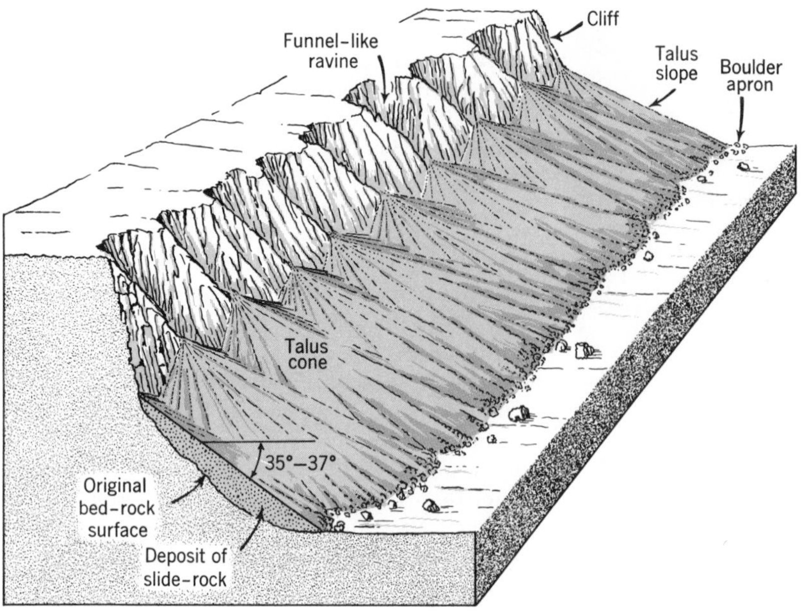

Figure 24.16 Idealized diagram of talus cones formed at the base of a cliff, which might be 200 to 500 ft (60 to 150 m) high.

Soil creep

On almost any moderately steep, soil-covered slope, some evidence may be found of extremely slow downslope movement of soil and overburden, a process called *soil creep*. Figure 24.17 shows some of the evidence that the process is going on. Joint blocks of distinctive rock types are found moved far downslope from the outcrop. In some layered rocks such as shales or slates, edges of the strata seem to bend in the downhill direction. This is not true plastic bending, but is the result of slight movement on many small joint cracks (Figure 24.18). Fence posts and telephone poles lean downslope and even shift measurably out of line. Retaining walls of road cuts lean and break outward under pressure of soil creep from above.

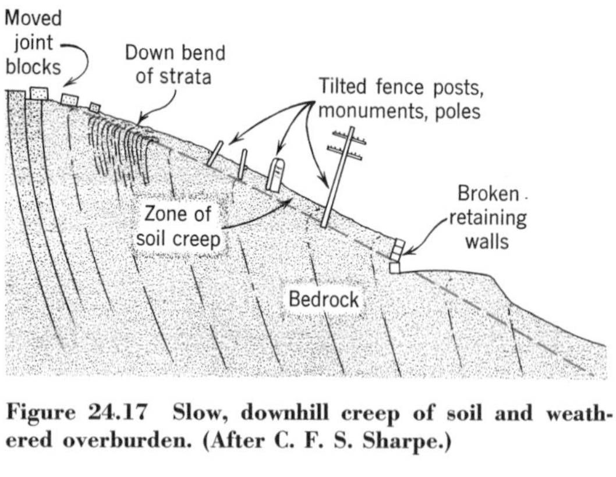

Figure 24.17 Slow, downhill creep of soil and weathered overburden. (After C. F. S. Sharpe.)

What causes soil creep? Heating and cooling of the soil, growth of frost needles, alternate drying and wetting of the soil, trampling and burrowing by animals, and shaking by earthquakes all produce some disturbance of the soil and mantle. Because gravity exerts a downhill pull on every such rearrangement that takes place, the particles are urged progressively downslope.

Creep affects rock masses enclosed in the soil or lying upon bare bedrock. Huge boulders which have gradually crept down a mountain side in large

Figure 24.18 Slow creep has caused this down hill bending of steeply dipping sandstone layers. (Photograph by courtesy of Ward's Natural Science Establishment.)

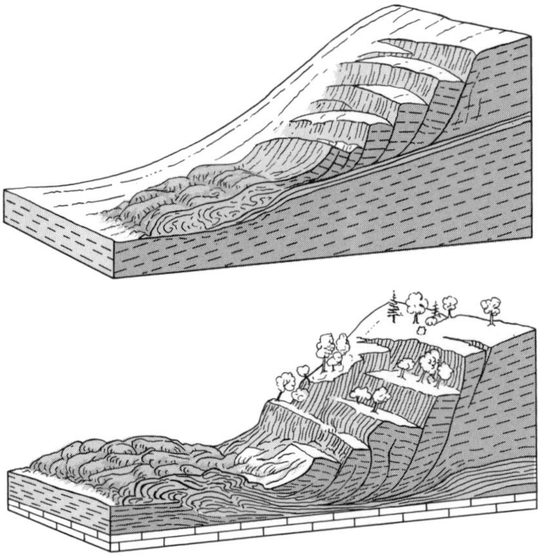

Figure 24.19 Two varieties of earthflow. (After E. Raisz.)

numbers may accumulate at the mountain base to produce a boulder field.

Earthflow

In humid climate regions, if slopes are steep, masses of water-saturated soil, overburden, or weak bedrock may slide downslope during a period of a few hours in the form of *earthflows.* Figures 24.19 and 24.20 are sketches of earthflows showing how the material slumps away from the top, leaving steplike terraces bounded by arcuate scarps, and flows down to form a bulging "toe" in which wrinkles are curved convexly downslope.

Shallow earthflows, affecting only the soil and residual overburden, are common on sod-covered slopes that have been saturated by heavy rains. An earthflow may affect a few square yards, or it may cover an area of several acres. If the bedrock is rich in clay (shale or deeply weathered igneous rocks), earthflow sometimes include millions of tons of bedrock, moving by plastic flowage like a great mass of thick mud.

A special variety of earth flowage characteristic of arctic regions is *solifluction* (from Latin words meaning *soil* and *to flow.* In late spring and early summer, when thawing has penetrated the upper

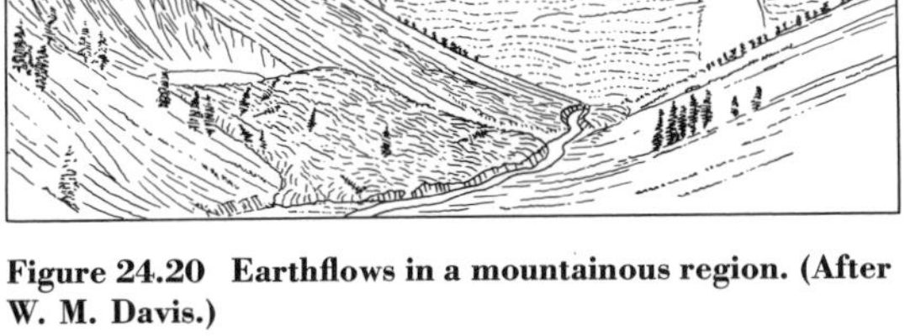

Figure 24.20 Earthflows in a mountainous region. (After W. M. Davis.)

Figure 24.21 Solifluction lobes cover this Alaskan mountain slope in the tundra climate region. (Photograph by P. S. Smith, U.S. Geologial Survey.)

few feet, soil is fully saturated with water which cannot escape downward because of the underlying impermeable frozen mass (permafrost). Flowing almost imperceptibly, this saturated soil forms terraces and lobes that give the mountain slope a stepped appearance (Figure 24.21).

Mudflow

One of the most spectacular forms of mass wasting is the *mudflow,* a mud stream of fluid consistency which pours down canyons in mountainous regions (Figure 24.22). In deserts, where vegetation does not protect the mountain soils,

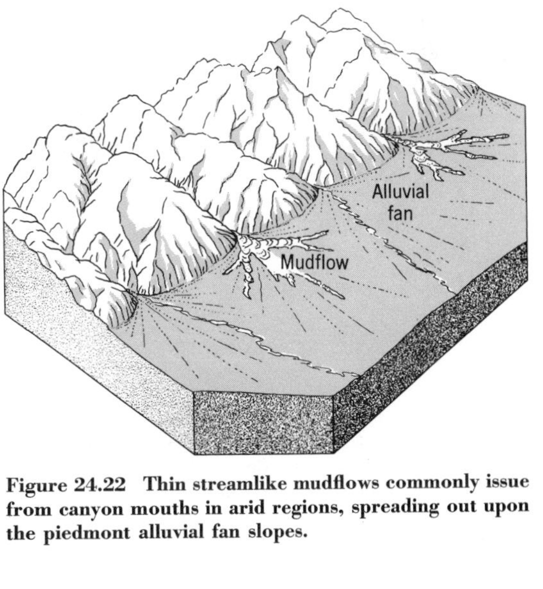

Figure 24.22 Thin streamlike mudflows commonly issue from canyon mouths in arid regions, spreading out upon the piedmont alluvial fan slopes.

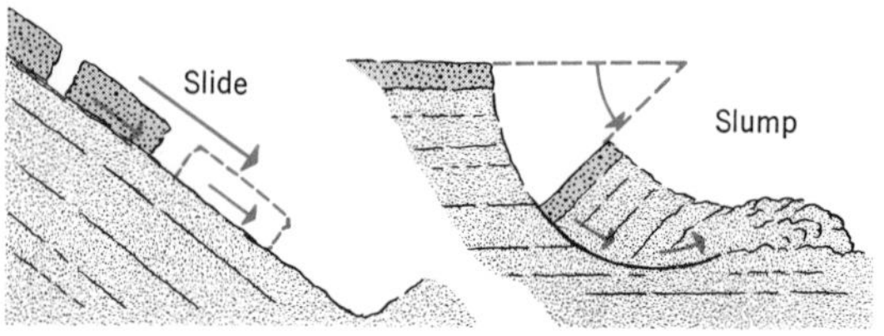

Figure 24.23 Landslides may involve (*A*) slip on a nearly plane surface or (*B*) slump with rotation on a curved plane.

violent local storms produce rain much faster than it can be absorbed by the soil. As the water runs down the slopes it forms a thin mud, which flows down to the canyon floors. Following stream courses, the mud continues to flow until it becomes so thickened that it must stop. Great boulders are carried along, buoyed up in the mud. Roads, bridges, and houses in the canyon floor are engulfed and destroyed. If the mudflow emerges from the canyon and spreads across a piedmont plain, property damage and loss of life can result, because in desert regions the plains lying at the foot of a mountain range which supplies irrigation water may be heavily populated.

Mudflows also occur on the slopes of erupting volcanoes. Freshly fallen volcanic ash and dust is turned into mud by heavy rains and flows down the slopes of the volcano. Herculaneum, a city at the base of Mt. Vesuvius, was destroyed by a mudflow during the eruption of 79 A.D., when the neighboring city of Pompeii was buried under volcanic ash.

Landslide

Landslide is the rapid sliding of large masses of rock with little or no flowage of the materials as in the previous types. Two basic forms of landslide are (*a*) *rockslide*, in which the bedrock mass slips on a relatively flat inclined rock plane, such as a fault or bedding plane, and (*b*) *slump*, in which there is backward rotation on a curved up-concave slip plane (Figure 24.23).

Wherever steep mountain slopes occur, there is a possibility of great and disastrous rockslides. In Switzerland, Norway, or the Canadian Rockies, for example, villages built on the floors of steep-sided valleys are sometimes destroyed and their inhabitants killed by the sliding of millions of cubic yards of rock, set loose without any warning.

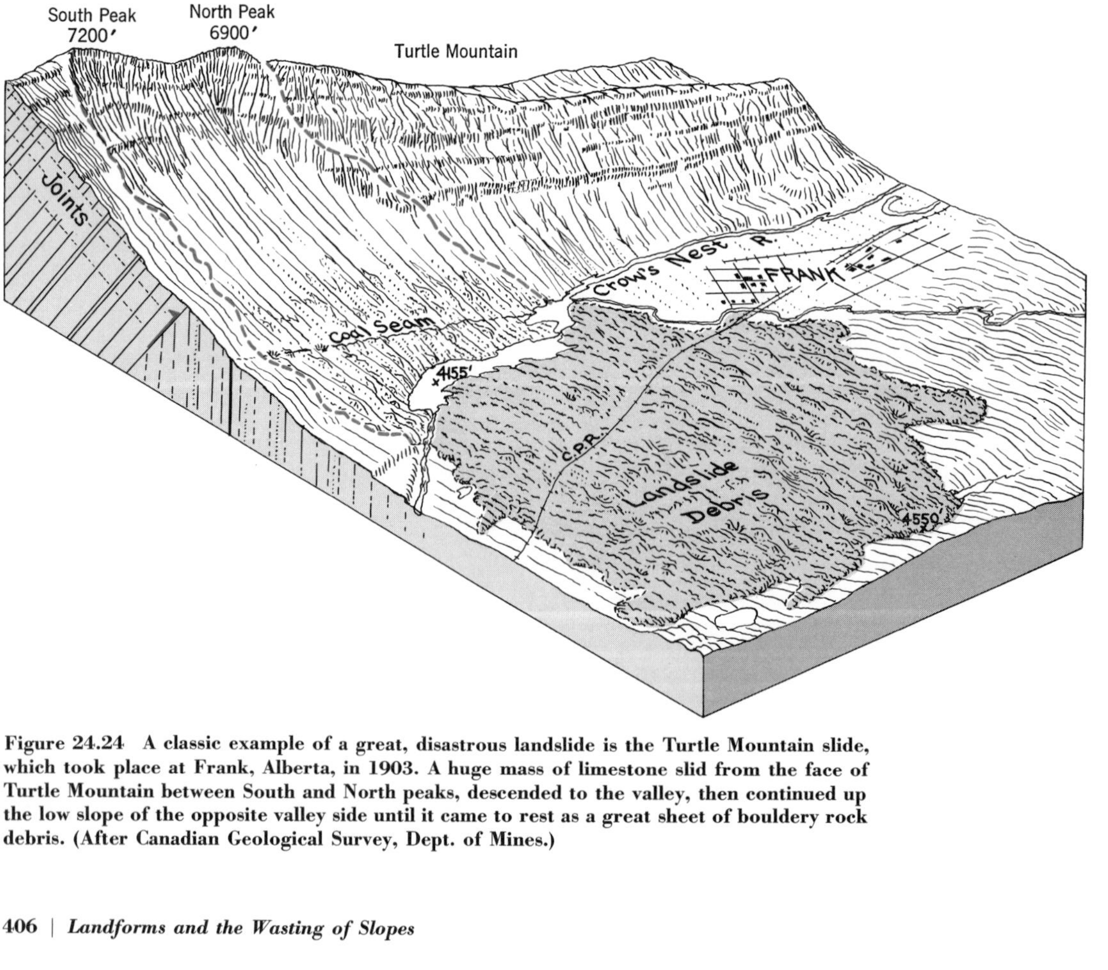

Figure 24.24 A classic example of a great, disastrous landslide is the Turtle Mountain slide, which took place at Frank, Alberta, in 1903. A huge mass of limestone slid from the face of Turtle Mountain between South and North peaks, descended to the valley, then continued up the low slope of the opposite valley side until it came to rest as a great sheet of bouldery rock debris. (After Canadian Geological Survey, Dept. of Mines.)

Figure 24.25 The Madison Slide, southwestern Montana, photographed shortly after it occurred. Water of the Madison River, impounded by the slide, is rising to form a new lake, named Earthquake Lake. Within three weeks the lake was nearly 200 ft (60 m) deep. (Photograph by N. R. Farbman, Life Magazine, © 1959, Time Inc.)

The Turtle Mountain landslide of 1903 in Alberta, shown in Figure 24.24, involved the sliding of an enormous mass of limestone, its volume estimated at 35 million cubic yards (27 million cu m), through a descent of 3000 ft (900 m). The debris buried a part of the town of Frank, with a loss of 70 persons. A similar disaster occurred in 1959 in Montana when a severe earthquake (Hebgen Lake earthquake, Richter scale 7.1) caused an entire mountainside to slide into the Madison River gorge, killing 27 persons (Figure 24.25). The Madison Slide, as it is known, formed a debris dam over 200 ft (60 m) high and produced a new lake. Volume of this slide was about the same as in the Turtle Mountain slide. The lake has now been made permanent by construction of a protected spillway. Severe earthquakes in mountainous regions are a major immediate cause of landslides, earthflows, and other forms of mass displacements.

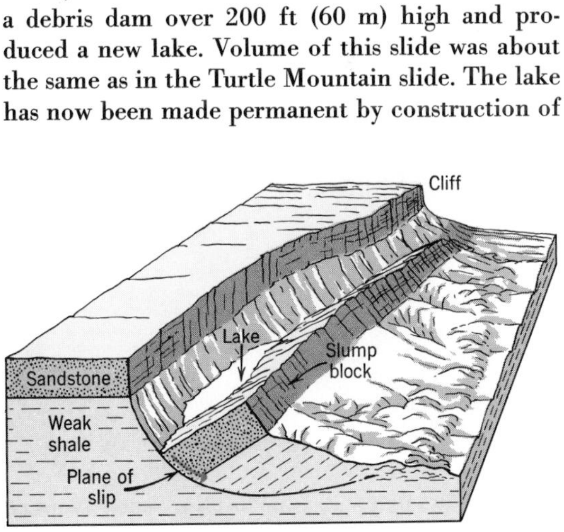

Figure 24.26 Slump blocks rotate backward as they slide from a cliff.

Aside from occasional great catastrophes, rockslides do not have strong environmental influence because of their sporadic occurrence in thinly populated mountainous regions. Small slides may, however, repeatedly block or break an important mountain highway or railway line.

The second form of landslide produces *slump blocks*, great masses of bedrock or overburden that slide downward from a cliff, at the same time rotating backward on a horizontal axis (Figure 24.26). Wherever massive sedimentary strata, usually sandstones or limestones, or lava beds, rest upon weak clay or shale formations, a steep cliff tends to be formed by erosion. As the weak rock is eroded from the cliff base, the cap rock is undermined. When a point of failure is reached, a large block breaks off, sliding down and tilting back along a curving plane of slip. Slump blocks may be as much as 1 to 2 mi (1.5 to 3 km) long and 500 ft (800 m) thick. A single block appears as a ridge at the base of the cliff. A closed depression or lake basin may lie between the block and the cliff.

Rockfall and debris avalanche

Most rapid of all mass wasting processes is *rockfall*, the free falling or rolling of single masses

Figure 24.27 This rockfall on the Palisades of the Hudson River, just north of the George Washington Bridge, took place in November 1955. About 1200 tons of broken rock fragments resulted from the fall of a large columnar joint block of diabase. (Photograph by Bergen Evening Record.)

of rock from a steep cliff. Individual fragments may be as small as boulders, or as large as a city block. Large blocks disintegrate upon falling, strewing the slope below with rubble and leaving a conspicuous scar on the cliff face (Figure 24.27).

A related phenomenon of high, alpine mountain chains, where glacial erosion has produced extremely steep valley gradients and where large quantities of glacial rock rubble (moraine) and relict glaciers are perched precariously at high positions, is the *alpine debris avalanche.* This sudden rolling of a mixture of rock waste and glacial ice can produce a tongue of debris traveling downvalley at a speed little less than that of a freely falling body. A recent disaster of this type occurred in 1970 in the high Andes of Peru. A severe earthquake (magnitude 7.7 on the Richter scale), which caused widespread death and destruction, set off the fall of a large snow cornice from a high peak, Huascaran. After a free fall of 3000 ft (900 m), the snow mass was partially melted by impact and incorporated a great quantity of loose rock to become a debris avalanche. Traveling downvalley at a speed calculated to have reached 300 mi (480 km) per hour, the avalanche wiped out the town of Yungay and several smaller villages. The death toll, which included earthquake casualties, was estimated in the thousands.

Man-induced mass wasting

Man's activities induce mass wasting in forms ranging from mudflow and earthflow to rockslide and slump. These activities include (1) piling up of waste soil and rock into unstable accumulations that fail spontaneously, and (2) removal of support by undermining natural masses of soil, overburden, and rock.

Spoil banks produced by strip mining of coal are unstable and a constant threat to the lower slope and valley bottom below. When saturated by heavy rains and melting snows, the spoil generates earthflows and mudflows that descend upon houses, roads, and forest.

At Aberfan, Wales, a major disaster occurred when a hill 600 ft (180 m) high, built of rock waste (culm) from a nearby coal mine, spontaneously began to move as an earthflow. The debris tongue overwhelmed part of the town below, destroying a school and taking over 150 lives (Figure 24.28).

Examples of both large and small earthflows induced or aggravated by man's activities are found in the Palos Verdes Hills of Los Angeles County, California. These movements occur in shales that tend to become plastic when water is added. The upper part of the earthflow undergoes a slump motion with backward rotation of the down-sinking mass. The interior and lower parts

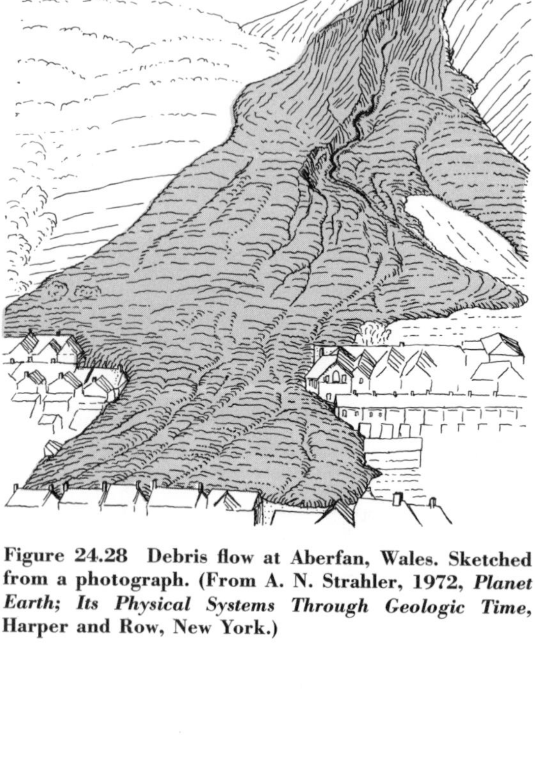

Figure 24.28 Debris flow at Aberfan, Wales. Sketched from a photograph. (From A. N. Strahler, 1972, *Planet Earth; Its Physical Systems Through Geologic Time,* Harper and Row, New York.)

of the mass move by slow flowage and a toe of extruded flowage material may be formed. Largest of the earthflows in this area is the Portuguese Bend "landslide," which affected an area of 300 to 400 acres. The total motion over a three-year period was about 70 ft (20 m). Damage to residential and other structures totaled some $10 million. The slide has been attributed by geologists to infiltration of water from cesspools and from irrigation water applied to lawns and gardens. A discharge of over 30,000 gallons of water per day from some 150 homes is believed to have sufficiently weakened the shale beneath to start and sustain the flowage.

REVIEW QUESTIONS

1. What are landforms? What is geomorphology? Of what environmental importance are landforms?

2. What is a genetic system of description? How does it differ from an empirical description?

3. What advantage is there to describing and classifying landforms according to structure, process, and stage? Explain these terms.

4. What is the basis for dividing all landforms into initial and sequential groups? What are the agents of land sculpture? Toward what goal do they work?

5. Explain what is meant by the terms *erosional landforms* and *depositional landforms.*

6. Explain how a landscape reflects the work of both internal and external earth forces and processes.

7. Define bedrock, overburden, and soil. What is an outcrop? What is the distinction between transported and residual overburden?

8. In what way is an understanding of rocks and their structures necessary in explaining landforms?

9. What is the meaning of the term slope?

10. Define and distinguish between weathering and mass wasting. Into what two large groups can the weathering processes be divided?

11. Describe four common geometrical forms of rock breakup.

12. List the physical processes of weathering. Of these, which require water to be present? Which do not? Which are controlled by changes in air temperature? Which can act deep in the soil or bedrock? Which cannot?

13. What is a felsenmeer? Under what climatic environment would it be formed?

14. How is a talus cone produced and what angle of slope does it normally have?

15. How can the growth of salt crystals cause rock disintegration? What climatic conditions favor this process? What forms result?

16. Under what conditions can temperature changes alone cause rock disintegration? What evidence is available that the process is effective?

17. Explain how sheeting structure and exfoliation domes result from unloading. What rock type is most likely to exhibit such structure?

18. Comment on the effectiveness of plant growth forces in producing the disintegration of rock.

19. What three groups of changes come under the general heading of chemical weathering processes?

20. How does hydrolysis promote weathering? What rocks are most susceptible to this type of decay? What are some of the visible forms resulting from hydrolysis in granitic rocks? In mafic rocks? How deep in the ground do the effects of hydrolysis extend?

21. How is carbonic acid formed and how does it act on rock? What type of rock is most susceptible? What surface forms are produced? What other acids are commonly found in soil water?

22. What role does mass wasting play in the denudation of the lands? List the evidences of soil creep. What general type of mechanism causes soil creep? Are large rock masses also affected by creep? Explain.

23. Describe an earthflow. How large are earthflows? Under what conditions of season and climate might earthflows be expected to take place? Describe and explain the special features of solifluction.

24. Under what topographic and climatic conditions do mudflows occur? Point out the similarities and differences between a mudflow and a stream.

25. To what kinds of movement and earth material is landslide restricted? Distinguish between rockslide and slump as two basic types of landslide.

26. What is rockfall? What sizes of rock masses may be included? What is a debris avalanche?

27. How can man's activity induce earthflows and landslides? Give examples.

Exercises

Exercise 1. ***Exfoliation Domes.*** (Source: Yosemite National Park, Calif., U.S. Geological Survey topographic map; scale 1:24:000.)

Explanatory Note: North Dome and Basket Dome, two great exfoliation domes of massive granitic intrusive rock rise above the northwest wall of Tenaya Canyon, a branch of Yosemite Valley. These domes are pictured in Figure 24.9, viewed from the south. In the southeastern part of this map is the floor of Tenaya Canyon, a deeply scoured glacial trough. Much rock was removed from the northwest wall of Tenaya Canyon by ice of the Wisconsin glacial stage, causing steep, irregular cliffs. North Dome and Basket Dome, lying above the level of this glacier, escaped modification.

The remarkable fidelity of contours on this map is a tribute to the great topographer-geologist, Francois E. Matthes, who not only surveyed the area and drew the contours, but who also published a comprehensive volume on the geologic history of Yosemite Valley, listed among the reading references.

QUESTIONS

1. **(*a*)** What contour interval is used on this map? **(*b*)** What is the difference in elevation between the summit of Basket Dome and the floor of Tenaya Canyon?

2. **(*a*)** Calculate the angle of slope of the southwest side of Basket Dome between 7300 and 7500 ft elevation. **(*b*)** Do the same for the steepest part of the northeast face of Basket Dome. **(*c*)** Can you give a geological reason for the fact that Basket Dome is highly unsymmetrical, whereas North Dome is highly symmetrical?

3. Explain the curious angular zigzag bends in the contours at 0.6–0.4[2] and 0.2–0.2. (Examine Figure 24.9 closely.)

4. Note that the topography of the wall of Tenaya Canyon below 6500 ft elevation is steep, rough, and blocky in comparison with the smooth, broadly rounded slopes at higher elevation. Can you explain this contrast?

Exercise 2. ***Landslide.*** (Source: Map 57*A*, Frank, Alta., Geological Survey of Canada; scale 1:9,600.)

Explanatory Note: The great Turtle Mountain landslide of April 29, 1903, is shown on this map, modified from a special large-scale map made during an investigation of the cause of the disastrous slide, which wiped out a part of the town of Frank, Alberta, taking the lives of 70 persons. Figure 24.24 is a block diagram of this area and gives a geological cross section. Between North Peak and South Peak, on Turtle Mountain, a great mass of limestone slid away, descending about 2500 ft to the Crow's Nest River. As the rock mass disintegrated into rubble, a flowage movement developed; the momentum of the 35 to 40 million cubic yards of material was so great as to carry some debris 400 ft above river level on the east side of the valley.

On this map, limits of the slide are shown by a dashed line. Contour interval is reduced within the area of debris east of the river in order to show details of the topography. Series of dashed lines on Turtle Mountain in the vicinity of 0.3–0.7 represent open cracks, or fissures, in the limestone, which are fractures produced at the time of the landslide. (For further information on this slide, see a report by R. A. Daly et al, listed among the reading references.)

QUESTIONS

1. **(*a*)** What two contour intervals are used on this map? **(*b*)** What is the length, in miles, of the landslide area, measured from South Peak (0.3–0.3) to 3.6–1.5?

2. **(*a*)** What is the elevation of the hachured contour located at 2.4–1.2? **(*b*)** at 3.3–1.4? **(*c*)** What is the summit elevation of the hill of landslide debris at 2.2–1.0?

3. Taking the volume of debris of the landslide east of the Crow's Nest River to be 30 million cubic yards, what average thickness has the landslide debris within the limits of the slide east of the river? To compute this roughly, draw in the 500-yd grid lines, then total the area of the squares and part squares included within the dashed line. For a more accurate answer, use 100-yd grid squares.

4. How steep is the east face of Turtle Mountain between the 6000-ft contour at 0.68–1.50 and the 5000-ft contour at 0.96–1.50? State this in feet per mile. **(*b*)** Draw a right triangle whose legs are scaled proportionately to the vertical and horizontal distances obtained in **(*a*)**. Measure the angle of slope from this triangle with a protractor and state the answer in degrees.

5. Draw a profile from 0.0–0.4 to 3.8–1.5. Use a vertical scale equal to the horizontal scale of the map. Draw in the geological cross section as shown in Figure 24.24. Notice that the joints in the limestone are inclined a little less than the slope of the mountain side and are probably the planes of fracture on which the block began to slide.

[2] On all maps used in the exercises of Chapters 24 through 33, a 1000-yard grid system is marked on the margins of the maps. Locations of places will be given in grid coordinates as explained in Chapter 3. The first number gives distance to right from lower left-hand corner; the second number gives distance up from bottom line.

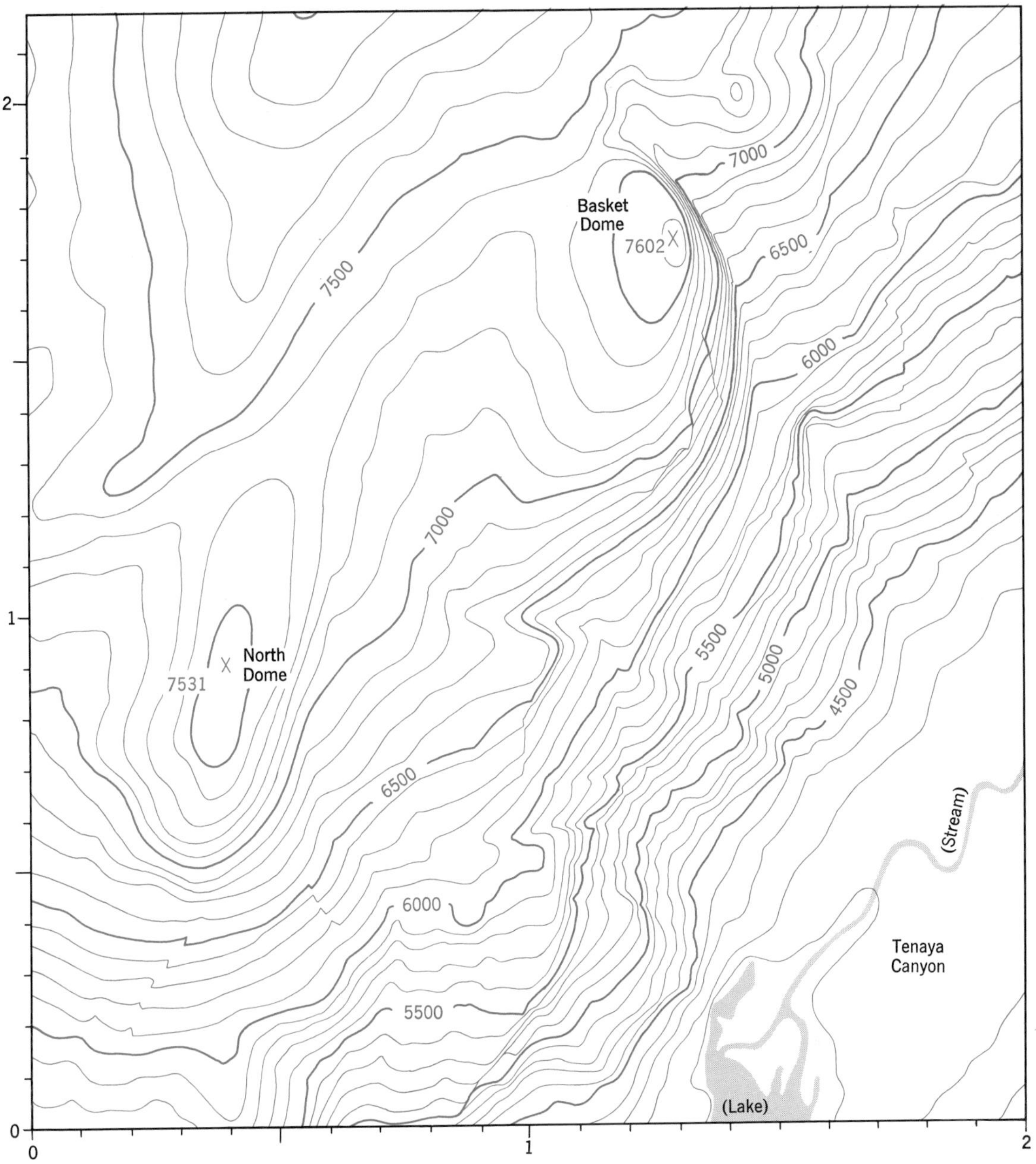
Basket Dome
7602
7000
6500
7500
6000
7000
North Dome
7531
5500
5000
4500
6500
(Stream)
6000
Tenaya Canyon
5500
(Lake)
2
1
0
0
1
2

Exercise 1

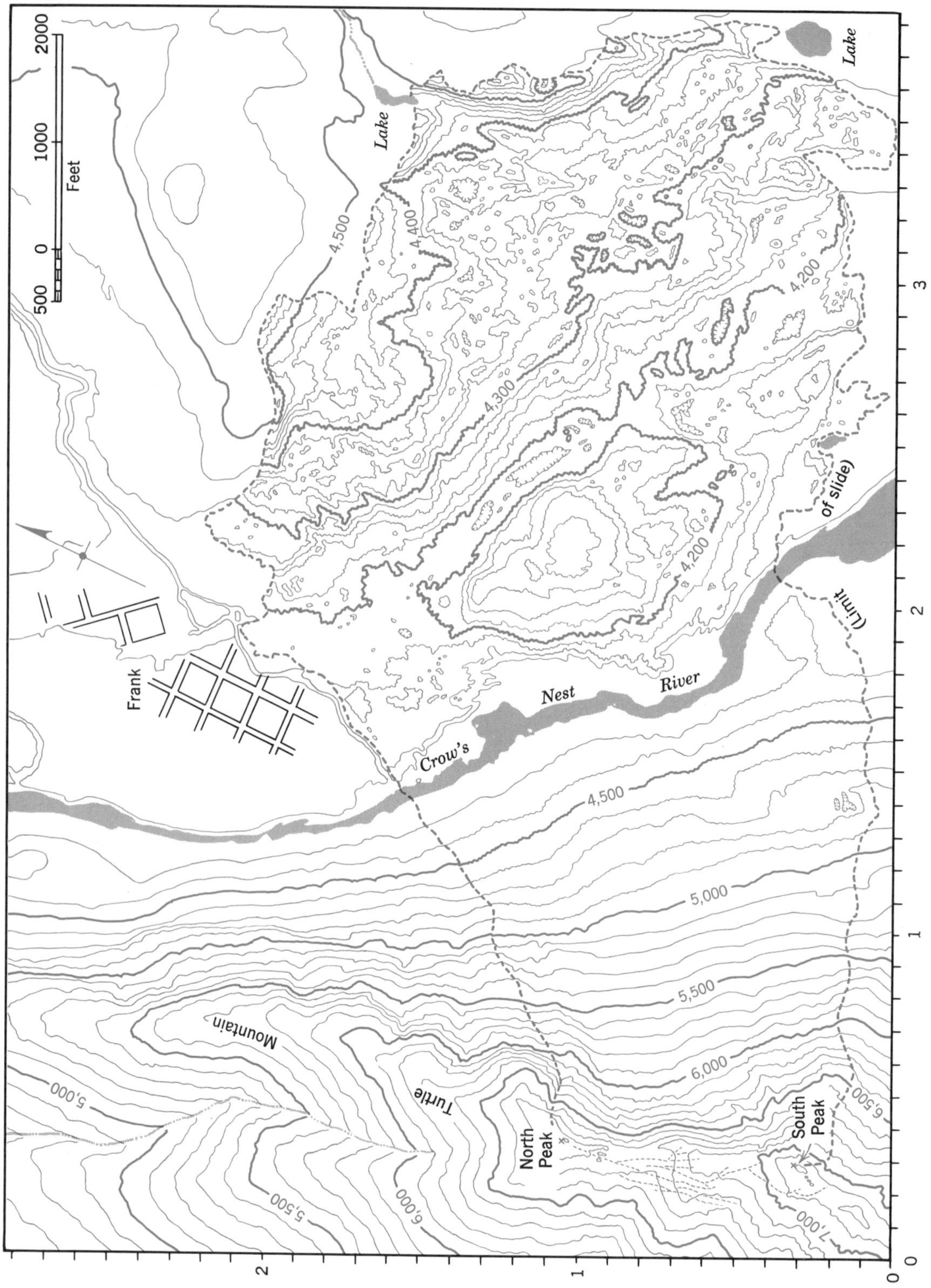
Lake
Lake
Frank
Crow's Nest River
(Limit of slide)
Turtle Mountain
North Peak
South Peak
Feet
500
0
1000
2000
4,200
4,300
4,400
4,500
5,000
5,500
6,000
6,500
7,000
0
1
2
3

Exercise 2

CHAPTER **25**

Landforms Made by Running Water

OVERLAND flow and streams have a geologic role as the principal agents of denudation. They not only constitute essential flow paths of the hydrologic cycle, but are also the carriers of mineral matter from the lands to the oceans. In this chapter we examine the work of running water on the lands as a fluid agent of both erosion and deposition. We start with the action of overland flow on land slopes, then turn to stream action in channel flow.

Erosion by overland flow

Overland flow, by exerting a dragging force over the soil surface, picks up particles of mineral matter ranging in size from fine clay to coarse sand or gravel, depending on the speed of the flow and the degree to which the particles are bound by plant rootlets or held down by a mat of leaves.

On bare soil surfaces, which may occur naturally in arid lands or in humid lands by man's activities in farming and grading of construction sites, rainfall plays a direct erosional role in conjunction with overland flow. The force of falling drops (Figure 25.1) causes a geyserlike splashing in which soil particles are lifted and then dropped into new positions, a process termed *splash erosion.* It is estimated that a violent rainstorm has the ability to disturb as much as 100 tons of soil per acre (225 metric tons per hectare). On a sloping ground surface, splash erosion tends to shift the soil slowly downhill. A more important effect is to cause the soil surface to become much less able to infiltrate water because the natural soil openings become sealed by particles shifted by raindrop splash. Reduced infiltration permits a much greater proportion of overland flow to occur from rain of given intensity and duration. The depth and velocity of overland flow then increase greatly, intensifying the rate of soil removal.

A cover of vegetation greatly increases the resistance of the ground surface to the force of erosion under overland flow. On a slope covered by grass sod, even a deep layer of overland flow causes little soil erosion because the energy of the moving water is dissipated in friction with the grass stems, which are tough and elastic. Similarly on a heavily forested slope, countless check dams made by leaves, twigs, roots, and fallen tree trunks take up the force of overland flow. Without such vegetative cover the eroding force is applied directly to the bare soil surface, easily dislodging the grains and sweeping them downslope.

Summarizing the variables involved, we find that eroding capacity of overland flow is directly proportional to the rate of precipitation and length of

Figure 25.1 A large raindrop (above) lands on a wet soil surface, producing a miniature crater (below). Grains of clay and silt are thrown into the air and the soil surface is disturbed. (Official U.S. Navy photograph.)

slope, but inversely proportional to both the infiltration capacity of the soil and the resistance of the surface. To complete this equation, we need only to add the effect of the steepness of ground slope. Obviously, the steeper the slope of ground, the faster is the flow and the more intense the erosion. We therefore add that the eroding capacity of overland flow increases directly with angle of slope. As the slope angle approaches the vertical, however, erosion will become less intense from overland flow because the ground surface intercepts much less of the vertically falling rain.

Figure 25.1 shows heavy runoff at the base of a long slope, where the accumulated overland flow has converged into broad shallow streams spreading across the slope.

Accelerated soil erosion

In humid climates having a permanent plant cover of forest trees, shrubs, or grasses, slow removal of soil is part of the natural geological process of landmass denudation and is both inevitable and universal. Under stable, natural conditions, the erosion rate in a humid climate is slow enough that a soil with distinct horizons is formed and maintained, enabling vegetation to maintain itself. Soil scientists refer to this state of activity as the *geologic norm.*

By contrast, the rate of soil erosion may be enormously speeded up through man-made activities or rare natural events to result in a state of *accelerated erosion,* removing the soil much faster than it can be formed. This condition comes about most commonly from a change in the conditions of vegetative cover and physical state of the ground surface. Destruction of vegetation by clearing of land for cultivation, or by forest fires, directly causes great changes in the relative proportions of infiltration to runoff. Interception of rain by foliage is ended; protection afforded by a ground cover of fallen leaves and stems is removed. Consequently the rain falls directly upon the mineral soil.

An important characteristic of soils in the undisturbed state is that the infiltration capacity is usually great at the start of a rain which has been preceded by a dry spell, but drops rapidly as the rain continues to fall and to soak into the soil. After several hours the soil's infiltration capacity becomes almost constant. The reason for the high starting value and its rapid drop is, of course, that the soil openings rapidly become clogged by particles brought from above, or tend to close up as the colloidal clays take up water and swell. From this we can easily reason that a sandy soil with little or no clay will not suffer so great a drop in infiltration capacity, but will continue to let the water through indefinitely at a generous rate. In contrast, the clayrich soil is quickly sealed to the point that it allows only a very slow rate of in-

Figure 25.2 Overland flow running down an 8 percent slope following a heavy thunderstorm. The ditch in the foreground receives the runoff and conducts it away as channel flow. (Soil Conservation Service photograph.)

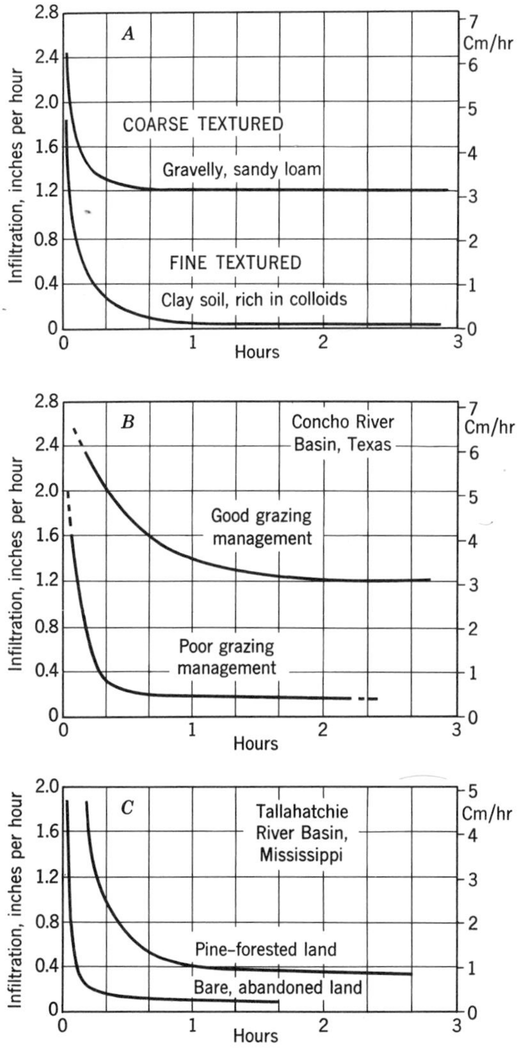

Figure 25.3 Infiltration rates vary greatly according to soil texture and land use. (Data from Sherman and Musgrave; Foster.)

filtration. This principle is illustrated by the graph in Figure 25.3*A* showing the infiltration curves of two soils, one sandy, one rich in clay.

It also follows that a sandy soil may be able to infiltrate even a heavy, long-continued rain without any surface runoff occurring, whereas the clay soil must divert much of the rain into overland flow, a process that may lead to erosion by gullies. Many forms of artificial disturbance of natural soils tend to decrease the infiltration capacity and to increase the amount of surface runoff (Figure 25.3*B*, *C*). Cultivation tends to leave the soil exposed so that rain beat quickly seals the soil pores. Fires, by destroying the protective vegetation and surface litter, also expose the soil to rain beat. Trampling by livestock will tamp the porous soil into a dense, hard layer. It is little wonder, then, that man has, through his farming and grazing practices, radically changed the original proportions of infiltration to runoff in such a way as to result in severe erosion damage and at the same time to decrease the reserves of soil moisture which might otherwise sustain plant growth and stream flow in droughts.

Forms of accelerated erosion

When a plot of ground is first cleared of forest and ploughed for cultivation, little erosion will occur until the action of rain splash has broken down the soil aggregates and sealed the larger opening. Following this, overland flow begins to remove the soil in rather uniform thin layers, a process termed *sheet erosion.* Because of seasonal cultivation, the effects of sheet erosion are often little noticed until the upper horizons of the soil (*A* and *B* horizons) are removed or greatly thinned. Reaching the base of the slope, where the angle of surface is rapidly reduced to meet the valley bottom, soil particles come to rest and accumulate in a thickening layer termed *colluvium,* or simply *slope wash.* This, too, has a sheetlike distribution and may be little noticed, except where it can be seen that fence posts or tree trunks are being slowly buried.

Material that continues to be carried by overland flow to reach a stream in the valley axis is then carried further down valley and may be built up into layers on the valley floor, where it becomes *alluvium,* a word applied generally to any stream-laid deposits. Colluvium and alluvium together are described as products of *sedimentation,* the opposite process from erosion. In many ways, sedimentation at the base of slopes and in valley bottoms is a process equally serious to erosion from the agricultural standpoint, because it results in burial of soil horizons under relatively infertile, sandy

Figure 25.4 Shoestring rills on a barren slope. (Soil Conservation Service photograph.)

Figure 25.5 This great gully, eroded into deeply weathered overburden, was typical of certain parts of the Piedmont region of South Carolina and Georgia before remedial measures were applied. (Soil Conservation Service photograph.)

layers and may choke the valleys of small streams, causing the water to flood broadly over the valley bottoms.

Where slopes are exceptionally steep and runoff from storms is exceptionally heavy, sheet erosion progresses into a more intense activity, that of *rill erosion*, or *rilling* (Figure 25.4), in which innumerable, closely spaced channels, referred to as *shoestring rills*, are scored into the soil in a system of long, parallel lines. In some cases, shoestring rills are merely seasonal features, developed during periods of torrential rain in the spring and summer, but healing over as ground frost heaves the soil during the winter season. Again, shoestring rills may actually represent a permanent change brought on by deforestation or cultivation, in which new stream channels are in the process of formation.

If rills are not destroyed by soil tillage, they may soon begin to integrate into still larger channels, termed *gullies*. This transformation comes about as the more active rills deepen more rapidly than their neighbors and incorporate the adjacent drainage areas. Erosive action thus is concentrated into a few large channels which deepen into steep-walled, canyonlike trenches whose upper ends grow progressively upslope (Figure 25.5).

Ultimately, a rugged, barren topography, resembling the badland forms of the arid climates, may result from accelerated soil erosion allowed to proceed unchecked. Curative measures developed by the Soil Conservation Service have proved effective in stopping accelerated soil erosion and permitting the return to slow erosion rates approaching the geologic norm. These measures include construction of terraces to reduce slope angle and distance of overland flow, permanent restoration of overly steep slope belts to dense vegetative cover, and the healing of gullies by placing check dams in the gully floors.

Figure 25.6 These potholes have been carved in granite in the channel of the James River, Henrico County, Virginia. (Photograph by C. K. Wentworth, U.S. Geological Survey.)

Geologic work of streams

The geologic work of streams consists of three closely interrelated activities: *erosion, transportation*, and *deposition*. Erosion by a stream is the progressive removal of mineral material from the floor and sides of the channel, whether this be carved in bedrock, or in residual or transported overburden.

Transportation consists of movement of the eroded particles by dragging along the bed, by suspension in the body of the stream, or in solution. Deposition is the progressive accumulation of transported particles upon the stream bed and floodplain, or on the floor of a standing body of water into which the stream empties. Obviously, erosion cannot occur without some transportation taking place, and the transported particles must eventually come to rest. Therefore, erosion, transportation, and deposition are simply three phases of a single activity.

Stream erosion

Streams erode in various ways, depending on the nature of the channel materials and the tools with which the current is armed. The force of the flowing water alone, exerting impact and a dragging action upon the bed, can erode poorly consolidated alluvial materials such as gravel, sand, silt, and clay, a process termed *hydraulic action*. Where rock particles carried by the swift current strike against bedrock channel walls, chips of rock are detached. The rolling of cobbles and boulders over the stream bed will further crush and grind smaller grains to produce an assortment of grain sizes. These processes of mechanical wear are combined under the general term *abrasion*, which is the principal means of erosion in bedrock too strong to be affected by simple hydraulic action. Finally, the chemical processes of rock weathering—acid reactions and solution—are effective in removal or rock from the stream channel and may be designated as *corrosion*. Effects of corrosion are most marked in limestone, which is a hard rock not easily carved by abrasion, but yielding readily to the action of carbonic acid in solution in the stream water.

One interesting form produced by stream abrasion is the *pothole*, a cylindrical hole carved into the hard bedrock of a swiftly moving stream (Figure 25.6). Potholes range in diameter from a few inches to several feet; the larger ones may be many feet deep. Often a spherical or discus-shaped stone is found in the pothole and is apparently the tool, or *grinder*, with which the pothole was deepened. A spiraling flow of water in the pothole causes the grinder to be rotated at the base of the hole, thus boring gradually into the rock. Many other features of abrasion, such as plunge pools, chutes, and troughs lend variety to the rock channel of a swift mountain stream.

Stream transportation

The load of a stream is carried in three forms. Dissolved matter is transported invisibly in the form of chemical ions. All streams carry some dissolved salts resulting from rock decomposition. Clay and silt are carried in *suspension*, that is, held up in the water by the upward elements of flow in turbulent eddies in the stream. This fraction of the transported matter is termed the *suspended load*. Sand, gravel, and still larger fragments move as *bed load* close to the channel floor by rolling or sliding and an occasional low leap.

The load carried by a stream varies enormously in the total quantity present and the size of the fragments, depending on the discharge and stage of the river. In flood, when velocities of 20 ft (6 m) per second or more are produced in large rivers, the water is turbid with suspended load. Boulders of great size may be moving over the stream bed, if the river gradient is steep. Frederick S. Dellenbaugh, a member of Major Powell's boat party which descended the Grand Canyon of the Colorado River in 1871 and 1872, wrote that as the men rested beside the river at night they could feel and hear the dull, thundering impacts of huge boulders rolled over and over on the channel bottom in the swift rapids.

Figure 25.7 A river in flood eroded this huge trench at Cavendish, Vermont, in November 1927. An area 1 mi (1.6 km) wide and 3 mi (4.8 km) long, once occupied by eight farms, was cut away by the flood waters. Damage was great because the material consisted of sand and gravel which offered little resistance. (Photograph by Wide World Photos.)

The hydraulic action of flood waters is capable of excavating enormous quantities of unconsolidated materials in a short time (Figure 25.7). Not only is the channel often greatly deepened in flood, but the undermining of the banks causes huge masses of alluvium to slump into the river where the particles are quickly separated and become a part of the stream's load. This process, known as *bank caving*, is an important source of sediment during high river stages, and is associated with rapid sidewise shifts in channel position on the outsides of river bends.

Suspended loads of large rivers

The load carried by a large river is of considerable importance in planning for construction of large storage dams and in the construction of canal systems for irrigation. Sediment will be trapped in the reservoir behind a dam, eventually filling the entire basin and ending the useful life of the reservoir as a storage body. At the same time, depriving the river of its sediment in the lower course below the dam may cause serious upsets in river activity. Resulting deep scour of the bed and lowering of river level may upset the grades of irrigation systems. In designing for canal systems, the forms of artificial channels must be adjusted to the size and quantity of sediment carried by the water, otherwise obstruction by deposition or abnormal scour may follow.

Table 25.1 gives comparative figures on the sediment loads of rivers in various stages.

TABLE 25.1 SUSPENDED LOADS OF SELECTED RIVERS.[a]

	Suspended Sediment Parts per Million	Fraction by Weight
Mississippi River		
Yearly average	500 to 600	1/1800 to 1/1660
Flood stage up to 2,000,000 cfs (56,600 cms)	2,600	1/400
Low stage 70,000 cfs (2,000 cms)	50 (water blue, clear)	1/2000
Missouri River		
Flood stage	20,000	1/50
Colorado River (before Hoover Dam)		
Flood stage 50,000 to 70,000 cfs (1400 to 2000 cms)	40,000	1/25
Yellow River, China		
Flood stage	Weight of solids may be greater than weight of water	

[a]Data from G. H. Matthes, "Paradoxes of the Mississippi," *Scientific American*, Vol. 184, pp. 19–23, 1951.

Although information is scanty on quantity of sediment moved as bed load, the proportion of suspended load is generally very high. For the Mississippi, 90 per cent of the total load is carried in suspension. Table 25.1 shows that rivers draining semi-arid or arid lands (Missouri and Colorado) have very high suspended loads because of the large expanses of barren soil from which sediment is easily swept into stream channels.

The Yellow River (Hwang Ho) of China heads the world list in annual suspended sediment load, while its watershed sediment yield is one of the highest known for a large river basin. The explanation lies in a high soil-erosion rate on intensively cultivated upland surfaces of wind-deposited silt (loess) in Shensi and Shansi provinces (see Chapter 33 and Figure 33.17). Much of the drainage area is in a semiarid climate with dry winters; vegetation is sparse and the runoff from heavy summer rains entrains a large amount of sediment.

Although it is difficult to assess the importance of man-induced soil disturbance upon the sediment load of major rivers, most investigators generally agree that land cultivation has greatly increased the sediment load of rivers of eastern and southeastern Asia, Europe, and North America. The increase due to man's activities is thought to be greater by a factor of $2\frac{1}{2}$ than the geologic norm for the entire world land area. For the more strongly affected river basins, the factor may be 10 or more times larger than the geologic norm. This form of environmental impact has attracted little attention because it has accumulated over many centuries of agricultural development.

How channels change in flood

We tend to think of a river in flood as changing largely through increase in height of water surface, which causes channel overflow and inundation of the adjoining floodplain. Because of the turbidity of the water we cannot see the changes taking place on the stream bed, but these can be determined by sounding the river depth during stream-gauging measurements (Figure 25.8). At first the bed may be built up by large amounts of bed load supplied to the stream during the first phases of heavy runoff. This is soon reversed, however, and the

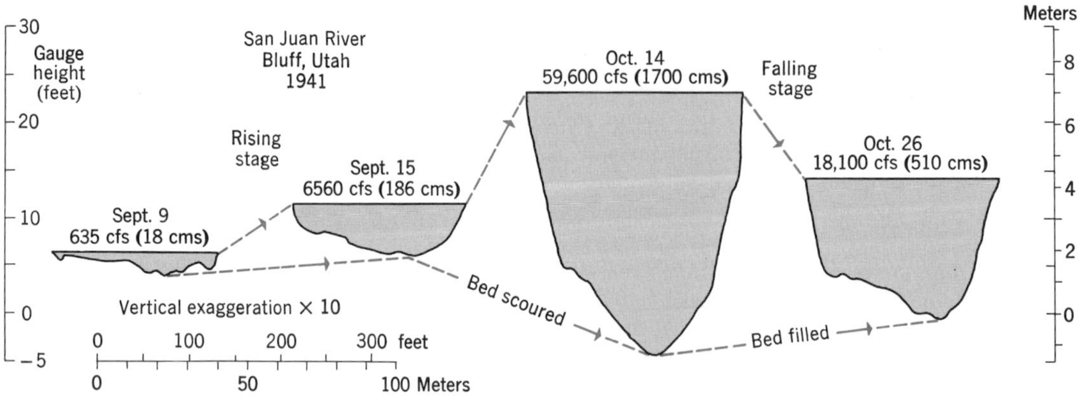

Figure 25.8 Changes in channel form of the San Juan River near Bluff, Utah, during the progress of a flood. (After Leopold and Maddock, U.S. Geological Survey.)

bed is actively deepened by scour as stream stage rises. Thus, in the period of highest stage, the river bed is at its lowest elevation. When the discharge then starts to decline, the level of the stream's surface drops and the bed is built back up by the deposition of alluvium. In the example shown in Figure 25.8, about 10 ft (3 m) of thickness of alluvium was *reworked*, that is, moved about in the complete cycle of rising and falling stages.

Alternate deepening by scour and shallowing by deposition of load are responses to changes in the stream's ability to transport its load. The maximum quantity, or load, of debris that can be carried by a stream is a measure of the stream's *capacity*. Load is usually stated in terms of the weight of material moved through a given stream cross section in a given unit of time, commonly in units of tons per day. Total load includes both the bed load and the suspended load.

If a stream is flowing in a channel of hard bedrock, it may not be able to pick up enough alluvial material to supply its full capacity for bed load. Such conditions exist in streams occupying deep gorges and having steep gradients, so that when flood occurs, the channel cannot be quickly deepened in response. In an alluvial river, however, where thick layers of silt, sand, gravel, and boulders underlie the channel, the rising river easily picks up and sets in motion all of the material that it is capable of moving. In other words, the increasing capacity of the stream for load is easily satisfied.

Capacity for load increases sharply with the stream's velocity, because the swifter the current the more intense is the turbulence and the stronger is the dragging force against the bed. Capacity to move bed load goes up about as the third to fourth power of the velocity. Thus, if a stream's velocity is doubled in flood, its ability to transport bed load is increased from eight to sixteen times. It is small wonder, then, that most of the conspicuous changes in the channel of a stream, such as sidewise shifting of the course, occur in flood stage, with few important changes occurring in low stages.

When the flood crest has passed and the discharge begins to decrease, the stream's capacity to transport load also declines. Therefore, some of the particles that are in motion must come to rest on the bed in the form of sand and gravel bars. First the largest boulders and cobblestones will cease to roll, then the pebbles and gravel, then the sand. Fine sand and silt carried in suspension can no longer be sustained, and settle to the bed. In this way the stream adjusts to its falling capacity. When restored to low stage the water may become quite clear, with only a few grains of sand rolling along the bed where the current threads are fastest.

Life history of a stream

Throughout its life history, a stream passes through a series of stages, each with certain definite characteristics (Figure 25.9). The initial stage occurs as soon as a new land surface is created by uplift and dislocation of a portion of the earth's crust. It is assumed here for simplicity of discussion that the surface was formerly under ocean level and has now become exposed for the first time. The landscape is thus composed entirely of initial landforms. Rain falling upon the land will produce overland flow. This must flow down the initial slopes, whatever their form. Water flow will be concentrated where slight depressions exist in the slopes, thus causing the development of stream channels, which are quickly deepened by erosive action of the water and any loose rock particles it carries. Depressions will fill up with water, making lakes. Overflow at the lowest points on the rims of these lake basins will serve to make a connected system of drainage from higher to lower lakes. Thus the initial stream system comes into existence. It is characterized by falls, rapids, and lakes along its course (Figure 25.9*A*).

Once formed, the stream enters upon the stage of *youth*. Deepening of the channel is the principal activity of a young stream, whose capacity for load exceeds the load available to it. Lake outlets are cut through, draining the lakes and extending the stream across the old lake floors. Waterfalls are cut down at the lip until they are nothing more than rapids. A deepening *gorge* or *canyon* is perhaps the most striking landform associated with a young stream. The gorge is steep-walled and has a V-shaped cross section. The stream occupies all the bottom of the gorge. From the steep walls much weathered rock material is shed into the stream. Landslides occur frequently, large fallen masses sometimes temporarily damming the stream. Talus slopes of loose rock fragments may here and there extend down into the water. Because of rapid denudation of the steep valley walls, bedrock outcrops are conspicuous, locally forming bold cliffs (Figure 25.9*B*).

The environmental importance of a young river valley can be readily imagined. There is no room for roads or railroads between the stream and the valley sides; hence road beds must be cut or blasted at great expense and hazard from the valley sides. Maintenance is expensive because of undercutting by the stream and the sliding and falling of rock, which can wipe out or damage the road bed. Yet a young gorge may afford the only passage through a mountain range. The Royal Gorge of the Arkansas River, in the Rocky Mountain Front Range of southern Colorado, is a striking example (Figure 25.10).

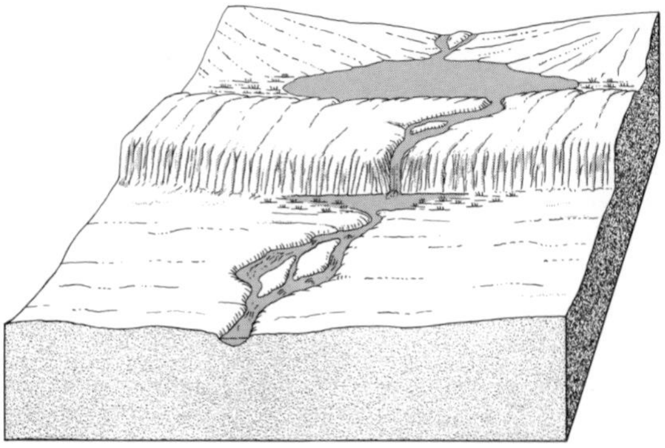

A. In the initial stage a stream has lakes, waterfalls, and rapids.

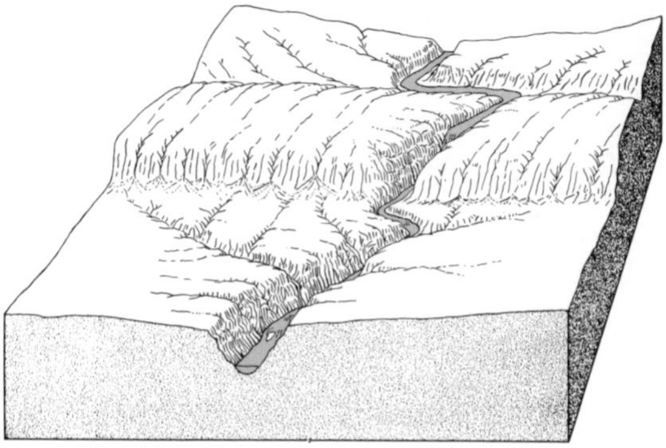

B. By middle youth the lakes are gone, but falls and rapids persist along the narrow incised gorge.

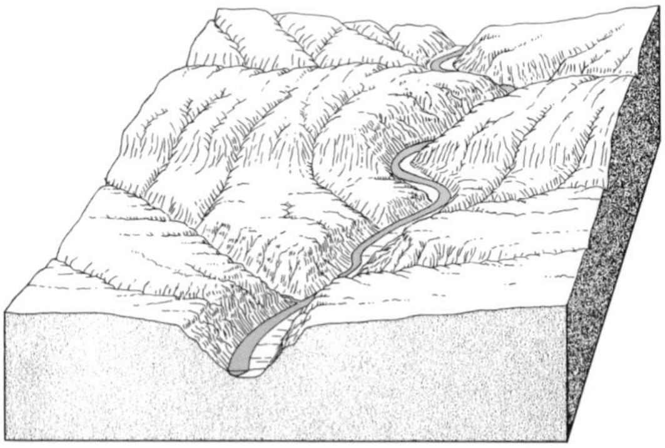

C. Early maturity brings a smoothly graded profile without rapids or falls, but with the beginnings of a flood plain.

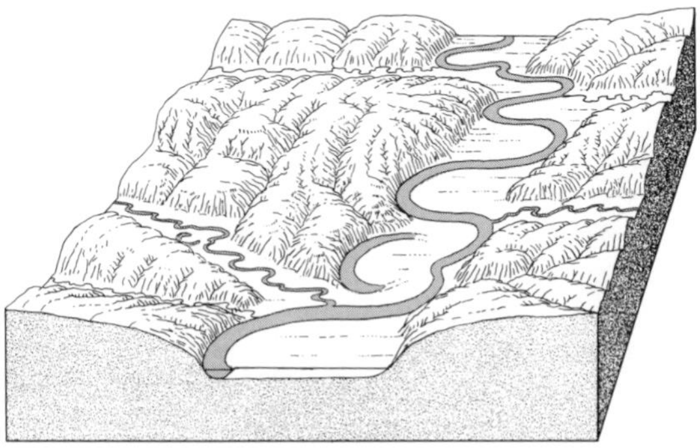

D. Approaching full maturity, the stream has a flood plain almost wide enough to accommodate its meanders.

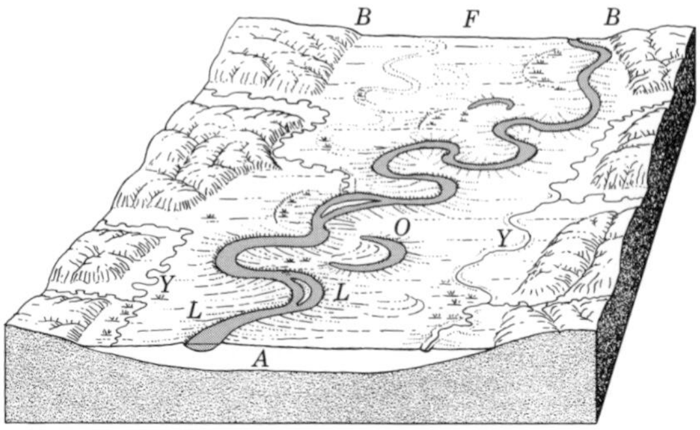

E. Full maturity is marked by a broad flood plain and freely developed meanders. *L* = levee; *O* = oxbow lake; *Y* = yazoo stream; *A* = alluvium; *B* = bluffs; *F* = flood plain.

Figure 25.9 Stages in the life history of a stream. (After E. Raisz.)

Another environmental consideration is that a young stream is not navigable, even though it might otherwise have a sufficient discharge.

The steep gradients of young streams, especially at waterfalls, sometimes make them important sources of hydroelectric power (Figures 25.11, 25.12). Most large young rivers, however, do not possess abrupt drops in gradient, and so it is necessary to build dams in order to create artificially the vertical drop necessary for turbine operation. An example is the Hoover Dam, behind which lies Lake Mead occupying the canyon of the young Colorado River.

As a stream progresses through the stage of youth it removes falls and rapids from its course, creating a smooth, even gradient. Deepening of the valley becomes greatly retarded, allowing the canyon or gorge walls to be worn down to more moderate slopes (Figure 25.13).

Stream equilibrium

The stage of *maturity* is reached when the stream has completed its phase of rapid downcutting and has prepared itself a smoothly graded

Figure 25.10 The Royal gorge of the Arkansas River in the Colorado Rockies illustrates the canyon of a young river with a steep gradient. Seen above is a suspension bridge, 1053 ft (321 m) above river level. (Photograph by Josef Muench.)

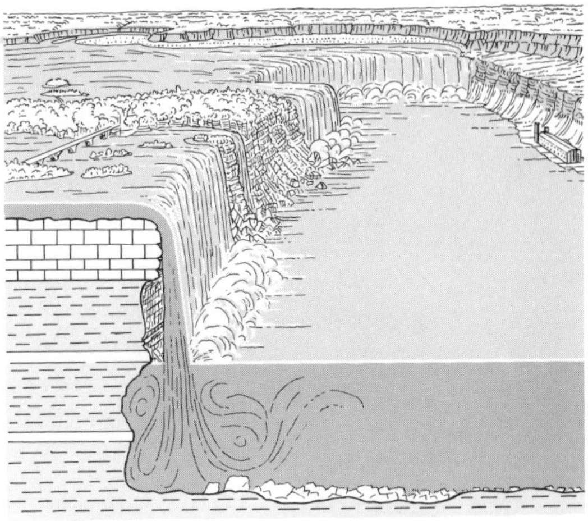

Figure 25.11 Niagara Falls is formed where the river passes over the eroded edge of a massive limestone layer. Continual undermining of weak shales at the base keeps the fall steep. (After G. K. Gilbert and E. Raisz.)

course. It is now in a state of balance, or *equilibrium*, in which the average rate of supply of rock waste to the stream from all its tributaries and their slopes is equal to the average rate at which the stream can transport the load. In other words, the stream's capacity is satisfied by the load supplied.

The longitudinal profile, representing the stream channel from upper to lower end, is a profile of equilibrium (Figure 25.14). It may also be said that the stream is *graded*, which simply means that it possesses a profile of equilibrium.

It is important to understand that the balance between load and a stream's capacity exists only as an average condition over periods of many years. As already explained, streams scour their channels in flood and deposit load when in low stage. Thus in terms of conditions of the moment, a stream is rarely in equilibrium; but over long periods of time, the graded stream maintains its level by restoring those channel deposits temporarily removed by the excessive energy of flood flows.

Having attained this state of balance, the stream continues to cut sidewise on the outsides of banks. It cannot continue to cut down without destroying the equilibrium condition, but the lateral cutting does not materially affect the equilibrium (Figure 25.9*C*).

Floodplain development

Immediate evidence of the earliest stage of maturity is the beginning of development of a flat valley floor. During enlargement of a bend, the river channel shifts toward the outer part of the bend, leaving a strip of relatively flat land, or *floodplain*, on the inner side of the bend (Figure 25.15). The floodplain is built of bars composed largely of sand and gravel brought as bed load scoured from the outsides of bends immediately upstream. Innundation of the floodplain approximately yearly in frequency allows finer silt and clay to settle out over the surface, adding to the floodplain height and covering the coarser alluvium beneath.

As lateral cutting by the stream continues, floodplain strips grow wider and presently join to form more or less continuous belts along either side of the stream (Figure 25.9*D*). The stream bends are now larger and more smoothly rounded. When the bends are developed into smooth, sinuous curves they are termed *meanders*. As valley development progresses the floodplain becomes wide enough to accommodate the meanders without cramping their form. The stream has then passed from *early maturity* to *full maturity* (Figure 25.9*E*).

Figure 25.12 This air view of Victoria Falls of the Zambezi River shows that the river has excavated a long cleft in the bedrock, probably along a fault or other zone of weakness. (Photographer not known.)

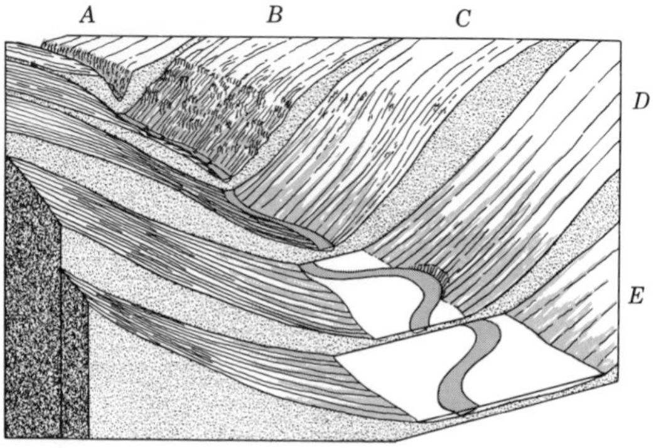

Figure 25.13 As a stream progresses from youth to maturity, its valley walls become more gentle in slope and the bedrock is covered by soil and weathered rock. (After W. M. Davis.)

The stage of early maturity of a stream is significant environmentally. The floodplain, though narrow, will accommodate roads or railroads. The graded stream profile assures low, smooth grades for the roadbeds. Agriculture, impossible on the steep walls of a young valley, can be practiced on the narrow floodplain strips. In advanced maturity, the floodplain valley has assumed more importance as a productive belt and is occupied by relatively greater numbers of persons than the upland areas between the valleys. Furthermore, the absence of rapids in the stream channels permits navigation, although streams in the early mature stage require locks to assure navigability.

After a stream has reached full maturity, its principal activity is to widen the floodplain. Eventually the floodplain attains a width several times as great as the meander belt. By *meander belt* is meant the area included by two lines, each one drawn on the side of the meandering stream in such a way as to connect the outermost points of the bends. It is common practice for the hydraulic engineer to apply the term *alluvial river* to a freely meandering stream with a broad floodplain.

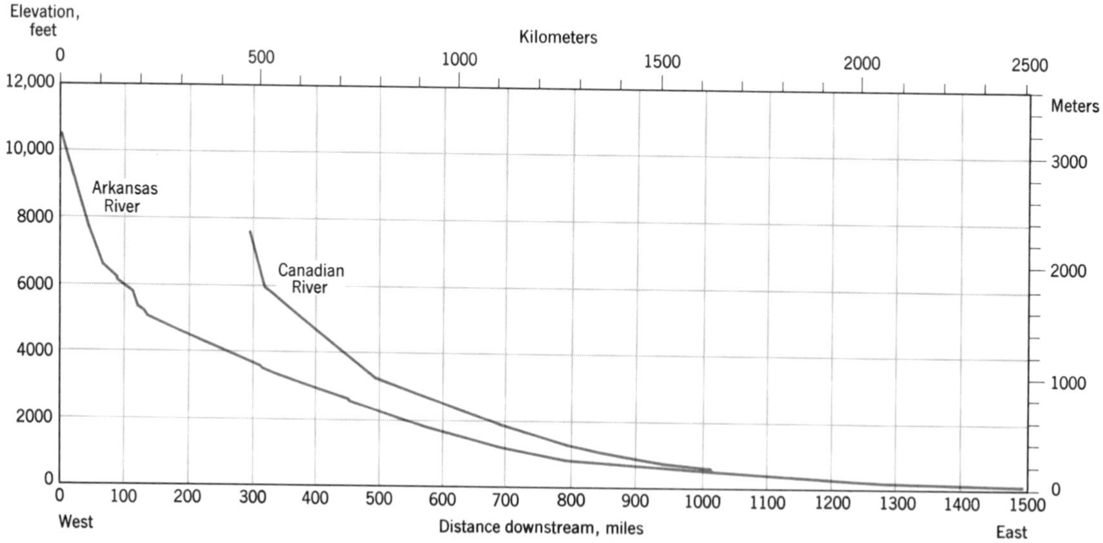

Figure 25.14 Longitudinal profiles of the Arkansas and Canadian rivers. The middle and lower parts of the profiles are for the most part smoothly graded, whereas the poorly graded upper parts reflect rock inequalities and glacial modifications within the Rocky Mountains. (From Gannett, U.S. Geological Survey.)

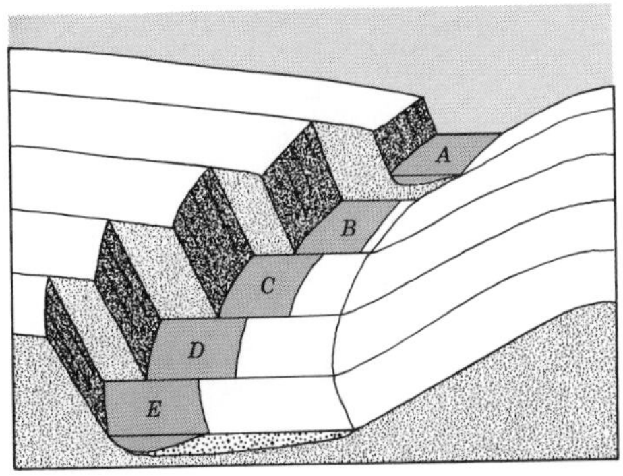

Figure 25.15 A graded stream cuts sidewise against the outside of a bend, leaving a flood-plain belt on the inside. (After W. M. Davis.)

The ultimate goal to which the stream profile is being lowered is an imaginary inland extension of a sea-level surface. Below this level, which the geologist calls *base level*, the valleys could not be deepened. The mouth of every stream that empties into the sea is at base level. Although theoretically the remainder of the stream might eventually reach base level, all alluvial rivers have a slight gradient, due to the fact that the lands within each drainage basin still stand well above sea level, and are shedding sediment which requires a sloping stream bed for transport.

Meanders of alluvial rivers

The floodplains of large alluvial streams have many special features of interest to the geographer (Figures 25.9*E* and 25.16). Meander bends grow as the stream undercuts the bank on the outside of the bend and deposits alluvium on the inside of the bend. These two sides of the bend are called the *undercut* and *slipoff slopes*, respectively. The bars of alluvium built on the slipoff slope are referred to as *point-bar deposits* (Figure 25.17).

The bends grow larger and larger until the channels meet, causing the intervening meander loop to be pinched off and abandoned. An occurrence of this type is called a *cutoff* (Figure 25.16). On a large river, such as the Mississippi, Missouri, or Arkansas, cutoffs are of considerable geographical importance. Where a state boundary is defined as the midline of a river, cutoffs cause portions of land within the cutoff bends to be transferred from one state to another, automatically altering the legal residences of persons who live on these lands and making them subject to laws and taxes of the other state. This difficulty can be overcome by fixing the boundary at a given time and maintaining it despite river changes. As a result, some maps show the old boundary meandering along a former river course (Figure 25.18).

Along the Mississippi, towns grew up at many places along the steep undercut banks of meander bends. Here a river steamer could safely come close to the bank, dropping its gangplank directly upon the shore. If the meander bend was cut off the river took a new channel and the town immediately felt the effects of economic strangulation. Lakes formed by sealing off of the cutoff bend are called *oxbow lakes* (Figure 25.9*E*). Several oxbow lakes and marshes can be identified in Figure 25.17.

Natural levees

A floodplain, as the name implies, is normally inundated by flood stages. A floodplain experiences inundation about once annually, at the season of highest runoff. Although the whole

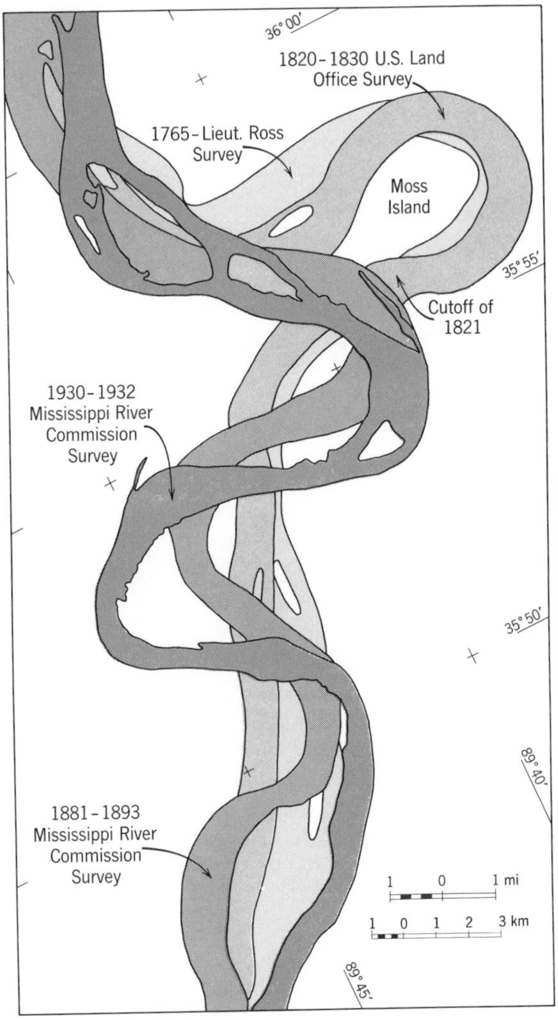

Figure 25.16 A series of four surveys of the Mississippi River shows considerable changes in the position of the channel and the form of the meander bends. Note that one meander cutoff has occurred (1821) and new bends are being formed. (After U.S. Army Corps of Engineers.)

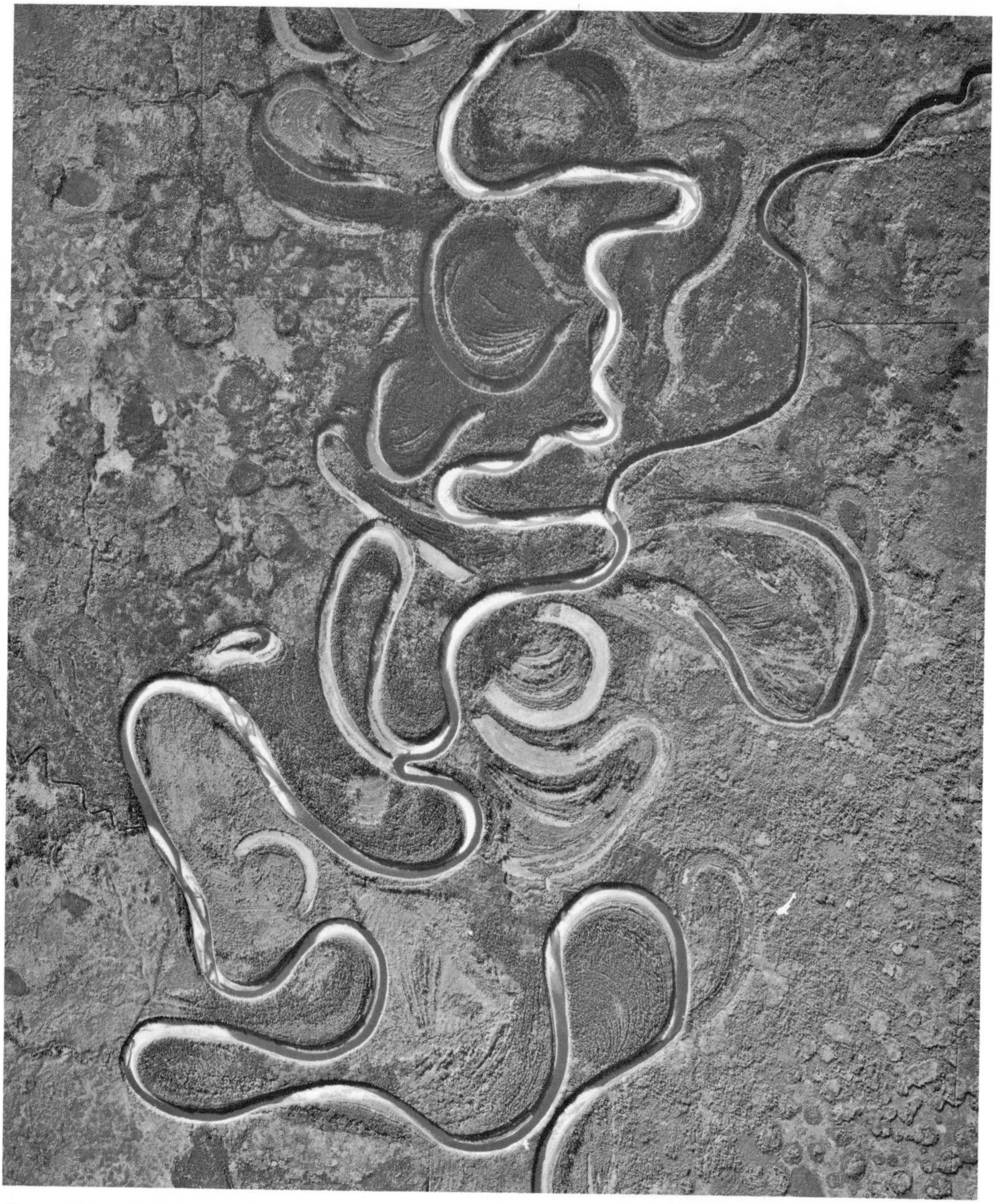

Figure 25.17 This vertical air photograph, taken from an altitude of about 20,000 ft (6100 m) shows meanders, cutoffs, oxbow lakes and swamps, and flood-plain of the Hay River, Alberta (lat. 58° 55′ N, long. 118° 10′ W). (National Air Photo Library, Surveys and Mapping Branch, Canada Department of Energy, Mines and Resources.)

plain, from one valley wall, or *bluff,* to the other, is under water at such times (Figure 25.19), the water current is most rapid along the deep line of the river channel. Silt-bearing water, which spreads out and mingles with shallow flood waters on either side, quickly loses velocity, and much of the silt and mud settles out. Because the greatest amount of sediment settles out adjacent to the river channel there is built up by many such floods a belt of slightly higher ground, known as a *natural levee,* on both sides of the river (Figure 25.9*E*). The levee surface slopes gently downward away from the river to lower portions of the flood-plain. Strangely enough, then, the highest ground

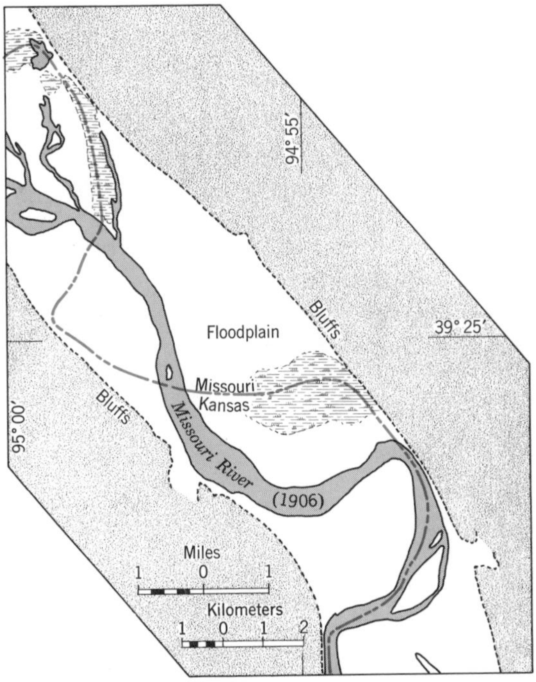

Figure 25.18 The Missouri-Kansas boundary was originally surveyed along the midline of the Missouri River, but the river has since shifted to a new course. (After U.S. Geological Survey.)

Figure 25.19 The Wabash River in flood near Delphi, Indiana, February 1954. An ice dam clogs the river channel, while lines of trees mark the crest of the bordering natural levees. The floodplain itself is inundated on both sides of the channel and reaches to the base of the bluff, at left. (U.P.I. Telephoto.)

on the floodplain is along the natural levees, immediately adjacent to the river. This narrow strip of ground may remain above water in all but the highest floods and is the safest place, aside from the floodplain bluffs themselves, of any on the floodplain.

Streams entering upon the floodplain cannot directly join the main river because of the levees; hence they flow down valley, parallel to the river, considerable distances until a point of entry can be found. Streams of this type are called *yazoo streams* (Figure 25.9*E*), after the Yazoo River of the Mississippi River floodplain.

The extensive, flat lands of a broad floodplain are usually highly productive and densely populated agricultural regions. The natural levee slopes are intensively cultivated; ditches assist in the natural surface water drainage away from the river bank. Farther from the river, in the lower parts of the floodplain and in oxbows and other sections of abandoned channels, are swamps that support a dense forest of trees adapted to the continually wet environment.

Flood abatement measures

In the face of repeated disastrous floods, vast sums of money have been spent on a wide variety of measures to reduce flood hazards. The economic, social, and political aspects of flood abatement are beyond the scope of this book; we discuss only the physical principles applied to the problem. Two basic forms of regulation are: (*a*) to detain and delay runoff by various means on the ground surfaces and in smaller tributaries of the watershed; (*b*) to modify the lower reaches of the river where floodplain inundation is expected.

The first form of regulation aims at treatment of watershed slopes, usually by reforestation or planting of other vegetative cover so as to increase the amount of infiltration and reduce the rate of overland flow. This type of treatment, together with construction of many small flood-storage dams in the valley bottoms, may greatly reduce the flood crests and allow the discharge to pass into the main

Figure 25.20 This old photograph shows the artificial levee of the Mississippi River near Greenville, Mississippi, during the great flood of March 1903. A crevasse, or break, at the distant point, *x*, is discharging flood water into the lower flood plain on the left. (Mississippi River Commission photograph.)

stream over a longer period of time. Treatment of watershed slopes is carried out in conjunction with soil erosion control measures, described in earlier pages of this chapter.

Under the second type of flood control, designed to protect the floodplain areas directly, two quite different theories can be practiced. First, the building of *levees*, or *dikes*, parallel with the river channel on both sides can function to contain the overbank flow and prevent inundation of the adjacent floodplain (Figure 25.20). Such levees are broad embankments built of earth and must be designed with great care, not only to possess the physical resistance to water pressures, but must be high enough to contain the greatest floods; otherwise they will be breached rapidly by great gaps, termed *crevasses*, at the points where water spills over (Figure 25.21). Under the control of the Mississippi River Commission, which began in 1879, a vast system of levees was built along the Mississippi River in the expectation of containing all floods. Figure 25.20 shows such a levee during the flood of 1903, when it was necessary in Louisiana to add to the top of the levee by means of planks and sand-filled bags for a distance of 71 mi (114 km) to prevent overflow. Levees have been continuously improved and now total more than 2500 mi (4000 km) in length and in places are as high as 30 ft (10 m). Additional levees on the lower, or deltaic, alluvial plain form floodways, by means of which excessive discharges can be diverted in time of flood and passed directly to the sea.

Because of natural and artificial levees, the river channel is often slowly built up to appreciably higher level than the floodplain, making these "bottom lands," as they are called, subject to repeated inundations and consequent heavy loss of life and property. Floodplains of many of the world's great rivers present serious problems of this type. In China, in 1887, the Hwang Ho inundated an area of 50,000 sq mi (130,000 sq km), causing the direct death of a million people and the indirect death of a still greater number through ensuing famine.

Figure 25.21 This air view taken in April 1952, shows a break in the artificial levee adjacent to the Missouri River in western Iowa. Water is spilling from the high river level at right to the lower floodplain level at left. (Photograph by Forsythe, U.S. Department of Agriculture.)

The second theory, practiced in more recent years on the Mississippi River by the U.S. Army Corps of Engineers, is to shorten the river course by cutting channels directly across the great meander loops to provide a more direct river flow. Shortening has the effect of increasing the river slope, which in turn increases the mean velocity. Greater velocity enables a given flood discharge to be moved through a channel of smaller cross-sectional area; the flood stage is correspondingly reduced. Channel improvement, begun in the 1930s, initially had a measurable effect in reducing flood crests along the lower Mississippi and the levees were thus not in such great danger of being overtopped. However, new meander bends grew rapidly and proved difficult to control. Certain parts of the floodplain are also set aside as temporary basins into which the river is to be diverted according to plan to reduce the flood crest. In the delta region, floodways are designed to conduct flow from river channels to the ocean by alternate routes.

Aggrading streams and braided channels

When a stream is supplied with more rock waste than it can carry, the excess material is spread along the channel bottom. Deposition of bed load serves to increase the channel gradient, which in turn increases the transporting power of the stream. The process of building up the channel is termed *aggradation;* it is the opposite of *degradation,* the normal downcutting process so marked in young streams.

Aggradation gives a distinctive broad, shallow form to stream channels. The stream divides and subdivides into two, three, four, or more threads, which come together and redivide in a manner suggestive of the braided strands of rope. The term *braided stream* is, in fact, used to describe the pattern of an aggrading stream. The reason for braiding and constant shifting of the channels is that deposition of sand and gravel bars on the channel floor causes the stream to split into two or more channels which shift sideways toward lower adjacent ground. The stream thus forces itself out of its own channel. Aggrading streams are commonest in dry regions where stream flow is small, but where large quantities of rock waste are swept into stream valleys from relatively bare, unprotected valley slopes.

Aggrading streams are also plentiful in association with glaciers. A remarkably fine example is shown in Figure 31.6. The stream is fed by meltwater, heavily charged with coarse rock debris, and this is spread downvalley in a braided channel.

Aggradation and sedimentation induced by man

Channel aggradation is a common form of environmental degradation brought on by man's activities. Accelerated soil erosion following cultivation, lumbering, and forest fires is the most widespread source of sediment for valley aggradation. On the other hand, this form of aggradation may be less conspicuous than other forms, since the land disturbance involves only the uppermost soil horizons. The results are typically seen in a gradual accumulation of sandy colluvium and alluvium in the smaller valleys, burying the finer-textured soils and lowering the agricultural quality of the valley bottoms. The silts, clays, and organic particles (humus) suspended in the runoff are carried far downvalley to distant sites of deposition. In localities of particularly severe soil erosion, such as that which affected the Piedmont Upland of the southeastern states and was manifested in deep gullying, aggradation was rapid in valley bottoms, inundating the surface with sediment and destroying the surface for productive use. Filling of reservoirs by

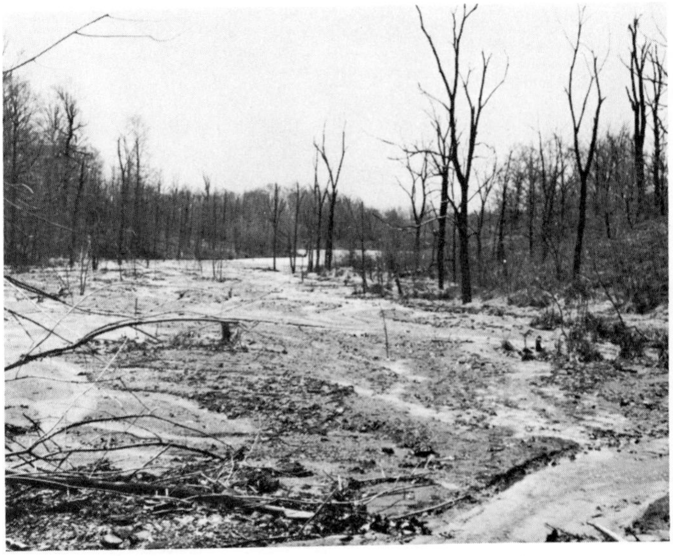

Figure 25.22 This valley bottom in Kentucky is choked with coarse debris from a nearby strip mine area. The natural channel has been completely buried under the rising alluvium. (Photograph by W. M. Spaulding, Jr., Fisheries and Wildlife, U.S. Dept. of the Interior.)

sediment is a major consequence of this type of valley sedimentation.

Aggradation of channels has also been a serious form of environmental degradation in coal-mining regions, along with water pollution (acid mine drainage) mentioned in Chapter 13. Throughout the Appalachian coal fields, channel aggradation is widespread because of the huge supplies of coarse sediment from mine wastes (Figure 25.22). Strip mining has enormously increased the aggradation of valley bottoms because of the vast surfaces of broken rock available to entrainment by runoff.

Urbanization and highway construction are also major sources of excessive sediment, causing channel aggradation. Major earthmoving projects are involved in creating highway grades and preparing sites for industrial plants and housing developments. While these surfaces are eventually stabilized, they are vulnerable to erosion for periods ranging from months to years. The regrading involved in these projects often diverts overland flow into different flow paths, further upsetting the regimen of streams in the area.

Mining, urbanization, and highway construction not only cause drastic increases in bed load, which cause channel aggradation close to the source, but also increase the suspended load of the same streams. Suspended load travels downstream and is eventually deposited in lakes, reservoirs, and estuaries far from the source areas. This sediment is particularly damaging to the bottom environments of aquatic life. Moreover, the accumulation

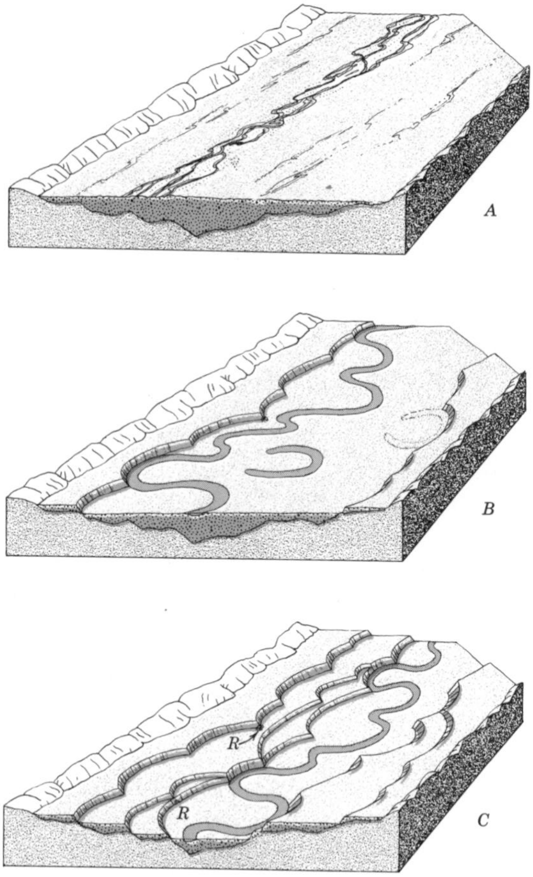

Figure 25.23 Alluvial terraces form when a graded stream slowly cuts away the alluvial fill in its valley.

and results in rapid filling of tidal estuaries, requiring increased dredging of channels.

Terraces

If a stream aggrades its valley for a long time the alluvial deposits may reach a thickness of many tens of feet, as Diagram *A* of Figure 25.23 shows. Among several possible causes of aggradation is the onset of a more arid climate, which reduces stream discharge and requires streams to steepen their gradients by building up their channels. Perhaps the commonest cause of aggradation in fairly recent geologic time in North America has been the advance and wasting away of ice sheets and valley glaciers (See Chapter 31). Melting water from the ice was heavily charged with rock debris, which caused virtually all streams near the ice front to fill their valleys with alluvium.

With return to normal conditions of reduced load, a stream will cut down through its alluvial deposit and eventually sweep most of it out of the valley. During this degradation a series of *alluvial terraces* is formed. A terrace is a relatively flat strip of ground bounded on one side by a steeply descending slope and on the other side by a steeply rising slope. A series of alluvial terraces resembles a flight of rather broad, low steps. As indicated in Diagrams *B* and *C* of Figure 25.23, terraces are made by the stream swinging from one side of the valley to the other as it slowly cuts down through the material. As each terrace is cut, the width of the next one above it is reduced.

All older terraces would be destroyed were it not that bedrock of the valley wall here and there projects through the alluvium and protects higher terraces. In Diagram *C*, the valley alluvium has been largely removed, but some *rock-defended terraces* remain on the valley sides, protected from stream attack by the rock which outcrops at the points labeled *R*. Notice that the scarps separating terraces are curved in broad arcs concave toward the valley. The curvature is easily explained as the result of cutting of the scarps by curved meander bends.

Terraces are of environmental importance similar to that of river floodplains. The relatively flat terrace surfaces are suitable for cultivation and make good sites for towns and cities, highways, and railroads. In all these utilizations terraces have one advantage over floodplains: their surfaces may be well above the level of even the highest floods, whereas floodplains are normally subject to frequent inundation.

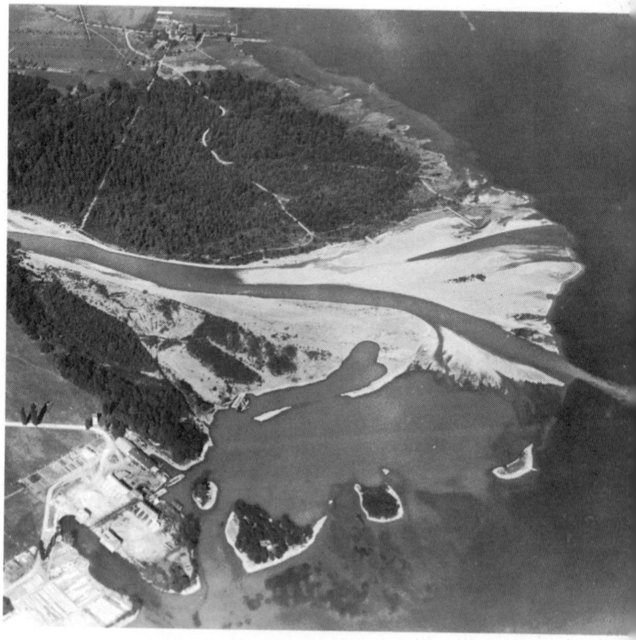

Figure 25.24 This air view of the Kander delta in Switzerland shows a tongue of silt-ladened water being projected into Lake Thun. (Swissair photo.)

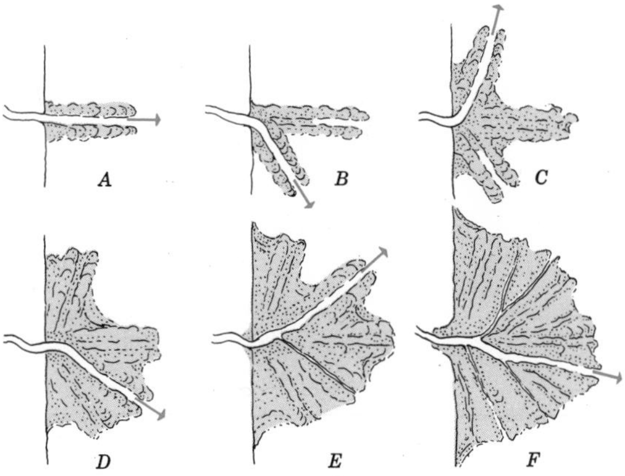

Figure 25.25 Stages in the formation of a simple delta. (After G. K. Gilbert.)

Deltas

The deposit of mud, silt, sand, or gravel made by a stream where it flows into a body of standing water is known as a *delta* (Figure 25.24). Deposition is caused by rapid reduction in velocity of the stream current as it pushes out into the standing water (Figure 25.25). The coarse particles settle out first; the fine clays continue out farthest and eventually come to rest in fairly deep water (Figure 25.26). Contact of fresh with salt water causes the finest clays to clot into larger aggregates which settle to the sea floor.

Deltas show a variety of shapes. The Nile delta, whose resemblance to the Greek letter "delta" suggested the name for this type of landform, has many *distributaries* which branch out in a radial arrangement (Figure 25.27*A*). Because of its broadly curving shoreline, causing it to resemble in outline an alluvial fan, this type may be described as an *arcuate* delta. The Mississippi River delta presents a very different sort of picture (Figure 25.27*B*). It is said to be of the *bird-foot* type because of the long, projecting fingers which grow far out into the water at the ends of each distributary. Where a river empties out upon a fairly straight shoreline along which wave attack is vigorous, the sediment brought out by the stream is spread along the shore in both directions from the river mouth, giving a pointed delta with curved sides. Because of its resemblance to a sharp tooth, this type is called a *cuspate* delta (Figure 25.27*C*). Where a river empties into a long, narrow estuary, the delta is confined to the shape of the estuary (Figure 25.27*D*). This type can be called an *estuarine* delta.

Deltas of large rivers have been of environmental importance from earliest historical times because their extensive flat areas support dense agricultural populations. Important coastal cities, linking ocean and river traffic, are often situated on or near deltas, as Alexandria on the Nile, Calcutta on the Ganges-Brahmaputra, Amsterdam and Rotterdam on the Rhine, Shanghai on the Yangtze, Marseilles on the Rhone, and New Orleans on the Mississippi, to mention but a few. Delta growth is often rapid, ranging from about 10 ft (3 m) per year for the Nile to 200 ft (60 m) per year for the Po and Mississippi rivers. Thus, some cities and towns that were at river mouths several hundred years ago are today several miles inland. An important engineering problem is to keep an open channel for ocean-going vessels which have to enter the delta distributaries to reach port. The ends of the Mississippi River delta distributaries, known as *passes*, have been extended by the construction of jetties, between which the narrowed stream is forced to move faster, thereby scouring a deep channel (Figure 25.27*B*).

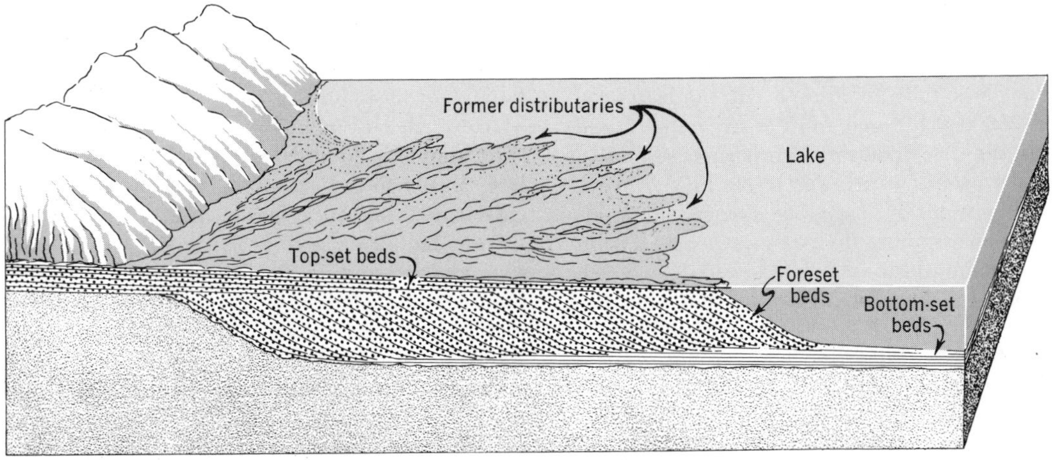

Figure 25.26 Structure of a simple delta shown in a vertical section. (After G. K. Gilbert.)

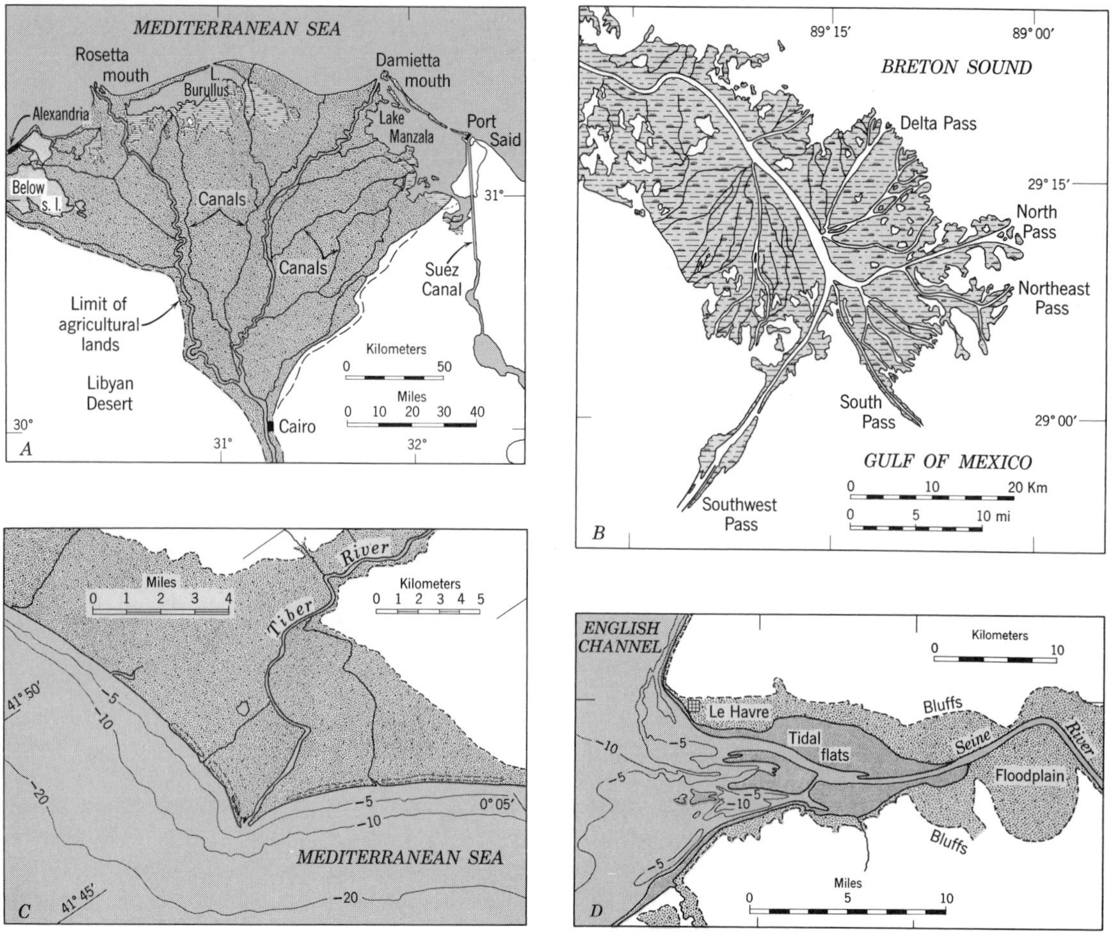

Figure 25.27 **Deltas.** ***A*****, The Nile delta has an arcuate shoreline and is triangular in plan.** ***B*****, The Mississippi delta is of the branching, bird-foot type with long passes.** ***C*****, The Tiber delta on the Italian coast is pointed, or cuspate, because of strong wave and current action.** ***D*****, The Seine delta is filling in a narrow estuary.**

Rejuvenated streams and entrenched meanders

A mature stream, which has developed a graded profile with respect to a fixed sea level at its mouth, may experience a marked change if the land rises or the sea level falls. In either event, the base level of the stream is lowered and the stream is caused to begin rapid downcutting in order to reestablish its graded profile at a lower level. This process, termed *rejuvenation,* begins as a series of rapids at the stream's mouth, where the water passes from the former mouth down to the lowered sea level. The rapids quickly shift upstream, and soon the entire stream valley is being trenched to form a new, youthful valley.

If rejuvenation occurs when a stream has reached maturity, the effect is to give a steep-walled inner gorge, on either side of which lies the former floodplain, now a flat terrace high above river level (Figures 25.28 and 25.29). Meanders which the river had formed on its floodplain have now become impressed into the bedrock and give the inner gorge a meandering pattern. These sinuous bends are termed *entrenched meanders* to distinguish them from the common floodplain meanders.

Although entrenched meanders are not free to shift about as floodplain meanders, they can enlarge slowly so as to produce cutoffs. Cutoff of an entrenched meander leaves a high, round hill surrounded on three sides by the deep abandoned river channel and on the fourth by the shortened river course. As might be guessed these hills form ideal natural fortifications. Many European fortresses of the Middle Ages were built on such cutoff meander spurs. A good example is Verdun, near the Meuse River.

Under unusual circumstances, where the bed rock includes a strong, massive sandstone forma-

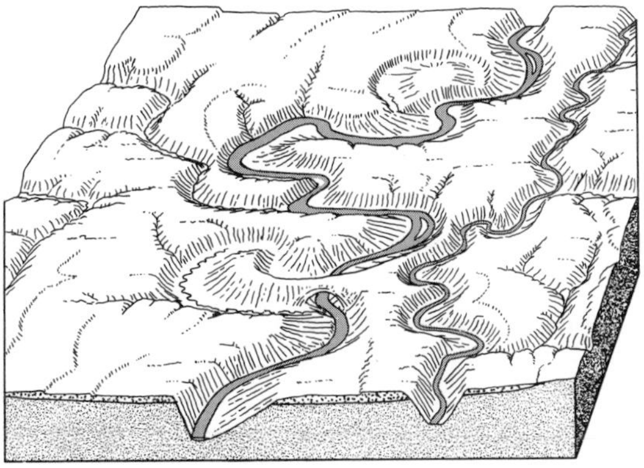

Figure 25.28 The winding valley of this stream resulted from entrenchment of the meandering river shown in Figure 25.9*E*. (After E. Raisz.)

tion, meander cutoff leaves a *natural bridge*, formed by the narrow meander neck (Figure 25.28). One well-known example is Rainbow Bridge at Navajo Mountain, in southeastern Utah; other fine examples can be seen in Natural Bridges National Monument at White Canyon in San Juan County, Utah.

Entrenched meanders do not offer ideal locations for railroads and highways, but in a few instances they have been the best available choices for arteries of travel. This point is well illustrated by the Moselle River, whose winding entrenched

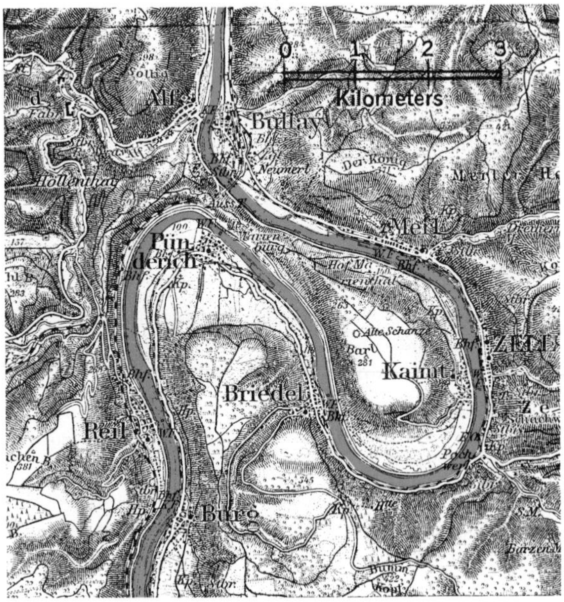

Figure 25.30 The Moselle River has a winding entrenched-meandering gorge through the Eifel district of Western Germany. (Portion of German 1:100,000 topographic map, 1890.)

meanders through the Ardennes mountain upland of Belgium and Western Germany have been utilized (Figure 25.30). Engineers have even cut tunnels through the narrow meander necks to shorten the distance of travel.

Figure 25.29 The Goose-Necks of the San Juan River in Utah are entrenched river meanders in horizontal sedimentary strata. (Spence Air Photos.)

REVIEW QUESTIONS

1. Describe the process of erosion by overland flow. What factors determine the intensity of erosion? What is splash erosion?

2. Distinguish between the geologic norm of soil erosion and accelerated soil erosion. How is infiltration capacity related to surface cover? To soil texture? To land use?

3. Describe the successively more severe forms of accelerated soil erosion. What happens to the eroded soil? Distinguish between *colluvium* and *alluvium*. In what ways is sedimentation harmful to valley floors?

4. What three geologic activities are carried on by a stream? Explain the ways in which stream erosion occurs. What is a pothole and how is it formed?

5. What are the modes of stream transportation? What sizes of particles are carried in each? When is bank caving of great importance?

6. How is the quantity of suspended sediment load of a stream stated? How does it vary in large rivers? Why?

7. Explain the changes in cross section of the channel of an alluvial river during a cycle of rising and falling discharge. Define stream capacity. What causes capacity to vary in a stream?

8. Describe the very first events in the formation of a new stream.

9. What features characterize the stage of youth of a stream? What is the environmental importance of young streams?

10. What event marks the attainment of the stage of maturity of streams? What is the profile of equilibrium? What is a graded stream? What is the form of the longitudinal profile of a graded stream?

11. Describe the floodplain and meanders of an alluvial river. What is the importance of a floodplain?

12. Explain the following terms: undercut and slipoff slopes, point-bar deposits, cutoffs, oxbow lakes, natural levees, yazoo streams.

13. What is the lower limit to which a stream can eventually degrade its channel? What stage of stream development would be found on a peneplain?

14. What basic forms of flood abatement are used? What principles lie behind each? What are the advantages and disadvantages of constructing extensive levee systems? How can channels be modified to reduce flood crests?

15. When does aggradation take place in a stream? Does aggradation change the stream gradient? Explain. What is a braided stream?

16. List the ways in which aggradation and sedimentation are induced or aggravated by man's activities. How can mining induce aggradation? What environmental impact does sedimentation have?

17. Explain how stream terraces are formed. How did the advance and retreat of the continental ice sheet cause terraces to form in the northeastern United States?

18. Describe the stages in growth of a simple delta. What internal structures has a simple delta? How is the sediment sorted according to particle size?

19. Describe and explain the various forms commonly assumed by deltas of large rivers. Cite examples of the rates of growth of deltas.

20. How can a mature, graded river become rejuvenated? How do entrenched meanders form?

Exercises

Exercise 1. ***Young Stream.*** (Source: Colfax, Calif., U.S. Geological Survey topographic map; scale 1:125,000)

Explanatory Note: The American River shown on this map is a young stream occupying a deep, V-shaped canyon in the west slope of the Sierra Nevada range.

QUESTIONS

1. (*a*) What contour interval is used on this map? **(*b*)** How deep is the canyon in the vicinity of 22–7?

2. (*a*) What is the average gradient in feet per mile of the river? **(*b*)** What is the gradient of the tributary stream flowing from 19–13 to 10–8?

3. (*a*) Measure the steepness of slope of the north wall of the main river canyon in the vicinity of 24–8. State this in feet per mile. **(*b*)** How many times steeper is this slope than the gradient of the river itself?

4. Make a topographic profile from 24–14 to 24–0. Use a vertical scale of 1 in. equals 5000 ft. What is the vertical exaggeration of this profile?

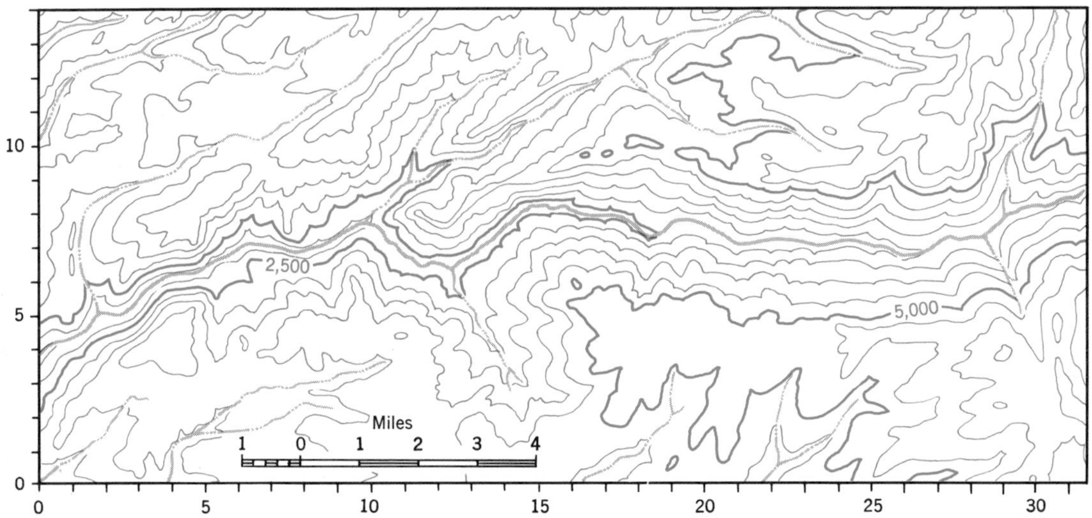

Exercise 1

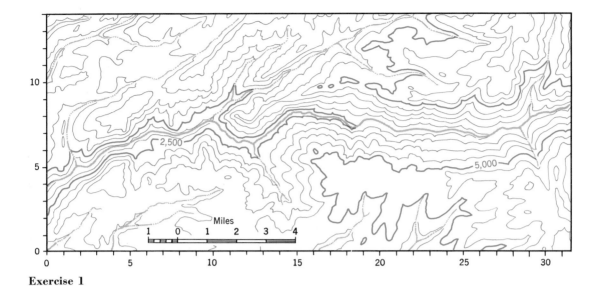

Exercise 2. *Stream in Early Maturity.* (Source: St. Albans, W. Va., U.S. Geological Survey topographic map; scale 1:62,500.)

Explanatory Note: The Kanawha River, shown on this map, has reached a stage of early maturity and has formed a flood plain between steep bluffs. (North is to the right on this map.) There is as yet insufficient width of flood plain to accommodate meanders.

QUESTIONS

1. **(*a*)** How wide is the river? **(*b*)** What is the contour interval of this map? **(*c*)** Give the elevation and location (in grid coordinates) of the highest point shown on this map.

2. **(*a*)** Is it possible to tell in which direction the Kanawha River is flowing? **(*b*)** Is it possible to determine the gradient of the river? **(*c*)** Draw in the 550-ft contour line as best you can to fit the assumption that the river flows from left to right.

3. Make a topographic profile from 7-0 to 7-7. Use a vertical scale of 1000 ft equals 1 in. What is the vertical exaggeration of this profile?

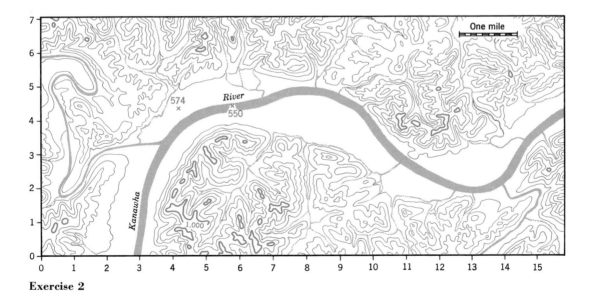

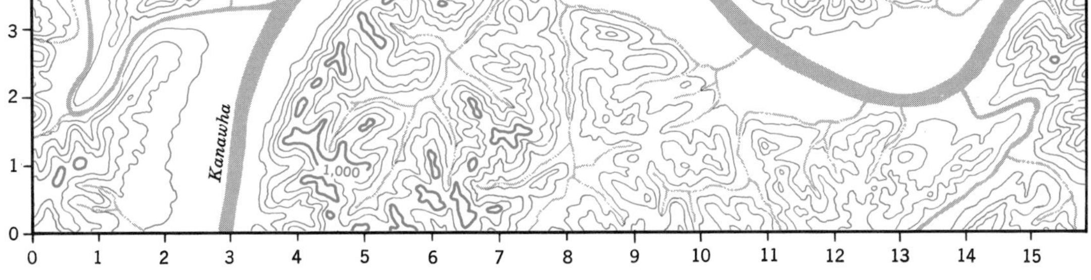

Exercise 2

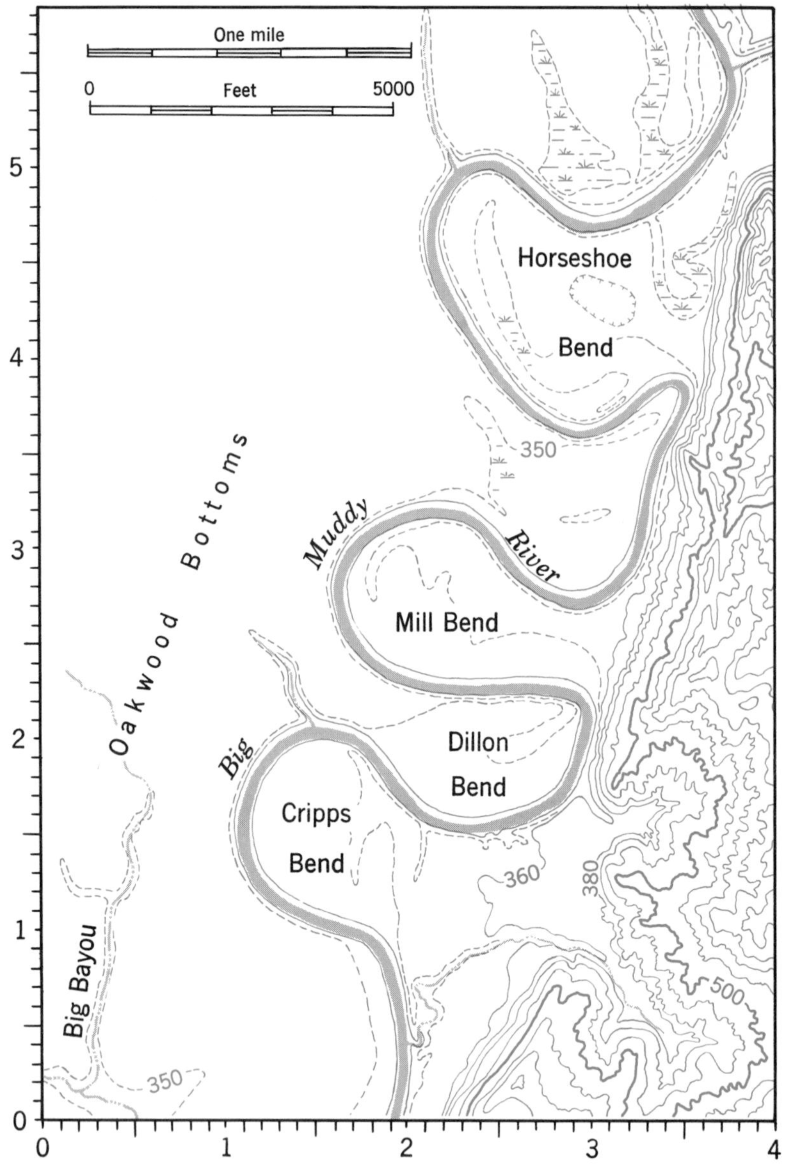

Exercise 3. ***Meandering Stream on Flood Plain.*** (Source: Gorham, Ill., U.S. Geological Survey topographic map; scale 1:24,000.)

Explanatory Note: The Big Muddy River meander belt lies close to valley bluffs on the east. To the west extends a broad flat flood plain, Oakwood Bottoms. The 350-ft contour line (dashed) reveals the existence of low natural levees bordering the channel. Three contours are omitted between 380 and 500 ft.

QUESTIONS

1. **(*a*)** What is the fractional scale of this map? **(*b*)** How many times greater (or smaller) is this scale than that of the map in Exercise 2?

2. **(*a*)** Measure the radius of curvature (in feet) of Mill Bend and Cripps Bend. This is simply the radius of a circle which best fits the meander curve along the midline of the river channel. **(*b*)** Measure the width of the stream channel (in feet), taking the average of several measurements along the stream. **(*c*)** How does the ratio of radius of curvature to stream width for the Big Muddy River compare with the corresponding ratio for the Mississippi River, computed from the cutoff of 1821 in Figure 25.16? State both ratios as whole numbers.

3. Place a sheet of thin tracing paper over the page and redraw a portion of the map to show the river as having cut off Dillon Bend, leaving an oxbow lake in its stead. Take care to see that all contours are correctly redrawn.

4. The gradient of Big Muddy River cannot be computed from this map alone. Suppose, however, that one regular contour crosses the stream just after it leaves the map area; that the next consecutive regular contour crosses the stream just before it enters the map area. What then is the river gradient in feet per mile?

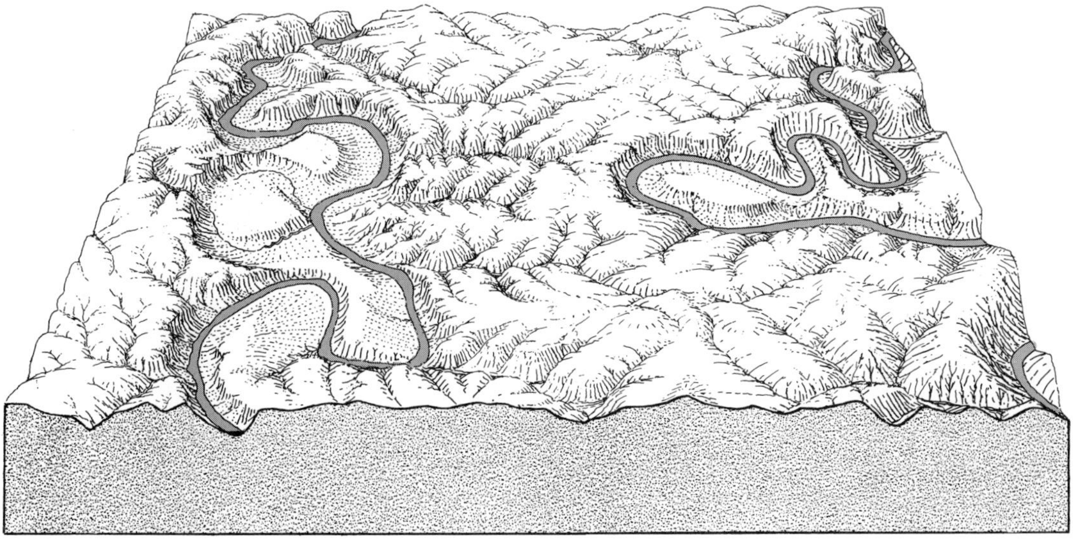

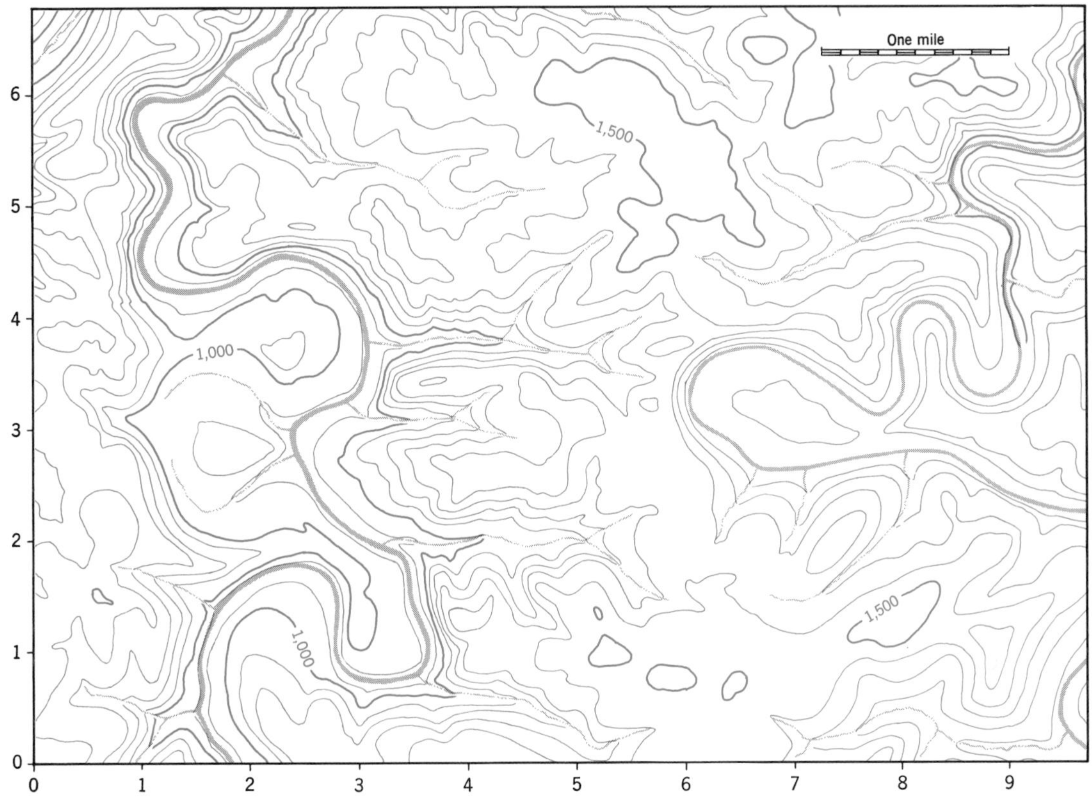

Exercise 4. *Entrenched Meanders.* (Source: Rural Valley, Pa., U.S. Geological Survey topographic maps; scale 1:62,500.)

Explanatory Note: Mahoning Creek (left) and Redbank Creek (right) are entrenched meandering streams in the Appalachian Plateau of western central Pennsylvania. On this map north is to the right.

QUESTIONS

1. **(*a*)** What contour interval is used on this map? **(*b*)** What is the fractional scale of this map? **(*c*)** What is the difference in elevation between the highest and lowest points shown on this map? Locate these points by grid coordinates.

2. The circular valley of an abandoned meander loop of Mahoning Creek is located in the vicinity of 1.7–2.8. **(*a*)** Draw in the former course of the stream around this bend. **(*b*)** By how many yards was the river shortened as a result of this cutoff?

3. **(*a*)** Make a profile from the point 2.3–3.7 to the point 2.3–4.8. **(*b*)** Why are the slopes of unequal steepness on the two sides of the stream at this place?

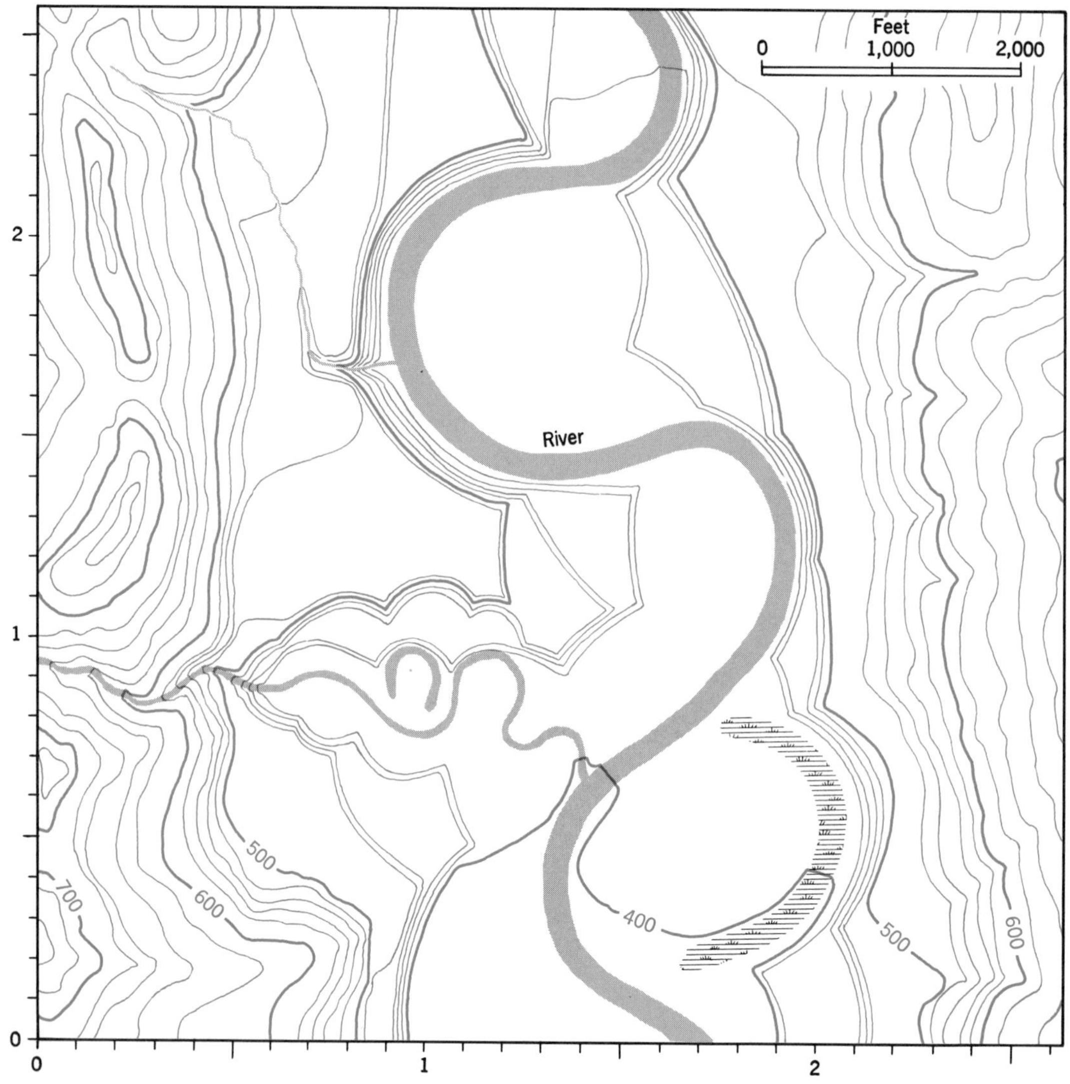

Exercise 5. ***Alluvial Terraces.***

Explanatory Note: This is a synthetic map, not representing any real area, and should be thought of as a diagram on which features are idealized. A large stream has cut down through alluvial material (with which a bedrock valley had been filled during the glacial period), leaving terraces at different levels. Curved terrace scarps are of various heights and have a radius of curvature similar to that of the meander bends of the present stream.

QUESTIONS

1. **(*a*)** What is the fractional scale of this map? **(*b*)** How many times larger is this scale than that of the map in Exercise 3? **(*c*)** What contour interval is used on this map?

2. How high is the terrace scarp **(*a*)** at 1.1–2.2? **(*b*)** at 1.6–1.7? **(*c*)** at 1.4–1.0?

3. **(*a*)** Measure the radius of curvature of the large stream meander at 1.3–1.8. (This is the same as the radius of a circle which best fits the meander bend along the midline of the stream.) **(*b*)** Measure the radius of curvature of the small stream meander at 1.2–0.9. **(*c*)** Of these two streams, which one cut the curved terrace scarp at 1.2–1.05?

4. What is the origin of the semicircular swamp extending from 1.7–0.2 to 1.8–0.8?

5. Why does the small stream have such a steep gradient at 0.5–0.9 but such a low gradient at 0.9–0.8?

6. Why does the contour line at 1.95–1.9 bulge westward in this place instead of bending east in a sharp V as the higher contours do just east of it?

7. Make a topographic profile from 2.6–1.2 to 0.0–1.2. Use a vertical scale of 1 in. equals 200 ft.

CHAPTER **26**

The Cycle of Landmass Denudation

THUS far, individual types of landforms produced by weathering, mass wasting, overland flow, and streams have been examined. Consider now the entire aspect of denudation of a large region under the combined attack of these agents. Imagine an area, such as a continent or a large portion of a continent, which is elevated by internal earth movements to provide a new landmass. This elevation will constitute the *initial* stage of a grand *cycle of landmass denudation,* in which the region passes through young, mature, and old stages. For the sake of simplicity, it must be assumed that the elevation of the landmass occurs rapidly and that further crustal deformation then ceases, leaving the denudational agents a long period of uninterrupted activity.

A single, ideal cycle of landmass denudation will not cover all occurrences. There is a difference between landform development in humid climates and that in arid climates. It is consequently necessary to describe two cycles, one for each climate. Furthermore, some initial landmasses are relatively smooth surfaced, representing an even sea floor broadly uparched by epeirogenic crustal movement. Others are mountainous because of breaking and bending of the rock during orogeny. For our purposes, two combinations will be discussed: (1) a smooth-surfaced landmass in a region of humid climate; (2) a rugged, mountainous landscape in an arid climate.

Cycle of landmass denudation in a humid climate

A landmass formed by uparching of a relatively smooth sea floor would have gentle slopes inclined seaward from the high central area. A portion of this slope is shown in Block *A* of Figure 26.1; it is said to be in the *initial stage.* Overland flow upon the new surface would drain off in the most convenient downslope direction and would soon develop initial streams. In the manner already explained in connection with the life history of a stream, these would begin to trench youthful V-shaped valleys into the initial landmass. Marshes and lakes occupying shallow depressions in the initial surface would soon be drained.

Block *B* shows a stage of *early youth* in the cycle of landmass denudation. The *relief* of the area—that is, the difference in elevation between valley bottoms and divide summits—is now increasing rapidly because the streams are cutting down rapidly, whereas between the streams there remain relatively flat portions of the initial land surface (Figure 26.2). As the valleys deepen they also widen, because rock waste is swept down the valley sides into the streams. The unconsumed areas between valleys thus are reduced in proportionate area, while the steep valley slopes increase in extent. Small tributary valleys branch out from the larger streams, further cutting into the initial landmass.

Block *C* of Figure 26.1 shows remnants of the initial land surface, between which are well-developed valley systems, but the remnants shrink in area until the greater proportion of the region consists of steep valley slopes, a stage that may be termed *late youth.* Relief has been steadily increasing as the streams have been actively downcutting. Next, however, conditions show a marked change. When the larger streams become graded and begin to form their floodplains, the increase in relief is halted. The remaining flat remnants of the initial surface are finally consumed, and the valley slopes intersect in narrow divides.

When relief has reached the maximum, the stage of *maturity* is attained (Block *D*, Figure 26.1 and Figure 26.3). From this time on, the valley floors are lowered with extreme slowness, whereas the interstream divides are rapidly lowered. Thus the relief of the region decreases steadily. Slopes become progressively lower in angle (Block *E*, Figure 26.1). Slope erosion and mass wasting no longer are so active as in previous stages.

Base level and peneplain

After a period of time much longer than was required for maturity to be reached, the landscape is reduced to a low rolling surface and may be said to have entered the *old stage* (Block *F*, Figure 26.1). By this time most of the streams have extremely low gradients and extensive floodplains. The ultimate goal, which would be reached if infinite time were available for its accomplishment, would be the reduction of the land to

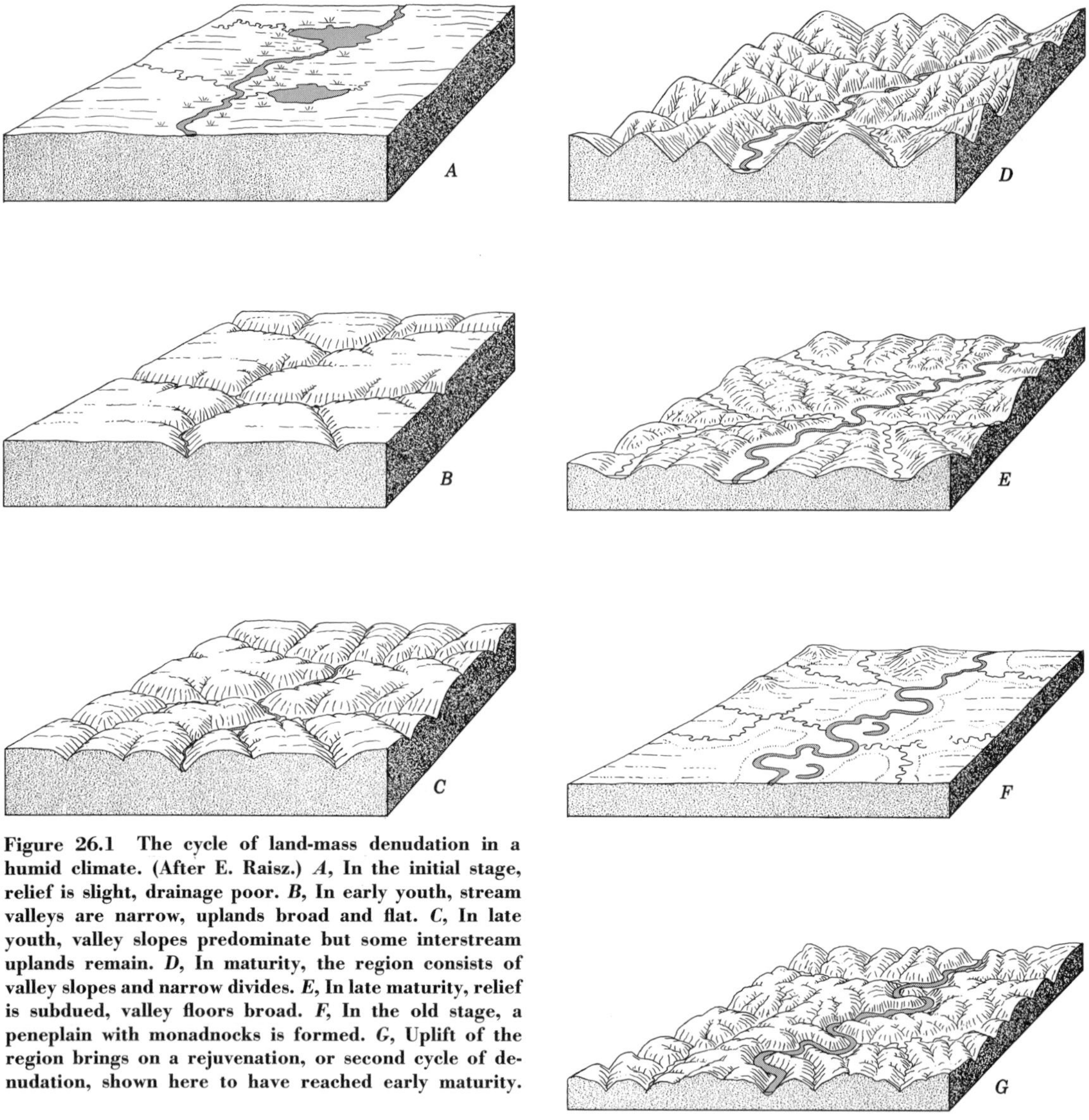

Figure 26.1 The cycle of land-mass denudation in a humid climate. (After E. Raisz.) *A,* In the initial stage, relief is slight, drainage poor. *B,* In early youth, stream valleys are narrow, uplands broad and flat. *C,* In late youth, valley slopes predominate but some interstream uplands remain. *D,* In maturity, the region consists of valley slopes and narrow divides. *E,* In late maturity, relief is subdued, valley floors broad. *F,* In the old stage, a peneplain with monadnocks is formed. *G,* Uplift of the region brings on a rejuvenation, or second cycle of denudation, shown here to have reached early maturity.

a surface coinciding with sea level projected inland. To this imaginary surface is applied the word *base level,* for it is both a base toward which denudation is progressing and a water-level surface. Although the base level is attainable only in theory, a fairly close approach to it has been made in various parts of the world throughout eras of the geologic past. The word *peneplain* is given to a land surface of faint relief produced in the old stage of a cycle of denudation.

A peneplain is not perfectly flat but has gentle slopes. Because the streams are sluggish and the land slopes low, further denudation is extremely slow. It is not easy to set a particular figure for the number of years required for a region to pass from initial stage to old stage, because this depends upon how high the landmass was elevated to begin with and how resistant the rocks are to weathering and erosion. Perhaps it would be safe to say that in known cases in the geologic record, several million years have been required to reduce a mountain mass to a peneplain.

Sometimes a region contains zones or patches of rock far more resistant to weathering and erosion than the rock of the region as a whole. As the cycle progresses through maturity into old age, these harder rocks are left standing in prominent hills or isolated mountains, which rise conspicuously above the surrounding peneplain (Figure 26.4). To a residual hill or mountain of this type is given the name *monadnock,* named for Mount Monadnock in southern New Hampshire.

Figure 26.2 The Grand Canyon of the Yellowstone River, viewed from over Inspiration Point, illustrates a youthful canyon carved into the initial surface of a lava plateau. (U.S. Army Air Service Photograph.)

Rejuvenation

Once formed, a peneplain is usually elevated again by crustal movement. This follows the principle, stated earlier, that the internal earth forces act spasmodically, and that periods of extreme stability of a particular part of the world are ended by uparching or severe deformation.

A peneplain that is badly folded or fractured during orogeny is quickly obliterated in the erosion cycle that follows. If, however, the region is merely elevated by epeirogenic movements of a few hundred or a few thousand feet, remains of the peneplain persist for a considerable period. In Block *G* of Figure 26.1 this occurrence is illustrated. Similarity between this and the young stage illustrated in Block *C* is marked. The principal difference between the two is that the initial land surface in Block *C* is a former sea floor, whereas the "initial" surface in Block *G* is the uplifted peneplain. Drainage on the uplifted peneplain is already well established, so that the streams are merely rejuvenated and cut deep V-shaped valleys into their old shallow courses. When maturity of the second cycle is reached, the former peneplain is completely consumed, but its influence is seen generally in the accordant summits of hilltops over the region as a whole. There is no particular limit to the number of cycles a region can undergo. In some regions, such as the Appalachians of eastern North America (Figure 23.12), three or four previous cycles can be interpreted from a study of the topography.

Environmental aspects of the erosion cycle

The environmental importance of stage in the cycle of landmass denudation is very great. Regions in the initial stage of the cycle are relatively flat plains on which drainage is poor and marsh lands often extensive. Sandy beach deposits left by waves as the land surface emerged from beneath the sea usually produce infertile soils. An excellent example is the coastal plain region of Georgia and northern Florida. The great Okefenokee Swamp occupies a shallow depression in the former sea floor whereas long sandy strips of higher land parallel the coast. The porous sands permit plant nutrients to be leached out of the soil.

Not all regions in an initial stage have emerged from the sea; some, such as the High Plains of eastern Colorado, western Kansas, and northern Texas, were built by aggrading streams and, although remarkably flat, are well drained and do not have extensive marshes. The High Plains possess a high productivity in wheat, not only because the soil and climate are favorable but also because the flatness of the land permits enormous grain fields to be cultivated and harvested by machines.

Still other areas of initial land surface are formed by lava flows, poured out profusely to inundate the previous topography and produce a high, undulating lava plateau. This is the origin of the plateau into which the Yellowstone River has carved its gorge (Figure 26.2). The Snake River Plain of southern Idaho is another example. A region in youth of the cycle supports its population on the relatively flat areas between deep V-shaped valleys. Because these valleys are in a young stage they have no floodplains; hence, roads, railroads, cities, and farms are situated on the uplands. A mature region, on the other hand, has no flat uplands remaining, hence is not favorable to habitation, agriculture, or transportation. Many of the world's mountain regions are in the stage of maturity in the erosion cycle. Extremely great relief and steep slopes are the result of recent, high uplift of those portions of the earth's crust. A coastal region, on the other hand, elevated only 200 or 300 ft (60 to 90 m) above sea level, can never attain really mountainous relief because the streams can cut into the mass no more deeply than the mass itself rises above sea level. In some maturely dissected regions, the larger streams have already reached full maturity in their own individual life cycles and have sizable floodplains at the same time that the surrounding region is extremely rugged. Under such circumstances, human activity is concentrated in the valley floors.

Regions with a humid climate in late mature

Figure 26.3 This air view of a maturely dissected region shows a complex of stream channels and small drainage basins. (Spence Air Photos.)

or old stages of the cycle are usually favorable to agriculture. Slopes are moderate or low and are well drained. Soils tend to be thick. Destructive soil erosion can be held in check. Roads and railroads cross the rolling surface without great difficulty or follow extensively developed floodplains.

Much of the Amazon Basin is a peneplain, but because of the heavy rainfall a dense forest vegetation renders the region passable only at the cost of great labor. Low, floodplain areas bordering the mature rivers are forbidding morasses.

Peneplains that have been uplifted and trenched by valleys in a new cycle are comparable environmentally to regions in young stages of the cycle of landmass denudation.

Drainage networks

Much of the earth's land surface consists of landforms produced in the cycle of denudation in a humid climate, although little is in the initial and early youth stages of the cycle because of the short duration of those stages. Thus, well-developed drainage systems characterize vast areas of the continents. Even in deserts, where a somewhat different cycle is followed by the landmass as a whole, the mountains and plateaus are dominated by drainage systems quite like those of the cycle in humid climates. It is therefore important to devote some attention to the form development, or *morphology*, of stream networks and the associated systems of slopes and drainage basins.

Figure 26.4 Stone Mountain, on the Piedmont upland near Atlanta, Georgia, is a striking monadnock about 1.5 mi (2.4 km) long and rising 650 ft (193 m) above the surrounding Piedmont peneplain surface. The rock is a light-gray granite, almost entirely free of joints, and has been rounded into a smooth dome by weathering processes. (Photograph by U.S. Army Air Service.)

Figure 26.5 is a map on which the permanent stream channels are drawn, together with divides of the major drainage basins. This is a typical example of the drainage network of a region in the stage of maturity.

Ground slopes converge upon the head of each channel, bringing overland flow in sufficient quantities during heavy rains to produce channel flow and keep the channel well scoured, free of soil and vegetation which might otherwise tend to obstruct and obliterate the channel (Figure 26.6). Over a long period of time, as the landmass cycle evolved through youth into maturity, the available ground has been apportioned to the individual

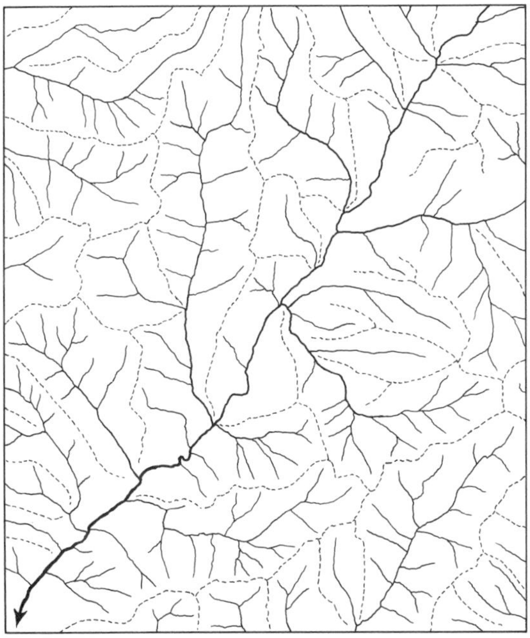

Figure 26.5 A drainage system consists of many small basins, each adjusted in size and shape to the magnitude of the stream it serves. Streams are shown by solid lines; divides by dashed lines.

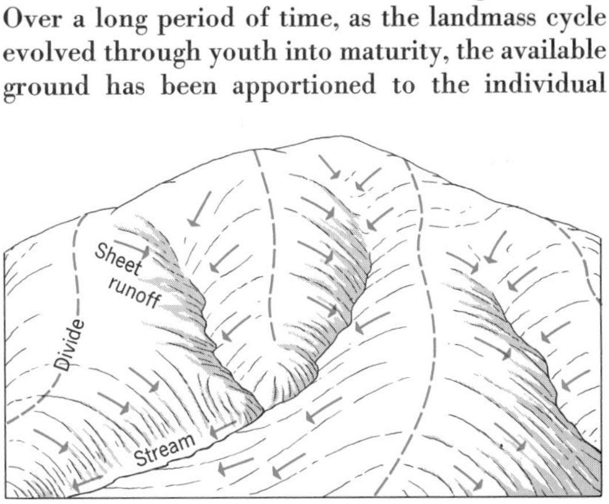

Figure 26.6 Overland flow from slopes in the headwater area of a stream system supplies water and rock debris to the smallest elements of the channel network.

small channels so that each receives the amount of runoff required to sustain it. Ground slopes and channel gradients have become mutually adjusted so that the rates of erosion are neither so rapid that a channel lengthens at the expense of its neighbor, nor is it so deficient in area of runoff that it becomes choked with debris of which it cannot dispose. Slow down-cutting of channels and slow lowering of slopes and divides continues, but with a general uniformity over the entire region.

Each finger-tip channel joins another, or enters a larger channel, in such a way that watersheds of increasing size are formed; exit channels are progressively larger in dimensions and discharge. Most stream junctions form acute angles, so that the discharge is carried in the general direction of the larger trunk streams in as direct a manner as possible and with as few channels as possible while, at the same time, providing each element of the stream network with an adequate area of surface runoff.

Playfair's law

Prior to about 1800 there was a widespread belief among naturalists that stream valleys and other landforms were the products of a great cataclysmic upheaval, or rending, of the earth's crust. It was supposed that streams later occupied the valleys produced in this manner, hence, that the valleys came first and were not the work of the streams themselves.

To John Playfair, an English geologist, is attributed the first clear and convincing statement of beliefs universally accepted today concerning streams and their valleys. In 1802, he published the following statement, now known as *Playfair's law:*

> Every river appears to consist of a main trunk, fed from a variety of branches, each running in a valley proportioned to its size, and all of them together forming a system of valleys connecting with one another, and having such a nice adjustment of their declivities that none of them join the principal valley on too high or too low a level; a circumstance which would be infinitely improbable if each of these valleys were not the work of the stream which flows in it.

The three main points of this law, restated in abstract form, are (1) valleys are proportioned in size to streams flowing in them; (2) stream junctions are accordant in level; (3) therefore, valleys are carved by the streams flowing in them, because both (1) and (2) would be "infinitely improbable" on the basis of chance alone. Exceptions occur in nature. Sometimes a tributary stream passes over a waterfall to reach the level of the master stream. Sometimes small streams occupy very large valleys. But for each of these exceptions, closer study reveals unusual or special conditions that have locally prevailed. Regions of homogeneous bedrock, which have been eroded by streams over a long period of time, invariably bear out the truth of Playfair's law.

From purely qualitative descriptions of drainage systems and their related land slopes, the science of geomorphology has advanced to quantitative analysis in which measurements of landscape elements are taken and subjected to mathematical treatment. These newer developments, discussed in Chapter 27, serve to support and to express more rigorously than words alone the concepts of the cycle of land-mass denudation and Playfair's law.

Fluvial processes in an arid climate

The general appearance of desert regions is strikingly different from that of humid regions, reflecting differences in both vegetation and landforms. It should be emphasized that rain falls in dry climates as well as in moist and that most landforms of desert regions are formed by running water. A particular locality in a dry desert may experience heavy rain only once in several years, but when it does fall, stream channels carry water and perform the same work as the constantly flowing streams of moist regions. Excess water runs off from the valley slopes into the streams, washing down rock particles into the channels, just as in the moister regions. We may even go further and say that, although running water is a rather rare phenomenon in dry deserts, it works with more spectacular effectiveness on the fewer occasions when it does act. This is explained by the meagerness of vegetation in dry deserts. The few small shrubs and herbs that survive offer little or no protection to soil or bedrock. Without a thick vegetative cover to protect the ground and hold back the swift downslope flow of water, excessive quantities of coarse rock debris are swept into the streams. A dry channel is transformed in a few minutes into a raging flood of muddy water, heavily charged with rock fragments.

From these statements it might be inferred that rainfall in desert regions is heavier and more violent than in moist regions. This is not true. In dry deserts, almost all rainfall is of the thunderstorm type, which is localized and intense, affecting only a small area directly under the storm. In humid middle-latitude climates there are even more frequent and heavier thunderstorm downpours, but also many prolonged periods of steady, light rain during which the soil takes in moisture and becomes fully saturated. Furthermore, the air of humid regions tends to be moist, thus reducing evaporation from the ground. Under such conditions, vegetation can grow densely and maintain its protective hold upon the ground. It is true that in the wet equatorial climate most of the rainfall is of the thunderstorm type, but this occurs so frequently that dense vegetation can flourish. In a dry desert the periods between rains are so prolonged that only a few widely spaced plants can survive.

Effluent and influent streams

An important contrast between regions of arid and humid climates lies in the nature of stream flow. An important consideration is the manner in which the water enters and leaves a stream channel.

In a humid region with a high water table sloping toward the stream channels, ground water moves steadily toward the channels, into which it seeps, producing permanent, or *perennial* streams. Such streams are described as *effluent* (Figure 26.7*A*). Surplus water in a humid region—both that which flows over the ground in wet weather, and that which seeps into the channel from the ground water table—escapes by through-flowing channels to regions of lower elevation and eventually to the sea. In the course of such drainage, dissolved substances (ions and colloids) that have been removed from the soil are taken out of the region, permanently removing them from the soil (Chapter 18).

In arid regions, where streams flow across plains of gravel and sand, water is lost from the channels by seepage to a water table that lies below the level of the streams. Such streams are designated as *influent* (Figure 26.7*B*). The water table beneath the channel tends to rise, forming a *water-table mound.*

Loss of discharge by influent seepage and evaporation strongly affects streams in alluvium-filled valleys of arid and semiarid regions. Aggradation occurs and the braided channel is conspicuous. One of the most important generalizations made about streams in desert regions is that "the streams are shorter than the slopes." Instead of long, continuously flowing streams extending to the sea, streams of desert regions are often short and terminate in alluvial deposits or on shallow, dry lake floors.

One very common landform built by braided, aggrading streams is the *alluvial fan,* a low cone of alluvial sands and gravels resembling in outline an open Japanese fan (Figure 26.8). The apex, or central point of the fan, lies at the mouth of a canyon, ravine, or gully and is built out upon an adjacent plain. Alluvial fans are of many sizes, ranging from tiny miniature fans a foot or two across, such as the ones seen alongside a roadcut, to huge fans many miles across (Figure 26.9).

Fans are built by young streams carrying heavy loads of coarse rock waste out from a mountain or upland region. The braided channel shifts constantly but, because it is firmly fixed in position at the canyon mouth, must sweep back and forth. This fixed apex accounts for the semicircular form and the downward slope in all radial directions from the apex.

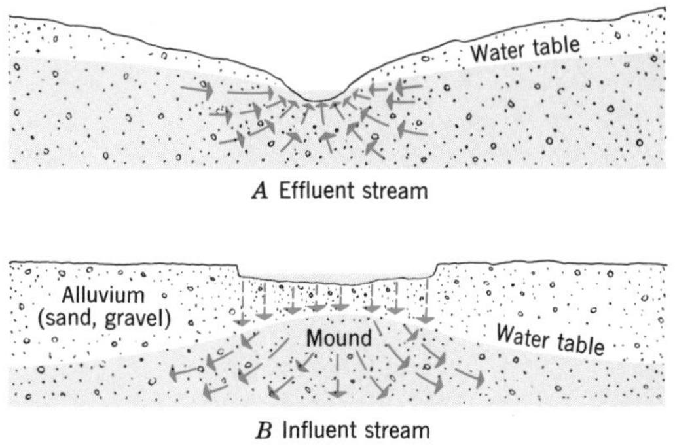

Figure 26.7 Effluent and influent streams. (From A. N. Strahler, 1971, *The Earth Sciences*, 2nd ed., Harper and Row, New York.)

Ground water in alluvial fans

Large, complex alluvial fans also include mudflows (Figure 24.22). As shown in Figure 26.10, mud layers are interbedded with sand and gravel layers. Water infiltrates the fan at its head, making its way to lower levels along sand layers that serve as water conductors, or *aquifers.* The mudflow layers serve as barriers to ground water movement, or *aquicludes.* Ground water trapped beneath an aquiclude is under hydraulic pressure from the load of water at higher elevation toward the fan head. When a well is drilled into the lower slopes of the fan, water rises spontaneously and may emerge at the well head, a phenomenon known as *artesian flow.* The well is designated an *artesian well.*

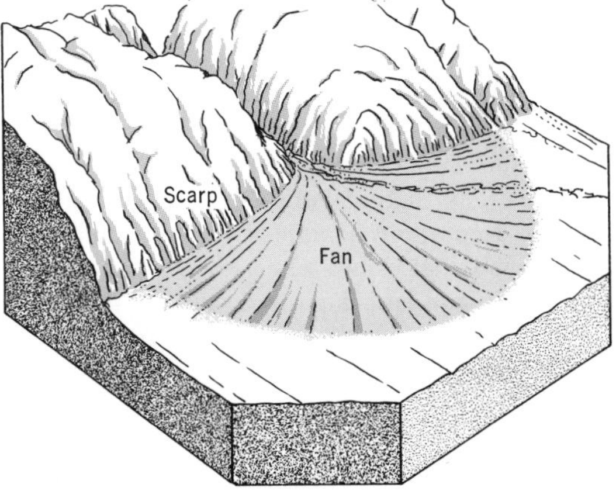

Figure 26.8 A simple alluvial fan.

Figure 26.9 A great alluvial fan in Death Valley, built of debris swept out of a large canyon. Observe the braided stream channels. (Copyrighted Spence Air Photos.)

Alluvial fans are the dominant class of ground-water reservoirs in the southwestern United States. Sustained heavy pumping of these reserves for irrigation has lowered the water table severely in many fan areas. Rate of recharge is extremely slow in comparison. However, efforts are made to increase this recharge by means of water-spreading structures and infiltrating basins on the fan surfaces.

A serious side effect of excessive ground water withdrawal is that of subsidence of the ground surface. Important examples of subsidence are found in alluvial valleys filled to great depth with alluvial fan and lake-deposited sediments. For example, in one locality in the San Joaquin Valley of California, where the water table has been drawn down over 100 ft (30 m), ground subsidence has amounted to about 10 ft (3 m) over a 35-year period. As a result, wells in the area have been damaged. Withdrawal of ground water from alluvium beneath Mexico City has caused ground subsidence exceeding 20 ft (6 m).

Alluvial fan environments

Fans are of major environmental importance. In mountainous regions of arid climates, populations and cities are concentrated on the gentler outer fringes of alluvial fans near the base of a mountain range. Water for irrigation is diverted from the streams that issue from the canyon mouths or can be had from wells.

A striking example of fan utilization is found in the Los Angeles basin of southern California. Flash floods have been a serious menace to such communities as Burbank, Glendale, Montrose, and Pasadena, whose favored residential districts are on the higher, inner slopes of these fans. These *debris floods*, as they are called by engineers, are distinctive in that the sediment load is very great

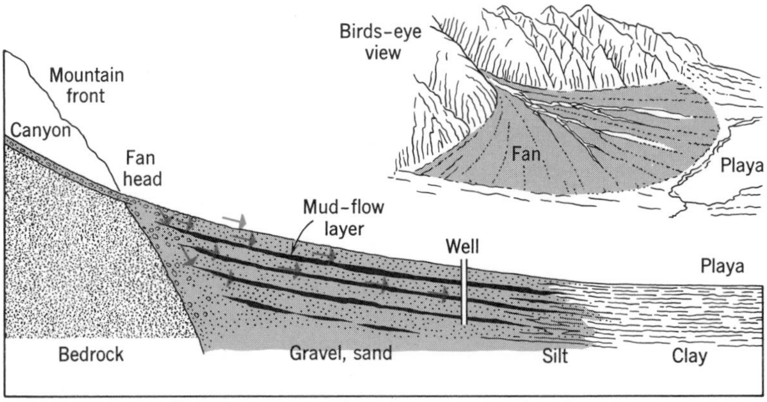

Figure 26.10 An idealized cross section of an alluvial fan showing mudflow layers (aquicludes) interbedded with sand layers (aquifers). (From A. N. Strahler, 1972, *Planet Earth; Its Physical Systems Through Geologic Time,* Harper and Row, New York.)

and may include large boulders as well as sand, silt, and clay. The floods arise on the steep slopes of mountain watersheds, where brush fires have destroyed protective vegetation (Figure 26.11).

Cycle of landmass denudation in an arid climate

A mountainous desert such as that shown in Figure 26.12 undergoes a gradual evolution of its landforms quite different in certain respects from the evolution of a landscape under a humid climate having a large water surplus.

A cycle of landmass denudation in an arid region is illustrated in Figure 26.13. In this ideal cycle we imagine a mountainous region formed by folding or fracturing of the earth's crust and lying in an interior part of the continent (Block *A*). Relief is at the maximum in the initial stage and diminishes throughout successive stages. Numerous large depressions exist between mountain ranges. These do not fill up with water to form lakes, as in a moist climate, but remain dry because of excessive evaporation in the hot, dry climate. The flat central parts of such depressions provide the beds of temporary lakes and are known as *playas.* Playas lakes are shallow and fluctuate considerably in level, often disappearing entirely for long periods. Because they have no outlets, playa lakes contain salt water often more strongly saline than ocean water.

Figure 27.11 Debris deposits at the mouth of Benedict Canyon, Ballona Creek basin, California, following a flood in January 1952. (Photograph by U.S. Army Corps of Engineers.)

Throughout the erosion cycle, the intermontane depressions become filled with rock waste as alluvial fans are built out from the adjoining mountain masses (Figure 26.12). When the basins are filled with alluvium and the mountains masses are cut up into an intricate set of canyons, divides, and peaks, the region is said to be in the *mature stage* (Block *B*, Figure 26.13). As maturity progresses, the mountains are worn lower, at the same time shrinking in size as the alluvium of the fans encroaches progressively farther inward upon the mountain base.

When the *old stage* is reached, mountains are represented by small islandlike remnants of their former selves (Block *C*). Eventually even these remnants, which may be compared to monadnocks on a peneplain, are eroded away and a vast plain remains. This surface is a type of peneplain, but it has not been developed with reference to sea level as a base because no streams drain out to the sea, and it may, therefore, even lie many hundreds of feet above sea level. It contains shallow depressions occupied by playas rather than by floodplains of meandering rivers. Wind action in a dry climate is effective in eroding shallow depressions and in making dunes of shifting sand.

In most deserts, the sloping surfaces of boulders, gravel, and sand which extend from the abrupt base of steep mountain faces to the flat ground of the playas are alluvial fans, underlain by thick deposits of alluvium shed by the mountains. In some places, however, this alluvium, although

Figure 26.12 This air view of Death Valley, California, shows a mature desert landscape comparable to that shown in Figure 26.13*B*. (Copyrighted Spence Air Photos.)

outwardly taking the form alluvial fans, is nothing more than a veneer perhaps 10 to 20 ft (3 to 6 m) thick, overlying a smooth sloping floor of solid bedrock. To such a rock surface, fringing a desert mountain range or cliff line, the term *pediment* is applied. On the righthand cross section of Blocks *B* and *C*, Figure 26.13, pediment surfaces are shown in profile in a narrow zone between the thick basin alluvium and the rugged mountain masses.

Because a pediment is normally veneered by alluvium while it is being formed, the only way to be certain that a pediment exists is to see the bedrock widely exposed to view by later erosion, as in the case of the pediment shown in Figure 26.14.

Because the old stage surface of the arid climate landmass cycle consists in part of pediments, the term *pediplain* has been introduced as equivalent to the term "peneplain" in the humid climate landmass cycle. A pediplain consists of alluvial fan and playa surfaces as well as pediments; it is thus partly erosional and partly depositional.

The environmental aspects of mountainous deserts are well illustrated by the vast basin-and-range region which includes Nevada, southern California, western Utah, and the southern halves of Arizona and New Mexico. Here, excellent examples of various stages in the cycle can be observed as one travels through the desert on the main transcontinental highways, railroads, and airways. Sparseness of population goes hand in hand with sparseness of vegetation. In 1970, the state of Nevada, which lies wholly within this region, had only about 500,000 inhabitants in its 110,000 sq mi (286,000 sq km). Contrast this with Pennsylvania, whose 45,000 sq mi (117,000 sq km) contained nearly 12,000,000 persons. The vastness of waste land in this region can be partly appreciated when we realize that the explosion of the first test atom bomb was kept a secret within a single intermontane basin in New Mexico.

To the traveler, this landscape of mountainous deserts seems to be composed of three distinctive elements: (1) intensely rugged and inhospitable mountains; (2) broad sloping pediments and alluvial fans scored with innumerable shallow dry stream channels, or *washes;* and (3) perfectly flat playa lake floors, covered with shallow water or

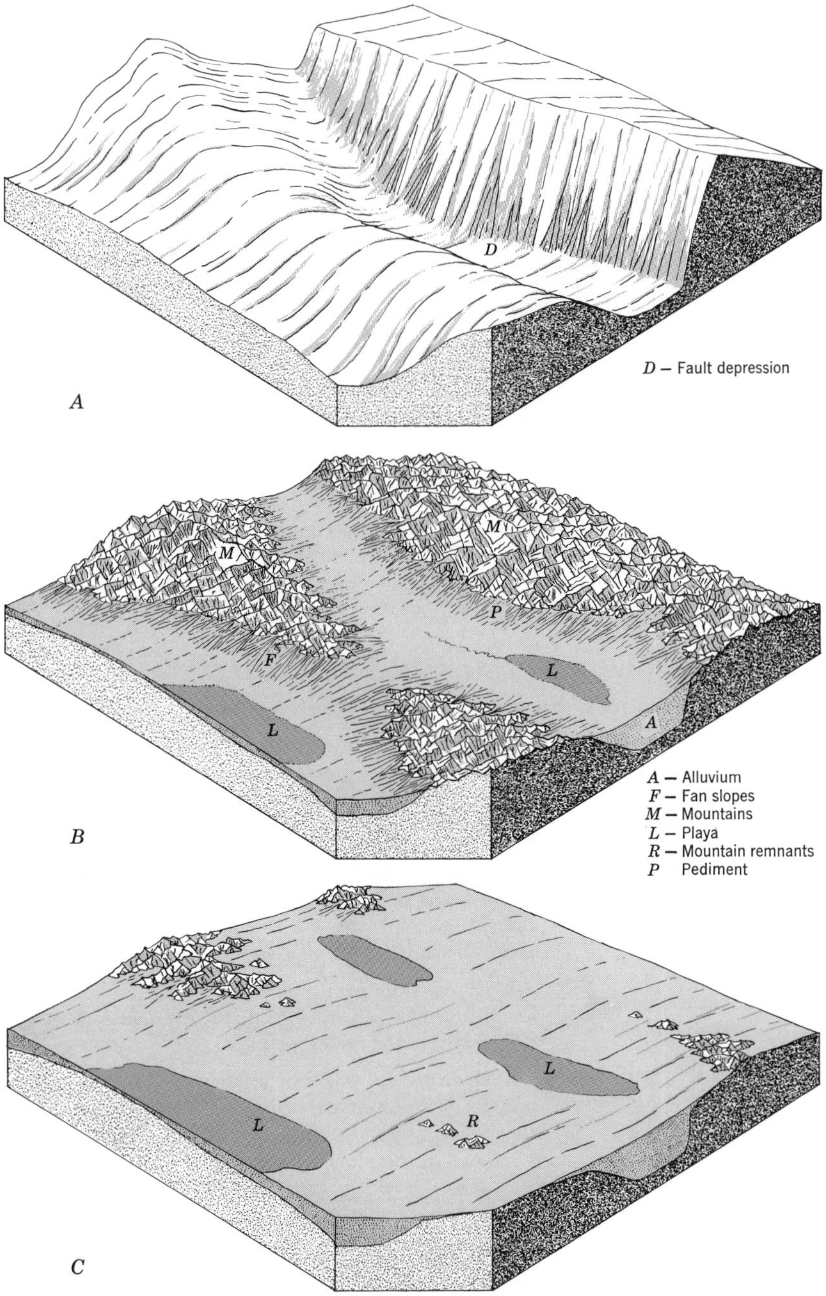

Figure 26.13 The cycle of land-mass denudation in a mountainous desert. *A*, In the initial stage, relief made by crustal deformation is at the maximum. *B*, In the mature stage, the mountains are completely dissected and the basins are filled with alluvial fan material and playa deposits. *C*, In the old stage, relief is low and alluvial deposits have largely buried the eroded mountain masses, whose remnants project here and there as islandlike groups.

white salt deposits. In the rugged mountains occur valuable mineral deposits, which are an outstanding source of economic wealth. The alluvial fan slopes are virtually worthless except in the few places where wells bring in a flow of water for the needs of isolated communities. The playas provide mineral wealth of a different sort, including salts of calcium, sodium, and potassium.

Figure 26.14 Bedrock is widely exposed over this pediment surface at the foot of the Dragoon Mountains near Benson, Arizona. (Photograph by Douglas Johnson.)

REVIEW QUESTIONS

1. Explain the concept of the cycle of landmass denudation. What possible combinations of initial relief and climate may be considered?

2. Describe the various stages in the landmass denudation cycle in a humid climate. Include discussion of the development of the drainage system, valley slopes and divides, and topographic relief. Which stage has greatest relief?

3. Describe the appearance of a peneplain. What is a monadnock? If elevation of the land causes a new cycle to begin, what happens to the peneplain?

4. Discuss the environmental aspects of the cycle of landmass denudation in a humid climate. Give examples from regions in different stages of the cycle.

5. Describe the appearance of a drainage network. How is it adapted to efficient discharge of runoff and sediment load?

6. State Playfair's law. What other view was held in Playfair's time? Does the law have any exceptions? Explain.

7. How does the role of running water in arid regions differ from that in humid climates? Contrast the forms of stream channels in humid and arid climates.

8. Describe effluent and influent streams and their relation to ground water.

9. Describe and explain the development of an alluvial fan. Where and why are alluvial fans of great importance to man?

10. Explain the occurrence of artesian ground water conditions beneath an alluvial fan. What side effect has sustained ground water withdrawal?

11. Describe the stages of a landmass cycle in mountainous deserts. Compare this cycle, stage by stage, with the cycle of landmass denudation in a humid climate. Compare a peneplain with a pediplain.

12. Describe a pediment. How does it resemble an alluvial fan? How does it differ from an alluvial fan?

13. Discuss the economic and environmental aspects of mountainous deserts. What physical factors in the environment tend to limit human occupation of such areas?

Exercises

Exercise 1. ***Region in Stage of Youth.*** (portion of Yellowstone National Park topographic map, U.S. Geological Survey; scale 1:125,000.)

Explanatory Note: Much of the area of this map represents an undulating plateau surface at 8000 to 8500 ft elevation, which may be considered as an initial land surface built by repeated outpourings of lavas. Hot springs and geysers suggest the geologic recency of volcanic activity. Low stream gradients and marshes are typical of streams flowing across the undissected upland. In strong contrast with the initial upland is the deep, youthful Grand Canyon of the Yellowstone River downstream from the Upper and Lower Falls. Figure 26.2 is an oblique air photograph taken from a point just above the canyon, looking southwestward toward the falls.

QUESTIONS

1. **(*a*)** What contour interval is used on this map? **(*b*)** What is the fractional scale of the map? **(*c*)** Construct a graphic scale in miles for the map.

2. **(*a*)** Compute the gradient in feet per mile of the Yellowstone River in the Grand Canyon. **(*b*)** How does this gradient compare with the gradient of the Yellowstone River upstream from the Upper and Lower Falls?

3. Construct a topographic profile along the line from 15.0–0.0 to 15.0–13.3. Show the position of sea level with respect to the profile.

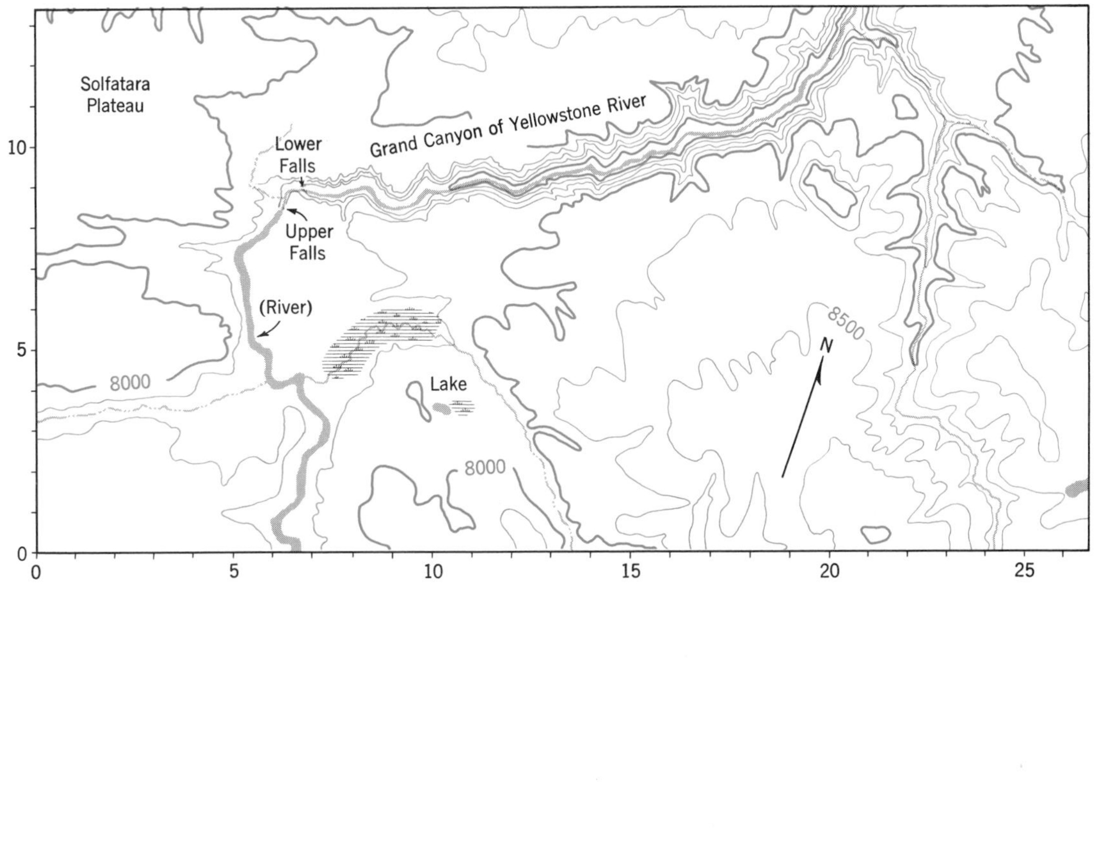

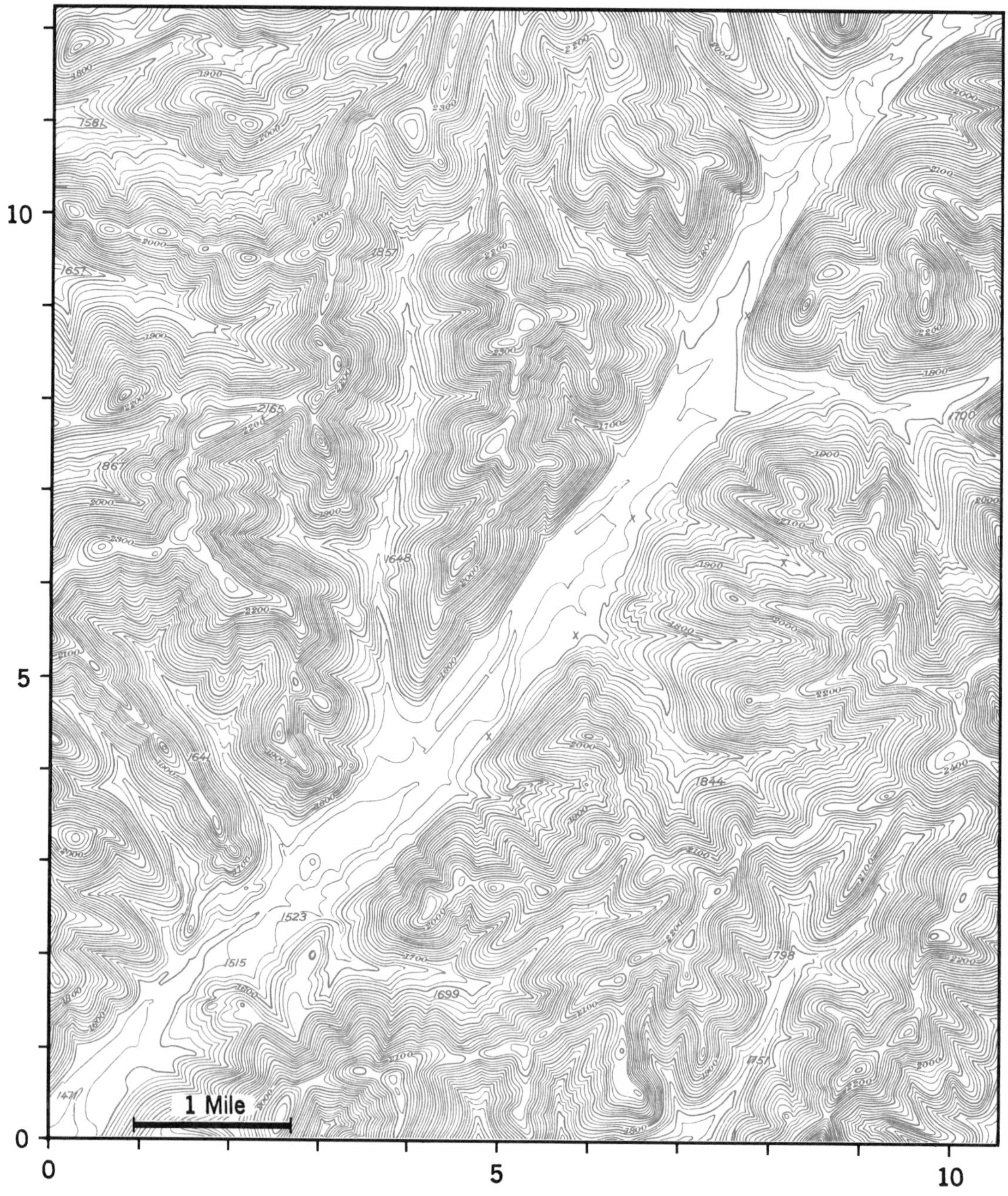

Exercise 2. *Region in Stage of Maturity.* (Portion of Belmont, N.Y., U.S. Geological Survey topographic map; scale 1:62,500.)

Explanatory Note: A maturely dissected region in a humid continental climate is shown without the drainage network. All the region has been formed into closely fitting drainage basins separated by narrow, rounded divides.

QUESTIONS

1. **(*a*)** What is the fractional scale of this map? **(*b*)** What contour interval is used? **(*c*)** What is the approximate average relief in this area? (Difference between elevations of divides and adjacent valley bottoms.)

2. Using a blue pencil, draw in the drainage network for this area. Indicate a stream channel wherever the contour lines form a series of sharply defined V's. (Use a tracing paper overlay for this exercise.)

3. Using a red pencil, outline all divides on this map. Carry the divide lines down the spurs to the valley bottoms in such a way as to enclose a single basin for each finger-tip tributary of the drainage system. (Do this exercise on the same sheet of tracing paper used in Question 2.)

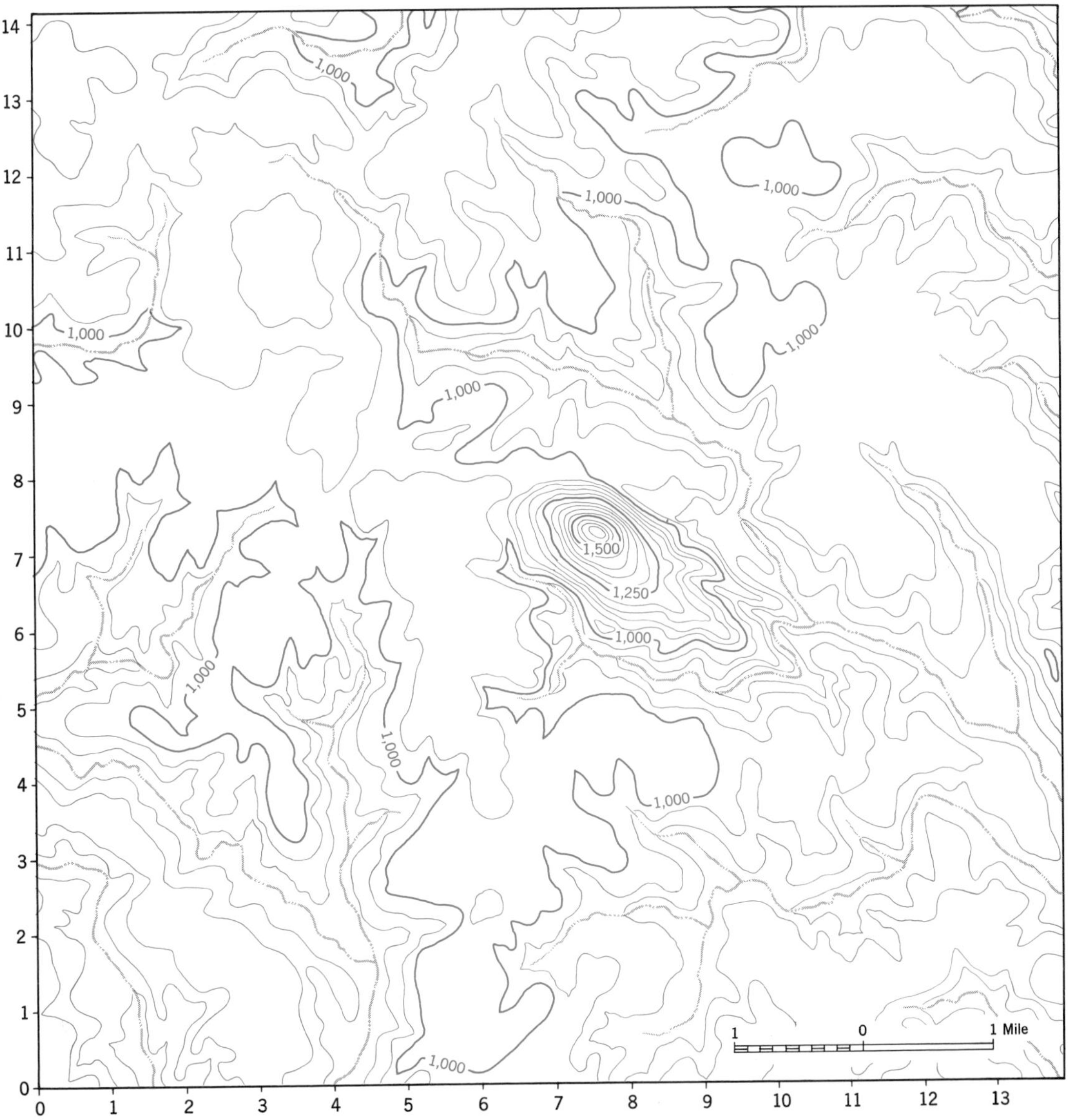

Exercise 3. *Monadnock on Peneplain.* (Source: Atlanta, Ga., U.S. Geological Survey topographic map; scale 1:125,000.)

Explanatory Note: The prominent knob in the center of the map is Stone Mountain, a small intrusive body of granite surrounded by ancient gneisses and schists. Because of its greater resistance to weathering and erosion, the Stone Mountain granite formed a conspicuous monadnock while the surrounding region was reduced to a peneplain, seen now in the broad, relatively flat divide areas between 950 and 1050 ft elevation. In a second erosion cycle, stream valleys were cut into the peneplain surface, producing the present topography. Figure 26.4 is an air photo of Stone Mountain.

QUESTIONS

1. **(*a*)** What contour interval is used on this map? **(*b*)** How high does the summit of Stone Mountain rise above the peneplain surface one mile west of the summit?

2. Make a topographic profile through the summit of Stone Mountain, from 13.9–12.0 to 0.0–1.5. Use a vertical scale of 1 in. equals 1000 ft.

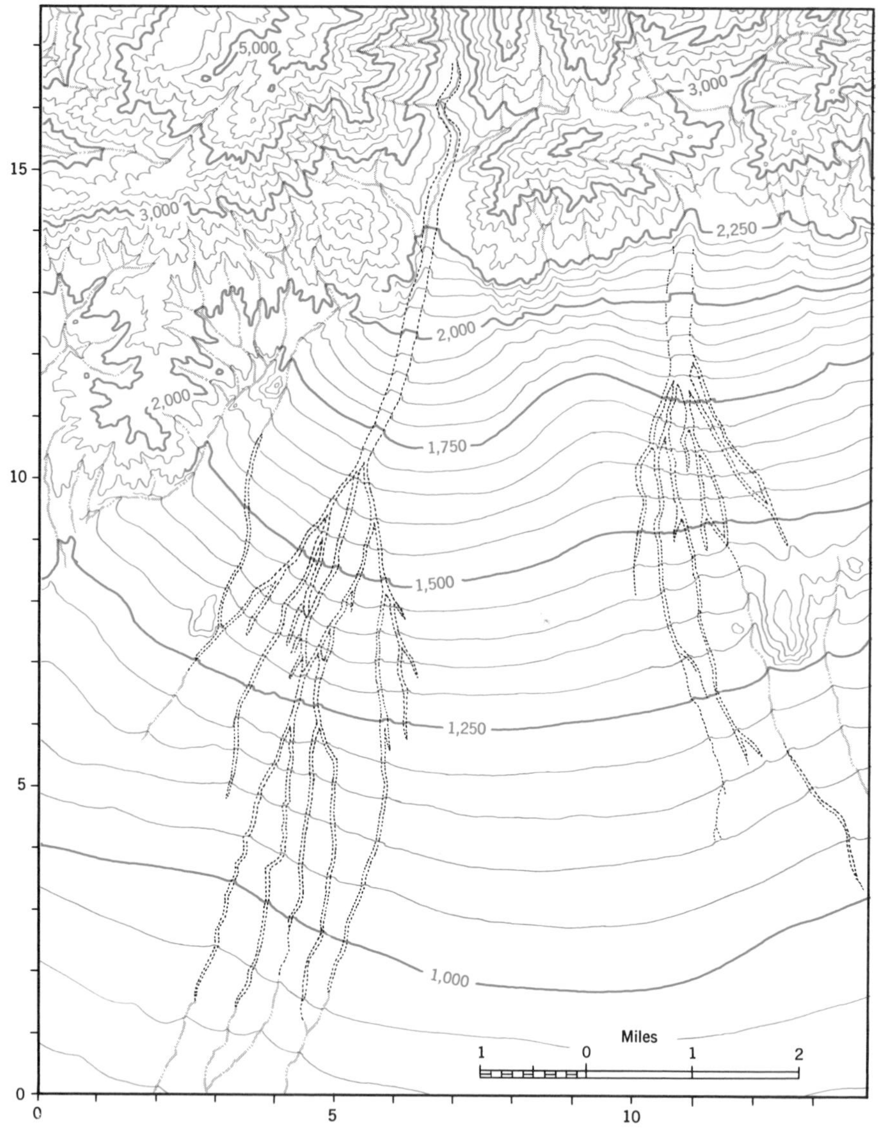

Exercise 4. ***Alluvial Fans.*** (Source: Cucamonga, Calif., U.S. Geological Survey topographic map; scale 1:62,500.)

Explanatory Note: The northern third of this map shows the southern slopes of the San Gabriel Mountain range of southern California, from which issue debris-laden streams of steep gradient. Large alluvial fans, covering the remainder of the map, have been built of gravels and bouldery debris by flood waters of these south-flowing streams. The sloping piedmont alluvial plain, therefore, consists of alluvial fans of various sizes arranged side by side. The contour interval on this map changes at the contact of the fans and mountain slopes because the difference in relief of the two types of topography is very great. For further details of these fans and the history of their development, see the paper by Rollin Eckis, listed among the reading references.

QUESTIONS

1. **(*a*)** What contour interval is used on the fan areas? **(*b*)** What contour interval is used in the mountain areas? **(*c*)** What is the fractional scale of this map?

2. The main stream located just west of the center line of the map is San Antonio Canyon. **(*a*)** What is the gradient (feet per mile) of this stream in the mountain canyon measured between the top edge of the map and the 2250-ft contour? **(*b*)** Why does the stream disappear at the canyon mouth, to be replaced by a broad channel shown by rows of dots? (The dotted lines mark boulder-strewn flood channels, cut somewhat below the level of the fan surface and normally dry.)

3. **(*a*)** What is the average gradient (feet per mile) of the fan surface between the 2000– and 1750–ft contours near 5.5–11.5? **(*b*)** What is the average gradient between the 1250– and 1000–ft contours near 7.0–4.0? **(*c*)** Make a profile of the fan of the San Antonio Wash from 6.5–14.0 to 2.0–0.0, using a vertical scale of 1 in. equals 1000 ft. **(*d*)** Why is the fan profile concave-up, steepening toward the fan apex?

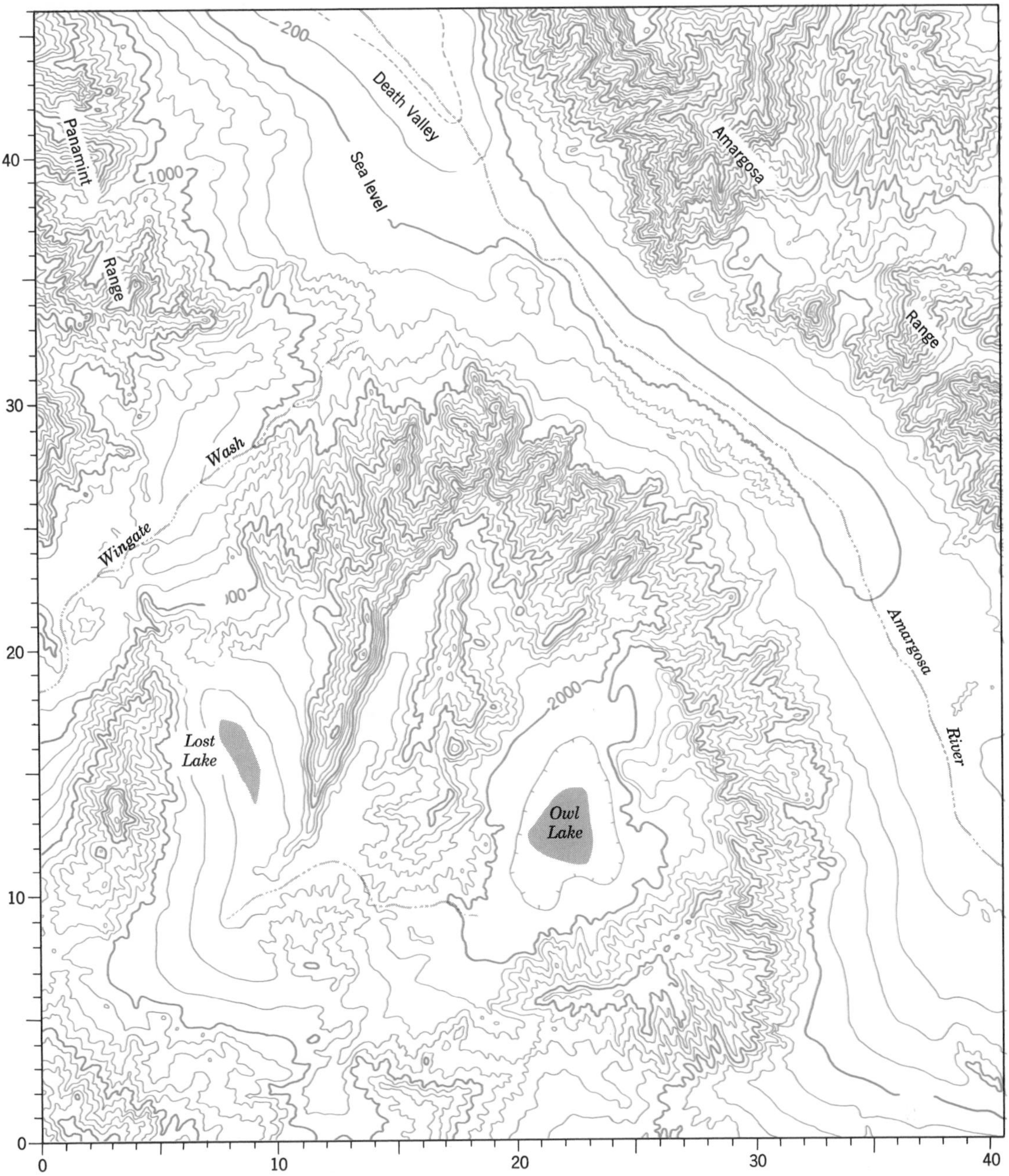

Exercise 5. ***Mountainous Desert Landscape.*** (Source: Avawatz Mountains, Calif., U.S. Geological Survey topographic map; scale 1:250,000.)

Explanatory Note: This map shows a portion of the Basin and Range physiographic province, the region used by Professor W. M. Davis to typify the land-mass denudation cycle in an arid climate. Death Valley, a downfaulted block, lies well below sea level despite filling with vast quantities of alluvium swept down from the Amargosa and Panamint Ranges, upfaulted blocks which bound the valley on either side. Figure 26.12 is an air photograph taken from a point above Death Valley, looking southeastward up the Amargosa River. Owl Lake and Lost Lake are ephemeral playa lakes occupying independent closed depressions considerably higher than Death Valley.

QUESTIONS

1. **(*a*)** How many square miles are covered by this map? **(*b*)** Give grid coordinate location and elevation of the highest and lowest points on this map.

2. By study of Figure 26.12, determine the position of the plane when the air photograph was taken. **(*a*)** Give grid coordinates of this position. **(*b*)** Give the azimuth of the line along which the camera axis was pointed.

3. **(*a*)** What is the maximum depth of the closed depression in which Owl Lake lies? **(*b*)** Give the grid coordinates of the lowest point on the rim of this depression. **(*c*)** What drainage changes are in progress there? **(*d*)** What is the relationship between Lost Lake and Owl Lake basins?

CHAPTER 27

Quantitative Analysis of Erosional Landforms

IN the preceding chapters, landscapes shaped by the combined action of weathering, mass wasting, overland flow, and stream flow have been treated largely in a qualitative and descriptive style. Such an approach to the understanding of landforms is in keeping with the explanatory-descriptive system introduced by Professor William Morris Davis about 1890 (Chapter 24). Most geomorphologists followed Davis' lead. For several decades, the classification and description of landforms was verbal and almost totally devoid of measurement, other than some descriptive parameters such as the approximate length, width, height, and areal extent of the landscape elements. Comparisons were in terms of such adjectives as "steeper" or "gentler," "faster" or "slower," "well adjusted" or "poorly adjusted."

Notice that the sciences of climatology and hydrology, closely linked with geomorphology, progressed from the very outset by quantitative statements. Measurements of air temperature, pressure, winds, humidity, cloudiness, and precipitation were painstakingly collected and averaged over decades. Flow of streams was gauged routinely to secure discharge data. All subsequent study of runoff, whether by overland flow or in channels, has been treated by hydraulic engineers in rigorously quantitative terms. Historically, then, geomorphology until recently has followed a course of action quite different from related sciences of climatology and hydrology. Soil science, so far as it has concerned the classification and distribution of major soil types, has also until recently been largely qualitative and descriptive. But the scene is changing in all of these classical fields of physical geography, just as it is changing in the fields of human geography. Quantification has come to stay. Methods of mathematical statistics are now routine. In all branches of geography, investigators seek to state the relationships that they observe in the form of mathematical models. A student's first introduction to physical geography therefore should include a brief look at the newer approaches to geomorphology.

Fluvial erosion systems

Although quantitative methods of study can be applied to any group of sequential landforms, produced by any process of erosion or deposition, this chapter is limited to considering a landscape dominated by processes of erosion and transportation by running water. Starting at a watershed, overland flow of runoff contributes both water and rock disintegration products to stream channels, which in turn carry these materials out of the system. Such a system is here referred to as a *fluvial erosion system*, in recognition of the dominant role of running water. Those parts played by rock weathering, soil creep, and various other forms of mass wasting (Chapter 24), although important, are considered here as secondary to running water.

The ideal fluvial erosion system used for analysis is that described in Chapter 26 as being in the mature and later stages of the cycle of denudation in a humid climate. Refer to Figures 26.1*D* and 26.3 for illustrations of mature landscapes. In such areas, the denudation process has acted long enough that all available surface area is fully occupied by drainage basins, closely nested together. Except for relatively small areas of floodplain along the larger streams, the landscape is dominantly one of removal and export of water and debris by means of the trunk streams of the region. Such landforms as alluvial fans, deltas, and broad alluvial valley floors are not included within this system; they are depositional landforms.

The fluvial system described above also has its place in arid and semiarid environments. For example, the maturely dissected mountain masses shown in Figure 26.13*B* are composed of fluvial erosion systems. Providing that analysis is limited to such areas, excluding the pediment, alluvial fan, and playa areas of the intermontane basins, the same principles will apply as hold in a humid region with a large yearly water surplus. It is true that the mountain slopes within a desert are largely barren of plant cover and have poorly developed, coarse-textured soils. It is also true that ground-water seepage into stream channels is of minor importance in arid mountains, and hence that base flow is usually absent. Nevertheless, despite obvious surface differences, fluvial systems in both humid and arid climates are remarkably similar in geometry, where mature areas are compared.

As later chapters will show, differences in resistance of rock bodies to weathering and water erosion exert strong controls over landforms. Geologic structure in many places controls landforms. For purposes of presenting the principles of an ideal fluvial erosion system, the assumption is usually made that the bedrock underlying the area is of uniform composition and structure throughout. Such has also been assumed in describing the progress of the cycle of land-mass denudation in Chapter 26. In fact, many areas can be found that exhibit such uniformity to a high degree. An example is an area underlain by a granite batholith (Chapter 30).

Elements of fluvial morphometry

Measurement of the shape, or geometry, of any natural form—be it plant, animal, or relief feature—is termed *morphometry*. We shall use the term *fluvial morphometry* to denote the measurement of geometrical properties of the land surface of a fluvial erosion system.

Fluvial morphometry might seem, at first glance, to be a frightfully complex undertaking. We all realize that no two landforms are exactly alike and that every vista differs in some detail from the next. Similarly, no two human faces are exactly alike and no two city blocks are identical. It is the unique configuration of a given assemblage of forms that makes possible recognition of people and places. Yet the basic components that go into each complex form are essentially alike and can be described and classified in a systematic way.

What are the basic form elements of a fluvial erosion landscape? First and simplest are the *linear properties* of the stream channel system. A branching system of lines is analyzed. Regardless of differences in channel widths, all streams are considered as pure lines, having infinitely small width. Linear properties are therefore limited to the numbers, lengths, and arrangements of sets of line segments. Although these lines, in fact, do have inclinations from the horizontal (all streams must have a gradient), the analysis of linear properties is made on the basis of a projection of the stream channel system to a horizontal plane. Such study is described by the word *planimetric*, meaning "measurement in a single plane."

The second class of form elements of a fluvial erosion system deals with *areal properties* of drainage basins. Again, the ground surface is projected to a horizontal plane; hence, the study is planimetric. Areal properties include the surface areas of drainage basins, as well as descriptions of the shapes (outlines) of those basins. Area is a two-dimensional property, the product of both length and width; whereas lines have only the one dimension of length. As a generalization, it can be stated that areas perform best the function of intercepting precipitation and of supplying rock debris; whereas lines (channels) perform best the function of transporting water and debris out of the area.

Third of the classes of form elements are the *relief properties* of the fluvial system. Relief refers to the relative heights of points on surfaces and lines with respect to the horizontal base of reference. Relief properties can be thought of as relating to the third dimension, perpendicular to the horizontal base on which the planimetric measurements are made. One group of form elements is a study of relief itself, defined as the elevation of a given point above the planimetric base, or difference in elevation of any two given points. In other words, relief expresses the magnitude of the vertical dimension of the landform. Another group of form elements in this class consists of the gradients, or slopes, of ground surfaces and stream channels. Such measurements tell the rate of drop of the runoff and are measures of the intensity of the processes of erosion and transportation.

In the pages that follow, we shall merely sample from among the more important morphometric studies of fluvial systems. A set of laws or generalizations will be considered. Furthermore, since a statement of form is in itself a static configuration, it will be desirable to take note of the way in which forms change with time. If the fluvial landscape changes with time, as the denudation cycle described by Professor Davis requires, it should be possible to measure such changes and define the stages in rigorous quantitative terms, using the data of morphometry gathered from various areas. Morphometry becomes of further scientific value when form is related to hydrologic process. For example, we shall inquire into the relationship between area of drainage basin and the trunk stream discharge.

Stream orders

Commencing with the linear properties of a stream system, the first consideration is to analyze the composition of the branching systems of channels, treating them as lines on a plane.

Given a map of a complete stream channel network, we can subdivide the network into individual lengths of channel, or *channel segments*, according to a hierarchy of orders of magnitude, assigning a sequence of numbers to the orders as

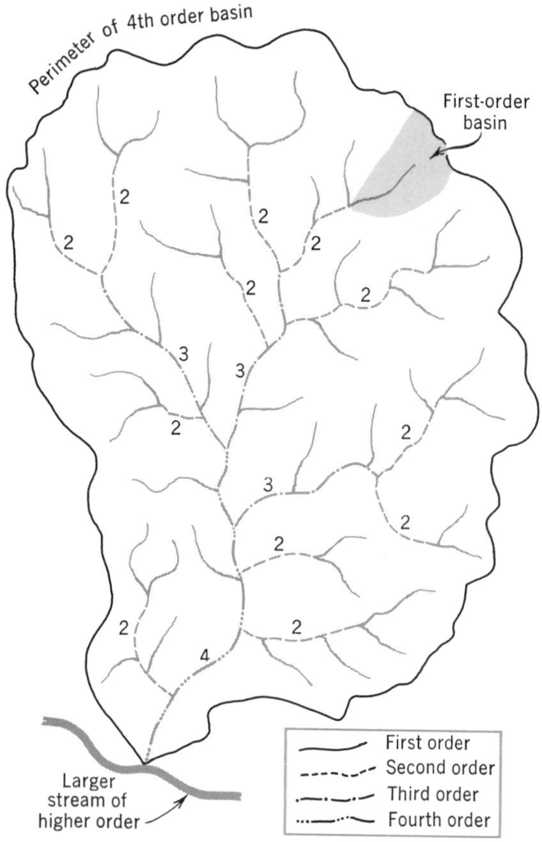

Figure 27.1 Orders of magnitude may be assigned to the segments of a branching stream system.

shown in Figure 27.1. Each finger-tip channel is designated as a segment of the *first order.* At the junction of any two first-order segments, a channel of the *second order* is produced and extends down to the point where it joins another second-order channel, whereupon a segment of *third order* results, and so forth. However, should a segment of the first order join a second- or third-order segment, no increase in order occurs at that point of junction. The trunk stream of any watershed bears the highest order number of the entire system. Channels of the first and second order usually carry flowing streams only in wet weather.

If large numbers of channel networks in a given region are divided into segments, each assigned an order according to the above rules, it will be possible to make some generalizations about the form and dimensions of the drainage network characteristic of that region. Consider first the distribution of numbers of segments of each order in a single watershed. In a carefully surveyed, large-scale map of a single drainage basin in the Big Badlands of South Dakota, all stream segments were assigned orders. The numbers of segments of each order were then counted to yield the figures in Table 27.1. The order of a stream segment is designated by the symbol, u; the number of segments of a given order by the symbol N_u. Consider the ratio between the number of stream segments of any given order to the number of segments of the next higher order, a proportion designated the *bifurcation ratio* (symbol R_b). The bifurcation ratio between successive orders is then defined as

$$R_b = \frac{N_u}{N_{u+1}}$$

In the example from the Big Badlands there are just over three times as many first-order segments as second-order; over four times as many second-order segments as third-order; three and two-thirds times as many third-order as fourth-order; and three times as many fourth-order as fifth-order. The differences in these bifurcation ratios can be attributed to chance variations in the shape of any stream network. An average of the four bifurcation ratios is close to 3.5, which is a good representative value for the series.

Studies of many stream networks confirm the principle that in a region of uniform climate, rock type, and stage of development, the bifurcation ratio tends to remain constant from one order to the next. Values of bifurcation ratio between 3 and 5 are characteristic of natural stream systems.

A noted hydraulic engineer, Robert E. Horton, is credited with formulating a *law of stream numbers,* which can be stated as follows. *The numbers of stream segments of successively lower orders in a given basin tend to form a geometric series, beginning with a single segment of the highest order and increasing according to a constant bifurcation ratio.* For example, if the bifurcation ratio is 3, and the trunk segment is of the sixth order, the numbers of segments will be 1, 3, 9, 27, 81, and 243.

TABLE 27.1*

Stream Order, u	Number of Stream Segments, N_u	Bifurcation Ratio, R_b
1	139	
		3.02
2	46	
		4.18
3	11	
		3.66
4	3	
		3.00
5	1	

* Data from K. G. Smith, 1958.

A geometric progression of numbers (such as 1, 3, 9, 27, 81, and 243) represents a constant ratio of increase. This is to say that each number above 1 is a threefold increase over the preceding lower number. As explained in Appendix III, when a geometric progression is plotted on a constant-ratio (logarithmic) scale, the numbers of the series position themselves at equal distance intervals on the scale. Let us then compose a graph on which the numbers of streams (N_u) are plotted on a vertical constant-ratio scale, against the stream order (u), on a horizontal arithmetic scale (Figure 27.2). Although the points do not form a perfectly straight line, the departures from that line are very small. Figure 27.2 also shows a plot of a similar sequence of number-order data from the Allegheny River drainage basin (Table 27.2). This larger basin has seven orders. The numbers of streams of each order are therefore very much greater than in the first example. Moreover, the points show larger departures from a straight line. Even so, a straight line is a good description of the sequence of points.

The relationship between orders and numbers that follow a geometric progression conforms to a mathematical model known as a *negative exponential function.* (This form of regression of two related variables is explained in Appendix III.) The formalized statement of Horton's law of stream numbers is then as follows:

$$N_u = R_b^{(k-u)}$$

The symbols N_u, u, and R_b are as defined in an earlier paragraph. The symbol k is defined as the order of the main trunk stream; it designates the segment of highest order. There is, of course, only one trunk segment. In Figure 27.2, the value of k for the Big Badlands example is 5; for the Allegheny River basin it is 7. It should be kept in mind that the above equation has meaning only for integer values of u (for example, 1, 2, 3, 4, etc.) because there are no intermediate values of order under the definition that has been adopted for order of a stream segment.

A simple test of the exponential equation can be made as follows. Assume an ideal stream net with a bifurcation ratio of exactly 3. Let the highest order, k, be 5.

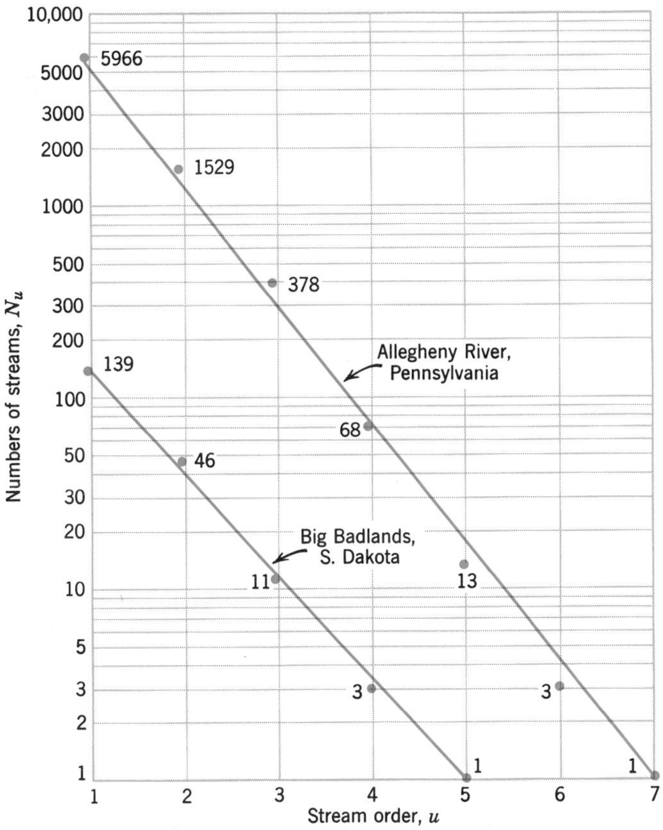

Figure 27.2 Numbers of stream segments in each order, plotted against order, produce a straight-line regression of negative exponential form. (Data from M. E. Morisawa, 1959, and K. G. Smith, 1958.)

Order, u	Number of Segments, N_u	Bifurcation Ratio, R_b
1	81	
		3.0
2	27	
		3.0
3	9	
		3.0
4	3	
		3.0
$k = 5$	1	
	$N_u = 121$	

Suppose that we wish to determine the number of segments of the second order (N_2), knowing only that the bifurcation ratio is 3 and that $k = 5$. Substituting in the equation:

$$N_2 = 3^{(5-2)}$$
$$N_2 = 3^{(3)}$$
$$N_2 = 27$$

Horton further observed that the total number of stream segments of the entire drainage basin can be expressed as follows:

$$\Sigma N_u = \frac{R_b{}^k - 1}{R_b - 1}$$

The symbol ΣN_u means "the sum of segments within each order" (the Greek letter *sigma* has been adopted to instruct the reader to sum all terms). Testing this equation against the ideal set of numbers in which R_b equals 3, we obtain

$$\Sigma N_u = \frac{3^5 - 1}{3 - 1}$$

$$\Sigma N_u = \frac{243 - 1}{2} = \frac{242}{2}$$

$$\Sigma N_u = 121$$

Stream lengths

Referring again to the drainage network map, Figure 27.1, it is apparent that the first-order channel segments have, on the average, the shortest length, and that segments become longer as order increases. Table 27.2 gives measurements for a part of the Allegheny River basin in McKean County, Pennsylvania.

The master stream of this basin is of the seventh order, but its entire length was not completely measured above the gauging station arbitrarily selected as the basin mouth. Therefore, only the first six orders should be examined.

The mean length of stream segments, in miles, increases by a ratio of roughly three times with each increase in stream order. This proportion of length increase is known as the *length ratio* (symbol, R_L), and tends to be approximately constant for a given drainage system. Chance variations to be expected in the configuration of any drainage system will produce inequalities of observed length ratio from one order to the next.

Definition of length ratio resembles that of bifurcation ratio and is as follows:

$$R_L = \frac{\bar{L}_u}{\bar{L}_{u-1}}$$

The symbol $\bar{L}_u$ represents the mean length of all stream segments of order u. In the practice of morphometry, a distance-measuring instrument (map-measurer) is run over all segments of a given order on the map and their total distance read off. This total length is then divided by the number of segments of that order, yielding the mean length. Stated in rigorous fashion,

$$\bar{L}_u = \frac{\Sigma L_u}{N_u}$$

where ΣL_u means "the sum of lengths of all stream segments of order u."

Study of many drainage systems led Horton to formulate a *law of stream lengths,* which with necessary modifications may be stated as follows: *the cumulative mean lengths of stream segments of successive orders tend to form a geometric series beginning with the mean length of the first-order segments and increasing according to a constant length ratio.* The word "cumulative" in this law indicates that the mean lengths are progressively added (cumulated) starting with the second order. For order 2, mean lengths of first and second orders are summed; for order 3, mean lengths of orders 1, 2, and 3 are summed; and so forth. On Table 27.2, a column of cumulative mean lengths is given immediately to the right of the column of mean lengths.

As with the law of stream orders, the law of stream lengths can be given mathematical expression by an exponential regression equation. Figure 27.3 is a graph in which cumulative mean stream lengths are plotted on a constant-ratio (logarithmic) scale on the vertical axis; stream order is plotted on an arithmetic scale on the horizontal axis. If the plotted points fall on a nearly straight line, the validity of Horton's law of stream lengths can be regarded as strongly supported. In the case of the Allegheny River basin data of Table 27.2, points for orders 2 through 6 fall on a remarkably straight course, whereas the point for order 1 deviates substantially. Because each first-order stream segment has a free terminus that is not easy to determine with confidence, it is a possibility that many of the first-order segments were not measured in their full lengths. For data of Fern Canyon, a drainage basin in California, also shown on Figure 27.3, the straight-line relationship of points is

TABLE 27.2 ALLEGHENY RIVER DRAINAGE BASIN CHARACTERISTICS*

Stream Order u	Number of Segments N_u	Bifurcation Ratio R_b	Mean Length of Segments, Miles $\bar{L}_u$	Cumulative Mean Length, Miles $\bar{L}_u$	Length Ratio R_L	Average Watershed Area, Square Miles $\bar{A}_u$
1	5966		0.09	0.09		0.05
		3.9			3.3	
2	1529		0.3	0.4		0.15
		4.0			2.7	
3	378		0.8	1.2		0.86
		5.7			3.1	
4	68		2.5	3.9		6.1
		5.3			2.8	
5	13		7	11		34
		4.3			2.9	
6	3		20	31		242
		3.0				
7	1		8 + (not complete)			550 (not complete)

* Data by Marie E. Morisawa, 1959.

excellent. Notice that mean lengths differ greatly for the same orders between one basin and the other. This fact leads to the conclusion that the segments of a stream system cover a wide range of dimensions. Observe that the lines of points slope upward from left to right, whereas those on the number-order plot (Figure 27.2) slope downward from left to right.

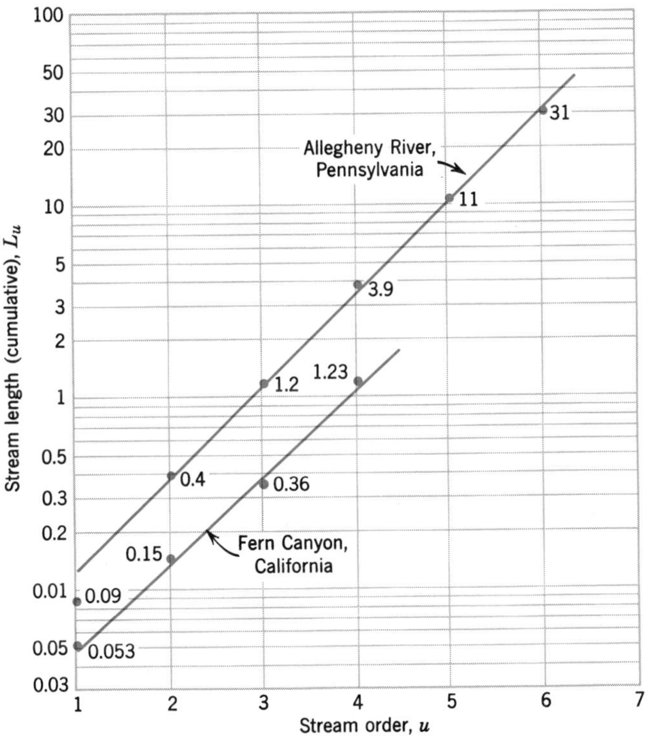

Figure 27.3 Mean length of streams of each order (cumulative), plotted against order, produces a straight-line regression of positive exponential form. (Data from M. E. Morisawa, 1959; and J. C. Maxwell, 1960.)

Horton's law of stream lengths is stated mathematically by the following equation

$$\bar{L}_u = \bar{L}_1 R_L^{(u-1)}$$

where $\bar{L}_1$ is the mean length of the first-order segments, the other symbols having been previously defined.

Basin areas

Turning next to the areas of drainage basins, we can study the relationship between mean area of basin of a given order (symbol, $\bar{A}_u$) and the order itself. In most respects, this relationship is of the same form as that between mean stream lengths and orders. First, it is necessary to examine the way in which surface areas contribute to basins of each order.

Figure 27.4 shows a nested group of basins of orders 1 and 2. There are 4 basins of first-order; 2 basins of second-order. The second-order basins are shown to contribute to a stream channel of the third-order. Orders of channels are indicated by numbers; the direction of overland flow by short arrows. In each first-order basin, all ground surface of the basin contributes directly to the first-order channel. In each second-order basin, considered in its entirety, only part of the overland flow enters the first-order channels directly. There are shown, in addition, two triangular or trapezoidal patches of ground on which overland flow passes down-slope directly into the second-order channel. These surfaces are known as *interbasin areas.* We see, then, that the area of an entire second-order basin is the sum of the first-order basins which it contains plus all interbasin areas within its perimeter.

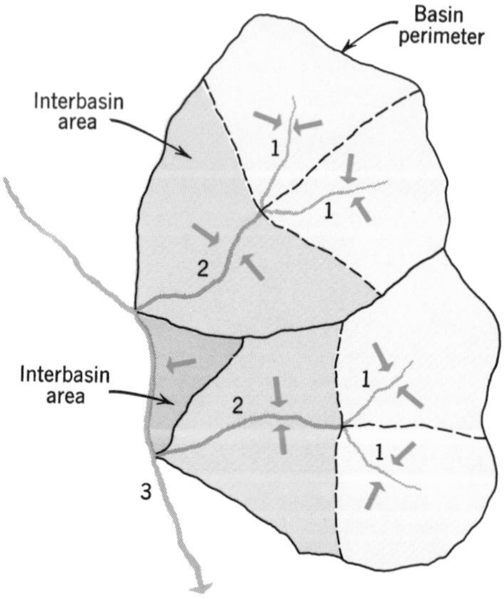

Figure 27.4 Nested basins of first and second orders, with contributing interbasin areas.

For each higher order basin, there will be interbasin areas contributing directly to the highest order channel. One such interbasin area, shown in Figure 27.4, contributes to the third-order channel. In summary, the area of a basin of order u is defined as the total area of surface contributing to all first-order channels plus all included interbasin areas. In practice, it is only required that a single perimeter be located for a basin of a given order, and that this area be measured with an instrument known as a *planimeter.* Basin area is therefore automatically cumulative in summing all nested basins of lower orders within it.

Horton's law of stream length has been paraphrased into a *law of basin areas,* stated as follows: *the mean basin areas of successive stream orders tend to form a geometric series beginning with mean area of the first-order basins and increasing according to a constant area ratio.* The definition of area ratio, R_a, is

$$R_a = \frac{\bar{A}_u}{\bar{A}_{u-1}}$$

where $\bar{A}_u$ is mean area of basins of order u. By analogy with the law of stream lengths, the law of basin areas is as follows:

$$\bar{A}_u = \bar{A}_1 R_a^{(u-1)}$$

The symbol $\bar{A}_1$ denotes mean area of the first-order basins. Figure 27.5 shows basin areas plotted on the same type of graph as used in Figure 27.3. Data for the Allegheny River basin will be found in Table 27.2.

Stream flow and basin area

One of the purposes of fluvial morphometry is to derive information in quantitative form about the geometry of the fluvial system that can be correlated with hydrologic information.

One example is the relationship of stream discharge, Q, to area of watershed. (Refer to Chapter 13 for explanation of stream discharge and its measurement.) Common sense tells us that the discharge of a stream increases with increasing drainage basin area. It remains to be determined what mathematical model applies to such an increase.

If stream systems were fitted with gauges at the lower end of each channel segment of each order, the investigation could proceed according to basin areas by order. In practice, gauges are situated at various points on streams. Therefore, we can only relate stream discharge to the total contributing area of watershed above the gauge.

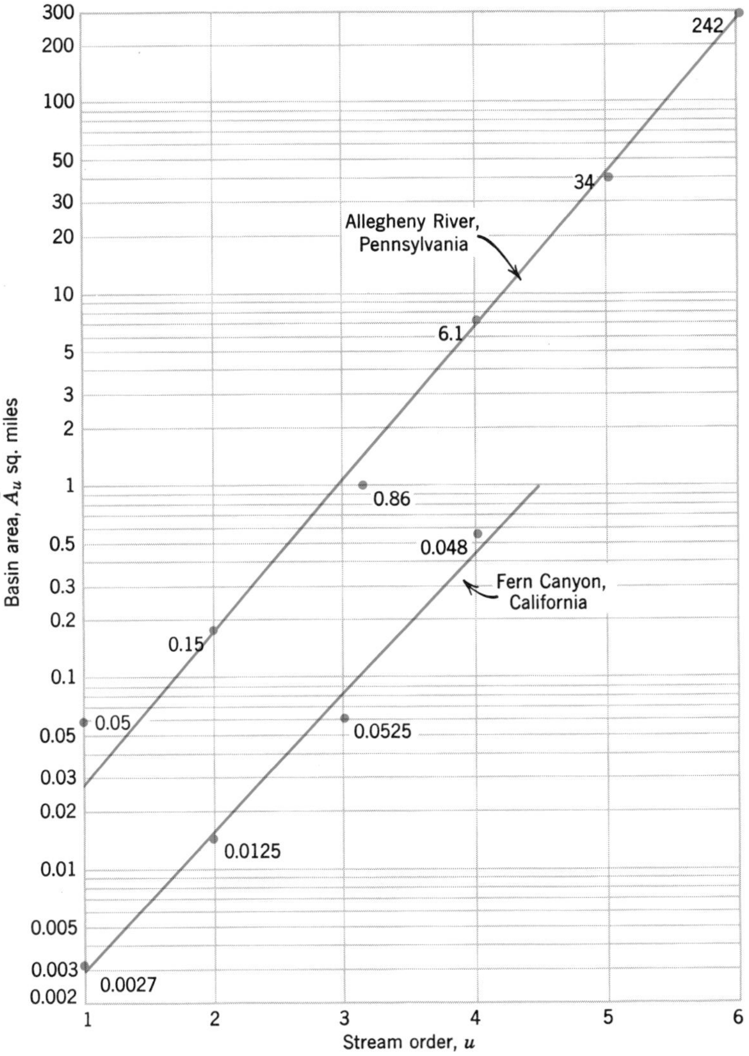

Figure 27.5 Mean area of basins of each order, plotted against order, produces a straight-line regression of positive exponential form. (Data from M. E. Morisawa, 1959, and J. C. Maxwell, 1960.)

Figure 27.6 shows the observed relationship of *average discharge*, $\overline{Q}$, to drainage area, A, for the Potomac River basin. Each point represents one gauge. Obviously, gauges in the headwater regions show as points at the lower left; those far downstream lie at the upper right. While the individual points show marked departures from the fitted straight line, the trend is obvious.

Stated mathematically, the relationship between average (mean) discharge and drainage area is

$$\overline{Q} = aA^b$$

where a is a numerical constant and b is an exponent. As explained in Appendix III, this regression equation represents a *power function.*

Because the fitted straight line runs at 45° across the graph, it can be said that the value of b is exactly 1. The meaning of the exponent of 1 is that the discharge increases in direct proportion to the area. Actually, values of b differing somewhat from 1 are observed.

One practical use of the mathematical equation relating stream discharge to basin area is that it enables the hydrologist to estimate mean discharge at any point in the system by measuring the watershed area lying above that point. Such knowledge would be essential in designing hydraulic structures, such as dams, bridges, and irrigation diversions.

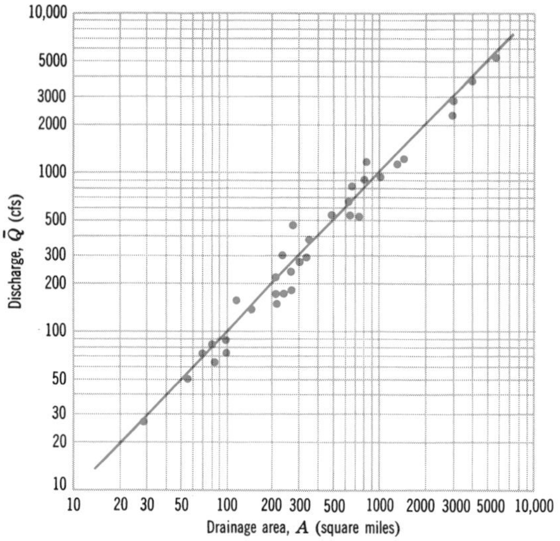

Figure 27.6 Relation of mean stream discharge to drainage area for all gauging stations in the Potomac River basin. Each point represents a gauge. (Data from John T. Hack and U.S. Geological Survey, 1957.)

Drainage density and texture of topography

If we study an area of badlands, the intricately eroded forms that develop in barren areas of soft clays in arid climates (Figure 28.19), we cannot fail to be impressed with the way that the landforms resemble miniature mountains. Innumerable tiny channels carve tiny valleys to reproduce on a small scale the same great canyon and ridge forms seen in a rugged mountain range, such as the San Gabriel Range of California, or the Great Smoky Mountains of North Carolina. Evidently nature follows the laws of stream numbers, lengths, and areas, regardless of whether the first-order drainage basin is so small that one can stand astride it, or whether it is a full mile across. Because such similarity of geometry prevails in maturely eroded land masses, it is necessary to have some means of describing and measuring the scale of magnitude of the forms.

If, for the drainage network map of Figure 26.5, we should measure the total length in miles of all channels, and divide this figure by the total area in square miles of the entire map or watershed, the *drainage density* is found:

$$\text{Drainage density} = \frac{\text{total length of streams (miles)}}{\text{area (square miles)}}$$

Stated in symbols

$$D = \frac{\Sigma L_k}{A_k}$$

where D represents drainage density in miles per square mile, ΣL_k represents the total length of all channels of all orders, and A_k is the total basin area.[1]

Suppose that a drainage density value of 12 is obtained; this number is interpreted as meaning that there are 12 miles of channel for every square mile of land surface. Because area and length are measured from a map, which projects the sloping surfaces and channels upon a horizontal plane, the measured values are somewhat less than the true values, to a degree depending upon the amount of slope.

Figure 27.7 shows four topographic maps of the U.S. Geological Survey, each covering one square mile. A great range in drainage density is shown. Map A is from a region of *low drainage density*, averaging three to four miles of channel per square mile. This example comes from a region underlain by massive, hard sandstone beds and is under heavy forest cover. Such a region of low drainage density may be described as having *coarse texture*, since the individual elements of the topography are very large, or gross.

Map B shows a region of *medium drainage density*, averaging 12 to 16. This area is underlain by thin-bedded sandstones and thick shales, relatively easily eroded, but developed under a heavy deciduous forest cover. It is typical of large parts of the humid eastern United States where the stage of land-mass erosion is mature. This area may be described as of *medium texture*.

Map C shows *high drainage density*, or *fine texture*, developed in easily eroded, weak sedimentary strata in southern California, where vegetative cover is sparse. Drainage density runs from 30 to 40 under such conditions. Much higher values of drainage density are found in badlands, where there may be from 200 to 500 or more miles of channel per square mile. Such topography would be described as *ultrafine texture*.

Map D, taken from the badlands region of South Dakota, illustrates the appearance of badlands on maps of the same scale as the preceding three textures, but much intricate detail is lacking because it is impossible to draw the minute crenulations of the contours on a map of this scale. For this reason no drainage lines have been drawn in, but they can be seen in Figure 27.8, a portion of an air photograph covering one square mile at a nearby location within the Badlands National Monument, South Dakota.

What factors control drainage density? One highly important control is rock type. Hard, resistant rocks, such as intrusive granitic rock, gneiss, sandstone, and quartzite, tend to give low drainage density (coarse texture). This is because stream

[1] Metric units may also be used. To convert to kilometers per square kilometer, multiply drainage density by 0.62.

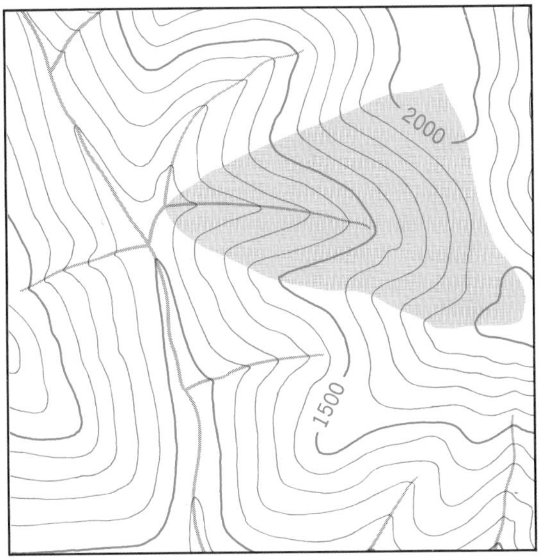

A. Low drainage density or coarse texture, Driftwood. Pennsylvania, Quadrangle.

B. Medium drainage density or medium texture, Nashville, Indiana, Quadrangle.

C. High drainage density or fine texture, Little Tujunga, California, Quadrangle.

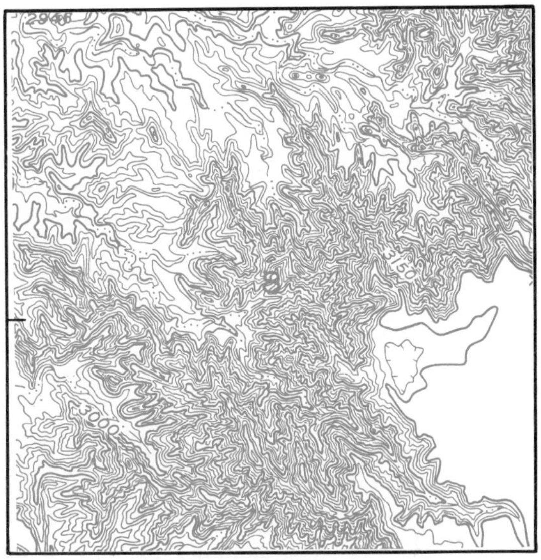

D. Extremely high drainage density or ultrafine texture, Cuny Table West, South Dakota, Quadrangle.

Figure 27.7 Four areas of one square mile each serve as representative examples of the natural range in drainage density. Colored areas are representative drainage basins of the first order. (From maps of the U.S. Geological Survey.)

erosion is difficult and only a relatively large channel can maintain itself. Therefore, the first-order basins are large and provide large amounts of runoff to the channels. In weak rocks, such as shales and clays, even a small watershed can supply enough runoff for channel erosion.

A second factor is the relative ease of infiltration of precipitation into the ground surface and downward to the water table (see Chapter 13). Highly permeable materials, such as sand or gravel, tend to give low drainage density because infiltration is great and little water is available as surface runoff to maintain channels. Clays and shales, on the other hand, have a high proportion of surface runoff and this combines with their weakness to give high drainage density.

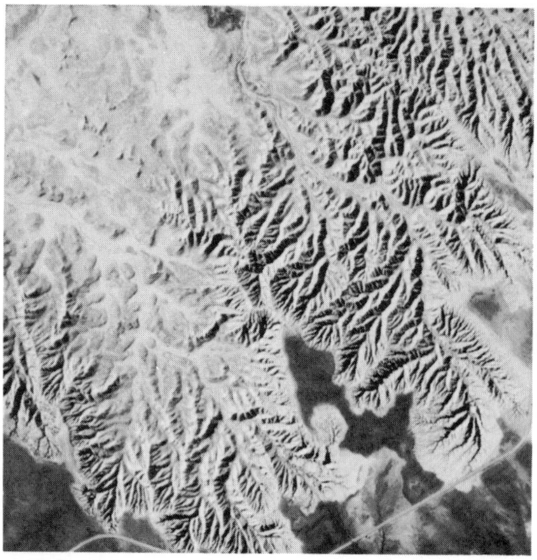

Figure 27.8 This vertical air photograph of an area of one square mile in the Big Badlands of South Dakota illustrates ultrafine texture. The topography is very similar, but not identical to that shown in Figure 27.7*D*. North is toward the bottom of the page. (U.S. Dept. of Agriculture.)

Figure 27.9 Seen from the air, the maturely dissected sected Allegheny Plateau of West Virginia appears largely forested. Relief of 700 to 800 ft (210 to 240 m) is here developed on shales of Devonian age. (Photograph by J. L. Rich, Courtesy of the *Geographical Review*.)

A third major factor is the presence or absence of vegetative cover. A weak rock will have much lower drainage density in a humid climate, where a strong, dense cover of forest or grass protects the underlying material, than in an arid region where no protective cover exists (Figure 27.9). For this reason, badlands are characteristic of arid climates, and the drainage density there tends to be markedly higher on all rock types.

Stream slopes

As stated in Chapter 25, the typical profile of a graded river is upwardly concave and shows a progressive flattening of slope (gradient) in the downstream direction (Figure 25.14). This observation leads to a consideration of the relationship of channel slope to stream order. For this purpose, an average slope value is obtained for all stream segments of a given order within the drainage basin.

Channel slope is defined here as the ratio of vertical drop to horizontal distance, measured from the upper end to the lower end of a single stream segment of given order. Slope is given the symbol S. Slope is stated as a ratio, or proportion, and is without dimension. Thus, a slope of 0.01 is a ratio of 1:100, for example, a drop of 1 ft vertically in 100 ft horizontally.

If the slopes of all channel segments of the first order are measured and averaged, a mean slope is obtained for that order, and is denoted by the symbol $\bar{S}_1$. The same operation is carried out for slopes of orders 2, 3, 4, and so on ($\bar{S}_2$, $\bar{S}_3$, $\bar{S}_4$ etc.). Figure 27.10 shows diagrammatically what has been done. Each of the triangles on the graph represents an order. The vertical leg of the triangle is the average vertical drop ($\bar{H}_u$) of that order; the horizontal leg is the average horizontal distance of that order and is identical with mean stream length, $\bar{L}_u$. The hypotenuse of the triangle shows the average slope, $\bar{S}_u$. Values of $\bar{S}_u$ are given for each order.

The slope segments in Figure 27.10 tend to approximate a curve which is upwardly concave and which flattens in the downstream direction. To this degree it resembles the continuous river profile in Figure 25.14. Let us now plot the mean slope of each order, $\bar{S}_u$, against order, u, using the constant-ratio (logarithmic) scale for slope, as in previous graphs of stream numbers, stream lengths, and basin areas. Figure 27.11 shows the plotted data of Tables 27.3 and 27.4. The points show minor departures from the fitted straight lines, but the agreement is generally good.

On the basis of data, such as shown in Figure 27.11, Horton formulated a *law of stream slopes*, stated as follows: *the mean slopes of stream segments of successively higher orders in a given basin tend to form an inverse geometric series, decreasing according to a constant slope ratio.* Stated as an equation, the law of stream slopes is as follows:

$$\bar{S}_u = \bar{S}_1 R_s^{(u-1)}$$

The symbol R_s represents the slope ratio, and is defined as $R_s = [\bar{S}_u/(\bar{S}_{u-1})]$. Slope ratios must be less than 1. Values from 0.3 to 0.6 are typical.

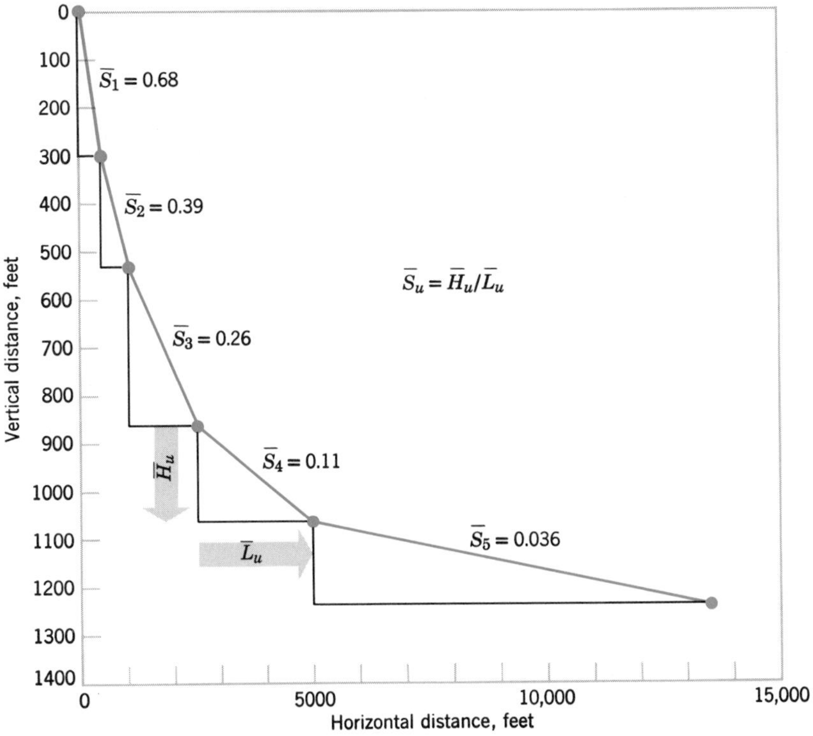

Figure 27.10 Mean channel slope for each order can be represented by a descending chain of triangles. Data are for Chileno Canyon watershed, California. (From A. J. Broscoe, 1959.)

Individual slope ratios differ from order to order, as Tables 27.3 and 27.4 show, due to variations in the resistance of the rock beneath the stream channel. In Horton's law of stream slopes a single, constant value of slope ratio is assumed. Although the data of stream slopes show a rather

TABLE 27.3 Home Creek, Ohio*

Order u	Mean Channel Slope, $\bar{S}_u$	Slope Ratio, R_s
1	0.181	
		0.48
2	0.087	
		0.32
3	0.028	
		0.32
4	0.009	
		0.56
5	0.005	

* Data from Marie E. Morisawa, 1959.

TABLE 27.4 Perth Amboy Badlands, New Jersey*

Order u	Mean Channel Slope, $\bar{S}_u$	Slope Ratio, R_s
1	0.60	
		0.68
2	0.41	
		0.83
3	0.34	
		0.53
4	0.18	
		0.61
5	0.11	

* Data from S. A. Schumm, 1956.

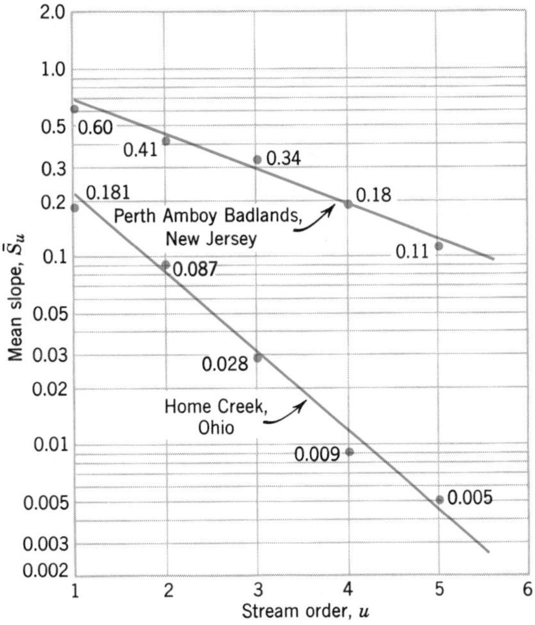

Figure 27.11 Mean slope of channel of each order, plotted against order, produces a straight-line regression of negative exponential form. (Data from S. A. Schumm, 1956, and M. E. Morisawa, 1959.)

high degree of variability within large drainage basins, the law appears to be generally valid.

The mathematical laws of Horton, governing stream numbers, lengths, areas, and slopes, taken together, form a modern extension of Playfair's law of streams (Chapter 26). Playfair's law, a purely qualitative expression, states that the branches of a stream run in valleys proportioned to their sizes and that they have "a nice adjustment of their declivities (slopes)."

Valley-side slopes

Complementing the slopes of stream channels are the slopes of the valley sides that enclose each channel. Together, channel slopes and valley-side slopes provide the gradient for water flow and debris transport within the fluvial system.

It is obvious to any casual observer that the characteristic slope of valley sides differs from one region to another. In a rugged mountainous region—such as the San Gabriel Mountains of southern California—valley-side slopes are very steep, so steep in fact that they are difficult to ascend or descend on foot. Loose debris rolls and slides freely down such slopes in dry weather. In contrast, valley-side slopes found in the Piedmont region of Virginia, the Carolinas, and Georgia are comparatively gentle. They offer little resistance to cross-country traverse and are widely cultivated. As noted in Chapter 26, steep valley-side slopes are typical of the youthful and early-mature stages of the cycle of denudation; gentle slopes of late-mature and old stages.

In recent years, geomorphologists have undertaken to make systematic and detailed measurements of valley-side slopes and to relate these data to other form elements of the fluvial system. Going directly into the field, or using detailed maps of high quality, the investigator measures the slope angle (inclination from the horizontal) at its maximum point along a given profile line from divide to adjacent stream channel. These measurements are repeated at regular intervals along the valley sides of a region until a sufficiently large sample has been obtained. The mean value of the measurements is computed. The symbol S_g is used to denote valley-side slope, in distinction to the symbol S_c for channel slope, where both are in ratio units previously explained. However, we may also measure these slopes in degrees of arc, as is customary in the geologist's measurement of dip of strata. The Greek letter *theta* is used for slope angles in degrees. Hence $\bar{\theta}_c$ and $\bar{\theta}_g$ are symbols for mean channel slope and mean ground slope respectively. (The bar over a symbol always denotes a mean value derived from a sample of observations.)

Channel slope and ground slope cannot be wholly independent of one another. Steep valley sides shed water and coarse debris at high rates; a steep channel gradient is necessary to transport this flow and prevent the debris from choking the valley. Gentle valley sides contribute little debris and it is of a fine size grade; hence, streams can function on low gradients.

Observations of mean valley-side slopes and mean stream gradients from a wide range of localities can be plotted on a single graph to study these relationships. Figure 27.12 is such a graph. Each point represents a locality. On the vertical scale is plotted the mean value of valley-side slope angles; on the horizontal scale is plotted the mean slope of the adjacent channel. For a realistic comparison from one region to another, all samples have been limited to second-order channels and their immediately adjacent slopes. Constant-ratio (logarithmic) scales are used. The points are well represented by a straight line sloping upward to the right, although moderate deviations from the line exist. The relationship is considered as meaningful and conforms generally to the expected relationship based on a consideration of the activities of a fluvial system.

The mathematical equation that represents the straight line in Figure 27.12 is of the power form

$$\bar{\theta}_g = a\bar{\theta}_c^{\,b}$$

where a, a numerical constant, has a value of 0.6. The exponent, b, is approximately 0.8. The equation is strictly *empirical,* meaning that it is based upon the observed data and not upon physical theory. When data of other localities are introduced, the values of a and b in this power equation will change. Much of quantitative geomorphology is a continuing search for new and improved data to provide a basis for better empirical equations.

It is now possible to relate the stages of the cycle of land-mass denudation to the information shown in Figure 27.12. A region that has undergone strong uplift and has subsequently reached early maturity will have both steep valley-side slopes and steep channel slopes. Such conditions are represented by the upper-right end of the fitted line in Figure 27.12. The examples plotted in this position are of badlands. These are, in effect, miniature mountain ranges that may be thought of as scale-models of full-sized mountain ranges. Erosion rates are high at this stage; the land surface is being lowered rapidly.

As relief diminishes with time, the point on the graph representing a given region will migrate down the sloping line, toward the lower left. Both valley-side slopes and stream channel slopes will decrease in constant ratio as time passes and stage moves into late maturity. Old age, a time of low slopes,

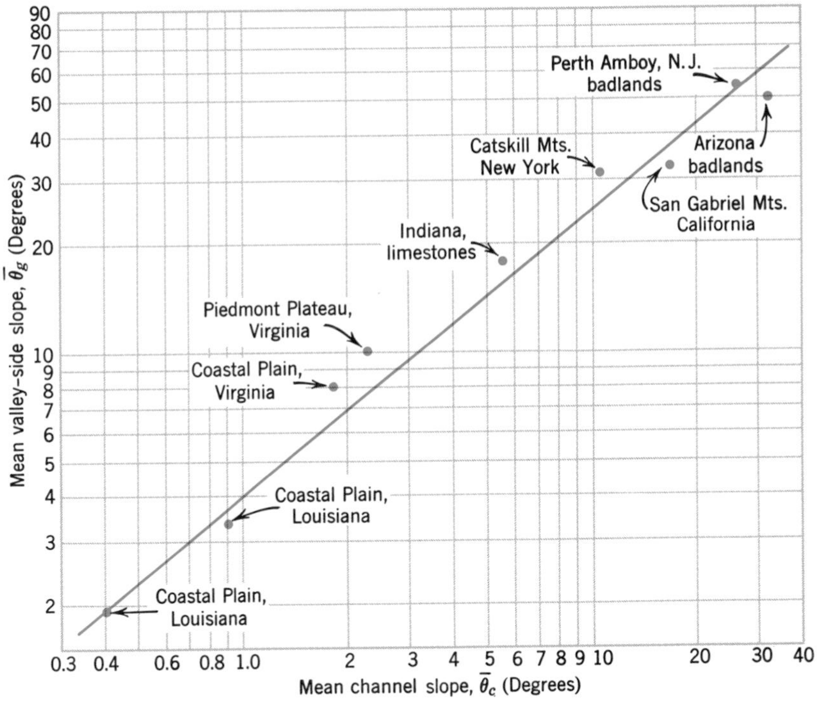

Figure 27.12 A regression of the power form describes the relationship between mean valley-side slopes and adjacent channel slopes over a wide range of geologic and climatic environments. (From A. N. Strahler, 1950.)

is represented by a point at the lower-left end of the sloping line. As constant-ratio scales have no zero point, there is no provision in this model for reduction of the land surface to a condition of zero slope, that is, perfect horizontality at base level. Instead, the peneplain is represented, its gentle slopes declining only with extreme slowness.

Quantitative geomorphology in review

In this chapter only a few topics have been introduced from the large body of information presently available within the full scope of quantitative analysis of fluvial erosion systems. There are, in addition, substantial bodies of information relating to quantitative analysis of the landforms produced by glaciers, waves and currents, and winds. Fundamentally, quantitative studies in all of these fields proceed along the same basic lines. Relationships between form elements and between form and process are expressed mathematically in equations of the types described in this chapter. Field observations as well as laboratory model studies furnish data for empirical equations. Ideal mathematical models are generated from established laws of physics.

Within the limited space of a single chapter, we have observed that landforms of fluvial erosion systems, despite their complexity and infinite variety, can be systematically analyzed in terms of their linear, areal, and relief properties. We have seen that the relationships of stream numbers, stream lengths, basin areas, and channel slopes to orders follow a group of related laws of similar form. These laws of Horton are expressed by exponential equations. We have seen that purely qualitative and descriptive laws and concepts, such as Playfair's law and the Davis model of the denudation cycle, can be restated in modern quantitative forms at increased levels of understanding.

Review Questions

1. Discuss the role that quantitative investigations have played in the past in the various fields of physical geography, including climatology, hydrology, soil science, and geomorphology.

2. What are the basic characteristics of a fluvial erosion system? Under what climatic and geologic environments are such systems studied?

3. What is fluvial morphometry? What are the basic form elements of a fluvial erosion landscape? Explain the term *planimetric.*

4. Explain how the segments of a channel system are classified by orders. Which segment has the highest order? Which the lowest?

5. Discuss the relationships between stream numbers and orders. What is the bifurcation ratio? What range of values has the bifurcation ratio in natural stream systems?

6. State Horton's law of stream numbers. Explain a geometric series. What form of regression equation represents the law of stream numbers?

7. State Horton's law of stream lengths. What is the length ratio? What form of regression equation represents the law of stream lengths?

8. State Horton's law of basin areas. Compare this law with the law of stream lengths. How is basin area defined? What is an interbasin area?

9. What empirical relationship has been observed between stream discharge and basin area? Of what practical value is this knowledge?

10. Define drainage density. What range of values of drainage density is observed in nature? How is the term *texture* used in association with drainage density?

11. What physical factors—geologic and otherwise—control the drainage density in a given region? Explain how each factor exerts its control.

12. What is meant by the mean channel slope of stream segments of a given order? How is this quantity measured? State Horton's law of stream slopes? What regression equation applies to this law?

13. How are valley-side slopes measured in a fluvial system? How are the measurements analyzed? Relate valley-side slope to stream-channel slope.

14. How are stages in the cycle of land-mass denudation reflected in the changes in valley-side slopes and channel slopes?

15. Discuss the role that quantitative and statistical analysis will play in future developments in geography. What is to be gained by the application of such methods?

Exercises

Exercise 1. *Stream Orders and Numbers.* The accompanying map shows the stream pattern of a drainage basin in the Appalachian Plateau region of Kentucky. The basin perimeter is also shown. The region is in a stage of maturity in the cycle of land-mass denudation. The drainage pattern is of a pure dendritic type.

QUESTIONS

1. Assign orders to all segments of the stream system (see Figure 27.1). Use a sheet of thin tracing paper placed over the page. Select several pencils of different colors and assign a color to each order. Find and trace in color all first-order segments. Then find and color the second-order segments; followed by the third, and so forth. Prepare a key to your color code.

2. **(*a*)** Count the numbers of segments of each order. Check your count at least twice. Record the data in a table such as Table 27.1. **(*b*)** Determine the bifurcation ratios between successive orders and record in the table. **(*c*)** Using semilogarithmic paper, prepare a graph similar to Figure 27.1 showing relationship of stream numbers to orders. Fit the points with a straight line. **(*d*)** Evaluate the results in terms of the degree of agreement with Horton's law of stream numbers.

3. **(*a*)** Using a map measurer, measure the total length of first-order segments. Measure total lengths of segments of each of the remaining orders. Convert to miles and record in the same table used in Question 2. **(*b*)** Divide the total length of each order by the number of segments of that order, obtaining mean stream length. Record these values in the table. **(*c*)** Calculate and record length ratios between each pair of successive orders. **(*d*)** Cumulate the mean lengths and record in a column beside the mean stream lengths. **(*e*)** Using semilogarithmic graph paper, plot the cumulative mean stream lengths against order. Fit the points with a straight line. **(*f*)** Comment on the degree to which these data conform with Horton's law of streams.

4. **(*a*)** Using a *polar planimeter,* measure the area of the drainage basin within indicated perimeter. If no planimeter is available, estimate the area by counting squares of a sheet of graph paper placed under the traced map. Convert the measured area into square miles. **(*b*)** Calculate drainage density. (Divide basin area by total stream length of all orders.) **(*c*)** Comment upon the measured drainage density of this area in comparison with the known range of drainage density values in natural fluvial systems.

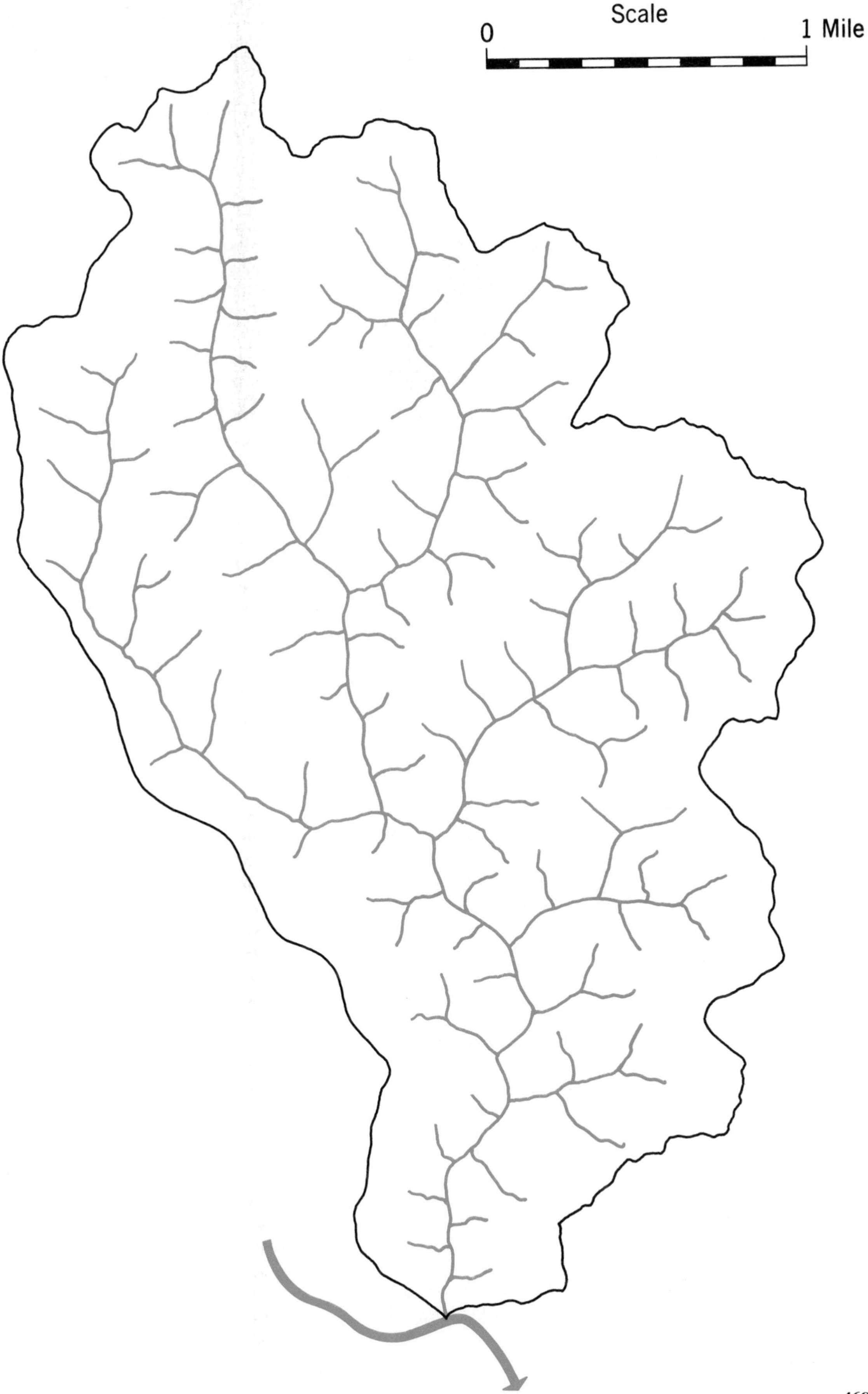
Scale
0
1 Mile

Exercise 2. *Horton's Laws.* The accompanying table gives morphometric data for Little Mill Creek, Ohio, a basin of the fourth order with an area of 2.7 square miles. The fourth-order basin is not complete, hence values in parentheses should not be plotted. The data can be regarded as a combined sample consisting of five complete third-order basins.

QUESTIONS

1. (*a*) Compute the bifurcation ratios between orders 1 and 2 and between orders 2 and 3. Compute an average bifurcation ratio. **(*b*)** Using semilogarithmic graph paper, prepare a graph similar to Figure 27.2 showing the relationship of numbers to orders. Label fully. Fit the points with a straight line. (Plot only orders 1, 2, and 3.)

2. (*a*) Compute the length ratios between orders 1 and 2 and between orders 2 and 3. **(*b*)** Calculate the cumulative mean stream length for orders 1, 2, and 3. **(*c*)** Using semilogarithmic graph paper, prepare a graph similar to Figure 27.3, showing the relationship of cumulative mean stream length to order. Label fully. (Plot only orders 1, 2, and 3.) Fit the points with a straight line.

3. (*a*) Compute the area ratios between orders 1 and 2 and between 2 and 3. **(*b*)** Using semilogarithmic graph paper, prepare a graph similar to Figure 27.5 showing the relationship of mean basin area to order. (Plot only orders 1, 2, and 3.) Label fully. Fit a straight line to the points.

4. (*a*) Compute the slope ratios for orders 1 and 2 and for orders 2 and 3. **(*b*)** Using semilogarithmic graph paper, prepare a graph similar to Figure 27.11 showing relationship of mean channel slope to order. (Plot only orders 1, 2, and 3.) Fit the points with a straight line. Label fully.

5. Evaluate the plotted data in terms of degree to which these data conform with Horton's laws of stream numbers, stream lengths, basin areas, and channel slopes. Is the agreement good or poor? Is the sample adequate in size to sustain such an evaluation?

Little Mill Creek, Coshocton County, Ohio*

Order	Number of Streams	Mean Stream Length (Miles)	Mean Basin Area (Square Miles)	Mean Channel Slope
u	N_u	$\bar{L}_u$	$\bar{A}_u$	$\bar{S}_c$
1	104	0.07	0.025	0.37
2	22	0.19	0.12	0.12
3	5	0.65	0.58	0.04
4	1	(1.2)	(2.7)	(0.01)

* Data from M. E. Morisawa, 1959.

CHAPTER 28

Coastal Plains, Horizontal Strata, Domes

THE foregoing chapters on landforms produced by weathering, mass wasting, and streams, have given little or no account of the manner in which variations in rock composition and structure are capable of exerting a powerful control upon the shapes and sizes of landforms. Instead, by assuming that all bedrock is of uniform composition throughout, it has been possible to describe the simple, ideal erosional landforms produced by the agents of denudation. There are, it is true, large land areas where bedrock is fairly uniform throughout, and it is in such areas that the denudational agents can produce the ideal forms. Elsewhere, sedimentary rocks are tilted, folded, domed, or faulted; metamorphic rocks are arranged in belted patterns; intrusive igneous rocks have solidified in a variety of bodies. It is with such structures and their topographic expression that these chapters deal.

Classification of landmasses

As illustrated in Figure 28.1, landmasses fall into several groups, distinguished according to the structure and composition of the bedrock comprising the mass.

A. **Undisturbed structures.**
   1. **Coastal plains.** Recently emerged coastal belts underlain by sedimentary rock layers which lap over older rocks of the continents.
   2. **Horizontal strata.** Sedimentary strata, essentially horizontal in attitude, which have been raised over a large area, but not otherwise seriously disturbed. Horizontal lava flows of great thickness and extent may be included in this group.

B. **Disturbed structures.**
   3. **Domes and basins.** Circular or oval zones of uplift or depression causing sedimentary layers to be convexly or concavely bent.
   4. **Folds.** Sedimentary strata that have been deformed by mountain-making crustal movements into long belts of wavelike folds. The folds may be broad and open or tightly compressed, depending on the degree of crustal compression.
   5. **Fault blocks.** Crustal masses of any rock type or structure broken by faulting into sharply cut blocks that have been displaced in relation to one another. Some tilting usually accompanies the faulting.
   6. **Homogeneous crystallines.** Masses of intrusive igneous rock or metamorphic rock which are essentially uniform throughout as regards their resistance to weathering and erosion.
   7. **Belted metamorphics.** Narrow zones of metamorphic rocks forming parallel mountain and valley belts.
   8. **Complex structures.** Crustal masses that have suffered a combination of folding, faulting, or intrusion by igneous rocks so as to make a mass of irregular and complicated structures.

C. **Volcanoes and related forms.**
   All types of rock masses resulting from the extrusion of molten rock. These include various types of volcanoes and lava flows.

There is a significant difference in the initial appearance of the undisturbed and the disturbed structures. The former have surfaces of low relief (plains or plateaus) before erosional modification sets in. The disturbed structures and volcanic forms, on the other hand, usually have bold, mountainous relief in the initial stage. Relief is greatest at the beginning of their life history, and the masses are ultimately reduced to surfaces of faint relief.

Each one of the structural types described above passes through an orderly series of erosional stages, patterned after the general cycle of landmass denudation already explained. For regions of horizontal strata or homogenous crystalline rocks, this cycle is very similar to the ideal general cycle because these rock masses are of uniform composition and structure in every direction horizontally. Folds, fault blocks, domes, and volcanoes, however, have life cycles quite different, not only from the ideal cycle, but also from one another.

Rock structure as a landform control

As explained in Chapter 23, in the description of evolution of the Appalachian Mountains, denudation acts to lower surfaces underlain by weak rocks more rapidly than surfaces underlain by resistant rocks. As a result, landscape relief de-

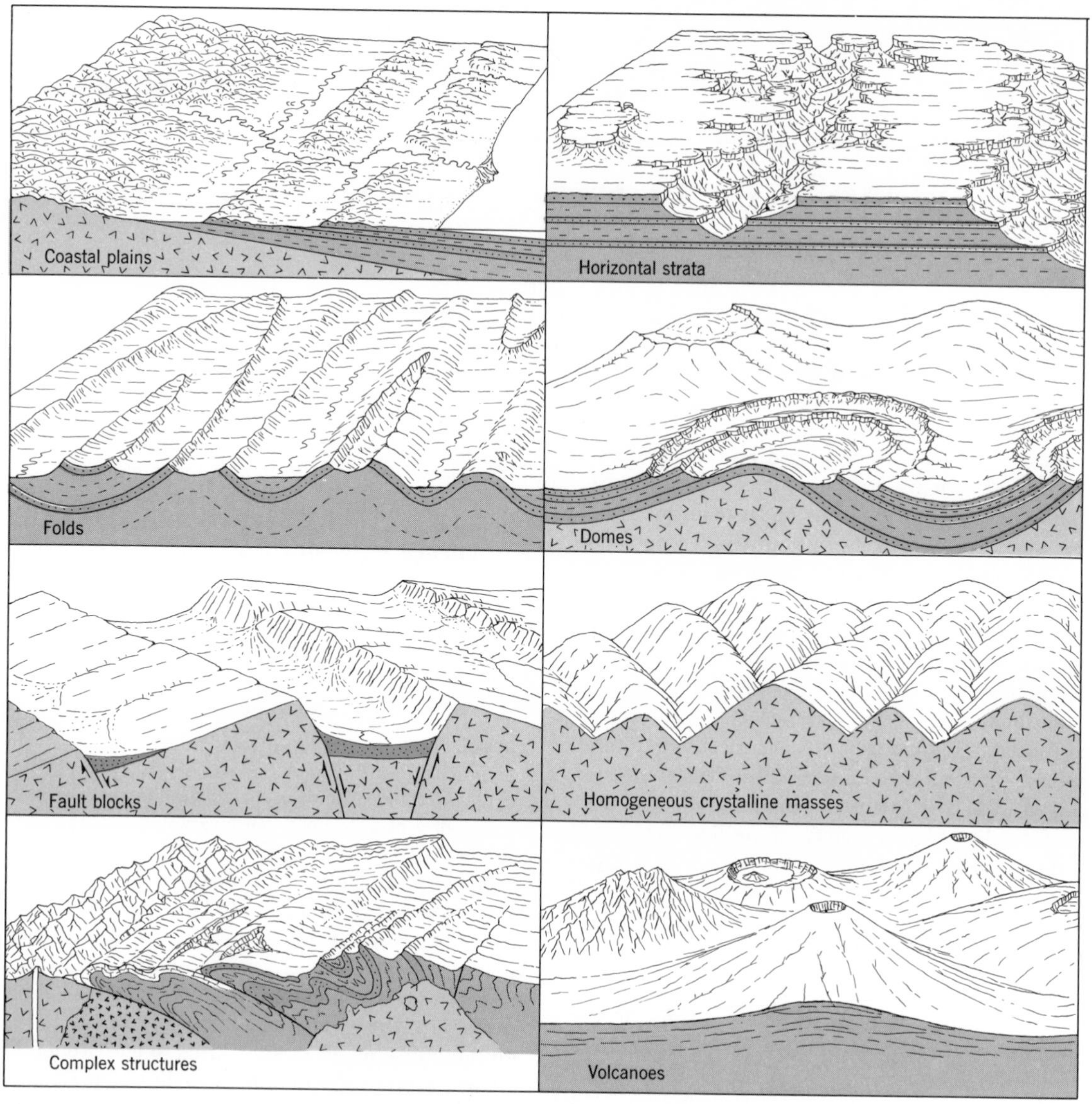

Figure 28.1 Landmasses can be classified according to the groups illustrated here. Belted metamorphics are shown in Figure 28.3.

velops in close conformity with patterns of bedrock composition and structure.

Figure 28.2 shows five types of sedimentary rock together with a mass of much older igneous rock upon which the sediments were laid. Their usual landform habit, whether to form valleys or mountains, is indicated, together with the conventional symbols used on cross sections by geologists. These rock strata have been strongly tilted and deeply eroded.

Shale is usually a weak rock and is reduced to the lowest valley surfaces of the region. Limestone, easily subject to carbonic acid reaction, is also a valley former in humid climates. On the other hand, limestone is highly resistant and usually stands high in arid climates. Sandstone and conglomerate are typically resistant rocks and usually form ridges and uplands.

The metamorphic rocks are, as a group, more resistant to denudation than their sedimentary parent types. However, as shown in Figure 28.3, there are conspicuous differences among the types of metamorphic rocks.

Slates and schists are relatively resistant and tend to form hills and uplands. In comparison with granite, however, these rocks are less resistant, so that granites will usually form markedly higher mountain masses.

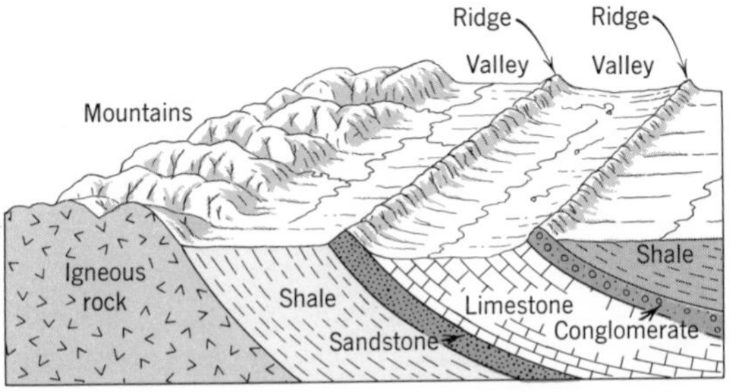

Figure 28.2 Many landscape features originate through the slow erosional removal of weaker rock, leaving the more resistant rock standing as ridges or mountains.

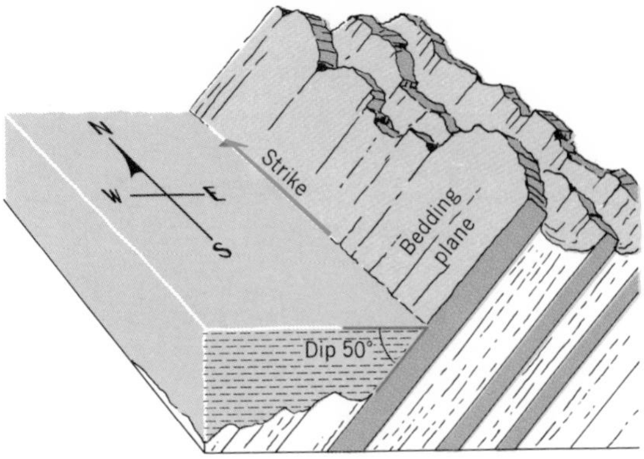

Figure 28.4 Strike and dip.

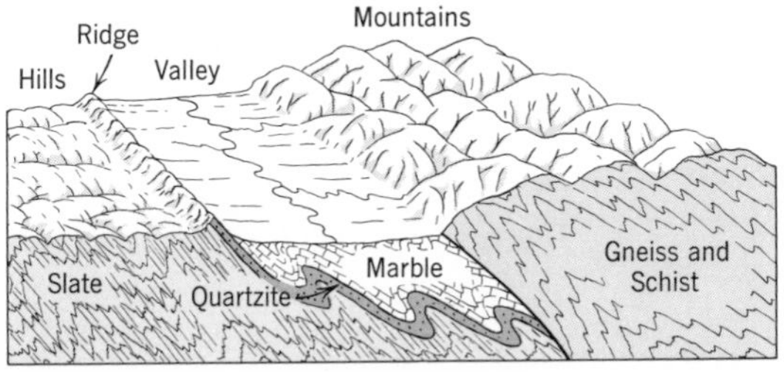

Figure 28.3 Metamorphic rocks tend to form elongate, parallel belts of valleys and mountains.

Like limestone, marble is easily decomposed by carbonic acid and is usually found occupying valleys and lowlands. The extreme hardness of quartzite, combined with its high immunity to chemical decay, makes it the most resistant of all rocks. Many prominent ridge crests and peaks in a region of metamorphic rock will be found to be composed of quartzite Gneisses are strong, resistant rocks, which, like granite, generally form bold highlands or mountain chains.

Dip and strike

Because natural planes are characteristic of the structure of each type of rock, the geologist requires a system of geometry to enable him to measure and describe the attitude of these natural planes and to indicate them on maps. Examples of such planes are the bedding layers of sedimentary strata, the sides of a dike, the upper and lower surfaces of a sill, slaty cleavage of slates, and the joints in a granite. Rarely are these planes truly horizontal.

The acute angle formed between a natural rock plane and an imaginary horizontal plane is termed the *dip*, and is stated in degrees ranging from 0° for a horizontal plane to 90° for a vertical plane. Figure 28.4 shows the dip angle for an outcropping layer of sandstone, against which rests a horizontal water surface.

The compass direction, or bearing, of the line of intersection between the inclined rock plane and an imaginary horizontal plane is the *strike*. In Figure 28.4 the strike is north.

Coastal plains

Coastal plains evolve through erosional stages illustrated in Figure 28.5. In Block *A* the region has recently emerged from beneath the sea, where it was formerly a shallow continental shelf accumulating successive layers of sediment brought from the land and distributed by currents. On the initial surface, streams flow directly seaward, down the slope of the new surface. A stream of this origin is a *consequent stream*, defined as any stream whose course is controlled by the initial slope of land surface. Consequent streams occur on many landforms, such as volcanoes, fault blocks, or beds of drained lakes. Streams that formerly drained the land surface inland from the coastal plain, but that now have become extended across it to reach the new shoreline, are called *extended consequent streams*. The term *oldland* is applied to the area of older rock lying inland from the coastal plain.

In the mature stage of coastal-plain development a new series of streams and topographic features has developed (Block *B*). Where more easily eroded strata (usually clay or shale) are exposed, denudation is rapid, making *lowlands*. Between them rise broad low ridges or belts of hills comprising *cuestas*. The lowland lying between the oldland and the first cuesta is called the *inner lowland*. Cuestas are commonly underlain by sand, sandstone, limestone, or chalk. They have a fairly steep slope on the landward side, or *inface*, because the edge of the eroded layer is exposed on this side. The seaward slope, or *backslope*, of the cuesta is gentle because it follows the top surface of the gently inclined harder layer. Where the resistant

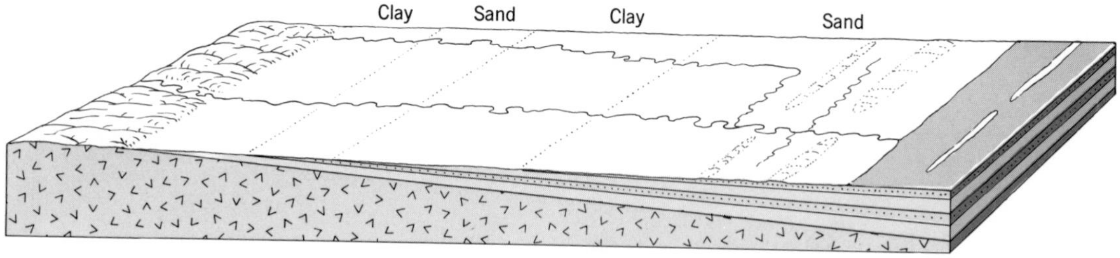

A. **Initial stage; plain recently emerged.**

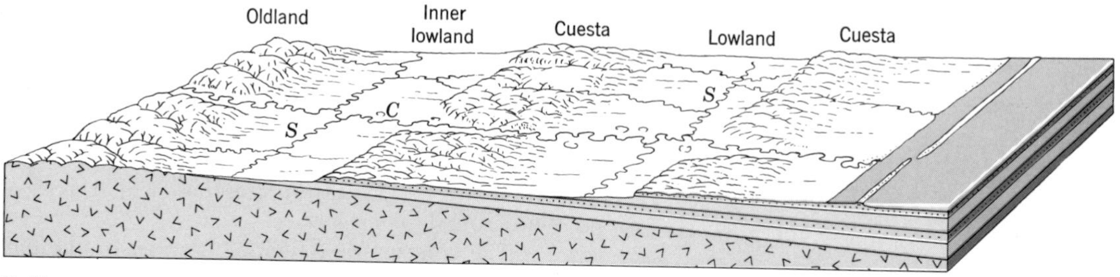

***B*. Mature stage; cuestas and lowlands developed. *S* = subsequent: *C* = consequent.**

Figure 28.5 Development of a broad coastal plain. (After A. K. Lobeck.)

Figure 28.6 This sharply defined cuesta in the Paris basin of northern France has its steep face to the east (left), a very gentle slope westward from the crest. (Photograph by Douglas Johnson.)

layer is very hard and is underlain by a weak layer, the cuesta face is often steep with occasional rock cliffs, as in the limestone cuesta near Rheims, France (Figure 28.6). More commonly, however, the cuesta is merely a belt of low hills.

Streams that develop along the trend of the lowlands, parallel with the shoreline, are of a class known as *subsequent* streams. They take their position along any belt or zone of weak rock and therefore follow closely the pattern of rock exposure. Subsequent streams occur in many regions and will be mentioned frequently in the discussion of folds, domes, and fault blocks. The drainage lines on a maturely dissected coastal plain combine to form a *trellis* pattern (See Figure 29.2).

Coastal plains of the United States and England

Splendid examples of coastal plains are present along the Atlantic and Gulf coasts of the United States, in southeastern England, and in the Paris Basin region of north-central France.

The coastal plain of the United States is by far the largest of these, ranging in width from 100 to 300 mi (160 to 500 km) and extending for 2000 mi (3000 km) along the Atlantic and Gulf coasts. Strata are of Cretaceous and Tertiary age, the former being exposed nearest to the inner margin of the coastal plain because they lie directly upon the oldland rocks of Paleozoic and Precambrian age. The coastal plain starts at Long Island, which

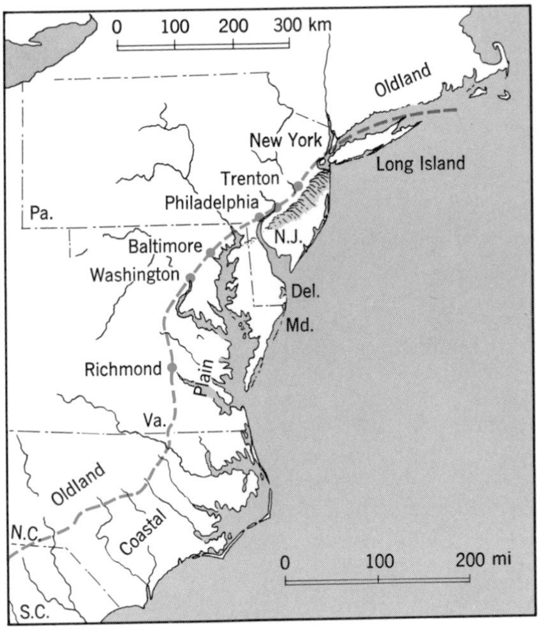

Figure 28.7 The coastal plain of the Atlantic seaboard states shows little cuesta development except in New Jersey. The inner limit of the coastal plain is marked by a series of fall-line cities. (After A. K. Lobeck.)

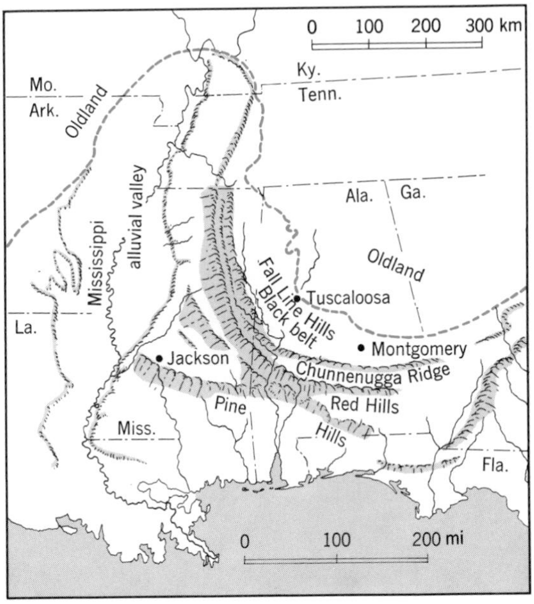

Figure 28.8 The Alabama-Mississippi coastal plain is belted by a series of sandy cuestas and shale or marl lowlands. (After A. K. Lobeck.)

is a partly submerged cuesta, and widens rapidly southward so as to include much of New Jersey, Delaware, Maryland, and Virginia (Figure 28.7). Throughout this portion the coastal plain has but one cuesta, that which forms the Atlantic Highlands, Mt. Laurel, Pine Hills, and similar hill groups. The cuesta is underlain by a porous sand formation of Tertiary and Cretaceous age, which resists erosion by absorbing rainwater rapidly and thereby minimizing overland flow. The inner lowland is a continuous broad valley developed on a weak clay formation of Cretaceous age.

In Alabama and Mississippi the coastal plain is maturely dissected in all but the coastal area, which has recently emerged from the sea. Cuestas and lowlands run in belts roughly parallel with the coast (Figure 28.8). Hence the term *belted coastal plain* is applied to these regions. The cuestas tend to be underlain by sandy formations and support a growth of pines. Limestone forms fertile lowlands such as the Black Belt in Alabama.

The entire southeastern portion of England is a former coastal plain (Figure 28.9). Two cuestas dominate the topography. The innermost is of Jurassic limestone and is locally named the Cotswold Hills. In England, the term *wold* is applied to a cuesta, *vale* to an intercuesta lowland. The outer or southeastern cuesta is of white chalk of Cretaceous age and includes the Chiltern Hills. Between cuestas is a lowland in which lie Oxford and Cambridge. An extensive inner lowland runs between the inner cuesta and the old-land rock masses of Cornwall, Wales, and the Pennine Range. In the inner lowlands are the important cities of Bristol, Gloucester, Birmingham, Nottingham, Lincoln, and York, as well as extensive farm lands. This lowland is drained by the Severn, Avon, Trent, and Ouse rivers.

Environmental aspects of coastal plains

Broad coastal plains, such as those of the eastern United States and southeastern England, show intensive agricultural development because of the fertility and easy cultivation of broad lowlands. Although important seaport cities have developed on coastal plains there was not the same impelling necessity toward marine occupations that was induced by mountainous coastal belts bordered by shorelines of submergence.

Cuestas provide valuable forests, as in England and Europe and in the southern United States. Where excessively porous sands occur, as in the New Jersey coastal plain, pine and oak are supported.

Transportation tends to follow the lowlands and to connect the larger cities located there. For example, important roads and railroads connect New York with Trenton, Philadelphia, Baltimore and Washington, all of which are situated in an inner lowland. Cuesta topography, however, is rarely so rugged as to interfere seriously with the location of communication lines.

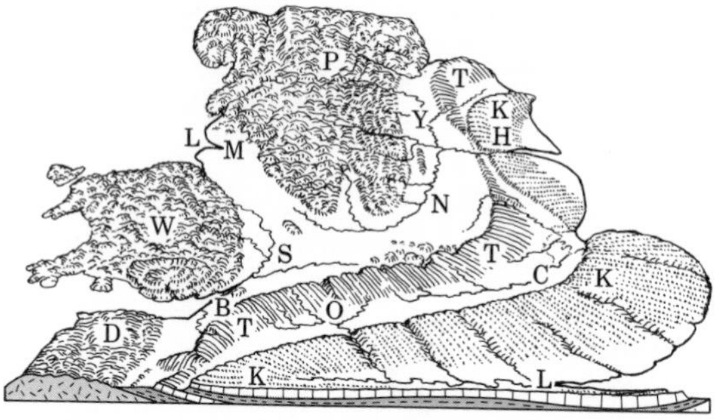

Figure 28.9 Southeastern England is a broadly curved mature coastal plain. L = London; K = Cretaceous chalk cuesta; C = Cambridge; O = Oxford; H = Humber River; Y = York; N = Nottingham; S = Severn River; B = Bristol; D = Dartmoor; W = Wales; M = Manchester; L = Liverpool; P = Pennine Range: T = Jurassic limestone cuesta. (After W. M. Davis.)

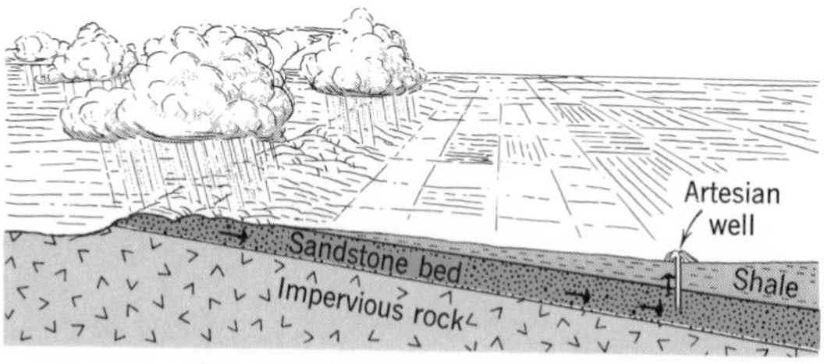

Figure 28.10 An artesian well requires a dipping sandstone layer. (After E. Raisz.)

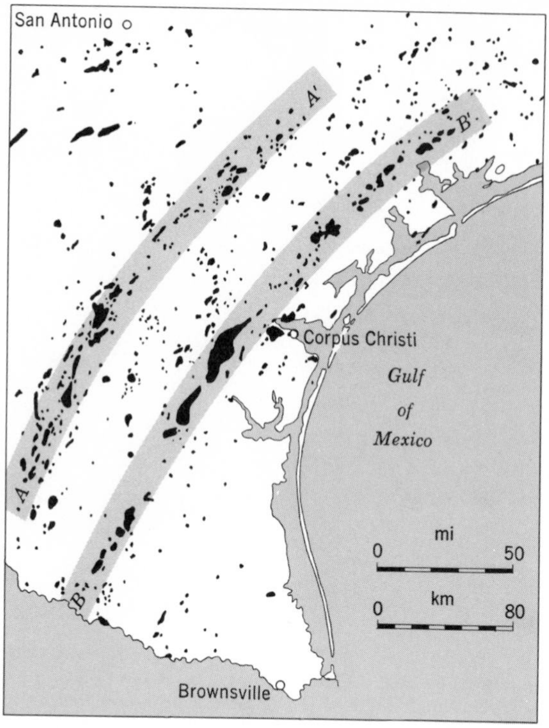

Figure 28.12 Two zones of oil pools on up-dip pinchouts of sands of Eocene age (AA') and Oligocene age (BB'). (After A. I. Levorsen.)

The seaward dip of sedimentary strata in a coastal plain provides a structure favorable to the development of artesian water wells. Water penetrates deeply into a sandy cuesta stratum, which is overlain by shales or clays impervious to the flow of underground waters. When a well is drilled into the sand formation considerably seaward of its surface exposure, water under hydraulic pressure reaches the surface (Figure 28.10). Artesian water in large quantities is available in many parts of the Atlantic and gulf coastal plains, although it is no longer sufficient to supply the demands of densely populated and industrialized localities.

The Gulf Coastal Plain of the United States contains petroleum and natural gas accumulations of enormous economic value. Oil occurs in *stratigraphic traps,* which are layers or lenses of permeable sand or sandstone capped by impermeable shales or clays. One kind of stratigraphic trap is the *pinch out,* illustrated in Figure 28.11. A sand-

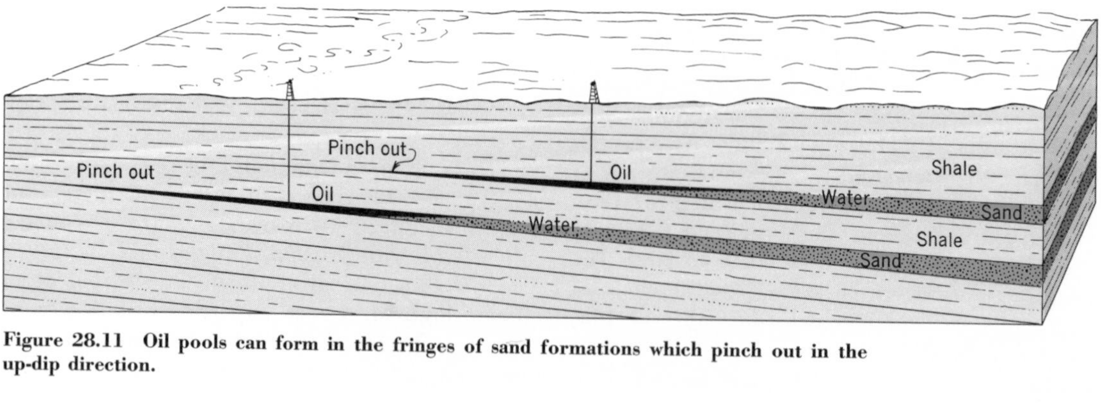

Figure 28.11 Oil pools can form in the fringes of sand formations which pinch out in the up-dip direction.

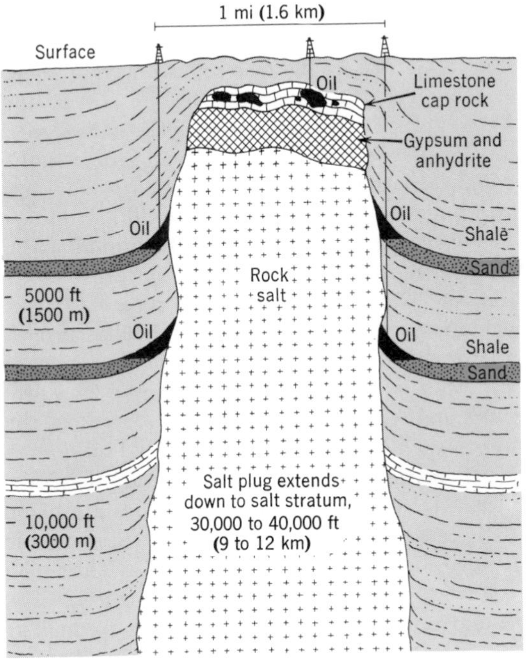

Figure 28.13 Idealized structure section of a salt dome.

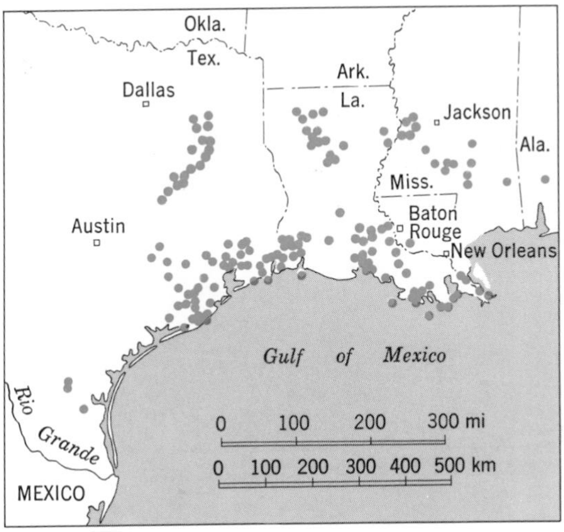

Figure 28.14 Distribution of salt domes of the Gulf Coast region is indicated by dots. (After K. K. Landes.)

stone formation in the coastal plain sequence of strata thins in the updip (landward) direction to a feather edge, where it disappears, whether through lack of deposition, or by later erosion that preceded deposition of the next younger beds. Capped above by impermeable beds, the sandstone wedge forms a trap for oil migrating updip. Figure 28.12 show two curving bands of oil *pools* of this type in the Gulf Coast of Texas.

Another quite different type of oil pool common in coastal plains and other regions of thick sedimentary strata occurs on *salt domes*, or *salt plugs* (Figure 28.13). These strange, stalklike bodies of rock salt project upward through coastal plain strata. Apparently they were forced up by slow plastic flowage from thick salt formations lying in deep lower layers of the coastal plain. Surrounding strata are sharply bent up and faulted against the side of the salt plug, making traps for petroleum. Salt plugs commonly have a cap rock of limestone resting upon a plate of gypsum and anhydrite. Oil may collect in cavities in the limestone. Distribution of salt domes of the Gulf Coast is shown in Figure 28.14. The salt dome should not be confused with sedimentary domes discussed later in this chapter.

Other mineral deposits of economic importance in coastal plains include: *sulfur*, occuring in the coastal plain of Louisiana and Texas; *phosphate* beds, found in Florida; *lignite* (a low-grade, woody coal), used as a fuel in Alabama, Mississippi, and Texas; and *clays*, used in manufacture of pottery, tile, and brick in New Jersey and the Carolinas.

Horizontal strata

Considerable areas of the continental shields are covered by thick sequences of sedimentary-rock layers, which at one time in the geologic past were the bottom deposits of shallow inland seas or were stream deposits spread over vast alluvial plains. Strata of all three post-Cambrian geologic eras are represented. When uplifted with little disturbance other than a faint warping or minor faulting, these sedimentary strata pass through a series of stages of erosion such as those illustrated in the series of block diagrams of Figure 28.15.

In the initial stage the land is fairly smooth and is drained by consequent streams following the gentle slope of the surface. If the elevation of the initial surface is high, these streams soon cut deep canyons, leaving plateau surfaces between. Should the region have initially low elevation above sea level, the streams are prevented by the base level from cutting deeply, and hence strong relief can never develop in the region. Throughout the young stage, the region, whether of great or small relief, is dissected by streams whose valley network develops at the expense of the initial land surface.

When the initial land surface is entirely or almost entirely consumed and the region has reached its most rugged character, the stage of maturity has been reached. Throughout the remainder of the erosion cycle the relief diminishes and the slopes perhaps tend to become more gentle. In the old stage the region is reduced to a rolling plain upon which the larger streams have broad, flat floodplains.

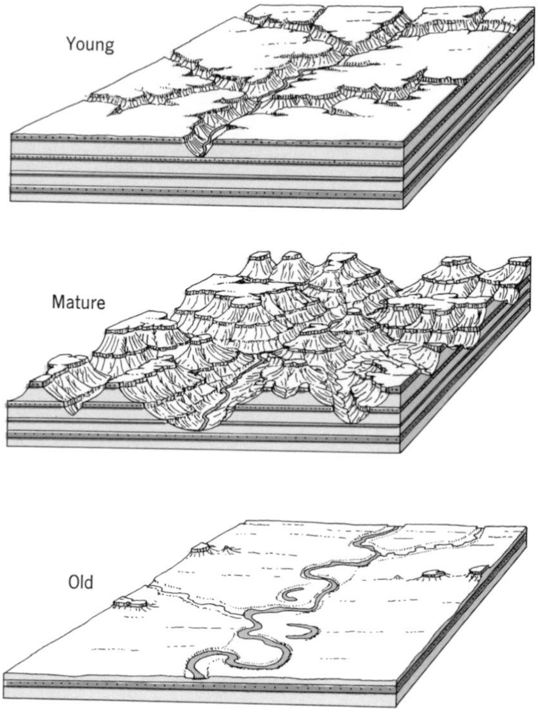

Figure 28.15 Erosional development of horizontal sedimentary strata. The development of cliffs is accentuated here, as typical of an arid climate. (After E. Raisz.)

The horizontal attitude of the rock layers gives rise to distinctive landforms where the layers are of alternately weak and resistant nature (Figure 28.16). The resistant layers, usually of sandstone and limestone (the latter particularly in arid climates), form *cliffs* or steep slopes. The weak layers, usually of shale, clay, or marl, are easily washed away from beneath the lower edges of the resistant layers, hence serving to accentuate the cliffs above them and form smoothly descending slopes at each cliff base. In dry climates, where vegetation is scant and the action of rainwash especially effective, sharply defined topographic forms develop. They comprise what may be described as *scarp-slope-shelf* topography, because the normal sequence of forms is a cliff, or scarp, at the base of which is a smooth slope. This in turn flattens out to make a shelf, terminated at the outer edge by the cliff of the next lower set of forms. In the walls of the great canyons of the Colorado Plateau region in Colorado, Utah, Arizona, and New Mexico, these forms are wonderfully displayed (Figure 28.17).

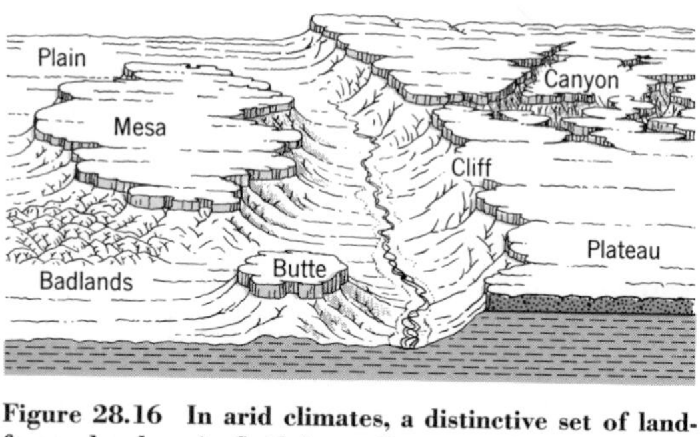

Figure 28.16 In arid climates, a distinctive set of landforms develops in flat-lying sedimentary formations.

In plateau regions underlain by horizontal strata, the erosion processes tend to strip successive layers from the plateau surface. Cliffs, capped by hard rock layers, retreat as near-perpendicular surfaces because the weak clay or shale formations exposed at the cliff base are rapidly washed away by storm runoff and channel erosion. Thus undermined, the rock in the upper cliff face repeatedly breaks away along vertical fractures. Where a cliff has thus retreated a considerable distance from a canyon, there remains a broad, flat bench which is the exposed surface of the next layer below. The bench so formed is termed an *esplanade* (Figure 28.17). Should the entire plateau surface be formed by the complete or almost complete removal of a rock series, leaving a plateau capped by a resistant layer, the plateau is termed a *stripped surface* or *stratum plain*.

In the later stages of erosion the landscape in an arid region has many *mesas* (Figure 28.16), tabletopped hills or mountains bordered on all sides by cliffs and representing the remnant of a formerly extensive layer of resistant rock. Often a mesa is capped by a lava flow, which is generally more resistant than the sedimentary rocks over which it once flowed. As a mesa is reduced in area by retreat of the cliffs that border it, it maintains its flat top and altitude. Before its complete consumption the final stage is a small, steep-sided hill or peak known as a *butte* (Figure 28.18).

Where extremely weak clays or shales, lacking a protective vegetative cover, are exposed to rainwash and gully erosion in dry regions, a very rugged topography resembling miniature mountains develops. Such areas are termed *badlands* (Figures 28.16 and 28.19).

In humid climates the elements described above, namely, scarp-slope-shelf topography, stripped surfaces, mesas, and buttes, are present only in greatly subdued aspect. This is due to the thick cover of vegetation that protects the ground from rapid rainwash and permits a layer of soil and residual overburden to cloak the bedrock. Nevertheless, occasional lines of cliffs do form, and mesalike mountains occur. Vast areas in Pennsylvania, New York, Ohio, West Virginia, Kentucky, Tennessee, and Alabama consist of maturely dissected horizontal strata. Much of this land is mountainous and forested (Figure 27.9).

In a maturely dissected region of horizontal strata the stream system forms a *dendritic drainage pattern,* in which the smaller streams show no predominant directional orientation or control (Figure 26.5). This pattern has been likened to the branching of an apple tree.

In addition to regions underlain by sedimentary rocks, the regions of horizontal strata may be made to include thick accumulations of lava flows. In some parts of the world, such as the Columbia Plateau region of eastern Washington and Oregon, or the Deccan Plateau of western India, the vast outpourings of highly fluid basalt lavas now cover thousands of square miles and are several thousand feet thick. Interbedded with the lavas are lake and stream deposits of sands, gravels, and clays. Consequently, the landforms are very similar to those of sedimentary strata.

Figure 28.18 This early photograph shows a butte of horizontal red sandstones capped by a gypsum layer, near Cambria, Wyoming. (Photograph by N. H. Darton, U.S. Geological Survey.)

Environmental aspects of horizontal strata

Generalizations cannot readily be made about the utilization of areas underlain by horizontal strata because of the great variations in surface relief that exist. On initial and old surfaces, where the topography is plainlike, agriculture is widely developed. On the high plains of western Kansas and Nebraska, eastern Wyoming and Colorado, New Mexico, and Texas, wheat farming is the predominant activity. Here the plain is in its initial or very early stage of erosion. Despite elevations of 3000 to 5000 ft (900 to 1500 m), the plain is trenched only by a few major through-flowing streams.

Some regions of horizontal strata in the interior United States, including much of Illinois, Indiana, Ohio, Missouri, Montana, Kansas and Iowa, are maturely dissected but a mantle of glacial drift has reduced relief so that slopes are gentle and are highly cultivated. Where relief is strong, as in the

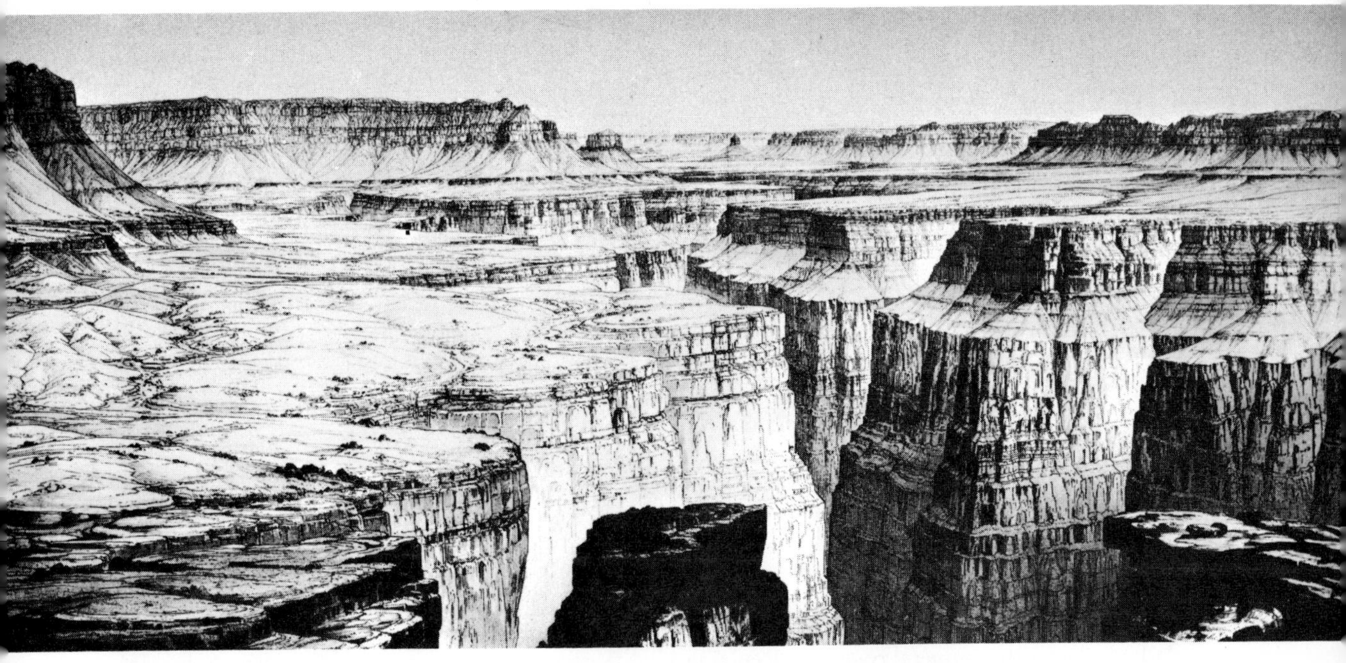

Figure 28.17 This panoramic drawing by noted geologist-artist, W. H. Holmes, published in 1882, shows the Grand Canyon at the mouth of the Toroweap. In this part of the canyon, rarely seen by tourists, a broad bench called The Esplanade is well developed. (From Dutton, *Atlas to accompany Monograph II*, U.S. Geological Survey.)

Figure 28.19 Badlands, such as these in the Petrified Forest Nationai Monument, Arizona, are like minature mountain topography on bare clay formations. (Photograph by B. Mears, Jr.)

mountain areas of the Alleghenies or the Cumberland Mountains, cultivation is limited to a few small tracts, such as the floodplain belts of larger streams, despite the favorable humid climate (Figure 27.9). In the canyon lands of the Colorado Plateau an extremely low population density exists, not only because the high relief and aridity do not favor agriculture, but also because human access is virtually impossible across the network of sheer-walled canyons.

Mineral resources of horizontal strata

As with coastal plains, regions of horizontal strata have only those minerals and rocks of economic value that are associated with sedimentary rocks (or lavas). Building stone, such as the Bedford limestone in Indiana or the Berea sandstone in Ohio, is a valuable product. Limestone may be quarried for use in manufacture of portland cement or as flux in iron smelting. Some important deposits of lead, zinc, and iron ores occur in sedimentary rocks. For example, the lead and zinc mines of the Tristate district (Missouri, Kansas, Oklahoma) are in horizontal limestones. Uranium ores are important in the Colorado Plateau.

Perhaps the greatest mineral resources occurring in sedimentary strata are coal and petroleum. Where the strata are undisturbed, coal is of the *bituminous*, or soft, variety and lies in horizontal layers from a few inches to several feet thick and hundreds of square miles in extent. Where the relief of the region is great, coal seams outcrop along the valley walls into which mine openings termed *drifts* can be tunneled to obtain the coal. This is common practice in the bituminous field of western Pennsylvania, eastern Ohio, and West Virginia. Where the coal seams are not exposed at the surface they must be reached by vertical shafts, as in the Illinois coal fields.

Where the coal seams lie close to the surface or actually outcrop along hillsides the *strip mining* method is used. Here, earthmoving equipment removes the covering strata (*overburden*) to bare the coal, which is lifted out by power shovels. There are two kinds of strip mining, each adapted to the given relationship between ground surface and coal seam. *Area strip mining* is used in regions of nearly flat land surface under which the coal seam lies horizontally (Figure 28.20*A*). After the first trench is made and the coal removed, a parallel trench is made, the overburden of which is piled as a spoil ridge into the first trench. Thus the entire seam is gradually uncovered and there remains a series of parallel soil ridges. In this connection, it is interesting from the aspect of environmental impact to know that phosphate beds are mined extensively by the area strip mining method in Florida, and that the method is also used for mining clay layers.

The *contour strip mining* method is used where a coal seam outcrops along a steep hillside (Figure 28.20*B*). The coal is uncovered as far back into the hillside as possible and the overburden dumped on the downhill side. There results a bench bounded on one side by a freshly cut rock wall and upon the other by a ridge of loose spoil with a steep outer slope leading down into the

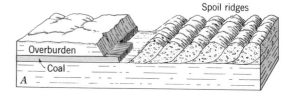

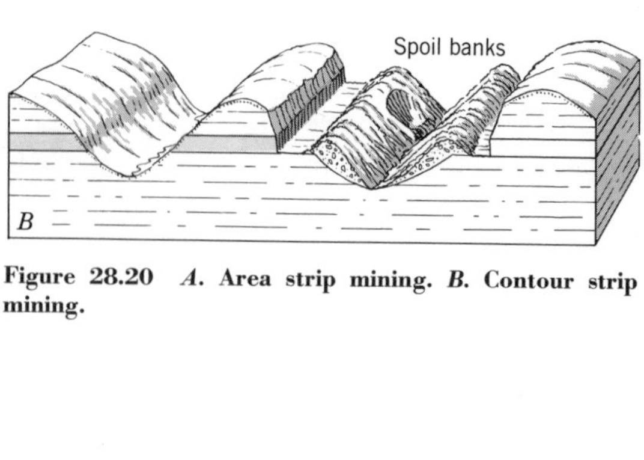

Figure 28.20 *A*. Area strip mining. *B*. Contour strip mining.

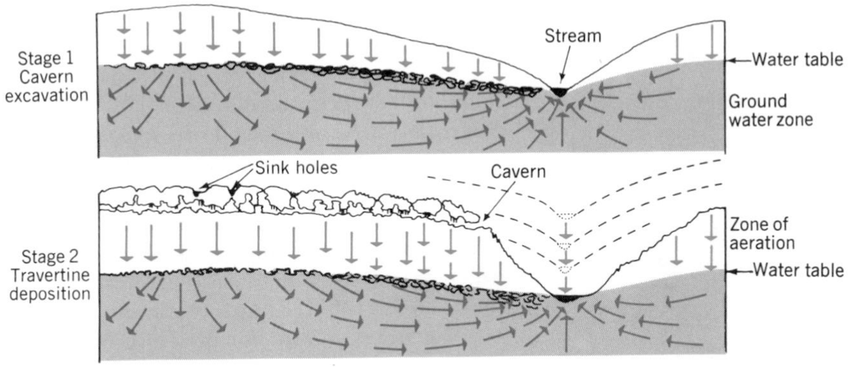

Figure 28.21 Cavern development in the ground-water zone, followed by deposition in the zone of aeration. (From A. N. Strahler, 1971, *The Earth Sciences*, 2nd ed. Harper and Row, New York.)

valley bottom. The benches form sinuous patterns following the plan of the outcrop. Strip mining is carried to depths as great as 100 feet below the surface. Associated with contour strip mining is *auger mining* in which enormous auger drills are run horizontally into the exposed face of the coal seam after the initial strip mining is completed. Augers with cutting heads several feet in diameter are used and are capable of penetrating as far as 200 feet into the seam.

Petroleum occurs within permeable sandstone layers, in which the oil is trapped by overlying impermeable shales. Because the strata are not perfectly horizontal, but are affected by minor warpings and faults as well as by changes in thickness and character of the sandstone layers, there are many structures in which petroleum is localized into pools. Stratigraphic traps, similar in principle to those of coastal plains, form important pools. Traps also result from faulting, which offsets the edges of the strata (Figure 29.14).

Limestone caverns

Most persons are familiar with the names of famous caverns, such as Mammoth Cave or Carlsbad Caverns, and many Americans have visited one or more of these famous tourist attractions. Although caverns may develop in folded, faulted, and steeply dipping limestone layers, most caverns occur in areas of flat-lying strata.

The consensus of opinion among geologists is that most cavern systems were opened out in the ground-water zone. Figure 28.21 suggests how caverns may develop. In the upper diagram the action of carbonic acid upon calcium carbonate is shown to be particularly concentrated just below the water table. Products of solution are carried along in the ground-water flow paths to emerge in streams and leave the region in stream flow. In a later stage, shown in the lower diagram, the stream has deepened its valley and the water table has been correspondingly lowered to a new position. The cavern system previously excavated is now in the zone of aeration. Evaporation of percolating water on exposed rock surfaces in the caverns now begins the deposition of carbonate matter, known as *travertine.* Encrustations of travertine take many beautiful forms—stalactites, stalagmites, columns, drip curtains, and terraces (Figures 28.22 and 28.23).

Figure 28.22 A drip curtain of travertine, Jenolan Caves, N.S.W., Australia. (N.S.W. Government Printer.)

Figure 28.23 A travertine terrace, Jenolan Caves, N.S.W., Australia. (N.S.W. Government Printer.)

The environmental importance of caves is felt in several ways. Throughout man's early development, caves were an important habitation. Now we find the skeletal remains of these people, together with their implements and cave drawings, preserved through the centuries in caves in many parts of the world. Today, with increasing destructiveness of weapons of warfare, caverns are achieving importance as possible sites for storage of valuable materials, as living quarters, and as factories for important types of production.

Caverns have provided some valuable deposits of *guano*, the excrement of birds or bats, which is rich in nitrates and is used in the manufacture of fertilizers and explosives. Bat guano was taken from Mammoth Cave for making gunpowder during the war of 1812. Much more recently a valuable guano deposit in a limestone cavern in the wall of the Grand Canyon of the Colorado River has been mined and lifted to the canyon rim by cable car.

Karst landscapes

Where limestone solution has been especially active there results a landscape with many unique landforms. This is especially true along the Dalmatian coastal area of Yugoslavia, where the topography is termed *karst*. The term may be applied to the topography of any limestone area where sinkholes are numerous and small surface streams nonexistent. A *sinkhole* is a depression, often steep-sided, in limestone of a cavernous region (Figure 28.24). It may represent a collapse structure of the bedrock.

Four stages of development of a karst landscape are shown in Figure 28.25. Exposed limestone surfaces are deeply grooved and fluted into *lapiés* (*A*). Deep, steep-walled funnel-like sinkholes, called *dolines* (*B*), are numerous. In places, these have coalesced to make open, flat-floored valleys termed *poljes* (*C*). Here surface streams flow and the soil may be suitable for agriculture.

Some other regions of karst or karstlike topography are the Mammoth Cave region of Kentucky, the Causses region of France, the Yucatan Peninsula, and parts of Cuba and Puerto Rico.

Domes and basins

Sedimentary layers in many places show a warping into broad domelike or basinlike structures, which may range from 100 to 200 mi (160 to 325

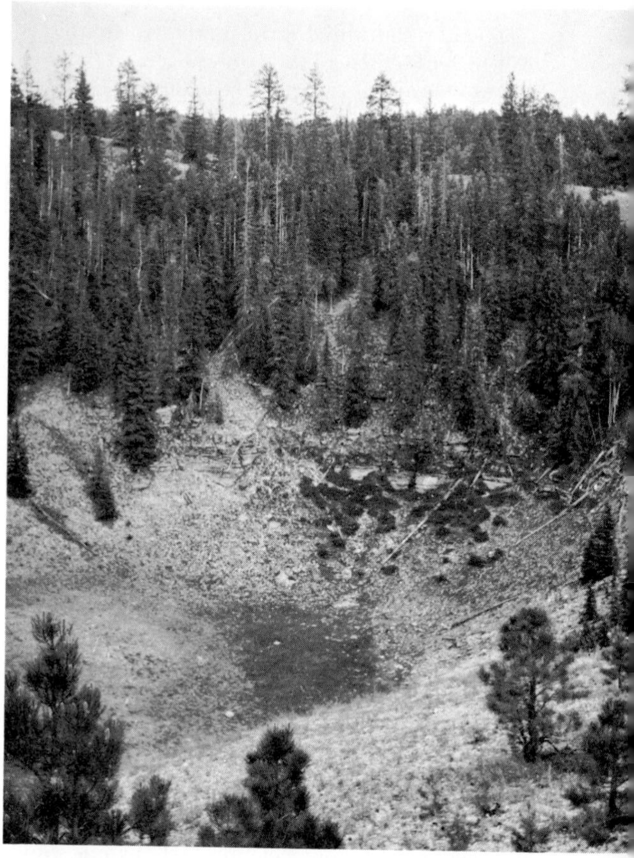

Figure 28.24 Outcrops of horizontal limestone strata show in the walls of this deep sinkhole on the Kaibab Plateau of northern Arizona. Because of high elevation, 8500 ft (2500 m), the climate here is cool and humid, favoring limestone solution. (Photograph by A. N. Strahler.)

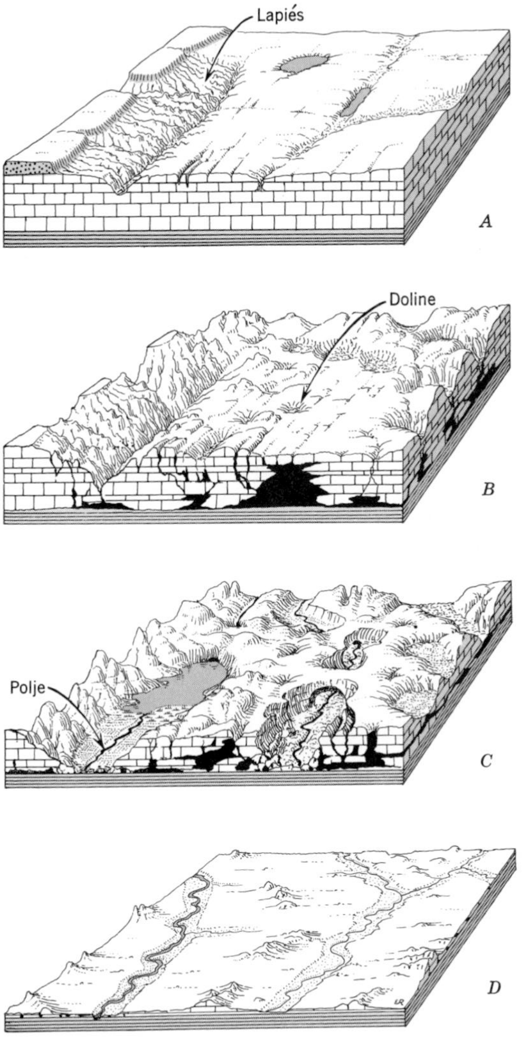

Figure 28.25 Stages of evolution of a karst landscape show increased relief and cavern development followed by decreasing relief and the removal of the limestone mass. (Drawn by E. Raisz.)

km) in diameter, but in which the strata are nowhere inclined more than 1° or 2° from the horizontal. Such domes and basins tend to form concentric cuestas when dissected. An example is the Weald Uplift of southeastern England (Figure 28.26). As the central part of the dome is eroded away, older rock layers are exposed. As each layer is cut through, it is eroded back from the center. Thus the cuestas have their steep inface toward the center of the dome and retreat away from the center. An example of a broad, shallow basin structure is the Paris Basin. The cuestas have their steep infaces outward, and as erosion progresses the cuestas retreat toward the center of the basin.

Domes may also be steep-sided and high, differing from the broad, low type in that the strata on the flanks dip outward at angles up to 25° or

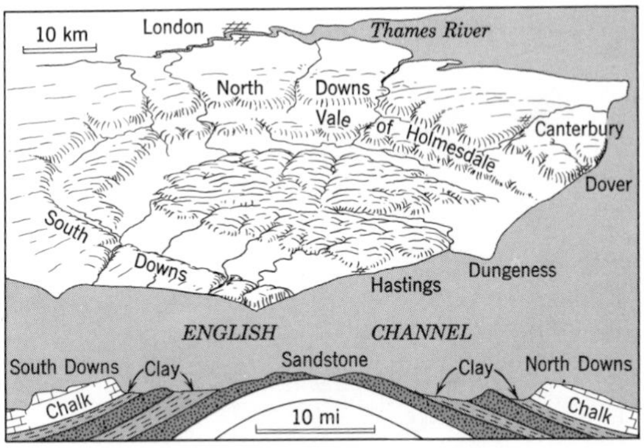

Figure 28.26 The Weald region of southeastern England is a broad dome of sedimentary strata from which the top has been removed. (After A. K. Lobeck.)

more. Thus the dome forms a conspicuous hill or mountain and, when maturely dissected, may constitute a truly rugged mass of peaks. To distinguish this type of dome it will be referred to as a *mountainous dome.*

The eroded edges of steeply inclined strata form sharp-crested *hogback ridges* alternating with narrow valleys (Figure 28.27). The ridges are formed of resistant strata, such as sandstone, whereas the valleys are developed on shales. Streams occupying the valleys are of the subsequent class. In combination with short tributaries that drain the ridge flanks, the subsequent streams form a *trellis drainage pattern* (See Figure 29.2).

The erosion of a mountainous dome is shown in the diagrams of Figure 28.28. In the stage of early youth a series of consequent streams drains outward in a *radial* drainage pattern. These streams entrench the flanks of the dome and quickly expose the underlying layers. As erosion progresses, the resistant strata begin to stand out as sharp-crested ridges, or hogbacks, encircling the

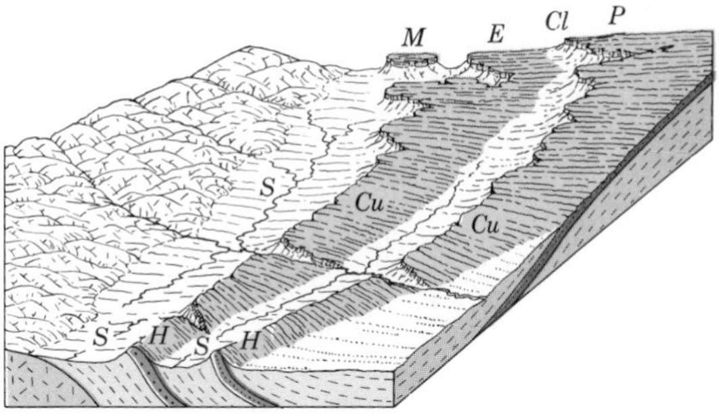

Figure 28.27 Hogbacks may gradually merge into cuestas, the cuestas into plateaus and esplanades, if the dip of the strata becomes less from one place to another. *S* = subsequent stream; *H* = hogback ridge; *Cu* = cuesta; *M* = mesa; *E* = esplanade; *Cl* = cliff; *P* = plateau. (After W. M. Davis.)

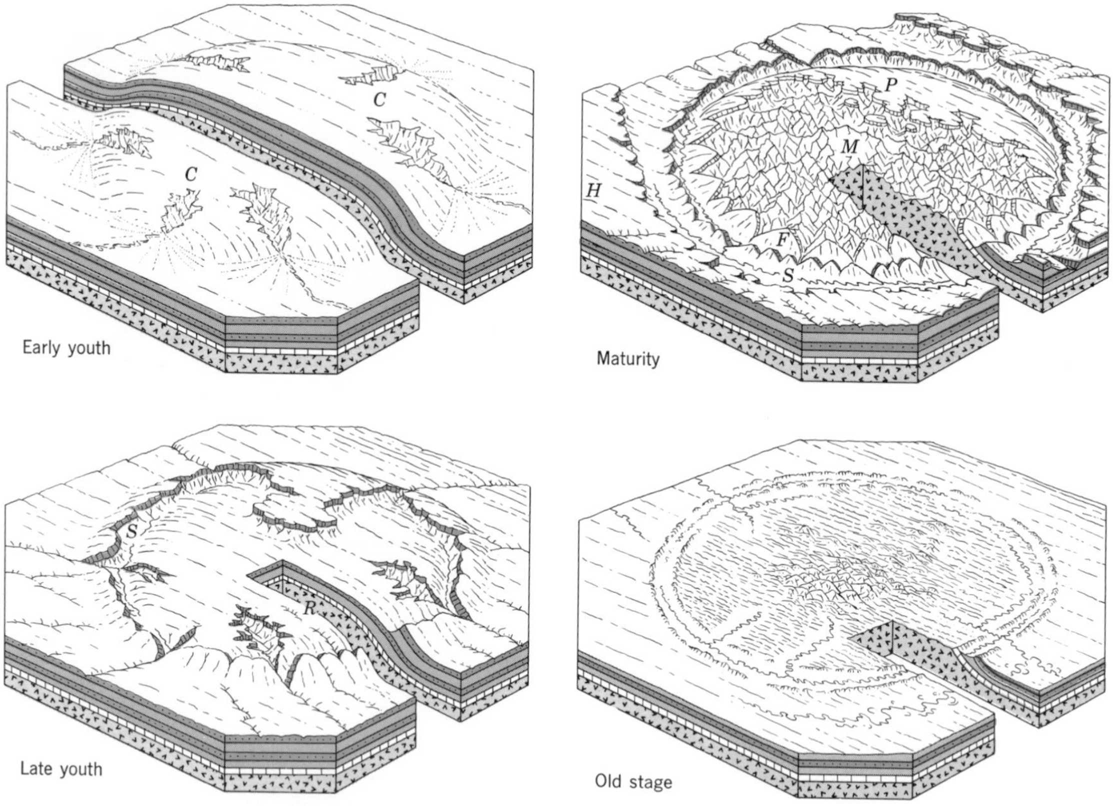

Figure 28.28 Stages in the development of a mountainous dome. C = consequent stream; S = subsequent stream; F = flatiron; P = plateau in center of dome; M = mountains of crystalline rock; H = horizontal strata surrounding dome; R = resequent stream.

dome (Figure 28.29). A concentric arrangement of alternate hogback ridges and valleys develops on the mature dome. Streams occupying the weak rock valleys are subsequent in origin, forming a concentric, or *annular*, drainage pattern (Figure 28.30).

As dome erosion progresses, older and deeper rocks are exposed in the center. If geologic conditions are favorable, erosion may reveal in the center a core of intrusive igneous rock representing material that was forced up to produce the dome. In this event the igneous rock is younger than the sedimentary rock of the dome.

In other domes the central core is of ancient rock, much older than even the sedimentary layers, and represents the rock upon which those sediments were deposited. It shows through in the dome core because the strata are not thick enough to cover it when the dome is fully dissected.

The last sedimentary rock layer to be stripped from the central core of crystalline rock clings to the sides of the core in triangular patches known as *flatirons*. The flatirons cap the ends of mountain spurs and are separated by V-shaped canyons.

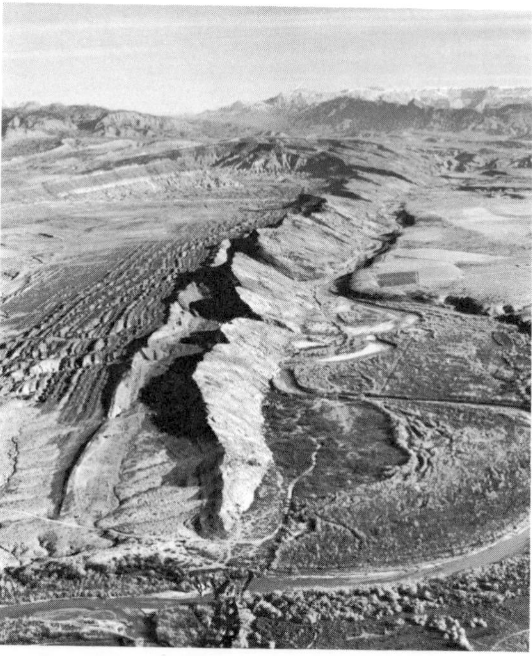

Figure 28.29 A hogback on the flank of the Virgin anticline in southwestern Utah. (Photograph by Frank Jensen.)

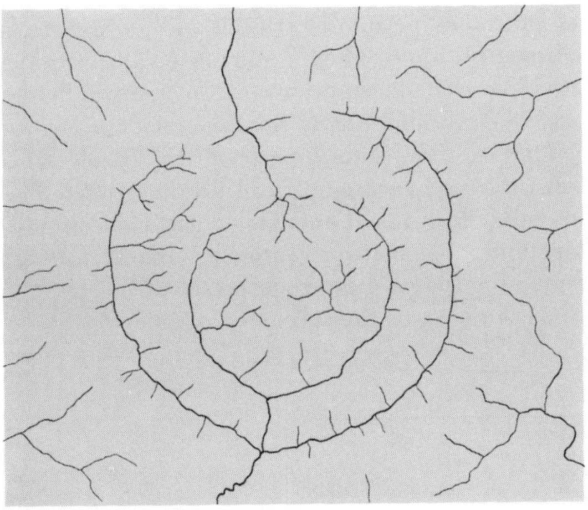

Figure 28.30 The drainage pattern on a maturely eroded dome combines annular and radial elements. It resembles a trellis pattern bent into a circle.

In the old stage, a mountainous dome has been reduced to a peneplain on which the hogback ridges are represented by faint rows of hills. In the central core, a few monadnock masses may rise conspicuously above the peneplain level.

Environmental and resource aspects of domes

Important accumulations of petroleum and natural gas occur in domes of sedimentary strata. Many domes of the Rocky Mountain region have been important petroleum producers, for example, the Rock Springs Dome and Teapot Dome in Wyoming. Oil tends to accumulate in the domed sandstone layers which are overlain by impervious shales. An example of a very low dome in the initial stage of development, which is a valuable producer of oil, is the Dominguez Hills dome in the Los Angeles Basin (Figure 28.31). Figure 28.32 shows the arrangement of oil, gas, and ground water in a sedimentary dome.

Mountainous domes have some unique environmental features. These are illustrated by the great Black Hills dome of western South Dakota and eastern Wyoming (Figure 28.33).

The annular subsequent valleys that encircle the dome are splendid locations for railroads and highways. Thus it is natural that towns and cities should grow in these valleys. In the Black Hills dome, one annular valley in particular, the Red Valley, is continuously developed around the entire dome and has been termed the Race Track because of its shape. It is underlain by a weak shale which is easily washed away. In the Red Valley lie such towns as Rapid City, Spearfish, and Sturgis. On the outer side of the Red Valley is a high, sharp hogback of Dakota sandstone, known simply as the Hogback Ridge. It rises some 400 to 500 ft (120

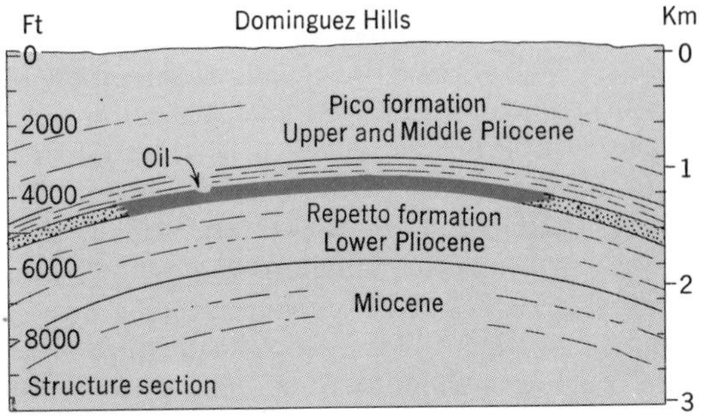

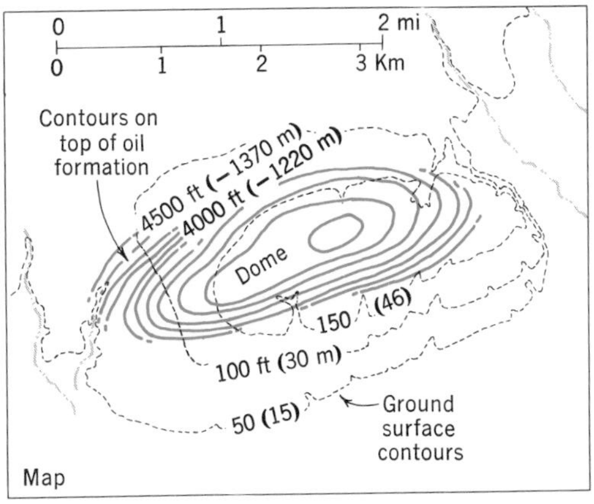

Figure 28.31 The Dominguez Hills, a low dome in an early stage of erosion, has beneath it a valuable oil pool. (After H. W. Hoots and U.S. Geological Survey.)

or 150 m) above the level of the Red Valley. Farther out toward the margins of the dome the strata are less steeply inclined and form a series of cuestas. Artesian water is obtained from wells drilled in the surrounding plain.

The eastern central part of the Black Hills consists of a mountainous core of intrusive and metamorphic rocks. These mountains are richly

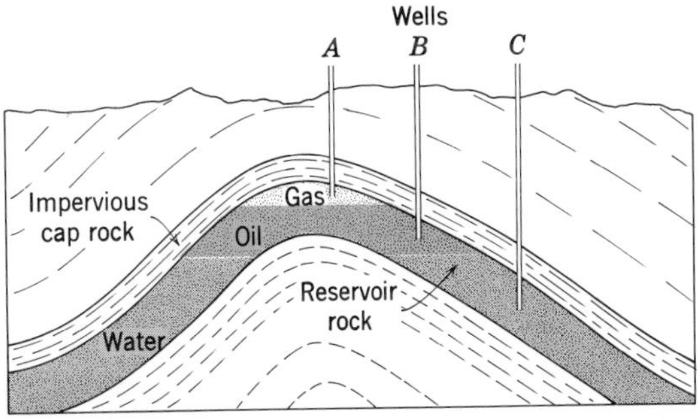

Figure 28.32 Idealized cross section of an oil pool on a dome structure in sedimentary strata. Well *A* will draw gas; well *B* will draw oil; and well *C* will draw water. The caprock is shale; the reservoir rock is sandstone. (From A. N. Strahler, 1972, *Planet Earth; Its Physical Systems Through Geologic Time*, Harper and Row, New York.)

forested, whereas the intervening valleys are beautiful open parks. Thus the region is attractive as a summer resort area. Harney Park, elevation 7242 ft (2207 m), is highest of the peaks of the core. In the northern part of the central core, in the vicinity of Lead and Deadwood, are valuable ore deposits. At Lead is the fabulous Homestake Mine, one of the world's richest gold-producing mines. In the southern part of the central crystalline area, at Pennington, is the Etta Mine, known widely for its enormous pegmatite crystals of spodumene, a source of lithium. These occurrences illustrate the principle that the interior cores of mature domes may be favorable places for mineral deposits.

The western central part of the Black Hills consists of a limestone plateau deeply dissected by streams. The original dome has a flattened summit. The limestone plateau represents one of the last remaining sedimentary rock layers to be stripped from the core of the dome.

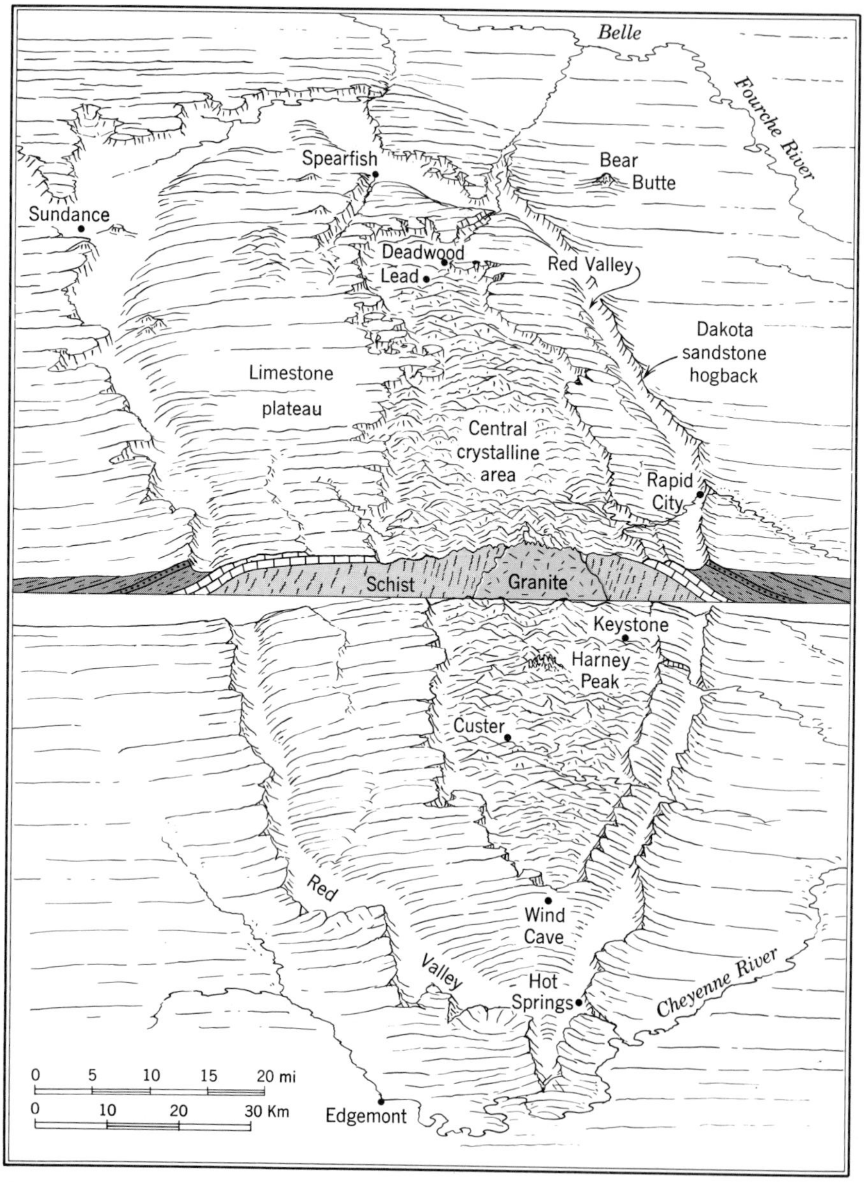

Figure 28.33 Black Hills consist of a broad, flat-topped dome, deeply eroded to expose a core of crystalline rock.

REVIEW QUESTIONS

1. What groups of geologic structures make up the bedrock landmasses of the earth? Name and describe briefly each variety.

2. How does bedrock composition and structure exert a control upon landforms? Compare sedimentary, metamorphic, and igneous rocks in this respect.

3. Define *dip* and *strike*. To what kinds of rock structures do these terms apply?

4. How is a coastal plain formed? By what rock types is it underlain? What is a consequent stream? Where and how does it form on a new coastal plain?

5. Describe the topography of a mature coastal plain. Name the component parts as they appear along a profile line beginning on the oldland and extending to the sea. What is a subsequent stream?

6. Describe the coastal plain of the Atlantic and Gulf coast states. What is notable about the drainage pattern of the Gulf coastal plain? What kind of shoreline is associated with the Gulf coastal plain?

7. Briefly state the salient features of the coastal plain topography of southeastern England.

8. How are agriculture and transportation influenced by coastal-plain topography? Explain how artesian wells can function on a coastal plain.

9. What are some of the mineral products of coastal plains? Which of these is most important in the United States? In what structures does petroleum occur beneath coastal plains? What is a pinch out? A salt dome?

10. Describe the stages of erosional development of a region of horizontal strata. What determines the maximum relief that can occur?

11. What special features develop during erosion of horizontal layers of shale and sandstone in an arid climate? Explain the following: mesa, butte, esplanade, stripped surface (stratum plain), badlands.

12. What is a dendritic drainage pattern? On what kind of rock and structure is it developed?

13. How do the landforms of a region of horizontal strata in the various stages of the erosion cycle influence agriculture and human settlements? What are the principal economic mineral products of regions of horizontal strata?

14. How are limestone caverns formed? What are sinkholes? What is travertine? What is guano?

15. What is karst? From what region does this term originate? Describe lapiés, dolines, and poljes. Name some regions of karst topography.

16. Describe the various kinds of dome and basin structures that occur in regions underlain by sedimentary rock formations.

17. Outline the systematic changes in topography that occur as a mountainous dome is eroded through the young, maturity, and old stages. What type of drainage pattern is formed? Compare this pattern with that of a coastal plain.

18. What is a hogback? How is it formed? What are flatirons?

19. How can domes act as traps for petroleum? Give examples of domes that have become valuable oil producers.

20. Describe briefly the important structural and topographical features of the Black Hills. In what ways are these features related to the economic products of the Black Hills?

Exercises

Exercise 1. ***Cuestas of the Paris Basin.*** (Source: France; scale 1:200,000; Chalons Sheet.)

Explanatory Note: This map shows a part of two cuestas of the Paris Basin in the region east of Paris, France. The Argonne cuesta, supporting the Argonne Forest of World War I fame, runs down along the eastern side of the map. It is a deeply dissected, rugged sandstone cuesta with steep scarps on the east side. The Hills of Champagne (Monts de Champagne) comprise a lower but well-defined cuesta running through the western part of the map, with divide summits rising to slightly over 200 meters. Between the two cuestas is a lowland on weak marls and clays, the Wet Champagne (Champagne Humide) in which flows the Aisne River. West from the Hills of Champagne the cuesta backslope descends gradually to form the Dry Champagne (Champagne Pouilleuse), a chalk plain with few surface streams. For a detailed discussion of the topography, geology, and military geography of this region, see the volume by Douglas Johnson (1921) listed among the reading references.

QUESTIONS

1. **(*a*)** State the scale of this map in miles to the inch, and construct a graphic scale in miles. **(*b*)** State the contour interval of this map in feet.

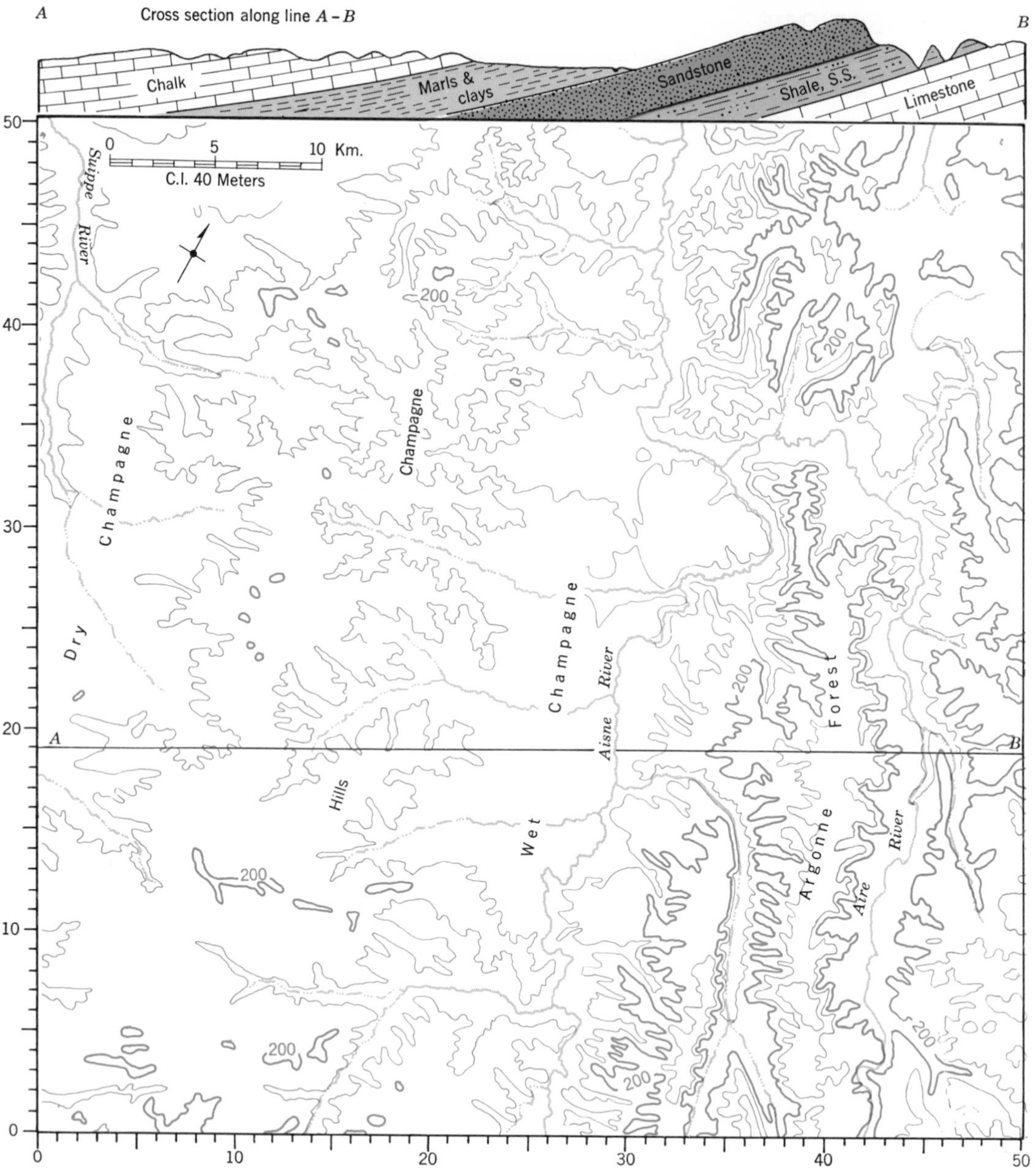

2. On a sheet of thin tracing paper laid over this map, copy off the principal streams, whether indicated on the map by a line or merely by contour indentations. Use the following color scheme: subsequent streams, blue; resequent streams, red; and obsequent streams, green. The result will be a trellis pattern typical of a mature coastal plain.

3. On a sheet of thin tracing paper laid over the map, draw the boundaries separating the zones underlain by various types of rock as shown in the structure section above. First, mark off the boundaries along the line of section *AB* on the map, then extend the lines north and south as best fits the topography. Color chalk and limestone areas blue; clay and marl, green; and sandstone, yellow.

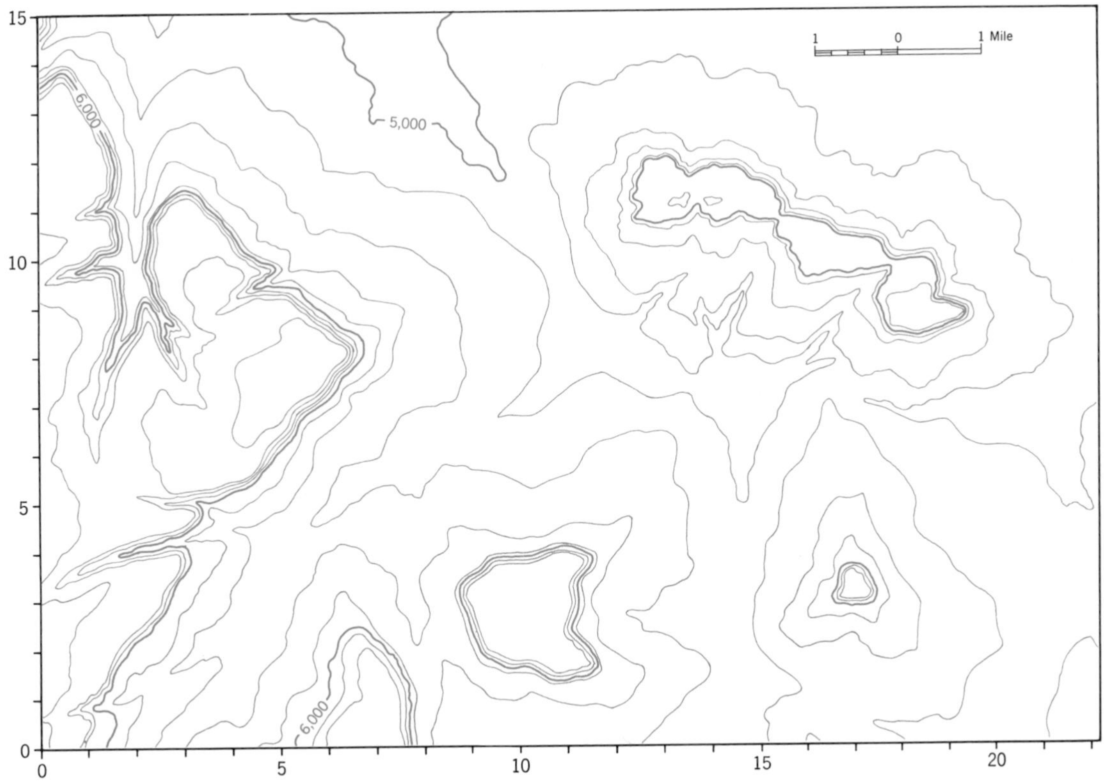

Exercise 2. *Mesas, Buttes, and Canyons.* (Source: Marsh Pass, Ariz., U.S. Geological Survey topographic map; scale 1:250,000.)

Explanatory Note: Erosion of a resistant sandstone layer underlain by a weak shale formation has produced the steep-sided flat-topped mesas, buttes, and plateaus shown on this map of a semiarid part of northern Arizona. Refer to Figure 28.16 for terminology.

QUESTIONS

1. **(*a*)** What contour interval is used on this map? **(*b*)** Locate by grid coordinates and give the elevation of the highest point on the map. **(*c*)** How high is the cliff at 4.5–5.5?

2. On a sheet of thin tracing paper laid over this map, label one good example of each of the following forms: mesa, butte, plateau, canyon, and cliff.

3. On the same sheet of paper used in Question 2, draw in a complete drainage system for this area, showing channels wherever indicated by the contours.

4. Make a topographic profile from 22.0–3.0 to 0.0–3.0, using a vertical scale of 1 in. equals 2000 ft. Then show on the profile the sandstone and shale formations in a manner similar to that in Figure 28.16. For those who wish to do so, a block diagram of the map area may be extended behind the completed profile.

5. Assuming the sandstone cap rock to be of uniform thickness wherever it now remains, have we any reason to think that the layer is not horizontal, but instead that it is slightly tilted or bent? Cite evidence bearing on this question.

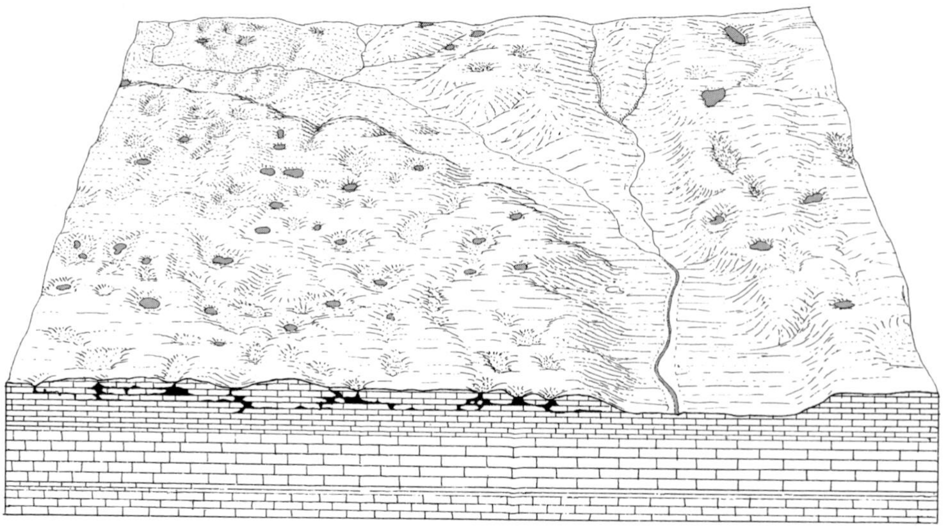

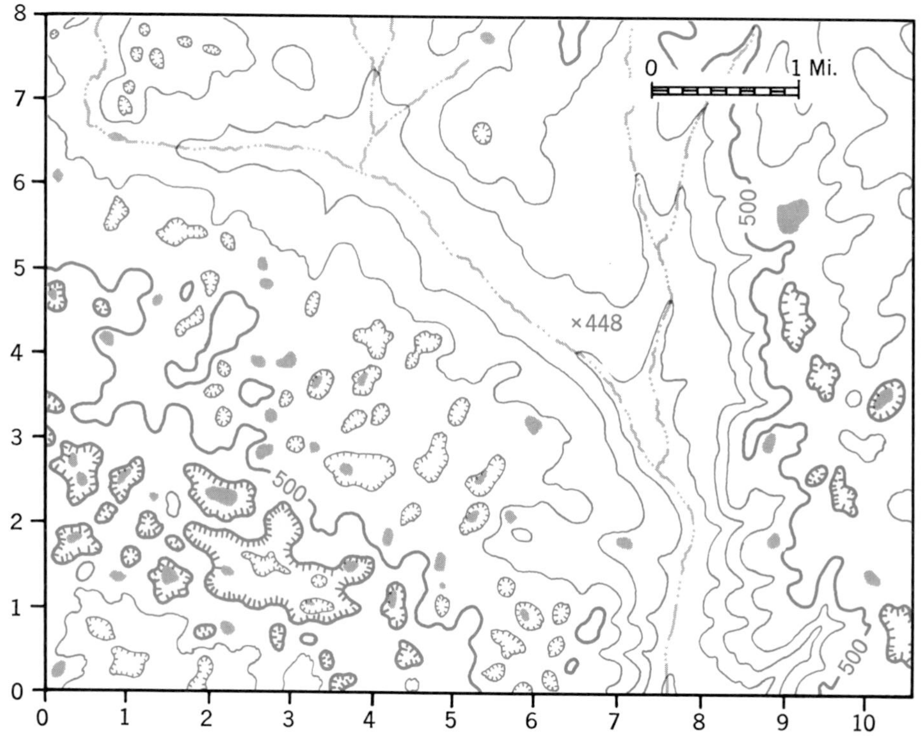

Exercise 3. ***Sinkholes.*** (Source: Princeton, Ky., U.S. Geological Survey topographic map; scale 1 : 62,500.)

Explanatory Note: Numerous small sinkholes, some containing lakes (solid black), are represented by the closed hachured contours. This topography indicates a limestone formation beneath the surface.

QUESTIONS

1. What contour interval has been used on this map? (Refer to the numbered heavy contour and to the spot height located at 6.5–4.4.)

2. Estimate the depth of each of the sinkholes whose location is given by the grid coordinates listed below. By "depth" is meant the difference in elevation between the lowest outlet point on the rim of the depression and the deepest point on the bottom of the depression. **(*a*)** 2.7–1.5. **(*b*)** 2.2–2.3. **(*b*)** 9.1–4.5.

3. Draw an east-west topographic profile across the map from 10.6–1.6 to 0.0–1.6. Use a vertical scale of 100 ft equals 1 in.

4. On the broad divide located at 6.5–5.5, draw in contours to show a sinkhole whose depth is more than 40 but less than 80 ft. Label the contours with their correct elevations.

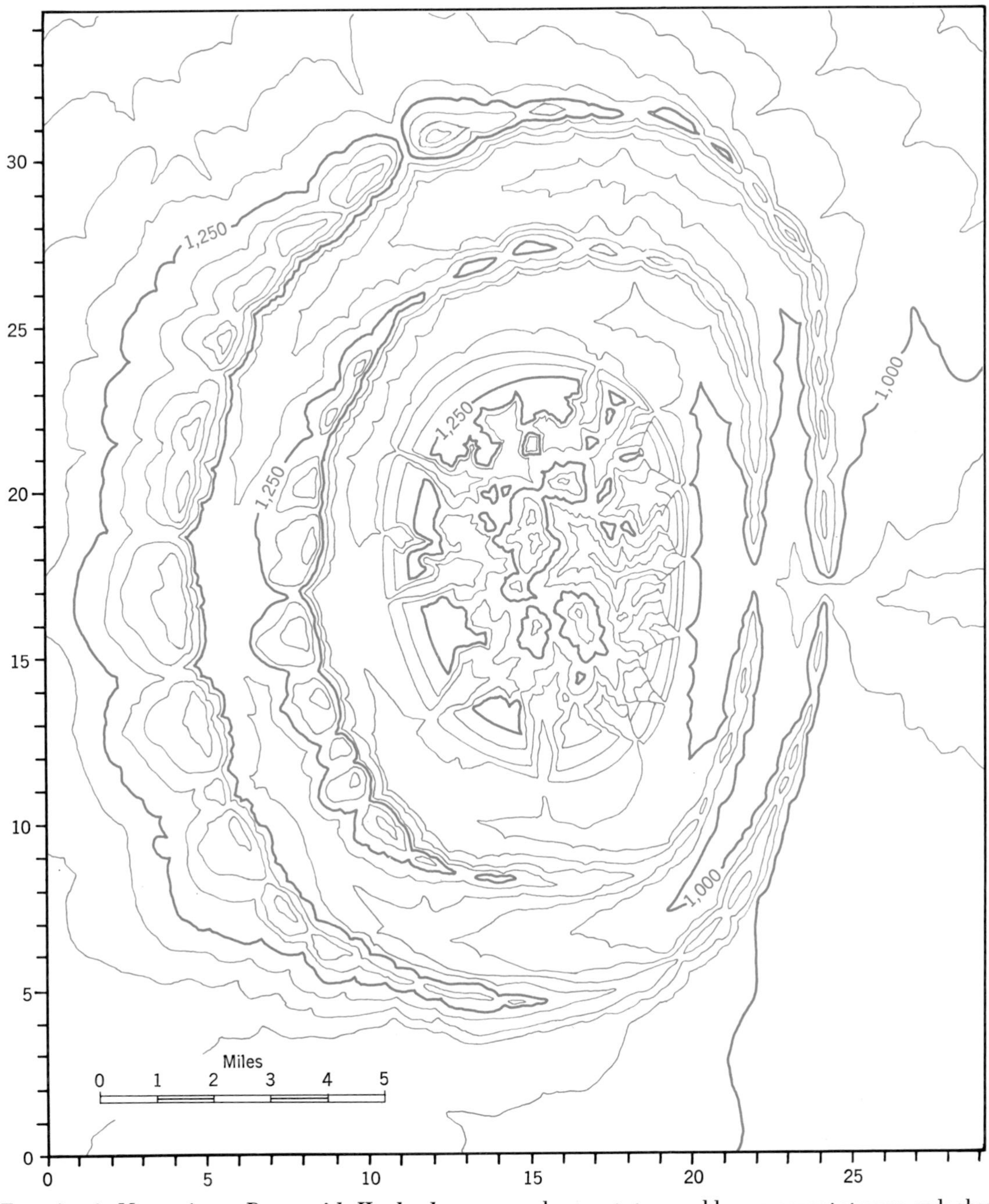

Exercise 4. *Mountainous Dome with Hogbacks.*

Explanatory Note: This is a synthetic map, not representing any real area, and should be regarded as an idealized diagram for illustrating landforms typical of a maturely dissected dome, such as those shown in Figures 28.28 and 28.33.

QUESTIONS

1. On a sheet of thin tracing paper laid over this map, label good examples of the following: hogback, flatiron, subsequent valley, central crystalline area, and watergap.

2. On the same sheet used in Question 1, draw in a complete drainage system, showing all streams indicated by the contours. Use the following color scheme: subsequent streams, blue; resequent streams, red; obsequent streams, green; and insequent streams, black.

3. Make a topographic profile from 29–19 to 0–19, using a vertical scale of 1 in. equals 1000 ft. Then draw in sandstone and shale layers to fit the topography, using Figures 28.28 and 28.33 as a guide to your interpretation. Indicate crystalline rocks in the core of the dome.

4. Assuming the sandstone formations that make the two hogback ridges to be of uniform thickness where present over this region, why are the hogback ridges broad and cuestalike on the west side? Does your answer also explain why the major streams drain out through gaps in the east side of the dome?

CHAPTER **29**

Folds, Faults, and Fault Blocks

REGIONS of sedimentary strata that have been compressed into sets of parallel, wavelike folds pass through a series of stages illustrated in the diagrams of Figure 29.1. Short consequent streams drain the flanks of the folds to join major consequent streams that follow the axes of the troughs.

In geological terminology, a downfold is known as a *syncline;* an upfold, an *anticline* (Block *A*). It may help in remembering these terms to know that the root word *clino* means "lean," as, for example, in the word *incline.* The prefix *syn* means "together." Hence, a syncline is a structure in which strata dip toward the center line of the trough. *Anti,* meaning against or opposite, implies that in an anticline the layers dip away from the center line. In the initial stages of the development of folds, anticlines are identical with the mountains, or ridges; synclines, with the valleys.

Block *A* (Figure 29.1) shows erosion of anticlines occurring during the last stages of folding. Synclines are being filled with alluvial fan materials swept down from the adjacent anticlines. After folding has ceased the upper layers of soft, unconsolidated rock are removed until, as shown in Block *B*, a hard, well-cemented sandstone layer is exposed, reflecting the full amplitude of the folding. Coinciding with synclines are *synclinal valleys;* with anticlines, *anticlinal mountains.*

Streams that drain the flanks of the anticlines quickly cut deep ravines, exposing the underlying layers. The breaching spreads rapidly to the crest of the anticline, where a long, narrow valley is opened out along the summit (Block *B*). This valley is occupied by a subsequent stream excavating a belt of weak rock and is known as an *anticlinal valley*, because it lies upon the center line of the anticline. As this valley grows in length, depth, and breadth it replaces the original anticlinal mountain. A reversal of topography thus occurs. The synclinal valley, which originally contained the major stream, is now shrunken between the growing anticlinal valleys on either side. Moreover, the anticlinal valleys are the more rapidly deepened because of the core of weak rock exposed to attack, so that eventually the syncline becomes a mountain ridge, termed a *synclinal mountain* (Block *C*). This change of landform might well bring to mind the words of the prophet, Isaiah, "Every valley shall be exalted, and every mountain and hill shall be made low." At this stage, which is that of maturity of the folds, the original topography has been completely reversed. The drainage pattern is a trellis type, similar in most respects to the trellis pattern of a mature coastal plain, but different in that the major subsequent streams are more closely spaced and their tributaries are shorter (Figure 29.2).

Here and there in a belt of folds, a principal stream crosses several folds at nearly right angles, passing through the sharply defined ridges by narrow *watergaps.* These streams are likely to have existed previously to the folding and to have maintained themselves as the folds were formed. The term *antecedent* has been applied to such streams. Some are illustrated in Block *B* of Figure 23.12, crossing the series of folds produced in the Appalachian revolution of Permian time.

As the dissection of the folded strata progresses, there is a continuous change in the form and position of the various types of ridges and valleys. Following the reversal of topography shown in Block *C,* Figure 29.1, the synclinal ridges will be completely removed by erosion. Meanwhile, new ridges are appearing in the centers of the anticlinal valleys. These form as a result of the uncovering of still older resistant strata which were folded along with the rest, but which previously lay below the general level of the land surface. The new ridges, which may be thought of as second-generation *anticlinal ridges*, grow in height as the weak rock is stripped from both sides. In time the anticlinal ridges, like the original anticlines of the initial land surface, are breached by streams and finally are transformed into anticlinal valleys. Thus an inversion of topography is again accomplished. The hard sandstone layers stand as narrow, sharp ridges separated by long, parallel valleys. Where the strata in a ridge dip in one direction only, representing one flank of an anticline or syncline, the ridge is termed a *homoclinal ridge* (Block *C*). Likewise, a valley of weak shales or limestones in which the layers all dip in one direction is termed a *homoclinal valley*.

In summary, mature topography developed on alternately resistant and weak sedimentary strata may have three types of ridges, or mountains: anticlinal, synclinal, and homoclinal; and three types of valleys: anticlinal, synclinal, and homoclinal.

Ultimately the belt of folds is reduced to a peneplain (Figure 29.1, Block *D*). Even here, the ridges rise as rows of low hills and the streams maintain their trellis drainage pattern.

Zigzag ridges and plunging folds

The folds illustrated in Figure 29.1 are continuous and even-crested, hence they produce ridges that are approximately parallel in trend and continue for great distances. In some fold regions, however, the folds are not continuous and level-crested but instead have crests that rise or descend from place to place. When maturely dissected, such folds give rise to a zigzag line of ridges (Figure 29.3).

The topographic form of a syncline whose trough descends, or *plunges*, differs from that of an anticline which plunges in the same direction, when both are maturely dissected. Figure 29.3 compares the two forms. The plunging syncline is represented by a ridge with a slightly concave summit but steeply descending cliffs on the end

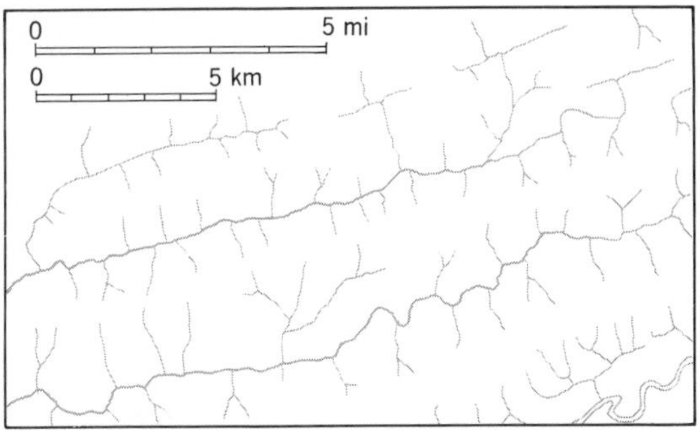

Figure 29.2 A trellis drainage pattern on folds.

and sides. Along the direction of plunge of the fold center line, or *axis*, this ridge develops an increasing concavity, then separates into two diverging homoclinal ridges. The plunging anticline is represented by a ridge that points in a direction opposite to the plunging synclinal mountain. The end is smoothly rounded and descends gradually to the level of the valley in the direction of plunge. In the opposite direction the mountain splits into

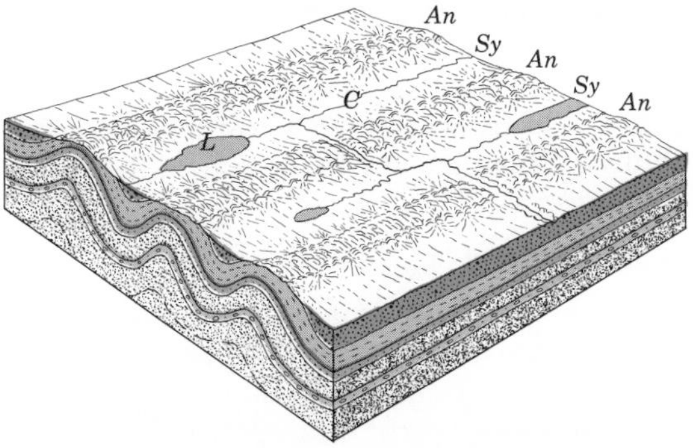

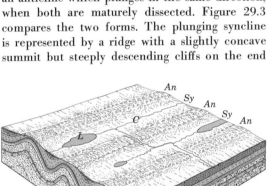

***A.* While folding is still in progress, erosion cuts down the anticlines; alluvium fills the synclines, keeping relief low. *An* = anticline; *Sy* = syncline; *C* = consequent stream; *L* = lake.**

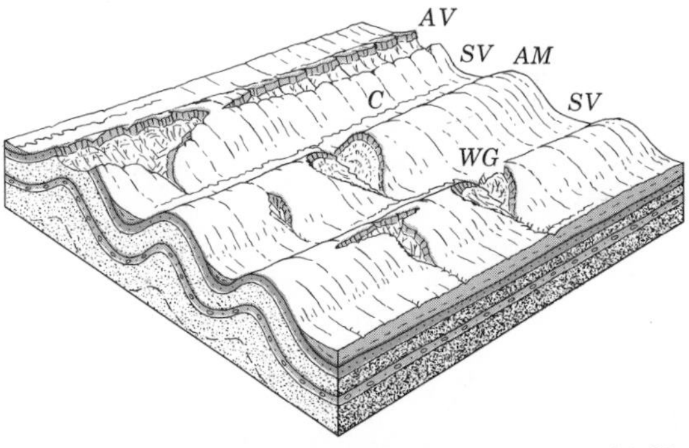

***B.* Long after folding has ceased, erosion exposes a highly resistant layer of sandstone or quartzite. *AV* = anticlinal valley; *SV* = synclinal valley; *C* = consequent stream; *WG* = water-gap.**

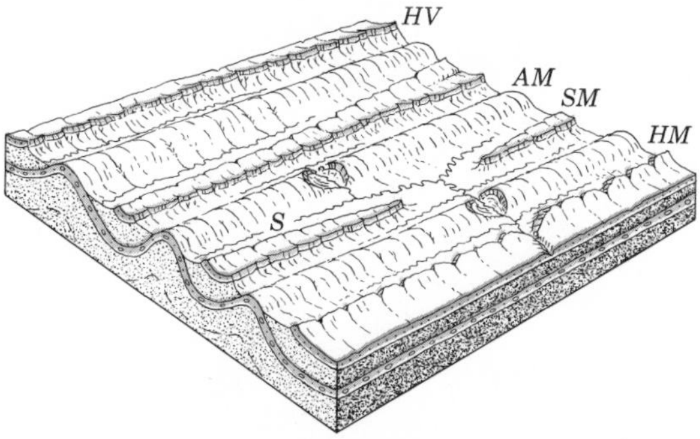

***C.* Continued erosion partly removes the resistant formation but reveals another below it. *AM* = anticlinal mountain; *SM* = synclinal mountain; *HM* = homoclinal ridge; *HV* = homoclinal valley: *S* = subsequent stream.**

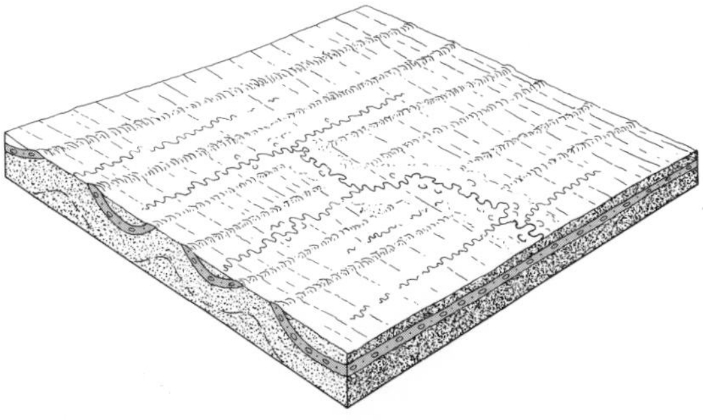

***D.* Peneplanation reduces the fold belt to low relief, but the hard-rock ridges still show.**

Figure 29.1 Stages in the erosional development of folded strata.

two homoclinal ridges. Enclosed is a valley bounded by steep cliffs (Figure 29.3). The end of this valley, where the cliff line swings around, is often termed an *anticlinal cove.* In comparing the two forms it is apparent that the cliff slope faces outward on a plunging syncline, but faces inward in a plunging anticline.

Environmental and resource aspects of fold regions

Some of the environmental and economic aspects of maturely dissected fold regions are illustrated by the Appalachians of south-central and eastern Pennsylvania (Figure 29.4). The ridges, of resistant sandstones and conglomerates, rise boldly to heights of 500 to 2000 ft (150 to 600 m) above broad lowlands underlain by weak shales and limestones. Major highways run in the valleys, crossing from one valley to another through the watergaps of streams that have cut through the ridges. Important cities may be situated near the watergaps of major streams. An example is Harrisburg, where the Susquehanna River issues from a series of watergaps cut in Blue Mountain, Second Mountain, and Peters Mountain (Figure 29.4). Where no watergaps are conveniently located, the roads must climb in long, steep grades over the ridge crests. The ridges are heavily forested; the valleys are rich agricultural belts. Fire towers are situated upon the ridge crests and command splendid views of distant ridges.

In various fold regions of the world, for example, in the Pennsylvania Appalachians, an important resource is *anthracite,* or hard coal (Figure 29.5). This occurs in strata that have been folded and squeezed. Pressure has converted the coal from bituminous into anthracite. Because of extensive erosion, all coal has been removed except that which lies in the central parts of synclines. The coal seams dip steeply; workings penetrate deeply to reach the coal that lies in the bottoms of the synclines. Seams near the surface are worked by strip mining.

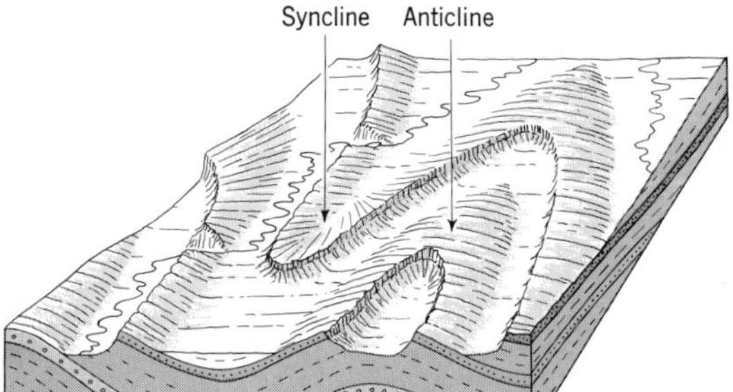

Figure 29.3 Folds with crests that plunge downward give zigzag ridges when maturely eroded. (After E. Raisz.)

Broad, gentle anticlinal folds may form important traps for the accumulation of petroleum. The principle is the same as in low domes of sedimentary strata (Figure 28.32). Oil migrates in permeable sandstone beds to the anticlinal crest, where it is trapped by an impervious cap rock of shale. Many of the oil and gas pools of western Pennsylvania, where petroleum production first succeeded, are on low anticlines.

Pressures which created folds also changed shales into slates of considerable commercial value. The slate quarries of Pennsylvania occur in rows paralleling the base of a great sandstone ridge. Limestone, if of satisfactory quality for the manufacture of portland cement, is quarried along narrow belts where a steeply dipping bed appears at the surface.

Not all fold regions contain as extensive agricultural valleys as the Pennsylvania Appalachians. In parts of Maryland, West Virginia, and Virginia the ridges are predominant and form a great rugged mountain belt difficult to cross and very thinly populated.

Other fold regions somewhat like the Appalachians are the Ouachita Mountains of Arkansas and the Jura Mountains of the Swiss and French Alps region. The Jura Mountains consist almost entirely of anticlinal limestone ridges. Good illustrations of fold belts likewise occur in North Africa, principally in Tunisia and Algeria, and in the Union of South Africa, not far north of Capetown.

Faults and fault blocks

A *fault* is a break in the brittle surficial rocks of the earth's crust as a result of unequal stresses. Faulting is accompanied by a slippage or displacement along the plane of breakage. Faults are often of great horizontal extent, so that the *fault line* can be traced along the ground for many miles,

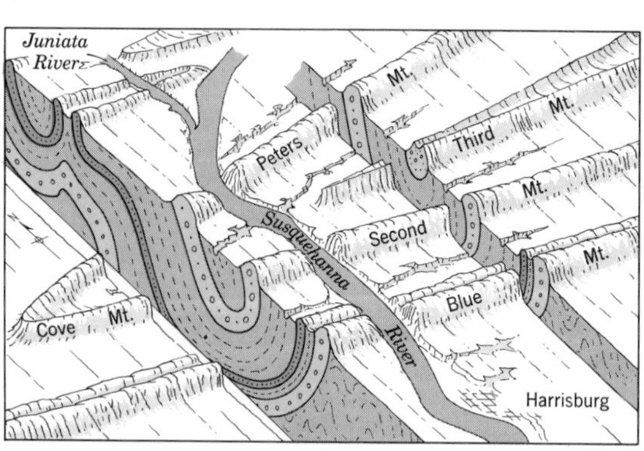

Figure 29.4 A great synclinal fold involving three resistant quartzite-conglomerate formations and thick intervening shales has been eroded to form bold ridges through which the Susquehanna River has cut a series of watergaps. (After A. K. Lobeck.)

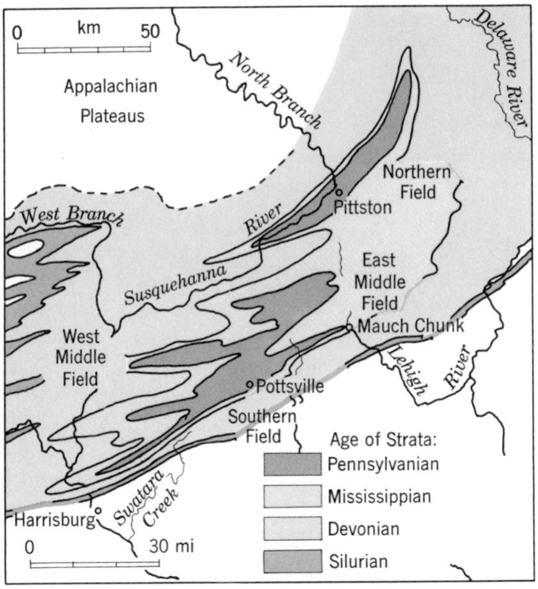

Figure 29.5 Anthracite coal basins of central Pennsylvania correspond with areas of Pennsylvania strata, downfolded into long synclinal troughs.

sometimes even 100 mi (160 km) or more. Little is known of what happens to faults at depth, but in all probability most extend down for at least several thousands of feet.

Faulting occurs in sudden slippage movements which generate earthquakes, the wavelike ground tremors that start in the zone of maximum movement. A particular fault movement may result in a slippage of as little as an inch (2.5 cm) or as much as 25 or 50 ft (8 or 15 m). Successive movements may occur many years apart, even many tens or hundreds of years apart, but aggregate total displacements of hundreds or thousands of feet. In some places clearly recognizable sedimentary rock layers are offset on opposite sides of a fault and the amount of displacement can be accurately measured.

According to the nature and relative direction of the displacement, several types of faults can be recognized (Figure 29.6). A *normal fault* has a steep or nearly vertical *fault plane.* Movement is predominantly in a vertical direction, so that one side is raised or *upthrown* relative to the other, which is *downthrown.* A normal fault results in a steep, straight *fault scarp,* whose height is an approximate measure of the vertical element of displacement (Figure 29.7). Fault scarps range in height from a few feet to a few thousand feet. Their length is measurable in miles; often they attain lengths of 100 to 200 mi (160 to 320 km). Normal faulting is an expression of tension in the earth's outer crust. It is an evident geometric observation that sliding upon an inclined surface of the type indicated in Figure 29.6 *A* must result in a spreading apart of points situated on opposite sides of the fault.

In a *reverse fault* the inclination of the fault plane is such that one side rides up over the other and a crustal shortening occurs (Figure 29.6*B*). Reverse faults produce fault scarps similar to those of normal faults, but the possibility of landsliding is greater because an overhanging scarp tends to be formed.

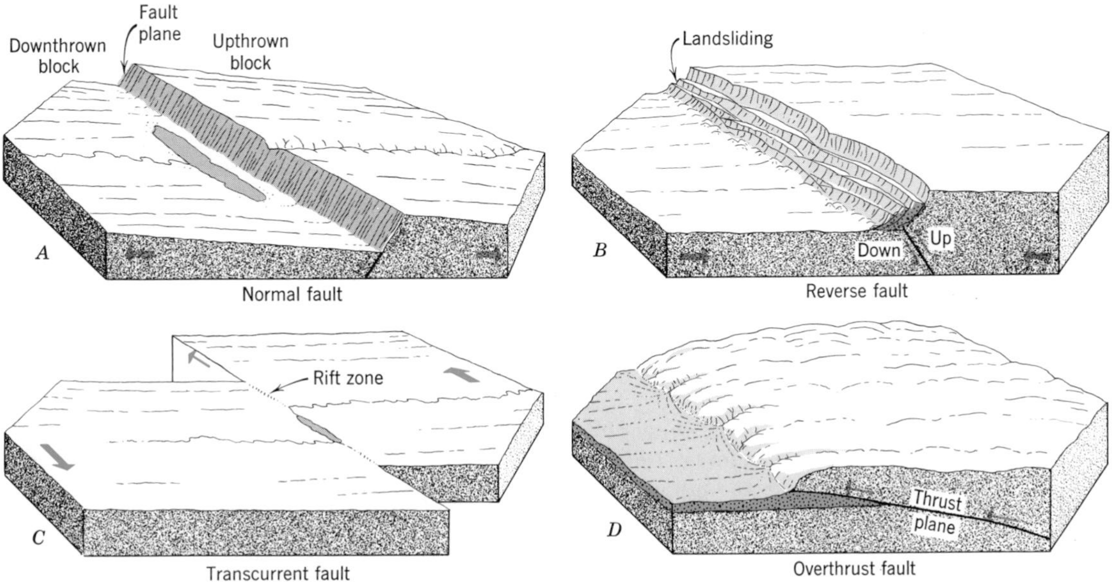

Figure 29.6 Four types of faults and their topographic expression.

Figure 29.7 Formation of this fresh fault scarp in alluvial materials accompanied the Hebgen Lake earthquake of August 17, 1959, in Gallatin County, Montana. Displacement was about 19 ft (6 m) at the maximum point. Vehicle stands on the upthrown side of the fault. (Photograph by J. G. Stacy, U.S. Geological Survey.)

A *transcurrent fault* is unique in that the movement is predominantly in a horizontal direction (Figure 29.6*C*). Hence no scarp results, or a very low one at most. Instead, only a thin line is traceable across the surface. Streams sometimes turn and follow the fault line for a short distance. Sometimes a narrow trench, or *rift*, marks the fault line (Figure 29.15).

A *low-angle overthrust fault* (Figure 29.6*D*) likewise involves predominantly horizontal movement, but the fault plane is in a horizontal position and one slice of rock rides up over the adjacent ground surface. A thrust slice may be a few hundred or thousand feet thick but up to 25 or 50 mi (40 to 80 km) wide. Overthrusting of this type is generally associated with strong crustal compression in which intense folding also occurs. The scarp produced by low-angle overthrusting is not straight or smooth, as in normal and reverse faults; instead in may be irregular in plan.

Faults rarely are isolated features. More often they occur in multiple arrangements, commonly as a parallel series of faults. This gives rise to a grain or pattern of rock structure and topography. A narrow block dropped down between two normal faults is a *graben* (Figure 29.8). A narrow block elevated between two normal faults is a *horst*. Grabens make conspicuous topographic trenches, with straight, parallel walls. Horsts make blocklike plateaus or mountains, often with a fairly flat top, but steep, straight sides.

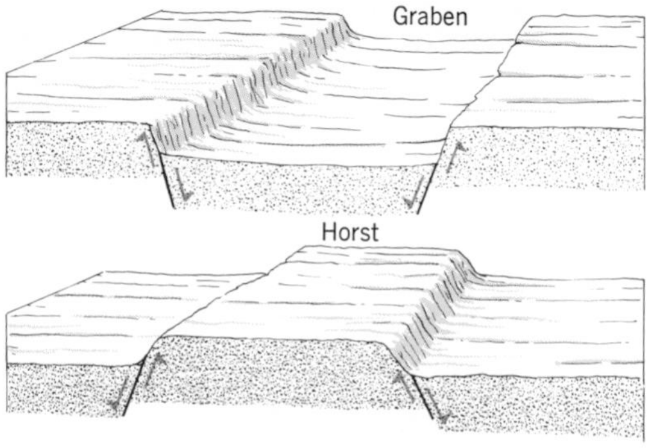

Figure 29.8 Graben and horst.

Closely related to normal faulting is *monoclinal flexing* (Figure 29.9), in which sedimentary layers are sharply bent between the upthrown and downthrown sides instead of being fractured. A *monocline* passes through a series of stages of erosion quite similar to an anticline, except that only half the anticline is represented.

Erosional development of a fault scarp

Forms attained by normal faults throughout the erosional period that follows their formation are illustrated in Figure 29.10. At the rear part of the block is the original fault scarp, produced directly by crustal movement, and therefore belonging to the group of initial landforms. Form *A* is a *fault scarp;* the scarp is still straight and smooth, with only a few stream-cut canyons and some talus cones and alluvial fans built along the scarp base.

Form *B* is a scarp resulting entirely through erosion and is designated a *fault-line scarp*. It appears because the weak shale formation has been stripped from both sides of the fault, revealing a basement of resistant igneous rocks upon which the shale was once deposited (Figure 29.11).

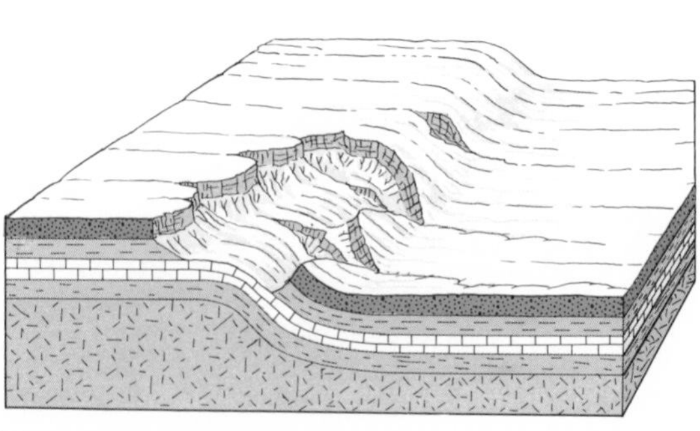

Figure 29.9 A monocline.

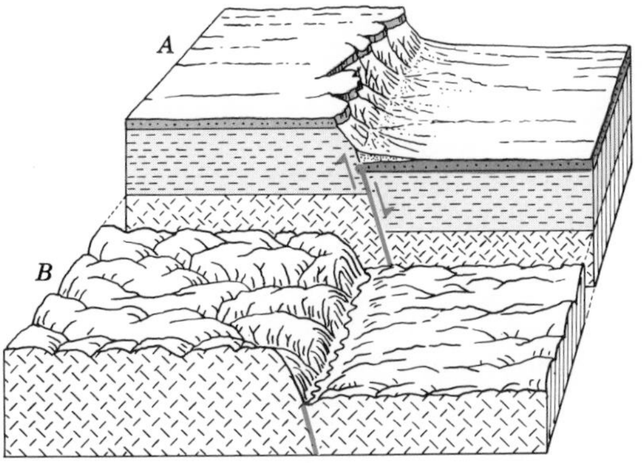

Figure 29.10 *A.* Fault scarp. *B.* Fault-line scarp. (From A. N. Strahler, 1971, *The Earth Sciences*, 2nd ed., Harper and Row, New York.)

Figure 29.11 Fault-line scarp, MacDonald Lake, near Great Slave Lake, Northwest Territories, Canada. (Canadian Forces Photograph No. 5120-105R.)

Fault-block mountains

In regions where normal faulting is on a grand scale, with displacements up to several thousands of feet, huge mountain masses are produced. In general, these faulted mountain blocks can be classed as *tilted* and *lifted* (Figure 29.12). A tilted block has one steep face, the fault scarp, and one gently sloping side. The initial divide lies near the top of the fault scarp and is hence situated over to one side of the block. A lifted block, which is a type of horst, is bound by steep slopes on both sides.

Figure 29.13 shows stages in the life history of a tilted fault block. Most numerous good examples of fault-block ranges are in the desert of the western United States. This region is known as the Basin-and-Range Province. Geomorphic processes and landforms of this province have been described in Chapter 26 (See Figures 26.12 and 26.13). It should not be supposed, however, that block mountains and arid climates necessarily go hand in hand.

In youth, the fault block is asymmetrical and has generally even sides, despite the presence of numerous small stream valleys that have been developing as the fault block was elevated.

In maturity (Figure 29.13), the range is dissected into a great number of divides, spurs, and peaks separated by deep canyons. The main crest line of the range is now pushed back to a more nearly central position and the simple blocklike aspect of the mountain range has disappeared. Along the base of the fault scarp, between canyon mouths, remain some parts of the original fault scarp. These make *triangular facets*, aligned nicely along the base line (Figure 29.13). Fans are more extensive than in early youth. The adjoining basins are filled higher with alluvium. In a humid climate, a rolling landscape of moderate relief develops in late maturity, whereas in the old stage the range is reduced to a subdued peneplain surface.

Environmental and resource aspects of faults and block mountains

Faults are of environmental and economic significance for both geologic and topographic reasons. Fault planes are usually zones along which the rock has been pulverized, or at least considerably fractured. This has the effect of permitting ore-forming chemical solutions to rise along fault planes. Many important ore deposits lie in fault planes or in rocks that faults have broken across.

Another related phenomenon is the easy rise of underground water along fault planes. Springs, both cold and hot, are commonly situated along fault lines. They occur along the bases of young fault-block mountains, as, for example, Arrowhead Springs along the base of the San Bernardino Range and Palm Springs along the foot of the San Jacinto Mountains, both in southern California.

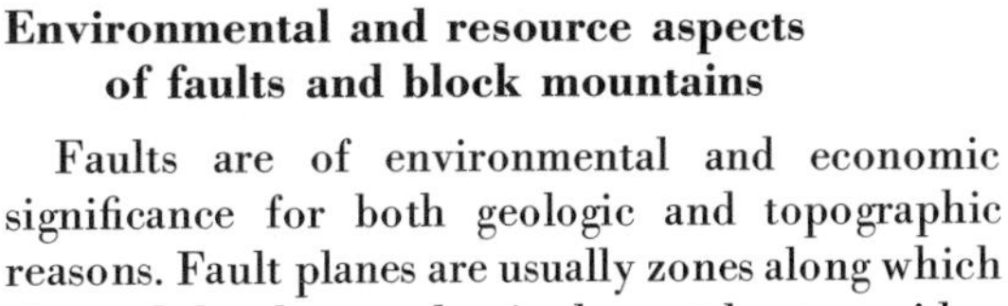

Figure 29.12 Fault block mountains may be of tilted type (left) or lifted type (right). (After W. M. Davis.)

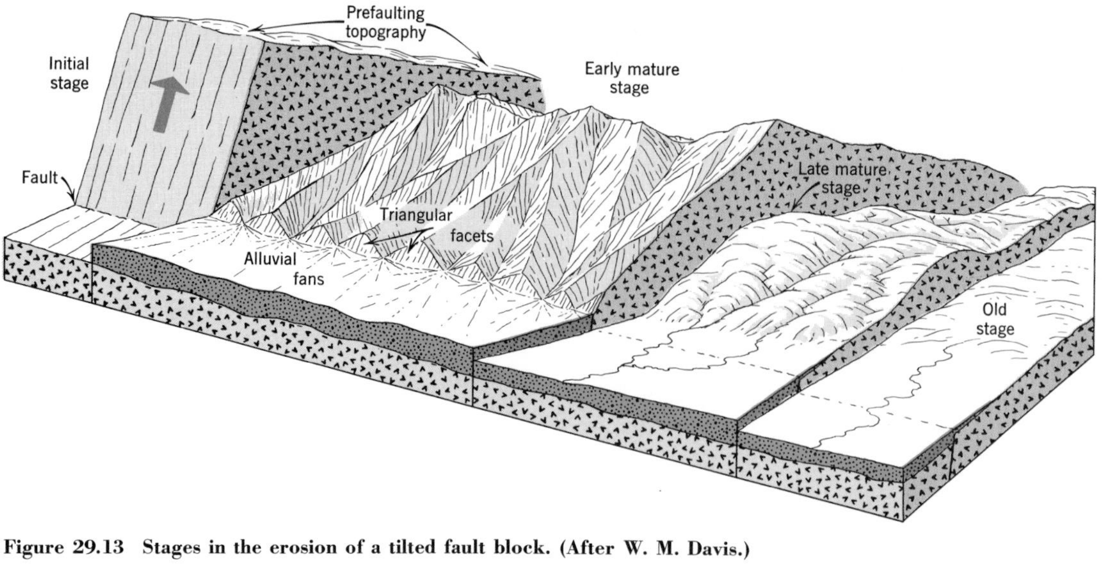

Figure 29.13 Stages in the erosion of a tilted fault block. (After W. M. Davis.)

Petroleum, too, finds its way along fault planes where the rocks have been rendered permeable by crushing, or it becomes trapped in porous beds that have been faulted against impervious shale beds (Figure 29.16). Some of the most intensive searches for oil center about areas of faulted sedimentary strata because of the great production that has been achieved from pools of this type.

Fault scarps and fault-line scarps may form imposing topographic barriers across which it is difficult to build roads and railroads. The great Hurricane Ledge of southern Utah is a feature of this type, in places a steep wall 2500 ft (760 m) high.

Grabens may be of such size as to form broad lowlands. An illustration is the Rhine graben of Western Germany. Here a belt of rich agricultural land, 20 mi (32 km) wide and 150 mi (240 km) long, lies between the Vosges and Black Forest ranges, both of which are block mountains faulted up in contrast with the downdropped Rhine graben block.

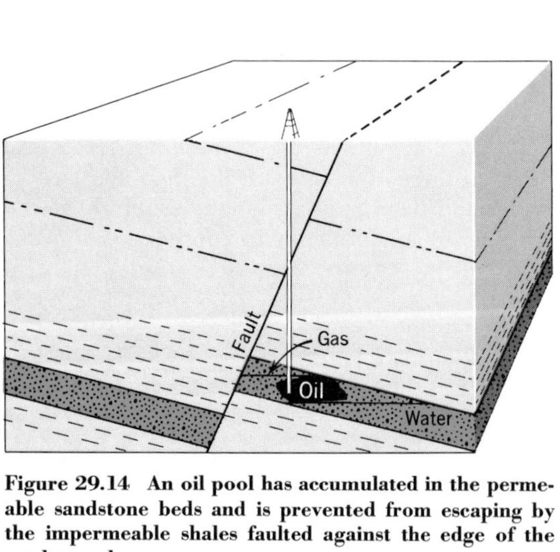

Figure 29.14 An oil pool has accumulated in the permeable sandstone beds and is prevented from escaping by the impermeable shales faulted against the edge of the sandstone layer.

Earthquake—an environmental hazard

Everyone has read many news accounts about disastrous earthquakes and has seen pictures of their destructive effects. Californians know about severe earthquakes from first-hand experience, but many other areas in North America have experienced earthquakes, and of these a few have been severe. The *earthquake* is a motion of the ground surface, ranging from a faint tremor to a wild motion capable of shaking buildings apart and causing gaping fissures to open up in the ground.

The earthquake is a form of kinetic energy of wave motion transmitted through the surface layer of the earth in widening circles from a point of sudden energy release—the *focus.* Like ripples produced when a pebble is thrown into a quiet pond, the waves travel outward in all directions, gradually losing energy through frictional resistance encountered within the rock that is flexed by the passing waves.

As already noted, earthquakes are produced by sudden movements along faults; commonly these are normal faults or transcurrent faults. Through the San Francisco Bay area passes the famed *San Andreas fault.* The devastating earthquake of 1906 resulted from slippage along this fault, which is of the transcurrent type. The fault is 600 mi (965 km) long and extends into southern California, passing about 40 mi (60 km) inland of the Los Angeles metropolitan area. In places the fault is expressed as a *rift valley* (Figure 29.15). Associated with the San Andreas fault are several impor-

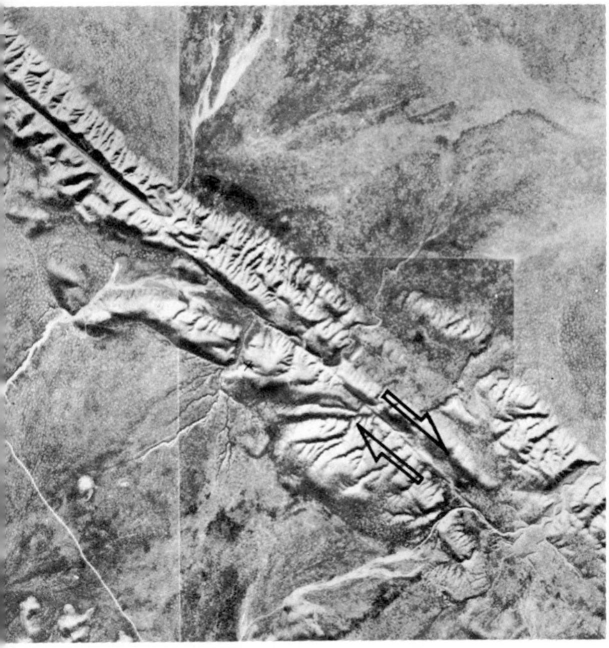

Figure 29.15 In this vertical air view, the nearly straight trace of the San Andreas fault contrasts sharply with the sinuous lines of stream channels. San Bernardino County, California. (Photograph by Litton Industries, Aero Service Division.)

Figure 29.16 This road in the Santa Cruz Mountains of California was offset by fault movement during the San Francisco earthquake of 1906. Lateral displacement was about 4 ft (1.2 m). (Sketched from a photograph by E. P. Carey.)

tant transcurrent branch faults, all capable of generating severe earthquakes.

We shall not go into the details of mechanics of faults and how they produce earthquakes. It must be enough to say that rock on both sides of the fault is slowly bent over many years as horizontal forces are applied in the movement of lithospheric plates. Energy accumulates in the bent rock, just as it does in a bent crossbow. When a critical point is reached, the strain is relieved by slippage on the fault and a large quantity of energy is instantaneously released in the form of *seismic waves.* Slow bending of the rock takes place over many decades. Its release then causes offsetting of features that formerly crossed the fault in straight lines, for example, a roadway or fence (Figure 29.16). Faults of this type can also show a slow, steady displacement known as *fault creep,* which tends to reduce the accumulation of stored strain.

The *Richter scale* of earthquake magnitudes was devised in 1935 by the distinguished seismologist, Charles F. Richter, to indicate the quantity of energy released by a single earthquake. Scale numbers range from 0 to 9, but there is no upper limit except for nature's own limit of energy release. A value of 8.6 was the largest observed between 1900 and 1950.

One of the great earthquakes of recent times was the Good Friday Earthquake of March 27, 1964, with the surface point of origin located about 75 mi (120 km) from Anchorage, Alaska. Its magnitude was 8.4 to 8.6 on the Richter scale, approaching the maximum known. Of particular interest in connection with earthquakes as environmental hazards are the *secondary effects.* At Anchorage most of the damage was from secondary effects, since buildings of wooden frame construction often experience little damage when on solid rock areas. Damage was largely from earth movements in weak clays underlying the city (Figure 29.17). These clays developed liquid properties upon being shaken (they are called *quick clays*) and allowed great segments of ground to subside and to pull apart in a succession of steps, tilting and rending houses. Other secondary effects were from the rise of water level and landward movement of large waves, destroying shipping and low-lying structures.

Another important secondary effect of a major earthquake, nicely illustrated by the Good Friday Earthquake, is the *seismic sea wave,* or *tsunami,* as it is known to the Japanese. A train of these waves is often generated in the ocean at a point near the earthquake source by a sudden movement of the sea floor. The waves travel over the ocean in ever-widening circles, but they are not perceptible at sea in deep water. However, when a wave arrives at a distant coastline, the effect is to cause a slow rise of water level over a period of 10 to 15 minutes. This rise is reinforced by a favorable configuration of the bottom offshore. Wind-driven waves, superimposed upon the heightened water level allow the surf to attack places inland that are normally above the reach of waves. For example, the particularly destructive seismic sea wave of 1933 in the Pacific Ocean caused waves to attack ground as high as 30 ft (9 m) above normal

Figure 29.17 Slumping and flowage of unconsolidated sediments, resulting in property destruction at Anchorage, Alaska, Good Friday earthquake of March 27, 1964. (U.S. Army Corps of Engineers photograph.)

tide level, causing widespread destruction and many deaths by drowning in low-lying coastal areas. It is thought that coastal flooding that occurred in Japan in 1703, with an estimated life loss of 100,000 persons, may have been caused by seismic sea waves.

Earthquakes and urban planning—the San Fernando earthquake

The toll in human lives and severe structural effects of the San Fernando earthquake of February 9, 1971 shocked the entire Los Angeles community into renewed awareness of the need for urban planning to minimize or forestall the damaging effects of a major earthquake. Although the earthquake was not in the really severe category by the Richter scale (it measured 6.6, which is moderate in severity), local areas experienced a ground motion as intense as any previously measured in an earthquake. Fortunately, the ground shaking was of brief duration; had it persisted for a longer time, the structural damage would have been much more severe than it was.

Particularly disconcerting was the collapse of the Olive View Hospital in Sylmar, a new structure supposedly conforming with earthquake-resistant standards. The Veterans Hospital in Sylmar also suffered severe damage and the collapse of several buildings (Figure 29.18). A crack produced in the Van Norman Dam caused authorities to drain that reservoir to prevent dam collapse and disastrous flooding of a densely built-up area. The Sylmar Converter Station, one of the key elements in the electrical power transmission system of the Los Angeles area, was severely damaged. Collapse of a freeway overpass blocked the highway beneath, and freeway pavements were cracked and dislocated (Figure 29.19). Fortunately, the time of the quake was 6 A.M., when most persons were at home and few were traveling the major arteries.

Yet the fault movement that set off the San Fernando earthquake was not on the great San Andreas fault, but rather from a locality some 15 mi (25 km) from that fault, along a system of relatively minor faults. This section of the San Andreas fault is believed capable of producing an earthquake of far greater intensity than the 1971 San Fernando earthquake, and although the year of this event is not predictable within decades, the progress of urbanization will have greatly expanded the structures and population subject to devastation.

Therefore, soon after the earthquake had occurred, the National Academy of Sciences and the National Academy of Engineering set up a joint panel of experts to study the earthquake effects and to draw up recommendations. The panel concluded that existing building codes do not provide adequate damage control features, and should be revised. The panel further recommended that public buildings, such as hospitals, schools, and buildings housing police and fire departments and other emergency services should be so constructed as to withstand the most severe shaking to be anticipated. Fortunately, most school buildings constructed following the Long Beach earthquake of the 1930s showed no structural damage, but many of the older school buildings were rendered unfit for use. Damage from the San Fernando earthquake of 1971 has been estimated at $500 million, but experts think that an earthquake as severe as that of 1906 at San Franciso would cause damage on the order of $20 billion if it occurred now in a large metropolitan area.

Figure 29.18 Severe structural damage and collapse of buildings of the Veterans Administration Hospital, Sylmar, Los Angeles County, caused by the San Fernando earthquake of February 1971. (Wide World Photos.)

Figure 29.19 Collapsed pavement and overpass on the Golden State Freeway at the northern end of the San Fernando Valley, California, resulting from the earthquake of February 1971. (Wide World Photos.)

REVIEW QUESTIONS

1. What is an anticline? A syncline? How are these features related in a series of folds?

2. What is an anticlinal valley? A synclinal valley? How does reversal of topography normally occur in the process of erosion of a fold region?

3. How are watergaps formed in a fold region? Of what environmental importance are watergaps? What is an antecedent stream? How might antecedent streams develop in a fold region?

4. What type of drainage pattern is developed in a maturely dissected region of folds?

5. How do homoclinal ridges and valleys differ from anticlinal and synclinal ridges and valleys? Show all these types by simple cross-sectional diagrams.

6. What effect does a downplunge of fold axes have upon the forms of ridges and valleys? How can a plunging synclinal mountain be distinguished from a plunging anticlinal mountain?

7. Discuss the environmental aspects of fold regions of rugged ridge-and-valley topography. What economic mineral products are important in fold regions?

8. What is a fault? What is a fault line?

9. Describe a normal fault, and explain how it differs from a reverse fault. What kind of topographic feature is produced by these types of faulting?

10. What is a transcurrent fault? How does it differ in topographic expression from a normal fault? What is a rift?

11. Describe a low-angle overthrust fault. With what kind of crustal deformation is it associated? What topographic expression does an overthrust fault have?

12. Distinguish between a graben and a horst. Briefly describe the Rhine graben region as an illustration of graben and horst forms.

13. How does monoclinal flexing differ from faulting? Are the two structures basically related?

14. Why is it necessary to distinguish between a fault scarp and a fault-line scarp? Which form comes first?

15. What kinds of block mountains can be formed? Compare the stages of youth and maturity of a large, tilted fault-block mountain. What are the triangular facets? How do they differ from flatirons?

16. In what ways do faults influence the occurrence of ground water and springs? How is petroleum concentrated by fault structures? Why are fault zones favorable for occurrences of ore minerals?

17. What causes an earthquake? What is the focus? What are seismic waves? What is fault creep? Describe the Richter scale.

18. Describe the San Andreas fault and its earthquake history.

19. What side effects often accompany an earthquake? Give examples. What is a seismic sea wave?

20. What lessons can be learned from the San Fernando earthquake of February 9, 1971?

Exercises

Exercise 1. ***Mountains Developed on Folded Strata.***

Explanatory Note: This is a synthetic map, not representing any real topography, and should be regarded as an idealized diagram illustrating the ridge and valley forms typical of a maturely dissected region of folded strata. Refer to Figures 29.1 and 29.3 for aid in understanding the relation between structure and topography on this map.

QUESTIONS

1. On a sheet of thin tracing paper laid over this map, label each ridge and valley with the names *anticlinal ridge, synclinal ridge, homoclinal ridge, anticlinal valley, synclinal valley*, or *homoclinal valley*. To help you do this, study Figures 29.1 and 29.3, comparing the forms shown on the map with the labeled forms on the block diagrams.

2. On the same sheet of paper used in Question 1, draw in a complete drainage system using the following color scheme: subsequent streams, blue; resequent streams, red; and obsequent streams, green.

3. Make a topographic profile from the upper left-hand corner of the map to the lower right-hand corner. Use a vertical scale of 1 in. equals 2500 ft. Draw in sandstone and shale formations in such a way as to fit the interpretation of topography in your answer to Question 1. Be guided by the cross sections in Figures 29.1 and 29.3.

4. Explain the fact that the deep gap in the ridge at 2.5–6.5 now contains no through-flowing stream, although it resembles the watergap at 11.0–2.0. What drainage change seems to have occurred here?

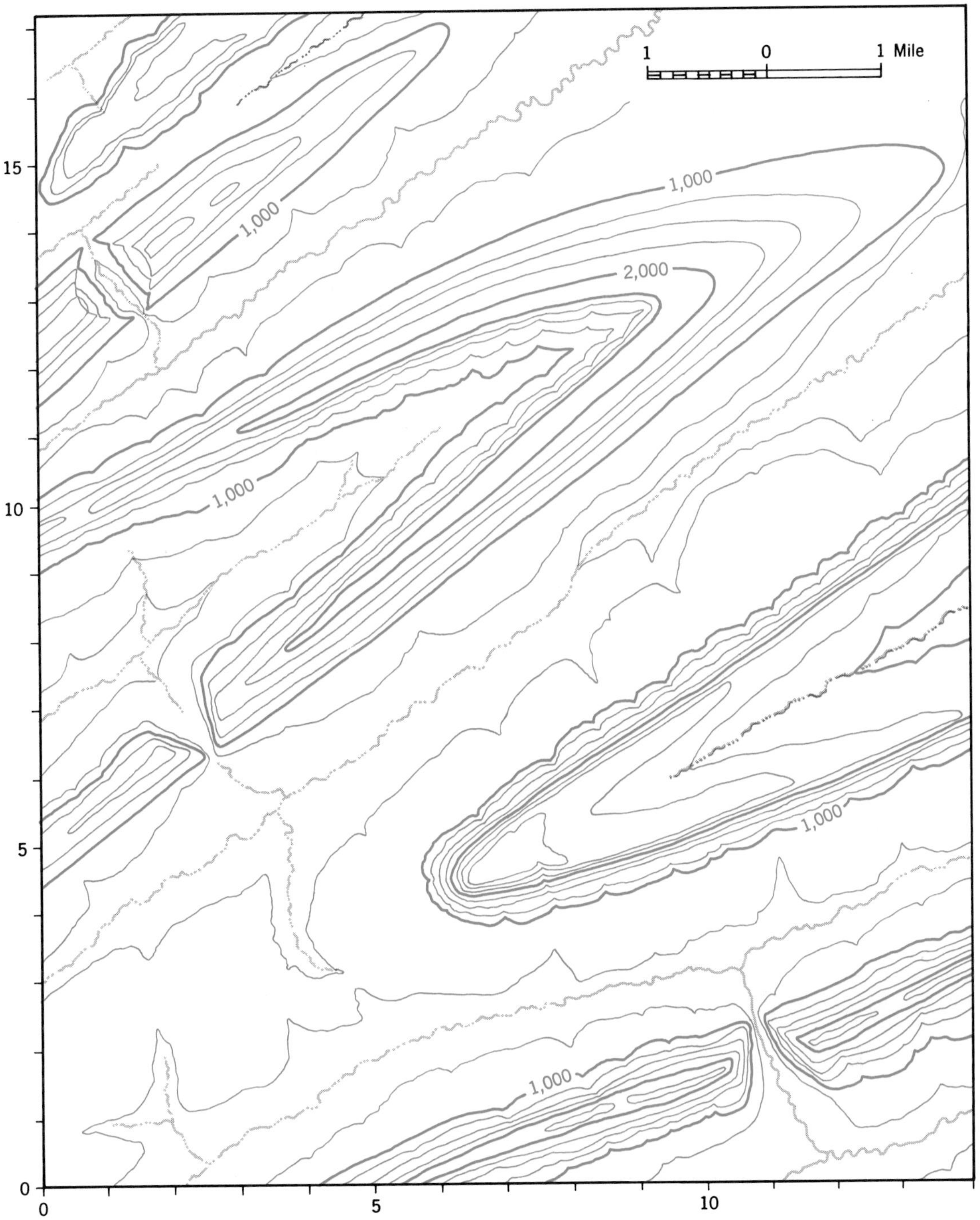

Exercise 1

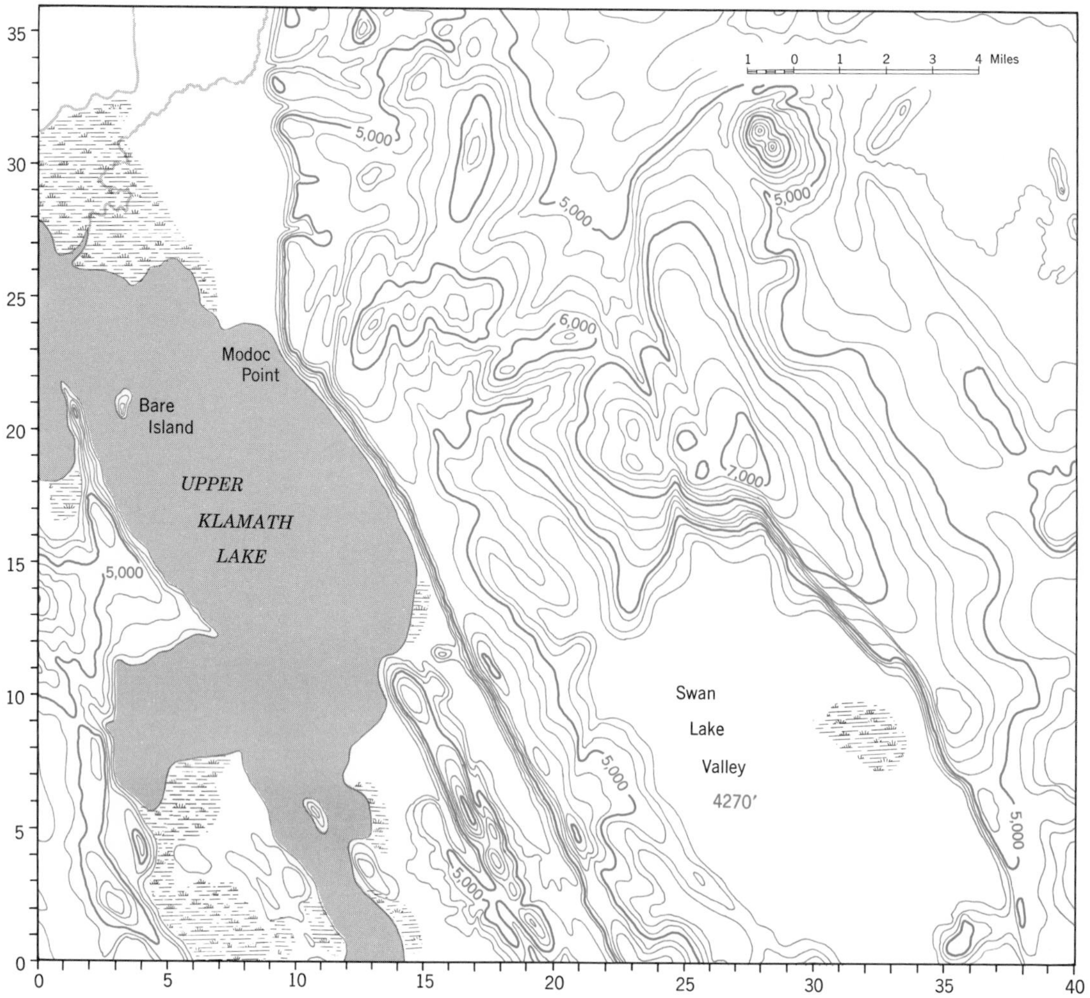

Exercise 2. ***Fault scarps and graben.*** (Source: Klamath, Oregon, U.S. Geological Survey topographic map; scale 1:250,000.)

Explanatory Note: The steep, simple scarps trending northwest to southeast on this map are fault scarps. Upper Klamath Lake occupies a graben and is bounded by fault scarps on both sides. Swan Lake Valley is a downtilted block with a fault scarp on the east side only.

QUESTIONS

1. On a sheet of thin tracing paper laid over this map, mark all fault lines which you can safely interpret from the topographic forms. Write the letter *D* and *U* on opposite sides of each fault line to indicate which side went down, which went up.

2. Make a topographic profile across the map from 0.0–8.0 to 40.0–8.0, using a vertical scale of 1 in. equals 2000 ft. Then indicate the positions of the faults as in Figure AI.24.

3. Why is the floor of Swan Lake Valley so flat?

CHAPTER 30

Crystalline Masses and Volcanic Forms

THE term *crystalline rock* is useful in referring collectively to intrusive igneous rock and the metamorphic rocks, such as schists and gneisses. It is of value in understanding regional landform assemblages to recognize the range of landmass types possible under the general class of crystalline rocks, even though distinct subgroups cannot always be separated.

Homogeneous crystallines

Intrusive igneous rocks, such as granites, generally occur in enormous batholiths (Figure 30.1). One batholith in Idaho is exposed over an area of 16,000 sq mi (40,000 sq km), a region almost as large as New Hampshire and Vermont combined. The rock seems to extend down many thousands of feet, and for practical purposes may be considered bottomless. A smaller body of igneous rock less than 40 sq mi (100 sq km) in surface extent is termed a *stock*.

Batholiths and stocks do not reach the surface of the earth when formed, hence produce no initial landforms and have no initial stage in the erosional development cycle. They appear only after prolonged erosion has stripped away the older, overlying rock when the landmass is in a mature or old stage of denudation (Figure 30.2). An additional illustration of the process of exposure of deep-seated rocks was given in the discussion of the manner in which the central core of a dome becomes exposed (Chapter 28).

Landforms developed on batholiths vary somewhat according to texture and composition of the rock and whether or not the mass has been faulted. Where the rock is quite uniform and free of strong faults, it is eroded into a maze of canyons and ravines which follow no predominant trend (Figure 30.3). The drainage pattern is dendritic and resembles that developed on horizontal sedimentary strata. In fact, the two patterns may be indistinguishable.

Where faulting has occurred, making a series of intersecting zones of crushed and weakened rock, the drainage follows the fault lines and may form a linear or a rectangular pattern (Figure 30.4). The streams are of subsequent type because they developed in the zones of weakness.

Certain areas of metamorphic rocks, such as gneisses and schists, also develop a dendritic drainage pattern of insequent streams because the variations in rock texture and composition seem to have little influence upon valley development. The topography of such areas may be identical with that of batholiths. Therefore it is convenient to use the term *homogeneous crystallines* to include both the igneous intrusive and metamorphic rock masses.

Belted metamorphics

Regions of metamorphic rock normally show a strong grain in the topography. Ridges tend to be elongate in one direction and to be separated by long, roughly parallel valleys (see Figure 28.3). Neither ridges nor valleys have the sharpness of folded sedimentary strata, but the drainage pattern is clearly of trellis or rectangular form. Regions

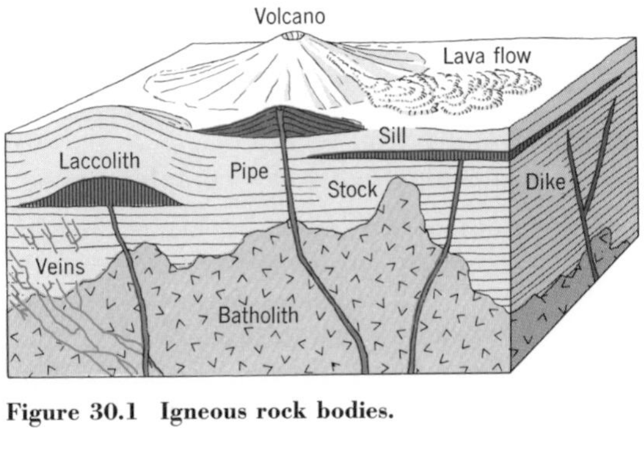

Figure 30.1 Igneous rock bodies.

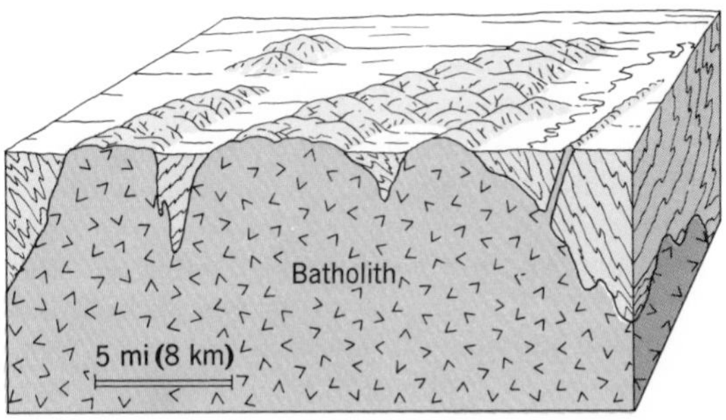

Figure 30.2 Deep-seated igneous rocks appear at the surface only after long-continued erosion has removed thousands of feet of overlying rocks. (After Longwell, Knopf, and Flint.)

Figure 30.3 This dendritic drainage pattern is developed on the maturely dissected Idaho batholith.

of this type may be classed as *belted metamorphics* because the topography is a reflection of different rates of denudation of parallel belts of metamorphic rocks, such as schist, slate, quartzite, and marble. Marble tends to form distinctive valleys; slate and schist make belts of medium to strong relief; quartzite usually stands out boldly and may produce conspicuous narrow hogback ridges. Furthermore, most metamorphic rocks have been broken by reverse and overthrust faults which run parallel with the different belts and often separate one rock type from another. Subsequent stream valleys occupying these fault lines help bring out the grain of the topography. Much of New England, particularly the Taconic and Green Mountains, illustrates these principles well. The larger valleys trend north and south and are underlain by marble. These are flanked by ridges of gneiss, schist, slate, or quartzite. The highlands of the Hudson and of northern New Jersey continue this belted pattern southward where it joins the Blue Ridge. Near Harpers Ferry, Maryland, quartzite ridges rise prominently above broad valley belts of schist.

Areas of complex structure

Some parts of the earth's crust, particularly the continental shields, have undergone several periods

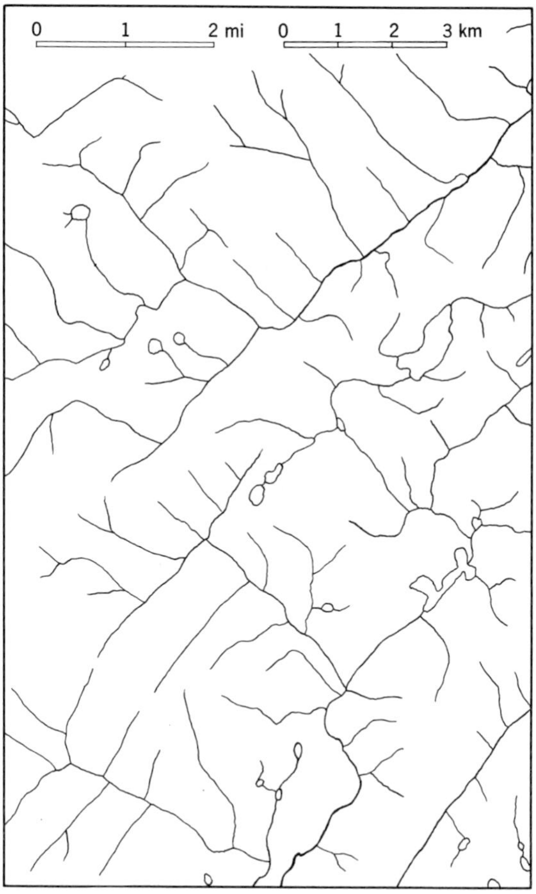

Figure 30.4 A rectangular drainage pattern, developed on faulted crystalline rocks of the Adirondack Mountains, New York.

of folding, faulting, intrusion, and volcanism. As each event occurred, new rocks or new structures were added to the mass, so that it may appear today as a region of *complex structure* (Figure 30.5). Because of the variety of rock types and structures, the landforms show a variety of forms.

Figure 30.6 diagram is a schematic block of a geologically complex region following long continued denudation. A highly irregular drainage pattern has developed. The rugged topography consists of fault and fault-line scarps, hogbacks, old volcanoes, and other landform types. Intrusive and sedimentary rocks of various ages and shapes are exposed.

Natural resources of batholiths, metamorphic belts, and complex areas

Regions of intrusive igneous rock, metamorphic rock, and complex structure are often rich in mineral wealth. Where igneous intrusion occurred repeatedly, metallic ores were deposited. Examples from the Rocky Mountains are the copper, silver, gold, and lead of Butte, Montana; the silver and

Figure 30.5 Complexly folded and faulted strata of Precambrian age, northern Rockies of Glacier National Park, Montana. (Photograph by Chapman, U.S. Geological Survey.)

the lead, zinc, and silver of Leadville, Colorado.

Metamorphic rocks, such as slates, quartzites, marbles, and schists, are not likely to contain metallic ores of importance, unless intrusive rocks have penetrated them. They do, however, have economic value for the slate or marble which can be quarried from them. Vermont has these rocks, along with intrusive granites which are quarried as well.

Among the foreign examples of valuable metallic mineral deposits occurring in igneous, metamorphic, and complex rocks might be cited the tin deposits of the Katanga district, Republic of the Congo.

Much of the batholithic, metamorphic, and complex areas of the world are mountainous, hence, heavily forested and thinly populated. Examples are the Salmon River Mountains, underlain by the great Idaho batholith; the Great Smoky Mountains of North Carolina, Georgia, and Tennessee underlain by intrusive and metamorphic rocks in generally complex arrangement; or the Taconic and Green mountain ridges of belted metamorphics in Vermont and Massachusetts. Lumber is thus an important resource of these regions, in addition to their mineral wealth. At the same time, transportation is often difficult, and in the absence of agricultural land the population is thinly distributed if not actually absent from large areas.

Other extensive regions of intrusive or metamorphic rocks are of low relief—the continental shields—having been reduced to peneplains and only slightly dissected in the present erosion cycle. Much of eastern Canada is of this type of topography, as are parts of Sweden and Finland. Not only are they similar in rock conditions, but these regions because of glaciation have countless lakes, and because of similar climate support a needleleaf evergreen forest.

In the United States, the outstanding example of a peneplain on intrusive and metamorphic rocks is the Piedmont Upland of Virginia, the Carolinas, and Georgia. A rolling landscape of monotonously uniform hill-top level extends in a vast belt between the Blue Ridge Mountains on the west and the Coastal Plain on the east. Above this surface rise a few monadnocks.

Volcanoes and associated landforms

Volcanoes are built by the eruption of molten rock and heated gases under pressure from a relatively small pipe, or *vent,* leading from a magma reservoir at depth (Figure 30.1.) Both explosive and quiet types of eruption occur, the forms built differing for the two types.

Volcanoes of explosive eruption are *cinder cones* and *composite cones;* those formed by relatively quiet outflow of lava are *lava domes.* Quiet eruption of lava, if issuing from extensive cracks, or *fissures,* in sufficient quantities may make great plains or plateaus of lava, classified with horizontal strata (Chapter 28).

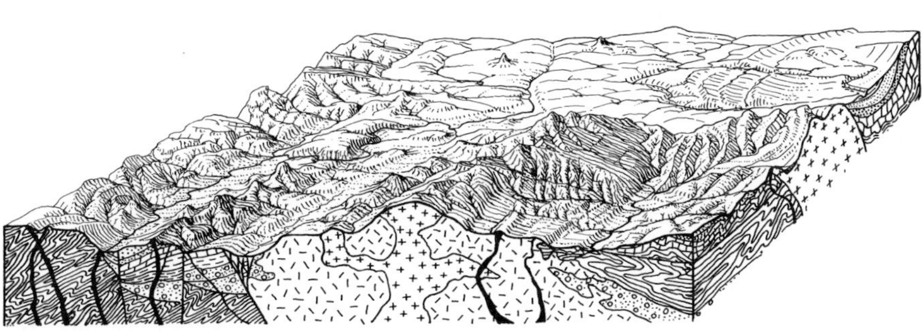

Figure 30.6 A region of highly complex structure. (Drawn by E. Raisz.)

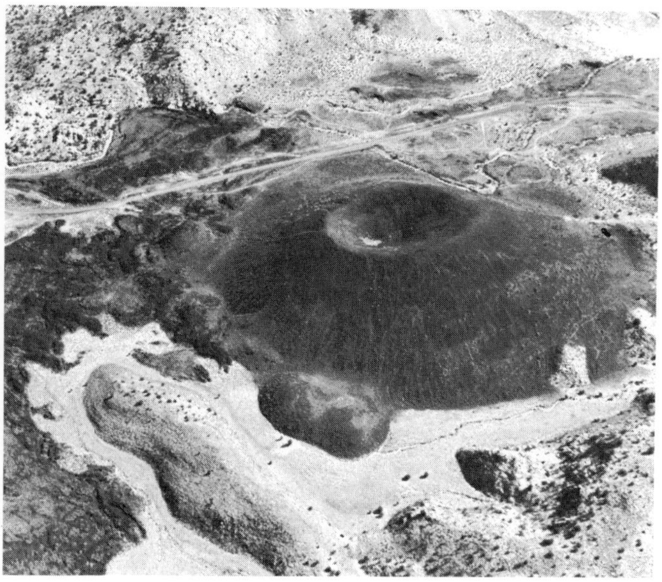

Figure 30.7 A fresh cinder cone and its associated basaltic lava flow (left) have partially blocked a valley. Dixie State Park, about 17 mi (27 km) northwest of St. George, Utah. (Photograph by Frank Jensen.)

Cinder cones

Smallest of the volcanoes are the cinder cones, built entirely of pieces of solidified lava thrown from a central vent. They form where a high proportion of gas in the molten rock causes it to froth into a bubbly mass and to be ejected from a vent with great violence. The froth breaks up into small fragments which solidify as they are ejected and fall as solid particles near the vent (Figure 30.7). The fragments resemble clinkers and ash taken from a coal furnace. Large pieces up to several tons in weight are *volcanic bombs;* they may be somewhat plastic when ejected. Smaller pieces, a fraction of an inch up to an inch or two (1 to 5 cm) in size, are *cinders;* these make up the bulk of the cinder cone. Still finer particles are termed *ash* and *volcanic dust.* The ash falls like snow upon the ground within a few miles of the eruption (see Figure 30.10). Finer dust is carried by winds to distant regions and may settle out only after years of drifting in the atmosphere.

Cinder cones rarely grow to more than 500 or 1000 ft (150 to 300 m) in height. Growth is rapid. Monte Nuovo, near Naples, Italy, grew to a height of 400 ft (120 m) in the first week of its existence. Paricutin, in Mexico, started as a cinder cone and reached a height of 1000 ft (300 m) in the first three months. The angle of slope of a recently formed cinder cone ranges between 26° and 30°. So loose is the material that it absorbs heavy rain without permitting surface runoff. Erosion is thus delayed until weathering produces a soil which fills the interstices.

Lava flows sometimes issue from the same vent as a cinder cone. They may burst apart the side of the cone but more commonly do not alter its form. Cinder cones may erupt in almost any conceivable topographic location, on ridges, on slopes, and in valleys (Figure 30.7). Cinder cones usually occur in groups, often many dozens in an area of a few tens of square miles. They sometimes show an alignment parallel with fault lines in the underlying rock.

Composite volcanoes

Most of the world's great volcanoes are composite cones. They are built of layers of cinder and ash alternating with layers of lava, and for this reason have been called *strato-volcanoes* by some writers. The steep-sided form is governed by the angle at which the cinder and ash stands, whereas the lava layers provide strength and bulk to the volcano (Figure 30.9). Among the outstanding examples of recently formed composite volcanoes are Fujiyama in Japan, Mayon in the Philippines, Mt. Hood in Oregon, and Shishaldin in the

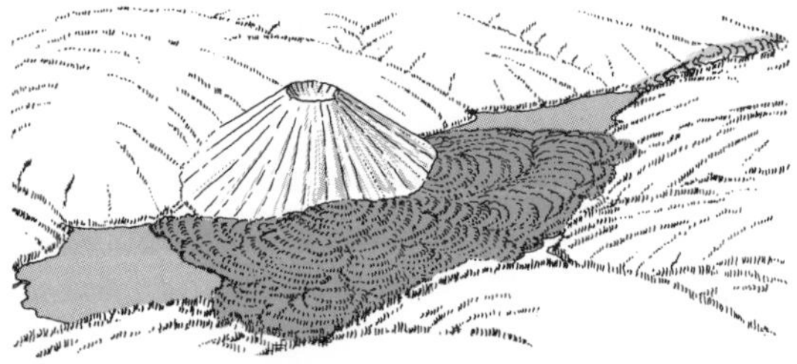

Figure 30.8 A cinder cone with its lava flows has dammed a valley, making a lake. Farther down valley, in the distance, another lava mass has made a second dam. (After W. M. Davis.)

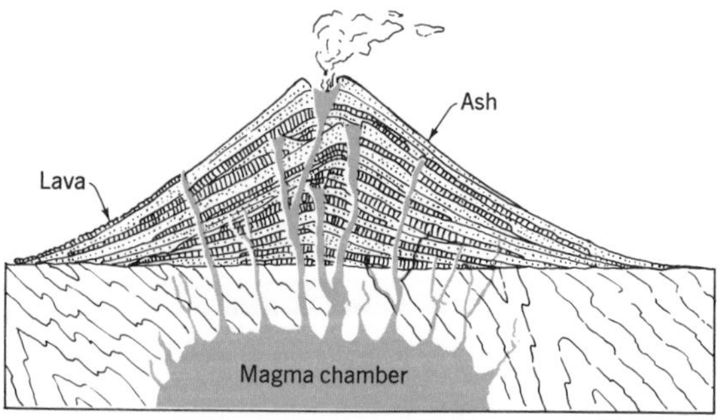

Figure 30.9 Idealized cross section of a composite volcanic cone with feeders from magma chamber beneath. (From A. N. Strahler, 1972, *Planet Earth; Its Physical Systems Through Geologic Time*, Harper and Row, New York.)

A. **This distant view of Sakurajima shows the great cauliflower cloud of volcanic gases and condensed steam.**

C. **Reaching the sea, the hot lava makes clouds of steam.**

B. **A blocky lava flow is advancing slowly over a ground surface littered with volcanic bombs and ash.**

D. **Volcanic ash has buried this village.**

Figure 30.10 Sakurajima, a Japanese volcano, erupted violently in 1914. These pictures show various scenes from the eruption. (Photograph by T. Nakasa.)

Aleutians. Other famous ones, less perfectly formed, are Vesuvius, Etna, and Stromboli in Italy and Sicily. Heights of several thousand feet and slopes of 20° to 30° are characteristic.

Many composite volcanoes lie in a great belt, the *circum-Pacific ring*, extending from the Andes in South America, through the Cascades and the Aleutians, into Japan; thence south into the East Indies and New Zealand. There is also an important Mediterranean group, mentioned above, which includes active volcanoes of Italy and Sicily. Otherwise, Europe has no active volcanoes. Island and mountain arcs bearing composite volcanoes are shown by a solid line in Figure 23.8.

The eruption of large composite volcanoes is usually accompanied by explosive issue of steam, cinders, bombs, and ash, and by lava flows (Figure 30.10). The crater may change form rapidly, both from demolition of the upper part and from new accumulation.

Calderas

One of the most catastrophic of natural phenomena is a volcanic explosion so violent as to destroy the entire central portion of the volcano. There remains only a great central depression, termed a *caldera*. A portion of the upper part of the volcano is blown outward in fragments, whereas most of the mass subsides into the ground beneath the volcano. Although calderas have been formed in historic time, conditions near the volcano do not permit observation of the process. Vast quantities of ash and dust are emitted and fill the atmosphere for many hundreds of square miles around.

Krakatoa, a volcanic island in Indonesia, exploded in 1883, leaving a great caldera. It is estimated that 18 cu mi (75 cu km) of rock disappeared during the explosion. Great seismic sea waves, or *tsunamis*, generated by the explosion killed many thousands of persons living on low coastal areas of Java and Sumatra. Another historic explosion was that of Katmai, on the Alaskan Peninsula, in 1912. A caldera more than 2 mi (3 km) wide and 2000 to 3700 ft (600 to 1100 m) deep was produced at this time. The explosion was heard at Juneau, 750 mi (1200 km) distant, while at Kodiak, 100 mi (160 km) away, the ash formed a layer 10 in (25 cm) deep.[1]

A classic example of a caldera produced in prehistoric time is Crater Lake, Oregon (Figure 30.11). Mt. Mazama, the former volcano, is estimated to have risen 4000 ft (1200 m) higher than the present rim. Valleys previously cut by streams and glaciers into the flanks of Mt. Mazama were beheaded by the explosive subsidence of the central portion and now form distinctive notches in the rim. The event occurred about 6600 years ago. Wizard Island, a recent volcano with associated flows, has since grown in the floor of the caldera.

Erosion cycle of volcanoes

Figure 30.12 shows successive stages in the erosion of volcanoes, lava flows, and a caldera. In the first block are active volcanoes in the process of building. These are in their initial stage. Lava flows issuing from the volcanoes have spread down into a stream valley, following the downward grade of the valley and forming a lake behind the lava dam.

In the next block some changes have taken place, the most conspicuous of which is the destruction of the largest volcano to produce a caldera. A lake occupies the caldera, and a small cone has been built inside. One of the other volcanoes, formed earlier, has become extinct. It has been dissected by streams, losing the initial form, and may be said to be in a stage of late youth. Smaller, neighboring volcanoes are still active, and the contrast in form is marked. The examples of large, beautifully formed volcanoes cited above are all in their initial or very young stage.

The drainage pattern of streams upon a volcanic cone is of necessity *radial* in pattern. Because these streams take their positions upon a slope of an initial land surface they are of consequent origin. It is often possible to recognize volcanoes from a drainage map alone (Figure 30.13) because of the perfection of the radial pattern. Where a well-formed crater exists, small streams flow from the crater rim toward the bottom of the crater. Here the water is absorbed in the porous layers of ash within the cone, or is conducted outward by means of a single gap in the crater rim. This inward

[1] A. K. Lobeck, *Geomorphology*, McGraw-Hill Book Co., New York, 1939, p. 685.

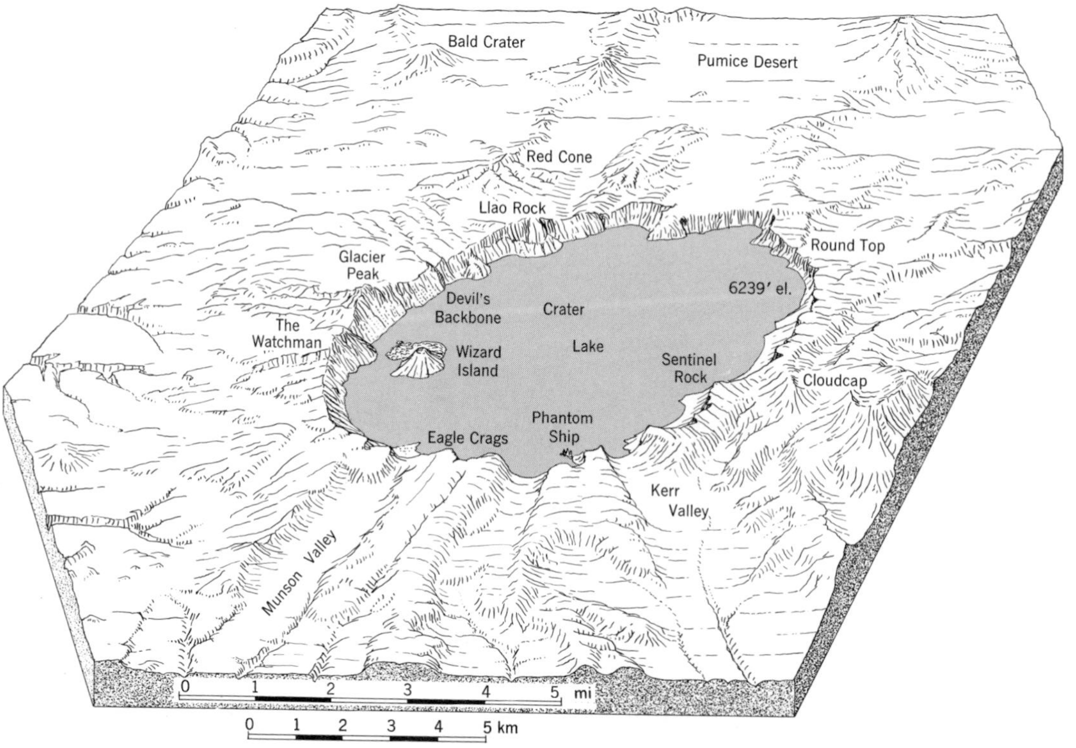

Figure 30.11 Crater Lake, Oregon, is an outstanding illustration of a caldera, now holding a lake. (After E. Raisz.)

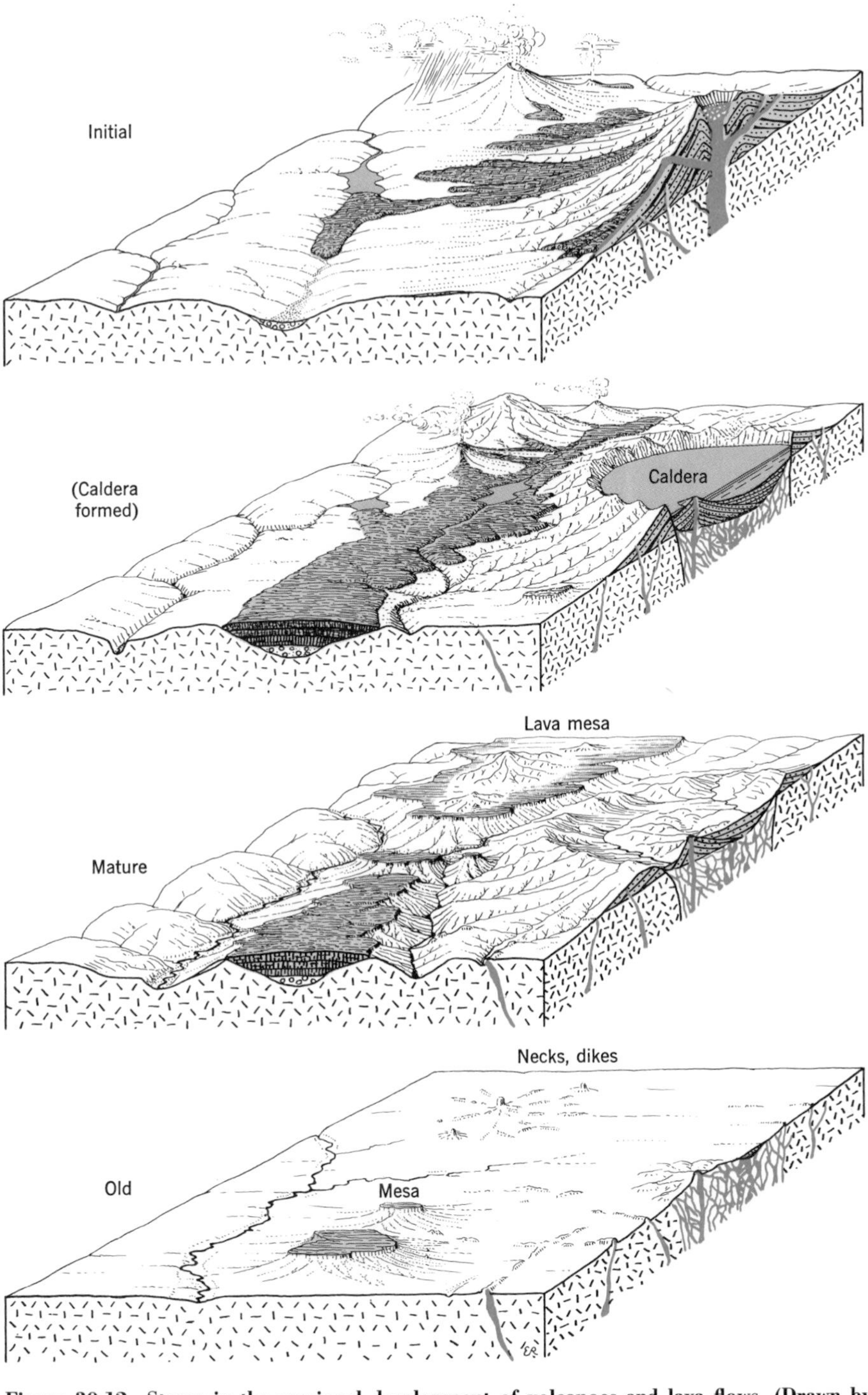

Figure 30.12 Stages in the erosional development of volcanoes and lava flows. (Drawn by E. Raisz.)

drainage, described as *centripetal*, often adds to the certainty of interpreting a volcano from a drainage pattern.

In the third block, Figure 30.12, all volcanoes are extinct and have been eroded into the stage of maturity. The caldera lake has been drained and the rim worn to a low, circular ridge. The lava flows which formerly flowed down stream valleys have been able to resist erosion far better than the rock of the surrounding area and have come to stand as mesas high above the general level of the region.

An example of a maturely dissected volcano is Mt. Shasta in the Cascade Range (Figure 30.14). A smaller subsidiary cone of more recent date, named Shastina, is attached to the side.

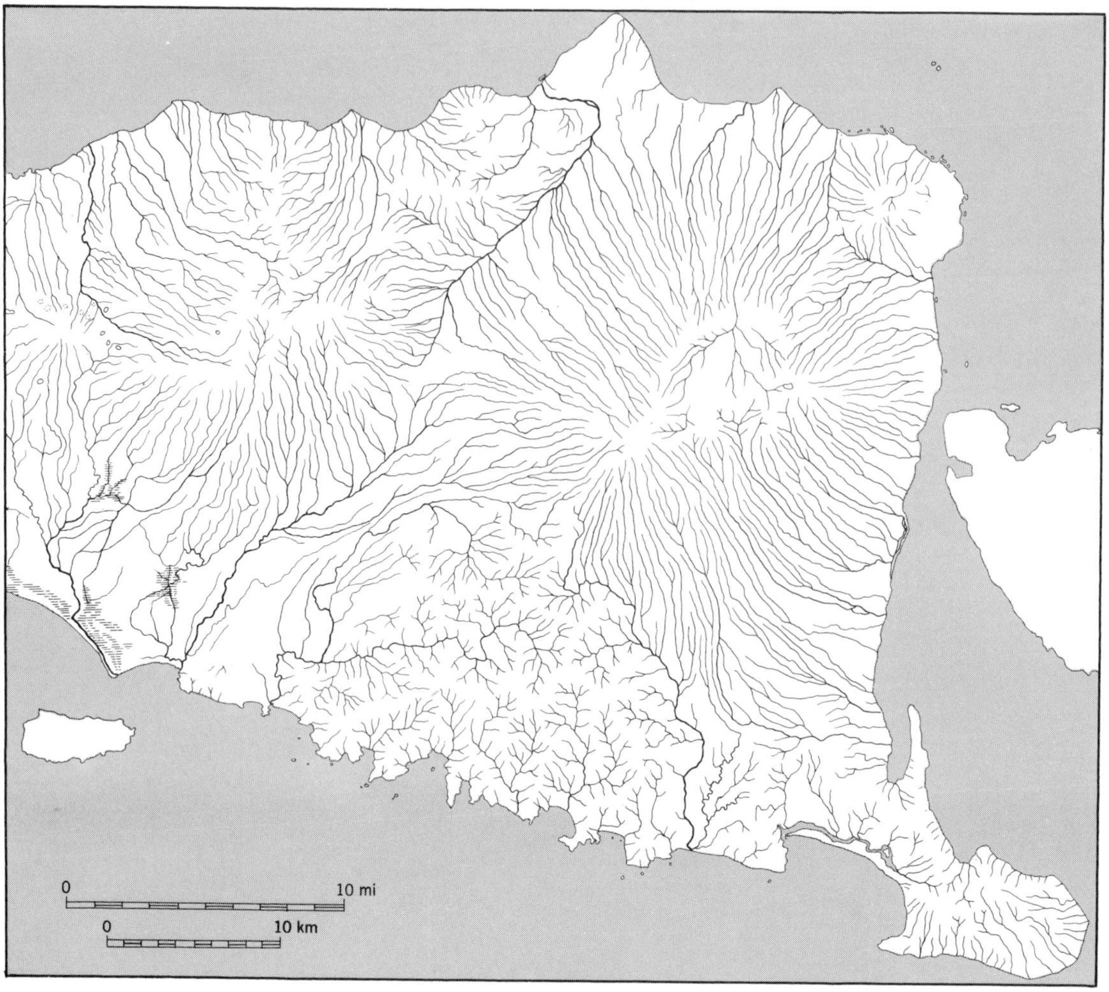

Figure 30.13 Radial drainage patterns of volcanoes in the East Indies. (After Verbeck and Fennema.)

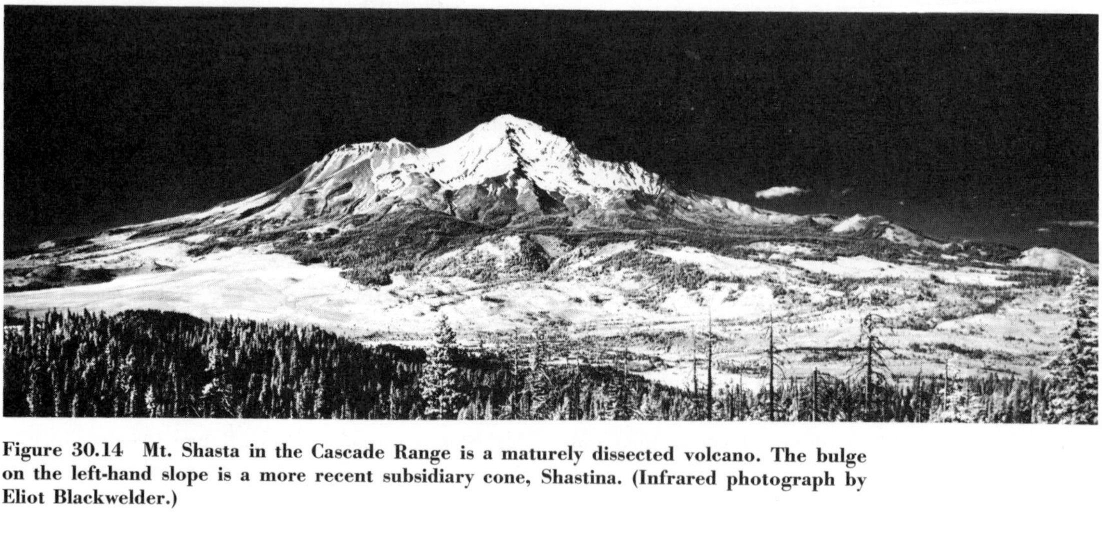

Figure 30.14 Mt. Shasta in the Cascade Range is a maturely dissected volcano. The bulge on the left-hand slope is a more recent subsidiary cone, Shastina. (Infrared photograph by Eliot Blackwelder.)

Figure 30.15 Shiprock, New Mexico, is a volcanic neck. Radiating from it are dikes. (Spence Air Photos.)

Figure 30.12 shows the old stage of erosion of volcanoes. There remains now only a small sharp peak, or *volcanic neck*, representing the solidified lava in the pipe, or neck, of the volcano. Radiating from this are wall-like *dikes*, formed of lava, which previously filled fractures around the base of the volcano. Perhaps the finest illustration of a volcanic neck with radial dikes is Ship Rock, New Mexico (Figure 30.15). Because the central neck and radial dikes extend to great depths in the rock below the base of the volcano, they may persist as landforms long after the cone and its associated flows have been removed.

Lava domes or shield volcanoes

A very important type of volcano, differing greatly in form from those already discussed, is the *lava dome* or *shield volcano*. The best examples are from the Hawaiian Islands, which consist entirely of lava domes (Figure 30.16).

Lava domes are characterized by gently rising, smooth slopes which tend to flatten near the top, producing a broad-topped volcano. The Hawaiian domes range to elevations up to 13,000 ft (4,000 m) above sea level, but including the basal portion lying below sea level they are more than twice that high. In width they range from 10 to 50 mi (16 to 80 km) at sea level and up to 100 mi (160 km) wide at the submerged base.

Lava domes, as the name implies, are built by repeated outpourings of lava. Explosive behavior and emission of fragments are not important, as they are for cinder cones and composite cones. The lava, which in the Hawaiian lava domes is of a dark basaltic type, is highly fluid and travels far down the low slopes, which do not usually exceed 4° or 5°.

Instead of the explosion crater, lava domes have a wide, steep-sided *central depression*, or *sink*, which may be 2 mi (3.2 km) or more wide and several hundred feet deep. These large depressions are a type of caldera produced by subsidence accompanying the removal of molten lava from beneath. Molten basalt is actually seen in the floors of deep *pit craters*, steep-walled depressions 0.25 to 0.5 mi (0.4 to 0.8 km) wide or smaller, which occur on the floor of the sink or elsewhere over the surface of the lava dome (Figure 30.17). Most lava flows issue from cracks, or *fissures*, on the sides of the volcano.

Lava domes of the Hawaiian islands are in various stages of erosion (Figure 30.16). Active volcanoes such as Kilauea and Mauna Loa are in the initial stage and have smooth slopes. Others, such as East Maui, are partly dissected by deep canyons but still possess sizable parts of the original surface. Still others, such as West Maui, are fully dissected. Rising from the sea are some steep-walled stacks representing the last vestiges of old domes; and there exist submarine banks some 250 ft (75 m) below sea level, representing the final stage in destruction.

Environmental aspects of volcanoes

Volcanic eruptions count among the earth's great natural disasters. Wholesale loss of life and destruction of towns and cities are frequent in the history

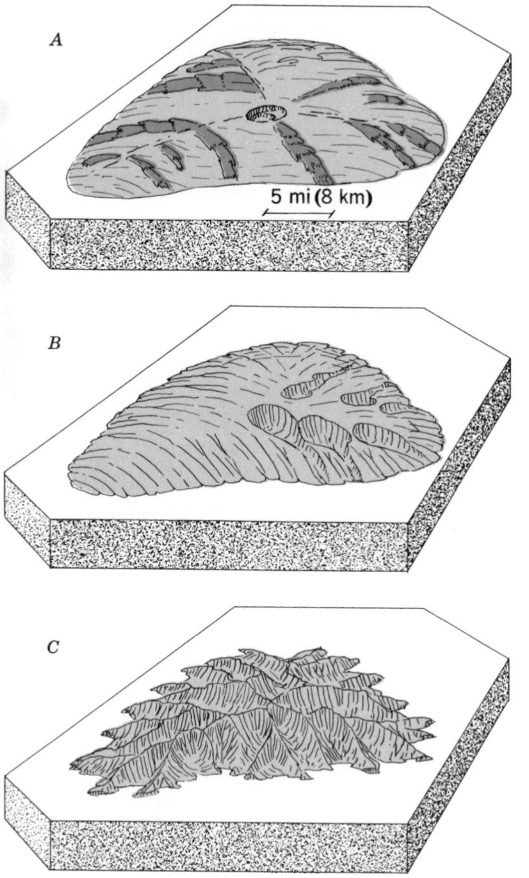

Figure 30.16 Lava domes in various stages of erosion make up the Hawaiian Islands. (Data from Stearns and Macdonald.) *A*. Initial dome with central depression and fresh flows issuing from radial fissure lines. *B*, Young stage with deeply eroded valley heads. *C*, Mature stage with steep slopes and great relief.

Figure 30.17 Halemaumau, a pit crater on Mauna Ioa, with a fire fountain of molten lava on its floor, 1952. (National Park Service, U.S. Dept. of the Interior.)

of peoples who live near active volcanoes. Loss occurs principally from sweeping clouds of incandescent gases that descend the volcano slopes like great avalanches; from lava flows whose relentless advance engulfs whole cities; from the descent of showers of ash, cinders, and bombs; from violent earthquakes associated with the volcanic activity; and from mudflows of volcanic ash saturated by heavy rain. For habitations along lowlying coasts there is the additional peril of great seismic sea waves, generated by submarine earth faults. These do not necessarily accompany volcanic activity and may occur without warning in the ocean basins which are bordered by belts of active mountain-making.

The surfaces of volcanoes and lava flows remain barren and sterile for long periods after their formation. Certain types of lava surfaces are extremely rough and difficult to traverse; the Spaniards who encountered such terrain in the southwestern United States named it *malpais* (bad ground). Most volcanic rocks in time produce highly fertile soils that are extensively cultivated.

Volcanic ash may have a remarkably beneficial effect upon productivity of soil where the ash fall is relatively light. The eruption of Sunset Crater, near Flagstaff, Arizona, in 1065 A.D., spread a layer of sandy volcanic ash over the barren reddish soil of the surrounding region and caused it to become highly productive because of the moisture-conserving effect of the ash, which acted as a mulch in the semi-arid climate. Because Hopi Indian corn grows well in sand, this development attracted Indians, who settled the area thickly. As the ash was gradually washed off of the slopes by heavy summer rains or blown into thick dunes by wind, the fertility declined and after about 200 years of occupation the region was abandoned to its previous state.

Young and mature volcanoes possess most of the natural resources of rugged mountains of other types. Steep slopes prevent extensive agriculture, although providing valuable timber resources. Thus the San Francisco Mountains, a group of maturely dissected volcanoes in northern Arizona, are clothed in what is perhaps the finest known western yellow pine forest (ponderosa pine). A lumber industry centered about the towns of Flagstaff and Williams has flourished for many years.

As scenic features of great beauty, attracting a heavy tourist trade, few landforms outrank volcanoes. National parks have been made of Mt. Rainier, Mt. Lassen, and Crater Lake in the Cascade Range. Mt. Vesuvius and Fujiyama also attract many visitors.

Mineral resources, particularly the metallic ores, are conspicuously lacking in volcanoes and lava flows, unless later geologic events have resulted in the injection or diffusion of ore minerals into the volcanic rocks. The gas-bubble cavities in some ancient lavas have become filled with copper or other ores. The famed *kimberlite* rock of South Africa, source of diamonds, is the pipe of an ancient volcano.

As a source of crushed rock for concrete aggregate or railroad ballast, and other engineering purposes, lava rock is often extensively used. Thus the ancient lava layers that make up the Watchung ridges of northern New Jersey have in places been virtually leveled in quarrying operations continued over several decades.

Geomorphic provinces of North America

Using their knowledge of control of landforms by rock structure, together with information on the age and geologic history of the bedrock, geographers have made extended efforts to divide the continental surfaces into landform regions. These efforts have paralleled the work of soil scientists and plant geographers in defining regions of great soil groups and natural vegetation classes. Thus, *geomorphic provinces* have been established as distinctive regions of landform assemblages. Within each province the landforms share many characteristics in common, but are on the whole of a different character from landforms of adjacent provinces.

Geologic structure is the primary basis of defining geomorphic provinces. Groundwork for a classification has already been laid in Chapter 28 through the recognition of landmasses of three major classes: undisturbed structures, disturbed structures, and volcanic forms. Within the first two classes eight structural types are identified, and these we have treated systematically in Chapters 28 to 30. What remains to be done is to delineate those regions of a continent belonging largely to a single structural type; each region is thus a geomorphic province.

A secondary basis for distinguishing between geomorphic provinces is that of process. So far, we have concentrated upon the fluvial denudation process; it is the dominant process of the earth's land surfaces. However, continental glaciation, treated in Chapter 31, is sufficiently important as an agent of landform development to define certain geomorphic provinces and subprovinces. Further subdivisions within provinces can be made on the basis of stage in the fluvial denudation cycle, differentiating regions in the young, mature, and old stages of the cycle.

The concept of geomorphic provinces was introduced near the close of the nineteenth century by the famed geologist-explorer, Major John Wesley Powell, and a number of his coworkers. However, it remained for a geographer, Professor Nevin M. Fenneman, to work out in detail the definitions and boundaries of geomorphic provinces of the United States and southeastern Canada. His map and descriptions of 25 provinces and over 75 subdivisions (the latter known as *sections*) were published in 1928. Another American geographer, Professor Armin K. Lobeck, later extended Fenneman's work to all continents of the globe. Lobeck's classification system is presented here for North America.

Lobeck recognized five major geomorphic divisions of North America lying north of the 20th parallel; these are shown in Figure 30.18 and listed in Table 30.1. Within the five major divisions are a total of 28 geomorphic provinces. Table 30.1 gives generalized information on rock ages, predominant rock type and structure, characteristic relief, and typical upland and summit elevations of the North American provinces. A map, Figure 30.19, shows province boundaries for the 48 contiguous United States, southern Canada, and northern Mexico. Figure 30.20 is a corresponding *physiographic diagram* of the same region, using a conventionalized pictorial process to represent the relief features. Both the table and maps are greatly generalized and simplified; they do not reveal numerous secondary features within many of the provinces. Further subdivision into sections would be required to sort out and explain these secondary landform groups.

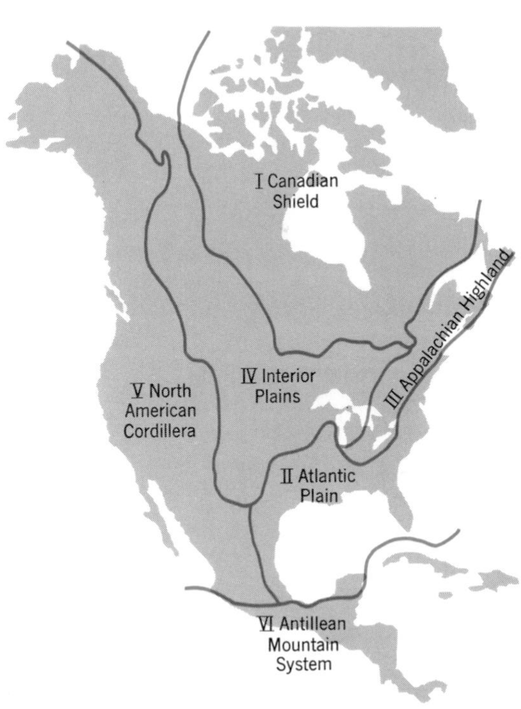

Figure 30.18 The major geomorphic subdivisions of North America. (After A. K. Lobeck, 1948, *Physiographic Diagram of North America*, Hammond Incorporated, New York.)

TABLE 30.1 GEOMORPHIC DIVISIONS OF NORTH AMERICA[a]

Major Division		Geomorphic Province	Predominant Rock Ages
I. CANADIAN SHIELD		1. Laurentian Upland	Precambrian
		(2. Arctic archipelago)[b] (3. Greenland)	
II. ATLANTIC PLAIN		4. Atlantic and Gulf coastal plain	Cretaceous, Cenozoic
		5. Continental shelf	Cenozoic
III. APPALACHIAN HIGHLANDS		6. New England-Maritime	Precambrian, Paleozoic
		7. Older Appalachians	
		8. Triassic lowlands	Triassic
		9. Newer (Folded) Appalachians	Paleozoic
		10. Appalachian Plateau	Paleozoic
IV. INTERIOR PLAINS		11. Interior low plateaus	Paleozoic
		12. Central lowland	Paleozoic, Cretaceous, Pleistocene
		13. Ozark Plateau	Paleozoic, minor Precambrian
		14. Ouachita Mountains	Paleozoic
		15. Great Plains	Cretaceous, Cenozoic
V. NORTH AMERICAN CORDILLERA	A. Rocky Mountain system	16. Southern Rocky Mountains 17. Middle Rocky Mountains	Precambrian–Cenozoic
		18. Northern Rockies	Precambrian–Cenozoic
	B. Intermontane plateau system	(19. Arctic Rockies)	
		(20. Central Alaska uplands & plains)	
		(21. Interior plateaus of Canada)	
		22. Columbia Plateau	Cenozoic
		23. Colorado Plateau	Paleozoic–Cenozoic, minor Precambrian
		24. Basin and Range 25. Mexican highlands	Precambrian–Cenozoic
	C. Pacific mountain system	26. Sierra–Cascade–Coast mountains	Mesozoic–Cenozoic
		27. Pacific troughs	Cenozoic
		28. Pacific Coast ranges	Precambrian–Cenozoic

[a]Data of A. K. Lobeck, 1948. From A. N. Strahler, 1971, *The Earth Sciences*, 2nd Ed., Harper & Row, New York.
[b]Provinces in parentheses are not shown in Figures 30.19 and 30.20.

Predominant Rock Type and Structure	Characteristic Relief	Typical Upland and Summit Elevations feet	meters
Metamorphosed sediments, volcanics, and intrusives. Local thin sedimentary cover	Low hills to low mountains; many lakes	1000–2500	300–760
Unsolidated sedimentary strata, low seaward dip	Undulating plains, floodplains, hill belts (cuestas)	0–800	0–240
Unconsolidated sediments	Gently undulating ocean floor	to–600	to–180
Folded overthrust sediments, metamorphics, igneous instrusives	Rolling hills to subdued mountains	1000–5000	300–1500
	Rolling hills (Piedmont) Subdued mountains (Blue Ridge)	300–1500 3000–6000	90–460 900–1800
Redbeds and basaltic lavas in isolated basins	Undulating lowlands, narrow ridges	500–800	150–240
Open folds, overturned folds, thrust sheets	Parallel and zigzag ridges and valleys	2000–3000	600–900
Horizontal to gently dipping strata	Rugged hills and low mountains	1200–4000	360–1200
Horizontal to gently dipping strata; broad domes and basins	Undulating plains to rugged hills; escarpments	500–1000	150–300
Horizontal strata. Pleistocene cover in parts	Plains, low hills; moraine belts	600–1200	180–360
Horizontal strata	Rugged hills, low mountains	1200–2500	360–760
Folded and overthrust strata	Parallel and zigzag ridges and valleys	500–2500	150–760
Horizontal strata; isolated domes	Plains, low hills and escarpments, badlands; isolated mountains	2000–5000	600–1500
Anticlinal uplifts and domes; basin sediments; volcanics	High, rugged mountains; intermontane plains	10,000–14,000	3000–4300
Sedimentary strata; batholiths; faulted mountain blocks	High, rugged mountains; narrow intermontane basins	9000–12,000	2700–3600
Basaltic lavas; fault blocks	Rolling plateaus; isolated mountains	2000–5000	600–1500
Horizontal strata broken by faults and monoclines; minor volcanics	High plateaus, steep escarpments, deep canyons	6000–11,000	1800–3300
Fault blocks of complex structure; alluvium-filled basins	High rugged mountains; broad intermontane plains	4000–12,000	1200–3600
Metasediments, batholiths, volcanics, upfaulted blocks (Sierra Nevada)	High rugged mountains; deep narrow valleys	8000–14,000	2400–4300
Thick sediments in subsiding basins	Extensive plains; low hills	200–500	60–150
Upfaulted blocks of complex structure	Rugged mountains; narrow intermontane lowlands	8000–10,000	2400–3000

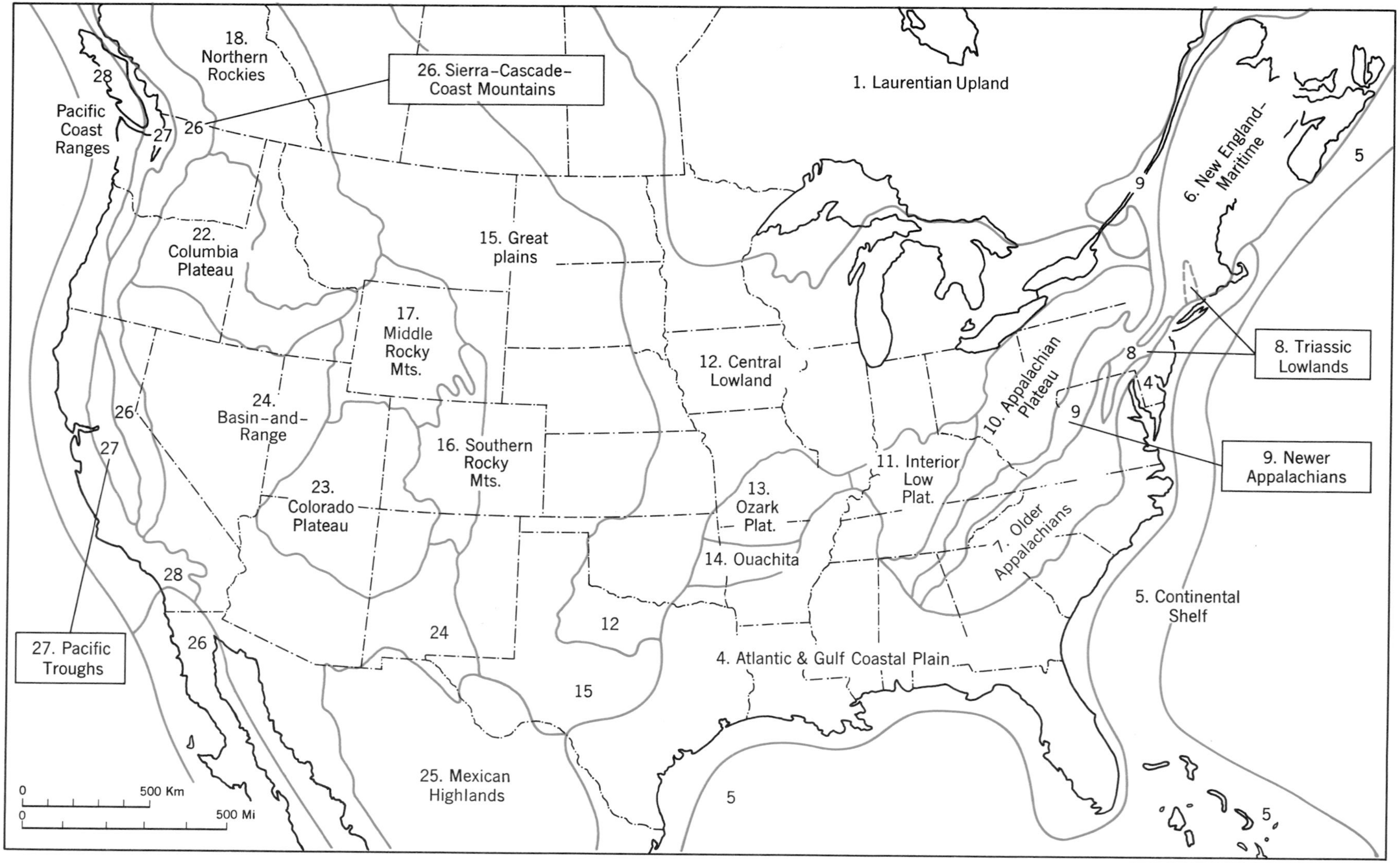

Figure 30.19 Geomorphic provinces of the 48 contiguous United States, southern Canada, and northern Mexico. (Data of A. K. Lobeck, 1948, *Physiographic Diagram of North America*, Hammond Incorporated, New York.)

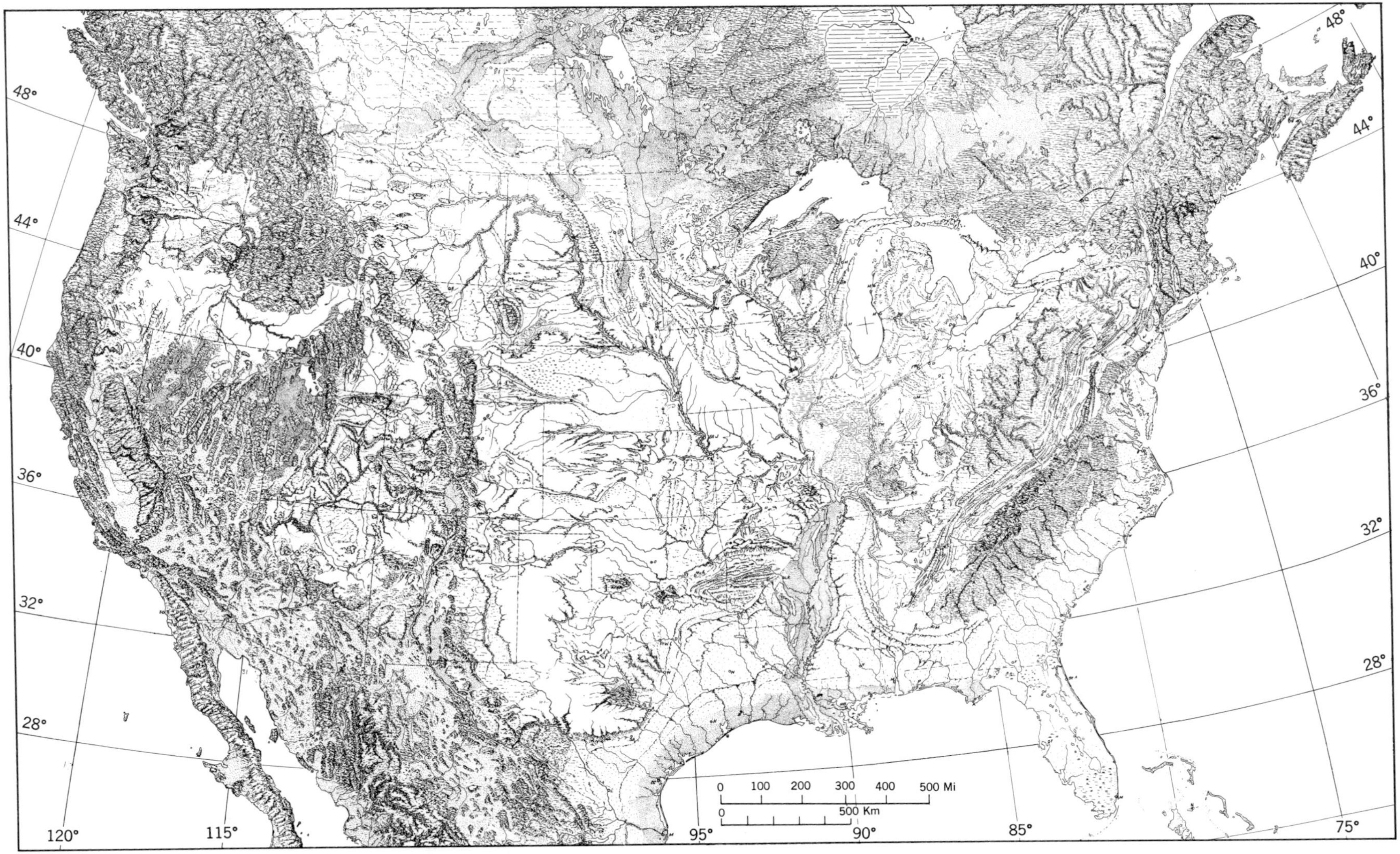

Figure 30.20 Physiographic diagram of the 48 contiguous United States, southern Canada, and northern Mexico. (From A. K. Lobeck, 1948, *Physiographic Diagram of North America*, Hammond Incorporated, New York.)

A new world-wide system of landform classification

The most recent comprehensive treatment of landforms on a global scope is that by Professor Richard E. Murphy. It was evolved by him during the 1960s and is based in part upon the same concepts used by Powell, Fenneman, and Lobeck. The Murphy system of landform classification is explained in Appendix IV; his world map of landforms is reproduced as Plate 5 (end of book).

REVIEW QUESTIONS

1. What is a batholith? A stock? How large are these features? Of what kind of rock is a batholith composed? What kind of topography does it produce in a mature stage?

2. What is the meaning of a rectangular drainage pattern in a region underlain by homogeneous crystalline rock of a batholith?

3. What kind of topography develops in regions of belted metamorphic rocks such as gneiss, schist, slate, marble, and quartzite? Which of these rock types would form valleys? Which would form narrow ridges; which broad ridges?

4. What are some of the varieties of rocks and structure that might be found in a region of complex structure? From the standpoint of geologic time would a complex region be ancient or comparatively recent in development? Would it be a likely region for mineral deposits of economic value?

5. What types of volcanoes are recognized? How do they differ?

6. Describe a cinder cone. Of what material is it formed? What size is normally attained by cinder cones? How fast do they form? How are groups of cinder cones arranged in plan?

7. How is a composite type of volcano formed? Name several famous large composite volcanoes. Along what belts are most of the world's active, or recently active, large volcanoes located?

8. What is a caldera? What volcanic activity occurred at Krakatoa (1883) and at Katmai (1912)? Describe Crater Lake as an illustration of a caldera.

9. Describe the stages of youth, maturity, and old age in the erosion of a large volcano. What form of drainage pattern is typically present on volcanoes?

10. Describe a volcanic neck with radial dikes. Give an example.

11. Discuss the form and development of large lava domes, or shield volcanoes, as exemplified by the Hawaiian Islands. What feature of a lava dome corresponds to the crater of a composite volcano?

12. Discuss the environmental and resource aspects of volcanic landforms. In what way are soils and vegetation affected by volcanic forms?

13. What is the basis for defining geomorphic provinces? What individuals made important contributions to description and mapping of geomorphic provinces of North America?

14. Give a general description of A. K. Lobeck's system of geomorphic divisions of North America. Describe four strikingly unlike geomorphic provinces in terms of rock structure, rock age, and characteristic relief. Evaluate each province in terms of its mineral resources.

Exercises

Exercise 1. ***Volcano.*** (Source: Dunsmuir, Calif., U.S. Geological Survey topographic maps; scale 1:125,000.)

Explanatory Note: This map shows Mount Shasta, a great composite volcano of the Cascade Range. Figure 30.14 is a photograph of the same volcano, taken from the southwest. The main part of the volcano is dissected by streams and glaciers, and no longer shows a crater. Several small glaciers, shown by the cross-hatched pattern, remain on the mountain. Shastina, a subsidiary cone, is seen on the western slope of the main cone. Shastina is relatively young in date of formation and still shows a crater rim.

QUESTIONS

1. (a) Determine the summit elevation of Mt. Shasta. **(*b*)** Determine the summit elevation of Shastina (13.5–17.5). **(*c*)** How wide is the volcano at its base, assuming the 5000-ft contour to represent the base?

2. Make a topographic profile across this volcano, from 30.0–28.0 to 0.0–4.0, using a vertical scale twice that of the horizontal scale of the map.

3. Compute the angle of the volcano slopes between the 10,000 and 12,500-ft contours. Use the map for this purpose.

4. The serrate, or sawtooth, contours near 9.0–26.0 mean that a rough, blocky lava flow of recent date is present. **(*a*)** Give grid coordinates of the source of this lava, as closely as you can locate it from the contour indications. **(*b*)** Locate by grid coordinates two similar lava flows.

5. What is the origin of the round hill at 2.0–12.0?

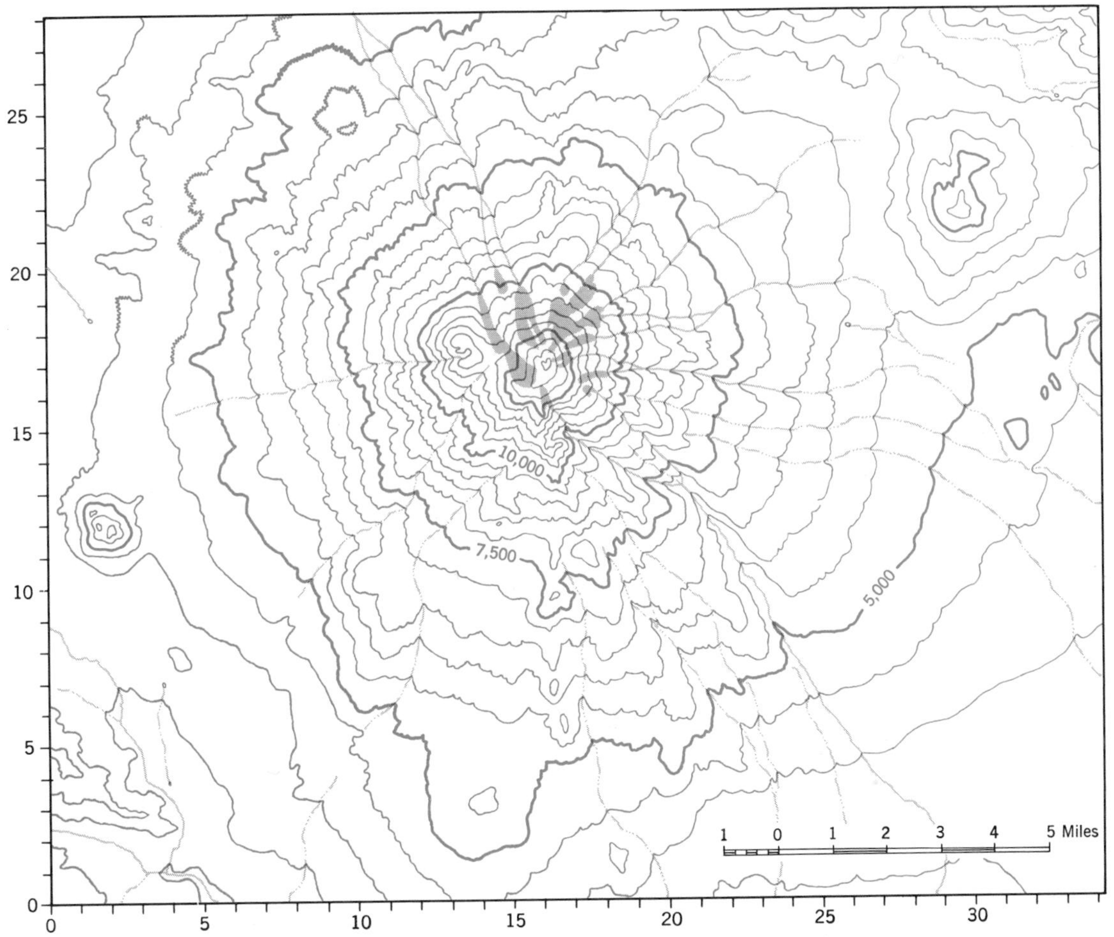

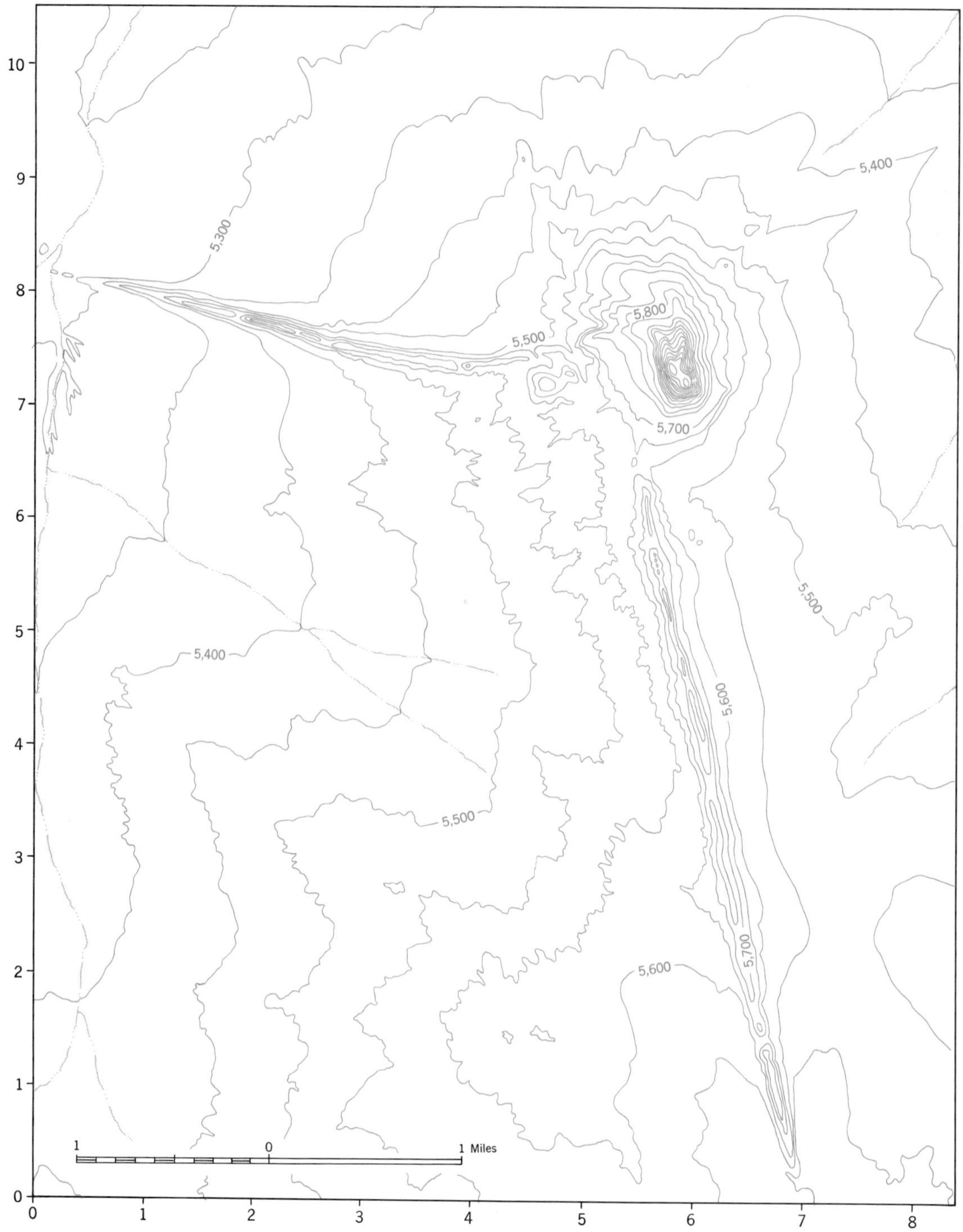

Exercise 2. *Volcanic Neck and Dikes.* (Source: Ship Rock, N.M., U.S. Geological Survey topographic map; scale 1:62,500.)

Explanatory Note: Ship Rock, a steep-sided volcanic neck, rises above a broad plain, underlain by soft shales which have been easily removed by running water, leaving the harder volcanic rocks standing in bold relief. Figure 30.15 is a photograph of the map area taken from the northeast. Details shown on the photograph can easily be found on the map. Two great dikes extend radially outward, southward and westward from the neck.

QUESTIONS

1. **(*a*)** What contour interval is used on this map? **(*b*)** What is the summit elevation of Ship Rock (5.8–7.3)? **(*c*)** How high is the dike at 6.7–1.2?

2. On a sheet of thin tracing paper laid over this map, color red all volcanic rock on this map. The dikes should be shown as thin red lines. In addition to the two great dikes, at least two small dikes are indicated by sharply pointed contours. Locate three small pipes of volcanic rock which show on the photograph (Figure 30.15).

CHAPTER **31**

Landforms Made by Glaciers

MOST of us know ice only as a brittle, crystalline solid because we are accustomed to seeing it only in small quantities. Where a great thickness of ice exists, let us say 200 to 300 ft (60 to 90 m) or more, the ice at the bottom behaves as a plastic material and will slowly flow in such a way as to spread out the mass over a larger area, or to cause it to move downhill, as the case may be. This behavior characterizes *glaciers*, which may be defined as any large natural accumulations of land ice affected by present or past motion.

Figure 31.1 The Eklutna Glacier, Chugach Mountains, Alaska, seen from the air. A deeply crevassed ablation zone with a conspicuous medial moraine (foreground) contrasts with the smooth-surfaced firn zone (background). (Photograph by Steve McCutcheon, Alaska Pictorial Service.)

Conditions requisite to the accumulation of glacial ice are simply that snowfall of the winter shall, on the average, exceed the amount of melting and evaporation of snow that occurs in summer. (The term *ablation* is used by glaciologists to include both evaporation and melting of snow and ice.) Thus, each year a layer of snow is added to what has already accumulated. As the snow compacts, by surface melting and refreezing, it turns into a granular ice, then is compressed by overlying layers into hard crystalline ice. When the ice becomes so thick that the lower layers become plastic, outward or downhill flow commences, and an active glacier has come into being.

At sufficiently high altitudes, whether in high or low latitudes, glaciers form both because air temperature is low and mountains receive heavy orographic precipitation (Chapter 10). Glaciers that form in high mountains are characteristically long and narrow because they occupy previously formed valleys and bring the plastic ice from small collecting grounds high upon the range down to lower elevations, and consequently warmer temperatures, where the ice disappears by ablation (Figure 31.1). Such *valley glaciers*, or *alpine glaciers*, are distinctive types.

In arctic and polar regions, prevailing temperatures are low enough that ice can accumulate over broad areas, wherever uplands exist to intercept heavy snowfall. As a result, areas of many thousands of square miles may become buried under gigantic plates of ice whose thickness may reach several thousand feet. The term *icecap* is usually applied to an ice plate limited to high mountain and plateau areas. During glacial periods an icecap spreads over surrounding lowlands, enveloping all landforms it encounters and ceasing its spread only when the rate of ablation at its outer edge balances the rate at which it is spreading. This extensive type of ice mass is called a *continental glacier* or *ice sheet.*

Figure 31.2 illustrates a number of features of alpine glaciers. The center illustration is of a simple glacier occupying a sloping valley between steep rock walls. Snow is collected at the upper

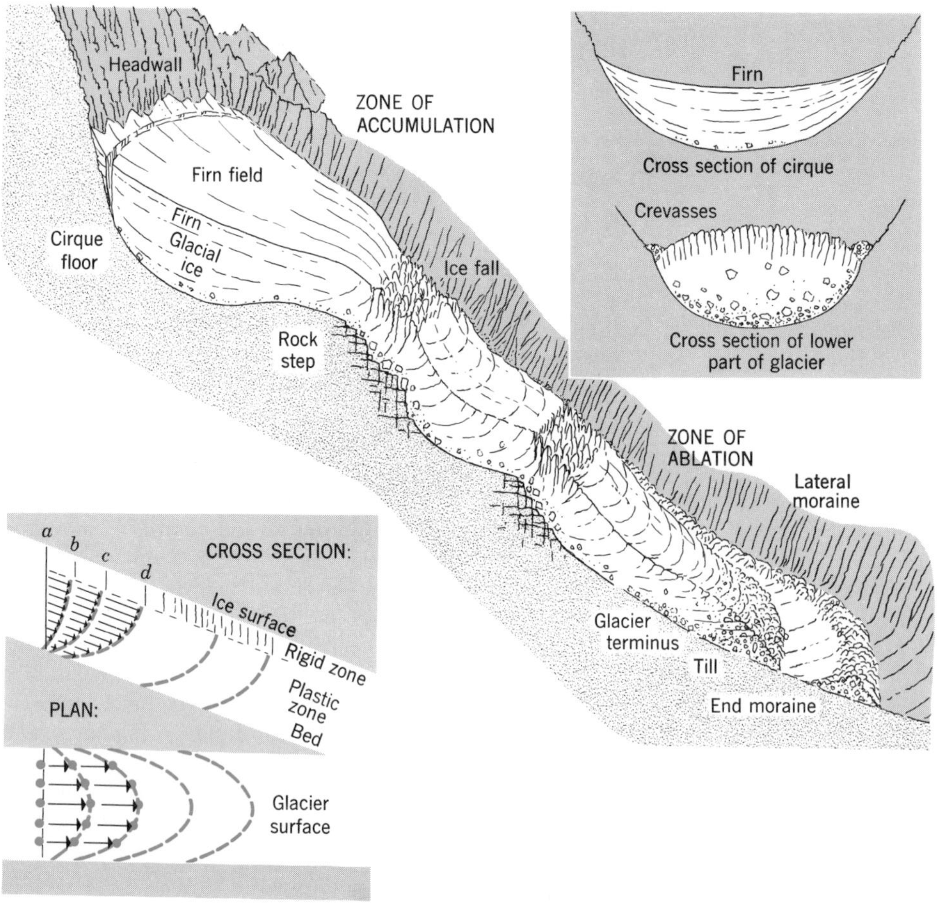

Figure 31.2 Structure and flowage of a simple alpine glacier.

end in a bowl-shaped depression, the *cirque*. The upper end constitutes the *zone of accumulation*. Layers of snow in the process of compaction and recrystallization constitute the *firn* (or *névé*). The smooth *firn field* is slightly concave up in profile (upper right). Flowage in the glacial ice beneath carries the excess ice out of the cirque, down valley. An abrupt steepening of the grade comprises a *rock step*, over which the rate of ice flow is accelerated and produces deep crevasses (gaping fractures) which form an *ice fall*. The lower part of the glacier lies in the *zone of ablation*. Here the rate of ice wastage is rapid and old ice is exposed at the glacier surface, which is extremely rough and deeply crevassed. The glacier *terminus*, or *snout* is heavily charged with rock debris. The lower part of the glacier is usually of upwardly convex cross-profile, the center being higher than the sides (upper right).

The uppermost layer of a glacier, perhaps 200 ft (60 m) in thickness is brittle and fractures readily into crevasses, whereas the ice beneath behaves as a plastic substance and moves by slow flowage (lower left). If one were to place a line of stakes across the glacier surface, the glacier flow would gradually deform the line into a parabolic curve, indicating that rate of movement is fastest in the center and diminishes toward the sides.

Altogether, a simple glacier forms a system that readily establishes a dynamic equilibrium in which the rate of accumulation at the upper end balances the rate of ablation at the lower end. Ice is transferred by flowage to maintain the glacier at approximately a given length and cross-sectional area. Equilibrium is easily upset by changes in the average annual rates of nourishment or wastage.

Rate of flowage of both alpine and continental glaciers is very slow indeed, amounting to a few inches per day for large ice sheets and for the more sluggish valley glaciers, up to several feet per day for an active valley glacier.

Glacial erosion

Most glacial ice is heavily charged with rock fragments, ranging from pulverized rock flour to huge angular boulders of fresh rock. This material is derived from the rock floor upon which the ice moves, or in alpine glaciers, from material that slides or falls from valley walls. Glaciers are capable of great erosive work, both by *abrasion*, erosion caused by ice-held rock fragments that scrape and grind against the bedrock, and by *plucking*, in which the moving ice lifts out blocks of bedrock that have been loosened by freezing of water in joint fractures. The process of plucking out of joint blocks is illustrated in Figure 31.2, at the rock steps.

The debris thus obtained must eventually be left stranded at the outer edge or lower end of a glacier when the ice is dissipated. Thus there are two glacial activities to consider, erosion and deposition. Both result in distinctive landforms, which in some cases are further differentiated according to the type of glacier, whether alpine or continental.

Landforms made by alpine glaciers

Landforms made by alpine glaciers can best be studied by a series of diagrams (Figure 31.3), in which a previously unglaciated mountainous

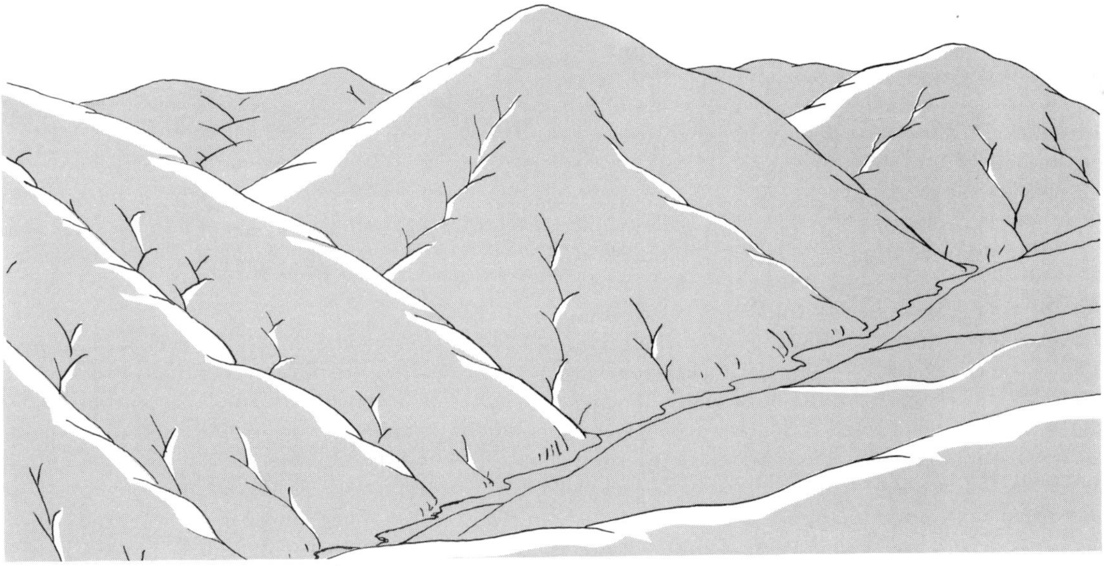

A. Before glaciation sets in, the region has smoothly rounded divides and narrow, V-shaped stream valleys.

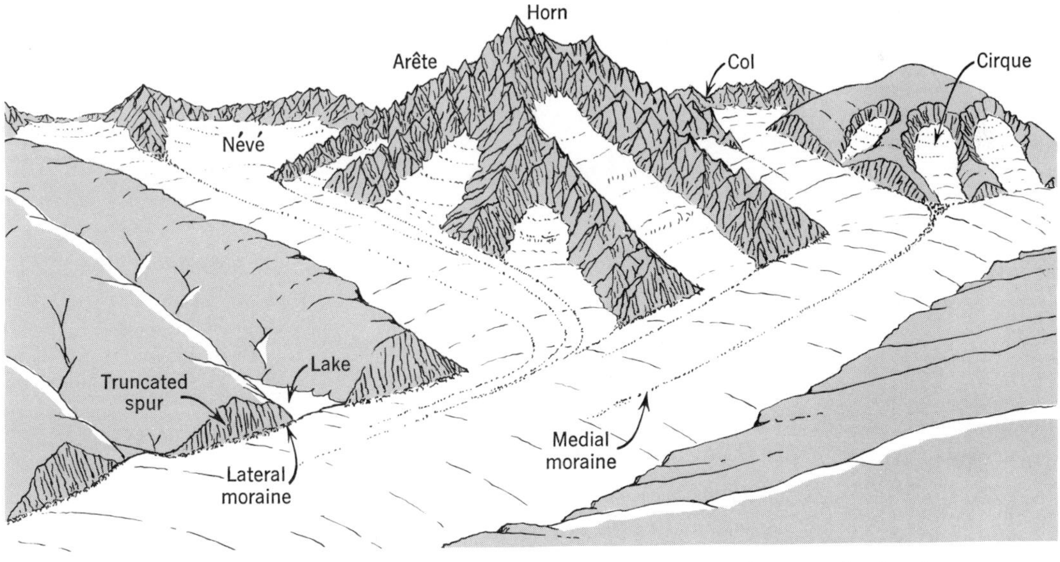

B. After glaciation has been in progress for thousands of years new erosional forms are developed.

Figure 31.3 Landforms produced by alpine glaciers. (After W. M. Davis and A. K. Lobeck.)

region is imagined to be attacked and modified by glaciers, after which the glaciers disappear and the remaining landforms are exposed to view.

Block *A* shows a mountainous region sculptured entirely by weathering, mass wasting, and streams. The mountains have a smooth, full-bodied appearance, with rather rounded divides. Although this is not always true, it is typical of the appearance of mountains in humid regions, for example, the Great Smoky Mountains of the southern Appalachians. Imagine now a climatic change in which the average annual temperature becomes several degrees lower and results in the accumulation of snow in the heads of most of the valleys high upon the mountain sides. An early stage of glaciation is shown at the right hand side of Block *B*, where snow is collecting and cirques are being carved by the outward motion of the ice and by intensive frost shattering of the rock near the masses of compacted snow.

In Block *B*, glaciers have filled the valleys and are integrated into a system of tributaries that feed a trunk glacier just as in a stream system. Glaciers are, of course, enormously thicker than streams, because the extremely slow rate of ice motion requires a great cross section if a glacier is to maintain a discharge equivalent to a swiftly flowing stream. Tributary glaciers join the main glacier with smooth, accordant junctions, but, as we shall see later, the bottoms of their channels are quite discordant in level.

Vigorous freezing and thawing of melt water from snows lodged in crevices high upon the walls of the cirque shatters the bare rock into angular fragments, which fall or creep down upon the snowfield and are incorporated into the glacier.

Figure 31.4 The Swiss Alps appear from the air as a sea of sharp arêtes and toothlike horns. In the foreground is a cirque. (Swissair Photo.)

Frost shattering also affects the rock walls against which the ice rests. The cirques thus grow steadily larger. Their rough, steep walls soon replace the smooth, rounded slopes of the original mountain mass. Where two cirque walls intersect from opposite sides, a jagged, knifelike ridge, called an *arête*, results. Where three or more cirques grow together, a sharp-pointed peak is formed by the intersection of the arêtes. The name *horn* is applied to such peaks in the Swiss Alps (Figure 31.4). One of the best known is the striking Matterhorn. Where the intersection of opposed cirques has

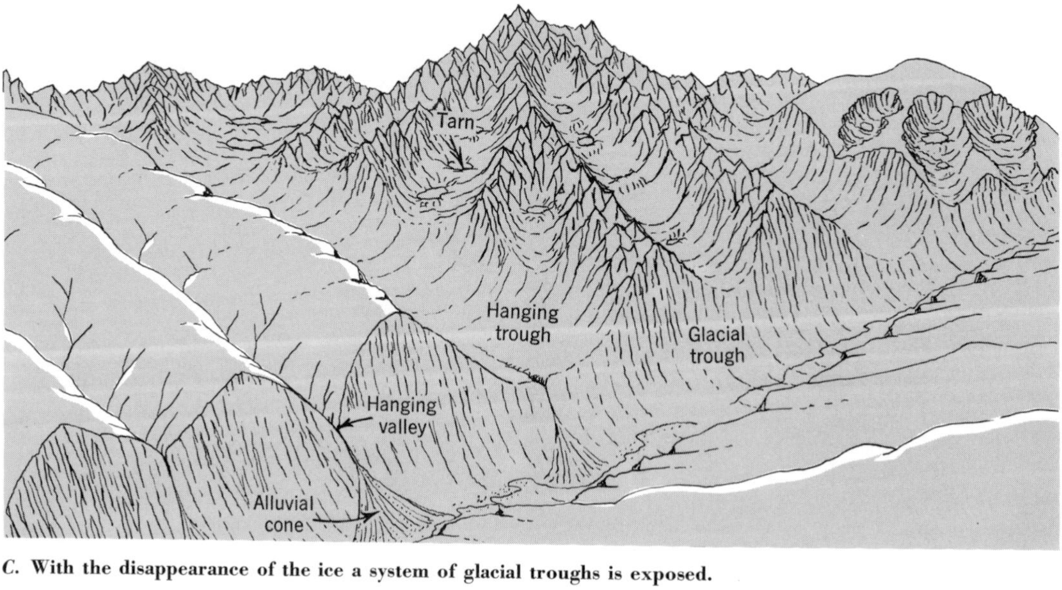

***C.* With the disappearance of the ice a system of glacial troughs is exposed.**

been excessive, a pass or notch, called a *col*, is formed.

Glacier flow constantly deepens and widens its channel so that after the ice has finally disappeared there remains a deep, steep-walled *glacial trough*, characterized by a relatively straight or direct course and by the U-shape of its transverse profile (Block *C*, Figure 31.3). Tributary glaciers likewise carve U-shaped troughs, but they are smaller in cross section, with floors lying high above the floor level of the main trough, so are called *hanging troughs*. Streams, which later occupy the abandoned trough systems, form scenic waterfalls and cascades where they pass down from the lip of a hanging trough to the floor of the main trough. These streams quickly cut a small V-shaped notch in the trough bottom.

Valley spurs that formerly extended down to the main stream before glaciation occurred have been beveled off by ice abrasion and are termed *truncated spurs* (Block *B*). Under a glacier the bedrock is not always evenly excavated, so that the floors of troughs and cirques may contain *rock basins* and *rock steps*. Cirques and upper parts of troughs thus are occupied by numerous small lakes, called *tarns* (Figure 31.5). The major troughs frequently contain large, elongate *trough lakes*, sometimes referred to as *finger lakes*. Landslides are numerous because glaciation leaves oversteepened trough walls. In glaciated countries such as Switzerland and Norway slides are a major type of natural disaster, because most towns and cities lie in the trough floors where they are readily destroyed by mudflows and landslides (Chapter 24).

Figure 31.6 The braided stream in the foreground is aggrading the floor of a glacial trough. A shrunken glacier in the distance provides the meltwater and debris. Peters Creek, Chugach Mountains, Alaska. (Photograph by Steve McCutcheon. Alaska Pictorial Service.)

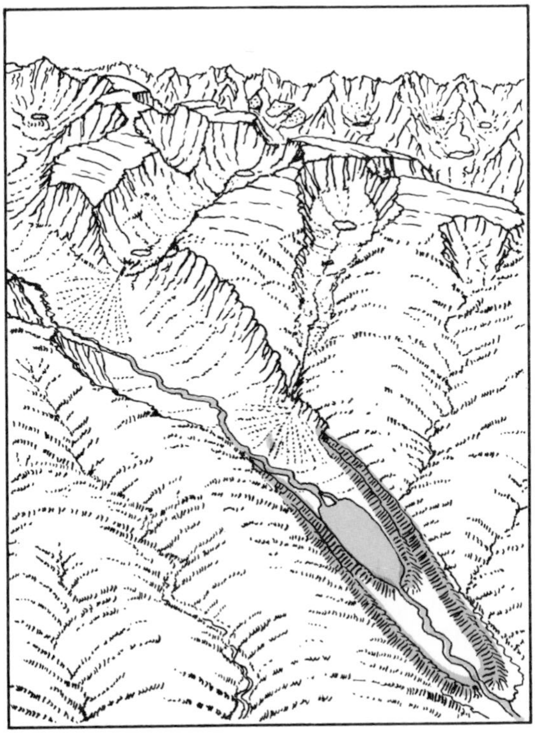

Figure 31.5 Moraines of a former valley glacier appear as looped embankments marking successive positions of the ice margins. (After W. M. Davis.)

Debris may be carried by an alpine glacier within the ice, or it may be dragged along between the ice and the valley wall as a *lateral moraine* (Figures 31.2 and 31.3*B*). Where two ice streams join, this marginal debris is dragged along to form a *medial moraine*, riding upon the ice in midstream (Figure 31.1). At the terminus of a glacier debris accumulates in a heap known as a *terminal moraine*, or *end moraine*. This heap is usually in the form of a curved embankment lying across the valley floor and bending upvalley along each wall of the trough to merge with the lateral moraines (Figures 31.2 and 31.5). As the end of the glacier wastes back, scattered debris is left behind. Successive halts in ice retreat produce successive moraines, termed *recessional moraines*.

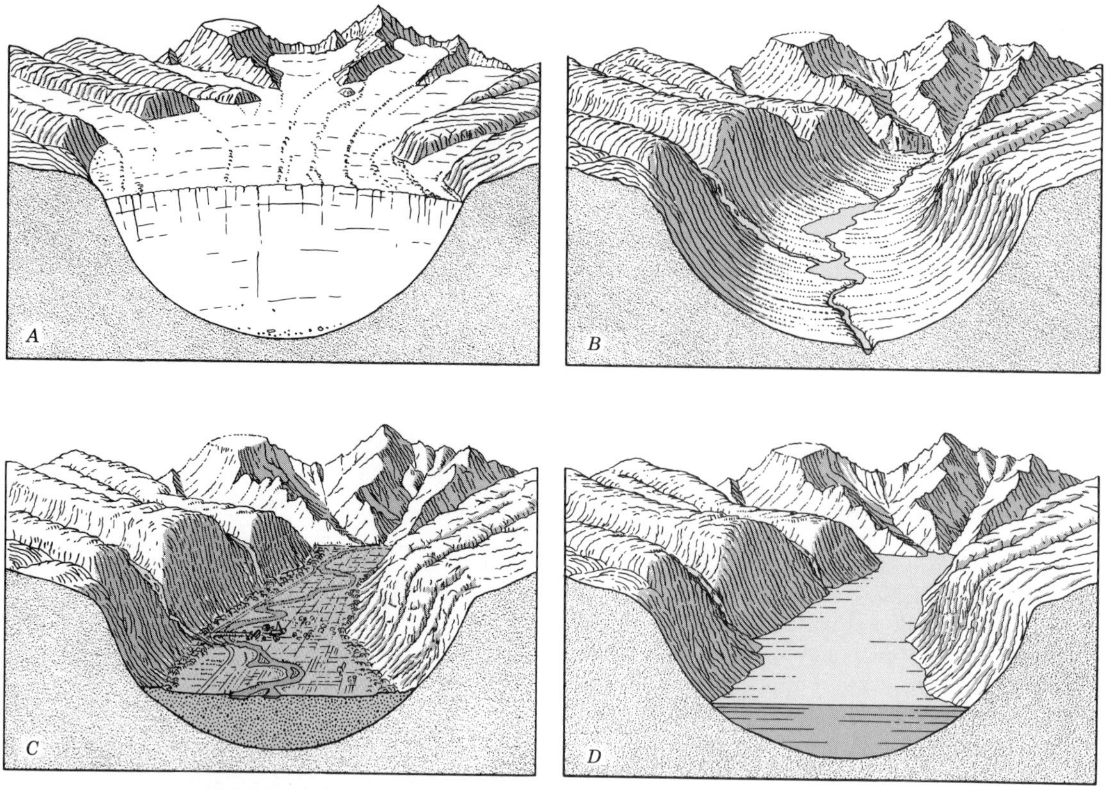

Figure 31.7 Development of a glacial trough. (After E. Raisz.) *A*, During maximum glaciation, the U-shaped trough is filled by ice to the level of the small tributaries. *B*, After glaciation, the trough floor may be occupied by a stream and lakes. *C*, If the main stream is heavily loaded, it may fill the trough floor with alluvium. *D*, Should the glacial trough have been deepened below sea level, it will be occupied by an arm of the sea, or fiord.

Glacial troughs and fiords

Many large glacial troughs now are nearly flat-floored because aggrading streams that issued from the receding ice front were heavily laden with rock fragments. The deposit of alluvium extending downvalley from a melting glacier is the *valley train.* Figure 31.6 shows a braided meltwater stream in the process of constructing a valley train. Figure 31.7 shows a comparison between a trough with little or no fill and another with an alluvial-filled bottom.

When the floor of a trough open to the sea lies below sea level, the sea water will enter as the ice front recedes, thus producing a narrow estuary known as a *fiord* (Figure 31.8). Fiords may originate either by submergence of the coast or by glacial erosion to a depth below sea level. Most fiords are explained in the second way because ice is of such a density that, when floating, from three-fourths to nine-tenths of its mass lies below water level. Therefore, a glacier several hundred feet thick could erode to considerable depth below sea level before the buoyancy of the water reduced its erosive power where it entered the open water. Fiords are observed to be opening up today along the Alaskan coast, where some glaciers are melting back rapidly and the fiord waters are being extended along the troughs. Fiords are found largely along mountainous coasts in latitudes 50° to 70° N and S. The explanation of this distribution lies in climate and is treated under the discussion of marine west coast climate (Chapter 16).

Environmental aspects of alpine glaciation

In general, the ruggedness of fully glaciated mountains such as the Alps, Pyrenees, Himalayas, or Sierra Nevada makes for sparseness of population and difficulty of access. Land above timber line is useless for any purpose except summer pastures and the extraction of such minerals as may lie in the rocks. Locally, recreational uses—mountain climbing and winter sports—are intensively developed.

Below timber line are rich forests. U-shaped glacial troughs provide broad, accessible strips of land at relatively low levels. These are utilized for town sites, for winter pasture, and as arteries of transportation. In the Italian Alps several great flat-floored glacial troughs extend from the heart

Figure 31.8 This Norwegian fiord has the steep rock walls of a deep glacial trough. (Photograph by Mittet and Co.)

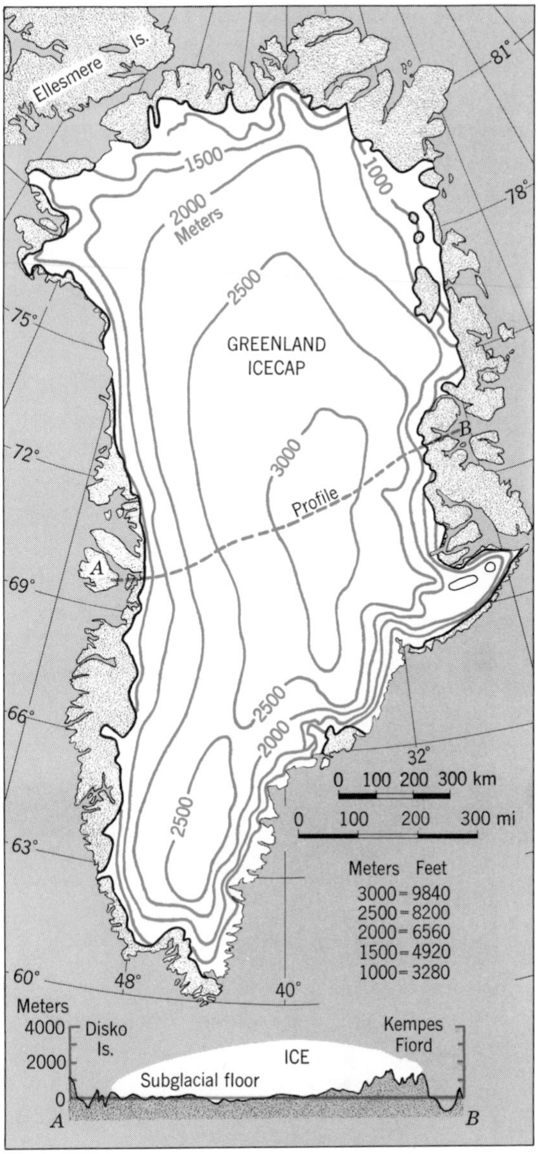

Figure 31.9 Generalized map of Greenland. (After R. F. Flint, *Glacial and Pleistocene Geology*.)

of the Alps southward to the plain of northern Italy. These are important geographic controls because they provide smooth and easy access into the heart of the Alps and to the principal Alpine passes. The Brenner Pass lies at the head of a magnificent trough of this type, the Adige River valley. The steep-walled troughs contain many waterfalls and rapids readily used for hydroelectric plants.

Ice sheets of the present

Two enormous accumulations of glacial ice are the Greenland and Antarctic ice sheets. These are huge plates of ice, several thousand feet thick in the central areas, resting upon landmasses of subcontinental size. The Greenland ice sheet has an area of 670,000 sq mi (1,740,000 sq km) and occupies about seven-eighths of the entire island of Greenland (Figure 31.9). Only a narrow, mountainous coastal strip of land is exposed.

The Antarctic ice sheet covers 5,000,000 sq mi (13,000,000 sq km) and in places spreads out into the ocean to form floating ice layers (Figure 31.10). A significant point of difference between the two ice sheets is their position with reference to the poles. Whereas the antarctic ice rests almost squarely upon the south pole, the Greenland ice sheet is considerably offset from the north pole, with its center about at 75° N lat. This position illustrates a fundamental principle; that a large area of high land is essential to the accumulation of a great ice sheet. No land exists near the north pole; ice accumulates there only as sea ice.

Contours drawn upon the surface of the Greenland ice sheet show it to have the form of a very broad, smooth dome. From a high point of about 10,000 ft (3000 m) elevation east of the center there is a gradual slope outwards in all directions. The rock floor of the ice sheet lies near sea level under the central region, but is higher near the edges. Accumulating snows add layers of ice to the surface, while at great depth the plastic ice slowly flows outward toward the edges. At the outer edge of the sheet the ice thins down to a few hundreds of feet. Continual loss through ablation keeps the position of the ice margin relatively steady where it is bordered by a coastal belt of land. Elsewhere the ice extends in long tongues, called *outlet glaciers,* to reach the sea at the heads of fiords. From the floating glacier edge huge masses of ice break off and drift out to open sea with tidal currents. The breakup of the ice front is known as *calving* and is brought about by strains caused by the rise and fall of tide level as well as by the undercutting

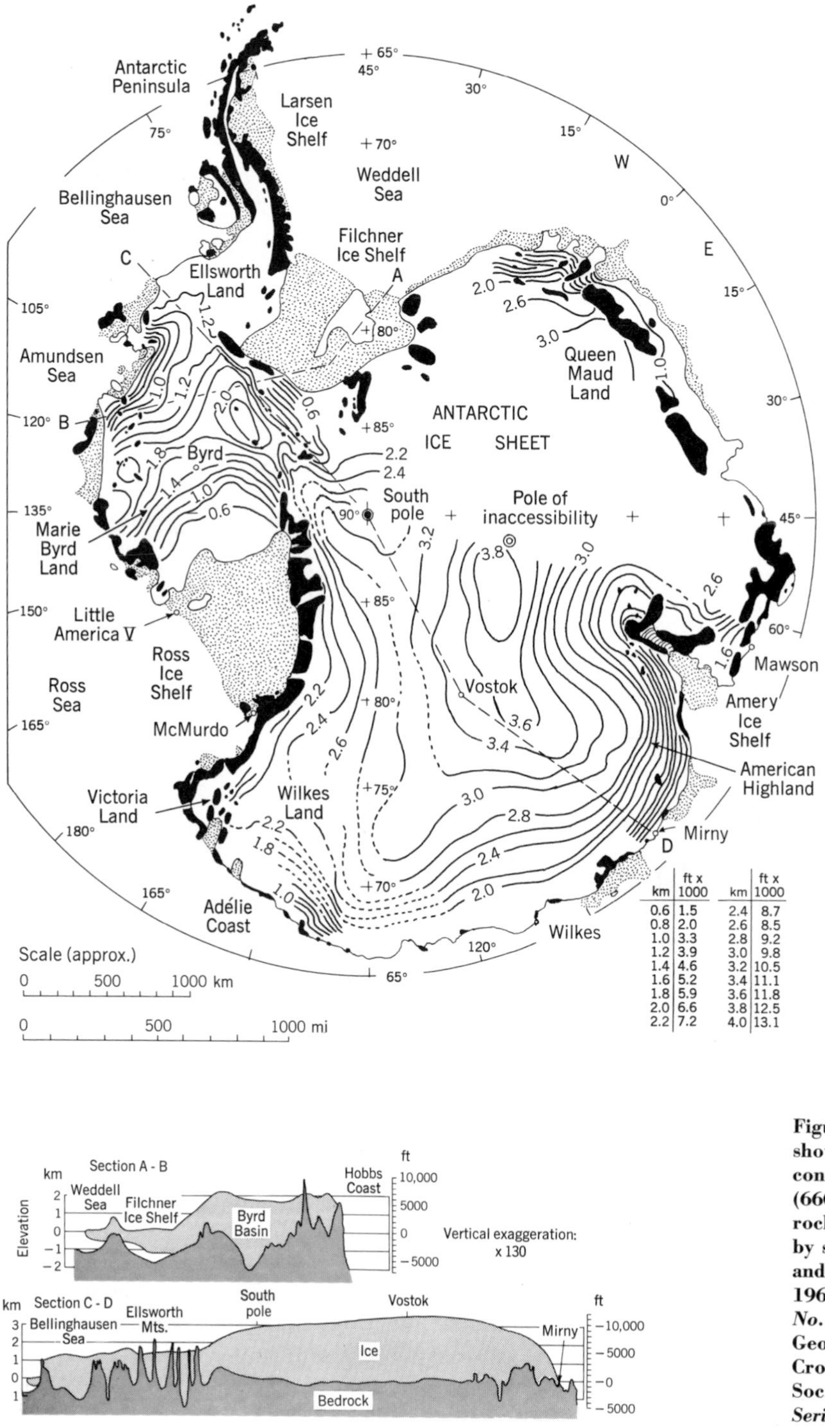

km	ft x 1000	km	ft x 1000
0.6	1.5	2.4	8.7
0.8	2.0	2.6	8.5
1.0	3.3	2.8	9.2
1.2	3.9	3.0	9.8
1.4	4.6	3.2	10.5
1.6	5.2	3.4	11.1
1.8	5.9	3.6	11.8
2.0	6.6	3.8	12.5
2.2	7.2	4.0	13.1

Figure 31.10 Map of Antarctica showing ice surface elevations by contours with interval of 0.2 km (660 ft). Areas of exposed bedrock shown in black; ice shelves by stipple pattern. (Map redrawn and simplified from C. R. Bentley, 1962, *Geophysical Monograph No. 7,* Washington, D.C., Amer. Geophys. Union, p. 14, Figure 2. Cross sections after Amer. Geog. Soc. (1964), *Antarctic Map Folio Series,* Folio 2, Plate 2.)

and melting at and below the water line. The calving of floating glacier fronts is an extremely rapid process compared to ablation of ice fronts on land. Consequently, icecaps are limited in their seaward extent and rarely extend far into the ocean beyond the limits of bays and shallow continental shelves.

Ice thickness in Antarctica is even greater than that of Greenland. For example, on Marie Byrd Land a thickness of 13,000 ft (4000 m) was measured, the rock floor lying 6500 ft (2000 m) below sea level.

An important glacial feature of Antarctica is the presence of great plates of floating glacial ice, termed *ice shelves* (Figure 31.10). The largest of these is the Ross Ice Shelf with an area of about 200,000 sq mi (520,000 sq km) and a surface elevation averaging about 225 ft (70 m) above the sea. Ice shelves are fed by the ice sheet, but also accumulate new ice through the compaction of snow.

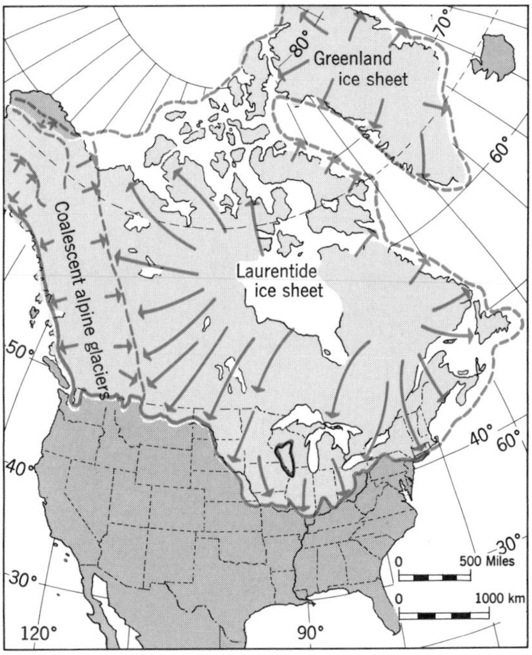

Figure 31.11 Pleistocene ice sheets of North America at their maximum spread reached as far south as the present Ohio and Missouri rivers. (After R. F. Flint.)

Ice sheets of the Pleistocene Epoch

Much of North America and Europe and parts of northern Asia and southern South America were covered by enormous ice sheets in a period of time designated the Pleistocene Epoch. This ice age ended 10,000 to 15,000 years ago with the rapid wasting away of the ice sheets. Landforms made by the last ice advance and recession are very little modified by erosional agents.

Figures 31.11 and 31.12 show the extent to which North America and Europe were covered at the maximum known spread of the last advance of the ice. In the United States, most of the land lying north of the Missouri and Ohio rivers was covered, as well as northern Pennsylvania and all of New York and New England. In Europe, the ice sheet centered upon the Baltic Sea, covering the Scandinavian countries and spreading as far south as central Germany. The British Isles were almost covered by an icecap that had several centers on highland areas and spread outward to coalesce with the Scandinavian ice sheet. The Alps at the same time were heavily inundated by enlarged alpine glaciers, fused into a single icecap. All high mountain areas of the world underwent greatly intensified alpine glaciation at the time of maximum ice-sheet advance. Today, only small remnant alpine glaciers. In less favorable mountain regions no glaciers remain.

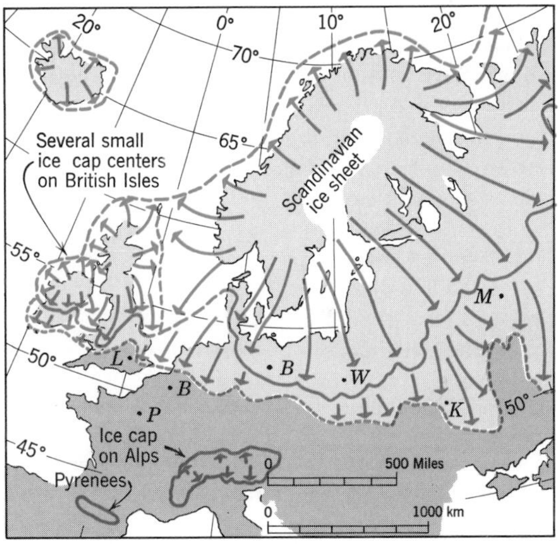

Figure 31.12 The Scandinavian ice sheet dominated northern Europe during the Pleistocene glaciations. Solid line shows limits of ice in the last glacial stage; dotted line on land shows maximum extent at any time. (After R. F. Flint.)

Proof of the former great extent of ice sheets has been carefully accumulated since the middle of the nineteenth century when the great naturalist Louis Agassiz first announced the bold theory. Careful study of the deposits left by the ice has led to the knowledge that not one but four major advances and retreats occurred, spaced over a total period of about a million years. It is the deposits of the last advance, known as the *Wisconsin Stage*, that form fresh and conspicuous landforms. For the most part, then, the discussion of glacial landforms will concern these most recent deposits.

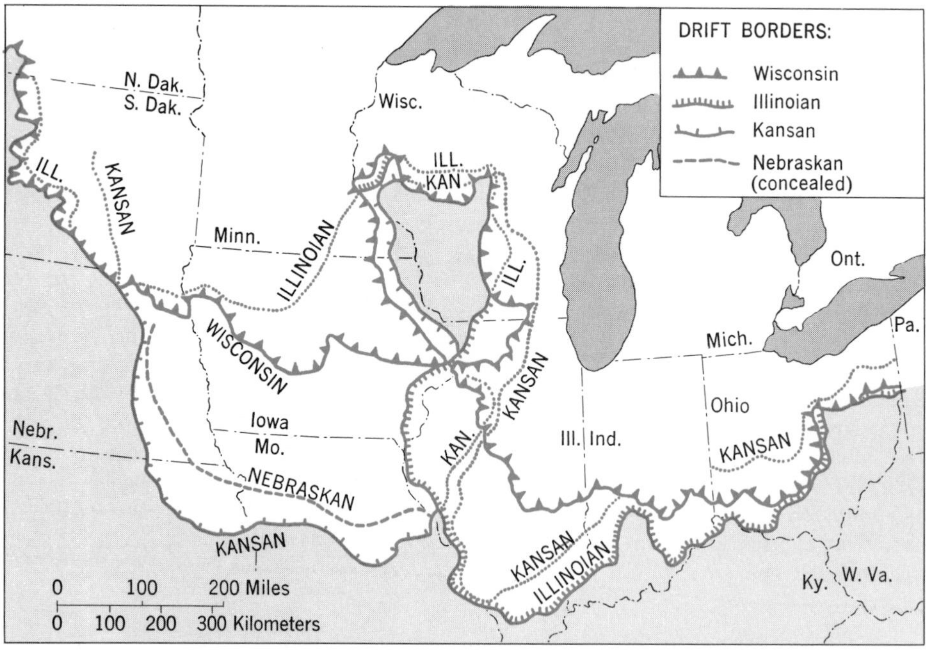

Figure 31.13 In each glacial stage, the ice sheet reached a different line of maximum advance. (After R. F. Flint, *Glacial and Pleistocene Geology*.)

Glacial stages

The four principal glacial stages of North America are matched by corresponding stages of Europe, as deciphered in the Alps. Together with the interglacial stages (periods of mild climate during which the ice sheets disappeared), these glacial stages in order of increasing age are:

North American Stages (Northcentral U.S.)	European Stages (Alps)
Wisconsin glacial	Würm glacial
Sangamon interglacial	Riss-Würm interglacial
Illinoian glacial	Riss glacial
Yarmouth interglacial	Mindel-Riss interglacial
Kansan glacial	Mindel glacial
Aftonian interglacial	Günz-Mindel interglacial
Nebraskan glacial	Günz glacial

Figure 31.13 shows the limit of southerly spread of ice of each glacial stage in the north-central United States. Where older limits were overridden by younger, the boundaries are conjectural. Notice that an area in Wisconsin (the *Driftless Area*) was surrounded by ice but never inundated.

The absolute age and duration in years of the Pleistocene Epoch and its stages are extremely difficult to fix, even though the relative order of events is well established. The last ice disappeared 10,000 to 15,000 years ago from the north-central United States. The Wisconsin Stage may have endured 60,000 years. The earliest onset of glacial advance (Nebraskan) may have occurred 300,000 to 600,000 years ago, although the beginning of the Pleistocene Epoch of geologic time is commonly set by geologists at about one million years.

Cause of continental glaciation

It should not be inferred that the occurrence of an ice age as the last event of geologic history means that the earth as a planet is cooling off. There is excellent evidence in the form of rock deposits of similar periods of glaciation in the early part of the earth's geologic history. But, beyond the fact that glaciations are occasional events in geologic history, knowledge concerning the cause of glaciation is speculative. Obviously a prolonged period of colder climate with ample snowfall brings on the growth of ice sheets.

One possible contributing cause of glaciation is a decrease in the quantity of solar radiation intercepted by the earth. It has been reasoned that, should this quantity of energy have diminished somewhat at the beginning of the Pleistocene epoch, the average temperature of the earth's atmosphere would have dropped to a lower level. Providing that this change did not also diminish the quantity of snowfall, the reduced ablation of snowfields would increase the quantity of snow turned into glacial ice, with resultant growth and spread of icecaps. A second factor, possibly working in harmony with the first, is the known increase

Figure 31.14 Glacial striations and fracture marks, mostly crescentic gouges, cover the smoothly rounded surface of this rock knob. These marks were made by the East Twin Glacier, Alaska. The ice moved in a direction away from the photographer. (Photograph by Maynard Miller, Foundation for Glacier Research.)

in elevation of large parts of the continents during the Pliocene and early Pleistocene epochs as a result of widespread mountain making (orogeny), as well as broad uparching of continental interiors (epeirogenic uplift). That mountains intercept large quantities of precipitation has already been explained (Chapter 10) in the discussion of orographic precipitation. The combined effects of reduced solar energy and increased altitude of continents would bring on colder climates with increased snowfall over favorable areas of the continents, such as the Laurentian Upland of eastern Canada and the Scandinavian peninsula. This composite theory of cause of glaciation may be called the *solar-topographic theory*.

An important and widely held theory attributes glaciation to a reduction of the carbon dioxide content of the atmosphere. The role of carbon dioxide in absorbing long-wave radiation and thus warming the atmosphere has been explained in Chapter 8. It is estimated that if the carbon dioxide content of the atmosphere, which is now 0.03 percent by volume, were reduced by half that amount, the earth's average surface temperature would drop about 7 F° (4 C°). Such a reduction in carbon dioxide content might be postulated in combination with increased continental altitude to bring on the growth of ice sheets.

Other theories invoke quite different mechanisms. It has been postulated that increased quantities of volcanic dust in the atmosphere might bring about a glacial period because more solar energy would be reflected back into space, permitting less to enter the lower atmosphere. Along with the reduced air temperature would be the increase in numbers of tiny dust particles to serve as nuclei for the condensation of moisture, thus favoring increased precipitation. Another group of theories propose shifts in the positions of the continents with respect to the poles, bringing various parts of the landmasses into favorable geographical positions for the growth of ice sheets. Still another theory requires that changes in oceanic currents, specifically the diversion or blocking of such warm currents as the Gulf Stream, would have brought colder climates to the subarctic regions. Variations in the earth's orbit, causing changes in the amounts of solar energy received by the earth, have also been considered as the cause of glacial periods.

Erosion by ice sheets

Like alpine glaciers, ice sheets are effective eroding agents. The slowly moving ice may scrape and grind away much solid bedrock, leaving behind smoothly rounded rock masses bearing countless minute abrasion marks. Scratches, or *striations*, trend in the general direction of ice movement (Figure 31.14), but variations in ice direction from time to time often result in intersecting lines. Certain kinds of rock were susceptible to deep grooving (Figure 31.15).

Where a strong, sharp-pointed piece of rock was held by the ice and dragged over the bedrock surface, there resulted a series of curved cracks fitted together along the line of ice movement. These *chatter marks*, and closely related *crescentic gouges*, whose curvature is the opposite, are good indicators of the direction of ice movement (Figure 31.14). Some very hard rocks have acquired highly polished surfaces from the rubbing of fine clay particles against the rock. The evidences of ice erosion described here are common throughout the northeastern United States. They may be seen on almost any exposed hard rock surface. Once understood, they may be recognized by any alert observer.

Commonly bearing the above abrasion marks is a type of conspicuous knob of solid bedrock that has been shaped by the moving ice (Figure 31.16). One side, that from which the ice was approaching, is characteristically smoothly rounded and shows a striated and grooved surface. This is termed the *stoss* side. The other, or *lee* side, where the ice plucked out angular joint blocks, is irregular, blocky, and steeper than the stoss side. The quaint term *roches moutonnées* has long been applied by glaciologists to such glaciated rock knobs.

Figure 31.15 Unusually smooth, deep glacial grooves were carved in limestone by the ice action on Kelley's Island near the south shore of Lake Erie. (Photograph by State of Ohio, Department of Industrial and Economic Development.)

Figure 31.17 Seen from the air, this esker in the Canadian shield area appears as a narrow embankment crossing the terrain of glacially eroded lake basins. (Photograph by Canadian Department of Mines, Geological Survey.)

Vastly more important than the minor abrasion forms are enormous excavations that the ice sheets made in some localities where the bedrock is of a weak variety and the ice current was accentuated by the presence of a valley paralleling the direction of ice flow. Under such conditions the ice sheet behaved much as a valley glacier, scooping out a deep, U-shaped trough. As fine examples may be cited the Finger Lakes of western New York State. Here a set of former stream valleys lay parallel to southward spread of the ice, which scooped out a series of deep troughs. Blocked at the north ends by glacial debris the basins now hold elongated lakes. Many hundreds of lake basins were created in a similar manner all over the glaciated portion of North America and Europe. Countless small lakes of Minnesota, Canada, and Finland occupy rock basins scooped out by ice action (Figure 31.17). Irregular debris deposits left by the ice are also important in causing lake basins.

Deposits left by ice sheets

The term *glacial drift* has long been applied to include all varieties of rock debris deposited in close association with glaciers. Drift is of two major types: (1) *Stratified drift* consists of layers of sorted and stratified clays, silts, sands, or gravels deposited by melt-water streams or in bodies of water adjacent to the ice. (2) *Till* is a heterogeneous mixture of rock fragments ranging in size from clay to boulders and is deposited directly from the ice without water transport. Moraines of valley glaciers, previously described, are composed largely of till, whereas the valley train is composed of stratified drift.

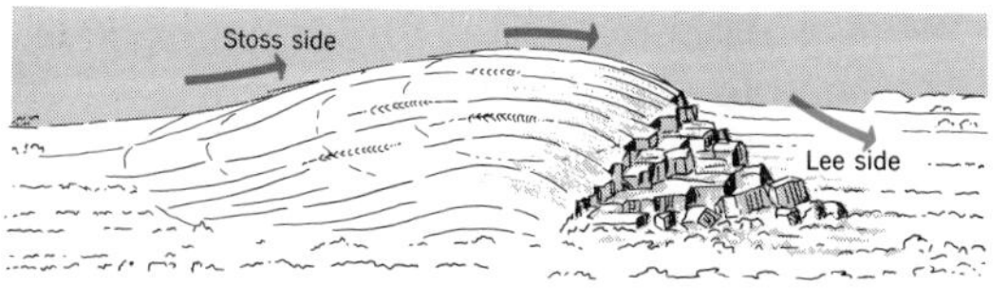

Figure 31.16 A glacially abraded rock knob. (From A. N. Strahler, 1971, *The Earth Sciences*, 2nd ed., Harper and Row, New York.)

Over those parts of the United States formerly covered by Pleistocene ice sheets, glacial drift averages from 20 ft (6 m) thick over mountainous terrain such as New England, to 50 ft (15 m) and more thick over the lowlands of the north-central United States. Over Iowa, drift is from 150 to 200 ft (45 to 60 m) thick; over Illinois, it averages more than 100 ft (30 m) thick. Locally, where deep stream valleys existed prior to glacial advance, as in Ohio, drift may be several hundred feet deep.

In order to understand the form and composition of deposits left by ice sheets, it is desirable to consider the conditions prevailing at the time of existence of the ice, as shown in Figure 31.18. Block *A* shows a region partly covered by an ice sheet with a relatively stationary front edge. This condition occurs when the rate of ice ablation balances the amount of ice brought forward by spreading of the ice sheet. Any increase in ice movement would cause the ice to shove forward to cover more ground; an increase in the rate of wasting would cause the edge to recede and the ice surface to become lowered. Although the Pleistocene ice fronts did advance and recede in many minor and major fluctuations, there were considerable periods when the front was essentially stable. This condition is represented in Block *A*.

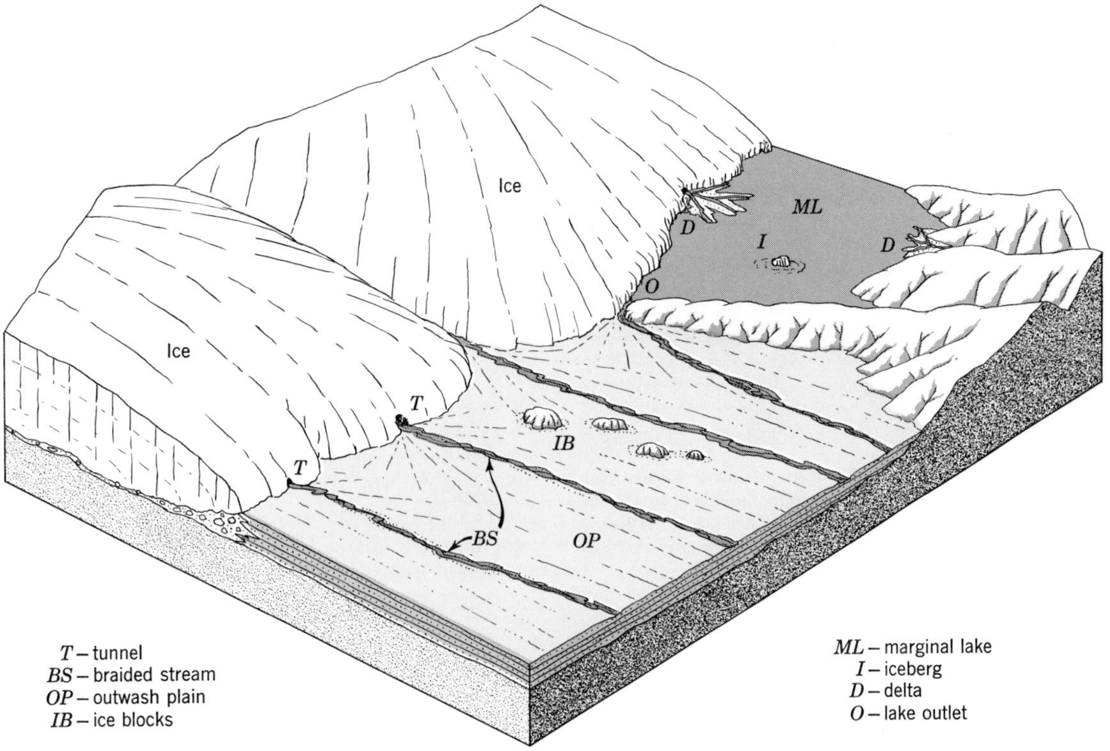

A. With the ice front stabilized and the ice in a wasting, stagnant condition, various depositional features are built by melt water.

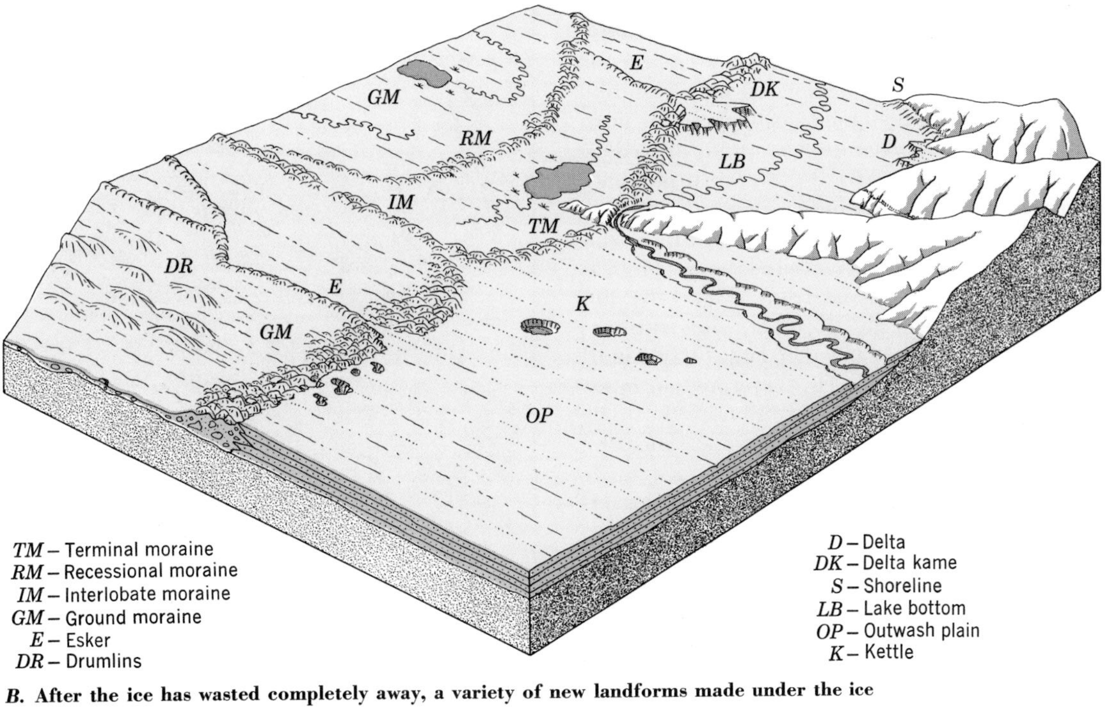

B. After the ice has wasted completely away, a variety of new landforms made under the ice is exposed to view.

Figure 31.18 Marginal landforms of continental glaciers.

Figure 31.19 Rugged topography of small knobs and kettles characterizes this interlobate moraine northeast of Elkhart Lake, Sheboygan County, Wisconsin. (Photograph by W. C. Alden, U.S. Geological Survey.)

The transportational work of an ice sheet may be likened to that of a great conveyor belt. Anything carried on the belt is dumped off at the end and if not constantly removed will pile up in increasing quantity. Rock fragments brought within the ice are deposited at the edge as the ice evaporates or melts. There is no possibility of return transportation.

Glacial till that accumulates at the immediate ice edge forms an irregular, rubbly heap known as a *terminal moraine.* After the ice has disappeared, as in Diagram *B*, the moraine appears as a belt of knobby hills interspersed with basinlike hollows, some of which hold small lakes. The term *knob and kettle topography* is often applied to morainal belts (Figure 31.19). Terminal moraines tend to form great curving patterns, the convex form of curvature being directed southward and indicating that the ice advanced as a series of great *lobes,* each with a curved front (Figure 31.20). Where two lobes came together, the moraines curved back and fused together into a single moraine pointed northward. This is termed an *interlobate moraine* (Figure 31.18, Block *B*). In its general recession accompanying disappearance, the ice front paused for some time along a number of lines, causing morainal belts similar to the terminal moraine belt to be formed. These belts, known as *recessional moraines* (Figures 31.18 and 31.20), run roughly parallel with the terminal moraine but are often thin and discontinuous.

Figure 31.18, Block *A*, shows a smooth, sloping plain lying in front of the ice margin. This is the *outwash plain,* formed of stratified drift left by braided streams issuing from the ice. Their deposits are in reality great alluvial fans upon which are spread layer upon layer of sands and gravels. The adjective *glaciofluvial* is often applied to stream-laid stratified drift.

Large streams issue from tunnels in the ice, particularly when the ice for many miles back from the front has become stagnant, without forward movement. Tunnels then develop throughout the ice mass, serving to carry off the melt water. After the ice has gone (Block *B*, Figure 31.18) the outwash plain remains in its original form, but may be bounded on the iceward side by a steep slope which is the mold of the ice against which the outwash was built. Such a slope is called an *ice-contact slope.* Farther back, behind the terminal moraine, the position of a former ice tunnel is marked by a long, sinuous ridge known as an *esker.* The esker is the deposit of sand and gravel formerly laid upon the floor of the ice tunnel. Inasmuch as ice formed the sides and roof of the tunnel, its disappearance left merely the stream-bed deposit, which now forms a ridge (Figure 31.21). Eskers are often many miles long; in parts of Maine a few are more than 100 mi (160 km) long. Some have branches just as streams do. Because the esker is made of highly porous sand and gravel, the rapid draining away of water from the crest may prevent

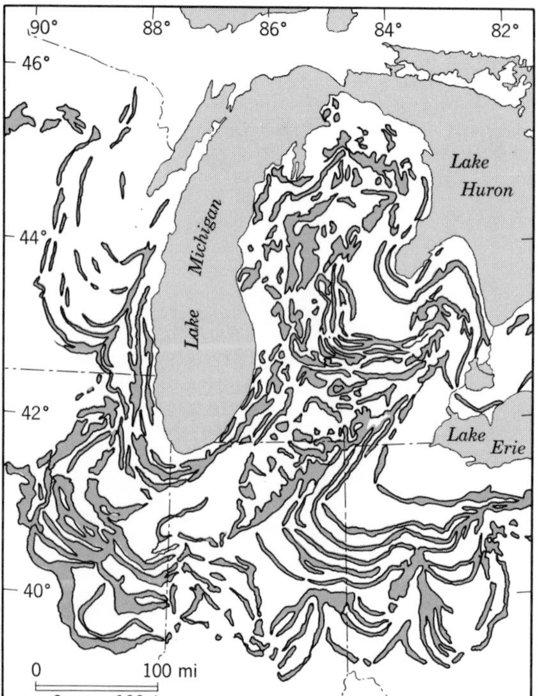

Figure 31.20 Moraine belts of the north-central United States have a festooned pattern left by ice lobes. (After R. F. Flint and others, *Glacial Map of North America,* 1945.)

the growth of trees along the top of some eskers, which look as if artificially cleared of forest (Figure 31.17).

Another curious glacial form is the *drumlin*, a smoothly rounded, oval hill resembling the bowl of an inverted teaspoon. It consists of glacial till (Figure 31.22). Drumlins invariably lie in a zone behind the terminal or recessional moraines. They commonly occur in groups or swarms, which may number in the hundreds. The long axis of each drumlin parallels the direction of ice movement, and the drumlins thus point toward the terminal moraines and serve as indicators of direction of ice movement. From a study of the composition and structure of drumlins, it has been generally agreed that they were formed under moving ice by a kind of plastering action in which layer upon layer of bouldery clay was spread upon the drumlin. This would have been possible only if the ice were so heavily choked with debris that the excess had to be left behind.

Between the terminal, recessional, and interlobate moraines, the surface left by the ice is usually overspread by a cover of glacial till known as *ground moraine*. This cover is often inconspicuous because it forms no prominent or recognizable topographic feature. Nevertheless, the ground moraine may be thick and may obscure or entirely bury the hills and valleys that existed before glaciation. Where thick and smoothly spread, the ground moraine forms an extensive, level *till plain*, but this condition is likely only in regions already fairly flat to start with. In more dissected regions, the preglacial valleys and hills retain their same general outlines despite glaciation.

Deposits built into standing water

Where the general land slope is toward the front of an ice sheet, a natural topographic basin is formed between the ice front and the rising ground. Valleys that may have opened out northward are blocked by ice. Under such conditions, *marginal glacial lakes* form along the ice front (Figure 31.18, Block *A*). These lakes overflow along the lowest available channel, which lies between the ice and the ground slope or over some low pass along a divide. Into marginal lakes streams of melt water from the ice build *glacial deltas*, similar in most respects to deltas formed by any stream flowing into a lake. Streams from the land likewise build deltas into the lake. When the ice has disappeared the lake drains away, exposing the bottom upon which layers of fine clay and silt have been laid. These fine-grained strata, which have settled out from suspension in turbid lake waters, are called *glaciolacustrine* sediments and are a variety of stratified drift. The layers are commonly of banded appearance, with alternating dark and light layers, termed *varves*. Glacial lake plains are extremely flat, with meandering streams and extensive areas of marshland.

Figure 31.21 This esker has developed a cover of soil and vegetation, concealing the coarse gravel that lies within it. Dodge County, Wisconsin. (Photograph by W. C. Alden, U.S. Geological Survey.)

Deltas, built with a flat top at what was formerly the lake level, are now curiously isolated, flat-topped landforms known as *delta kames*. Delta and stream channel deposits built between a stagnant ice mass and the wall of a valley become *kame terraces*, whose steep scarps are ice-contact slopes (Figure 31.23). Kame terraces are difficult to distinguish from the uppermost member of a series of alluvial terraces, but most kames have undrained depressions or pits produced by the melting of enclosed ice blocks. Built of very well-washed and sorted sands and gravels, kames commonly show the steeply dipping foreset beds characteristic of deltas (Figure 31.24).

Figure 31.22 This small drumlin, located south of Sodus, New York, shows a tapered form from upper right to lower left, indicating that the ice moved in that direction (north to south). (Photograph by Ward's Natural Science Establishment, Inc., Rochester, N.Y.)

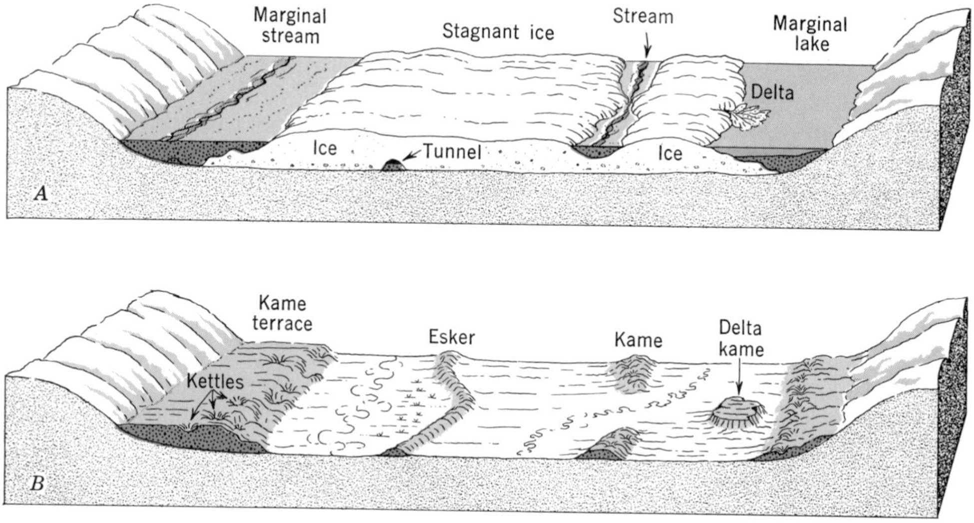

Figure 31.23 Kames may originate as stream or lake deposits laid between a stagnant ice mass and the valley sides. (After R. F. Flint.)

Environmental aspects of glacial landforms

Because much of Europe and North America was glaciated by the Pleistocene ice sheets, landforms associated with the ice are of fundamental environmental importance in influencing activities of human beings. Agricultural influences of glaciation are both favorable and unfavorable, depending on preglacial topography and whether the ice eroded or deposited heavily. In hilly or mountainous regions, such as New England, the glacial till is thinly distributed and extremely stony. Cultivation is made difficult by countless boulders and stones in the soil. Along morainal belts the steep slopes, irregularity of topography, and abundance of boulders and stones in the till are unfavorable to cultivation. Hills and lakes, however, make morainal belts extremely desirable as suburban residential areas.

Extensive till plains, outwash plains, and lake plains, on the other hand, make for some of the

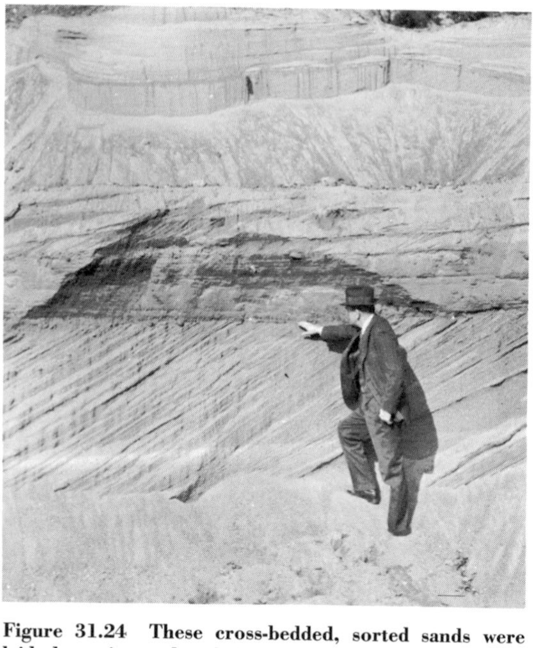

Figure 31.24 These cross-bedded, sorted sands were laid down in a glacial delta near North Haven, Connecticut. (Photograph by R. J. Lougee.)

Figure 31.25 Thick layers of outwash sands and gravels such as these on the north shore of Long Island are excavated in great quantities for use in highway and building construction. The dark layer at the top is a bed of glacial till, left by a glacial advance. Boulders in the foreground are glacial erratics that have rolled down from the till bed. (Photograph by A. K. Lobeck.)

most productive agricultural land in the world. In this class belong the prairie lands of Indiana, Illinois, Iowa, Nebraska, and Minnesota.

Glaciofluvial deposits are of great economic value. The sands and gravels of outwash plains, kames, and eskers provide the aggregate necessary for concrete and other building purposes (Figure 31.25). The purest sands may be used for molds needed for metal castings. Huge quantities of ground water are contained in glaciofluvial deposits. Where deep, preglacial valleys were filled with such materials, large quantities of water can be pumped from wells penetrating the deposit. In this way, many large cities and industrial plants in Ohio, Pennsylvania, and New York obtain their water supplies.

REVIEW QUESTIONS

1. What is a glacier? What conditions are necessary for the formation of glaciers? Define ablation. What is firn? What is the distinction between alpine and continental glaciers?

2. Describe a simple alpine glacier and explain how it operates. What is glacier equilibrium? How fast does a glacier flow?

3. How do glaciers erode their channels? Compare alpine glaciers with streams in regard to erosion and transportation activities.

4. How do alpine glaciers modify mountain topography? Describe and explain the following features: cirque, arête, horn, col, glacial trough, hanging trough, rock basin, rock step, tarn, finger lake.

5. What is the form of a glacial trough? How does this form differ from that of a normally eroded stream valley? Explain how a fiord is formed. Where are fiords found?

6. Describe the various kinds of deposits made by a valley glacier. What are the location and form of the following types of moraines: medial, lateral, terminal, and recessional? What is a valley train?

7. Where are the ice sheets of the world today? How thick is the ice in these ice sheets? Explain how an ice sheet is fed and how the ice moves. What are ice shelves? Outlet glaciers? What is calving?

8. Describe the extent of ice sheets of the Pleistocene Epoch in North America and Europe. Describe the southern glacial limit in the United States. What great scientist was largely responsible for convincing the scientific world that the Pleistocene glaciation actually occurred?

9. Name in order the glacial and interglacial stages of North America. How long ago did the last glacial ice disappear from the United States?

10. Review the principal theories that attempt to explain the occurrence of glacial periods.

11. What erosional features on rock surfaces give evidence of the former presence of an ice sheet? How can the direction of ice movement be inferred?

12. How can the direction of ice movement be inferred from the shapes of a glaciated rock knobs (roches moutonnées)? How were the Finger Lakes of western New York State formed?

13. What is glacial drift? What is the distinction between stratified drift and till?

14. Describe and explain the various depositional forms associated with the margin of an ice sheet.

15. What kinds of moraines are left by ice sheets? What is a glacial outwash plain? Describe the surface topography of an outwash plain.

16. How is an esker formed? How long are eskers? Of what material are they composed?

17. What is a drumlin? Of what material is it composed? How is it formed? Where are most drumlins found in relation to the terminal moraine?

18. What is a till plain? How is it formed?

19. Explain how marginal glacial lakes are formed. What types of deposits are formed in them? What is a delta kame? A kame terrace? What are varves?

20. Discuss the environmental and resource aspects of glacial landforms.

Exercises

Exercise 1. ***Young Stage of Alpine Glaciation.*** (Source: Cloud Peak, Wyo., U.S. Geological Survey topographic map; scale 1:125,000.)

Explanatory Note: Steep-walled cirques have been cut into the broadly rounded summit of the Big Horn range in northern Wyoming, but much of the preglacial surface remains. The cirques contain lakes which form chainlike groups extending down the glacial troughs which lead from the cirques.

QUESTIONS

1. **(*a*)** Determine the fractional scale of this map. **(*b*)** What is the contour interval? **(*c*)** Give location (grid coordinates) and the approximate elevation of the highest point on this map.

2. What is the depth of the cirque located at 3.1–2.3, measured from the lake at this point to the upland surface at 2.9–2.0?

3. Cirques heading on the east side of the range seem to have eaten back closer to the main divide running down the center of the map then those heading on the west side of the range. Can you give an explanation for this?

4. Lay out the shortest possible route for a trail over this range from 5.5–2.3 to 0.0–1.5 in such a way that the gradient nowhere exceeds 1600 ft per mile. Avoid all cirques and troughs.

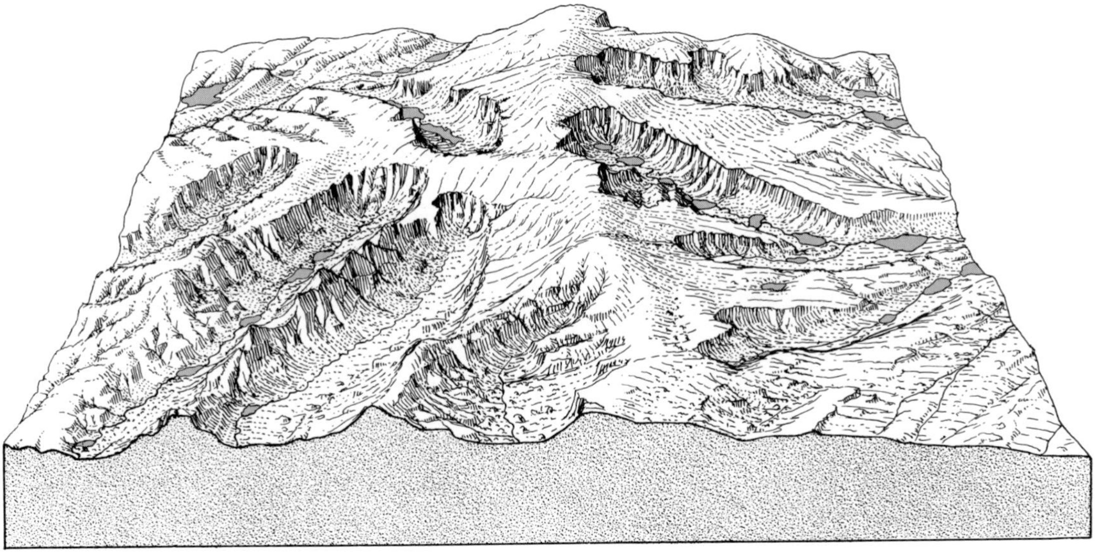

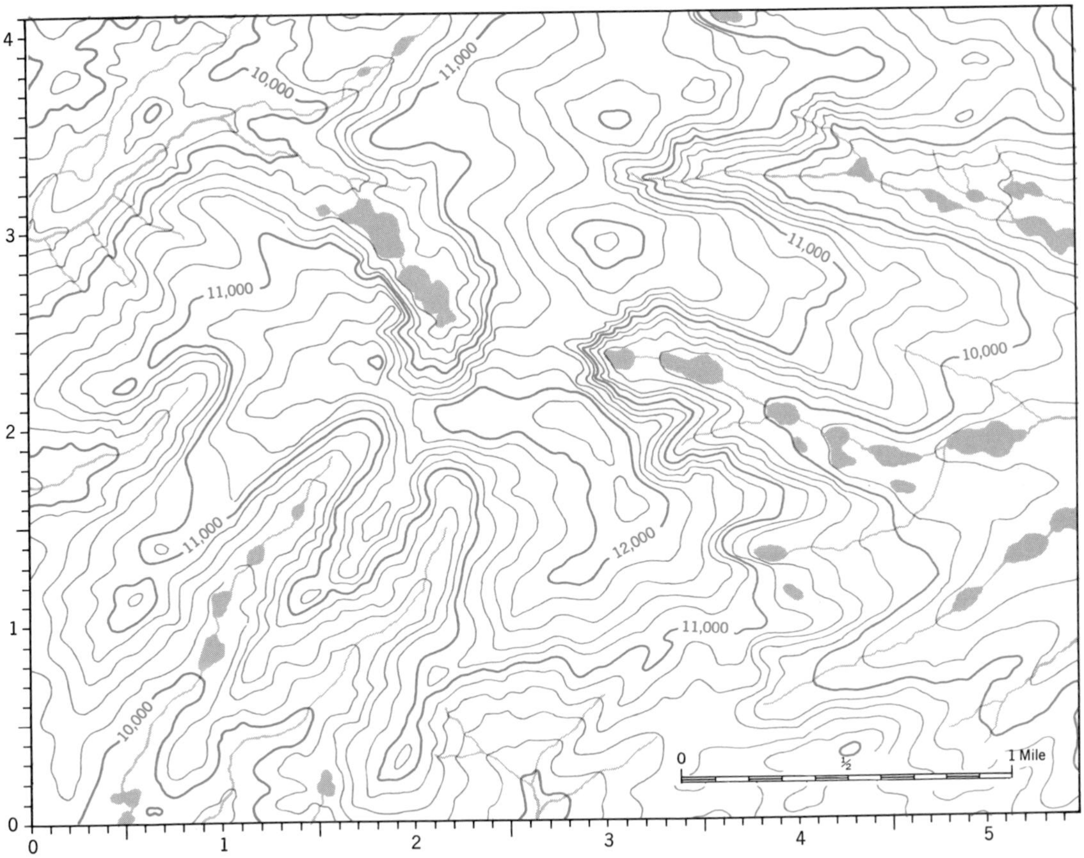

10,000
11,000
11,000
11,000
11,000
12,000
11,000
10,000
10,000
0
½
1 Mile
0
1
2
3
4
0
1
2
3
4
5

Exercise 1

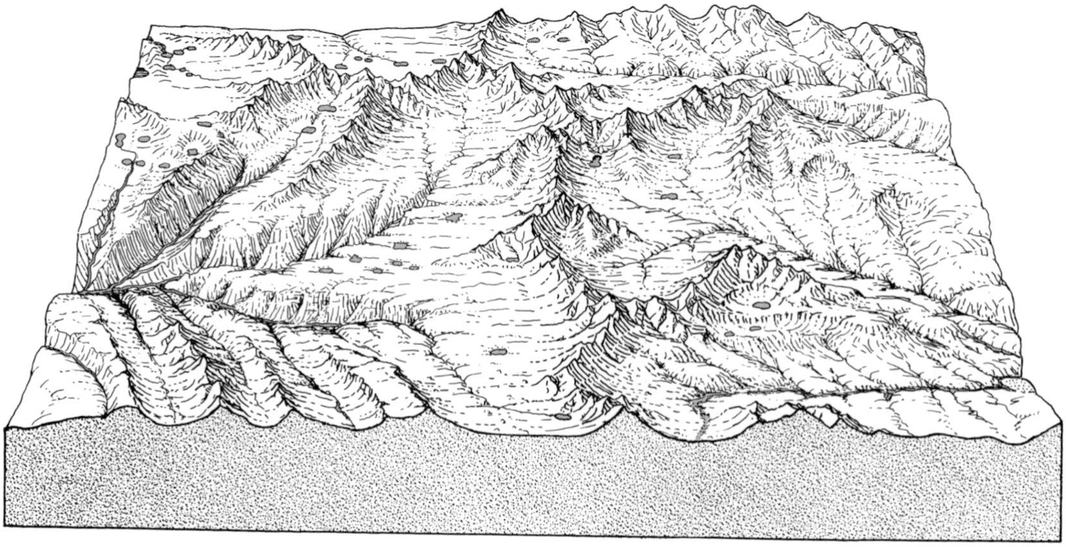

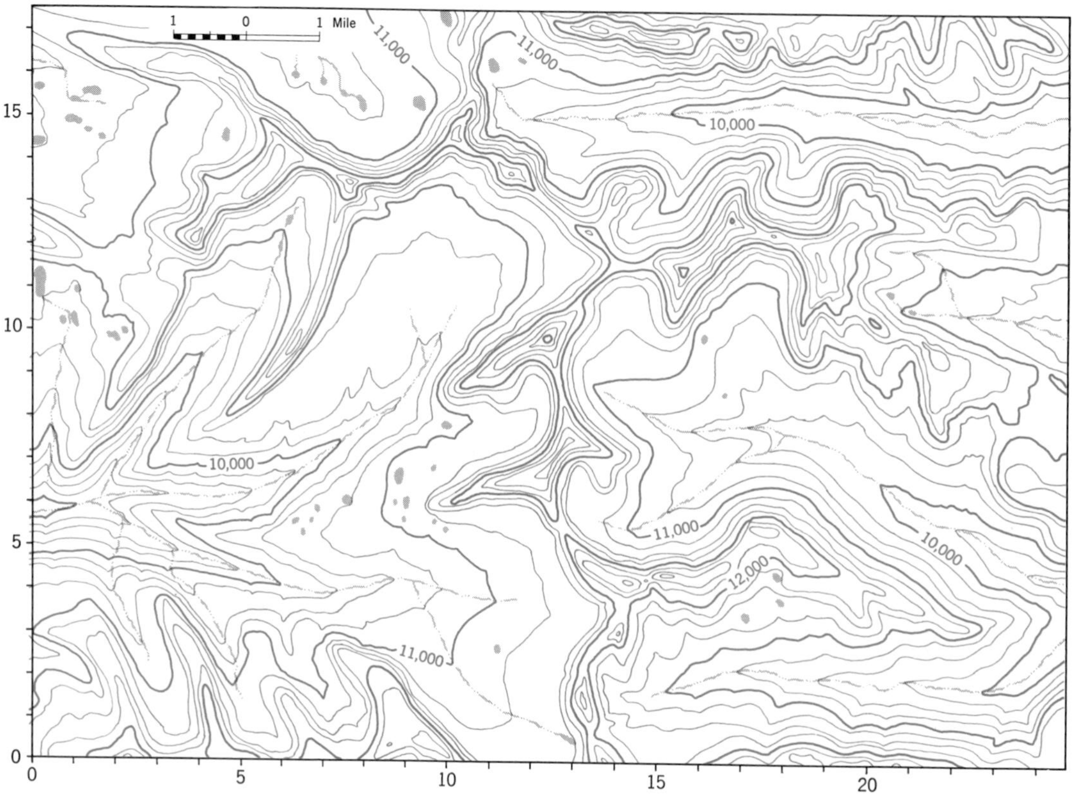

Exercise 2. ***Mature Stage of Alpine Glaciation.*** (Source: Hayden Peak, Utah, U.S. Geological Survey topographic map; scale 1:125,000.)

Explanatory Notes: The map and block diagram show a maturely glaciated portion of the great Uinta Mountain range of northern Utah. North is toward the right. Broad, relatively flat-floored cirques are separated by narrow, steep-walled divides, consisting of pointed horns connected by sharp arêtes. Streams have cut deep V-shaped canyons into the cirque mouths. For further details on the glacial features of this area, see a report by W. W. Atwood (1909), listed among the reading references.

QUESTIONS

1. Make a topographic profile from 14.0–17.5 to 2.0–0.0, using a vertical scale of 1 in. equals 2000 ft.

2. On a sheet of thin tracing paper laid over this map, write the word *cirque* on all cirques that you can identify. Label all good examples of arêtes, horns, and cols.

3. On another sheet of tracing paper laid over this map, color blue all areas which you would imagine to be covered by glaciers at a stage of maximum glaciation. Add arrows to show directions of ice movement. In another color, draw in all medial moraines which you might expect.

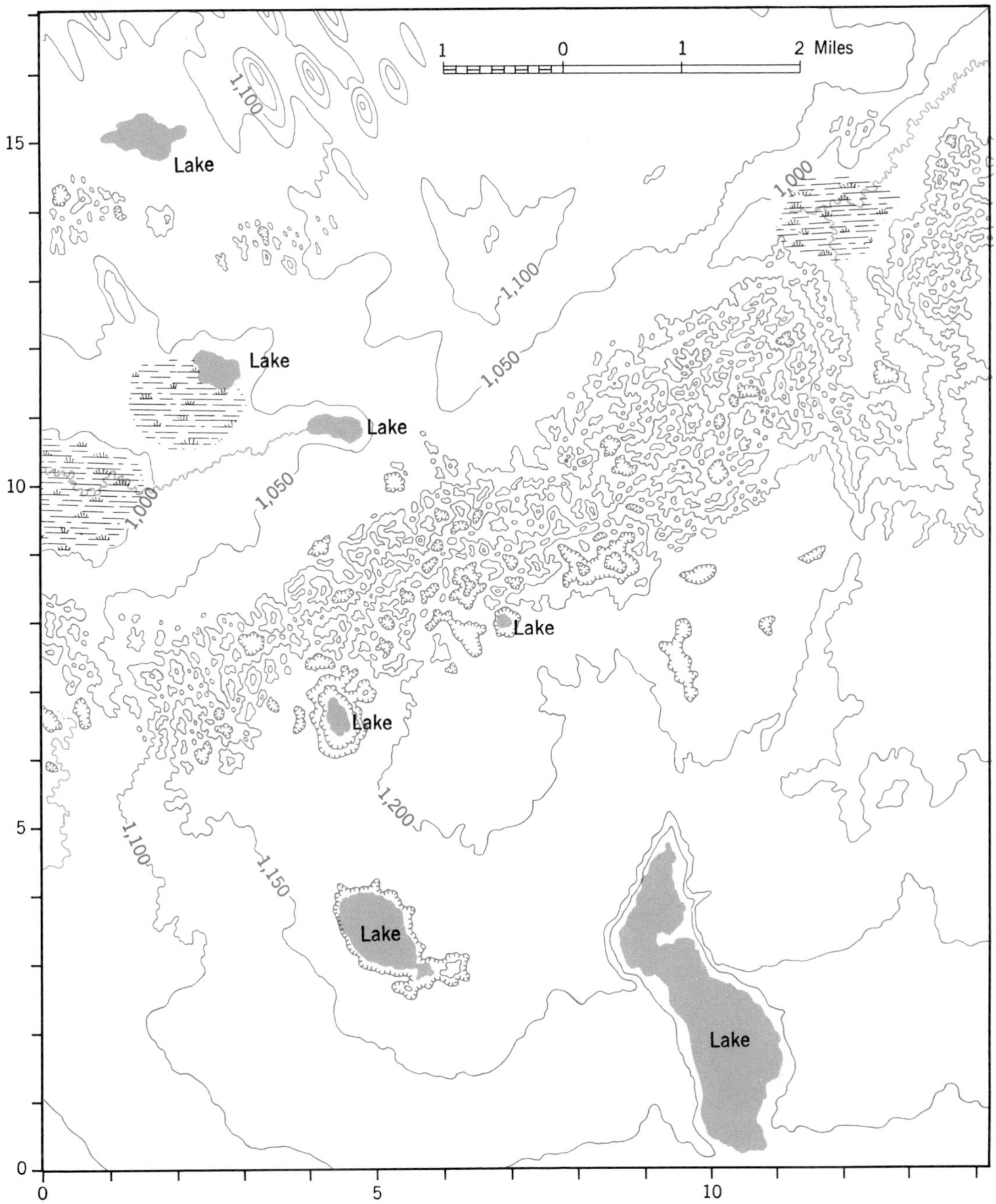

Exercise 3. ***Moraine and Pitted Outwash Plain.*** (Source: St. Croix Dalles, Wis.-Minn., U.S. Geological Survey topographic map; scale 1:62,500.)

Explanatory Note: Running across this map from northeast to southwest is a terminal moraine belt composed of innumerable small hills and depressions. To the southeast of this moraine is a sloping outwash plain containing ice-block lakes. North of the moraine is a lower area with marshes and drumlins, which was beneath the ice and has a cover of thin ground moraine.

QUESTIONS

1. **(*a*)** What is the contour interval used on this map? **(*b*)** How wide is the terminal moraine belt in the center of the map?

2. **(*a*)** How deep is the closed depression at 8.5–8.8? **(*b*)** What is the elevation of the surface of the lake at 4.5–6.5? **(*c*)** What is the slope (feet per mile) of the outwash plain surface between 6.0–7.5 and 9.0–0.0?

3. Give the compass bearing or azimuth on which the ice was moving, as indicated by the orientation of the drumlins at 4.0–16.0 and by the trend of the moraine belt.

4. Did the ice advance over the area now covered by the outwash plain? What is the evidence?

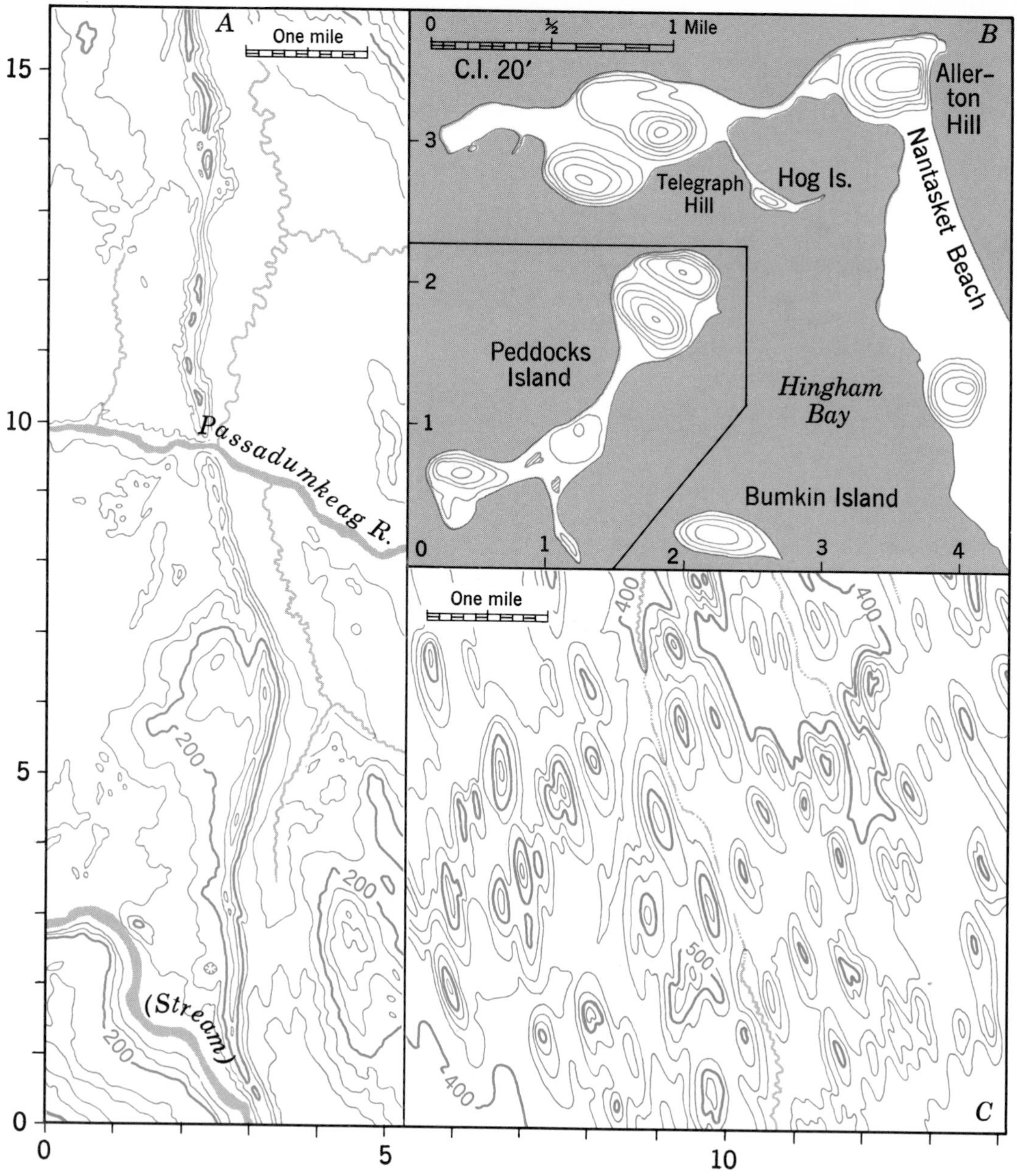

Exercise 4. ***Esker and Drumlins.*** (Source: Map *A*, Passadumkeag, Maine, 1:62,500. Map *B*, Hull, Mass., 1:31,680, Map *C*, Weedsport, N.Y., 1:62,500 U.S. Geological Survey topographic maps.)

Explanatory Note: Map *A* shows an esker ridge, named Enfield Horseback, running from north to south down the center of the map. Maps *B* and *C* show drumlins from two well-known drumlin localities, Boston Bay and western New York State, respectively. Note that the scale of Map *B* differs from that of the other two and therefore has a grid system of its own.

QUESTIONS

Map A: **1.** **(*a*)** How wide (yards) is this esker, approximately? **(*b*)** How high (feet) is the esker crest above the surrounding lowland, just north of the Passadumkeag River?

2. **(*a*)** Why does the esker crest rise and fall in elevation? **(*b*)** Explain the closed depressions at 2.3–13.9.

Map B: **3.** Give the length, width, and height of four drumlins on this map, locating each by grid coordinates.

4. Why has Allerton Hill such a steep slope on the east?

Map C: **5.** How many drumlins are shown on this map? Count all hills represented by one or more closed contours.

6. Which drumlins are higher, those on Map *C* or those on Map *B*? Compare the highest three on each map.

7. In which direction did the ice move when forming the drumlins on Map *C*? Give compass bearing. Compare with direction of movement indicated by drumlins in Map *B*. Compare the shapes of drumlins on the two maps.

CHAPTER **32**

Landforms Made by Waves and Currents

OCEAN waves work unceasingly to transform the shores of continents and islands. Waves travel across the deep ocean with only gradual loss of energy, but when shallow water is reached, the wave form changes radically and new water motions are developed. These motions take the form of powerful surges and currents capable of performing great erosive and transportational work.

Most shorelines have a rather smooth, sloping bottom extending out beneath the water level. As waves approach this shallow zone, wave velocity becomes less and wave crests become more closely spaced (Figure 32.1). Wave height and steepness increase rapidly until the crest rolls forward to make a *breaker* (Figure 32.2). After the breaker has collapsed, a foamy, turbulent sheet of water rides up the beach slope. This *swash*, or *uprush*, is a powerful surge causing a landward movement of sand and gravel on the beach. When the force of the swash has been spent against the slope of the beach, the return flow, or *backwash*, pours down the beach, but much disappears by infiltration into the permeable beach sand. Sands and gravels are swept seaward by the backwash.

Wave erosion

In times of relative calm or moderate winds, waves do little erosional work but tend instead to build beaches and sand bars out of sand and gravel. In times of storm, when enormous waves break

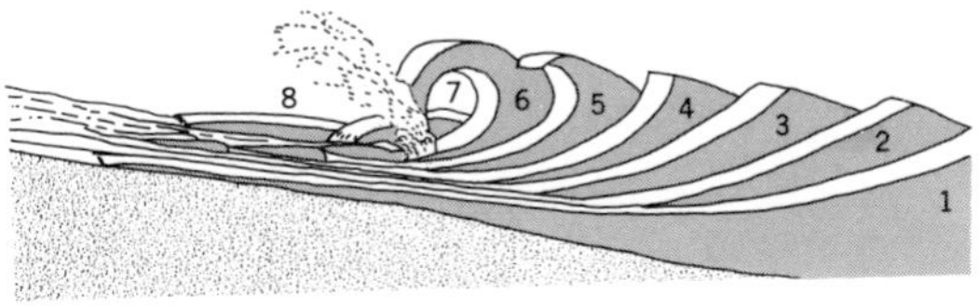

Figure 32.2 A breaking wave.

and throw tons of water against the shoreline, erosion is rapid (Figure 32.3). The violent uprush hurls cobbles and boulders against exposed bedrock along the shore, causing fragments to be broken free. A continual crushing and grinding action goes on as the stones are jostled together. The products of this breaking up are sorted according to size. Fine particles are carried out to sea, where they eventually come to rest in deep, quiet water to form silt and clay layers. Sands and gravels remain close to shore, forming beaches and bars.

Should the shoreline consist of hard rock, erosion will be slow and the storms of many years' time will make little visible change. Where soft materials, such as glacial moraines or outwash plains, form the shoreline, erosion is very rapid. The force of the water alone is sufficient to erode such deposits, and the sea cliff may recede many feet in a single storm (Figure 32.4). Some of the particular forms produced by wave erosion are described and explained in a later paragraph.

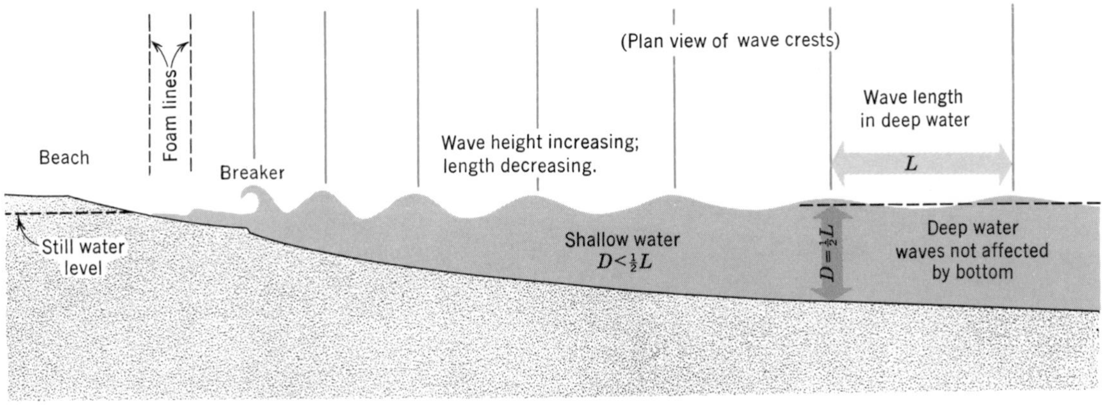

Figure 32.1 As waves enter shallow water the form changes until breaking occurs.

Figure 32.3 Tremendous forward thrust is evident in these storm waves breaking against a sea wall at Hastings, England. (Photographer not known.)

Development of sea cliffs

Where a steeply sloping land surface descends below the water the development of steep cliffs bordering the shoreline is especially favored. Such a condition may come about by a sinking of the land or a rise of ocean level, bringing the water line against what were formerly the steep slopes of mountains or hills.

Figure 32.5 illustrates the development of sea cliffs. Block *A* shows an early stage, termed the *nip stage*, in which wave attack has carved out a small cliff in the hard bedrock. At the base of the cliff is a small rock platform sloping seaward and lying just below water level. Disintegrated rock

(Figure 32.4 Storm waves breaking against a coast underlain by weak sand quickly undermined this shore home at Seabright, New Jersey. The barrier of wooden pilings (right) proved ineffective in preventing cutting back of the cliff. (Photograph by Douglas Johnson.)

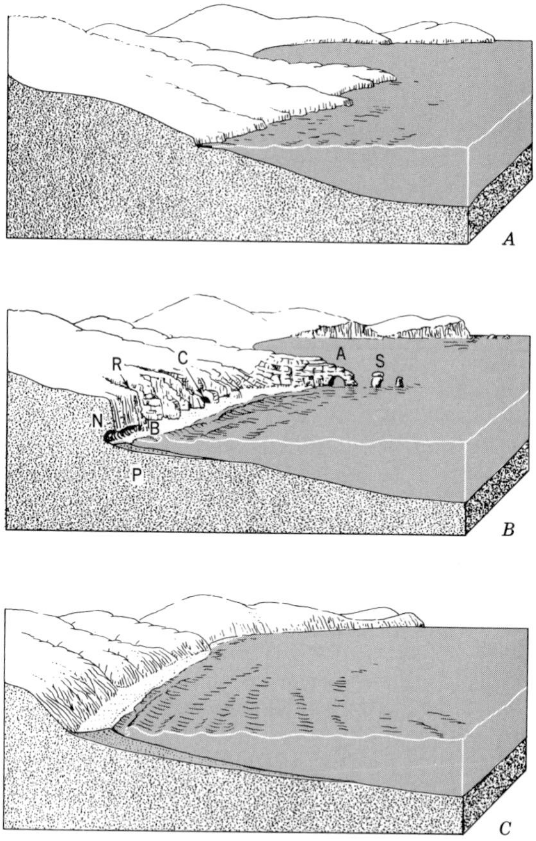

Figure 32.5 Development of sea cliffs. (After E. Raisz.) *A* = arch: *S* = stack: *C* = cave: *N* = notch: *P* = abrasion platform: *B* = beach: and *R* = crevice.

fragments are swept seaward because wave energy is excessive and will not permit sand and gravel to remain as a beach.

Block *B* shows a cliff developed to considerable height, because, as the cliff is cut back, the rise of land slope causes its height to increase. The waves have sought out points of comparative weakness in the rock, penetrating to form crevices and *sea caves.* Some more solid portions of rock project seaward. Where attached to the mainland they may form *sea arches;* where detached, they rise as *stacks* (Figure 32.6). At this stage, the cliff line has reached its greatest degree of irregularity. It is still being vigorously undercut, as evidenced by a wave-cut *notch* at the cliff base. The sloping rock-floored platform at the cliff base, known as the *abrasion platform* is now relatively wide. The inner edge is covered by water only at times of high tide or storm. A beach of sand or cobblestones may be present, but is transitory and may disappear during a single storm, to be built back very slowly.

Streams that formerly flowed to the sea in

Figure 32.6 The chalk cliffs of Normandy, along the French channel coast, show stacks, arches, and sea caves. (Photographer not known.)

valleys whose lower ends were at sea level may now be shortened and left as *hanging valleys* (Figure 32.7), having been unable to deepen their valleys with sufficient rapidity to keep pace with the retreating sea cliff. If the rock material is of an unstable variety, large masses may slump or slide from the cliff (Figure 32.8).

A late stage in the development of sea cliffs, shown in Block *C* of Figure 32.5, may be considered the *mature stage.* The abrasion platform has been so greatly widened that all, or almost all, wave energy is expended in friction as the waves travel over the shallow platform, and in shifting sand across the beach. Consequently, wave attack upon the cliff base is greatly reduced. Weathering and rainbeat acting upon the cliff face wear it down to a lower slope. Irregularities such as sea caves and crevices disappear. The beach may now be broad and deep with little bedrock appearing at the surface.

The environmental influence of marine cliffs is felt in several ways. The coast may be inaccessible because of the high cliffs. In military operations this is especially significant, because a landing is hazardous and a beachhead difficult to expand where a sheer cliff parallels the beach. If the stream valleys are hanging, there are few points where access may be had to the coastal region. Along such coasts, which tend to be fairly straight, the only natural harbors are at the mouths of large streams that have been able to cut down to an accordant junction with the sea (Figures 32.6 and

Figure 32.7 When a marine cliff is cut back rapidly, hanging valleys appear as notches in the cliff. The large stream at the right has been able to maintain an accordant junction with the sea level and provides a small harbor. (After W. M. Davis.)

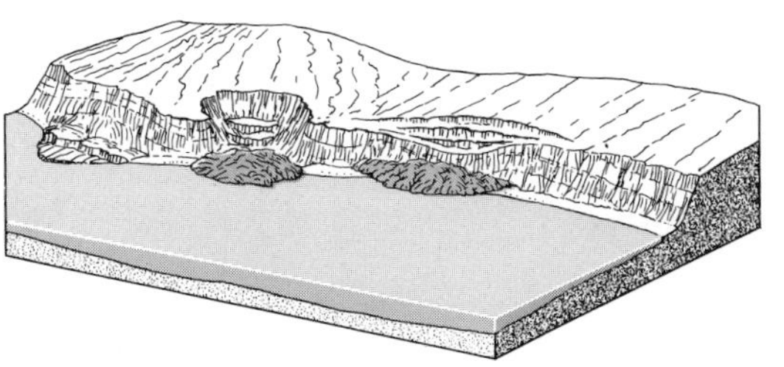

Figure 32.8 Coastal landslides in weak sedimentary strata are the result of oversteepening of a marine cliff by wave attack. (After W. M. Davis.)

32.7). This condition prevails along parts of the French channel coast.

A shore bordered by a marine cliff may be of limited value for summer beach use because of the inaccessibility of the shore from the land back of the cliff and because the beach may be thin or rocky even if present. At high tide, the abrasion platform is sometimes inundated to the cliff base, making the shore a dangerous place where bathers can be trapped by rising tide.

Wave refraction

The phenomenon of change in direction, or bending, of wave fronts as they approach the shore is known as *wave refraction.* Figure 32.9 shows a shoreline with bays and promontories. Successive positions of a wave are indicated by the lines numbered 1, 2, 3, etc. In deep water the wave fronts are parallel. As the shore is neared, the retarding influence of shallow water is felt first in the areas in front of the promontories. Shallowing of water reduces speed of wave travel at those places, but in the deeper water in front of the bays the retarding action has not yet occurred. Consequently, the wave front is bent, or *refracted,* in rough conformity with the shoreline. If the shoreline pattern consists of broad, open curves, the waves may break everywhere along the shore at the same time, but this is unusual. The wave ordinarily will break first upon the promontory and on the bay head last, as indicated in Figure 32.9.

Particularly important in understanding the development of embayed shorelines is the distribution of wave energy along the shore. On Figure 32.9, dashed lines (lettered *a, b, c, d,* etc.) divide the wave at position 1 into equal parts, which may be taken to include equal amounts of energy. Along the headlands the energy is concentrated into a short piece of shoreline; along the bays it is spread over a much greater length of shoreline. Consequently, the breaking waves act as powerful erosional agents on the promontories, but are relatively weak and ineffective at the bay heads. The important principle is that headlands and promontories are rapidly eroded back, tending to produce a simple, broadly curving shoreline as a stable form.

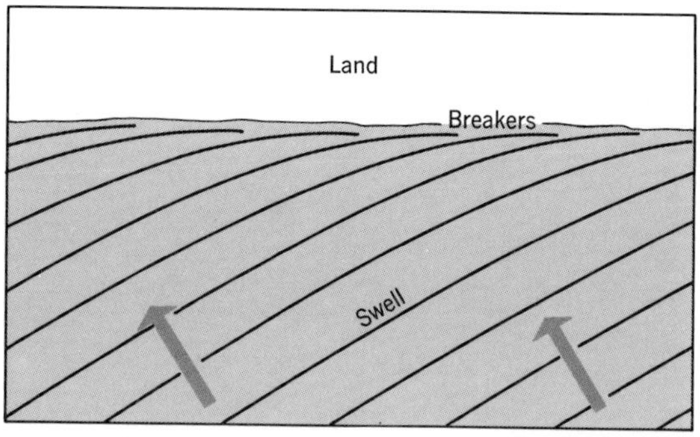

Figure 32.10 Wave refraction on a straight shoreline.

Wave refraction also occurs where oblique waves approach a perfectly straight shore (Figure 32.10). They are turned so as to break almost parallel with the beach. Wave-refraction patterns can be studied from air photographs and may provide valuable information regarding the form of the bottom in the vicinity of the shoreline.

Littoral drift

The unceasing alternate shifting of materials with swash and backwash of breaking waves results not only in movements of rock fragments in seaward and landward directions on the beach, but also in a sidewise movement known as *beach drifting* (Figure 32.11). Wave fronts usually approach the shore with a slight obliquity rather

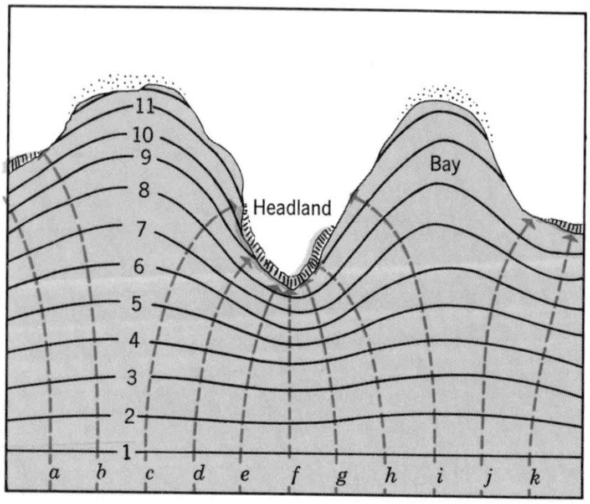

Figure 32.9 Wave refraction along an embayed coast.

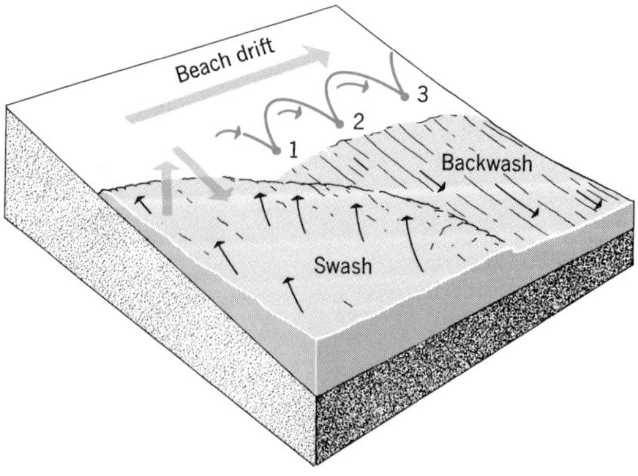

Figure 32.11 Swash and backwash.

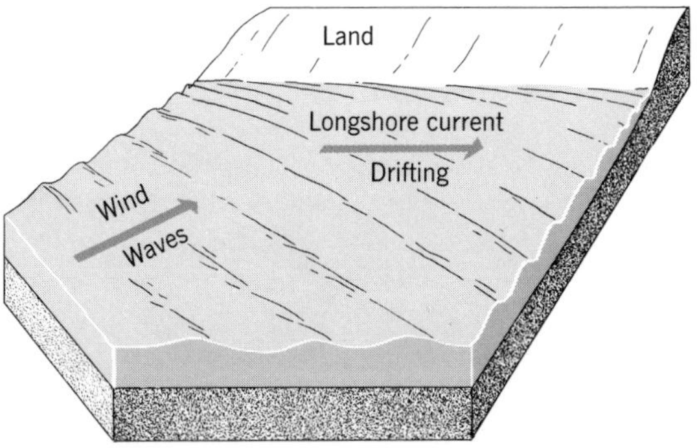

Figure 32.12 Longshore current drifting.

than directly head on. The swash is therefore directed obliquely up the beach, and the sand, gravel, and cobbles are consequently moved obliquely up the slope. After the water has spent its energy, the backwash flows down the slope of the beach, being controlled by the pull of gravity which urges it in the most direct downhill direction. The particles are therefore dragged directly seaward and come to rest at a position to one side of the starting place. Because, on a particular day, wave fronts approach consistently from the same direction this movement is repeated many times. Individual rock particles thus travel a considerable distance along the shore. Multiplied many thousands of times to include the numberless particles of the beach, this form of mass transport is a major process in shoreline development.

Although the word *beach* is in common language use, it would be well at this point to define a beach as a deposit of sand, gravel, or cobbles formed inshore from the zone of breaking waves by the action of swash and backwash. If the sand is arriving at a particular section of the beach more rapidly than it can be carried away, the beach is widened and built shoreward, a change called *progradation.* If sand is leaving a section of beach more rapidly than it is being brought in, the beach is narrowed and the shoreline moves landward, a change called *retrogradation.*

A process related to beach drifting is *longshore drifting.* When waves approach a shoreline under the influence of strong winds, the water level is slightly raised near shore by a slow shoreward drift of water. There is thus an excess of water pushed shoreward, which must escape. A *longshore current* is set up parallel to shore in a direction away from the wind (Figure 32.12). When wave and wind conditions are favorable, this current is capable of moving sand along the sea bottom in a direction parallel to the shore.

Both beach drifting and longshore drifting move particles in the same direction for a given set of onshore winds and oblique wave fronts and therefore supplement each other's influence. The combined processes can be referred to as *littoral drift.*

Sand bars, sand spits, and pocket beaches

Where the continued abrasion of a marine cliff has produced an ample supply of sand, or where the delta of a stream furnishes a sand supply, the sand is moved by littoral drift away from the source toward regions of less intense wave action in more sheltered locations.

In the case of the embayed shoreline (Figure 32.13), littoral drift carries sediment from the cliffed headlands landward along the bay sides to the head of the bay. Here the sediment accumulates as a *pocket beach.* An example is seen in Figure 32.6. Pocket beaches are often formed of cobbles, and are referred to as *shingle beaches.*

In the case of a straight shoreline indented by a deep bay, as shown in Figure 32.13, shore drifting extends the beach in a more or less direct line into the open water of the bay to produce a *sand spit.* Spits are usually curved toward the land at their extremities, and are described as *recurved spits.* Ultimately, a sand spit will be extended to the opposite side of the bay and will form a continuous sand bar, or *baymouth bar,* separating the bay from the open water body (Figure 32.14).

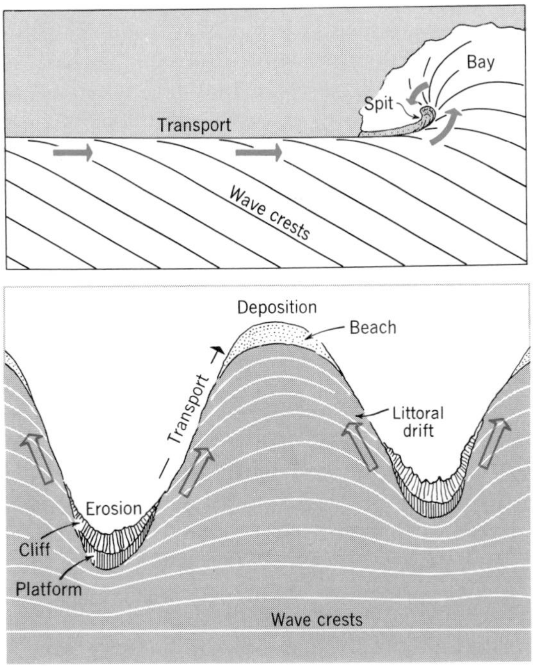

Figure 32.13 Littoral drift along a straight section of coastline, ending in a bay (*below*); along an embayed coast (*above*). (From A. N. Strahler, 1971, *The Earth Sciences*, 2nd ed., Harper and Row, New York.)

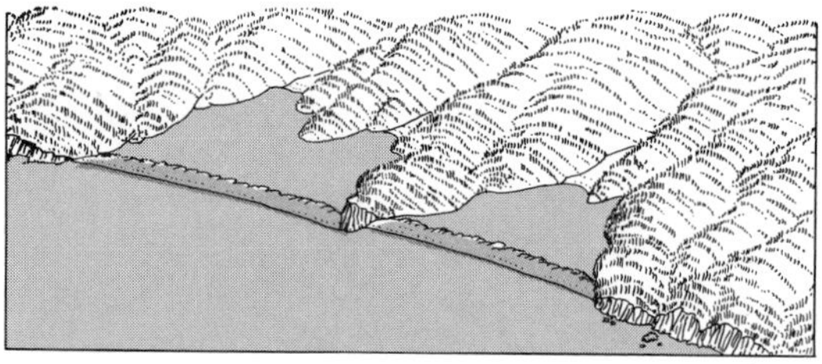

Figure 32.14 These baymouth bars have sealed off two bays. (After W. M. Davis.)

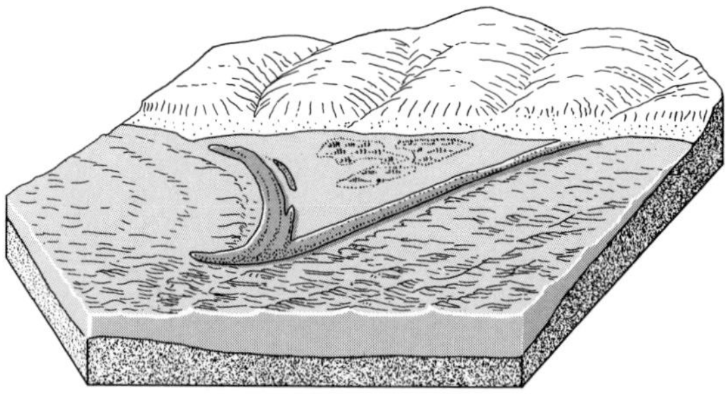

Figure 32.16 This cuspate bar, which has enclosed a triangular lagoon, receives drifted beach materials from both sides. (After E. Raisz.)

Figure 32.15 Two tombolos have connected this island with the mainland. (After W. M. Davis.)

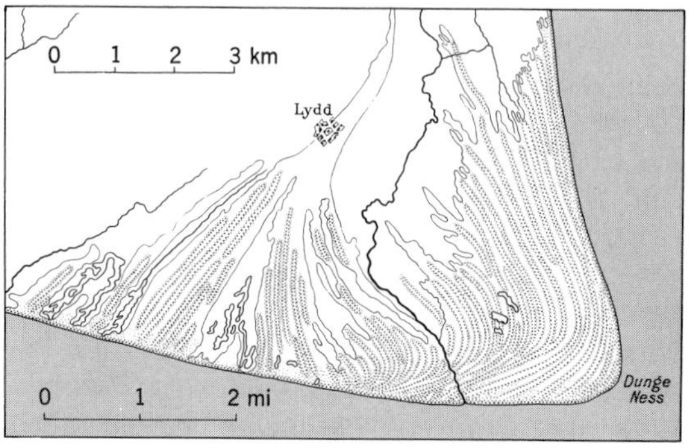

Figure 32.17 Dungeness Foreland, on the Dover Straits of southeastern England, is a large cuspate foreland with curving beach ridges.

Littoral drift from an island commonly forms a *tombolo*, which is a sand bar connecting the island with the mainland (Figure 32.15). Drift of sand along a shore from opposing directions results in the building of a pointed sand-bar deposit known as a *cuspate bar* (Figure 32.16). Long-continued deposition, adding one beach upon another in a ribbed arrangement, produces a prominent *cuspate foreland* (Figure 32.17).

Littoral drift and shore protection

Stretches of shoreline from which sediment is being removed by littoral drift faster than it is supplied are affected by retrogradation. The beach may be seriously depleted or even entirely destroyed. When this occurs, cutting back of the marine scarp can be rapid in weak materials, destroying valuable shore property. Protective engineering structures, such as sea walls, designed for direct resistance to frontal wave attack, are prone to failure and are extremely expensive, as well. In some circumstances, a successful alternative strategy is to install structures that will cause progradation, building a broad protective beach. The principle here is that the excess energy of storm waves will be dissipated in reworking the beach deposits. Cutting back of the beach in a single storm will be restored by beach-building between storms.

Progradation requires that sediment moving as littoral drift be trapped by the placement of baffles across the path of mass transport. To accomplish this result, groins are installed at close intervals along the beach. A *groin* is simply a wall or embankment built at right angles to the shoreline; it may be constructed of huge rock masses (*rip-rap*), of concrete, or of wooden pilings. Figure 32.18 shows the shoreline changes induced by groins. Sand accumulates on the up-drift side of the groin, developing a curved shoreline. On the

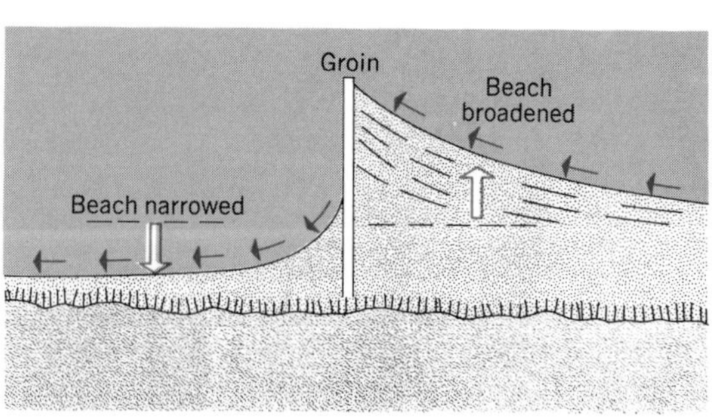

Figure 32.18 Construction of a groin causes marked changes in configuration of the sand beach. (From A. N. Strahler, 1972, *Planet Earth: Its Physical Systems Through Geologic Time*, Harper and Row, New York.)

down-drift side of the groin, the beach will be depleted because of the cutting off of the normal supply of drift sand. The result may be harmful retrogradation and cutting back of the scarp. For this reason groins must be closely spaced so that the trapping effect of one groin will extend to the next. Ideally, when the groins have trapped the maximum quantity of sediment, beach drift will be restored to its original rate for the shoreline as a whole. However, there have been many instances of damaging retrogradation induced by groin or pier construction on the up-drift side of an unprotected shoreline.

In some instances the source of beach sand is from the mouth of a river. Construction of dams on the river may drastically reduce the sediment load and therefore also cut off the source of sand for littoral drift. Retrogradation may then occur on a long stretch of shoreline. The Mediterranean shoreline of the Nile Delta has suffered retrogradation because of reduction in sediment supply following dam construction far up the Nile.

Tidal current deposits

Ebb and flood currents generated by tides perform several important functions. (See Chapter 6.) First, the currents that flow into and out of bays through narrow inlets are very swift and can scour the inlet strongly to keep it open despite the tendency of shore drifting processes to close the inlet with sand. Second, the tidal currents carry much fine silt and clay in suspension, derived from streams which enter the bays or from bottom muds agitated by storm wave action. This fine sediment tends to become clotted into small aggregates (process of *flocculation*) where fresh water mixes with salt water. The sediment then settles to the floors of the bays and estuaries where it accumulates in layers and gradually fills the bays. Much organic matter is normally present in such sediment.

In time, tidal sediments fill the bays and produce *mud flats*, which are barren expanses of silt and clay exposed at low tide but covered at high tide (Figure 32.19). Next, there takes hold upon the mud flat a growth of salt-tolerant plants (such as the genus *Spartina*). The plant stems entrap more sediment and the flat is built up to approximately the level of high tide, becoming a *salt marsh* (Figure 32.20). Tidal currents maintain their flow through the salt marsh by means of a highly complex network of sinuous *tidal streams* in which the water alternately flows seaward and landward (Figure 32.21).

Salt marsh is of interest to geographers because it is land that can be drained and made agriculturally productive. The salt marsh is first cut off

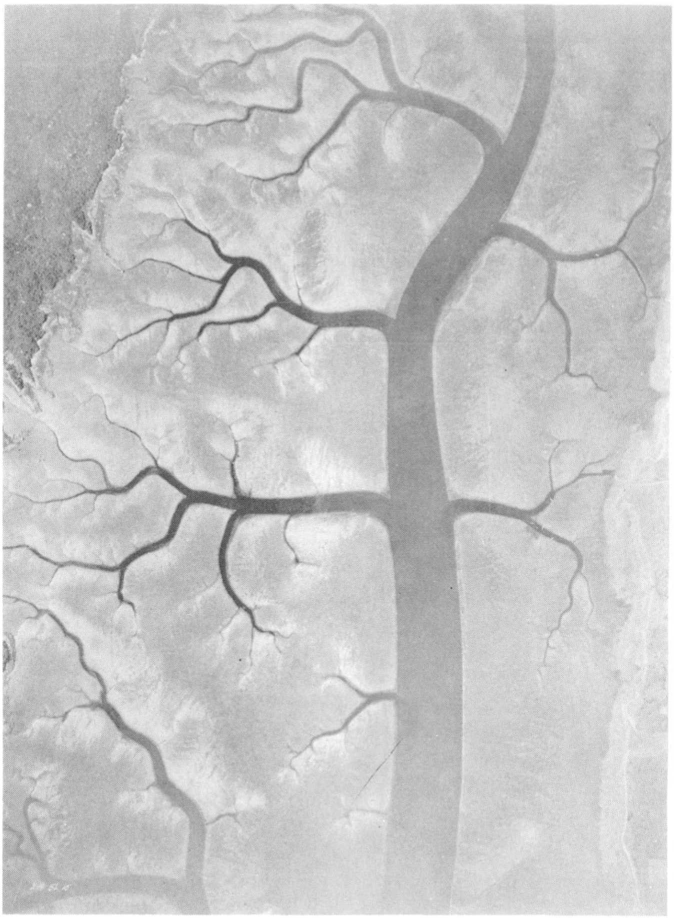

Figure 32.19 Viewed from the air at low tide these tidal mud flats near Yarmouth, Nova Scotia, show a well-adjusted branching system of tidal streams. The area shown is 1.5 mi (2.4 km) wide. (Canadian Forces photograph, No. K.A. 51-15.)

Figure 32.20 Salt marsh at low tide, Green Harbor, Boston Bay region, Massachusetts. Distant houses are on a higher sand ridge separating the salt marsh from the sea. (Photograph by Douglas Johnson.)

from the sea by construction of an embankment of earth, a *dike*, in which gates are installed to allow the fresh water drainage of the land to exit during flood flow. Gradually the salt water is excluded and soil water of the diked land becomes fresh. Such diked lands are intensively developed in Holland (*polders*) and southeast England (*fenlands*). Over many decades the surface of reclaimed salt marsh subsides because of compaction of the underlying peat layers and may come to lie well below mean sea level. The threat of flooding by salt water, when storm waves breach the dikes, hangs constantly over the inhabitants of such low areas. The reclamation of salt marsh by dike construction and drainage was practiced by New World settlers in New England and Nova Scotia. In the industrial era, large expanses of salt marsh have been destroyed by land fill, a practice only recently inhibited by new legislation.

Figure 32.21 This broad tidal marsh along the east coast of Florida is laced with serpentine tidal channels. (Aerial photograph by Laurence Lowry.)

Shoreline classification

Shorelines are classified according to their origin and development. Five major classes of shorelines have been found to include most known types (Figure 32.22).

1. *Shorelines of submergence.* Wherever a sinking down of the earth's crust occurs near the border of a land area, or there is a rise of sea level, the new shoreline takes a position along what was approximately a former contour of the land surface. Below this level all the surface formerly exposed to the air is now submerged beneath the sea. In this way, the term *shoreline of submergence* is applicable.

It is possible to subdivide shorelines of submergence into subgroups, depending on the type of topography that existed before the submergence. Wherever a region was dissected by streams into a system of valleys and divides, submergence produces highly irregular, embayed shoreline termed a *ria shoreline*. Former valleys become deep embayments; former hilltops produce islands; former divides between valleys produce promontories or peninsulas. Variations in ria shoreline form depend on the stage of dissection and relief of the land immediately before submergence. Two possibilities, one a mountainous region (*1A*), the other a coastal plain of very low relief (*1B*), are illustrated in Figure 32.22.

Some coastal regions have been heavily eroded by valley glaciers, whose troughs were excavated below sea level. After the glaciers have disappeared, a *fiord shoreline* results (Figure 32.22, *1C*). (See also Chapter 31.) Such shorelines are distinctive because of the steepness of the fiord walls, the great depth of water, and the great inland extent of the fiords.

Other subtypes of shorelines of submergence result from the submergence of landscapes formed by continental glaciation (Figure 32.22, *1D*).

2. *Shorelines of emergence.* Wherever a rising of the earth's crust has occurred near the border of a continent, or the sea level has fallen, a *shoreline of emergence* is created. The water line takes a position against what was formerly a slope of the sea floor. Above the new shoreline lies a new coastal land belt which has emerged from the sea. Withdrawal of water to form extensive continental ice sheets is an effective cause of sea-level lowering. It is likely that shorelines of emergence were widely distributed during the time of maximum glacier advance in the Pleistocene epoch.

Most sea floors that have been submerged for a long period of geologic time near the margins of continents have been receiving layered deposits of muds, sands, and gravels derived from erosion of the lands and distributed by ocean currents. These continental shelves have a relatively smooth

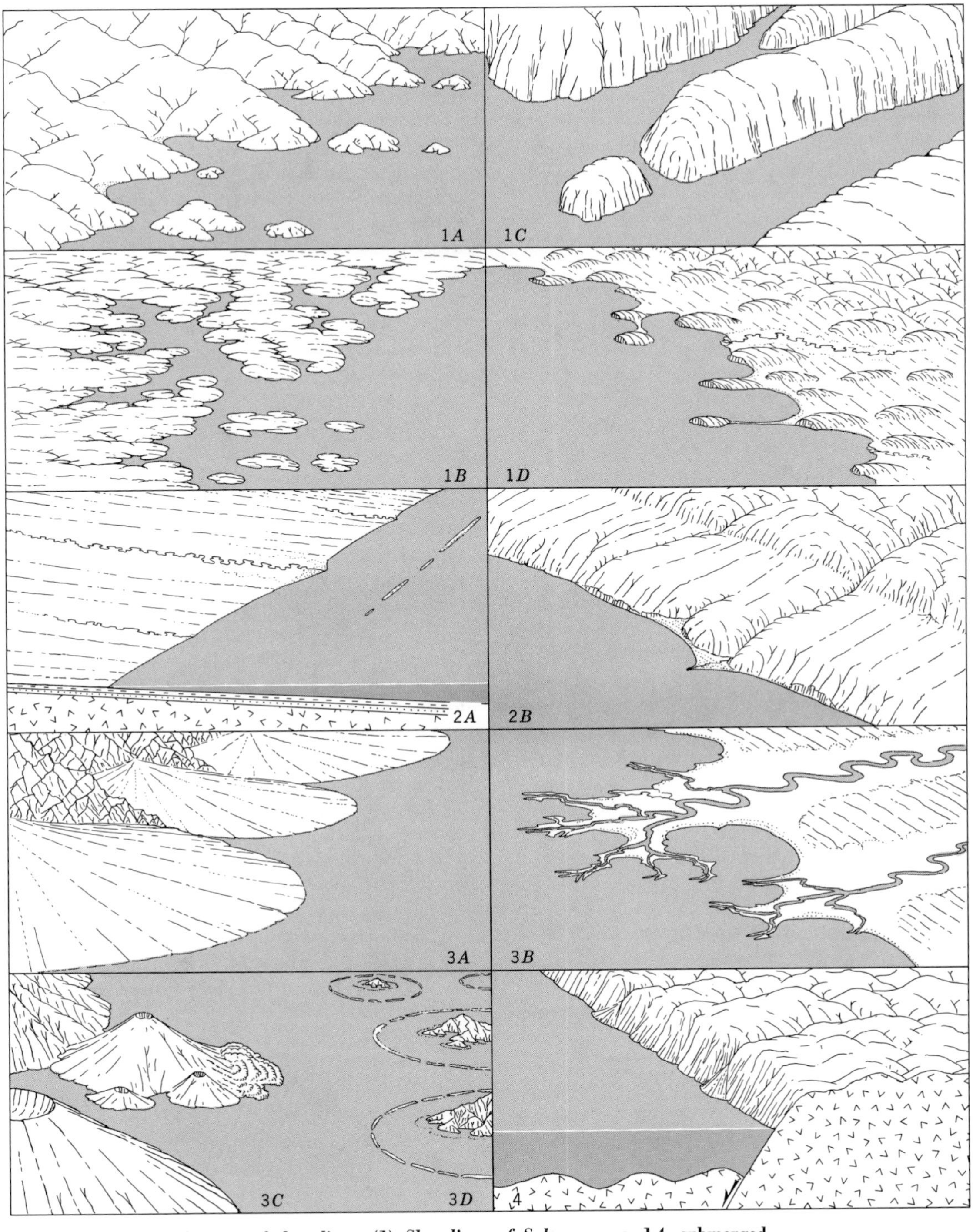

Figure 32.22 Classification of shorelines. (1) *Shorelines of Submergence:* 1*A*, submerged mountainous coast; 1*B*, submerged coastal plain, low relief; 1*C*, fiord coast; 1*D*, submerged glacial deposits (drumlins). (2) *Shorelines of Emergence:* 2*A*, coastal plain type, low relief; 2*B*, steeply sloping type, strong relief. (3) *Neutral Shorelines:* 3*A*, alluvial fan shoreline; 3*B*, delta shoreline; 3*C*, volcano shoreline; 3*D*, coral reef shoreline. (4) *Fault Shorelines.*

surface and gentle slope away from the continents. When a continental shelf is exposed by emergence it produces a low, smooth, gently sloping coastal plain, bounded by a simple, even shoreline—a *coastal-plain shoreline* (Figure 32.22, 2*A*).

Along some coastal belts, the submarine topography contains steep slopes. The shoreline of emergence here differs from that of the coastal-plain shoreline in that deep water lies close offshore and the coastal belt may be relatively mountainous to within a short distance of the shore (Figure 32.22, *2B*). No simple name has been given to this subtype, but it may be designated a *steeply sloping shoreline of emergence.* Old wave-cut cliffs and benches at various levels above the sea indicate that emergence was spasmodic.

3. *Neutral shorelines.* Wherever a shoreline has been formed as a result of new materials being built out into the water, it is classified as a *neutral shoreline.* The word "neutral" implies that there need be no relative change between the level of the sea and the coastal region of the continent.

Several types of neutral shorelines are known. Each type is easily understood when referred to the agent responsible for building out the material into the water (Figure 32.22, *3A, 3B, 3C, 3D*). An *alluvial fan shoreline,* curved in outline, is built by braided streams building a fan in the manner already explained in Chapter 26. Very similar are shorelines formed by building of a glacial outwash plain into the sea where the ice front stood near the shore. A *delta shoreline* is built of material brought out by a stream system (Chapter 25). Where a volcanic eruption has occurred, the slope of a volcano or the edge of a lava flow may compose the shore, which is then classified as a *volcano shoreline.* A *coral-reef shoreline* is built by marine organisms in the shallow-water zones of tropical seas.

4. *Fault shorelines.* An unusual type of shoreline is produced by faulting of the earth's crust in such a way that there is a dropping down of a segment of the crust on the seaward side of the break and corresponding rising up of the landward side. Should the downdropped block subside below sea level, the shoreline will come to rest against the steeply sloping land surface which is the surface of the plane of faulting, and can be termed a *fault shoreline* (Figure 32.22, *4*). A similar result would be obtained where a crustal mass bounded by fault planes arises from beneath the sea.

5. *Compound shorelines* are those that show the forms of two of the previous classes combined, for example, submergence followed by emergence, or vice versa.

Life history of a shoreline of submergence

A shoreline of submergence of the ria type passes through a series of developmental stages, as illustrated in Figure 32.23. In the initial stage, submergence has just occurred to modify the shoreline. The coast is deeply bayed, with long peninsulas or promontories. Numerous islands lie offshore.

In the stage of early youth, wave attack is vigorous upon the headlands and upon the seaward sides of islands. Wave refraction brings maximum attack to these points. Wave-cut cliffs form such minor features as sea caves, stacks, and arches. The term *cliffed headland* is applied to the beveled peninsulas. Rock abrasion platforms develop at the cliff base, but beaches are thin or absent.

In the stage of late youth, the cliffs have increased in height and have been cut back considerably. Some of the smaller islands have been entirely planed off by wave abrasion and the larger ones have lost considerable area. In this stage, large numbers of depositional forms begin to appear. They are built of sand and gravel by the processes of beach and longshore drifting, described earlier, and represent a series of sequential landforms built out of the waste products of wave abrasion against the cliffs.

Fronting the cliffed headlands are *headland beaches.* From these sources sands have drifted out across the bay mouths to produce spits extending into open water. These spits will normally be recurved, or bent toward the land. Continued growth may add new curved portions to the end of a recurved spit, forming a *compound recurved spit.* Should waves generated within the waters of the bay break upon the spit on its landward side, a secondary spit may be built by shore drifting, forming a complex spit.

As shoreline development continues the spits join to produce baymouth bars, which cut off the bays from the open ocean. A variety of other types of bars also forms along the coast. Tombolos connect islands with the mainland. Between an island and mainland may lie a harbor suitable for small craft. Sometimes a double tombolo is formed, enclosing a lagoon of quiet water between island and mainland (Figure 32.15). A *looped bar* may grow along the landward side of an island as a result of the drift of materials around the lee side of the island from the cliffed portion on the seaward side.

Along the sides and ends of the bays are formed *bayside beaches* and *bayhead beaches.* These grow as a result of the drift of sand along the shore from the headlands. Because the bay heads are places of minimum wave attack, the sand tends to accumulate there. If the bays are long and narrow, curving bars will be built across the bay. If located in the middle portion of the bay, they are called *midbay bars;* if near the bay heads, *bayhead bars.* These bars are always smoothly curved, the concave part of the curve facing seaward and merging smoothly with the line of the bayside beaches. A cuspate bar, described in an earlier paragraph, sometimes forms along the side of a bay or on the outer shoreline (Figure 32.16). At bay heads, deltas may be built into water, thus aiding in the process of filling the bays.

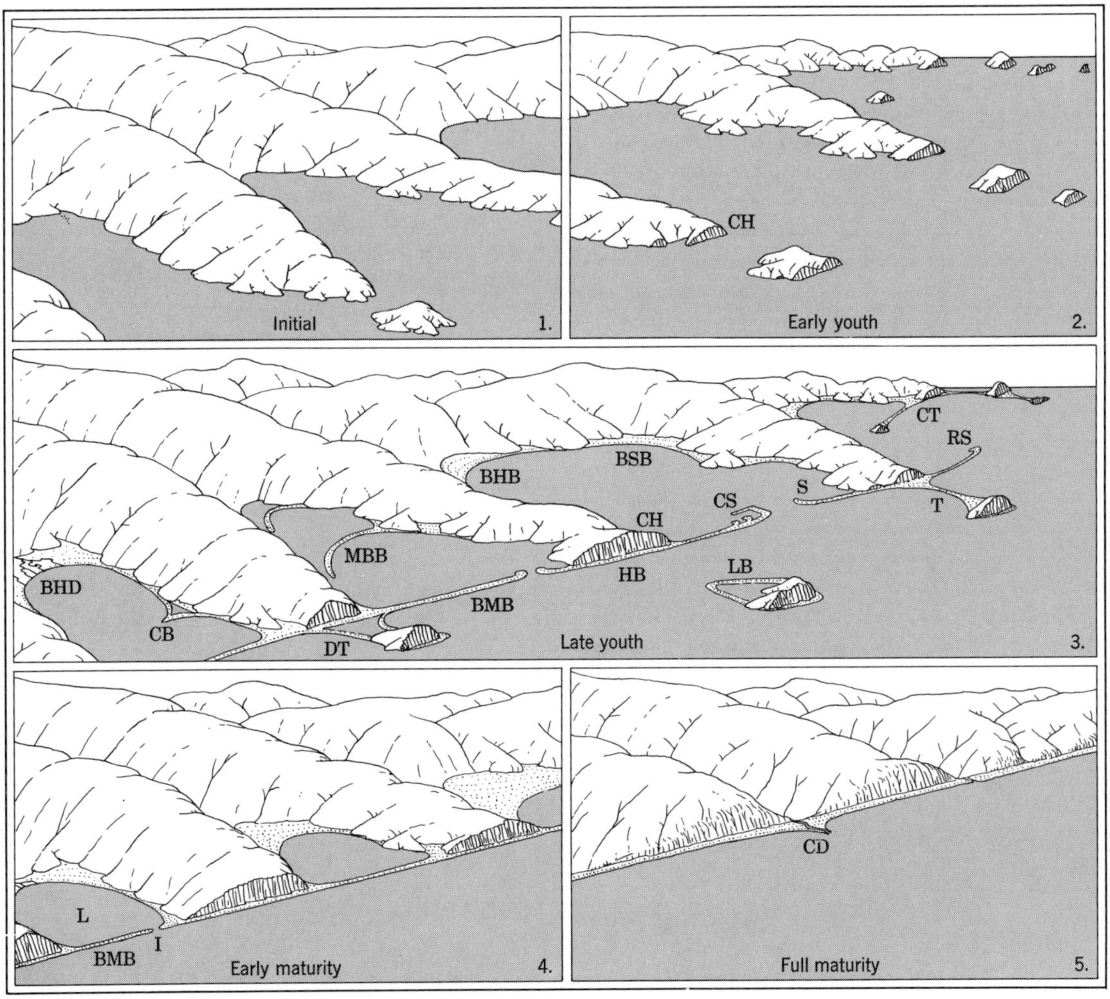

Figure 32.23 Development of the shoreline of submergence. T = tombolo; S = spit; RS = recurved spit: CS = complex spit; CT = complex tombolo; LB = looped bar; CH = cliffed headland; DT = double tombolo; HB = headland beach; BMB = baymouth bar; CB = cuspate bar; BHB = bayhead beach; BSB = bayside beach; BHD = bayhead delta; L = lagoon; I = inlet; CD = cuspate delta.

As the stage of youth draws to a close, the outlying islands are completely consumed and the cliffed headlands begin to form a fairly straight line. Baymouth bars carry the smooth line of the shore from headland to headland.

With the attainment of a simple, smooth shoreline the stage of early maturity is reached. This coast consists alternately of cliffed headlands and baymouth bars. The point of fundamental importance is that throughout the life cycle thus far described, a highly irregular shoreline has been replaced by a nearly straight shoreline. The bays now become filled in by delta materials supplied by streams and by deposition of silts as the tide rises and falls, and the tide-induced currents flow in and out through narrow passes in the bars.

Throughout early maturity the shoreline continues to retreat landward. The sea cliffs are progressively cut back while the baymouth bars are pushed back to keep in a straight line. Eventually a position is reached where the shoreline coincides with the original line of the bay heads. The baymouth bars and all other depositional features except the outer beach disappear, and the cliff of bedrock extends along the entire shoreline, which is then said to have attained full maturity (Figure 32.23). No further major development occurs, aside from the continued landward retreat of the shoreline. This retrogression will, however, become a very slow process, as the increased relief of the land causes the cliffs to be heightened and to supply more detritus to the shore.

Environmental aspects of shorelines of submergence

The influence of shorelines on human activity is strong, especially that of shorelines of submergence. The deep embayments of the youthful shoreline make splendid natural harbors. Much of the shoreline of Scandinavia, France, and the British Isles is thus provided with harbor facilities. Consequently, these peoples have a strong tradition of fishing, ship-building, ocean commerce, and marine activity generally. Mountainous relief of ria and fiord coasts makes agriculture difficult or impossible, forcing the people to turn to the sea for a livelihood. Rich forests and cheap hydroelectric power have, however, stimulated lumbering and manufacturing. New England and the Maritime Provinces of Canada have a youthful shoreline of submergence with abundant good harbors. The influence of this environment has been to foster the same development of fishing, whaling, ocean commerce, shipbuilding, and manufacturing seen in the British Isles and Scandinavian countries.

Development of barrier-island coasts

In contrast to ria and fiord coasts, with their bold relief and deeply embayed outlines, are coasts of low relief from which the land slopes gently beneath the sea. The coastal plain of the Atlantic and Gulf coasts of the United States presents a particularly fine example of such a gently-sloping surface. As explained in Chapter 28, this coastal plain is a belt of relatively young sedimentary strata, formerly accumulated beneath the sea as deposits on the continental shelf. Emergence as a result of repeated crustal uplifts of epeirogenic nature has characterized this coastal plain during the latter part of the Cenozoic era and into recent time. There exist various elevated marine features, such as wave-cut scarps and platforms, extensive sand beaches, and lagoonal deposits of tide-water origin, lying many miles inland and at elevations of many tens of feet.

Such evidence points toward the conclusion that the shoreline of the Atlantic and gulf coastal plain is one of emergence, but on closer examination we find that the lower portions of stream valleys along the entire coast are drowned by a rise of sea level and that tidal channels and tidal marshes extend in many places for miles inland in the valley bottoms. (See Figure 28.12.) Clearly the last event has been one of a rise of sea level, occasioned by the melting of the Pleistocene ice sheets, and has produced many characteristics of a shoreline of submergence.

Along with the rise of sea level there has developed along much of the Atlantic and Gulf coast shoreline a *barrier island*, which is a low ridge of sand built by waves and further increased in height by the growth of dunes shaped from beach

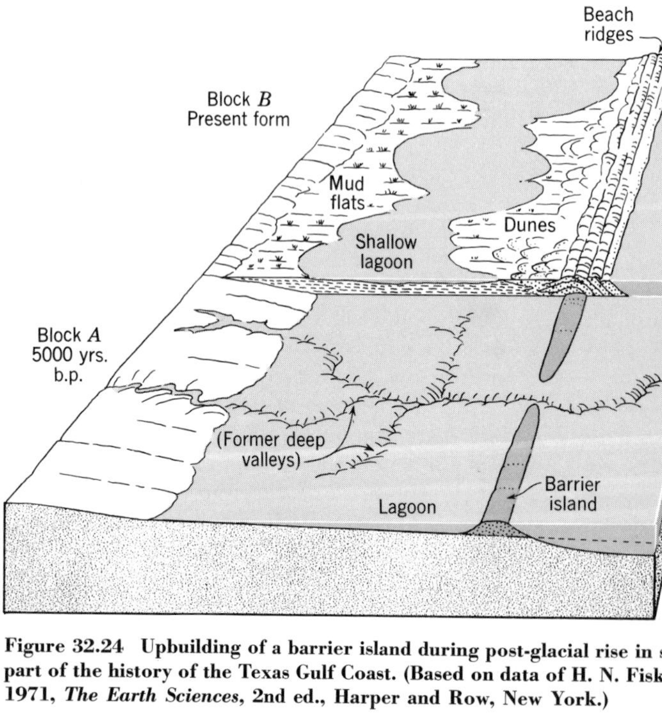

Figure 32.24 Upbuilding of a barrier island during post-glacial rise in sea level is an essential part of the history of the Texas Gulf Coast. (Based on data of H. N. Fisk, From A. N. Strahler, 1971, *The Earth Sciences*, 2nd ed., Harper and Row, New York.)

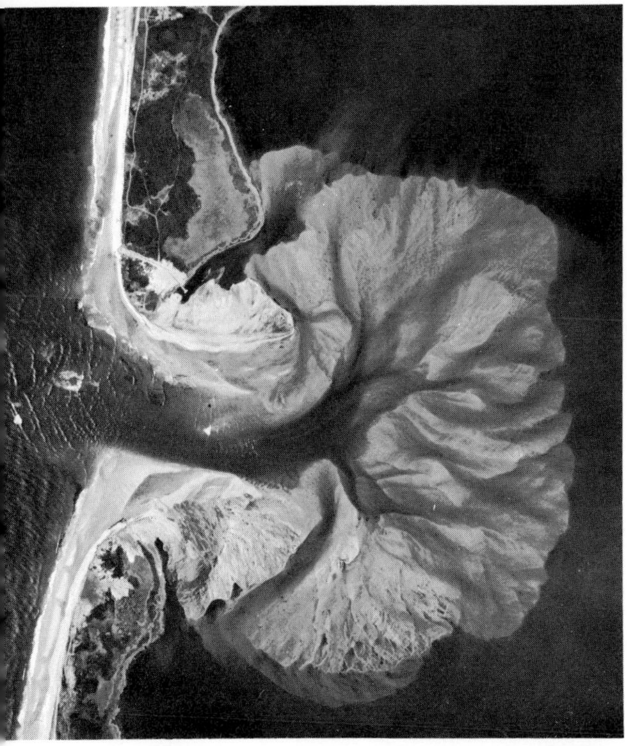

Figure 32.25 East Moriches Inlet was cut through Fire Island, a barrier island off the Long Island shoreline, during a severe storm in March 1931. This aerial photograph, taken a few days after the breach occurred, shows the underwater tidal delta being built out into the lagoon (right) by currents. The entire area shown is about 1 mi (1.6 km) long. North is to the right; the open Atlantic Ocean on the left. (U.S. Army Air Forces Photograph.)

sands by wind action. Behind the barrier island lies a *lagoon,* which is a broad expanse of shallow water, often several miles wide, and in places largely filled with tidal deposits. (See Figure 28.12.)

The manner in which barrier islands may have come into existence along gently sloping coasts of the world since the end of the last glacial stage is illustrated in Figure 32.24 which represents the Gulf coast of Texas. At the time when ice sheets were at their maximum extent over the continents, sea level was drawn down to perhaps as much as 330 ft (100 m) below present sea level. The shoreline then was many miles farther seaward than today and a broad sloping plain lay exposed. Streams draining the land were extended across this plain and carved deep trenches into it. Then, as the ice began to melt rapidly some 10,000 to 12,000 years ago, the sea level began its rise and the shoreline rapidly shifted landward. The forward part of the block diagram in Figure 32.24 shows conditions about 5000 years ago. A low barrier island was formed of beach sands derived from the shallow sea floor and from longshore drift. As the sea level rose, the waves continued to add material to the crest of the barrier, building it up to keep pace with rise of water level. Correspondingly, the lagoon widened and the inner shoreline encroached farther upon the gently sloping land surface. Today, as shown in the rear part of the block in Figure 32.24, the barrier island is a complex structure consisting of several wave-built beach ridges and of dunes which have widened the island on the landward side. The lagoon is partly filled by tidal muds.

Other examples of barrier-island coasts are found where glacial outwash plains were built out into the sea to produce a gently sloping plain. Post-glacial rise of sea level has partly submerged these outwash plains while at the same time a barrier beach has formed and produced a lagoon. A particularly striking example is the south shore of Long Island, New York, along which Fire Island separates Great South Bay, a lagoon over 5 mi (8 km) wide, from the Atlantic Ocean.

Tidal inlets and tidal deltas

A characteristic feature of most barrier islands—and of many baymouth bars as well—is the presence of gaps, known as *tidal inlets,* through which strong currents flow alternately seaward and landward as the tide rises and falls, building *tidal deltas.*

The spacing of tidal inlets in a barrier beach depends in part on the average range of tides along the coast, the spacing being closer where the range is greater. In heavy storms, the barrier may be breached by new inlets (Figure 32.25). Tidal currents will subsequently tend to keep a new inlet open, but it may be closed by shore drifting of sand. Among these opposing activities a sort of balance is maintained, so that neither too many nor too few inlets exist.

Environmental aspects of barrier-island shorelines

Shallow water results in generally poor natural harbors along barrier-island shorelines. The lagoon itself may serve as a harbor if channels and dock areas are dredged to sufficient depths. Ships enter and leave through one of the passes in the barrier island, but artificial sea walls and jetties are required to confine the current and thereby keep sufficient channel depth. Frequently the major port cities are located where a large river empties into the lagoon. Drowning of the lower courses of large rivers provides tidal channels which may be dredged to accommodate large vessels and thus make seaports of cities many miles inland.

One of the finest examples of a barrier island and lagoon is along the Gulf coast of Texas. Here

Figure 32.26 The horizontal step seen in the mountain base is a wave-cut bench with associated gravel deposits. It represents a high stand of ancient Lake Bonneville, which occupied the Great Salt Lake basin during the glacial period. (Photograph by Hal Rumel. Utah Travel Council.)

the island is unbroken for as much as 100 mi (160 km) at a stretch and passes are few. The lagoon is 5 to 10 mi (8 to 16 km) wide, indicating that the original slope of the sea bottom was very slight. Galveston is built upon the barrier island adjacent to an inlet connecting Galveston Bay with the sea. Most other Texas ports, however, are located on the mainland shore. Submergence has caused an embayed inner shoreline with extensive estuaries marking the mouths of the larger streams. Corpus Christi, Rockport, Texas City, Lavaca, and other ports are located along the shores of these embayments. (See Figure 28.12.)

Still another good illustration of the geographical aspects of a barrier-island shoreline is seen in the Atlantic coast of New Jersey, Delaware, Maryland, Virginia, and North Carolina. (See Figure 28.7.) Virtually the entire coast from Sandy Hook to Cape Lookout is bordered by a barrier beach. Extensive post-glacial submergence has caused vast embayments such as Chesapeake Bay and Delaware Bay. Great port cities, such as Baltimore, Wilmington, and Philadelphia are situated at the heads of these tidal estuaries. Along the New Jersey coast, however, the old inner shoreline of the lagoon is remarkably straight and shows vestiges of earlier wave cutting. The lagoon is largely filled now and forms a flat plain through which wander sinuous tidal creeks. The barrier beach makes a splendid resort and has been fully utilized with the building of such cities as Atlantic City and Asbury Park.

Elevated shorelines

A shoreline may, at any stage in its development, be raised above water level so as to become a land feature. At the same time, a new shoreline, which is a true shoreline of emergence, is produced at the new position of the water line. The raised shoreline, or *elevated shoreline*, is not a shoreline of emergence; in fact, it is not a true shoreline at all because it is no longer associated with wave and current action. Having once been elevated, it is attacked by weathering, mass wasting, and streams and will eventually be destroyed.

Elevated shorelines result either from crustal

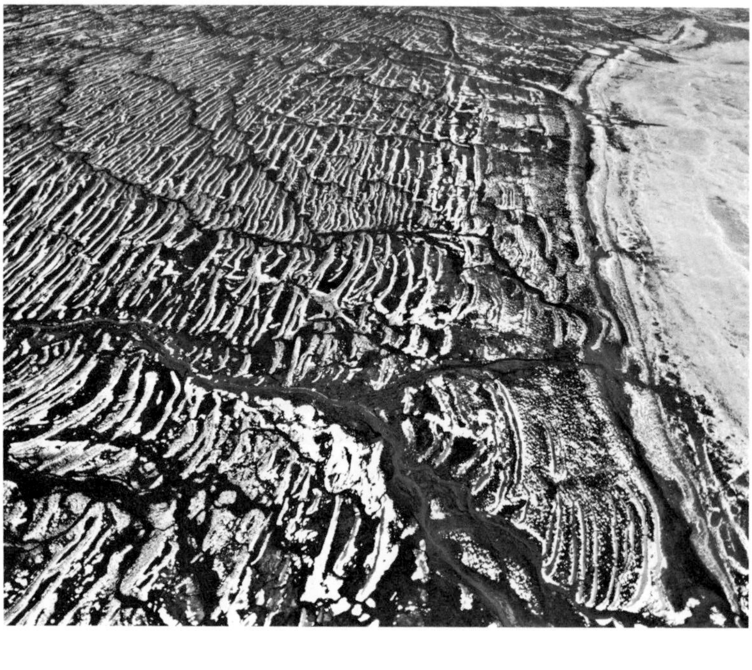

Figure 32.27 This great succession of elevated strand lines bordering the shore of Hudson Bay documents the almost-continuous post-glacial rise of the earth's crust which has followed the removal of ice load. (Photograph by Canada Department of Energy, Mines, and Resources; National Library Air Photo.)

Figure 32.28 A fringing reef on the south coast of Java forms a broad bench between surf zone (left) and a white coral-sand beach. Inland is rainforest. (Photograph by Luchtvaart-Afdeeling, Bandung.)

uplift along a coastal belt such as that of Alaska or California coast, which are subject to faulting and earthquakes, or from a falling of sea or lake level.

Lake Bonneville, the ancestor of the present Salt Lake in Utah, rose to a maximum level in the Pleistocene epoch when rainfall was greater and evaporation less than now. Excellent illustrations of elevated shorelines are to be found on the lower slopes of the mountain ranges in the Salt Lake region (Figure 32.26). From a study of these ancient wave-cut benches with their associated beaches, spits, and bars, it has been possible to reconstruct the history of the old lake and to make inferences as to climates of the past.

Where lake level falls steadily, or coastal regions rise steadily, the elevated shorelines become *strand lines* resembling natural contours of the land (Figure 32.27). Crustal rise of this continual and rather uniform nature has been occurring in the Baltic Sea area and along the Arctic coast of North America as a result of the recovery of the crust following its depression under the load of Pleistocene ice sheets.

Coral-reef shorelines

As a variety of neutral shoreline, coral-reef shorelines are unique in that the addition of new land is made by organisms: *corals*, which secrete lime to form their skeletons, and *algae*, plants that also make limy encrustations. Corals are colonial types of animals, that is, they occur in large colonies of individuals. As coral colonies die, new ones are built upon them, thus developing a coral limestone made up of the strongly cemented limy skeletons. Coral fragments torn free by wave attack and pulverized may be deposited to form beaches, spits, and bars, which later are cemented into a limestone.

Coral-reef shorelines occur in warm, tropical water between the limits 30° N and 25° S lat. Water temperatures above 68°F (20°C) are necessary for dense reef coral growth. Furthermore, reef corals live near the water surface, down to limiting depths of about 200 ft (60 m). Water must be free of suspended sediment and well aerated for vigorous coral growth; hence corals thrive in positions exposed to wave attack from the open sea. Because muddy water prevents coral growth, reefs are missing opposite the mouths of muddy streams. Coral reefs are remarkably flat on top (Figure 32.28) and have a level approximately equal to the upper one-third mark of the range of tide. Thus they are exposed at low tide and covered at high tide.

Three general types of coral reefs may be recognized: (1) fringing reefs, (2) barrier reefs, and (3) atolls. *Fringing reefs* are built as platforms attached to shore (Figure 32.29). They are widest in front of headlands where wave attack is strongest, and the corals receive clean water with abundant food supply. Fringing reefs are usually absent near the mouths and deltas of streams, where the water is muddy. This is a fact of great military importance where the problem is to find reef-free places for landing of troops and supplies. Fringing reefs may be from 0.25 to 1.5 mi (0.4 to 2.5 km) wide, depending on the length of time that the reef has been developing.

Barrier reefs lie out from shore and are separated from the mainland by a lagoon which may range from 0.5 to 10 mi (2.5 to 16 km) or more in width (Figure 32.30). The reef itself may be from 20 to 3000 ft (6 to 900 m) wide. The lagoon is shallow and flat-floored, usually 120 to 240 ft (35 to 75 m) deep. There are, however, many towerlike columns of coral in the lagoon. *Passes*, which occur at intervals in barrier reefs, are narrow gaps through which excess water from breaking waves is returned from the lagoon to the open sea. They sometimes occur opposite deltas on the mainland shore, because of the inhibiting effect of mud on coral growth. Passes are of environmental and military importance because they provide the only means of entrance by ship into the lagoon.

Atolls are more or less circular coral reefs enclosing a lagoon, but without any land inside (Figure 32.31). In all other respects they are

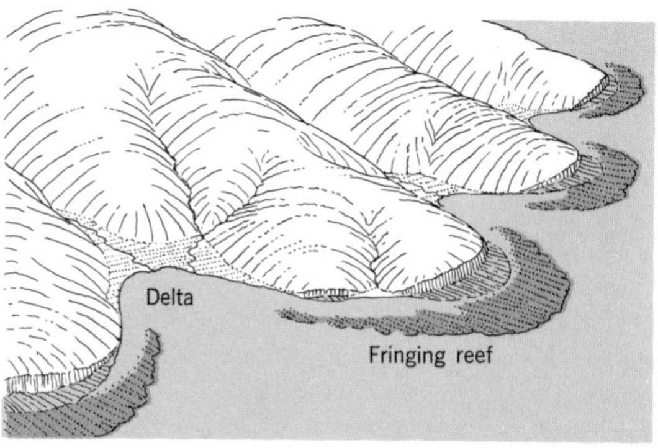

Figure 32.29 Fringing reefs are widest in front of headlands and may be absent near the mouths of streams. (After W. M. Davis.)

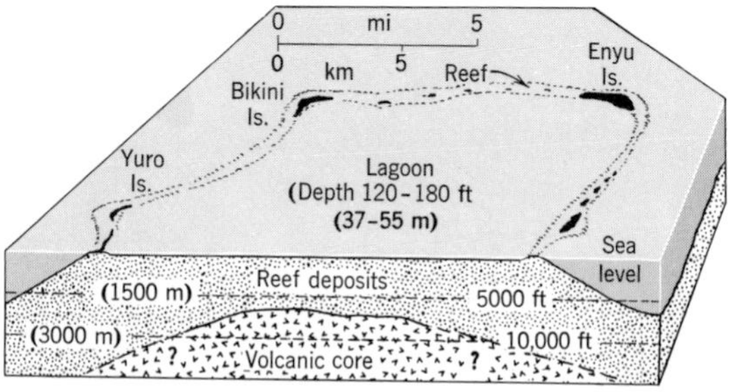

Figure 32.31 Bikini Atoll in the Pacific, scene of early atom bomb tests, is thought to consist of a great thickness of reef deposits resting on a sea mountain of volcanic rock. (After M. Dobrin, et al.)

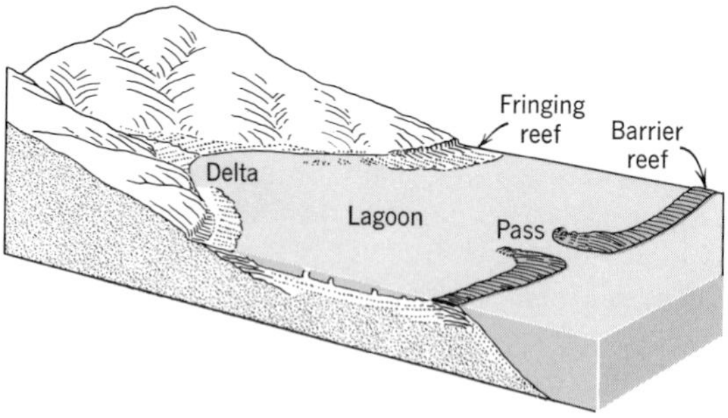

Figure 32.30 A barrier reef is separated from the mainland by a shallow lagoon. (After W. M. Davis.)

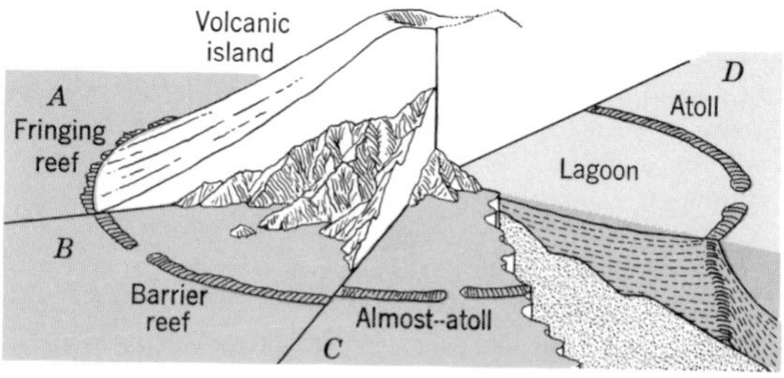

Figure 32.32 The subsidence theory of barrier-reef and atoll development is shown in four stages, beginning with a fringing reef attached to a volcanic island and ending with a circular reef. (After W. M. Davis.)

similar to barrier reefs. On large atolls, parts of the reef have been built up by wave action and wind to form low island chains, connected by the reef. A cross section of an atoll shows that the lagoon is flat-floored and shallow, and that the outer slopes are steep, often descending thousands of feet to great ocean depths.

Several plausible theories have been advanced for the origin of atolls and barrier reefs. To explain each one and discuss the advantages and disadvantages of each would take many pages. One interesting theory of origin, which has been popular since it was first outlined by a great scientist, Charles Darwin, in 1842, may be called the *subsidence theory* (Figure 32.32). He supposed that small islands, such as volcanoes, slowly subsided in a general downwarping of the earth's crust over parts of the ocean basin. Coral reefs, which were originally fringing reefs attached to the island shores, continued to build upward as the island subsided. Thus the area of the island shrank and a lagoon formed, creating a barrier reef. Finally the island sank out of sight, but the reef persisted, maintained at sea level by vigorous coral and algal growth.

The environmental aspects of atoll islands are unique in some respects. First, there is no rock other than coral limestone, composed of calcium carbonate. This means that trees requiring other minerals, such as silica, cannot be cultivated without the aid of fertilizers or some outside source of rock from a larger island composed of volcanic or other igneous rock. The palm tree is native to atoll islands because it thrives on brackish water, and the seed, or palm nut, is distributed widely by floating from one island to another. Native inhabitants have cultivated the cocoanut palm to provide food, clothing, fibers, and building materials. Fresh water is scarce on small atoll islands because there is not enough surface area for the collection of rainfall, and the land is so low that a high water table of fresh water is not present to supply springs, streams, and wells. Rainfall must be caught in open vessels or catchment basins and carefully conserved. Fish and other marine animals are an important part of the human diet on atoll islands. Calm waters of the lagoon make a good place for fishing and for beaching canoes. Coral islands of the western Pacific stand in continual danger of devastation by tropical cyclones (typhoons). Breaking waves wash over the low-lying ground, sweeping away palm trees and houses and drowning the inhabitants. There is no high ground for refuge. In the same way, great seismic sea waves of unpredictable occurrence may inundate atoll islands.

Mangrove coasts are of great extent in tropical and equatorial regions, where an abundance of fine sediment in suspension prevents coral reef growth. Mangrove is discussed in Chapter 21.

Review Questions

1. Why do breakers form in shallow water? What do the terms *swash* and *backwash* mean? How are beach materials moved by these currents?

2. Under what conditions do waves do the greatest amount of shore erosion? When are beaches built?

3. Describe the development of sea cliffs, beginning with a newly submerged slope and continuing through to a mature sea cliff. Explain the following terms: nip, sea cave, sea arch, stack, wave-cut notch, abrasion platform, beach, shore-face terrace hanging valley.

4. Describe the phenomenon of wave refraction. Why does refraction occur? When waves are refracted along an embayed coast where is the energy of wave erosion concentrated? Where is it least? What important result does this have on the form of the shoreline?

5. Explain beach drifting and longshore drifting. What forms are built by these processes? What is a longshore current, and how is it caused?

6. What are sand spits? How and where do they form? What shape characterizes the end of a spit? How are pocket beaches formed?

7. How can littoral drift lead to depletion of a beach? In what way are groins used to control retrogradation of the shore?

8. Explain the formation of baymouth bars, tombolos, and cuspate bars. How does a cuspate foreland differ from a cuspate bar?

9. What deposits of sediment are associated with tidal currents in bays and estuaries? Describe these deposits as to form and composition. How are salt marshes utilized by man?

10. How can shorelines be classified? Name the five principal classes.

11. What kinds of shorelines of submergence are recognized? Name and explain the subdivisions.

12. Describe the changing coastal landforms of a shoreline of submergence as it passes through its development cycle.

13. What is a barrier-island coast? How are barrier islands formed? How are tidal inlets related to barrier islands?

14. How have shorelines of submergence exerted an influence on human activities?

15. Discuss the environmental aspects of barrier-island shorelines. Compare harbor facilities of a barrier-island shoreline with those of a young shoreline of submergence.

16. Name several kinds of neutral shorelines.

17. What is meant by an elevated shoreline? How does it differ from a shoreline of emergence? Where may some excellent elevated shorelines be seen today?

18. How are coral reefs formed? Under what conditions do reef-building corals flourish?

19. What three general types of reefs are formed? Describe each type. Explain Darwin's subsidence theory of atolls.

Exercises

Exercise 1. ***Shoreline of Submergence, Stage of Early Youth.*** (Source: Brest, France, topographic sheet No. 21; scale 1:200,000.)

Explanatory Note: The coastal region around the port of Brest lies in the peninsula of Brittany, which projects westward into the Atlantic Ocean. The coast here is deeply embayed and represents a shoreline of submergence modified appreciably by wave erosion only where the shore is exposed to waves of the open sea. This coastline resembles in many ways the Maine coast of the United States, but lies much farther north (48° N) and was never modified by intensive glacial action, as was the Maine coast. Submarine contours are shown as dashed lines for depths of 1, 5, 10, 20, 30, 40, and 50 meters. Contour interval on land is 20 meters.

QUESTIONS

1. Place a sheet of tracing paper over the map. In red pencil line, mark all parts of the shoreline where a marine cliff is well developed. In blue pencil, shade all probable sand beaches.

2. **(*a*)** Why are there few prominent cliffs along the shoreline of the Harbor of Brest? **(*b*)** Study the peninsula ending at 11–4. Which side seems to have undergone the greatest marine erosion? What is the topographic evidence? Is this what you would expect, knowing that the open Atlantic lies to the west, a bay 15 mi wide to the southeast?

3. On the same tracing sheet used in Question 1, draw in black lines a reconstruction of the drowned-stream system, which may have occupied the Harbor of Brest before the region was submerged.

4. To what stage of development would you assign the shoreline between 25–22 and 31–21? Between 22–10 and 34–0?

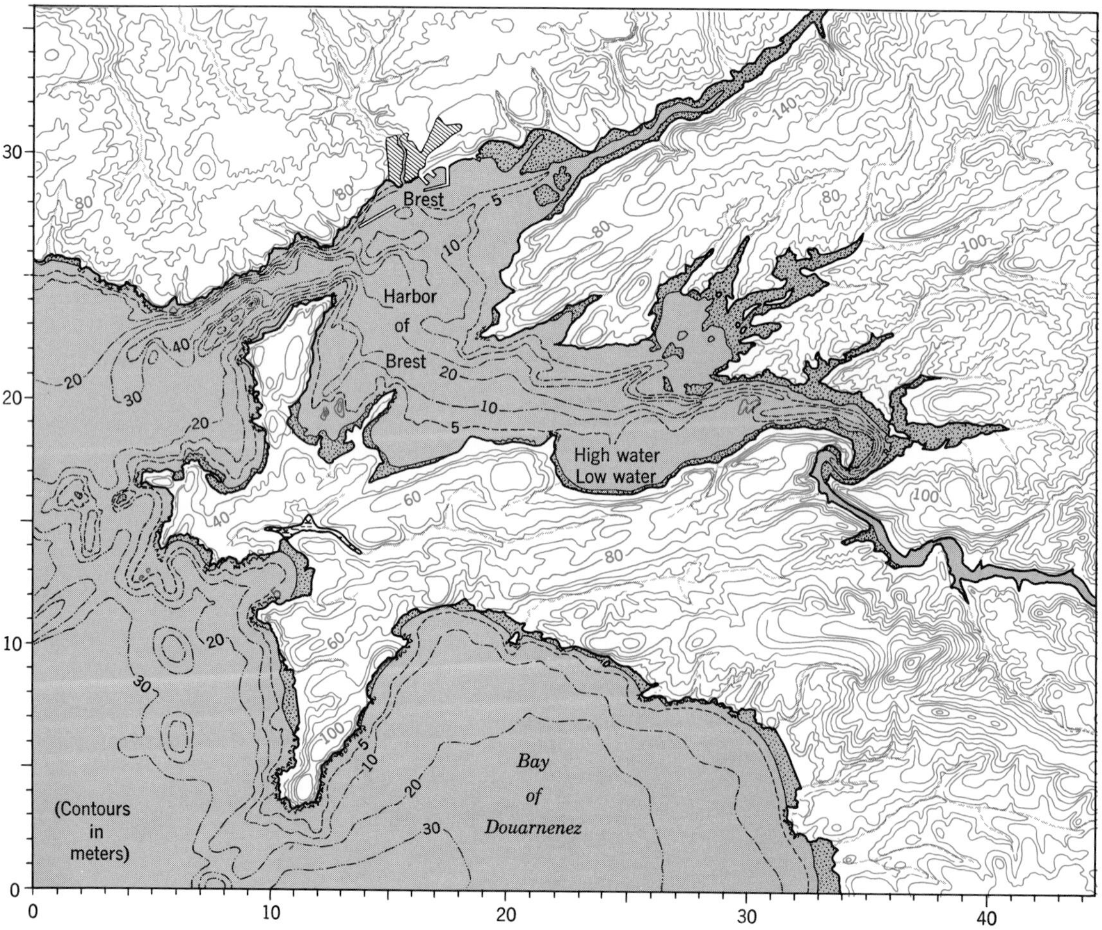

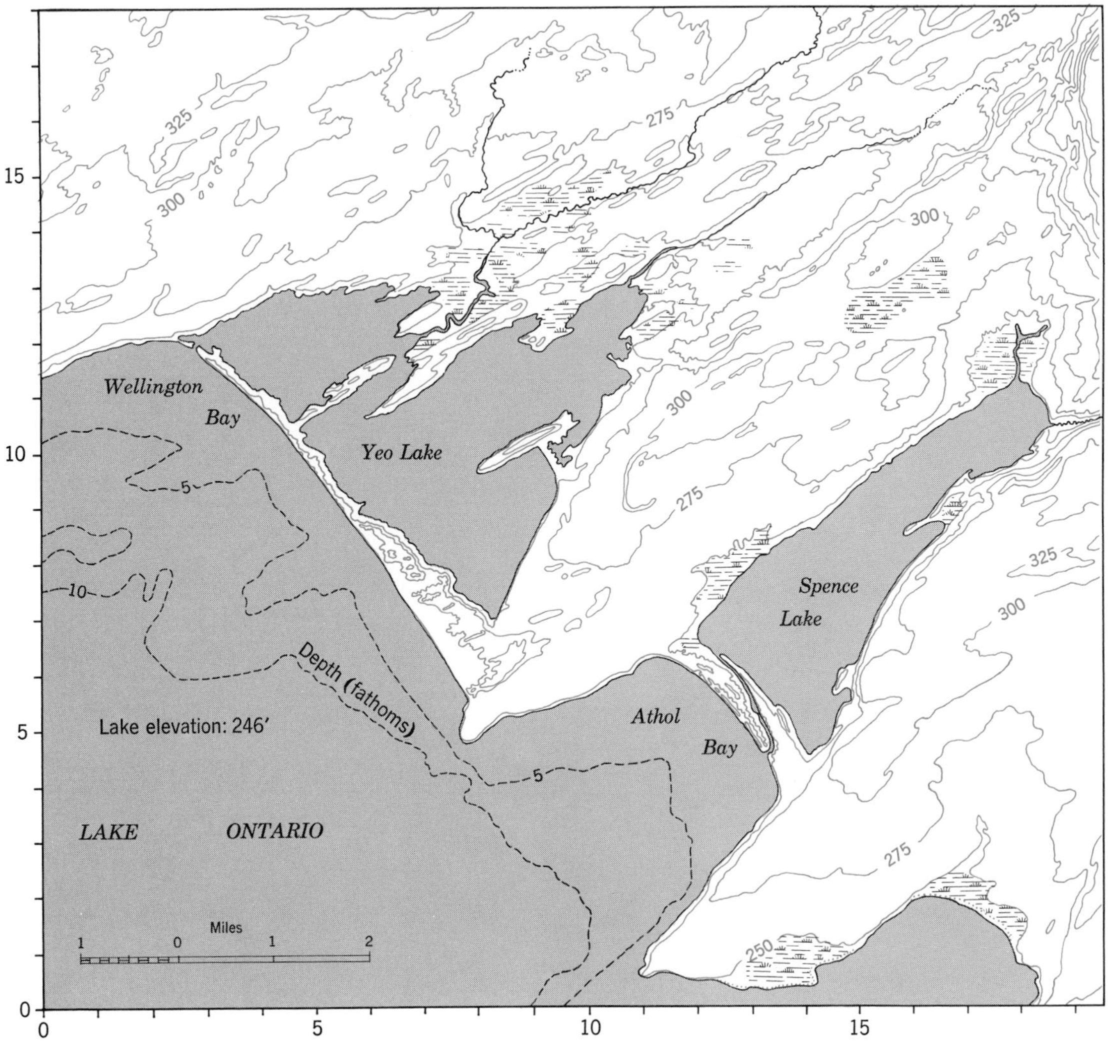

Exercise 2. *Young Shoreline of Submergence with Bars.* (Source: Wellington, Ont., topographic maps; scale 1:63,360. Geological Survey of Canada.)

Explanatory Note: The north shore of Lake Ontario is a shoreline of submergence on a topography developed first by normal stream erosion and associated weathering and mass wasting processes, then heavily glaciated by the Pleistocene ice sheets. A baymouth bar separates Yeo Lake from Wellington Bay, and a midbay bar has cut off the inner part of Athol Bay to produce Spence Lake.

QUESTIONS

1. **(*a*)** What fractional scale has this map? **(*b*)** What is the height (feet) above lake level of the small hills at 6.8–8.0? **(*c*)** Is it more likely that these hills are sand dunes or that they are beach-ridge deposits thrown up by storm waves? Explain.

2. **(*a*)** What is peculiar about the form of the outlet channel through the bar between Spence Lake and Athol Bay? Explain in terms of the beach drifting process. **(*b*)** Are tidal currents responsible for keeping this channel clear?

3. On a thin sheet of tracing paper laid over this map, redraw the shoreline as a smooth, simple shoreline of early maturity passing approximately through 0.0–14.0, 6.0–11.0, 10.0–8.0, 16.0–4.0 and 19.0–0.0. Imagine all headlands to be cut back to this line. Redraw the contours to show the cliffed headlands and baymouth bars.

Exercise 3. *Barrier Island Shoreline.* (Source: Accomac, Va., U.S. Geological Survey, topographic map; scale 1:62,500.)

Explanatory Note: A barrier island represented by Metomkin Island and Cedar Island is separated from the mainland by a lagoonal belt consisting of expanses of open water, Metomkin Bay and Floyd's Bay, and of salt marsh with sinuous tidal creeks. At some stage prior to the appearance of the barrier island, the shoreline lay just east of the 10-ft contour line on the seaward side of Parker Neck, Bailey Neck, Joynes Neck, and Custis Neck. The age of this earlier shoreline is difficult to determine.

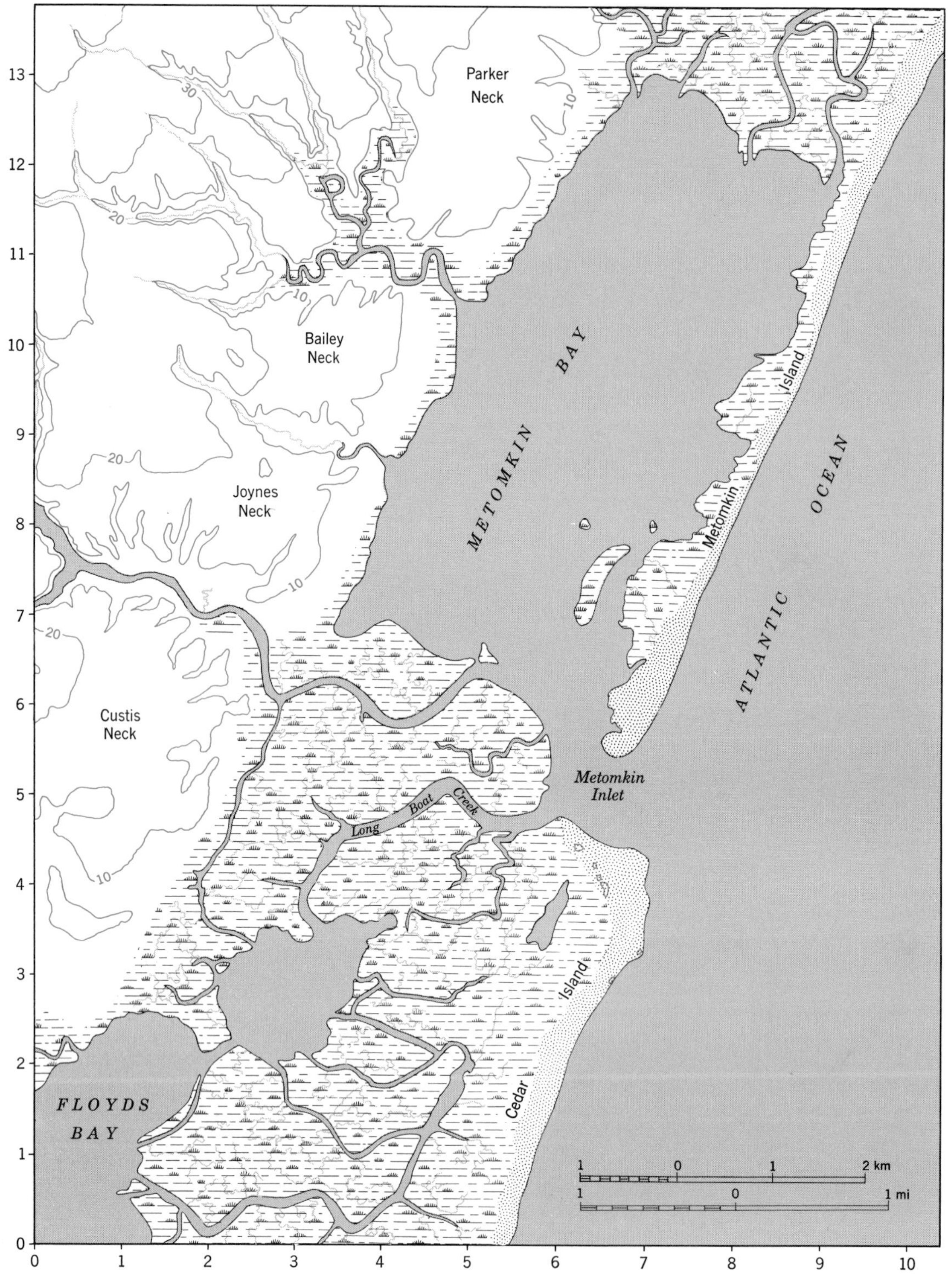

Exercise 3

QUESTIONS

1. **(*a*)** How far (miles) seaward of the earlier shoreline does the barrier island lie in the vicinity of the northern end of Metomkin Bay? **(*b*)** in the vicinity of Cedar Island? **(*c*)** Why does this distance increase toward the the south? Give two possible explanations.

2. **(*a*)** Why are the ends of Cedar Island and Metomkin Island offset at Metomkin Inlet? **(*b*)** In which direction do you think materials are being moved along this coast by shore drifting processes? What is the evidence?

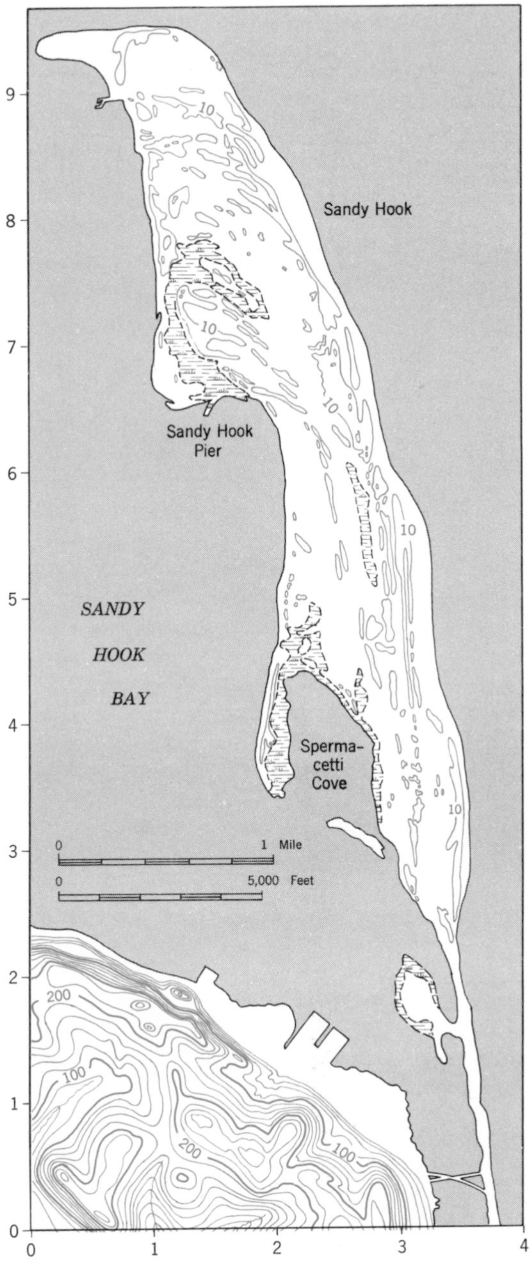

Exercise 4. ***Sandy Hook Spit.*** (Source: Navesink, N.J., topographic map; scale 1:24,000. State of New Jersey, Department of Conservation and Development.)

Explanatory Note: Sandy Hook is a large spit extending north from Navesink Highlands into the Atlantic Ocean and forming a part of the enclosure of Lower Bay of New York Harbor. It is formed, in part, of sand carried northward along the New Jersey coast by shore drifting processes. Numerous beach ridges show various stages in the growth of the spit. The tendency of these ridges to recurve westward or landward is typical of complex spits. The tip of the spit is said to have grown about one mile since 1764, one-half mile since 1865.

QUESTIONS

1. **(*a*)** What contour interval is used on the spit? **(*b*)** On the mainland?

2. **(*a*)** What is the origin of the small spit enclosing Spermacetti Cove on the west? **(*b*)** In what direction do beach materials generally drift on the west shore of Sandy Hook? **(*c*)** Is this the same direction as prevailing shore drifting on the eastern (Atlantic Ocean) shore? Explain.

3. Would you expect to find sand dunes on Sandy Hook spit? If so, give the grid coordinates of possible dune forms shown by contour lines.

4. What is the origin of the marsh in the vicinity of 1.4–7.6?

5. Do the Navesink Highlands (mainland) show any topographic forms produced by wave erosion? If so, describe and locate the forms.

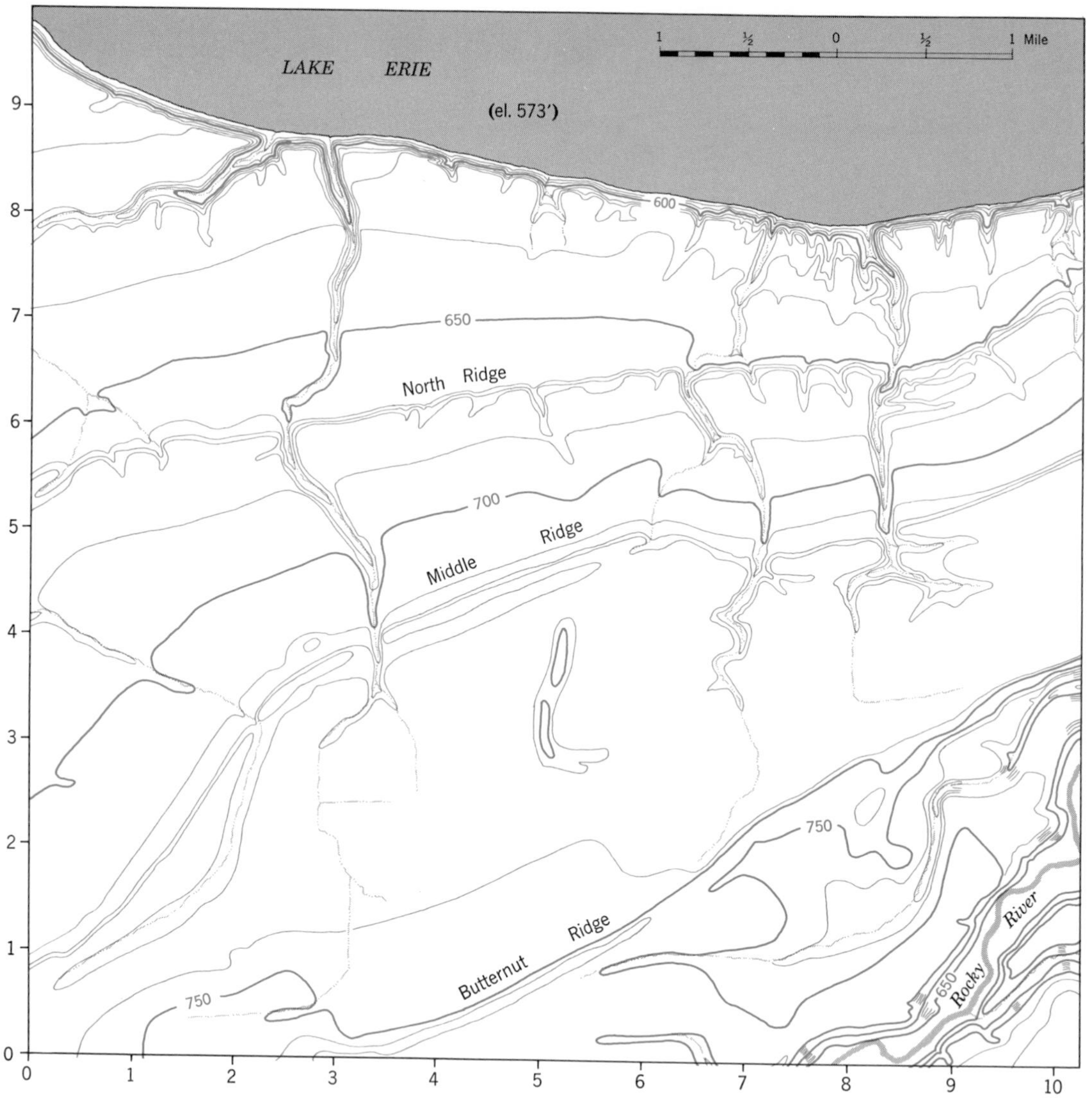

Exercise 5. ***Elevated Shorelines and Beach Ridges.*** (Source: Berea, Ohio, U.S. Geological Survey topographic map; scale 1:62,500.)

Explanatory Note: Lake Erie, whose present shoreline is located in the northern part of this map, formerly stood higher than at present. Three ridges mark previous higher stands of the water level; Butternut Ridge, the highest, is also the oldest; Middle Ridge and North Ridge mark progressively lower stages. On both Middle Ridge and Butternut Ridge, a well-developed beach ridge is indicated by the contours, whereas North Ridge is simply a low, wave-cut escarpment. The present shoreline of Lake Erie is a good example of a mature shoreline with a continuous wave-cut scarp and numerous hanging valleys.

QUESTIONS

1. Make a topographic profile across the map from north to south, starting at 4.9–9.0. Use a vertical scale of 1 in. equals 100 ft. Label the three ridges.

2. On the profile, draw horizontal lines to show the lake level at the time each of the three elevated shorelines was formed. Note that Middle Ridge and Butternut Ridge may have been barrier beaches typical of a shoreline of emergence.

3. At 5.2–3.5 is a curious curved ridge. Could this be some type of depositional shore feature? Explain.

4. **(*a*)** How high is the present scarp along the shore of Lake Erie? **(*b*)** Make an enlarged contour sketch map of a sea cliff with a hanging valley similar to those shown on the map.

CHAPTER **33**

Landforms Made by Wind

WIND, the fourth of the agents of erosion thus far discussed, produces a variety of interesting sequential landforms, both erosional and depositional. In terms of total mass of material thereby removed or deposited, however, it ranks below mass wasting, running water, waves, and ice, except in certain desert regions especially favorable to its action. In humid regions, with ample soil moisture and dense vegetative cover, there are few evidences of the work of wind. These usually are coastal sand dunes. Elsewhere, vegetation holds the ground in place unless man has laid it bare.

Erosion by wind

Wind performs two kinds of erosional work. Loose particles lying upon the ground surface may be lifted into the air or rolled along the ground. This process is *deflation.* Where the wind drives sand and dust particles against an exposed rock or soil surface, causing it to be worn away by the impact of the particles, the process is *abrasion.* Abrasion requires cutting tools carried by the wind; deflation is accomplished by air currents alone.

Deflation acts wherever the ground surface is thoroughly dried out and is littered with small, loose particles derived by rock weathering or previously deposited by running water, ice, or waves.

Thus, dry river courses, beaches, and areas of recently formed glacial deposits are highly susceptible to deflation. In dry climates, virtually the entire ground surface is subject to deflation because the soil or rock is everywhere bare. Wind is selective in its deflational action. The finest particles, those which constitute clay and silt, are lifted most easily and raised high into the air. Sand grains are moved only by moderately strong winds and tend to travel close to the ground. Gravel fragments and rounded pebbles up to 2 or 3 in (5 to 8 cm) in diameter may be rolled over flat ground by strong winds but do not travel far. They become easily lodged in hollows or between their fellows. Consequently, where a mixture of sizes of

Figure 33.1 This blowout hollow on the plains of Nebraska contains a remnant column of the original material, thus providing a natural yardstick for the depth of material removed by deflation. (Photograph by N. H. Darton, U.S. Geological Survey.)

particles is present on the ground, the finer sizes are removed; the coarser particles remain behind.

The principal landform produced by deflation is a shallow depression termed a *blowout*, or *deflation hollow*. This depression may be from a few yards to a mile or more in diameter, but is usually only a few feet deep. Blowouts form in plains regions in dry climates. Any small depression in the surface of the plain, particularly where the grass cover is broken through, may develop into a blowout. Rains fill the depression, creating a shallow pond or lake. As the water evaporates the mud bottom dries out and cracks, forming small scales or pellets of dried mud which are lifted out by the wind. In grazing lands, cattle may trample the margins of the depression into a mass of mud, breaking down the protective grass-root structure and facilitating removal when dry. Thus the depression is progressively enlarged (Figure 33.1). Blowouts are also found on rock surfaces where the rock is being disintegrated by weathering.

In the great deserts of southeastern California, Arizona, and New Mexico, the floors of intermontane basins are subject to deflation. The flat floors of the vast, shallow playas have in some places been reduced by deflation as much as several feet over areas of many square miles.

Where deflation has been active on a ground surface littered with loose fragments of a wide range of sizes, the pebbles that remain behind tend to accumulate until they cover the entire surface (Figure 33.2). By rolling or jostling about as the fine particles are blown away, the pebbles may become closely fitted together, forming a *desert pavement*. In North Africa such a pebble-covered surface is called a *reg*. The precipitation of calcium carbonate, gypsum, and other salts near the surface, as ground water is drawn to the surface and evaporated in dry weather, tends to cement the pebbles together, forming a highly effective protection against further deflation.

The sandblast action of wind against exposed rock surfaces is limited to the basal few feet of a cliff, hill, or other rock mass rising above a relatively flat plain, because sand grains do not rise high into the air. Wind abrasion produces pits, grooves, and hollows in the rock. Where a small rock mass projects above the plain it may be cut away at the base to make a *pedestal rock*, delicately balanced upon a thin stem. Most pedestal rocks, or *mushroom rocks*, are, however, produced by weathering processes.

Figure 33.2 This desert pavement of quartzite fragments was formed by action of both wind and water on the surface of an alluvial fan in the desert of southeastern California. Fragments range in size from 1 to 12 in (2.5 to 33 cm). Silt underlies the layer of stones. (Photograph by C. S. Denny, U.S. Geological Survey.)

Dust storms and sand storms

In dry seasons over plains regions, strong, turbulent winds lift great quantities of fine dust into the air, forming a dense, high cloud called a *dust storm*. The dust storm is generated where ground surfaces have been stripped of protective vegetative cover by cultivation or grazing, or where they naturally carry no vegetation cover because of extreme aridity of the climate. A dust storm approaches as a great dark cloud extending from the ground surface to heights of several thousand feet (Figure 33.3). Within the dust cloud deep gloom or even total darkness prevails. visibility is cut to a few yards, and a fine choking dust penetrates everywhere.

It has been estimated that as much as 4000 tons of dust may be suspended in a cubic mile of air (875 metric tons per cubic kilometer). On this basis, a dust storm 300 mi (500 km) in diameter might be carrying more than 100 million tons (90 million metric tons) of dust—enough to make a

Figure 33.3 Front of an approaching dust storm, Coconino Plateau, Arizona. (Photograph by D. L. Babenroth.)

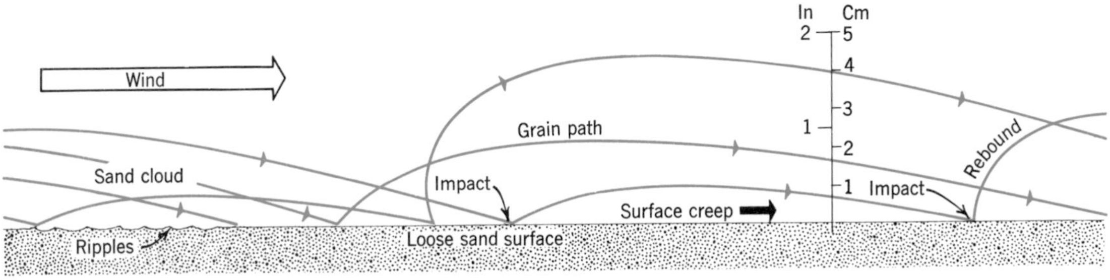

Figure 33.4 Sand particles travel in a series of long jumps. (After R. A. Bagnold.)

hill 100 ft (30 km) high and 2 mi (3 km) across the base.[1] A region that supplied the dust for thousands of such storms would thus lose a considerable mass over a span of thousands of years. Whether during the same period streams would remove more material from the same area is difficult to say, but in all probability they would remove much more.

Dust travels enormous distances in the air. That of individual dust storms is often traceable as far as 2500 mi (4000 km). Volcanoes erupt much extremely fine dust into the air. The renowned eruption of the volcano Krakatoa in Indonesia in the year 1883 cast out an enormous quantity of dust, some of which was caught by atmospheric circulation at high levels and carried around the entire earth. It is said that unusually brilliantly colored sunsets occurred in the British Isles in the years following 1883 as a result of the presence of the Krakatoa dust in the atmosphere. These were referred to as the "Chelsea sunsets," and were a favorite subject for paintings by English artists of the period.

The true desert *sandstorm* is a low cloud of moving sand that rises usually only a few inches

[1]A. K. Lobeck, *Geomorphology*, McGraw-Hill Book Co., New York. 1939, 731 pp., p. 380.

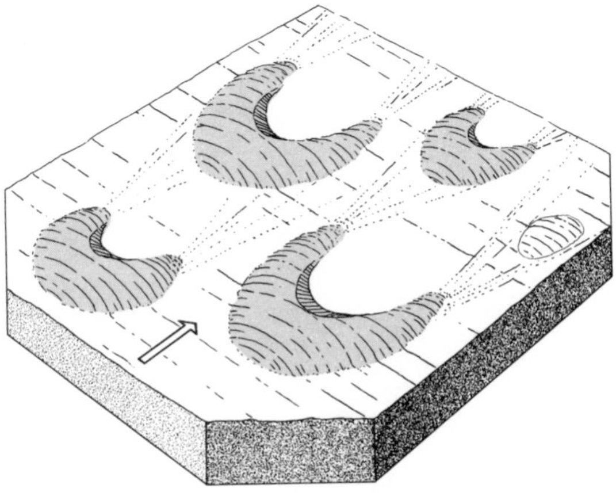

Figure 33.5 Barchans, or crescentic dunes. Arrow indicates wind direction.

and at most 6 ft (2 m) above the ground. It consists of sand particles driven by a strong wind. Those who have experienced sandstorms report that a man standing upright may have his head and shoulders entirely above the limits of the sand cloud. The reason why the sand does not rise higher is that the individual particles are engaged in a leaping motion, termed *saltation* (Figure 33.4). Grains describe a curved path of travel and strike the ground with considerable force but at a low angle. The impact causes the grain to rebound into the air. At the same time, the surface layer of sand grains creeps downwind as the result of the constant impact of the bouncing grains.

The erosional effect of blown sand is thus concentrated on surfaces exposed less than a foot or two (0.3 to 0.6 m) above the flat ground surface. Telephone poles on wind-swept sandy plains are quickly cut through at the base unless a protective metal sheathing or heap of large stones is placed around the base.

Sand dunes

A *dune* is any hill or accumulation of sand shaped by the wind. Dunes may be active, or *live*, when bare of vegetation and constantly changing form under wind currents. They may be inactive, or *fixed*, dunes, covered by vegetation that has taken root and serves to prevent further shifting of the sand.

Several common varieties of dunes are treated here. The *crescentic dune*, or *barchan* (also spelled barcan, barkhan, or barchane), is an isolated dune, which in plan view resembles a blunted crescent (Figure 33.5). The broadly rounded ends of the crescent point downwind and indicate the direction of dune motion and prevailing winds. On the windward side of the crest the sand slope is gentle, being the slope up which the sand grains move. On the lee side of the dune, within the crescent, is a steep curving dune slope, the *slip face*, which maintains an angle of about 35° from the horizontal (Figure 33.6). Sand grains fall or slide down the steep face after being blown free of the crest. When a strong wind is blowing, the flying sand makes a perceptible cloud at the crest. The term

Figure 33.6 Barchan dunes at Biggs, Oregon. (Photograph by G. K. Gilbert, U.S. Geological Survey.)

smoking crest has been used for this feature. Crescentic dunes rest upon a flat, pebble-covered ground surface. The sand may originate as a drift in the lee of some obstacle, such as a small hill, rock, or clump of brush. Once a sufficient mass of sand has formed it begins to move downwind, taking the form of a crescent dune. Thus the dunes are commonly arranged in chains extending downwind from the source drifts.

Where sand is so abundant that it completely covers the ground, dunes take the form of wavelike ridges separated by troughlike furrows. The dunes are called *transverse dunes* because their crests trend at right angles to direction of wind (Figure 33.7). The entire area may be called a *sand sea*, for it resembles a storm-tossed sea suddenly frozen to immobility. The term *erg*, referring to any large expanse of dunes in the Sahara Desert, has been

Figure 33.7 This air photograph of a sand-dune field between Yuma, Arizona, and Calexico, California, shows a sand sea of transverse dunes in the background and a field of crescentic barchan dunes in the foreground. (Copyrighted Spence Air Photos.)

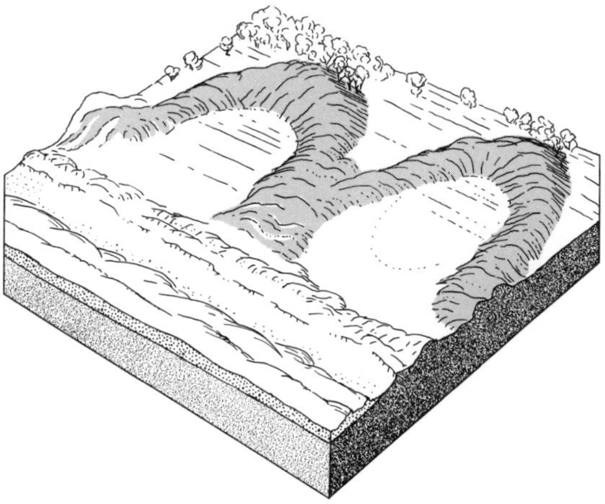

A. **Coastal blowout dunes with saucerlike depressions.**

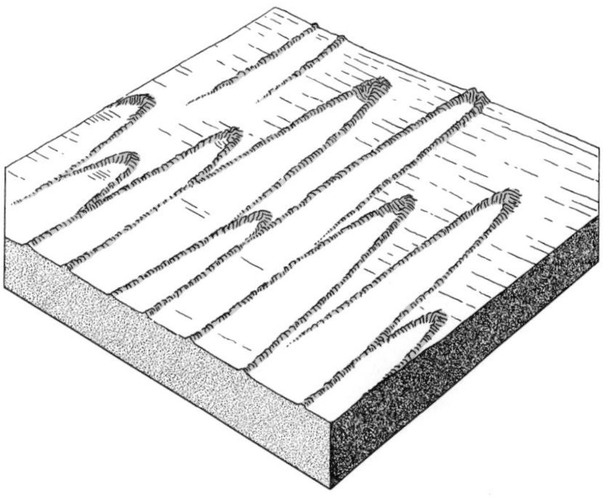

C. **Parabolic dunes of hairpin form.**

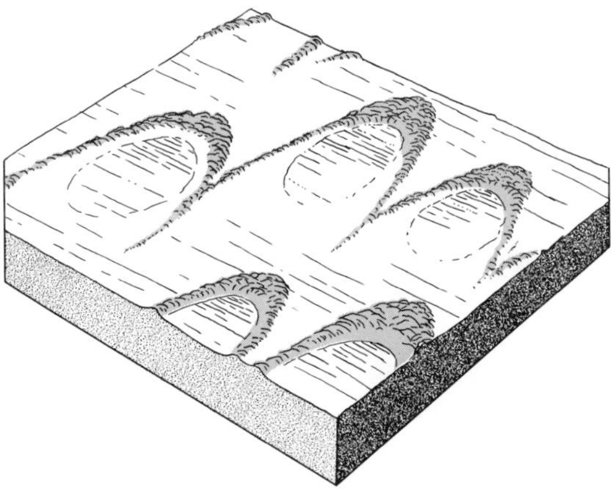

B. **Parabolic blowout dunes on an arid plain.**

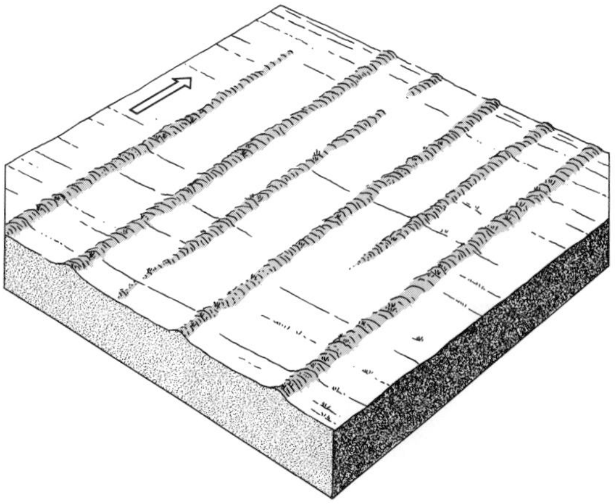

D. **Longitudinal dune ridges on a desert plain.**

Figure 33.8 Four types of dunes. Prevailing wind, shown by arrow, is the same for all diagrams.

Figure 33.9 At Beaufort Harbor, near Cape Lookout, North Carolina, coastal sand dunes are inundating a forest of live oak. (Photograph by J. A. Holmes, U.S. Geological Survey.)

adopted by geographers for this type of landscape. Individual sand ridges have sharp crests and are asymmetrical, the gentle slope being on the windward, the steep slope on the lee side. Deep depressions lie between the dune ridges. Sand seas require huge quantities of sand, often derived from weathering of a sandstone formation underlying the ground surface, or from adjacent alluvial plains. Still other transverse dune belts form adjacent to beaches which supply abundant sand and have strong onshore winds (Figure 33.10).

Another group of dunes belongs to a family in which the curve of the dune crest is bowed convexly downwind, the opposite of the curvature of crests in the barchan and transverse dunes. These may be described as *parabolic* in form. A common representative of this class, the *coastal blowout dune,* is formed adjacent to beaches, where large supplies of sand are available and are blown landward by prevailing winds (Figure 33.8*A*). A saucer-shaped depression is formed by deflation; the sand is heaped in a great curving ridge resembling a

Figure 33.10 The arrows on this photograph point to elongate blowout dunes of hairpin form, which once advanced from the beach and have since become stabilized by vegetation. Active transverse dunes are overriding the blowout dunes in a fresh wave, San Luis Obispo Bay, California. (Spence Air Photos.)

horseshoe in plan. On the landward side is a steep slip face which advances over the lower ground and buries forests, killing the trees (Figure 33.9). Coastal blowout dunes are well displayed along the southern and eastern shore of Lake Michigan, those of the southern shore being set aside for public use as the Indiana Dunes State Park.

In arid plains and plateaus, where vegetation is sparse and winds strong, groups of *parabolic blowout dunes* develop to the lee of shallow deflation hollows (Figure 33.8*B*). Sand is caught by low bushes and accumulates in a broad, low ridge. These dunes have no steep slip faces, and may remain relatively immobile. In some cases, however, the dune ridge migrates downwind, drawing the parabola into a long, narrow form with parallel sides (Figure 33.8*C*). This form resembles a hairpin in plan; hence has been named a *hairpin dune*, although it is a member of the parabolic family. Hairpin dunes stabilized by vegetative growth are seen in Figure 33.10.

Still another class of dunes is described as *longitudinal* because the dune ridges run parallel with the wind direction. On desert plains and plateaus, where sand supply is meager but winds are strong from one direction, *longitudinal dune ridges* are formed (Figure 33.8*D*). These are usually only a few feet high, but may be several miles long. In some areas, at least, the longitudinal dune is produced by extreme development of the hairpin dune, such that the parallel side ridges become the dominant form.

Figure 33.11 This oblique air photograph shows longitudinal sand-dune ridges reaching as far as the eye can see. Simpson Desert, southeast of Alice Springs, Australia. (Photograph by George Silk, *Life Magazine*, © Time, Inc.)

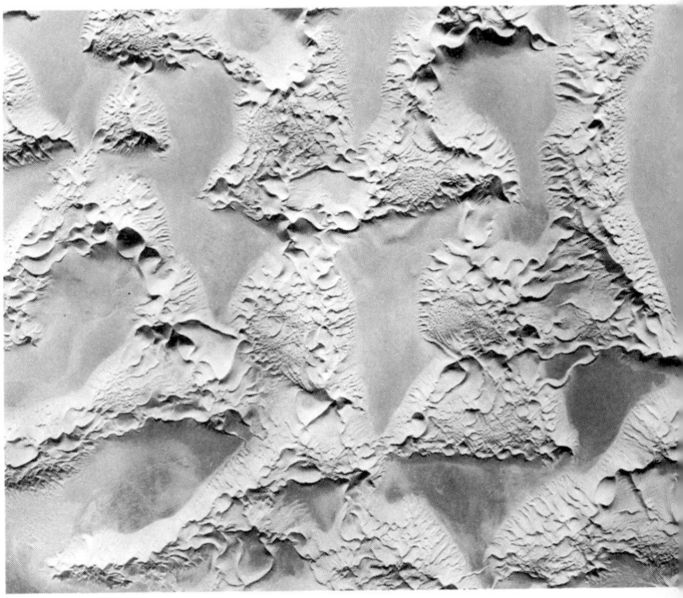

Figure 33.12 Seen from an altitude of 6 mi (10 km), these sand dunes of the Libyan desert appear as irregular patches which rise to star-shaped central peaks 300 to 600 ft (90 to 180 m) higher than the intervening flat ground. Width of the photograph represents about 7 mi (11 km). (Aero Service Corporation. Division of Litton Industries.)

Longitudinal dune ridges, oriented parallel with dominant winds, occupy vast areas of central Australia referred to as *sand-ridge deserts* (Figure 33.11). Ridges average 30 to 50 ft (10 to 15 m) in height, are spaced 0.25 to 1.5 mi (0.4 to 2.4 km) apart, and run in continuous length as much as 25 to 50 mi (40 to 80 km).

In the vast deserts of North Africa, Arabia, and southern Iran are large, complex dune forms apparently not represented in the United States. One of these is the *seif dune*, or *sword dune*, which is a huge tapering sand ridge whose crestline rises and falls in alternate peaks and saddles and whose side slopes are indented by crescentic slip faces. Seif dunes may be a few hundred feet high and tens of miles long. Another Saharan type is the *star dune*, *pyramidal dune*, or *heaped dune*, a great hill of sand whose base resembles a many-pointed star in plan (Figure 33.12). Radial ridges of sand rise toward the dune center, culminating in sharp peaks as high as 300 ft (100 m) or more above the base. Star dunes seem to remain fixed in position for centuries and can serve as reliable landmarks for desert travelers.

Coastal dunes and man

Landward of sand beaches there is usually present a narrow belt of dunes in the form of irregularly shaped hills and depressions; these constitute the *foredunes*. They normally bear a cover of beachgrass and a few other species of plants capable of survival in the severe environment. Dunes

Figure 33.13 Foredunes protected by beachgrass, Provincelands of Cape Cod. (Photograph by A. N. Strahler.)

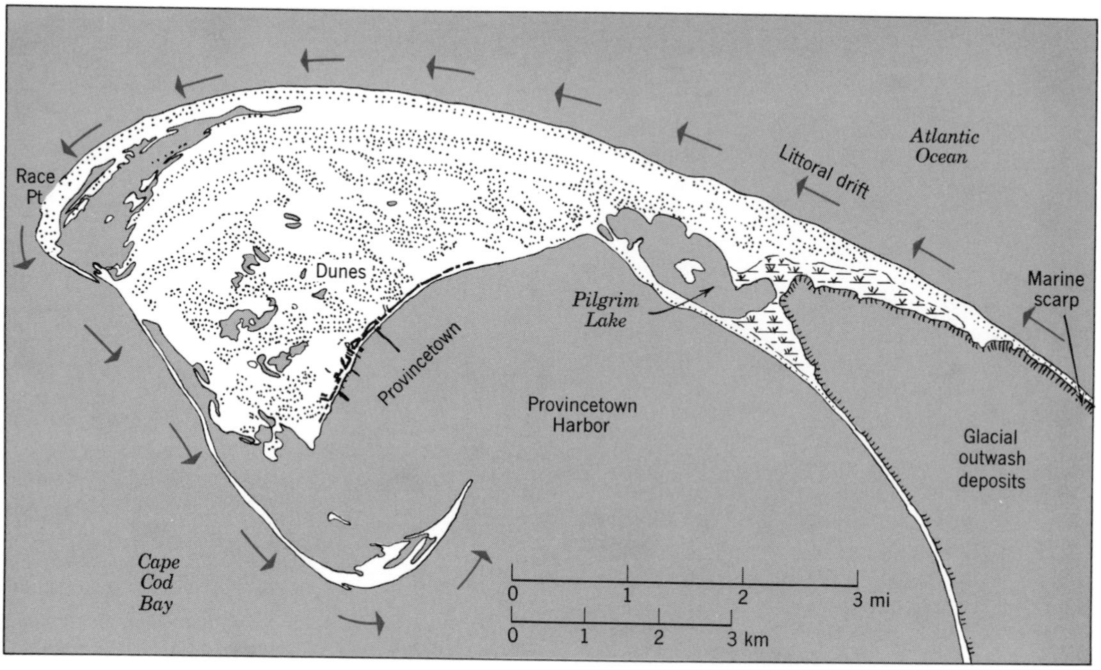

Figure 33.14 Sketch map of the Provincelands of Cape Cod, Massachusetts, as mapped in 1887. Dunes show as stippled pattern. (Data of U.S. Geological Survey.)

that evolve in association with a partial cover of vegetation are classed as *phytogenic* dunes. (Parabolic dunes are also in the phytogenic class.)

On coastal foredunes, the cover of beachgrass and other small plants, sparse as it seems to be, acts as a baffle to trap sand in saltation moving landward from the adjacent beach (Figure 33.13). As a result, the foredune ridge is built up as a barrier rising many feet above high tide level. For example, dune summits of the Landes coast of France reach elevations of 250 to 300 ft (80 to 90 m) and span a belt 2 to 6 mi (3 to 10 km) wide.

The swash of storm waves, acting at high water of tide and under conditions of raised water level due to lowered barometric pressure and the onshore drift of surface water, cuts away the upper part of the beach and the dune barrier is attacked. Undermining of the dunes supplies sand to the swash and this is spread over a wide area of beach slope. Wave energy is then largely expended in the riding of water up a long, sloping ramp, and the force of storm attack is dissipated. Between storms the beach is rebuilt and in due time the dune ridge is also restored if plants are maintained. In this way the foredunes form a protective barrier for tidal lands lying on the landward side.

If, now, the plant cover of the dune ridge is depleted by vehicular and foot traffic, or by bulldozing of sites for approach roads and buildings, a blowout will rapidly develop. The cavity thus formed may extend as a trench across the dune ridge. With the onset of a storm with high water levels, swash is funneled through the gap and spreads out upon the tidal marsh or tidal lagoon behind the ridge. Sand swept through the gap is spread over the tidal deposits, a phenomenon called *overwash.* A new tidal inlet may be created in this way, but few of these persist.

For many coastal communities of the eastern United States seaboard, the breaching of a dune ridge with its accompanying overwash brings a certain measure of environmental damage to the tidal marsh or estuary, and this may cause extensive property damage as well. However, this effect is minor compared with the effects of breaching of dune barriers along the North Sea Coast, which are part of the system to exclude seawater from reclaimed polders and fenlands. Protection of the dunes of the Netherlands coast assumes vital importance in view of the loss of life and property that a series of storm breaches can bring.

Another form of environmental damage related to dunes is the rapid downwind movement of sand when the dune status is changed from one of fixed or plant-controlled forms to that of live dunes of free sand. When plant cover is depleted, saltation rapidly reshapes the dunes to produce crests and slip faces. Blowout dunes are developed, and as already noted, the free sand slopes advance upon forests, roads, buildings, and agricultural lands (Figures 33.9 and 33.10). In the Landes region of

Figure 33.15 This perpendicular road cut in loess south of Vicksburg is typical of thick glacial loess accumulations on the eastern bluffs of the Mississippi River. (Photograph by Orlo Childs.)

coastal dunes on the southwestern coast of France, landward dune advance has overwhelmed houses and churches and even caused entire towns to be abandoned.

A striking case of man's interference with a dune environment is that of the Provincelands of Cape Cod, located at the northern tip of that peninsula, making up the fist of the armlike outline of the Cape (Figure 33.14). The Provincelands has been constructed of beach sand carried by northward littoral drift from a fast-eroding marine scarp in glacial deposits of the arm of the Cape. The structure consists of a succession of beach ridges, and these have been modified in form and increased in height by dune building. When the first settlers arrived in Provincetown, a city now occupying the south shore of the Provincelands, the dunes were naturally stabilized by grasses and other small plants covering the dune summits and by pitch pine and other forest trees on lower surfaces and in low swales between dune ridges, although dunes were probably active then, as now, on ridges close to the northern shore. Inhabitants grazed their livestock on the dune summits and rapidly cut the forests for fuel, with the result that the dunes were activated and began to move southward.

By 1725, Provincetown was being overwhelmed by drifting sand. Some buildings were partially buried and some had to be abandoned. Sand was carted away from the streets in large volumes. By the early 1800s the major dune ridges were moving southward at a rate estimated to be 90 ft (27 m) per year. Fortunately, beachgrass plantings, begun in 1825, and the rigorous enforcement of laws forbidding grazing and tree cutting, resulted finally in stabilization of all but the northernmost dunes. Even today, high slip faces advance upon a major highway and into Pilgrim Lake (Figure 33.14). The area is today a part of the Cape Cod National Seashore. Authorities have made extensive new plantings of dunegrass and have minimized vehicular traffic in an attempt to bring further control to sand movement.

Loess

In several parts of the world the ground is underlain by deposits of wind-transported silt, which has settled out from dust storms over many thousands of years. The material thus formed is known as *loess.* It generally has a uniform buff color and lacks any visible layering or other banding. Loess has a tendency to break, or *cleave,* along vertical cliffs wherever it is exposed by the cutting of a stream or by man (Figure 33.15). The cleavage is possibly produced by a slight shrinkage of the entire mass as it has compacted after being laid down.

Perhaps the greatest deposits of loess are in China, where thicknesses over 100 ft (30 m) are common and a maximum of 300 ft (90 m) has been measured. It covers many hundreds of square miles in northern China and appears to have been derived from the interior of Asia, out of which blow dry winter winds. Loess deposits are also important in central Europe, Argentina, and New

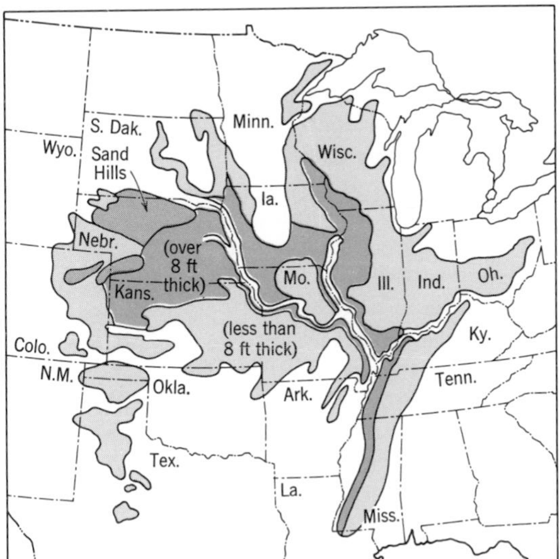

Figure 33.16 Map of loess distribution in the central United States. (Data from Map of Pleistocene Eolian Deposits of the United States, Geol. Soc. Amer., 1952.)

Zealand, but not so extensive or thick as in China.

In the United States, important loess deposits lie in the Missouri-Mississippi valley (Figure 33.16). Much of the prairie plains region of Indiana, Illinois, Iowa, Missouri, Nebraska, and Kansas is underlain by a loess layer ranging in thickness from 3 to 100 ft (1 to 30 m). There are also extensive deposits along the lands bordering the lower Mississippi River floodplain on its east side, throughout Tennessee and Mississippi. Still other loess deposits are in northeast Washington and western Idaho. The American and European loess deposits are directly related to the continental glaciers of the Pleistocene Epoch. At the time when the ice covered much of North America and Europe, it is possible that a generally dry winter climate prevailed in the land bordering the ice sheets and that strong winds blew southward and eastward over the bare ground, picking up silt from the floodplains of braided streams which discharged the meltwater from the ice. This dust settled upon the ground between streams, gradually building up to produce a smooth, level ground surface. The loess is particularly thick along the eastern sides of the valleys because of prevailing westerly winds, and is well exposed along the bluffs of most streams flowing through the region today.

The importance of loess in world agricultural resources cannot be easily overestimated. Loess plains and plateaus have developed rich, black soils especially suited to cultivation of grains. These are the prairie, chernozem, chestnut, and brown soils described in Chapter 19. The highly productive plains of southern Russia, the Argentine Pampa, and the rich grain region of north China are underlain by loess. In the United States, corn is extensively cultivated on the loess plains in those states, such as Iowa and Illinois, where rainfall is sufficient; wheat is grown farther west on loess plains of Kansas and Nebraska and in the Palouse region of eastern Washington.

Because loess forms vertical walls along valley sides and is able to resist sliding or flowage, but at the same time is easily dug into, it has been widely used for cave dwellings both in China and in Central Europe. In China, old trails and roads in the loess have become deeply sunken into the ground as a result of the pulverization of the loess of the road bed and its removal by wind and water (Figure 33.17).

Figure 33.17 Road sunken deeply into loess, Shensi, China. (Photograph by Frederick G. Clapp, courtesy of *The Geographical Review*.)

Man as an agent in inducing deflation

Cultivation of vast areas of plains under a climatic regime of substantial seasonal water deficiency, where only sparse grasses are normally sustained, is a practice inviting deflation of soil surfaces. Much of the Great Plains region, including all or parts of the states of New Mexico, Texas, Oklahoma, Kansas, Colorado, Nebraska, and the Dakotas is such a marginal region and has in past centuries experienced many dust storms generated by turbulent winds. Strong cold fronts frequently sweep over this area, lifting dust high into the troposphere at times when soil moisture is low. However, deflation and soil drifting reached disastrous proportions during a series of drought years in the middle 1930s, following a great expansion of wheat cultivation. These former grasslands are underlain by humus-rich brown and chestnut soils. During the drought a sequence of exceptionally intense dust storms occurred. Within their formidable black clouds visibility declined to nighttime darkness, even at noonday. The area affected became known as the *Dust Bowl.* Many inches of topsoil were removed from fields and transported out of the region as suspended dust, while the coarser silt and sand particles accumulated in drifts along fence lines and around buildings. The combination of environmental degradation and repeated crop failures caused widespread abandonment of farms and a general exodus of farm families. Well known through popular novels and screen plays is the migration of the impoverished "Okies" westward to new homes and occupations in California.

Among scientists who have studied the Dust Bowl phenomenon, there is a difference of opinion as to how great a role soil cultivation and livestock grazing played in augmenting deflation. The drought was a natural event over which man had no control, but it seems reasonable that the natural grassland would have sustained far less soil loss and drifting, had it not been destroyed by the plow.

Although man cannot prevent cyclic occurrences of drought over the Great Plains, measures can be taken to minimize the deflation and soil drifting occurring in periods of dry soil conditions. Improved farming practices include use of *listed furrows* (deeply carved furrows) that act as traps to soil movement. Stubble mulching will reduce deflation when land is lying fallow. Tree belts may have significant effect in reducing the intensity of wind stress at ground level.

Man's activities in the very dry, hot deserts have contributed measurably to raising of dust clouds. In the desert of northwest India and West Pakistan (the Thar Desert bordering the Indus River) the continued trampling of fine-textured soils by hooves of grazing animals, and by human feet as well, produces a blanket of dusty hot air that hangs over the region for long periods and extends to a height of 30,000 ft (9 km). In other deserts, such as those of North Africa and the southwestern United States, ground surfaces in the natural state contribute comparatively little dust because of the presence of desert pavements and sheets of coarse sand from which fines have already been winnowed. This protective layer is easily destroyed by wheeled vehicles, exposing finer textured materials and allowing deflation to raise dust clouds. It is said that the disturbance of large expanses of North African desert by tank battles during World War II caused great dust clouds, and that dust from this source was traced as far away as the Caribbean region.

Review Questions

1. Explain the processes of deflation and abrasion by wind.

2. What conditions favor deflation? What topographic forms are produced?

3. What is a desert pavement? How does it form? What is a reg?

4. What forms does wind abrasion produce?

5. How do dust storms originate? How much material might be carried in a single dust storm? How far does the dust travel?

6. How do sand grains travel in a sandstorm? How high do the grains rise? At what level is abrasion concentrated?

7. What are sand dunes? What is the distinction between live dunes and fixed dunes?

8. Describe a crescentic, or barchan, dune. In which direction does it move?

9. What are transverse dunes? What is a sand sea? What is the source of the sand? What places would be most favorable for the development of dune areas? What is an erg landscape?

10. Describe a coastal blowout dune, a parabolic blowout dune and a hairpin dune. How does each develop?

11. Describe longitudinal sand dunes. What do they indicate as to the direction of prevailing winds?

12. Where are seif dunes and star dunes found?

13. How do man's activities result in environmental damage to coastal foredunes and their bordering tidal marshes? How can the migration of live coastal dunes be controlled?

14. What is loess? How is it formed? What structure has loess?

15. Describe the distribution of loess throughout the world. What origin has the loess of northern China?

16. What states of the Mississippi-Missouri river region have loess deposits? Of what economic importance is loess?

17. In what ways do man's activities induce deflation on a large scale in arid lands?

Exercises

Exercise 1. *Crescentic (Barchan) Dunes.* (Source: Sieler, Wash., U.S. Geological Survey topographic map; scale 1:24,000.)

Explanatory Note: The isolated, crescent-shaped hills on this map are barchan dunes similar in appearance to those shown in Figures 33.5 and 33.6 and in the lower right hand corner of Figure 33.7. They lie near the easternmost fringe of a large dune field in the vicinity of Moses Lake, Washington. On many of these barchan dunes, a low ridge curves westward from each end of the crescent, reversing the normal curvature of the barchan. This results from a tendency of the dunes to take the parabolic form (Figure 33.8*B*) which has become superimposed on the barchan form. The parabolic form is well shown in dunes at 0.1–1.3, 0.6–1.2, and 0.1–0.6.

QUESTIONS

1. (*a*) What contour interval is used on this map? **(*b*)** What is the fractional scale of this map?

2. (*a*) On a sheet of tracing paper placed over the map, outline in pencil the base of at least twelve well-developed barchan dunes. **(*b*)** Indicate by an arrow the direction of prevailing wind. **(*c*)** Choosing the three highest, best-formed dunes, measure the length and height of each in feet and write this information directly beside the dune outline. **(*d*)** Mark the slip faces of these three selected dunes with hachures to show the extent and form of the slip face.

3. What evidence is there that the ground-water table lies close to the surface between dunes? What effect might this have on dune development?

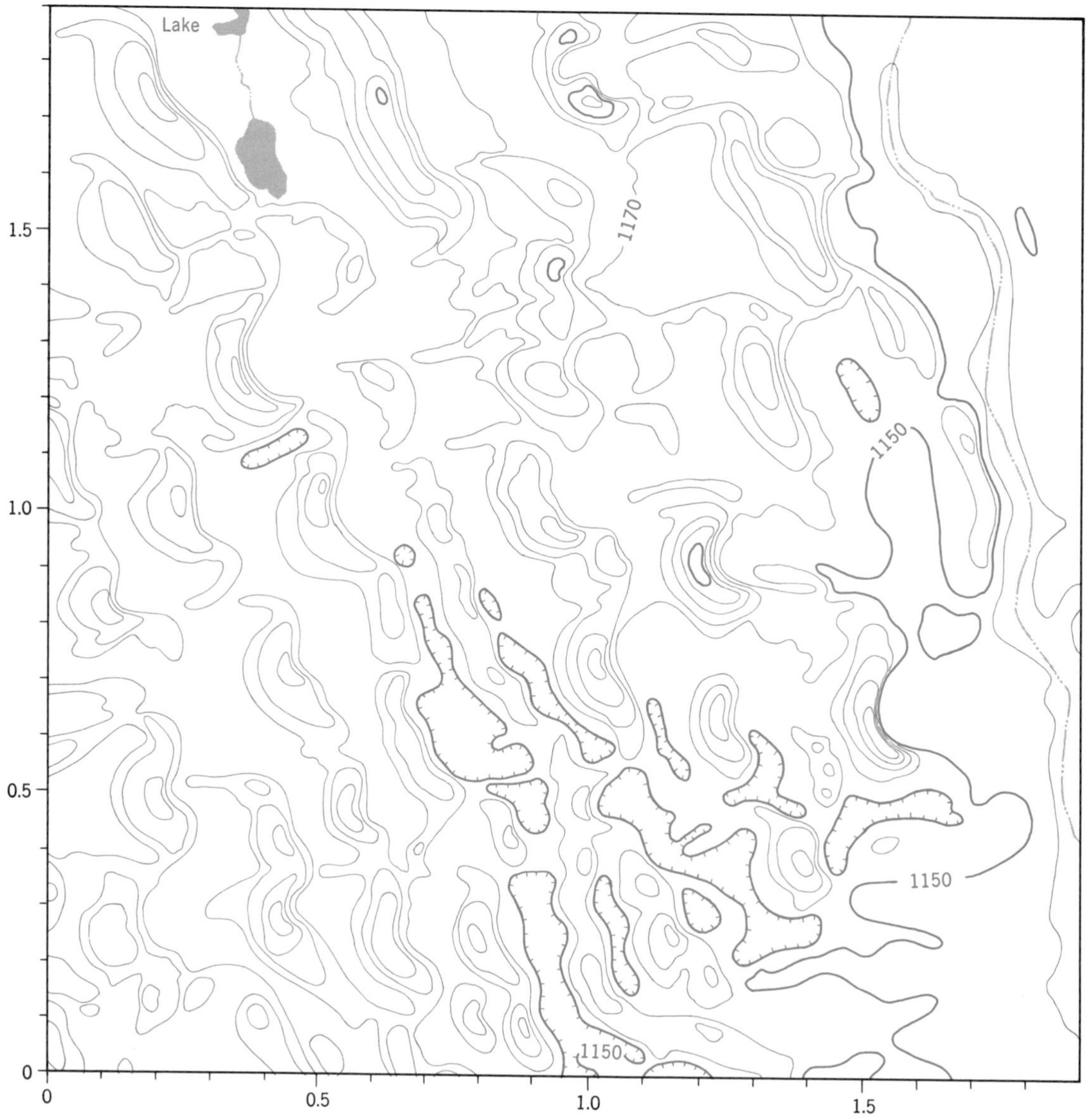

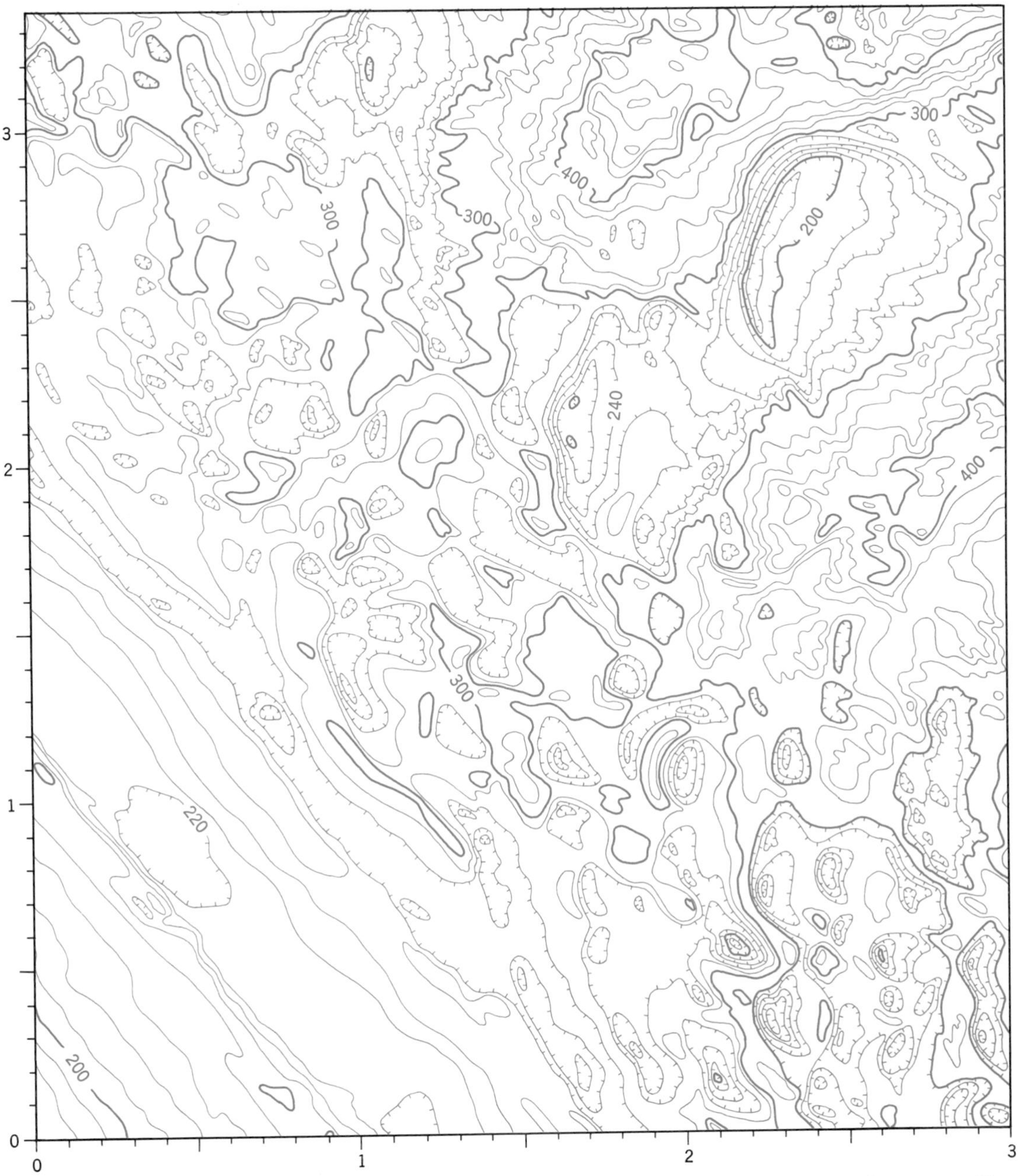

Exercise 2. ***Sand Sea of Transverse Dunes.*** (Source: Glamis Southeast, Calif., U.S. Geological Survey topographic map; scale 1:24,000).

Explanatory Note: Near Yuma, Arizona, in the dry, hot Sonoran Desert, is the largest active sand dune belt of the United States. The portion shown here consists of barren dunes of loose, pale-yellow sand separated by irregular depressions. An air view of this dune belt is shown in Figure 33.7. Although the area of the map lies north of the photo area, the dune topography is very similar to that pictured in the middle distance of the photograph, which is a view toward the west. A series of parallel dune ridges, seen cutting diagonally across the southwest corner of the map, forms the southwestern border of the dune belt. Over most of the map, however, the sand is formed into great ridges separated by deep hollows. High steep slip faces dip toward the southeast and east.

QUESTIONS

1. Compare map scale and contour interval with those of the map of Exercise 1. **(*a*)** What is the maximum relief of this dune belt, measuring from the highest dune peak to the deepest point in a dune depression? **(*b*)** Give elevations and grid coordinates of the highest dune peak and lowest point in the deepest closed depression.

2. In what way do the shapes of the dune ridges and depressions indicate direction of prevailing winds? What is this direction?

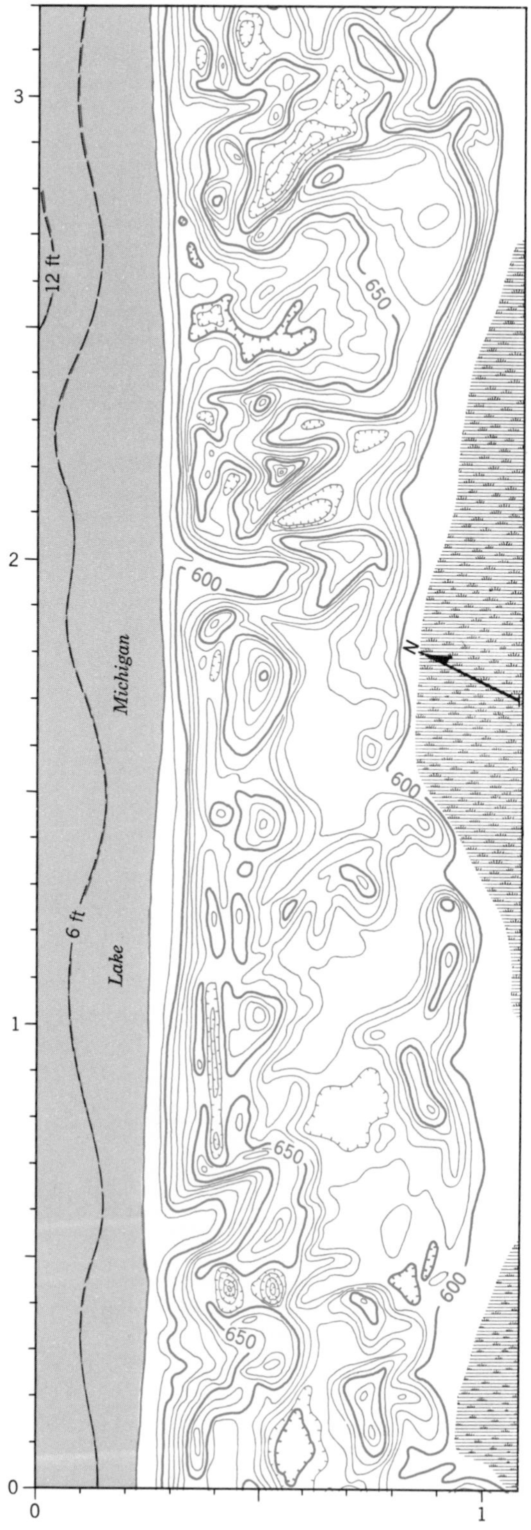

Exercise 3. *Coastal Blow-Out Dunes.* (Source: Dune Acres, Ind., U.S. Geological Survey topographic map; scale 1:24,000.)

Explanatory Note: This map shows a part of the southern coast of Lake Michigan, set aside for public use as the Indiana Dunes State Park. Abundant supplies of beach sand have accumulated here by shore drifting toward the southern end of Lake Michigan. Strong onshore winds, from northerly and northwesterly directions, have produced a series of large coastal blowout dunes of the type shown in Figure 33.8*A*.

QUESTIONS

1. **(*a*)** State the contour interval on this map. **(*b*)** Give the grid coordinates of all points of elevation greater than 700 ft. **(*c*)** What is ground distance in miles, represented by the long dimension of this map? **(*d*)** Give the width in miles of the dune belt at its widest point.

2. Place a tracing sheet over this map. Outline in pencil the well-developed blowout dunes at 0.8–2.7, 0.7–2.0, and 0.5–0.6. Show the form of each of these dunes by hachures drawn on the slopes.

3. What type and distribution of vegetation would you expect over this coastal dune belt?

Exercise 4. ***Hairpin Dunes.*** (Source: Idaho Falls South, Idaho, U.S. Geological Survey topographic map; scale 1:24,000.)

Explanatory Note: Long narrow dune ridges on this map trend southwest to northeast. Here and there a ridge is doubled back on itself to produce the characteristic hairpin form of highly elongated parabolic blowout dunes, such as those illustrated in Figures 33.8*C* and 33.10. The ends of several of the hairpin dunes have been built up into prominent dune masses.

QUESTIONS

1. **(*a*)** Compare the contour interval of this map with that of Exercise 3. **(*b*)** Which of the two exercise maps has the larger scale?

2. On a sheet of tracing paper laid over this map, outline the base of all dunes, connecting the individual ridge and hill elements so as to produce and emphasize the outlines of a few simple hairpin dunes. Show by arrow the direction of north on the map. Use another arrow to show prevailing wind direction required for development of these dune forms.

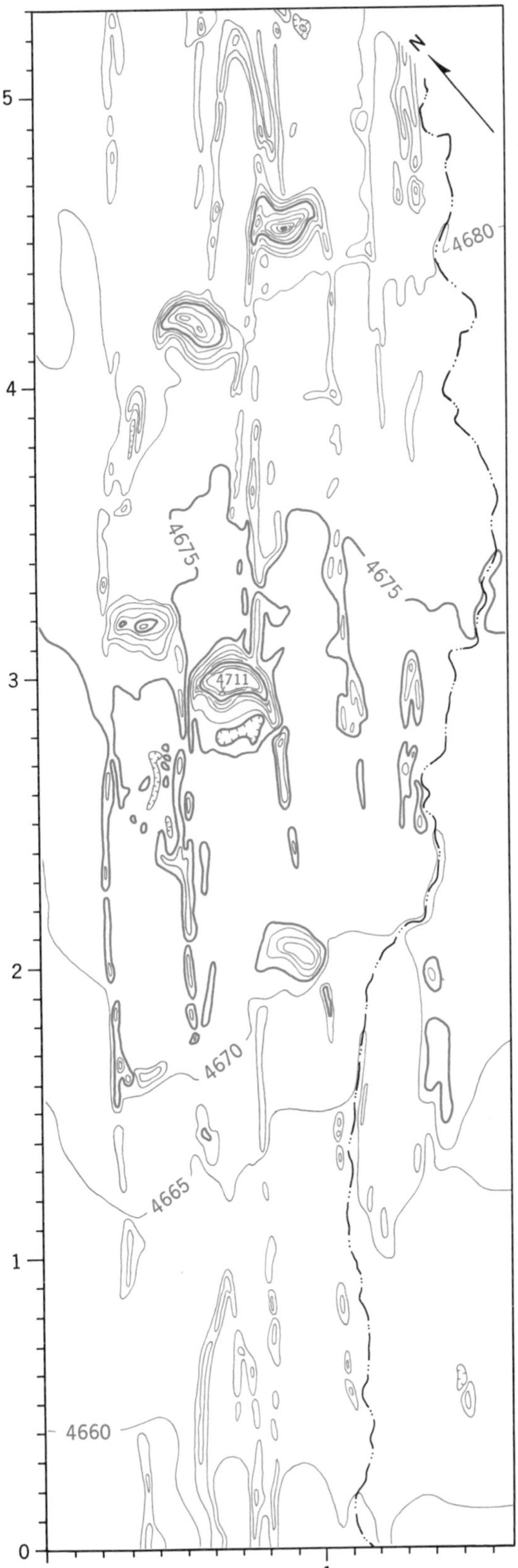

APPENDIX A

APPENDIX I

Map Reading

THE techniques of reading planimetric and topographic maps include application of principles of map scales, azimuths of lines, location of points by coordinate systems, and the representation of relief features. In this appendix a number of the principles treated in Chapters 1, 2, and 3 are applied and developed in further detail.

Map scale

Scale is the ratio between map distance and the actual ground distance that the map represents (Chapter 2). Given a fractional scale (representative fraction, or R.F.) on which to draw his map, the cartographer must convert this fraction to units of measurement, for he must use a ruler scaled in inches or centimeters to measure lengths of line on the map to represent ground distances in miles or kilometers.

For example, if the map scale is given as 1/63,360, or 1:63,360, as explained in Chapter 2, this fractional scale may be interpreted as "one inch on the map represents one mile on the ground." A map scale of 1:100,000 could be read as "one centimeter represents one kilometer."

Most maps of small areas carry a fractional scale printed on the map margin. Conversion to equivalent ratios of inches to miles or centimeters to kilometers is left to the reader to compute. For practical map use, however, a *graphic scale* is printed on the map margin. This is a length of line divided off into numbered segments (Figure AI.1). The units are in conventional terms of measurement, such as feet, yards, and miles, or meters and kilometers. To use the graphic scale, a piece of paper with a straight edge is held along the line to be measured on the map and the distance marked on the edge of the paper. The paper is then placed along the graphic scale and the length of the line read directly. Where many measurements are to be made it will save time to copy the graphic scale onto the edge of a piece of paper and apply it directly between points on the map (Figure AI.1).

In the study of map projections (Chapter 2) it was emphasized that no global map on a flat sheet of paper can have a constant scale in all parts and in all directions. Within the limits of maps of large scale, showing only a very small part of the earth's surface, the scale changes are so slight as to be dismissed. A graphic scale will be true in terms of the extent to which the eye of the user can distinguish the width of a finely printed line.

Conversion of scale from one form to another

An important skill in cartography and map reading is the conversion of scale from one form to another. These manipulations are best explained by examples.

Example A. ***Fractional form to length units.*** *Problem:* How may the scale 1:100,000 be stated in familiar units, such as inches to the mile? To avoid any mistake, go back to the definition of scale and work by easy steps to the final answer. The following sequence is suggested:

$$\frac{\text{Distance on map}}{\text{Distance on ground}} = \frac{\text{1 in. on map}}{\text{100,000 in. on ground}}$$

Now, in order to have the ground distance in miles, we must divide 100,000 in. by 63,360, because there are 63,360 in. to a mile. Therefore

$$\frac{\text{1 in. on map}}{100{,}000 \div 63{,}360} = \frac{\text{1 in. on map}}{\text{1.57 mi on ground}}$$

or "1 in. to 1.57 mi."

Example B. ***Length units to fractional form.*** *Problem:* How may the scale "1 in. to 1 mi" be written as a fraction? Following the general rule, write the information as follows:

$$\frac{\text{1 in. on map}}{\text{1 mi on ground}}$$

Convert the denominator into inches by multiplying by 63,360 (because there are 63,360 in. in a mile). Then

$$\frac{\text{1 in. on map}}{\text{63,360 in. on ground}} \quad \text{or} \quad \frac{1}{63{,}360}$$

Example C. *Problem:* Construct a graphic scale for the scale "1 in. to 1 mi." This is the simplest of all scale operations. Lay off a line and divide it into segments 1 in. long. Each unit represents 1 mi and can be so labeled.

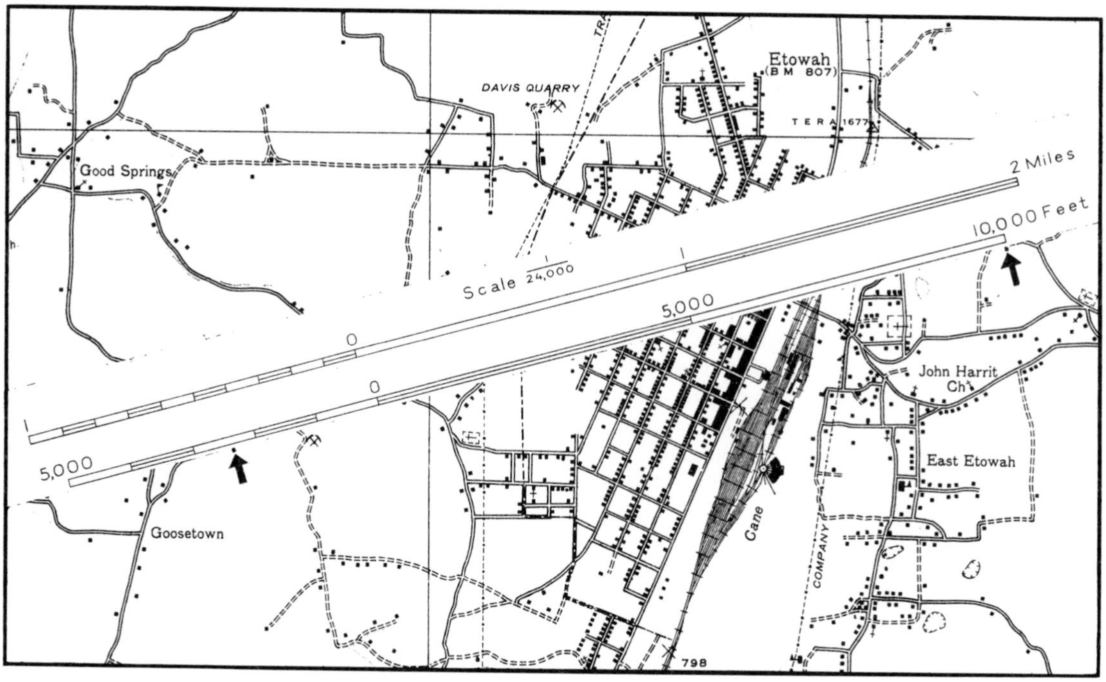

Figure A I.1 The distance between two points on a map can be read directly from a graphic scale. (Portion of U.S. Geological Survey map.)

Example D. *Problem:* Construct a graphic scale for the scale "1 in. to 11.3 mi." Here, although the line could be divided into 1-in. parts, marking them "0,11.3,22.6,33.9, etc.," such a graphic scale would be of little use to anyone. The problem is to make the graphic scale consist of some even-numbered unit, preferably some power of 10, such as 1, 10, 100, or 1000 miles or kilometers. Therefore, perform the following operation:

$$\frac{1 \text{ inch}}{11.3 \text{ miles}} = \frac{x \text{ in.}}{10 \text{ mi}}$$

Solving for x,

$$11.3x = 10$$
$$x = \frac{10}{11.3}$$
$$x = 0.885 \text{ in.}$$

Now lay off segments 0.885 in. apart on the line. Each unit represents 10 mi.

Example E. *Problem:* Given the fractional scale 1:50,000, make a graphic scale using miles as units of measure. This problem can be broken down to two problems already treated. First, convert the fractional scale to length units of inches and miles, as explained in Example A, then construct the graphic scale, as explained in Example D.

Example F. *Problem:* Given the graphic scale shown below, determine the fractional scale:

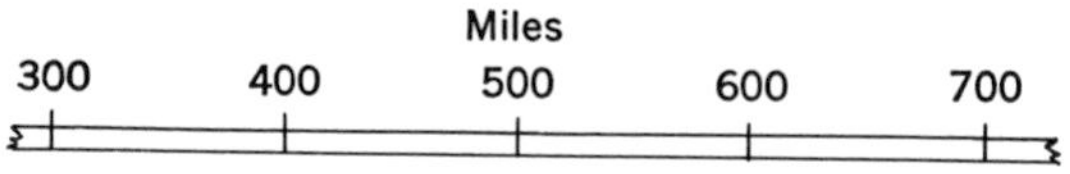

Solution of this problem requires a ruler, preferably one scaled off in tenths of inches. Measure the exact length of several units on the graphic scale, recording your results in the following form:

$$\frac{1.97 \text{ in.}}{300 \text{ mi}}$$

First reduce the 300 mi to inches by multiplying by 63,360 (because there are 63,360 in. in a mile); then reduce to a simple fraction:

$$\frac{1.97 \text{ in.}}{300 \times 63{,}360 \text{ in.}} = \frac{1.97}{19{,}008{,}000}$$
$$= \frac{1}{9{,}650{,}000 \text{ approx.}}$$

Large-scale and small-scale maps

The relative size of two different scales is determined according to which fraction is the larger and which the smaller. For example, a scale of 1:10,000 is twice as large as a scale of 1:20,000. Many students are confused about this because they unthinkingly suppose that the fraction with the larger denominator represents the larger scale. If in doubt, ask yourself the question: "Which fraction is larger: $\frac{1}{4}$ or $\frac{1}{2}$?"

Maps with scales ranging from 1:600,000 down to 1:100,000,000 or smaller are known as *small-scale maps.* Those of scale 1:600,000 to 1:75,000 are *medium-scale* maps; those of scale greater than 1:75,000 are *large-scale* maps.

Starting at the small-scale end, the following are examples of the use of various scales. A 6-in. globe or a Mercator map of the world measuring about 12 by 18 in. has an equatorial scale of about 1:85,000,000 and is classed as a very small-scale map. A large map of the world suitable for hanging on the wall of a classroom has an equatorial scale of about 1:15,000,000 if the earth's equator is represented on the map as a line 8 ft long. One inch on this small-scale map represents a distance of 240 mi. A wall map of the United States measuring about 7 ft across has a fractional scale of 1:2,500,000. Each inch represents a distance of about 40 mi. A wall map of Wyoming measuring 3 by 4 ft is on a medium scale of 1:500,000 and represents distances on the ratio of 1 in. to about 8 mi.

For representing details of the earth's surface configuration, or relief, large-scale maps are needed and the area of land surface shown by an individual map sheet must necessarily be small. A topographic sheet 10 by 20 in., on a scale of 1:63,360 (1 in. to 1 mi) would, of course, include an area 10 by 20 mi, or 200 sq mi. Of the common sets of topographic maps published by national governments for general distribution, most fall within the scale range of 1:20,000 to 1:250,000.

Relation between scales and areas

Assuming that two maps, each on a different scale, have the same dimensions, what is the relation between the ground areas shown by each? In Figure A1.2 are shown three maps, each having the same dimensions, but representing scales of 1:20,000, 1:10,000, and 1:5,000, respectively, from left to right. Although map *B* is on twice the scale of map *A*, it shows a ground area only one-fourth as great. Map *C* is on four times as large a scale as map *A*, yet it covers a ground area only one-sixteenth as much. From this example can be deduced the following rule: the ground area that is represented by a map of given outside dimensions varies inversely with the square of the change in scale. Thus, if the scale is reduced to one-third its original value, the area that can be shown on a map of fixed dimensions increases to nine times the original value.

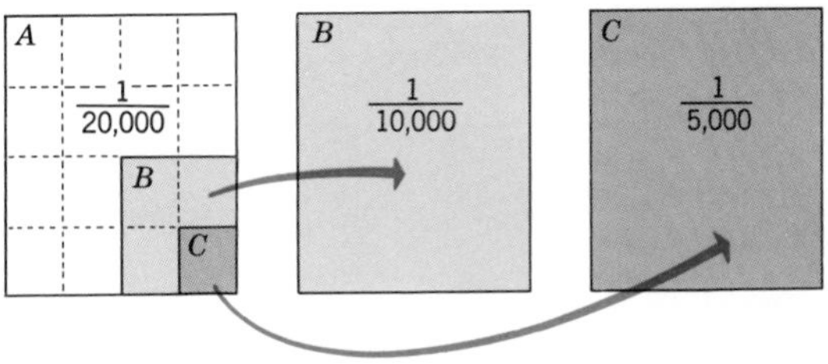

Figure A I.2 Area shown on a map decreases as scale increases.

Map quadrangle systems

As explained in Chapter 3, parallels and meridians usually form the boundaries of individual map sheets of a series and are known as *quadrangles* (Figure 3.13).

Good examples of the geographic grid system used on large-scale maps are found on the U.S. Geological Survey's topographic maps of the United States. Their relations to one another are shown in Figures A1.3 and A1.4. Most 15-minute and 30-minute quadrangles measure 17.5 in (44.5 cm) from top to bottom, a figure that remains constant over the entire country because the length of a degree of latitude, as shown in an earlier chapter, is everywhere nearly the same. Width of these quadrangles varies from about 15 in (38 cm) for Texas to about 12 in (30.5 cm) for North Dakota because of the northward convergence of meridians.

Seven standard scales comprise the National Topographic Map Series:

Series	R.F.	Unit Equivalents
7.5 minute	1:24,000	1 in. to 2000 ft
7.5 minute	1:31,680	1 in. to $\frac{1}{2}$ mi
15 minute	1:62,500	1 in. to about 1 mi
Alaska	1:63,360	1 in. to 1 mi
30 minute	1:125,000	1 in. to about 2 mi
1:250,000	1:250,000	1 in. to about 4 mi
1:1,000,000	1:1,000,000	1 in. to about 16 mi (1 cm to 10 km)

Figures A1.3 and A1.4 compare the coverages and sizes of standard quadrangles on several of these series.

Alaska provides an illustration of the geographic grid applied to map quadrangles in high latitudes, where the length of a degree of longitude is much shorter than a degree of latitude (see Figure 3.13). Two scales are used. The map series on 1:250,000 consists of quadrangles of one degree of latitude by three degrees of longitude. The large-scale series, 1:63,360, consists of quadrangles all covering 15

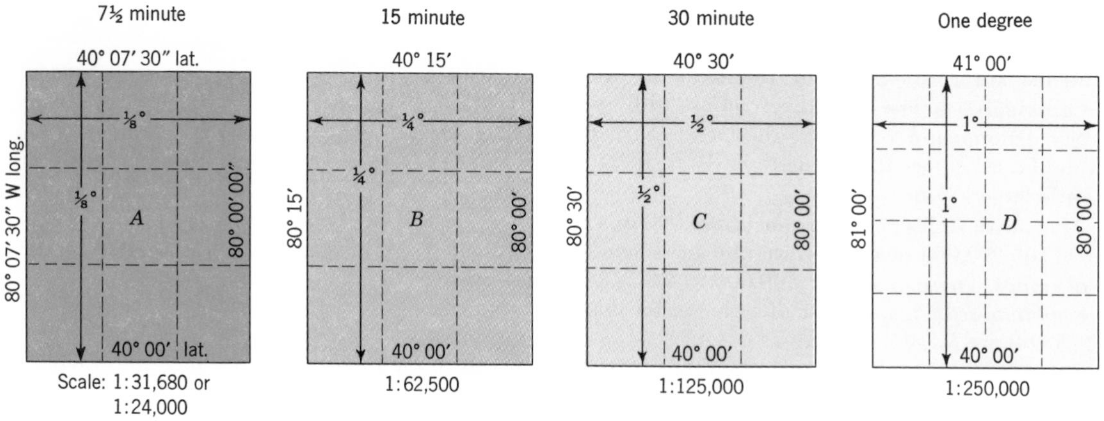

Figure A I.3 Large-scale maps of the U.S. Geological Survey are bounded by parallels and meridians to form quadrangles. The four quadrangles shown here represent the scales and areas commonly used in the United States, exclusive of Alaska.

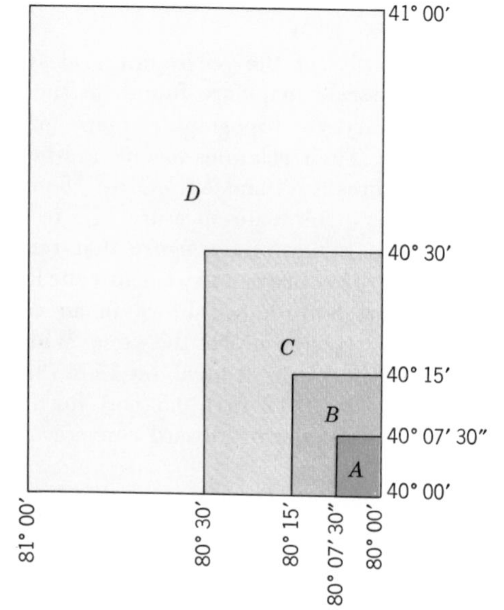

Figure A I.4 If the four quadrangles of Figure I.3 were reduced to the same scale, their areas would compare as shown here.

minutes of latitude, but including 20, $22\frac{1}{2}$, 30, or 36 minutes of longitude according to latitude, so as to provide sheets of about equal map dimensions.

Another example of the geographic grid applied to map sheets is the series of U.S. World Aeronautical Charts issued by the U.S. Coast and Geodetic Survey on the scale of 1:1,000,000. These maps are for use in military and civilian air navigation the world over. Figure A1.5 is a portion of the index map of World Aeronautical Charts and shows how parallels and meridians form the boundaries of individual map sheets. All sheets have a latitude coverage of 4°, but the longitude span ranges from 6°, in low latitudes, to 12° at 60° latitude. Poleward of 60° the longitude span increases rapidly.

Scale factors and the military grids

The military grid and its coordinate designations are treated in Chapter 3. The lines of the UTM and UPS military grid are exactly uniform in spacing, so that the printed squares are exactly the same in dimension wherever printed on maps of a given stated scale, such as 1:100,000. In the study of map projections, it was emphasized that no map printed on a flat sheet can preserve a truly constant scale in all parts of the map and in all directions. Consequently, we must reckon with the fact that the so-called "1000-meter grid square" does not everywhere represent a ground square exactly 1000 m wide. On the UTM grid, the map scale is exactly true with respect to the grid along two lines that lie parallel with the central meridian at a distance of 180,000 m on each side (Figure A1.6). Thus the scale is true at eastings of 320,000 m and 680,000 m. Between these grid lines, the map scale is smaller than the true scale by a *scale factor* of 0.9996. This means that a straight horizontal 1000 m line on the ground at sea level will appear to be 999.6 m long with respect to the grid scale. The scale factor at the outer edge of the grid zone is 1.0010, so that a 1000-m line on the ground will appear to be 1001 m long with respect to the grid scale. These scale errors are trivial in most practical problems of map reading, but the principle is nevertheless important.

On the UPS grid, the scale factor is set arbitrarily as 0.994 at the pole. The scale is true to the grid (scale factor 1000) at latitude 81°06′. At 80° latitude, the limit of the UPS system, the scale is too large by a factor of 1.0016 (Figure A1.6).

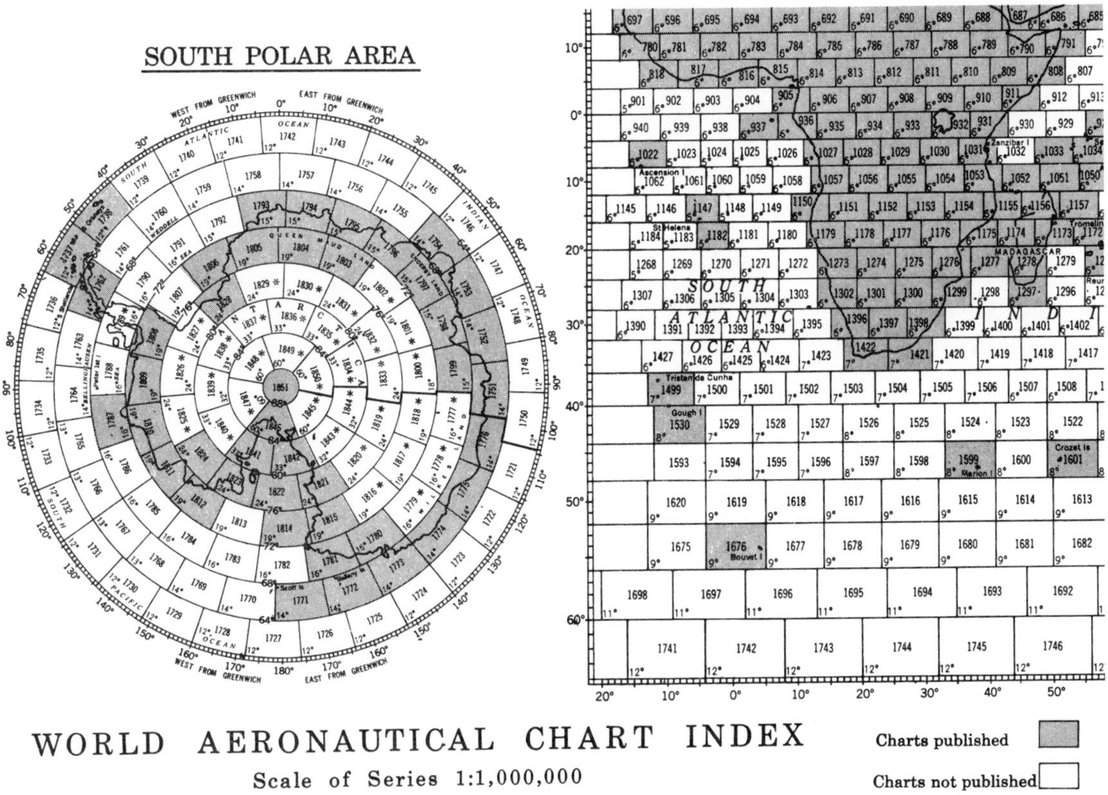

Figure A I.5 A part of the index map of World Aeronautical Charts. (U.S. Coast and Geodetic Survey.)

The military grid referencing system

The large number of grid zones of the UTM grid system requires that some orderly system be available to designate approximate global position, not only as to the correct grid zone, but also the part of the zone, because these are long narrow strips of great latitudinal extent. The *Military Grid Referencing System* provides such information by means of numerals and letters (Figure A1.7). The world between the 80th parallels north and south is considered divided into geographical areas, or quadrilaterals, extending 6° in longitude and 8° in latitude. The north-south boundaries of these areas conform to the 60 UTM grid zones, so that the east-west position of the area columns is designated by the grid zone number, which begins with Zone 1, 180° to 174° W, and increases eastward by integers to Zone 60. The 8° rows of areas are designated by letters, starting with *C*, 80° to 72° S, and extending to *X* at 72° to 80° N. Letters *A*, *B*, *Y*, and *Z* are reserved for the UPS grid system. The *grid zone designation* of a quadrilateral is thus given by a numeral and a letter, as for example, *3P* in Figure A1.7.

To give location within a 6° by 8° quadrilateral, a secondary order of lettered columns and rows is used, applying to *100,000 m squares* (Figure A1.8). Starting at the 180th meridian and proceeding eastward along the equator for 18°, the 100,000 m columns are lettered consecutively *A* through *Z*, omitting *I* and *O*, including the partial columns along grid zone junctions. This alphabetical sequence is repeated at 18° intervals of longitude and at two-million meter intervals of latitude. Other details of the letter arrangement are omitted here. A particular 100,000-m grid square shown in Figure A1.8 is designated as *3PWN*. To this may be added the grid coordinates in terms of 10,000-m and smaller units, as described previously. Thus a complete military grid reference to the nearest 100 m might read: *3PWN539544*.

For polar areas in the UPS grid system, the letters *A*, *B*, *Y*, and *Z* designate the largest grid zones, while a system of letters by columns and rows designates the 100,000-m grid squares (Figure A1.8). Complete details of the military grid reference system are available in manuals prepared by the Departments of the Army and Air Force and published by the U.S. Government Printing Office. The geographer will find a knowledge of the universal grid systems a valuable asset because of their global application.

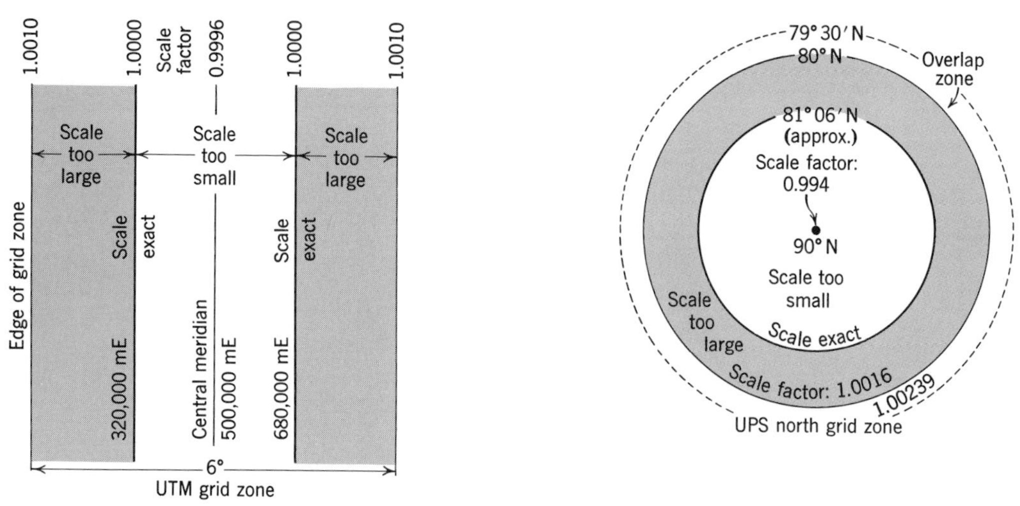

Figure A I.6 Scale factors in the UTM and UPS grids. (After U.S. Departments of the Army and Air Force, TM 5-241, 1951.)

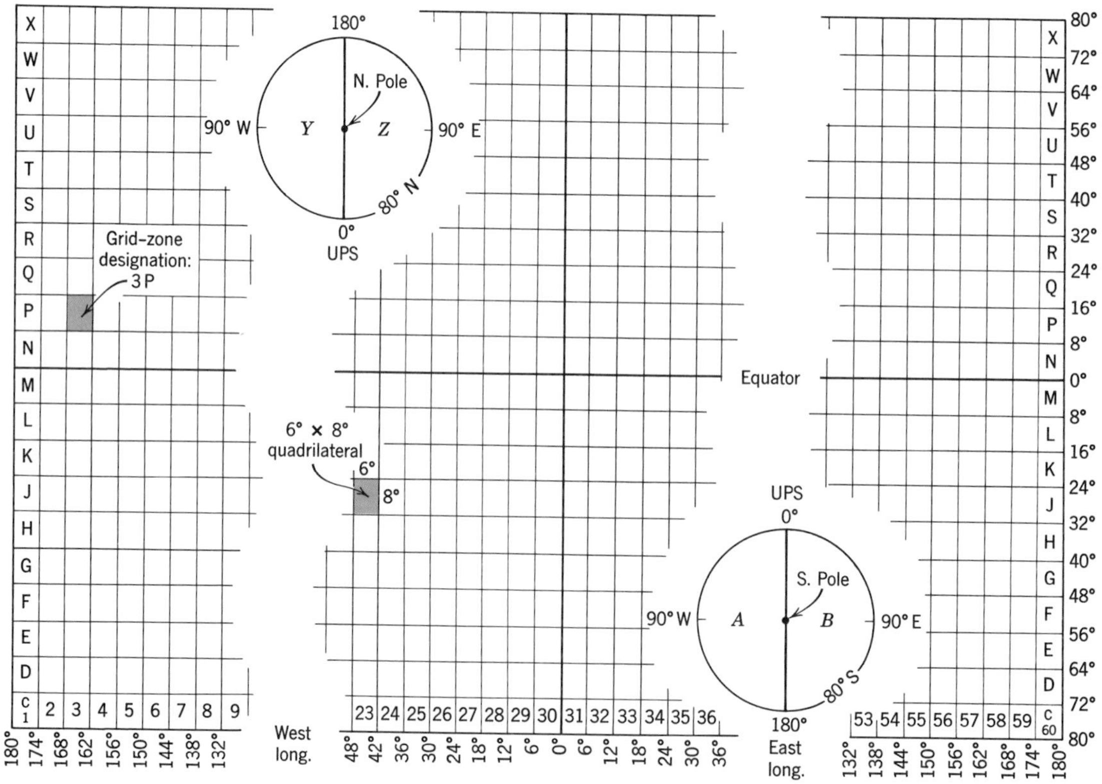

Figure A I.7 Identification of 6° x 8° quadrilaterals in the grid referencing system. (After U.S. Departments of the Army and Air Force, TM 5-241, 1951.)

Figure A I.8 Identification of 100,000-meter squares in the grid referencing system. (After U.S. Departments of the Army and Air Force, TM 5-241, 1951.)

100,000 METER SQUARE IDENTIFICATIONS
FOR THE
MILITARY GRID REFERENCE SYSTEM

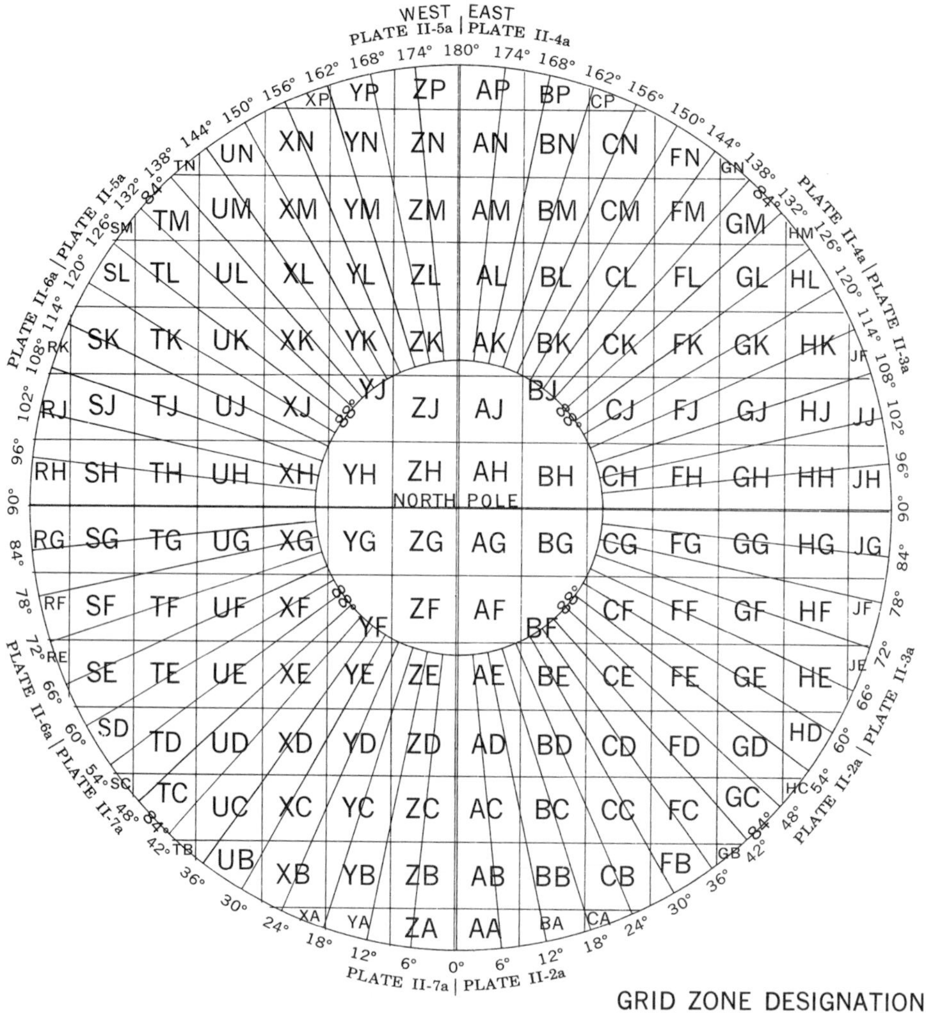

GRID ZONE DESIGNATIONS

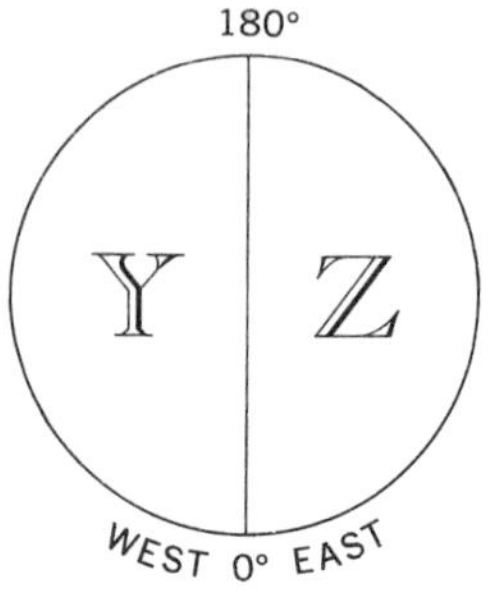

INTERNATIONAL SPHEROID

Figure A I.8 (Continued from previous page.)

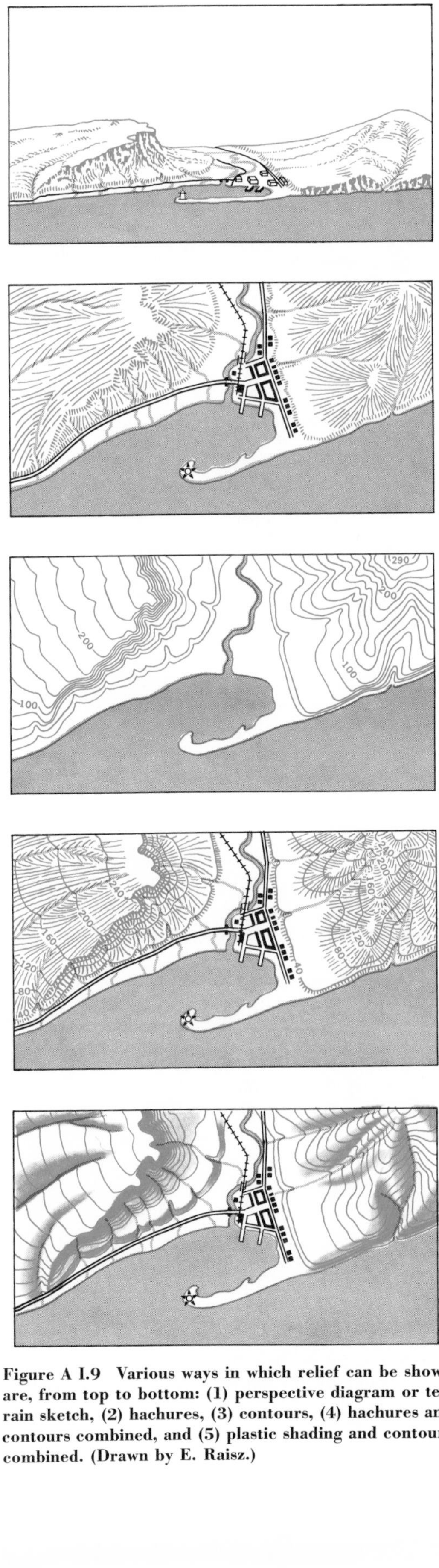

Figure A I.9 Various ways in which relief can be shown are, from top to bottom: (1) perspective diagram or terrain sketch, (2) hachures, (3) contours, (4) hachures and contours combined, and (5) plastic shading and contours combined. (Drawn by E. Raisz.)

Topographic maps

The most important tools available for laboratory study of landforms are topographic maps and air photographs. The student of geomorphology must be able to read and interpret topographic maps quickly and accurately, translating the characteristic forms he sees on maps into verbal statements covering classifications, descriptions, and evolutionary aspects of each.

Several methods have been used to show accurately the configuration of the land surface on topographic maps. Methods described below are *plastic shading, altitude tints, hachures,* and *contours* (Figure A1.9). The first three processes give a strong visual effect of three dimensions so that even untutored persons can grasp the essential character of the landscape features without preliminary explanation. But, as compensation for ease of understanding, such methods of showing relief are inadequate because they do not tell the reader the elevation above sea level of all points on the map, or how steep the slopes are. The method of topographic contours, however, gives this information and makes the most useful type of topographic map.

Plastic shading

Maps using plastic shading to show relief look very much like photographs taken down upon a plaster relief model of the land surface illuminated from directly above or from an oblique angle (Figure A1.10). They may also be likened to air photographs taken when the sun's rays are striking the earth at a fairly low angle (Figure A1.11). The effect of relief is produced by gray or brown tones applied according to one of two principles. In the *oblique illumination* method, light rays are imagined as coming from a point in the northwestern sky somewhere intermediate between the horizon and zenith. Thus all slopes facing southeast receive the heaviest shades and are darkest where the slopes are steepest (Figure A1.10). Northwest-facing slopes are light in tone and lightest where the slope most nearly approaches a right angle to the imaginary rays. Although it is true that, in the northern hemisphere, the sun's rays illuminate the ground from the southeast, south, and southwest, it has been found that an imaginary light source from the south side of a map gives an inverted or "negative" effect to the relief. To demonstrate this, turn the page upside down and look at Figures A1.10 and A1.11. For many persons the relief will seem to reverse itself, so that the stream valleys look like ridges, and vice versa.

An alternative method of applying plastic shading is that of *vertical illumination,* in which a light source is imagined as located directly above the map. This might seem at first thought to produce

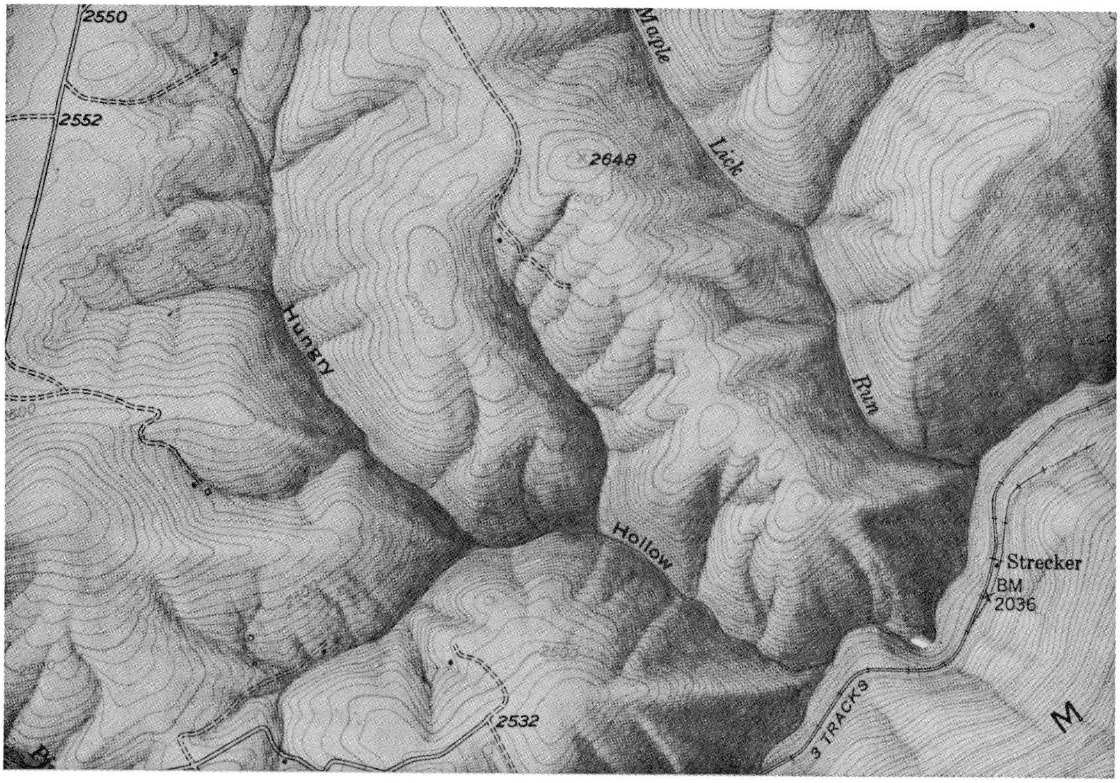

Figure A I.10 Plastic shading combined with contours greatly enhances the visual effect of relief. (Portion of U.S. Geological Survey, Kitzmiller, Md.–W. Va., topographic quadrangle, scale 1:24,000.)

no relief effect at all, because no shadows would be cast, but on a sloping surface the illumination must be spread over a larger area than on a horizontal surface. Consequently, sloping surfaces are somewhat darker than horizontal ones and the darkness increases where the slopes steepen. One advantage possessed by vertical illumination is that surfaces of equal slope have the same tone and, by the use of a standard scale of tones, the steepness of slope can be approximately determined. On some maps, shading of both kinds is used simultaneously and seems to enhance the effect of three dimensions.

Altitude tints

In its simplest form, altitude tinting consists of assigning a certain color, or a certain depth of tone of a color, to all areas on the map lying within a specified range of elevation. Schoolroom wall maps commonly show low areas by deep green, intermediate ranges of elevations by successive shades of buff and light brown, and high mountain elevations in darker shades of brown or red. The method is effective for small-scale maps that are viewed at some distance. A psychological disadvantage is that green suggests a fertile plain with luxuriant vegetation whereas the area may actually be a dry desert. One application of altitude tinting has been in air navigation maps. At high speed a plane passes so rapidly over topographic features that little time is available for detailed map study. Altitude tinting in which livid color shades are applied to high mountain masses brings the dangerous areas forcibly to the attention of pilots or navigators.

Two methods of altitude tinting are used. The change from one solid color tone to the next may be made abruptly along the line where two elevations zones meet. This method is easy to produce, but may appear unreal and give some erroneous ideas about the topography. The second method employs a gradual merging of color tones with rising elevations. Where colors are tastefully selected and lightly applied, a pleasing effect results. Merged altitude tints have been successfully used in combination with other relief methods on recent sheets of the British Ordnance Survey on a scale of one inch to one mile.

Hachures

Hachures are tiny, short lines arranged in such a way as to look as if someone had placed thousands of match sticks side by side into roughly parallel rows. Each hachure line lies along the direction

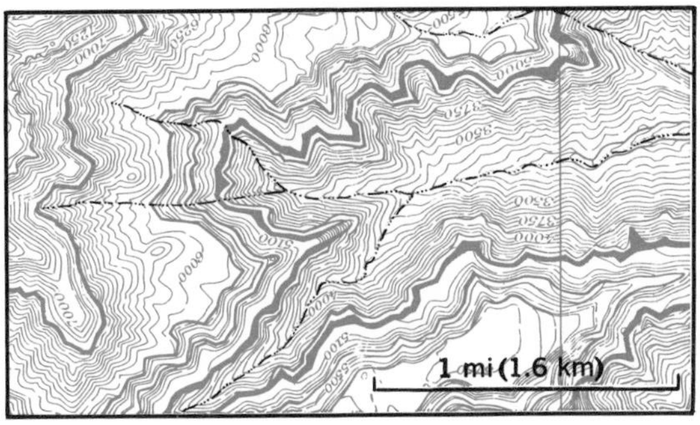

Figure A I.11 A vertical air photograph (above) is a kind of topographic map, showing all relief details but lacking elevation information. The contour topographic map (below) covers the same area and shows a small side canyon of the Grand Canyon. North is toward the bottom of these maps to give the proper effect of relief in the photograph. (U.S. Forest Service and U.S. Geological Survey.)

of the steepest slope and represents the direction that would be taken by water flowing down the surface.

A precise system of hachures, adapted to representation of detailed topographic features on accurate, large-scale maps, was invented by a Saxon army officer, Major J. G. Lehmann (1765–1811), and was widely used on military maps of European countries throughout the nineteenth century. In the Lehmann system, steepness of slope is indicated by thickness of the hachure line. For steep slopes, the hachures are heavy, giving the appearance of dark shading on the map. For gentle slopes, hachures are thin, hairlike lines, giving a light shading on the map. A hachure map following the Lehmann system is shown in Figure A1.12, together with a greatly enlarged diagram illustrating the Lehmann principle of determining width of the line. The Lehmann system of hachures attained wide popularity in Europe during the nineteenth century, but died out rapidly when multicolor printing made possible the production of clear contour lines.

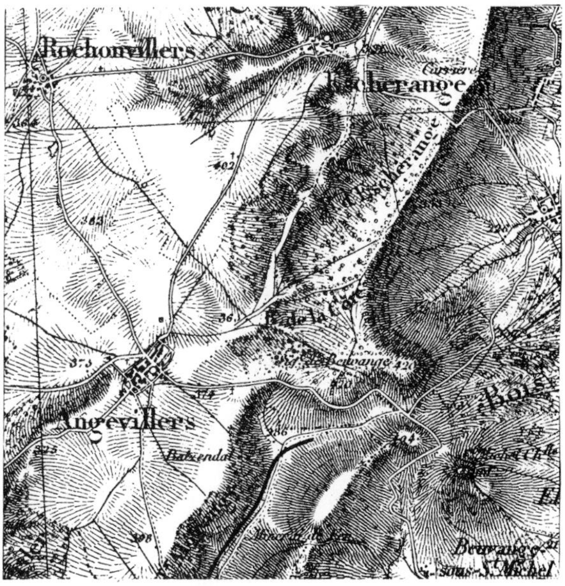

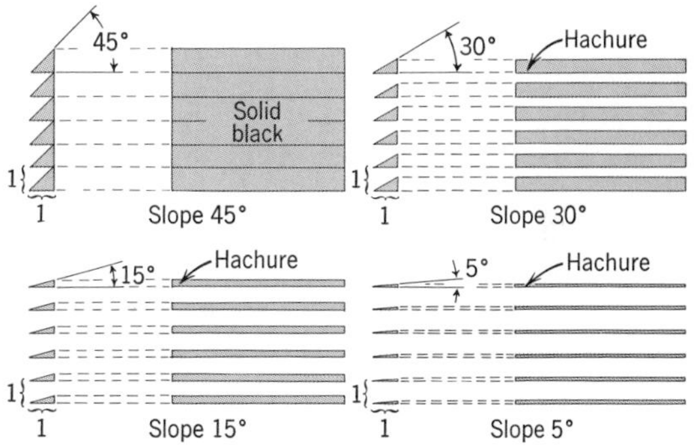

Figure A I.12 A portion of the Metz sheet is shown above to correct scale. This map is one of the French 1:80,000 topographic series using black hachures and spot heights. The Lehmann system of hachuring, shown below, varies the thickness of hachure line according to ground slope.

Because hachures do not tell the elevation of surface points, it is necessary to print numerous elevation figures on hilltops, road intersections, towns and other strategic locations. These numbers are known as *spot heights.* Without them a hachure map would be of little practical value.

A variation of the Lehmann system carries hachures to a very specialized degree of development. If hachures are made thicker on southeast slopes and thinner on northwest slopes, an effect of plastic shading under oblique illumination can be obtained. The method is most effective in rugged mountain regions and was used in the Dufour maps of Switzerland, famous as remarkable examples of the art of cartography (Figure A1.13).

Although hachure maps have now been superseded by other types, the hachure principle still finds occasional use. Where steep cliffs and rock crags occur, they may be missed completely or partially by contour lines. A line of hachures, inserted in addition to contour lines, thus adds important information to the map. Some European cartographers go so far as actually to draw in miniature the form of cliffs and rocky crags. The method is sometimes referred to as *rock drawing* (Figure A1.14).

Contours

A *contour* may be defined as an imaginary line on the ground, every point of which is at the same altitude, or elevation, above sea level. *Contour lines* on a map are simply the graphic representations of ground contours, drawn for each of a series of specified elevations such as 10, 20, 30, 40, or 50 ft or meters above sea level or any other chosen base, known as a *datum plane.* The resulting line pattern not only gives a visual impression of topography to the experienced student of maps but also supplies accurate information about true elevations and slopes.

In order to clarify the contour principle, various commonplace things can be used for illustration. Imagine, for example, a small island as shown in Figure A1.15. The shoreline is a natural contour line because it is a line connecting all points having zero elevation. Suppose that sea level could be made to rise exactly 10 ft (or that the island could be made to sink exactly 10 ft); the water would come to rest along the line labeled "10." This would be the 10-ft contour because it connects all points on the island that are exactly 10 ft higher than the original shoreline. By successive rises in water level, each exactly 10 ft more than the last, the positions of the remaining contours would be fixed.

Although contours are almost never obtained in this way, a very similar procedure was followed in the mapping of some parts of the valley that Lake Mead occupies behind Boulder Dam. As the lake level slowly rose, airplane photographs were taken. With each rise of 2 ft (0.6 m) in lake level, the successive shorelines accurately showed positions of the contours.

Contour interval

Contour interval is the vertical distance separating successive contours. The interval remains constant over the entire map, except in special cases where two or more intervals are used on the map sheet. It is essential then that full information be present on the map margin, describing the areas in which each interval is used.

Because the vertical contour interval is fixed, horizontal spacing of contours on a given map

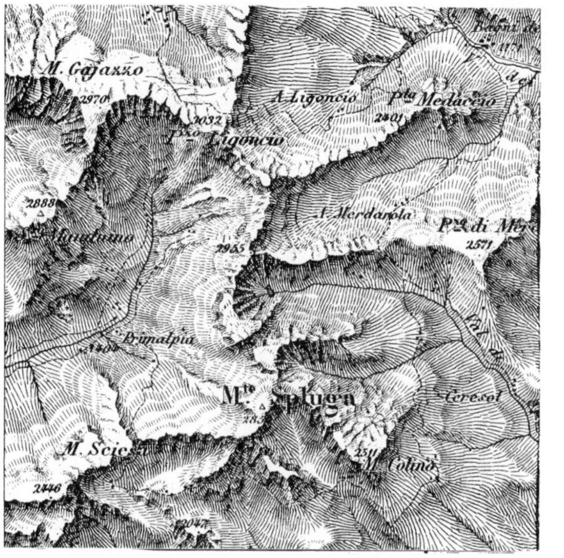

Figure A I.13 The Dufour map of Switzerland, a portion of which is shown here to correct scale (1:100,000), adds the effect of oblique shading to the Lehmann system of hachuring.

Figure A I.14 Rocky cliffs and crags may be shown by rock drawing superimposed on a contour map. Portion of topographic map of Switzerland, Biasca Sheet, reproduced to correct scale, 1:50,000.

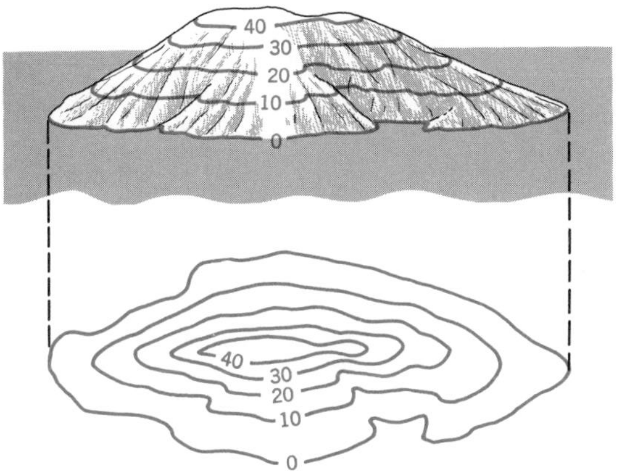

Figure A I.15 Contours on a small island.

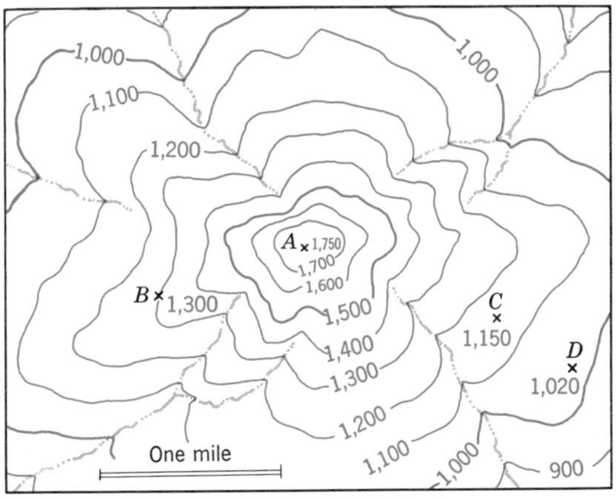

Figure A I.17 Stream valleys produce V-shaped indentations of the contours.

varies with changes in land slope. The general rule is: close crowding of contour lines represents a steep slope; wide spacing represents a gentle slope. Figure A1.16 shows a small island, one side of which is a steep, clifflike slope. From the summit point *B* to the cliff base at *A*, the contours are crossed within a short horizontal distance and therefore appear closely spaced on the map. From *B* to the shore at *C* the same total vertical descent is made, but because the slope is gentle, the horizontal distance is much greater. Hence, the contours between *B* and *C* are widely spaced on the map.

Selection of contour interval depends both on relief of the land and on scale of the map. Topographic maps showing regions of strong relief require a large interval, such as 50, 100, or 200 ft (15, 30, or 60 m); regions of moderate relief, intervals such as 10, 20, 25 ft (2, 5, or 10 m). In flat country, and interval of 5 ft (1 m) or less may

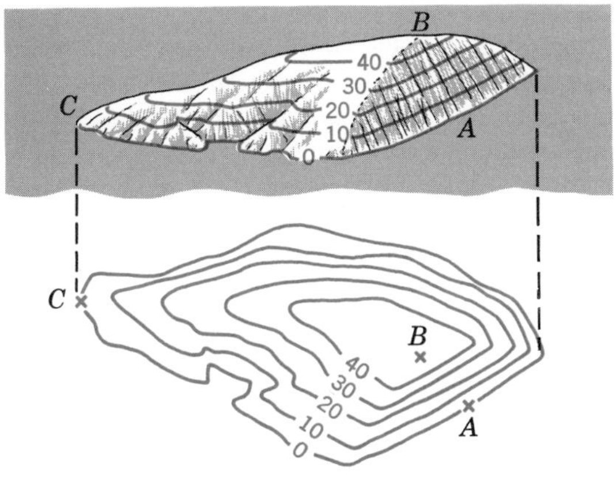

Figure A I.16 On the steep side of this island the contours appear more closely spaced.

be required. Large intervals are used on small-scale maps both because a greater range of elevation is likely to be included and because there is a limit to how closely contours can be printed without fusing into a dense mass.

Because by far the greatest parts of the earth's land surfaces are sculptured by streams flowing in valleys, special note should be made of how contours behave when crossing a stream valley. Figure A1.17 is a small contour sketch map illustrating some stream valleys. Notice that each contour is bent into a V whose apex lies on the stream and points in an upstream direction. The reason for this deflection is that the contour must maintain the same elevation, hence must follow the valley side upstream to a point where the stream gradient brings the stream to the same elevation as the contour. On most topographic maps only the larger streams are actually shown by any line, but the positions of numerous small channels can be deduced from V indentations of the contours.

Determing elevations by means of contours

Although each contour stands for a certain precise elevation, it is not practical to number all the lines, or to place the numbers so close together as to be always close at hand. A common practice is to make every fifth contour line much heavier than the rest, and to number the heavy lines at frequent intervals. This not only makes it easier to grasp essential features of the topography but also facilitates finding the elevation numbers.

Figure A1.17 can be used to illustrate the determination of elevations. Point *B* is easy to determine because it lies exactly on the 1300-ft contour. Point *C* requires interpolation. Because it lies

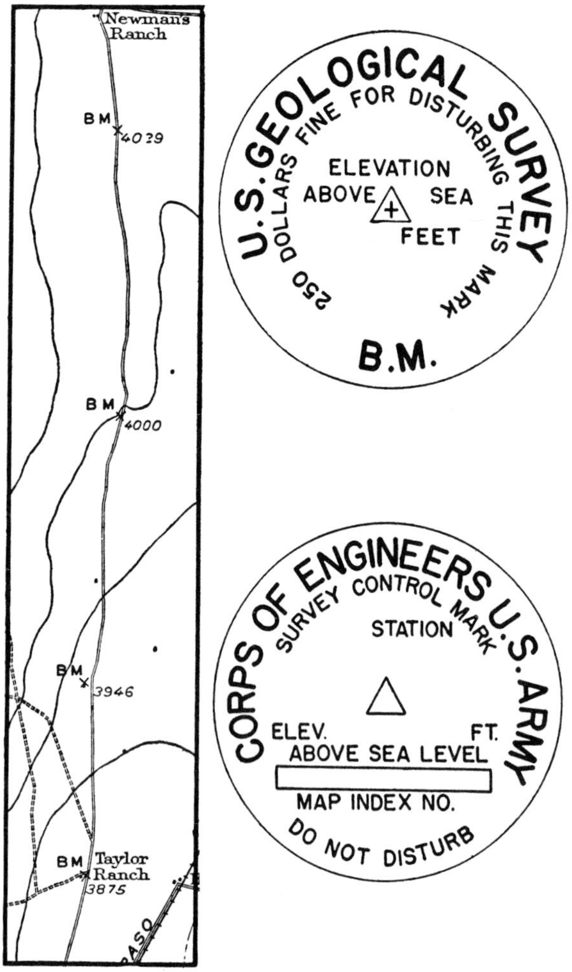

Figure A I.18 A bench mark is a permanently fixed brass plate whose elevation has been carefully surveyed. (U.S. Geological Survey.)

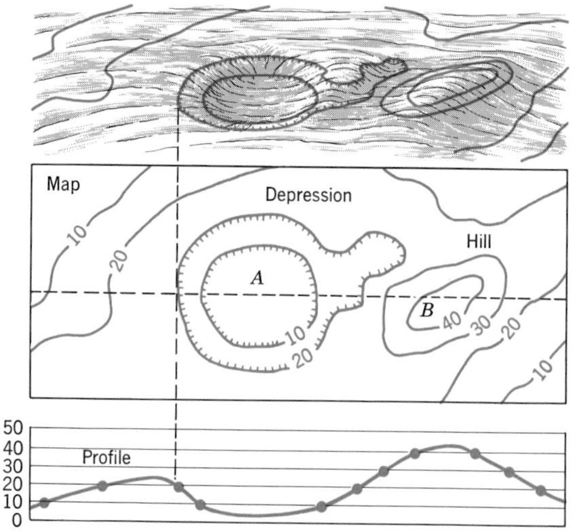

Figure A I.19 Contours which close in a circular manner show either closed depressions or hills.

midway between the 1100- and 1200-ft lines, its elevation is likewise the midvalue of the vertical interval, or 1150 ft. Point *D* lies about one-fifth of the distance from the 1000 to the 1100-ft contours. Because one-fifth of the contour interval is 20 ft, point *D* has an approximate elevation of 1020 ft. For the last two points only a guess has been made as to the true elevation, but if the ground is not too irregular the error will probably be small. Determination of the summit elevation, point *A*, involves still more uncertainty. It is certain that the summit point is more than 1700 ft and less than 1800 because the 1700-ft contour is the highest one shown. Because a sizable area is included within the 1700-ft contour, it may be supposed that the summit rises appreciably higher than 1700. A guess would place the true elevation at about 1750 ft.

On many topographic maps the elevation of hilltops, road intersections, bridges, lakes, and so forth, is printed on the map to the nearest foot or meter. These spot heights do away with the need for estimating elevations at key points.

Federal agencies, such as the U.S. Coast and Geodetic Survey and the U.S. Geological Survey, in mapping an area, carefully determine elevation and position of convenient reference points. These are known as *bench marks.* On the map they are designated by the letters B.M., together with the elevation stated to the nearest foot (Figure A1.18). The brass discs are firmly embedded in rock or concrete, or in the masonry of buildings, and are inscribed with the elevation.

Depression contours

A special type of contour is used where the land surface has basinlike hollows, or *closed depressions,* which would make small lakes if they could be filled with water. This is the *hachured contour,* or *depression contour.* Figure A1.19 is a sketch of a depression in a gently sloping plain. Below it is the corresponding contour map. Hachured contours have the same elevations and contour intervals as regular contours on the same map. If, as often happens, the closed depressions are so small that the hachured contours cannot be numbered, the hachured contour allows the reader to distinguish a closed depression from a small hill of similar size. Not all closed depressions are shown this way, however. Very large basins, such as the floor of Death Valley, are shown by regular contours, because such a large number of hachured lines would make a bad appearance on the map.

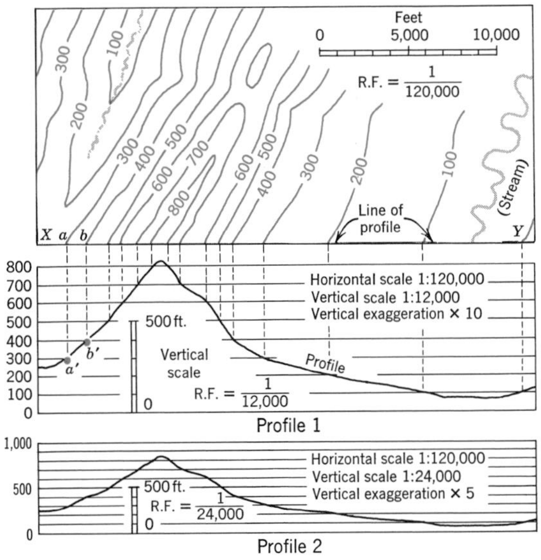

Figure A I.20 A topographic profile can be constructed from a contour map along any desired line.

Topographic profiles

In order to get a better idea of the nature of the relief, *topographic profiles* are sometimes drawn. These are lines that show the rise and fall of the land surface along a selected line crossing the map. Figure A1.20 illustrates the construction of a profile. A line, *XY*, is lightly drawn across the map at the desired location. A piece of paper, ruled with horizontal lines, is placed so that its top edge lies along the line *XY*. Each horizontal line represents a contour level and is so numbered along the left-hand side. Starting at the left, a perpendicular is dropped from the point *a* where the map contour intersects the profile line, *XY*, down to the corresponding horizontal level. A point *a* is marked on the horizontal line. Next, the procedure is repeated for the 400-ft contour at point *b*, and so on, until all points have been plotted. A smooth line is then drawn through all the points, completing the profile. Where contours are widely spaced, some judgment is required in drawing of the profile.

Figure A1.20 shows two profiles, both of which are drawn along the same line *XY*. The difference is one of degree of exaggeration of the vertical scale. In this illustration, horizontal map scale is 1 in. to 10,000 ft, or 1 : 120,000, whereas the vertical scale of the upper profile is 1 in. to 1000 ft, or 1 : 12,000. The vertical scale is thus ten times as large as the horizontal map scale, and the profile is said to have a *vertical exaggeration* of ten times. In the lower profile, the horizontal scale remains the same, of course, but the vertical scale is 1 in. to 2000 ft, or 1:24,000. The vertical exaggeration is therefore five times. Some degree of vertical exaggeration is usually needed to bring out the nature of the topography. A *natural scale* profile, one in which the vertical and horizontal scales are the same, would give a profile with such tiny fluctuations as to be not only difficult to read but also difficult to draw or reproduce. The human eye exaggerates the height of topographic features and the steepness of slopes when they are seen from ground level. For this reason, a moderate degree of profile exaggeration may seem perfectly natural. Excessive exaggeration should be avoided. In Figure A1.20 the upper profile is excessively exaggerated, but the lower one is more suitable for general purposes.

Topographic profiles are used in highway and railroad planning to estimate the degree of cutting or filling needed to establish a smooth grade. In military operations, profiles are required to determine the limits of visibility from key observation points.

How contour maps are made

An understanding of the methods of making contour maps will help the student to judge what can and cannot be safely interpreted from the contours. A topographic surveying party takes into the field a large map board called a *plane table*, which is set up firmly on a tripod at a point commanding a good view of the surrounding ground (Figure A1.21). A telescope, called an *alidade*, mounted on a flat brass plate and containing cross hairs in the eyepiece, is placed on the plane table so that it can be moved around freely (Figure A1.22). Because the straight metal edge of the telescope base is exactly parallel to the telescope tube, lines drawn along the straight edge onto a sheet of heavy paper tacked to the plane table board represent true sight lines.

One of the party takes a *rod*, a flat strip of wood about 12 ft long marked off plainly in 1-ft and 0.1-ft divisions. This he sets up vertically on a distant point. The man at the plane table sights the rod through his telescope and, by means of a *stadia system* of cross hairs, is able to calculate the distance to the rod. The stadia principle is simply that the segment of rod seen between two horizontal hairs in the eyepiece is proportioned to the distance between telescope and rod. Using the straight edge of the brass plate under the telescope as a guide, he draws a line from the point representing his own position toward the point where the rod is located. On this line, he scales off the distance and thus locates exactly on his map the position of the distant point. At the same time, he computes the difference in elevation between himself and the distant point. This he can do because he has determined the distance and can read the vertical angle of his line of sight.

After many points have thus been located and their elevations determined, the topographer draws in the contours. Since very few of the points actually fall on exact contour elevations, it is necessary to interpolate the position of the contour line; and because the points may be scattered widely, the topographer must use his best judgment in sketching in the contours so as to show various minor details of the topography. Usually a topographer who understands geology and knows the typical form that contours take on certain types of rocks will make a better map than a topographer who is not acquainted with geology.

In view of what has been said above, it is obvious that the accuracy of a map depends on various things not evident from the map itself. The speed with which topographic parties work determines how much detail they can put into a map. The speed is itself influenced by the appropriation of money available for the job. The purpose for which

Figure A I.21 The instrument man at his plane table is sighting through a telescopic alidade to the stadia rod held by a distant rod man. (U.S. Geological Survey photograph.)

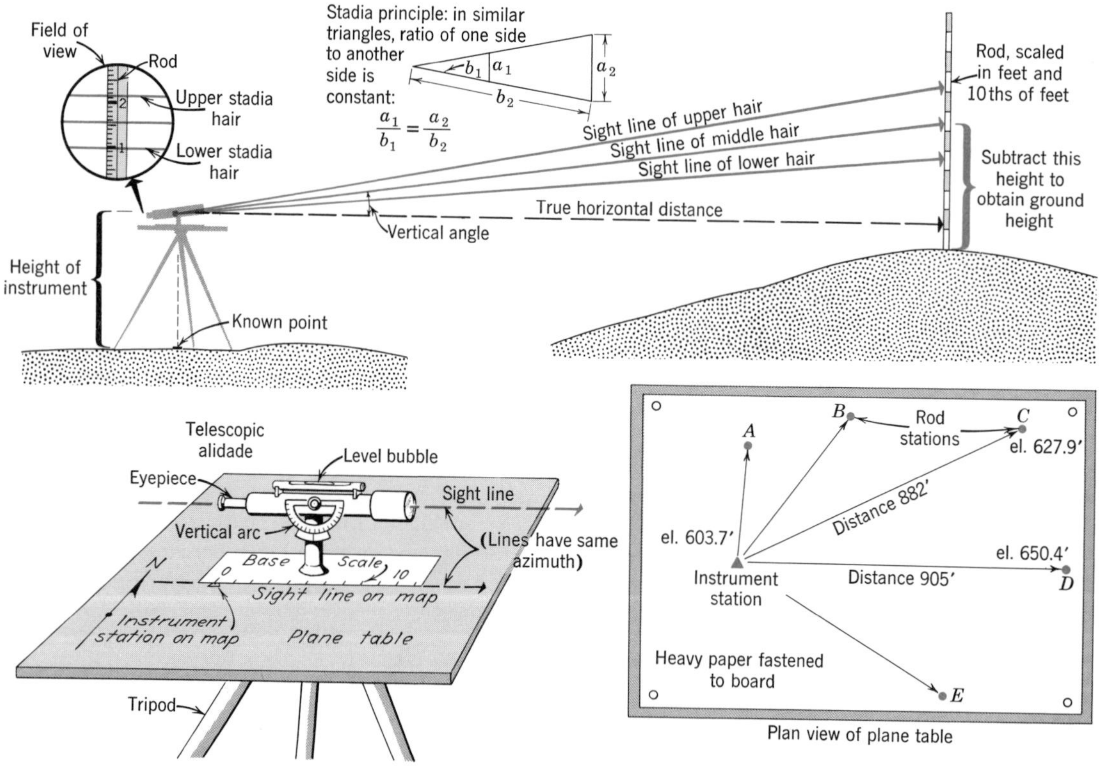

Figure A I.22 Topographic mapping in the field requires telescopic alidade, plane table, and rod. The stadia principle is used to determine distances.

Figure A I.23 This *multiplex*, a precision optical device, enables contours to be drawn directly from sets of overlapping air photographs. (U.S. Geological Survey photograph.)

the map was made is a factor. A reconnaissance contour map to accompany a geographic report cannot be expected to be as accurate as a large-scale highway planning map from which estimates of volume of cut and fill are to be derived. The geographer must be alert for any clues as to the reliability of the map and must not place more confidence on data derived from a map than the accuracy of the map itself permits.

Contour mapping from air photographs

Shortly before World War II, the Topographic Division of the U.S. Geological Survey began to use precision optical instruments by means of which a trained operator working in a laboratory can draw contours from series of overlapping aerial photographs taken with cameras mounted in aircraft (Figure A1.23). In general, the science of mapping and surveying by means of photographs taken from the air or ground is termed *photogrammetry*, a specialized field of cartography based on mathematical and geometrical principles that accurately relate the forms seen on the photograph to the true forms on the ground surface. The science of photogrammetry developed rapidly during World War II and has since provided the principal means of constructing large-scale contour maps. The rate of contour map production by government agencies has therefore increased enormously. Whereas formerly a small, but steady, stream of topographic quadrangles on the scale of 1:62,500 (15-minute quadrangles) was being distributed by the U.S. Geological Survey, there is now a veritable flood of quadrangles on the larger scale of 1:24,000, each sheet covering $7\frac{1}{2}$ minutes of latitude and longitude. Other United States governmental agencies, notably the Army Map Service of the Department of Defense, have contributed heavily to the mapping program, but all topographic maps are distributed to the public through the Geological Survey, which edits and publishes most of the maps.

The use of air photo plotting instruments has by no means done away with the need for ground surveys. It is still necessary to survey control points on the ground and to check the laboratory map in the field by direct comparison with landforms and cultural features. Furthermore, a heavy forest cover may so obscure the ground from photography that field surveys by plane table and alidade are necessary.

National standards for contour map accuracy were agreed upon in 1942 and include the provision that (1) horizontally, 90 percent of the well-defined planimetric features shall be plotted in correct position on the published map sheet within a tolerance of one-fiftieth inch; and (2) vertically, 90 percent of the elevations interpolated from the contours shall be correct within a tolerance of one-half contour interval. (On a map of scale 1:24,000, one-fiftieth inch is approximately equivalent to 40 ft on the ground.) Maps conforming to these specifications bear a statement to that effect printed in the lower map margin.

In Chapters 24 through 33, small portions of topographic maps have been reproduced for landform study laboratory exercises. To simplify these maps, most of the cultural features have been deleted. Changes of scale and contour interval have been required in some cases to emphasize the landforms of interest. In all cases the source maps are described; most are available by direct purchase from the U.S. Geological Survey, which will supply information on availability and price. Index maps of each state are available to show the name, scale, location, and year of publication of all quadrangles.

The front end papers of this book reproduce to exact color and scale a representative portion of a modern topographical map quadrangle of the U.S. Geological Survey, together with a set of map symbols.

Geologic maps and structure sections

The *areal geologic map* shows by means of colors or patterns the surface distribution of each rock unit with special emphasis upon the lines of contact of rocks unlike in variety or age. Faults are shown as lines. Strikes and dips of strata are added by special symbols.

Figure AI.24 is a simple geologic map of the same area shown by a perspective diagram in

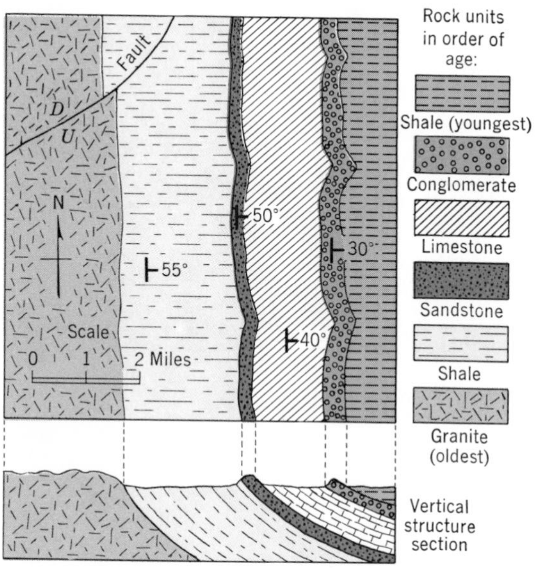

Figure A I.24 A geologic map shows the surface distribution of rocks and structures. The structure section shows rocks at depth.

Figure 28.2. If map reproduction is limited to black and white, patterns are applied to differentiate the rock units. Shorthand letter combinations may be added as a code to set apart formations of different ages. Small T-shaped symbols, seen on the map, tell strike-and-dip. The long bar gives direction of strike; the short bar which abuts it at right angles show direction of dip. Amount of dip in degrees is given by a figure beside the symbol. A small fault, cutting across the northwest corner of the map, is shown by a solid line. The letters D and U indicate which side slipped downward, which upward.

To show the internal geologic structure of an area the geologist resorts to the *structure section*, an imaginary vertical slice through the rocks. A structure section is illustrated in the lower part of Figure AII.21. The upper line of the section is a topographic profile.

Review Questions

1. Explain the three ways of expressing map scale and the uses of each. Explain the procedures used in conversion of a fractional scale (R.F.) to length units; of a fractional scale to a graphic scale.

2. What is the distinction between large-scale, medium-scale, and small-scale maps? What rule expresses the way in which ground area shown on a map varies with map scale, assuming a map of fixed dimensions?

3. Describe in detail the quadrangle system used by the U.S. Geological Survey. What standard scales are used?

4. Explain the relation of map scale to the UTM and UPS military grids. What is meant by *scale factor?*

5. What is the military grid referencing system? Tell how the location of a particular point on the earth's surface is uniquely given by this system.

6. List the methods used to show relief features on topographic maps.

7. What two kinds of plastic shading can be used? What are the advantages of each method?

8. For what kinds of maps are altitude tints most effective?

9. What are hachures? Explain how the Lehmann system of hachures is applied. How are differences in steepness of slope shown? What disadvantages does the hachure system have in the making and using of maps?

10. How do Dufour maps differ from maps using the Lehmann hachure system? What is rock drawing?

11. What is a contour line? What is contour interval? What factors determine the selection of contour interval?

12. How do contours give indication of a stream valley? How is a hilltop shown by contours?

13. How can the elevations of points be determined if they fall in the spaces between contours?

14. What are spot heights? What are bench marks? Of what value are bench marks?

15. How are closed depressions shown by contours? How can the depth of the bottom of the depression below the lowest point in the rim be estimated from the contour?

16. What is a topographic profile? How is it constructed? What is vertical exaggeration? When should it be used?

17. Explain the general principles of topographic field surveying. What instruments are needed? How can distances be determined without measuring with a tape? Where should points be located to provide most effective aid in drawing contours?

18. What is photogrammetry? How have air photo methods influenced the production of topographic quadrangles? Why are field surveys still necessary?

19. State the accuracy standards set for horizontal and vertical features on a modern map.

20. What is an areal geologic map? A structure section? How are faults indicated?

Exercises

1. Refer to Figure A1.1. **(*a*)** This map was originally on a scale of 1:24,000, but this has been reduced for use as an illustration. What is the fractional scale of this map as printed here? **(*b*)** Determine the width and length in miles of the area shown by the map. **(*c*)** Determine the distance in yards from Davis Quarry to John Harrit Church. **(*d*)** Prepare a graphic scale in kilometers for this map. **(*e*)** State the scale of this map in inches to miles.

2. On an air photograph taken from a vertical position, a measured distance between two points is 7.36 in. This value represents 2.37 mi on the ground. **(*a*)** What is the fractional scale of the photograph? **(*b*)** Construct an equivalent graphic scale in units of 1000 yd. **(*c*)** Construct a graphic scale in units of meters.

3. Refer to Figure A1.1. **(*a*)** Determine the bearing of a line from John Harrit Church to Davis Quarry. (See Chapter 3 for explanation of bearings and azimuths.) **(*b*)** State the azimuth of this same line. **(*c*)** Determine the azimuth and distance of a line between the two points marked by the heavy black arrows below the graphic scale.

4. Show by contours alone each of the following geometric forms. Construct these carefully with compass and straightedge. **(*a*)** A straight-sided cone. **(*b*)** A cone with sides concave-up. **(*c*)** A hemisphere. **(*d*)** A cube.

5. Make a contour map of a semicircular island, the straight side of which has a very steep slope as compared with the rest of the island. The summit of the island is 105 ft above sea level. The contour interval is 10 ft. Two stream valleys extend down the gently sloping sides of the island. The island is about 4 mi wide. Use a map scale of 1 in. to 1 mi.

6. The map on the next page shows contouring about half completed. **(*a*)** Using the elevations printed on the left-hand side of the map, complete the drawing of the 100-ft contours. (The contours may be drawn on a sheet of thin tracing paper laid over the page.) **(*b*)** Make a profile along the line *AB*, using a vertical exaggeration of five times. **(*c*)** Determine the fractional scale of the map. **(*d*)** How high is the cliff at the point *C* in the southeastern part of the map? **(*e*)** What is the amount of the magnetic declination in this area?

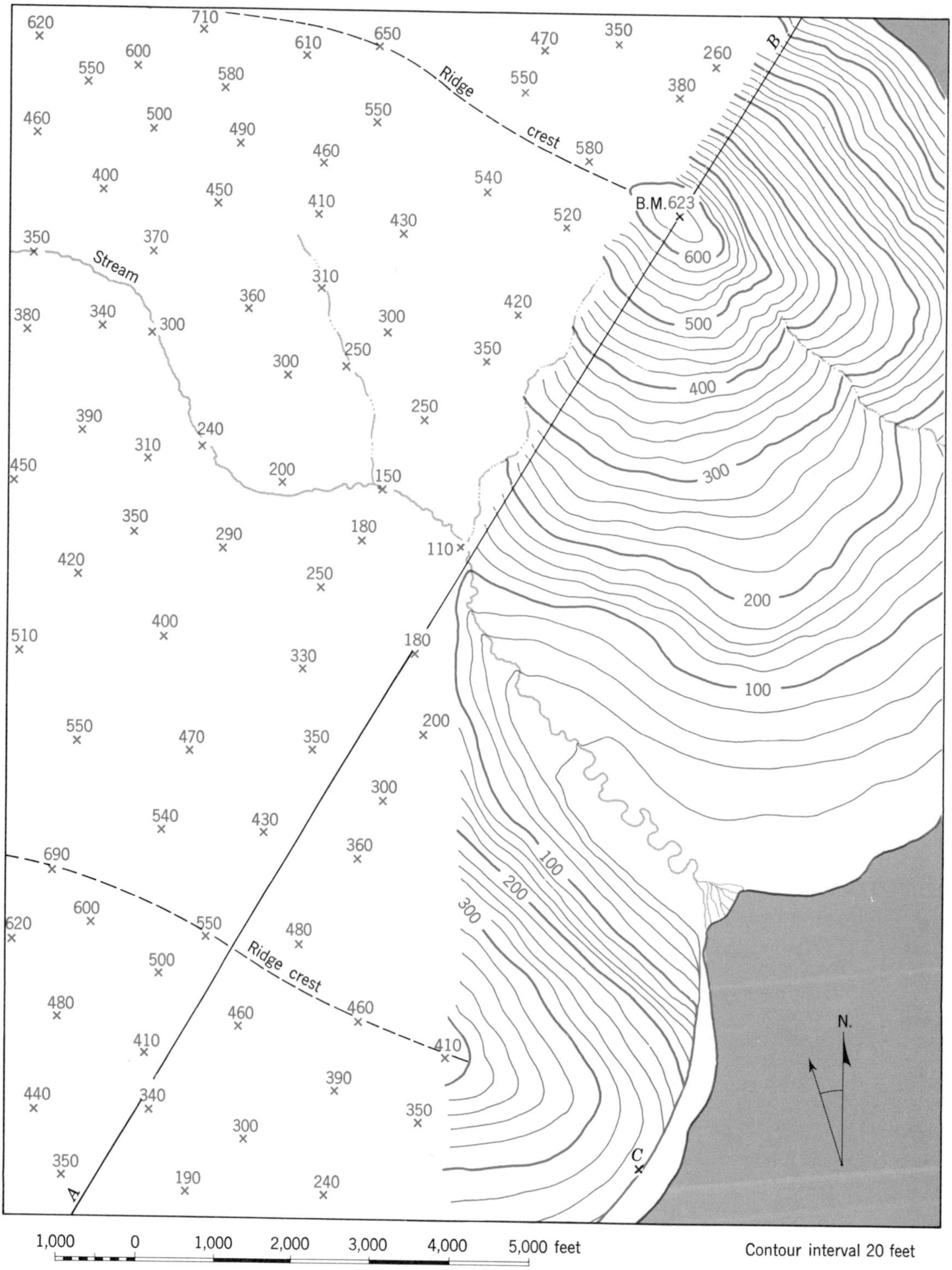

Ridge crest
Stream
B.M. 623
Ridge crest
A
B
C
N.
1,000 0 1,000 2,000 3,000 4,000 5,000 feet
Contour interval 20 feet

Exercise 6

APPENDIX **II**

World Climate Data

MONTHLY temperature and precipitation data of 183 representative stations are given below. The upper line of figures gives mean daily air temperature in degrees Fahrenheit, derived as the average of the monthly mean of daily maximum temperatures and the monthly mean of daily minimum temperatures. Temperatures are taken in standard shelters, 4 to 6 ft (1.2 to 1.8 m) above ground level. The lower line of figures (boldface type) gives mean monthly precipitation in inches. In the last column are given mean annual temperature and mean annual total precipitation.

Station data are from *Tables of temperature, relative humidity and precipitation for the world* (1958) Parts I through VI, (M.O.617 a–f), Meteorological Office, Air Ministry of Great Britain. These volumes are published by Her Majesty's Stationery Office, London, and are obtainable in the United States from the British Information Services, 845 Third Avenue, New York, N.Y. Contained in the description of each station in the data below is a roman numeral and arabic numeral in parentheses, indicating the volume and page respectively of the M.O. table from which the data have been taken. The Meteorological Office tables also give mean and absolute highest and lowest temperatures, mean relative humidities, maximum precipitation of 24 hours, and average number of days with 0.01 inch of precipitation or more. Location maps and bibliographies are included in each volume of the M.O. tables.

The stations for which data are given below are arranged by climate groups. The Köppen symbol is given at the left. Sequence of stations within each climate group follows a set geographical pattern, starting with northern hemisphere at maximum west longitude and proceeding eastward around the globe. The southern hemisphere sequence follows in like manner. An alphabetical finding list of stations follows the data.

Appendix I Temperature and Precipitation Data For Selected Stations

WET EQUATORIAL CLIMATE

Köppen code	Station, location, elevation, data source		J	F	M	A	M	J	J	A	S	O	N	D	Year
Af	1. Georgetown, British Guiana	°F	79	79	79.5	80.5	80	80	80	80.5	81.5	81.5	81	79.5	80
	06°50′N, 58°12′W, 6 ft. (II-29)	In	8.0	4.5	6.9	5.5	11.4	11.9	10.0	6.9	3.2	3.0	6.1	11.3	88.7
Af	2. Barumbu, Republic of the Congo	°F	77.5	78	78.5	79	78	77.5	76.5	76.5	77	76.5	76.5	76.5	77
	01°15′N, 23°29′E, 1378 ft (IV-25)	In	2.8	3.5	5.6	7.0	6.1	5.7	6.9	6.5	7.3	8.1	7.2	4.1	70.8
Af	3. Colombo, Ceylon	°F	79	79.5	81	82	82.5	81	81	81	81	80	79	78.5	80.5
	06°54′N, 79°52′E, 24 ft (V-10)	In	3.5	2.7	5.8	9.1	14.6	8.8	5.3	4.3	6.3	13.7	12.4	5.8	93.1
Af	4. Sandakan, North Borneo	°F	79.5	80	81	82.5	82.5	82	82	82	82	81.5	81	80	81.5
	05°50′N, 118°07′E, 152 ft (V-45)	In	19.0	10.9	8.6	4.5	6.2	7.4	6.7	7.9	9.3	10.2	14.5	18.5	123.7
Af	5. Jaluit, Marshall Islands	°F	81.5	82.5	83	82.5	82	82	82.5	82.5	82.5	83.5	82.5	82	82.5
	05°55′N, 169°38′E, 20 ft (VI-41)	In	10.2	8.5	14.2	15.8	16.6	15.3	15.4	12.0	13.1	12.2	11.9	13.6	158.8
Af	6. Uaupés, Brazil	°F	80	80.5	80	80	79.5	78.5	77.5	79	80	80	81	80	79.5
	00°08′S, 67°05′W, 272 ft (II-17)	In	10.3	7.7	10.0	10.6	12.0	9.2	8.8	7.2	5.1	6.9	7.2	10.4	105.4
Af	7. Padang, Sumatra, Indonesia	°F	80.5	80.5	80.5	81	81.5	80.5	80.5	80.5	80	80	80	80	80.5
	00°56′S, 100°22′E, 22 ft (V-47)	In	13.8	10.2	12.1	14.3	12.4	12.1	10.9	13.7	6.0	19.5	20.4	18.9	174.3
Af	8. Amboina, Moluccas, Indonesia	°F	82	82	82	82	79.5	78	77.5	77.5	78.5	79.5	81.5	82	80
	03°42′S, 128°10′E, 14 ft (V-45)	In	5.0	4.7	5.3	11.0	20.3	25.1	23.7	15.8	9.5	6.1	4.5	5.2	136.2
Af	9. Madang, New Guinea	°F	81	80.5	80.5	81	81.5	81	81	81	81	81.5	81.5	81.5	81
	05°14′S, 145°45′E, 20 ft (VI-27)	In	12.1	11.9	14.9	16.9	15.1	10.8	7.6	4.8	5.3	10.0	13.3	14.5	137.2

MONSOON CLIMATES

Köppen code	Station, location, elevation, data source		J	F	M	A	M	J	J	A	S	O	N	D	Year
Am	10. Cristobal, Panama	°F	80	80.5	81	81.5	81	81	80.5	80.5	80.5	80.5	79.5	80.5	80.5
	09°21′N, 79°54′W, 35 ft (II-30)	In	3.4	1.5	1.5	4.1	12.5	13.9	15.6	15.3	12.7	15.8	22.3	11.7	130.3
Am	11. Cayenne, French Guiana	°F	79	79.5	79.5	80.5	79.5	80	80.5	81.5	82.5	82.5	81.5	80	80.5
	04°56′N, 52°27′W, 20 ft (II-29)	In	14.4	12.3	15.8	18.9	21.7	15.5	6.9	2.8	1.2	1.3	4.6	10.7	126.4
Am	12. Freetown, Sierra Leone	°F	80	81	81.5	82	81	79	77.5	77	78	78.5	79.5	80	80
	08°37′N, 13°12′W, 92 ft (IV-135)	In	0.4	0.2	1.2	3.1	9.5	14.3	29.2	36.5	22.3	14.2	5.5	1.2	137.6
Am	13. Douala, Cameroon	°F	79.5	80	79.5	79.5	79.5	77.5	75.5	75.5	76.5	76	78.5	79	78
	04°03′N, 09°41′E, 26 ft (IV-42)	In	1.8	3.7	8.0	9.1	11.8	21.2	29.2	27.3	20.9	16.9	6.1	2.5	158.5

Station		J	F	M	A	M	J	J	A	S	O	N	D	Year
Am 14. Cochin, India	°F	80.5	82	84	85.5	84	80	79	79.5	80	80	81.5	81	81.5
09°58′N, 76°14′E, 10 ft (V-27)	In	**0.9**	**0.8**	**2.0**	**4.9**	**11.7**	**28.5**	**23.3**	**13.9**	**7.7**	**13.4**	**6.7**	**1.6**	**115.3**
Am 15. Akyab, Burma	°F	70	72.5	78	82.5	84	81.5	80.5	80.5	81.5	81.5	78	72	79
20°08′N, 92°55′E, 29 ft (V-7)	In	**0.1**	**0.2**	**0.4**	**2.0**	**15.4**	**45.3**	**55.1**	**44.6**	**22.7**	**11.3**	**5.1**	**0.7**	**202.9**
Am 16. Rangoon, Burma	°F	77	79.5	83.5	86.5	84.5	81	80.5	80.5	81	82	80.5	77.5	81
16°46′N, 96°11′E, 18 ft (V-9)	In	**0.1**	**0.2**	**0.3**	**2.0**	**12.1**	**18.9**	**22.9**	**20.8**	**15.5**	**7.1**	**2.7**	**0.4**	**103.0**
Am 17. Quang-Tri, Vietnam	°F	68	69.5	73	79.5	83.5	85	85	85	81.5	77.5	73.5	69.5	77.5
16°44′N, 107°11′E, 23 ft (V-40)	In	**6.7**	**2.2**	**2.7**	**2.2**	**3.9**	**3.0**	**3.5**	**3.8**	**15.6**	**22.1**	**22.3**	**12.0**	**100.0**
Am 18. Manaus, Brazil	°F	81.5	81.5	81.5	81	81.5	81.5	82	83	83.5	84	83.5	82.5	82
03°08′S, 60°01′W, 144 ft (II-12)	In	**9.8**	**9.1**	**10.3**	**8.7**	**6.7**	**3.3**	**2.3**	**1.5**	**1.8**	**4.2**	**5.6**	**8.0**	**71.3**
Am 19. Djakarta, Indonesia	°F	79	79	80	81	81	80.5	80	80	81	80.5	80	79.5	80
06°11′S, 106°50′E, 26 ft (V-44)	In	**11.8**	**11.8**	**8.3**	**5.8**	**4.5**	**3.8**	**2.5**	**1.7**	**2.6**	**4.4**	**5.6**	**8.0**	**70.8**
Am 20. Papeete, Tahiti, Society Islands	°F	80.5	80.5	80.5	80.5	78.5	77.5	77	77	77.5	78.5	79.5	80	79
17°32′S, 149°34′W, 302 ft (VI-45)	In	**9.9**	**9.6**	**16.9**	**5.6**	**4.0**	**3.0**	**2.1**	**1.7**	**2.1**	**3.5**	**5.9**	**9.8**	**64.1**

TRADE-WIND LITTORAL CLIMATE

Station		J	F	M	A	M	J	J	A	S	O	N	D	Year
Am 21. Belize, British Honduras	°F	74	75.5	77.5	80	81	81	81	81.5	80.5	79	75.5	74.5	78.5
17°31′N, 88°11′W, 17 ft (II-18)	In	**5.4**	**2.4**	**1.5**	**2.2**	**4.3**	**7.7**	**6.4**	**6.7**	**9.6**	**12.0**	**8.9**	**7.3**	**74.4**
Am 22. Aparri, Philippines	°F	74.5	76	78.5	81.5	84	84.5	83.5	83.5	82.5	80	78	75.5	80
18°22′N, 121°38′E, 17 ft (V-80)	In	**6.0**	**3.3**	**2.4**	**1.8**	**4.9**	**6.7**	**8.5**	**9.5**	**11.3**	**14.1**	**12.8**	**8.2**	**89.5**
Af 23. Salvador, Bahia, Brazil	°F	80	80	80	80	77	75.5	74	74	75.5	77	78	78.5	77.5
13°00′S, 38°30′W, 154 ft (II-14)	In	**2.6**	**5.3**	**6.1**	**11.2**	**10.8**	**9.4**	**7.2**	**4.8**	**3.3**	**4.0**	**4.5**	**5.6**	**74.8**
Af 24. Santos, São Paulo, Brazil	°F	78	78.5	77	74.5	70.5	68	67	66.5	69	70.5	73.5	76	72.5
23°56′S, 46°19′W, 10 ft (II-15)	In	**11.0**	**9.8**	**12.3**	**7.3**	**6.1**	**5.7**	**4.3**	**4.1**	**5.7**	**6.4**	**7.7**	**7.8**	**88.1**
Af 25. Farafangana, Malagasy Republic	°F	79.5	80.5	79	76	73.5	69.5	68.5	68.5	70.5	73.5	75.5	78	74
22°49′S, 47°47′E, 10 ft (IV-99)	In	**12.8**	**15.8**	**14.1**	**11.1**	**8.9**	**7.3**	**8.7**	**5.9**	**3.8**	**3.2**	**6.7**	**10.4**	**108.8**
Am 26. Cairns, Queensland, Australia	°F	82	81.5	80	77.5	73.5	71.5	69.5	71	73.5	77	79	81.5	76.5
16°55′S, 145°47′E, 16 ft (VI-9)	In	**16.6**	**15.7**	**18.1**	**11.3**	**4.4**	**2.9**	**1.6**	**1.7**	**1.7**	**2.1**	**3.9**	**8.7**	**88.7**

TROPICAL WET-DRY CLIMATES		J	F	M	A	M	J	J	A	S	O	N	D	Year
Aw 27. Manzanillo, Mexico 19°04′N, 104°20′W, 26 ft (I-41)	°F	77	76	76	77	80	83.5	84.5	84.5	83	83.5	81	78.5	80.5
	In	0.1	0.2	<0.1	0.0	0.1	4.7	5.7	6.4	14.5	5.1	0.9	1.8	39.5
Aw 28. San José, Costa Rica 09°56′N, 84°08′W, 3760 ft (II-24)	°F	66.5	67	69	70.5	71	70.5	69.5	69.5	70	68.5	68.5	66.5	68.5
	In	0.6	0.2	0.8	1.8	9.0	9.5	8.3	9.5	12.0	11.8	5.7	1.6	70.8
Aw 29. Key West, Florida, U.S.A. 24°33′N, 81°48′W, 22 ft (I-51)	°F	70	71	73	76	79	82	83.5	83.5	83.5	79	74	70.5	77
	In	2.0	1.3	1.4	1.3	3.5	4.2	3.3	4.5	6.7	6.0	2.2	1.7	38.1
Aw 30. Santo Domingo, Dominican Republic 18°29′N, 69°54′W, 57 ft (II-40)	°F	75	75.5	75.5	77	78.5	79.5	80	80.5	80	79.5	78	76	78
	In	2.4	1.4	1.9	3.9	6.8	6.2	6.4	6.3	7.3	6.0	4.8	2.4	55.8
Aw 31. Calabozo, Venezuela 08°56′N, 67°20′W, 348 ft (II-36)	°F	82.5	82	84.5	83	81.5	80	80	80.5	81	81.5	82.5	82.5	82
	In	<0.1	0.1	0.3	2.8	6.7	7.4	9.1	8.6	7.2	5.1	3.3	0.6	51.3
Aw 32. Bathurst, Gambia 13°21′N, 16°40′W, 90 ft (IV-74)	°F	73.5	75.5	78.5	78	78	81	80	79	80	80.5	77	74.5	78
	In	0.1	0.1	<0.1	<0.1	0.4	2.3	11.1	19.7	12.2	4.3	0.7	0.1	51.0
Aw 33. Ibadan, Nigeria 07°26′N, 03°54′E, 656 ft (IV-121)	°F	80.5	82	83	82	80.5	78	76	75	77.5	79	80	80	79.5
	In	0.3	0.9	3.0	4.9	5.7	6.4	5.2	2.9	6.7	6.0	1.7	0.4	44.1
Aw 34. Wau, Sudan 07°42′N, 28°03′E, 1443 ft (IV-155)	°F	80	82	85	85.5	83	81	79	79	80	81	81.5	80	81.5
	In	<0.1	0.2	0.9	2.6	5.3	6.5	7.5	8.2	6.6	4.9	0.6	<0.1	43.3
Aw 35. Bangalore, India 12°57′N, 77°37′E, 3021 ft (V-25)	°F	69	73	78	81	80.5	76	74	74	73.5	73.5	71	69	74.5
	In	0.2	0.3	0.4	1.6	4.2	2.9	3.9	5.0	6.7	5.9	2.7	0.4	34.2
Aw 36. Mandalay, Burma 21°59′N, 96°06′E, 252 ft (V-8)	°F	68.5	73.5	81.5	89	88.5	85.5	85.5	84.5	83.5	81	75.5	68.5	80.5
	In	0.1	0.1	0.2	1.2	5.8	6.3	2.7	4.1	5.4	4.3	2.0	0.4	32.6
Aw 37. Saigon, Vietnam 10°47′N, 106°42′E, 30 ft (V-40)	°F	79.5	81	83.5	85.5	84	82	81.5	81.5	81	81	80	79	82
	In	0.6	0.1	0.5	1.7	8.7	13.0	12.4	10.6	13.2	10.6	4.5	2.2	78.1
Aw 38. Guayaquil, Ecuador 02°10′S, 79°53′W, 20 ft (II-25)	°F	79	79	80	80	78	77.5	75.5	75.5	76.5	77	78	79	78
	In	9.4	9.8	10.9	4.6	1.1	0.3	0.2	0.0	0.1	0.3	0.1	2.0	38.8
Aw 39. Catalão, Brazil 18°10′S, 47°32′W, 2723 ft (II-9)	°F	73	73.5	73.5	72.5	69.5	67	68	70	73	74	73.5	72.5	71.5
	In	11.8	10.2	8.8	3.8	1.1	0.3	0.5	0.3	2.3	6.1	8.3	14.9	68.1
Aw 40. Iguatu, Brazil 06°24′S, 39°35′W, 685 ft (II-12)	°F	84	82	81.5	80.5	79.5	79.5	79.5	81	83.5	84.5	85	85	82
	In	3.5	6.8	7.3	6.3	2.4	1.4	0.2	0.1	0.7	0.7	0.4	1.3	31.0

Station		J	F	M	A	M	J	J	A	S	O	N	D	Year
Cwa 41. Nova Lisboa, Angola 12°48′S, 15°45′E, 5577 ft (IV-18)	°F	68	68	68	67.5	64.5	61	62	66	69.5	69.5	68	68	67
	In	8.7	7.8	9.8	5.7	0.4	0.0	<0.1	<0.1	0.6	5.5	9.6	8.9	57.0
Aw 42. Beira, Mozambique 19°50′S, 38°51′E, 28 ft (IV-111)	°F	82	82	80.5	78.5	73.5	70	69	70	73.5	79	79.5	80.5	76.5
	In	10.9	8.4	10.1	4.2	2.2	1.3	1.2	1.1	0.8	5.2	5.3	9.2	59.9
Aw 43. Port Darwin, Northern Territory, Australia 12°28′S, 130°51′E, 97 ft (VI-7)	°F	83.5	83.5	84	84	82	78.5	77	79.5	87.5	85	86	85	82.5
	In	15.2	12.3	10.0	3.8	0.6	0.1	<0.1	0.1	0.5	2.0	4.7	9.4	58.7
TROPICAL DESERT CLIMATES (WARM)														
BWh 44. Yuma, Arizona, U.S.A. 32°45′N, 114°36′W, 141 ft (I-46)	°F	54.5	59	64.5	70	76.5	85	91.5	90.5	85	73	62.5	55.5	72.5
	In	0.4	0.4	0.3	0.1	<0.1	<0.1	0.2	0.6	0.4	0.3	0.2	0.5	3.4
BWh 45. Cape Juby, Morocco 27°56′N, 12°55′W, 20 ft (IV-147)	°F	61.5	62	63	65	66	68	69	69.5	69.5	69	67	63	66
	In	0.3	0.2	0.2	<0.1	<0.1	<0.1	<0.1	<0.1	0.3	<0.1	0.6	0.3	1.9
BWh 46. In Salah, Algeria 27°12′N, 02°28′E, 919 ft (IV-7)	°F	56	61	68	77	84	95	97.5	96.5	91	80	66.5	58	77.5
	In	0.1	0.1	<0.1	<0.1	<0.1	<0.1	0.0	0.1	<0.1	<0.1	0.2	0.1	0.6
BWh 47. Cairo, U.A.R. 29°52′N, 31°20′E, 381 ft (IV-32)	°F	56	58.5	63.5	70	77	81.5	83	83	79	75.5	68	59	71.5
	In	0.2	0.2	0.2	0.1	0.1	<0.1	0.0	0.0	<0.1	<0.1	0.1	0.2	1.1
BWh 48. Aswân, U.A.R. 24°02′N, 32°53′E, 366 ft (IV-31)	°F	62	65	72.5	81	88.5	92.5	92.5	92.5	89	84.5	74.5	65	80.5
	In	<0.1	<0.1	<0.1	<0.1	<0.1	<0.1	0.0	0.0	0.0	<0.1	<0.1	<0.1	<0.1
BWh 49. Baghdad, Iraq 33°20′N, 44°24′E, 111 ft (V-49)	°F	49.5	53	59.5	71	82	89	93	93	87	76.5	64	53	73
	In	0.9	1.0	1.1	0.5	0.1	<0.1	<0.1	<0.1	<0.1	0.1	0.8	1.0	5.5
BWh 50. Karachi, Pakistan 24°48′N, 66°59′E, 13 ft (V-73)	°F	66	68.5	76	81.5	86	87.5	86	83.5	82.5	81.5	75.5	66.5	78
	In	0.5	0.4	0.3	0.1	0.1	0.7	3.2	1.6	0.5	<0.1	0.1	0.2	7.7
BWh 51. Lima, Peru 12°05′S, 77°03′W, 394 ft (II-33)	°F	74	75	74.5	76.5	67	63	62	61	62.5	64.5	67	70	68
	In	0.1	<0.1	<0.1	<0.1	0.2	0.2	0.3	0.3	0.3	0.1	0.1	<0.1	1.6
BWh 52. Upington, Union of South Africa 28°27′S, 21°15′E, 2641 (IV-184)	°F	81	80.5	76.5	69.5	60.5	55.5	55	59	64	70.5	75	79.5	69
	In	0.6	1.0	1.7	0.7	0.5	0.1	0.1	0.1	0.1	0.3	0.4	0.5	6.1
BWh 53. William Creek, South Australia 28°55′S, 136°21′E, 247 ft (VI-16)	°F	82.5	83	76.5	67.5	59	54	53	56	62.5	70	77	81	69
	In	0.5	0.6	0.3	0.3	0.3	0.5	0.2	0.3	0.3	0.5	0.5	0.7	5.0

TROPICAL STEPPE CLIMATES (WARM)		J	F	M	A	M	J	J	A	S	O	N	D	Year
BSh 54. Monterrey, Mexico	°F	58	62	66.5	73	77.5	81	80.5	82	78	72	63	57.5	71
25°40′N, 100°18′W, 1732 ft (I-43)	In	0.6	0.7	0.8	1.3	1.3	3.0	2.3	2.4	5.2	3.0	1.5	0.8	22.8
BSh 55. Kayes, Mali	°F	78	82	88.5	93.5	95.5	89.5	83.5	81	82.5	85	84	83.5	85
14°26′N, 11°26′W, 98 ft (IV-60)	In	0.1	0.0	0.0	<0.1	1.0	3.8	6.3	9.5	7.4	1.7	<0.1	<0.1	29.8
BSh 56. Mogadiscio, Somalia	°F	79.5	80	82	84	83	79.5	78	78	79	81	81	80.5	80.5
02°02′N, 45°21′E, 39 ft (IV-138)	In	<0.1	<0.1	<0.1	2.3	2.3	3.8	2.5	1.9	1.0	0.9	1.6	0.5	16.9
BSh 57. Bushire, Iran	°F	57.5	59	66	74	82.5	86.5	89.5	90.5	86.5	80	70.5	61.5	75.5
28°59′N, 50°49′E, 14 ft (V-76)	In	2.9	1.8	0.8	0.4	<0.1	0.0	0.0	<0.1	0.0	0.1	1.6	3.2	10.8
BSh 58. Jaipur, India	°F	60.5	64.5	74.5	84.5	91.5	92	86	83	83	79.5	69.5	62	78
26°55′N, 75°50′E, 1431 ft (V-30)	In	0.4	0.3	0.3	0.2	0.6	2.2	7.7	8.1	3.2	0.5	0.1	0.3	23.9
BSh 59. Poona, India	°F	70	73	79.5	84.5	85.5	81	77.5	76.5	77	77.5	73	69	77
18°32′N, 73°51′E, 1834 ft (V-34)	In	<0.1	<0.1	<0.1	0.6	1.1	4.5	6.6	3.5	5.3	3.5	1.1	0.1	26.3
BSh 60. Santiago del Estero, Argentina	°F	83	81	77	70.5	63	56.5	57	60.5	67.5	73	78	80.5	71
27°46′S, 64°18′W, 653 ft (II-5)	In	3.4	3.0	3.0	1.3	0.6	0.3	0.2	0.2	0.5	1.4	2.5	4.1	20.4
BSh 61. Francistown, Bechuanaland	°F	76.5	75	73	69.5	63.5	57.5	58	62	70	75.5	76.5	76.5	69
21°13′S, 27°30′E, 3294 ft (IV-22)	In	4.2	3.1	2.8	0.7	0.2	0.1	<0.1	<0.1	<0.1	0.9	2.2	3.4	17.7
BSh 62. Broome, Western Australia	°F	85.5	85.5	85	82.5	76.5	71	70	72.5	77	81.5	84.5	86	79.5
17°57′S, 122°15′E, 63 ft (VI-19)	In	6.3	5.8	3.9	1.2	0.6	0.9	0.2	0.1	<0.1	<0.1	0.6	3.3	22.9
BSh 63. Bourke, New South Wales, Australia	°F	84.5	83	77.5	68.5	60	53.5	52.5	56.5	63	70.5	78	82	69
30°05′S, 145°58′E, 361 ft (VI-3)	In	1.4	1.5	1.1	1.1	1.0	1.1	0.9	0.8	0.8	0.9	1.2	1.4	13.2

TROPICAL WEST-COAST DESERT CLIMATES (COOL)		J	F	M	A	M	J	J	A	S	O	N	D	Year
BWk 64. Antofagasta, Chile	°F	69.5	69.5	67.5	64	61	58.5	57	57	58.5	60.5	63.5	66	63
23°42′S, 70°24′W, 308 ft (II-19)	In	0.0	0.0	0.0	<0.1	<0.1	0.1	0.2	0.1	<0.1	0.1	<0.1	0.0	0.5
BWk 65. Walvis Bay, Southwest Africa	°F	66	67	66.5	65	63	61	58.5	57	57	59	62.5	64.5	62.5
22°56′S, 14°30′E, 24 ft (IV-144)	In	<0.1	0.2	0.3	0.1	0.1	<0.1	<0.1	0.1	<0.1	<0.1	<0.1	<0.1	0.9
BWk 66. Port Nolloth, Union of South Africa	°F	60	60.5	60	58	57	55.5	53.5	54	55	56.5	58.5	59.5	57
29°14′S, 16°52′E, 23 ft (IV-182)	In	0.1	0.1	0.2	0.2	0.3	0.3	0.3	0.3	0.2	0.1	0.1	0.1	2.3

MIDDLE-LATITUDE DESERT CLIMATES (COOL)		J	F	M	A	M	J	J	A	S	O	N	D	Year
BWk 67. Las Vegas, Nevada, U.S.A.	°F	44.5	50.5	55.5	63	70.5	80	85.5	84	76	65.5	53.5	45.5	64.5
36°12′N, 115°10′W, 2075 ft (I-63)	In	**0.7**	**0.5**	**0.3**	**0.3**	**0.2**	**0.2**	**0.5**	**0.5**	**0.3**	**0.3**	**0.2**	**0.4**	**4.4**
BWk 68. Kazalinsk, Kazak, U.S.S.R.	°F	10.5	13	26	42.5	64	73.5	77.5	73	61.5	46	30	19.5	45
45°46′N, 62°06′E, 207 ft (V-97)	In	**0.4**	**0.4**	**0.5**	**0.5**	**0.6**	**0.2**	**0.2**	**0.3**	**0.3**	**0.4**	**0.5**	**0.6**	**4.9**
BWk 69. Sufu (Kashgar), China	°F	22.5	31	45.5	59.5	69.5	76.5	80	78	70	57	41.5	27.5	55
39°24′N, 76°07′E, 4296 ft (V-16)	In	**0.6**	**0.1**	**0.5**	**0.2**	**0.3**	**0.2**	**0.4**	**0.3**	**0.1**	**0.1**	**0.2**	**0.3**	**3.2**
BWk 70. Santa Cruz, Argentina	°F	59	59	54.5	48	42	35.5	34.5	39	43.5	48.5	52	55.5	48
50°01′S, 68°32′W, 39 ft (II-5)	In	**0.6**	**0.3**	**0.3**	**0.6**	**0.4**	**0.5**	**0.4**	**0.5**	**0.3**	**0.3**	**0.4**	**0.7**	**5.3**

MIDDLE-LATITUDE STEPPE CLIMATES (COOL)		J	F	M	A	M	J	J	A	S	O	N	D	Year
BSk 71. Reno, Nevada, U.S.A.	°F	32	36.5	41.5	48	55	62.5	70.5	69.5	61	51.5	41.5	33.5	50.5
39°30′N, 119°47′W, 4397 ft (I-63)	In	**1.5**	**1.1**	**0.8**	**0.5**	**0.5**	**0.3**	**0.2**	**0.2**	**0.2**	**0.3**	**0.6**	**0.9**	**7.1**
BSk 72. Medicine Hat, Alberta, Canada	°F	12	14.5	28	45	55	63	69.5	66.5	56	45.5	28	19.5	42
50°01′N, 110°37′W, 2144 ft (I-10)	In	**0.6**	**0.6**	**0.6**	**0.8**	**1.6**	**2.4**	**1.7**	**1.4**	**1.1**	**0.6**	**0.7**	**0.7**	**12.8**
BSk 73. Cheyenne, Wyoming, U.S.A.	°F	25.5	27	33	41	50	60.5	66.5	65.5	57	45	34.5	28.5	44.5
41°09′N, 104°49′W, 6139 ft (I-78)	In	**0.4**	**0.6**	**1.0**	**1.9**	**2.4**	**1.6**	**2.1**	**1.6**	**1.2**	**1.0**	**0.5**	**0.5**	**14.8**
BSk 74. Odessa, Ukrainian S.S.R., U.S.S.R.	°F	25	28.5	35.5	46.5	61	68	72	76.5	62	52	39	30	49
46°29′N, 30°44′E, 214 ft (III-135)	In	**1.0**	**0.7**	**0.7**	**1.1**	**1.1**	**1.9**	**1.6**	**1.4**	**1.1**	**1.4**	**1.1**	**1.1**	**14.3**
BSk 75. Ankara, Turkey	°F	31.5	34	41	51.5	61	65.5	72.5	73	65	56.5	47	36	53
39°57′N, 32°53′E, 2825 ft (V-89)	In	**1.3**	**1.2**	**1.3**	**1.3**	**1.9**	**1.0**	**0.5**	**0.4**	**0.7**	**0.9**	**1.2**	**1.9**	**13.6**
BSk 76. Tehrān, Iran	°F	36	41	49	60	70	79.5	85.5	84	77	64.4	53	42	62
35°41′N, 51°25′E, 4002 ft (V-79)	In	**1.8**	**1.5**	**1.8**	**1.4**	**0.5**	**0.1**	**0.1**	**0.1**	**0.1**	**0.3**	**0.8**	**1.2**	**9.7**
BSk 77. Akmolinsk, Kazak S.S.R., U.S.S.R.	°F	−2	0.5	10.5	32	53	62	66.5	58	51.5	34.5	17	5	33
51°12′N, 71°23′E, 1004 ft (V-96)	In	**0.7**	**0.5**	**0.5**	**0.6**	**1.1**	**1.7**	**1.6**	**1.5**	**1.0**	**1.1**	**0.7**	**0.6**	**11.7**
BSk 78. Sian (Hsian), China	°F	31.5	37.5	51	62.5	73.5	82.5	85.5	83.5	72.5	61.5	47	36	60.5
34°03′N, 109°01′E, 1197 ft (V-19)	In	**0.3**	**0.3**	**0.7**	**1.8**	**1.9**	**1.8**	**3.9**	**3.9**	**2.3**	**1.6**	**0.5**	**0.3**	**19.3**

SUBTROPICAL WET-DRY AND MONSOON CLIMATES		J	F	M	A	M	J	J	A	S	O	N	D	Year
Cwa 79. Allahabad, India	°F	61	65	76.5	87	93.5	93	86	84	84	78.5	68.5	61.5	78
25°17′N, 81°44′E, 322 ft (V-24)	In	0.9	0.6	0.6	0.2	0.6	5.0	12.6	10.0	8.4	2.3	0.3	0.3	41.8
Cwa 80. Lashio, Burma	°F	60	63.5	71	75.5	77	77	76.5	76.5	76	73	66.5	61	71.5
22°58′N, 97°51′E, 2802 ft (V-8)	In	0.3	0.3	0.6	2.2	6.9	9.8	12.0	12.7	7.8	5.7	2.7	0.9	61.9
Cwa 81. Hanoi, Vietnam	°F	62	63.5	68.5	75.5	82	85	84.5	84	82	77.5	71	65.5	75.5
21°02′N, 105°52′E, 53 ft (V-39)	In	0.7	1.1	1.5	3.2	7.7	9.4	12.7	13.5	10.0	3.9	1.7	0.8	66.2
Cwa 82. Asunción, Paraguay	°F	83	82.5	80.5	79.5	67.5	62.5	63.5	67.5	71.5	74	77.5	82	74
25°17′S, 57°30′W, 456 ft (II-31)	In	5.5	5.1	4.3	5.2	4.6	2.7	2.2	1.5	3.1	5.5	5.9	6.2	51.8
Cwa 83. Kasempa, Northern Rhodesia	°F	69.5	69.5	69.5	67	63	58	58	63	70	72	71	70	66.5
13°27′S, 25°50′E, 4439 ft (IV-129)	In	12.5	8.4	6.4	1.4	0.1	<0.1	<0.1	<0.1	<0.1	1.3	5.7	10.7	46.5
Cwa 84. Newcastle, Union of South Africa	°F	71	70.5	68	63	56.5	51.5	51.5	57.5	62.5	67	69	71	63
27°45′S, 29°56′E, 3934 ft (IV-180)	In	6.6	5.5	5.1	1.7	1.0	0.5	0.6	0.5	1.3	3.0	5.1	5.3	36.2
Cwa 85. Tananarive, Malagasy Republic	°F	70	69.5	69.5	67	63.5	59.5	58	59	62.5	67	69.5	70	65.5
18°55′S, 47°33′E, 4500 ft (IV-103)	In	11.8	11.0	7.0	2.1	0.7	0.3	0.3	0.4	0.7	2.4	5.3	11.3	53.4
Cwa 86. Rockhampton, Queensland, Australia	°F	80.5	80.5	78.5	74.5	68.5	63.5	61.5	65	70.5	75	78	80.5	73
23°24′S, 150°30′E, 37 ft (VI-11)	In	7.5	7.6	4.4	2.6	1.6	2.6	1.8	0.8	1.3	1.8	2.4	4.7	39.1

HUMID SUBTROPICAL CLIMATES		J	F	M	A	M	J	J	A	S	O	N	D	Year
Cfa 87. Nashville, Tennessee, U.S.A.	°F	39	41.5	49.5	59	68	76.5	79.5	78	72	61	49	41	59.5
36°10′N, 87°47′W, 546 ft (I-71)	In	4.6	4.1	5.1	4.3	3.8	4.1	4.0	3.6	3.4	2.6	3.5	4.0	47.1
Cfa 88. New Orleans, Louisiana, U.S.A.	°F	54.5	57.5	63	69	75.5	81	83	83	79.5	71.5	62.5	56	69.5
29°57′N, 90°04′W, 8 ft (I-56)	In	4.6	4.2	4.7	4.8	4.5	5.5	6.6	5.8	4.8	3.5	3.8	4.6	57.4
Cfa 89. Wilmington, North Carolina, U.S.A.	°F	47.5	48.5	54.5	61.5	70	77	79.5	79.5	74	65	55.5	48.5	63.5
34°14′N, 77°57′W, 72 ft (I-66)	In	3.3	3.3	3.2	2.7	3.4	5.1	7.1	6.4	4.5	3.3	2.0	2.8	47.1
Cfa 90. Varna, Bulgaria	°F	34	35.5	42.5	51.5	61.5	67	74	73.5	67	59	49	38.5	54.5
43°12′N, 27°55′E, 115 ft (III-8)	In	1.4	0.9	1.2	1.1	1.6	2.6	1.8	1.0	1.5	2.0	2.0	2.0	19.1
Cfa 91. Amoy, China	°F	57	56.5	60	67.5	75	81	84.5	84.5	82.5	76.5	69	62.5	71.5
24°26′N, 118°04′E, 16 ft (V-11)	In	1.3	3.0	3.5	4.9	6.2	7.0	5.2	6.6	4.3	1.9	1.2	1.3	46.4

			J	F	M	A	M	J	J	A	S	O	N	D	Year
Cfa	92. Nagasaki, Japan 32°44′N, 129°53′E, 436 ft (V-56)	°F	42.5	43	49	58	65	71.5	79	81	74.5	65	56	46.5	61
		In	**2.8**	**3.3**	**4.9**	**7.3**	**6.7**	**12.3**	**10.1**	**6.9**	**9.8**	**4.5**	**3.7**	**3.2**	**75.5**
Cfa	93. Buenos Aires, Argentina 34°35′S, 58°29′W, 89 ft (II-1)	°F	74	73	69.5	62.5	55.5	49	49.5	51.5	55	59.5	66	71.5	61.5
		In	**3.1**	**2.8**	**4.3**	**3.5**	**3.0**	**2.4**	**2.2**	**2.4**	**3.1**	**3.4**	**3.3**	**3.9**	**37.4**
Cfa	94. Brisbane, Queensland, Australia 27°28′S, 153°02′E, 137 ft (VI-8)	°F	77	76.5	74	70	65	60	58.5	60.5	65.5	70	73	76	69
		In	**6.4**	**6.3**	**5.7**	**3.7**	**2.8**	**2.6**	**2.2**	**1.9**	**1.9**	**2.5**	**3.7**	**5.0**	**44.7**

MARINE WEST COAST CLIMATES (WARM SUMMER)

			J	F	M	A	M	J	J	A	S	O	N	D	Year
Cfb	95. Vancouver, British Columbia, Canada 49°17′N, 123°05′W, 45 ft (I-13)	°F	36.5	39	43.5	49	55	60.5	64	63.5	57	50.5	43.5	39	50
		In	**8.6**	**5.8**	**5.0**	**3.3**	**2.8**	**2.5**	**1.2**	**1.7**	**3.6**	**5.8**	**8.3**	**8.8**	**57.4**
Cfb	96. Shannon Airport, Ireland 52°41′N, 08°55′W, 8 ft (III-50)	°F	41	42	45.5	48	52.5	58	59.5	60.5	57.5	51.5	46	42.5	50.5
		In	**3.8**	**3.0**	**2.0**	**2.2**	**2.4**	**2.1**	**3.1**	**3.0**	**3.0**	**3.4**	**4.2**	**4.3**	**36.5**
Cfb	97. Aberdeen, Scotland, U.K. 57°10′N, 02°06′W, 79 ft (III-109)	°F	39	39	41	42.5	48	54	57.5	56.5	53.5	48	43	40	47
		In	**3.0**	**2.2**	**2.0**	**2.2**	**2.7**	**2.1**	**3.3**	**2.8**	**2.7**	**3.5**	**3.4**	**3.1**	**33.0**
Cfb	98. Bordeaux, France 44°50′N, 00°43′W, 157 ft (III-20)	°F	41.5	44	49	53.5	59	64.5	69	68.5	64.5	56.5	48	43	55
		In	**2.7**	**2.8**	**2.9**	**2.6**	**2.5**	**2.3**	**2.0**	**1.9**	**2.2**	**3.0**	**3.9**	**3.9**	**32.7**
Cfb	99. Bergen, Norway 60°24′N, 05°19′E, 141 ft (III-66)	°F	35	35	37.5	44.5	52.5	58	61.5	60	54.5	47.5	41	36.5	47
		In	**7.9**	**6.0**	**5.4**	**4.4**	**3.9**	**4.2**	**5.2**	**7.3**	**9.2**	**9.2**	**8.0**	**8.1**	**78.8**
Cfb	100. Bremen, Germany 53°05′N, 08°47′E, 52 ft (III-31)	°F	33.5	35	39.5	45.5	54.5	59.5	63	62	57	48.5	40.5	35.5	48
		In	**1.9**	**1.6**	**1.8**	**1.5**	**2.1**	**2.6**	**3.2**	**2.8**	**2.1**	**2.2**	**2.0**	**2.2**	**26.0**
Cfb	101. Valdivia, Chile 39°48′S, 73°14′W, 16 ft (II-22)	°F	62.5	62	59	54	49.5	47	46.5	47	49.5	53.5	55.5	59.5	53.5
		In	**2.6**	**2.9**	**5.2**	**9.2**	**14.2**	**17.7**	**15.5**	**12.9**	**8.2**	**5.0**	**4.9**	**4.1**	**102.4**
Cfb	102. Port Elizabeth, Union of South Africa 33°59′S, 25°36′E, 190 ft (IV-181)	°F	69.5	70	68	64	60.5	56.5	56	57.5	59	62	64.5	67	63
		In	**1.2**	**1.3**	**1.9**	**1.8**	**2.4**	**1.8**	**1.9**	**2.0**	**2.3**	**2.2**	**2.2**	**1.7**	**22.7**
Cfb	103. Hobart, Tasmania, Australia 42°53′S, 147°20′E, 177 ft (VI-17)	°F	62	62	59.5	55.5	51	47	46	48	51	54.5	57	60	54.5
		In	**1.9**	**1.5**	**1.8**	**1.9**	**1.8**	**2.2**	**2.1**	**1.9**	**2.1**	**2.3**	**2.4**	**2.1**	**24.0**
Cfb	104. Auckland, New Zealand 36°47′S, 174°39′E, 85 ft (VI-29)	°F	66.5	66.5	63	61.5	56.5	53	53	52	54.5	57.5	60	63.5	59
		In	**3.1**	**3.7**	**3.2**	**3.8**	**5.0**	**5.4**	**5.7**	**4.6**	**4.0**	**4.0**	**3.5**	**3.1**	**49.1**

Station		J	F	M	A	M	J	J	A	S	O	N	D	Year
Cfb 105. Invercargill, New Zealand 46°26′S, 168°21′E, 12 ft (VI-30)	°F	57	57	55	51	46.5	43	41.5	44.5	47.5	51	52.5	55.5	50
	In	**4.2**	**3.3**	**4.0**	**4.1**	**4.4**	**3.6**	**3.2**	**3.2**	**3.2**	**4.1**	**4.2**	**4.0**	**45.5**
MARINE WEST COAST CLIMATES (COOL SUMMER)														
Cfc 106. Dutch Harbor, Alaska, U.S.A. 53°53′N, 166°32′W, 13 ft (I-8)	°F	32.5	33.5	34	36	40.5	46.5	51	53	48.5	42	36.5	33.5	40.5
	In	**5.9**	**5.6**	**4.9**	**3.9**	**4.2**	**2.7**	**1.9**	**2.4**	**5.3**	**7.3**	**5.7**	**7.0**	**56.8**
Cfc 107. Kodiak, Alaska, U.S.A. 57°48′N, 152°22′W, 152 ft (I-4)	°F	31.5	31.5	33	38	43	49.5	54.5	55.5	50.5	43	36	32.5	41.5
	In	**4.7**	**4.7**	**3.8**	**3.8**	**5.9**	**4.7**	**3.4**	**5.1**	**5.3**	**7.2**	**5.8**	**6.0**	**60.4**
Cfc 108. Reykjavik, Iceland 64°09′N, 21°56′W, 92 ft (III-46)	°F	32	32.5	34.5	38	44.5	49.5	53	52	46.5	40	35.5	34	41.5
	In	**4.0**	**3.1**	**3.0**	**2.1**	**1.6**	**1.7**	**2.0**	**2.6**	**3.1**	**3.4**	**3.6**	**3.7**	**33.9**
Cfc 109. Thorshavn, Faeroes 62°02′N, 06°45′W, 82 ft (III-14)	°F	37.5	37.5	37.5	40.5	44	48.5	51.5	51	48.5	44	40	38	43.5
	In	**6.6**	**5.2**	**4.8**	**3.6**	**3.4**	**2.5**	**3.1**	**3.5**	**4.7**	**5.9**	**6.3**	**6.6**	**56.2**
Cfc 110. Trondheim, Norway 63°25′N, 10°27′E, 417 ft (III-72)	°F	26.5	27.5	31.5	38.5	46.5	53	58.5	56	49	41	33	28.5	41
	In	**3.1**	**2.7**	**2.6**	**2.0**	**1.7**	**1.9**	**2.4**	**3.0**	**3.4**	**3.7**	**2.8**	**2.8**	**32.1**
Cfc 111. Punta Arenas, Chile 53°10′S, 70°54′W, 26 ft (II-21)	°F	51.5	51	47.5	44.5	40	37	35.5	37.5	40.5	44.5	47	50	44
	In	**1.5**	**0.9**	**1.3**	**1.4**	**1.3**	**1.6**	**1.1**	**1.2**	**0.9**	**1.1**	**0.7**	**1.4**	**14.4**
MEDITERRANEAN CLIMATES (WARM SUMMER)														
Csa 112. Sacramento, California, U.S.A. 38°35′N, 121°30′W, 69 ft (I-48)	°F	45.5	50.5	55	58.5	64	70	74	73	70	62.5	54	46.5	60.5
	In	**3.8**	**2.8**	**2.8**	**1.5**	**0.8**	**0.1**	**<0.1**	**<0.1**	**0.3**	**0.8**	**1.9**	**3.8**	**18.6**
Csa 113. Sevilla, Spain 37°29′N, 05°59′W, 98 ft (III-89)	°F	50	53	57.5	62	68.5	76	81.5	82.5	76.5	67.5	58	52	65.5
	In	**2.2**	**2.9**	**3.3**	**2.3**	**1.3**	**0.9**	**0.1**	**0.1**	**1.1**	**2.6**	**3.7**	**2.8**	**23.3**
Csa 114. Algiers, Algeria 36°46′N, 03°03′E, 194 ft (IV-1)	°F	54	55	57.5	61.5	66	71.6	76.5	78	75	68.5	61	55.5	65
	In	**4.4**	**3.3**	**2.9**	**1.6**	**1.8**	**0.6**	**<0.1**	**0.2**	**1.6**	**3.1**	**5.1**	**5.4**	**30.0**
Csa 115. Marseille, France 43°18′N, 05°23′E, 246 ft (III-26)	°F	45.5	44.5	46.5	50	55.5	62	68	72	71.5	66.5	58.5	51	55
	In	**1.9**	**1.5**	**1.8**	**2.0**	**1.9**	**1.0**	**0.6**	**0.9**	**2.6**	**3.7**	**3.1**	**2.2**	**23.2**
Csa 116. Naples, Italy 40°53′N, 14°18′E, 220 ft (III-55)	°F	48	49	53	58.5	64.5	71.5	76.5	76.5	72	64	56	51	62
	In	**4.8**	**3.5**	**1.7**	**1.8**	**2.2**	**0.7**	**0.6**	**1.3**	**4.3**	**4.6**	**4.1**	**4.7**	**34.3**

Station		J	F	M	A	M	J	J	A	S	O	N	D	Year
Csa 117. Cirene, Libya	°F	47	48	52	58	64.5	71	72	71.5	69	66	58.5	51.5	61
32°49′N, 21°52′E, 2037 ft (IV-86)	In	**6.1**	**3.0**	**2.3**	**0.7**	**0.6**	**<0.1**	**<0.1**	**<0.1**	**0.3**	**1.7**	**2.9**	**6.0**	**23.6**
Csa 118. Jerusalem, Israel	°F	48	49	55.5	61.5	69	72.5	75	75.5	73.5	70	61.5	52	63
31°47′N, 35°13′E, 2485 ft (V-52)	In	**5.2**	**5.2**	**2.5**	**1.1**	**0.1**	**<0.1**	**0.0**	**0.0**	**<0.1**	**0.5**	**2.8**	**3.4**	**20.8**
Csa 119. Perth, Western Australia	°F	74	74	71	61.5	61	57	55.5	56	58.5	61.5	66.5	71	64
31°57′S, 115°51′E, 197 ft (VI-24)	In	**0.3**	**0.4**	**0.8**	**1.7**	**5.1**	**7.1**	**6.7**	**5.7**	**3.4**	**2.2**	**0.8**	**0.5**	**34.7**
Csa 120. Adelaide, South Australia	°F	73.5	74	70	64	58	54	52	54	57	62	67	71	63
34°56′S, 138°35′E, 140 ft (VI-13)	In	**0.8**	**0.7**	**1.0**	**1.8**	**2.7**	**3.0**	**2.6**	**2.6**	**2.1**	**1.7**	**1.1**	**1.0**	**21.1**
MEDITERRANEAN CLIMATES (COOL SUMMER)														
Csb 121. Portland, Oregon, U.S.A.	°F	39	42	46.5	52	56.5	62.5	66.5	66.5	61.5	54.5	47	41.5	53
45°32′N, 122°40′W, 154 ft (I-68)	In	**6.1**	**5.2**	**4.6**	**2.8**	**2.1**	**1.6**	**0.5**	**0.6**	**1.8**	**3.3**	**6.2**	**7.0**	**41.8**
Csb 122. San Francisco, California, U.S.A.	°F	50	53	54.5	55.5	57	59	59	59	62	61	57	52	56.5
37°47′N, 122°25′W, 52 ft (I-49)	In	**4.7**	**3.8**	**3.1**	**1.5**	**0.7**	**0.1**	**<0.1**	**<0.1**	**0.3**	**1.0**	**2.5**	**4.4**	**22.1**
Csb 123. Pôrto, Portugal	°F	48	49.5	53	56	58.5	64.5	67	67.5	65.5	60	53.5	48.5	57.5
41°08′N, 08°36′W, 328 ft (III-78)	In	**6.0**	**4.6**	**5.4**	**4.1**	**3.3**	**1.7**	**0.9**	**0.7**	**2.1**	**4.3**	**5.9**	**6.5**	**45.5**
Csb 124. Santiago, Chile	°F	69	68	64.5	59.5	52.5	47.5	48	50.5	54	58.5	63	67	58.5
33°27′S, 70°42′W, 1706 ft (II-21)	In	**0.1**	**0.1**	**0.2**	**0.5**	**2.5**	**3.3**	**3.0**	**2.2**	**1.2**	**0.6**	**0.3**	**0.2**	**14.1**
Csb 125. Capetown, Union of South Africa	°F	69	69.5	67.5	62.5	58	55.5	54	55	57	61	64	67	62
33°54′S, 18°32′E, 56 ft (IV-176)	In	**0.6**	**0.3**	**0.7**	**1.9**	**3.1**	**3.3**	**3.5**	**2.6**	**1.7**	**1.2**	**0.7**	**0.4**	**20.0**
Csb 126. Cape Leeuwin, Western Australia	°F	67.5	68.5	67.5	65	61	58.5	56.5	56.5	57.5	59	62.5	65.5	62
34°22′S, 115°08′E, 163 ft (VI-19)	In	**0.6**	**0.7**	**1.2**	**2.3**	**5.9**	**7.3**	**7.3**	**5.4**	**3.4**	**2.8**	**1.2**	**0.8**	**38.9**
HUMID CONTINENTAL CLIMATES (WARM SUMMER)														
Dfa 127. Lincoln, Nebraska, U.S.A.	°F	24	27	38.5	52	62	72	77.5	75.5	67.5	55	40	28.5	51.5
40°49′N, 96°42′W, 1181 ft (I-62)	In	**0.6**	**0.9**	**1.2**	**2.5**	**4.0**	**4.2**	**3.9**	**3.5**	**2.9**	**2.0**	**1.2**	**0.8**	**27.7**
Dfa 128. Chicago, Illinois, U.S.A.	°F	25	27	36	47.5	57.5	67.5	73.5	72	65.5	54	40.5	29.5	49.5
41°53′N, 87°38′W, 823 ft (I-54)	In	**2.0**	**2.0**	**2.6**	**2.8**	**3.4**	**3.5**	**3.3**	**3.2**	**3.1**	**2.6**	**2.4**	**2.0**	**32.9**

Station		J	F	M	A	M	J	J	A	S	O	N	D	Year
Cfa 129. St. Louis, Missouri, U.S.A. 38°38′N, 90°12′W, 568 ft (I–60)	°F	32	34.5	45	56	66	75	79.5	78	71	59	46	35.5	56.5
	In	**2.3**	**2.5**	**3.5**	**3.8**	**4.5**	**4.5**	**3.5**	**3.4**	**3.2**	**2.9**	**2.8**	**2.5**	**39.4**
Cfa 130. New Haven, Connecticut, U.S.A. 41°18′N, 72°56′W, 107 ft (I–50)	°F	28.5	28.5	37	47.5	58	67	72.5	70.5	64.5	53.5	42.5	32.5	50
	In	**4.0**	**3.8**	**4.4**	**3.8**	**3.6**	**3.4**	**4.2**	**4.4**	**3.7**	**3.6**	**3.7**	**3.8**	**46.2**
Dfa 131. Rostov-Na-Donu, R.S.F.S.R., U.S.S.R. 47°13′N, 39°43′E, 159 ft (III–127)	°F	19	25	32.5	47	61.5	68.5	73	72	60.5	49.5	34.5	25.5	48
	In	**1.3**	**1.6**	**1.2**	**1.4**	**1.6**	**2.4**	**2.3**	**1.0**	**1.0**	**1.3**	**1.4**	**1.5**	**18.1**
Dwa 132. Mukden, Manchuria 41°48′N, 123°23′E, 141 ft (V–69)	°F	11.5	17	32	49	62	72.5	78	76	68.5	49.5	31.5	15	46.5
	In	**0.3**	**0.3**	**0.7**	**1.1**	**2.7**	**3.3**	**7.2**	**6.7**	**2.5**	**1.4**	**1.1**	**0.6**	**27.9**
Dwa 133. Inchon, Korea 37°29′N, 126°38′E, 231 ft (V–62)	°F	26	29	38	50	59.5	68.5	75	77.5	68.5	58	43.5	30.5	52
	In	**0.8**	**0.7**	**1.2**	**2.6**	**3.3**	**3.9**	**10.9**	**8.8**	**4.3**	**1.6**	**1.6**	**1.1**	**40.8**

HUMID CONTINENTAL CLIMATES (COOL SUMMER)

Station		J	F	M	A	M	J	J	A	S	O	N	D	Year
Dfb 134. Calgary, Alberta, Canada 51°06′N, 114°01′W, 3540 ft (I–9)	°F	13	17.5	25.5	40	49.5	56	61.5	59.5	50.5	41.5	27.5	19	38.5
	In	**0.5**	**0.5**	**0.8**	**1.0**	**2.3**	**3.1**	**2.5**	**2.3**	**1.5**	**0.7**	**0.7**	**0.6**	**16.7**
Dfb 135. Winnipeg, Manitoba, Canada 49°54′N, 97°14′W, 786 ft (I–16)	°F	−3	0.5	16	37.5	52	62	67	63.5	54	41	21.5	6	35.5
	In	**0.9**	**0.9**	**1.2**	**1.4**	**2.3**	**3.1**	**3.1**	**2.5**	**2.3**	**1.5**	**1.1**	**0.9**	**21.2**
Dfb 136. Escanaba, Michigan, U.S.A. 45°48′N, 87°05′W, 594 ft (I–58)	°F	15	15	24.5	38	50.5	61	67	64	57	46	33.5	22	41
	In	**1.4**	**1.4**	**1.9**	**2.2**	**3.0**	**3.3**	**3.3**	**3.3**	**3.3**	**2.7**	**2.1**	**1.6**	**29.5**
Dfb 137. Toronto, Ontario, Canada 43°40′N, 79°24′W, 379 ft (I–30)	°F	23	22.5	25	42	53.5	63.5	69	67.5	60	48	37	27	45
	In	**2.7**	**2.4**	**2.6**	**2.5**	**2.9**	**2.7**	**2.9**	**2.7**	**2.9**	**2.4**	**2.8**	**2.6**	**32.1**
Dfb 138. St. Johns, Newfoundland, Canada 47°34′N, 52°42′W, 243 ft (I–20)	°F	23.5	22	27.5	35.5	47.5	52.5	59.5	61	54.5	46.5	37	29	40.5
	In	**5.3**	**4.9**	**4.6**	**4.2**	**3.6**	**3.5**	**3.5**	**3.7**	**3.8**	**5.3**	**5.9**	**5.5**	**53.8**
Cfb 139. Prague, Czechoslovakia 50°05′N, 14°25′E, 662 ft (III–10)	°F	29.5	33	39	47.5	57	63.5	66	65	58.5	49	38	31.5	48
	In	**0.9**	**0.8**	**1.1**	**1.5**	**2.4**	**2.8**	**2.6**	**2.2**	**1.7**	**1.2**	**1.2**	**0.9**	**19.3**
Dfb 140. Leningrad, R.S.F.S.R., U.S.S.R. 59°56′N, 30°16′E, 16 ft (III–122)	°F	17.5	18	25.5	38	50	58.5	64	59.5	51	41	30.5	22	39.5
	In	**1.0**	**0.9**	**0.9**	**1.0**	**1.6**	**2.0**	**2.5**	**2.8**	**2.1**	**1.8**	**1.4**	**1.2**	**19.2**
Dfb 141. Kiev, Ukranian S.S.R., U.S.S.R. 50°27′N, 30°30′E, 600 ft (III–134)	°F	21.5	24	32	44.5	59.5	66	68	66	57.5	46	38.5	25.5	45.5
	In	**1.3**	**1.0**	**1.6**	**1.7**	**1.9**	**2.6**	**3.1**	**2.3**	**1.8**	**1.8**	**1.5**	**1.5**	**22.1**

		J	F	M	A	M	J	J	A	S	O	N	D	Year
Dfb 142. Kazan, R.S.F.S.R., U.S.S.R.	°F	4.5	9	19	37	54	61.5	67	62.5	51	37.5	22	10	36
55°47′N, 49°08′E, 262 ft (III-121)	In	**0.8**	**0.6**	**0.7**	**0.8**	**1.3**	**2.4**	**2.3**	**2.0**	**1.7**	**1.6**	**1.2**	**0.9**	**16.3**
Dfb 143. Omsk, R.S.F.S.R., U.S.S.R.	°F	−7.5	−1.5	9.5	30	49.5	60	65	61	50.5	33.5	13	−1	30
54°58′N, 73°20′E, 279 ft (V-106)	In	**0.6**	**0.3**	**0.3**	**0.5**	**1.2**	**2.0**	**2.0**	**2.0**	**1.1**	**1.0**	**0.7**	**0.8**	**12.5**
Dwb 144. Vladivostok, R.S.F.S.R., U.S.S.R.	°F	6.5	14	26	40	49	57.5	65.5	69.5	66.5	48	30	14	40
43°07′N, 131°55′E, 94 ft (V-112)	In	**0.3**	**0.4**	**0.7**	**1.2**	**2.1**	**2.9**	**3.3**	**4.7**	**4.3**	**1.9**	**1.2**	**0.6**	**23.6**
Dwb 145. Khabarovsk, R.S.F.S.R., U.S.S.R.	°F	−7.5	−1.5	14	34.5	50	61.5	69	69	56.5	41	16.5	−3	33.5
48°28′N, 135°03′E, 165 ft (V-102)	In	**0.3**	**0.2**	**0.3**	**0.7**	**2.0**	**3.5**	**4.1**	**3.3**	**3.0**	**0.7**	**0.6**	**0.5**	**19.2**
Dfb 146. Nemuro, Japan	°F	22.5	21.5	27.5	37.5	44	50.5	58.5	64	61	51.5	40	29	42.5
43°20′N, 145°35′E, 84 ft (V-57)	In	**1.5**	**1.2**	**2.4**	**3.0**	**3.7**	**3.7**	**3.9**	**4.1**	**5.8**	**4.1**	**3.5**	**2.3**	**39.2**

CONTINENTAL SUBARCTIC CLIMATES

		J	F	M	A	M	J	J	A	S	O	N	D	Year
Dfc 147. Dawson, Yukon Territory, Canada	°F	−21	−12	4	28.5	46.5	56.5	59.5	54.5	42	26	1.5	−14	23
64°04′N, 139°29′W, 1062 ft (I-34)	In	**0.9**	**0.7**	**0.5**	**0.5**	**1.0**	**1.2**	**1.5**	**1.5**	**1.4**	**1.2**	**1.1**	**1.0**	**12.6**
Dfc 148. Churchill, Manitoba, Canada	°F	−19	−16.5	−6	14	30	43	53.5	52.5	41.5	27	5.5	−11	17.5
58°47′N, 94°11′W, 43 ft (I-14)	In	**0.5**	**0.6**	**0.9**	**0.9**	**0.9**	**1.9**	**2.2**	**2.7**	**2.3**	**1.4**	**1.0**	**0.7**	**16.0**
Dfc 149. Chibougamau, Quebec, Canada	°F	−1.5	2.5	10	28	44	55.5	61.5	59	50	37.5	22	5.5	31
49°54′N, 74°18′W, 1234 ft (I-31)	In	**2.3**	**2.0**	**2.7**	**2.6**	**2.9**	**4.3**	**4.5**	**5.1**	**5.4**	**4.0**	**2.7**	**3.4**	**41.9**
Dfc 150. Arkhangel'sk, R.S.F.S.R., U.S.S.R.	°F	5.5	6	16	29.5	40.5	51	57.5	54.5	45	33	20	9.5	31
64°33′N, 40°32′E, 22 ft (III-118)	In	**1.2**	**1.1**	**1.1**	**0.7**	**1.3**	**1.9**	**2.6**	**2.7**	**2.2**	**1.9**	**1.6**	**1.3**	**19.8**
Dfc 151. Turukhansk, R.S.F.S.R., U.S.S.R.	°F	−24.5	−12.5	−1	11	29	46	58.5	53.5	40.5	19	−9	−22	16.5
65°55′N, 87°38′E, 131 ft (V-110)	In	**0.3**	**0.3**	**0.3**	**0.4**	**0.7**	**1.5**	**1.7**	**2.2**	**2.0**	**1.0**	**0.5**	**0.4**	**11.3**
Dwc 152. Irkutsk, R.S.F.S.R., U.S.S.R.	°F	−6	−1.5	13.5	31	44.5	56	60	58	46	31	11	−4	28.5
52°16′N, 104°19′E, 1532 ft (V-101)	In	**0.5**	**0.4**	**0.3**	**0.6**	**1.3**	**2.2**	**3.1**	**2.8**	**1.7**	**0.7**	**0.6**	**0.6**	**14.9**
Dwd 153. Yakutsk, R.S.F.S.R., U.S.S.R.	°F	−49	−33.5	−10	16.5	40	57.5	63.5	57.5	42	17	−19.5	−42.5	12.5
62°01′N, 129°43′E, 535 ft (V-112)	In	**0.3**	**0.2**	**0.1**	**0.3**	**0.4**	**1.1**	**1.6**	**1.3**	**1.0**	**0.5**	**0.4**	**0.3**	**7.4**
Dwd 154. Verkhoyansk, R.S.F.S.R., U.S.S.R.	°F	−58.5	−48.5	−26	4.5	32.5	54	56.5	49	35	4.5	−35.5	−54	1
67°34′N, 133°51′E, 328 ft (V-111)	In	**0.2**	**0.2**	**0.1**	**0.2**	**0.3**	**0.9**	**1.1**	**1.0**	**0.5**	**0.3**	**0.3**	**0.2**	**5.3**

Station		J	F	M	A	M	J	J	A	S	O	N	D	Year
Dwc 155. Okhotsk, R.S.F.S.R., U.S.S.R. 59°21′N, 143°17′E, 18 ft (V-106)	°F	−11.5	−8	10	19.5	32.5	42	52.5	54	46	27	5	−6	21.5
	In	**0.1**	**0.1**	**0.2**	**0.4**	**0.9**	**1.6**	**2.2**	**2.6**	**2.4**	**1.0**	**0.2**	**0.1**	**11.8**
Dfc 156. Petropavlovsk, R.S.F.S.R., U.S.S.R. 52°53′N, 158°42′E, 286 ft (V-107)	°F	17	16.5	21.5	30	37.5	45.5	51.5	55.5	49.5	40	28	21	34.5
	In	**3.0**	**2.2**	**3.4**	**2.5**	**2.2**	**2.0**	**3.1**	**3.2**	**3.8**	**3.9**	**3.6**	**3.0**	**35.9**
Dfd 157. Nizhne-Kolymsk, R.S.F.S.R., U.S.S.R. 68°32′N, 160°59′E, 16 ft (V-105)	°F	−41	−30.5	−18	−0.5	23.5	46.5	52	45.5	34	7.5	−13	−31.5	6
	In	**0.5**	**0.3**	**0.2**	**0.2**	**0.2**	**0.7**	**1.5**	**1.2**	**0.9**	**0.5**	**0.4**	**0.3**	**6.9**
MARINE SUBARCTIC CLIMATES														
EM 158. St. Paul Island, Alaska, U.S.A. 57°09′N, 170°13′W, 22 ft (I-6)	°F	25	23.5	24.5	29	35	41.5	45.5	47.5	45.5	39.5	33.5	28.5	35
	In	**1.7**	**1.2**	**1.2**	**1.0**	**1.2**	**1.2**	**2.4**	**3.2**	**3.4**	**2.9**	**2.5**	**2.0**	**23.9**
EM 159. Ivigtut, Greenland 61°12′N, 48°10′W, 82 ft (I-36)	°F	18	19	23.5	31	40	46.5	49.5	48	41.5	34.5	27	21.5	33.5
	In	**3.3**	**2.6**	**3.4**	**2.5**	**3.5**	**3.2**	**3.1**	**3.7**	**5.9**	**5.7**	**4.6**	**3.1**	**44.6**
EM 160. Vardø, Norway 70°22′N, 31°06′E, 43 ft (III-72)	°F	23	22	24.5	30	36	42.5	48.5	48.5	43.5	35	29.5	26	34
	In	**2.5**	**2.5**	**2.3**	**1.5**	**1.3**	**1.3**	**1.5**	**1.7**	**1.9**	**2.5**	**2.1**	**2.4**	**23.5**
EM 161. Stanley, Falkland Is. 51°42′S, 57°51′W, 6 ft (II-27)	°F	49	48	46.5	43	39	36	35.5	36	39	41.5	44.5	46.5	42
	In	**2.8**	**2.3**	**2.5**	**2.6**	**2.6**	**2.1**	**2.0**	**2.0**	**1.5**	**1.6**	**2.0**	**2.8**	**26.8**
EM 162. Cumberland Bay, South Georgia Is. 54°16′S, 36°30′W, 8 ft (II-26)	°F	41.5	42	40.5	35.5	32.5	28.5	28.5	32	32	34.5	38	39	35
	In	**3.3**	**4.3**	**5.3**	**5.4**	**5.2**	**4.9**	**5.5**	**5.3**	**3.5**	**2.6**	**3.4**	**3.0**	**51.7**
TUNDRA CLIMATES														
ET 163. Wrangel Island, R.S.F.S.R., U.S.S.R. 70°58′N, 178°33′W, 10 ft (V-112)	°F	−10.5	−14	−10	1	16.5	32.5	37	35.5	28.5	16.5	1	−5.5	11
	In	**0.2**	**0.2**	**0.2**	**0.2**	**0.2**	**0.4**	**0.6**	**0.9**	**0.5**	**0.4**	**0.1**	**0.2**	**4.1**
ET 164. Barrow Point, Alaska, U.S.A. 71°18′N, 156°47′W, 22 ft (I-1)	°F	−15.5	−18.5	−15	−0.5	18.5	34	39.5	38.5	30.5	17	1	−10.5	10
	In	**0.2**	**0.1**	**0.1**	**0.1**	**0.1**	**0.3**	**0.9**	**0.8**	**0.5**	**0.5**	**0.3**	**0.2**	**4.1**
ET 165. Upernivik, Greenland 72°47′N, 56°07′W, 59 ft (I-39)	°F	−7	−9.5	−5.5	7	25.5	36	41.5	41.5	33.5	25	14.5	2	17
	In	**0.4**	**0.5**	**0.7**	**0.6**	**0.6**	**0.5**	**0.9**	**1.1**	**1.1**	**1.1**	**1.1**	**0.6**	**9.2**
ET 166. Grønfjorden, Spitsbergen 78°02′N, 14°15′E, 23 ft (III-73)	°F	3	0.5	−3.5	6	22.5	35.5	42	40.5	32	21	12.5	7.5	18
	In	**1.4**	**1.3**	**1.1**	**0.9**	**0.5**	**0.4**	**0.6**	**0.9**	**1.0**	**1.2**	**0.9**	**1.5**	**11.7**

			J	F	M	A	M	J	J	A	S	O	N	D	Year
ET	167. Laurie Island, South Orkneys	°F	32	32	31	26	19	13	12	13	19	24.5	28	33	23.5
	60°44′S, 44°44′W, 13 ft (II-27)	In	1.4	1.5	1.9	1.6	1.2	1.0	1.3	1.3	1.1	1.1	1.3	1.0	15.7

ICECAP CLIMATES

			J	F	M	A	M	J	J	A	S	O	N	D	Year
EF	168. Eismitte, Greenland	°F	−43	−53	−40	−25.5	−6	2	10	−1	−8	−32.5	−45	−37	−23
	70°53′N, 40°42′W, 9843 ft (I-35)	In	0.6	0.2	0.3	0.2	0.1	0.1	0.1	0.4	0.3	0.5	0.5	1.0	4.3
EF	169. Little America, Antarctica	°F	20	3	−6.5	−19.5	−24.5	−16.5	−36	−33	−40	−15	1.5	18.5	−12.5
	78°34′S, 163°56′W, 30 ft (VI-1)	In													
EF	170. McMurdo Sound, Antarctica	°F	23	14.5	3.5	−8.5	−13	−13.5	−15	−15.5	−14	−5	13	23.5	−0.5
	77°40′S, 166°30′E, 0 ft (VI-2)	In													

HIGHLAND CLIMATES

		J	F	M	A	M	J	J	A	S	O	N	D	Year
171. Mt. Wilson, California, U.S.A.	°F	42.5	43.5	45	49.5	56.5	67	73	72	68	57.5	50.5	45	55.5
34°14′N, 118°04′W, 5850 ft (I-48)	In	6.3	6.7	6.1	2.6	1.2	0.2	<0.1	0.1	0.5	1.1	1.9	4.4	31.1
172. Flagstaff, Arizona, U.S.A.	°F	27.5	31	36	43.5	50	56	65	63.5	56.5	46.5	37	29	45.5
35°12′N, 111°40′W, 6903 ft (I-45)	In	2.5	2.1	2.4	1.3	1.0	0.4	3.1	2.7	1.6	1.4	1.4	2.0	21.9
173. Mexico City, Mexico	°F	54	56	61	64	66	65.5	63	63.5	63.5	60	57	54.5	60.5
19°24′N, 99°12′W, 7575 ft (I-42)	In	0.5	0.2	0.4	0.8	2.1	4.7	6.7	6.0	5.1	2.0	0.7	0.3	29.4
174. Bogotá, Colombia	°F	57.5	58.5	58.5	59	58.5	58	57	57.5	57.5	58	58	57.5	58
04°36′N, 74°05′W, 8678 ft (II-23)	In	2.3	2.6	4.0	5.8	4.5	2.4	2.0	2.2	2.4	6.3	4.7	2.6	41.5
175. Fanaråken, Norway	°F	9	10	12	17	26	31	36.5	36.5	29.5	23	16.5	13	21.5
61°31′N, 07°54′E, 6647 ft (III-67)	In	3.8	4.6	3.5	4.2	1.9	3.2	4.4	5.0	4.7	3.4	3.4	3.8	45.9
176. Sonnblick, Austria	°F	8.5	8.5	11.5	16.5	24.5	30	34	34	30.5	24	16.5	11	21
57°03′N, 12°57′E, 10190 ft (III-5)	In	4.3	4.4	5.2	6.0	5.9	4.9	5.4	4.8	4.1	4.6	4.2	4.7	58.5
177. Addis Ababa, Ethiopia	°F	59	61.5	63	63.5	63.5	61.5	59.5	59.5	60.5	60	58	57	60.5
09°20′N, 38°45′E, 8038 ft (IV-37)	In	0.5	1.5	2.6	3.4	3.4	5.4	11.0	11.8	7.5	0.8	0.6	0.2	48.7
178. Lhasa, Tibet	°F	29	34	40.5	46.5	54	62	61.5	60	57.5	48	39	32	47
29°40′N, 91°07′E, 12090 ft (V-88)	In	<0.1	0.5	0.3	0.2	1.0	2.5	4.8	3.5	2.6	0.5	0.1	0.0	16.0

		J	F	M	A	M	J	J	A	S	O	N	D	Year
179. Quito, Ecuador	°F	59	59	59	58.5	58.5	58	58	59	59	59	58.5	59	59
0°13′S, 78°32′W, 9446 ft (II-25)	In	3.9	4.4	5.6	6.9	5.4	1.7	0.8	1.2	2.7	4.4	3.8	3.1	43.9
180. LaPaz, Bolivia	°F	53	53	53	52.5	50.5	48	47.5	49	51	53	54.5	53.5	51.5
16°30′S, 68°08′W, 12001 ft (II-7)	In	4.5	4.2	2.6	1.3	0.5	0.3	0.4	0.5	1.1	1.6	1.9	3.7	22.6
181. LaQuiaca, Argentina	°F	55.5	55	54.5	50.5	42.5	38	38	42	47	51.5	55	56	48.5
22°06′S, 65°36′W, 11345 ft (II-3)	In	3.5	2.6	1.8	0.3	<0.1	0.0	<0.1	<0.1	0.1	0.3	1.0	2.7	12.3
182. Tshibinda, Republic of the Congo	°F	61.5	61	61.5	61	60.5	57	58.5	60	60.5	61	60.5	60.5	60.5
02°19′S, 28°45′E, 6939 ft (IV-28)	In	6.7	7.0	7.5	9.2	6.7	2.3	1.3	2.3	5.2	8.7	8.5	8.1	73.5
183. Equator, Kenya	°F	57	58	58	57.5	56.5	55	53.5	53.5	55	56	56	55.5	56
0°00′, 35°33′E, 9062 ft (IV-78)	In	0.7	1.6	3.5	5.8	5.7	5.0	5.9	7.6	3.9	1.3	3.4	1.5	45.9

Alphabetical List of Climate Stations

97 Aberdeen, Scotland, U.K.
177 Addis Ababa, Ethiopia
120 Adelaide, South Australia
77 Akmolinsk, Kazak S.S.R., U.S.S.R.
15 Akyab, Burma
114 Algiers, Algeria
79 Allahabad, India
8 Amboina, Moluccas, Indonesia
91 Amoy, China
75 Ankara, Turkey
64 Antofagasta, Chile
22 Aparri, Philippines
150 Arkhangel'sk, R.S.F.S.R., U.S.S.R.
82 Ascunción, Paraguay
48 Aswân, U.A.R.
104 Auckland, New Zealand
49 Baghdad, Iraq
35 Bangalore, India
164 Barrow Point, Alaska, U.S.A.
2 Barumbu, Republic of the Congo
32 Bathurst, Gambia
42 Beira, Mozambique
21 Belize, British Honduras
99 Bergen, Norway
174 Bogotá, Colombia
98 Bordeaux, France
63 Bourke, New South Wales, Australia
100 Bremen, Germany
94 Brisbane, Queensland, Australia
62 Broome, Western Australia
93 Buenos Aires, Argentina
57 Bushire, Iran
26 Cairns, Queensland, Australia
47 Cairo, U.A.R.
31 Calabozo, Venezuela
134 Calgary, Alberta, Canada
45 Cape Juby, Morocco
126 Cape Leeuwin, Western Australia
125 Capetown, Union of South Africa
39 Catalaõ, Brazil
11 Cayenne, French Guiana
73 Cheyenne, Wyoming, U.S.A.
149 Chibougamau, Quebec, Canada
128 Chicago, Illinois, U.S.A.
148 Churchill, Manitoba, Canada
117 Cirene, Libya
14 Cochin, India
3 Colombo, Ceylon
10 Cristobal, Panama
162 Cumberland Bay, South Georgia Is.
147 Dawson, Yukon Territory, Canada
19 Djakarta, Indonesia
13 Douala, Cameroon
106 Dutch Harbor, Alaska, U.S.A.
168 Eismitte, Greenland
183 Equator, Kenya
136 Escanaba, Michigan, U.S.A.
175 Fanaråken, Norway
25 Farafangana, Malagasy Republic
172 Flagstaff, Arizona, U.S.A.
61 Francistown, Bechuanaland
12 Freetown, Sierra Leone
1 Georgetown, British Guiana
166 Grønfjorden, Spitzbergen
38 Guayaquil, Ecuador
81 Hanoi, Vietnam
103 Hobart, Tasmania, Australia
33 Ibadan, Nigeria
40 Iguatu, Brazil
133 Inchon, Korea
46 In Salah, Algeria
105 Invercargill, New Zealand
152 Irkutsk, R.S.F.S.R., U.S.S.R.
159 Ivigtut, Greenland
58 Jaipur, India
5 Jaluit, Marshall Islands
118 Jerusalem, Israel
50 Karachi, Pakistan
83 Kasempa, Northern Rhodesia
55 Kayes, Mali
68 Kazalinsk, Kazak, U.S.S.R.
142 Kazan, R.S.F.S.R., U.S.S.R.
29 Key West, Florida, U.S.A.
145 Khabarovsk, R.S.F.S.R., U.S.S.R.
141 Kiev, Ukranian S.S.R., U.S.S.R.
107 Kodiak, Alaska, U.S.A.
180 La Paz, Bolivia
181 La Quiaca, Argentina
80 Lashio, Burma
67 Las Vegas, Nevada, U.S.A.
167 Laurie Island, South Orkneys
140 Leningrad, R.S.F.S.R., U.S.S.R.
178 Lhasa, Tibet
51 Lima, Peru
127 Lincoln, Nebraska, U.S.A.
169 Little America, Antarctica
170 McMurdo Sound, Antarctica
9 Madang, New Guinea
18 Manaus, Brazil
36 Mandalay, Burma
27 Manzanillo, Mexico
115 Marseille, France
72 Medicine Hat, Alberta, Canada
173 Mexico City, Mexico
56 Mogadiscio, Somalia
54 Monterrey, Mexico
171 Mt. Wilson, California, U.S.A.
132 Mukden, Manchuria
92 Nagasaki, Japan
116 Naples, Italy
87 Nashville, Tennessee, U.S.A.
146 Nemuro, Japan
84 Newcastle, Union of South Africa
130 New Haven, Connecticut, U.S.A.
88 New Orleans, Louisiana, U.S.A.
157 Nizhne-Kolymsk, R.S.F.S.R., U.S.S.R.
41 Nova Lisboa, Angola
74 Odessa, Ukranian S.S.R., U.S.S.R.
155 Okhotsk, R.S.F.S.R., U.S.S.R.
143 Omsk, R.S.F.S.R., U.S.S.R.
7 Padang, Sumatra, Indonesia
20 Papeete, Tahiti, Society Is.
119 Perth, Western Australia
156 Petropavlosk, R.S.F.S.R., U.S.S.R.
59 Poona, India
43 Port Darwin, Northern Territory, Australia
102 Port Elizabeth, Union of South Africa
121 Portland, Oregon, U.S.A.
66 Port Nolloth, Union of South Africa
123 Pôrto, Portugal
139 Prague, Czechoslovakia
111 Punta Arenas, Chile
17 Quang-Tri, Vietnam
179 Quito, Ecuador
16 Rangoon, Burma
71 Reno, Nevada, U.S.A.
108 Reykjavik, Iceland
86 Rockhampton, Queensland, Australia
131 Rostov-Na-Donu, R.S.F.S.R., U.S.S.R.
112 Sacramento, California, U.S.A.
37 Saigon, Vietnam
138 Saint Johns, Newfoundland, Canada
129 Saint Louis, Missouri, R.S.A.
158 Saint Paul Island, Alaska, U.S.A.
23 Salvador, Bahia, Brazil
4 Sandaken, North Borneo
122 San Francisco, California, U.S.A.
28 San José, Costa Rica
70 Santa Cruz, Argentina
124 Santiago, Chile
60 Santiago del Estero, Argentina
30 Santo Domingo, Dominican Republic
24 Santos, São Paulo, Brazil
113 Sevilla, Spain
96 Shannon Airport, Ireland
78 Sian (Hsian), China
176 Sonnblick, Austria
161 Stanley, Falkland Islands
69 Sufu (Kashgar), China
85 Tananarive, Malagasy Republic
76 Tehrān, Iran
109 Thorshavn, Faeroes
137 Toronto, Ontario, Canada
110 Trondheim, Norway
182 Tshibinda, Republic of the Congo
151 Turukhansk, R.S.F.S.R., U.S.S.R.
6 Uaupés, Brazil
165 Upernivik, Greenland
52 Upington, Union of South Africa
101 Valdivia, Chile
95 Vancouver, British Columbia, Canada
160 Vardø, Norway
90 Varna, Bulgaria
154 Verkhoyansk, R.S.F.S.R., U.S.S.R.
144 Vladivostok, R.S.F.S.R., U.S.S.R.
65 Walvis Bay, Southwest Africa
34 Wau, Sudan
53 William Creek, South Australia
89 Wilmington, North Carolina, U.S.A.
135 Winnipeg, Manitoba, Canada
163 Wrangel Island, R.S.F.S.R., U.S.S.R.
153 Yakutsk, R.S.F.S.R., U.S.S.R.
44 Yuma, Arizona, U.S.A.

APPENDIX III

Quantitative Analysis of Geographic Variables

ONE of the most useful of a geographer's tools in deriving the laws that relate the change of one quantity with respect to change in another related quantity is known as *bivariate analysis.* The term bivariate means simply "two-variable"; that is, two measurable physical properties engage in simultaneous variation. Reaching back into weather science for an example, refer to Figure 7.6, a graph showing how atmospheric pressure decreases with altitude. Pressure in millibars is scaled on one axis of the graph; altitude in kilometers on the other. We can say: "Pressure depends on altitude." Pressure is one variable quantity; altitude is the other. Their simultaneous variation is systematic and is represented by a smooth curve. A rather simple mathematical equation describes the simultaneous variation of pressure and altitude; it conforms to a well-known law of physics.

Independent and dependent variables

In bivariate analysis it is important to assign, where possible, the role of *cause* to one variable, of *effect* to the other. In the example cited above, altitude is considered the "cause", and pressure the "effect." Turning to another example, the rating curve of a stream (Figure 13.15), there exists a bivariate relationship between gauge height and stream discharge. Discharge is the cause; gauge height the effect. This conclusion not only seems reasonable, but the reverse relationship would be absurd, for there is no physical means whereby the gauge height can rise except in response to an increase in stream discharge.

In the rigorous language of mathematical statistics, the cause is referred to as the *independent variable* (it varies independently), while the effect constitutes the *dependent variable* (it depends upon the other variable). In mathematical statements, the independent variable is designated by the symbol X, and the dependent variable by Y. The mathematician states: "Y is some function of X." The procedure of establishing that equation which best describes the observed relationship between Y and X is referred to as *regression analysis.* The equation selected becomes a *mathematical model.*

Role of time in regression analysis

In a large proportion of the bivariate relationships studied in physical geography, the role of independent variable is played by *time.* That is, some physical property changes in magnitude as time passes. Time is scaled on the horizontal axis of the graph. For example, in a hydrograph (Figure 13.16), runoff, the dependent variable, changes with time. Obviously, it would be absurd to say that time depends upon runoff. In this example, time begins with a zero value at an arbitrary starting point. Where X represents time and Y represents runoff, both X and Y have zero values at the *origin* of the graphic coordinates, located at the lower left-hand corner of the graph.

Simple harmonic motion

Where time is referred to the calendar, the horizontal scale of the graph is laid off into months and years. In the climate graphs of Chapters 15 through 17, and in the water-budget graphs of Chapter 12, time spans one year only and represents an indefinitely repetitive cycle. The dependent quantities of temperature, precipitation, and evapotranspiration represent the averages of many yearly cycles superimposed. Such a cyclic relationship requires yet another form of mathematical analysis involving angular relationships of a full circle. Tide curves, exhibited in Chapter 6, also fall into this category.

In the ideal mathematical model for the simplest of tide curves, the lunar half-day of about $12^h\,24^m$ is considered as equivalent to the sweep of a radius through 360°, a full circle. Time in hours and minutes is replaced by a *phase angle* in degrees (Figure AIII.1). Height of water, scaled on the vertical axis, is determined by the *sine* (or *cosine*) of the phase angle and constitutes the *amplitude.* Within a full lunar day, there will be two low points and two high points, equally spaced throughout the lunar day. This type of regression model of amplitude on time is known as *simple harmonic motion,* and is found widely throughout those fields of physics dealing with wave motions.

Actual tide curves represent the superimposition

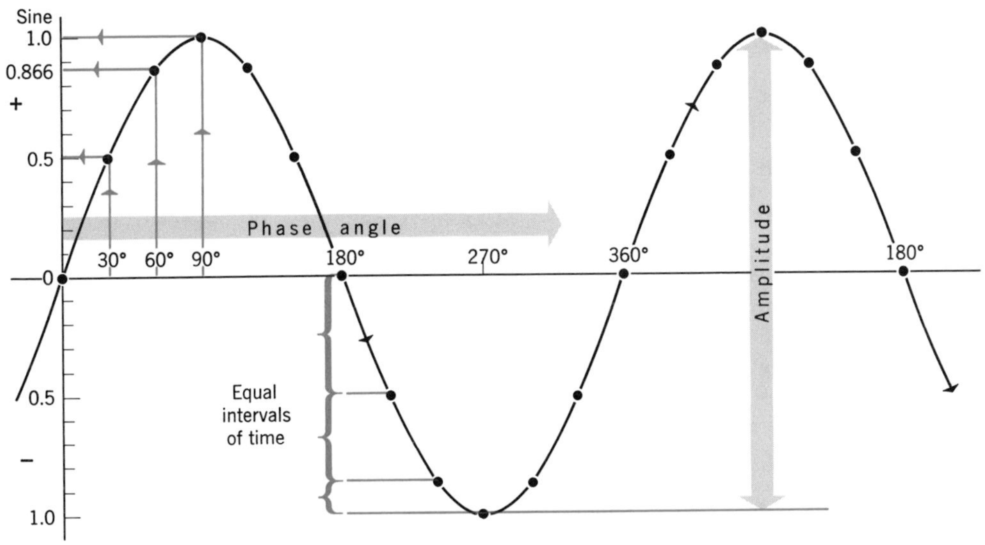

Figure A III.1 An ideal tide curve illustrates simple harmonic motion. Compare with Figure 6.8.

of two or more sine curves, differing in both phase and amplitude. Thus the distinctive curve of tropic tides (Figure 6.10) results from the curve of a full lunar day being superimposed on the curve of a half lunar day. Ideal tidal curves for a given place are predicted by machines that generate the combined curve for a number of constituent astonomical cycles.

Examples of the curve of simple harmonic motion will be found in the annual cycle of incoming solar radiation (Figure 8.5).

The annual cycles of air temperature variation shown in climate graphs throughout Chapter 15, 16, and 17, in some instances, are well approximated by the curve of simple harmonic motion.

Correlation of two dependent variables

It should be further recognized that two physical quantities can vary simultaneously in a meaningful and systematic way without either quantity having the role of cause. Take, for example, the data of Table 1.1, in which divergence of a light ray from the earth's surface is related to distance along the earth's surface. Divergence and distance are obviously two related variables, and their strict mathematical relationship is obvious. Yet both variables are merely geometrical properties and there is no physical influence of one upon the other.

In general, where two simple geometrical properties are involved in a relationship, no roles of cause or effect can be assigned. It can be said only that a *correlation* exists. Moreover, both properties are likely to be controlled by yet another variable property. For example, in the hydraulic geometry of a stream channel (Chapter 13), both the depth, d, and the width, w, normally increase as discharge increases and decrease as discharge falls (see Figure 25.8). Yet a graph of depth plotted against width (or *vice versa*) would show a simple and meaningful bivariate relationship. In fact, both depth and width are dependent variables of a third, independent variable (discharge) that controls them both.

Methods of mathematical statistics used in correlation analysis are different in a number of respects from regression analysis and are not discussed here.

Constant-ratio scales

Only three of the simplest and most important mathematical relationships between an independent variable, X, and its dependent variable, Y, need be considered to give understanding to a surprisingly large number of bivariate relationships that will be found in physical geography. Most of the remaining cases fall into special cyclic functions of time, discussed in earlier paragraphs.

Our approach is through arithmetic of numbers and presupposes no mathematical competence beyond the simplest expressions of algebra.

Consider first that there exist two forms of graphic scales, each entirely different in meaning from the other. The first scale is familiar to all; it is found on the ordinary yardstick (meter stick) marked off into inches or centimeters. All of our lives we have been accustomed to measuring the lengths of things by this scale, which has exactly equal units throughout its entire length. This *arithmetic scale* is the simplest form of scale by which quantities can be plotted on a graph.

The second graphic scale is a *constant-ratio scale.*

The marks that designate successive integers (1, 2, 3, 4, 5, etc.) are not equally spaced on this scale, as they are on the arithmetic scale. Instead, the successive spacing between integers decreases rather rapidly as the numbers increase in value. Figure AIII.2 shows both arithmetic and constant-ratio scales from 1 through 100. The total distance span from 1 to 100 on this page is exactly the same for both.

To understand the significance of the constant-ratio scale, undertake the following operation. Obtain a pair of dividers, or a drawing compass. Set one point of the dividers on mark *1* on the constant-ratio scale; set the other point to mark *2*. Note that the *ratio of increase* from *1* to *2* is a two-fold increase (factor of two, or doubling). Next, move the divider up the scale, keeping one point on mark *2*. Observe that the other divider point falls on mark *4*. Again, the ratio of increase is two (doubling). Successively, the divider points will fall upon marks 8, 16, 32, and 64. Repeat the experiment, starting with divider points on marks *1* and *3*. It will be found that the successive points fall on marks 9, 27, and 81. Here the ratio of increase is *three* (factor of three, or tripling). Experiment with any other initial span of the dividers, for example, points on marks *1* and *1.5*. Here the successive points fall on *2.25, 3.37, 5.06, 7.50,* and so on, the ratio of increase being 50 percent.

On a constant-ratio scale, therefore, equal distances on the scale represent equal proportions (ratios) of increase. Worked in reverse, equal distances of descent on the constant-ratio scale result in equal proportions of decrease. For example, starting with one divider point on mark *100,* set the other point on mark *50,* a decrease of 50 percent. The divider point will fall next on mark *25*; thereafter on *12.5, 6.25, 3.125, 1.5625,* and so on.

Students of mathematics will recognize the constant-ratio scale as a *logarithmic scale.* Although a knowledge of logarithms and exponents is essential for serious study of bivariate regression analysis, this discussion attempts to explain the basic concepts without the need for understanding these mathematical tools.

When the same procedure with dividers is carried out on the arithemetic scale, it is found that ratio of increase changes as equal increments of length are taken. For example, using the dividers, place one point on mark *10* of the arithmetic scale, and the other point on mark *40.* The ratio of increase is fourfold, or 400 percent. Moving the dividers up to the next segment, the point falls on mark *70,* for a 75 percent increase. The next point falls on mark *100,* for a 43 percent increase. The ratio of increase falls off rapidly with larger numbers on the scale. Similarly, when one descends the arithmetic scale in equal-distance intervals, the ratio of decrease becomes rapidly larger.

Given two kinds of scales and two coordinates to use on a rectangular graph, four graphic combinations are possible. Each constitutes a model of behavior of one variable with respect to the other.

X-Axis Scale (Independent Variable)	Y-Axis Scale (Dependent Variable)	Regression Model
1. Arithmetic	Arithmetic	Simple arithmetic
2. Arithmetic	Constant-ratio	Exponential
3. Constant-ratio	Arithmetic	Logarithmic
4. Constant-ratio	Constant-ratio	Power

Each of the four models is described mathematically by a simple equation, representing an ideal relationship between the two variables.

In terms of Y, the dependent variable, and X, the independent variable, regression equations 1, 2, and 4 are as follows. (Equation 3 is rarely needed in analyzing the data of physical geography.)

1. Simple arithmetic

$$Y = a + bX \quad \text{(positive form)}$$
$$\text{and } Y = a - bX \quad \text{(negative form)}$$

2. Exponential

$$Y = ae^{bX}$$
$$\text{or } \log Y = \log a + bX \quad \text{(positive form)}$$
$$\text{and } Y = ae^{-bX}$$
$$\text{or } \log Y = \log a - bX \quad \text{(negative form)}$$

4. Power

$$Y = aX^b$$
$$\text{or } \log Y = \log a + b \log X \quad \text{(positive form)}$$
$$\text{and } Y = aX^{-b}$$
$$\text{or } \log Y = \log a - b \log X \quad \text{(negative form)}$$

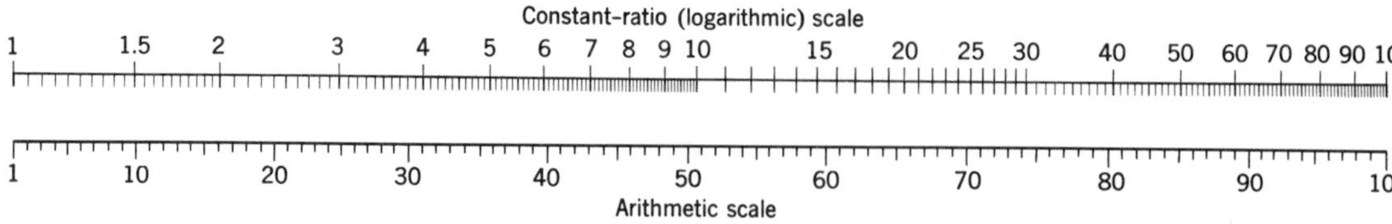

Figure A III.2 Arithmetic and constant-ratio (logarithmic) scales compared.

The symbol a denotes a numerical constant. The symbol b is also a constant for a given set of data and is known as the *regression coefficient.* The symbol e in the exponential equation is a constant (base of natural logarithms) with a value of 2.71828. The terms *log Y* and *log X* denote the *logarithm of Y* and *logarithm of X*[1]. For those unfamiliar with logarithms, the expression "log Y" can be taken as an instruction to use a constant-ratio (logarithmic) scale for plotting values of Y.

Figure AIII.3 shows graphs, representing regression equations 1, 2, and 4. Sloping straight lines on each graph represent the plotted forms of the equations. Where the line slopes upward to the right, the positive form is shown; where the line slopes downward to the right, the negative form is shown.

The constant a in each equation controls the position of the regression line, whereas the constant b controls the inclination, or *slope,* of the regression line. By controlling both a and b, the regression line can be made to take any desired position and orientation on the coordinate field of the graph.

Selection of a regression equation

It is the goal of the geographer, when working with two related variables, to ascertain which regression model best describes the numerical data that he has derived by observation. Although his data will show irregularities and variations that depart from any simple mathematical model, the plot often shows that one model is distinctly superior to others in correspondence with the observations.

Without using sophisticated techniques of mathematical statistics, some elementary procedures can be followed in selecting a regression model to fit observed data. If there is no prior information or experience for guidance, the simplest procedure is to plot the same data on each of the three kinds of graphs. If the plotted points fall on a strongly curved line, the model is rejected. If the points form an essentially straight line, the model is accepted. If extreme irregularity (scatter) of points appears, the relationship between the two variables may be of questionable significance and no model can be selected.

In Figure AIII.4 the procedures are applied to the data of Table 1.1. Because the figures for divergence and distance are themselves calculated from the mathematical equation given on page 8, we can be assured that the plotted points will form either a simple curve or a true straight line. From

[1] In equation 2, the natural logarithm to the base e is correct for the equivalent logarithmic form. In practice, logarithms to the base 10 are generally used throughout in regression analysis.

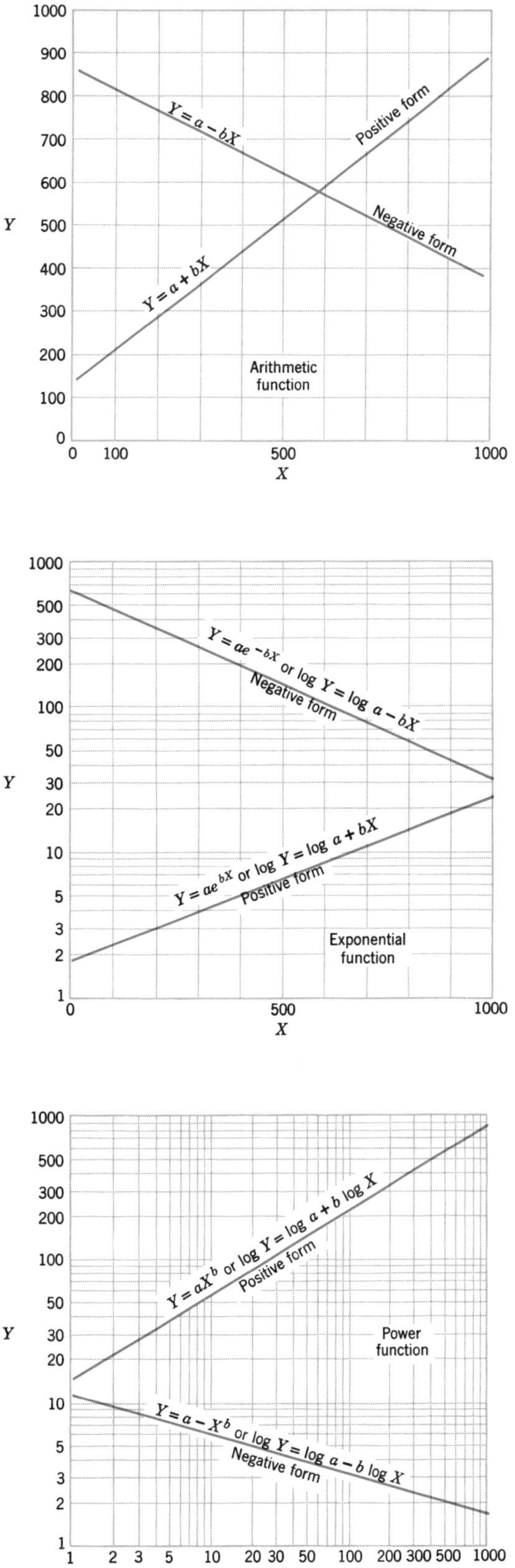

Figure A III.3 Graphs of arithmetic, exponential, and power regression equations.

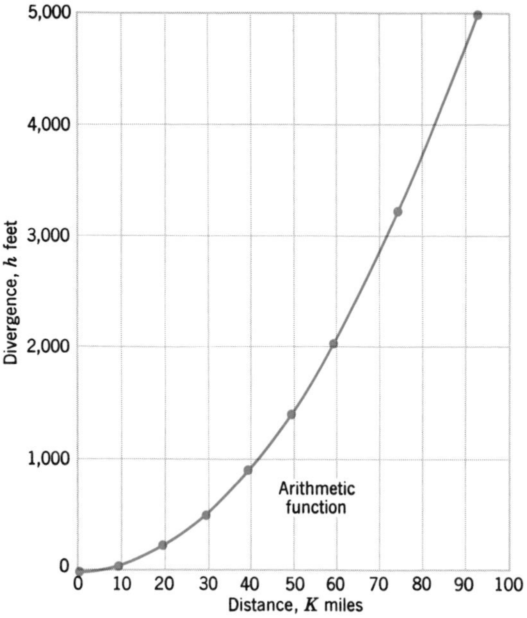

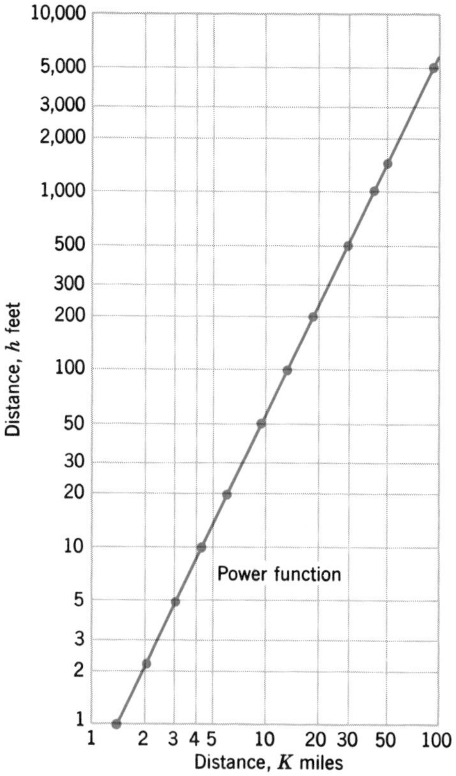

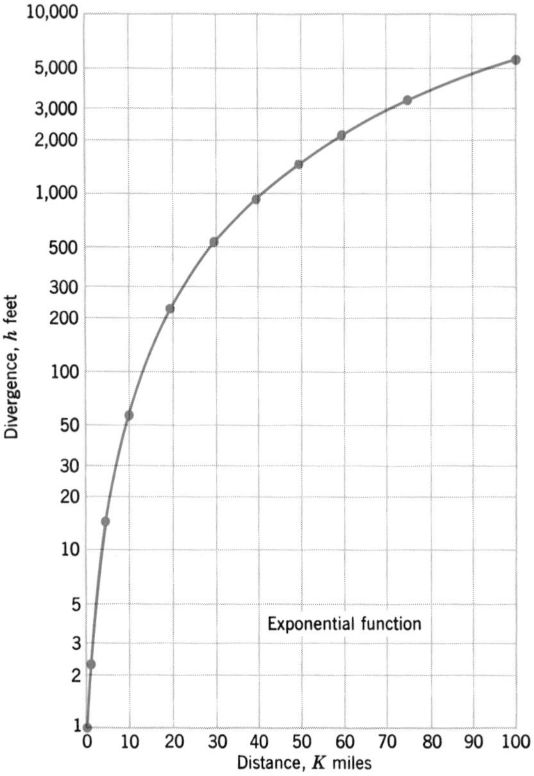

Figure A III.4 Graphs of divergence of a light ray with horizontal distance plotted in three forms. Refer to Table 1.1 for data.

the fact that strongly curved lines result for models 1 and 2, but a straight line results for model 4, the last, or power regression model, is chosen. (A strongly curved line also results with model 3, not shown.)

Applications in geography

The same general principles of quantitative data analysis apply not only throughout the branches of physical geography, but to many phases of economic and urban geography as well. Not only can the same basic methods of regression analysis and model selection be transferred from one field to another, they also provide a means for relating a variable quantity from one field of study with a variable quantity in quite a different field. For example, some measure of economic development might prove to be related through a causative mechanism to some variable quantity in the physical environment, such as soil moisture, or land slopes.

Most problems in geography concern more than two variables. Typically, several independent variables contribute to the observed value of a single dependent variable. Although far more complex forms of regression analysis and other techniques of multiple-variable analysis are required to cope with such investigations, all are built upon elementary concepts, some of which have been suggested here.

APPENDIX IV

Landforms of the World—the Murphy System

The Murphy system of landform classification[1]

The Murphy system of landform classification uses three levels, or categories, of information in successive application to identify a landform type in terms of its geologic origin and rock composition (structural regions), the configuration of its surface (topography), and finally the nature of the geomorphic process by which it has been shaped (erosional or depositional landscapes). The threefold basis of the genetic approach to landform study—structure, process, and stage—is included in the first and third categories of the classification system. A particularly valuable attribute of the genetic approach is that a highly trained and experienced geomorphologist can use his background of knowledge to predict or anticipate many characteristic details of the landforms that are not implicitly stated in the definitions of the classes.

The empirical ingredient of the Murphy system is found in the second level of classification, namely topography. Here subdivision of geometrical properties of the land surface follows strict numerical definitions. Elevation above sea level and local relief (difference in elevation between highest and lowest points in adjacent locations) form the basis for defining classes. The empirical approach to description of the topography lends an element of useful and unambiguous information to the classification system.

The entire system uses three sets of letter symbols, the first to represent structural regions, the second to represent topographical classes, and the third to indicate the kinds of erosional or depositional landscapes.

[1] Richard E. Murphy, 1968, *Landforms of the world*, Annals Map Supplement Number Nine, Annals, Association of American Geographers, 58(1). Richard E. Murphy, 1967, *A spatial classification of landforms based on both genetic and empirical factors—a revision*, Annals, Association of American Geographers, Vol. 57, No. 1, pp. 185–186, 1967. Reproduced by permission of the author and the Association of American Geographers.

Structural regions

Seven structural regions are recognized and are designated by the capital letters *A, C, G, L, R, S,* and *V*, defined as follows.

A *Alpine system.* World-girdling system of mountain chains formed since the Jurassic period. Faulted areas, plateaus, basins, and coastal plains enclosed by such ranges are included in the system.

C *Caledonian (or Hercynian or Appalachian) remnants.* Remains of mountain chains and ranges formed during the Paleozoic and Mesozoic eras, prior to the Cretaceous period and experiencing no orogeny since then, although epeirogenic movements may (and often have) occurred. In some cases, only worn-down roots remain. Faulted areas, plateaus, basins, and coastal plains enclosed by these remnants are included with them.

G *Gondwana shields.* Areas of stable, massive blocks of the earth's sialic crust, lying south of the great east-west portion of the Alpine system, where Precambrian rocks form either the entire surface rock or where Precambrian rocks form an encircling enclosure with no gap of more than 200 mi (320 km) between outcroppings or covering extrusives and within which crystalline rocks form more than 50 percent of the surface rock. These shields have not been subject to orogeny since the Cambrian period.

L *Laurasian shields.* Areas of stable, massive blocks of the earth's sialic crust lying north of the great east-west portion of the Alpine system. (Remainder of definition same as in *G*, above.)

R *Rifted shield areas.* Block-faulted areas of shields forming grabens together with associated horsts and volcanic features.

S *Sedimentary covers.* Areas of sedimentary layers which have not been subjected to orogeny and which lie outside either the crystalline rock enclosures of the shields or the enclosing mountains and hills of the Alpine or older orogenic systems. These areas of sedimentary rock form continuous covers over underlying structures.

V *Isolated volcanic areas.* Areas of volcanoes, active or extinct, with associated volcanic features, lying outside the Alpine or older mountain systems and outside the rifted shield areas.

Figure AIV.1 shows the distribution of the seven structural regions, together with major oceanic rift and fault lines, undersea axial connections of the Alpine system, and continental shelf areas.

Topographical regions

Six classes of topography, represented by the capital letters *P, H, T, M, W*, and *D*, are defined as follows (metric equivalents are approximate).

P *Plains.* Surfaces with local relief less than 325 ft (100 m). On the marine side the surface slopes gently to the sea. Plains rising continuously inland may attain elevations of high plains, over 2000 ft (600 m).

H *Hills and low tablelands.* Hill areas have local relief more than 325 ft (100 m) but less than 2000 ft (600 m). At the oceanic shoreline, however, local relief may be as low as 200 ft (60 m). A low tableland is an area less than 5000 ft (1500 m) in elevation, with local relief less than 325 ft (100 m), but which (unlike plains) either does not reach the sea or, where it does, terminates in a bluff at least 200 ft (60 m) high. A tableland may also terminate in a similar bluff overlooking a low coastal plain.

T *High tablelands.* Upland surfaces over 5000 ft (1500 m) in elevation having local relief less than 1000 ft (300 m), except where cut by widely separated canyons.

M *Mountains.* Areas of steep slopes with local relief more than 2000 ft (600 m).

W *Widely spaced mountains.* Mountains which are discontinuous and stand in isolation with intervening areas having local relief less than 500 ft (150 m).

D *Depressions.* Basins surrounded by mountains, hills, or tablelands which abruptly delimit the basins.

Figure AIV.2 shows the distribution of the six topographical regions.

Erosional and depositional landforms

The kind of geomorphic process acting currently, or relatively recently in geologic time, to shape the landscape into its present form provides the basis for five classes of areas, indicated by lower-case letters *h, d, g, w*, and *i*, and defined as follows.

h *Humid landform areas.* Areas in which the pattern of permanent streams has a density of at least one stream in every 10 mi (16 km) traverse distance, and which have not been subject to glaciation since the beginning of the Pleistocene epoch.

d *Dry landform areas.* Areas in which the pattern of stream density is more sparse than one stream in 10 mi (16 km), and which have not been subject to glaciation since the beginning of the Pleistocene epoch. Some karst as well as arid areas are included in this category.

g *Glaciated areas.* Areas covered by glacial ice at some time since the beginning of the Pleistocene epoch, but earlier than the Wisconsin and Würm glaciations. The symbol *g* is also used for undifferentiated glaciated areas.

w *Wisconsin and Würm glaciated areas.* Areas covered by glacial ice during or since the Wisconsin and Würm glaciations but now free of glacial ice.

i *Icecaps.* Areas covered by glacial ice at present.

Figure AIV.3 shows the distribution of the five classes of erosional and depositional landforms.

Combined landform classes

Under the Murphy system, the three categories of classification are superimposed to yield a complete world landform map (Plate 5). To designate a particular area within the complete system, the code letter of each category is stated in sequence. For example, the Colorado Plateau is symbolized as *ATd*, meaning that it is an elevated tableland over 5000 ft (1500 m) in elevation, enclosed within the Alpine system, and characterized by landforms developed under conditions more arid than humid. The Congo basin is symbolized as *GDh*, indicating a depression in the Gondwana shield subject to geomorphic processes of a humid climate. Central Poland is designated as *SPg*, a sedimentary plain which has been subjected to Pleistocene glaciation.

As with all attempted classifications of climates, soils, and natural vegetation types, any classification of landforms is difficult to implement in the form of a world map because of lack of sufficient information. Over vast areas of the continents information on bedrock geology and surface topography is only generalized and of questionable reliability because of the reconnaissance nature of field surveys. Consequently, as new information of increasing detail of scale and accuracy becomes available, the world map must be modified. In the light of new information, minor changes in the categories and their definitions may also be desirable. Nevertheless, the existing world landform map (Plate 5) represents a vast quantity of soundly established geologic and topographic information, organized into a logical and meaningful classification system.

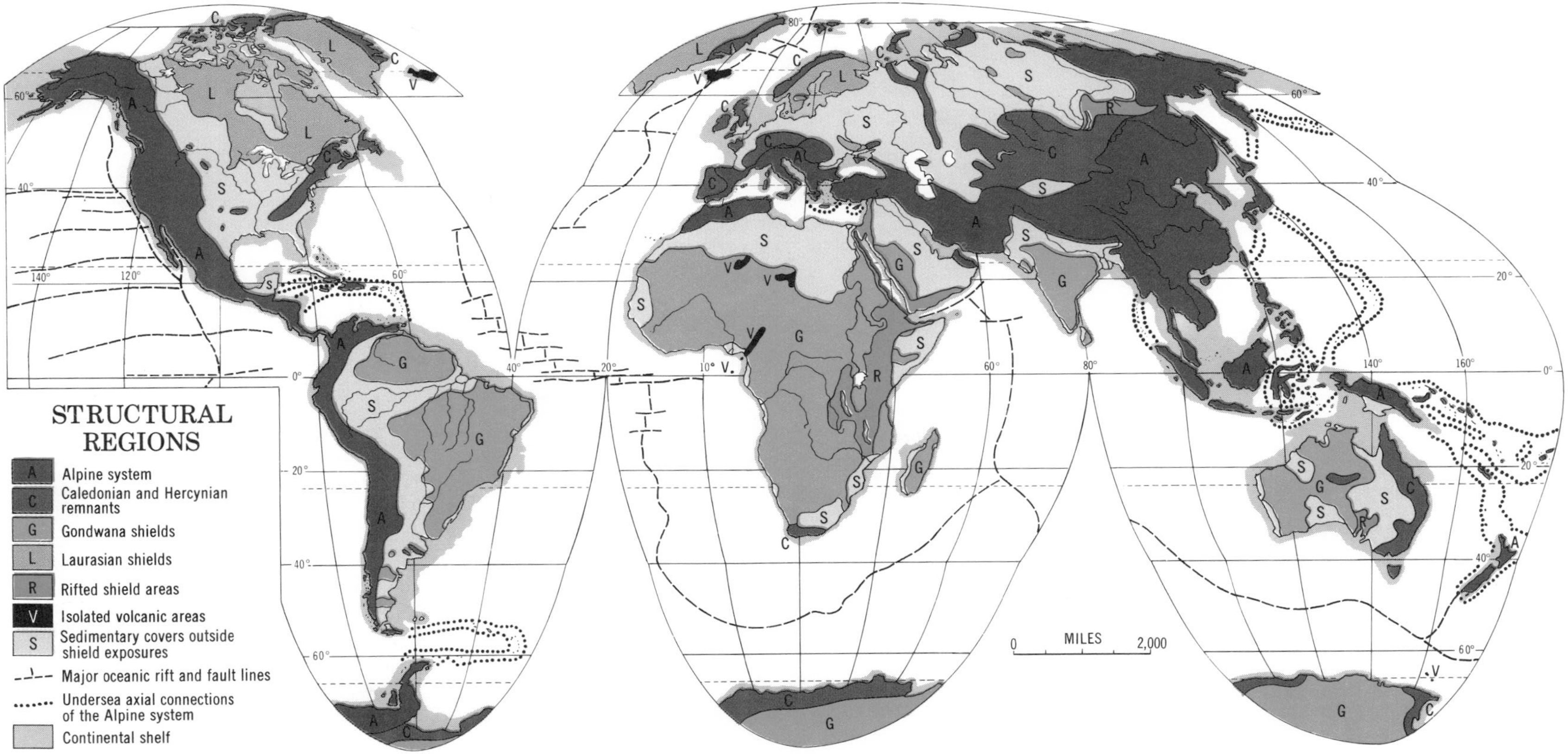

Figure A IV.1 World structural regions. (From R. E. Murphy, 1968, Annals, A.A.G., Map Supplement No. 9. Based on Goode Base Map. Copyright by the University of Chicago. Used by permission of the University of Chicago Press.)

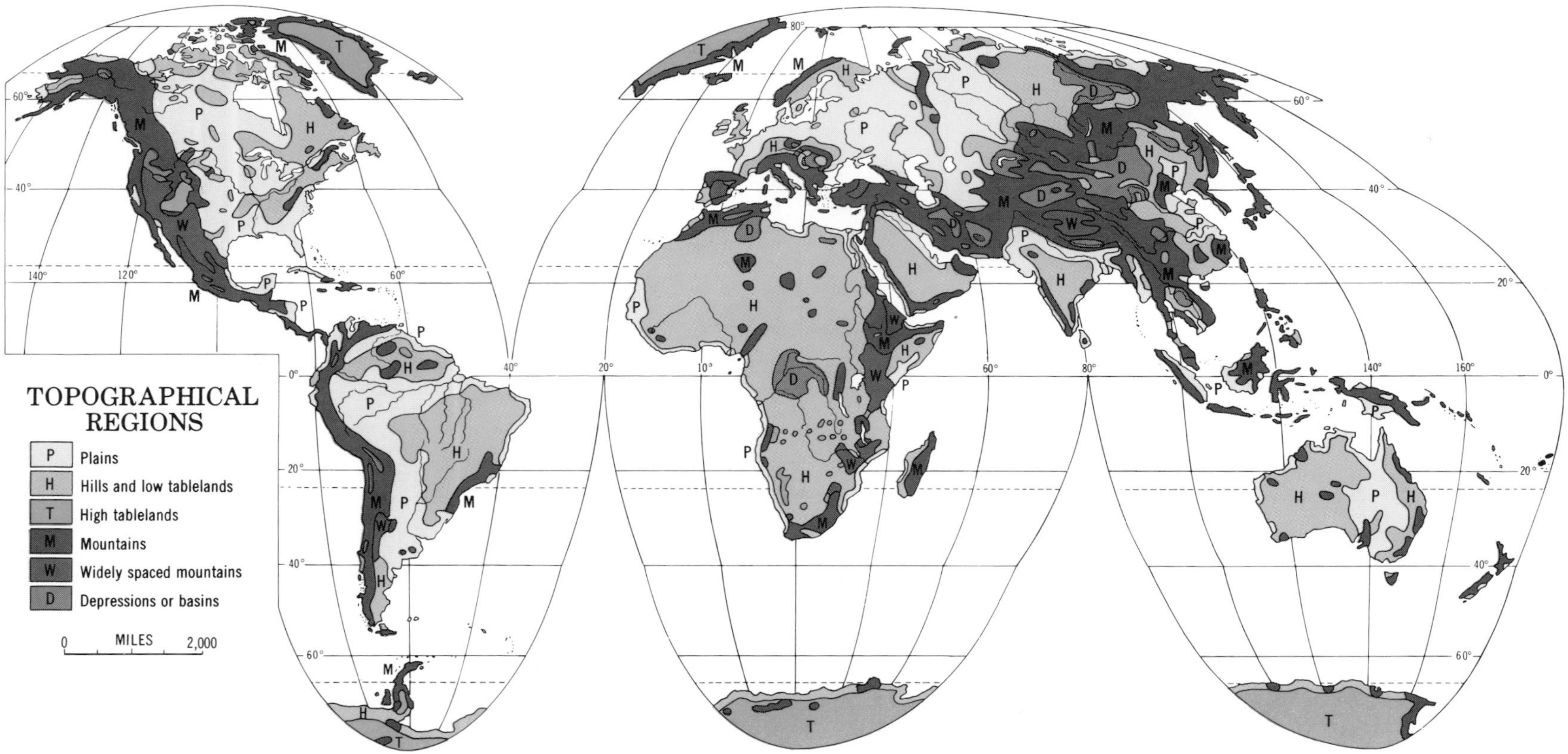

Figure A IV.2 World topographical regions. (From R. E. Murphy, 1968, Annals, A.A.G., Map Supplement No. 9. Based on Goode Base Map. Copyright by the University of Chicago. Used by permission of the University of Chicago Press.)

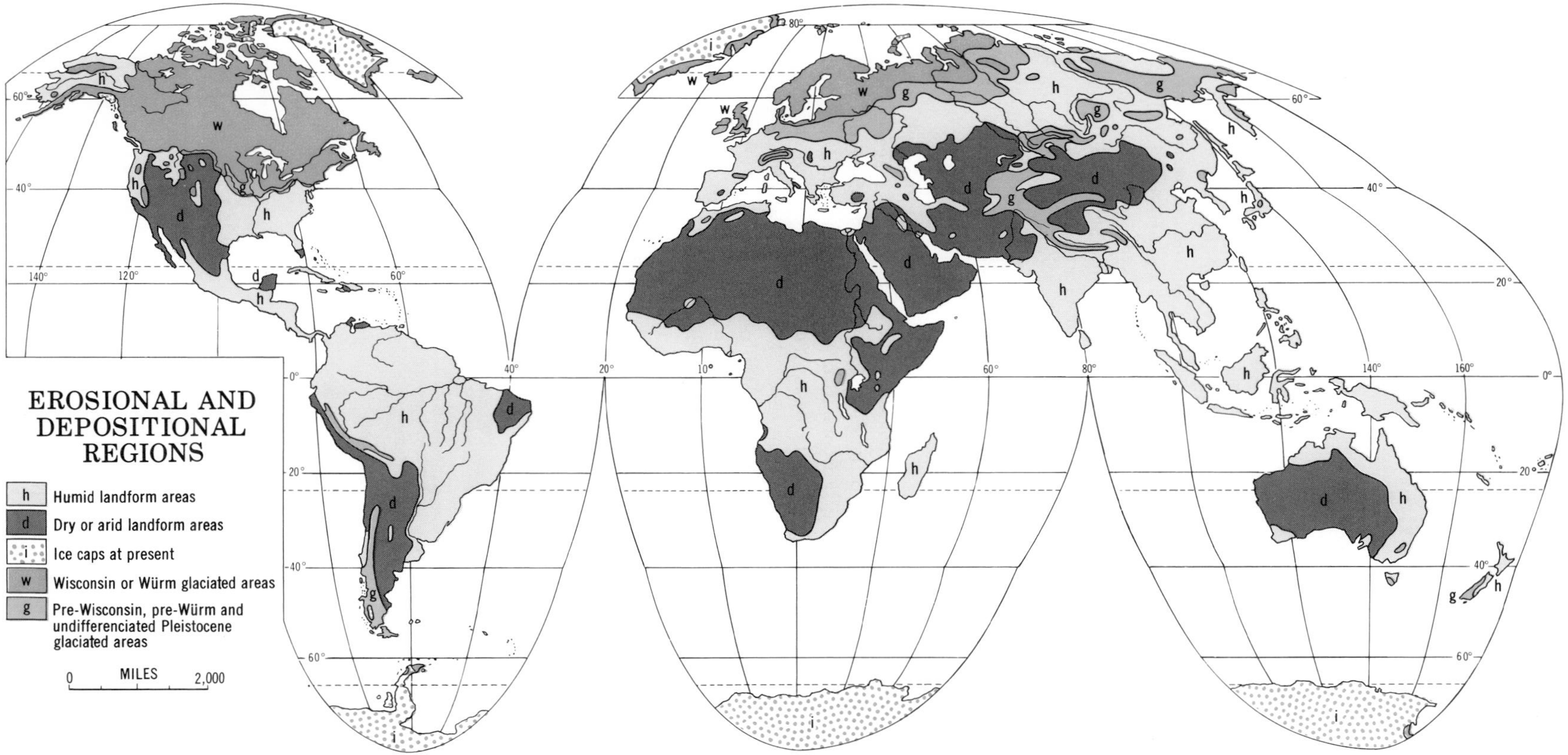

Figure A IV.3 World distribution of erosional and depositional landforms. (From R. E. Murphy, 1968, Annals, A.A.G., Map Supplement No. 9. Based on Goode Base Map. Copyright by the University of Chicago. Used by permission of the University of Chicago Press.)

APPENDIX V

Remote Sensing Techniques in Geographical Research

IN various branches of geography, as in other sectors of the earth sciences, a new technical discipline called *remote sensing* has expanded rapidly within the past decade and is adding greatly to our ability to perceive and analyze the physical, chemical, biological, and cultural character of the earth's surface. In its broadest sense, remote sensing is the measurement of some property of an object by using means other than direct contact through the observer's senses of touch, taste, or smell. Hearing and seeing are remote sensing activities of organisms and depend upon reception of wave forms of energy transmitted from the object to the observer. Use of an ordinary camera represents a form of remote sensing in which a detecting instrument senses and records the light waves reflected from objects. In its more restricted operational meaning, remote sensing refers to gathering information from great distances and over broad areas, usually through instruments mounted on aircraft or orbiting space vehicles.

All substances, whether naturally occurring or synthesized by man, are capable of reflecting, absorbing, and emitting energy in forms that can be detected by instruments known collectively as *remote sensors.* We can recognize four basic classes of remote sensors: (1) *Radiometers* measure the energy level of portions of the electromagnetic spectrum. (2) *Audiometers* measure intensities of sound waves. (3) *Magnetometers* measure minute variations in the strength of the earth's magnetic field. (4) *Gravimeters* measure minute variations in the acceleration of gravity.

Although the first class of sensors, the radiometers, are by far the most important tool of remote sensing for geographical purposes, the last three are of very great importance in research and exploration in geology and geophysics. For example, the microphone, a form of audiometer, is used to receive sound waves emitted from the hull of a research vessel and reflected back from the ocean floor. By this device, the configuration of the ocean floor has been mapped in great detail (See Chapter 23). Magnetometers towed behind research vessels detect magnetic "stripes" in the basalts of the ocean floor, enabling the amount and rate of sea-floor spreading to be determined (See Chapter 23). Gravimeters used over land surfaces can detect the presence of rock masses of varying density and thus assist in the search for ore bodies or petroleum-bearing structures. Gravimeters carried on orbiting satellites allow refined determinations of the earth's oblateness and geodetic data of the moon and planets. However, in the remainder of this explanation of remote sensing we shall deal only with the use of radiometers to detect and record incoming energy from portions of the electromagnetic spectrum.

The electromagnetic spectrum

For those not familiar with basic physics, it will be necessary for us to review a few of the basic principles governing electromagnetic radiation. For practical purposes, electromagnetic radiation can be described as a *wave form* of energy. Simple ocean waves (Figure 32.1) as well as tide curves (Chapter 6) illustrate some of the principles of wave motion. In Appendix III, under the discussion of simple harmonic motion, you will find a description of the ideal *sine wave* (See Figure AIII.1). Wave motion of the electromagnetic spectrum can be visualized as taking the sine-wave form.

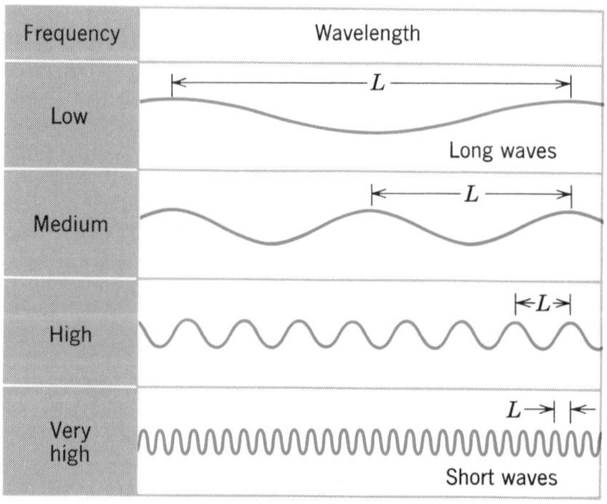

Figure AV.1 Relationship of wavelength to frequency. (From A. N. Strahler, 1971, *The Earth Sciences*, Harper and Row, New York, Figure 6.5.)

Figure AV.1 illustrates how electromagnetic waves differ in dimensions throughout their entire range, or *spectrum*. Waves are described in terms of either wavelength or frequency. *Wavelength*, *L*, is the actual distance between two successive wave crests (or two successive wave troughs). Metric units of length are always used, for example, centimeters or meters. Next, we know that all waves of the electromagnetic spectrum travel at the same velocity, a quantity known as the *speed of light*. This speed is approximately 300,000 km per second (about 186,000 mi per second). The number of waves passing a fixed point in a unit of time (one second) is known as the *frequency*. Frequency depends upon the wavelength. Long waves have *low frequency;* short waves will have *high frequency*. Frequency is stated in terms of *cycles per second;* the unit of measure is the *hertz*. One hertz is a frequency of one cycle per second; one *megahertz* is a frequency of one million cycles per second. Knowing the speed of light to be 300,000 km per sec, it is easy to calculate that a frequency of one megahertz is associated with a wavelength of 300 m. Equivalents between wavelength and frequency can be read directly on the scales of Figure AV.2, which shows the divisions of the electromagnetic spectrum, ranging from the shortest waves at the left to the longest waves at the right. The scale is a logarithmic scale (constant-ratio scale). Refer to Appendix III and Figure AIII.2 for an explanation of the logarithmic scale.

At the short end of the spectrum are *gamma rays* of extremely high frequency and short wavelength. It is customary to describe these short wavelengths in units of *Angstroms*. One Angstrom is equal to 0.000,000,01 cm (10^{-8} cm). Gamma rays are shorter than 0.03 Angstroms. Next come the *X-rays*. The shorter X-rays, described as "hard," fall in the range 0.03 to about 0.6 Angstroms. The longer X-rays, described as "soft," range from about 0.6 to about 100 Angstroms; they grade into *ultraviolet rays*, which extend to about 4000 Angstroms. For the visible light spectrum it is customary to switch to a longer unit of length, the *micron*. One micron equals 0.0001 cm (10^{-4} cm); and one micron thus equals ten thousand Angstroms. The term *micrometer* can be used in place of micron. The *visible light* portion of the spectrum begins with violet at 0.4 microns. Colors then grade successively through blue, green, yellow, orange, and red, reaching the end of the visible spectrum at about 0.7 microns. In Figure AV.2 the visible light spectrum has been expanded to show the limits of the various colors.

Next in the spectrum comes the *infrared region*, consisting of wavelengths from about 0.7 microns to about 300 microns. Infrared rays are referred to in Chapter 8 as comprising *longwave radiation* emitted by the earth's surface (See Table 8.2 and

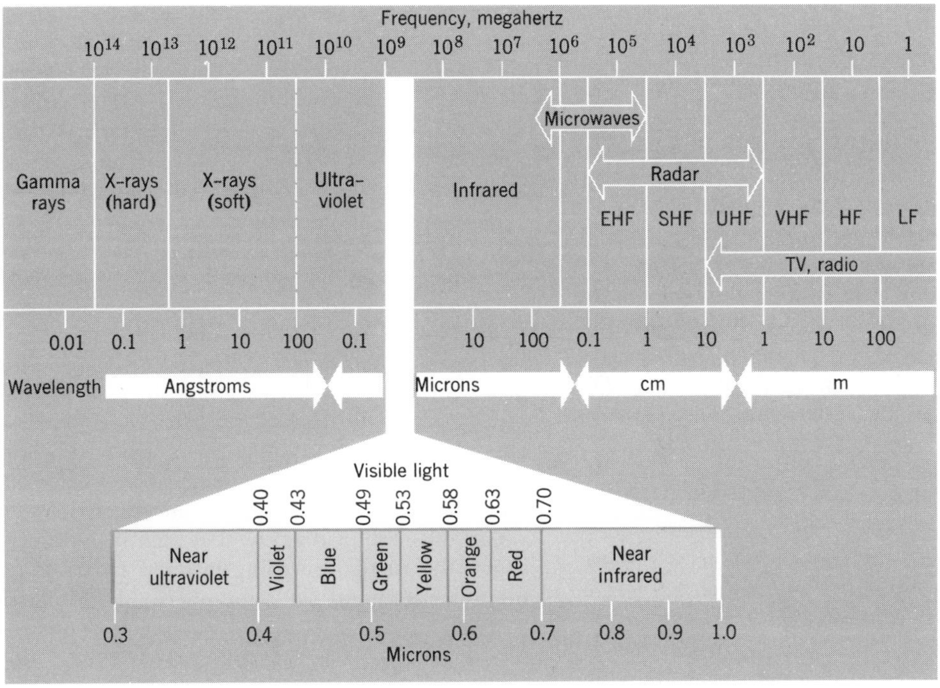

Figure AV.2 The electromagnetic radiation spectrum.

Figure 8.8). These rays are not visible, but can sometimes be felt as "heat rays" from a hot object.

Infrared rays grade into yet longer wavelengths referred to as *microwaves.* Shifting to centimeters, we can place the microwave region as between about 0.03 cm and about 1 cm. Most persons are familiar with microwaves as the form of energy used in the microwave oven for very rapid heating or cooking of foods. Microwaves are also used in direct-line transmission of messages from one tower to another across country. Within the microwave region is the *radar* region beginning at about 0.1 cm and extending through to about 100 cm (1 m). Radar systems are actually microwave sensor systems. Frequencies at which radar systems operate grade into television and radio frequencies, the latter extending into wavelengths exceeding 300 m.

As shown in Figure AV.2, the radar-TV-radio spectrum is divided into equal logarithmic units according to wavelength. Thus each division point is greater by one power of ten than the next lower division point. The abbreviations shown in Figure AV.2 have the following meanings:

EHF	Extremely high frequency	0.1 to 1 cm
SHF	Superhigh frequency	1 to 10 cm
UHF	Ultrahigh frequency	10 to 100 cm (1 m)
VHF	Very high frequency	1 to 10 m
HF	High frequency	10 to 100 m
LF	Low frequency	Longer than 100 m

Remote sensing makes extensive use of the ultraviolet, visible, infrared, microwave, and radar, portions of the spectrum. The longer wavelengths are largely used in TV and radio communications systems.

Radiation and temperature

To understand how electromagnetic radiation is useful in remote sensing you need to know certain of the underlying physical laws governing the radiation of energy in relation to the temperature of a radiation-emitting object. First, any substance whose temperature is above absolute zero (−273°C) emits electromagnetic radiation. The total energy emitted for each unit area of surface increases with increased temperature. Thus one square centimeter of the sun's surface at the extremely high temperature of several thousand degrees will emit a vastly greater quantity of energy each second than a square centimeter of the surface of the distant planet Pluto, which has a temperature close to absolute zero. Moreover, the radiation spectrum of emission of a cold surface differs greatly from that of a hot surface. For the cold surface, most of the energy emitted is in the long wavelengths; for the hot surface most is concentrated in the short (visible and ultraviolet) wavelengths. Comparing the radiation properties of the sun with that of the earth shows how this principle applies.

Figure AV.3 is a graph comparing the intensity of emission of energy of sun and earth. A logarithmic scale of wavelengths is used, as in Figure AV.2. The vertical scale, which is also logarithmic, shows intensity of energy emission within each micron width of spectrum. For both sun (left) and earth (right) smooth, ideal energy curves are shown by dashed lines of arched form. These represent the curves for ideal bodies which are perfect radiators of energy; they are called *black bodies* by the physicist. A black body will not only absorb all radiation falling upon it, but it will also radiate energy in a manner solely dependent upon its surface temperature. The sun closely resembles a black body with a temperature of about 6000°K.[1] This ideal body has a peak intensity of energy output in the visible light region. The actual radiation intensity curve of the sun is shown by the solid line just below the dashed line; it has minor irregularities and reduced intensity in the ultraviolet region. For the earth, the ideal black-body radiation curve is for a planet with average surface temperature of 300°K (27°C, 80°F). This curve lies entirely within the infrared region and peaks at about 10 microns. Notice that the peak is much lower than for the sun (about one-fifth as intense). The area under the earth curve is much smaller, meaning that the total energy emitted by a square centimeter of area is very much less than from a corresponding area of the sun's surface. This relationship is stated in strict terms by the *Stefan-Boltzmann law:* the total energy radiated by each unit of surface per unit of time varies as the fourth power of the absolute temperature (°K).

Referring again to Figure AV.3, we find beneath the arched black-body curves of both sun and earth deeply notched curves showing the intensity of energy measured after passage through the earth's atmosphere. The left-hand curve shows the solar spectrum as depleted by passage of the solar beam through the earth's atmosphere. (This depletion process is described in Chapter 8.) Practically all ultraviolet rays have been removed, while certain bands of the infrared region have been cut out through absorption by atmospheric water vapor and carbon dioxide. The right-hand curve shows the earth's outgoing infrared radiation at the outer limits of the atmosphere. Much of the

[1] *K* stands for *absolute temperature* in *degrees Kelvin.* The degrees are of the same unit value as in the Celsius scale, but the zero point of the Kelvin scale is at −273°C, a value called *absolute zero.*

outgoing infrared energy has been absorbed by water vapor and carbon dioxide in the atmosphere. However, in certain wavelength bands, known as *windows*, energy escapes to outer space. Important windows occur between 3 and 5 microns and between 8 and 20 microns. These windows represent bands within which energy can best be received by a sensor operating from an orbiting space vehicle. Sensor systems must then be adjusted to be sensitive in those wavelengths capable of passing freely through the atmosphere.

Sensing systems

Two classes of electromagnetic sensor systems are recognized: passive systems and active systems. *Passive systems* measure radiant energy reflected or emitted by an object. Mostly, this energy falls in the visible light region (reflected) and the infrared and microwave regions (emitted). The most familiar instrument of this class is the camera, using film sensitive to reflected energy at wavelengths in the visible range.

Active systems use a man-made beam of wave energy as a source. This type of sensor system sends a beam of energy to an object. A part of the energy is reflected back to the source, where it is recorded by a detector. A simple analogy would be the use of a spotlight on a dark night to illuminate a target and reflect light back to the eye. Active systems used in remote sensing operate mostly in the radar region.

Short wave radiation sensors

Of the passive sensor systems, camera photography in the visual portion of the spectrum is most familiar to the average person. Black-and-white aerial photographs, taken by cameras from aircraft, have been in wide use by geographers, geologists, and other environmental scientists since before World War II. Commonly, the field of one photograph overlaps the next along the plane's flight path, so that the photographs can be viewed stereoscopically for three-dimensional effect. Color film can be used to increase the level of information on the aerial photographs. Because of its high resolution (degree of sharpness), aerial photography remains one of the most valuable of the older remote-sensing techniques. Photography has been extended to greater distances through the

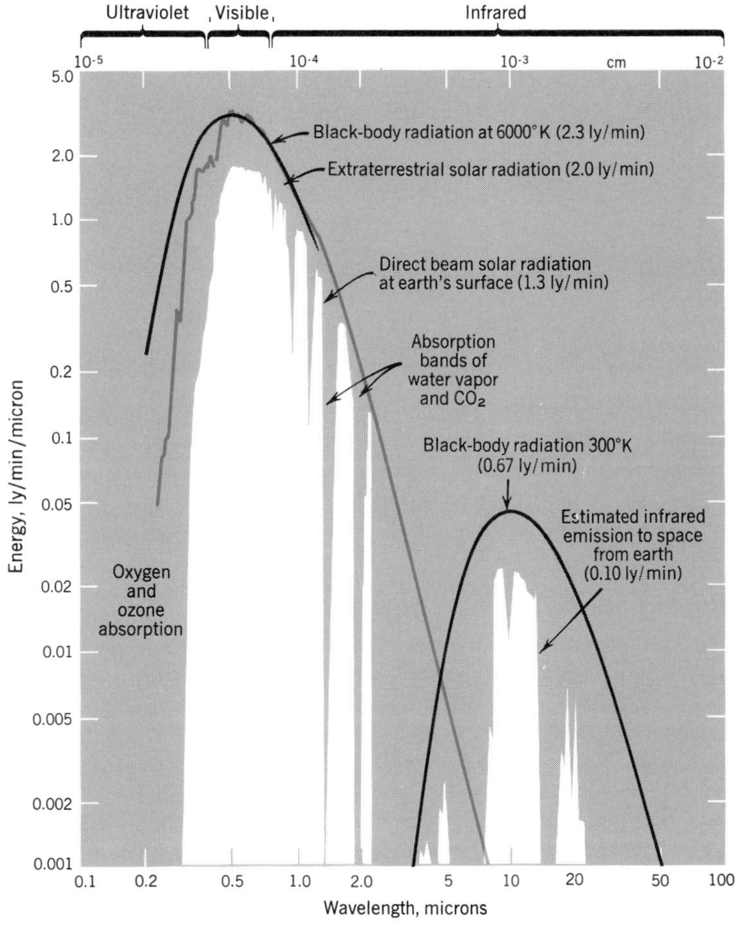

Figure AV.3 Spectra of solar and earth radiation. (From W. D. Sellers, 1965, *Physical Climatology*, Univ. of Chicago Press, Chicago, p. 20, Figure 6.)

use of cameras operated by astronauts in orbiting space vehicles. Most persons are familiar with striking color photographs obtained during the early 1960s by the several Gemini missions. Examples of these photographs have been published by NASA in widely distributed albums.

Photography using reflected electromagnetic radiation also extends into the ultraviolet and infrared wavelengths. Conventional cameras equipped with suitable filters and film can be used in the near-ultraviolet region between 3000 and 4000 Angstroms (0.3–0.4 microns) (See Figure AV.2). The application of *absorption spectrophotometry* to the ultraviolet wavelengths reflected from the earth's surface allows some degree of analysis of the element composition of surface rocks and soils. Spectrophotometry is a major tool of astronomy. Light entering a telescope from the sun or other stars is passed through a *spectroscope*, a device to spread out the radiation band for examination. An explanation of the principles of spectroscopy is beyond the scope of our discussion, but is found in elementary textbooks of physics and astronomy.

There is a small region of the infrared spectrum, the *near-infrared* region immediately adjacent to the visible red region, in which reflected rays can be recorded by cameras with suitable film and filter combinations. Figure 30.14, a photograph of Mt. Shasta, is an example of infrared photography in this region of reflected wavelengths. Notice that the blue sky, which emits little longwave radiation, appears black (blue rays have been excluded by a dense red filter).

Besides conventional cameras using films, images in the reflected radiation region can be generated by *scanning systems*. The television camera and receiver illustrate one type of scanning system. Scanning is the process of receiving information instantaneously from only a very small area of the frame under surveillance. A receiving beam runs very rapidly across the field of view, shifting position after each traverse so as to obtain information along a set of closely spaced parallel lines. After the entire frame is scanned the process can be immediately repeated. Reflected light is scanned by a means of a rapidly rotating *scanning mirror* which deflects the incoming rays to form a set of lines across a detecting field. This field may be the face of an electron tube, on which the energy of the beam is transformed into luminescence and thus generates an image of the field being scanned. Then, an electron beam within the electron tube scans the face of the tube and converts the luminous image into a succession of signals of varying intensities, and these in turn can be stored on tape. Taped data can be reconverted into *imagery*, a graphic form resembling a conventional photograph. Other forms of detectors besides the electron tube can receive the energy of the scanning beam and convert it into electronic signals. Some additional information on scanning systems is given in a later paragraph describing the equipment used on orbiting earth satellites.

Photography using films or plates is greatly improved as a remote-sensing tool by use of selected narrow frequency bands. A filter placed over the camera lens allows only the desired wavelengths to pass through to the film. For example, one filter passes rays of the blue-green portion of the visible spectrum; another the visible red rays; and a third the near-infrared rays. Because some surfaces reflect one portion of the spectrum much better than another, a comparison of the photographs under different wavelengths allows the type of surface to be identified. The term *multiband spectral photography* applies to this technique; it requires simultaneous photography by several cameras or by a multiple-lens camera.

One of the most recently developed and sophisticated devices for remote sensing in the visible and near-infrared regions is the *Multispectral Scanning System* (MSS). The MSS system used on the Earth Resources Technology Satellite (ERTS) was developed by the Hughes Aircraft Company; it simultaneously gathers data from four spectral bands in the range from about 0.4 to 1.1 microns (Figure AV.4). A scanner feeds the radiation to detectors and the information is digitized for continuous transmission to a ground receiving station. As the satellite travels the MSS continuously scans a belt of the earth's surface 185 km (115 mi) wide. On the ground the data are transformed into square-framed images 185 km on a side. There is a 10 percent forward lap between frames.

As shown in Figure AV.4 various types of earth surfaces have varying degrees of reflectivity over the spectral range covered by the MSS. Water surfaces have their highest reflectivity in the blue-green region. Green vegetation reflects much more strongly in the red and near-infrared region than in the blue-green region. Consequently, a comparison of data from the four bands allows the nature of the surface to be identified.

As a further refinement of the multiband scanning system, each one of three bands of the MSS imagery is reproduced in a different color and the three colors superimposed in a single print. Although at first glance this print may seem to be a color photograph of the type we get from an ordinary camera, it is by no means the same thing. The colors derived in the MSS imagery are "false" colors. The colors blue, green, and red are selected for use in reproducing bands 1, 2, and 3 (or bands 1, 3, or 4) respectively, of the MSS imagery. Thus the resulting color imagery combines the intensi-

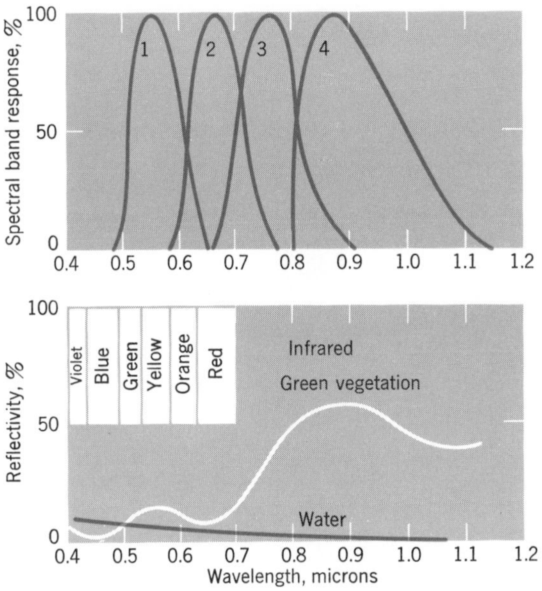

Figure AV.4 Schematic diagram of sensitivities of four filters in the MSS on ERTS-1, compared with reflectivity of water and green vegetation surfaces.

ties of reflected energy in the three bands at a given point within the frame. Typically, urbanized areas show as blue-gray colors. In contrast, healthy, dense, actively growing vegetation shows as deep red. Various agricultural crops appear as color shades ranging from pink to orange-red to deep red. Mature crops and dried vegetation (dormant grasses) appear yellow or brown. Shallow water areas appear blue; deep water appears dark blue to blue-black.

Examples of multispectral false-color imagery are shown in the color plate accompanying this appendix.

Thermal infrared sensors

We turn next to the longer infrared wavelengths emitted by objects because they possess sensible heat. This radiation, in the range approximately from 1 to 20 microns, is described as *thermal infrared* to distinguish it from reflected radiation in the near-infrared region. Thermal infrared radiation is sampled by a scanning system, usually of the rotating-mirror type already described. However, the detector consists of a surface coated with a material sensitive to infrared wavelengths, rather than to light.

Infrared radiation can be recorded during day or night, since it is emitted, rather than reflected energy. The rays pass readily through haze and smoke—atmospheric contaminants that tend to obscure radiation in the visual range. Because a warmer object emits greater energy than a cooler object, the latter appears lighter in tone on infrared imagery.

Besides absolute temperature, the intensity of infrared emission depends upon an intrinsic property of the object, known as *infrared emissivity.* Whereas the black body, mentioned earlier, is an ideal perfect radiator of energy, most substances are imperfect radiators, or *gray bodies.* Infrared emissivity is the ratio of emission of a gray body to that of a black body at the same absolute temperature; it ranges from zero for a body with no emission to unity (100 percent) for a black body. For most natural terrestrial surfaces the infrared emissivity is comparatively high—in the range from 85 to 99 percent. Differences in emissivity can be important in determining the patterns of the infrared image.

Two examples of infrared imagery acquired at night are shown in Figure AV.5. Keep in mind while examining these images that the differences in tone, by means of which objects are delineated, are caused by differences in the level of emission of infrared energy. Differences in temperature are the most important variable in causing the differences in tone, although emissivity plays an important role in the emissions from certain man-made materials. In the upper image, taken during the early morning hours (about 2 to 4 A.M.) pavements and water emit stronger infrared radiation and appear light in tone. Trees lining many of the roadways also appear bright in tone. Agricultural areas, moist soil surfaces, and buildings are "cooler" and appear darker. The lower image shows the shore of the Salton Sea. Currents in the shallow water containing varying sediment loads appear as swirling bands of lighter and darker tones, allowing the dynamics of water motion to be studied.

Passive microwave sensors

Microwaves in the range from 0.1 to 1.5 cm (EHF band) resemble infrared radiation in that they are emitted from an object because of the presence of sensible heat. Microwaves can be detected by *passive microwave sensors;* these use slow-scanning radiometers of the type used in radio astronomy. Because microwaves can pass through substantial thicknesses of solid and unconsolidated earth materials, it is possible to detect microwaves originating from beneath covers of ice, snow, or soil. However, microwave sensing faces disadvantages in that the resolution is generally low and the systems have poor sensitivity.

A passive microwave sensor, known as the *Electrically Scanning Microwave Radiometer* (ESMR), installed in 1972 on a meteorological satellite (Nimbus 5), operates on the wavelength

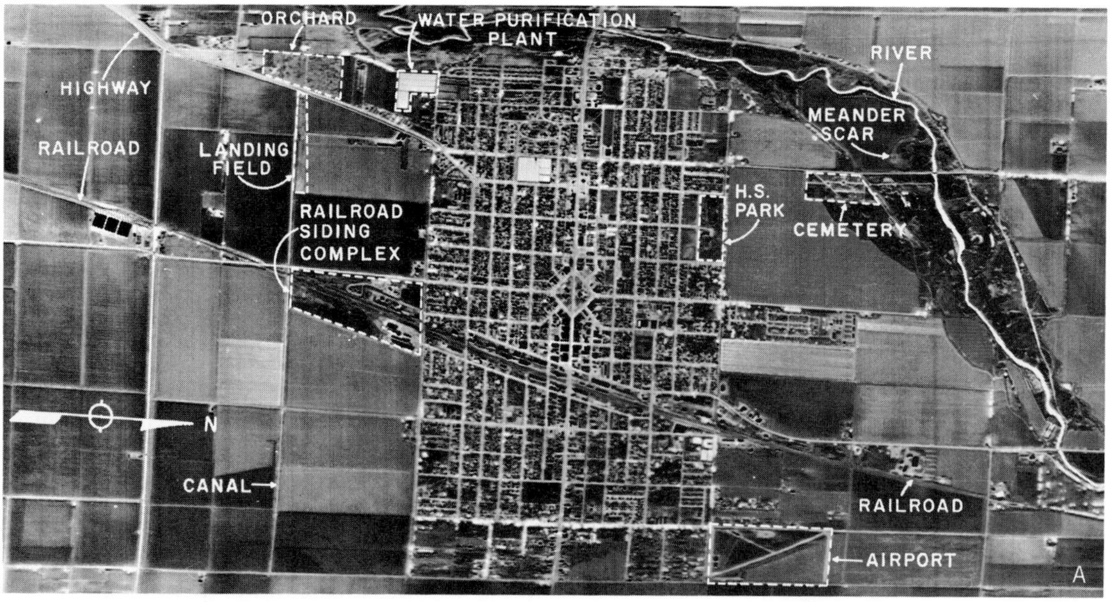

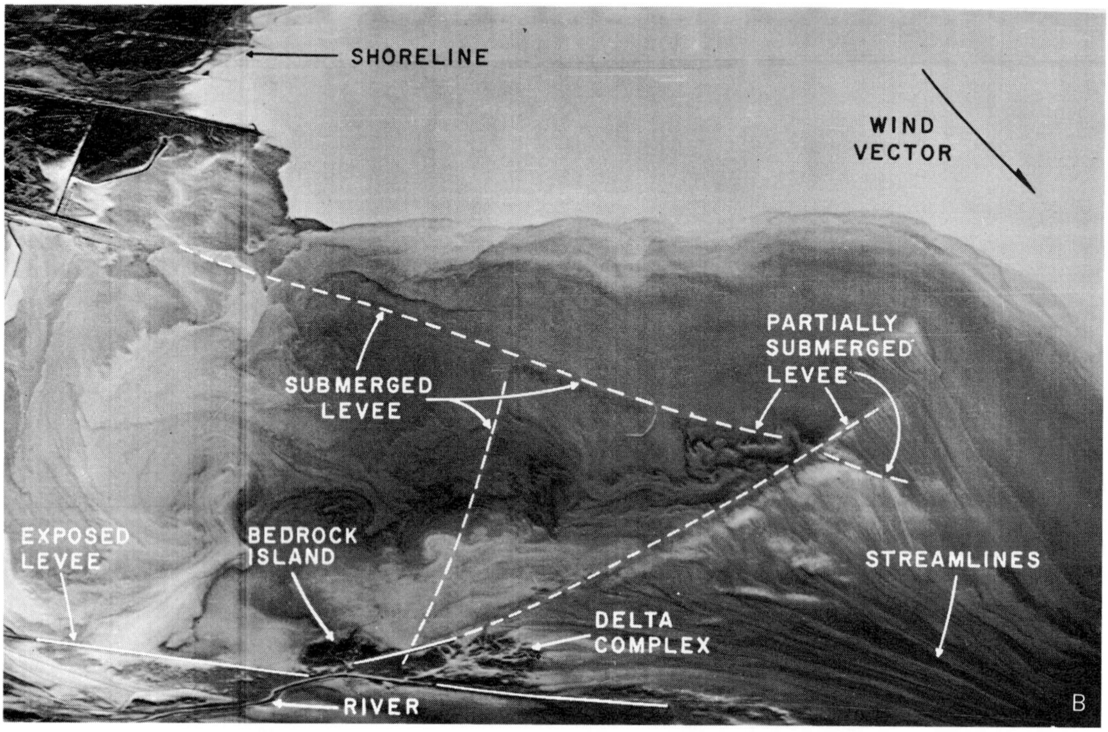

Figure AV.5 Examples of infrared imagery. (Courtesy of Environmental Analysis Department HRB—Singer, Inc.) *A.* Imagery of Brawley, California, a small farm community within the Imperial Valley farming region of southernmost California. *B.* A portion of the shoreline of the Salton Sea, California. Water appears lighter than land, indicating that this is a nighttime image.

of 1.5 cm. Together with a microwave spectrometer sensing system, ESMR determines atmospheric temperature structure and maps sea ice cover and rainfall patterns over oceans, regardless of cloud cover. On the front cover of this book is reproduced a microwave false-color image of Antarctica. The low resolution of the image is evident from the large size of the unit squares (dots) comprising the information.

Radar sensing systems

We turn next to the active mode of remote sensing within the radar portion of the electromagnetic spectrum (0.1 to 100 cm). The active sending system uses pulses of energy emitted by sources mounted on aircraft or space vehicles. A beam of such pulses is directed at the ground and a portion of the energy is returned as an echo signal. A scanner directs the beam back and forth across the target surface. Radar pulses pass readily through all but the densest of clouds and precipitation.

Radar-sensing systems are effectively used on aircraft. The types most often used to produce imagery are known as *side-looking airborne radar* (SLAR) systems. These systems send their impulses toward either side of the aircraft from a rectangular antenna. A scanning beam is used. The rectangular antenna produces a *polarized* beam. Polarization is a process of restricting the wave motion to a single direction. Similarly, the receiving antenna is rectangular in outline and can select incoming wave motion in a single direction. When the transmitting antenna is oriented with the long axis horizontal (*H*), horizontally polarized pulses are emitted. Vertical orientation (*V*) produces a pulse polarized in the vertical direction. The echo can be received with a horizontally oriented (*H*) or vertically oriented (*V*) antenna. Consequently a horizontally polarized pulse can be received with a horizontally oriented antenna (*HH*), or with a vertically oriented antenna (*HV*). The two combinations yield different image tones, depending upon the target surface.

SLAR images show terrain features with remarkable sharpness and contrast (Figure AV.6). Surfaces oriented most nearly at right angles to the radar beam appear lightest in tone; those turned away from the beam are darkest. The effect is to produce an image resembling a relief map using oblique illumination (See Appendix I). Various types of surfaces, such as forest land, rangeland, and agricultural land can also be identified by variations in image tone and pattern.

We should also be aware that radar imagery is produced by ground-based stations, using atmospheric phenomena as targets. An example is seen in Figure 11.3, the photograph of a radar screen on which thunderstorms show as bright patches. Although this form of radar sensing is not usually included within the conventional meaning of remote sensing, the data can be processed and interpreted in much the same way as for SLAR imagery.

Lasers represent yet another probe for remote sensing systems. The importance of laser systems is not yet sufficiently great in geographical research to warrant inclusion in this brief review.

Orbiting earth satellites

It is only since the advent of orbiting earth satellites that remote sensing has burgeoned into a major branch of geographical research, going far beyond the limitations of conventional aerial photography. The first significant orbiting satellite with major capabilities in remote sensing of geographical phenomena was *Tiros I*, a meteorological observation satellite launched in 1960. This satellite orbited at an average height of about 460 mi (740 km) and was equipped with TV cameras for observing cloud systems and associated weather phenomena, including cyclonic storms and fronts. However, the plane of the Tiros I orbit was oriented at an angle of 48° with respect to the plane of the earth's equator, so that the highest terrestrial latitude covered by its earth track was 48°N and S. With each revolution, Tiros I shifted the point of its equatorial crossing westward by about 25° of longitude.

To cover the entire globe, an earth satellite was needed that would orbit over the earth's polar region. An orbit exactly perpendicular to the earth's equator, passing over the poles, would remain fixed in position with respect to the stars. However, because remote sensing relies upon reflected solar radiation, such a truly polar orbit would be undesirable, since the sun shifts its position continually with respect to the stars. What is needed is a *sun-synchronous* orbit in which the satellite's plane remains fixed with respect to the sun. Only in this way can the solar illumination over the satellite's earth track be maintained uniform in direction and intensity. A sun-synchronous orbit is illustrated in Figure AV.7. The plane of this orbit makes an angle of 80° with the plane of the earth's equator and the satellite's earth track makes a tangent contact with the 80th parallels of latitude north and south. With this orbit, the torque exerted by the earth's equatorial bulge upon the satellite's motion is just sufficient to shift the orbit westward at a rate matching the shift in angle between the sun and stars. In the example illustrated in Figure AV.7 the satellite completes its orbit in a period such that the point of equatorial crossing shifts westward $28\frac{1}{2}°$ per orbit. This shift is at the rate of 4 minutes of time per

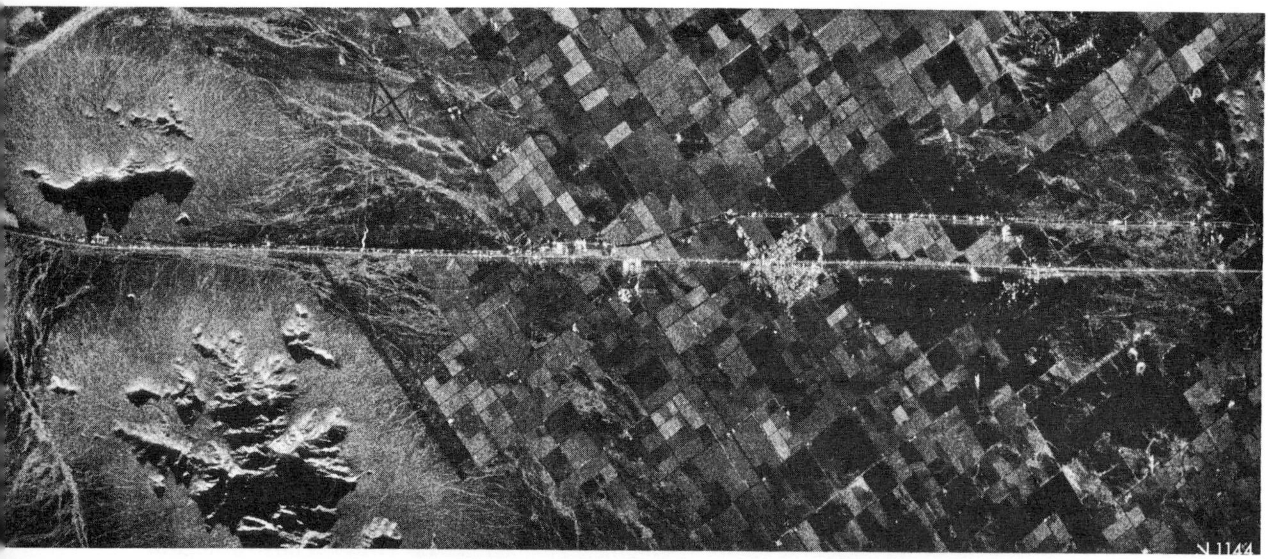

Radar imagery of a desert landscape in southern Arizona. The flight line parallels the main highway between Phoenix and Tucson. North is toward the upper left. At the left are remnant mountain masses surrounded by pediment and fan slopes. Washes appear as anastomosing white lines. The "shadows" on southwest slopes show that the radar beam was directed from the northeast. At the right are cultivated lands, laid off into sections and quarter-sections of the Land office Grid. The larger dark squares are 1 mi. (1.6 km) on a side. Built-up areas along the highway are brightly reflective.

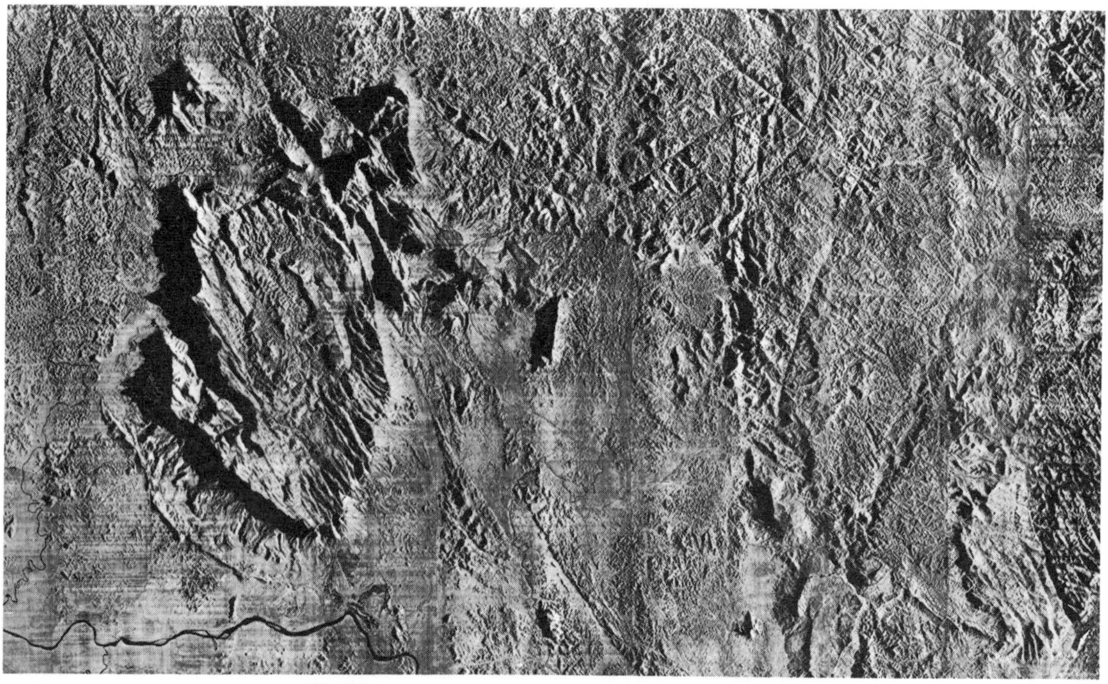

Radar imagery of an area in southern Venezuela, in the headwater region of the Orinoco River, latitude 3° to 4° N. The entire region is clothed in dense equatorial rainforest. A massive synclinal remnant of sedimentary rock forms a highland at the left. Closely-set fractures, trending northwest-southeast, control the trends of monor stream valleys carved into the highland escarpment. Other intersecting lineaments in the right-hand portion of the area suggest fault control of major drainage lines. Horizontal dimension of the area shown is about 100 mi (160 km).

Figure AV.6 Side-Looking Radar (SLAR) imagery. The upper illustration shows imagery along a single flight line; whereas the lower illustration shows a mosaic and covers a much larger area. (Courtesy of Goodyear Aerospace Corporation and Aero Service Corporation.)

degree of longitude, or 15° per hour. As explained in Chapter 5, this rate is the same as the rate of earth rotation with respect to the sun (i.e., mean solar time).

The requirements stated above were first met by a new generation of weather-observing satellites, of the *Nimbus* series. *Nimbus I* was launched in 1964. Although the orbit was far from perfect, the satellite was successful in transmitting TV photographs as well as high-resolution infrared radiometer information. Later Nimbus satellites and the subsequent NOAA satellite series have permitted greatly improved data collecting performance. Besides photography and infrared radiometry, weather satellites record the vertical profile of temperatures in the atmosphere between the satellite and the earth's surface. They also record global distributions of ozone, water vapor, albedo, cloud cover, and precipitation. The satellite photographs have become an important adjunct to weather forecasting, and are particularly useful in identification and monitoring of tropical cyclones. The period of orbit has been adjusted to 107 minutes, so that the satellite can scan the entire globe twice daily.

A major step forward in remote sensing of the environment was taken by NASA in July 1972 with the launching of the first *Earth Resources Technology Satellite* (ERTS-1). It is the first of two orbiting observatories designed to monitor the natural resources of the planet. ERTS-1 is a one-ton (900-kg) satellite orbiting at a height of 567 mi (614 km). It completes one orbit each 103 minutes, crossing the equator at about 9:30 A.M. local time on its north-south leg. This timing is associated with the best conditions of sunlight and shadow for terrain observation. Each day ERTS-1 completes 14 orbits. It photographs a strip 115 mi (185 km) wide; three such strips cover North America daily. There is an overlap of 14 percent in photograph strips near the equator; more overlap near the poles. At any given point on the earth's surface, the satellite passes overhead at the same position once every 18 days, giving opportunity to match conditions of illumination exactly at 18-day intervals. In this way changes that occur with time can be most readily detected.

In addition to the MSS scanner system described in an earlier paragraph, ERTS-1 carries an imaging system called *Return Beam Vidicon* (RBV), using three individual high-resolution television cameras. Each camera is equipped with a different spectral filter: one senses the visible blue-green region, another the visible red, the third the near infrared. As the cameras are simultaneously shuttered, the image is received and stored on the face plates of vidicon tubes. Scanners then read out the information on each face plate

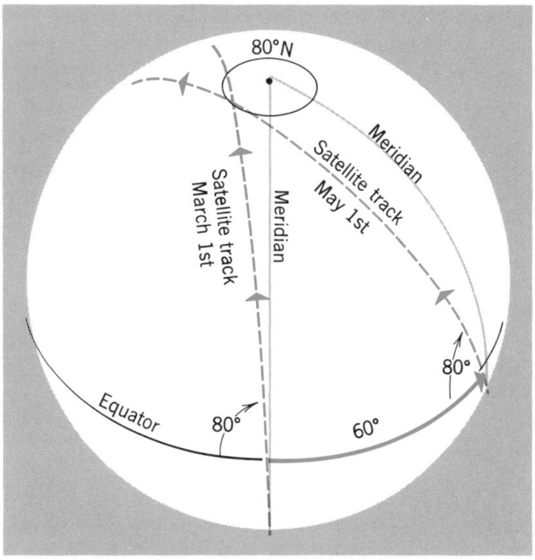

Figure AV.7 Earth-track of a sun-synchronous satellite. In the period from March 1 to May 1 the orbit has retrogressed eastward about 60° with respect to space coordinates. (From A. N. Strahler, 1971, *The Earth Sciences*, Harper and Row, New York, Figure 8.17.)

by electron beam, transmitting the data to earth by microwave. The cameras are reshuttered every 25 seconds, during which time the satellite has traveled some 102 mi (165 km). Thus sequential images overlap by about 10 percent.

ERTS-1 also carries a *data collection system* (DCS) that picks up and relays data sent by radio transmission from about 100 ground stations situated in remote locations in North America. These ground stations may monitor such diverse environmental parameters as water quality, rainfall, snow depth, seismic activity, animal migrations, and other earth phenomena.

Imagery obtained by MSS and RBV is transmitted directly to ground stations in the United States and Canada as the satellite moves over that continent. Imagery gathered over other parts of the world is stored in videotape recorders and replayed when the satellite passes over the United States at night. Other ground receiving stations are being constructed in Europe and Australia. Whereas all possible scenes are photographed as the satellite passes over North America, photography over other world areas is selective, depending upon reports of favorable observing conditions. In the first 10 months of operation, ERTS-1 generated imagery for more than $1\frac{1}{2}$ million photographs and mapped more than 75 percent of the earth's land areas. Imagery is processed by NASA and sent to cooperating federal agencies and to over 300 United States and foreign scientific investigators working on various applications of remote sensing. Images are offered for sale to the public through the Earth Resources Observations Systems Data Center of the U.S. Department of the Interior in Sioux Falls, South Dakota. Some investigators claim that they can identify on the photographs features as small as 300 ft (90 m) in diameter; linear features, such as roads, as narrow as 50 ft (15 m). Use of false-color photographic processing techniques, described earlier, greatly increases the amount of information available from the imagery. When designed, the useful life of ERTS-1 was estimated to be one year, but it may continue to operate for a second or even a third year. A second ERTS satellite is scheduled to be launched early in 1976.

Skylab

A new dimension was added to remote sensing by the launching in May 1973 of *Skylab*, a manned orbiting space laboratory. Skylab travels in an orbit oblique to the plane of the earth's equator and covers the region between 50°N and 50°S. Traveling at a height of 270 mi (430 km) Skylab orbits the earth once every 93 minutes. It is occupied by three-man crews for work periods of from 4 to 8 weeks. The Skylab workshop is a cabin 48 ft (15 m) long and 21 ft (6.5 m) in diameter; the total volume is about 10,000 cu ft (280 cu m). In addition, the multiple docking adapter houses the control panel for a telescope array and has a large window through which the earth's surface can be examined. One of the major scientific installations is the Apollo Telescope Mount with instruments for observing the sun.

Observation of earth resources is another major objective of the Skylab missions. Multispectral imagery is obtained by a package of six instruments housed in the *Earth Resources Experimental Package* (EREP). In addition, cameras allow photographs of high resolution to be taken at will and the films returned to earth. A wide variety of terrestrial phenomena covering atmosphere, oceans, and land surfaces has been observed visually by Skylab crew and photographed for detailed study. During the second Skylab mission, lasting over 59 days, astronauts brought back 17 miles of magnetic tape bearing stored data of the multispectral EREP cameras, as well as nearly 17,000 frames of film of the earth's surface. Most of the United States was covered by these high-resolution photographs. Subjects included drainage patterns, crustal fracture patterns, and mineral deposits. The third mission, ending early in 1974, greatly augmented the store of EREP data.

Contributions of remote sensing to physical geography

Table V.1 lists some of the classes of geographical information derived by remote sensing techniques. Opportunities for reconnaissance studies covering vast areas of the globe are particularly promising. Changes with time can be studied through examination of imagery repeated at intervals. Of all the forms of space technology, remote sensing promises to contribute the most useful knowledge of man's environment and resources in ratio to its development costs.

TABLE AV.1 EXAMPLES OF APPLICATIONS OF REMOTE SENSING IN GEOGRAPHY

METEOROLOGY, CLIMATOLOGY

Monitoring and mapping of atmospheric temperature, surface temperature, water vapor, cloud cover, albedo, precipitation, ozone, and turbidity (dust).
Assessment of energy balance through measurements of reflected shortwave energy, longwave emission, net radiation, soil-heat flux, turbulent heat flux, and latent heat flux.
Mapping seasonal changes and extent of snow cover, lake ice, sea ice.
River flood surveillance and impact analysis.
Assessing disaster effects of storms (hurricanes, floods).

OCEANOGRAPHY

Measurement of sea-surface temperatures, mapping of water masses.
Measurement of ocean currents; identification of upwellings.
Measurements of waves and swells, sea surface roughness.
Observing seasonal changes in sea ice; ice breakup.
Mapping of bottom topography in shallow water.
Observation of sediment transport and accumulation.
Mapping of plant life of oceans (kelp, phytoplankton).
Surveillance of fishing fleets.

VEGETATION, SOILS, AGRICULTURE

Mapping of large-scale patterns of vegetation and soils.
Mapping distribution of biomass, ecosystem energetics.
Observing annual cycles of vegetation growth and dormancy.
Mapping soil-moisture content, annual changes.
Mapping soil erosion intensity.
Mapping of forest-grasslands boundaries.
Assessing ecological effects of fires.
Mapping abandonment of land, shifting tropical agriculture.
Determination of growth and vigor of crops.
Time-series study of harvests.
Monitoring and assessment of crop disease and insect infestations.
Survey of agricultural resources in developing countries.

GEOMORPHOLOGY, GLACIOLOGY

Mapping of landform types, drainage patterns.
Identification of ancient dune fields and wind directions.
Mapping mineral composition of surfaces: rocks, alluvium.
Mapping extent of glacial ice.
Calculating mass budgets of ice sheets and glaciers.
Observing relation of ground water to landforms, effluent ground water.
Detection of lineaments, including active faults.
Measurement of energy fluxes along shorelines.
Monitoring of volcanic eruptions.
Earthquake disaster surveillance.

URBAN GEOGRAPHY, TRANSPORTATION, LAND USE

Monitoring urban air pollution.
Measuring heat emissions of cities.
Mapping categories of land use.
Monitoring changes in regional land use patterns.
Mapping traffic densities.
Assessing local and regional effects of major new urban developments.
Identifying remains of ancient settlements and caravan routes.

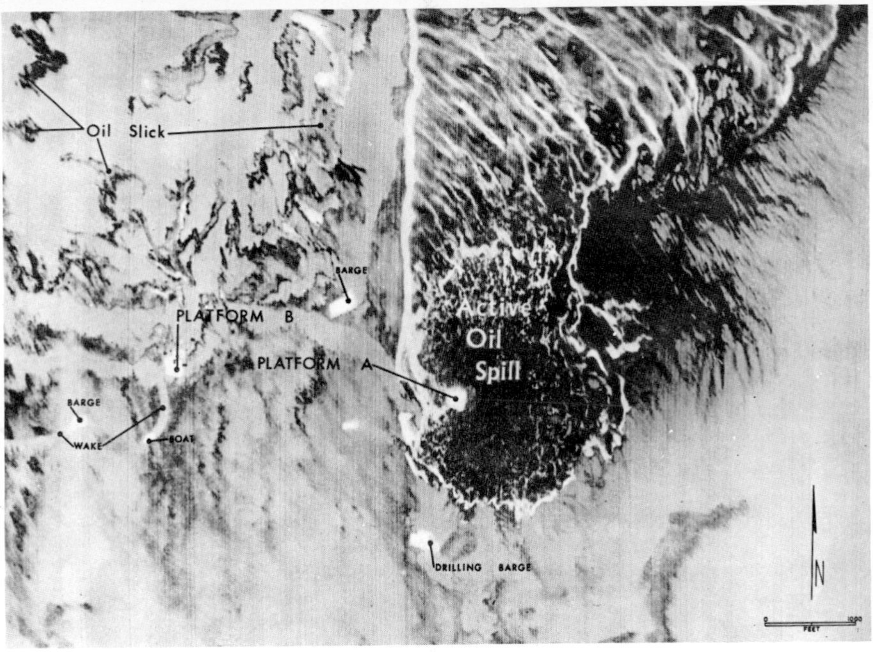

Figure 1 Thermal infrared image of the source area of the Santa Barbara Oil Spill, taken in early February, 1969. This image shows the patterns of emitted thermal infrared energy from the sea surface. Imagery such as this can be acquired day or night. (Photo by courtesy of North American Rockwell Corporation; labels by Geography Remote Sensing Unit, University of California, Santa Barbara.)

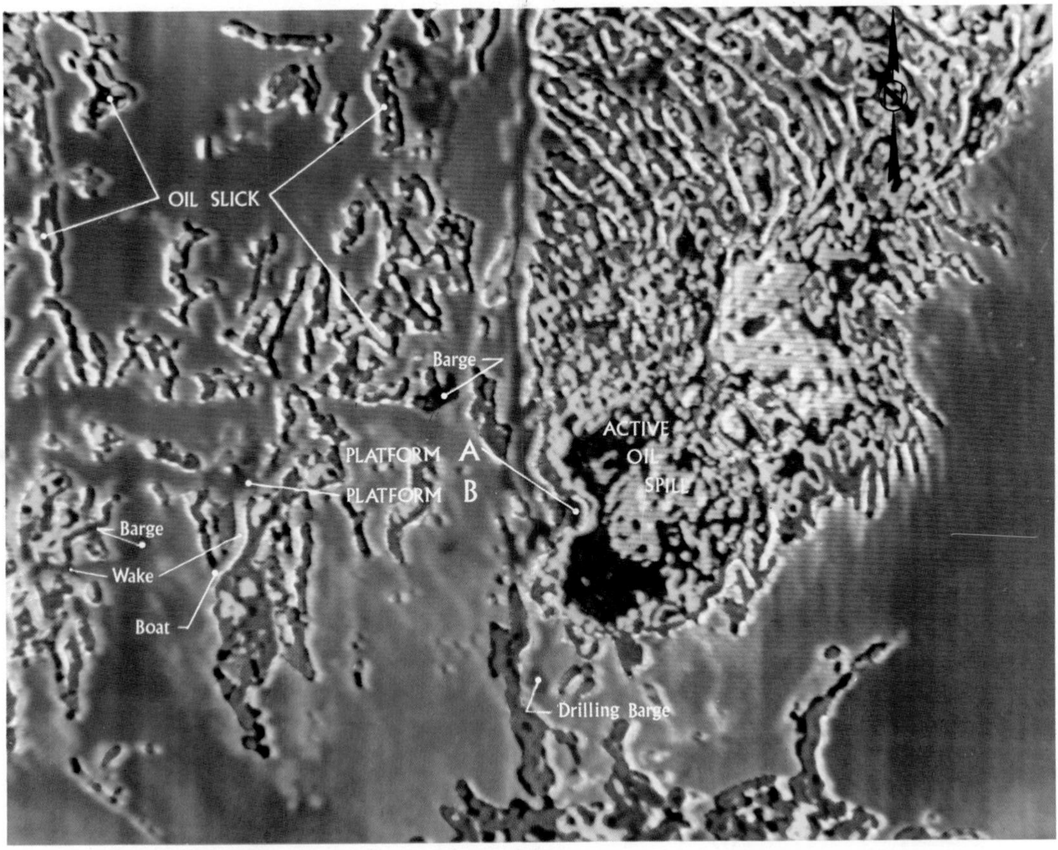

Figure 2 An electronic enhancement of Figure 1. Image enhancement systems, such as the one which generated this image, expand the image interpreter's ability to study the meaning of subtle variations in an images' optical density. This effect is achieved by assigning colors to specific gray levels on the original image. In this instance researchers were attempting to ascertain whether or not the patterns of emitted thermal energy emanating from this scene were related to variations in thickness of the oil film on the seawater. (Photo by courtesy of Spatial Data Systems Inc.; labels by Geography Remote Sensing Unit, University of California, Santa Barbara.)

This color brochure illustrating applications of remote sensing was compiled and annotated by John E. Estes and Leslie W. Senger.

Figure 3 This Skylab 4 color infrared photo of the Flagstaff, Arizona, area was imaged by the Earth Resources Experiment Package (EREP) S-1908 (five-inch terrain camera). Humphrey's Peak, the Sunset Crater volcanic field, and Flagstaff are at the left of the photo. Meteor Crater, seen in the right center of the photo, may be used for scale comparison; the crater is approximately one mile in diameter. In this photo living vegetation appears in red, snow in white, water in blue. Photos such as this are used for snow mapping in an attempt to gain a better understanding of an area's total water resource capability. (Photo courtesy National Aeronautics and Space Administration.)

Figure 4 Portion of a color infrared vertical photo of the eastern coast of Sicily taken during the Skylab 3 mission with the Earth Resources Experiment Package (EREP) S-1908B (five-inch earth terrain camera). Mt. Etna, the highest volcano in Europe, is still active, as evident by the thin plume of smoke emanating from its crest. On Etna's flanks recent lava flows appear black, while older lava flows and volcanic debris, which now support vegetation, appear in shades of red. Photos such as this can be used not only to study the localized patterns of geologic phenomena, such as volcanic activity or faulting, but regional or global patterns and distributions as well. (Photo by courtesy of National Aeronautics and Space Administration.)

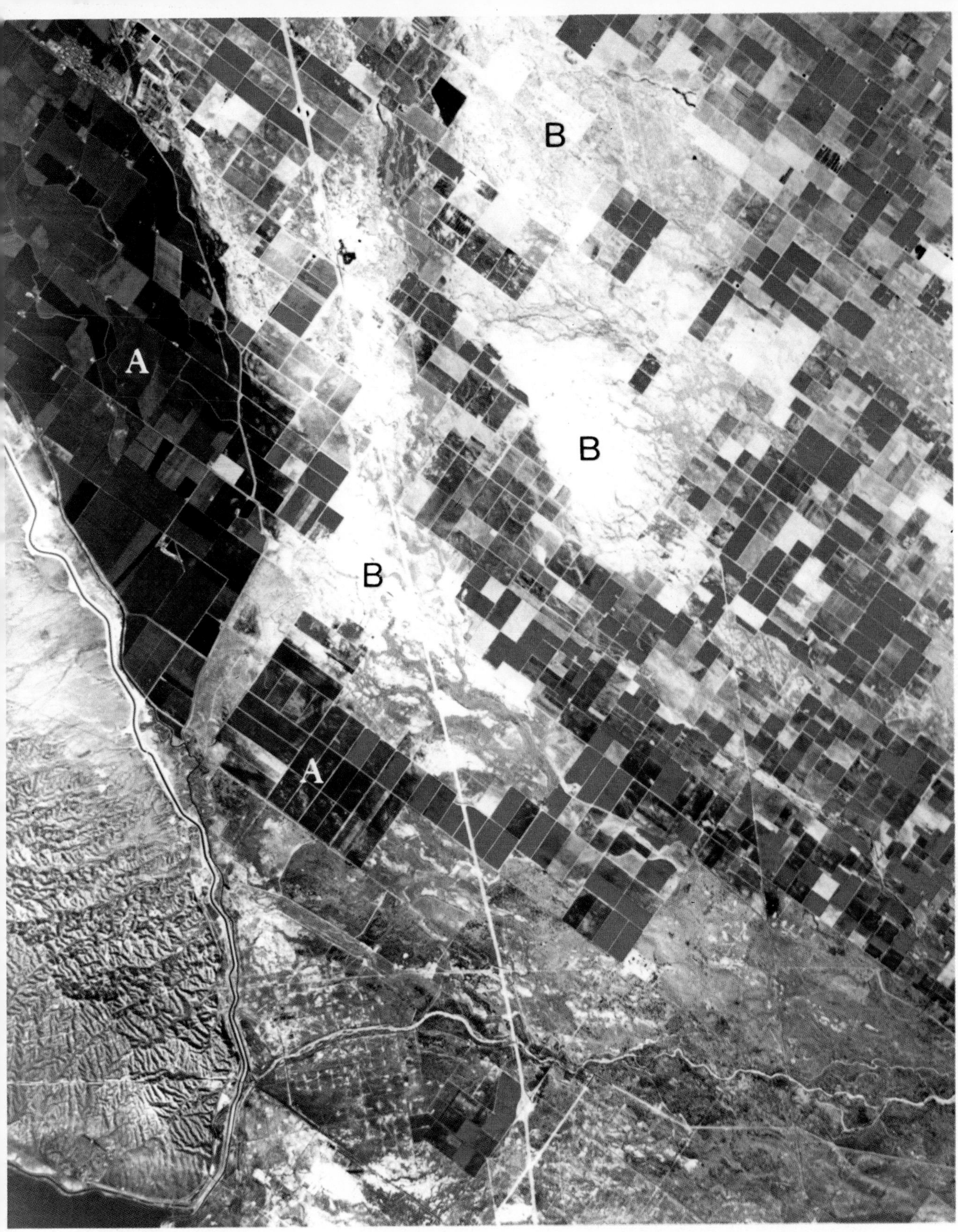

Figure 5 High-altitude color infrared aerial photograph of an area near Bakersfield, California, in the Southern San Joaquin Valley, taken by a National Aeronautics and Space Administration U-2 aircraft flying at approximately 60,000 ft. The original photo scale was about 1 : 120,000. Photos such as this are used to study problems associated with agriculture. In this area perched ground water areas seen at (A) and areas of high soil salinity (B) create problems which can effect crop yields. The various red tones in the fields indicate that various types of crops are being grown. By knowing which crops are typically grown in this area and their growing cycle, determinations can be made of the amount and areal extent of individual crops being cultivated. Estimates of areal, regional, or global yield may be made by adding this type of information to a stratified statistical model with appropriate sampling at the local, regional, or global level and which has as an integral part the monitoring of the growing cycle of the crop or crops of interest and with the addition of information concerning significant meteorological, hydrological, and pedological variables. (Photo by courtesy of National Aeronautics and Space Administration.)

Figure 6 A high-altitude color infrared photograph of Goleta California taken in April 1971, by a National Aeronautics and Space Administration RG-57 aircraft from approximately 60,000 ft, with an orginal scale of 1 : 120,000. 1971 was a normal rainfall year for this area and, at this time, grassy natural vegetation (A) appears light pink while chaparral slopes (B) appear darker red. (Photo by courtesy of National Aeronautics and Space Administration.)

Figure 7 A second National Aeronautics and Space Administration color infrared photograph of the Goleta Valley imaged in January 1973 by a U-2 aircraft from about 60,000 ft (again with an original contact scale of 1 : 120,000). This image, taken after a fall period of heavier than normal rainfall, shows grassy vegetation (A) in bright red with the chaparral slopes (B) again slighter darker. (Photo by courtesy of National Aeronautics and Space Administration.)

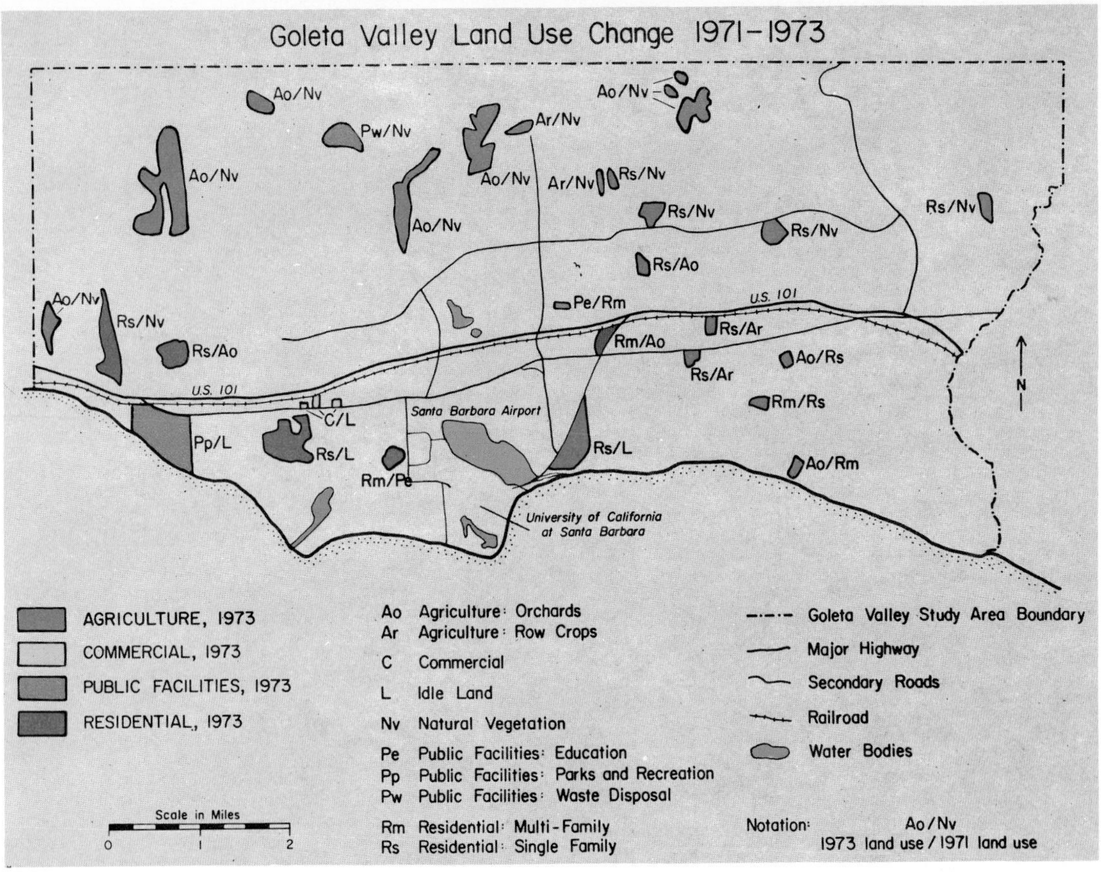

Figure 8 Land use change map of the Goleta Valley, California. compiled from information interpreted from Figures 6 and 7 on the facing page. Many types of environmental information are perishable. Remotely sensed images, such as those seen here, provide a data bank of information concerning an area at a point in time, which can be used by many types of environmental specialists. By acquiring imagery of an area through time, users can gain insights into the types, rates, and amounts of changes which may be occurring within a given area. Information obtained in this manner may then be analyzed; the data thus extracted can be utilized for a variety of applications. These can range from the testing of theoretical models to the making of policy decisions. Information contained in the photographs of Figures 6 and 7 could be used by rangeland managers in their decision-making process. Decisions concerning the readiness of the area's range grasses could be made so that the animals could be moved on and off the range at the optimum time; thus, neither underuse or overuse of the resource occurs. Land use planners and policy makers can use the information shown here to monitor the changes occurring, and whether or not those changes conform to general planning objectives and requirements. As a final example, it may be possible to relate the land use data, which can be extracted from images such as these, to the development of water consumption equations for the land use categories depicted and the modeling of water demand in the area shown. (Map by courtesy of Geography Remote Sensing Unit, University of California, Santa Barbara.)

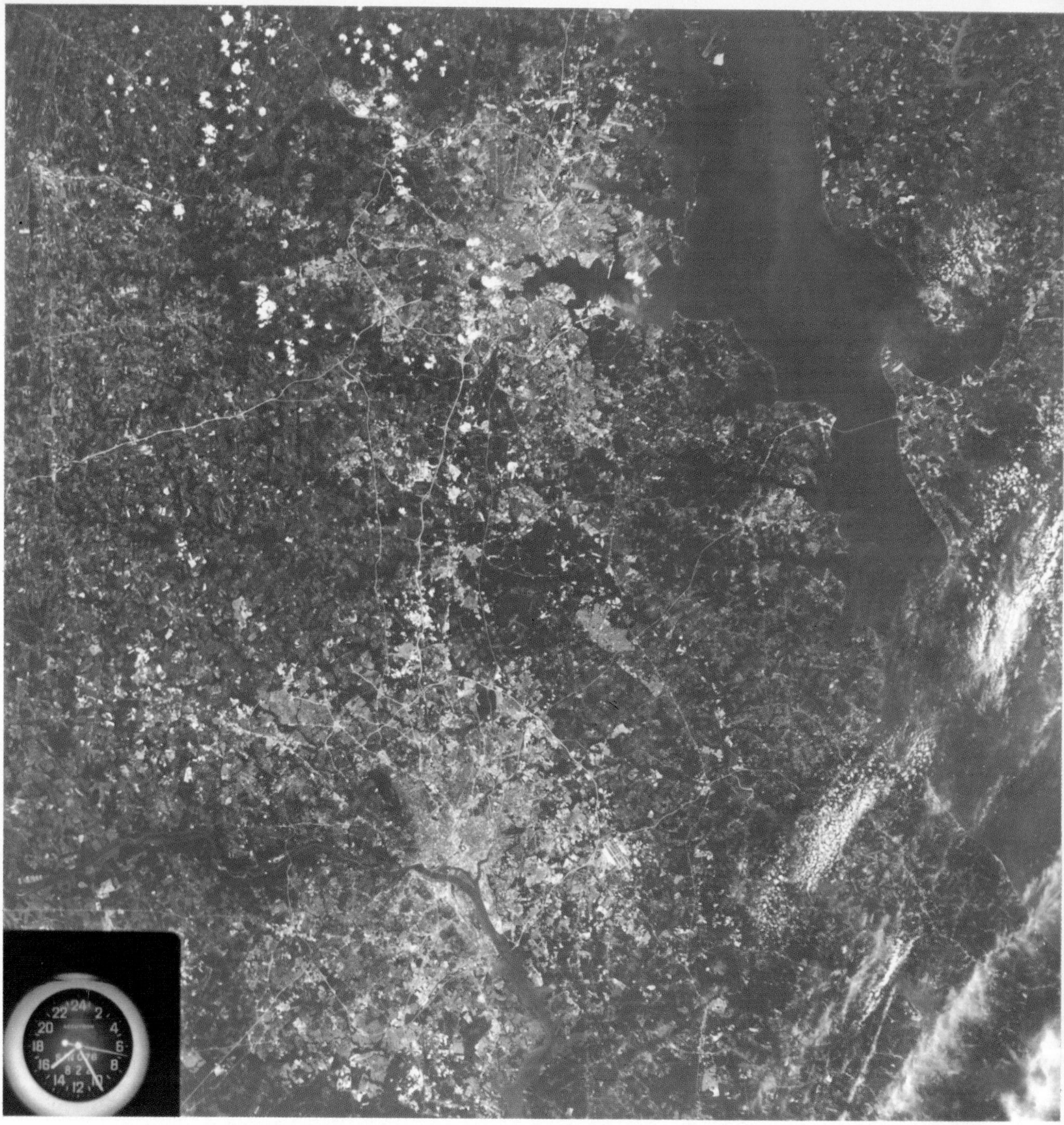

Figure 9 Color photograph of the Washington, D.C. and Baltimore, Maryland, area imaged from the Skylab 3 Earth Resources Experimental Package (EREP) S-1908 (five-inch terrain camera). From the regional perspective provided by photos such as this, the extent of urbanization is readily apparent, with suburban housing showing up as a pink-gray tone. Beltways around the cities, Interstate 95 between Baltimore and Washington, Interstate 70N leading west (left) from Baltimore, harbor facilities, bridges across the Potomac, and airfields are easily identifiable. Sedimentation patterns are detectable as lighter tones in the water. (Photo by courtesy of National Aeronautics and Space Administration and Hughes Aircraft Company.)

Figure 10 Portion of an Earth Resources Technology Satellite (ERTS) color infrared composite (Multispectral Scanner Bands 4, 5 and 7) image of essentially the same area seen in Figure 9. Urban patterns and transportation networks can be readily interpreted. Vegetated areas stand out more clearly than other features in this photography because of their high reflectance in the infrared portion of the electromagnetic (E-M) spectrum and because of the increased haze penetrating ability of this type of photography, which images longer wavelength energy than conventional color photography. Both Figures 9 and 10 have applications for land use analysis, sediment studies, and general resource surveys. (Photo by courtesy of National Aeronautics and Space Administration and Hughes Aircraft Company.)

Figure 11 The upper frame is a portion of a false color Earth Resources Technology Satellite (ERTS) photograph of the Los Angeles basin. A standard ERTS photograph covers an area approximately 100 $\times$ 100 nautical miles. Multispectral Scanner Bands 5, 6, and 7 have been color-combined to show vegetation as yellow and commercial concentrations as a bluish gray (purplish) color. The lower frame is an enlargement of a portion of an image of the Los Angeles area, which has been processed from ERTS computer compatible digital tapes. The image has been electronically enhanced by having a computer use statistical techniques to correlate and assign colors to combinations of gray levels from several ERTS multispectral bands. This technique generates a thematic map showing heavy commercial areas as yellow. Enhancement systems, such as the one whose product is shown here, are designed to aid image interpreters in extracting the greatest amount of information possible from each image. (Photos by courtesy of National Aeronautics and Space Administration, Hughes Aircraft Company, and General Electric Company.)

Bibliography

THIS selective bibliography emphasizes the contributions of North American geographers to the literature of physical geography. Entries have been selected from the full life-span of the *Annals of the Association of American Geographers* and *The Geographical Review*, both beginning in the year 1910. Thus contemporary students of geography can become acquainted with leading physical geographers of the past and can trace changes in their research methods and concepts over the past half-century. Nearly all of the remaining journal entries are taken from the following publications:

American Journal of Science
American Scientist
Arctic
Bulletin of the American Meteorological Association
Bulletin of the Geological Society of America
Journal of Geology
Weatherwise
Journal of Geomorphology
Professional Geographer
Science
Scientific American
Soil Science
Technology Review

In addition, the bibliography includes basic textbooks, monographs, and volumes of collected works. Of these, most are recent works in print, but the list includes some classic works dating back as far as the late nineteenth century. Taken as a whole, this bibliography represents a basic study collection of English-language works in physical geography suitable for colleges in the United States and Canada.

CHAPTER 1

FORM OF THE EARTH; GEOGRAPHIC GRID

Pratt, J. H. **(1942)**, American prime meridians, *Geog. Rev.*, 32:233–244.

Depts. of the Army and the Air force **(1951)**, *The Universal Grid Systems*, TM 5-241, TO 16-1-233, U.S. Govt. Printing Office, Washington, D.C. 324 pp.

Heiskanen, W. A. **(1955)**, The earth's gravity, *Scientific American*, 193:164–174.

Knox, R. W. **(1957)**, Precise determination of longitude in the United States, *Geog. Rev.*, 47:555–563.

Garland, G. D. **(1965)**, *The earth's shape and gravity*, Pergamon Press, Oxford, 183 pp.

King-Hele, D. **(1967)**, The shape of the earth, *Scientific American*, 217:67–76.

Jacobs, J. A., R. D. Russell, and J. T. Wilson **(1974)**, *Physics and geology*, McGraw-Hill Book Co., New York, 622 pp. See Chapter 4.

CHAPTER 2

PRINCIPLES OF MAP PROJECTION

Stewart, John Q. **(1943)**, The use and abuse of map projections, *Geog. Rev.*, 33:589–604.

Marschner, F. J. **(1944)**, Structural properties of medium- and small-scale maps, *Annals A. A. G.*, 34:1–46.

Deetz, C. H., and O. S. Adams **(1945)**, Elements of map projection, *Special Publ.* 68, U.S. Dept. Commerce, U.S. Govt. Printing Office, Washington D.C., 226 pp.

Powell, L. H. **(1945)**, New uses for globes and spherical maps, *Geog. Rev.*, 35:49–58.

Chamberlin, W. **(1947)**, *The round earth on flat paper*, Nat. Geog. Soc., Washington, D.C., 126 pp.

Raisz, E. **(1948)**, *General cartography*, McGraw-Hill Book Co., New York, 354 pp.

Kellaway, G. P. **(1949)**, *Map projections*, Methuen and Co., London, 127 pp.

Robinson, A. H. **(1949)**, An analytical approach to map projections, *Annals A. A. G.*, 39:283–290.

Robinson, A. H. **(1953)**, Interrupting a map projection; A partial analysis of its value, *Annals A. A. G.*, 43:216–225.

Bailey, H. P. **(1956)**, A grid formed of meridians and parallels for comparison and measurement of area, *Geog. Rev.*, 46:239–245.

Steers, J. A. **(1962)**, *An introduction to the study of map projections*, 13th edition, Univ. of London Press, London, 288 pp.

Tobler, W. R. **(1962)**, A classification of map projections, *Annals A. A. G.*, 52:167–175.

Tobler, W. R. **(1963)**, Geographic area and map projections, *Geog. Rev.*, 53:59–78.

Mackay, J. R. **(1969)**, The perception of conformality of some map projections, *Geog. Rev.*, 59:373–387.

Robinson, A. H. **(1969)**, *Elements of cartography*, 3rd edition, John Wiley and Sons, New York, 415 pp.

Richardus, P., and R. K. Adler **(1972)**, *Map projections: an introduction*, American Elsevier Publ. Co., New York, 200 pp.

SPECIAL PROJECTIONS DESCRIBED

Goode, J. P. **(1925)**, The homolosine projection; A new device for portraying the earth's surface entire, *Annals A. A. G.*, 15:119–125.

Goode, J. P. **(1929)**, The polar equal area: A new projection for the world map, *Annals A. A. G.*, 19:157–161.

Jefferson, M. **(1930)**, The six-six world map, giving larger, better continents, *Annals A. A. G.*, 20:1–6.

Miller, O. M. **(1941)**, A conformal map projection for the Americas, *Geog. Rev.*, 31:100–104.

Miller, O. M. *(1942)*, Notes on cylindrical world map projections, *Geog. Rev.*, 32:424–430.

Spilhaus, A. F. **(1942)**, Maps of the whole world ocean, *Geog. Rev.*, 32:431–435.

Fisher, I. **(1943)**, A world map on a regular icosahedron by gnomonic projection, *Geog. Rev.*, 33:605–619.

Philbrick, A. K. **(1953)**, An oblique equal area map for world distributions, *Annals A. A. G.*, 43:202–215.

Leighly, J. **(1956)**, Extended uses of polyconic projection tables, *Annals A. A. G.*, 46:150–173.

Lee, L. P. **(1965)**, Some conformal projections based on elliptic functions, *Geog. Rev.*, 55:563–580.

CHAPTER 3

MAP MAKING AND MAP READING

Deetz, C. H. **(1943)**, *Cartography*, U.S. Govt. Printing Office, Washington, D.C., 85 pp.

Lobeck, A. K., and W. J. Tellington **(1944)**, *Military maps and air photographs*, McGraw-Hill Book Co., New York, 256 pp.

Espenshade, E. B., Jr. **(1951)**, Mathematical scale problems, *J. Geog.*, 50:107–113.

Monkhouse, F. J., and H. R. Wilkinson **(1952)**, *Maps and diagrams: Their compilation and construction*, Methuen and Co., London, 330 pp.

Dept. of the Army **(1953)**, *Topographic surveying*, TM 5-234, U.S. Govt. Printing Office, Washington, D.C., 280 pp.

Dept. of the Army **(1956)**, *Map reading*, FM 21-26, U.S. Govt. Printing office, Washington, D.C., 253 pp.

Military Service Publishing Co. **(1957)**, *Map and aerial photograph reading*, Harrisburg, Pa., 177 pp.

Rude, G. T. **(1957)**, Our last frontier: the Coast and Geodetic Survey's work in Alaska, *Geog. Rev.*, 47:349–364.

Raisz, E. **(1962)**, *Principles of cartography*, McGraw-Hill Book Co., New York, 315 pp.

Robinson, A. H. **(1969)**, *Elements of cartography*, 3rd edition, John Wiley and Sons, New York, 415 pp.

GEOMAGNETISM

Nelson, J. H., L. Hurwitz, and D. G. Knapp **(1962)**, *Magnetism of the earth,* Coast and Geodetic Survey, Publ. 40-1, U.S. Govt. Printing Office, Washington, 79 pp.

Jacobs, J. A. **(1963)**, *The earth's core and geomagnetism,* The Macmillan Company, New York, 137 pp.

Takeuchi, H., S. Uyeda, and H. Kanamori **(1970)**, *Debate about the earth,* revised edition, Freeman, Cooper and Co., San Francisco. See Chapter 3.

Jacobs, J. A., R. D. Russell, and J. T. Wilson **(1974)**, *Physics and geology,* McGraw-Hill Book Co., New York, 622 pp. See Chapter 8.

MILITARY GRID

Depts. of the Army and the Air Force **(1951)**, *The universal grid systems,* TM 5-241, TO 16-1-233, U.S. Govt. Printing Office, Washington, D.C. 324 pp.

O'Keefe, J. A. **(1952)**, The universal transverse Mercator grid and projection, *Professional Geographer,* 4:19–24.

Dept. of the Army **(1958)**, *Grids and grid references,* TM 5–241–1, U.S. Govt. Printing Office, Washington, D.C., 185 pp.

U.S. LAND OFFICE SURVEY

U.S. Govt. Printing Office **(1934)**, *Manual of instructions for the survey of the public lands of the United States,* Washington, D.C., 530 pp.

Johnson, H. B. **(1957)**, Rational and ecological aspects of the quarter section, *Geog. Rev.,* 47:330–348.

CHAPTER 4

ILLUMINATION OF THE GLOBE

Boggs, S. W. **(1931)**, Seasonal variations in daylight, twilight, and darkness, *Geog. Rev.,* 21:656–659.

Beckinsale, R. P. **(1945)**, The altitude of the zenithal sun: A geographic approach to determination and climatic significance, *Geog. Rev.,* 35:596–600.

Hosmer, G. L., and J. M. Robbins **(1948)**, *Practical astronomy,* 4th edition, John Wiley and Sons, New York, 355 pp.

Mehlin, T. G. **(1959)**, *Astronomy,* John Wiley and Sons, New York, 391 pp.

CURRENT ASTRONOMICAL DATA

Old Farmer's Almanac, issued annually, Yankee, Incorporated, Dublin, N. H.

Royal Astronomical Society of Canada, *The observer's handbook,* issued annually, Univ. of Toronto Press.

U.S. Naval Observatory, *The American ephemeris and nautical almanac,* issued annually, U.S. Govt. Printing Office, Washington, D.C.

U.S. Naval Observatory, *The Air Almanac,* issued annually, U.S. Govt. Printing Office, Washington, D.C.

World Journal Tribune, *The world almanac and book of facts,* issued annually, Newspaper Enterprise Association, New York. Hard cover edition: Doubleday and Co., Inc. New York.

CHAPTER 5

TIME

Strong, H. M. **(1935)**, Universal world time, *Geog. Rev.,* 25:479–484.

Pyne, T. E. **(1958)**, *Standard time,* Interstate Commerce Commission, U.S. Govt. Printing Office, Washington, D.C., 8 pp.

U.S. Naval Observatory, *The air almanac,* issued annually, U.S. Govt. Printing Office, Washington, D.C. Tables give standard times of most countries.

Strahler, A. N. **(1971)**, *The earth sciences,* 2nd edition, Harper and Row, New York, 824 pp. See Chapter 3.

North, J. D. **(1974)** The astrolabe, *Scientific American,* 230, No. 1:96–106.

CHAPTER 6

THE MOON

Mehlin, T. G. **(1959)**, *Astronomy,* John Wiley and Sons, New York, 391 pp.

Glasstone, S. **(1965)**, *Sourcebook on the space sciences,* Van Nostrand Co., Princeton, N.J., 937 pp. See Chapter 9.

Whipple, F. L. **(1968)**, *Earth, moon, and planets,* 3rd edition, Harvard Univ. Press, Cambridge, Mass., 297 pp. See Chapters 7, 8, and 9.

Carpenter, D. G., ed. **(1972)**, *Environmental space sciences,* Whitehall Company, Northbrook, Ill., 719 pp. See Chapter 26.

Hartmann, W. K. **(1972)**, *Moons and planets: an introduction to planetary science,* Wadsworth Publ. Co., Belmont, Calif., 404 pp. See Chapter 10, 11.

Marvin, U. B. **(1973)**, The moon after Apollo, *Technology Review,* 75, No. 8:13–23.

Muehlberger, W. R., and E. W., Wolfe **(1973)**, The challenge of Apollo 17, *Amer. Scientist,* 61:660–668.

TIDES

Macgowan, D. J. **(1855)**, On the eagre of the Tsien-Tang, *Am. J. Sci.,* 2nd ser., 20:305–314, Eyewitness account of the tidal bore at Hangchow.

Marmer, H. A. **(1921)**, Tide tables, *Geog. Rev.,* 11:406–413.

Marmer, H. A. **(1922)**, Tides in the Bay of Fundy, *Geog. Rev.,* 12:195–205.

Marmer, H. A. **(1923)**, Flood and ebb in New York Harbor, *Geog. Rev.,* 12:413–444.

Marmer, H. A. **(1925)**, Tides and currents in New York Harbor, U.S. Dept. Commerce, Coast and Geodetic

Survey, *Special Publ.* 111. Washington, D.C., 174 pp.

Marmer, H. A. **(1926)**, *The tide*, Appleton and Co., New York, 282 pp.

Marmer, H. A. **(1928)**, On cotidal maps, *Geog. Rev.*, 18:129–143.

Marmer, H. A. **(1932)**, Tides and tidal currents, Chapter 7 of Physics of the earth, Vol. 5: *Bull. Nat Research Council* 85, pp. 229–309.

Bauer, H. A. **(1933)**, A world map of tides, *Geog. Rev.*, 23:259–270.

Marmer, H. A. **(1948)**, Is the Atlantic Coast sinking? The evidence from the tide, *Geog. Rev.*, 38:652–657.

Defant, A. **(1958)**, *Ebb and flow*, Univ. of Mich. Press, Ann Arbor, 121 pp.

Darwin, G. H. **(1962)**, *The tides, and kindred phenomena of the solar system*, W. H. Freeman and Co., San Francisco, 378 pp. (Reprint. Originally published in 1898.)

Macmillan, D. H. **(1966)**, *Tides*, American Elsevier Publ. Co., New York, 240 pp.

Clancy, E. P. **(1968)**, *The tides; pulse of the earth*, Doubleday and Co., Garden City, N.Y., 228 pp.

Coast and Geodetic Survey, published annually, *Tide tables*, U.S. Govt. Printing Office, Washington, D.C. (Gives times and heights of high and low waters for principal ports.)

CHAPTER 7

GENERAL METEOROLOGY

Fairbridge, R. W., ed. **(1967)**, *The encyclopedia of atmospheric sciences and astrogeology*, Reinhold Publ. Co., New York, 1200 pp.

Petterssen, S. **(1969)**, *Introduction to meteorology*, 3rd edition, McGraw-Hill Book Co., New York, 333 pp.

Riehl, H. **(1972)**, *Introduction to the atmosphere*, 2nd edition, McGraw-Hill Book Co., New York, 516 pp.

Byers, H. R. **(1974)**, *General meteorology*, 4th edition, McGraw-Hill Book Co., New York, 461 pp.

UPPER ATMOSPHERE

Korff, S. A. **(1960)**, Geographical aspects of cosmic-ray studies, *Geog. Rev.*, 50:504–522.

Hare, F. K. **(1962)**, The stratosphere, *Geog. Rev.*, 52:525–547.

Bates, D. R. **(1964)**, *The planet earth*, 2nd edition, Pergamon Press, Oxford, 370 pp.

Craig, R. A. **(1968)**, *The edge of space; exploring the upper atmosphere*, Doubleday and Co., New York, 150 pp.

Carpenter, D. G., ed. **(1972)**, *Environmental space sciences*, Whitehall Co., Northbrook, Ill., 719 pp. See Chapters 13, 14, 15, and 19.

Jacobs, J. A., R. D. Russell, and J. T. Wilson **(1974)**, *Physics and geology*, 2nd edition, McGraw-Hill Book Co., New York, 622 pp. See Chapter 9.

GENERAL OCEANOGRAPHY

Fairbridge, R. W. **(1966)**, *The encyclopedia of oceanography*, Reinhold Publ. Corp., New York, 1021 pp.

Multiple authorship **(1969)**, *The ocean*, A Scientific American Book, W. H. Freeman, San Francisco, 140 pp.

Weyl, P. K. **(1970)**, *Oceanography*, John Wiley and Sons, New York, 535 pp.

Davis, R. A. **(1972)**, *Principles of oceanography*, Addison-Wesley Publ. Co., Reading, Mass., 434 pp.

Pirie, R. G., ed. **(1973)**, *Oceanography: contemporary readings in ocean sciences*, Oxford Univ. Press, New York, London, 530 pp.

CHAPTER 8

RADIATION AND HEAT BALANCE

Gates, D. M. **(1962)**, *Energy exchange in the biosphere*, Harper and Row, New York, 151 pp.

Larsson, P. **(1963)**, The distribution of albedo over arctic surfaces, *Geog. Rev.*, 53:572–579.

Geiger, R. **(1965)**, *The climate near the ground*, 4th edition, Harvard Univ. Press, Cambridge, Mass., 611 pp.

Miller, D. H. **(1965)**, The heat and water budget of the earth's surface, *Advances in Geophysics*, Vol. II, p. 176–277., Academic Press, New York.

Sellers, W. D. **(1965)**, *Physical climatology*, Univ. of Chicago Press, 272 pp.

Miller, D. H. **(1968)**, *A survey course: the energy and mass budget at the surface of the earth*, Commission on College Geography, Publ. No. 7, Assoc. of Amer. Geographers, 142 pp. Extensive bibliography.

Chang, J-H. **(1970)**, Global distribution of net radiation according to a new formula, *Annals A. A. G.*, 60:340–351.

Chang, J-H. **(1970)**, Potential photosynthesis and crop productivity, *Annals A. A. G.*, 60:92–101.

Leighly, J. **(1970)**, Graphic derivation of elements of the solar climate, *Annals A. A. G.*, 60:174–184.

Gates, D. M. **(1972)**, *Man and his environment: Climate*, Harper and Row, New York, 175 pp.

Terjung, W. H., and S. S-F. Louie **(1973)**, Energy budget and photosynthesis of canopy leaves, *Annals A. A. G.*, 63:109–130.

SOLAR ENERGY

Daniels, F. **(1964)**, *Direct use of the sun's energy*, Yale Univ. Press, New Haven, 374 pp.

Brinkworth, B. J. **(1972)**, Solar energy for man, John Wiley and Sons, New York, 251 pp.

Halacy, D. S., Jr. **(1973)**, *The coming age of solar energy*, revised edition, Harper and Row, New York, 231 pp.

AIR AND SOIL TEMPERATURES

Jefferson, M. **(1918)**, The real temperatures throughout North and South America, *Geog. Rev.*, 6:240–267.

Visher, S. S. **(1922)**, Laws of temperature. *Annals A. A. G.*, 13:15–40.

Jefferson, M. **(1926)**, Actual temperatures of South America, *Geog. Rev.*, 16:443–466.

Parkins, A. E. **(1926)**, The temperature region map, *Annals, A. A. G.*, 16:151–165.

Hartshorne, R. **(1938)**, Six standard seasons of the year, *Annals A. A. G.*, 28:165–178.

Jefferson, M. **(1938)**, Standard seasons, *Annals A. A. G.*, 28:1–12.

Jones, S. B. **(1942)**, Lags and ranges of temperature in Hawaii, *Annals A. A. G.*, 32:68–97.

Leighly, J. B. **(1947)**, Profiles of air temperatures normal to coast lines, *Annals A. A. G.*, 37:75–80.

Court, A. **(1949)**, How hot is Death Valley? *Geog. Rev.*, 39:214–220.

Court, A. **(1953)**, Temperature extremes in the United States, *Geog. Rev.*, 43:39–49.

Sumner, A. R. **(1953)**. Standard deviation of mean monthly temperatures in Anglo-America, *Geog. Rev.*, 43:50–59.

Miller, D. H. **(1956)**, The influence of open pine forest on daytime temperature in the Sierra Nevada, *Geog. Rev.*, 46:209–218.

Chang, Jen-Hu **(1957)**, World patterns of monthly soil temperature distribution, *Annals A. A. G.*, 47:241-249.

Fluker, B. J. **(1958)**, Soil temperatures, *Soil Sci.*, 86:35–46.

Ludlum, D. M. **(1963)**, Extremes of heat in the United States, *Weatherwise*, 16:108–129.

Ludlum, D. M. **(1963)**, Extremes of cold in the United States, *Weatherwise*, 16:275–291.

Dunbar, G. S. **(1966)**, Thermal belts in North Carolina. *Geog. Rev.*, 56:516-526.

Leighton, P. A. **(1966)**, Geographical aspects of air pollution, *Geog. Rev.*, 56:151–174.

VanLoon, H. **(1966)**, On the annual temperature range over the Southern oceans, *Geog. Rev.*, 56:497–515.

Kopeck, R. J. **(1967)**, Aerial patterns of seasonal temperature anomalies in the vicinity of the Great Lakes, *Bull. Am. Met. Soc.*, 48:884–889.

Williams, L. **(1967)**, Occurrence of high temperatures at Yuma Proving Ground, Arizona, *Annals A. A. G.*, 57:579–592.

Dury, G. H. **(1972)**, High temperature extremes in Australia, *Annals A. A. G.*, 62:388–400.

CARBON DIOXIDE, DUST, AND GLOBAL CLIMATE CHANGE

Bryson, R. A., and D. A. Baerreis **(1967)**, Possibilities of major climatic modification and their implications, *Bull. Am. Met. Soc.*, 48:136–142.

Davitaya, F. F. **(1969)**, Atmospheric dust content as a factor affecting glaciation and climatic change, *Annals A. A. G.*, 59:552–560.

Council on Environmental Quality **(1970)**, Man's inadvertent modification of weather and climate, *Bull. Am. Met. Soc.*, 51:1043–1047.

Landsberg, H. E. **(1970)**, Man-made climatic changes, *Science*, 170:1265–1274.

Singer, S. F., ed. **(1970)**, *Global effects of environmental pollution*, Springer-Verlag, New York, 218 pp.

Frisken, W. R. **(1971)**, Extended industrial revolution and climate change, *EOS*, 52:500–508.

Hare, F. K. **(1971)**, Future climates and future environments, *Bull. Am. Met. Soc.*, 52:451–456.

Massachusetts Institute of Technology **(1971)**, *Inadvertent climate modification*, Report of the Study of Man's Impact on Climate (SMIC), The MIT Press, Cambridge, Mass., 308 pp.

Rasool, S. I., and S. H. Schneider **(1971)**, Atmospheric carbon dioxide and aerosols: effects of large increases on global climates, *Science*, 173:138–141.

Budyko, M. I. **(1972)**, The future climate, *EOS*, 53:868–874.

Machta, L. **(1972)** Mauna Loa and global trends in air quality, *Bull. Am. Met. Soc.*, 53:402–420.

Hobbs, P. V., H. Harrison, and E. Robinson **(1974)**, Atmospheric effects of pollutants, *Science*, 183:909–915.

Kalnicky, R. A. **(1974)**, Climate change since 1950, *Annals A. A. G.*, 64:100–112.

Kukla, G. J., and H. J. Kukla **(1974)**, Increased surface albedo in the Northern Hemisphere, *Science*, 183:709–714.

URBANIZATION AND THE RADIATION-HEAT BALANCE

Kopec, J. **(1970)**, Further observations on the urban heat island in a small city, *Bull. Am. Met. Soc.*, 51:602–606.

Landsberg, H. **(1970)**, Man-made climatic changes, *Science*, 170:1265–1274.

Terjung, W. H., et al. **(1970)**, The energy balance climatology of a city-man system, *Annals A. A. G.*, 60:466–492.

Terjung, W. H. **(1970)**, Urban energy balance climatology, *Geog. Rev.*, 60:31–53.

Bach, W. **(1972)**, *Atmospheric pollution*, McGraw-Hill Book Co., New York, 144 pp. See Chapter 2.

CHAPTER 9

WINDS, GLOBAL CIRCULATION

Borchert, J. R. **(1953)**, Regional differences in world atmospheric circulation, *Annals, A. A. G.*, 43:14–26.

Hare, F. K., **(1960)**, The westerlies, *Geog. Rev.*, 50:345–367.

Warntz, W. **(1961)**, Transatlantic flights and pressure patterns, *Geog. Rev.*, 51:187–212.

Newell, R. E. **(1964)**, The circulation of the upper atmosphere, *Scientific American*, 210:62–74.

Slusser, W. **(1965)**, Wind rose maps of the United States, *Weatherwise*, 18:260–263.

Lorenz, E. N. **(1966)**, The circulation of the atmosphere, *American Scientist*, 54:402–420.

Reiter, E. R. **(1967)**, *Jet streams*, Doubleday and Co., New York, 189 pp.

Miller, W. H. **(1968)**, Santa Ana winds and crime, *Professional Geographer*, 20:23–37.

Ashwell, I. (**1971**), Warm blast across the snow-covered prairie, (Chinook winds) *Geographical Magazine*, 43:858–863.

Chang, J-H. (**1971**), The Chinese monsoon, *Geog. Rev.*, 61:370–395.

Newell, R. E. (**1971**), The global circulation of atmospheric pollutants, *Sci. American*, 224, No. 1:32–42.

Chang, J-H. (**1972**), *Atmospheric circulation systems and climates*, Oriental Publ. Co., Honolulu, Hawaii, 328 pp.

Harman, J. R., and J. G. Hehr (**1972**), Lake breezes and summer rainfall, *Annals A. A. G.*, 62:375–387.

Sonu, C. J., et al. (**1973**), Sea breeze and coastal processes, *EOS*, 54:820–833.

Kalnicky, R. A. (**1974**), Climatic change since 1950, *Annals A. A. G.*, 64:100–112.

OCEAN CURRENTS

Murphy, R. C. (**1923**), The oceanography of the Peruvian littoral with reference to the abundance and distribution of marine life, *Geog. Rev.*, 13:64–85.

Marmer, H. A. (**1929**), The Gulf Stream and its problems, *Geog. Rev.*, 19:457–478.

Pettersson, O. (**1929**), Changes in oceanic circulation and their climatic consequences, *Geog. Rev.*, 19:121–131.

Church, P. E. (**1932**), Surface temperatures of the Gulf Stream and its bordering waters, *Geog. Rev.*, 22:286–293.

Isaac, W. E. (**1937**), South African coastal waters in relation to ocean currents, *Geog. Rev.*, 27:651–664.

Munk, W. (**1955**), The circulation of the oceans, *Scientific American*, 193:96–104.

Defant, A. (**1961**), *Physical Oceanography*, Volume I, Pergamon Press, Oxford, 729 pp. See Part II. Dynamical oceanography.

Fairbridge, R. W. (**1966**), *Encyclopedia of oceanography*, Reinhold Publ. Co., New York, 1021 pp. See Ocean currents, Oceanic circulation.

Davis, R. A., Jr. (**1972**), *Principles of oceanography*, Addison-Wesley Publ. Co., Reading, Mass. 434 pp. See Chapter 5.

CHAPTER 10

HUMIDITY, CLOUDS, FOG, DEW

McAdie, A. (**1931**), The commercial importance of fog control, *Annals A. A. G.*, 21:91–100.

Stone, R. G. (**1935**), Fog in the United States and adjacent regions, *Geog. Rev.*, 25:111–134.

Ives, R. L. (**1941**), Colorado Front Range crest clouds and related phenomena, *Geog. Rev.*, 31:23–45.

Ashbel, D. (**1949**), Frequency and distribution of dew in Palestine, *Geog. Rev.*, 39:291–297.

Went, F. W. (**1955**), Fog, mist, dew, and other sources of water, *Yearbook of Agriculture, 1955*, U.S. Dept. Agriculture, pp. 103–109.

Patton, C. P. (**1956**), Climatology of summer fogs in the San Francisco Bay area, *Univ. of California Publ. in Geography*, 10, No. 3:113–200.

Ludlam, F. H., and R. S. Scorer (**1957**), *Cloud study; a pictorial guide*, John Murray, London, 80 pp.

World Meteorological Organization (**1956**), *International cloud atlas*, Geneva, Switzerland, 2 vols., English language edition.

Battan, L. J. (**1962**), *Cloud physics and cloud seeding*, Doubleday and Co., New York, 144 pp.

Schloss, M. (**1962**), Cloud cover of the Soviet Union, *Geog. Rev.*, 52:389–399.

Malkus, J. S. (**1963**), Cloud patterns over tropical oceans, *Science*, 141:767–778.

Duvdevani, S. (**1964**), Dew in Israel and its effects on plants, *Soil Sci.*, 98:14–21.

Court, A., and R. D. Gerston (**1966**), Fog frequency in the United States, *Geog. Rev.*, 56:543–550.

Myers, J. N. (**1968**), Fog, *Scientific Amer.* 219, No. 6:75–81.

Knight, C., and N. Knight (**1973**), Snow crystals, *Scientific Amer.* 228, No. 1:100–107.

PRECIPITATION

McAdie, A. (**1922**), Monsoon and trade winds as rain makers and desert makers, *Geog. Rev.*, 12:412–419.

Visher, S. S. (**1923**), The laws of winds and moisture, *Annals A. A. G.*, 13:169–207.

Sykes, G. (**1931**), Rainfall investigations in Arizona and Sonora by means of long-period rain gauges, *Geog. Rev.*, 21:229–233.

Church, J. E. (**1933**), Snow surveying, *Geog. Rev.*, 23:529–563.

Leighly, J. (**1935**), Continental precipitation on a rotating earth, *Geog. Rev.*, 25:657–666.

Visher, S. S. (**1938**), Rainfall-intensity contrasts in Indiana, *Geog. Rev.*, 28:627–637.

Tannehill, I. R. (**1947**), *Drought, its causes and effects*, Princeton Univ. Press, Princeton, N.J., 264 pp.

Thornthwaite, C. W. (**1947**), Climate and moisture conservation, *Annals A. A. G.*, 37:87–100.

Visher, S. S. (**1947**), Precipitation seasons in the United States, *Geog. Rev.*, 37:106–111.

Foster, E. E. (**1949**), *Rainfall and runoff*, The Macmillan Co., New York, 487 pp. See Chapters 2–7.

Gardner, C., Jr. (**1955**), Hauling down more water from the sky, *Yearbook of Agriculture, 1955*, U.S. Dept. Agriculture, pp. 91–95.

Hiatt, W. E., and R. W. Schloemer (**1955**), How we measure the variations in precipitation, *Yearbook of Agriculture, 1955*, U.S. Dept. Agriculture, pp. 78–84.

Tannehill, I. R. (**1955**), Is weather subject to cycles? *Yearbook of Agriculture, 1955*, U.S. Dept. Agriculture, pp. 84–90.

Work, R. A. (**1955**), Measuring snow to forecast water supplies. *Yearbook of Agriculture, 1955*, U.S. Dept. Agriculture, pp. 94–102.

Chow, V. T., ed. (**1964**), *Handbook of applied hydrology*, McGraw-Hill Book Co., New York. See Sections 9 and 10.

Small, R. T. (**1966**), Terrain effects on precipitation in Washington State, *Weatherwise,* 19:204–207.

Eichenlaub, V. L. (**1970**), Lake effect snowfall to the lee of the Great Lakes: its role in Michigan, *Bull. Am. Met. Soc.,* 51:403–412.

Williams, A., Jr. (**1973**), *The use of radar imagery in climatological research,* Resource Paper No. 21, Commission on College Geography, Association of American Geographers Washington, D.C., 29 pp.

THUNDERSTORMS, HAIL

Lemons, H. (**1942**), Hail as a factor in the regional climatology of the United States, *Geog. Rev.,* 32:471–475.

Byers, H. R. (*1949*), *The thunderstorm,* U.S. Dept. Commerce, Weather Bureau, U.S. Govt. Printing Office, Washington, D.C., 287 pp.

Flora, S. D. (**1956**), *Hailstorms of the United States,* Univ. of Oklahoma Press, Norman, 201 pp.

Battan, L. J. (**1961**), *The nature of violent storms,* Doubleday and Co., Garden City, N.Y., 158 pp.

Frisby, E. M. (**1964**), A study of hailstorms of the upper Great Plains of the North American continent, *Weatherwise,* 17:68–74.

Fankhauser, J. C., and C. W. Newton (**1965**), The migration of thunderstorms as related to wind and to moisture supply, *Weatherwise,* 18:68–73.

Brown, R. M. (**1967**), Hail, hail—A case history, Weatherwise, 20:254–258.

Knight, C., and N. Knight (**1971**), Hailstones, *Sci. American,* 224, No. 4:97–103.

Tyson, P. D. (**1971**), Spatial variation of rainfall spectra in South Africa, *Annals A. A. G.,* 61:711–720.

AIR POLLUTION

Bryson, R. A., and J. E. Kutzbach (**1968**) Air pollution, Resource Paper No. 2, Commission on College Geography, Assoc. of Amer. Geographers, Washington, D.C., 52 pp.

Leinwald, G. (**1969**), *Air and water pollution,* Washington Square Press, New York, 160 pp.

Neiburger, M. (**1969**), The role of meteorology in the study and control of air pollution, *Bull. Am. Met. Soc.,* 50:957–965.

Bach, W. (**1972**), *Atmospheric pollution,* McGraw-Hill Book Co., New York, 144 pp.

Bates, D. V. (**1972**), *A citizen's guide to air pollution,* McGill Queens Univ. Press, Montreal and London, 140 pp.

Brodine, V. (**1973**), *Air pollution,* Harcourt Brace Jovanovich, New York, 205 pp. Extensive bibliography.

de Nevers, N. (**1973**), Enforcing the Clean Air Act of 1970, *Sci. American,* 228, No. 6:14–21.

Kromm, D. E. (**1973**), Response to air pollution in Ljubljana, Yugoslavia, *Annals A. A. G.,* 63:208–217.

EFFECTS OF CITIES ON CLIMATE

Chandler, T. J. (**1967**), Absolute and relative humidities in towns, *Bull. Am. Met. Soc.,* 48:394–399.

Changnon, S. A. (**1968**), The LaPorte weather anomaly: fact or fiction? *Bull. Am. Met. Soc.,* 49:4–11.

Changnon, S. A., Jr. (**1969**), Recent studies of urban effects on precipitation in the United States, *Bull. Am. Met. Soc.,* 50:411–421.

Holzmann, B. G., and H. C. Thom (**1970**), The LaPorte precipitation anomaly, *Bull. Am. Met. Soc.,* 51:335–337.

Bach, W. (**1971**), Atmospheric turbidity and air pollution in greater Cincinnati, *Geog. Rev.,* 61:573–594.

Harman, J. R., and W. M. Elton (**1971**), The LaPorte, Indiana, recipitation anomaly, *Annals A. A. G.,* 61:468–480.

Bach, W. (**1972**), *Atmospheric pollution,* McGraw-Hill Book Co., New York, 144 pp. See Chapter 2.

Bryson, R. A., and J. E. Ross (**1972**), The climate of the city, pp. 51–68 in *Urbanization and environment,* T. R. Detwyler and M. G. Marcus, eds., Duxbury Press, Wadsworth Publ. Co., Belmont, Calif., 287 pp.

Huff, F. A., and S. A. Changnon (**1973**), Precipitation modification by major urban areas, *Bull. Am. Met. Soc.,* 54:1220–1232.

Landsberg, H. (**1973**), The meteorologically utopian city, *Bull. Am. Met. Soc.,* 54:86–89.

Terjung, W. H., and S. S-F. Louie (**1973**), Solar radiation and urban heat islands, *Annals A. G. G.,* 63:181–207.

Terjung, W. H. (**1974**), Climatic modification, pp. 105–151 in *Perspectives on environment,* Assoc. of Amer. Geographers, Washington, D.C., 395 pp.

CHAPTER 11

FRONTS, CYCLONES

Visher, S. S. (**1925**), Effects of tropical cyclones upon the weather of middle latitudes, *Geog. Rev.,* 15:106–114.

Thornthwaite, C. W. (**1937**), The life history of rainstorms, *Geog. Rev.,* 27:92–111.

Brooks, C. F. (**1939**), Hurricanes into New England, *Geog. Rev.,* 29:119–127.

James, P. E. (**1939**), Air masses and fronts of South America, *Geog. Rev.,* 29:132–134.

Brooks, C. F., and C. Chapman (**1945**), The New England hurricane of September, 1944, *Geog. Rev.,* 35:132–136.

Riehl, H. (**1954**), *Tropical meteorology,* McGraw-Hill Book Co., New York, 392 pp.

Tannehill, I. R. (**1956**), *Hurricanes,* Princeton Univ. Press, Princeton, N.J., 308 pp.

Rosendal, H. E. (**1963**), Mexican west coast tropical cyclones, 1947–1961, *Weatherwise,* 16:226–229.

Dunn, G. E., and B. I. Miller (**1964**), *Atlantic hurricanes,* Louisiana State Univ. Press, Baton Rouge, 377 pp.

Cullen, W. C., and L. W. Crow (**1967**), A case history of residential property damage as a hurricane moves inland, *Bull. Amer. Met. Soc.,* 48:10–12.

Helm, T. (**1967**), *Hurricanes: Weather at its worst,* Dodd, Mead and Co., New York, 234 pp.

Hovey, W., K. Sirinek, and F. Storer (**1967**), A synop-

tic analysis of New England's "backdoor" cold fronts, *Weatherwise*, 20:264–267.
Miller, B. I. **(1967)**, Characteristics of hurricanes, *Science*, 157:1389–1399.
Cressman, G. P. **(1969)**, Killer storms, *Bull. Am. Met. Soc.*, 50:850–855.
Frank, N. L., and S. A. Husain **(1971)**, The deadliest tropical cyclone in history?, *Bull. Am. Met. Soc.*, 52:438–444.
White, R. M. **(1972)**, The national hurricane warning program, *Bull. Am. Met. Soc.*, 53:631–533.

TORNADOES

Carey, J. P. **(1917)**, The central Illinois tornado of May 26, 1917, *Geog. Rev.*, 4:122–130.
Flora, S. D. **(1954)**, *Tornadoes of the United States*, Univ. of Oklahoma Press, Norman, 221 pp.
Battan, L. J. **(1961)**, *The nature of violent storms*, Doubleday and Co., Garden City, N.Y., 158 pp.
Changnon, S. A. Jr., and R. G. Semonin **(1966)**, A great tornado disaster in retrospect, *Weatherwise*, 19:56–65.
Kessler, E. **(1970)**, Tornadoes, *Bull. Am. Met. Soc.*, 51:926–936.
Kessler, E. **(1972)**, On tornadoes and their modification, *Technology Review*, 74, No. 6:48–55.
Sims, J. H., and D. B. Baumann **(1972)**, The tornado threat: coping styles of the North and South, *Science*, 176:1386–1392.
Fujita, T. T. **(1973)**, Tornadoes around the world, *Weatherwise*, 26:56–62, 78–83.

PLANNED WEATHER MODIFICATION

Woodley, W. L. **(1970)**, Rainfall enhancement by dynamic cloud modification, *Science*, 170:127–132.
Agee, E. M. **(1971)**, An artificially induced local snowfall, *Bull. Am. Met. Soc.*, 52:557–560.
Hammond, A. L. **(1971)**, Weather modification: a technology coming of age, *Science*, 172:548–549.
Meyer, J. W. **(1971)**, Toward hurricane surveillance and control, *Technology Rev.*, 74, No. 1:59–66.
Simpson, J., and W. L. Woodley **(1971)**, Seeding cumulus in Florida: new 1970 results, *Science*, 172:117–126.
Droessler, E. G. **(1972)**, Weather modification; review and perspective, *Bull. Am. Met. Soc.*, 53:345–348.
Kessler, E. **(1972)**, On tornadoes and their modification, *Technology Rev.*, 74, No. 6:48–55.
Sewell, W. R. D., et al **(1973)**, *Modifying the weather: a social assessment*, Western Geographical Series, Volume 9, Department of Geography, Univ. of Victoria, Victoria, B. C., 349 pp. Extensive bibliography.

CHAPTER 12

GENERAL HYDROLOGY: HYDROLOGIC CYCLE

Foster, E. E. (1949), *Rainfall and runoff*, The Macmillan Co., New York, 487 pp.
Wisler, C. O., and E. F. Brater **(1949)**, *Hydrology*, John Wiley and Sons, New York, 419 pp.
Ackerman, W. C., E. A. Colman, and H. O. Ogrosky **(1955)**, From ocean to sky to land to ocean, *Yearbook of Agriculture, 1955*, U.S. Dept. Agriculture, pp. 41–51.
Thornthwaite, C. W., and J. R. Mather **(1955)**, The water balance, *Publications in climatology*, 8, No. 1, Laboratory of Climatology, Centerton, N.J., 104 pp.
Chow, V. T., ed. **(1964)**, *Handbook of applied hydrology*, McGraw-Hill Book Co., New York. See Sections 1, 11, 12, 20–24.
Miller, D. H. **(1965)**, The heat and water budget of the earth's surface, *Advances in Geophysics*, Volume 11, Academic Press, New York, pp. 175–302. Extensive bibliography.
Bruce, J. P., and R. H. Clark **(1966)**, *Introduction to hydrometeorology*, Pergamon Press, Oxford, 319 pp.
Miller, D. H. **(1968)**, *A survey course: the energy and mass budget at the surface of the earth*, Commission on College Geography, Publ. No. 7, Assn. Amer. Geog., Washington, D.C., 142 pp.
Chorley, R. J., ed. **(1971)**, *Introduction to physical hydrology*, Methuen and Co., London, 211 pp.
Kazmann, R. G. **(1972)**, *Modern hydrology*, 2nd. edition, Harper and Row, New York, 365 pp.
Lvovitch, M. I. **(1973)**, The global water balance, United States I. H. D. Bull. No. 23, in *EOS*, 54:28–42.

THE SOIL MOISTURE BALANCE

Thornthwaite, C. W. **(1948)**, An approach toward a rational classification of climate, *Geog. Rev.*, 38:55–94.
Bernstein, L. **(1955)**, The needs and uses of water by plants, *Yearbook of Agriculture, 1955*, U.S. Dept. Agriculture, pp. 18–25.
Fletcher, H. C., and H. B. Elmendorf **(1955)**, Phreatophytes—a serious problem in the West, *Yearbook of Agriculture, 1955*, U.S. Dept. Agriculture, pp. 423–429.
Haise, H. R. **(1955)**, How to measure the moisture in the soil, *Yearbook of Agriculture, 1955*, U.S. Dept. Agriculture, pp. 362–371.
Musgrave, G. W. **(1955)**, How much rain enters the soil? *Yearbook of Agriculture, 1955*, U.S. Dept. Agriculture, pp. 151–159.
Richards, L. A. **(1955)**, Retention and transmission of water in soil, *Yearbook of Agriculture, 1955*, U.S. Dept. of Agriculture, pp. 144–151.
Thornthwaite, C. W., and J. R. Mather **(1955)**, *The water balance*, Drexel Inst. of Technology, Laboratory of Climatology, Publ. in Climatology, Vol. 8, No. 1, Centerton, N.J., 86 pp.
Thornthwaite, C. W., and J. R. Mather **(1955)**, The water budget and its use in irrigation, *Yearbook of Agriculture, 1955*, U.S. Dept. Agriculture, pp. 346–358.
Wadleigh, C. H. **(1955)**, Soil moisture in relation to plant growth. *Yearbook of Agriculture, 1955*, U.S. Dept. Agriculture, pp. 358–361.

Chang, J-H. (**1959**), An evaluation of the 1948 Thornthwaite classification, *Annals, A. A. G.*, 49:24–30.

Horton, J. H., and R. H. Hawkins (**1965**), Flow path of rain from the soil surface to the water table, *Soil Sci.*, 100:377–383.

Hillel, D. (**1971**), *Soil and water; physical principles and processes*, Academic Press, New York, 288 pp.

GROUND WATER

Bryan, K. (**1919**), Classification of springs, *J. Geol.*, 27:522–561.

Meinzer, O. E. (**1923**), The occurrence of ground water in the United States with a discussion of principles, U.S. Geol. Survey, *Water Supply Paper 489*, 321 pp.

Stearns, H. T. (**1936**), Origin of large springs and their alcoves along the Snake River in southern Idaho, *J. Geol.*, 44:429–450.

Hubbert, M. K. (**1940**), Theory of ground water motion, *J. Geol.*, 43:708–728.

Thomas, H. E. (**1951**), *The conservation of ground water*, McGraw-Hill Book Co., New York, 321 pp.

Gregor, H. F. (**1952**), The Southern California water problem in the Oxnard area, *Geog. Rev.*, 42:16–36.

Meigs, P. (**1952**), Water problems in the United States, *Geog. Rev.*, 42:346–366.

Garver, H. L. (**1955**), Water supplies for homes in the country, *Yearbook of Agriculture, 1955*, U.S. Dept. Agriculture, pp. 655–663.

Muckel, D. C. (**1955**), Pumping ground water so as to avoid overdraft, *Yearbook of Agriculture, 1955*, U.S. Dept. Agriculture, pp. 294–301.

Muckel, D. C., and L. Schiff (**1955**), Replenishing ground water by spreading, *Yearbook of Agriculture, 1955*, U.S. Dept. Agriculture, pp. 302–310.

Parker, G. G. (**1955**), The encroachment of salt water into fresh, *Yearbook of Agriculture, 1955*, U.S. Dept. Agriculture, pp. 615–635.

Rohwer, C. (*1955*), Wells and pumps for irrigated lands, *Yearbook of Agriculture, 1955*, U.S. Dept. Agriculture, pp. 285–295.

Thomas, H. E. (**1956**), Changes in quantities and qualities of ground and surface waters, pp. 542–563 of *Man's role in changing the face of the earth*, Univ. of Chicago Press, Chicago, 1193 pp.

Cressey, G. B. (**1957**), Water in the desert, *Annals A. A. G.*, 47:105–124.

Todd, D. K. (**1959**), *Ground water hydrology*, John Wiley and Sons, New York, 336 pp.

GROUND WATER POLLUTION, SALT WATER INTRUSION

LeGrand, H. E. (**1965**), Environmental framework of ground-water contamination, *Ground Water*, 3, No. 2:11–15.

Heath, R. C., B. L. Foxworthy, and P. Cohen (**1966**), *The changing patterns of groundwater development on Long Island, New York*, Circular 524, U.S. Geological Survey, Washington, D.C., 10 pp.

McGauhey, P. H. (**1968**), Man-made contamination hazards, *Ground Water*, 6, No. 3:10–13.

Emrich, G. H., and G. L. Merritt (**1969**), Effects of mine drainage on ground water, *Ground Water*, 7, No. 7:27–32.

Parizek, R. R. (**1971**), Impact of highways on the hydrogeologic environment, pp. 151–199 in *Environmental Geomorphology*, D. R. Coates, ed., Publ. in Geomorphology, State Univ. of New York, Binghamton, N.Y., 262 pp.

Franke, O. L. and N. E. McClymonds (**1973**), *Summary of the hydrologic situation on Long Island, New York, as a guide to water-management alternatives*, Professional Paper 627-F, U.S. Geol. Survey, Washington, D.C., 59 pp.

RUNOFF

Goldthwait, J. W. (**1928**), The gathering of floods in the Connecticut River system, *Geog. Rev.*, 18:428–445.

Corbett, D. M., and others (**1943**), Stream-gauging procedure, U.S. Geol. Survey, *Water Supply Paper*, 888, 245 pp.

Foster, E. E. (**1948**), *Rainfall and runoff*, The Macmillan Co., New York, 487 pp.

Colman, E. A. (**1953**), *Vegetation and watershed management*, The Ronald Press Co., New York, 412 pp.

Langbein, W. B., and J. V. B. Wells (**1955**), The water in the rivers and creeks, *Yearbook of Agriculture, 1955*, U.S. Dept. Agriculture, pp. 52–62.

Wisler, C. O., and E. F. Brater (**1959**), *Hydrology*, 2nd edition, John Wiley and Sons, New York, 408 pp.

Chow, V. T. (**1964**), *Handbook of applied hydrology*, McGraw-Hill Book Co., New York. Sections 14 and 15.

Grover, N. C., and A. W. Harrington (**1966**), *Stream flow measurements, records and their uses*, Dover Publications, New York, 363 pp.

Kazmann, R. G. (1972), *Modern hydrology*, 2nd edition, Harper and Row, New York, 365 pp.

FLOODS

Brooks, C. F., and A. H. Thiessen (**1937**), The meteorology of the great floods in the eastern United States, *Geog. Rev.*, 27:269–290.

Leopold, L. B., and T. Maddock, Jr. (**1954**), *Big dams, little dams, and land management*, The Ronald Press Co., New York, 278 pp.

Leopold, L., and T. Maddock, Jr. (**1954**), *The flood control controversy*, The Ronald Press Co., New York, 255 pp.

Brown, C. B., and W. T. Murphy (**1955**), Conservation begins on the watersheds, *Yearbook of Agriculture, 1955*, U.S. Dept. Agriculture, pp. 161–165.

Ford, E. C., and W. L. Cowan, and H. N. Holtan (**1955**), Floods—And a program to alleviate them, *Yearbook of Agriculture, 1955*, U.S. Dept. Agriculture, pp. 170–176.

Heard, W. L., and V. B. MacNaughton (**1955**), The Yazoo-Little Tallahatchie flood prevention project, *Yearbook of Agriculture, 1955*, U.S. Dept. Agriculture, pp. 199–205.

Hoyt, W. G., and W. B. Langbein (**1955**), *Floods*, Princeton Univ. Press, Princeton, N.J., 469 pp.
Kautz, H. M. (**1955**), The story of Sandstone Creek watershed, *Yearbook of Agriculture, 1955*, U.S. Dept. Agriculture, pp. 210–218.
Matson, H. O., W. L. Heard, G. E. Lamp, and D. M. Ilch (**1955**), The possibilities of land treatment in flood prevention, *Yearbook of Agriculture, 1955*, U.S. Dept. Agriculture, pp. 176–179.
Storey, H. C. (**1955**), Frozen soil and spring and winter floods, *Yearbook of Agriculture, 1955*, U.S. Dept. Agriculture, pp. 179–184.
Bordne, E. (**1957**), Some hydrologic aspects of the flood of August, 1955, in a Connecticut valley, *Geog. Rev.*, 47:211–223.
White, G. F., et al (**1958**), *Changes in urban occupance of flood plains in the United States*, Research Paper 57, Dept. of Geog., University of Chicago, 235 pp.
Burton, I., and R. W. Kates (**1964**), The floodplain and the seashore: A comparative analysis of hazard-zone occupance, *Geog. Rev.*, 54:366–385.
White, G. H. (**1964**), Choice of adjustment to floods, Research Paper 93, Dept. of Geog. Univ. of Chicago, 150 pp.
Haley, R. J. (**1965**), The spring flood of April–May 1965 in the upper Mississippi Valley, *Weatherwise*, 18:115–119.
Henry, W. K. (**1965**), The ice jam floods of the Yukon River, *Weatherwise*, 18:81–85.
Bue, C. D. (**1967**), *Flood information for flood-plain planning*, Circular 539, U.S. Geol. Survey, Washington, D.C., 10 pp.
Morgan, A. E. (**1971**), *Dams and other disasters*, Porter Sargent Publisher, Boston, Mass., 422 pp.
Thompson, H. J. (**1972**), The Black Hills flood, *Weatherwise*, 25:162–167,173.

HYDROLOGIC EFFECTS OF URBANIZATION

Wolman, M. G., and P. A. Schick (**1967**), Effects of construction on fluvial sediment, urban and suburban areas of Maryland, *Water. Resources Res.*, 3:451–462.
Leopold, L. B. (**1968**), *Hydrology for urban planning*, Circular 554, U.S. Geol. Survey, Washington, D.C., 18 pp.
Emerson, J. W. (**1971**), Channelization: a case study, *Science*, 173:325–326.
Mrowka, J. P. (**1974**), Man's impact on stream regimen and quality, pp. 79–104 in *Perspectives on Environment*, Assoc. of Amer. Geographers, Washington, D.C., 395 pp.

WATER POLLUTION

Boughey, A. S. (**1971**), *Man and the environment*, The Macmillan Co., New York, 472 pp. See Chapter 11.
Edmonson, W. T. (**1971**), Fresh water pollution, pp. 213–229 in *Environment; resources, pollution and society*, W. W. Murdoch, ed., Sinauer Assoc. Inc., Stamford, Conn., 440 pp.
Hubschman, J. H. (**1971**), Lake Erie: pollution abatement, then what? *Science*, 171:536–540.
Nelkin, D. (**1971**), *The Cayuga Lake controversy*, Cornell Univ. Press, Ithaca, N.Y., 128 pp.
Strobbe, M. A., ed. (**1971**), *Understanding environmental pollution*, C. V. Mosby Co., Saint Louis, Mo., 357 pp.
Wolman, M. G. (**1971**), The nation's rivers, *Science*, 174:905–818.
Bohn, H. L., and R. C. Cauthorn (**1972**), Pollution: the problem of misplaced waste, *Amer. Scientist*, 60:561–565.
Westman, W. F. (*1972*), Some basic issues in water pollution control legislation, *Amer. Scientist*, 60:767–773.
Strahler, A. N., and A. H. Strahler (**1974**), *Introduction to environmental science*, Hamilton Publ. Co., Santa Barbara, Calif., 632 pp. See Chapters 13, 21, and 23.

FRESH WATER RESOURCES

Fowler, F. J. (**1950**), Some problems of water distribution between East and West Punjab, *Geog. Rev.*, 40:583–599.
Borchert, J. R. (**1954**), The surface water supply of American municipalities, *Annals A. A. G.*, 44:15–32.
Van Burkalow, A. (**1959**), The geography of New York City's water supply; a study of interactions, *Geog. Rev.*, 49:369–386.
Quinn, F. (**1968**), Water transfers: must the American west be won again? *Geog. Rev.*, 58:108–132.
Schneider, W. J., and A. M. Spieker (**1969**), *Water for the Cities—the outlook*, Circular 601-A, U.S. Geol. Survey, Washington, D.C., 6 pp.
Loeffler, M. J. (**1970**), Australian-American interbasin water transfer, *Annals A. A. G.*, 60:493–516.
Thomas, H. E. (**1970**), *Water as an urban resource and nuisance*, Circular 601D, U.S. Geological Survey, Washington, D.C., 9 pp.
Todd, D. K., ed. (**1970**), *The water encyclopedia*, Water Information Center, Port Washington, N.Y., 559 pp.
van Hylckama, T. E. A. (**1971**), Water resources, pp. 135–155 in *Environment: resources, pollution and society*, W. W. Murdoch, ed., Sinauer Assoc. Inc., Stamford, Conn., 440 pp.
Wollman, N., and G. W. Bonem (**1971**), *The outlook for water: quality, quantity, and national growth*, The Johns Hopkins Press, Baltimore, Md., 286 pp.
Hobbs, J., and E. Woolmington (**1972**), Water and urban decentralization in New South Wales, *Annals A. A. G.*, 62:37–41.
Murray, C. R. (**1972**), *Estimated use of water in the United States*, Circular 676, U.S. Geol. Survey, Washington, D.C., 37 pp.

CHAPTER 14

GENERAL CLIMATOLOGY, CLIMATE CLASSIFICATION

Russell, R. J. (**1934**), Climate years, *Geog. Rev.*, 24:92–103.
Brooks, C. F. (**1948**), The climatic record: its con-

tent, limitations and geographic value, *Annals A. A. G.*, 38:153–168.
Brooks, C. E. P. **(1949)**, *Climate through the ages*, McGraw-Hill Book Co., New York, 395 pp.
Kendrew, W. G. **(1953)**, *The climates of the continents*, Oxford Univ. Press, London, 607 pp.
Hare, F. K. **(1955)**, Dynamic and synoptic climatology, *Annals A. A. G.*, 45:152–162.
Court, A. **(1957)**, Climatology: Complex, dynamic, and synoptic, *Annals A. A. G.*, 47:125–136.
Landsberg, H. E. **(1958)**, Trends in climatology, *Science*, 128:749–758.
Lydolph, P. E. **(1959)**, Fedorov's complex method in climatology, *Annals A. A. G.*, 49:120–144.
Trewartha, G. T. **(1961)**, *The earth's problem climates*, Univ. of Wisconsin Press, Madison, 334 pp.
Bailey, H. P. **(1964)**, Toward a unified concept of the temperate climate, *Geog. Rev.*, 54:516–545.
Hare, F. K. **(1966)**, The concept of climate, *Geography*, 51:99–110.
Trewartha, G. T. **(1968)**, *An introduction to climate*, McGraw-Hill Book Co., New York, 408 pp.
Oliver, J. E. **(1970)**, A genetic approach to climatic classification, *Annals A. A. G.*, 60:615–637.
Chang, J-H. **(1972)**, *Atmospheric circulation systems and climates*, The Oriental Publ. Co., Honolulu, Hawaii, 328 pp.
Lamb, H. H. **(1972)**, *Climate: present, past and future*, volume I., Methuen, London, 613 pp.
McBoyle, G. **(1973)**, *Climate in review*, Houghton Mifflin, Boston, 314 pp.
Oliver, J. E. **(1973)**, *Climate and Man's environment; an introduction to applied climatology*, John Wiley & Sons, New York, 517 pp. Extensive bibliography.
Barrett, E. C. **(1974)**, *Climatology from satellites*, Methuen, London, 424 pp.
Critchfield, H. J. **(1974)**, *General climatology*, 3rd edition, Prentice-Hall, Englewood Cliffs, N. J., 446 pp.

CLIMATE DATA

Clayton, H. H. **(1927, 1934)**, World weather records, *Smithsonian Miscellaneous Collections*, 79, 1927, and 90, 1934. Data of many stations throughout the world.
U.S. Dept. Agriculture **(1928)**, *Atlas of American agriculture*, U.S. Govt. Printing Office, Washington, D.C.
Brooks, C. F., A. J. Connor, and others **(1936)**, *Climatic maps of North America*, Harvard Univ. Press, Cambridge, Mass.
Thomas, M. K. **(1953)**, *Climatological atlas of Canada*, Nat. Research Council, Ottawa, Canada, 253 pp.
Visher, S. S. **(1954)**, *Climatic atlas of the United States*, Harvard Univ. Press, Cambridge, Mass., 403 pp.
U.S. Navy **(1955** and later), *Marine climatic atlas of the world*, 4 vols., U.S. Govt. Printing Office, Washington, D.C.
Meteorological Office of Great Britain **(1958)**, *Tables of temperature, relative humidity and precipitation for the world*, Her Majesty's Stationery Office, London.

APPLIED CLIMATOLOGY

U.S. Dept. Agriculture **(1941)**, Climate and man, *Yearbook of Agriculture, 1941*, U.S. Govt. Printing Office, Washington, D.C. Many papers on climate topics. United States climate data.
Curry, L. **(1952)**, Climate and economic life: a new approach with examples from the United States, *Geog. Rev.*, 42:367–383.
Lee, D. H. K. **(1953)**, Physiological climatology as a field of study, *Annals A. A. G.*, 43:127–137.
Griffiths, J. F. **(1966)**, *Applied climatology, an introduction*, Oxford University Press, London, 118 pp.
Terjung, W. H. **(1966)**, Physiological climates of the conterminous United States: A bioclimatic classification based on man, *Annals A. A. G.*, 56:141–179.
Shaw, R. H., ed. **(1967)**, *Ground level climatology*, Publ. No. 86, Amer. Assoc. for the Advancement of Sci., Washington, D. C., 395 pp.
Terjung, W. H. **(1967)**, Annual physioclimatic stresses and regimes in the United States, *Geog. Rev.*, 57:225–240.
Chang, J-H. **(1968)**, *Climate and agriculture; an ecological survey*, Aldine Publ. Co., Chicago, Ill., 304 pp. Extensive bibliography.
Sewell, W. R., R. W. Kates, and L. E. Phillips **(1968)**, Human response to weather and climate, *Geog. Rev.*, 58:262–280.
Gates, D. M. **(1972)**, *Man and his environment: climate*, Harper and Row, New York, 175 pp.
Wilken, G. C. **(1972)**, Microclimate management by traditional methods, *Geog. Rev.*, 62:544–560.
Oliver, J. E. **(1973)**, *Climate and man's environment; an introduction to applied climatology*, John Wiley & Sons, New York, 517, pp.

ENERGY-BALANCE CLIMATOLOGY

Terjung, W. H. **(1969)**, A global classification of solar radiation, *Solar Energy*, 13:67–81.
Terjung, W. H. et al **(1970)**, The energy balance climatology of a city-man system, *Annals, A. A. G.* 60:466–492.
Terjung, W. H. **(1970)**, Urban energy balance climatology, *Geog. Rev.*, 60:31–53.
Terjung, W. H., and S. S-F. Louis **(1971)**, Potential solar radiation climates of man, *Annals A. A. G.*, 61:481–500.
Terjung, W. H., and S. S-F. Louie **(1972)**, Energy input-output climates of the world: a preliminary attempt, *Arch. Met. Geoph. Biokl.*, Ser. B, 20:129–166.

KÖPPEN SYSTEM

Köppen, W. **(1923)**, *Die Klimate der Erde; Grundriss der Klimakunde*, Walter de Gruyter Co., Berlin, 369 pp.
Köppen, W., and R. Geiger **(1930** and later), *Hand-*

buch der Klimatologie, 5 vols., Gebrüder Borntraeger, Berlin. See Vol. 1, Part C (1936) for general analysis of Köppen system of climatology.

Ackerman, E. A. (**1941**), The Köppen classification of climates in North America, *Geog. Rev.,* 31:105–111.

Kesseli, J. E. (**1942**), The climates of California according to the Köppen Classification, *Geog. Rev.,* 32:476–480.

Köppen-Geiger (**1954**), Klima der Erde (map), Justus Perthe, Darmstadt, Germany. American distributor, A. J. Nystrom and Co., Chicago.

Lewis, P. F. (**1961**), Dichotomous keys to the Köppen system, *Professional Geographer,* 13:25–31.

Shear, J. A. (**1966**), A set-theoretic view of the Köppen dry climates, *Annals A. A. G.,* 56:508–515.

Wilcock, A. A. (**1968**), Köppen after fifty years *Annals, A. A. G.,* 58:12–28.

THORNTHWAITE SYSTEM

Thornthwaite, C. W. (**1931**), The climates of North America, according to a new classification, *Geog. Rev.,* 21:633–655.

Thornthwaite, C. W. (**1933**), The climates of the earth, *Geog. Rev.,* 23:433–440, with world map.

Thornthwaite, C. W. (**1943**), Problems in the classification of Climates, *Geog. Rev.,* 33:233–255.

Thornthwaite, C. W. (**1948**), An approach toward a rational classification of climate, *Geog. Rev.,* 38:55–94.

Wilcock, A. A. (**1951**), Potential evapotranspiration: a simplification of Thornthwaite's method, *Proc. Royal Soc. Victoria,* 63:25–30.

Carter, D. B. (**1954**), Climates of Africa and India according to Thornthwaite's 1948 classification, The Johns Hopkins University, Laboratory of Climatology, Centerton, N.J., Publ. in Climatology, vol. 7, No. 4:453–470.

Carter, D. B. (**1956**), The water balance of the Mediterranean and Black seas, Drexel Inst. of Technology, Laboratory of Climatology, Publ. in Climatology, vol. 9, No. 3:123–174.

van Hylckama (**1956**), The water balance of the earth, Drexel Institute of Climatology, Laboratory of Climatology, Centerton, N.J., Publ. in Climatology, vol. 9, No. 2, 117 pp.

Thornthwaite Associates (**1957**), Instructions and tables for computing potential evapotranspiration and the water balance, 5th printing, C. W. Thornthwaite Associates, Laboratory of Climatology, Elmer, N.J., vol. 10, No. 3:311 pp.

Chang, Jen-Hu (**1959**), An evaluation of the 1948 Thornthwaite classification, *Annals A. A. G.,* 49:24–30.

Thornthwaite Associates (**1962–1965**), Average climatic water balance data of the continents, Parts I–VIII, C. W. Thornthwaite Associates, Laboratory of Climatology, Elmer, N.J., Volumes 15 through 18.

Mather, J. R., and G. A. Yoshioka (**1968**), The role of climate in the distribution of vegetation, *Annals A. A. G.,* 58:29–41.

Basile, R. M., and S. W. Corbin (**1969**), A graphical method for determining Thornthwaite climate classifications, *Annals A. A. G.,* 59:561–572.

CHAPTER 15

WET EQUATORIAL CLIMATE

Ross, E. (**1919**), The climate of Liberia and its effect on man, *Geog. Rev.,* 7:387–402.

Scrivenor, J. B. (**1921**), The physical geography of the southern part of the Malay Peninsula, *Geog. Rev.,* 11:351–371.

van Valkenburg, S. (**1925**), Java: the economic geography of a tropical island, *Geog. Rev.,* 15:563–583.

Stamp, L. D. (**1930**), Burma: an undeveloped monsoon country, *Geog. Rev.,* 20:86–109.

Harley, G. W. (**1939**), Roads and trails in Liberia, *Geog. Rev.,* 29:447–460.

Brass, L. J. (**1941**), Stone age agriculture in New Guinea, *Geog. Rev.,* 31:555–569.

Pendleton, R. L. (**1942**), Land utilization and agriculture of Mindanao, Philippine Islands, *Geog. Rev.,* 32:180–210.

Schneeberger, W. F. (**1945**), The Kerayan-Kalabit highland of central northeast Borneo, *Geog. Rev.,* 35:544–562.

Lee, D. H. K. (**1951**), Thoughts on housing for the humid tropics, *Geog. Rev.,* 41:124–147.

Lee, D. H. K. (**1957**), *Climate and economic development in the tropics,* Harper and Bros., New York, 182 pp.

Spencer, J. E. (**1959**), Seasonality in the tropics: the supply of fruit to Singapore, *Geog. Rev.,* 49:475–484.

Fosberg, F. R., B. J. Garnier, and A. W. Küchler (**1961**), Delimitation of the humid tropics, *Geog. Rev.,* 51:333–347.

Ignatieff, V., and Lemos, P. (**1963**), Some management aspects of more important tropical soils, *Soil Sci.,* 95:243–249.

Chang, J-H. (**1968**), The agricultural potential of the humid tropics, *Geog. Rev.,* 58:333–361.

Nieuwolt, S. (**1968**), Diurnal rainfall variation in Malaya, *Annals A. A. G.,* 58:313–326.

Harris, D. R. (**1971**), The ecology of swidden cultivation in the upper Orinoco rain forest, Venezuela, *Geog. Rev.,* 61:475–495.

TRADE WIND LITTORAL CLIMATE

Durland, W. D. (**1922**), The forests of the Dominican Republic, *Geog, Rev.,* 12:206–222.

James, P. E. (**1932**), The coffee lands of southeastern Brazil, *Geog, Rev.,* 22:225–244.

van Royen, W. (**1938**), A geographical reconnaissance of the Cibao of Santo Domingo, *Geog. Rev.,* 28:556–572.

von Hagen, V. W. (**1940**), The Mosquito Coast of Honduras and its inhabitants, *Geog. Rev.,* 30:238–259.

Portig, W. H. (**1965**), Central American rainfall, *Geog. Rev.,* 55:68–90.

TROPICAL DESERT AND STEPPE CLIMATES

Hobbs, W. H. **(1917)**, A pilgrimage in northeastern Africa, with studies of desert conditions, *Geog. Rev.*, 3:337–355.

MacDougal, D. T. **(1917)**, A decade of the Salton Sea, *Geog. Rev.*, 3:457–473.

Gautier, E. F. **(1925)**, The trans-Saharan railway, *Geog. Rev.*, 15:51–69.

Clapp, F. F. **(1926)**, In the northwest of the Australian desert, *Geog. Rev.*, 16:206–231.

Taylor, G. **(1926)**, The frontiers of settlement in Australia, *Geog. Rev.*, 16:1–25.

de Martonne, E. **(1927)**, Regions of interior-basin drainage, *Geog. Rev.*, 17:397–414.

Sykes, G. **(1927)**, The Camino del Diablo: with notes on a journey in 1925, *Geog. Rev.*, 17:62–74.

Hoover, J. W. **(1929)**, The Indian country of southern Arizona, *Geog. Rev.*, 19:38–60.

Wilson, E. D. **(1931)**, New mountains in the Yuma Desert, *Geog. Rev.*, 21:221–228.

Shreve, F. **(1934)**, Rainfall, runoff and soil moisture under desert conditions, *Annals A. A. G.*, 24: 131–156.

Gautier, E. F., tr. by D. F. Mayhen **(1935)**, *Sahara, the great desert,* Columbia Univ. Press, New York, 264 pp.

Forbes, R. H. **(1942)**, Egyptian-Libyan borderlands, *Geog. Rev.*, 32:294–302.

Twitchell, K. S. **(1944)**, Water resources of Saudi Arabia, *Geog. Rev.*, 34:365–386.

Capot-Rey, R. **(1945)**, Dry and humid morphology in the western Erg, *Geog. Rev.*, 35:391–407.

Moolman, J. H. **(1946)**, The Orange River, South Africa, *Geog. Rev.*, 36:653–676.

Gorrie, R. M. **(1948)**, Countering desiccation in the Punjab, *Geog. Rev.*, 38:30–40.

Ives, R. L. **(1949)**, Climate of the Sonoran Desert, *Annals A. A. G.*, 39:143–187.

Bunker, D. G. **(1953)**, The southwest borderlands of the Rub Al Khali, *Geog. Jour.*, 119:418–430.

Awad, H. **(1963)**, Some aspects of the geomorphology of Morocco as related to the Quaternary climate, *Geog. Jour.*, 129:129–139.

Gulick, L. H., Jr. **(1963)**, Irrigation systems of the former Sind Province, West Pakistan, *Geog. Rev.*, 53:79–99.

Ebert, C. H. V. **(1965)**, Water resources and land use in the Qatif Oasis of Saudi Arabia, *Geog. Rev.*, 55:496–509.

Hills, E. S., ed. **(1966)**, *Arid lands; a geographical appraisal,* Methuen and Co., London, 461 pp.

Bryson, R. A., and D. A. Baerreis **(1967)**, Possibilities of major climatic modification and their implications: Northwest India, a case for study, *Bull. Amer. Met. Soc.*, 48:136–142.

McGuiness, W. G., B. J. Goldman, and P. Paylore, eds. **(1968)**, *Deserts of the world,* Univ. of Arizona Press, Tucson, 788 pp.

WEST COAST DESERT CLIMATE

Rich, J. L. **(1941)**, The nitrate district of Tarapacá, Chile: An aerial traverse, *Geog. Rev.*, 31:1–22.

Light, M., and R. Light **(1946)**, Atacama revisited: "Desert trails" seen from the air, *Geog. Rev.*, 36:525–545.

Rudolph, W. E. **(1951)**, Chuquicamata revisited twenty years later, *Geog. Rev.*, 41:88–113.

Hammon, E. H. **(1954)**, A geomorphic study of the Cape region of Baja California, *Univ. of California Publ. in Geography*, 10, No. 2:45–111.

Lydolph, P. E. **(1957)**, A comparative analysis of the dry western littorals, *Annals A. A. G.*, 47:211–230.

Meigs, P. **(1966)**, *Geography of coastal deserts,* UNESCO Publications Center, New York, 140 pp.

Amiran, D. H. K. and A. W. Wilson, eds. **(1973)**, *Coastal deserts—their natural and human environments,* Univ. of Arizona Press, Tucson, 207 pp.

TROPICAL WET-DRY CLIMATE

Heller, E. **(1918)**, The geographical barriers to the distribution of big game animals in Africa, *Geog. Rev.*, 6:297–319.

Taylor, G. **(1919)**, The settlement of tropical Australia, *Geog. Rev.*, 8:84–115.

Shantz, H. L. **(1922)**, Urundi, territory and people, *Geog. Rev.*, 12:329–357.

Bennett, H. H. **(1925)**, Some geographic aspects of western Ecuador, *Annals A. A. G.*, 15:126–147.

Sheppard, G. **(1930)**, Notes on the climate and physiography of southwestern Ecuador, *Geog. Rev.*, 20:445–453.

Darby, H. C. **(1931)**, Settlement in northern Rhodesia, *Geog. Rev.*, 21:559–573.

Crist, R. **(1932)**, Along the Llanos-Andes border in Zamora, Venezuela, *Geog. Rev.*, 22:411–422.

Forbes, R. H. **(1932)**, The desiccation problem in West Africa: the capture of the Sourou by the Black Volta, *Geog., Rev.*, 22:97–106.

Gautier, E. F. **(1933)**, Climatic and physiographic notes on French Guinea, *Geog. Rev.*, 23:248–258.

Price, A. G. **(1933)**, Pioneer reactions to a poor tropical environment, *Geog. Rev.*, 23:353–371.

Pendleton, R. L. **(1943)**, Land use in northeastern Thailand, *Geog. Rev.*, 33:15–41.

Rudolph, W. E. **(1944)**, Agricultural possibilities in northwestern Venezuela, *Geog. Rev.*, 34:36–56.

Dobby, E. H. G. **(1945)**, Winds and fronts over Southeast Asia, *Geog. Rev.*, 35:204–218.

Waibel, L. **(1948)**, Vegetation and land use in the Planalto Central of Brazil, *Geog. Rev.*, 38:529–554.

James, P. E. **(1952)**, Observations on the physical geography of northeast Brazil, *Annals A. A. G.*, 42:153–176.

Whittlesey, D. **(1956)**, Southern Rhodesia—an African compage, *Annals A. A. G.*, 46:1–97.

Church, R. J. H. **(1961)**, *West Africa, a study of environment and man's use of it,* John Wiley and Sons, New York, 547 pp.

Mahew, W. A., Jr. **(1965)** The climate pattern of North and South Vietnam, *Weatherwise*, 18: 162–165.

Chang, J.-H. **(1967)**, The Indian summer monsoon, *Geog. Rev.*, 57:373–396.

Gleave, M. B., and H. P. White **(1969)**, The West

African Middle Belt: environmental fact or geographer's fiction? *Geog. Rev.*, 59: 123–139.

CHAPTER 16

HUMID SUBTROPICAL CLIMATE

Tower, W. S. **(1918)**, The Pampa of Argentina, *Geog. Rev.*, 5:293–315.

Emerson, F. V. **(1919)**, The southern long-leaf pine belt, *Geog. Rev.*, 7:81–90.

Durland, W. D. **(1924)**, The quebracho region of Argentina, *Geog. Rev.*, 14:227–241.

Trewartha, G. T. **(1928)**, A geographic study in Shizuoko Prefecture, Japan, *Annals A. A. G.*, 18:127–259.

Frothingham, E. H. **(1931)**, Timber growing and logging practice in the southern Appalachian region, *U.S. Dept. Agriculture Tech. Bull*, 250, 93 pp.

Visher, S. S. **(1941)**, Torrential rains as a serious handicap in the South, *Geog. Rev.*, 31:644–652.

MARINE WEST COAST CLIMATE

Hoover, J. W. **(1933)**, The littoral of northern California as a geographic province, *Geog. Rev.*, 23:217–229.

Bilham, E. G. (*1938*), *The Climate of the British Isles*, Macmillan and Co., London, 347 pp.

Cumberland, K. B. **(1941)**, A century's change: natural to cultural vegetation in New Zealand, *Geog. Rev.*, 31:529–554.

Küchler, A. W. **(1946)**, The broadleaf deciduous forests of the Pacific Northwest, *Annals A. A. G.*, 36:122–147.

Taylor, J. A., and R. A. Yates **(1958)**, *British weather in maps*, St. Martin's Press, New York, 256 pp.

MEDITERRANEAN CLIMATE

Smith, J. R. **(1916)**, The oak tree and man's environment, *Geog. Rev.*, 1:3–19.

Semple, E. C. **(1919)**, Climatic and geographic influences on ancient Mediterranean forests and the lumber trade, *Annals A. A. G.*, 9:13–37.

Coulter, J. W. **(1930)**, Land utilization in the Santa Lucia region, *Geog. Rev.*, 20:469–479.

Raup, H. F. **(1935)**, Land use and water-supply problems in southern California: market gardens of the Palos Verdes Hills, *Geog. Rev.*, 25:264–269.

Torbet, E. N. **(1935)**, The specialized commercial agriculture of the northern Santa Clara Valley, *Geog. Rev.*, 25:247–263.

Stotz, C. L. **(1939)**, The Bursa region of Turkey, *Geog. Rev.*, 29:81–100.

Leighly, J. **(1941)**, Settlement and cultivation in the summer-dry climates, *Yearbook of Agriculture, 1941*, U.S. Dept. Agriculture, pp. 197–204.

Fish, W. B. **(1944)**, The Lebanon, *Geog. Rev.*, 34:235–258.

Nuttonson, M. Y. **(1947)**, Agroclimatology and crop ecology of Palestine and Trans-Jordan and climatic analogues in the United States, *Geog. Rev.*, 37:436–456.

Whyte, R. O. **(1950)**, The phytogeographical zones of Palestine, *Geog. Rev.*, 40:600–614.

MIDDLE-LATITUDE DESERT AND STEPPE CLIMATES

Gregory, H. E. **(1915)**, The oasis of Tuba City, Arizona, *Annals A. A. G.*, 5:107–119.

Jefferson, M. **(1916)**, The oasis at the foot of the Wasatch, *Geog. Rev.*, 1:346–358.

Smith, J. W. **(1920)**, Rainfall of the Great Plains in relation to cultivation, *Annals A. A. G.*, 10:69–74.

Stein, A. **(1920)**, Explorations in the Lop Desert, *Geog. Rev.*, 9:1–34.

Bryan, K. **(1929)**, Flood-water farming, *Geog. Rev.*, 19:444–456.

Hoover, J. W. **(1930)**, Tusayan: The Hopi Indian country of Arizona, *Geog. Rev.*, 20:425–444.

Bowman, I. **(1931)**, Jordan country, *Geog. Rev.*, 21:22–55.

Bowman, R. H. **(1933)**, Belle Fourche valleys and uplands, *Annals A. A. G.*, 23:127–164.

Brown, R. H. **(1934)**, Irrigation in a dry-farming region, *Geog. Rev.*, 24:596–604.

Bowman, I. **(1935)**, Our expanding and contracting desert, *Geog. Rev.*, 25:43–61.

Crowe, P. R. **(1936)**, The rainfall regime of the Western Plains, *Geog. Rev.*, 26:463–484.

Lackey, E. E. **(1937)**, Annual-variability rainfall maps of the Great Plains, *Geog. Rev.*, 27:665–670.

Thornthwaite, C. W. **(1941)**, Climate and settlement in the Great Plains, *Yearbook of Agriculture, 1941*, U.S. Dept. Agriculture, pp. 177–196.

Bryan, K., and C. C. Albritton **(1943)**, Soil phenomena as evidence of climate change, *Am. J. Sci.*, 241:469–490.

Nuttonson, M. Y. **(1947)**, Agroclimatology and crop ecology of the Ukraine and climatic analogues in North America, *Geog. Rev.*, 37:216–232.

Bretz, J. H., and L. Horberg **(1949)**, Caliche in southeastern New Mexico, *J. Geol.*, 57:491–511.

Borchert, J. R. **(1950)**, The climate of the central North American grassland, *Annals A. A. G.*, 40 pp. 1–39.

Villmow, J. R. **(1956)**, The nature and origin of the Canadian dry belt, *Annals A. A. G.*, 46:211–232.

White, G. F., ed. **(1956)**, The future of arid lands, *Am. Assoc. Advancement of Sci. Publ.* 43, Washington, D.C., 453 pp.

Zierer, C. M., ed. **(1956)**, *California and the southwest*, John Wiley and Sons, New York, 376 pp.

Jaeger, E. C. **(1957)**, *The North American deserts*, Standford Univ. Press, Stanford, Cal., 308 pp.

Fonaroff, L. S. **(1963)**, Conservation and stock reduction on the Navajo tribal range, *Geog. Rev.*, 53:200–223.

Chappell, J. E., Jr. **(1970)**, Climatic change reconsidered: another look at the "pulse of Asia", *Geog. Rev.*, 60:347–373.

HUMID CONTINENTAL CLIMATE

Kincer, J. B. **(1923)**, The climate of the Great Plains

as a factor in their utilization, *Annals A. A. G.*, 13:67–80.
Durand, L., Jr., and K. Bertrand **(1935)**, The forest and woodland regions of Wisconsin, *Geog. Rev.*, 25:264–271.
Church, P. E. **(1936)**, A geographical study of New England temperatures, *Geog. Rev.*, 26:283–292.
Rose, J. K. **(1936)**, Corn yield and climate in the corn belt, *Geog. Rev.*, 26:88–102.
McCune, S. **(1941)**, Climatic regions of Korea and their economy, *Geog. Rev.*, 31:95–99.
Sauer, C. O. **(1941)**, The settlement of the humid East, *Yearbook of Agriculture, 1941,* U.S. Dept. Agriculture, pp. 157–176.
Forbes, C. B. **(1942)**, Snowfall in Maine, *Geog. Rev.*, 32:245–251.
Manley, G. **(1945)**, The effective rate of altitudinal change in temperature Atlantic climates, *Geog.* Rev., 35:408–417.
Calef, W. **(1950)**, The winter of 1948–49 in the Great Plains, *Annals A. A. G.*, 40:267–292.
Hare, F. K. **(1952)**, The climate of the island of Newfoundland: A geographical analysis, *Geog. Bull.*, 2:36–88.
Brumback, J. J. **(1965)**, *The Climate of Connecticut,* Conn. Geol. and Nat. Hist. Survey, Bull. 99, 215 pp.
Kopec, R. J. **(1965)**, Continentality around the Great Lakes, *Bull. Amer. Met. Soc.*, 46:54–57.
Muller, R. A. **(1966)**, Snowbelts of the Great Lakes, *Weatherwise,* 19:248–255.
Stommel, H. G. **(1966)**, The great blizzard of '66 on the northern Great Plains, *Weatherwise,* 19:189–193.
Kopec, R. J. **(1967)**, Areal patterns of seasonal temperature anomalies in the vicinity of the Great Lakes, *Bull, Amer. Met. Soc.*, 48:884–889.
Landsberg, H. E. **(1967)**, Two centuries of New England climate, *Weatherwise*, 20:52–57.

CHAPTER 17

POLAR AND ARCTIC CLIMATES; NORTH AMERICA, ASIA

Alcock, F. J. **(1916)**, The Churchill River, *Geog. Rev.*, 2:433–448.
Flaherty, R. J. **(1918)**, Two traverses across Ungava Peninsula, Labrador, *Geog. Rev.*, 6:116–132.
Hall, H. U. **(1918)**, A Siberian wilderness: native life on the Lower Yenisei, *Geog. Rev.*, 5:1–21.
Kindle, E. M. **(1920)**, Arrival and departure of winter conditions in the Mackenzie River basin, *Geog. Rev.*, 10:388–399.
Holtedahl, O. **(1922)**, Novaya Zemlya, a Russian arctic land, *Geog. Rev.*, 12:521–531.
Novakovsky, S. **(1922)**, Climatic provinces of the Russian far East in relation to human activities, *Geog. Rev.*, 12:100–115.
Kindle, E. M. **(1925)**, The James Bay coastal plain, *Geog. Rev.*, 15:226–236.
Transehe, N. A. **(1925)**, The Siberian Sea road, *Geog. Rev.*, 15:367–398.
Ekblaw, W. E. **(1927, 1928)**, The material response of the polar Eskimo to their far arctic environment, *Annals A. A. G.*, 17:147–198; 18:1–24.
Bell, J. M. **(1929)**, Great Slave Lake, *Geog. Rev.*, 19:556–580.
Soper, J. D. **(1930)**, Explorations in Foxe Peninsula and along the west coast of Baffin Island, *Geog. Rev.*, 20:397–424.
Albright, W. D. **(1933)**, Crop growth at high latitudes, *Geog. Rev.*, 23:608–620.
Albright, W. D. **(1933)**, Gardens of the Mackenzie, *Geog. Rev.*, 23:1–22.
Zubov, N. N. **(1933)**, The circumnavigation of Franz Josef Land, *Geog. Rev.*, 23:394–401.
Wheeler, E. P. **(1935)**, The Nain-Okak section of Labrador, *Geog. Rev.*, 25:240–254.
Soper, J. D. **(1939)**, Wood Buffalo Park, *Geog. Rev.*, 29:383–399.
Adams, J. Q. **(1941)**, Settlements of the northeastern Canadian arctic, *Geog. Rev.*, 31:112–126.
Rockie, W. A. **(1942)**, A picture of Matanuska, *Geog. Rev.*, 32:353–371.
Department of Transport, Air Services Branch, Meteorological Division **(1944)**, *Meteorology of the Canadian arctic*, Ottawa, Canada, 85 pp.
Cabot, E. C. **(1947)**, The northern Alaska coastal plain interpreted from aerial photographs, *Geog. Rev.*, 37:639–648.
Sanderson, M. **(1948)**, Drought in the Canadian Northwest, *Geog. Rev.*, 38:289–299.
Hare, F. K. **(1952)**, The Labrador frontier, *Geog. Rev.*, 42:405–424.
Arctic Institute of North America **(1955)**, Arctic research, edited by D. Rowley, *Arctic*, 7:117–375; also reprinted as *Special Publ.* 2. Status of research in the North America arctic and subarctic, including physical, biological, and social sciences.
Kimble, G. H. T., and D. Good, eds. **(1955)**, Geography of the Northlands, *Amer. Geog. Soc., Special Publ.* No. 32, John Wiley and Sons, New York, 534 pp.
Jenness, J. L., **(1957)**, *Dawn in arctic Alaska*, University of Minnesota Press, Minneapolis, 222 pp.
Sim, V. W. **(1957)**, Geographical aspects of weather and climate at Eureka, Northwest Territories, *Geog. Bull.*, 10:34–53.
Sonnenfeld, J. **(1959)**, An arctic reindeer industry: Growth and decline, *Geog. Rev.*, 49:76–94.
Malmström, V. H. **(1960)**, Influence of the arctic front on the climate and crops of Iceland, *Annals A. A. G.*, 50:117–122.
Ives, J. D., and J. T. Andrews **(1963)**, Studies in physical geography of north-central Baffin Island, N. W. T., *Geog. Bull.*, 19:5–48.
Shear, J. A. **(1964)**, The polar marine climate, *Annals A. A. G.*, 54:310–317.
Bird, J. B. **(1967)**, *The physiography of arctic Canada,* The Johns Hopkins Press, Baltimore, Md., 336 pp.
Hare, F. K., and J. C. Ritchie **(1972)**, The boreal bioclimates, *Geog. Rev.*, 62:333–365.

ARCTIC GEOMORPHOLOGY; PERMAFROST

Porsild, A. E. **(1938)**, Earth mounds in unglaciated

arctic northwestern America, *Geog. Rev.*, 28:46–58.
Sharp, R. P. **(1942)**, Ground-ice mounds in tundra, *Geog. Rev.*, 32:417–423.
Sharp, R. P. **(1942)**, Soil structures in the St. Elias Range, Yukon Territory, *J. Geomorphology*, 5:274–301.
Taber, S. **(1943)**, Perennially frozen ground in Alaska: its origin and history, *Geol. Soc. Am. Bull.*, 54:1433–1548.
Washburn, A. L. **(1947)**, Reconnaissance geology of portions of Victoria Island and adjacent regions, arctic Canada, *Geol. Soc. Am. Memoir*, 22, 142 pp.
Black, R. F. **(1950)**, Permafrost, Chapter 14 of *Applied sedimentation*, edited by P. D. Trask, John Wiley and Sons, New York, pp. 247–275.
Ray, L. L. **(1951)**, Permafrost, *Arctic*, 4:196–203.
Flint, R. F. **(1952)**, The Ice Age in the North American arctic, *Arctic*, 3:135–152.
Jenness, J. L. **(1952)**, Erosive forces in the physiography of western arctic Canada, *Geog. Rev.*, 42:238–252.
Black, R. F. **(1954)**, Permafrost—a review, *Geol. Soc. Am. Bull.*, 65:839–856.
Hopkins, D. M., and T. N. V. Karlstrom **(1954)**, Permafrost and ground water in Alaska, *U.S. Geol. Survey Prof. Paper* 264-F, 34 pp.
Bird, J. B. **(1955)**, Terrain conditions in the central Canadian arctic, *Geog. Bull.*, 7:1–16.
Washburn, A. L. **(1956)**, Classification of patterned ground and review of suggested origins, *Geol. Soc. Am. Bull.*, 67:823–866.
Fraser, J. K. **(1959)**, Freeze-thaw frequencies and mechanical weathering in Canada, *Arctic*, 12:40–53.
Brown, R. J. E. **(1960)**, The distribution of permafrost and its relation to air temperature in Canada and the U.S.S.R., *Arctic*, 13:163–177.
Carson, C. E., and K. M. Hussey **(1962)**, The oriented lakes of arctic Alaska, *J. Geol.*, 70:417–439.
Drew, J. V., and J. C. F. Tedrow **(1962)**, Arctic soil classification and patterned ground, *Arctic*, 15:109–116.
Mackay, J. R. **(1962)**, Pingos of the Pleistocene Mackenzie delta area, *Geog. Bull.*, 18:21–63.
Péwé, T. L. **(1969)**, *The periglacial environment*, McGill-Queens Univ. Press, Montreal, 487 pp.
Mackay, J. R. **(1972)**, The world of underground ice, *Annals A. A. G.*, 62:1–22.
Price, L. W. **(1972)**, *The periglacial environment, permafrost, and man*, Resource Paper No. 14, Commission on College Geography, Assn. of Amer. Geog., Washington, D.C., 88 pp. Extensive bibliography.
Nat. Res. Council of Canada and U.S. National Acad. Sci. **(1973)**, *Permafrost*, North American Contribution, Second Internat. Conference, Yakutsk, U.S.S.R., 1973, Nat. Acad. Sci., Washington, D.C., 783 pp.
Washburn, A. L. **(1973)**, *Periglacial processes and environments*, St. Martin's Press, New York, 320 pp.

ANTARCTIC CLIMATES

Mawson, D. **(1930)**, The antarctic cruise of the "Discovery," *Geog. Rev.*, 20:535–554.
Gould, L. M. **(1931)**, Some geographical results of the Byrd antarctic expedition, *Geog. Rev.*, 21:177–200.
Ronne, F. **(1949)**, *Antarctic conquest*, G. P. Putnam's Sons, New York, 299 pp.
American Geophysical Union **(1956)**, Antarctica in the International Geophysical Year, Nat. Acad. Sci., Nat. Research Council, Washington, D. C., *Publ.* 462, 133 pp.
Dufek, G. J. **(1957)**, *Operation Deepfreeze*, Harcourt, Brace and Co., New York, 243 pp.
Gould, L. M. **(1957)**, Antarctic prospect, *Geog. Rev.*, 47:1–28.
Nichols, R. L. **(1963)**, Geologic features demonstrating aridity of McMurdo Sound area, Antarctica, *Am. J. Sci.*, 261:20–31.
Stepanove, N. A. **(1963)**, The world's lowest temperature record, *Weatherwise*, 16:268–269.
Haurwitz, B. **(1966)**, Antarctic exploration, *Bull. Amer. Met. Soc.*, 47:258–274.
Quam, L. O., ed. **(1971)**, *Research in the Antarctic*, Publ. No. 93, Assn. for the Advancement of Sci., Washington, D.C., 768 pp. See Part IV.

SEA ICE; ICE ISLANDS

Weaver, J. C. **(1946)**, Ice atlas of the northern hemisphere, *U.S. Navy Oceanographic Office Publ.* 550, Washington, D.C., 105 pp.
Crary, A. P., R. D. Cotell, and T. F. Sexton **(1952)**, Preliminary report on scientific work on "Fletcher's Ice Island, T 3," *Arctic*, 5:211–223.
Koenig, L. S., K. R. Greenaway, M. Dunbar, and G. Hattersley-Smith **(1952)**, Arctic ice islands, *Arctic*, 5:67–103.
Helk, J. V., and M. Dunbar **(1953)**, Ice islands: Evidence from North Greenland, *Arctic*, 6:263–271.
U.S. Navy **(1957)**, Oceanographic Atlas of the Polar Seas, Part I: Antarctic, *Oceanographic Office Publication* 705, *U.S. Navy*, Washington, D.C., 70 pp.
Crary, A. P. **(1958)**, Arctic ice island and ice shelf studies, *Arctic*, 11:3–42; 13:32–50.
Bilello, M. A. **(1961)**, Formation, growth, and decay of sea-ice, *Arctic*, 14:3–24.
Robin, G. deQ. **(1962)**, The ice of the Antarctic, *Scientific American*, 207:132–146.
Nutt, D. C. **(1966)**, The drift of ice island WH-5, Arctic, 19:244–262.

HIGHLAND CLIMATES, EFFECTS OF ALTITUDE

Bowman, I. **(1916)**, *The Andes of Peru*, Henry Holt and Co., New York, 336 pp.
Bowman, I. **(1916)**, The country of the shepherds, *Geog. Rev.*, 1:419–442.
Harshberger, J. W. **(1919)**, Alpine fell-fields of eastern North America, *Geog. Rev.*, 7:233–255.
Peattie, R. **(1929)**, Andorra: A study in mountain geography, *Geog, Rev.*, 19:218–233.

Pearson, G. A. (**1931**), Forest types in the southwest as determined by climate and soil, *U.S. Dept. Agriculture Tech. Bull.* 247, 144 pp.

Antevs, E. (**1932**), *Alpine zone of Mt. Washington Range*, Merrill and Webber Co., Auburn., Me., 118 pp.

Platt, R. S. (**1932**), Six farms in the Central Andes, *Geog. Rev.*, 22:245–259.

Garnett, A. (**1935**), Insolation, topography, and settlement in the Alps, *Geog. Rev.*, 25:601–617.

Peattie, R. (**1936**), *Mountain geography. A critique and field study*, Harvard Univ. Press, Cambridge, Mass., 257 pp.

Seifriz, W. (**1936**), Vegetation zones in the Caucasus, *Geog. Rev.*, 26:59–66.

Deffontaines, P. (**1937**), Mountain settlement in the central Brazilian Plateau, *Geog. Rev.*, 27:394–413.

Hanson-Lowe, J. (**1941**), Notes on the climate of the South Chinese-Tibetan borderland, *Geog. Rev.*, 31:444–453.

Ives, R. L. (**1942**), The beaver-meadow complex, *J. Geomorphology*, 5:191–203.

Spencer, J. E., and W. L. Thomas (**1948**), The hill stations and summer resorts of the Orient, *Geog. Rev.*, 38:637–651.

Miller, D. H. (**1955**), Snow cover and climate in the Sierra Nevada, California, *Univ. of California Publ. in Geography*, Vol. 11, Berkeley, 218 pp.

Drewes, W. U., and A. T. Drewes (**1957**), *Climate and related phenomena of the eastern Andean slopes of central Peru*, Syracuse Univ. Research Institute, Syracuse, N.Y., 85 pp.

Logan, R. F. (**1961**), Winter temperatures of a mid-latitude desert mountain range, *Geog. Rev.*, 51:236–252.

Miller, M. M. (**1964**), Glacio-meteorology on Mt. Everest in 1963, *Weatherwise*, 17:167–179.

Thompson, W. F. (**1968**), New observations on alpine accordances in the western United States, *Annals A. A. G.*, 58:650–669.

Terjung, W. H., et al (**1969**), Energy and moisture balances of an alpine tundra in mid July, *Arctic and Alpine Research*, 1:247–266.

Terjung, W. H., et al (**1969**), Terrestrial, atmospheric and solar radiation fluxes on a high desert mountain in mid-July: White Mountain Peak, California, *Solar Energy*, 12:363–375.

Terjung, W. H. (**1970**), The energy budget of man at high altitudes, *Internat. Jour. Biometeor.*, 14, No. 1:13–43.

CHAPTER 18

SOIL-FORMING PROCESSES AND FACTORS

U.S. Dept. Agriculture (**1938**), Soils and men, *Department of Agriculture Yearbook, 1938*, U.S. Govt. Printing Office, Washington, D.C., 1232 pp.

Nikiforoff, C. C. (**1939**), Weathering and soil evolution, *Soil Science*, 67:219–230.

Jenny, H. (**1941**), *Factors of soil formation*, McGraw-Hill Book Co., New York, 281 pp.

Nikiforoff, C. C. (**1943**), Introduction to paleopedology, *Am. J. Sci.*, 241:194–200.

Nikiforoff, C. C. (**1948**), Stony soils and their classification, *Soil Science*, 66:347–363.

Pendleton, R. L., and D. Nickerson (**1951**), Soil colors and special Munsell soil color charts, *Soil Sci.*, 71:35–43.

Baver, L. D. (**1956**), *Soil physics*, 3rd edition, John Wiley and Sons, New York, 489 pp.

Carter, G. F., and R. L. Pendleton (**1956**), The humid soil: process and time, *Geog. Rev.*, 46:488–507.

Albrecht, W. A. (**1957**), Soil fertility and biotic geography, *Geog. Rev.*, 47:86–105.

Cooper, A. W. (**1960**), An example of the role of microclimate in soil genesis, *Soil Sci.*, 90:109–120.

Bidwell, O. W., and F. D. Hole (**1965**), Man as a factor in soil formation, *Soil Sci.*, 99:65–72.

Bunting, B. T. (**1965**), *The geography of soil*, Aldine Publ. Co., Chicago, 213 pp.

Buol, S. W. (**1965**), Present soil-forming factors and processes in arid and semiarid regions, *Soil Sci.*, 99:45–49.

de Villiers, J. M. (**1965**), Present soil-forming factors and processes in tropical and subtropical regions, *Soil Sci.*, 99:50–57.

Millar, C. E., L. M. Turk, and H. D. Foth (**1965**), *Fundamentals of soil science*, 4th edition, John Wiley and Sons, New York, 491 pp.

Money, D. C. (**1965**), *Climate, soils and vegetation*, University Tutorial Press, Ltd., London, 272 pp.

Retzer, J. L. (**1965**), Present soil-forming factors and processes in arctic and alpine regions, *Soil Sci.*, 99:38–44.

Riecken, F. F. (**1965**), Present soil-forming factors and processes in temperate regions, *Soil Sci.*, 99:58–64.

Stephens, C. G. (**1965**), Climate as a factor of soil formation through the Quaternary, *Soil Sci.*, 99:9–14.

Thorp, J. (**1965**), The nature of the pedological record in the Quaternary, *Soil Sci.*, 99:1–8.

Gile, L. H., F. F. Peterson, and R. B. Grossman (**1966**), Morphological and genetic sequences of carbonate accumulation in desert soils, *Soil Sci.*, 101:347–360.

Yaalon, D. H., and B. Yaron (**1966**), Framework for man-made soil changes—an outline of metapedogenesis, *Soil Sci.*, 102:272–277.

Legget, R. F. (**1967**), Soil: Its geology and use, *Geol. Soc. Am. Bull.*, 78:1433–1460.

Eyre, S. R. (**1968**), *Vegetation and soils; a world picture*, 2nd edition, Aldine Publ. Co., Chicago, 328 pp. See Part I.

Loughnan, F. C. (**1969**), *Chemical weathering of the silicate minerals*, Amer. Elsevier Publ. Co., New York, 154 pp.

Gibson, J. S., and J. W. Batten (**1970**), *Soils: their nature, classes, distribution, uses and care*, Univ. of Alabama Press, University, Ala., 296 pp.

Wilde, S. A. (**1971**), Forest humus: its classification on a genetic basis, *Soil Sci.*, 111:1–12.

Cruickshank, J. G. (**1972**), *Soil geography*, Halsted Press Div., John Wiley and Sons, New York, 265 pp.

Birkeland, P. W. **(1974)**, *Pedology, weathering and geomorphological research,* Oxford Univ. Press, New York, 304 pp.

CHAPTER 19

SOIL CLASSIFICATION PRINCIPLES

Marbut, C. F. **(1925)**, The rise, decline and revival of Malthusianism in relation to geography and character of soils, *Annals A. A. G.,* 15:1–29.

Glinka, K. D. **(1927)**, *The great soil groups of the world and their development* (tr. by C. F. Marbut), Edwards Brothers, Ann Arbor, Mich., 235 pp.

Wolfanger, L. A. **(1929)**, Major soil groups and some of their geographic implications, *Geog. Rev.,* 19:94–113.

Baldwin, M., C. E. Kellogg, and J. Thorp **(1938)**, Soil classification, *U.S. Dept. Agriculture Yearbook of Agriculture, 1938,* pp. 979–1001.

Kellogg, C. E. **(1941)**, Climate and soil, *Yearbook of Agriculture, 1941,* U.S. Dept. Agriculture, pp. 265–291.

Thorp, J., and G. D. Smith **(1949)**, Higher categories of soil classification: order, suborder and great soil groups, *Soil Science,* 67:117–126.

Edelman, C. H. **(1952)**, Soils, *Soil Sci.,* 74:15–20.

Pierre, W. H. **(1958)**, Relationship of soil classification to other branches of soil science, *Soil Sci. Soc. Amer. Proc.,* 22:167–170.

Bunting, B. T. **(1965)**, *The geography of soil,* Aldine Publ. Co., Chicago, 213 pp. See Chapter 9.

Bridges, E. M. **(1970)**, *World soils,* Cambridge University Press, 89 pp.

SOIL TAXONOMY (SEVENTH APPROXIMATION)

Soil Survey Staff **(1960)**, *Soil classification, A comprehensive System—7th approximation,* U.S. Dept. Agriculture, Soil Conservation Service, U.S. Govt. Printing Office, Washington, D.C.

Simonson, R. **(1962)**, Soil classification in the United States, *Science,* 137:1027–1034.

Cline, M. G. **(1963)**, Logic of the new system of soil classification, *Soil Sci.,* 96:17–22.

Johnson, W. N. **(1963)**, Relation of the new comprehensive soil classification system to soil mapping, *Soil Sci.,* 96:31–34.

Kellogg, C. E. **(1963)**, Why a new system of soil classification? *Soil Sci.,* 96:1–5.

Simonson, R. W. **(1963)**, Soil correlation and the new classification system, *Soil Sci.,* 96:23–30.

Smith, Guy D. **(1963)**, Objectives and basic assumptions of the new soil classification system, *Soil Sci.,* 96:6–16.

Stephens, C. G. **(1963)**, The seventh approximation—A symposium, *Soil Sci. Soc. Amer. Proc.,* 27:212–228. (Seven authors discuss various aspects of the 7th approximation.)

Tavernier, R. **(1963)**, The 7th approximation: Its application in western Europe, *Soil Sci.,* 96:35–39.

Soil Survey Staff **(1974)** *Soil taxonomy: a basic system of soil classification for making and interpreting soil surveys,* U.S. Dept. Agriculture, Handbook No. 436, U.S. Govt, Printing Office, Washington, D.C.

SOILS OF MIDDLE LATITUDES

Marbut, C. F. **(1923)**, Soils of the Great Plains, *Annals A. A. G.,* 13:41–66.

Strahorn, A. T. **(1929)**, Agrieulture and soils of Palestine, *Geog. Rev.,* 19:581–602.

Marbut, C. F. **(1931)**, Russia and the United States in the world's wheat market, *Geog. Rev.,* 21:1–21.

Thorp. J. **(1931)**, The effects of vegetation and climate upon soil profiles in northern and northeastern Wyoming, *Soil Science,* 32:283–302.

Wolfanger, L. A. **(1931)**, Economic geography of the gray-brownerths of the eastern United States, *Geog. Rev.,* 21:276–296.

Taylor, G. **(1933)**, The soils of Australia in relation to topography and climate, *Geog. Rev.,* 23:108–113.

Dachnowski-Stokes, A. P. **(1934)**, Peat-land utilization, *Geog. Rev.,* 24:238–250.

Marbut, C. F. **(1935)**, Soils of the United States, Part III of *Atlas of American Agriculture,* U.S. Govt. Printing Office, Washington, D.C., 29 pp.

Kellogg, C. E. **(1936)**, Development and significance of the great soil groups of the United States, *U.S. Dept. Agriculture Misc. Publ.* 229, 40 pp.

Moyer, R. T. **(1936)**, Agricultural soils in a loess region of North China, *Geog. Rev.,* 26:414–425.

Nikiforoff, C. C. **(1937)**, The inversion of great soil zones in western Washington, *Geog. Rev.,* 27:200–213.

Foscue, E. J. **(1938)**, Influence of contrasted soil types upon changing land values near Grapevine, Texas, *Annals A. A. G.,* 28:137–144.

Strong, H. M. **(1938)**, A land use record in Blackland prairies of Texas, *Annals A. A. G.,* 28:128–236.

U.S. Dept. Agriculture **(1938)**, Soils and man, *Department of Agriculture Yearbook, 1938,* U.S. Govt. Printing Office, Washington, D.C., 1232 pp. See Part V. Soils of the United States.

Thorp, J. **(1948)**, How soils develop under grass, *Yearbook of American Agriculture, 1948,* U.S. Dept. Agriculture, pp 55–66.

Smith, R. **(1949)**, A comparison of the reddish chestnut soils of the United States with the redbrown earth of Australia, *Soil Science,* 67:209–218.

Hurst, F. B. **(1951)**, Climates prevailing in the yellow-gray earth and yellow-brown earth zones in New Zealand, *Soil Sci.,* 72:1–19.

Putnam, D. F. **(1951)**, Pedogeography of Canada, *Geog. Bull.,* 1:57–85.

Krusekopf, H. H. **(1958)**, Soils of Missouri—Genesis of great soil groups, *Soil Sci.,* 85:19–27.

Storie, R. E., and F. Harradine **(1958)**, Soils of California, *Soil Sci.,* 85:207–227.

Matelski, R. P. **(1959)**, Great soil groups of Nebraska, *Soil Sci.,* 88:228–239.

Giddens, J., H. F. Perkins, and R. L. Carter **(1960)**, Soils of Georgia, *Soil Sci.,* 89:229–238.

Harradine, F. (**1963**), Morphology and genesis of noncalcic brown soils in California, *Soil Sci.*, 96:277–287.

Papadakis, J. (**1963**), Soils of Argentine, *Soil Sci.*, 95:356–366.

Retzer, J. L. (**1963**), Soil formation and classification of forested mountain lands in the United States, *Soil Sci.*, 96:68–74.

Harradine, F. (**1966**), Comparative morphology of lateritic and podzolic soils in California, *Soil Sci.*, 101:142–151.

ARCTIC SOILS

Tedrow, J. C. F., and D. E. Hill (**1955**), Arctic brown soil, *Soil Sci.*, 80:265–275.

Tedrow, J. C. F., and J. E. Cantlon (**1958**), Concepts of soil formation and classification in arctic regions, *Arctic*, 11:166–179.

Hill, D. E., and J. C. F. Tedrow (**1961**), Weathering and soil formation in the Arctic environment, *Am. J. Sci.*, 259:84–101.

Tedrow, J. C. F., and L. A. Douglas (**1964**), Soil investigations on Banks Island, *Soil Sci.*, 98:53–65.

Bliss, L. C., and G. M. Woodwell (**1965**), An alpine podzol on Mount Katahdin, Maine, *Soil Sci.*, 100:274–297.

Tedrow, J. C. F. (**1966**), Polar desert soils, *Soil Sci. Soc. Amer. Proc.*, 30:381–387.

Brown, J. (**1967**), Tundra soils formed over ice wedges, northern Alaska, *Soil Sci. Soc. Amer. Proc.*, 31: 686–691.

Charlier, R. H. (**1969**), The geographic distribution of polar desert soils in the Northern Hemisphere, *Geol. Soc. Am. Bull.*, 80:1985–1996.

SOILS OF TROPICAL, EQUATORIAL REGIONS

Marbut, C. F., and C. B. Manifold (**1926**), The soils of the Amazon Basin in relation to agricultural possibilities, *Geog. Rev.*, 16:414–442.

Bennett, H. H. (**1928**), Some geographic aspects of Cuban soils, *Geog. Rev.*, 18:62–82.

Thorp, J., and M. Baldwin (**1940**), Laterite in relation to soils of the tropics, *Annals A. A. G.*, 30:163–194.

Pendleton, R. L. (**1941**), Laterite and its structural uses in Thailand and Cambodia, *Geog. Rev.*, 31:177–202.

Powers, W. L. (**1945**), Soil development and land use in northern Venezuela, *Geog. Rev.*, 35:273–285.

Goldich, S. S., and H. R. Bergquist (**1948**), Aluminous lateritic soil of the Republic of Haiti, W. I., *U.S. Geol. Survey Bull.* 954-C, pp. 99–109.

Vermaat, J. G., and C. F. Bentley (**1955**), The age and channeling of Ceylon laterite, *Soil. Sci.*, 79:239–247.

Bennema, J. (**1963**), The red and yellow soils of the tropical and subtropical uplands, *Soil Sci.*, 95:250–257.

Dudal, R. (**1963**), Dark clay soils of tropical and subtropical regions, *Soil Sci.*, 95:264–270.

Edelman, C. H., and P. K. J. Van der Voorde (**1963**), Important characteristics of alluvial soils in the tropics, *Soil Sci.*, 95:258–263.

Harris, S. A. (**1963**), On the classification of latosols and tropical brown earths of high-rainfall areas, *Soil Sci.*, 96:210–216.

Mukerjee, H. N. (**1963**), Determination of nutrient needs of tropical soils, *Soil Sci.*, 95:276–280.

Vann, J. H. (**1963**), Developmental processes in laterite terrain in Amapá, *Geog. Rev.*, 53:406–417.

McNeil, M. (**1964**), Laterite soils, *Scientific American*, 211:96–102.

Tan, K. H. (**1965**), The andosols in Indonesia, *Soil Sci.*, 99:375–378.

Chang, J-H. (**1968**), The agricultural potential of the humid tropics, *Geog. Rev.*, 58:333–361.

Rutherford, G. K. (**1968**), Observations on a succession of soils on Mt. Giluwe, eastern New Guinea, *Annals A. A. G.*, 58:304–312.

UNESCO (**1971**), *Soils and tropical weathering*, Symposium, Bandung, Indonesia, November 1969. Natural Resources Research 11, UNESCO Publ. Center, New York, 150 pp.

Paton, T. R., and M. A. J. Williams (**1972**), The concept of laterite, *Annals A. A. G.*, 62:42–56.

SALINIZATION AND WATERLOGGING OF SOILS

Ebert, C. H. V. (**1971**), Irrigation and salt problems in Renmark, South Australia, *Geog. Rev.*, 61:355–369.

Achi, K. (**1972**), Salinization and water problems in the Algerian northeast Sahara, pp. 276–287 in *The careless technology*, M. T. Farvar and J. P. Milton, eds., Natural History Press, Garden City, New York, 1030 pp.

Michel, A. (**1972**), The impact of modern irrigation technology in the Indus and Helmand basins of southwest Asia, pp. 257–275 in *The careless technology*, M. T. Farvar and J. P. Milton, eds., Natural History Press, Garden City, New York, 1030 pp.

CHAPTER 20

BIOGEOGRAPHY; PLANT ECOLOGY

Carter, G. F. (**1946**), The role of plants in geography, *Geog. Rev.*, 36:121–131.

Newbigin, Marion I. (**1948**), *Plant and animal geography*, 2nd edition, E. P. Dutton and Co., New York, 298 pp.

Oosting, H. J. (**1956**), *The study of plant communities; an introduction to plant ecology*, 2nd edition, W. H. Freeman, San Francisco, 440 pp.

Dansereau, Pierre (**1957**), *Biogeography; An ecological perspective*, The Ronald Press, New York, 394 pp.

Daubenmire, R. F. (**1959**), *Plants and environment: A textbook of plant autecology*, 2nd edition, John Wiley and Sons, New York, 422 pp.

Polunin, Nicholas (**1960**), *Introduction to plant geography and some related sciences*, McGraw-Hill Book Co., New York, 640 pp.

Wagner, P. L. (**1962**), Natural and artificial zonation in a vegetation cover: Chiapas, Mexico, *Geog. Rev.*, 52:253–274.
Gleason, H. A., and A. Cronquist (**1964**), *The natural geography of plants*, Columbia University Press, New York, 420 pp.
Mather, J. R., and G. A. Yoshioka (**1968**), The role of climate in the distribution of vegetation, *Annals A. A. G.*, 58:29–41.
Watts, D. (**1971**), *Principles of biogeography*, McGraw-Hill Book Co., New York, 402 pp. Extensive bibliography.
Jensen, W. A., and F. B. Salisbury (**1972**), *Botany: an ecological approach*, Wadsworth Publ. Co., Belmont, Calif., 748 pp.
Collier, B. D., et al (**1973**), *Dynamic ecology*, Prentice-Hall, Englewood Cliffs, N. J., 563 pp.
Cox, C. B., I. N. Healey, and P. D. Moore (**1973**), *Biogeography; an ecological and evolutionary approach*, John Wiley and Sons, New York, 288 pp.
Seddon, B. (**1973**), *Introduction to biogeography*, Methuen, London, 220 pp.

VEGETATION CLASSIFICATION SYSTEMS

Küchler, A. W. (**1947**), A geographic system of vegetation, *Geog. Rev.*, 37:233–240.
Küchler, A. W. (**1949**), A physiognomic classification of vegetation, *Annals A. A. G.*, 39:201–210.
Küchler, A. W. (**1955**), A comprehensive method of mapping vegetation, *Annals A. A. G.*, 45:404–415.
Küchler, A. W. (**1956**), Classification and purpose in vegetation maps, *Geog. Rev.*, 46:155–167.
deLaubenfels, D. J. (**1957**), The status of "conifers" in vegetation classifications, *Annals A. A. G.*, 47:145–149.
Wagner, P. L. (**1957**), A contribution to structural vegetation mapping, *Annals A. A. G.*, 47:363–369.
Küchler, A. W. (**1964**), Potential natural vegetation of the coterminus United States, Map and Manual, Amer. Geog. Soc., *Special Publ.* No. 36, 116 pp.
Küchler, A. W. (**1966**), Analyzing the physiognomy and structure of vegetation, *Annals A. A. G.*, 56:112–127.
Eyre, S. R. (**1968**), *Vegetation and soils: a world picture*, 2nd edition, Aldine Publ. Co., Chicago, 328 pp.
Shimwell, D. W. (**1971**), *The description and classification of vegetation*, Univ. of Washington Press, Seattle, 322 pp.

CHAPTER 21

WORLD VEGETATION DISTRIBUTION

Shantz, H. L., and R. Zon (**1924**), *Natural vegetation*, Atlas of American Agriculture, Section E, U.S. Department of Agriculture, U.S. Govt. Printing Office, Washington, D.C., 29 pp.
Brockmann-Jerosch, H. (**1951**), *Vegetation of the earth* (Wall map), Justus Perthes, Gotha.
Dansereau, P. (**1957**), *Biogeography; An ecological perspective*, The Ronald Press Co., New York, 394 pp.
Polunin, N. (**1960**), *Introduction to plant geography*, McGraw-Hill Book Co., New York, 640 pp.
Gleason, H. A., and A. Cronquist (**1964**), *The natural geography of plants*, Columbia University Press, New York, 420 pp.
Eyre, S. R. (**1968**), *Vegetation and soils*, 2nd edition Aldine Publishing Co., Chicago, 328 pp.
Eyre, S. R., ed. (**1971**), *World vegetation types*, Columbia Univ. Press, New York, 264 pp.
Hammond, A. L. (**1972**), Ecosystem analysis: biome approach to environmental research, *Science*, 175:46–48.

FOREST BIOCHORE

Matoon, W. R. (**1936**), Forest trees and forest regions of the United States, *U.S. Dept. Agriculture Misc. Publ.* 217, 54 pp.
Hodge, W. B. (**1943**), The vegetation of Dominica, *Geog. Rev.*, 33:349–375.
Raup, H. M. (**1945**), Forests and gardens along the Alaska Highway, *Geog. Rev.*, 35:22–48.
U.S. Department of Agriculture (**1949**), *Trees*, Yearbook of Agriculture, 1949, U.S. Govt. Printing Office, Washington, D.C., 944 pp.
Hare, F. K. (**1950**), Climate and zonal divisions of the boreal forest formation in eastern Canada, *Geog. Rev.*, 40:615–635.
Hustich, I. (**1953**), The boreal limits of the conifers, *Arctic*, 6:149–162.
Amer. Geog. Society (**1956**), *A world geography of forest resources*, Edited by Haden-Guest, Wright, and Teclaff, The Ronald Press Co., New York, 736 pp.
Hare, F. K., and R. G. Taylor (**1956**), The position of certain forest boundaries in southern Labrador-Ungava, *Geog. Bull.*, 8:51–73.
Ritchie, J. C. (**1958**), Vegetation map from the southern spruce forest zone of Manitoba, *Geog. Bull.*, 12:39–46.
Hopkins, D. M. (**1959**), Some characteristics of the climate in forest and tundra regions in Alaska, *Arctic*, 12:215–220.
Sjörs, H. (**1959**), Bogs and fens in the Hudson Bay lowlands, *Arctic*, 12:3–19.
Vann, J. H. (**1959**), Landform-vegetation relationships in the Atrato delta, *Annals A. A. G.*, 49:345–360.
Thompson, K. (**1961**), Riparian forests of the Sacramento Valley, California, *Annals A. A. G.*, 51:294–315.
Parsons, J. J. (**1962**), The acorn-hog economy of the oak woodlands of southwestern Spain, *Geog. Rev.*, 52:211–235.
Kellogg, C. E. (**1963**), Shifting cultivation, *Soil Sci.*, 95:221–230.
Platt, R. (**1965**), *The great American forest*, Prentice-Hall, Englewood Cliffs, N.J., 271 pp.
Chang, J-H. (**1968**), The agricultural potential of the humid tropics, *Geog. Rev.*, 58:333–361.
Reichle, D. E., ed. (**1970**), *Analysis of temperate forest ecosystems*, Springer-Verlag, New York, 304 pp.
Harris, D. R. (**1971**), The ecology of swidden culti-

vation in the upper Orinoco rain forest, Venezuela, *Geog. Rev.*, 61:475–495.
Igbozuirke, M. U. (**1971**), Ecological balance in tropical agriculture, *Geog. Rev.*, 61:519–529.
Gómez-Pompa, C. Vázquez-Yanes, and S. Guevara (**1972**), The tropical rain forest: a nonrenewable resource, *Science*, 177:762–765.
Hare, F. K., and J. C. Ritchie (**1972**), The boreal bioclimates, *Geog. Rev.*, 62:333–365.
Harris, D. R. (**1972**), The origins of agriculture in the tropics, *Amer. Scientist*, 60:180–193.
Strahler, A. H. (**1972**), Forests of the Fairfax line, *Annals A. A. G.*, 62:664–684.
Janzen, D. H. (**1973**), Tropical agroecosystems, *Science*, 182:1212–1219.
Mitchell, A. L. (**1973**), A theoretical tree line in central Canada, *Annals A. A. G.*, 63:296–301.
Richards, P. W. (**1973**), The tropical rain forest, *Sci. American*, 229, No. 6:58–67.

SAVANNA BIOCHORE

Williams, L. (**1941**), The Caura Valley and its forests, *Geog. Rev.*, 31:414–429.
Bates, M. (**1948**), Climate and vegetation in the Villavicencio region of eastern Colombia, *Geog. Rev.*, 38:555–574.
Gillman, C. (**1949**), A vegetation-types map of Tanganyika Territory, *Geog. Rev.*, 39:7–31.
Parsons, J. J. (**1955**), The Miskito pine savanna of Nicaragua and Honduras, *Annals A. A. G.*, 45:36–63.
Shantz, H. L., and B. L. Turner (**1958**), *Photographic documentation of vegetational changes in Africa over a third of a century*, Univ. of Arizona, College of Agriculture, Tucson, 158 pp.
Davis, C. M. (**1959**), Fire as a land-use tool in northeastern Australia, *Geog. Rev.*, 49:552–560.
Bell, R. H. V. (**1971**), A grazing ecosystem in the Serengeti, *Sci. American*, 225, No. 1:86–93.
Myers, N. (**1972**), National parks in Savannah Africa, *Science*, 178:1255–1263.
Hills, T. L. (**1974**), The savanna biome: a case study of human impact on biotic communities, pp. 342–373 in *Perspectives on environment*, I. R. Manners and M. W. Mikesell, eds., Association of Amer. Geographers, Washington, D.C., 395 pp.

GRASSLAND BIOCHORE

Visher, S. S. (**1916**), The biogeography of the northern Great Plains, *Geog. Rev.*, 2:89–115.
Shantz, H. L. (**1923**), The natural vegetation of the Great Plains, *Annals A. A. G.*, 13:81–107.
U.S. Department of Agriculture (**1948**), *Grass*, Yearbook of Agriculture, 1948, U.S. Govt. Printing Office, Washington, D.C., 892 pp.
Leopold, L. B. (**1951**), Vegetation of southwestern watersheds in the nineteenth century, *Geog. Rev.*, 41:295–316.
Curtis, J. T. (**1956**), The modification of mid-latitude grasslands and forests by man, pp. 721–736 in *Man's Role in Changing the Face of the earth*, W. L. Thomas, ed., Univ. of Chicago Press, 1193 pp.
Malin, J. C. (**1956**), The grassland of North America: its occupance and the challenge of continuous reappraisal, pp. 350–366 in *Man's role in changing the face of the earth*, W. L. Thomas, ed., Univ. of Chicago Press, 1193 pp.
Weaver, J. E., and F. W. Albertson (**1957**), *Grasslands of the Great Plains*, Johnsen Publ. Co., Lincoln, Nebr., 395 pp.
Watts, F. B. (**1960**), The natural vegetation of the southern Great Plains of Canada, *Geog. Bull.*, 14:25–43.
Wells, P. V. (**1965**), Scarp woodlands, transported grassland soils, and concept of grassland climate in the Great Plains region, *Science*, 148:246–249.
Wells, P. V. (**1969**), Postglacial vegetational history of the Great Plains, *Science*, 167:1574–1582.
Love, R. M. (**1970**), The rangelands of the western U.S., *Sci. American*, 222, No. 2:89–96.

DESERT BIOCHORE, TUNDRA

Benninghoff, W. S. (**1952**), Interaction of vegetation and soil frost phenomena, *Arctic*, 5:34–44.
Aleksandrova, V. D. (**1960**), Some regularities in the distribution of the vegetation in the arctic tundra, *Arctic*, 13:147–162.
Henoch, W. E. S. (**1960**), String-bogs in the arctic 400 miles north of the tree-line, *Geog. Jour.*, 126:335–339.
Bliss, L. C. (**1962**), Adaptations of arctic and alpine plants to environmental conditions, *Arctic*, 15:117–144.
Thomson, D. F. (**1962**), The Bibindu expedition: Exploration among the desert aborigines of Western Australia, *Geog. Jour.*, 128:1–14, 143–157, 262–278.
Maycock, P. F. (**1966**), An *Arctic forest* in the tundra of northern Ungava, Quebec, *Arctic*, 19:114–144.
Britton, M. (**1957**), Vegetation of the arctic tundra, *Proc. 18th Biology Colloquium*, Oregon State College, Corvallis, pp. 26–61.
Krebs, J. S., and R. G. Barry (**1970**), The arctic front and the tundra-taiga boundary in Eurasia, *Geog. Rev.*, 60:548–554.
Carter, L. G. (**1974**), Off-road vehicles: a compromise plan for the California desert, *Science*, 183:396–399.

CHAPTER 22

MINERALS AND ROCKS

Spock, L. E. (**1962**), *Guide to the study of rocks*, 2nd edition, Harper and Row., New York, 298 pp.
Krumbein, W. C., and L. L. Sloss (**1963**), *Stratigraphy and sedimentation*, 2nd edition, W. H. Freeman, San Francisco, 660 pp.
Deer, W. A., R. A. Howie, and J. Zussman (**1966**), *An introduction to the rock-forming minerals*, John Wiley and Sons, New York, 528 pp.

Mason, B. (**1966**), *Principles of geochemistry*, 3rd edition, John Wiley and Sons, New York, 329 pp.
Simpson, B. (**1966**), *Rocks and minerals*, Pergamon Press, Oxford, 302 pp.
Ernst, W. G. (**1969**), *Earth materials*, Prentice-Hall, Englewood Cliffs, N.J., 149 pp.
Turekian, K. K. (**1972**), *Chemistry of the earth*, Holt, Rinehart and Winston, New York, 131 pp.

ORE DEPOSITS, MINERAL RESOURCES

Bateman, A. M. (**1950**), *Economic mineral deposits*, 2nd edition, John Wiley and Sons, New York, 916 pp.
Mero, J. L. (**1964**), *The mineral resources of the sea*, Elsevier Publ. Co., Amsterdam, 312 pp.
Park, C. F., Jr. (**1968**), *Affluence in jeopardy: minerals and the political economy*, Freeman, Cooper and Co., San Francisco, 368 pp.
Committee on Resources and Man (**1969**), *Resources and man*, Nat. Acad. of Sci.—Nat. Res. Council, W. H. Freeman and Co., San Francisco, 259 pp.
Earney, F. C. (**1969**), New ores for old furnaces: pelletized iron, *Annals A. A. G.*, 59:512–534.
Mason, P. F. (**1969**), Some changes in domestic iron mining as a result of pelletization, *Annals A. A. G.*, 59:535–551.
Skinner, B. J. (**1969**), *Earth resources*, Prentice-Hall, Englewood Cliffs, N.J., 149 pp.
Brobst, D. A., and W. P. Pratt, eds. (**1973**), *United States mineral resources*, Professional Paper 820, U.S. Geol. Survey, U.S. Govt. Printing Office, Washington, D.C., 722 pp.
Warren, K. (**1973**), *Mineral resources*, Halsted Press, John Wiley and Sons, New York, 272 pp.
McKelvey, V. E. (**1974**), Approaches to the mineral supply problem, *Technology Review*, 76; No. 5:13–23.

CHAPTER 23

EARTH'S INTERIOR AND CRUST

Philips, O. M. (**1968**), *The heart of the earth*, Freeman, Cooper and Co., San Francisco, 236 pp.
Gordon, R. B. (**1972**), *Physics of the earth*, Holt, Rinehart and Winston, New York, 207 pp. See Chapters 6, 7, and 8.
Jacobs, J. A., R. D. Russell, and J. T. Wilson (**1974**), *Physics and geology*, 2nd edition, McGraw-Hill Book Co., New York, 622 pp. See Chapters 2, 3, and 10.

SUBMARINE GEOLOGY AND TOPOGRAPHY

Heezen, B. C., M. Tharp, and M. Ewing (**1959**), The floors of the oceans, Geol. Soc. Am., *Special Paper* 65, 122 pp.
Shepard, F. P. (**1963**), *Submarine geology*, 2nd edition, Harper and Row, New York, 557 pp.
Fairbridge, R. W., ed. (**1966**), *Encyclopedia of oceanography*, Reinhold Publ. Corp., New York, 1021 pp.
Winslow, J. H. (**1966**), Raised submarine canyons: An exploratory hypothesis, *Annals A. A. G.*, 56:634–672.
Shepard, F. P. (**1967**) *The earth beneath the sea*, revised edition, The Johns Hopkins Press, Baltimore, Md., 242 pp.
Starke, G. W. and A. D. Howard (**1968**), Polygenetic origin of Monterey submarine canyon, *Geol. Soc. Am. Bull.*, 79:813–126.
Turekian, K. K. (**1968**), *Oceans*, Prentice-Hall, Englewood Cliffs, N.J., 120 pp.
Weyl, P. K. (**1970**), *Oceanography*, John Wiley and Sons, New York, 535 pp. See Part III.
Multiple authorship (**1971**), *Oceanography*, Readings from Scientific American, W. H. Freeman and Co., San Francisco, 417 pp. See Part III.
Davis, R. A., Jr. (**1972**), *Principles of oceanography*, Addison-Wesley Publ. Co., Reading, Mass., 434 pp. See Chapters 1–3, 18–22.

PLATE TECTONICS, CONTINENTAL DRIFT

Takeuchi, H., S. Uyeda, and H. Kanamori (**1970**), *Debate about the earth*, revised edition, Freeman, Cooper, San Francisco, 281 pp.
Wilson, J. T., ed. (**1970**), *Continents adrift*, Readings from Scientific American, W. H. Freeman, San Francisco, 172 pp.
Clark, S. P., Jr. (**1971**), *Structure of the earth*, Prentice-Hall, Englewood Cliffs, N.J., 131 pp.
Wyllie, P. J. (**1971**), *The dynamic earth: textbook in geosciences*, John Wiley and Sons, New York, 416 pp.
Bird, J. M., and B. Isacks, eds. (**1972**), *Plate tectonics, selected papers from the Journal of Geophysical Research*, Amer. Geophys. Union, Washington, D.C., 951 pp.
Cox, A., ed. (**1973**), *Plate tectonics and geomagnetic reversals*, W. H. Freeman, San Francisco, 702 pp. Extensive bibliography.
Hallam, A. (**1973**), *A revolution in the earth sciences*, Clarendon Press, Oxford, 127 pp.
Seyfert, C. K., and L. A. Sirkin (**1973**), *Earth history and plate tectonics*, Harper and Row, New York, 504 pp.
Jacobs, J. A., R. D. Russell, and J. T. Wilson (**1974**), *Physics and geology*, 2nd edition, McGraw-Hill Book Co., New York, 622 pp. See Chapters 12–17.

CHAPTER 24

PRINCIPLES OF GEOMORPHOLOGY

Davis, W. M. (**1909**), *Geographical essays*, Ginn and Co., Boston, 777 pp. Reprinted in 1954, Dover Publications, New York.
Davis, W. M. (**1915**), The principles of geographical description, *Annals A. A. G.*, 5:61–105.
Lobeck, A. K. (**1939**), *Geomorphology*, McGraw-Hill Book Co., New York, 731 pp.
Cotton, C. A. (**1941**), *Landscape as developed by processes of normal erosion*, Cambridge Univ. Press, Cambridge, England, 301 pp.

Cotton, C. A. **(1942)**, *Climatic accidents in landscape-making*, Whitcombe and Tombs, Christchurch, New Zealand, 354 pp.
Bryan, K. **(1950)**, The place of geomorphology in the geographic sciences, *Annals A. A. G.*, 40:196–208.
Strahler, A. N. **(1952)**, Dynamic basis of geomorphology, *Geol. Soc. Amer. Bull.*, 63:923–938.
Chorley, R. J., A. J. Dunn, and R. P. Beckinsale **(1964)**, *The history of the study of landforms, or the development of geomorphology;* Vol. One *Geomorphology before Davis*, Methuen and Co., London, 678 pp.
Birot, P. **(1966)**, *General physical geography*, John Wiley and Sons, New York, 360 pp. See Book II, The landscapes of the continents, pp. 111–343.
Zakrzewska, B. **(1967)**, Trends and methods in landform geography, *Annals A. A. G.*, 57:128–165.
Scheidegger, A. E. **(1970)**, *Theoretical geomorphology* 2nd edition, Prentice-Hall, Englewood Cliffs, N.J., 435 pp.
Fairbridge, R. W. ed. **(1968)**, *The encyclopedia of geomorphology*, Reinhold Book Co., New York, 1295 pp.
Bloom, A. L. **(1969)**, *The surface of the earth*, Prentice-Hall, Englewood Cliffs, N.J., 152 pp.
Thornbury, W. D. **(1969)**, *Principles of geomorphology*, 2nd edition, John Wiley and Sons, New York, 594 pp.

WEATHERING PROCESSES AND FORMS

Gilbert, G. K. **(1904)**, Domes and dome structures of the High Sierras, *Geol. Soc. Am. Bull.*, 15:29–36.
Cvijic, J. **(1924)**, The evolution of lapiés, *Geog. Rev.*, 14:26–49.
Blackwelder, E. **(1925)**, Exfoliation as a phase of rock weathering, *J. Geol.*, 33:793–806.
Blackwelder, E. **(1927)**, Fire as an agent in rock weathering, *J. Geol.*, 35:134–140.
Blackwelder, E. **(1929)**, Cavernous rock surfaces of the desert, *Am. J. Sci.*, 17:393–399.
Taber, S. F. **(1929)**, Frost heaving, *J. Geol.*, 37:428–461.
Matthes, F. E. **(1930)**, Geologic history of Yosemite Valley, *U.S. Geol. Survey Prof. Paper* 160, 137 pp. See pp. 114–116.
Taber, S. **(1930)**, The mechanics of frost heaving, *J. Geol.*, 38:303–317.
Blackwelder, E. **(1933)**, The insolation hypothesis of rock weathering, *Am J. Sci.*, 26:97–113.
Balk, R. **(1939)**, Disintegration of glaciated cliffs, *J. Geomorphology*, 2:305–334.
Chapman, R. W. **(1940)**, Monoliths in the White Mountains of New Hampshire, *J. Geomorphology*, 3:302–310.
Smith, L. L. **(1941)**, Weather pits in granite of the southern Piedmont, *J. Geomorphology*, 4:117–127.
Jahns, R. H. **(1943)**, Sheet structure in granites: its origin and uses as a measure of glacial erosion, *J. Geol.*, 51:71–98.
White, W. A. **(1945)**, Origin of granite domes in the southeastern Piedmont, *J. Geol.*, 53:276–282.
Ruhe, R. V. **(1959)**, Stone lines in soils, *Soil Sci.*, 87:223–231.
Ollier, C. D. **(1963)**, Insolation weathering: examples from central Australia, *Am. J. Sci.*, 261:376–381.
Williams, L. **(1964)**, Regionalization of freeze-thaw activity, *Annals A. A. G.*, 54:597–611.
Roth, E. S. **(1965)**, Temperature and water content as factors in desert weathering, *J. Geol.*, 73:454–468.
Leopold, L. B., M. G. Wolman, and J. P. Miller **(1964)**, *Fluvial processes in geomorphology*, W. H. Freeman and Company, San Francisco, 522 pp. See Chapter 4.
Loughnan, F. C. **(1969)**, *Chemical weathering of the silicate minerals*, American Elsevier Publ. Co., New York, 154 pp.
Carroll, D. **(1970)**, *Rock weathering*, Plenum Press, 203 pp.
Chapman, R. W. **(1974)**, Calcareous duricrust in Al-Hasa, Saudia Arabia, *Geol. Soc. Am. Bull.*, 85:119–130.

MASS WASTING

Andersson, J. G. **(1906)**, Solifluction, a component of subaerial denudation, *J. Geol.*, 14:91–112.
Howe, E. **(1909)**, Landslides in the San Juan Mountains, *U.S. Geol., Survey Prof. Paper* 67; 58 pp.
Capps, S. R., Jr. **(1910)**, Rock glaciers in Alaska, *J. Geol.*, 18:359–375.
Blackwelder, E. **(1912)**, The Gros Ventre slide, an active earth-flow, *Geol. Soc. Am Bull.*, 23:487–492.
Daly, R. A., W. G. Miller, and G. S. Rice **(1912)**, Report of the Commission appointed to investigate Turtle Mountain, Frank, Alberta, Canada; Dept, Mines, Geol. Survey Branch, Ottawa, *Memoir* 27, 34 pp.
Blackwelder, E. **(1928)**, Mudflows as a geologic agent in semi-arid mountains, *Geol. Soc. Am. Bull.*, 39:465–480.
Sharpe, C. F. S. **(1938)**, *Landslides and related phenomena*, Columbia Univ. Press, New York, 137 pp.
Putnam, W. C., and R. P. Sharp **(1940)**, Landslides and earthflows near Ventura, Southern California, *Geog. Rev.*, 30:591–600.
Strahler, A. N. **(1940)**, Landslides of the Vermilion and Echo Cliffs, northern Arizona, *J. Geomorphology*, 3:285–296.
Ives, R. L. **(1941)**, Vegetative indications of solifluction, *J. Geomorphology*, 4:128–132.
Blackwelder, E. **(1942)**, The process of mountain sculpture by rolling debris, *J. Geomorphology*, 5:325–328.
Sharp, R. P. **(1942)**, Mudflow levees, *J. Geomorphology*, 5:222–227.
Sharpe, C. F. S. **(1942)**, Relation of soil-creep to earthflow in the Appalachian Plateaus, *J. Geomorphology*, 5:312–324.
Wentworth, C. K. **(1943)**, Soil avalanches on Oahu, Hawaii, *Geol. Soc. Am. Bull.*, 54:53–64.
Terzaghi, K. **(1950)**, Mechanism of landslides, *Geol. Soc. Am., Berkey Vol.*, pp. 83–123.

Sharp, R. P., and L. H. Nobles (**1953**), Mudflow of 1941 at Wrightwood, Southern California, *Geol. Soc. Am. Bull.*, 64:547–560.

Brice, J. C. (**1958**), Origin of steps on loess-mantled slopes, U.S. Geol. Survey, *Bull.*, 1071-C, 85 pp.

Highway Research Board (**1958**), *Landslides and engineering practice*, Special Report 29, N. A. S.-N.R.C. Publication 544, Washington, D.C., 232 pp.

Wahrhaftig, C., and A. Cox (**1959**), Rock glaciers in the Alaska Range, *Geol. Soc. Am. Bull.*, 70:383–436.

Williams, P. J. (**1959**), An investigation into processes occurring in solifluction, *Am. J. Sci.*, 257:481–490.

Lutz, H. J. (**1960**), Movement of rocks by uprooting of forest trees, *Am. J. Sci.*, 258:752–756.

Merriam, R. (**1960**), Portuguese Bend landslide, Palos Verdes Hills, California *J. Geol.*, 68:140–153.

Kerr, P. F. (**1963**), Quick clay, *Scientific American*, 209:132–142.

Watson, R. A., and H. E. Wright, Jr. (**1963**), Landslides on the east flank of the Chuska Mountains, Northwestern New Mexico, *Am. J. Sci.*, 261:525–548.

Schumm, S. A., and R. J. Chorley (**1964**), The fall of a threatening rock, *Am. J. Sci.*, 262:1041–1054.

Shreve, R. L. (**1966**), Sherman landslide, Alaska, *Science*, 154:1639–1643.

Kirkby, M. J. (**1967**), Measurement and theory of soil creep, *J. Geol.*, 75:359–378.

Marangunic, C., and C. Bull (**1968**), The landslide on the Sherman Glacier, pp. 383–394 in *The Great Alaska Earthquake*, vol. 3, Nat. Acad. Sci., Washington, D.C.

Zaruba, Q. (**1969**), *Landslides and their control*, Amer. Elsevier Publ. Co., New York, 212 pp.

CHAPTER 25

SOIL EROSION, SOIL CONSERVATION

Bennett, H. H. (**1928**), The geographical relation of soil erosion to land productivity, *Geog. Rev.*, 18:579–605.

Shaw, C. F. (**1929**), Erosion pavement, *Geog. Rev.*, 19:638–641.

Bennett, H. H. (**1931**), The problem of soil erosion in the United States, *Annals A. A. G.*, 21:147–170.

Bennett, H. H. (**1933**), The quantitative study of erosion technique and some preliminary results, *Geog. Rev.*, 23:423–432.

Sharpe, C. F. S. (**1938**), What is soil erosion? *U.S. Dept. Agriculture Misc. Publ.* 286, 85 pp.

Rockie, W. A. (**1939**), Man's effects on the Palouse, *Geog. Rev.*, 29:34–45.

Tieh, T. M. (**1941**), Soil erosion in China, *Geog. Rev.*, 31:570–590.

Bennett, H. H. (**1943**), Adjustment of agriculture to its environment, *Annals A. A. G.*, 33:163–198.

Bennett, H. H. (**1944**), Food comes from soil, *Geog. Rev.*, 34:57–76.

Cumberland, K. B. (**1944**), Contrasting regional morphology of soil erosion in New Zealand, *Geog. Rev.*, 34:77–95.

Gottschalk, L. C., and V. H. Jones (**1955**), Valleys and hills, erosion and sedimentation, *Yearbook of Agriculture, 1955*, U.S. Dept. Agriculture, pp. 135–143.

Osborn, B. (**1955**), How rainfall and runoff erode soil, *Yearbook of Agriculture, 1955*, U.S. Dept. Agriculture, pp. 127–135.

Leopold, L. B. (**1956**), Land use and sediment yield, pp. 639–647 of *Man's role in changing the face of the earth*, Univ. of Chicago Press, Chicago, 1193 pp.

Strahler, A. N. (**1956**), The nature of induced erosion and aggradation, pp. 621–638 of *Man's role in changing the face of the earth*, Univ. of Chicago Press, Chicago, 1193 pp.

Stallings, J. H. (**1957**), *Soil conservation*, Prentice-Hall, Englewood Cliffs, N.J., 575 pp.

Bennett, H. H. (**1960**), Soil erosion in Spain, *Geog. Rev.*, 50:59–72.

Smith, R. M., and W. L. Stamey (**1965**), Determining the range of tolerable erosion, *Soil Sci.*, 100:414–424.

Rockie, W. A. (**1971**), Soil conservation, pp. 99–132 in *Conservation of natural resources*, G.-H. Smith, ed., 4th edition, John Wiley and Sons, New York, 685 pp.

Beasley, R. P. (**1972**), *Erosion and sediment pollution control*, Iowa State Univ. Press, Ames, 320 pp.

Mosley, M. P. (**1972**), Evolution of a discontinuous gully system, *Annals A. A. G.*, 62:655–663.

Coates, D. R., ed. (**1973**), *Environmental Geomorphology and landscape conservation*, Vol. III: Non-Urban, Dowden, Hutchinson, and Ross, Stroudsburg, Pa., 483 pp.

Butzer, K. W. (**1974**), Accelerated soil erosion: a problem of man-land relationships, pp. 57–78 in *Perspectives on Environment*, I. R. Manners, ed., Assoc. of Amer. Geographers, Washington, D.C. 395 pp.

STREAM CHANNELS; STREAM PROFILES

Gilbert, G. K. (**1914**), The transportation of debris by running water, *U.S. Geol. Survey Prof. Paper* 86, 263 pp.

Johnson, D. (**1929**), Baselevel, *J. Geol.*, 37:575–582.

Johnson, D. (**1932**), Streams and their significance, *J. Geol.*, 40:481–497.

Mackin, J. H. (**1948**), Concept of the graded river, *Geol. Soc. Am. Bull.*, 59:463–512.

Holmes, C. D. (**1952**), Stream competence and the graded stream profile, *Am. J. Sci.*, 250:899–906.

Leopold, L. B., and T. Maddock, Jr. (**1953**), The hydraulic geometry of stream channels and some physiographic implications, *U.S. Geol. Survey Prof. Paper* 252, 57 pp.

Wolman, M. G. (**1955**), The natural channel of Brandywine Creek, Pa., *U.S. Geol. Survey Prof. Paper* 271, 56 pp.

Leopold, L. B., and J. P. Miller (**1956**), Ephemeral streams—Hydraulic factors and their relation to the drainage net, *U.S. Geol. Survey Prof. Paper* 282-*A*, 37 pp.

Hack, J. T. (**1957**), Studies of longitudinal stream

profiles in Virginia and Maryland, *U.S. Geol. Survey Prof. Paper* 294-*B*, 97 pp.

Wolman, M. G. (**1959**), Factors influencing erosion of a cohesive river bank, *Am. J. Sci.*, 257:204–216.

Brush, L. M., Jr., and M. G. Wolman (**1960**), Knickpoint behavior in noncohesive material: A laboratory study, *Geol. Soc. Am. Bull.*, 71:59–74.

Schumm, S. A. (**1960**), The shape of alluvial channels in relation to sediment type, *U.S. Geol. Survey Prof. Paper*, 352-*B*, 30 pp.

Wolman, M. G., and J. P. Miller (**1960**), Magnitude and frequency of forces in geomorphic processes, *J. Geol.*, 68:54–74.

Schumm, S. A. (**1961**), Effect of sediment characteristics on erosion and deposition in ephemeral-stream channels, *U.S. Geol. Survey Prof. Paper*, 352-C, 70 pp.

Carlston, C. W. (**1963**), Drainage density and streamflow, *U.S. Geol. Survey Prof. Paper*, 422-*C*, 8 pp.

Brice, J. C. (**1964**), Channel patterns and terraces of the Loup Rivers in Nebraska, *U.S. Geol. Survey Prof. Paper*, 422-*D*, 41 pp.

Leopold, L. B., M. G. Wolman, and J. P. Miller (**1964**), *Fluvial processes in geomorphology*, W. H. Freeman, San Francisco, 522 pp.

Schumm, S. A., and R. W. Lichty (**1965**), Time, space, and causality in geomorphology, *Am. J. Sci.*, 263:110–119.

Leopold, L. B., W. W. Emmett, and R. M. Myrick (**1966**), Channel and hillslope processes in a semiarid area, New Mexico, *U.S. Geol. Survey Prof. Paper*, 352-*G*, 253 pp.

Tuan, Y.-F. (**1966**), New Mexican gullies: A critical review and some recent observations, *Annals A. A. G.*, 56:573–597.

Wertz, J. B. (**1966**), The flood cycle of ephemeral mountain streams in the southwestern United States, *Annals A. A. G.*, 56:598–633.

Morisawa, M. (**1968**), *Streams: Their dynamics and morphology*, McGraw-Hill Book Co., New York, 175 pp.

Chorley, R. J., ed. (**1971**), *Introduction to fluvial processes*, Methuen and Co., London, 218 pp.

Schumm, S. A., ed. (**1972**), *River morphology*, Benchmark papers in geology, Dowden, Hutchinson and Ross, Stroudsburg, Pa., 429 pp.

GORGES, CANYONS, FALLS

Powell, J. W. (**1875**), *Exploration of the Colorado River of the West and its tributaries*, U.S. Govt. Printing Office, Washington, D.C. 291 pp.

Gilbert, G. K. (**1895**), Niagara Falls and their history, *Nat. Geog. Mag., Monograph* 1, pp. 203–236. Also in *The Physiography of the United States* (1896), American Book Co., New York, pp. 203–236.

Darton, N. H. (**1896**), Examples of stream robbing in the Catskill Mountains, *Geol. Soc. Am. Bull.*, 7:505–507.

Dellenbaugh, F. S. (**1908**), *A canyon voyage*, G. P. Putnam's Sons, New York, 277 pp.

Freeman, O. W. (**1938**), The Snake River Canyon, *Geog. Rev.*, 28:597–608.

Philbrick, S. S. (**1974**), What future for Niagara Falls? *Geol. Soc. Am. Bull.*, 85:91–98.

ALLUVIAL RIVERS; FLOOD PLAINS; MEANDERS

Macar, P. F. (**1934**), Effects of cut-off meanders on the longitudinal profiles of streams, *J. Geol.*, 42:523–536.

Melton, F. A. (**1936**), An empirical classification of flood-plain streams, *Geog. Rev.*, 26:593–609.

Happ, S. C., G. Rittenhouse, and G. C. Dobson (**1940**), Some principles of accelerated stream and valley sedimentation, *U.S. Dept. Agriculture Tech. Bull.* 695, 134 pp.

Matthes, G. H. (**1951**), Paradoxes of the Mississippi, *Scientific American*, 184:19–23.

Russell, R. J. (**1954**), Alluvial morphology of Anatolian rivers, *Annals A. A. G.*, 44:363–391.

Leopold, L. B., and M. G. Wolman (**1957**), River channel patterns: braided, meandering and straight, *U.S. Geol. Survey, Prof. Paper* 282-*B*, Washington, D.C., 47 pp.

Wolman, M. G., and L. B. Leopold (**1957**), River flood plains: some observations on their formation, *U.S. Geol. Survey Prof. Paper* 282-*C*, 109 pp.

Bagnold, R. A. (**1960**), Some aspects of the shape of river meanders, *U.S. Geol. Survey Prof. Paper*, 282-*E*, pp. 135–144.

Crickmay, C. H. (**1960**), Lateral activity in a river of northwestern Canada, *J. Geol.*, 68:377–391.

Leopold, L., and M. G. Wolman (**1960**), River meanders, *Geol. Soc. Am. Bull.*, 71:769–794.

Schmudde, T. H. (**1963**), Some aspects of land forms of the lower Missouri River floodplain, *Annals A. A. G.*, 53:60–73.

White, G. F. (**1963**), The Mekong River plan, *Scientific American*, 208:49–60.

Sigafoos, R. S. (**1964**), Botanical evidence of floods and flood-plain deposition, *U.S. Geol. Survey Prof. Paper*, 485-*A*, 35 pp.

Langbein, W. B., and L. B. Leopold (**1966**), River meanders—Theory of minimum variance, *U.S. Geol. Survey Prof. Paper*, 422-H, 15 pp.

Leopold, L. B., and W. B. Langbein, (**1966**), River meanders, *Scientific American*, 214:60–70.

Alexander, C. A., and N. R. Nunnally (**1972**), Channel stability on the lower Ohio River, *Annals A. A. G.*, 62:411–417.

Brice, J. C. (**1974**), Evolution of meander loops, *Geol. Soc. Am. Bull.*, 85:581–586.

MAN-INDUCED CHANGES IN STREAM CHANNELS AND SEDIMENT TRANSPORT

Nelson, J. G. (**1966**), Man and geomorphic process in the Chemung River valley, New York and Pennsylvania, *Annals A. A. G.*, 56:24–32.

Judson, S. (**1968**), Erosion of the land, or what's happening to our continents? *Amer. Scientist*, 56:356–374.

Meade, R. H. (**1969**), Errors in using modern stream-load data to estimate natural rates of denudation, *Geol. Soc. Am. Bull.*, 80:1265–1274.

Guy, H. P. (**1970**), *Sediment problems in urban areas*,

Circular 601E, U.S. Geological Survey, Washington, D.C., 8 pp.
Emerson, J. W. **(1971)**, Channelization: a case study, *Science*, 173:325–326.
Wolman, M. G. **(1971)**, The nation's rivers, *Science*, 174:905–918.
Gillette, R. **(1972)**, Stream channelization: conflict between ditchers, conservationists, *Science*, 176:890–894.
Knox, J. C. **(1972)**, Valley alluviation in southwestern Wisconsin, *Annals A. A. G.*, 62:401–410.
Wilson, L. **(1972)**, Seasonal sediment yield patterns of U.S. rivers, *Water Res. Research*, 8:1470–1479.
Leopold, L. B. **(1973)**, River channel change with time: an example, *Geol. Soc. Am. Bull.*, 84:1845–1860.

TERRACES

Davis, W. M. **(1909)**, River terraces in New England, *Geographical essays*, Ginn and Co., Boston, pp. 514–586. Reprinted in 1954, Dover Publications, New York.
Trewartha, G. T. **(1932)**, The Prairie du Chien Terrace: geography of a confluence site, *Annals A. A. G.*, 22:119–158.
Mackin, J. H. **(1937)**, Erosional history of the Big Horn Basin, *Wyoming, Geol. Soc. Am. Bull.*, 48:813–894.
Cotton, C. A. **(1940)**, Classification and correlation of river terraces, *J. Geomorphology*, 3:27–37.
Frye, J. C., and A. R. Leonard **(1954)**, Some problems of alluvial terrace mapping, *Am. J. Sci.*, 252:242–251.
Leopold, L. B., and J. P. Miller **(1954)**, A postglacial chronology for some alluvial valleys in Wyoming, *U.S. Geol. Survey Water-Supply Paper*, 1261, 90 pp.
Moss, J. G., and W. E. Bonini **(1961)**, Seismic evidence supporting a new interpretation of the Cody Terrace near Cody, Wyoming, *Geol. Soc. Am. Bull.*, 72:547–556.

DELTAS

Sykes, G. **(1926)**, The delta and estuary of the Colorado River, *Geog. Rev.*, 16:232–255.
Cressey, G. B. **(1935)**, The Fenghsien landscape: a fragment of the Yangtze delta, *Geog. Rev.*, 25:396–413.
Russell, R. J. **(1936)**, Physiography of the Lower Mississippi delta, Louisiana Conservation Dept., *Bull.* 8; 3–199.
Sykes, G. **(1937)**, The Colorado delta, *Am. Geog. Soc. Special Publ.* 19, 193 pp.
Russell, R. J. **(1942)**, Flotant, *Geog. Rev.*, 32:74–98.
Russell, R. J. **(1942)**, Geomorphology of the Rhone delta, *Annals A. A. G.*, 32:149–254.
Dobby, E. H. G. **(1951)**, The Kelantan delta, *Geog. Rev.*, 41:226–255.
Mackay, J. R. **(1963)**, *The Mackenzie delta area, N. W. T.*, Memoir 8, Geographical Br., Mines and Tech. Surveys, Ottawa, 202 pp.
Houston Geological Society **(1966)**, *Deltas in their geologic framework*, Houston, Texas, 251 pp.
Holtz, R. K. **(1969)**, Man-made landforms in the Nile delta, *Geog. Rev.*, 59:253–269.
Morgan, J. P. **(1970)**, Deltas—a résume, *Jour. of Geol. Education*, 18:107–117.
Shlemon, R. J. **(1971)**, The Quaternary deltaic and channel system in the central Great Valley, California, *Annals A. A. G.*, 61:427–440.
Kassas, M. **(1972)**, Impact of river control schemes on the shoreline of the Nile Delta, pp. 179–188 in *The careless technology*, M. T. Farvar and J. P. Milton, eds., Natural History Press, Garden City, New York, 1030 pp.

ENTRENCHED MEANDERS

Davis, W. M. **(1913)**, Meandering valleys and underfit rivers, *Annals A. A. G.*, 3:3–28.
Rich, J. L. **(1914)**, Certain types of stream valleys and their meaning, *J. Geol.*, 22:469–497.
Miser, H. D., K. W. Trimble, and S. Paige **(1923)**, Rainbow Bridge, Utah, *Geog. Rev.*, 13:518–531.
Moore, R. C. **(1926)**, Origin of enclosed meanders on streams of the Colorado Plateau, *J. Geol.*, 34:29–57, 97–130.
Mahard, R. H. **(1942)**, Origin and significance of intrenched meanders, *J. Geomorphology*, 5:32–44.
Strahler, A. N. **(1946)**, Elongate intrenched meanders of Conodoguinet Creek, Pa., *Am. J. Sci.*, 244:31–40.
Dury, G. H. **(1954)**, Contribution to the general theory of meandering valleys, *Am. J. Sci.*, 252:193–224.
Hack, J. T., and R. S. Young **(1959)**, Intrenched meanders of the North Fork of the Shenandoah River, Virginia, *U.S. Geol. Survey Prof. Paper*, 354-A, 10 pp.
Dury, G. H. **(1960)**, Misfit streams: problems in interpretation, discharge, and distribution, *Geog. Rev.*, 50:219–242.
Dury, G. H. **(1963)**, Underfit streams in relation to capture: A reassessment of the idea of W. M. Davis, *Inst. of British Geographers*, Publ. No. 32, pp. 83–94.

CHAPTER 26

CYCLE OF DENUDATION

Gilbert, G. K. **(1877)**, *Geology of the Henry Mountains*, U.S. Geog. and Geol. Survey, Rocky Mt. Region (Powell). See Land sculpture, pp. 99–150.
Davis, W. M. **(1909)**, The geographical cycle, *Geographical essays*, Ginn and Co., Boston, pp. 249–278. Reprinted in 1954, Dover Publications, New York.
Rich, J. L. **(1917)**, Cultural features and the physiographic cycle, *Geog. Rev.*, 4:297–308.
Davis, W. M. **(1923)**, The scheme of the erosion cycle, *J. Geol.*, 31:10–25.
Johnson, D. W. **(1929)**, Baselevel, *J. Geol.*, 37:775–782.

Glock, W. S. (**1931**), The development of drainage systems; A synoptic view, *Geog. Rev.*, 21:475–482.
Johnson, D. (**1933**), Development of drainage systems and the dynamic cycle, *Geog. Rev.*, 23:114–121.
Fenneman, N. M. (**1936**), Cyclic and non-cyclic aspects of erosion, *Geol. Soc. Am. Bull.*, 47:173–186.
Peltier, L. (**1950**), The geographic cycle in periglacial regions as it is related to climatic geomorphology, *Annals A. A. G.*, 40:214–236.
Melton, F. A. (**1959**), Aerial photographs and structural geomorphology, *J. Geol.*, 67:351–370.
Hack, J. T. (**1960**), Interpretation of erosional topography in humid temperate regions, *Am. J. Sci.*, 258:80–97.
Menard, H. W. (**1961**), Some rates of regional erosion, *J. Geol.*, 69:154–161.
Bretz, J. H. (**1962**), Dynamic equilibrium and the Ozark land forms, *Am. J. Sci.*, 260:427–438.
Leopold, L. B., and W. B. Langbein (**1962**), The concept of entropy in landscape evolution, *U.S. Geol. Survey Prof. Paper*, 500-*A*, 20 pp.
Schumm, S. A. (**1963**), The disparity between present rates of denudation and orogeny, *U.S. Geol. Survey Prof. Paper*, 454-*H*, 13 pp.
Holmes, C. D. (**1964**), Equilibrium in humid-climate physiographic processes, *Am. J. Sci.*, 262:436-445.
Judson, S., and D. F. Ritter (**1964**), Rates of regional denudation in the United States, *Jour. Geophysical Res.* 69:3395–3401.
Holzner, L., and Weaver, G. D. (**1965**), Geographic evaluation of climatic and climo-genetic geomorphology, *Annals A. A. G.*, 55:592–602.
Howard, A. D. (**1965**), Geomorphological systems—equilibrium and dynamics, *Am. J. Sci.*, 263:302–312.
Marchand, D. E. (**1971**), Rates and modes of denudation, White Mountains, eastern California, *Amer. Jour. Sci.*, 270:109–135.
Derbyshire, E., ed. (**1974**), *Climatic geomorphology*, Barnes and Noble, New York, 296 pp.
Garner, H. F. (**1974**), *The origin of landscapes: a synthesis of geomorphological research*, Oxford University Press, New York, 750 pp.

SLOPE DEVELOPMENT

Gilbert, G. K. (**1909**), The convexity of hilltops, *J. Geol.*, 17:344–350.
Bryan, K. (**1940**), Gully gravure—A method of slope retreat, *Jour. Geomorphology*, 3:89–107.
Von Engeln, O. D., et al (**1940**), Symposium: Walther Penck's contribution to geomorphology, *Annals A. A. G.*, 30:219–284.
White, S. E. (**1949**), Processes of erosion on steep slopes of Oahu, Hawaii, *Am. J. Sci.*, 247:168–186.
Strahler, A. N. (**1950**), Equilibrium theory of erosional slopes approached by frequency distribution analysis, *Am. J. Sci.*, 248:673–696, 800–814.
Cotton, C. A. (**1952**), The erosional grading of convex and concave slopes, *Geog. Jour.*, 143:197–204.
Penck, W. (**1953**), *Morphological analysis of land forms*, Macmillan and Co., London, 429 pp.
Holmes, C. D. (**1955**), Geomorphic development in humid and arid regions: A synthesis, *Am. J. Sci.*, 253:377–390.
Schumm, S. A. (**1956**), The role of creep and rainwash on the retreat of badland slopes, *Am. J. Sci.*, 254:693–706.
Strahler, A. N. (**1956**), Quantitative slope analysis, *Geol. Soc. Am. Bull.*, 67:571–596.
Melton, M. A. (**1960**), Intravalley variation in slope angles related to microclimate and erosional environment, *Geol. Soc. Am. Bull.*, 71:133–144.
Scheidegger, A. E. (**1961**), Mathematical models of slope development, *Geol. Soc. Am. Bull.*, 72:37–50.
Savigear, R. A. G. (**1962**), Some observations on slope development in North Devon and North Cornwall, *Inst. British Geographers, Publ.*, 31:23–42.
Culling, W. E. H. (**1965**), Theory of erosion on soil-covered slopes, *J. Geol.*, 73:230–254.
Melton, M. A. (**1965**), Debris-covered hillslopes of the southern Arizona desert—Consideration of their stability and sediment contribution, *J. Geol.*, 73:715–729.
Schumm, S. A. (**1966**), The development and evolution of hillslopes, *Jour. Geol. Education*, 14:98–104.
LaMarche, V. C., Jr. (**1968**), Rates of slope degradation as determined from botanical evidence, White Mountains, California, *U.S. Geol. Survey Prof. Paper*, 352-*I*, 377 pp.
Carson, M. A., and M. J. Kirkby (**1972**), *Hillslope form and process*, Cambridge Univ. Press, 483 pp.
Clayton, K. (**1972**), *Slopes*, Oliver and Boyd, Edinburgh, 288 pp.
Schumm, S. A., and M. P. Mosley, eds. (**1973**), *Slope morphology*, Benchmark papers in geology, Dowden, Hutchinson and Ross, Stroudsburg, Pa., 454 pp.

PENEPLAINS

Davis, W. M. (**1909**), Base-level, grade, and peneplain, *Geographical essays*, Ginn and Co., Boston, pp. 381–412. Reprinted in 1954, Dover publications, New York.
Davis, W. M. (**1909**), Plains of marine and subaerial denudation, *Geographical essays*, Ginn and Co., Boston, pp. 323–349. Reprinted in 1954, Dover Publications, New York.
Davis, W. M. (**1909**), The peneplain, *Geographical essays*, Ginn and Co., Boston, pp. 350–380. Reprinted in 1954, Dover Publications, New York.
Davis, W. M. (**1911**), The Colorado Front Range, *Annals A. A. G.*, 1:21–83.
Johnson, D. (**1916**), Plains, planes, and peneplanes, *Geog. Rev.*, 1:443–447.
Davis, W. M. (**1922**), Peneplains and the geographic cycle, *Geol. Soc. Am. Bull.*, 23:587–598.
Ward, F. (**1930**), The role of solution in peneplanation, *J. Geol.*, 38:262–270.
Rich, J. L. (**1938**), Recognition and significance of multiple erosion surfaces, *Geol. Soc. Am. Bull.*, 49:1695–1722.
Meyerhoff, H. A. (**1940**), Migration of erosion surfaces, *Annals A. A. G.*, 30:247–254.

Dixey, F. **(1944)**, African landscape, *Geog. Rev.*, 34:457–465.
Geyl, W. F. **(1960)**, Geophysical speculations on the origin of stepped erosion surface, *J. Geol.*, 68:154–176.

CYCLE IN ARID CLIMATE; PEDIMENTS

McGee, W J **(1897)**, Sheetflood erosion, *Geol. Soc. Am. Bull.*, 8:87–112.
Davis, W. M. **(1909)**, The geographical cycle in an arid climate, *Geographical essays*, Ginn and Co., Boston, pp. 296–322. Reprinted in 1954, Dover Publications, New York.
Paige, S. **(1912)**, Rock-cut surfaces of the desert ranges, *J. Geol.*, 20:442–450.
Davis, W. M. **(1930)**, Rock floors in arid and humid climates, *J. Geol.*, 38:1–27, 136–158.
Blackwelder, E. **(1931)**, Desert plains, *J. Geol.*, 39:133–140.
Fenneman, N. M. **(1931)**, *Physiography of Western United States*, McGraw-Hill Book Co., New York, See pp. 326–333, 340–348.
Johnson, D. **(1931)**, Planes of lateral corrosion, *Science*, n. ser. 73:174–177.
Johnson, D. **(1932)**, Rock plains of arid regions, *Geog. Rev.*, 22:656–665.
Bagnold, R. A. **(1933)**, A further journey through the Libyan Desert, *Geog. J.*, 82:103–129, 211–235.
Willis, B. **(1934)**, Inselbergs, *Annals A. A. G.*, 24:123–129.
Rich, J. L. **(1935)**, Origin and evolution of rock fans and pediments, *Geol. Soc., Am. Bull.*, 46:999–1024.
Davis, W. M. **(1938)**, Sheetfloods and streamfloods, *Geol. Soc. Am. Bull.*, 49:1337–1416.
Peel, R. F. **(1941)**, Denudational landforms of the central Libyan Desert, *J. Geomorphology*, 4:3–23.
Howard, A. D. **(1942)**, Pediments and the pediment pass problem, *J. Geomorphology*, 5:3–31, 95–136.
Childs. O. E. **(1948)**, Geomorphology of the Little Colorado River, Arizona, *Geol. Soc. Am. Bull.*, 59:353–388.
Tator, B. A. **(1952)**, Pediment characteristics and terminology, *Annals A. A. G.*, 42:295–317 and 43:47–53.
Tuan, Y.-F. **(1962)**, Structure, climate, and basin land forms in Arizona and New Mexico, *Annals A. A. G.*, 52:51–68.
Mabbutt, J. A. **(1966)**, Mantle-controlled planation of pediments, *Am. J. Sci.*, 264:78–91.
Denny, C. S. **(1967)**, Fans and pediments, *Am. J. Sci.*, 265:81–105.
Hadley, R. F. **(1967)**, Pediments and pediment-forming processes, *Jour. of Geol. Education*, 15:83–89.
Rahn, P. H. **(1967)**, Sheetfloods, streamfloods, and the formation of pediments, *Annals A. A. G.*, 57:593–604.
Twidale, C. R. **(1967)**, Origin of the piedmont angle as evidenced in South Australia, *J. Geol.*, 75:393–411.
Cooke, R. U. **(1970)**, Morphometric analysis of pediments and associated landforms in the western Mojave Desert, California, *Amer. Jour. Sci.*, 269:26–38.
Cooke, R. U. **(1970)**, Stone pavements in deserts, *Annals A. A. G.*, 60:560–577.
Twidale, C. R. **(1972)**, Landform development in the Lake Eyre region, Australia, *Geog. Rev.*, 62:40–70.

ALLUVIAL FANS

Eckis, R. **(1928)**, Alluvial fans of the Cucamonga district, southern California, *J. Geol.*, 36:224–247.
Chawner, W. D. **(1955)**, Alluvial fan flooding. The Montrose, California, flood of 1934, *Geog. Rev.*, 25:255–263.
Beaty, C. B. **(1963)**, Origin of alluvial fans, White Mountains, California and Nevada, *Annals A. A. G.*, 53:516–535.
Bull, W. B. **(1964)**, Geomorphology of segmented alluvial fans in western Fresno County, California, *U.S. Geol. Survey Prof. Paper*, *352-E*, 129 pp.
Ruhe, R. V. **(1964)**, Landscape morphology and alluvial deposits in southern New Mexico, *Annals A. A. G.*, 54:147–159.
Denny, C. S. **(1965)**, Alluvial fans in the Death Valley region, California and Nevada, *U.S. Geol. Survey Prof. Paper*, 466, 62 pp.
Melton, M. A. **(1965)**, The geomorphic and paleoclimatic significance of alluvial deposits in southern Arizona, *J. Geol.*, 73:1–38.
Denny, C. S. **(1967)**, Fans and pediments, *Am. J. Sci.*, 265:81–105.
Hook, R. LeB. **(1967)**, Processes on arid-region alluvial fans, *J. Geol.*, 75:438–460.
Shlemon, R. J. **(1971)**, The Quaternary deltaic and channel system in the central Great Valley, California, *Annals A. A. G.*, 61:427–440.

CHAPTER 27

QUANTITATIVE ANALYSIS OF FLUVIALLY ERODED LANDSCAPES

Horton, R. E. **(1945)**, Erosional development of streams and their drainage basins; hydrophysical approach to quantitative morphology, *Geol. Soc. Am. Bull.*, 56:275–370.
Smith, K. G. **(1950)**, Standards of grading texture of erosional topography, *Am. Jour. Sci.*, 248:655–668.
Strahler, A. N. **(1952)**, Hypsometric (area-altitude) analysis of erosional topography, *Geol. Soc. Am. Bull.*, 63:1117–1142.
Strahler, A. N. **(1954)**, Statistical analysis in geomorphic research, *J. Geol.*, 62:1–25.
Schumm, S. A. **(1956)**, Evolution of drainage systems and slopes in badlands at Perth Amboy, N.J., *Geol. Soc. Am. Bull.*, 67:597–646.
Strahler, A. N. **(1956)**, Quantitative slope analysis, *Geol. Soc. Am. Bull.*, 67:571–596.
Chorley, R. J. **(1957)**, Climate and morphometry, *J. Geol.*, 65:628–638.
Melton, M. A. **(1957)**, Geometric properties of mature

drainage basins and their representation in a E_4 phase space, *J. Geol.*, 66:35–54.
Strahler, A. N. **(1957)**, Quantitative analysis of watershed geomorphology, *Trans. Am. Geophysical Union*, 38:913–920.
Melton, M. A. **(1958)**, Correlation structure of morphometric properties of drainage systems and their controlling agents, *J. Geol.*, 66:442–460.
Morisawa, M. **(1958)**, Measurement of drainage-basin outline form, *J. Geol.*, 66:587–591.
Strahler, A. N. **(1958)**, Dimensional analysis applied to fluvially eroded landforms, *Geol. Soc. Am. Bull.*, 69:279–300.
Geyl, W. F. **(1961)**, Morphometric analysis and the world-wide occurrence of stepped erosion surfaces, *J. Geol.*, 69:388–416.
Chorley, R. J., and M. A. Morgan **(1962)**, Comparison of morphometric features, Unaka Mountains, Tennessee and North Carolina, and Dartmoor, England, *Geol. Soc. Am. Bull.*, 73:17–34.
Peltier, L. C. **(1962)**, Area sampling for terrain analysis, *Professional Geographer*, 14:24–28.
Lubowe, J. K. **(1964)**, Stream junction angles in the dendritic drainage pattern, *Am. J. Sci.*, 262:325–339.
Morisawa, M. **(1964)**, Development of drainage systems on an upraised lake floor, *Am. J. Sci.*, 262:340–354.
Strahler, A. N. **(1964)**, Quantitative geomorphology of drainage basins and channel networks, Section 4-II of *Handbook of Applied Hydrology*, McGraw-Hill Book Co., New York.
Woodruff, J. F. **(1964)**, A comparative analysis of selected drainage basins, *Professional Geographer*, 16:15–19.
Chorley, R. J. **(1966)**, The application of statistical methods to geomorphology, pp. 275–387, of *Essays in Geomorphology*, G. H. Dury, ed., Amer. Elsevier Publ. Co., New York.
Clarke, J. I. **(1966)**, Morphometry from maps, pp. 235–274 of *Essays in Geomorphology*, G. H. Dury, ed., Amer. Elsevier Publ. Co., New York.
McConnell, H. **(1966)**, A statistical analysis of spatial variability of mean topographic slope on stream-dissected glacial materials, *Annals A. A. G.*, 56:712–728.
Shreve, R. L. **(1966)**, Statistical law of stream numbers, *J. Geol.*, 74:17–37.
Woldenberg, M. J. **(1966)**, Horton's laws justified in terms of allometric growth and steady state in open systems, *Geol. Soc. Am. Bull.*, 77:431–434.
Shreve, R. L. **(1967)**, Infinite topologically random channel networks, *J. Geol.*, 75:178–186.
Woldenberg, M. J., and B. J. L. Berry **(1967)**, Rivers and central places: Analogous systems? *Jour. of Regional Sci.*, 7:129–139.
Eyles, R. J. **(1968)**, Stream net ratios in West Malaysia, *Geol. Soc. Am. Bull.*, 79:701–712.
Hooke, R. Le B. **(1968)**, Steady-state relationships on arid region alluvial fans in closed basins, *Amer. Jour. Sci.*, 266:609–629.
Smart, J. S. **(1969)**, Topological properties of channel networks, *Geol. Soc. Am. Bull.*, 80:1757–1774.
Woldenberg, M. J. **(1969)**, Spatial order in fluvial systems: Horton's laws derived from mixed hexagonal hierarchies of drainage basin areas, *Geol. Soc. Am. Bull.*, 80:97–112.
Swan, S. B. St. C. **(1970)**, Analysis of residual terrain: Johor, Malaya, *Annals A. A. G.*, 60:124–133.
Chorley, R. J., ed. **(1971)**, *Introduction to fluvial processes*, Methuen and Co., London, 218 pp.
Chorley, R. J., and B. A. Kennedy **(1971)**, *Physical geography; a systems approach*, Prentice-Hall International, London, 370 pp.
Eyles, R. J. **(1971)**, A classification of west Malaysian drainage basins, *Annals A. A. G.*, 61:460–467.
Morisawa, M., ed. **(1971)**, *Quantitative geomorphology: some aspects and applications*, Proc. Second Annual Geomorphology Symposia Series, Publ. in Geomorphology, State University of New York, Binghamton, 315 pp.
Smart, J. S., and V. L. Moruzzi **(1971)**, Computer simulation of Clinch Mountain drainage networks, *Jour. Geol.*, 79:572–584.
Gregory, K. J., and D. E. Walling **(1974)**, *Drainage basin form and process*, John Wiley and Sons, New York, 456 pp.

CHAPTER 28

REGIONAL LANDFORM CLASSIFICATION AND DESCRIPTION

Powell, J. W., and others **(1896)**, *The physiography of the United States*, American Book Co., New York, 345 pp.
Fenneman, N. M. **(1928)**, Physiographic divisions of the United States, *Annals A. A. G.*, 18:261–353.
Fenneman, N. M. **(1931)**, *Physiography of the Western United States*, McGraw-Hill Book Co., New York, 534 pp.
Zernitz, E. R. **(1932)**, Drainage patterns and their significance, *J. Geol.*, 40:498–521.
Fenneman, N. M. **(1938)**, *Physiography of the Eastern United States*, McGraw-Hill Book Co., New York, 691 pp.
Atwood, W. W. **(1940)**, *The physiographic provinces of North America*, Ginn and Co., Boston, 536 pp.
Strahler, A. N. **(1946)**, Geomorphic terminology and classification of landmasses, *J. Geol.*, 54:32–42.
Thornbury, W. D. **(1965)**, *Regional geomorphology of the United States*, John Wiley and Sons, New York, 609 pp.
Hunt, C. B. **(1967)**, *Physiography of the United States*, W. H. Freeman and Co., San Francisco, 480 pp.
Hunt, C. B. **(1974)**, *Natural regions of the United States and Canada*, W. H. Freeman, San Francisco, 725 pp.

COASTAL PLAINS

Davis, W. M. **(1895)**, The development of certain English rivers, *Geog. J.*, 5:127–146.
Davis, W. M. **(1909)**, The Seine, the Meuse, and the Moselle, *Geographical essays*, Ginn and Co., Bos-

ton, pp. 587–616. Reprinted in 1954, Dover Publ., New York.

Davis, W. M. **(1918)**, *Handbook of northern France*, Harvard Univ. Press, Cambridge, Mass., 174 pp.

Cleland, H. F. **(1920)**, The Black Belt of Alabama, *Geog. Rev.*, 10:375–387.

Grabau, A. W. **(1920)**, The Niagara cuesta from a new viewpoint, *Geog. Rev.*, 9:264–276.

Johnson, D. W. **(1921)**, Battlefields of the World War, *Am. Geog. Soc. Research Ser.*, No. 3, 648 pp.

Renner, G. T., Jr. **(1927)**, The physiographic interpretation of the Fall Line, *Geog. Rev.*, 17:278–286.

Johnson, D. W. **(1931)**, A theory of Appalachian geomorphic evolution, *J. Geol.*, 39:497–508.

Lobeck, A. K. **(1939)**, *Geomorphology*, McGraw-Hill Book Co., New York, 731 pp. Chapter 13, Coastal plains, pp. 439–468.

Stokes, G. A. **(1957)**, Lumbering and western Louisiana cultural landscapes, *Annals A. A. G.*, 47:250–266.

Landes, K. K. **(1959)**, *Petroleum geology*, 2nd edition, John Wiley and Sons, New York, 433 pp.

Prince, H. C. **(1959)**, Parkland in the Chilterns, *Geog. Rev.*, 49:18–31.

Levorsen, A. I. **(1967)**, *Geology of petroleum*, 2nd edition, W. H. Freeman, San Francisco, 724 pp.

CAVERNS, KARST

Eigenmann, C. H. **(1917)**, The homes of blindfishes, *Geog. Rev.*, 4:171–182.

Sanders, E. M. **(1921)**, The cycle of erosion in a karst region (after Cvijić), *Geog. Rev.*, 11:593–604.

Lobeck, A. K. **(1929)**, The geology and physiography of the Mammoth Cave National Park, *Ky. Geol. Survey*, 6th ser., 31:327–399.

Davis, W. M. **(1930)**, Origin of limestone caverns, *Geol. Soc. Am Bull.*, 41:475–628.

Thorp, J. **(1934)**, The asymmetry of the "Pepino Hills" of Puerto Rico in relation to the trade winds, *J. Geol.*, 42:537–545.

Dicken, S. N. **(1935)**, Kentucky karst landscapes, *J. Geol.*, 43:708–728.

Bretz, J. H. **(1938)**, Caves of the Galena formation, *J. Geol.*, 46:828–841.

Meyerhoff, H. A. **(1938)**, The texture of karst topography in Cuba and Puerto Rico, *J. Geomorphology*, 1:279–295.

Bretz, J. H. **(1942)**, Vadose and phreatic features of limestone caverns, *J. Geol.*, 50:675–811.

Bretz, J. H. **(1949)**, Carlsbad Caverns and other caves of the Guadalupe block, New Mexico, *J. Geol.*, 57:447–463.

Jordan, R. H. **(1950)**, An interpretation of Floridan karst, *J. Geol.*, 58:261–268.

Bretz, J. H. **(1953)**, Genetic relations of caves to peneplains and big springs in the Ozarks, *Am. J. Sci.*, 251:1–24.

Doerr, A. H., and D. R. Hoy **(1957)**, Karst landscapes of Cuba, Puerto Rico, and Jamaica, *Scientific Monthly*, 81:178–187.

Sweeting, M. M. **(1958)**, The karstlands of Jamaica, *Geog. Jour.*, 124:184–199.

Sweeting, M. M., and others **(1965)**, Denudation in limestone regions; A symposium, *Geog. Jour.*, 131:34–56.

LaValle, P. **(1967)**, Some aspects of linear karst depression developments in south central Kentucky, *Annals A. A. G.*, 57:49–71.

Poulson, T. L., and W. B. White **(1969)**, The cave environment, *Science*, 165:971–981.

White, W. B., et al **(1970)**, The central Kentucky karst, *Geog. Rev.*, 165:88–115.

Herak, M., and V. T. Stringfield, eds. **(1972)**, *Karst, American Elsevier Publ. Co.*, New York, 565 pp.

LeGrand, H. E. **(1973)**, Hydrological and ecological problems of karst regions, *Science*, 179:859–864.

HORIZONTAL STRATA

Dutton, C. E. **(1882)**, Tertiary history of the Grand Canyon district, *U.S. Geol. Survey Monograph* 2, 264 pp.

Atwood, W. W. **(1911)**, A geographic study of the Mesa Verde, *Annals A. A. G.*, 1:95–100.

Lee, W. T. **(1921)**, The Raton mesas of New Mexico and Colorado, *Geog. Rev.*, 11:384–397.

Birdseye, C., and R. C. Moore **(1924)**, A boat voyage through the Grand Canyon of the Colorado, *Geog. Rev.*, 14:177–196.

Haas, W. H. **(1926)**, The Cliff-dweller and his habitat, *Annals A. A. G.*, 16:167–215.

Pike, R. W. **(1930)**, Land and peoples of the Hadhramant, Aden Protectorate, *Geog. Rev.*, 30:627–648.

Smith, G. H. **(1935)**, The relative relief of Ohio, *Geog. Rev.*, 25:272–284.

Lobeck, A. K. **(1939)**, *Geomorphology*, McGraw-Hill Book Co., New York, 731 pp. Chapter 14, Plains and plateaus, pp. 469–502.

Ives, R. **(1947)**, Reconnaissance of the Zion hinterland, *Geog. Rev.*, 37:618–638.

Koons, D. **(1955)**, Cliff retreat in the southwestern United States, *Am. J. Sci.*, 253:53–60.

Doerr, A., and L. Guernsey **(1956)**, Man as a geomorphological agent: the example of coal mining, *Annals A. A. G.*, 46:197–210.

Schumm, S. A. **(1956)**, The role of creep and rainwash on the retreat of badland slopes, *Am. J. Sci.*, 254:693–706.

Smith, K. G. **(1958)**, Erosional processes and landforms in Badlands National Monument, South Dakota, *Geol. Soc. Am. Bull.*, 69:975–1008.

Ahnert, F. **(1960)**, The influence of Pleistocene climates upon the morphology of cuesta scarps on the Colorado Plateau, *Annals A. A. G.*, 50:139–156.

Zakrzewska, B. **(1963)**, An analysis of landforms in a part of the Central Great Plains, *Annals A. A. G.*, 53:536–568.

Butzer, K. W. **(1965)**, Desert landforms at the Kurkur Oasis, *Egypt, Annals A. A. G.*, 55:578–591.

Aghassy, J. **(1970)**, Jointing, drainage, and slopes in a west African epeirogenic savanna landscape, *Annals A. A. G.*, 60:286–298.

Nephew, E. A. **(1972)**, Healing wounds (strip mining), *Environment*, 14, No. 1:12–21.

Greenburg, W. (**1973**), Chewing it up at 200 tons a bite: strip mining, *Technology Review*, 75, No. 4:46–55.

DOMES

Gilbert, G. K. (**1877**), *Report on the geology of the Henry Mountains*, U.S. Geol. and Geol. Survey Rocky Mt. Region (Powell), pp. 18–98.
Newton, H., and W. P. Jenny (**1880**), *Geology of the Black Hills*, U.S. Geog. and Geol. Survey Rocky Mt. Region (Powell), 566 pp.
Cross, C. W. (**1894**), The laccolithic mountain groups of Colorado, Utah and Arizona, U.S. Geol. Survey, *14th Ann. Rept.*, Part 2, pp. 157–241.
Darton, N. H., and S. Paige (**1925**), *Central Black Hills*, U.S. Geol. Survey, Folio 219.
Lobeck, A. K. (**1939**), *Geomorphology*, McGraw-Hill Book Co., New York, 731 pp. Chapter 15, Dome mountains, pp. 503–542.

CHAPTER 29

FOLDS

Willis, B. (**1895**), The northern Appalachians, *Nat. Geog. Soc. Monograph*, 1:169–202.
Davis, W. M. (**1906**), The mountains of southernmost Africa, *Am. Geog. Soc. Bull.*, 38:593–623.
Davis, W. M. (**1909**), The rivers and valleys of Pennsylvania, *Geographical essays*, Ginn and Co., Boston, pp. 413–484, Reprinted in 1954, Dover Publ., New York.
Chamberlin, R. T. (**1910**), The Appalachian folds of central Pennsylvania, *J. Geol.*, 18:228–251.
Johnson, D. (**1931**), *Stream sculpture on the Atlantic slope*, Columbia Univ. Press, New York, 142 pp. Reprinted in 1967, Hafner Publ. Co., New York.
Meyerhoff, H. A., and E. W. Olmsted (**1936**), The origins of Appalachian drainage, *Am. J. Sci.*, 232:21–41.
Fenneman, N. M. (**1938**), *Physiography of the Eastern United States*, McGraw-Hill Book Co., New York, pp. 195–278.
Mackin, J. H. (**1938**), The origin of Appalachian drainage—A reply, *Am. J. Sci.*, 236:27–53.
Lobeck, A. K. (**1939**), *Geomorphology*, McGraw-Hill Book Co., New York, 731 pp. Chapter 17, Folded mountains, pp. 581–612.
Rich, J. L. (**1939**), A bird's-eye cross section of the central Appalachian Mountains and Plateau: Washington to Cincinnati, *Geog. Rev.*, 29:561–586.
Thompson, H. D. (**1939**), Drainage evolution in the southern Appalachians, *Geol. Soc. Am. Bull.*, 50:1323–1356.
Strahler, A. N. (**1945**), Hypotheses of stream development in the folded Appalachians of Pennsylvania, *Geol. Soc. Am. Bull.*, 56:45–88.
Bethune, P. de (**1948**), Geomorphic studies in the Appalachians of Pennsylvania, *Am. J. Sci.*, 246:1–22.

FAULTING AND EARTHQUAKES

Davis, W. M. (**1934**), The Long Beach earthquake, *Geog. Rev.*, 24:1–11.
Heck, N. H. (**1935**), A new map of earthquake distribution, *Geog. Rev.*, 25:125–130.
Page, B. M. (**1935**), Basin-range faulting of 1915 in Pleasant Valley, Nevada, *J. Geol.*, 43:690–707.
Macelwane, J. B. (**1947**), *When the earth quakes*, The Bruce Publ. Co., Milwaukee, Wisc., 288 pp.
Kingdon-Ward, F. (**1953**), The Assam earthquake of 1950, *Geog. Jour.*, 119:169–182.
Kingdon-Ward, F. (**1955**), Aftermath of the great Assam earthquake of 1950, *Geog. Jour.*, 121:290–303.
Richter, C. F. (**1958**), *Elementary seismology*, W. H. Freeman and Co., San Francisco, 768 pp.
Whitkind, I. J., et al. (**1962**), Geologic features of the earthquake at Hebgen Lake, Montana, August 17, 1959, *Bull. Seismological Soc. Am.*, 52:163–180.
Hodgson, J. H. (**1964**), *Earthquakes and earth structure*, Prentice-Hall, Englewood Cliffs, N.J., 166 pp.
Eckel, E. B. (**1970**), *The Alaska Earthquake, March 27, 1964: lessons and conclusions*, Professional Paper 564, U.S. Geological Survey, U.S. Govt. Printing Office, Washington, D.C., 57 pp.
Wallace, R. E. (**1970**), Earthquake recurrence intervals on the San Andreas Fault, *Geol. Soc. Am. Bull.*, 81:2875–2890.
Iacopi, R. (**1971**), *Earthquake country*, 3rd edition, Lane Books, Menlo Park, Calif., 160 pp.
Multiple authors (**1971**), *The San Fernando, California, earthquake of February 9, 1971*, Professional Paper 733, U.S. Geological Survey, U.S. Govt. Printing Office, Washington, D.C., 254 pp.
Coffman, J. L., and C. A. von Hake, eds. (**1973**), *Earthquake history of the United States*, rev. edition, NOAA, U.S. Govt. Printing Office, Washington, D.C. 208 pp.

FAULT FORMS AND BLOCK MOUNTAINS

Davis, W. M. (**1909**), Mountain ranges of the Great Basin, *Geographical essays*, Ginn and Co., Boston, pp. 725–772. Reprinted in 1954, Dover Publ., New York.
Cushing, S. W. (**1913**), Coastal plains and block mountains in Japan, *Annals A. A. G.*, 3:43–61.
Johnson, D. W. (**1918**), Block faulting in the Klamath Lakes region, *J. Geol.*, 26:229–236.
Blackwelder, E. (**1928**), The recognition of fault scarps, *J. Geol.*, 36:289–311.
Fuller, R. E., and A. A. Waters (**1929**), The nature and origin of the horst and graben structure of southern Oregon, *J. Geol.*, 37:204–338.
Davis, W. M. (**1930**), The Peacock Range, Arizona, *Geol. Soc. Am. Bull.*, 41:293–313.
Longwell, C. R. (**1930**), Faulted fans west of the Sheep Range, southern Nevada, *Am. J. Sci.*, 220:1–13.
Fenneman, N. M. (**1931**), *Physiography of the Western United States*, McGraw-Hill Book Co., New York, 691 pp. See pp. 326–395.
Teale, E. O., and H. E. Teale (**1933**), A physiographical

map of Tanganiyka Territory, *Geog. Rev.*, 23:402–413.

Willis, B. **(1938)**, San Andreas Rift, California, *J. Geol.*, 46:793–827.

Johnson, D. **(1939)**, Fault scarps and fault-line scarps, *J. Geomorphology*, 2:174–177.

Lobeck, A. K. (1939), *Geomorphology*, McGraw-Hill Book Co., New York, 731 pp. Chapter 16, Block Mountains, pp. 543–580.

Sharp, R. P. **(1939)**, Basin-range structure of the Ruby-East Humboldt Range, northeastern Nevada, *Geol. Soc. Am. Bull.*, 50:881–920.

Dixey, F. **(1941)**, Geomorphic development of the Shire Valley, Nyasaland, *J. Geomorphology*, 4:97–116.

Gardner, L. S. **(1941)**, The Hurricane fault in southwestern Utah and northwestern Arizona, *Am. J. Sci.*, 239:241–260.

Strahler, A. N. **(1948)**, Geomorphology and structure of the West Kaibab fault zone and Kaibab Plateau, Arizona, *Geol. Soc. Am. Bull.*, 61:717–757.

Cotton, C. A. **(1950)**, Tectonic scarps and fault valleys, *Geol. Soc. Am. Bull.*, 61:717–757.

Cotton, C. A. **(1953)**, Tectonic relief: with illustrations from New Zealand, *Geog. Jour.*, 119:213–222.

Beaty, C. B. **(1961)**, Topographic effects of faulting: Death Valley, California, *Annals A. A. G.*, 51:234–240.

Higgins, C. G. **(1961)**, San Andreas fault north of San Francisco, California, *Geol. Soc. Am. Bull.*, 72:51–68.

Donath, F. A. **(1962)**, Analysis of Basin-Range structure, south-central Oregon, *Geol. Soc. Am. Bull.*, 73:1–16.

Stewart, J. H. **(1971)**, Basin and Range structure: a system of horsts and grabens produced by deep-seated extension, *Geol. Soc. Am. Bull.*, 82:1019–1044.

CHAPTER 30

CRYSTALLINE AND COMPLEX MASSES

Davis, W. M. **(1911)**, Colorado Front Range, *Annals A. A. G.*, 1:21–83.

Lobeck, A. K. **(1917)**, Position of the New England peneplain in the White Mountains region, *Geog. Rev.*, 3:53–60.

Fenneman, N. M. **(1931)**, *Physiography of Western United States*, McGraw-Hill Book Co., New York, 534 pp. Chapters 2, 4, 5, and 9.

Fenneman, N. M. **(1938)**, *Physiography of the Eastern United States*, McGraw-Hill Book Co., New York, 691 pp. Chapters 3, 6, 7, and 13.

Taylor, G. **(1942)**, British Columbia, A study in topographic control, *Geog. Rev.*, 32:372–402.

Woodruff, J. F., and E. J. Parizek **(1956)**, Influence of underlying rock structures on stream courses and valley profiles in the Georgia Piedmont, *Annals A. A. G.*, 46:129–139.

Hack, J. T. **(1966)**, Circular patterns and exfoliation in crystalline terrane, Grandfather Mountain area, North Carolina, *Geol. Soc. Am. Bull.*, 77:975–986.

VOLCANOES

Atwood, W. W. **(1906)**, Red mountain: a dissected volcanic cone, *J. Geol.*, 14:138–146.

Hobbs, W. H. **(1906)**, The grand eruption of Vesuvius in 1906, *J. Geol.*, 14:636–655.

Johnson, D. W. **(1907)**, Volcanic necks of the Mount Taylor region, New Mexico, *Geol. Soc. Am. Bull.*, 18:303–324.

Stearns, H. T. **(1924)**, Craters of the Moon National Monument, Idaho, *Geog. Rev.*, 14:362–372.

Peacock, M. A. **(1931)**, The Modoc Lava field, northern California, *Geog. Rev.*, 21:68–82.

Colton, H. S. **(1932)**, Sunset Crater: the effects of a volcanic eruption on an ancient Pueblo people, *Geog. Rev.*, 22:582–590.

Shippee, R. **(1932)**, Lost valleys of Peru, *Geog. Rev.*, 22:562–581.

Atwood, W. W., Jr., **(1935)**, The glacial history of an extinct volcano, Crater Lake National Park, *J. Geol.*, 43:142–168.

Putnam, W. C., **(1938)**, The Mono Crater, California, *Geog. Rev.*, 28:68–82.

Lobeck, A. K. **(1939)**, *Geomorphology*, McGraw-Hill Book Co., New York, 731 pp. Chapter 19, Volcanoes, pp. 647–704.

Williams, H. **(1941)**, *Crater Lake, the story of its origin*, Univ. of California Press, Berkeley and Los Angeles, 97 pp.

Stearns, H. T., and G. A. Macdonald **(1942)**, Geology and ground-water resources of the island of Maui, Hawaii, *Bull.* 7, Div. of Hydrography, Terr. of Hawaii, Honolulu, 344 pp.

Cotton, C. A. **(1944)**, *Volcanoes as landscape forms*, Whitcombe and Tombs, 416 pp.

Nichols, R. L. **(1946)**, McCartys basalt flow, Valencia County, New Mexico, *Geol. Soc. Am. Bull.*, 57:1049–1086.

Williams, H. **(1961)**, The floor of Crater Lake, *Am. J. Sci.*, 259:81–83.

Bullard, F. M. **(1962)**, *Volcanoes: In history, in theory, in eruption*, Univ. of Texas Press, Austin, 441 pp.

Hopson, C. A., et al. **(1962)**, The latest eruptions from Mt. Rainier volcano, *J. Geol.*, 70:635–647.

MacDonald, G. A., and A. T. Abbott **(1970)**, *Volcanoes in the sea: the geology of Hawaii*, Univ. of Hawaii Press, Honolulu, 441 pp.

Schmidt, R. G., and H. R. Shaw **(1971)**, *Atlas of volcanic phenomena*, U.S. Geological Survey, U.S. Govt. Printing Office, Washington D.C., Sheets 1–20.

CHAPTER 31

ALPINE GLACIERS

Boyd, L. A. **(1932)**, Fiords of East Greenland, *Geog. Rev.*, 22:529–561.

Cooper, W. S. **(1937)**, The problem of Glacier Bay, Alaska, A study of glacier variations, *Geog. Rev.*, 27:37–62.

Lewis, W. V. **(1940)**, The function of meltwater in cirque formation, *Geog. Rev.*, 30:64–83.

Sharp, R. P. **(1947)**, The Wolf Creek glaciers, St. Elias Range, Yukon Territory, *Geog. Rev.*, 37:26–52.
Dyson, J. L. **(1948)**, Shrinkage of Sperry and Grinnell glaciers, Glacier National Park, Montana, *Geog. Rev.*, 38:95–103.
Sharp, R. P. **(1948)**, The constitution of valley glaciers, *J. Glaciology*, 1:182–189.
Lawrence, D. B. **(1950)**, Glacier fluctuation for six centuries in southeastern Alaska and its relation to solar activity, *Geog. Rev.*, 40:191–223.
Sharp, R. P. **(1951)**, Accumulation and ablation on the Seward-Malaspina glacier system, Canada-Alaska, *Geol. Soc. Am. Bull.*, 62:725–744.
Field, W. O., Jr., and C. J. Heusser **(1952)**, Glaciers—Historians of climate, *Geog. Rev.*, 42:337–345.
Sharp, R. P. **(1954)**, Glacier flow: a review, *Geol. Soc. Am. Bull.*, 65:821–838.
Baird, P. D. **(1955)**, Glaciological research in the Canadian arctic, *Arctic*, 8:96–108.
Field, W. O., Jr. **(1955)**, Glaciers, *Scientific American*, 193:84–92.
Sharp, R. P. **(1956)**, Glaciers in the arctic, *Arctic*, 9:78–117.
Sharp, R. P. **(1958)**, The latest major advance of Malaspina Glacier, Alaska, *Geog. Rev.*, 48:16–26.
Marcus, M. G. **(1960)**, Periodic drainage of glacier-dammed Tulsequah Lake, British Columbia, *Geog. Rev.*, 50:89–106.
Sharp, R. P. **(1960)**, *Glaciers*, Univ. of Oregon Press, Eugene, 78 pp.
Dyson, J. L. **(1962)**, *The World of Ice*, Alfred A. Knopf, New York, 292 pp.
Stone, K. H. **(1963)**, The annual emptying of Lake George, Alaska, *Arctic*, 16:27–40.
Ives, J. D. **(1967)**, Glacier terminal features in northeast Baffin Island, *Geog. Bull.*, 9:62–70.
Paterson, W. S. B. **(1969)**, *The physics of glaciers*, Pergamon Press, Oxford, 250 pp.
Wood, W. A. **(1970)**, Recent glacier fluctuations in the Sierra Nevada de Santa Marta, Colombia, *Geog. Rev.*, 60:374–392.

LANDFORMS OF ALPINE GLACIATION

Atwood, W. W. **(1907)**, The glaciation of the Uinta Mountains, *J. Geol.*, 15:790–804.
Davis, W. M. **(1909)**, The sculpture of mountains by glaciers, *Geographical essays*, Ginn and Co., Boston, pp. 617–634. Reprinted in 1954, Dover Publications, New York.
Atwood, W. W. **(1909)**, Glaciation of the Uinta and Wasatch Mountains, *U.S. Geol. Survey Prof. Paper* 61, 96 pp. Describes the area shown in Exercise 2.
Davis, W. M. **(1920)**, Features of glacial origin in Montana and Idaho, *Annals A. A. G.*, 10:75–148.
Martin, L., and F. E. Williams **(1924)**, An ice-eroded fiord, *Geog. Rev.*, 14:576–596.
Matthes, F. E. **(1930)**, Geologic history of the Yosemite Valley, *U.S. Geol. Survey Prof. Paper* 160. See pp. 45–97.
Hubbard, G. C. **(1932)**, The geography of residence in Norway fiord areas, *Annals A. A. G.*, 22:109–118.
Matthes, F. E. **(1950)**, *The incomparable valley; a geological interpretation of the Yosemite*, edited by F. Fryxell, Univ. of California Press, Berkeley, 168 pp.
Nichols, R. L., and M. M. Miller **11951)**, Glacial geology of the Ameghino Valley, Lago Argentino, Patagonia, *Geog. Rev.*, 41:274–294.
Dyson, J. L. **(1952)**, Ice-ridged moraines and their relation to glaciers, *Am. J. Sci.*, 250:204–211.

ICE SHEETS OF ANTARCTICA AND GREENLAND

Demorest, M. **(1943)**, Ice sheets, *Geol. Soc. Am. Bull.*, 54:363–400.
Katz, H. R. **(1953)**, Journey across the Nunataks of central East Greenland, *Arctic*, 6:3–14.
Swithinbank, C. **(1955)**, Ice Shelves, *Geog. Jour.*, 121:64–76.
Neuburg, H. A. C., et al. **(1959)**, The Filchner Ice Shelf, *Annals A. A. G.*, 49:110–119.
Crary, A. P. **(1962)**, The Antarctic, *Scientific American*, 207:60–73.
Robin, G. deQ. **(1962)**, The ice of the Antarctic, *Scientific American*, 207:132–146.
Wexler, H., M. J. Rubin, and J. E. Caskey, Jr. eds. **(1962)**, *Antarctic research*, Geophysical Monograph No. 7, Amer. Geophysical Union, Washington, D.C., 228 pp.
Woolard, G. P. **(1962)**, The land of the Antarctic, *Scientific American*, 207:151–166.
Swithinbank, C. **(1964)**, To the valley glaciers that fed the Rosa Ice Shelf, *Geog. Jour.*, 130:32–48.
Hatherton, T., ed. **(1965)**, *Antarctica*, Frederick A. Praeger, New York-Washington, 511 pp.
Committee on Polar Research **(1970)**, *Polar research; a survey*, Nat. Academy of Sciences, Washington, D.C., 204 pp. See Chapter 4.
Quam, L. O., ed. **(1971)**, *Research in the Antarctic*, Publ. No. 93, Amer. Assoc. for the Advancement of Sci., Washington, D. C., 768 pp. See Part III.

PLEISTOCENE GLACIATION

Flint, R. F. **(1943)**, Origin of the former North American ice sheet, *Geog. Rev.*, 33:479–481.
Flint, R. F., and others **(1945)**, Glacial map of North America, Geol. Soc. Am., *Special Paper* 60, 37 pp. Pt. 1, Glacial map; Pt. 2, Explanatory notes.
Upson, J. E. **(1949)**, Late Pleistocene and recent changes of sea level along the coast of Santa Barbara County, California, *Am. J. Sci.*, 247:94–115.
Carter, G. F. **(1950)**, Evidence for Pleistocene man in Southern California, *Geog. Rev.*, 40:84–102.
Flint, R. F. **(1952)**, The ice age in the North American Arctic, *Arctic*, 5:135–152.
Ewing, M., and W. L. Donn **(1956)**, A theory of ice ages, *Science*, 123:1061–1066.
Plass, G. N. **(1956)**, Carbon dioxide and the climate, *American Scientist*, 4:302–316.
Carter, G. F. **(1957)**, *Pleistocene man at San Diego*, The Johns Hopkins Press, Baltimore, Md., 400 pp.
Sauer, C. O. **(1957)**, The end of the ice age and its witnesses, *Geog. Rev.*, 47:29–43.
Geol. Assn. of Canada **(1958)**, Glacial map of Canada,

Toronto, Ontario, scale 1:3,881,600, 50 × 62 inches.
Hough, J. L. **(1958)**, Geology of the Great Lakes, Univ. of Illinois Press, Urbana, 313 pp.
Flint, R. F. **(1959)**, Pleistocene climates in eastern and southern Africa, *Geol. Soc. Am. Bull.*, 70:343–373.
Wright, H. E., Jr. **(1961)**, Late Pleistocene climate of Europe: A review, *Geol. Soc. Am. Bull.*, 72:933–984.
Donn, W. L., W. R. Farrand, and M. Ewing **(1962)**, Pleistocene ice volumes and sea-level lowering, *J. Geol.*, 70:206–214.
Farrand, W. R. **(1962)**, Postglacial uplift in North America, *Am. J. Sci.*, 260:181–199.
Schultz, G. **(1963)**, *Glaciers and the Ice Age*, Holt, Rinehart and Winston, New York, 128 pp.
Tanner, W. F. **(1965)**, Cause and development of an ice age, *J. Geol.*, 73:413–430.
Wright, H. E., and D. G. Frey, eds. **(1965)**, *The Quaternary of the United States*, Princeton University Press, Princeton, N.J., 922 pp.
Galloway, R. W. **(1970)**, The full-glacial climate in the southwestern United States, *Annals A. A. G.*, 50:245–256.
Flint, R. F. **(1971)**, *Glacial and quaternary geology*, John Wiley and Sons, New York, 892 pp.
Black, R. F., R. P. Goldthwait, and H. B. Willman, eds. **(1973)**, *The Wisconsin Stage*, Memoir 136, Geol. Soc. of Amer., Boulder, Colo., 334 pp.

LANDFORMS OF CONTINENTAL GLACIATION

Alden, W. C. **(1905)**, The drumlins of southeastern Wisconsin, *U.S. Geol. Survey Bull.*, 273, 46 pp.
Whitbeck, R. H. **(1913)**, Economic aspects of glaciation in Wisconsin, *Annals A. A. G.*, 3:62–87.
Thwaites, F. T. **(1926)**, The origin and significance of pitted outwash, *J. Geol.*, 34:308–319.
Flint, R. F. **(1928)**, Eskers and crevasse fillings, *Am. J. Sci.*, 5th ser., 15:410–416.
Flint, R. F. **(1929)**, The stagnation and dissipation of the last ice sheet, *Geog. Rev.*, 19:256–289.
Flint, R. F. **(1930)**, The classification of glacial deposits, *Am. J. Sci.*, 19:169–176.
Brown, T. C. **(1931)**, Kames and kame terraces of central Massachusetts, *Geol. Soc. Am. Bull.*, 42:467–479.
Cotton, C. A. **(1942)**, *Climatic accidents in landscape making*, Whitcombe and Tombs, Christchurch, N.Z., 354 pp.
von Engeln, O. D. **(1945)**, Glacial diversion of drainage, *Annals A. A. G.*, 35:79–120.
Gravenor, C. P. **(1951)**, Bedrock source of tills in southwestern Ontario, *Am. J. Sci.*, 249:66–71.
Gravenor, C. P. **(1953)**, The origin of drumlins, *Am. J. Sci.*, 251:674–681.
Horberg, L., and R. C. Anderson **(1956)**, Bedrock topography and Pleistocene glacial lobes in central United States, *J. Geol.*, 64:101–116.
Aronow, S. **(1959)**, Drumlins and related streamlines features in the Warwick-Tokio Area, North Dakota, *Am. J. Sci.*, 257:191–203.
Gravenor, C. P., and W. O. Kupsch **(1959)**, Ice-disintegration features in western Canada, *J. Geol.*, 67:48–64.
von Engeln, O. D. **(1961)**, *The Finger Lakes Region: Its origin and nature*, Cornell University Press, Ithaca, N.Y., 156 pp.
Kupsch, W. O. **(1962)**, Ice-thrust ridges in western Canada, *J. Geol.*, 70:582–594.
Reed, B., C. J. Galvin, Jr., and J. P. Miller **(1962)**, Some aspects of drumlin geometry, *Am. J. Sci.*, 260:200–210.
Zakrzewska, B. **(1971)**, Valleys of driftless areas, *Annals A. A. G.*, 61:441–459.
Donahue, J. J. **(1972)**, Drainage intensity in western New York, *Annals A. A. G.*, 62:23–36.
Miller, J. W., Jr. **(1972)**, Variations in New York drumlins, *Annals A. A. G.*, 62:418–423.
White, W. A. **(1972)**, Deep erosion by continental ice sheets, *Geol. Soc. Am. Bull.*, 83:1037–1056.
Hill, A. R. **(1973)**, The distribution of drumlins in County Down, Ireland, *Annals A. A. G.*, 63:226–240.

CHAPTER 32

OCEAN WAVES

Johnson, D. W. **(1919)**, *Shore processes and shoreline development*, John Wiley and Sons, New York, 584 pp. Reprinted, 1965, Hafner Publ. Co., New York.
Cornish, V. **(1934)**, *Ocean waves and kindred geophysical phenomena*, Cambridge Univ. Press, New York, 164 pp.
Bigelow, H. B., and W. T. Edmondson **(1947)**, Wind waves at sea; Breakers and surf, *U.S. Navy Oceanographic Office Publ.* 602, Washington, D.C., 177 pp.
Macdonald, G. A., F. P. Shepard, and D. Cox **(1947)**, The tsunami of April 1, 1946, in the Hawaiian Islands, *Pacific Science*, 1:21–37.
Steers, J. A. **(1953)**, The east coast floods; January 31–February 1, 1953, *Geog. Jour.*, 119:280–298.
Russell, R. C. H., and D. H. Macmillan **(1954)**, *Waves and tides*, Hutchinson's Sci. and Tech. Publ., London, 348 pp.
Bacomb, W. **(1964)**, *Waves and beaches*, Doubleday and Co., New York, 260 pp.

SHORELINE PROCESSES AND CLASSIFICATION

Gilbert, G. K. **(1890)**, Lake Bonneville, *U.S. Geol. Survey Monograph* 1, See pp. 29–65.
Cotton, C. A. **(1916)**, Fault coasts of New Zealand, *Geog. Rev.*, 1:20–33.
Cotton, C. A. **(1918)**, The outline of New Zealand, *Geog. Rev.*, 6:320–340.
Johnson, D. W. **(1919)**, *Shore processes and shoreline development*, John Wiley and Sons, New York, 584 pp. Reprinted in 1965, Hafner Publ., New York.
Putnam, W. C. **(1937)**, The marine cycle of erosion for a steeply sloping shoreline of emergence, *J. Geol.*, 45:844–850.

Shepard, F. P. **(1937)**, Revised classification of marine shorelines, *J. Geol.*, 45:602–624.
Lucke, J. **(1938)**, Marine shorelines reviewed, *J. Geol.*, 46:985–995.
Brown, C. W. **(1939)**, Hurricanes and shoreline changes in Rhode Island, *Geog. Rev.*, 29:416–430.
Steers, J. A. **(1946)**, *The coastline of England and Wales*, Cambridge Univ. Press, New York, 644 pp.
Kuenen, P. H. **(1950)**, *Marine geology*, John Wiley and Sons, New York, 568 pp.
Cotton, C. A. **(1954)**, Deductive morphology and genetic classification of coasts, *Scientific Monthly*, 78:163–181.
Cotton, C. A. **(1955)**, The theory of secular marine planation, *Am. J. Sci.*, 253:580–589.
Davis, J. H. **(1956)**, Influences of man upon coast lines, pp. 504–521 of *Man's role in changing the face of the earth*, Univ. of Chicago Press, Chicago, 1193 pp.
Guilcher, A. **(1958)**, *Coastal and submarine morphology*, John Wiley and Sons, New York, 274 pp.
McGill, J. T. **(1958)**, Map of coastal landforms of the world (with separate map), *Geog. Rev.*, 48:402–405.
King, C. A. M. **(1959)**, *Beaches and coasts*, Edward Arnold, London, 403 pp.
Dietz, R. S. **(1963)**, Wave-base, marine profile of equilibrium, and wave-built terraces: A critical appraisal, *Geol. Soc. Am. Bull.*, 74:971–990.
Shepard, F. P. **(1963)**, *Submarine geology*, 2nd edition, Harper and Row, New York, 557 pp.
Bascom, W. **(1964)**, *Waves and beaches*, Doubleday and Co., New York, 260, 260 pp.
Alexander, C. S. **(1966)**, A method of descriptive shore classification and mapping as applied to the northeast coast of Tanganyika, *Annals A. A. G.*, 56:128–140.
Dolan, R., et al **(1972)**, *Classification of the coastal environments of the world*, Part I, The Americas, Office of Naval Research, Geography Programs, Dept. of Environmental Sciences, Univ. of Virginia, Charlottesville, 163 pp.

MARINE CLIFFS; ELEVATED SHORELINES

Davis, W. M. **(1923)**, The Halligs, vanishing islands of the North Sea, *Geog. Rev.*, 13:99–106.
Stearns, H. T. **(1935)**, Shore benches on the Island of Oahu, Hawaii, *Geol. Soc. Am. Bull.* 46:1467–1482.
Jutson, J. T. **(1939)**, Shore platforms near Sydney, N.S.W., *J. Geomorphology*, 2:236–250.
Edwards, A. B. **(1941)**, Storm-wave platforms, *J. Geomorphology*, 4:223–236.
Wengerd, S. **(1951)**, Elevated strandlines of Frobisher Bay, Baffin Island, Canadian Arctic, *Geog. Rev.*, 41:622–637.
Fairbridge, R. W. **(1960)**, The changing level of the sea, *Scientific American*, 202:69–79.
Alexander, C. S. **(1961)**, The marine terraces of Aruba, Bonaire, and Curacao, Netherlands Antilles, *Annals A. A. G.*, 51:102–123.
Butzer, K. W. **(1962)**, Coastal geomorphology of Majorca, *Annals A. A. G.*, 52:191–212.
Russell, R. J. **(1963)**, Recent recession of tropical cliffy coasts, *Science*, 139:9–15.
Byrne, J. V. **(1964)**, An erosional classification for the northern Oregon coast, *Annals A. A. G.*, 54:329–335.
McIntyre, W. G., and H. J. Walker **(1964)**, Tropical cyclones and coastal morphology in Mauritius, *Annals A. A. G.*, 54:582–596
Sugden, D. E., and B. S. John **(1965)**, The raised marine features of Kjove Land, East Greenland, *Geog. Jour.*, 131:235–247.
Snead, R. E. **(1967)**, Recent morphological changes along the coast of West Pakistan, *Annals A. A. G.*, 57:550–565.
Nir, D. **(1971)**, Marine terraces of southern Sinai, *Geog. Rev.*, 61:32–50.
Chappell, J. **(1974)**, Geology of coral terraces, Huon Peninsula, New Guinea: a study of Quaternary tectonic movements and sea-level changes, *Geol. Soc. Am. Bull.*, 85:553–570.

BEACHES, SPITS, COASTAL ENGINEERING

Davis, W. M. **(1909)**, The outline of Cape Cod, *Geographical essays*, Ginn and Co., Boston, pp. 690–724. Reprinted in 1954, Dover Publications, New York.
Johnson, D. W., and W. G. Reed **(1910)**, The form of Nantasket Beach, Mass., *J. Geol.*, 18:162–189.
Evans, O. F. **(1942)**, The origin of spits, bars, and related structures, *J. Geol.*, 50:846–865.
Grant, U. S. **(1943)**, Waves as a sand-transporting agent, *Am. J. Sci.*, 241:117–123.
Krumbein, W. C. **(1950)**, Geological aspects of beach engineering, Geol. Soc. Am., *Berkey Vol.*, pp. 195–223.
Davis, J. H. **(1956)**, Influences of man upon coast lines, pp. 504–521 in *Man's role in changing the face of the earth*, W. L. Thomas, ed., Univ. of Chicago Press, 1193 pp. (Reprinted 1971 in two volumes.)
Zeigler, J. M., C. R. Hayes, and S. D. Tuttle **(1959)**, Beach changes during storms on outer Cape Cod, Massachusetts, *J. Geol.*, 67:318–336.
Bird, E. C. F. **(1960)**, The coastal barriers of East Gippsland, Australia, *Geog. Jour.*, 127:460–468.
Nichols, R. L. **(1961)**, Characteristics of beaches formed in polar climates, *Am. J. Sci.*, 259:694–708.
Tanner, W. F. **(1962)**, Reorientation of convex shores, *Am. J. Sci.*, 260:37–43.
Russell, R. J., and W. G. McIntyre **(1965)**, Southern hemisphere beach rock, *Geog. Rev.*, 55:17–45.
Yasso, W. E. **(1965)**, Plan geometry of headland-bay beaches, *J. Geol.*, 73:702–714.
Dolan, R. **(1966)**, Beach changes on the outer banks of North Carolina, *Annals A. A. G.*, 56:699–711.
Strahler, A. N. **(1966)**, Tidal cycle of changes in an equilibrium beach, Sandy Hook, New Jersey, *J. Geol.*, 74:247–268.
Dolan, R., and J. C. Ferm **(1968)**, Crescentic landforms along the Atlantic Coast of the United States, *Science*, 159:627–629.
Burton, I., and R. W. Kates **(1964)**, The floodplain

and the seashore; a comparative analysis of hazard-zone occupance, *Geog. Rev.*, 54:366–385.
Alexander, C. S. (**1969**), Beach ridges in northeastern Tanzania, *Geog. Rev.*, 59:104–122.
Burton, I., R. W. Kates, and R. E. Snead (**1969**), *The human ecology of coastal flood hazard in Megalopolis*, Research Paper 115, Dept. of Geography, University of Chicago., 196 pp.
Corps of Engineers (**1971**), *Shore Protection guidelines*, Dept. of the Army, U.S. Govt. Printing Office, Washington, D.C., 59 pp.
Hoyt, J. G., and V. J. Henry, Jr. (**1971**), Origins of capes and shoals along the southeastern coast of the United States, *Geol. Soc. Am. Bull.*, 82:59–66.
Dolan, R., and K. Bosserman (**1972**), Shoreline erosion and the Lost Colony, *Annals A. A. G.*, 62:424–426.

ESTUARIES; TIDAL DEPOSITS

Lucke, J. B. (**1934**), A theory of evolution of lagoon deposits on shorelines of emergence, *J. Geol.*, 42:561–584.
Hellinga, F. (**1952**), Water control, *Soil Sci.*, 74:21–33.
Zuur, A. J. (**1952**), Drainage and reclamation of lakes and of the Zuiderzee, *Soil Sci.*, 74:75–89.
West, C. (**1956**), Mangrove swamps of the Pacific coast of Colombia, *Annals A. A. G.*, 46:98–121.
Thompson, K. (**1957**), Origin and use of the English peat fens, *Scientific Monthly*, 81:68–76.
Ahnert, F. (**1960**), Estuarine meanders in the Chesapeake Bay area, *Geog. Rev.*, 50:390–401.
Myrick, R. M., and L. B. Leopold (**1963**), Hydraulic geometry of a small tidal estuary, *U.S. Geol. Survey Prof. Paper*, 422-*B*, 18 pp.
El-Ashry, M. T., and H. R. Wanless (**1965**), Birth and early growth of a tidal delta, *J. Geol.*, 73:404–406.
Pestrong, R. (**1965**), The development of drainage patterns on tidal marshes, *Stanford Univ. Publ., Geol. Sciences*, 10, No. 2, 87 pp.
McCrone, A. W. (**1966**), The Hudson River estuary: hydrology, sediments, and pollution, *Geog. Rev.*, 56:175–189.
Lauff, G. H., ed. (**1967**), *Estuaries*, Amer. Assn. Advancement of Science, Publication No. 83, Washington, 757 pp.
Schubel, J. R., and D. W. Pritchard (**1972**), The estuarine environment, *Jour. Geol. Education*, 20:179–188.

BARRIER BEACHES AND ISLANDS; INLETS

Patton, R. S. (**1931**), Moriches Inlet: a problem in beach evolution, *Geog. Rev.*, 21:627–632.
Hitchcock, C. B. (**1934**), The evolution of tidal inlets, *Geog. Rev.*, 24:653–654.
Russell, R. J., and H. V. Howe (**1935**), Cheniers of southwestern Louisiana, *Geog. Rev.*, 25:449–461.
Howard, A. D. (**1939**), Hurricane modification of the offshore bar of Long Island, New York, *Geog. Rev.*, 29:400–415.
Zeigler, J. M. (**1959**), Origin of the Sea Islands of the southeastern United States, *Geog. Rev.*, 49:222–237.
Hoyt, J. H. (**1967**), Barrier island formation, *Geol. Soc. Am. Bull.*, 78:1125–1136.
Dolan, R., P. J. Godfrey, and W. E. Odum (**1973**), Man's impact on the barrier islands of North Carolina, *Amer. Scientist*, 61:152–162.
Schwartz, M., ed. (**1973**), *Barrier islands*, Dowden, Hutchinson and Ross, Stroudsburg, Pa, 451 pp.

CORAL REEFS, ATOLLS

Dana, J. D. (**1874**), *Corals and coral islands*, Dodd and Mead, New York, 406 pp.
Darwin, C. (**1898**), *The structure and distribution of coral reefs*, 3rd edition, Appleton and Co., New York, 344 pp.
Davis, W. M. (**1922**), The barren reef of Tagula, New Guinea, *Annals A. A. G.*, 12:97–151.
Davis, W. M. (**1928**), The coral reef problem, *Am. Geog. Soc. Special Publ.* 9, 596 pp.
Steers, J. S. (**1929**), The Queensland coast and the Great Barrier Reefs, *Geog. Jour.*, 74:232–370.
Steers, J. A. (**1937**), The coral islands and associated features of the Great Barrier Reefs, *Geog.* Jour., 89:1–146.
Steers, J. A. (**1940**), The cays and palisadoes, Port Royal, Jamaica, *Geog. Rev.*, 30:279–296.
Stearns, H. T. (**1946**), An integration of coral-reef hypotheses, *Am. J. Sci.*, 244:772–791.
Emery, K. O. (**1948**), Submarine geology of Bikini atoll, *Geol. Soc. Am. Bull.*, 59:855–860.
Teichert, C., and R. W. Fairbridge (**1948**), Some coral reefs of the Sahul Shelf, *Geog. Rev.*, 38:222–249.
Fairbridge, R. W. (**1950**), Recent and Pleistocene coral reefs of Australia, *J. Geol.*, 58:330–401.
Murphy, R. E. (**1950**), The economic geography of a Micronesian atoll, *Annals A. A. G.*, 40:58–83.
Cloud, P. E., Jr. (**1954**), Surficial aspects of modern organic reefs, *Scientific Monthly*, 78:195–208.
McKee, E. D. (**1958**), Geology of Kapingamaraangi Atoll, Caroline Islands, *Geol Soc. Am. Bull.*, 69:241–278.
Wiens, H. J. (**1959**), Atoll development and morphology, *Annals A. A. G.*, 49:31–54.
Walker, H. J. (**1962**), Coral and lime industry of Mauritius, *Geog. Rev.*, 52:325–336.
Newell, N. D., and A. L. Bloom (**1970**), The reef flat and 'two-meter eustatic terrace' of some Pacific atolls, *Geol. Soc. Am. Bull.*, 81:1881–1894.
Maragos, J. E., G. B. K. Baines, and P. J. Beveridge (**1973**), Tropical cyclone Bebe creates a new land formation on Funafuti Atoll, *Science*, 181:1161–1164.

CHAPTER 33

DEFLATION; ABRASION

Hobbs, W. H. (**1917**), The erosional and degradational processes of desert depressions, *Annals A. A. G.*, 7:25–60.
Bryan, K. (**1923**), Wind erosion near Lee's Ferry, *Am. J. Sci.*, 5th ser., 6:291–307.

Blackwelder, E. **(1931)**, The lowering of playas by deflation, *Am. J. Sci.*, 5th ser., 21:140–144.

Page, L. R., and R. W. Chapman **(1933)**, The dust fall of Dec. 15-16, 1933, *Am. J. Sci.*, 5th ser., 28:288–297.

Blackwelder, E. **(1934)**, Yardangs, *Geol. Soc. Am. Bull.*, 45:159–166.

Kellogg, C. E. **(1935)**, Soil blowing and dust storms, *U.S. Dept. Agriculture Misc. Publ.* 221, 11 pp.

King, L. C. **(1936)**, Wind-faceted stones from Marlborough, New Zealand, *J. Geol.*, 44:201–213.

Maxson, J. H. **(1940)**, Fluting and faceting of rock fragments, *J. Geol.*, 48:717–751.

Leverett, F. **(1942)**, Wind work accompanying or following the Iowan glaciation, *J. Geol.*, 50:548–559.

Sharp, R. P. **(1949)**, Pleistocene ventifacts east of the Big Horn Mountains, Wyoming, *J. Geol.*, 57:175–195.

Higgins, C. G. **(1956)**, Formation of small ventifacts, J. Geol., 64:506–516.

Borchert, J. R. **(1971)**, The dust bowl in the 1970s, *Annals A. A. G.*, 61:1–22.

Beasley, R. P. **(1972)**, *Erosion and sediment pollution control*, Iowa State Univ. Press, Ames, 320 pp. See Chapter 3.

Idso, S. B., and J. M. Pritchard **(1972)**, An American haboob, *Bull. Am. Met. Soc.*, 53: 930–935.

DUNES

Madigan, C. T. **(1936)**, The Australian sand-ridge deserts, *Geog. Rev.*, 26:205–227.

Smith, H. T. U. **(1939)**, Sand dune cycle in western Kansas, *Geol. Soc. Am. Bull.*, 50:1934–1935.

Melton, F. A. **(1940)**, A tentative classification of dunes, *J. Geol.*, 48:113–173.

Bagnold, R. A. **(1941)**, *The physics of blown sand and desert dunes*, Methuen and Co., London, 265 pp.

Hack, J. T. **(1941)**, Dunes of the Navajo country, *Geog. Rev.*, 31:240–263.

Capot-Rey, R. **(1945)**, Dry and humid morphology in the western Erg., *Geog. Rev.*, 35:391–407.

Black, R. F. **(1951)**, Eolian deposits of Alaska, *Arctic*, 3:89–111.

Cooper, W. S. **(1958)**, Coastal sand dunes of Oregon and Washington, *Geol. Soc. Am. Memoir* 72, 169 pp.

Olson, J. S. **(1958)**, Lake Michigan dune development, *J. Geol.*, 66:254–263, 345–351, 473–483.

Finkel, H. J. **(1959)**, The barchans of southern Peru, *J. Geol.*, 67:614–647.

Ives, R. L. **(1959)**, Shell dunes of the Sonoran shore, *Am. J. Sci.*, 257:449–457.

Simonett, D. S. **(1960)**, Development and grading of dunes in western Kansas, *Annals A. A. G.*, 50:216–214.

Norris, R. M., and K. S. Norris **(1961)**, Algodones dunes of southeastern California, *Geol. Soc. Am. Bull.*, 72:605–620.

Sharp, R. P. **(1963)**, Wind ripples, *J. Geol.*, 71:617–636.

Stokes, W. L. **(1964)**, Incised, wind-aligned stream patterns of the Colorado Plateau, *Am. J. Sci.*, 262:808–816.

Smith, H. T. U. **(1965)**, Dune morphology and chronology in central and western Nebraska, *J. Geol.*, 73:557–578.

Norris, R. M. **(1966)**, Barchan dunes of Imperial Valley, California, *J. Geol.*, 74:292–306.

Sharp, R. P. **(1966)**, Kelso dunes, Mojave Desert, California, *Geol. Soc. Am. Bull.*, 77:1045–1074.

Beheiry, S. A. **(1967)**, Sand forms in the Coachella Valley, southern California, *Annals A. A. G.*, 57:25–48.

Flint, R. F., and G. Bond **(1968)**, Pleistocene sand ridges and pans in Western Rhodesia, *Geol. Soc. Am. Bull.*, 79:299–314.

Calkin, P. E., and R. H. Rutford **(1974)**, The sand dunes of Victoria Valley, Antarctica, *Geog. Rev.*, 64:189–216.

LOESS

Fuller, M. L. **(1922)**, Some unusual features of the loess of China, *Geog. Rev.*, 12:570–584.

Fuller, M. L., and F. G. Clapp **(1924)**, Loess and rock dwellings of Shensi, China, *Geog. Rev.*, 14:215–226.

Moyer, R. T. **(1936)**, Agricultural soils in a loess region of North China, *Geog. Rev.*, 26:414–425.

Russell, R. J. **(1940)**, Lower Mississippi Valley loess, *Geol. Soc. Am. Bull.*, 55:1–40.

Bryan, K. **(1945)**, Glacial versus desert origin of loess, *Am. J. Sci.*, 243:245–248.

Symposium on Loess **(1945)**, *Am. J. Sci.*, 243:225–303. Ten papers on loess.

Leighton, M. M., and H. B. Willman **(1950)**, Loess formations of the Mississippi Valley, *J. Geol.*, 58:599–623.

Fisk, H. N. **(1951)**, Loess and Quaternary geology of the lower Mississippi Valley, *J. Geol.*, 59:333–356.

Péwé, T. L. **(1951)**, An observation on wind-blown silt, *J. Geol.*, 59:399–401.

Thorp. J., and others **(1952)**, Map of Pleistocene eolian deposits of the United States, Alaska, and parts of Canada. Scale 1:2,500,000, *Geol. Soc. Am., New York.*

Péwé, T. L. **(1955)**, Origin of the upland silt near Fairbanks, Alaska, *Geol. Soc. Am. Bull.*, 66:699–724.

Lewis, P. F. **(1960)**, Linear topography in the southwestern Palouse, Washington-Oregon, *Annals A. A. G.*, 50:98–111.

Lugn, A. L. **(1962)**, The origin and sources of loess, *University of Nebraska Studies*, New series No. 26, 105 pp.

Flint, R. F. **(1971)**, *Glacial and Quaternary geology*, John Wiley and Sons, New York, 892 pp. See Eolian features, pp. 243–266.

APPENDIX I

TOPOGRAPHIC MAPS

Salisbury, R. D., and W. W. Atwood **(1908)**, The interpretation of topographic maps, *U.S. Geol. Survey*

Prof. Paper 60, U.S. Govt. Printing Office, Washington, D.C.

Raisz, E., and J. Henry (**1937**), An average slope map of southern New England, *Geog. Rev.*, 27:467–472.

Cozzens, A. B. (**1940**), An angle of slope scale, *J. Geomorphology*, 3:52–56.

Brown, C. B. (**1941**), Mapping Lake Mead, *Geog. Rev.*, 31:385–405.

Wright, J. K. (**1942**), Map makers are human, *Geog. Rev.*, 32:527–544.

Lobeck, A. K., and W. J. Tellington (**1944**), *Military maps and air photographs*, McGraw-Hill Book Co., New York, 256 pp.

Platt, R. R. (**1945**), Official topographic maps: a world index, *Geog. Rev.*, 35:175–181.

Am. Geog. Soc. (**1946**), The Map of Hispanic America on the scale of 1:1,000,000, *Geog. Rev.*, 36:1–28.

Tanaka (**1950**), The relief contour method of representing topography on maps, *Geog. Rev.*, 40:444–456.

Low, J. W. (**1952**), *Plane table mapping*, Harper and Bros., New York, 365 pp.

Calef, W., and R. Newcomb (**1953**), An average Slope map of Illinois, *Annals A. A. G.*, 43:305–316.

Lobeck, A. K. (**1956**), *Things maps don't tell us*, The Macmillan Co., New York, 159 pp.

Robinson, A. H. (**1956**), Mapping the land, *Scientific Monthly*, 82:294–303.

Jenks, G. F., and D. S. Knos (**1961**), The use of shading patterns in graded series, *Annals A. A. G.*, 51:316–334.

Washburn, B. (**1961**), A new map of Mt. McKinley, Alaska, the life story of a cartographic project, *Geog. Rev.*, 51:159–186.

Greenhood, D. (**1964**), *Mapping*, Univ. of Chicago Press, Chicago, 289 pp.

Upton, W. B. (**1970**), *Landforms and topographic maps*, John Wiley and Sons, New York, 108 pp.

Dept. of Earth, Space and Graphic Sciences, U.S. Military Academy (**1974**), *Atlas of Landforms*, 2nd edition John Wiley and Sons, New York.

Riffel, P. (**1974**), *Reading maps*, Hubbard Press, Northbrook, Illinois, 72 pp.

PERSPECTIVE DIAGRAMS, RELIEF MODELS

Lobeck, A. K. (**1924**), *Block diagrams*, John Wiley and Sons, New York, 206 pp. Reprinted 1958, with revisions, by Emerson Trussell Book Co., Amherst, Mass.

Raisz, E. (**1931**), The physiographic method of representing scenery on maps, *Geog. Rev.*, 21:297–304.

Reed, H. P. (**1946**), The development of the terrain model in the war, *Geog. Rev.*, 36:632–652.

King, P. B., and E. M. McKee (**1949**), Terrain diagrams of the Philippine Islands, *Geol. Soc. Am. Bull.*, 60:1829–1836.

Hammond, E. H. (**1954**), Small-scale continental landform maps, *Annals A. A. G.*, 44:33–42.

Robinson, A. H., and N. J. W. Thrower (**1957**), A new method of terrain representation, *Geog. Rev.*, 47:507–520.

Stacy, J. R. (**1958**), Terrain diagrams in isometric projection—simplified, *Annals A. A. G.*, 48:232–236.

Ridd, M. K. (**1963**), The proportional relief landform map, *Annals A. A. G.*, 53:569–576.

AIR PHOTOGRAPHS, AERIAL MAPPING

Matthes, G. H. (**1926**), Oblique aerial surveying in Canada, *Geog. Rev.*, 16:568–582.

Birdseye, C. H. (**1940**), Stereoscopic phototopographic mapping, *Annals A. A. G.*, 30:1–24.

Smith, H. T. U. (**1941**), Aerial photographs in geomorphic studies, *J. Geomorphology*, 4:172–205.

Smith, H. T. U. (**1943**), *Aerial photographs and their applications*, Appleton-Century Co., New York, 372 pp.

Raisz, E. (**1951**), The use of air photos for landform maps, *Annals A. A. G.*, 41:324–330.

Walker, F. (**1953**), *Geography from the air*, Methuen and Co., London, 111 pp.

Institut Géographique National (**1956**), *Relief form atlas*, Paris, France, 179 pp. Maps and air photographs of representative geomorphic types in France and North Africa.

Monkhouse, F. J. (**1959**), *Landscape from the air*, Cambridge Univ. Press, New York, 53 pp.

Miller, V. C. (**1961**), *Photogeology*, McGraw-Hill Book Co., New York, 248 pp.

Strandberg, C. H. (**1967**), *Aerial discovery manual*, John Wiley and Sons, New York, 249 pp.

APPENDIX V

REMOTE SENSING TECHNIQUES IN GEOGRAPHICAL RESEARCH

Proceedings of international symposium on remote sensing of the environment (First Symposium, 1962, through Ninth Symposium, 1974) Institute of Science and Technology, Willow Run Laboratories, Environmental Research Institute of Michigan, University of Michigan, Ann Arbor.

Bird, J. B., and A. Morrison (**1964**), Space photography and its geographical applications, *Geog. Rev.*, 54:463–486.

Committee on Geography, O.N.R., Division of Earth Sciences (**1966**) *Spacecraft in geographic research*, National Academy of Sciences National Research Council, Washington, D.C., 107 pp.

Lancaster, J. (**1968**), *Geographers and remote sensing, Jour. Geog.*, 67:301–310.

McCoy, R. M. (**1969**), Drainage network analysis with K-band imagery, *Geog. Rev.*, 59:493–512.

NASA (**1970**), *Ecological surveys from space*, NASA SP-230 U.S. Govt. Printing Office, Washington, D.C., 75 pp.

NASA (**1970**), *This island earth*, NASA Sp-250, U.S. Govt. Printing Office, Washington, D.C., 182 pp.

Macdonald, H. C., A. J. Lewis, and R. S. Wing (**1971**), Mapping and landform analysis of coastal regions with radar, *Geol. Soc. Am. Bull.*, 82:345–358.

Lathram, E. H. (**1972**), Nimbus IV view of the major

structural features of Alaska, *Science,* 175:1423–1427.

Colwell, N. **(1973)**, Remote sensing as an aid to the management of earth resources, *American Scientist,* 61:175–183.

Fink, D. J. **(1973)**, *Monitoring earth's resources from space, Technology Review,* 75; No. 7:32–41.

Holz, R. K., ed. **(1973)**, *The surveillant science: remote sensing of the environment,* Houghton Mifflin, Boston, 390 pp.

Magoon, O. T. **(1973)**, Use of earth resources technology satellite (ERTS-1) in coastal studies. *NASA/Goddard Space Flight Center ERTS-1 Symposium 29 September 1973, Symposium Proceedings,* 108–116.

Thomson, K. P. B., R. K. Lane, and S. C. Csallany, eds. **(1973)**, *Remote sensing and water resources management,* Amer. Water Resources Assn., Urbana, Ill. 437 pp.

Estes, J. E., and L. W. Senger, eds. **(1974)**, *Remote sensing techniques for environmental analysis,* Hamilton Publ. Co., Santa Barbara, Calif., 340 pp. Extensive bibliography.

Rich, E. I., and W. C. Steele **(1974)**, Geologic structures in Northern California as detected from ERTS-1 satellite imagery, *Geology,* 2:165–169.

Index

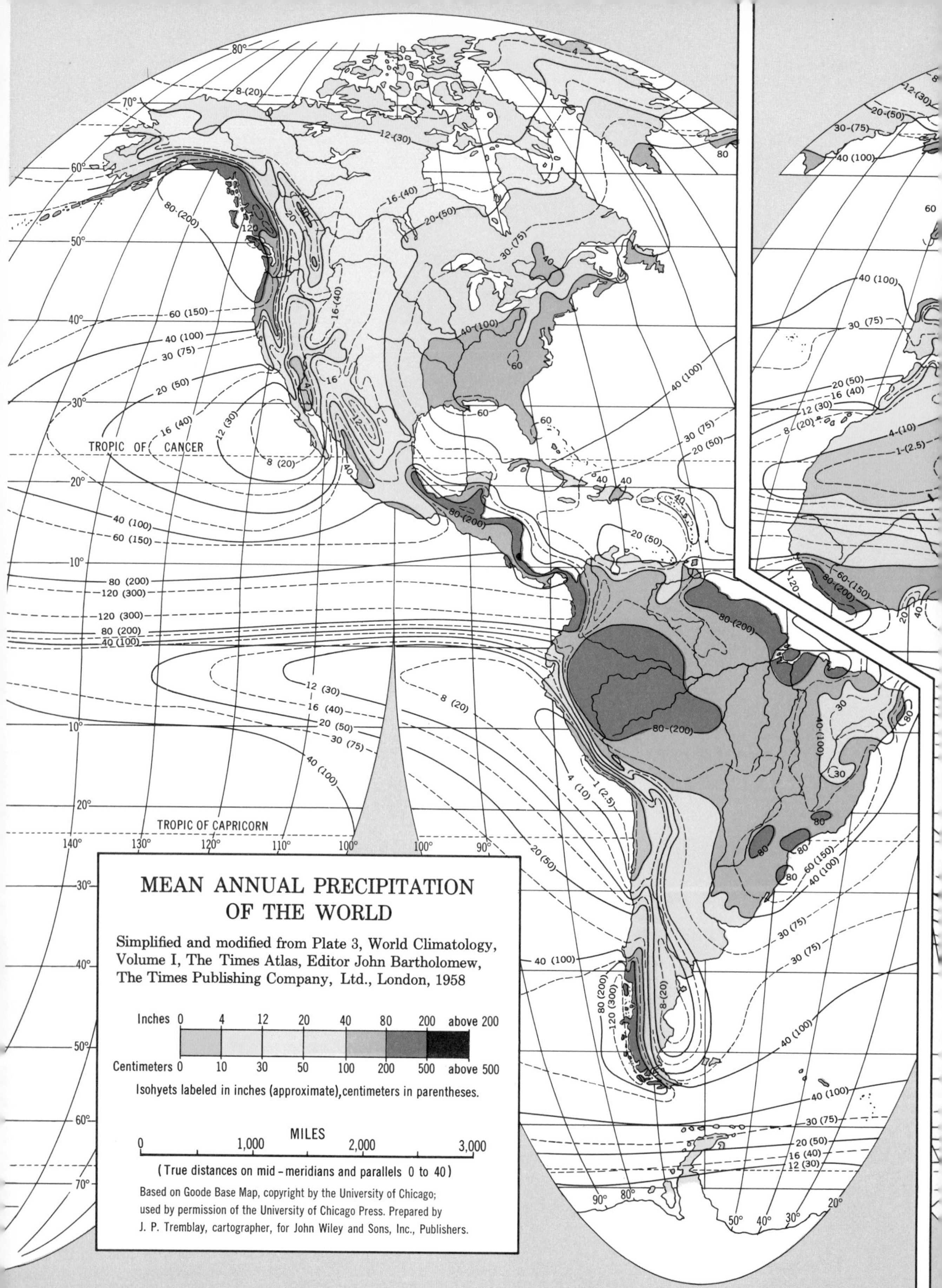
MEAN ANNUAL PRECIPITATION OF THE WORLD
Simplified and modified from Plate 3, World Climatology, Volume I, The Times Atlas, Editor John Bartholomew, The Times Publishing Company, Ltd., London, 1958
Inches 0 4 12 20 40 80 200 above 200
Centimeters 0 10 30 50 100 200 500 above 500
Isohyets labeled in inches (approximate), centimeters in parentheses.
MILES
0 1,000 2,000 3,000
(True distances on mid-meridians and parallels 0 to 40)
Based on Goode Base Map, copyright by the University of Chicago; used by permission of the University of Chicago Press. Prepared by J. P. Tremblay, cartographer, for John Wiley and Sons, Inc., Publishers.
TROPIC OF CANCER
TROPIC OF CAPRICORN

Plate 1
MEAN ANNUAL PRECIPITATION
OF THE WORLD

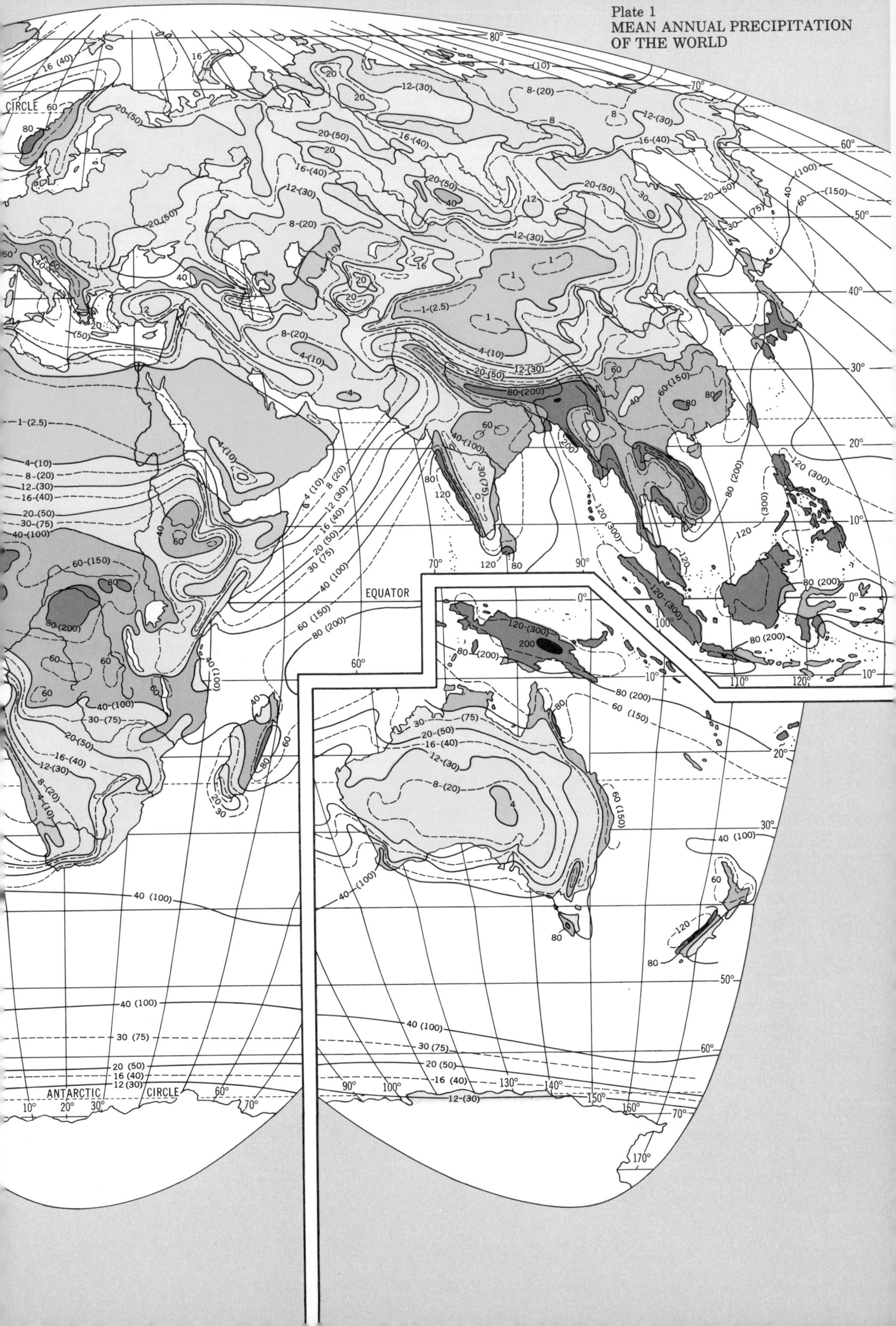

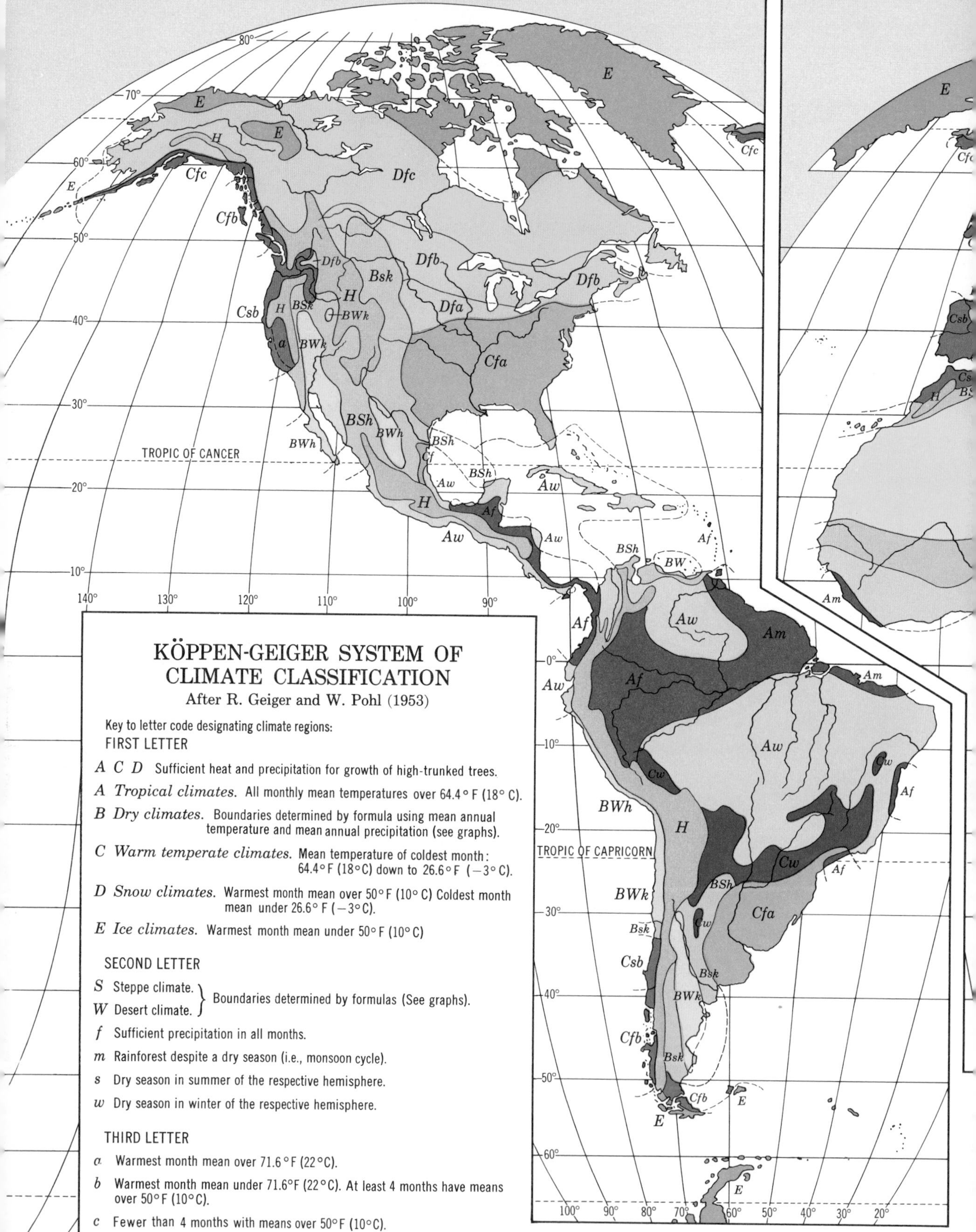

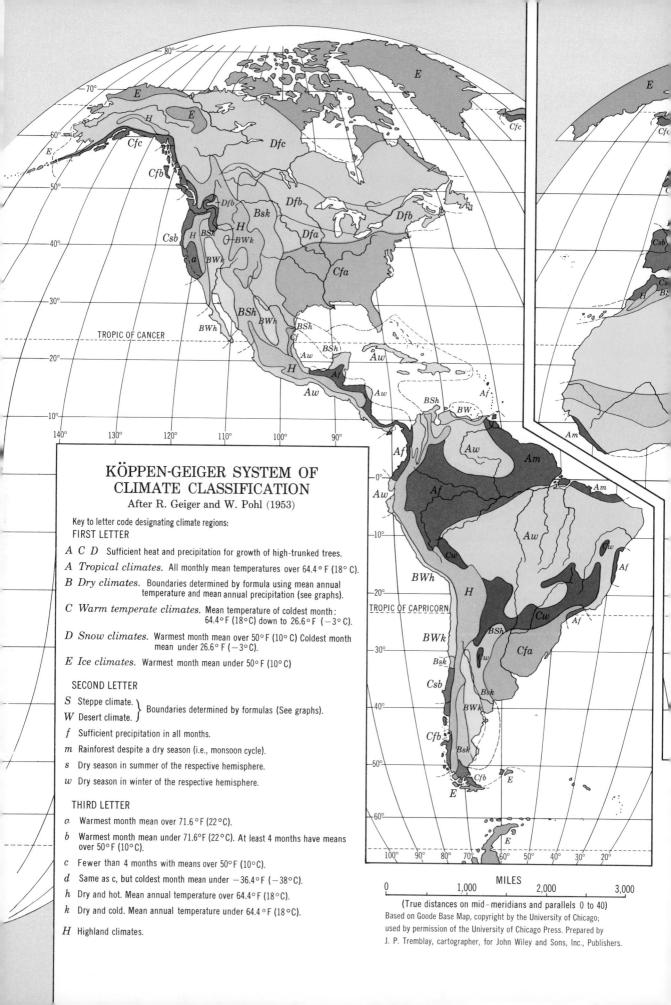

KÖPPEN-GEIGER SYSTEM OF CLIMATE CLASSIFICATION

After R. Geiger and W. Pohl (1953)

Key to letter code designating climate regions:

FIRST LETTER

A C D Sufficient heat and precipitation for growth of high-trunked trees.

A Tropical climates. All monthly mean temperatures over 64.4° F (18° C).

B Dry climates. Boundaries determined by formula using mean annual temperature and mean annual precipitation (see graphs).

C Warm temperate climates. Mean temperature of coldest month: 64.4°F (18°C) down to 26.6°F (−3°C).

D Snow climates. Warmest month mean over 50°F (10° C) Coldest month mean under 26.6° F (−3°C).

E Ice climates. Warmest month mean under 50°F (10° C)

SECOND LETTER

S Steppe climate. }
W Desert climate. } Boundaries determined by formulas (See graphs).

f Sufficient precipitation in all months.

m Rainforest despite a dry season (i.e., monsoon cycle).

s Dry season in summer of the respective hemisphere.

w Dry season in winter of the respective hemisphere.

THIRD LETTER

a Warmest month mean over 71.6°F (22°C).

b Warmest month mean under 71.6°F (22°C). At least 4 months have means over 50°F (10°C).

c Fewer than 4 months with means over 50°F (10°C).

d Same as c, but coldest month mean under −36.4°F (−38°C).

h Dry and hot. Mean annual temperature over 64.4°F (18°C).

k Dry and cold. Mean annual temperature under 64.4°F (18°C).

H Highland climates.

MILES

0 1,000 2,000 3,000

(True distances on mid-meridians and parallels 0 to 40)

 Prepared by J. P. Tremblay, cartographer, for John Wiley and Sons, Inc., Publishers.

Plate 2
CLIMATES OF THE WORLD

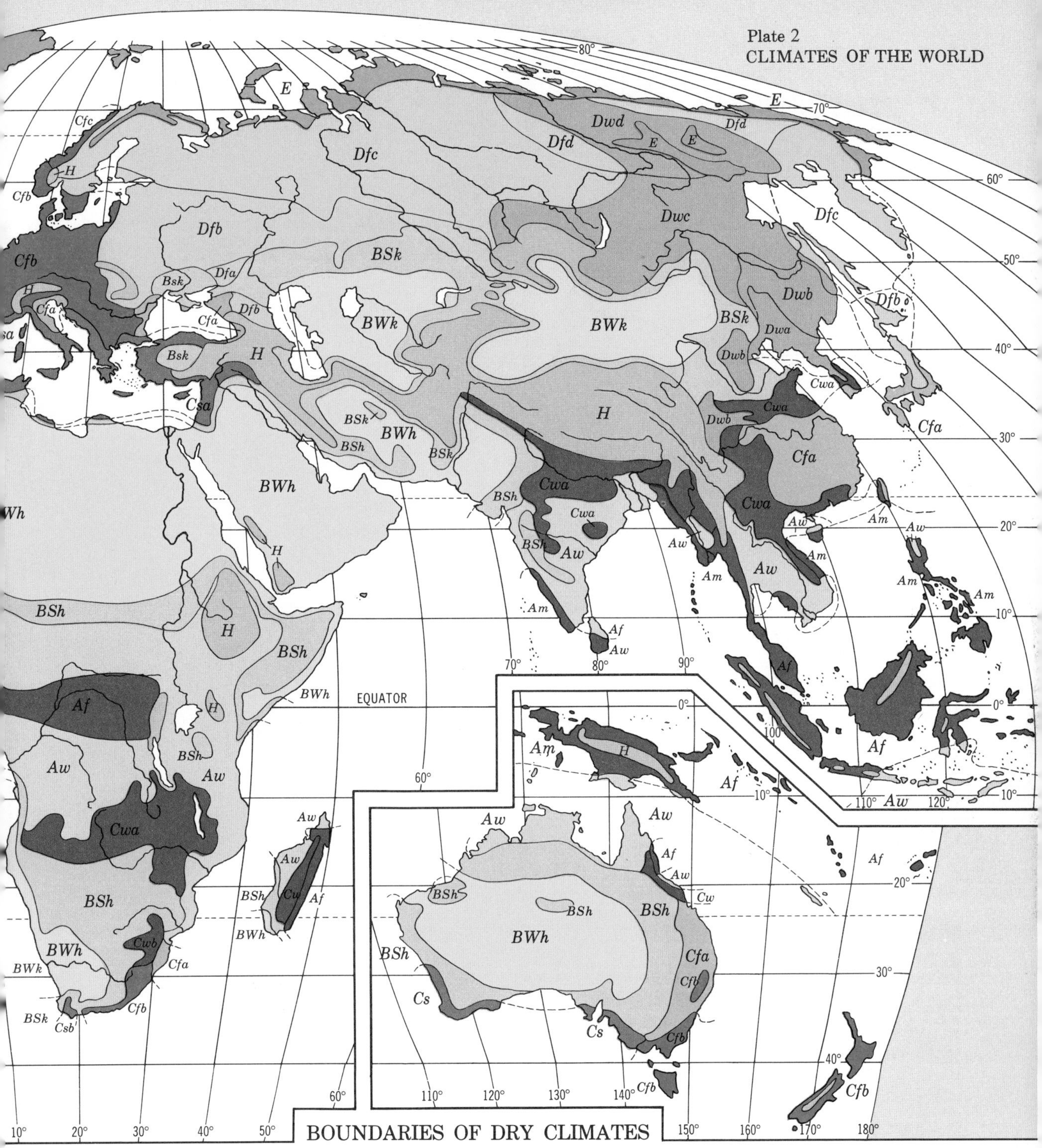

BOUNDARIES OF DRY CLIMATES

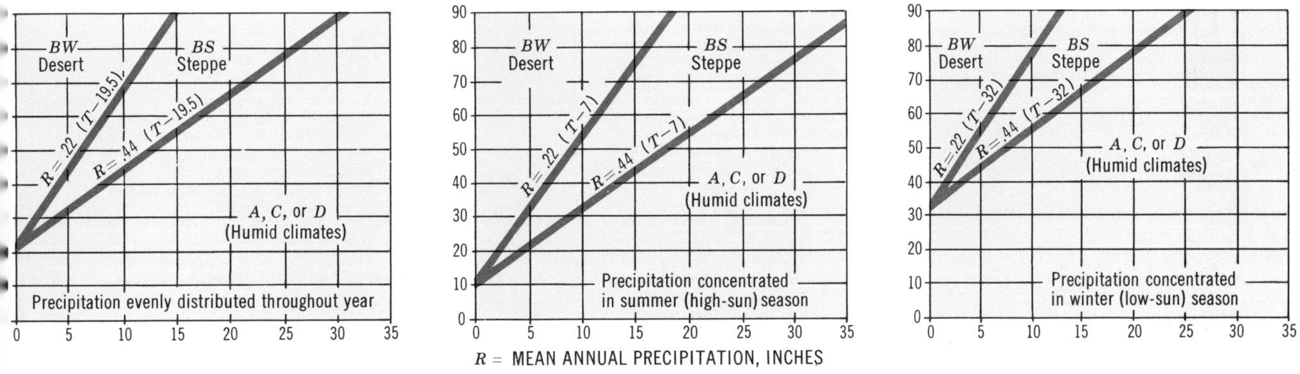

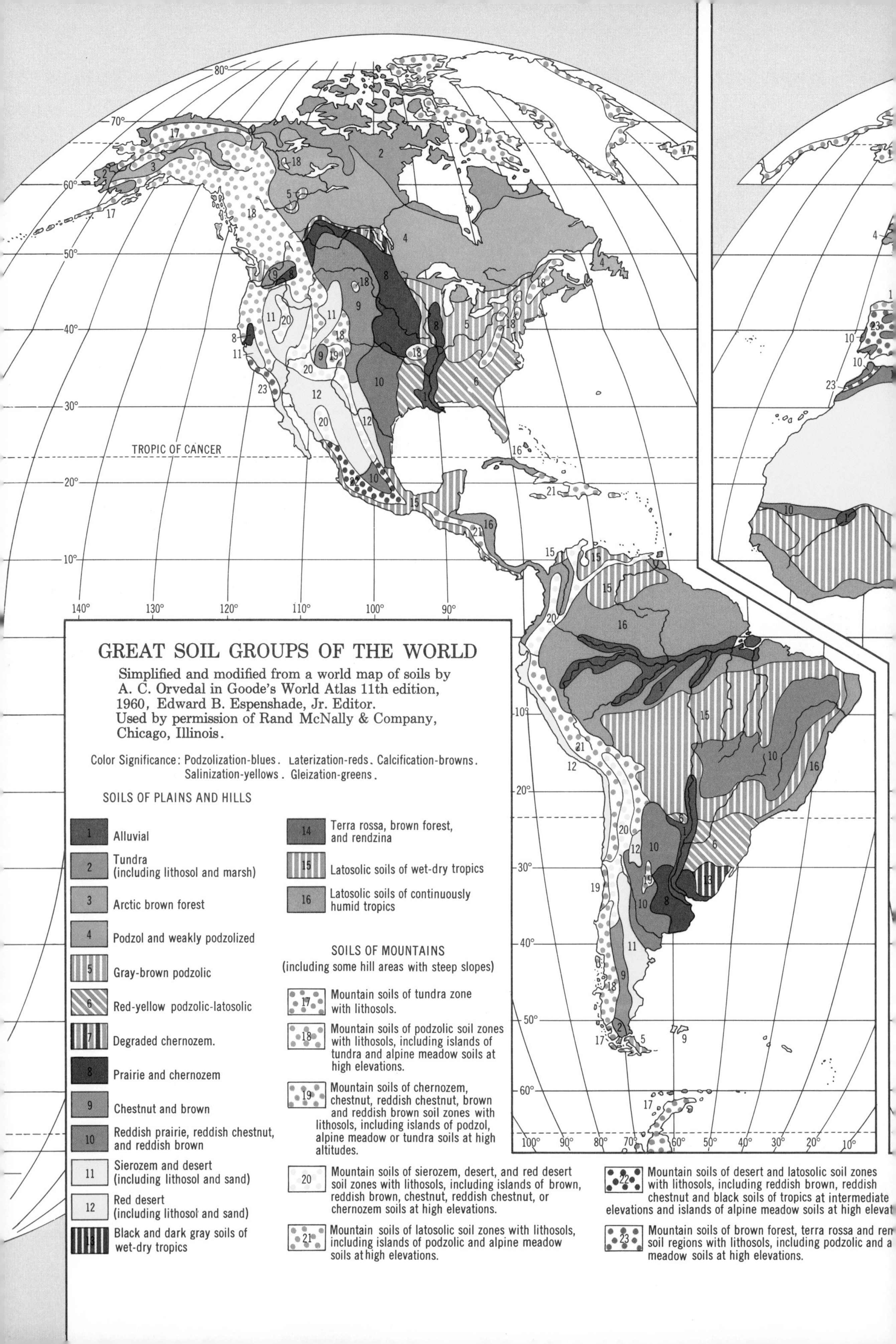
GREAT SOIL GROUPS OF THE WORLD
Simplified and modified from a world map of soils by A. C. Orvedal in Goode's World Atlas 11th edition, 1960, Edward B. Espenshade, Jr. Editor.
Used by permission of Rand McNally & Company, Chicago, Illinois.
Color Significance: Podzolization-blues. Laterization-reds. Calcification-browns. Salinization-yellows. Gleization-greens.
SOILS OF PLAINS AND HILLS
1 Alluvial
2 Tundra (including lithosol and marsh)
3 Arctic brown forest
4 Podzol and weakly podzolized
5 Gray-brown podzolic
6 Red-yellow podzolic-latosolic
7 Degraded chernozem.
8 Prairie and chernozem
9 Chestnut and brown
10 Reddish prairie, reddish chestnut, and reddish brown
11 Sierozem and desert (including lithosol and sand)
12 Red desert (including lithosol and sand)
13 Black and dark gray soils of wet-dry tropics
14 Terra rossa, brown forest, and rendzina
15 Latosolic soils of wet-dry tropics
16 Latosolic soils of continuously humid tropics
SOILS OF MOUNTAINS
(including some hill areas with steep slopes)
17 Mountain soils of tundra zone with lithosols.
18 Mountain soils of podzolic soil zones with lithosols, including islands of tundra and alpine meadow soils at high elevations.
19 Mountain soils of chernozem, chestnut, reddish chestnut, brown and reddish brown soil zones with lithosols, including islands of podzol, alpine meadow or tundra soils at high altitudes.
20 Mountain soils of sierozem, desert, and red desert soil zones with lithosols, including islands of brown, reddish brown, chestnut, reddish chestnut, or chernozem soils at high elevations.
21 Mountain soils of latosolic soil zones with lithosols, including islands of podzolic and alpine meadow soils at high elevations.
22 Mountain soils of desert and latosolic soil zones with lithosols, including reddish brown, reddish chestnut and black soils of tropics at intermediate elevations and islands of alpine meadow soils at high elevat
23 Mountain soils of brown forest, terra rossa and ren soil regions with lithosols, including podzolic and a meadow soils at high elevations.
TROPIC OF CANCER
80°
70°
60°
50°
40°
30°
20°
10°
140°
130°
120°
110°
100°
90°
-10°
-20°
-30°
-40°
-50°
-60°
100°
90°
80°
70°
60°
50°
40°
30°
20°
10°

Plate 3
GREAT SOIL GROUPS OF THE WORLD

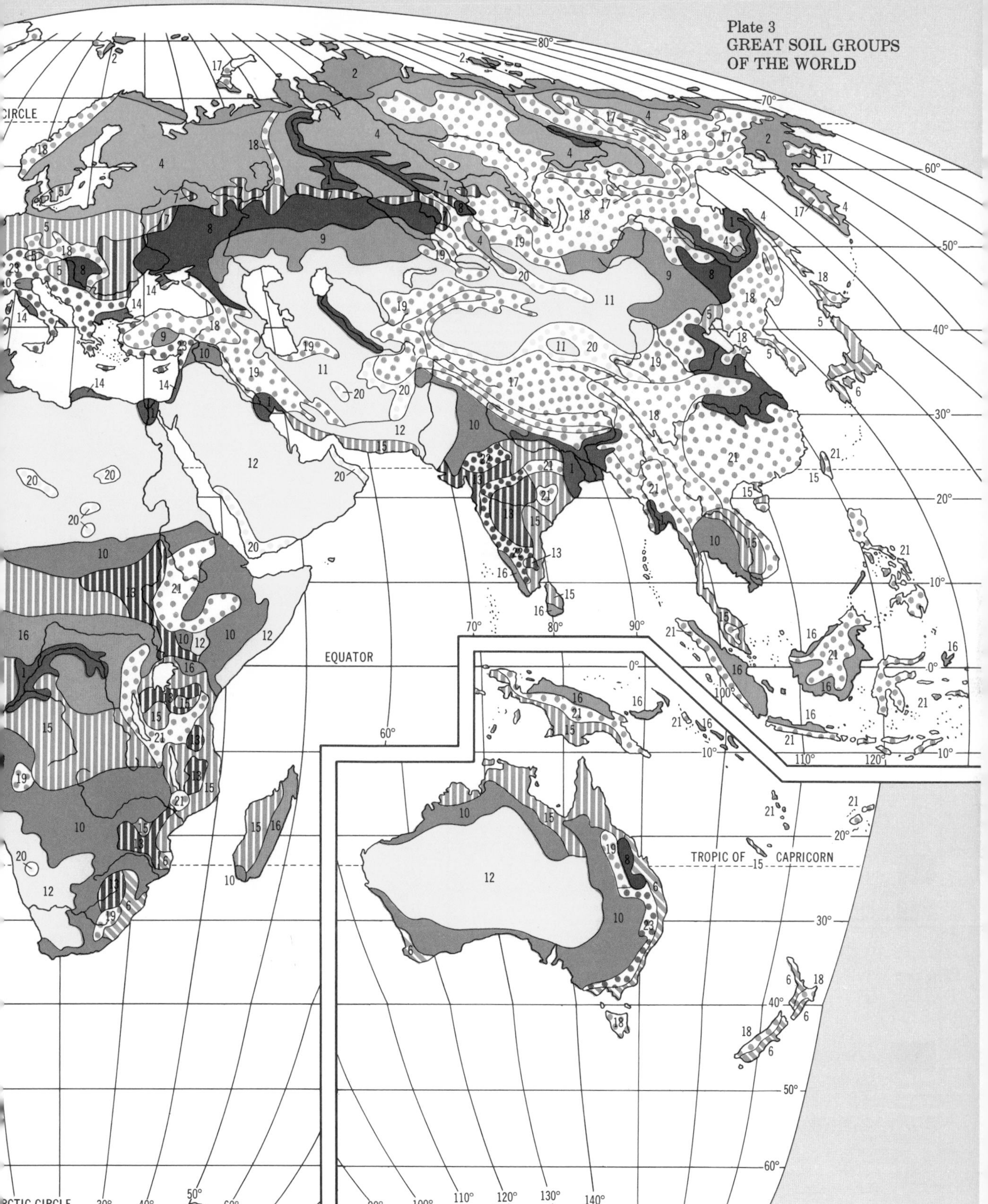

MILES
1,000
3,000
(True distances on mid-meridians and parallels 0 to 40)
sed on Goode Base Map, copyright by the University of Chicago;
d by permission of the University of Chicago Press. Prepared by
P. Tremblay, cartographer, for John Wiley and Sons, Inc., Publishers.

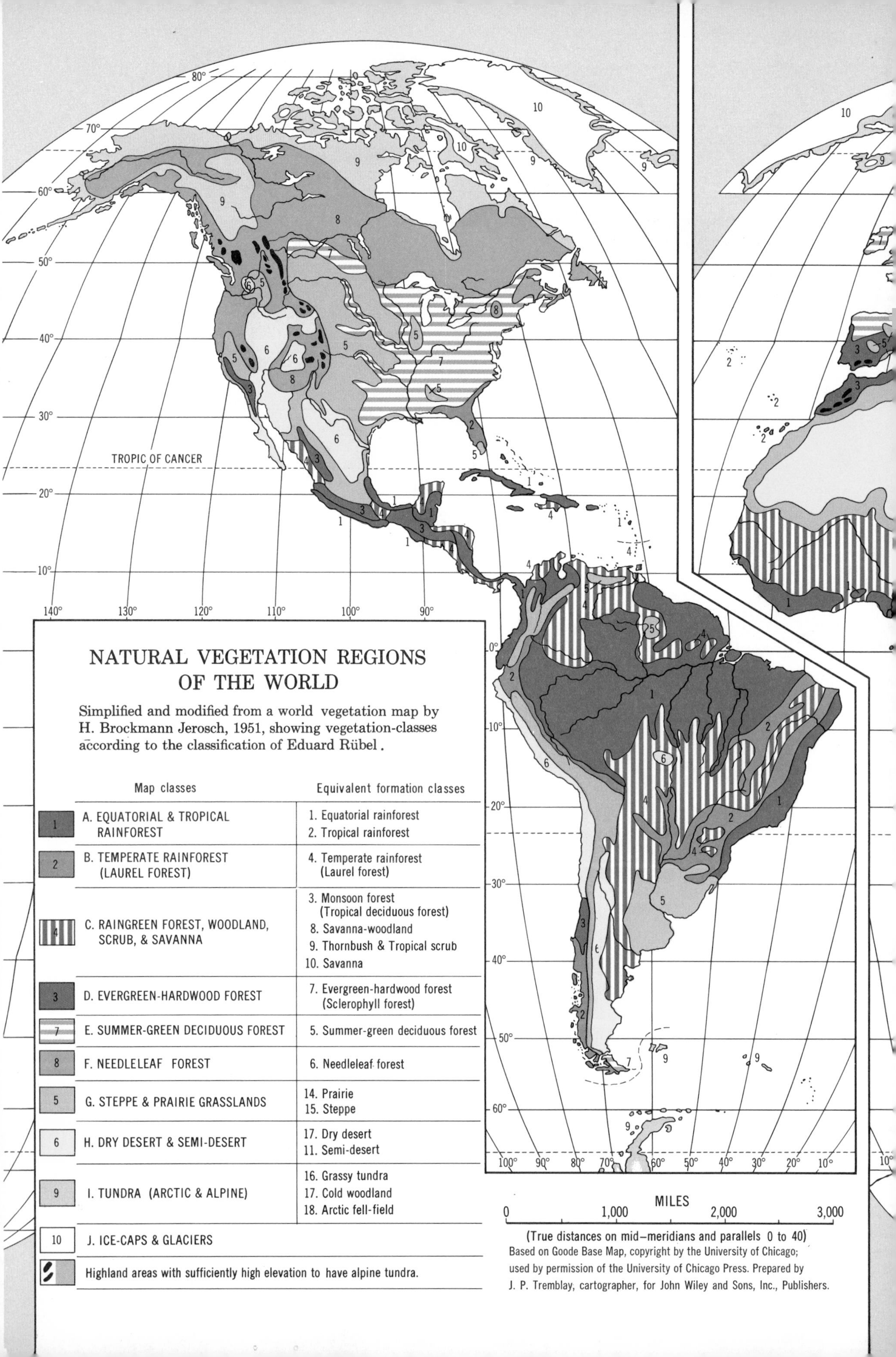

NATURAL VEGETATION REGIONS OF THE WORLD

Simplified and modified from a world vegetation map by H. Brockmann Jerosch, 1951, showing vegetation-classes according to the classification of Eduard Rübel.

	Map classes	Equivalent formation classes
1	A. EQUATORIAL & TROPICAL RAINFOREST	1. Equatorial rainforest 2. Tropical rainforest
2	B. TEMPERATE RAINFOREST (LAUREL FOREST)	4. Temperate rainforest (Laurel forest)
4	C. RAINGREEN FOREST, WOODLAND, SCRUB, & SAVANNA	3. Monsoon forest (Tropical deciduous forest) 8. Savanna-woodland 9. Thornbush & Tropical scrub 10. Savanna
3	D. EVERGREEN-HARDWOOD FOREST	7. Evergreen-hardwood forest (Sclerophyll forest)
7	E. SUMMER-GREEN DECIDUOUS FOREST	5. Summer-green deciduous forest
8	F. NEEDLELEAF FOREST	6. Needleleaf forest
5	G. STEPPE & PRAIRIE GRASSLANDS	14. Prairie 15. Steppe
6	H. DRY DESERT & SEMI-DESERT	17. Dry desert 11. Semi-desert
9	I. TUNDRA (ARCTIC & ALPINE)	16. Grassy tundra 17. Cold woodland 18. Arctic fell-field
10	J. ICE-CAPS & GLACIERS	
	Highland areas with sufficiently high elevation to have alpine tundra.	

(True distances on mid–meridians and parallels 0 to 40)

Based on Goode Base Map, copyright by the University of Chicago; used by permission of the University of Chicago Press. Prepared by J. P. Tremblay, cartographer, for John Wiley and Sons, Inc., Publishers.

Plate 4
NATURAL VEGETATION REGIONS
OF THE WORLD

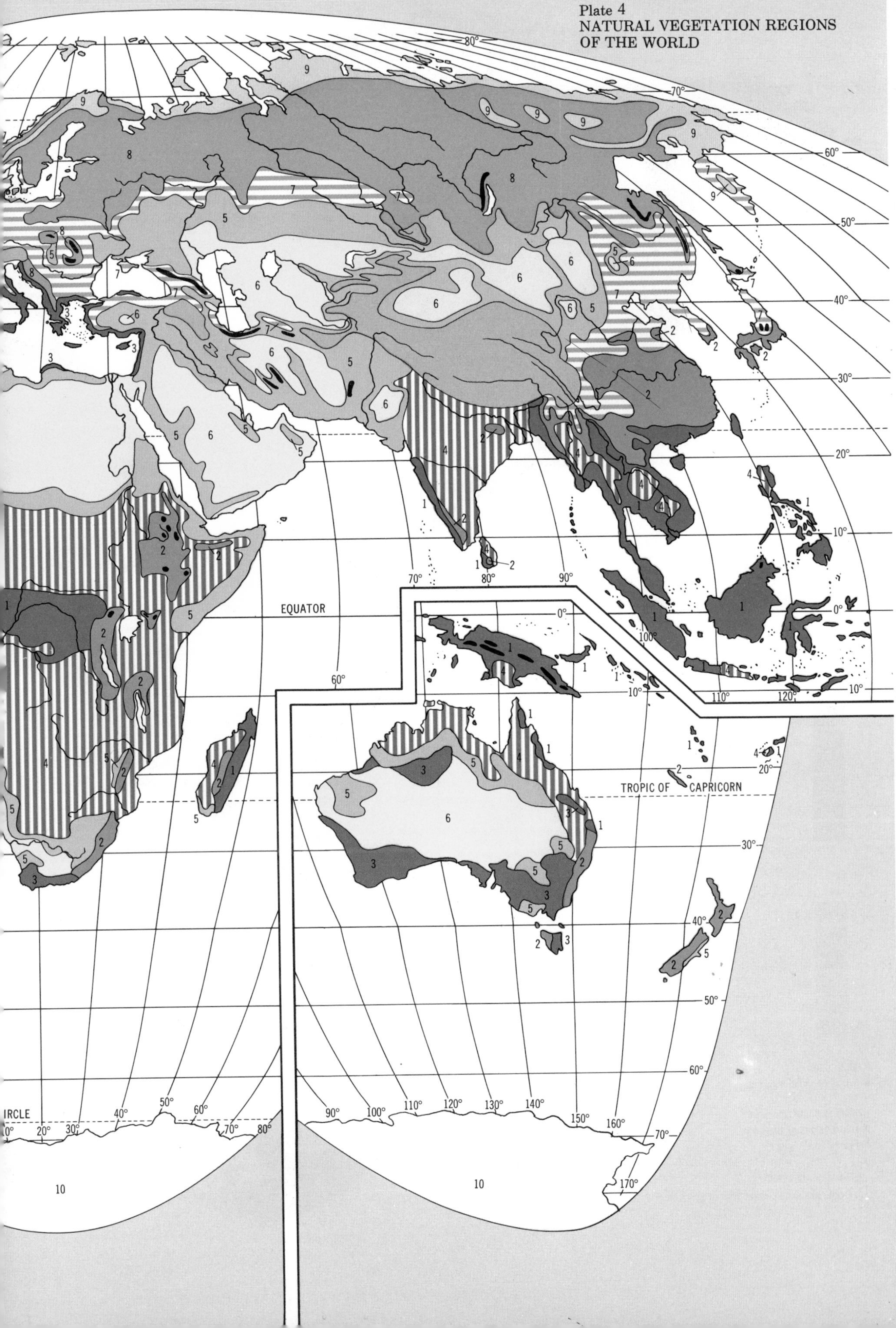

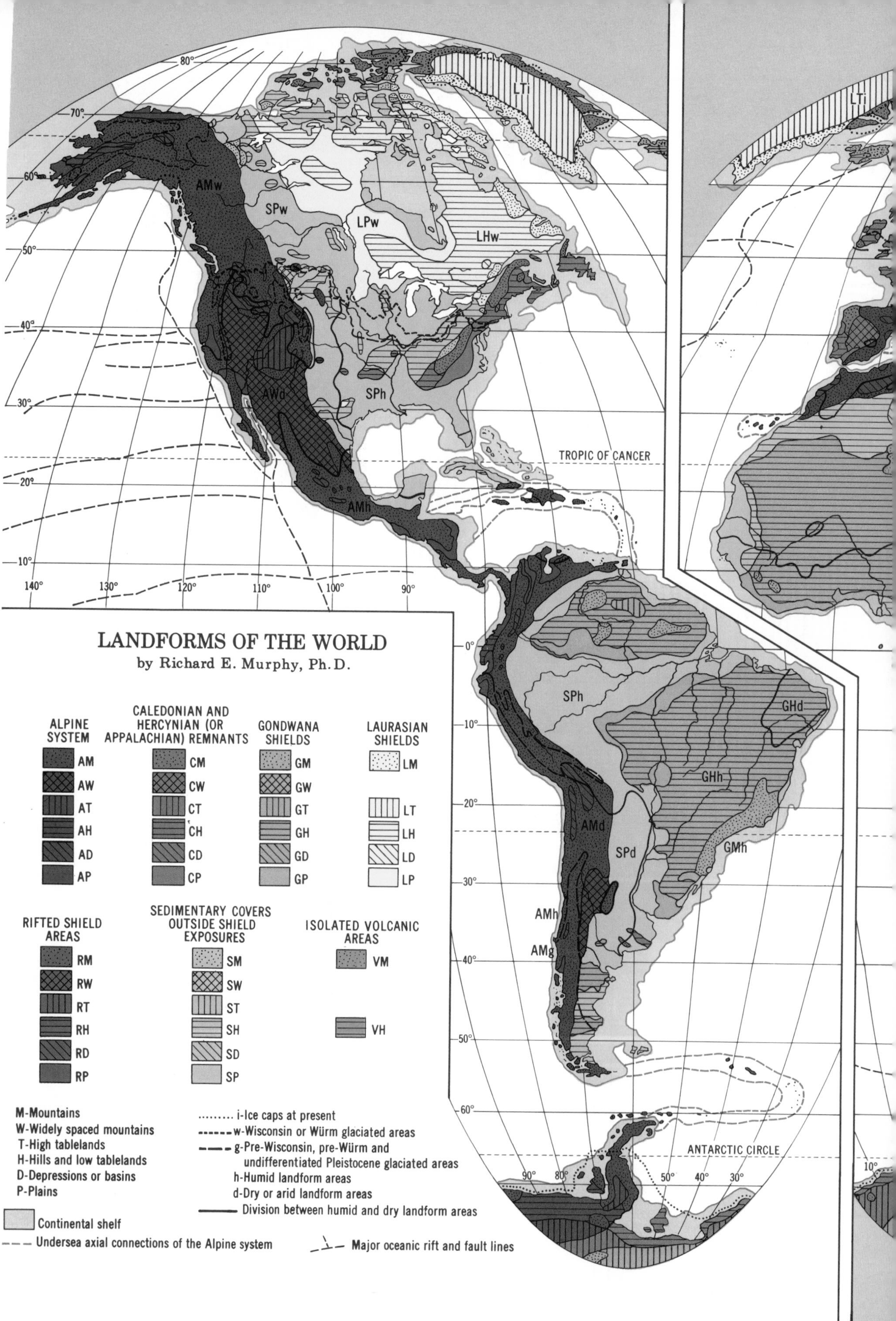

LANDFORMS OF THE WORLD
by Richard E. Murphy, Ph.D.
ALPINE SYSTEM
AM
AW
AT
AH
AD
AP
CALEDONIAN AND HERCYNIAN (OR APPALACHIAN) REMNANTS
CM
CW
CT
CH
CD
CP
GONDWANA SHIELDS
GM
GW
GT
GH
GD
GP
LAURASIAN SHIELDS
LM
LT
LH
LD
LP
RIFTED SHIELD AREAS
RM
RW
RT
RH
RD
RP
SEDIMENTARY COVERS OUTSIDE SHIELD EXPOSURES
SM
SW
ST
SH
SD
SP
ISOLATED VOLCANIC AREAS
VM
VH
M-Mountains
W-Widely spaced mountains
T-High tablelands
H-Hills and low tablelands
D-Depressions or basins
P-Plains
Continental shelf
Undersea axial connections of the Alpine system
i-Ice caps at present
w-Wisconsin or Würm glaciated areas
g-Pre-Wisconsin, pre-Würm and undifferentiated Pleistocene glaciated areas
h-Humid landform areas
d-Dry or arid landform areas
Division between humid and dry landform areas
Major oceanic rift and fault lines
AMw
SPw
LPw
LHw
LTi
AWd
SPh
AMh
TROPIC OF CANCER
SPh
GHd
GHh
AMd
SPd
GMh
AMh
AMg
ANTARCTIC CIRCLE
80°
70°
60°
50°
40°
30°
20°
10°
140°
130°
120°
110°
100°
90°
0°
10°
20°
30°
40°
50°
60°
90°
80°
50°
40°
30°
10°

Plate 5
LANDFORMS OF THE WORLD

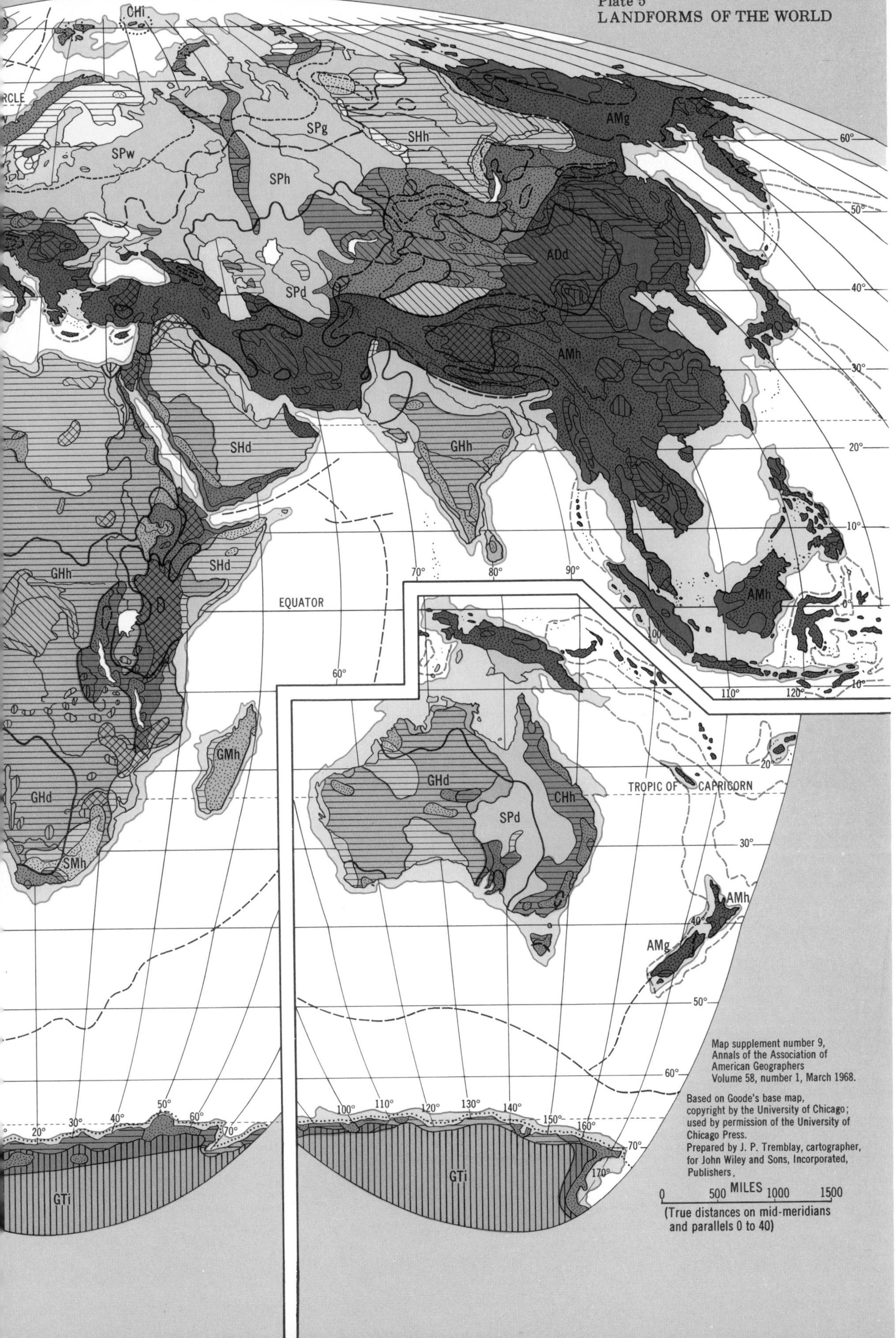

Map supplement number 9,
Annals of the Association of
American Geographers
Volume 58, number 1, March 1968.

Based on Goode's base map,
copyright by the University of Chicago;
used by permission of the University of
Chicago Press.
Prepared by J. P. Tremblay, cartographer,
for John Wiley and Sons, Incorporated,
Publishers,

(True distances on mid-meridians
and parallels 0 to 40)

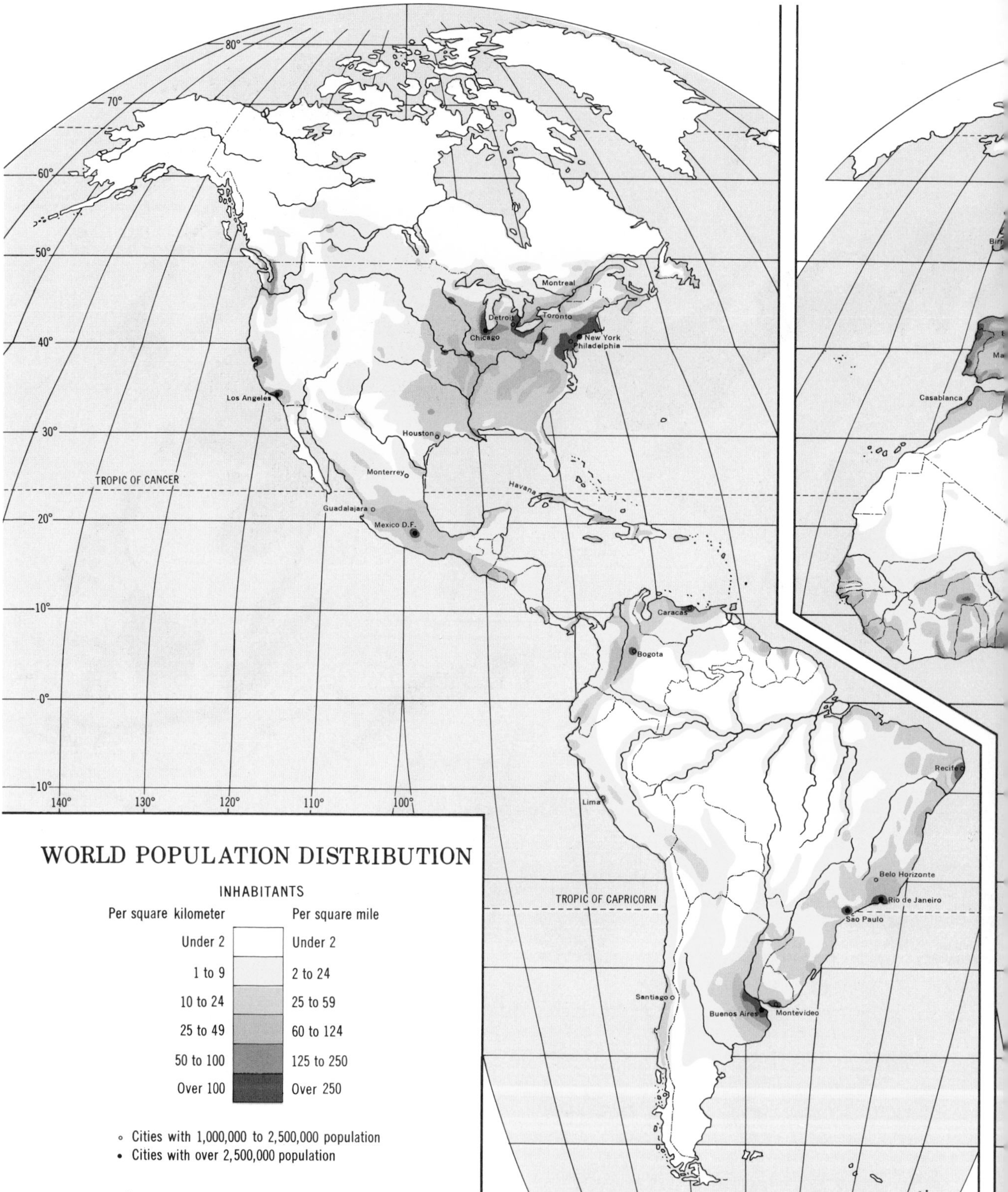
WORLD POPULATION DISTRIBUTION
INHABITANTS
Per square kilometer
Per square mile
Under 2
Under 2
1 to 9
2 to 24
10 to 24
25 to 59
25 to 49
60 to 124
50 to 100
125 to 250
Over 100
Over 250
◦ Cities with 1,000,000 to 2,500,000 population
• Cities with over 2,500,000 population
0
1,000
2,000
MILES
True distances on mid-meridians and parallels 0 to 40
TROPIC OF CANCER
TROPIC OF CAPRICORN
Montreal
Toronto
Detroit
Chicago
New York
Philadelphia
Los Angeles
Houston
Monterrey
Havana
Guadalajara
Mexico D.F.
Caracas
Bogota
Lima
Recife
Belo Horizonte
Rio de Janeiro
Sao Paulo
Santiago
Buenos Aires
Montevideo
Casablanca
80°
70°
60°
50°
40°
30°
20°
10°
0°
-10°
140°
130°
120°
110°
100°
90°
80°
70°
60°
50°
40°
30°

WORLD POPULATION DISTRIBUTION

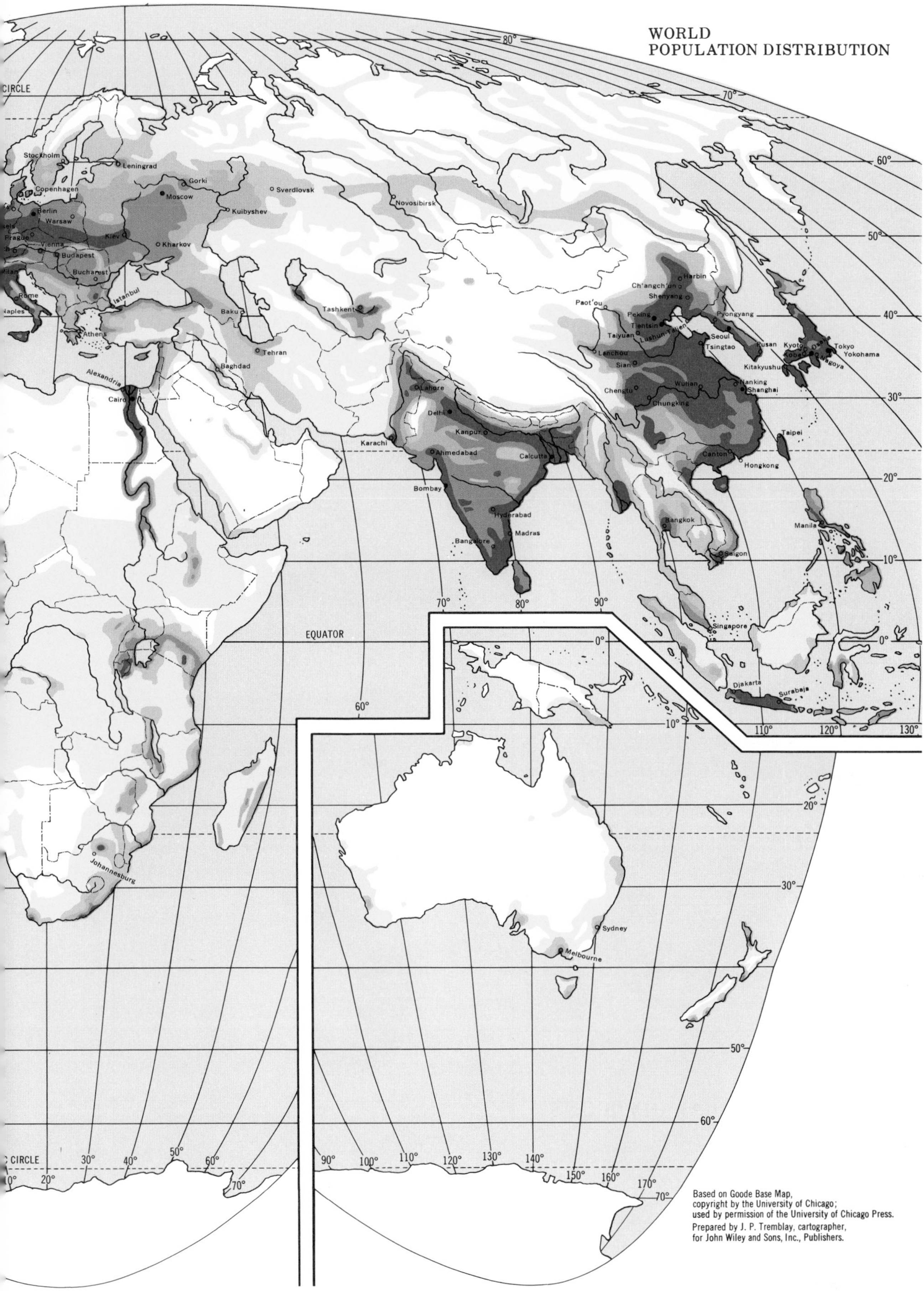

Based on Goode Base Map,
copyright by the University of Chicago;
used by permission of the University of Chicago Press.
Prepared by J. P. Tremblay, cartographer,
for John Wiley and Sons, Inc., Publishers.

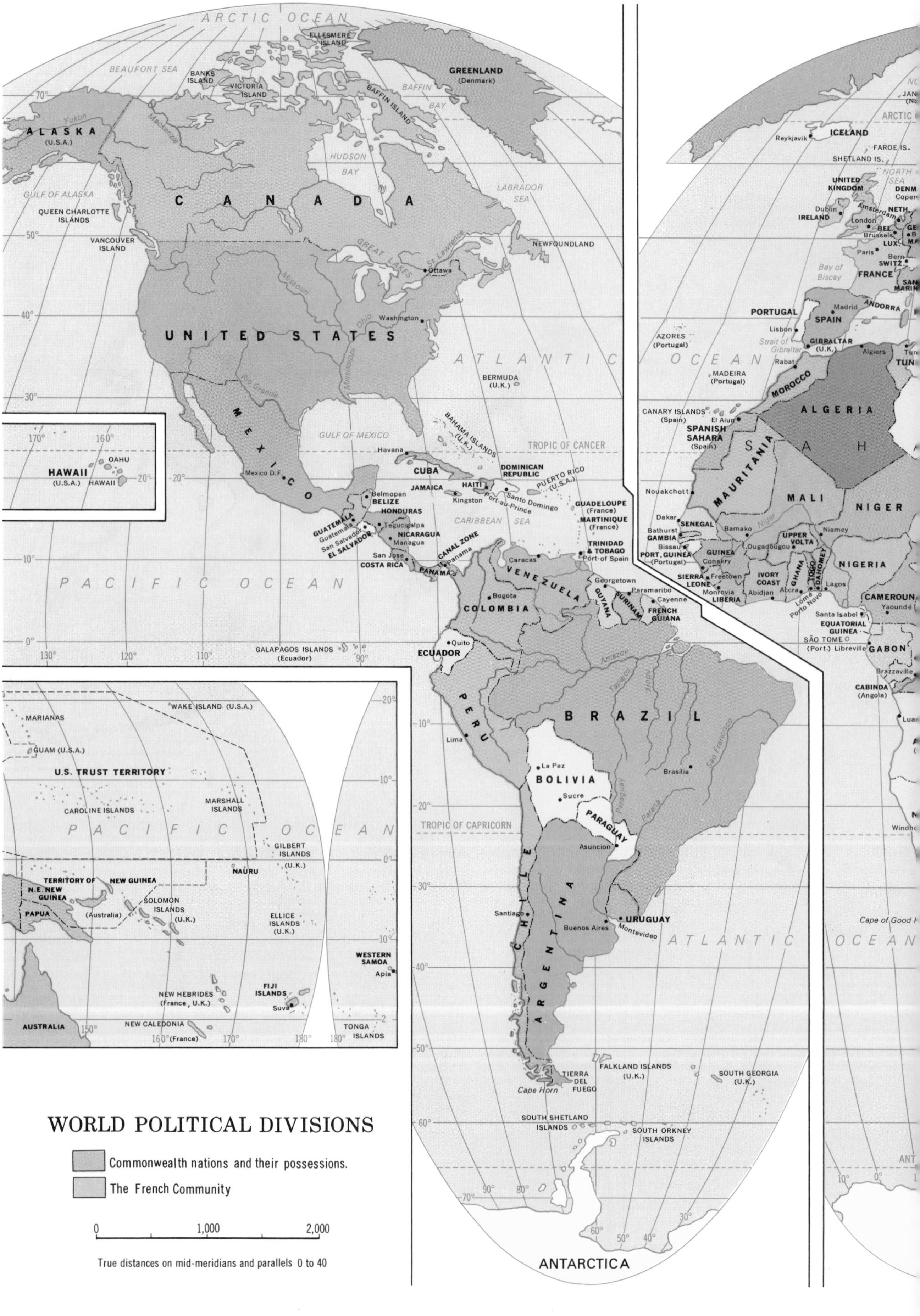

WORLD POLITICAL DIVISIONS

- Commonwealth nations and their possessions.
- The French Community

0 1,000 2,000

True distances on mid-meridians and parallels 0 to 40

WORLD POLITICAL DIVISIONS

Based on Goode Base Map, copyright by the University of Chicago; used by permission of the University of Chicago Press.
Prepared by J. P. Tremblay, cartographer, for John Wiley and Sons, Inc., Publishers.